Readings from *Conservation Biology*

EDITED BY DAVID EHRENFELD

Genes, Populations, and Species

A Joint Publication of
The Society for Conservation Biology
and Blackwell Science, Inc.

BLACKWELL SCIENCE

EDITORIAL OFFICES:
238 Main Street, Cambridge, Massachusetts 02142, USA
Osney Mead, Oxford OX2 0EL, England
25 John Street, London WC1N 2BL, England
23 Ainslie Place, Edinburgh EH3 6AJ, Scotland
54 University Street, Carlton, Victoria 3053, Australia
Arnette Blackwell SA, 1 rue de Lille, 75007 Paris, France
Blackwell Wissenschafts-Verlag GmbH, Kurfürstendam 57, 10707 Berlin, Germany
Blackwell MZV, Feldgasse 13, A-1238 Vienna, Austria

DISTRIBUTORS:
USA
Blackwell Science, Inc.
238 Main Street
Cambridge, Massachusetts 02142
(Telephone orders: 800-215-1000 or 617-876-7000)

Canada
Oxford University Press
70 Wynford Drive
Don Mills, Ontario M3C 1J9
(Telephone orders: 416-441-2941)

Australia
Blackwell Science Pty Ltd
54 University Street
Carlton, Victoria 3053
(Telephone orders: 03-347-5552)

Outside North America and Australia
Blackwell Science, Ltd.
c/o Marston Book Services, Ltd.
P.O. Box 87
Oxford OX2 0DT
England
(Telephone orders: 44-865-791155)

Printed and bound by The Sheridan Press, Hanover, PA, on acid-free recycled stock.

Printed in the United States of America

95 96 97 98 5 4 3 2 1

Library of Congress Cataloging in Publication Data

Genes, populations, and species/edited by David Ehrenfeld.
p. cm.—(Readings from Conservation biology)
Includes bibliographical references.
ISBN 0-86542-452-7
1. Nature conservation. 2. Biological diversity conservation. 3. Population biology. 4. Genetics. I. Ehrenfeld, David. II. Society for Conservation Biology. III. Series. QH75.G44 1994
333.95'16—dc20

94-46719
CIP

Contents

Preface

Conservation biology has always been a broad discipline, welcoming contributions from physiologists and philosophers, economists and engineers. The science of preserving diversity encourages diversity in its own ranks as a source of strength. But even diverse disciplines have to be held together and directed by a core, and that core is modern population ecology, a powerful field at the intersection point of genetics, evolution, ecology, and systematics. This point is where conservation biology started, and it remains at the center of our work.

I have chosen 32 articles from the 30 issues of *Conservation Biology* published at the time of this writing, to provide a sample of the best and most creative science that species-oriented biology can offer to conservation. The papers in this collection were selected because of their quality, their general interest, and the lasting value of the approaches that they take or the points that they make. The volume is intended to be of use both as an aid to teaching and as an important part of the working libraries of all conservation researchers and managers. A few of the papers employ advanced statistical techniques, but even these offer lessons that are clearly accessible to any interested reader. In a field as new and fast-paced as this one, there can be no "last word"—the controversies that are presented here, such as the debate about the possible hybrid origin of the red wolf, are proof of the vitality and excitement of this dynamic branch of conservation biology.

The first paper in this collection, by Martha Rojas, raises a critical question: if a principal aim of conservation biology is to protect species, what exactly is it that we are protecting? As Rojas explains, different approaches to delimiting species produce different results. Conservation biologists will have to use incomplete information to determine the units of protection, and will have to involve systematists in their work.

Once the target species or population has been determined, the next step is to assess the extent to which it is threatened. Georgina Mace and Russell Lande redefine the original, largely subjective, IUCN categories of threat by applying population viability theory to species evaluated for a projected time period of 100 years.

Systematists are in short supply for all taxa of organisms. Ian Oliver and Andrew Beattie outline a new method, using "biodiversity technicians," for the rapid assessment of biodiversity at sites proposed for conservation. The method estimates species richness of field samples of spiders, ants, polychaetes, and mosses.

Our detailed consideration of populations begins with a paper by Richard Harris and Fred Allendorf. The authors consider various methods for estimating the genetically effective population sizes of large mammals and propose a simulation model that reduces the need for extensive demographic data. Lamberson et al., entering the world of politics, courts, and economic interests, devise a population model for Northern Spotted Owls that will help design a reserve that the authors believe will protect the species with a minimal impact on competing economic resources.

Eric Menges, studying a prairie perennial plant, the royal catchfly, shows that seed germination in the laboratory is directly dependent on the size of the population from which the seeds were taken. Thus habitat fragmentation may act, through several possible mechanisms, to decrease the viability of plants in residual populations. Are small populations of plants even worth preserving? Peter Lesica and Fred Allendorf examine the data from eight studies and conclude that small, stressed populations may lose heterozygosity more slowly than similar-sized populations in benign environments. Such populations may be especially valuable for conservation.

R. W. Braithwaite and W. M. Lonsdale look at a shrew-like, marsupial insectivore, *Sminthopsis virginiae,* in the Northern Territory of Australia, and find that determining the cause of rarity is not always a simple matter; it requires a detailed knowledge of the species' natural history and ecological interactions. Hori et al. use the littoral fish communities in Lake Tanganyika to reach a similar conclusion. Cichlid species diversity is dependent upon food webs and inter-species behavioral interactions that are almost too complex to represent graphically.

Insects provide excellent subjects for population studies in conservation biology. To determine the size of reserves needed to preserve the checkerspot butterfly on California's serpentine-based grasslands, Paul Ehrlich and Dennis Murphy recommend a "minimum viable metapopulation" approach which would take into account the crucial roles of habitat diversity and environmental stochasticity in formulating a conservation strategy. Oskar Kindvall and Ingemar Ahlén present a metapopulation study of a species of bush cricket living in the sandy Vomb Valley in Sweden. They quantify the processes of extinction and recolonization that affect the fate of this species in a highly fragmented landscape. Their results show that it is appropriate to include geometrical factors in metapopulation models.

Useful as the metapopulation approach may be in species conservation, it has its limits. Wilson et al., citing the example of the highly endangered Puerto Rican Parrot, claim that further division of the parrot population into captive subgroups, with exchange of individuals between metapopulations, carries substantial risks of disease transmission and interference with the dynamics of mate selection.

Organisms that take a long time to reach sexual maturity and that live many years require different conservation strategies than the majority of species. Congdon, Dunham, and van Loben Sels analyze life tables for the Blanding's turtle and discover that long-lived organisms have severe constraints on their abilities to respond to chronic disturbances. In long-lived species with already reduced populations, sustained mortality of adults is unlikely to be compensated by efforts to increase juvenile survival.

Island biogeography is a cornerstone of population studies. Michael Rosenzweig and Colin Clark show that regular censuses conducted over a long time allow the calculation of extinction and immigration rates for island populations. Using bird data for three different British islands, their method helps to elucidate the impact of secondary effects on extinction rate.

Genetics is the tool that permits fine-tuning of population management in conservation. Gary Meffe and Robert Vrijenhoek, addressing the problem of the decline of North American desert fishes, discuss the need to incorporate experimental studies of population genetics and fitness into conservation management. They recommend experiments to test the relationship between heterozygosity and individual fitness, to evaluate the effects of mild outcrossing in order to provide information for restocking efforts, and to separate true genetic differentiation from environmental induction of phenotypic differences between populations. Robin Waples and David Teel have examined allele frequencies of Pacific salmon and have detected recent changes in hatchery, but not wild, populations. Their findings underline the importance of ongoing genetic monitoring of large-scale artificial propagation programs.

Ashley, Melnick, and Western use mitochondrial DNA to compare the level of genetic differences among remaining populations of black rhinoceros. The data indicate a close genetic relationship among all the black rhinos in the survey, suggesting that within national boundaries separate populations may be combined for breeding purposes. On the other hand, Walters, Decker-Walters, and Gordon test for 25 enzyme loci in wiregrass and find that genetic diversity both within and among populations is high. They recommend that *ex situ* collections and restocking efforts be organized to reflect this diversity. Bowen et al. analyze mitochondrial DNA variation among loggerhead sea turtles from five nesting populations in the southeastern U.S. and Greece. They report that there is little gene flow between populations, reinforcing prior management strategies that treat the separate nesting populations as demographically independent units.

Some of the most interesting and productive applications of genetics in conservation concern wild and captive wolf populations. Wayne et al. examined mitochondrial DNA variability among 26 wild populations of gray wolves from throughout their geographical range. Genetic differentiation among wolf populations is consistent with a recent history of population declines and habitat fragmentation, especially in the Old World. Because differences among gray wolf populations are comparatively small, conservation programs should concentrate more on behavioral and demographic factors than on maintenance of genetic purity of supposed subspecies. Wayne and others (in articles not reprinted here) have applied similar genetic criteria to the closely related, southern U.S. species known as the red wolf. They claim that the red wolf is merely a recent hybrid between wolves and coyotes, created by human-induced habitat disruption, and is not deserving of special conservation efforts. Dowling et al. defend the validity of the red wolf as a taxonomic entity, stating that failure to detect divergence between red and gray wolf mitochondrial DNAs is not evidence of either hybridization or conspecific status, but would suggest instead recent common ancestry. They reject genetic purity arguments as a sole guide to conservation decision-making. Robert Wayne replies using morphologic analyses to support his genetic evidence, but his conclusions are in turn disputed by Ronald Nowak and by Michael Phillips and Gary Henry, on the basis of historical, taxonomic, behavioral, and ecological criteria. Dowling et al. respond with a discussion of the nature of hybrids and their place in conservation. I invite the reader to form his or her own conclusions.

In the small populations characteristic of endangered species, inbreeding is a potentially serious problem. Ralls, Ballou, and Templeton study the pedigrees of 40 captive mammalian populations belonging to 38 species to estimate the effect of inbreeding on juvenile survival. Highly inbred matings produced offspring with a 33 percent higher mortality rate than matings of unrelated parents. Laikre, Ryman, and Thompson consider a difficult genetic management problem concerning inbreeding in very small zoo populations of animals that are rare or extinct in the wild: can the frequency of a deleterious allele in a population be reduced without significant loss of the remaining genetic variability? Using the case of hereditary blindness in captive wolves of Fennoscandic origin, the authors show that the frequency of the blindness allele can be reduced significantly without compromising the survival of other founder alleles.

Captive maintenance of endangered species and subspecies is a costly enterprise, especially for large animals with expensive food and housing requirements. Lynn

Maguire and Robert Lacy evaluate the question of allocating zoo space among the four surviving subspecies of tigers. Their analysis shows that by assigning half of the spaces to the most genetically variable (and numerous) subspecies, with the three other subspecies sharing the other half, a satisfactory compromise is reached between survival of each subspecies and overall maintenance of gene diversity.

Many varieties of domesticated species are also endangered. S. J. G. Hall presents registration details for seven rare British breeds of pig. He finds that successful conservation depends on the registration of stock of each breed with its breed society. But the preservation of populations of rare varieties of feral livestock is a different matter: here a balance must be struck between the conservation of potentially useful genes and gene constellations and the minimization of environmental damage done by the feral livestock. Susan Bratton offers a set of criteria for use by managers of nature reserves, national parks, and public lands to determine whether feral populations should be kept or eliminated. She feels that the burden of proof should rest on the advocates of feral species not on the advocates for native species. Dirk Van Vuren and Philip Hedrick discuss management alternatives and genetic goals in maintaining feral livestock populations. For captive populations they stress the importance of retention of genetic variation and unusual characteristics, plus continued selection for feral traits and the avoidance of selection for non-feral traits.

In the final paper in the collection, Klinger, Elam, and Ellstrand introduce discussion of a possible new danger to the conservation of biodiversity—the escape of engineered genes from transgenic crops into wild populations via crop-weed mating. Measuring the mating between bioengineered and wild radishes, the authors trace a marker allele as it moves from crop to weed across separations of up to 1000 m. They conclude that at least for insect-pollinated, outcrossing crops, containment strategies other than distance from wild populations will be needed to keep engineered genes out of the ambient environment.

This is the last of six volumes currently planned for this series—others may follow later. The first volume was *To Preserve Biodiversity: An Overview,* a general survey of the field of conservation biology. The next four volumes were: *The Social Dimension—Ethics, Policy, Law, Management, Development, Economics, Education; Wildlife and Forests; Plant Conservation;* and *The Landscape Perspective.* Overlap of papers between volumes is small—approximately 10 to 20 percent. Each volume is intended to stand by itself with a minimum of duplication. I had made every effort to select papers that will not age rapidly; nevertheless, the contents of each volume will be reviewed and revised at regular intervals.

I thank Jim Krosschell, Jane Humphreys, and Kathleen Grimes, of Blackwell Scientific Publications, for their assistance and for their effort to produce attractive, durable volumes printed on acid-free, recycled paper and published at an affordable price. I am also grateful to the officers of the Society for Conservation Biology, especially Peter Brussard, Steven Humphrey, and Reed Noss, for their support.

David Ehrenfeld

The Species Problem and Conservation: What are We Protecting?

MARTHA ROJAS*
Fundación Natura
AA. 55402
Bogotá, Colombia

Abstract: *Although conservation biologists base most of their activities on species taxa, they have participated little in the debate of systematists and evolutionary biologists about the recognition and nature of species. It is clear from this debate that there is no agreement on what species are, how they should be delimited, or what they represent. But in conservation science, and particularly in the selection, design and management of protected areas, species are either treated as types or as evolutionary units. A typological view of species may lead to a failure to protect the real diversity of life in the long term, if conservation priorities are set on the basis of species numbers or on circumscriptions of endangered, threatened, or endemic species taxa that result from the use of different species concepts. Considering species as typological entities may also lead workers to disregard geographic variation and to neglect the problem of deciding which level of variability to protect. It may even affect the goal of maintaining the evolutionary potential of organisms, if the focus is placed on the preservation of a sample of the existing types in populations that will persist as little changed as possible. When species are considered as evolutionary units, attention is shifted from the level of species to that of populations and metapopulations. Problems may still arise if these populations are considered in the context of species taxa, particularly if their composition is managed or if indicator species are used. A closer interaction between systematists and conservation biologists is suggested. Although the pressures of time and the difficulties faced by conservation biologists cannot be denied, considering the species problem more critically may result in recognizing the limitations of the taxonomic information used; it may also contribute to the refinement of the concepts and methods involved in their activities, making them more congruent with the nature of biological diversity.*

Resumen: *A pesar de que los biólogos de la conservación basan gran parte de sus actividades en taxa al nivel de especie, su participación en la discusión de sistemáticos y evolucionistas sobre el reconocimiento y naturaleza de las especies ha sido limitada. Este debate ha puesto en evidencia que existe total desacuerdo sobre qué son las especies, cómo deben delimitarse y qué representan. Sin embargo, en biología de la conservación y, en particular, en la selección, diseño y manejo de áreas protegidas, las especies son consideradas como tipos o como unidades evolutivas. Un enfoque tipológico de las especies puede implicar que no se logre proteger la diversidad biológica en el largo plazo, si las prioridades de conservación se establecen a partir de números de especies, o de especies en peligro, amenazadas o endémicas, cuya delimitación resulte del uso de diferentes conceptos de especie. Considerar a las especies como entidades tipológicas también puede llevar a que se ignore la variación geográfica y a que no se enfrente el problema de decidir el nivel de variabilidad a proteger. Puede aún afectar el objetivo de mantener el potencial evolutivo de los organismos protegidos, en la medida en que la conservación se centre en preservar una muestra de los tipos existentes en poblaciones que persistan con un mínimo de cambio. Cuando las especies son consideradas como unidades evolutivas, el foco de atención se desplaza del nivel de especies al de poblaciones y metapoblaciones. Pero aún pueden presentarse problemas, si las poblaciones se consideran en el contexto de especies taxonómicas y, en particular, si se maneja su composición, o se utilizan especies indicadoras. Por lo tanto, se recomienda una mayor interacción entre sistemáticos y biólogos de la conservación. Aunque la presión del tiempo y las dificultades que enfrentan los biólogos de la conservación son innegables, el considerar el problema de las especies de una forma más crítica puede resultar en el reconocimiento de las limitaciones de la información taxonómica empleada. Igualmente, puede contribuir al refinamiento de los conceptos y métodos utilizados en sus actividades, para hacerlos más acordes con la naturaleza de la diversidad biológica.*

* *Address correspondence to Martha Rojas c/o Juan Pablo Ruiz, Director, Fundación Natura, AA. 55402, Bogotá, Colombia.*
Paper submitted May 23, 1991; revised manuscript accepted February 5, 1992.

Introduction

The recognition and nature of species have been central subjects in the history of natural sciences. A variety of species concepts has been proposed and each of them has led to extensive discussions and debates within the scientific community. Issues such as the noncorrespondence of species as circumscribed using different concepts, claims about the "reality" of species either as groups or ranks, the nature of species in relation to taxa at other levels, and the role of species as evolutionary units are just some of the subjects that have been widely discussed. Unfortunately, agreement on the species problem is still far from being attained.

Surprisingly, all this discussion is largely restricted to systematists and evolutionary biologists. The users of the information provided by these scientists have participated little in this debate. This is particularly true of conservation biologists who, although discussing species-level taxa, have hardly questioned what these species represent. However, most if not all of their activities are affected by how species are circumscribed. For instance, strategies for conservation are developed on the basis of data on biological diversity, as well as on information concerning threatened, endangered or endemic species.

In this essay I will consider some of the implications that the species problem may have for conservation aims and activities. The goal is to present possible conflicts that conservation biologists may find useful to address, rather than to offer solutions to the species problem. I will concentrate on one aspect of conservation, namely the protection of natural areas such as national parks and reserves. This subject should, however, be considered in the context of other issues—for instance, the preservation of diversity in botanical gardens, zoological parks, and germplasm banks, and the legal and economic aspects relating to conservation, such as property rights of species (see publications of the Botanic Gardens Conservation Secretariat, the Centre for Plant Conservation, the Captive Breeding Specialist Group, and the International Board for Plant Genetic Resources of the International Union for Conservation of Nature and Natural Resources [IUCN]; the Food and Agriculture Organization [FAO] Commission on Plant Genetic Resources; Conway 1989; Kloppenburg 1988). I will focus on the implications of the species problem on the selection, design, and management of protected areas.

The Species Problem

Various criteria have been used to delimit species, including morphological discontinuity, interbreeding ability and reproductive isolation, relationships of ancestry and descent, ecological adaptation, and genetic cohesion. The actual criteria used depends on the purposes of the researcher, on "what is to be investigated and explained" (Endler 1989:632). This has resulted in a proliferation of species concepts, including "morphological," "biological" (see Mayr 1969, 1982), "ecological" (Van Valen 1969), "evolutionary" (Simpson 1961; Wiley 1978), "phylogenetic" (Eldredge & Cracraft 1980), "cladistic" (Nelson & Platnick 1981), "cohesion" (Templeton 1989), and "taxonomic" (Blackwelder 1967). Variations of these, such as "polytypic," "recognition" (Ghiselin 1974; Patterson 1985), and "isolation" (Patterson 1985), have also been proposed. Some of these concepts have been formulated as more general and overarching approaches. The phylogenetic and cohesion concepts, for instance, can satisfactorily deal with phenomena such as asexual reproduction and hybridization that earlier concepts failed to consider. They also subsume other species concepts; for example, the biological species concept is a subset of the cohesion concept. However, there is controversy over these unifying concepts as species criteria and over their applicability.

The main problem is that there may be noncongruence between species delimited using different approaches (Mishler & Donoghue 1982). This has been particularly true among botanists, for whom phenomena such as reticulate evolution, natural hybridization, polyploidy, apomixis, and interspecific introgression further complicate the delimitation of species taxa (see Grant 1981). However, noncongruence is also common in animal species (see Cracraft 1989), and there is growing evidence that these species too may be imperfectly isolated taxa (Woodruff 1989).

There is another tension central to the species debate, namely the relationship between taxonomic species—pattern—and what Stevens (1992) calls "the active units in the natural world"—process. Species have been considered to be simply constructions of taxonomists that permit the convenient cataloguing of the variety of nature (see Regan 1926; Dobzhansky 1937; Gilmour 1940; Camp 1951; Levin 1979); the extreme version of this view results in a species being defined as "what a competent taxonomist says it is" (see Stevens 1992). On the other hand, it has been claimed by both evolutionary biologists (see Mayr 1969) and phylogenetic systematists (see Cracraft 1989; Nelson 1989) that species are the units of evolution.

Even when pattern analysis is concerned with relations of homology and reconstruction of phylogeny, some authors argue that pattern is incompatible with process, although the two are complementary (see Lidén & Oxelman 1989; Rieppel 1988). In this view, pattern analysis—the taxic approach—is considered to be ahistorical, indicating nothing more than character distribution. No assumptions are made here about the processes that give rise to patterns of variation. Process analysis is considered to be a complementary activity that provides a causal explanation of the pattern observed. Taxa, including species, are thus seen as "the products of evolution but not as its determinants" (Lidén & Oxelman 1989:231), and evolutionary mechanisms are said to operate at the level of localized populations rather than within taxa.

Pattern is also seen as being compatible with process (see Cracraft 1989). In the phylogenetic approach, for instance, pattern is considered to be historical. Species are seen as irreducible, distinct and diagnosable clusters of organisms within which is a parental pattern of ancestry and descent. Phylogenetic species are equivalent to basal evolutionary taxa, and their use is thought to "unify the notion of species as it is applied to evolutionary theory, and as it is used in taxonomic practice" (Cracraft 1989:35). Isolated populations, independent of their size, may thus be delimited as individual species taxa.

The issue of the "reality" of species, either as groups or as a rank, has also been discussed in this context, and some biologists have argued that species are "real," objectively-defined units, different from other taxa (see Simpson 1961; Mayr 1969; Cracraft 1989). Considerations as to whether species are "individuals" or "classes" (Hull 1980), monophyletic or paraphyletic (de Queiroz & Donoghue 1988; Cracraft 1989; Nelson 1989), "theoretical" or "operational" (Hull 1968, 1970; Cracraft 1989) further obscure and confuse the species issue for nonexperts—and probably also for experts.

The nature of the species problem has led to alternative approaches. For example, de Queiroz and Donoghue (1988) advocate neither a new species concept nor any existing one; rather, they propose a change in the way of looking at natural diversity. As "no one species concept can meet the needs of all comparative biologists" (de Queiroz & Donoghue 1988:317), the use of the term "species" becomes less relevant, and it may as well "become restricted to one of the classes of real biological entities . . . or continue to be used as a general term referring to an assemblage of several classes sharing nothing more than having been conflated historically" (de Queiroz & Donoghue 1988:335). This notion of a species would seem to have little utility. Other authors have suggested that species should be treated in the same way as other taxa (see Lidén & Oxelman 1989; Nelson 1989; Stevens 1990), and that a pluralistic approach should be adopted over their delimitation (Mishler & Donoghue 1982). Emphasis has been placed on an understanding of patterns of variation rather than on tying them to a particular rank (Stevens 1990).

Nomenclatural stability has been a major preoccupation in this whole debate. It has been argued that this could also be attained very simply if no particular significance was attached to taxonomic rank (Stevens 1990). This would imply that users of species names should be aware that species may be "only equivalent by designation, and not by virtue of the nature or extent of their evolutionary differentiation" (Davis & Heywood 1963:92).

Some Implications of the Species Problem in Conservation Science

As users of species taxa, conservation biologists may be affected by the species problem if their activities depend on the way species are circumscribed or in how they are interpreted.

Use of Species Taxa in the Selection, Design, and Management of Protected Areas

Conservation biologists are one of the main groups of users of species taxa. In fact, species concepts are integral to the major objective of the conservation of wildlife areas, that of "maintaining the diversity of species and ecosystems" (Harrison et al. 1984; Walker 1989).

The level of biological diversity—biodiversity—is a major parameter for setting priorities for conservation, as areas of high "richness" are usually targeted for the establishment of protected areas. Biodiversity here is usually focused at the ecosystem and species levels, although considerations of genetic diversity within species are also included in its definition (see McNeely et al. 1990). In practice, the criterion of ecosystem diversity is the principle used for setting conservation priorities, both because of lack of knowledge of variation at the species level and below, and because entire regions are under threat. It is assumed that protecting ecosystem diversity will guarantee the protection of most of the species present in the ecosystem.

Diversity at the species level is also used as a criterion on its own, when data are available. "Hotspots" of diversity are identified by the comparison of species lists, and conservation measures such as the establishment of protected areas are applied (McNeely et al. 1990).

Areas of biogeographic importance that are based on the distribution of taxa, largely at the species level, are also used for setting conservation priorities. For example, centers of endemism and Pleistocene refugia identified in the Amazon region (Haffer 1969; Brown 1982; Prance 1982) have been used as a planning tool for the establishment of protected areas (Wettemberg et al. 1981).

Yet other criteria for the selection of protected areas involve the use of species taxa. For example, the presence of threatened species, especially of those in the endangered, vulnerable, and rare categories of the IUCN's Red Data Books (McNeely et al. 1990), or those included in international and national legislation such as the U.S. Endangered Species Act, may be an important factor in determining conservation strategies. Furthermore, data bases are being built all over the world to handle and make readily available existing information about species for conservation biologists.

How species are delimited is also central to the good design and management of protected areas, and therefore to the achievement of long term conservation goals, particularly if the viability in perpetuity and the maintenance of the "evolutionary potential" of organisms are considered. Thus, the theory of island biogeography, which has been used in the design of protected "refuges," is centered in part on species-area relationships. In this theory, sizes of communities and thus species numbers have played an important role, especially in the early stages of its development (Simberloff 1988). Concepts such as minimum viable populations (MVP) (see Wilcox 1984; Gilpin & Soulé 1986) and minimum critical size of ecosystems (see Lovejoy 1980), developed later by conservation biologists in response to the necessity of preserving in perpetuity the biological elements protected, while focusing on the population level, are also ultimately related to particular circumscriptions of species taxa. The MVP is defined as the "set of specifications concerning the size and structure of the populations comprising a species that is necessary to provide a margin of safety from extinction," and the minimum area requirements correspond to "the amount and type of habitat that will satisfy the MVP" (Wilcox 1984:645). The minimum viable density has also been stated as a correlate of the MVP (Shaffer 1981). More recently, a procedure called population vulnerability (or viability) analysis (PVA) has been introduced as a "comprehensive or systems-level approach to estimating MVP size" (Woodruff 1989:79). The use of the metapopulation concept—a metapopulation can be defined as a "set of local populations which interact via individuals moving among populations" (Hanski & Gilpin 1991:7)—and a model for conservation derived from it have also been involved in the design and management of reserves (Simberloff 1988; see also the Biological Journal of the Linnean Society 1991, 42:1–323).

Thus, many conservation activities, especially those related to the selection, design, and management of protected areas, depend on how species are delimited and hence on particular species concepts. But what do conservation biologists consider species to be? Two main approaches may be identified: species may be considered as types or as evolutionary units. Although these views cannot be totally separated in practice, I will focus on each of them individually to identify some of their implications for conservation activities.

Conserving Species as Types

Most of the current methods for the selection of protected areas are based on a typological approach to species, the aim of these activities being to preserve a representative sample of the existing "types" or "kinds" in nature. Conservation efforts are focused on the protection of highly representative areas, for instance, in Malaysia, in an area of 50 ha which is said to include one third of the known tree flora of the lowland dipterocarp forest (Ashton 1990).

Clearly, the number of existing "kinds" (the biodiversity at the species level) that is represented in the protected areas will depend on the way species taxa are delimited. The lack of sufficient knowledge concerning basic species-level diversity has already been identified as a major problem that will affect the feasibility of its preservation. It has been estimated that approximately between 1.5 and 1.6 million species have been described (Wilson 1989) out of a total of 5 to 30 million (see Erwin 1988; Wilson 1989). These numbers, however, are unlikely to be the same if we are considering biological species, cladistic species, or evolutionary species, to name just three species concepts. The species circumscribed using these concepts may have different limits, and different amounts of variability may be included within them. If taxa currently treated as subspecies, races, varieties, or breeding populations are considered as the basic units of diversity, we will get yet another much higher figure. On the other hand, if a polytypic species concept is used, the total species numbers will be reduced. An example from New Guinea illustrates this situation, as there may be one or thirty species of *Drimys,* depending on the species concept used (Stevens 1988). Hence, the "coverage" of conservation measures will vary. For example, if the total number of species taxa increases, the "coverage" of species diversity by the existing protected areas would be reduced, and the number of extinct, threatened, or endemic species would be increased, making the conservation task more difficult or less successful.

The typological approach has other implications in conservation, as the geographic variation of the species represented in the protected areas may be disregarded. In the example mentioned by Ashton (1990, see above), the issue is not simply that one third of the species of the lowland dipterocarp forest are in a 50 ha area, but that if this area is protected these species can be considered conserved even though they grow elsewhere, some being widely ranging and morphologically variable species. The species are treated as types, and infraspecific variation is discounted. Crucial decisions concerning the level of variability that should be protected, either within or between populations, are thus overlooked.

The same issues apply if conservation focuses on threatened or endemic species. There are good examples of the effect of taxonomic ambiguity on the fate of species. For instance, some distinctive populations of the tuatara of New Zealand (*Sphenodon guntheri*) are threatened, since legislation recognizes only a single, more widespread species (*S. punctatus*) (May 1990).

The different possible resolutions of taxonomic relationships can also affect the conservation of endangered species and subspecies such as the Florida panther (*Felis concolor coryi*), the gray wolf (*Canis lupus*), and the red wolf (*Canis rufus*); the possibility that these animals represent hybrids may exclude them from the protection of the U.S. Endangered Species Act (O'Brien & Mayr 1991). The "hybrid policy," as this hotly debated issue is known, clearly results from the use of the biological species concept. Other species concepts, such as the cohesion concept, consider hybridization in a way that would avoid the difficulties imposed by this policy.

A typological view of species also has implications for the design and management of reserves. For here the goal is just to guarantee that existing types are adequately represented and that they reproduce themselves, so that their populations can be maintained. But conservation in the long term requires particular characteristics of the reserves, such as size and shape, and special management techniques that can be defined only if species are seen as something more than types. It is hardly surprising, in this context, that conservation biologists consider species as evolutionary units (see Frankel & Soulé 1981; Allendorf & Leary 1986; Templeton 1986; Vrijenhoek 1989; Woodruff 1989), although the typological view is not totally abandoned.

Conserving Species as Evolutionary Units

Considering species to be evolutionary units, i.e. evolving entities, relates to a major goal of conservation science, that is, the maintenance of the evolutionary potential of the organisms protected. This issue is usually considered in terms of the maintenance of "the ability of species to respond to new evolutionary challenges" (Vrijenhoek 1989:90), these challenges being mainly presented by a changing biological or physical environment (Frankel & Soulé 1981). But, although the species are mentioned by conservation biologists as the units of evolution, the focus is now on the diversity within the species.

Patterns of variation at the species level and below have been taken into account mainly because of the influence of ideas from population genetics. This has helped to shift the focus from communities of species to individual populations (Simberloff 1988). Thus, concepts and methods developed for wildlife management, such as MVP, MDP, and PVA, have focused on populations (see Gilpin & Soulé 1986) or metapopulations (see Simberloff 1988; Soulé & Kohm 1989), and much attention has been paid to variation within species, particularly at the genetic level. The long term maintenance of genetic diversity has been considered to be a requirement for the survival of the organisms protected, especially because the loss of this diversity has been associated with reduced fitness in local populations (Vrijenhoek 1989). Conservation biology has thus focused on the loss of genetic variability, and its possible effects on population fitness, through genetic drift, inbreeding, colonization events (founder events), and population bottlenecks (Frankel & Soulé 1981, Simberloff 1988, Vrijenhoek 1989).

The focus on this level of variability makes the design and management of protected areas less vulnerable to idiosyncracies in the way in which species are circumscribed. But problems may still arise if, for instance, immediately unrelated populations are thought to belong to the same taxonomic species. There has been some concern expressed over this issue. According to Woodruff (1989:81), "if the goal of conservation efforts is to permit the continued evolution of species, then it is important to establish that the managed individuals or populations are conspecific. Currently, [however], many operational conservation units are poorly defined, and the true evolutionary relationships of isolated populations treated as belonging to the same species are unknown." The issue is not only that of "true evolutionary relationships," but that isolated and differentiated populations may be ignored because they are included in one species taxon.

There is in most cases, however, little information available on variability below the species level, although the need to concentrate on this issue was identified early in conservation biology (see Frankel & Soulé 1981; Prescott-Allen & Prescott-Allen 1984). This has led to the selection of particular species for detailed study to obtain more general information about the environmental quality of whole communities. These species, called "ecologically significant species for management and monitoring" by Soulé and Kohm (1989), include keystone species, indicator species, and mobile link species. Obviously, how these species are delimited may affect the results of management activities based on them, whether they are used as indicators of the biological or of the physical conditions of the environment. For example, a study by T. Parker showed that the delimitation of a species of worm (*Capitella capitata*) used as an indicator of water quality affected the validity of the monitoring and protecting activities of the U.S. Environmental Protection Agency. It was demonstrated that there was a complex of species differentiated by nonmorphological characters, with these species possibly showing different toxic responses (Parker 1990). A similar situation was reported for a Caribbean coral (*Montastraea annularis*) used in studies of environmental degradation and climate change (Knowlton et al. 1992).

The species problem is especially evident when the composition of a population is manipulated, whether by reintroductions from external stocks or by reestablishment of gene flow and migration patterns by the exchange of individuals from different populations. It has been proposed, for instance, that if habitats were fragmented, gene flow would have to be reestablished, "at least within the confines of major subspecific groups" (Vrijenhoek 1989:97). It seems to be necessary, therefore, to draw attention to the definition of these "confines." If these confines are wrongly defined, and if populations not closely related are mixed, the resulting hybridization may lead to an "outbreeding depression [that can] greatly increase the chances of population extinction if it is severe during the early generations, when population size is small" (Templeton 1986:114).

Other examples come from the management of remnant populations of endangered species. For instance, members of the last population of the now extinct dusky seaside sparrow (*Ammodramus maritimus nigrescens*) were crossed with a subspecies thought to be closely related (*A. m. peninsulae*). But recent molecular genetic studies have revealed that there is a fundamental phylogenetic distinction between the two groups, and that the dusky seaside sparrow had a closer affinity with other populations along the Atlantic coast of the United States (Avise & Nelson 1989). The example of the Tatra Mountain ibex (*Capra ibex ibex*) in Czechoslovakia, cited by Templeton (1986), also illustrates this problem.

Discussion

Clearly, the way in which species are interpreted by both systematists and conservation biologists affects the goals and activities of conservation. In particular, the use of a typological approach to species by conservation biologists may lead to a failure to protect the real diversity of life in the long term. This will become especially true if diversity at the species level is more widely used as a criterion for setting conservation priorities. A typological approach to diversity may also affect the design and management of the protected areas, if the study of variation at the population or metapopulation levels is carried out in the context of species taxa.

Considering species as dynamic evolutionary units is less affected by the species problem. The units discussed in conservation science are not tied to the criteria used by a particular evolutionary species concept, but rather relate to the goal of maintaining the evolutionary potential of the organisms protected. In this context, conservation biologists focus on the capacity of organisms to respond to environmental change. However, thinking of conservation in this manner suggests that this simple response is the only evolutionary path left to organisms and that evolution is now largely in response to the pressures of the human population. It has been proposed not unreasonably that all the diversity that makes up life cannot be maintained (Frankel & Soulé 1981), but it has also been suggested that the direction of evolution of virtually all surviving species will change, with little possibility of diversification and evolution into new taxa (Soulé 1989). If this situation is indeed the case, we do not seem to be far from the typological view discussed above. We may be trying to preserve in the long term a sample of the existing types, in healthy populations that persist as little changed as possible.

It is obvious that conservation biologists must act in the context of changes introduced by man, and still try and maintain in the long term—for as long as "this current interval of universal anthropogenic disturbance" lasts (Soulé 1989:303)—as much biological diversity as possible. We must therefore come to an understanding of the nature of this diversity. I emphasized earlier that the complexity of natural diversity is currently hardly amenable to the application of unified criteria and methods for the delimitation of species taxa. This suggests that there must be a change in the way conservation is approached. For instance, methods currently used for setting priorities that are based on species numbers and on measures of the representativity of protected areas may have to be reconsidered. In general, all activities based on the particular circumscription of species will need to be assessed critically.

A change may also be needed in the participation of systematics in conservation. There is some concern about this issue among systematists. They have proposed different areas where their efforts should concentrate, including the development and maintenance of expertise to facilitate the ready identification of organisms; the study of diversity at or below the level of conventional taxonomic species; the characterization of areas of endemism; and the definition of priorities for the study and conservation of species based on information from the phylogenetic position of these species (Humphries et al. 1990; Lefkovitch et al. 1990). An interesting contribution along these lines has recently been provided by Vane-Wright et al. (1991).

We may also identify consequences of a more practical nature. For example, if conservation biologists worry in detail about species concepts, there may be a problem of credibility and support among the public. It has been argued that "people are more likely to stand up for named, classified species, than for anonymous victims" (Soulé 1989), or even named subspecies or varieties (Stevens 1990). It is hardly surprising that species have become the "flag-bearers" of conservation (Lucas 1984). It may be more difficult to gain support for the conservation of things such as patterns of variation. Of course, it could be argued that it is just a matter of public education. But even with the "flags" being currently used, popular and governmental support is still difficult to get.

Also, if understanding the variation in nature and being aware of the species problem requires the reinterpretation of taxon boundaries, it will be necessary to treat each case individually (see also Woodruff 1989). Taxonomic information can be "standardized" only with great difficulty in the data bases that are being built for conservation purposes, and thus users cannot know what the species included represent, other than things called species by taxonomists. This does not mean that these data bases will not be usable. Rather, their limitations will have to be taken into account whenever they are used in decision making. And this will be true for all taxonomic information used in conservation strategies and actions. Concepts, not names, will have to be tracked.

But this implies that the pace at which decisions are made will have to be slowed down, However, conservation biologists have to act rapidly, and they have to make decisions with the information available now. Involving problems about species delimitation in their concerns will add to the difficulties that already exist over other issues such as ecological problems related to the definition and managing of reserves, as well as the socio-economic, political, and financial factors on which conservation depends. Nevertheless, at least in situations where there is more time to act, the additional effort involved in considering the species issue will result in sounder actions. A compromise must be attained, as it is very likely that, at least for the time being, conservation science will have to use incomplete information that is not easily interpretable. But simply being aware of the limitations of this information will be necessary for conservation science if it is to succeed in one of its goals, that of the long term protection of the diversity of life. Considering the species problem more critically will also contribute to a better understanding of the complex patterns of variability of nature, and hence to the refinement of the concepts and methods involved in the selection, design, and management of protected areas.

Conclusions

Although integrating the consequences of dealing with the species problem with conservation science will require additional effort, it may have a considerable effect on the soundness and efficiency of actions carried out. On the one hand, awareness of this issue may allow conservation biologists to consider the limitations of the information on which their actions are based, and thus to act in a more cautious way. It may also contribute to a better understanding of the patterns of variation present in nature and help conservation scientists to set priorities and to plan and implement their activities in light of this diversity.

The pressures of time cannot be denied, however, nor the complexity of conservation science itself. But the gradual inclusion of the species issue will only enrich the field. A closer interaction between academic researchers and applied biologists is required. Conservation biologists cannot afford to be out of the species debate if they are to guarantee the preservation of nature's diversity in the long term.

Acknowledgments

I am indebted to Peter F. Stevens, at the Harvard University Herbaria, for helpful discussion and comments on previous drafts of this paper, as well as for his encouragement and support during my stay at Harvard. I am also grateful to Dan Perlman, Santiago Madriñan, and three anonymous reviewers for very useful comments on the manuscript, and to the staff of the Botany Laboratories for their assistance.

Literature Cited

Allendorf, F. W., and R. B. Leary. 1986. Heterozygosity and fitness in natural populations of animals. Pages 57–76 in M. E. Soulé, editor, Conservation biology: the science of scarcity and diversity. Sinauer Associates, Inc., Sunderland, Massachusetts.

Ashton, P. S. 1990. Exploration expeditions in the tropics: what is no longer needed, what is still needed and what is urgently needed? A review of aims and goals. Flora Malesiana Bulletin, Special Volume 1:27–30.

Avise, J. C., and W. S. Nelson. 1989. Molecular genetic relationships of the extinct dusky seaside sparrow. Science 243:646–648.

Blackwelder, R. E. 1967. Taxonomy. J. Wiley & Sons, New York.

Brown, K. S., Jr. 1982. Paleoecology and regional patterns of evolution in neotropical forest butterflies. Pages 255–308 in G. Prance, editor. Biological diversification in the tropics. Columbia University Press, New York.

Camp, W. H. 1951. Biosystematy. Brittonia 7:113–127.

Conway, W. G. 1989. The prospects for sustaining species and their evolution. Pages 199–209 in D. Western, and M. C. Pearl, editors. Conservation for the twenty-first century. Oxford University Press, New York.

Cracraft, J. 1989. Speciation and its ontology: the empirical consequences of alternative species concepts for understanding patterns and processes of differentiation. Pages 28–59 in D. Otte and J. A. Endler, editors. Speciation and its consequences. Sinauer Associates, Inc., Sunderland, Massachusetts.

Davis, P. H., and V. H. Heywood. 1963. Principles of angiosperm taxonomy. Oliver and Boyd, Edinburgh, Scotland.

de Queiroz, K., and M. J. Donoghue. 1988. Phylogenetic systematics and the species problem. Cladistics 4:317–338.

Dobzhansky, T. 1937. Genetics and the origin of species. Columbia University Press, New York.

Eldredge, N., and J. Cracraft. 1980. Phylogenetic patterns and the evolutionary process. Columbia University Press, New York.

Endler, J. A. 1989. Conceptual and other problems in speciation. Pages 625–648 in D. Otte and J. A. Endler, editors. Speciation and its consequences. Sinauer Associates, Inc., Sunderland, Massachusetts.

Erwin, T. 1988. The tropical forest canopy: the heart of biotic diversity. Pages 123–129 in E. O. Wilson and F. M. Peter, editors. Biodiversity. National Academy Press. Washington, D.C.

Frankel, O. H., and M. E. Soulé. 1981. Conservation and evolution. Cambridge University Press, Cambridge, England.

Ghiselin, M. T. 1974. A radical solution to the species problem. Systematic Zoology **23**:536–544.

Gilmour, J. S. L. 1940. Taxonomy and philosophy. Pages 461–474 in J. Huxley, editor. The new systematics. Oxford University Press, Oxford, England.

Gilpin, M. E., and M. E. Soulé. 1986. Minimum viable populations: processes of species extinction. Pages 19–34 in M. E. Soulé, editor. Conservation biology: the science of scarcity and diversity. Sinauer Associates, Inc., Sunderland, Massachusetts.

Grant, V. 1981. Plant speciation. Columbia University Press, New York.

Haffer, J. 1969. Speciation in Amazonian forest birds. Science **165**:131–137.

Hanski, I., and M. Gilpin. 1991. Metapopulation dynamics: a brief history and conceptual domain. Biological Journal of the Linnean Society **42**:3–16.

Harrison, J., K. Miller, and J. McNeely. 1984. The world coverage of protected areas: development goals and environmental needs. Pages 24–33 in J. A. McNeely and K. R. Miller, editors. National parks, conservation, and development: the role of protected areas in sustaining society. Proceedings of the World Congress on National Parks, Bali, Indonesia, 11–22 October 1982. Smithsonian Institution Press, Washington, D.C.

Hull, D. L. 1968. The operational imperative: sense and nonsense of operationism. Systematic Zoology **17**:438–457.

Hull, D. L. 1970. Contemporary systematic philosophies. Annual Revue of Ecology and Systematics **1**:19–54.

Hull, D. L. 1980. Individuality and selection. Annual Revue of Ecology and Systematics **11**:311–332.

Humphries, J., W. Fink, and J. Cracraft. 1990. BBS task force testimony. Society of Systematic Zoology. Unpublished document.

Kloppenburg, J. R., Jr., editor. 1988. Seeds and sovereignty: the use and control of plant genetic resources. Duke University Press, Durham, North Carolina.

Knowlton, N., E. Weil, L. A. Weigt, and H. M. Guzmán. 1992. Sibling species in *Montastraea annularis,* coral bleaching, and the coral climate record. Science **255**:330–333.

Lefkovitch, L. P., F. Oberwinkler, C. Patterson, P. F. Stevens, A. R. Templeton, and A. Saura. 1990. International evaluation of systematics, December. Swedish Natural Science Research Council, Stockholm, Sweden.

Levin, D. A. 1979. The nature of plant species. Science **204**:381–384.

Lidén, M., and B. Oxelman. 1989. Species—pattern or process? Taxon **38(2)**:228–232.

Lovejoy, T. E. 1980. Discontinuous wilderness: minimum areas for conservation. Parks **5(2)**:13–15.

Lucas, G. L. 1984. The survival of species genetic diversity. Pages 56–57 in J. A. McNeely and K. R. Miller, editors. National parks, conservation, and development: the role of protected areas in sustaining society. Proceedings of The World Congress on National Parks, Bali, Indonesia, 11–22 October 1982. Smithsonian Institution Press, Washington, D.C.

May, R. M. 1990. Taxonomy as destiny. Nature **347**:129–130.

Mayr, E. 1969. The biological meaning of species. Biological Journal of the Linnean Society **1**:311–320.

Mayr, E. 1982. The growth of biological thought. Harvard University Press, Cambridge, Massachusetts.

McNeely, J. A., K. R. Miller, W. V. Reid, R. A. Mittermeier, and T. B. Werner. 1990. Conserving the world's biological diversity. International Union for Conservation of Nature and Natural Resources, World Resources Institute, Conservation International, World Wildlife Fund—U.S. and the World Bank, Gland.

Mishler, B. D., and M. J. Donoghue. 1982. Species concepts: a case for pluralism. Systamatic Zoology **31**:491–503.

Nelson, G. 1989. Species and taxa: systematics and evolution. Pages 60–81 in D. Otte and J. A. Endler, editors. Speciation and its consequences. Sinauer Associates, Inc., Sunderland, Massachusetts.

Nelson, G., and N. I. Platnick. 1981. Systematics and biogeography: cladistics and vicariance. Columbia University Press, New York.

O'Brien, S. J., and E. Mayr. 1991. Bureaucratic mischief: recognizing endangered species and subspecies. Science **251**:1187–1188.

Parker, T. 1990. A commentary on the Clean Water Act and taxonomy. Association of Systematics Collections Newsletter **18(6)**:81–82.

Patterson, H. E. H. 1985. The recognition concept of species. Pages 21–29 in E. Vrba, editor. Species and speciation. Transvaal Museum Monograph 4, Pretoria.

Prance, G. 1982. Forest refuges: evidence from woody angiosperms. Pages 137–157 in G. Prance, editor. Biological diversification in the tropics. Columbia University Press, New York.

Prescott-Allen, R., and C. Prescott-Allen. 1984. Park your genes: protected areas as in situ genebanks for the maintenance of wild genetic resources. Pages 634–638 in J. A. McNeely and K. R. Miller, editors. National parks, conservation, and development: the role of protected areas in sustaining society. Proceedings of The World Congress on National Parks, Bali, Indonesia, 11–22 October 1982. Smithsonian Institution Press, Washington, D.C.

Regan, C. T. 1926. Organic evolution. Report of the British Association for the Advancement of Science. 1925, 75–86.

Rieppel, O. C. 1988. Fundamentals of comparative biology. Birkhauser Verlag, Basel.

Shaffer, M. L. 1981. Minimum population sizes for species conservation. BioScience **31**:131–134.

Simberloff, D. 1988. The contribution of population and community biology to conservation science. Annual Review of Ecology and Systematics **19**:473–511.

Simpson, G. G. 1961. Principles of animal taxonomy. Columbia University Press, New York.

Soulé, M. E. 1989. Conservation biology in the twenty-first century: summary and outlook. Pages 297–303 in D. Western and M. C. Pearl, editors. Conservation for the twenty-first century. Oxford University Press, New York.

Soulé, M. E. and K. A. Kohm, editors. 1989. Research priorities for conservation biology. Island Press Critical Issues Series #1, Washington, D.C.

Stevens, P. F. 1988. New Guinea. Pages 120–132 in D. G. Campbell and H. D. Hammond, editors. Floristic inventory of tropical countries. New York Botanic Garden, New York.

Stevens, P. F. 1990. Nomenclatural stability, taxonomic instinct, and flora writing—a recipe for disaster? Pages 387–410 in P. Baas et al., editors. The plant diversity of Malesia. Kluwer Academic Publishers, The Netherlands.

Stevens, P. F. 1992. Species. In E. F. Keller and E. Lloyd, editors. Keywords in evolutionary biology. Harvard University Press, Cambridge, Massachusetts. In press.

Templeton, A. R. 1986. Coadaptation and outbreeding depression. Pages 105–116 in M. E. Soulé, editor. Conservation biology: the science of scarcity and diversity. Sinauer Associates, Inc., Sunderland, Massachusetts.

Templeton, A. R. 1989. The meaning of species and speciation: a genetic perspective. Pages 3–27 in D. Otte and J. A. Endler, editors. Speciation and its consequences. Sinauer Associates, Inc., Sunderland, Massachusetts.

Van Valen, L. 1969. Ecological species, multispecies and oaks. Taxon **25**:233–239.

Vane-Wright, R. I., C. J. Humphries, and P. H. Williams. 1991. What to protect?—systematics and the agony of choice. Biological Conservation **55**:235–254.

Vrijenhoek, R. C. 1989. Population genetics and conservation. Pages 89–98 in D. Western and M. C. Pearl, editors. Conservation for the twenty-first century. Oxford University Press, New York.

Walker, B. 1989. Diversity and stability in ecosystem conservation. Pages 121–130 in D. Western and M. C. Pearl, editors. Conservation for the twenty-first century. Oxford University Press, New York.

Wettemberg, G. B., G. T. Prance, and T. Lovejoy. 1981. Conservation progress in Amazonia: a structural review. Parks **6(2)**:5–10.

Wilcox, B. A. 1984. In situ conservation of genetic resources: determinants of minimum area requirements. Pages 639–647 in J. A. McNeely and K. R. Miller, editors. National parks, conservation, and development: the role of protected areas in sustaining society. Proceedings of The World Congress on National Parks, Bali, Indonesia, 11–22 October 1982. Smithsonian Institution Press, Washington, D.C.

Wiley, E. O. 1978. The evolutionary species concept reconsidered. Systematic Zoology **27**:17–26.

Wilson, E. O. 1989. Conservation: the next hundred years. Pages 3–7 in D. Western and M. C. Pearl, editors. Conservation for the twenty-first century. Oxford University Press, New York.

Woodruff, D. S. 1989. The problems of conserving genes and species. Pages 76–88 in D. Western and M. C. Pearl, editors. Conservation for the twenty-first century. Oxford University Press, New York.

Essay

Assessing Extinction Threats: Toward a Reevaluation of IUCN Threatened Species Categories

GEORGINA M. MACE

Institute of Zoology
Zoological Society of London
Regent's Park, London NW1 4RY, U.K.

RUSSELL LANDE

Department of Ecology and Evolution
University of Chicago
Chicago, Illinois 60637, U.S.A.

Abstract: *IUCN categories of threat (Endangered, Vulnerable, Rare, Indeterminate, and others) are widely used in 'Red lists' of endangered species and have become an important tool in conservation action at international, national, regional, and thematic levels. The existing definitions are largely subjective, and as a result, categorizations made by different authorities differ and may not accurately reflect actual extinction risks. We present proposals to redefine categories in terms of the probability of extinction within a specific time period, based on the theory of extinction times for single populations and on meaningful time scales for conservation action. Three categories are proposed (CRITICAL, ENDANGERED, VULNERABLE) with decreasing levels of threat over increasing time scales for species estimated to have at least a 10% probability of extinction within 100 years. The process of assigning species to categories may need to vary among different taxonomic groups, but we present some simple qualitative criteria based on population biology theory, which we suggest are appropriate at least for most large vertebrates. The process of assessing threat is clearly distinguished from that of setting priorities for conservation action, and only the former is discussed here.*

Resumen: *La categorización de la Unión Internacional para la Conservación de la Naturaleza (UICN) de las especies amenazadas (en peligro, vulnerables, raras, indeterminadas y otras) son ampliamente utilizadas en las Listas Rojas de especies en peligro y se han convertido en una herramienta importante para las acciones de conservación al nivel internacional, nacional, regional y temático. Las definiciones de las categorías existentes son muy subjetivas y, como resultado, las categorizaciones hechas por diferentes autores difieren y quizás no reflejen con certeza el riesgo real de extinción. Presentamos propuestas para re-definir las categorías en términos de la probabilidad de extinción dentro de un período de tiempo específico. Las propuestas están basadas en la teoría del tiempo de extinción para poblaciones individuales y en escalas de tiempo que tengan significado para las acciones de conservación. Se proponen tres categorías (CRITICA, EN PELIGRO, VULNERABLE) con niveles decrecientes de amenaza sobre escalas de tiempo en aumento para especies que se estima tengan cuando ménos un 10% de probabilidad de extinción en 100 años. El proceso de asignar especies a categorías puede que necesite variar dentro de los diferentes grupos taxonómicos pero nosotros presentamos algunos criterios cualitativos simples basados en la teoría de la biología de las poblaciones, las cuales sugerimos son apropiadas para cuando ménos la mayoría de los grandes vertebrados. El proceso de evaluar la amenaza se distingue claramente del de definir las prioridades para las acciones de conservación, sólamente el primero se discute aquí.*

Paper submitted February 12, 1990; revised manuscript accepted October 8, 1990.

Introduction

Background

The Steering Committee of the Species Survival Commission (SSC) of the IUCN has initiated a review of the overall functioning of the Red Data Books. The review will cover three elements: (1) the form, format, content, and publication of Red Data Books; (2) the categories of threat used in Red Data Books and the IUCN Red List (Extinct, Endangered, Vulnerable, Rare, and Indeterminate); and (3) the system for assigning species to categories. This paper is concerned with the second element and includes proposals to improve the objectivity and scientific basis for the threatened species categories currently used in Red Data Books (see IUCN 1988 for current definitions).

There are at least three reasons why a review of the categorization system is now appropriate: (1) the existing system is somewhat circular in nature and excessively subjective. When practiced by a few people who are experienced with its use in a variety of contexts it can be a robust and workable system, but increasingly, different groups with particular regional or taxonomic interests are using the Red Data Book format to develop local or specific publications. Although this is generally of great benefit, the interpretation and use of the present threatened species categories are now diverging widely. This leads to disputes and uncertainties over particular species that are not easily resolved and that ultimately may negatively affect species conservation. (2) Increasingly, the categories of threat are being used in setting priorities for action, for example, through specialist group action plans (e.g., Oates 1986; Eudey 1988; East 1988, 1989; Schreiber et al. 1989). If the categories are to be used for planning then it is essential that the system used to establish the level of threat be consistent and clearly understood, which at present it does not seem to be. (3) A variety of recent developments in the study of population viability have resulted in techniques that can be helpful in assessing extinction risks.

Assessing Threats Versus Setting Priorities

In the first place it is important to distinguish systems for assessing threats of extinction from systems designed to help set priorities for action. The categories of threat should simply provide an assessment of the likelihood that if current circumstances prevail the species will go extinct within a given period of time. This should be a scientific assessment, which ideally should be completely objective. In contrast, a system for setting priorities for action will include the likelihood of extinction, but will also embrace numerous other factors, such as the likelihood that restorative action will be successful; economic, political, and logistical considerations; and perhaps the taxonomic distinctiveness of the species under review. Various categorization systems used in the past, and proposed more recently, have confounded these two processes (see Fitter & Fitter 1987; Munton 1987). To devise a general system for setting priorities is not useful because different concerns predominate within different taxonomic, ecological, geographical, and political units. The process of setting priorities is therefore best left to specific plans developed by specialist bodies such as the national and international agencies, the specialist groups, and other regional bodies that can devise priority assessments in the appropriate regional or taxonomic context. An objective assessment of extinction risk may also then contribute to the decisions taken by governments on which among a variety of recommendations to implement. The present paper is therefore confined to a discussion of assessing threats.

Aims of the System of Categorization

For Whom?

Holt (1987) identifies three different groups whose needs from Red Data Books (and therefore categories of threat) may not be mutually compatible: the lay public, national and international legislators, and conservation professionals. In each case the purpose is to highlight taxa with a high extinction risk, but there are differences in the quality and quantity of information needed to support the assessment. Scott et al. (1987) make the point that in many cases simple inclusion in a Red Data Book has had as much effect on raising awareness as any of the supporting data (see also Fitter 1974). Legislators need a simple, but objective and soundly based system because this is most easily incorporated into legislation (Bean 1987). Legislators frequently require some statement about status for every case they consider, however weak the available information might be. Inevitably, therefore, there is a conflict between expediency and the desire for scientific credibility and objectivity. Conservationists generally require more precision, particularly if they are involved in planning conservation programs that aim to make maximal use of limited resources.

Characteristics of an Ideal System

With this multiplicity of purposes in mind it is appropriate to consider various characteristics of an ideal system:

(1) The system should be essentially simple, providing easily assimilated data on the risk of extinction. In terms of assessing risk, there seems to be little virtue in developing numerous categories, or in categorizing risk on the basis of a range of different parameters (e.g., abundance, nature of threat, likelihood of persistence of threat, etc.). The categories should be few in number,

should have a clear relationship to one another (Holt 1987; Munton 1987), and should be based around a probabilistic assessment of extinction risk.

(2) The system for categorization has to be flexible in terms of data required. The nature and amount of data available to assess extinction risks varies widely from almost none (in the vast majority of species) to highly detailed population data (in a very few cases). The categorization system should make maximum use of whatever data are available. One beneficial consequence of this process would be to identify key population data for field workers to collect that would be useful in assessing extinction risk.

(3) The categorization system also needs to be flexible in terms of the population unit to which it applies. Throughout this discussion, it is assumed that the system being developed will apply to any species, subspecies, or geographically separate population. The categorization system therefore needs to be equally applicable to limited lower taxonomic levels and to more limited geographical scope. Action planning will need to be focused on particular taxonomic groups or geographical areas, and can then incorporate an additional system for setting priorities that reflect taxonomic distinctiveness and extinction risks outside the local area (e.g., see East 1988, 1989; Schreiber et al. 1989).

(4) The terminology used in categorization should be appropriate, and the various terms used should have a clear relationship to each other. For example, among the current terms both 'endangered' and 'vulnerable' are readily comprehended, but 'rare' is confusing. It can be interpreted as a statement about distribution status, level of threat, or local population size, and the relationships between these factors are complex (Rabinowitz et al. 1986). Rare (i.e., low-density) species are not always at risk and many species at risk are not numerically rare (King 1987; Munton 1987; Heywood 1988). The relationship of 'rare' to 'endangered' and 'vulnerable' is also unclear.

(5) If the system is to be objectively based upon sound scientific principles, it should include some assessment of uncertainty. This might be in terms of confidence levels, sensitivity analyses, or, most simply, on an ordinal scale reflecting the adequacy of the data and models in any particular case.

(6) The categories should incorporate a time scale. On a geological time scale all species are doomed to extinction, so terms such as "in danger of extinction" are rather meaningless. The concern we are addressing here is the high background level of the current rates of extinction, and one aim is therefore preservation over the upcoming centuries (Soulé & Simberloff 1986). Therefore, the probability of extinction should be expressed in terms of a finite time scale, for example, 100 years. Munton (1987) suggests using a measure of number of years until extinction. However, since most models of population extinction times result in approximately exponential distributions, as in Goodman's (1987) model of density-dependent population growth in a fluctuating environment, mean extinction time may not accurately reflect the high probability that the species will go extinct within a time period considerably shorter than the mean (see Fig. 1). More useful are measures such as "95% likelihood of persistence for 100 years."

Population Viability Analysis and Extinction Factors

Various approaches to defining viable populations have been taken recently (Shaffer 1981, 1990; Gilpin & Soulé, 1986; Soulé 1987). These have emphasized that there is no simple solution to the question of what constitutes a viable population. Rather, through an analysis of extinction factors and their interactions it is possible to assess probabilities and time scales for population persistence for a particular taxon at a particular time and place. The development of population viability analyses has led to the definition of intrinsic and extrinsic factors that determine extinction risks (see Soulé 1983; Soulé 1987; Gilpin & Soulé 1986; see also King 1987). Briefly these can be summarized as population dynamics (number of individuals, life history and age or stage distribution, geographic structure, growth rate, variation in demographic parameters), population characteristics (morphology, physiology, genetic variation, behavior and dispersal patterns), and environmental effects (habitat quality and quantity, patterns and rates of environmental disturbance and change, interactions with other species including man).

Preliminary models are available to assess a population's expected persistence under various extinction pressures, for example, demographic variation (Goodman 1987*a*, *b*; Belovsky 1987; CBSG 1989), catastrophes (Shaffer 1987), inbreeding and loss of genetic diversity (Lande & Barrowclough 1987; Lacy 1987), metapopulation structure (Gilpin 1987; Quinn & Hastings 1987; Murphy et al. 1990). In addition, various approaches have been made to modeling extinction in populations threatened by habitat loss (e.g., Gutiérrez & Carey 1985; Maguire et al. 1987; Lande 1988), disease (e.g., Anderson & May 1979; Dobson & May 1986; Seal et al. 1989), parasites (e.g., May & Anderson 1979; May & Robinson 1985; Dobson & May 1986), competitors, poaching (e.g., Caughley 1988), and harvesting or hunting (e.g., Holt 1987).

So far, the development of these models has been rather limited, and in particular they often fail to successfully incorporate several different extinction factors and their interactions (Lande 1988). Nevertheless the approach has been applied in particular cases even with existing models (e.g., grizzly bear: Shaffer 1983; spotted owl: Gutiérrez & Carey 1985; Florida panther: CBSG 1989), and there is much potential for further development.

Although different extinction factors may be critical for different species, other, noncritical factors cannot be ignored. For example, it seems likely that for many species, habitat loss constitutes the most immediate threat. However, simply preserving habitats may not be sufficient to permit long term persistence if surviving populations are small and subdivided and therefore have a high probability of extinction from demographic or genetic causes. Extinction factors may also have cumulative or synergistic effects; for example, the hunting of a species may not have been a problem before the population was fragmented by habitat loss. In every case, therefore, all the various extinction factors and their interactions need to be considered. To this end more attention needs to be directed toward development of models that reflect the random influences that are significant to most populations, that incorporate the effects of many different factors, and that relate to the many plant, invertebrate, and lower vertebrate species whose population biology has only rarely been considered so far by these methods.

Viability analysis should suggest the appropriate kind of data for assigning extinction risks to species, though much additional effort will be needed to develop appropriate models and collect appropriate field data.

Proposal

Three Categories and Their Justification

We propose the recognition of three categories of threat (plus EXTINCT), defined as follows:

CRITICAL:	50% probability of extinction within 5 years or 2 generations, whichever is longer.
ENDANGERED:	20% probability of extinction within 20 years or 10 generations, whichever is longer.
VULNERABLE:	10% probability of extinction within 100 years.

These definitions are based on a consideration of the theory of extinction times for single populations as well as on meaningful time scales for conservation action. If biological diversity is to be maintained for the foreseeable future at anywhere near recent levels occurring in natural ecosystems, fairly stringent criteria must be adopted for the lowest level of extinction risk, which we call VULNERABLE. A 10% probability of extinction within 100 years has been suggested as the highest level of risk that is biologically acceptable (Shaffer 1981) and seems appropriate for this category. Furthermore, events more than about 100 years in the future are hard to foresee, and this may be the longest duration that legislative systems are capable of dealing with effectively.

It seems desirable to establish a CRITICAL category to emphasize that some species or populations have a very high risk of extinction in the immediate future. We propose that this category include species or populations with a 50% chance of extinction within 5 years or two generations, and which are clearly at very high risk.

An intermediate category, ENDANGERED, seems desirable to focus attention on species or populations that are in substantial danger of extinction within our lifetimes. A 20% chance of extinction within 20 years or 10 generations seems to be appropriate in this context.

For increasing levels of risk represented by the categories VULNERABLE, ENDANGERED, and CRITICAL, it is necessary to increase the probability of extinction or to decrease the time scale, or both. We have chosen to do both for the following reasons. First, as already mentioned, decreasing the time scale emphasizes the immediacy of the situation. Ideally, the time scale should be expressed in natural biological units of generation time of the species or population (Leslie 1966), but there is also a natural time scale for human activities such as conservation efforts, so we have given time scales in years and in generations for the CRITICAL and ENDANGERED categories.

Second, the uncertainty of estimates of extinction probabilities decreases with increasing risk levels. In population models incorporating fluctuating environments and catastrophes, the probability distribution of extinction times is approximately exponential (Nobile et al. 1985; Goodman 1987). In a fluctuating environment where a population can become extinct only through a series of unfavorable events, there is an initial, relatively brief period in which the chance of extinction is near zero, as in the inverse Gaussian distribution of extinction times for density-independent fluctuations (Ginzburg et al. 1982; Lande & Orzack 1988). If catastrophes that can extinguish the population occur with probability p per unit time, and are much more important than normal environmental fluctuations, the probability distribution of extinction times is approximately exponential, pe^{-pt}, and the cumulative probability of extinction up to time t is approximately $1 - e^{-pt}$. Thus, typical probability distributions of extinction times look like the curves in Figures 1A and 1B, and the cumulative probabilities of extinction up to any given time look like the curves in Figures 1C and 1D. Dashed curves represent different distributions of extinction times and cumulative extinction probabilities obtained by changing the model parameters in a formal population viability analysis (e.g., different amounts of environmental variation in demographic parameters). The uncertainty in an estimate of cumulative extinction probability up to a certain time can be measured by its coefficient of variation, that is, the standard deviation among different estimates of the cumulative extinction probability with respect to reasonable variation in model parameters, divided by the best estimate. It is apparent from Figures 1C and 1D that at least for small variations in the parameters (if the parameters are reasonably well known), the uncertainty of estimates of cumulative extinction probability at particular times decreases as the level of risk increases. Thus at times, t_1, t_2, and t_3 when the best estimates of the cumulative extinction probabilities are 10%, 20%, and 50% respectively, the corresponding ranges of extinction probabilities in Figure 1C are 6.5%–14.8%, 13.2%–28.6%, and 35.1%–65.0%, and in Figure 1D are 6.8%–13.1%, 13.9%–25.7%, and 37.2%–60.2%. Taking half the range as a rough approximation of the standard deviation in this simple illustration gives uncertainty measures of 0.41, 0.38, and 0.30 in Figure 1C, and 0.31, 0.29, and 0.23 in Figure 1D, corresponding to the three levels of risk. Given that for practical reasons we have chosen to shorten the time scales for the more threatened categories, these results suggest that to maintain low levels of uncertainty, we should also increase the probabilities of extinction in the definition of the ENDANGERED and CRITICAL categories.

These definitions are based on general principles of population biology with broad applicability, and we believe them to be appropriate across a wide range of life forms. Although we expect the process of assigning species to categories (see below) to be an evolving (though closely controlled and monitored) process, and one that might vary across broad taxonomic groups, we recommend that the definitions be constant both across taxonomic groups and over time.

Assigning Species or Populations to Categories

We recognize that in most cases, there are insufficient data and imperfect models on which to base a formal probabilistic analysis. Even when considerable information does exist there may be substantial uncertainties in the extinction risks obtained from population models containing many parameters that are difficult to estimate accurately. Parameters such as environmental stochasticity (temporal fluctuations in demographic parameters such as age- or developmental stage–specific mortality and fertility rates), rare catastrophic events, as well as inbreeding depression and genetic variability in particular characters required for adaptation are all difficult to estimate accurately. Therefore it may not be possible to do an accurate probabilistic viability analysis even for some very well studied species. We suggest that the categorization of many species should be based on more qualitative criteria derived from the same body of theory as the definitions above, which will broaden the scope and applicability of the categorization system. In these more qualitative criteria we use measures of effective population size (N_e) and give approximate equivalents in actual population size (N). It is important to recognize that the relationship between N_e and N depends upon a variety of interacting factors. Estimating N_e for a particular population will require quite extensive information on breeding structure and life history characteristics of the population and may then produce only an approximate figure (Lande & Barrowclough 1987). In addition, different methods of estimating N_e will give variable results (Harris & Allendorf 1989). N_e/N ratios vary widely across species, but are typically in the range 0.2 to 0.5. In the criteria below we give a value for N_e as well as an approximate value of N assuming that the N_e/N ratio is 0.2.

We suggest the following criteria for the three categories:

CRITICAL: 50% probability of extinction within 5 years or 2 generations, whichever is longer, or

(1) Any **two** of the following criteria:
- (a) Total population $N_e < 50$ (corresponding to actual $N < 250$).
- (b) Population fragmented: ≤2 subpopulations with $N_e > 25$ ($N > 125$) with immigration rates <1 per generation.
- (c) Census data of >20% annual decline in numbers over the past 2 years, or >50% decline in the last generation, or equivalent projected declines based on demographic projections after allowing for known cycles.
- (d) Population subject to catastrophic crashes (>50% reduction) per 5 to 10 years, or 2 to 4 generations, with subpopulations highly correlated in their fluctuations.

or (2) Observed, inferred, or projected habitat alteration (i.e., degradation, loss, or fragmentation) resulting in characteristics of (1).

or (3) Observed, inferred, or projected commercial exploitation or ecological interactions with introduced species (predators, competitors, pathogens, or parasites) resulting in characteristics of (1).

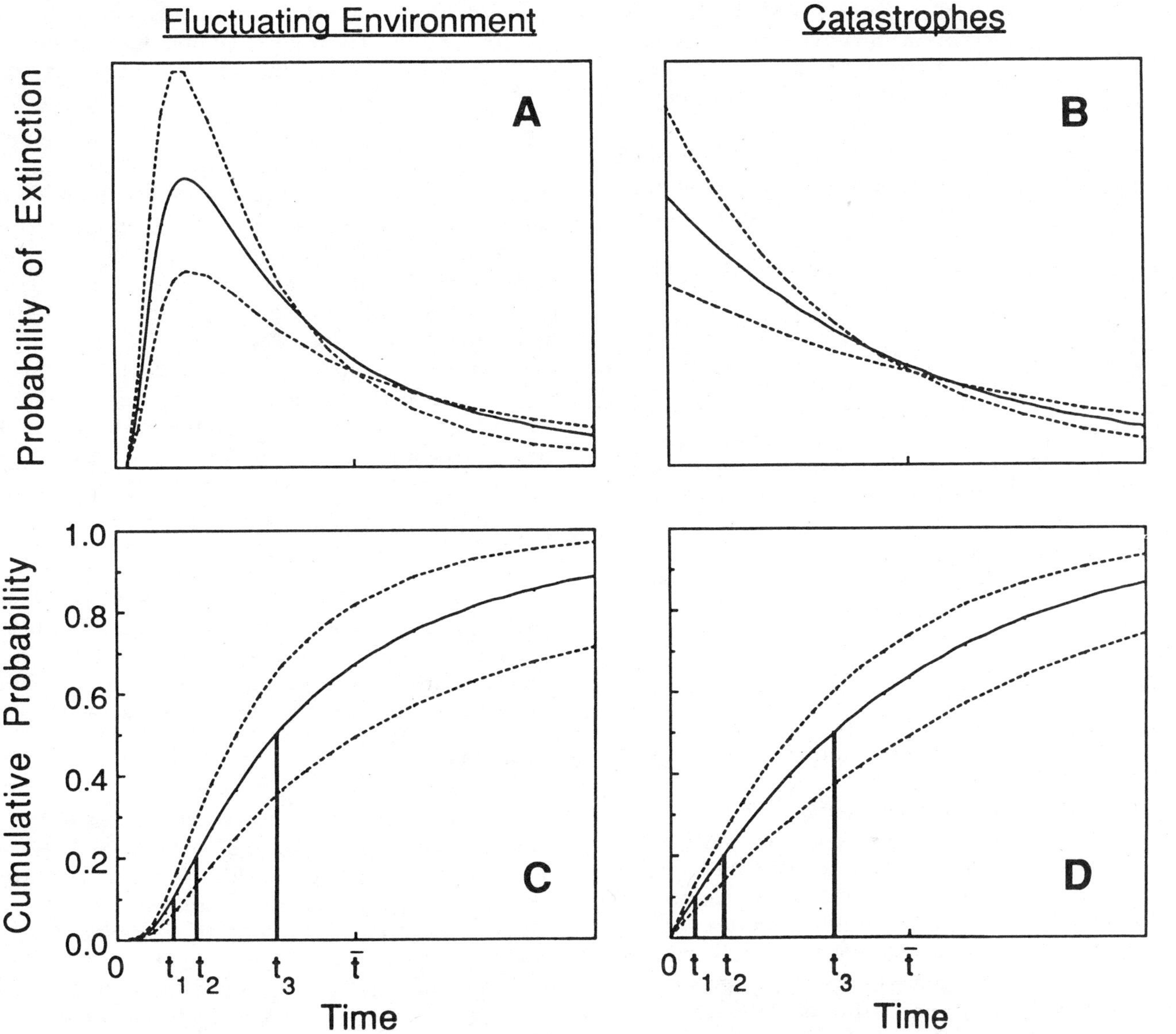

Figure 1. Probability distributions of time to extinction in a fluctuating environment, inverse Gaussian distributions (A), or with catastrophes, exponential distributions (B). Corresponding cumulative extinction probabilities of extinction up to any given time are shown below (C and D). Solid curves represent the best estimates from available data and dashed curves represent different estimates based upon the likely range of variation in the parameters. t_1, t_2, and t_3 are times at which the best estimates of cumulative extinction probabilities are 10%, 20%, and 50%. $\bar{t}$ is the expected time to extinction in the solid curves.

ENDANGERED:

20% probability of extinction within 20 years or 10 generations, whichever is longer, or

(1) Any **two** of the following or any **one** criterion under CRITICAL

(a) Total population $N_e < 500$ (corresponding to actual $N < 2{,}500$).

(b) Population fragmented:

(i) ≤ 5 subpopulations with $N_e > 100$ ($N > 500$) with immigration rates <1 per generation, or

(ii) ≤ 2 subpopulations with $N_e > 250$ ($N > 1{,}250$) with immigration rates <1 per generation.

(c) Census data of $>5\%$ annual decline in numbers over past 5 years, or $>10\%$ decline per generation over past 2 generations, or equivalent projected declines based on demographic data after allowing for known cycles.

(d) Population subject to catastrophic crashes: an average of >20% reduction per 5 to 10 years or 2 to 4 generations, or >50% reduction per 10 to 20 years or 5 to 10 generations, with subpopulations strongly correlated in their fluctuations.

or (2) Observed, inferred, or projected habitat alteration (i.e., degradation, loss, or fragmentation) resulting in characteristics of (1).

or (3) Observed, inferred, or projected commercial exploitation or ecological interactions with introduced species (predators, competitors, pathogens, or parasites) resulting in characteristics of (1).

VULNERABLE:

10% probability of extinction within 100 years, or

(1) Any **two** of the following criteria or any **one** criterion under ENDANGERED.

(a) Total population $N_e < 2{,}000$ (corresponding to actual $N < 10{,}000$).

(b) Population fragmented:
(i) ≤5 subpopulations with $N_e > 500$ ($N > 2{,}500$) with immigration rates <1 per generation, or
(ii) ≤2 subpopulations with $N_e > 1{,}000$ ($N > 5{,}000$) with immigration rates <1 per generation.

(c) Census data of >1% annual decline in numbers over past 10 years, or equivalent projected declines based on demographic data after allowing for known cycles.

(d) Population subject to catastrophic crashes: an average of >10% reduction per 5 to 10 years, >20% reduction per 10 to 20 years, or >50% reduction per 50 years, with subpopulations strongly correlated in their fluctuations.

or (2) Observed, inferred, or projected habitat alteration (i.e., degradation, loss, or fragmentation) resulting in characteristics of (1).

or (3) Observed, inferred, or projected commercial exploitation or ecological interactions with introduced species (predators, competitors, pathogens, or parasites) resulting in characteristics of (1).

Prior to any general acceptance, we recommend that these criteria be assessed by comparison of the categorizations they lead to in particular cases with the results of formal viability analyses, and categorizations based on existing methods. This process should help to resolve uncertainties about both the practice of, and results from, our proposals. We expect a system such as this to be relatively robust and of widespread applicability, at the very least for most higher vertebrates. For some invertebrate and plant taxa, different kinds of criteria will need to be developed within the framework of the definitions above. For example, many of these species have very high rates of population growth, short generation times, marked or episodic fluctuations in population size, and high habitat specificity. Under these circumstances, it will be more important to incorporate metapopulation characteristics such as subpopulation persistence times, colonization rates, and the distribution and persistence of suitable habitats into the analysis, which are less significant for most large vertebrate populations (Murphy et al. 1990; Menges 1990).

Change of Status

The status of a population or species with respect to risk of extinction should be up-listed (from unlisted to VULNERABLE, from VULNERABLE to ENDANGERED, or from ENDANGERED to CRITICAL) as soon as current information suggests that the criteria are met. The status of a population or species with respect to risk of extinction should be down-listed (from CRITICAL to ENDANGERED, from ENDANGERED to VULNERABLE, or from VULNERABLE to unlisted) only when the criteria of the lower risk category have been satisfied for a time period equal to that spent in the original category, or if it is shown that past data were inaccurate.

For example, if an isolated population is discovered consisting of 500 individuals and no other information is available on its demography, ecology, or the history of the population or its habitat, this population would initially be classified as ENDANGERED. If management efforts, natural events, or both caused the population to increase so that 10 years later it satisfied the criteria of the VULNERABLE category, the population would not be removed from the ENDANGERED category for a further period of 10 years. This time lag in down-listing prevents frequent up-listing and down-listing of a population or species.

Uncertain or Conflicting Results

Because of uncertainties in parameter estimates, especially those dealing with genetics and environmental variability and catastrophes, substantial differences may arise in the results from analyses of equal validity performed by different parties. In such cases, we recommend that the criteria for categorizing a species or population should revert to the more qualitative ones outlined above.

Reporting Categories of Threat

To objectively compare categorizations made by different investigators and at different times, we recommend that any published categorization also cite the method used, the source of the data, a date when the data were accurate, and the name of the investigator who made the categorization. If the method was by a formal viability model, then the name and version of the model used should also be included.

Conclusion

Any system of categorizing degrees of threat of extinction inevitably contains arbitrary elements. No single system can adequately cover every possibility for all species. The system we describe here has the advantage of being based on general principles from population biology and can be used to categorize species for which either very little or a great deal of information is available. Although this system may be improved in the future, we feel that its use will help to promote a more uniform recognition of species and populations at risk of premature extinction, and should thereby aid in setting priorities for conservation efforts.

Summary

1. Threatened species categories should highlight species vulnerable to extinction and focus appropriate reaction. They should therefore aim to provide objective, scientifically based assessments of extinction risks.
2. The audience for Red Data Books is diverse. Positive steps to raise public awareness and implement national and international legislation benefit from simple but soundly based categorization systems. More precise information is needed for planning by conservation bodies.
3. An ideal system needs to be simple but flexible in terms of data required. The category definitions should be based on a probabilistic assessment of extinction risk over a specified time interval, including an estimate of error.
4. Definitions of categories are appropriately based on extinction probabilities such as those arising from population viability analysis methods.
5. We recommend three categories, CRITICAL, ENDANGERED, and VULNERABLE, with decreasing probabilities of extinction risk over increasing time periods.
6. For most cases, we recommend development of more qualitative criteria for allocation to categories based on basic principles of population biology. We present some criteria that we believe to be appropriate for many taxa, but are appropriate at least for higher vertebrates.

Acknowledgments

We would like to acknowledge the support and encouragement of Simon Stuart, Steven Edwards, and Ulysses Seal in the preparation of this paper. We are also very grateful to the many members of the SSC network for the time they put into commenting upon earlier drafts of this paper, and only regret that they are too numerous to mention individually.

Literature Cited

Anderson, R. M., and R. M. May. 1979. Population biology of infectious diseases. Part I. Nature **280**:361–367.

Bean, M. J. 1987. Legal experience and implications. Pages 39–43 in R. Fitter and M. Fitter, editors. The road to extinction. IUCN, Gland, Switzerland.

Belovsky, G. E. 1987. Extinction models and mammalian persistence. Pages 35–57 in M. E. Soulé, editor. Viable populations for conservation. Cambridge University Press, Cambridge, England.

Caughley, G. 1988. A projection of ivory production and its implications for the conservation of African elephants. CSIRO consultancy report to CITES. CSIRO Division of Wildlife and Ecology.

CBSG. 1989. Florida panther: population viability analysis. IUCN/SSC/CBSG: Apple Valley, Minneapolis, Minnesota.

Cumming, D. H. M., R. F. du Toit, and S. N. Stuart. 1989. African elephants and rhinos: status, survey and conservation action plan. IUCN, Gland, Switzerland.

Dobson, A. P., and R. M. May. 1986. Disease and conservation. Pages 345–365 in M. Soulé, editor. Conservation biology—the science of scarcity and diversity. Sinauer Associates, Sunderland, Massachusetts.

Dobson, A. P., and D. Miller. 1989. Infectious diseases and endangered species management. Endangered Species Update 6(9):1–5.

East, R. 1988. Antelopes: global survey and regional action plans. Part 1. east and north east Africa. IUCN, Gland, Switzerland.

East, R. 1989. Antelopes: global survey and regional action plans. Part 2. southern and south central Africa. IUCN, Gland, Switzerland.

Eudey, A. 1988. Action plan for Asian primate conservation. IUCN/SSC, Gland, Switzerland.

Fitter, R. F. 1974. 25 years on: a look at endangered species. Oryx **12**:341–346.

Fitter, R., and M. Fitter, editors. 1987. The road to extinction. IUCN, Gland, Switzerland.

Fuller, W. A. 1987. Synthesis and recommendations. Pages 47–55 in R. Fitter and M. Fitter, editors. The road to extinction. IUCN, Gland, Switzerland.

Gilpin, M.E. 1987. Spatial structure and population vulnerability. Pages 125–139 in M. E. Soulé, editor. Viable populations for conservation. Cambridge University Press, Cambridge, England.

Gilpin, M. E., and M. E. Soulé. 1986. Minimum viable populations: processes of species extinctions. Pages 19–34 in M. E. Soulé, editor. Conservation biology—the science of scarcity and diversity. Sinauer Associates, Sunderland, Massachusetts.

Ginzburg, L. R., L. B. Slobodkin, K. Johnson, and A. G. Bindman. 1982. Quasiextinction probabilities as a measure of impact on population growth. Risk Analysis **2**:171–181.

Goodman, D. 1987*a*. The demography of chance extinction. Pages 11–34 in M. E. Soulé, editor. Viable populations for conservation. Cambridge University Press, Cambridge, England.

Goodman, D. 1987*b*. How do any species persist? Lessons for conservation biology. Conservation Biology **1**:59–62.

Gutiérrez, R. J., and A. B. Carey, editors. 1985. Ecology and management of the Spotted Owl in the Pacific Northwest. General Technical Report PNW-185, USDA Forest Service, Pacific Northwest Station, Portland, Oregon.

Harris, R. B., and F. W. Allendorf. 1989. Genetically effective population size of large mammals: an assessment of estimators. Conservation Biology **3**:181–191.

Heywood, V. H. 1988. Rarity: a privilege and a threat. Pages 277–290 in W. Greuter and B. Zimmer, editors. Proceedings of the XIV International Botanical Congress Koeltz, Konigstein/Taunus.

Holt, S. J. 1987. Categorization of threats to and status of wild populations. Pages 19–30 in R. Fitter and M. Fitter, editors. The road to extinction. IUCN, Gland, Switzerland.

IUCN. 1988. 1988 IUCN red list of threatened animals IUCN, Gland, Switzerland.

King, F. W. 1987. Thirteen milestones on the road to extinction. Pages 7–18 in R. Fitter and M. Fitter, editors. The road to extinction. IUCN, Gland, Switzerland.

Lacy, R. C. 1987. Loss of genetic diversity from managed populations: interacting effects of drift, mutation, immigration, selection and population subdivision. Conservation Biology **1**:143–157.

Lande, R. 1988. Genetics and demography in biological conservation. Science **241**:1455–1460.

Lande, R., and G. F. Barrowclough. 1987. Effective population size, genetic variation and their use in population management. Pages 87–123 in M. E. Soulé, editor. Viable populations for conservation. Cambridge University Press, Cambridge, England.

Lande, R., and S. H. Orzack. 1988. Extinction dynamics of age-structured populations in a fluctuating environment. PNAS **85**:7418–7421.

Leslie, P. H. 1966. Journal of Animal Ecology **25**:291.

Maguire, L. A., U. S. Seal, and P. F. Brussard. 1987. Managing critically endangered species: the Sumatran rhino as an example. Pages 141–158 in M. E. Soulé, editor. Viable populations for conservation. Cambridge University Press, Cambridge, England.

May, R. M., and R. M. Anderson. 1979. Population biology of infectious diseases. Part II. Nature **280**:455–461.

May, R. M., and S. K. Robinson. 1985. Population dynamics of avian brood parasitism. American Naturalist **126**:475–494.

Menges, E. S. 1990. Population viability analysis for an endangered plant. Conservation Biology **4**:52–62.

Munton, P. 1987. Concepts of threat to the survival of species used in Red Data books and similar compilations. Pages 72–95 in R. Fitter and M. Fitter, editors. The road to extinction. IUCN, Gland, Switzerland.

Murphy, D. D., K. E. Freas, and S. B. Weiss. 1990. An environment-metapopulation approach to population viability analysis for a threatened invertebrate. Conservation Biology **4**:41–51.

Nobile, A. G., L. M. Ricciardi, and L. Sacerdote. 1985. Exponential trends of first passage-time densities for a class of diffusion processes with steady-state distribution. J. Appl. Probab. **22**:611–618.

Oates, J. F. 1986. Action plan for African primate conservation: 1986–1990. IUCN/SSC, Gland, Switzerland.

Quinn, J. F., and A. Hastings. 1987. Extinction in subdivided habitats. Conservation Biology **1**:198–208.

Rabinowitz, D., S. Cairns, and T. Dillon. 1986. Seven forms of rarity and their frequency in the flora of the British Isles. Pages 182–204 in M. E. Soulé, editor. Conservation biology—the science of scarcity and diversity. Sinauer Associates, Sunderland, Massachusetts.

Schreiber, A., R. Wirth, M. Riffel, and H. von Rompaey. 1989. Weasels, civets, mongooses and their relations: an action plan for the conservation of mustelids and viverrids. IUCN, Gland, Switzerland.

Scott, P., J. A. Burton, and R. Fitter. 1987. Red Data Books: the historical background. Pages 1–5 in R. Fitter and M. Fitter, editors. The road to extinction. IUCN, Gland, Switzerland.

Seal, U. S., E. T. Thorne, M. A. Bogan, and S. H. Anderson. 1989. Conservation biology and the black-footed ferret. Yale University Press, New Haven, Connecticut.

Shaffer, M. L. 1981. Minimum population sizes for species conservation. Bioscience **31**:131–134.

Shaffer, M. L. 1983. Determining minimum viable population sizes for the grizzly bear. Int. Conf. Bear Res. Manag. **5**:133–139.

Shaffer, M. L. 1987. Minimum viable populations; coping with uncertainty. Pages 69–86 in M. E. Soulé, editor. Viable populations for Conservation. Cambridge University Press, Cambridge, England.

Shaffer, M. L. 1990. Population viability analysis. Conservation Biology **4**:39–40.

Soulé, M. E. 1983. What do we really know about extinction? Pages 111–124 in C. Schonewald-Cox, S. Chambers, B. MacBryde, and L. Thomas. Genetics and conservation. Benjamin/Cummings, Menlo Park, California.

Soulé, M. E., editor. 1987. Viable populations for conservation. Cambridge University Press, Cambridge, England.

Soulé, M. E., and D. Simberloff. 1986. What do ecology and genetics tell us about the design of nature reserves? Biological Conservation **35**:19–40.

A Possible Method for the Rapid Assessment of Biodiversity

IAN OLIVER
ANDREW J. BEATTIE

School of Biological Sciences
Macquarie University
New South Wales 2109, Australia

Abstract: *Inventories of vertebrate and flowering plants are frequently used as surrogates for estimates of total biodiversity. This is in part because the inclusion of invertebrates and nonflowering plants is perceived as being too time-consuming, costly, and difficult because of the shortage of specialists. Estimates of the species richness of field samples of spiders, ants, polychaetes, and mosses made by a biodiversity technician and by specialist taxonomists were compared. The biodiversity technician received a few hours training in the taxonomy of each group and separated specimens into recognizable taxonomic units (RTUs). The specialists sorted to species. For the three animal groups the biodiversity technician recorded 165 taxa and the specialists 147, with the error for the ants and spiders being 13% or less. A small amount of splitting and lumping of species was detected. The concordance of estimates remained very similar when small subsamples were used. The procedure was repeated by 13 undergraduates using a subsample of spiders. Their average error was 14.4%. The greatest similarity in estimates was for the mosses, but with high levels of splitting and lumping this result was entirely fortuitous. The results suggest that RTU estimates made by biodiversity technicians may be sufficiently close to formal taxonomic estimates of species richness to be useful for the rapid assessment of biodiversity. They also show, however, that the procedures outlined here should be used on invertebrate and nonflowering plant groups before they can be confidently included in biodiversity surveys.*

Paper submitted April 24, 1992; revised manuscript accepted July 29, 1992.

Un método posible para la evaluación rápida de la biodiversidad

Resumen: *Inventarios de vertebrados y de plantas fanerógamas son frecuentemente utilizados como substitutos de estimadores de biodiversidad total. Esto es en parte por que la inclusión de invertebrados y las plantas que no florecen es percibida como consumidora de tiempo, costosa y difícil por la escasez de especialistas. Estimaciones de riqueza de especies de muestras a campo de arañas, hormigas, poliquetos y musgos llevadas a cabo por un técnico en biodiversidad y un especialista taxónomo fueron comparadas. El técnico en biodiversidad recibió unas pocas horas de entrenamiento sobre la taxonomía para cada grupo y separó especímenes en Unidades Taxonómicas Reconocibles. El especialista separó por especies. Para los tres grupos de animales el técnico registró 165 taxa y el especialista 147, con un error para las hormigas y las arañas de un 13% o menos. Se detectó una pequeña cantidad de divisiones y de agrupamientos de especies. La concordancia de las estimaciones permaneció casi igual cuando pequeñas sub-muestras fueron utilizadas. El procedimiento fue repetido por 13 estudiantes utilizando submuestras de arañas. El error promedio que cometieron fue del 14.4%. La mayor similitud en las estimacione fue para los musgos, sin embargo, con los altos niveles de división y agrupamiento estos resultados fueron enteramente fortuitos. Los resultados sugieren que las estimaciones de las unidades taxonómicas reconocibles realizadas por técnicos en biodiversidad resultan lo suficientemente cercanas a las estimaciones taxonómicas formales para la riqueza de especies, por lo que son útiles para una rápida estimación de biodiversidad. Sin embargo, estas también muestran que los procedimientos discutidos aqui deben ser utilizados en invertebrados y plantas que no florecen antes de que sean incluídos con confianza en los relevamientos de biodiversidad.*

Introduction

Inventories of species at particular sites are essential data sets for conservation and resource management, but there are major problems with the deceptively simple task of producing them. First, biological surveys are becoming increasingly thorough and are revealing very large numbers of undescribed species that require major resources for taxonomic treatment. The formal determination of species names can take a long time and, in those groups where the formal taxonomy is poorly developed, may not be possible. For example Margules (in press) collected over 900 species of soil and litter arthropods in regrowth eucalyptus forest of southeastern Australia, but 80% of the taxa could not be named. Second, the time available for the construction of inventories is often very short. This may be the result of commercial or bureaucratic pressures to achieve quick results. Third, limitations on time and resources frequently restrict inventories, such as those of Conservation International's "RAP" team (Abate 1992), to vertebrates and flowering plants—groups that represent less than 20% of all named species (Wilson 1989). The weakness of this approach is that the relationships between vertebrate and flowering plant species richness to total biodiversity are usually unknown. In view of these problems, it is important to explore (1) other measures of the variety of taxa that may be useful for rapid biodiversity assessment, and (2) the use of invertebrates for biodiversity assessment: because of their abundance, their ubiquity and diversity should be at least as important to biodiversity surveys as that of angiosperms and vertebrates.

In this study we explored the first objective using mostly invertebrates as test organisms. The taxa used for measuring variety were recognizable taxonomic units (RTUs), and the goal was to discover whether or not they provided reasonable estimates of species richness, in comparison with formal Latin binomials, in a relatively short time. RTUs are taxa that are readily separable by morphological differences that are obvious to individuals with less training than professional taxonomists. These individuals are called biodiversity technicians in Australia (Kitching personal communication), and it is anticipated that they will, in association with specialist taxonomists, carry out all the procedures of rapid biodiversity assessment. In Costa Rica, nonprofessionals known as parataxonomists have been enlisted to collect and prepare arthropod specimens (Tangley 1990; Gamez 1991). To the authors' knowledge, however, the accuracy of estimates of species richness produced by them has not been critically examined.

At the start of the exercise it was thought that measurement by RTUs would be much faster than conventional methods. We used invertebrates to explore the first objective because (1) they are a major part of biodiversity, (2) they are rarely included in inventories, and (3) because there have been few studies that have explored relationships between invertebrate biodiversity and that of vertebrates and flowering plants.

There is some evidence that flowering plant and vertebrate biodiversity is not an accurate indicator of the species richness of invertebrates. Brooker (1982) showed that high invertebrate species richness correlated with low angiosperm species richness in some habitats. Majer (1983) showed that variation in ant species richness was explained by several factors and that plant species richness accounted for only 24% of the variation in the number of ant species. He also showed that ant species richness was significantly and positively correlated with that of several other invertebrate groups. Yen (1987) studied plant, vertebrate, and beetle communities at 32 sites in southeastern Australia and found that there was no correlation between the number of species of vertebrates and the number of species of beetles. Further, hierarchical classification did not separate beetle or vertebrate communities that corresponded to the vegetation communities. Burbidge et al. (in press) found no correlation between vertebrate and ant species richness at sites in SW Australia. As a result of these studies, Majer (1983), Yen (1987) and Burbidge et al. (in press) came to the conclusion that it is important to include invertebrates in biodiversity inventories because their patterns of diversity cannot be assumed to follow those of vertebrates and flowering plants.

While the inclusion of invertebrates in biodiversity surveys appears desirable, the demand on time and resources using conventional methods is immense. Disney (1986), McNeely et al. (1990), and Soulé (1990) have calculated that if the description of new species, using traditional methods, continues at the same rate with the same number of taxonomists as today, the cataloging of global biodiversity would take several thousand years to complete. Further, this process is slowing because funding for taxonomy continues to decline (McNeely et al. 1990; Stork & Gaston 1990; Wheeler 1990; Whitehead 1990; Gaston & May 1992; Richardson & McKenzie 1992). Therefore, it is likely that many species will not be named and described, or even collected, before they are extinct. If invertebrates are to be included on the agenda for biodiversity assessment, conventional naming to formal species will not be possible.

Classification of invertebrates into taxa other than formal species has been done previously. Rees (1983) first introduced the term "recognizable taxonomic unit" to sort Hemiptera from rainforest in Sulawesi. He defined RTUs as "taxa that differ from each other clearly" but accepted that they may not correspond to "good" species. Basset (1989) studied the canopy arthropods of one species of Australian rainforest tree using the RTU where species designations were not possible. Kitching (see Monteith 1990) classified Australian rainforest canopy insects by sorting them into "composite morphospecies which were then assigned to community guilds"; Yen (1987) and Hutcheson (1990) have used beetles sorted to morphospecies to characterize communities.

As far as we know, no study has attempted to compare estimates of species richness as determined by biodiversity technicians using RTUs with estimates made by professional taxonomists. The aim of this study was to determine whether a biodiversity technician with minimal training could produce accurate estimates of the number of species contained within samples, the accuracy being determined by taxonomic specialists. The emphasis was on the estimates of species richness and not on the naming of taxa.

Materials and Methods

We selected four groups of organisms for study: ants, spiders, marine polychaetes, and mosses. We included the mosses because problems relevant to invertebrates may apply equally to lower plants. The ants and spiders were collected in a study of the fauna of the Coolangubra State Forest, and the polychaetes were collected in a study of marine invertebrates off Barrenjoey Headland, both in New South Wales. The mosses were part of a plant survey of calcareous outcrops near the Jenolan Caves of New South Wales (Downing et al. 1991).

One of us (I. Oliver) was trained for three hours in the basic taxonomy of each group by experienced taxonomists. Prior to training Oliver had no experience in the taxonomy of spiders, polychaetes, or mosses, but some experience with the ants. The training included observation and utilization of the most important and easily identified taxonomic characters and an outline of possible problems such as sexual dimorphism and polymorphism.

Specimens were sorted into RTUs by Oliver with the use of a dissecting microscope with a maximum magnification of ×100. No scientific keys were used. A reference collection was compiled as new RTUs were encountered and illustrations of distinguishing characters made. Immature spiders and nonworker ants were not included. Following the first sorting, each RTU collection was examined for obvious misclassification. Then all specimens from every RTU were analyzed by specialists to obtain formal estimates of the number of species.

Comparison of the number of RTUs with the number of species for each group revealed the extent of overestimation and underestimation, and we carried out an analysis of the data to detect the effects of "lumping" and "splitting" on the RTU estimates. For each group we determined the number of cases in which one RTU contained one species, the number of cases in which one RTU contained more than one species, and the number of cases where a single species was split into different RTUs.

We then analyzed a subset of the data to determine whether or not smaller samples might estimate with reasonable accuracy the degree of overestimation and underestimation of species. If so, a smaller number of specimens per RTU sent for specialist analysis would greatly increase the speed of the process. As a first approximation, we took 10 specimens at random (or all specimens where there were fewer than 10 in the sample) from each RTU. Knowing the numbers of different species in RTUs and the number of specimens within species in the whole collection, we calculated the probability of representing all species in smaller samples. Splitting was detected by comparing RTU subsamples and identifying those species that appeared in more than one. We anticipated that in subsamples of 10 specimens there would be a proportion of cases in which species were not sampled by these procedures.

To examine variation between individuals, we asked 13 biology undergraduates to repeat the procedure for a subsample of spiders. The students had all completed a second-year course in invertebrate zoology and a third-year course on the biology of insects, but none were trained in taxonomy. Prior to sorting, Oliver trained the students in the relevant characters. The samples were the same as those used in the previous procedures. Every student was allocated a different pitfall catch, which he or she sorted to RTUs. This continued for three hours, with new pitfall catches being issued as the others were completed. It was not possible to have each student use the same sample due to deterioration of specimens from repeated handling.

Results

Table 1 compares the RTU and species estimates for each group of organisms. Column 4 shows that in the ants, spiders, and mosses the RTU estimates were within 13% of the formal species estimates. However, the data show that the RTU estimate for ants was an underestimate, and overestimates were apparent in the spiders and the polychaetes. These results indicated that some splitting and lumping had occurred.

Having had the entire sample of each group analyzed by specialists, we were able to determine how splitting and lumping affected the estimates. These data are shown in Figure 1. The greatest concordance between RTU and species estimates occurred in the spiders and ants (Fig. 1*a,b*). In both, the RTU estimates shown in Table 1 were relatively unaffected by splitting and lumping, with over 80% of cases represented by the "1 to 1" category—one RTU = one species. The greatest level of

Table 1. Comparison of RTU and species estimates.

TAXA	*Total # of Specimens* 1	*# of Species (A)* 2	*# of RTUs (B)* 3	*% Error* |A-B/A| 4	*Subsample* 5	*# Lumps Detected* 6
Spiders	1850	91	103	13	283	2 of 2
Ants	1233	35	33	6	219	2 of 3
Polychaetes	240	21	29	38	90	2 of 3
Mosses	220	86	87	1	—	—

overestimation of species occurred in the polychaetes (Table 1 and Fig. 1*c*). The right tail in Figure 1*c* shows the splitting of one species into 2, 3, or 4 RTUs on several occasions.

The RTU estimate for mosses shown in Table 1, with a 1% error, was the closest of the four groups to the formal number of species. However, Figure 1*d* shows the occurrence of splitting and lumping to be high, with pronounced tails and a low proportion of taxa in the 1-to-1 category. The result was a fortuitous symmetry in the graph in which the tails cancelled each other out. The problems of RTU estimates emerged clearly with the mosses, and the results emphasized the importance of the professional taxonomist to the process.

The subsampling procedure missed only two species. Overall, the specialists listed 147 species for the three animal groups, and of these 145 were represented in the subsamples (Table 1, columns 5 and 6). In the animal groups there were eight lumping events; in six of them two species occurred in approximately equal proportions. Therefore, the probability of selecting representatives of both species in the subsample was high. But in two cases one rare species was lumped with an abundant one and remained undetected (Table 1, column 6).

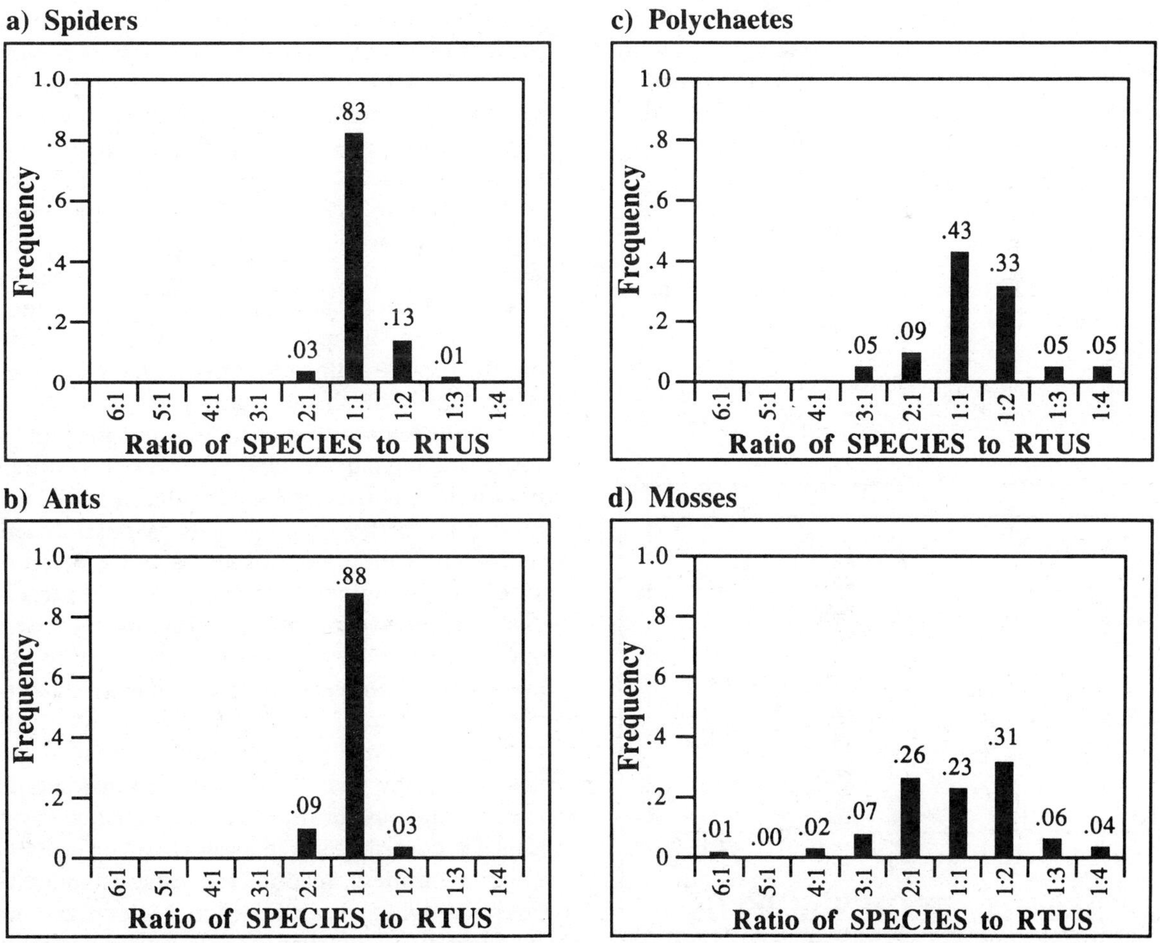

Figure 1. The frequency of lumping and splitting categories for four taxa. Categories on the x *axis to the left of 1:1 represent cases of lumping, and those to the right represent cases of splitting. (a)* n = *90, (b)* n = *32, (c)* n = *21, (d)* n = *84.*

All cases of splitting were represented in the subsamples.

The mosses were not included in the subsampling because they had an average of only 2.5 specimens per species, but the moss data illustrate the importance of the characters presented by a taxon. Figure 1*d* includes all 240 specimens, regardless of the developmental stages present. A very important and easily examined moss character used in classification is the sporophyte capsule. When only specimens exhibiting this character were analyzed, the number of taxa in the 1-to-1 category increased from 23% to 50%.

The undergraduates completed 20 pitfalls in three hours (mean = 1.5, S.D. = 0.8), and the average number of RTUs identified by them was 10.7 (S.D. = 2.9). Oliver analyzed 66 pitfalls in total and in the first 3 hours identified 13 RTUs, so his speed fell within one standard deviation of the mean of the undergraduates. The percentage error for the 13 students, as calculated for Table 1, averaged 14.4% (S.D. = 12.1%). This is similar to the 13.2% shown in Table 1 for the entire sample of spiders sorted by Oliver. However, the standard deviation shows the variation between students to be high. This error reflects underestimation of the number of species by the students, as shown in Figure 2 with a larger tail to the left of the graph. This is in contrast to Oliver's error of overestimation (Fig. 1*a*). Possible reasons for this difference are discussed later.

Discussion

The data show that RTU estimates for the ants, spiders, and mosses were close to the formal taxonomic estimates. However, the moss data also showed that RTU estimates can be misleading. Splitting and lumping produced an entirely fortuitous congruence between the RTU estimate and the formal estimate. This strongly suggests that different groups of organisms should be analyzed by the procedures described here before they can be confidently included in rapid biodiversity assessment using RTUs.

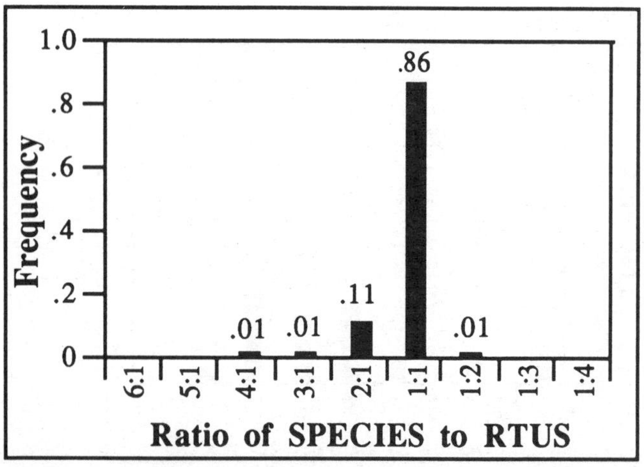

Figure 2. The frequency of lumping and splitting categories produced by 13 undergraduates sorting spiders (data pooled for all students), n = *121.*

The data also suggest that the use of RTUs requires a minimum of unambiguous morphological features that a relatively untrained individual can easily and quickly utilize. This was illustrated by the improvement in the moss RTU estimates when taxa with sporophytes were separated out. Ambiguity arises from sexual dimorphism and developmental polymorphism, and mistakes in the analysis of spiders and ants were probably avoided by including only sexually mature spiders and worker (nonalate) ants. Where taxa exhibit extreme sexual dimorphism, only one sex should be used. This approach may cause some species to be missed, but including all life stages or sexes in rapid biodiversity assessment may lead to large errors. This has been shown to occur in a similar study of freshwater invertebrates, where different larval instars of the same species were sorted into different RTUs (Cranston & Hillman 1992).

The very close estimate for the ants may be explained by the previous experience of Oliver. This suggests that estimates will improve as biodiversity technicians gain experience with the taxa they are sorting. It is also interesting to note that the ants were the only group where our estimate was an underestimate, where splitting occurred less often than lumping. A possible reason for this may be that we had a better understanding of intraspecific variation within the ants due to past experience.

Why did Oliver seem to be a "splitter" and the undergraduates "lumpers"? One reason may have been that the basic training Oliver received from the specialist was superior to the training he gave the students. Therefore, standardizing the training that future biodiversity technicians receive is essential. Different levels of training may have reduced lumping by Oliver but did not explain his tendency to split. A possible cause may be the larger numbers of species and specimens he analyzed: as the numbers of species sorted increased, the chance of not recognizing a previously recorded RTU may likewise increase. This may explain the higher frequency of splitting events shown in Figure 1*a* compared with Figure 2.

This experience strongly suggests that in large samples with many taxa, presorting into major groups may be advantageous for three reasons. First, several technicians can contribute specimens to major groups that are simply identified. Second, the establishment of major groups to which each additional specimen can be assigned reduces the chances of misclassification because it is not necessary to survey all RTUs every time, but rather the smaller number of major groups. Third, having assembled similar RTUs into major groups, the "problem" variation—the specimens that most resemble one another—can be assessed together. For example, all spiders up to 5 mm in length can be quickly examined to discover whether they are likely to be one or many RTUs. Keys may be useful for this phase.

Whether or not biodiversity assessment will include groups such as invertebrates in biological surveys depends on at least three factors: accuracy, time, and costs. Our data suggest that accuracy can be high for some invertebrate groups. With respect to time, the specialists were much faster than us at sorting the samples. Among the three animal groups studied, the specialists' time commitment was 30% to 75% of ours, with the greatest disparity for the polychaetes and the least for the spiders. However, as rapid biodiversity assessment is not concerned with the formal description and naming of species, its advantage as a time-saver will increase as a function of the learning curve of the biodiversity technician, who will be slower sorting to RTUs than specialists sorting to unnamed species at first, but then can be expected to speed up with the acquisition of skill and experience.

The final consideration of cost is difficult to analyze. Biodiversity technicians will receive less training than professional taxonomists, with concomitant savings in salaries and overheads. But with major biodiversity surveys already being carried out and others planned, the necessity for the establishment of sorting centers seems clear. Sorting centers might be seen as laboratories where biodiversity technicians, skilled in different groups of organisms, separate very large numbers of specimens, many with poorly-known taxonomic status, into units of variety such as RTUs. The reference collections built up by them would make important contributions by (1) revealing the spatial and temporal variation in species richness that has emerged as a major component of biodiversity; (2) estimating the degree of endemicity; (3) providing databases for conservation and natural resource management; and (4) providing resources for taxonomic research. The costs of sorting centers are hard to assess, as they will depend on existing scientific infrastructure.

Taxonomic specialists are vital at three stages of rapid biodiversity assessment: (1) to train the biodiversity technicians; (2) to provide the initial formal estimates for comparison with the RTU estimates; and (3) to check the efficacy of subsamples on a regular basis. In many situations the first two stages will be required only once, but we believe that if invertebrates are to be routinely incorporated into inventories the third stage should always be used. We have shown that with care the number of species missed when subsamples are analyzed by specialists can be small.

More research is needed to determine if the results presented in this paper can be reproduced by different biodiversity technicians studying different taxa. More efficient methods of rapid biodiversity assessment may be developed. For example, the automation of the sorting process may be possible in a few years with the use of laser interferometry (see Moldenke et al. 1991). The data presented here suggest that new methods of rapid biodiversity assessment are worth investigation and that invertebrates, which are a major part of all ecosystems, may provide much-needed and reliable estimates.

Acknowledgments

We are especially grateful to the taxonomic specialists Dr. Mike Gray of the Australian Museum, Sydney, Dr. M. A. O'Donnell of the Institute of Marine Ecology, University of Sydney, Alison Downing of the School of Biological Sciences, Macquarie University, and Dr. Alan York of the Forestry Commission of New South Wales, who examined the collections to produce formal estimates. Their support for the principles detailed in this paper was encouraging. We also thank the Forestry Commission of New South Wales, the Sydney Water Board, and Alison Downing for providing the samples on which this study was based, and Dave Briscoe, Jonathan Majer, and Hall Cushman for comments on earlier drafts of the manuscript. This is contribution number 144 to the Research Unit for Biodiversity and Bioresources.

Literature Cited

Abate, T. 1992. Environmental rapid-assessment programs have appeal and critics. BioScience **42**:486–489.

Basset, Y. 1989. The temporal and spatial distribution of arboreal arthropods associated with the rainforest tree *Argyrodendron actinophyllum*. Ph.D. dissertation. Griffith University, Australia.

Brooker, M. P., editor. 1982. Conservation of wildife in river corridors. Powys County Council, Wales.

Burbidge, A. H., K. Leichester, S. McDavitt, and J. D. Majer. 1992. Ants as indicators of disturbance at Yanchep National Park, Western Australia. Journal of the Royal Society of Western Australia. **75**: 89–95.

Cranston, P. S., and T. Hillman. 1992. Rapid assessment of biodiversity using "biological diversity technicians" (= "parataxonomists"). Australian Biologist **5**:144–154.

Disney, R. H. L. 1986. Inventory surveys of insect faunas: Discussion of a particular event. Antenna **10**:112–116.

Downing, A. J., H. P., Ramsay, and W. B. Schofield. 1991. Mosses in the vicinity of Jenolan Caves, New South Wales. Cunninghamia **2**:371–384.

Gamez, R. 1991. Biodiversity conservation through facilitation of its sustainable use: Costa Rica's National Biodiversity Institute. Trends in Ecology and Evolution **6**:377–378.

Gaston, K. J., and R. M. May. 1992. Taxonomy and taxonomists. Nature **356**:281–282.

Hutcheson, J. 1990. Characterization of terrestrial insect communities using quantified, malaise-trapped Coleoptera. Ecological Entomology **15**:143–151.

Majer, J. D. 1983. Ants: Bio-indicators of minesite rehabilitation, land-use, and land conservation. Environmental Management **7**:375–383.

Margules, C. R. The Wog Wog habitat fragmentation experiment. Environmental Conservation. In press.

McNeely, J. A., K. R. Miller, W. V. Ried, R. A. Mittermeier, and T. B. Werner. 1990. Conserving the world's biological diversity. International Union for the Conservation of Nature, Gland, Switzerland.

Moldenke, A., C. Shaw, and J. R. Boyle. 1991. Computer-driven image-based soil fauna taxonomy. Agriculture, Ecosystems and Environment **34**:177–185.

Monteith, G. 1990. Rainforest insects: Biodiversity, bioguesstimation, or just handwaving? News Bulletin of the Australian Entomological Society **26**:93–95.

Rees, C. J. C. 1983. Microclimate and the flying Hemiptera fauna of a primary lowland rainforest in Sulawesi. Pages 121–136 in S. L. Sutton, T. C. Whitmore, and A. C. Chadwick, editors. Tropical rainforest: Ecology and management. Blackwell Scientific Publications, Oxford, England.

Richardson, B. J., and A. M. McKenzie. 1992. Australia's biological collections and those who use them. Australian Biologist **5**:19–30.

Soulé, M. E. 1990. The real work of systematics. Annals of the Missouri Botanic Gardens **77**:4–12.

Stork, N. E., and K. Gaston. 1990. Counting species one by one. New Scientist **127**:31–35.

Tangley, L. 1990. Cataloging Costa Rica's diversity. BioScience **40**:633–636.

Wheeler, Q. D. 1990. Insect diversity and cladistic constraints. Annals of the Entomological Society of America **83**:1031–1047.

Whitehead, P. 1990. Systematics: An endangered species. Systematic Zoology **39**:179–184.

Wilson, E. O., editor. 1989. Biodiversity. National Academy Press, Washington, D.C.

Yen, A. L. 1987. A preliminary assessment of the correlation between plant, vertebrate and Coleoptera communities in the Victorian mallee. Pages 73–88 in J. D. Majer, editor. The role of invertebrates in conservation and biological survey. Department of Conservation and Land Management, Western Australia.

Genetically Effective Population Size of Large Mammals: An Assessment of Estimators

RICHARD B. HARRIS
Montana Cooperative Wildlife Research Unit
University of Montana
Missoula, MT 59812, U.S.A.

FRED W. ALLENDORF
Department of Zoology
University of Montana
Missoula, MT 59812, U.S.A.

Abstract: *We calculated genetically effective population sizes (N_e) for simulated grizzly bear (*Ursus arctos*) populations by tracing the loss of heterozygosity through time, and compared them with estimates of N_e produced by applying published formulas to demographic results from the simulation. Estimates of N_e using different formulas on identical data varied widely. Equations published by Hill (1972), and modifications of those used by Ryman et al. (1981) and Reed et al. (1986), provided the most accurate estimates. Minor population fluctuations had little effect on N_e, but variation in lifetime reproductive success among males (V_{km}) greatly reduced N_e from its expectation under random mating success. All methods to estimate N_e for populations with complex demographics require extensive data, but estimates for V_{km} in polygynous species are especially difficult to obtain. We suggest that simulation modeling may provide alternative methods to estimate V_{km} and N_e.*

Resumen: *Calculamos tamaños de poblaciones genéticamente efectivas (N_e) para poblaciones simuladas del oso gris (*Ursus arctos*) trazando la pérdida de heterozigasidad a tráves del tiempo, luego las comparamos con estimaciones de N_e producidas aplicando fórmulas publicadas a los resultados demográficos de la simulación. Los valores de N_e calculados usando diferentes fórmulas con datos idénticos, variaron mucho. Las ecuaciones publicadas por Hill (1972), y modificaciones de las usuadas por Ryman, et al. (1981) y Reed et al. (1986), proporcionaron los cálculos más precisos. Fluctuaciones menores en las poblacionales tuvieron poco efecto sobre N_e, pero la variación en el éxito reproductivo por vida entre los machos V_{km} redujo tremendamente el N_e comparado con el valor esperado bajo exito reproductivo al azar. Todos los métodos para calcular N_e para poblaciones con demografías complejas requieren datos extensos, pero estimaciones para V_{km} en especies polígamas son especialmente difíciles de obtener. Sugerimos que modelos de simulación pueden proveer métodos alternativos para calcular V_{km} y N_e.*

Introduction

A growing concern in management of endangered species is the increasing insularity of small populations. Especially worrisome is the deterioration in genetic variation of many species of large mammals, where population densities are normally low and spatial requirements high, such as tigers (*Panthera tigris*) and rhinos (Rhinocerotidae). The most useful concept in estimating the magnitude of genetic loss through time to be expected in small, isolated populations is that of effective population size (Wright 1969). While there is

Paper submitted June 21, 1988; revised manuscript accepted November 21, 1988.

universal agreement about the importance of estimating effective population size (N_e) for making informed management decisions (Soulé 1980), considerable confusion remains about exactly how it is best estimated, especially in the case of large mammals with their typically complex demographics and numerous departures from the genetically "ideal" population. The usual approach is to infer the rate of loss of genetic variation from the breeding structure of the population using various demographic statistics.

Wood (1987) described N_e as "something of a 'black box,' obscuring rather than illuminating the interactions and effects" of characteristics relevant to rate of loss of genetic diversity. Recently, numerous articles have attempted to provide guidance to managers of wild populations interested in estimating N_e (Ryman et al. 1981; Lemkuhl 1984; LaCava & Hughes 1984; Reed et al. 1986; Wood 1987; Lande & Barrowclough 1987; Crow & Denniston 1988). Unfortunately, they do not all use the same formulas, and applying their techniques results in differing estimates of N_e, and therefore of the rate of genetic loss to be expected over time.

Wright (1969) defined effective population number as "whatever must be substituted for N [population size] in the basic formula" (p. 211) whenever a population fails to meet genetically ideal conditions. Wright's "basic formula" defines the rate of loss of selectively neutral genetic variation (i.e., heterozygosity, H) as

$$H_t = H_0 [1 - 1/(2N)]^t \qquad (1)$$

where t is time in generations.

Estimates of N_e for populations of large mammals have addressed one or more of the following characteristics of non-ideal populations:

(1) unequal number of breeding males and females (the ideal population assumes monoecious individuals),
(2) fluctuations in population size,
(3) overlapping generations,
(4) non-Poisson (and nonbinomial) variance in distribution of offspring surviving to adulthood (the ideal population assumes all individuals have equal probability of contributing to subsequent generations),
(5) heritability of fertility (the ideal population assumes none), and
(6) geographic structure (the ideal population assumes random mating, i.e., panmixia).

In this paper, we compare approximations of N_e calculated from published demographic formulas to known N_e values from simulated grizzly bear populations. We focus particularly on those formulas that claim to be applicable to populations of large mammals. Thus, for example, we do not examine papers dealing with populations of a single sex, or those having discrete generations. Grizzly bears have many attributes (e.g., polygyny, delayed reproduction, extended parental care, between-sex differences in breeding success and survival of offspring) that make them an appropriate choice for assessing the application of published N_e formulas to other large mammals. We conclude with a discussion of the difficulties in obtaining data necessary to apply published N_e formulas to wild populations. We suggest that simulations may provide a method to estimate parameters that are otherwise difficult to gather in the field.

Model

The simulation model was a discrete-time, stochastic computer program that followed the history of each individual bear from birth until death.* The model retained information on each bear's age, sex, parents, mates, offspring, membership in a family group, and, if female, whether attended by offspring, pregnant, or solitary, thus allowing calculation of various demographic statistics required by N_e formulas. Five events affected each bear during the simulation year: breeding, hunting mortality, natural mortality, birth, and family breakup (weaning of juveniles). At each event, an individual bear's fate (dying, giving birth, etc.) was determined by comparing its age-specific probability with a random number from a uniform (0,1) probability distribution.

We limited our investigations to stable, stationary populations (i.e., those populations with similar numbers and age structures from year to year). To achieve equilibrium conditions in both hunted and unhunted populations, survival and natality were modeled as density-dependent functions of population size (N) relative to a hypothetical carrying capacity (K). Survival and natality rates increased slightly above mean values when a population fell below N/K = 1.0; rates decreased slightly when a population increased above K.

We used 30 age classes, and natality and survival rates typical of Rocky Mountain populations of grizzly bears (Appendix A). To assess published formulas for estimation of N_e it is not necessary that our simulation model emulate wild grizzly bear populations accurately (although elsewhere we make the claim that our model structure and parameters were reasonable approximations; Allendorf et al., in preparation). It is only necessary that the model provide detailed demographic data with which to compare N_e estimates using the published formulas.

A critical component for estimating effective popula-

---

* *Generalized Animal Population Projection System (GAPPS) is available for use on IBM-compatible microcomputers. For information about acquisition, contact the Montana Cooperative Wildlife Research Unit, University of Montana, Missoula, MT 59812.*

tion size, for which data are lacking, is the distribution of reproductive contribution among males. We simulated three different male mating systems (Table 1) with the intent of embracing the range of probable values: (1) a "random" male mating system, in which adult males were picked at random (with replacement) to sire the offspring of each pregnant female, (2) our "best estimate" of the likely situation in a wild grizzly bear population, in which adult males of a subjectively selected prime age group (ages 8–18) were assumed to have dominant status and given higher probabilities of fathering cubs than were males of other adult ages, and (3) an "extreme" age-selective system, similar to (2), but with heavier weighting toward the prime age group. In all three systems, males less than four years old were considered subadults and did not sire cubs. None of the three male mating systems allowed for variability of breeding success within age classes. Thus, variation within a particular cohort was Poisson. In nature, individuals of the same age class might be expected to display heterogeneity in reproductive success over and above Poisson variance (i.e., dominant males).

We simulated three alternative hunting strategies (*sensu* Ryman et al. 1981) that we termed "random," "chance," and "trophy" (Table 2). Each differed in its assumptions about interactions between hunters and bears and hence in the age and sex of bears killed. In each case, we limited the harvest to 4% of the standing population each year to prevent populations from declining due to overharvest (Harris 1986). Under the "random" hunting strategy, bears were killed by hunters in proportion to their abundance in the population. Under the "chance" hunting strategy, bears were killed in proportion to their likelihood of being encountered by a hunter, which was assumed to be related to the bears' amount of movement (Bunnell & Tait 1980). Subadult males, assumed to wander about in exploratory forays, were the most vulnerable to harvest. Mothers with young, assumed to constrain their movements to relatively small areas, were the least vulnerable. Under "trophy" hunting, bears were killed in proportion to their size and assumed desirability as a trophy. Adult males, the largest bears, were the most vulnerable; adult females and subadult males less vulnerable. Hunters employing this strategy were assumed to frequently pass up the opportunity to kill the more numerous younger (and smaller) bears, and avoid killing young or their attending mother in all cases.

Table 1. Coefficients given to males of different ages in the simulation model governing their probability (relative to each other) of siring the cubs of a pregnant female. The model builds a cumulative distribution, so that the sum of all males' probabilities = 1.0.

	Age			
Mating system	*0–3*	*4–7*	*8–18*	*19+*
Random	0.00	1.00	1.00	1.00
Estimated	0.00	0.25	1.00	0.33
Extreme	0.00	0.03	1.00	0.04

Table 2. Coefficients given to animals of each sex-age class governing their vulnerability (relative to each other) of being hunted, according to 3 different hunting regimes: random, chance, and trophy. The model builds a cumulative distribution, so that the sum of all individuals' vulnerabilities = 1.0. Animals are picked at random (with replacement) for removal by hunting, and the process continues until the specified number are removed.

	Relative vulnerability					
	Random		*Chance*		*Trophy*	
Category	*F*	*M*	*F*	*M*	*F*	*M*
Age 0 with mother	1.00	1.00	0.05	0.05	0.00	0.00
Age 0 without mother	1.00	1.00	2.00	2.00	0.00	0.00
Cub >0 with mother	1.00	1.00	0.20	0.20	0.00	0.00
Mother with young	1.00	—	0.20	—	0.00	—
Age 1–3	1.00	1.00	2.00	7.00	0.00	0.00
Age 4	1.00	1.00	2.00	7.00	0.25	0.25
Age 5–7	1.00	1.00	0.80	1.00	0.25	0.25
Age 8–30	1.00	1.00	0.80	1.00	0.25	8.00

Because our simulated grizzly bear populations exhibited relatively modest fluctuations, we also ran one set of simulations in which we deliberately inflated variances of vital rates to produce greater variation in population size.

Our simulation model did not include heritability of fertility, so we were unable to examine possible effects of this factor on the various estimators. With the exception of age-related mating probabilities for males as described above, the simulation also assumed complete panmixia.

Heterozygosity and N_e

We determined heterozygosity by calculating the inbreeding coefficient of individuals by the algorithm of Boyce (1983). This inbreeding coefficient (pedigree F; Wright 1969, p. 177) is equal to the probability that two genes at a locus in an individual are identical by descent from an allele in the foundation population. Thus, F is an exact measure of the loss in heterozygosity, H = (1 − F), over the entire genome relative to the founding populations (Wright 1969). The populations used to begin the simulations had age structures typical of grizzly bear populations; the initial heterozygosity (H_0) was taken to be 1.0 (F = 0) and all individuals were assumed to be unrelated.

We estimated the rate of loss of heterozygosity by regressing the natural logarithm of mean heterozygosity in newborns over time. From the basic formula (1),

$$H_t = H_0 [1 - (1/2N_e)]^t \qquad (2)$$

but because $H_0 = 1$,

$$H_t = [1 - (1/2N_e)]^t \quad (3)$$

or,

$$N_e = 1/(-2\,e^m + 2) \quad (4)$$

where t is the number of generations, and m is the slope of the natural log of mean heterozygosity of the newborn cohort regressed on time, measured in generations:

$$m = \frac{1}{H_t}\frac{dH_t}{dt} = \frac{d\ln(H_t)}{dt}$$

The generation interval is the mean age of parents contributing progeny, weighted by their number of progeny.

We note here that our procedure yields the *inbreeding* effective size (N_{eI}). Some of the demographic formulas we examined were clearly intended to yield this number, while others estimated the *variance* effective size (N_{eV}), and still others made no distinction between the two. In populations that are stable and stationary, as were our simulated populations, estimates of N_{eI} and N_{eV} converge (see Crow & Kimura 1970, p. 361; Wright 1969; Felsenstein 1971); thus no further distinction is made throughout this paper.

We performed 50 replicates for each combination of parameter values. Each replicate lasted 100 years; the first 24 years were omitted from the regression analysis to provide a delay of 2–3 generations in which the rate of increase in inbreeding reached an asymptote (Hill 1972, 1979; Johnson 1977). The mean inbreeding coefficient of newborns was calculated every five years, rather than each year, to save computing time. (We used population sizes of approximately 100, so that by the end of 100 years there were over 1,000 bears in each pedigree).

Demographic Estimation of N_e

We estimated N_e with demographic formulas by recording the relevant parameters during a simulation run of 76 years (years 25–100). Details of each demographic formula are presented in Appendix B. Estimators of N_e examined were:

(1) Hill (1972), who presented a general formula (Eq. B1) for calculating N_e for populations of constant size and sex ratio with overlapping generations.

(2) Ryman et al. (1981), who used formulas from Crow & Morton (1955) to calculate N_e (Eq. B2) from the demographic structure of the female component of simulated moose (*Alces alces*) populations in order to estimate effects of hunting on N_e.

(3) An alternative estimator using the same structure as Ryman et al. (1981), but utilizing data for both sexes, and calculating variances of offspring production directly from the simulated data rather than using the methods of Crow & Morton.

(4) Lemkuhl (1984), who suggested a process to estimate the census number of breeding animals needed to achieve $N_e = 50$ (Soulé 1980), and thereby an estimator of N_e from N (Eq. B8).

(5) LaCava & Hughes (1984), who asserted that the equation accounting for departures in equal sex ratio (Wright 1931; Crow & Kimura 1970) was sufficient to estimate N_e for polygynous species (Eq. B9).

(6) Reed et al. (1986), who modified Hill's (1972) formula for the case in which reproductive success of both sexes is Poisson (Eq. B10).

(7) An adjustment to Reed et al. (1986), in which Poisson reproductive success is not assumed (Eq. B11).

(8) Lacy & Clark (1989), who suggested an adjustment for population fluctuations (Eq. B12) in estimating N_e for black-footed ferrets (*Mustela nigripes*).

(9) An alternative adjustment for fluctuating population size (Eq. B13) of Crow & Kimura (1970, p. 360).

Results

Basic Simulation Runs and N_e

Mean N_e/N for hypothetical grizzly bear populations calculated from the 50 replicate runs by our pedigree analysis method varied from 0.241 (68% confidence limits 0.203–0.295) for a population with our estimated values for male mating, hunted under a "chance" hunting regime, to 0.324 (68% confidence limits 0.275–0.394) for a trophy-hunted population with our estimated values for male mating. As expected from theory, increasing the variance of reproductive success among males reduced N_e/N (Table 3).

Estimated N_e from Demographic Formulas

Table 4 summarizes demographic data used in estimating N_e using published formulas. The Hill (1972) formula predicted N_e among the six basic runs accurately, generally overestimating N_e by approximately 5% (Table 3). Hill made no allowance for fluctuating population size, while our simulated populations had standard deviations approximately 10% of their long-term average numbers. Covariances in production of male and female lifetime surviving offspring, used only in the Hill formula, were generally much lower than variances along any of the four pathways for gametes (mother-daughter, mother-son, father-daughter, father-son). Because polygyny affects reproductive heterogeneity among males more than among females (Wood 1987), variances along the male pathways, as well as male co-

Table 3. Effective population size as a function of total census number (N_e/N), as estimated by our simulation model, and using published formulas based on demographic data.

Simulation										
Hunt	*Male mating*	*Eq. 4*[a]	*Hill*[b]	*Ryman*[c]	*Ryman*[d]	*Ryman*[e]	*Lemkuhl*[f]	*LaCava*[g]	*Reed*[h]	*Reed*[i]
None	Random	0.311	0.334	0.431	0.372	0.344	0.055	0.974	0.514	0.322
None	Extreme	0.265	0.274	0.431	0.372	0.291	0.053	0.914	0.434	0.225
None	Estimated	0.285	0.304	0.431	0.372	0.326	0.054	0.961	0.494	0.283
Random	Estimated	0.324	0.308	0.425	0.363	0.317	0.045	0.970	0.520	0.291
Chance	Estimated	0.241	0.255	0.405	0.352	0.274	0.058	0.946	0.455	0.238
Trophy	Estimated	0.322	0.360	0.405	0.356	0.359	0.051	0.935	0.492	0.358

[a] *See methods section.*
[b] *Eq. B1.*
[c] *Ryman et al. 1981, assuming values used for moose: $N_{em}/N_m = 0.9 \times N_{ef}/N_f$, random suvival of subadults.*
[d] *Ryman et al. 1981, assuming values used for moose: $N_{em}/Nm = 0.9 \times N_{ef}/N_f$; 10% of subadult survival on family basis, 90% on random basis.*
[e] *Ryman et al. 1981; N_{em} calculated from male data; all variances of lifetime offspring production computed directly from data.*
[f] *Lemkuhl 1984; from Eq. B8.*
[g] *LaCava & Hughes 1984; Eq. B9.*
[h] *Reed et al. 1986; Eq. B10.*
[i] *Reed et al. 1986; adjusted for non-Poisson progeny distribution, Eq. B11.*

variance, were generally greater than along the female pathways.

Under the assumption that N_e/N for males was 0.9 N_e/N for females, used for moose in their paper, the procedure of Ryman et al. (1981) overestimated N_e/N, more so when none of the subadult survival was assumed to be on a family basis. However, when Ryman's procedure was performed assuming knowledge of the male, as well as the female demographic parameters, and when V_k was computed directly for each sex rather than with the equations of Crow & Morton (1955), results were closer to the actual N_e/N values. The average deviation from the true N_e among the six population variants was less than 10%. All values overestimated N_e/N, and all but one were higher than the corresponding estimate of Hill. The two formulas are similar, but Ryman's omits the covariance terms, which were always positive, leading to slight overestimation of N_e.

Lemkuhl's (1984) formula always produced substantial underestimation of the true N_e. LaCava & Hughes's (1984) formula produced similarly large overestimations of N_e. Some speculations on the causes of error in Lemkuhl's procedures are presented in the "Discussion" to follow.

Reed et al.'s (1986) estimation procedure overestimated N_e/N, usually by roughly 60%. However, when adjusted for non-Poisson distribution of surviving offspring, the modified Reed formula performed about as well as the Hill formula. We also computed Reed's formula using a single, pooled estimate of generation inter-

Table 4. Selected demographic statistics relevant to estimation of N_e from the 6 basic simulation runs.

	Type of Hunt					
	Male Mating System					
	None Random	*None Estimated*	*None Extreme*	*Random Estimated*	*Chance Estimated*	*Trophy Estimated*
Census population:						
Mean number	102.8	102.8	102.8	95.3	97.4	92.7
Coefficient of variation	9.9%	9.9%	9.9%	11.6%	8.0%	10.5%
Mean age, males	3.0	3.0	3.0	3.5	2.3	1.5
Mean age, females	7.4	7.4	7.4	6.6	7.7	7.6
Proportion Males:Females	43:57	43:57	43:57	46:54	43:57	42:58
Breeding population:						
Generation interval (L)						
Males	10.1	11.5	11.4	11.7	11.0	7.4
Females	10.1	10.1	10.1	10.0	10.1	9.7
Proportion Males:Females	42:58	40:60	35:65	41:59	38:62	38:62
Recruited yearly						
Males	2.2	1.8	1.3	1.8	1.6	2.5
Females	3.3	3.3	3.3	3.0	3.0	3.1
Variance in lifetime distribution of contributing progeny						
Males	6.3	8.2	10.3	7.4	11.0	4.2
Females	2.8	2.5	2.2	3.3	2.3	3.3

val (as in Hill), rather than the separate estimates. Differences in the resulting N_e estimates varied from none, under random male mating with no hunt, to 5% of the estimate under trophy hunting (in which the generation interval for females was 31% greater than for males).

Among those models most closely tracking N_e, overestimation was more common than underestimation. Our simulated populations fluctuated slightly, but only the Lemkuhl formula, which never gave reasonably correct results, included an explicit correction for nonconstant population size. In the experimental population with high fluctuations, the overestimation was more pronounced, as expected. The corrections suggested by Lacy & Clark (1989) and Crow & Kimura (1970) reduced N_e/N values estimated by Equations B1, B3, and B11 to much closer to the true N_e value (Table 5).

Discussion

N_e Estimators

For management purposes, it is probably unnecessary to strive for great precision in N_e estimates. Given the likely uncertainties in data necessary for its calculation by any method, excessively rigid dependence on even the best of estimations is unwarranted. Also, other factors may operate to change the rate of loss of genetic diversity from that predicted from N_e (e.g., selection, migration, mutation, avoidance of close inbreeding). At best, N_e estimates can provide managers with an approximation of the degree of genetic loss likely to take place. Nonetheless, the estimator of LaCava & Hughes (1984) and the unmodified estimator of Reed et al. (1986) are likely to paint an overly optimistic picture for most populations of large mammals. Similarly, the Lemkuhl (1984) procedure is likely to create more pessimism than is appropriate, possibly leading wildlife managers, many of whom already possess a great deal of skepticism regarding the importance of maintaining genetic diversity, to dismiss the issue entirely.

Table 5. N_e/N for a simulated grizzly bear population with high yearly population fluctuations, as estimated by 3 demographic formulas, with and without adjustments for nonconstant population size. The mean N_e value calculated using Eq. 4 was 0.234.

		Corrected for fluctuations in numbers	
	Uncorrected	*Lacy & Clark*[a]	*Crow & Kimura*[b]
Hill (1972)	0.286	0.189	0.206
Ryman et al. (1981)[c]	0.307	0.203	0.228
Reed et al. (1986)[d]	0.284	0.186	0.204

[a] *$N_e = N_e / (1 + CV^2)$, where CV = coefficient of variation in yearly numbers of breeders.*
[b] *$N_e = N_e - (\sigma^2/N)$, where σ^2 = variance in yearly numbers of breeders.*
[c] *All values computed directly (Eq. B3).*
[d] *Modified to adjust for non-Poisson variance in lifetime production of contributing offspring (Eq. B11).*

When accurate parameter estimation is possible, the Hill (1972), Ryman et al. (1981), and modified Reed et al. (1986) formulas all appear to predict N_e values well. When parameters for males are not available, the accuracy of the Ryman estimate depends greatly on the estimates used for the ratio of N_{em}/N_m to N_{ef}/N_f, and the ratio N_m/N_f. An estimate must also be made of the proportion of pre-adult mortality that occurs on a family basis. Ryman et al. (1981) used zero and 10%. Shull & Tipton (1987) in a recent application of Ryman's procedure for American bison (***Bison bison***), used zero and 30%. In our simulated bear populations, most mortality occurred after cubs had left their mother but before they became reproductively active, and was therefore random. However, a small portion was known to occur on a family basis, because litters that had lost their mother prior to the usual weaning age were immediately exposed to much higher mortality rates. We would have been hard-pressed to estimate the relative magnitudes of these two types of subadult mortality prior to running the simulations, despite our knowledge of the simulation's parameters. On the other hand, if heritability of fertility is thought to be present and can be estimated, the Ryman formula provides for this parameter. However, neither the Ryman nor Reed formulas includes the covariance terms, which could influence N_e in species with different subadult mortality patterns than our simulated bears.

Two additional estimators, not treated here, deserve mention. Recently, Crow & Denniston (1988) provided new derivations for both N_{eI} and N_{eV} of populations with discrete generations, and clarified differences between the two types of N_e. Both estimators included the covariance terms as defined by Hill (1972). They pointed out, however, that their estimator for N_{eV} reduced to approximately the Hill formula for overlapping generations if population size could be assumed to be constant. Although our simulated populations were not strictly constant (yearly numbers varied randomly about the long-term mean due to demographic stochasticity), they were intended to represent equilibrium conditions, and thus the constant size assumption was not grossly violated. Application of Crow & Denniston's formula (modified to reflect overlapping generations) to our simulated data yielded estimates within 2% of estimates using Hill's equation (Table 5).

Lande & Barrowclough (1987) also reviewed estimation and application of N_e for population management, suggesting methods to incorporate the various non-ideal population characteristics. We found their methods unwieldy when applied to simulated grizzly bear populations, and judged that some of the simpler equations we assessed provided estimates whose accuracy was well within the resolution of data likely to be collected for such populations in the field.

Variables Affecting N_e

Many authors have emphasized the importance of population fluctuations in reducing N_e from N (e.g., Hartl 1980). Our modeled grizzly bear populations fluctuated little, generally with a standard deviation of 10% or less of the mean population size. At this level of variation, population fluctuations were a minor contributor to the reduction of N_e relative to N. The adjustments of Crow & Kimura (1970) and that used by Lacy & Clark (1989) reduced the magnitude of underestimation slightly. We suspect that most populations of 'k-selected' large mammals show similarly modest yearly fluctuations.

The Hill (1972) and Crow & Denniston (1988) formulas included covariance terms for progeny distributions. In an analytical treatment, Wood (1987) found that covariances were potentially important components to the overall N_e estimate. However, our results suggest that, at least for our hypothetical grizzly bears, the covariances were relatively minor factors because they were small relative to the variances. Wood (1987) similarly found that covariance terms for a human population were considerably smaller than variance terms.

Some confusion arises in applying formulas for which an estimate of generation interval is required, over whether a pooled average or separate estimates for each sex are most appropriate. Our results with the Reed et al. (1986) formula suggest that, at least for grizzly bears, it makes little difference. Even when sex-specific generation intervals were maximally different, resultant estimates were within 5% of each other.

Another source of possible confusion surrounds the definition of adulthood, which is needed in most calculations. Ryman et al. (1981) specify adults as those attaining reproductive age, while Reed et al. (1986) require breeding as well. The analogous problem exists in defining k and V_k, that is, the mean and variance of lifetime reproductive output can be viewed as applying to all animals reaching a specified age, or only to animals that contribute at least one offspring. Our results lend support to the suggestion of N. Ryman (personal communication 1986) that the two approaches yield similar estimates as long as consistency in definition is maintained. That is, if adults are defined as those attaining a minimum breeding age, k and V_k of those adults must be defined as the mean and variance of numbers of their offspring attaining that age. If adults are defined as those actually contributing to subsequent generations, k and V_k must be defined as the mean and variance of their offspring who themselves become contributors.

In general, slight discrepancies in definition of terms were less important to the accuracy of a formula than was failure to include an important demographic consideration. Variance in lifetime production of surviving progeny was a crucial parameter; the unmodified Reed formula overestimated N_e because it ignored the possibility of non-Poisson variance. When corrected, the Reed formula gave results virtually indistinguishable from the Hill formula. Overlapping generations was an important consideration; the LaCava & Hughes formula produced wild overestimates because it ignored both overlapping generations and non-Poisson variance.

The formula of Lemkuhl (1984) included all these critical demographic parameters yet failed to produce accurate estimates. Lemkuhl's adjustment for non-Poisson variance of lifetime surviving offspring born to females (Eq. B4, Lemkuhl's Eq. 3) apparently underestimates N_e. The adjustment for nonconstant population size produces an additional underestimate (compare Eq. B7 with adjustments of Lacy & Clark [Eq. B12], and Crow & Kimura [Eq. B13]). Finally, to adjust for overlapping generations, Lemkuhl suggests that N_e is generally one-half the corresponding N with nonoverlapping generations. This rule of thumb appears reasonable when applied to our simulated grizzly bear populations (Table 5). However, Lemkuhl states that this adjustment must be made to the number of *breeding* animals (p. 172), thus requiring yet an additional adjustment to arrive at N_e/N if N is defined as the entire census population. The adjustments for overlapping generations made by the Hill, Crow & Denniston, Ryman, and Reed formulas implicitly account for the fact that not all animals are breeders. Thus Lemkuhl adjusts for overlapping generations twice, further contributing to the negative bias of his estimate.

Application of Results

To estimate N_e for a wild population using a published formula, reasonably accurate demographic data are needed. The likely sticking point with any formula is estimating the variance in progeny distribution terms from field data, to say nothing of the covariance terms. As Wood (1987) pointed out, these terms are also vulnerable to misinterpretation. Wood (1987) recently provided a method for estimating all these terms for use in the Hill formula, using data on a tribe in Papua New Guinea. However, he had access to detailed paternity records, data rarely available for wild populations. We agree with Wood (1987), however, that these variance terms can be exceedingly important in estimating N_e, and should not be omitted.

An alternative that may be useful in some cases is simulating population dynamics and calculating such parameters directly from the results for input into the accurate formulas (e.g., Hill). Our simulation model allowed us to derive estimates for the variances and covariances in progeny distributions given life table data and estimates of likely breeding success enjoyed by

Table 6. Mean reproductive parameters for female grizzly bears in the simulation model.

	Years			
Age at 1st reproduction:	*5*	*6*	*7*	*8*
Percent:	65%	26%	9%	1%
Mean age at 1st reproduction: 5.50 years				
	Cubs			
Litter size:	*1*	*2*	*3*	
Percent:	19%	55%	26%	
Mean litter size: 2.07 cubs				
	Years			
Breeding interval:	*2*	*3*	*4*	*5*
Percent:	14%	63%	20%	2%
Mean breeding interval: 3.09 years				
Mean cubs/reproductive female/year: 0.67				

males of differing ages. Alternative simulation structures might allow biologists to estimate the likely breeding success for males on the basis of different dominance rankings, how long dominance is maintained by individuals, and how dominance translates into breeding success. Admittedly, data with which to estimate these parameters are also rarely available. However, we have found that field biologists may be willing to make educated, insightful guesses for these parameters, whereas their ability to estimate such opaque parameters as, for example, the covariances needed in the Hill and Crow and Denniston formulas is questionable. Our procedure

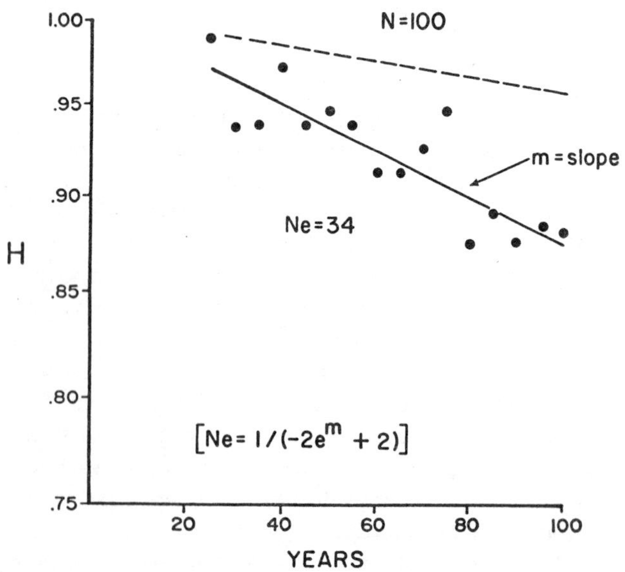

Figure 1. Least squares fit of the natural log of heterozygosity (H) on time (slope = m) from a randomly selected simulation with no hunting and random male mating. The dashed line represents the expected rate of loss of heterozygosity for an ideal population of 100 bears.

involved polling knowledgeable bear biologists about reproductive success of males, following a thorough discussion of the structure of our simulation model. Our simulation results then alleviated the need to make any additional assumptions regarding variances of reproductive success.

The simulation we developed also permitted estimation of N_e directly through a measure of the relatedness of all individuals. Our procedure gave an estimate that reflected the exact demographics of each particular run, and as such, required replicating a number of times to determine the expectation of N_e. The Hill estimate from a single run was a better predictor of the ultimate average N_e from all 50 runs than was the N_e from any single run, suggesting that it may use information more efficiently than our approach using pedigree analysis, or be less susceptible to sampling error. However, our approach also allowed the estimation of confidence limits for N_e, which themselves may have management utility. Confidence limits remind us that even the best estimate of N_e is only an expectation; stochastic variations in demographic events can easily alter the rate of genetic loss from that expected.

Acknowledgments

Funding for the senior author was provided by the U.S. Fish and Wildlife Service, through the office of grizzly bear coordinator C. W. Servheen. Administrative support was by the Montana Cooperative Wildlife Research Unit. The simulation language was written by Collin D. Bevins, Missoula, Montana. Anthony J. Boyce graciously allowed us to adopt his pedigree analysis algorithm into our program. For advice and clarification, we are greatly indebted to J. Ballou, J. Beecham, J. Felsenstein, M. Gilpin, R. Lacy, R. Lande, L. Metzgar, M. Reed, N. Ryman, and C. Servheen. We appreciate the improvements suggested by J. Crow and an anonymous reviewer.

Appendix A: Simulation Model Parameters

Natality and survival rates came from data on unhunted populations in Glacier National Park, British Columbia (Mundy & Flook 1973); Glacier National Park, Montana (Martinka 1974); and Yellowstone National Park (Craighead et al. 1974; Knight & Eberhardt 1984); as well as hunted populations in southeastern British Columbia (McClellan 1983) and northwestern Montana (Jonkel 1982; Aune & Stivers 1983). Our intent was not to mimic the behavior of a particular population, but rather to create a "representative" grizzly bear population of the North American Rocky Mountains by combining data from ecologically similar areas. The resulting model was similar in natality rates (Table 6) and age-structure characteristics to other published grizzly bear simulation models (McCullough 1981; Knight & Eberhardt 1985). Further descriptions of the stochastic model and rates used are available in Harris (1984) and Harris & Metzgar (1987).

Appendix B: Published Formulas for Estimation of N_e

Hill (1972)

$$1/N_e = 1/16ML[2 + \sigma^2_{mm} + 2(M/F)cov(mm,mf) + (M/F)^2\sigma^2_{mf}] + 1/16FL[2 + (F/M)^2\sigma^2_{fm} + 2(F/M)cov(fm,ff) + \sigma^2_{ff}] \quad (B1)$$

where M,F = number of males (or females) breeding yearly, L = average age of all parents, σ^2 = variance of number of progeny. Subscript mm denotes male progeny from fathers, mf female progeny from fathers, fm male progeny from mothers, and ff female progeny from mothers, cov(mm,mf) = covariance in male and female progeny from fathers, and cov(fm,ff) = covariance in male and female progeny from mothers. We interpreted M and F as the yearly number of new contributors to the subsequent generation (not merely those attaining reproductive age), and the variances and covariances of total progeny as pertaining only to those offspring who themselves become contributors.

Ryman et al. (1981)

$$\hat{N}_e = (N\hat{}L)/\{(1 + 3h^2)(V_{ka}/k^2_a) + (1/k_a)\} \quad (B2)$$

where $\hat{N}$ = annual recruitment of females of breeding age, L = mean age of mothers, h = heritability of fertility, and k_a and V_{ka} = mean and variance of individual lifetime production of offspring who themselves survive to reproductive age (a = adult).

Ryman et al.'s (1981) simulation model did not include males; they assumed for moose

$$N_e(♂)/N(♂)/N_e(♀)/N(♀) = 0.9$$

and

$$N(♂)/(N(♀) = 0.67$$

To calculate N_e for sexes combined,

$$N_e = 4\hat{N}_{e♂}\hat{N}_{e♀}/(\hat{N}_{e♂} + \hat{N}_{e♀}) \quad (B3)$$

In addition, their model did not provide data with which to estimate V_{ka} directly. Rather, they estimated the mean (k) and variance (V_k) of individual lifetime production of newborns, and converted to V_{ka} assuming alternatively random survival of offspring to adulthood, and 10% of survival being on a family basis (Crow & Morton 1955).

Lemkuhl (1984)

The number of breeding animals required to achieve N_e = 50:

$$N_1 = (2N_eV_k + 1)/k^2 \quad (B4)$$

where k and V_k are defined as by Ryman et al. (above). To account for imbalanced sex ratio:

$$N_2 = N_1 * \{[4N_mN_f/(N_m + N_f)]/N\} \quad (B5)$$

where N_m and N_f are numbers of breeding males and females, respectively. To account for overlapping generations:

$$N_3 = N_2 * 2 \quad (B6)$$

To adjust for population fluctuations:

$$N_4 = N_3 * (N_{max}/N_{min}) \quad (B7)$$

where N_{max} and N_{min} are maximum and minimum population levels. Finally, if the total number of animals (rather than just the number of breeders) is to be calculated:

$$N_5 = N_4 * (1/p) \quad (B8)$$

where p is the proportion of the total population made up by breeding animals.

LaCava and Hughes (1984)

$$N_e = 4m(1 - m)N \quad (B9)$$

where m = the proportion of males among adults. This is equivalent to the more common

$$1/N_e = 1/(4N_m) + 1/(4N_f)$$

originally put forward by Wright (1931).

Reed et al. (1986)

$$1/N_e = 1/(4L_mM_bk_ml_m) + 1/(4L_fF_bk_fl_f) \quad (B10)$$

where L = generation length, M_b and F_b = number of breeding males and females, respectively, k = number of young born to each sex each year, l = probability that a newborn of each age survives and reproduces, and subscripts m and f = male and female, respectively.

Reed et al. (1986), Adjustment

Although not in their paper, Reed (personal communication, 1987) suggested that their formula could be adjusted to account for non-Poisson variance in lifetime progeny distribution by multiplying the final expression by:

$$k/((k - 1) + (V_k/k)) \quad (B11)$$

where k and V_k take the values appropriate when offspring are considered only if they become contributors.

Lacy & Clark (In Press)

To adjust for population fluctuations,

$$N_e = N_e/(1 + cv^2) \quad (B12)$$

where N_e = effective population size calculated assuming constant population size, and cv = coefficient of variation of adult population size (which we interpreted as number actually contributing yearly).

Crow & Kimura (1970)

An alternative adjustment for population fluctuations,

$$N_e = N_e - (\sigma^2/N) \quad (B13)$$

Literature Cited

Allendorf, F. W., R. B. Harris, and L. H. Metzgar. In preparation. Estimation of the effective population size of grizzly bears by computer simulation.

Aune, K., and T. Stivers. 1983. Rocky Mountain front grizzly bear monitoring and investigation. Montana Department of Fish, Wildlife and Parks, Research Report, Helena, Montana. 180 pp.

Boyce, A. J. 1983. Computation of inbreeding and kinship coefficients on extended pedigrees. Journal of Heredity **74**:400–404.

Bunnell, F. L., and D. E. N. Tait. 1980. Bears in models and reality — implications to management. International Conference on Bear Research and Management **3**:15–23.

Craighead, J. J., J. R. Varney, and F. C. Craighead. 1974. A population analysis of the Yellowstone grizzly bears. Bulletin 40, Montana Conservation and Forest Experiment Station, University of Montana, Missoula.

Crow, J. F., and C. Denniston. 1988. Inbreeding and variance effective population numbers. Evolution **42**:482–495.

Crow, J. F., and M. Kimura. 1970. An introduction to population genetics theory. Harper and Row, New York.

Crow, J. F., and N. E. Morton. 1955. Measurement of gene frequency drift in small populations. Evolution **9**:202–214.

Felsenstein, J. 1971. Inbreeding and variance effective numbers in populations with overlapping generations. Genetics **68**:581–597.

Harris, R. B. 1984. Age-structure as an indicator of grizzly bear population status. Unpublished M.S. thesis. University of Montana, Missoula.

Harris, R. B. 1986. Modeling sustainable harvest rates for grizzly bears. Appendix K, pp. 268–279 in A. R. Dood, R. Brannon, and R. D. Mace, editors. The grizzly bear in Northwestern Montana. Final programmatic environmental impact statement. Montana Department of Fish, Wildlife and Parks, Helena.

Harris, R. B., and L. H. Metzgar, 1987. Harvest age structures as indicators of decline in small populations of grizzly bears. International Conference on Bear Biology and Management 7:109–116.

Hartl, D. J. 1980. Principles of population genetics. Sinauer Associates, Sunderland, Massachusetts.

Hill, W. G. 1972. Effective size of populations with overlapping generations. Theoretical Population Biology **3**:278–289.

Hill, W. G. 1979. A note on effective population size with overlapping generations. Genetics **92**:317–322.

Johnson, D. L. 1977. Inbreeding in populations with overlapping generations. Genetics **87**:581–591.

Jonkel, C. 1982. 5-year summary report, Border Grizzly Project, University of Montana, Missoula.

Knight, R. R., and L. L. Eberhardt. 1985. Population dynamics of Yellowstone grizzly bears. Ecology **66**:323–334.

LaCava, J., and J. Hughes. 1984. Determining minimum viable population levels. Wildlife Society Bulletin **12**:370–376.

Lacy, R. C., and T. W. Clark. 1989. Genetic variability in black-footed ferret populations: past, present and future. Pages 83–103 in Seal, U.S., E. T. Thorne, M. A. Bogan, and S. H. Anderson, editors. Conservation biology and the black-footed ferret. Yale Univesity Press, New Haven, Connecticut.

Lande, R., and G. F. Barrowclough. 1987. Effective population size, genetic variation, and their use in population management. Pages 87–123 in M. E. Soulé, editor. Viable populations for conservation. Cambridge University Press, Cambridge, England.

Lemkuhl, J. F. 1984. Determining size and dispersion of minimum viable populations for land management planning and species conservation. Environmental Management **8**:167–176.

Martinka, C. J. 1974. Population characteristics of grizzly bears in Glacier National Park, Montana. Journal of Mammalogy **55**:21–29.

McClellan, B. 1983. Akamina-Kishinena grizzly bear study annual report. British Columbia Wildlife Branch, Cranbrook, British Columbia. 22 pp.

McCullough, D. 1981. Population dynamics of the Yellowstone grizzly bear. Pages 173–196 in C. J. Fowler and T. W. Smith, editors. Dynamics of large mammal populations. John Wiley & Sons, New York.

Mundy, K. R. D., and D. R. Flook. 1973. Background for managing grizzly bears in the national parks of Canada. Canadian Wildlife Service Report Series, No. 22. Ottawa.

Reed, J. M., P. D. Doerr, and J. R. Walters. 1986. Determining minimum population sizes for birds and mammals. Wildlife Society Bulletin **14**:255–261.

Ryman, N., R. Baccus, C. Reuterwall, and M. H. Smith. 1981. Effective population size, generation interval, and potential loss of genetic variability in game species under different hunting regimes. Oikos **36**:257–266.

Shull, A. M., and A. R. Tipton. 1987. Effective population size of bison on the Wichita Mountains Wildlife Range. Conservation Biology **1**:35–41.

Soulé, M. E. 1980. Thresholds for survival: maintaining fitness and evolutionary potential. Pages 151–169 in M. E. Soulé and B. A. Wilcox, editors. Conservation Biology. Sinauer Associates, Sunderland, Massachusetts.

Wood, J. W. 1987. The genetic demography of the Gainj of Papua New Guinea. 2. Determinants of effective population size. American Naturalist **129**:165–187.

Wright, S. 1931. Evolution in Mendelian populations. Genetics **16**:97–159.

Wright, S. 1969. Evolution and the genetics of populations. Vol. 2. The theory of gene frequencies. University of Chicago Press, Chicago.

Reserve Design for Territorial Species: The Effects of Patch Size and Spacing on the Viability of the Northern Spotted Owl*

ROLAND H. LAMBERSON
Mathematics Department
Humboldt State University
Arcata, CA 95521, U.S.A.

BARRY R. NOON
CURTIS VOSS
KEVIN S. MCKELVEY
U.S.D.A. Forest Service
Redwood Sciences Laboratory
1700 Bayview Drive
Arcata, CA 95521, U.S.A.

Abstract: *Designing a reserve system for a threatened territorial species such as the Northern Spotted Owl requires the balancing of biological necessity against economic reality. The Spotted Owl conservation plan and the Pacific Northwest timber industry are in conflict because both demand large areas of mature and old-growth coniferous forests. The primary focus of this paper is the efficient design of a reserve system for territorial species with obligate juvenile dispersal. We examine the relationship between the degree of aggregation of suitable owl habitat, the level of occupancy of that habitat by pairs of Spotted Owls, and the likelihood of owl persistence given different amounts and spatial arrangements of suitable habitat across the landscape. We develop a population model for Spotted Owls that includes an abstracted forest landscape where suitable owl habitat is arrayed in clusters embedded in a matrix not suitable for owl habitation. We conclude from our study of this model that, for any given fraction of the landscape set aside in reserves, the level of occupancy (efficiency of use) of that area will increase as the aggregation of suitable habitat increases. After the reserves reach a size that includes territories for 20 to*

* *The model described in this paper is an extension of the territory-cluster model from the conservation plan for Northern Spotted Owls (Thomas et al. 1990).*

Paper submitted February 17, 1993; revised manuscript accepted July 19, 1993.

Diseño de reservas para especies territoriales: El efecto del tamaño de parche y del espacio en la viabilidad del Búho moteado del norte

Resumen: *El diseño de un sistema de reservas para especies territoriales en pelligro de extinción, tales como el búho moteado del norte, requiere del balance entre necesidad biológica y realidad económica. El plan de conservación del búho moteado y la industria maderera del Pacifico Noroeste estan en conflicto debido a la demanda de grandes áreas de bosques de coníferas maduros y de crecimiento antiguo. El principal foco de este trabajo es el diseño eficiente de un sistema de reservas para especies territoriales con dispersión juvenil obligatoria. A los efectos de alcanzar este logro nosotros examinamos la relación entre el grado de agregación de hábitats adecuados para el búho, el nivel de ocupación de tales hábitats por parejas de búhos moteados y la probabilidad de la persistencia de búhos bajo differentes cantidades y ordenamientos de hábitats adecuados a lo largo del paisaje. Desarrollamos un modelo poblacional para búhos moteados que incluye un paisaje forestal abstracto, donde hábitats adecuados para el búho se ordenan en grupos enclavados en una matriz que no es adecuada para el búho. A partir de nuestro estudio de este modelo concluímos que, para cualquier fracción del paisaje destinado a reservas, el nivel de ocupacion (eficiencia de uso) de dichas áreas aumentará*

25 owl pairs, however, there are diminishing returns from further increases. Preserving connectivity and increasing the geographical extent of the reserve begin to outweigh increased size in importance in insuring the long-term viability of the species.

a medida que la agregación de habitats adecuados incremente. Sin embargo, existe un rendimiento decreciente para aumentos que incrementen el tamaño de las reservas más allá de territorios que alberguen de 20 a 25 parejas de búhos. La preservación de la conectividad y el incremento de la extención geográfica de la reserva, comienza a superar el aumento en tamaño de la reserva en cuanto a su importancia para asegurar la viabilidad a largo plazo de la especie.

Introduction

To be both effective and acceptable for implementation, a conservation strategy must address two basic objectives: first, the strategy must provide for a high likelihood of long-term persistence of the species and, second, it must do so with a minimal impact on resources that have competing economic values. In this paper we address these requirements in the context of conservation planning for the Northern Spotted Owl (*Strix occidentalis caurina*). In many ways the timber industry and the Northern Spotted Owl are in direct conflict because they both demand incompatible access to the same habitat—mature and old-growth coniferous forests in the Pacific Northwest. The practical questions concern to what extent and in what spatial configuration can the timber resource be exploited without endangering the future existence of the Northern Spotted Owl.

It is apparent from earlier work on the viability of the Northern Spotted Owl that there may be sharp thresholds in habitat availability below which the long-term existence of this subspecies is threatened (Lande 1988*a*; Doak 1989; Lamberson et al. 1992; Carroll & Lamberson 1993). Further, it has recently become apparent that the distribution and shape of contiguous areas of suitable habitat can be nearly as important as the total amount of suitable habitat when it comes to insuring the stability of local populations of a threatened species (McKelvey et al. 1993; Carroll & Lamberson 1993).

In this paper we examine the relationship between the degree of aggregation of suitable habitat, the level of occupancy of that habitat by pairs of Spotted Owls, and the likelihood of owl persistence given different amounts and spatial arrangements of suitable habitat across the landscape. By suitable, we mean habitat utilized for nesting, with associated expected birth and death rates that allow for a stable or growing population. We are particularly interested in the efficient allocation of suitable habitat in preserving the owl, that is, an acceptable minimum number of hectares required to maintain a given number of owls with a high likelihood of persistence for the next century. We approach this problem by building a model that incorporates spotted owl population dynamics and, in addition, contains a stylized model of landscape pattern where the suitable habitat is arranged in clusters within a matrix of habitat that is not suitable. We believe the general inferences drawn from our analyses of Spotted Owl populations may be applicable to other territorial species with obligate juvenile dispersal.

Other Models

Most of the earlier attempts at modeling populations of Spotted Owls have either used models that contained no spatial structure (Boyce 1987; Marcot & Holthausen 1987; Noon & Biles 1990) or simply assumed that the suitable habitat was arranged uniformly or distributed randomly across the landscape (Lande 1987, 1988*a*; Lamberson et al. 1992). Other authors who have considered clusters of suitable habitat include Doak (1989), Carroll and Lamberson (1993), McKelvey et al. (1993), and ourselves in an earlier version of this model (described in Thomas et al. 1990: Appendix M).

The model by Doak considers dispersal both within and outside of clusters of suitable habitat, but both are treated somewhat differently than in ours. His model assumes that each cluster is equally accessible when the dispersing juvenile leaves its natal cluster. Doak's dispersal equations for successful dispersal within and between clusters do not incorporate specific geometry for the landscape, but instead he evaluates coefficients that reflect the ease with which dispersing owls may penetrate the landscape. Our model is more realistic in that between-cluster dispersal success is a function of the size of clusters, the distance between clusters, and the costs of locating them.

Carroll and Lamberson (1993) developed an analytic model in which they considered spatial structure, but they were only able to do an equilibrium analysis. Their results are completely compatible with ours even though their approach is entirely different. Both models establish a persistence threshold related to the amount of suitable habitat available and observe that the threshold is lowered (occurs at a lower density of suitable habitat) by increased aggregation of the suitable habitat.

Model Description

In the territory-cluster model, the landscape is portrayed as a rectangular array of identical circular clusters containing potential owl habitat (Fig. 1). Each cluster contains a variable number of owl home ranges (territories or sites), all of identical size (for example, 1000 ha). There is no interstitial space between sites within a cluster. The landscape matrix between clusters was assumed to be entirely unsuitable for owl territories, while the clusters were either partially or totally suitable. Thus, a cluster was a collection of territories with all sites, or only a subset of the sites, within a cluster being suitable owl habitat. The carrying capacity of a cluster, measured in terms of owl pairs, was equal to its number of suitable sites.

The life-history model was an all-female stage projection model with the stage-classes being juveniles, subadults, and adults. This structure was similar to that of Boyce (1987 unpublished data), Lande (1987, 1988*a*), and Noon and Biles (1990). Survival rates were treated as binomial random variables, and fecundity in our model was stochastic, with each year's fecundity drawn from a beta distribution. (The model also permits fecundity to be deterministic or to have a good year/bad year form.) Before dispersal, the number of juvenile owls was reduced by predispersal mortality (Table 1). These deaths were assigned at random to juvenile owls throughout the array of clusters. The mean values of the birth and survival rates used were chosen to approximate current estimates from field data (Table 1) but adjusted to allow the possibility of stable or growing populations.

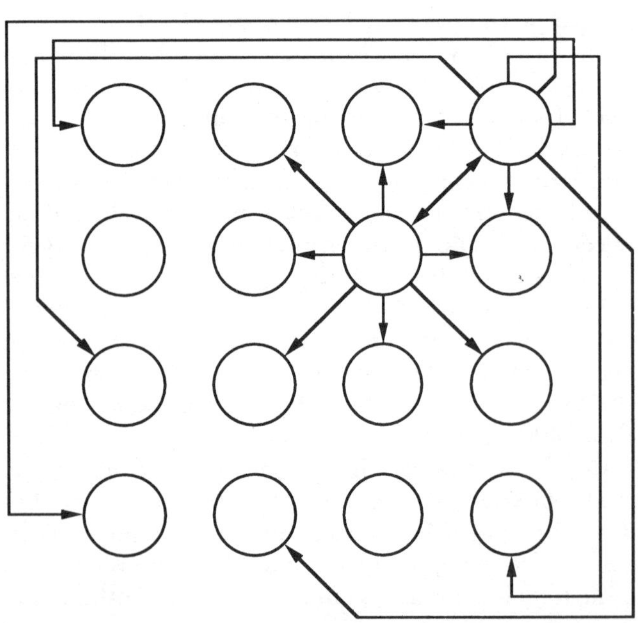

Figure 1. Example of a stylized landscape with a fixed array of circular clusters. Accessibility between clusters with wraparound effect is shown.

Table 1. Estimates of survival and fecundity based on Franklin et al. (1990), and values used in model simulations.

Parameter	*Franklin et al. Estimate*	*Model Values*
Juvenile survival	0.159	——
Predispersal survival	——	0.60, 0.69
Subadult survival	0.83	0.83
Adult survival	0.84	0.92
Fecundity*	0.38	0.38

* *Number of female young per territorial adult female.*

Each cluster was composed of an assigned number of sites, with a given number (or percentage) of them considered suitable habitat. The population status of a cluster was measured in terms of adult female owls (age ≥ 2 years old), subadult ($1 \leq$ age < 2 years), and postdispersal juveniles (age <1 year). The total population for a cluster (exclusive of predispersal juveniles) was never allowed to exceed the carrying capacity of that cluster.

Because little is known of the behavior of Spotted Owls as they search for a suitable site, we chose to model this process as simply as possible. The model distinguished between dispersal within a cluster and dispersal between clusters (compare with Doak 1989). The dispersal within a cluster followed Lande (1987), with each dispersing owl allowed to sample with replacement a given number, *m*, of sites within the cluster. The probability of successful dispersal within the cluster was computed as 1 minus the probability of failing to find a suitable, unoccupied site within the given number of trials:

$$Prob(success) = 1 - \left[1 - \frac{no.\ available\ suitable\ sites}{total\ no.\ sites}\right]^m \quad (1)$$

No further mortality was assigned to juveniles that succeeded in finding a site within their natal cluster (see Carroll and Lamberson 1993 for a treatment of the cost of dispersal both within and outside of the clusters).

To determine the allocation of search effort within a cluster, we assumed a random walk. The assumption was that an owl could traverse one site in a single time step. We estimated, by simulation, the expected number of steps taken before crossing the boundary of a circular cluster, given a random starting point within the cluster. Based on 10,000 simulations for various sized clusters and regressing the number of steps taken before crossing the boundary, we estimated the following relationship:

$$Expected\ (m) = 0.41 * number\ of\ sites\ cluster. \quad (2)$$

For a cluster of size 20 sites, for example, on the average 8 sites are searched before the cluster was exited. The total number of sites sampled (searched) across all clusters was based on the upper bound of the 90th percentile of the maximum straight-line dispersal distances of radio-tagged juvenile Spotted Owls (Thomas et al. 1990: 305) and the assumption that each site had a diameter of about 3.6 km.

If the juvenile female owl did not succeed in finding a territory within her natal cluster (based on a fixed number of searches), she was forced to disperse between clusters. Between-cluster dispersal was modeled as a straight-line path moving away from the natal cluster at a random azimuth. Two sources of mortality existed. First, death resulted if a juvenile female moved in a direction that did not intersect a neighboring cluster. As an example, when suitable habitat accounted for 31% of the landscape, juvenile mortality from this source occurred in about 45% of the cases where juveniles were forced to disperse between clusters. Restricting the likelihood of selecting the correct direction to the immediate neighborhood slightly underestimated juvenile survival for small clusters. For a cluster of size 5 with 60% suitable habitat, for example, the maximum bias was an underestimate of 0.02 in mean occupancy rate.

Second, if a correct direction was chosen, the likelihood of successful travel to the neighboring cluster lying a distance *d* km (edge-to-edge) away was modeled by a declining exponential:

$$Prob(success) = \exp(-k * d). \quad (3)$$

The value of *k* was estimated by arranging the maximum straight-line distances attained by the 56 radio-tagged juvenile owls in rank order, then regressing the natural log of the cumulative proportion represented by that distance (dependent variable) on the associated distance and forcing the regression through 0 ($r^2 = 0.97$).

The overall between-cluster dispersal success probability is given by

$$Prob\ (success) = Prob\ (intersects\ neighboring\ cluster) * \exp(-k * distance\ between\ clusters\ in\ km). \quad (4)$$

Based on an average cluster size of 20 sites occupying 31% of the landscape, when $k = 0.035$, about 53% of juveniles that chose a direction that intersected one of the nearest neighbor clusters successfully arrived at that cluster (about 6.7 km away, edge to edge), which gives an overall success rate of about 24%. Because search begins in the natal cluster and moves through adjacent clusters, the search success in the model, as in nature, is a function of the condition of the landscape near the bird's natal site, as well as the forest matrix.

We envision that most conservation plans will reflect a compromise between the size of reserve areas and the distance between them. Assuming a constant percentage of the landscape as potentially suitable owl habitat and restricting that habitat to the clusters had two important consequences. First, as average cluster size increased, the distance between adjacent clusters increased in a predictable way (Fig. 2). We believe this consequence accurately reflects real-world constraints encountered when developing a species' conservation strategy. Second, the dispersal angle and the probability associated with selecting this angle were constant regardless of the average cluster size (as long as the fraction of the landscape included in clusters remained fixed).

Once a juvenile had successfully traveled to an adjacent cluster, she repeated the within-cluster search process as in the *m* possible searches of the natal cluster (Equation 1). The number of clusters searched could be systematically varied and was based on the total number of sites in the cluster. If *m* searches were unsuccessful in the first cluster, travel to additional clusters was allowed until the maximum number of sites to be sampled was exhausted. Dispersing owls were chosen sequentially, and the occupancy data within each cluster were updated after each dispersal.

For the results reported here, the landscape simulated by the model had a "wrap-around" structure to exclude possible anomalous output that might result from edge effects (Fig. 1). The clusters on the right side of the grid were treated within the model as though they were immediately to the left of those on the left side of the grid. The top and bottom rows of clusters were treated in similar fashion. The effect is to equate immigration and emigration rates, and to assure that all clusters have exactly the same number of neighboring clusters.

Unless otherwise noted, the model was initialized at approximately 90% of a fixed carrying capacity with the age-structure near that expected at equilibrium (stable age-structure about 13% juveniles, 8% subadults, 79% adults). All simulations were initialized with approxi-

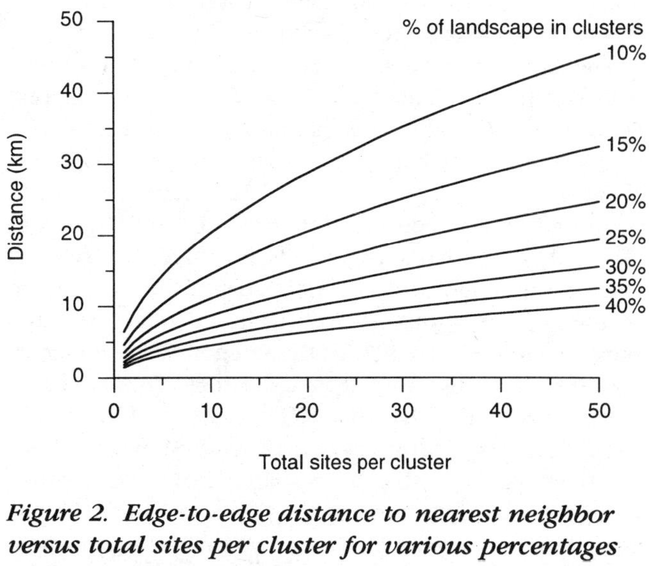

Figure 2. Edge-to-edge distance to nearest neighbor versus total sites per cluster for various percentages of landscape in clusters.

mately the same number of owls—thus, the array size (Fig. 1) was larger for smaller clusters than for larger clusters. The owls were distributed to the clusters at random, with care that none of the clusters exceeded its carrying capacity. All simulations were stochastic, with 50 replications per simulation.

Model Parameterization

The life-history component of the model was parameterized based on the demographic studies of Franklin et al. (1990), except that adult female survival was increased to allow for a stable or increasing population (Table 1). We assumed an annual adult and subadult survival probability of 0.92 and 0.83, a juvenile predispersal survival of 0.60 and 0.69 (both produce overall juvenile survival rates greater than those reported by field studies), fecundity (number of young fledged/adult female) of 0.76, and a 1:1 sex ratio at birth (see Table 1). Reproduction first occurred at age two years.

The standard search pattern used in these simulations was to have a dispersing owl choose, with replacement, a random sample of sites. The number of sites chosen (m) was equal to 0.4 times total sites within a cluster. If the sample chosen from the first cluster did not result in finding a suitable unoccupied site, the owl attempted to disperse to a randomly chosen adjacent cluster where the sampling process was repeated until successful, or until the total number of sites sampled equalled 22. For example, assuming successful movement between clusters, if the clusters were each 20 total sites, the sample from the first cluster would consist of 8 sites and the sample from the second cluster would again include 8 sites; if necessary, a final sample of 6 sites from a third cluster was chosen to complete the 22 allowed. The value for maximum number of searches (22) was based on the dispersal distances of radio-transmittered juveniles (above) and the fact that a 1000-ha site has a diameter of approximately 3.6 km.

The landscape parameters, based on the best estimates of the current status of lands of the U.S. Forest Service and Bureau of Land Management in the Northern Spotted Owl range, are variable and imprecise. Estimates of the percentage of the landscape within the proposed habitat conservation areas (HCAs; Thomas et al. 1990) range from 18% to 31% in different regions (Noon, unpublished data). Estimates of the percentage of the habitat within HCAs that is currently suitable range from less than 30% in the Oregon Coast Range to about 60% in northwestern California. Thomas et al. (1990) recommended that HCAs contain enough area to support at least 20 pairs of owls when suitable habitat has regrown. HCA sizes are variable, however, because among other things home range size is variable (Thomas et al. 1990: Appendix I). Given this variability, for most analyses we selected default settings for parameters of 25% of the forested landscape in clusters, of which 60% was suitable, with 20,000 ha (20 potential owl sites) in each cluster.

Results

The number of model parameters, their range of values, and their possible combinations were immense (Table 2). We examined model behavior using two different techniques. First, to incorporate the full range of potential variability in the model, we performed Monte Carlo sensitivity analyses. These allowed each parameter, for each simulation, to vary independently within some reasonable range of variation (Table 2). Because of the complete randomization of parameter values, we performed a large number of simulations (10,000), sampling from a uniform distribution over each parameter interval. For each 100 year simulation, we recorded three response variables: mean occupancy per suitable site, average population size in the last 30 years of the simulation, and whether the population had persisted or gone extinct (0,1 binary variable). The parameter values, selected during the randomization process for each simulation, served as predictor variables. We used two statistical models to examine response variable sensitivities: multiple regression (mean occupancy and average population size as dependent variables) and two-group discriminant function analysis (population persists or goes extinct). The interpretation of the sensitivities were very similar among the statistical models, so we report only the discriminant results.

Model output is a multi-dimensional, nonlinear surface. Sensitivity changes depending on the region of space over which parameters are allowed to vary. Knowledge of the reaction surface of each parameter when varied independently allowed us to determine those regions in which the model response was mostly linear. Thus, for the Monte Carlo simulations, we restricted the range of some parameter values in order to more clearly interpret the results of the linear statistical models.

We also performed a more restricted set of analyses varying selected model parameters one at a time, holding all others constant. This allowed us to explore in

Table 2. Values of model parameters used in the simulations.

Parameter	*Range of Variation*
Percentage of the landscape within the clusters	5–40
Cluster size	5–45
Percentage of sites within clusters that are suitable	20–100
Number of sites searched per cluster	0.2, 0.4, and 0.8 times cluster size
Dispersal coefficient	0.0004–0.30

Table 3. Two-group discriminant analysis contrasting populations that went extinct (E) or persisted (P) over 100-year simulations.*

Variable	Variable Range	Group Means (SD) E	Group Means (SD) P	Structure Coefficients
Cluster size (CS)	5–20	11.18 (4.46)	12.93 (3.99)	0.300
Percent of suitable sites per cluster	10–40	21.22 (8.48)	29.59 (7.62)	0.683
Total number of searches (TS)	3–20	9.71 (5.36)	13.60 (4.22)	0.552
Dispersal coefficient	0.0–0.1	0.05 (0.03)	0.04 (0.03)	−0.195
CS by TS interaction		102.52 (65.65)	178.52 (82.43)	0.680
Sample size		5332	4668	
Wilk's lambda		0.549		
Approximate *F*-statistic		1639.00**		

* *Summary statistics and group means and standard deviations of variables included in the model are presented. Proportion of the landscape within clusters = 0.20.*
** *Significant at* $p < 0.001$.

more detail the nonlinear relationships between occupancy rate and various parameter values. Specifically, we explored the effects of varying cluster size and spacing, key components common to any reserve design, under the assumption that the majority of sites within clusters were suitable and static.

Monte Carlo Sensitivities

Based on the randomized parameter values and their interactions, we observed significant discrimination between simulated populations that persisted and those that went extinct (Table 3). Interpreting the discriminant model on the basis of total structure coefficients (Green 1978:271–273), the most important variables were cluster size, total searches and their interaction, the proportion of the cluster that was suitable habitat, and the costs of dispersal between clusters (Table 3). Based on these variables, the a posteriori probability of correct classification of each simulation into either the persistant or extinct group exceeded 80%.

INDIVIDUAL PARAMETER SENSITIVITIES

The cluster size at which mean occupancy stabilized was affected by the number of sites that were searched within the natal cluster before dispersal into the surrounding matrix (Equation 1). To explore the sensitivity of mean occupancy to variation in search, we fixed the total number of sites sampled but systematically varied the number of sites searched in the natal and subsequent clusters (Fig. 3). Assuming 20 sites per cluster, 60% suitable, mean occupancy was moderately sensitive to variation of within-cluster search efficiency. For example, the difference in equilibrium occupancy between 4 sites and 16 sites searched per cluster was roughly 20 percent (Fig. 3). The sensitivity to within-cluster search efficiency increased as clusters became smaller. In contrast, increasing the total number of searches beyond about 22 had little effect on mean occupancy for any of the within-cluster search allocations we examined.

VARYING BETWEEN-CLUSTER DISPERSAL RISKS

We assumed that juvenile owls search systematically within their natal cluster for a vacant site before entering the forest matrix to search for another cluster. If dispersing juveniles allocate less search effort to their natal cluster, they will leave that cluster sooner and spend more time moving through the matrix. As the nature of the matrix changes, however, dispersal risks may increase and affect overall occupancy rates. To investigate the effect of increased dispersal costs, we varied the dispersal parameter ***k*** (Equation 4) but fixed the total number of searches. The general result was a reduction in equilibrium mean occupancy (Fig. 4). For clusters of size 20, for example, the reduction due to increasing ***k*** was about 20%. If less search is allocated to each cluster and dispersal risks are higher, mean occupancy is reduced even further.

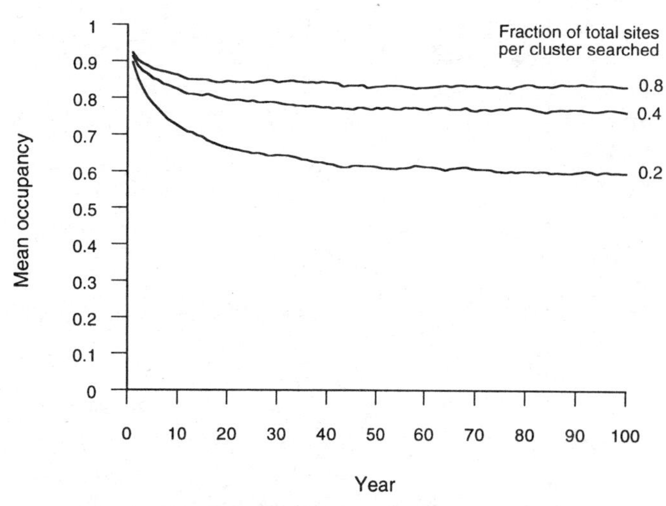

Figure 3. Mean occupancy versus time for various numbers of sites sampled per cluster during the search process. Sample sizes were 0.2, 0.4, and 0.8 times the total number of sites per cluster. All clusters had 20 sites, with 60% of the sites suitable.

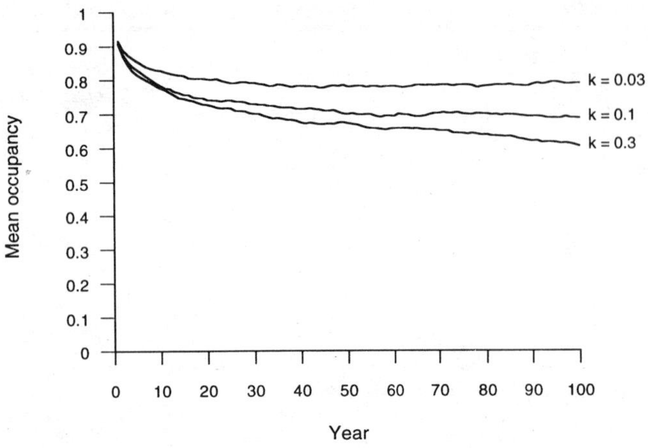

Figure 4. Mean occupancy versus time for various values of the exponential coefficient in Equation 3. All clusters had 20 sites, with 60% of the sites suitable.

VARIATION IN CLUSTER SIZE

The most significant results of this study pertain to the relationship between mean occupancy and the number of sites per cluster. Assuming 60% of the sites in each cluster to be suitable (approximately the current condition), the number of sites sampled per cluster to be 40% of cluster size, and a maximum of 22 sites searched, we did not observe a stabilization of mean occupancy within 100 years until each cluster contained at least 15 sites (Fig. 5). Clusters of 20 sites stabilized at approximately 77% occupancy, while clusters of 45 stabilized at near 90%. Once clusters contained 30 sites, however, increasing cluster size had little effect on the equilibrium level of mean occupancy (Fig. 5), with larger clusters stabilizing between 85% and 90% occupancy.

The sensitivity analyses suggest that the equilibrium mean occupancy would depend on the percentage of suitable sites within a cluster. Assuming that 80% of the sites within a cluster are suitable, mean occupancy increased beyond that observed at a level of 60% suitable (compare Figs. 5 and 6). Occupancy rate for clusters of 5–10 sites, however, remained well below those with 15 or more sites per cluster.

Another approach to choosing an optimal cluster size is to compute the number of owls expected to occupy a fixed amount of suitable habitat. Based on mean occupancies at 100 years (Fig. 5), we plotted mean occupancy verses cluster size as a measure of the efficiency of land use. For a fixed allocation of land in a reserve system, larger clusters support larger equilibrium population sizes (Fig. 7).

VARIATION IN THE FRACTION OF LANDSCAPE IN RESERVES

Variation in the percentage of the landscape within clusters interacted with cluster size to affect mean occupancy. We investigated the strength of this interaction through a series of simulations. Assuming that 60% of the sites in each cluster were suitable, we systematically varied the percentage of landscape in clusters and cluster size. For a given cluster size, increasing the percentage of the landscape in clusters increased mean occupancy (Fig. 8). The increase, however, was much more dramatic for small clusters (5–10 sites) than for large clusters (20 or more sites).

Given the relationship between the landscape frac-

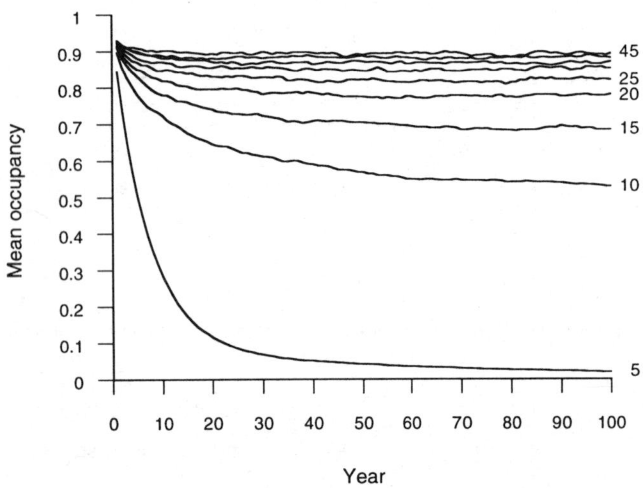

Figure 5. Mean occupancy versus time for various cluster sizes with 60% suitable within cluster. Total sites per cluster = 5, 10, 15, 20, 25, 30, 35, 40, and 45, with the lowest curve representing 5 sites per cluster and moving progressively upward to the top curve, which represents 45 sites per cluster.

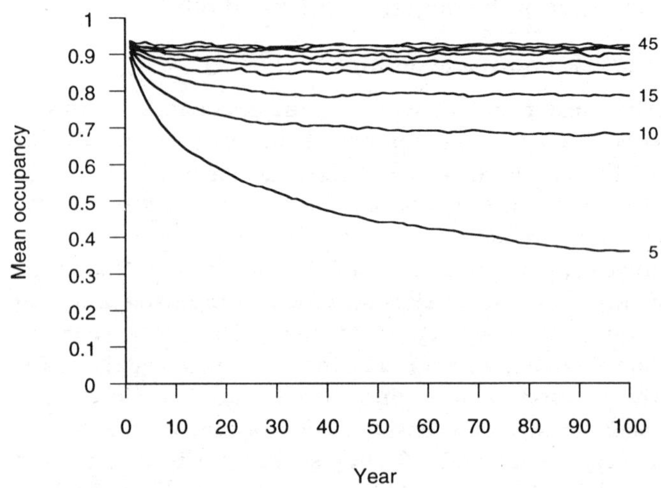

Figure 6. Mean occupancy versus total sites per cluster, with 80% suitable within cluster. Total sites per cluster = 5, 10, 15, 20, 25, 30, 35, 40, and 45, moving progressively upward with increasing sites per cluster.

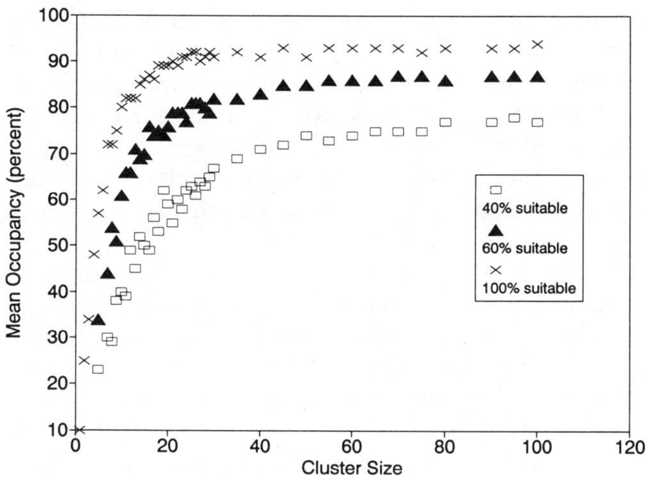

Figure 7. Mean occupancy versus total sites per cluster for 40%, 60%, and 100% of the sites suitable within each cluster. Occupancies are averaged over the last 10 years of each simulation.

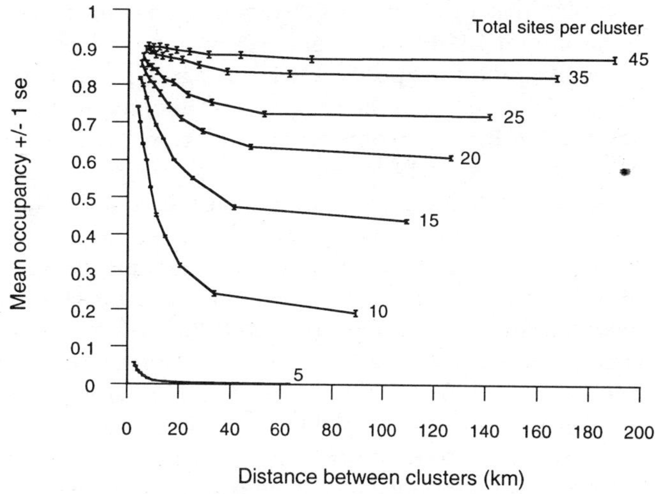

Figure 9. Mean occupancy ± 1 S.E. versus edge-to-edge distance to nearest cluster, for various total sites per cluster. Occupancy rates represent the last 10 years of each simulation. Total sites per cluster = 5, 10, 15, 20, 25, 35, and 45, of which 60% were suitable sites.

tion within clusters and cluster spacing (Fig. 2), different levels of allocation to a reserve system can also be viewed in terms of the distance between clusters. For clusters with less than 20 sites, we found mean occupancy to be strongly affected by increased spacing between clusters (Fig. 9). In contrast, clusters with more than 25 sites showed minimal distance effects beyond about 30 km spacing (Fig. 9). Including low levels of environmental variation in the survival rates lowered occupancy rates 3–7%. This suggests that in reality slightly larger cluster sizes are needed for the same levels of occupancy.

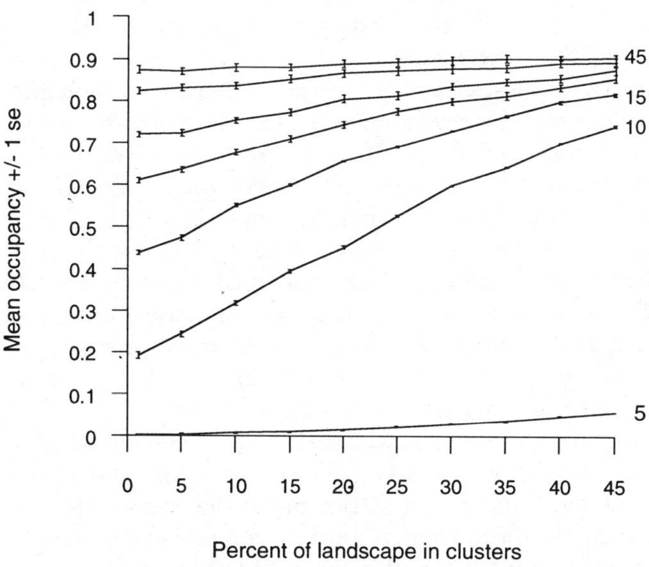

Figure 8. Mean occupancy ± 1 S.E. versus percentage of landscape in clusters, for various total sites per cluster. Occupancy rates represent the last 10 years of each simulation. Total sites per cluster = 5, 10, 15, 20, 25, 30, 35, and 45, of which 60% were suitable sites.

In general, mean occupancy increased with cluster size, percentage of suitable sites, and percentage of the landscape in clusters. With clusters of more than 25 sites, however, changes in landscape percentage had little effect on mean occupancy. This result was a consequence of our assumptions about the behavior of dispersing juveniles. With about 35 sites per cluster, a successful dispersal event usually occurred within the natal cluster. With fewer suitable sites, or smaller clusters, birds were frequently forced to leave their natal cluster and enter the surrounding matrix (Table 4). As a consequence, they had decreased dispersal success. Thus, to the extent that our within- and between-cluster dispersal equations reflect reality, cluster size is more important than cluster spacing.

Discussion

In keeping with the findings of Doak (1989) and Carroll and Lamberson (1993), we conclude that a conserva-

Table 4. Dispersal success rates for a range of cluster sizes.

Cluster Size	*Found Site in Natal Cluster (%)*	*Found Site Outside Natal Cluster (%)*	*Successfully Dispersed Between (%)*
5	43.0	13.6	37.3
10	52.1	11.6	29.4
15	55.0	6.1	25.0
20	55.8	5.4	21.6
25	59.6	2.3	18.6

tion plan that provides for clusters of territories above some minimal size should greatly increase the persistence likelihood of Spotted Owls, primarily by facilitating juvenile dispersal. Because of the uncertainties associated with environmental variation, precisely estimating a critical cluster size for a specified likelihood of persistence is very difficult at this time. Within the structure of our model simulations—60% of the sites suitable, moderate connectivity among clusters (Table 4), preferential search within the natal cluster before dispersal, equilibrium conditions, and no edge effects—clusters with 20 or more sites appear to support stable populations. As a member of a large, clustered population, an individual Spotted Owl becomes less susceptable to the uncertainties of between-cluster dispersal and the character of the landscape matrix.

Cluster size and the percentage of suitable sites were highly interdependent. On average, suitable territories are less contiguous in clusters below their potential, habitat-determined carrying capacities. Therefore, within-cluster dispersal becomes more uncertain, rates of territorial replacement when birds die are decreased, and birds spend more time traveling through the surrounding matrix. Collectively, these factors all contribute to lower survival and fecundity rates and increase the risks to long-term viability.

Of interest was the observation that once a minimum number of searches was allowed, the probability of finding a suitable site became largely independent of the number of searches. For the Spotted Owl life history and associated parameter values (Table 1), that value occurred at about 20 sites sampled (see Carroll and Lamberson 1993 for an analytical derivation).

Further, our results suggest that a reserve design that provides only for individual pairs or small clusters of pairs has a low likelihood of sustaining the species. In fact, even relatively large clusters for Spotted Owls have uncertain fates if they currently hold far less suitable habitat than their eventual carrying capacity. Such areas will need to be closely monitored until habitat within them has had time to recover. One way to increase their short-term occupancy rate is to increase their current effective size by preserving adjacent, suitable habitat.

We emphasize the fact that ours is an all-female model. This simplification eliminated the complications of mate finding, an additional source of uncertainty for dispersing birds. Consequences of mate finding difficulties are most pronounced for small clusters, or when the population is very diffuse relative to the amount of suitable habitat (Lamberson et al. 1992). Thus, the inclusion of search for mates will further reduce mean occupancies when cluster sizes are small, certainly when they contain less than 20 sites.

Models of the population dynamics of long-lived vertebrates are difficult to validate. Further, the output and inferences drawn from a model are always a reflection of the model's structure. The patterns we observed in our simulations reflect the model's structure and the assumptions we made about Spotted Owl behavior. We have structured our model to reflect our current understanding of owl behavior and life history. In some aspects of model structure, however, such as basic life-history patterns and the values of birth and death rates, we have more certainty than in others. For example, our model and its results are clearly the consequence of assumptions we have made about the dispersal behavior of juvenile owls within and between territory clusters. Unfortunately, little is known of Spotted Owl dispersal costs and how juvenile birds move through heterogeneous landscapes.

One inference drawn from our results—the positive effect of increasing cluster size—has much stronger support in both empirical and theoretical studies. Populations quickly escape from the dangers of demographic stochasticity with even slight increases in population size (see Goodman 1987). Populations also gain security from environmental uncertainty with increasing numbers, but at a slower rate than from demographic effects (Shaffer 1987). Therefore, this model result was not surprising. Of interest, however, was the fact that marginal gains in mean occupancy were not constant with incremental increases in cluster size. Rather, large gains occurred in moving from clusters of 5 to clusters of 20 sites; much smaller gains were made in moving from 20 to 45 sites per cluster.

In their individual territory model, Lamberson et al. (1992) assumed all sites were equally likely to be searched until a suitable site was located or search capabilities were exhausted (that is, the owl died). The probabilities of finding a suitable site, or an individual of the opposite sex, became insurmountable when suitable habitat was less than about 15% of the landscape, or the population density was too low. In the current model, we invoked the concept of clusters, groups of two or more adjacent suitable sites. If a single very large cluster is created with reflecting boundaries, however, model behavior is similar to the individual territory model. Clusters did not have reflecting boundaries, but owls searched preferentially within their natal cluster before searching the matrix for another cluster. With a finite number of searches and a sufficiently large cluster (of more than 40 sites), search seldom occurred outside the natal cluster. For very large clusters, the model behaves as if the cluster boundaries are reflecting—little between-cluster movement occurs, and occupancy rate is insensitive to dispersal costs or the distance between clusters. In general, large clusters become very stable if the proportion of suitable sites within the cluster lies above the threshold value derived in Lamberson et al. (1992).

If Spotted Owls do not preferentially search within clusters, the costs associated with moving through the matrix become considerably more important. Assuming a fixed search effort and less preference for within-cluster search, more effort is spent searching among clusters. Therefore, the necessary cluster size to achieve a given level of occupancy is increased proportional to the distance traveled in the matrix.

We assumed that the risks to owl viability were more pronounced for demographic than genetic factors (see Lande 1988*b*). There may be some long-term genetic consequences, however, if dispersing owls preferentially search their natal cluster before entering the matrix. The general effect of subdividing the owl population into a number of clusters may be to slightly increase the total genetic variation among clusters (Lande & Barrowclough 1987). Widely spaced but very large clusters would have local demographic stability but may eventually lead to a decline in genetic variation due to inbreeding. This is more likely if we have overestimated the vagility of dispersing owls. Smaller clusters more closely spaced promote higher dispersal rates between clusters but may lead to local demographic instability. Therefore, there may be an optimal size for a cluster that achieves a balance between these two processes.

The assumption of Thomas et al. (1990) that risks to owl viability were more pronounced from demographic than genetic factors could be significant if they have overestimated the vagility of dispersing owls in the forest matrix. If the assumption is false, there may be possible deleterious genetic effects from inbreeding within the large clusters at some future time. To have a higher degree of confidence in the generality of our model results, however, will require a considerably deeper understanding of how Spotted Owls move through their environment and respond to habitat heterogeneity at a variety of spatial scales.

The question of reserve design for Spotted Owls concerns both how big the clusters should be and how much of the landscape should be in clusters (thus controlling spacing). For an all female model, clusters with 15 or more sites appear stable but are still quite sensitive to cluster spacing (Fig. 9). To assure successful mate finding in a real reserve system subject to environmental variation and irregular spacing, this number should be increased to at least 20–25. With clusters of 20 sites, the number of owls that will be protected in the same amount of reserve land is more than three times that for clusters of 5 sites. Thus, to protect the same number of owls in smaller clusters would require a substantial increase in the amount of land that would have to be set aside.

Given some minimum criteria for cluster size and spacing, the next concern is the number of clusters required for a stable reserve system (the percentage of the landscape within the species range to include within clusters). To address this question requires knowledge of the frequency and magnitude of catastrophic events. Our model did not include these effects because, at this time, we do not know how to reliably model them. Several theoretical studies, however, have demonstrated that metapopulation extinction times increase exponentially with the number of local populations (see Nisbet & Gurney 1982), largely as a consequence of reducing the probability that local populations simultaneously experience major environmental perturbations (Goodman 1987; Harrison & Quinn 1989). Thus, the risk of stochastic failure of the reserve system is greatly diminished by increasing the number of large clusters and spreading them widely across the landscape.

The proposed cluster structure for the Northern Spotted Owl—the system of habitat conservation areas (HCAs) would reserve roughly 15–30% of the forested landscape on public lands in the owls' range (Thomas et al. 1990). Most of the conservation areas will have an eventual capacity of 15 to 25 pairs of owls, and they will be established throughout the species' current range (Thomas et al. 1990: Appendix Q). In addition, Thomas et al. (1990) proposed specific minimum habitat requirements for the forest matrix in order to facilitate owl dispersal. Our model results, assuming 60% suitable habitat within clusters with 20 or more sites, suggest a high likelihood for long-term stability. Given implementation of the their conservation strategy, and assuming rapid recovery of habitat within those HCAs that currently contain little suitable habitat, Thomas et al. (1990) speculated that the strategy would have a moderate to high likelihood of providing for long-term stability and at the same time the connectivity necessary to avert genetic difficulties in future owl populations.

The greatest uncertainty with the owl conservation strategy (Thomas et al. 1990), and a limitation of the equilibrium model used here, is our ignorance of the transient behavior of a recovering population while its habitat is regenerating. Several of the HCAs, particularly those on Bureau of Land Management lands in the northern coast range of Oregon, have far less than 60% suitable habitat. In these HCAs, we expect population instability in the near term. Given many alternative ways to attain the future steady-state reserve structure, preventing the population trajectory from passing through zero during the transition may depend critically on the pattern and rate of recovery of critical habitat within the reserve.

Acknowledgments

Professor Lamberson's work on this project was supported by U.S. Forest Service Cooperative Research Agreement PSW-91-000807. Helpful review comments were provided by Russ Lande, Barney Dunning, and Ron Pulliam.

Literature Cited

Boyce, M. S. 1987. A review of the U.S. Forest Service's viability analysis for the spotted owl. Final report to the National Council of the Paper Industry for Air and Stream Improvement. Unpublished.

Carroll, J. E., and R. H. Lamberson. 1993. A continuous model for the dispersal of territorial species, SIAM. Journal of Applied Mathematics 53(1):205–218.

Dennis, B. 1989. Allee effects: Population growth, critical density, and the chance of extinction. Natural Resources Modeling 3:481–538.

Doak, D. 1989. Spotted owls and old growth logging in the Pacific Northwest. Conservation Biology 3:389–396.

Franklin, A. B., J. A. Blakesley, and R. J. Gutierrez. 1990. Population ecology of the northern spotted owl in northwestern California: preliminary results. 1989. Final report submitted to the Forest Service and the California Department of Fish and Game. Unpublished.

Goodman, D. 1987. The demography of chance extinction. Pages 11–34 in M. E. Soule, editor. Viable populations for conservation. Cambridge University Press, Cambridge, England.

Green, P. E. 1978. Analyzing multivariate data. Dryden Press, Hinsdale, Illinois.

Harrison, S., and J. F. Quinn. 1989. Correlated environments and the persistence of metapopulations. Oikos 56:293–298.

Lamberson, R. H., R. McKelvey, B. R. Noon, and C. Voss. 1992. The effects of varying dispersal capabilities on the population dynamics of the northern spotted owl. Conservation Biology 6:505–512.

Lande, R. 1987. Extinction thresholds in demographic models of territorial populations. American Naturalist 130:624–635.

Lande, R. 1988*a*. Demographic models of the northern spotted owl (*Strix occidentalis caurina*). Oecologia 75:601–607.

Lande, R. 1988*b*. Genetics and demography in biological conservation. Science 241:1455–1460.

Lande, R. and G. F. Barrowclough. 1987. Effective population size, genetic variation, and their use in population management. Pages 87–123 in Viable populations for conservation. M. E. Soulé (ed.). Cambridge University Press, Cambridge. 189 pp.

Marcot, B. B., and R. Holthausen. 1987. Analyzing population viability of the spotted owl in the Pacific Northwest. Transactions of the Conference on North American Wildlife and Natural Resources 52:333–347.

McKelvey, K., B. R. Noon, and R. H. Lamberson. 1993. Conservation planning for species occupying fragmented landscapes: The case of the northern spotted owl. Pages 424–450 in P. Kareiva, J. Kingsolver, and R. Huey, editors. Biotic interactions and global change. Sinauer Associates, Sunderland, Massachusetts.

Nisbet, R. M., and W. S. C. Gurney. 1982. Modelling fluctuating populations. John Wiley & Sons, New York.

Noon, B. R., and C. M. Biles. 1990. The mathematical demography of the spotted owl in the Pacific Northwest. Journal of Wildlife Management 54:18–27.

Shaffer, M. 1987. Minimum viable populations: Coping with uncertainty. Pages 69–86 in M. E. Soule, editor. Viable populations for conservation. Cambridge University Press, Cambridge, England.

Thomas, J. W., E. D. Forsman, J. B. Lint, E. C. Meslow, B. R. Noon, and J. Verner. 1990. A conservation strategy for the northern spotted owl. 1990-791-171/20026. U.S. Government Printing Office, Washington, D.C.

USDA Forest Service. 1988. Final supplement to the environmental impact statement for an amendment to the Pacific Northwest regional guide. 2 vols. U.S. Department of Agriculture, Forest Service, Portland, Oregon.

Seed Germination Percentage Increases with Population Size in a Fragmented Prairie Species

ERIC S. MENGES
Archbold Biological Station
P.O. Box 2057
Lake Placid, FL 33852, U.S.A.

Abstract: *Laboratory germination percentages were relatively low in small populations of royal catchfly (*Silene regia*), a perennial prairie plant whose distribution has been severely reduced by habitat fragmentation. Populations with more than 150 individuals had uniformly high (>85%) germination, whereas smaller populations had great variation within and between populations. Germination percentage was related to the natural log of population size (r = .49, p = .017); in small populations the relationship was linear (r = .37, p = .156). Germination success was unrelated to population isolation. The overall dependence of germination percentage on population size has at least two possible explanations: (1) inbreeding depression in recently reduced populations, manifesting itself in lower fitness of seeds; or (2) increased proportion of inferior geitonogamously produced seed caused by reduced hummingbird visitation and reduced interplant movements in small populations.*

Resumen: *Los porcentajes de germinación en el laboratorio fueron relativamente bajos en las pequeñas poblaciones de* Silene regia, *una planta de pradera perenne cuya distribución ha sido severamente reducida por la fragmentación del hábitat. Las poblaciones con más de 150 individuos tuvieron una germinación uniformemementе alta (>85%) mientras que las poblaciones más pequeñas tuvieron gran variación dentro y entre las poblaciones. El porcentaje de germinaciones estaba relacionado con el logaritmo natural del tamaño de la población (r = .49, p = .017); en las poblaciones pequeñas la relación fue linear. (r = .37, p = .156). El éxito de la germinación no estuvo relaciondado con el aislamiento de la población. La total dependencia del porcentaje de la germinación en el tamaño de la población tiene cuando menos dos posibles explicaciones: (1) depresión por el entrecruzamiento consanguíneo en poblaciones recientemente reducidas, manifestándose como disminución en la viabilidad de las semillas; o (2) el aumento en la proporción de semillas inseriores producidas por polinización dentro de la misma planta causada por la redución de visitas de colibries y los reducidos movimientos entre planta y planta en las poblaciones pequeñas.*

Introduction

The theory of conservation biology holds that small and isolated populations may, for a number of reasons, be characterized by reduced population viability and persistence times (Shaffer 1981; Berger 1990; Menges 1991). The disadvantages suffered by small and isolated populations involve both reduced genetic variation and greater sensitivity to nongenetic forces such as environmental and demographic stochasticity. These processes may interact to promote high extinction probabilities (Gilpin & Soulé 1986).

Genetic variation is generally regarded as an important resource, allowing for short term adaptation to environmental fluctuation and long term evolutionary change (Wright 1948; Frankel & Soulé 1981; Beardmore 1983; Harris 1984). In small populations, we expect loss of additive genetic variance, increased inbreeding, increased homozygosity, and the potential expression of

Paper submitted July 6, 1990; revised manuscript accepted October 23, 1990.

deleterious alleles (Dobzhansky 1970; Wright 1977; Lande 1980; Ledig 1986). Because these negative consequences of small population size can be ameliorated through gene flow (Lewontin 1973), geographically isolated, small populations are most likely to suffer from depletion of genetic variation. In practice, the effects of population bottlenecks (periods of time spent at very low population size) on genetic variation appear complex (e.g., depending on the length of time a population stays in a bottleneck) and, in some cases, reversible (Rice & Jain 1985; Bryant et al. 1986). Also, populations recovered from previous bottlenecks may exhibit relatively low levels of inbreeding depression (Schemske & Lande 1985).

Large-scale destruction of natural habitat has reduced many species distributions to a scattering of small and isolated populations. In most parts of North America, these changes have occurred during the last 150 years (e.g., Curtis 1956; Burgess & Sharpe 1981). Habitat fragmentation can produce a pattern of small, isolated populations for which losses in genetic variation are likely. However, these potential effects are virtually undocumented for plant species.

Fitness reductions due to lowered genetic variation could be exposed early in the life of the sporophyte plant (e.g., fruit set, seed set, and germination). Most studies of inbreeding depression focus on fruit and seed set. However, germination differences among populations could well reflect inbreeding effects, and germination differences among individuals have important fitness consequences. Population differences in seed germination percentages and populations' responses to environmental conditions have been documented for many species (e.g., Farmer et al. 1986; Wu et al. 1987; Meyer et al. 1989), but to consider the possible effects of fragmentation-caused reduced genetic variation, these germination differences must be correlated with population size and isolation. To my knowledge, no studies make this link. This article reports such an effort using the prairie populations of *Silene regia* (Caryophyllaceae).

Silene regia (royal catchfly) is a perennial herb with hummingbird-pollinated, perfect flowers. All multiplication is through seed production as the plant does not spread clonally (Menges 1988). *Silene regia* is endemic to 12 states in midwestern and eastern North America, growing in prairies and dry woods in unglaciated areas of southern Missouri and neighboring states, and restricted to prairies in glaciated areas further east. Before and during the Pleistocene, repeated climatic and glacial fluctuations changed the distribution of midwestern prairies (Gleason 1923; Grimm 1983; Axelrod 1985). During prairie retreats, *S. regia* may have been restricted to Ozarkian refugia, where most populations are found today. In contrast, populations in Indiana and Ohio date from the last 10,000 years, before which ice sheets or conifer forests were present (Axelrod 1985).

In the last 100 years, prairies in eastern North America have been greatly reduced. Calculations comparing presettlement vegetation maps with current statewide heritage program data suggest that less than 0.1% of presettlement prairie remains in parts of Ohio and Indiana, while at least 4% remains in southwest Missouri (Menges 1988). *S. regia* currently occupies from 17%–33% of prairie remnants (by region), suggesting that it may have been fairly common in the landscape before European settlement. Now, however, most populations to the east of the Mississippi River (and many to the west) are small and isolated. Therefore, *S. regia* is a good candidate for understanding the effects of population size and isolation on individual fitness and population viability.

An experiment on seed germination as a function of population size and isolation was undertaken with *Silene regia*. The question addressed is: do small and isolated populations suffer reduced seed germination?

Methods

In 1987, I collected seeds from 23 prairie populations of *Silene regia* (eight from Ohio, two from Indiana, one from Illinois, ten from Missouri, two from Arkansas). Fruits were taken from 10–20 plants per population. Full-sized, healthy-looking seeds (40–50) from a single maternal plant were extracted and placed on filter paper within a petri dish. At least three dishes (i.e., seed families from three or more maternal plants) per population were created; eight populations had seeds from ten or more maternal plants. Seeds were moist stratified (5°C) for 48–60 days, conditions which produce high *S. regia* seed germination (Menges 1988). Subsequent germination under laboratory light conditions was monitored twice weekly for 4 weeks, at which point germination had ceased. Based on past laboratory experiments, I assumed nongerminated seeds were dead, not dormant (Menges 1988).

Sizes of Ohio and Indiana populations were assessed during 1987 peak flowering. Areas were carefully searched by several people walking parallel lines delimited by flagging or obvious landmarks. Within these areas, all plants were counted. Most populations were completely censused. The sizes of the largest Missouri and Arkansas populations were estimated based on partial counts in subsets of population areas. Counts could detect second-year plants but not seedlings. All population counts were complete before the germination experiment began.

Population distribution data were garnered over several years from sources indicated in the acknowledgments and from my own searches. Populations were classified as isolated (no populations of 50 or more individuals within 12 km) or not isolated. This arbitrary cutoff classified 11 populations as isolated and 12 as not isolated. Because the locations of undiscovered populations are unknown, this classification could overrate the isolation of certain study populations.

Analysis of covariance was used to examine the effects of isolation and population size. Germination data are reported as raw percentages, but analyses of variance were conducted on angular-transformed germination proportions. The relationship between population size and germination percentage was explored using a number of transformations and evaluated by proportion of variance explained and residual distribution. Nested analysis of variance was used to examine within-population versus between-population variation.

Results

Germination percentage, averaged within populations, increased with ln population size (Fig. 1). Only populations smaller than 100 plants had less than 50% germination. Larger populations (>150) had uniformly high rates of seed germination. The curvilinear regression fit implies that population size effects on germination operate mainly among smaller populations. The natural log of population size explained about one-quarter of the variance in mean germination rates ($r = 0.492$, p

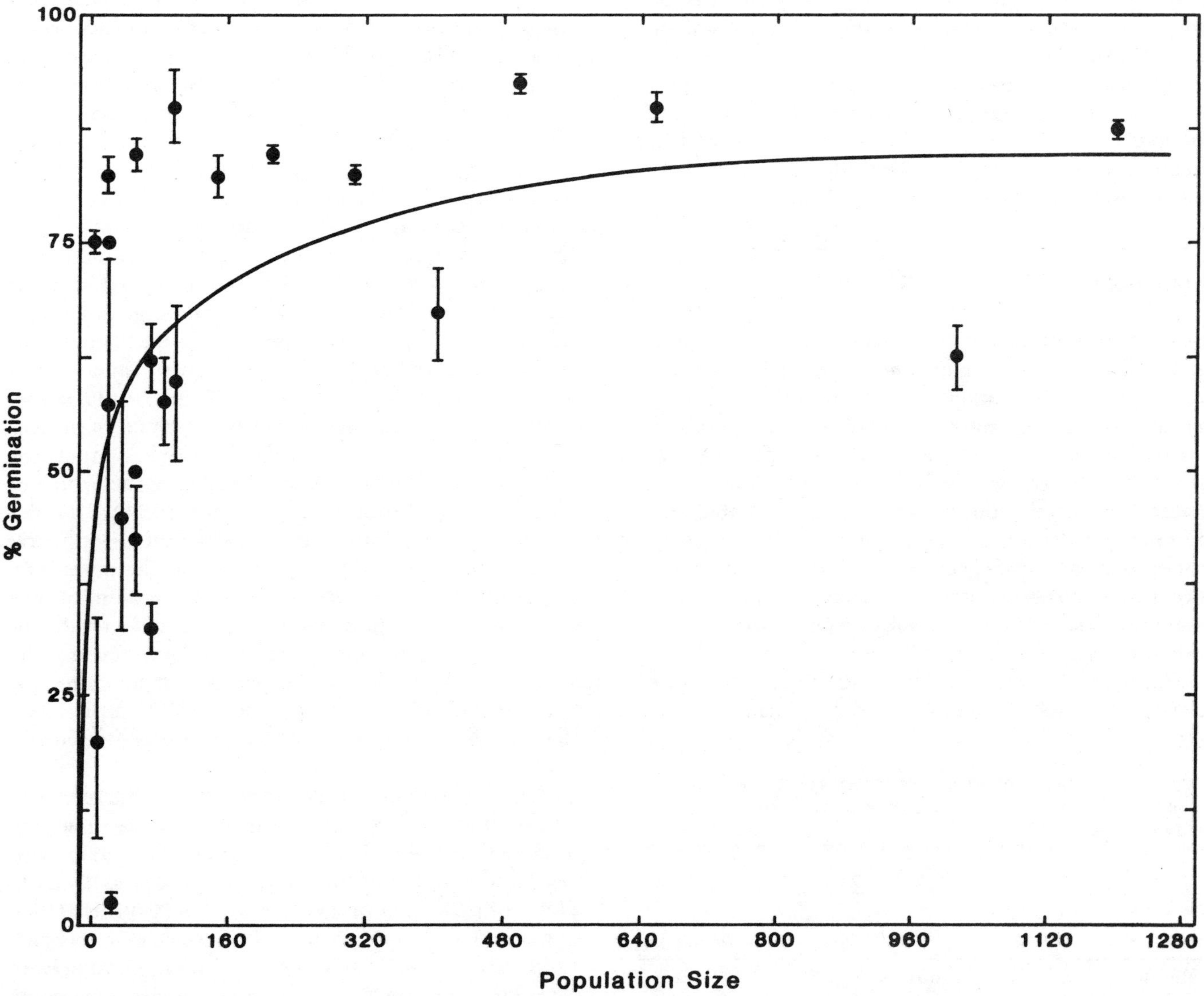

Figure 1. Average germination percentages for 23 populations (dots) as a function of population size. Error bars are one standard error. The equation for the fitted regression line is % germination = 7.42 (ln population size) + 32.6, $r = 0.492$, $n = 23$, $p = 0.017$.

= .017). For populations smaller than 150, germination percentage was linearly related to population size, but weakly ($r = .37$, $n = 15$, $p = .156$).

Population isolation was unimportant once population size was accounted for in analysis of covariance. The addition of a binary classification of population isolation did not increase the predictive power of population size, nor were population size and isolation interactions important (Table 1). Isolation was also not significant in analysis of covariance with size, for populations smaller than 150 (analysis not shown).

Germination among populations was highly variable (Fig. 1). This was due both to between-family variation (within populations) and between-population variation, as detected by nested analysis of variance (Table 2). Most of the variation in germination percentages exists between families within populations. Nevertheless, between-population patterns with regard to size are significant (Fig. 1).

Regional differences in germination were dramatic. Average germination in Indiana, Ohio, and Illinois populations (86.8%) was greater than in Missouri and Arkansas populations (65.7%). This difference was highly significant (one-way ANOVA, $p < 0.000$).

Discussion

Population size is linked to seed germination percentage in *Silene regia*. Larger populations had higher percentages of seed germination than smaller populations, regardless of region or isolation (Fig. 1). Overall, germination was related to the natural log of population size, but for small populations (<150) the relationship was linear. Small populations also had more variable seed germination percentages (the standard error of germination percentage was negatively related to population size, $r = -.26$). Laboratory germination of *S. regia* from three other small Illinois populations was also very low (Ketzner et al. 1989).

This study is the first to examine the effects of population size and isolation on seed germination within

Table 1. Analysis of covariance for percentage germination (population mean) as a function of ln population size and isolation.

Model	*Terms*	r^2	*F-value—adding additional term*	*Significance*
Saturated	Ln Size Isolation Interaction	.24661	.12	>.05
Additive	Ln Size Isolation	.24229	.01	>.05
Simple	Ln Size	.24175	6.70	.018

Table 2. Nested analysis of variance for population and maternal family effects on germination. The analysis uses angular-transformed germination percentages. Only populations with ten or more maternal families were included.

Source of variation	*Sum of squares*	*df*	*F*	*Significance*
Seed family (within population)	62,003	106	5.1	.0001
Population	14,789	7	3.6	.002
Residual	46,517	409		

one species. It is also the first to look directly at ecological differences among populations of once-widespread plant species suffering habitat loss and fragmentation. The most similar study known to me compared widespread and endemic congeneric *Astragalus* species for a number of components of inbreeding depression (Karron 1989). Germination percentages were similar in both species types, and germination was actually higher in selfed than in outcrossed seeds of the restricted species, suggesting a lack of inbreeding depression expressed at this stage. However, Karron (1989) did find inbreeding depression within populations of a restricted *Astragalus* species during the later stage of seedling growth.

Most studies considering population size and isolation have compared central and peripheral populations, and most are strictly genetic studies. Recent surveys of electrophoretic variation in widespread plant species have generally found decreased levels of genetic variation, especially in the number of alleles per locus, in peripheral populations (Shoumaker & Babbel 1980; Levin 1984; Furnier & Adams 1986), paralleling the results of classic *Drosophila* studies (Brussard 1984). This decrease is partly attributed to the restriction of peripheral populations to environmentally marginal, uniform habitats. Such populations are also likely to be founded by a small number of immigrants (founder effect) and to have high levels of inbreeding, resulting in fewer polymorphic loci and fewer alleles per locus than progenitor populations (Schwaegerle & Schaal 1979; Levin 1984; McClenaghan & Beaucamp 1986; Cwynar & MacDonald 1987).

The same pattern has been observed for differences in electrophoretically detectable genetic variation among congeners. Presumed derivative species (generally with restricted geographic ranges) contain a subset of the alleles of presumed progenitor species (Gottlieb 1973; Crawford et al. 1985; Karron 1987; Loveless & Hamrick 1988; Pleasants & Wendel 1989). Finally, the total level of genetic variation in endemic plant species is generally lower than that in regional or widespread species, and small local populations of endemics often show marked divergence due to drift and isolation (see Hamrick et al. 1979; Loveless & Hamrick 1984; and Kruckeberg & Rabinowitz 1985 for reviews). These differences among populations and species may have developed over long time periods.

In contrast, germination percentage patterns among *S. regia* populations suggest the importance of more recent effects of habitat fragmentation and destruction on fitness components. Today's configuration of *S. regia* in the agricultural landscape is primarily due to habitat fragmentation occurring over the past 150 years. Earlier patterns of population size and isolation were certainly much different. The generation time for *S. regia* is unknown, but demographic data suggests this perennial is long-lived (Menges 1988). Thus, important differences in population viability may have been created within a very few generations of habitat fragmentation. Nonetheless, population size explained only about one-quarter of total variation in population mean germination percentage. Other unknown factors, including past population sizes and distributional patterns, may account for other variation. The effects of selection during prolonged bottlenecks in removing deleterious recessive genes and thus mitigating inbreeding depression (Schemske & Lande 1985) might also account for variation in germination percentage.

Two non–mutually exclusive mechanisms could explain the pattern of lower seed germination percentages in smaller populations. First, small populations of the primarily outcrossing *S. regia* (especially if recently reduced) may be suffering from inbreeding depression. The occurrence of inbreeding depression in small populations is well known for animals. In plants, inbreeding depression could manifest itself early in the diploid stage of the life cycle. Inefficient germination of seeds is a likely result, and a laboratory seed germination experiment could detect differences.

Second, reduced populations may be disadvantaged because of reduced pollinator activity and changes in pollen sources. Plants in small and isolated *S. regia* populations may receive lower rates of hummingbird visitation. Because individual flowers cannot produce seed due to protandry (Menges 1988), lower fruit set will occur with fewer pollinator visits (Menges 1988). This situation also occurs in "mainland" versus "island" populations of the butterfly-pollinated *Dianthus deltoides* (Jennersten 1988). Also, in very small populations, most pollinator movements may occur within, rather than between, plants, so a greater proportion of seeds will be produced by within-plant crosses (geitonogamous pollinations are effective in producing seed in *S. regia* [Menges 1988]). The geitonogamous pollinations produce offspring that may suffer inbreeding depression.

Thus, the results of a lower proportion of pollinator movements among individuals could include both reduced seed set and reduced fitness of seed progeny. These differences among populations may have been exposed by the germination trials reported here.

Although confirmation of this pattern awaits genetic studies and more detailed demographic analyses of the entire life cycle (both underway), germination patterns across 23 populations of the prairie forb *Silene regia* suggest that habitat fragmentation may act through population size reductions to decrease fitness of plants in residual populations.

Acknowledgments

Thanks to Stacia Yoon, Leslie Barks, Rebecca Dolan, and Julie Alexander for assistance in field and laboratory work. Information on species distribution and individual sites came from many people and organizations, including Don Kurz and Tom Toney (Missouri Department of Conservation), Rick Thom, Rod Miller and Doug Ladd (The Nature Conservancy), Bill Shepherd (Arkansas Natural Heritage Commission), Terry Humble (Harrison School District), Bill McCloud (Arkansas State Parks), Jennifer Windus (Ohio Department of Natural Resources), Lee Casebere and Art Spingarn (Indiana Division of Natural Preserves), John Schwegman and Mary Kay Solecki (Illinois Department of Conservation), Jerry Baskin, Paul Somers, Steve Orzell, Robert Kral, Charles King, J. Jones, Tom Patrick, Ralph Brooks, R. McGregor, and J. Estes. Charles King, Jennifer Windus, Tom Toney, and Rebecca Dolan have participated in useful discussions of *S. regia* biology. I also acknowledge the thoughtful reviews of Jerry Baskin, Mark Deyrup, Rebecca Dolan, Laura Huenneke, Don Waller, and an anonymous reviewer. Marcia Moretto provided excellent word processing. Studies of *Silene regia* have been supported by the Ohio Department of Natural Resources Research Grants Program, the Indiana Academy of Sciences, Holcomb Research Institute, and the National Science Foundation (BSR 9000091).

Literature Cited

Axelrod, D. I. 1985. Rise of the grassland biome, North America. Botanical Review **51**:163–201.

Beardmore, J. A. 1983. Extinction, survival, and genetic variation. Pages in C. M. Schoenwald-Cox, S. M. Chambers, B. MacBryde, and L. Thomas, editors. Genetics and conservation. Benjamin/Cummings, Menlo Park, California. 722 pp.

Berger, J. 1990. Persistence of different-sized populations: an empirical assessment of rapid extinction in bighorn sheep. Conservation Biology **4**:91–98.

Brussard, P. F. 1984. Geographic patterns and environmental gradients: the central-marginal model in *Drosophila* revisited. Annual Review of Ecology and Systematics **15**:25–64.

Bryant, E. H. S. A. McCommas, and L. M. Combs. 1986. The effect of experimental bottleneck upon quantitative genetic variation in the housefly. Genetics **114**:1191.

Burgess, R. L., and D. M. Sharpe, editors. 1981. Forest dynamics in man-dominated landscapes. Springer-Verlag, New York. 309 pp.

Crawford, D. J., R. Ornduff, and M. C. Vasey. 1985. Allozyme variation within and between *Lasthenia minor* and its derivative species *L. maritima* (Asteraceae). American Journal of Botany **72**:1177–1184.

Curtis, J. T. 1956. The modifications of mid-latitude grasslands and forests by man. Pages 2721–2736 in W. L. Thomas, Jr., editor. Man's role in changing the face of the Earth. Volume 2. University of Chicago Press, Chicago, Illinois. 1159 pp.

Cwynar, L. C., and G. M. MacDonald. 1987. Geographic variation of lodgepole pine in relation to population history. American Naturalist **129**:463–469.

Dobzhansky, T. 1970. Genetics of the evolutionary process. Columbia University Press, New York. 505 pp.

Farmer, J. M., S. C. Price, and C. R. Bell. 1986. Population, temperature, and substrate influences on common milkweed (*Asclepias syriaca*) seed germination. Weed Science **34**:525–528.

Frankel, O. H., and M. E. Soulé. 1981. Conservation and evolution. Cambridge University Press, Cambridge, England.

Furnier, G. R., and W. T. Adams. 1986. Geographic patterns of allozyme variation in Jeffrey pine. American Journal of Botany **73**:1009–1015.

Gilpin, M. E., and M. E. Soulé. 1986. Minimum viable populations and processes of species extinctions. Pages 19–34 in M. E. Soulé, editor. Conservation biology: the science of scarcity and diversity. Sinauer Associates, Sunderland, Massachusetts.

Gleason, H. A. 1923. The vegetational history of the Middle West. Annals of the Association of American Geographers **12**:39–85.

Gottlieb, L. D. 1973. Genetic differentiation, sympatric speciation, and the origin of a diploid species of *Stephanomeria*. American Journal of Botany **60**:545–553.

Grimm, E. C. 1983. Chronology and dynamics of vegetation change in the prairie-woodland region of southern Minnesota, U.S.A. New Phytologist **93**:344–350.

Hamrick, J. L., Y. B. Linhart, and J. B. Milton. 1970. Relationships between life history characteristics and electrophoretically detectable genetic variation in plants. Annual Review of Ecology and Systematics **10**:173–200.

Harris, L. D., editor. 1984. The fragmented forest. University of Chicago Press, Chicago, Illinois. 211 pp.

Jennersten, O. 1988. Pollination in *Dianthus deltoides* (Caryophyllaceae): effects of habitat fragmentation on visitation and seed set. Conservation Biology **2**:359–366.

Karron, J. D. 1987. A comparison of levels of genetic polymorphism and self-compatibility in geographically restricted and widespread plant congeners. Evolutionary Ecology **1**:47–58.

Karron, J. D. 1989. Breeding systems and levels of inbreeding depression in geographically restricted and widespread species of *Astragalus* (Fabaceae). American Journal of Botany **76**:331–340.

Ketzner, D. M., E. F. Ulaszek, and L. R. Iverson. 1989. The status and ecology of *Silene regia* Sims, the royal catchfly, in Illinois. Unpublished report to Illinois Endangered Species Protection Board. 70 pp.

King, C. C. 1981. Distribution of royal catchfly (*Silene regia*) with special reference to Ohio populations. Pages 131–141 in R. L. Stuckey and K. J. Reese, editors. The prairie peninsula—in the "shadow" of Transeau: proceedings of the Sixth North American Prairie Conference. Ohio Biological Survey Biological Notes 15. 278 pp.

Kruckeberg, A. R., and D. Rabinowitz. 1985. Biological aspects of endemism in higher plants. Annual Review of Ecology and Systematics **16**:447–479.

Lande, R. 1980. Genetic variation and phenotypic evolution during allopatric speciation. American Naturalist **116**:463–479.

Ledig, F. T. 1986. Heterozygosity, heterosis, and fitness in outbreeding plants. Pages 77–104 in M. E. Soulé, editor. Conservation biology—the science of scarcity and diversity. Sinauer Associates, Sunderland, Massachusetts.

Levin, D. A. 1984. Genetic variation and divergence in a disjunct *Phlox*. Evolution **38**:223–225.

Lewontin, R. C. 1973. Population genetics. Annual Review of Genetics **7**:1–18.

Loveless, M. D., and J. L. Hamrick. 1984. Ecological determinants of genetic structure in plant populations. Annual Review of Ecology and Systematics **15**:65–95.

Loveless, M. D., and J. L. Hamrick. 1988. Genetic organization and evolutionary history in two North American species of *Cirsium*. Evolution **42**:254–265.

McClenaghan, L. R., and A. C. Beaucamp. 1986. Low genic differentiation among isolated populations of the California fan palm (*Washingtonia filifera*). Evolution **40**:315–322.

Menges, E. S. 1988. Population biology of a rare prairie forb, *Silene regia* 1985–1987. Report to the Ohio Department of Natural Resources and Indiana Academy of Sciences. Report 131, Holcomb Research Institute, Indianapolis, Indiana. 66 pp.

Menges, E. S. 1990. Population viability analysis for an endangered plant. Conservation Biology **4**:52–62.

Menges, E. S. 1991. Stochastic modeling of extinction in plant populations. Chapter in P. L. Fiedler and S. Jain, editors. Conservation biology: the theory and practice of nature conservation, population, preservation, and management. Chapman and Hall. In press.

Meyer, S. E., E. D. McArthur, and G. L. Jorgensen. 1989. Variation in germination response to temperature in rubber rabbitbrush *Chrysothamnus nauseosus* (Asteraceae) and its ecological implications. American Journal of Botany **76**:981–991.

Pleasants, J. M., and J. F. Wendel. 1989. Genetic diversity in a clonal narrow endemic, *Erythronium propullans,* and in its widespread progenitor, *Erythronium albidum.* American Journal of Botany **76:**1136–1151.

Rice, K., and S. Jain. 1985. Plant population genetics and evolution in disturbed environments. Pages 287–303 in S. T. A. Pickett and P. S. White, editors. The ecology of natural disturbance and patch dynamics. Academic Press, New York.

Schemske, D. W., and R. Lande. 1985. The evolution of self-fertilization and inbreeding depression in plants II. Empirical observations. Evolution **39:**41–52.

Schwaegerle, K. E., and B. A. Schaal. 1979. Genetic variability and founder effect in the pitcher plant *Sarracenia purpurea* L. Evolution **33:**1210–1218.

Shaffer, M. L. 1981. Minimal population sizes for species conservation. Bioscience **31:**131–134.

Shoumaker, K. M., and G. R. Babbel. 1980. Patterns of allozymic similarity in ecologically central and marginal populations of *Hordeum jubatum* in Utah. Evolution **34:**110–116.

Wright, S. 1948. On the roles of directed and random changes in gene frequency in the genetics of populations. Evolution **2:**279–294.

Wright, S. 1977. Evolution and genetics of populations. Volume 3. Inbreeding in animals: differentiation and depression. University of Chicago Press, Chicago, Illinois.

Wu, L., I. Till-Bottraud, and A. Torres. 1987. Genetic differentiation in temperature-enforced seed dormancy among golf course populations of *Poa annua* L. New Phytologist **107:** 623–631.

Are Small Populations of Plants Worth Preserving?

PETER LESICA

Division of Biological Sciences
University of Montana
Missoula, MT 59812, U.S.A.

FRED W. ALLENDORF

Division of Biological Sciences
University of Montana
Missoula, MT 59812, U.S.A.

Maintenance of genetic variation is considered essential for the long-term survival of a species (Frankel & Soulé 1981). Consequently, the rate at which genetic variation is lost in isolated populations is of great concern to conservation biologists who must decide how to apportion limited resources to protect rare or endangered species. Spending money and time to protect a population in which genetic variation will erode rapidly may be a poor investment. It is thought that small populations lose a large amount of genetic variability because of genetic drift and thus have reduced chance of long-term viability (Franklin 1980; Soulé 1980; Lande & Barrowclough 1987).

Most models of the rate of loss of genetic variation in small populations assume that all genotypes have equal fitness (i.e., selective neutrality). Neutral models of genetic drift assume that the amount of heterozygosity remaining after t generations is $(1-\frac{1}{2}N_e)^t$, where N_e is the effective population size (Wright 1969). Based on these models, it has been estimated that effective population sizes of at least 50 are needed to avoid harmful loss of genetic variation in the short term (Franklin 1980; Soulé 1980; Frankel & Soulé 1981; but see Simberloff 1988).

Some empirical studies have shown that selection against homozygotes occurs during early stages of growth in natural populations of plants. Consequently, models that assume selective neutrality may be misleading. In many species of plants fecundity is high, and mortality occurs principally in the seed and seedling stages (Burdon et al. 1983; Fenner 1987). Thus, only a small proportion of the progeny pool is recruited into the adult population, allowing selection to reduce the frequencies of homozygotes without reducing population growth.

We searched the literature for multilocus studies of plants in which genotypic frequencies at two or more stages in the life cycle were reported. We found eight such studies, which did this by following a single cohort through time (Clegg & Allard 1973; Burdon et al. 1983; Muona et al. 1987), by comparing genotypic frequencies in adults and their seedling progeny (Phillips & Brown 1977; Brotschol et al. 1986; Mitton 1990), or by comparing genotypic frequencies of different-age individuals sampled at a single time (Schaal & Levin 1976; Yazdani et al. 1985). Three tree species, one perennial herb, and three annual herbs were examined. We estimated the relative survival of homozygotes (1-s) relative to the survival of heterozygotes (1.0). In cases where data from more than one geographical location were reported, we arbitrarily used the data from only the first site presented so that each study was equally represented in the table. In many cases these values of s may be underestimates because the stages sampled do not cover the entire life span. We tested the statistical significance of these differences by contingency table analysis.

These data suggest that heterozygotes often have a survival advantage that could affect the rate of loss of

Paper submitted January 14, 1991; revised manuscript accepted July 8, 1991.

heterozygosity (Table 1). Of the 38 loci examined, 32 (89%) had a positive selection coefficient (s), indicating greater survival of heterozygotes. For 12 of these, the differences in survival between heterozygotes and homozygotes are significant ($P < 0.05$). However, an increase in heterozygosity in older life stages is not always observed (Mitton et al. 1989; Schnabel & Hamrick 1990).

In this paper, "heterozygous advantage" refers to an advantage of heterozygous genotypes regardless of the underlying genetic basis. There are three major possible explanations for an advantage of heterozygous genotypes: (1) inbreeding depression, (2) greater fitness of heterozygotes at the loci examined (i.e., overdominance), or (3) selection at loci that are linked or nonrandomly associated with the loci examined (i.e., associative overdominance, Ohta 1971). Whatever the cause, heterozygous advantage slows the loss of variation due to drift over what neutral models predict, regardless of the breeding system.

Table 1. Reduction in survival (s) of homozygotes relative to heterozygotes estimated from a single cohort sampled at different times or by sampling plants of different ages. The relative survival of heterozygotes is 1.0; the relative survival of homozygotes is 1-*s*. Probability values test for differences in survival between homozygotes and heterozygotes.

Species	*Locus*	*s*	*Life stages*
Avena barbata	Est4	0.14	seedling/adult
Clegg & Allard (1973)	Est9	0.64	
	Est10	0.30	
Echium plantagineum	Mdr	−0.04	seedling/adult
Burdon et al. (1983)	Pgi	0.41***	
	Pgm	0.30**	
	Sdh	0.11	
Eucalyptus pauciflora	Adh1	0.29	embryo/adult
Phillips & Brown (1977)	Got1	0.45	
	Got2	0.47	
	Lap2	−0.88	
	Mdh1	−1.27	
	Mdh2	0.15	
	Phos2	0.59	
Liatris cylindracea	Alk-P	0.41	6–10 yrs./
Schaal & Levin (1976)	Est2	−0.47	26–38 yrs.
	Est3	0.33	
	Pgi	0.29	
	Mdh	0.50**	
Liriodendron tulipifera	Acph1	0.02	seedling/adult
Brotschol et al. (1986)	Acph2	0.74**	
	Per	0.43	
	Est	0.41	
	Got1	0.55*	
Lolium multiflorum	Ap	0.36**	seedling/adult
Mitton (1990)	Got	−0.16	
	Per	0.15	
	Pgi	0.00	
	Pgm	0.34*	
Pinus sylvestris	Adh-B	0.84***	embryo/adult
Yazdani et al. (1985)	F-Est	0.55***	
	Gdh	0.43**	
	Got-B	0.37*	
	Mdh-B	−0.06	
Pinus sylvestris	F-Est	0.32*	embryo/juvenile
Muona et al. (1987)	Gdh	0.43	
	Got-B	0.13	
	Mdh-B	0.20	

*$P < 0.05$; **$P < 0.01$; ***$P < 0.001$.

No population genetic theory currently available predicts the effect of heterozygous advantage on the rate of loss of genetic variation in small populations. Therefore, we performed computer simulations to estimate the expected rate of loss of heterozygosity with different intensities of heterozygous advantage in populations of different sizes.

Simulation Model

We assumed random mating and a single locus with two alleles in which both homozygous genotypes had a relative survival of (1-s), in comparison to a relative survival of 1.0 for heterozygotes. The simulation model was a discrete-generation, stochastic computer program written in Pascal. The genotypes of zygotes and survival were determined by comparing two pseudorandom numbers drawn from a uniform (0,1) probability distribution with the expected genotypic proportions (based upon allele frequencies and binomial expectations) and the appropriate survival probabilities (1.0 for heterozygotes and 1-s for homozygotes). The value presented for each combination of selection intensity and population size is the mean of 1,000 repeats.

Figure 1 shows the proportion of heterozygosity retained after 25 generations for various combinations of population size and selection. Table 2 shows the population size with selectively neutral genotypes using the

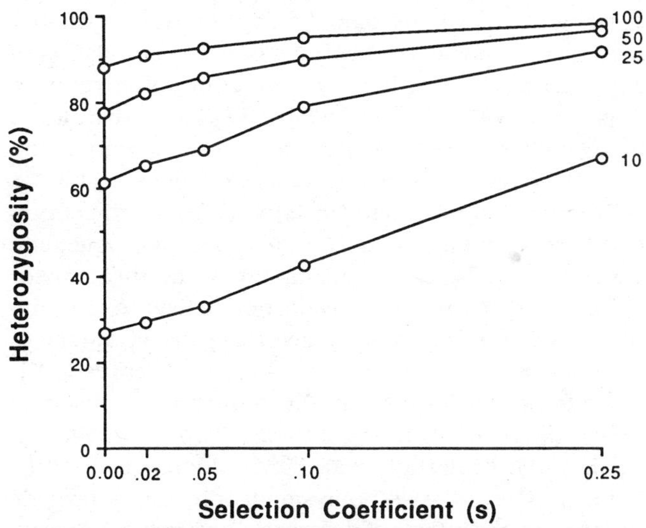

Figure 1. Simulation results showing mean proportion of heterozygosity remaining after 25 generations of genetic drift with overdominance in populations of 10, 25, 50, and 100 individuals. Heterozygotes have a fitness of 1.0, and both homozygotes have a relative fitness of (1-s). Values are the mean of 1000 repeats.

Table 2. Mean effective population sizes based upon 1000 simulations. Heterozygotes have a fitness of 1.0, and both homozygotes have a relative fitness of (1-*s*). Values are the population size that would lose genetic variation at a selectively neutral locus at the same rate as that observed in the simulations with overdominance.

	Population size			
s	*10*	*25*	*50*	*100*
0.00	9.7	25.9	49.6	100.4
0.02	10.4	29.5	63.4	129.6
0.05	11.4	34.1	81.0	162.6
0.10	14.8	52.7	116.2	242.9
0.25	31.3	145.1	361.6	718.7

basic formula $(1-\frac{1}{2}N_e)^t$ that would lose heterozygosity at the same rate as the simulated populations. Our results generally agree with those of Lacy (1987), who modeled loss of genetic variation under three intensities of overdominance with a population size of 120.

Discussion

Heterozygous advantage may be important in curtailing the loss of genetic variation in small populations of plants. The available empirical data suggest that 0.25 is a reasonable estimate of *s* at many loci (Table 1). Under selection of this intensity, over 92% of the original heterozygosity is maintained in populations with $N_e = 25$, and over 96% is retained with an N_e of 50 after 25 generations; this compares with values of 60% and 78% obtained with the neutral model. Even with a selection coefficient of 0.10, this model predicts retention of 30% more variation than the neutral model in a population of $N_e = 25$ after 25 generations. Stated in other terms, a population with $N_e = 25$ will lose genetic variation at a locus with $s = 0.25$ at the same rate as a population with $N_e = 145$ at a selectively neutral locus (Table 2).

We modeled overdominance to determine the effects of heterozygous advantage on the rate of loss of genetic variation because it is the simplest of the three major explanations to model. We do not believe overdominance is the major cause of heterozygous advantage in natural populations of plants; inbreeding depression is likely to be a major contributor to the greater observed survival of heterozygotes (Shaw & Allard 1982; Yazdani et al. 1985; Brotschol et al. 1986; Muona et al. 1987; Table 1). Nevertheless, our simulation results provide insight into the effects of differential survival of homozygotes and heterozygotes regardless of the underlying basis of heterozygous advantage. That is, the increase in the frequency of heterozygotes due to differential survival retards the rate of loss of genetic variation as modeled, whether or not the locus examined is the cause of differential survival.

Robertson (1962) has shown that overdominance is less effective in maintaining heterozygosity when the fitnesses of the homozygotes are unequal. Our model assumes equal fitness of homozygotes, and therefore represents the best case for overdominance retarding the loss of genetic variation for a given value of *s*. However, if inbreeding depression is a major contributor to heterozygous advantage, then most homozygous genotypes should be associated with approximately equal reduction in survival.

The effects of inbreeding depression in retarding the loss of genetic variation should be affected by breeding system and demographic history. Populations that have been purged of deleterious recessive alleles because of selfing or historically small population size are less likely to exhibit strong inbreeding depression (Lande & Schemske 1985). In contrast, inbreeding depression is expected to be most intense in outcrossing populations that have recently experienced a large reduction in size. Thus, we expect heterozygous advantage caused by inbreeding depression to be most effective in maintaining genetic variation in populations that are small because of a recent loss in habitat.

Studies on conifers (Libby et al. 1981; Neal & Adams 1985; Govindaraju & Dancik 1987; Stutz & Mitton 1988), grasses (McWilliams & Griffing 1965; McWilliams et al. 1969; Marshall & Allard 1970; Brown et al. 1974), and herbaceous dicots (Allard & Hansche 1964; Pederson 1968; Griffing & Zsiros 1971) have demonstrated that heterozygous advantage in plants is more pronounced under stressful conditions such as high temperatures or drought. Similarly, Shaw and Allard (1982) found that selection against heterozygotes between embryo and adult in Douglas fir, *Pseudotsuga menziesii,* was 0.12–0.21 in wild populations but only 0.05 in a nearby plantation with presumably less stressful conditions.

These findings and our results suggest that small populations subjected to stressful conditions may lose heterozygosity more slowly than those in more benign environments. Soulé (1973) has argued that benign environments do not promote loss of genetic variation, and Frankel and Soulé (1981) propose that populations in stressful environments exert strong directional selection and are likely to be genetically depauperate. Although these predictions may be correct, the occurrence of stress-induced heterozygous advantage in plants and the concomitant increased retention of variation suggests that small populations that occasionally undergo moderate levels of stress may retain higher levels of heterozygosity than similar-sized populations in benign environments.

The effects of heterozygous advantage have implications for conservation. First, caution should be exercised when measuring population heterozygosity from embryo or seedling samples because they may underestimate the level of heterozygosity in the adult population (Farris & Mitton 1984). In addition, small remnant populations of plants may retain a larger-than-expected proportion of their heterozygosity and consequently their capacity for quantitative adaptation for long periods of time. Moreover, small populations that occasionally undergo moderate levels of stress may be more appropriate for conservation than similar-sized populations in benign environments.

We are not recommending that small populations be considered equivalent to larger ones for conservation purposes. Small populations are subject to loss of rare alleles, and this may be important for the long-term response to selection and survival of populations (Denniston 1978; Allendorf 1986). Moreover, small populations are more prone to extinction from random environmental fluctuations (Goodman 1987). Nevertheless, our findings do suggest that small populations, particularly those subject to moderate levels of stress, may be much more valuable for conservation than predicted by models that assume selective neutrality.

Reznicek (1987) concluded that small reserves are more valuable for plants than for animals on the basis of demographic considerations (e.g., indefinite lifespan, persistent seed pools, and vegetative reproduction). He stressed the potential value of smaller preserves to provide a wider choice of sites for protection and expressed frustration that high-quality smaller sites may not be protected because of policies based upon theoretical grounds. Our results lend support to his conclusion that small preserves can play an important role in plant conservation.

Acknowledgments

We thank Jeff Mitton for sending us preprints of several papers and a reviewer for helpful comments. FWA was an employee of the National Science Foundation during the preparation of this paper.

Literature Cited

Allard, R. W., and P. E. Hansche. 1964. Some parameters of population variability and their implications in plant breeding. Advances in Agronomy **16**:281–325.

Allendorf, F. W. 1986. Genetic drift and the loss of alleles versus heterozygosity. Zoo Biology **5**:181–190.

Brotschol, J. V., J. H. Roberds, and G. Namkoong. 1986. Allozyme variation among North Carolina populations of *Liriodendron tulipifera* L. Silvae Genetica **35**:131–138.

Brown, A. D. H., D. R. Marshall, and L. Albrecht. 1974. The maintenance of alcohol dehydrogenase polymorphism in *Bromus mollis* L. Australian Journal of Biological Science **27**:545–559.

Burdon, J. J., D. R. Marshall, and A. H. D. Brown. 1983. Demographic and genetic changes in populations of *Echium plantagineum.* Journal of Ecology **71**:667–679.

Clegg, M. T., and R. W. Allard. 1973. Viability versus fecundity selection in the slender wild oat, *Avena barbata* L. Science **181**:667–668.

Denniston, C. 1978. Small population size and genetic diversity. Implications for endangered species. Pages 281–290 in S. A. Temple, editor. Endangered birds: management techniques for preserving threatened species. University of Wisconsin, Madison, Wisconsin.

Farris, M. A., and J. B. Mitton. 1984. Population density, outcrossing rate and heterozygote superiority in ponderosa pine. Evolution **38**:1151–1154.

Fenner, M. 1987. Seedlings. New Phytologist **106**(Supplement):35–47.

Frankel, O. H., and M. E. Soulé. 1981. Conservation and evolution. Cambridge University Press, Cambridge, England.

Franklin, I. R. 1980. Evolutionary change in small populations. Pages 135–150 in M. E. Soulé and B. A. Wilcox, editors. Conservation biology. Sinauer Associates, Sunderland, Massachusetts.

Goodman, D. 1987. The demography of chance extinction. Pages 11–34 in M. E. Soulé, editor. Viable populations for conservation. Cambridge University Press, Cambridge, England.

Govindaraju, D. R., and B. P. Dancik. 1987. Allozyme heterozygosity and homeostasis in germinating seeds of jack pine. Heredity **59**:279–283.

Griffing, B., and E. Zsiros. 1971. Heterosis associated with genotype–environment interactions. Genetics **68**:443–455.

Lacy, R. C. 1987. Loss of genetic diversity from managed populations: interacting effects of drift, mutation, immigration, selection, and population subdivision. Conservation Biology **1**:143–158.

Lande, R., and G. F. Barrowclough. 1987. Effective population size, genetic variation, and their use in population management. Pages 87–123 in M. E. Soulé, editor. Viable populations for conservation. Cambridge University Press, Cambridge, England.

Lande, R., and D. W. Schemske. 1985. The evolution of self–fertilization and inbreeding depression in plants. I. Genetic models. Evolution **39**:24–40.

Libby, W. J., B. G. McCutchan, and C. I. Millar. 1981. Inbreeding depression in selfs of redwood. Silvae Genetica **30**:15–25.

McWilliams, J. R., and B. Griffing. 1965. Temperature-dependent heterosis in maize. Australian Journal of Biological Science **18**:569–583.

McWilliams, J. R., B. D. H. Latter, and M. J. Mathison. 1969. Enhanced heterosis and stability in the growth of an interspecific *Phalaris* hybrid at high temperature. Australian Journal of Biological Science **22**:493–504.

Marshall, D. R., and R. W. Allard. 1970. Maintenance of isozyme polymorphisms in natural populations of *Avena barbata.* Genetics **66**:393–399.

Mitton, J. B. 1990. Physiological and demographic variation associated with allozyme variation. Pages 127–145 in D. E. Soltis and P. S. Soltis, editors. Isozymes in plant biology. Discorides Press, Portland, Oregon.

Mitton, J. B., H. P. Stutz, W. S. Schuster, and K. L. Shea. 1989. Genotypic differentiation at PGM in Engelmann spruce from wet and dry sites. Silvae Genetica **38**:217–221.

Muona, O., R. Yazdani, and D. Rudin. 1987. Genetic change between life stages in *Pinus sylvestris:* allozyme variation in seeds and planted seedlings. Silvae Genetica **36**:39–42.

Neal, D. B., and W. T. Adams. 1985. Allozyme and mating system variation in balsam fir (*Abies balsamea*) across a continuous elevational transect. Canadian Journal of Botany **63**:2448–2453.

Ohta, T. 1971. Associative overdominance caused by linked detrimental mutations. Genetical Research, Cambridge **18**:277–286.

Pederson, D. G. 1968. Environmental stress, heterozygote advantage and genotype-environment interaction in *Arabidopsis.* Heredity **23**:127–138.

Phillips, M. A., and A. H. D. Brown. 1977. Mating system and hybridity in *Eucalyptus pauciflora.* Australian Journal of Biological Science **30**:337–344.

Reznicek, A. A. 1987. Are small reserves worthwhile for plants? Endangered Species Update **5**(2):1–3.

Robertson, A. 1962. Selection for heterozygotes in small populations. Genetics **47**:1291–1300.

Schaal, B. A., and D. A. Levin. 1976. The demographic genetics of *Liatris cylindracea* Michx. (Compositae). American Naturalist **110**:191–206.

Schnabel, A., and J. L. Hamrick. 1990. Organization of genetic diversity within and among populations of *Gleditsia triacanthos* (Leguminosae). American Journal of Botany **77**:1060–1069.

Shaw, D. V., and R. W. Allard. 1982. Isozyme heterozygosity in adult and open-pollinated embryo samples of Douglas-fir. Silvae Fennica **16**:115–121.

Simberloff, D. 1988. The contribution of population and community biology to conservation science. Annual Review of Ecology and Systematics **19**:473–512.

Soulé, M. E. 1973. The epistasis cycle: a theory of marginal populations. Annual Review of Ecology and Systematics **4**:165–187.

Soulé, M. E. 1980. Thresholds for survival: maintaining fitness and evolutionary potential. Pages 135–50 in M. E. Soulé and B. A. Wilcox, editors. Conservation biology. Sinauer Associates, Sunderland, Massachusetts.

Stutz, H. P., and J. B. Mitton. 1988. Genetic variation in Engelmann spruce associated with variation in soil moisture. Arctic and Alpine Research **20**:461–465.

Yazdani, R., O. Muona, D. Rudin, and A. E. Szmidt. 1985. Genetic structure of a *Pinus sylvestris* L. seed–tree stand and naturally regenerated understory. Forest Science **31**:430–436.

Wright, S. 1969. Evolution and genetics of populations. Volume 2. The theory of gene frequencies. University of Chicago Press, Chicago, Illinois.

Note

The Rarity of *Sminthopsis virginiae* (Marsupialia: Dasyuridae) in Relation to Natural and Unnatural Habitats

R. W. BRAITHWAITE

Division of Wildlife and Rangelands Research
CSIRO Tropical Ecosystems Research Centre
PMB 44, Winnellie, NT 5789
Australia

W. M. LONSDALE*

Division of Entomology
CSIRO Tropical Ecosystems Research Centre
PMB 44, Winnellie, NT 5789
Australia

Species are rare because of natural ecological peculiarities or because populations are declining. It is important to be able to tell the difference (Slobodkin 1986), particularly from a conservation point of view. The shrew-like insectivore *Sminthopsis virginiae* (Marsupialia: Dasyuridae), like many of the 18 species in the genus, has a large geographic range but it appears to be very uncommon throughout that range. A compilation of the records of many workers in the Northern Territory of Australia, the major part of this species' range, produced a total of only 34 records before 1986 (Morton et al. 1987). During an exhaustive fauna survey of natural vegetation in Kakadu National Park in northern Australia in 1980–83 (Braithwaite 1985), only two individuals were caught (once each) in 21 600 trap-nights.

Then, over a three-day period (September 2–4) in 1986, a total of 28 captures of 19 individuals were recorded from 624 trap-nights in the wetlands of the Finniss River (13°S, 130°E). This represented a 56 percent increase in the number of records from the Northern Territory. This report concerns that trapping study. We will describe the results in detail and then speculate on the reasons behind them.

* *Correspondence and requests for reprints should be addressed to this author.*
Paper submitted May 28, 1987; revised manuscript accepted July 22, 1987.

One feature of the Finniss River area is its extensive stand of expanding populations of the invasive alien shrub mimosa (*Mimosa pigra*). Sampling focused on habitats with and without this shrub. Mimosa, a Central American invasive shrub of pantropical distribution (Lonsdale et al. 1985), arrived in northern Australia at some time in the 20 years prior to 1891 (Miller & Lonsdale 1987). It persisted as a minor weed for about 100 years, then underwent a massive population explosion in the late 1970s. It now infests some 450 square kilometers of wetlands in the central coast of northern Australia, much of it in dense thickets (Miller & Lonsdale 1987).

Eight sites were trapped for three nights with 26 Elliott traps set at each site at 5-meter intervals around the perimeter of a 55 m × 20 m rectangle. Two sites were in each of four habitat types: 1) *Hymenachne-Eleocharis-Oryza*–dominated sedgeland; 2) practically monospecific tall shrubland, formerly sedgeland, dominated by mimosa; 3) an open forest dominated by paperbark (*Melaleuca cajuputi*) with a graminoid understory; and 4) a paperbark open forest with a mimosa mid-story and virtually no ground flora. Thus, the four habitat types essentially consisted of sedgeland areas with and without mimosa (1 and 2) and paperbark forest areas with and without mimosa (3 and 4).

Totals of eight and 19 captures of five and 14 individuals of *S. virginiae* were recorded in the two mimosa shrubland sites (Table 1). A dense skirt of adventitious roots occurred at the base of most of the shrubs, providing a favored nest site for *S. virginiae,* which generally nests on the ground under cover (Taplin 1980; R. W. Braithwaite, unpublished) like most of its congeners (Hyett & Shaw 1980). The only other capture of *S. virginiae* was on a sedgeland site. By contrast, the rodents, *Rattus colletti* and *Melomys burtoni,* were apparently evenly distributed across the four types of habitat (Table 1). These two common rodent species were the only small mammals recorded besides *S. virginiae.* There are two remarkable features about these results. First, the captures were concentrated in only one of the four habitat types, and second, *S. virginiae* was the most commonly captured species. At ground level, the mimosa shrubland and the paperbark forest with a mimosa mid-story appear to be similar, and so the difference in *S. virginiae* capture rates between the two habitats—17 percent and zero, respectively—is surprising. The most obvious structural difference between mimosa-infested sites with and without paperbark is in the continuity of the ground and mid-story layers. Spaces 4–8 m wide occur under the wet-season canopy of *Nauclea orientalis* and *Barringtonia acutangula* in the paperbark forest sites. The seasonally dense canopy of these deciduous trees appears to shade out seedlings of mimosa and almost all other plants. These lacunae in the distribution of mimosa were common in the paperbark-mimosa sites; as a result, 31 percent of 1 m sq quadrats (n = 90) regularly distributed throughout the paperbark-mimosa sites were free of mimosa compared with only 14 percent of quadrats (n = 94) in the treeless mimosa sites ($X^2 = 7.93, P < 0.01$).

Predation is a possible reason for the difference in the abundance of *S. virginiae* between pure mimosa stands and paperbark forest with a mimosa mid-story. The discontinuities in the mimosa canopy in paperbark forest may promote greater predation than is possible in the dense thickets of pure mimosa. This idea is supported by the case of *Sminthopsis crassicaudata,* a common species in southern Australia that is generally found in low shrubland in arid and semiarid areas and is particularly common on the basalt plains of western Victoria (Morton 1978) in closely cropped short grassland with abundant small boulders for cover. Here barn owls (*Tyto alba*) prey heavily on *Mus musculus* but only occasionally manage to take the equally common *S. crassicaudata* (Morton 1975). It seems likely that *S. crassicaudata* requires good shelter from owl predation if it is to be common. This may also be true of other species of *Sminthopsis,* including *S. virginiae.* In paperbark forest the avian predators such as owls are able to roost in the trees and prey on ground-dwelling animals in the spaces. We have no specific information on this, but barking owls (*Ninox connivens*), southern boobooks (*Ninox novaeseelandiae*), eastern grass owls (*Tyto longimembris*), and rufous owls (*Ninox rufa*) are frequently recorded in the area (Blakers et al. 1984). Estbergs and Braithwaite (1985) have shown that *N. rufa* is a generalist predator on small mammals in northern Australia, taking animals as small as *Pseudomys delicatuus* (6–15 grams) and as agile as *Antechinus bellus.* Perhaps dingoes (*Canis familiaris*), too, are better able to hunt in the paperbark-mimosa sites than in pure mimosa.

Sminthopsis virginiae is found in a range of habitats across northern Australia (Braithwaite et al. 1985) but, before the present study, was generally associated with *Pandanus spiralis,* a palm-like tree that usually has a skirt of dead leaves and covers the nearby ground with prickly fronds. More than a century ago, pandanus was much more common and formed dense, impenetrable thickets in some areas (Leichhardt 1847). This changed with the arrival of water buffalo (*Bubalus bubalis*), which opened up the stands of pandanus. The proof of this is dramatically demonstrated with small exclosures—within five years an impenetrable tangle returns. We suggest that such areas were the primary habitat of *S. virginiae* and that the species was much more abundant in these areas before the intrusion of buffalo. It

Table 1. Total number of captures of three small mammal species on eight sites in the Finniss River area of the Northern Territory of Australia. Percentages in parentheses are captures per hundred trap-nights for each pair of sites.

		Habitat			
		Not mimosa dominated		*Mimosa dominated*	
Species	*Replicate*	*Sedgeland*	*Paperbark forest*	*Former sedgeland*	*Paperbark forest*
Sminthopsis virginiae	1	0	0	8	0
	2	(1%)	(0%)	(17%)	(0%)
		1	0	19	0
Melomys burtoni	1	1	7	5	7
	2	(3%)	(4%)	(4%)	(6%)
		4	0	1	3
Rattus colletti	1	3	5	7	0
	2	(7%)	(3%)	(6%)	(0%)
		0	0	5	0

should be possible to test this idea when buffalo are excluded from large areas over long periods.

Thus, *S. virginiae* stands in stark contrast to mimosa: the rapid expansion of the range of mimosa in the late 1970s may be partly attributed to the overgrazing of the wetlands by buffalo, whereas the increase in buffalo probably reduced the abundance of *S. virginiae* indirectly through vegetation change. Today, where mimosa is present it simulates the protective features of pandanus and thus is excellent habitat for *S. virginiae,* as long as suitable foraging areas with a graminoid ground layer are nearby. As these, too, are taken over by mimosa, the habitat would decrease in value for *S. virginiae.*

One of two scenarios, derived from policies of land management, will probably describe the future abundance of *S. virginiae.* In the first, the Brucellosis and Tuberculosis Eradication Campaign continues to maintain buffalo density at a low level in order to eliminate the diseases from these feral animals, while mimosa is brought under control (probably by biological control); the pandanus regenerates, and *S. virginiae* consequently may become much more abundant. In the second, mimosa is not controlled, and *S. virginiae* may become more abundant in the short term (say 10 years) but will ultimately decline again as the graminoid foraging areas disappear. In this case, buffalo being controlled or not being controlled would affect the outcome little. Thus, future governmental policies, related to the independent issues of buffalo and mimosa control and dictated by the requirements of the pastoral industry rather than conservation needs, could determine the future abundance of *S. virginiae.*

We have argued here that a species that has appeared to be rare because of some unknown ecological peculiarity may in fact be rare because of the decline of its natural habitat. A need for tangled thickets, which now appears an idiosyncrasy and only able to be supplied by alien vegetation, would not have seemed at all peculiar before introduced water buffalo destroyed the native pandanus thickets.

Acknowledgments

We thank C. M. Beurle, J. A. Estbergs, N. Jenkins, and C. C. Wanganeen for assistance with field work and G. R. Friend, S. R. Morton, and P. A. Werner for constructive comments on the manuscript.

Literature Cited

Blakers, M., S. J. J. F. Davies, and P. N. Reilly. 1984. The atlas of Australian birds. Melbourne University Press, Melbourne, Australia.

Braithwaite, R. W. 1985. The Kakadu fauna survey: an ecological survey of Kakadu National Park. Report to Australian National Parks and Wildlife Service, Canberra, Australia.

Braithwaite, R. W., J. W. Winter, J. A. Taylor, and B. S. Parker. 1985. Patterns of diversity and structure of mammalian assemblages in the Australian tropics. Australian Mammalogy **8**:171–186.

Estbergs, J. A., and R. W. Braithwaite. 1985. The diet of the rufous owl *Ninox rufa* near Cooinda in the Northern Territory. Emu **85**:202–205.

Hyett, J., and N. Shaw. 1980. Australian mammals. Nelson, Melbourne, Australia.

Leichhardt, L. 1847. Journal of an overland expedition in Australia. T & W Boone, London, England.

Lonsdale, W. M., K. L. S. Harley, and I. L. Miller. 1985. The biology of *Mimosa pigra* L. Proceedings (2) of the 10th Asian-Pacific weed science society conference. Chiang Mai, Thailand.

Miller, I. L., and W. M. Lonsdale. 1987. Early records of *Mimosa pigra* in the Northern Territory. Plant Protection Quarterly **2**:140–142.

Morton, S. R. 1975. The diet of the barn owl in southern Victoria. Emu **75**:31–34.

Morton, S. R. 1978. An ecological study of *Sminthopsis crassicaudata* (Marsupialia: Dasyuridae) I. Distribution, study areas and methods. Australian Wildlife Research **5**:151–162.

Morton, S. R., M. A. Armstrong, and R. W. Braithwaite. 1987. The breeding season of *Sminthopsis virginiae* (Marsupialia: Dasyuridae) in the Northern Territory. Australian Mammalogy **10**:41–42.

Slobodkin, L. B. 1986. On the susceptibility of different species to extinction: elementary instructions for owners of a world. Pages 226–242 *in* B. Norton, editor. The preservation of species. Princeton University Press, Princeton, New Jersey, USA.

Taplin, L. E. 1980. Some observations on the reproductive biology of *Sminthopsis virginiae* (Tarragon) [Marsupialia: Dasyuridae]. Australian Zoology **20**:407–418.

Littoral Fish Communities in Lake Tanganyika: Irreplaceable Diversity Supported by Intricate Interactions among Species

MICHIO HORI

Department of Biology
Wakayama Medical College
Hironishi 651
Wakayama, 649-63 Japan

MASTA MUKWAYA GASHAGAZA

Uvira Station/Research Center of Natural Sciences
Sud-Kivu, Zaire

MUDERHWA NSHOMBO

Uvira Station/Research Center of Natural Sciences
Sud-Kivu, Zaire

HIROYA KAWANABE

Department of Zoology and Center for Ecological Research
Kyoto University
Sakyo-ku
606-01 Kyoto, Japan

Abstract: *The rocky littoral areas of Lake Tanganyika harbor diverse fish communities, mainly composed of cichlids. Their stability, structure, and organizing mechanism were examined at three locations by census, behavioral observations, and dietary analyses. These fish communities were stable, characterized by their persistence during a 10-year period and resilience after a perturbation. Partitioning of spawning sites among substrate spawners was evident, which might be a factor in their ability to coexist. The fish communities were consistently composed of 12 food-habit groups, but composition of species in each group differed from location to location. Among species of different food-habit groups, facultative commensalisms were prevalent. Although aggressive interactions are common among fishes of the same food-habit group, each predatory fish obtained an advantage in feeding efficiency from the different feeding behaviors of other species of the same group. This mutualism and the facultative commensalism should both increase spe-*

Comunidades de peces litorales en el lago Tanganyika: Diversidad irreemplazable sustentada por intrincadas interacciones entre especies

Resumen: *Las áreas litorales rocosas del lago Tanganyika albergan diversas comunidades de peces compuestas principalmente por cíclidos. Su estabilidad, estructura, y mecanismos de organización fueron examinados en tres lugares por medio de censos, observaciones del comportamiento y análisis de la dieta. Estas comunidades de peces son estables, caracterizadas por su persistencia durante un período de 10 años y por su resiliencia luego de perturbaciones. Fue evidente la partición de los sitios de desove entre los desovantes de sustrato, esto podría ser un factor que contribuye a su coexistencia. Las comunidades de peces estaban compuestas, en forma consistente, por 12 grupos con distintos hábitos alimenticios, pero la composición de especies dentro de cada* *cies richness in different parts of the communities. These intricate interactions among species may be a base of stability of the fish communities. We assert that maintenance of ecological networks among species are most important for conservation of biodiversity in tropical regions.*

grupo varió de un lugar a otro. Entre las especies de grupos de hábitos alimenticios diferentes, prevaleció el comensalismo facultativo. Si bien interacciones agresivas son comunes entre los peces del mismo grupo de hábitos alimenticios, cada predador obtuvo una ventaja en la eficiencia alimenticia a partir de comportamientos alimenticios diferentes de las otras especies del mismo grupo. Este mutualismo y el comensalismo facultativo deben incrementar la riqueza de especies operando en diferentes partes de las comunidades. Estas intrincadas interacciones entre especies podrían ser la base de estabilidad de las comunidades de peces. Nosotros afirmamos que el mantenimiento de redes ecológicas entre especies son de gran importancia para la conservación de la biodiversidad en regiones tropicales.

Introduction

The Great Lakes in eastern Africa are one of the most speciose freshwater systems in the world (Lowe-McConnell 1987). Among them, Lake Tanganyika is the oldest, with an estimated age of 20 million years (Tiercelin & Mondeguer 1991), and it harbors the most morphologically diversified fish fauna (Fryer & Iles 1972; Greenwood 1991; Nishida 1991). Nearly 300 fish species are known in the lake, of which 60% are cichlids (Brichard 1989). The vast majority (97%) of cichlids are endemic, and they are considered to have evolved within the lake from more than 11 ancestral lineages (Nishida 1991). About 65% of the cichlids are restricted to rocky littoral habitats. These fish communities in rocky shores are mainly composed of cichlids allied to each other to various degrees (Hori et al. 1983; Kuwamura 1987; Brichard 1989).

During the last decade, intensive underwater studies of fish behavior began to throw light on the direct and indirect interactions among species, and to reveal that such interactions are an indispensable factor for maintaining the diversity in littoral areas of this lake (Hori 1983, 1987, 1991; Takamura 1983, 1984; Kawanabe et al. 1985; Kawanabe 1986; Gashagaza 1991; Yamaoka 1991).

Here we delineate the characteristics of community structure in the rocky shores of the lake based on a long-term census and a removal experiment. We make the point that the littoral fish community is very stable and in a state of equilibrium, hence indicating some stabilizing mechanisms or processes. Next, we review the interspecific relationships among fishes in their breeding and feeding activities and define the factors responsible for the community structure. Then we discuss what aspects of diversity need to be preserved and how we might best go about achieving such a preservation.

Study Areas and General Methods

Three rocky littoral fish communities located in the northern (Luhanga and Bemba, Zaire), central eastern (Mahale, Tanzania), and southern rocky shores (Mpulungu, Zambia) of Lake Tanganyika (Fig. 1) were studied from 1979 to 1989. The bottom of all three sites was gently sloping and composed mainly of stone and rubble, inlaid with rocks and boulders. At each site, a permanent quadrat of 400 m^2 was placed along the rocky bottom, extending from near shore to around 10 m deep.

In the northern study site, censuses were made seven times during a 10-year period. During this period, the water level of the lake went down 2 m, which caused about 15% of the quadrat bottom to be exposed. In the central site, census was done only twice in one month. In the southern site, a removal experiment and eight censuses to monitor the results were carried out. The diversity of the fish community in each site was evaluated by Simpson's (1949) index of diversity ($1/\lambda$). To compare the community structure between two census times, Pianka's (1973) index of overlap (α) was used.

Feeding behavior, breeding habit, and behavioral interactions of each species were observed using SCUBA in and around the quadrats. Fish samples were collected with gillnets in each study site. Stomach contents were analyzed by the points method (Hynes 1950). The detailed study methods and substrate condition in the study areas have been reported elsewhere (Hori 1983, 1987, 1991; Hori et al. 1983; Gashagaza 1991).

Community Structure and Interspecific Relationships

Diversity and Stability of Rocky Littoral Fish Communities

Figure 1 compares the census results in each quadrat from the three study sites and shows the pairwise sim-

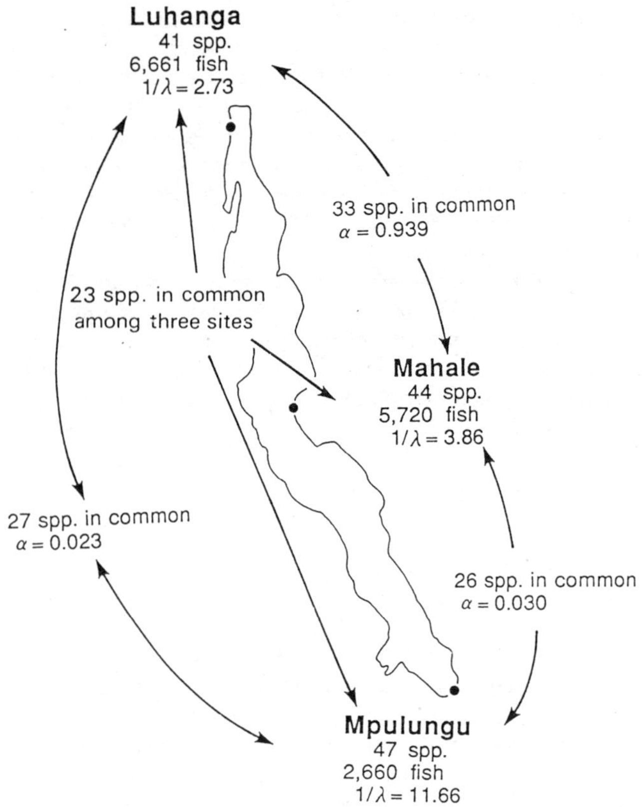

Figure 1. Comparison of three rocky littoral fish communities in Lake Tanganyika. Number of species, number of fish, and Simpson's index of diversity in the quadrat (400 m²) of each study site are shown. Pairwise comparison by number of species in common and Pianka's index of overlap between the three sites, and number of species in common among the three sites are also shown. Classification of species follows Poll (1986).

ilarity of species composition between them. In each quadrat, 2600 to 6600 fish belonging to some 40 species were found, most of which were cichlids. The fish density in Mpulungu was much lower than those in Mahale and Luhanga, while the number of species was greater than at the other two sites. The species composition and community structure in Mpulungu were also much different from those in the Luhanga and Mahale communities; these two communities had a similar structure, though about one quarter of the component species was specific to each site.

Stability of the fish community was examined from two aspects, the persistence and the resilience of structure, which are defined respectively as the ability to remain in a relatively constant state over time and the ability to return to an original state after a perturbation (Harrison 1979; Grossman 1982; Connel & Sousa 1983). By examining these attributes of a community, it is also possible to determine whether the system is regulated primarily through stochastic or deterministic processes (Grossman 1982).

Census results over 10 years in the quadrat of the northern site provide data about the persistence of the community structure (Fig. 2). The number of fish decreased gradually to less than half of the initial level, which should be partly attributed to the descending of the water level. The number of species, however, went virtually unchanged, though some species were not found and several species appeared after the first census. Examination of census data indicated that such alternations were mainly brought about by the coming and going of some rare and/or nomadic species; it did not suggest competitive replacements among species. Moreover, the Pianka's indices between the first and each of the other census times were maintained close to unity during the period. So composition of the community was highly persistent. Consequently, stochastic processes seem to be unimportant for this community.

The resilience of the fish community was examined by a removal experiment in the quadrat of the southern study site (Hori 1991). From a small part (70 m^2) of the quadrat, nearly all the territorial fish, about 70% of the initial count of fish, were removed, but some small fish escaped the removal. This procedure made the Pianka's index between censuses before and just after removal less than 0.5. Within two weeks, however, the number of fish had recovered, and the number of species also increased gradually toward the initial level. A census after 10 months revealed that the community composition returned to the initial state represented by Pianka's index at 0.97.

We conclude that littoral fish communities in this lake are very stable and in a state of equilibrium, indicating that some mechanism or process to maintain the community structure exists. The recovery from the removal was mainly achieved by recolonization of adult and subadult fishes invading from surrounding areas. So there is a possibility that the fish are competing with each other for available space.

Segregation in Space Utilization

Substantial evidence has been accumulated that feeding and/or breeding space is limited for many epilithic algal feeders (Nagoshi 1983; Takamura 1984; Kuwamura 1986, 1992; Mbomba 1986; Kohda 1991), some benthos feeders (Kondo 1986; Yanagisawa 1987; Yuma 1989), and rock-dwelling plankton feeders (Hori 1983), most of which exhibit some interspecific territoriality. But no comprehensive study has been done on space utilization of all the fish in these communities.

Gashagaza (1991) investigated the breeding habits of 12 cohabiting substrate breeders in the northern site, all belonging to genus *Lamprologus* (in a broad sense),

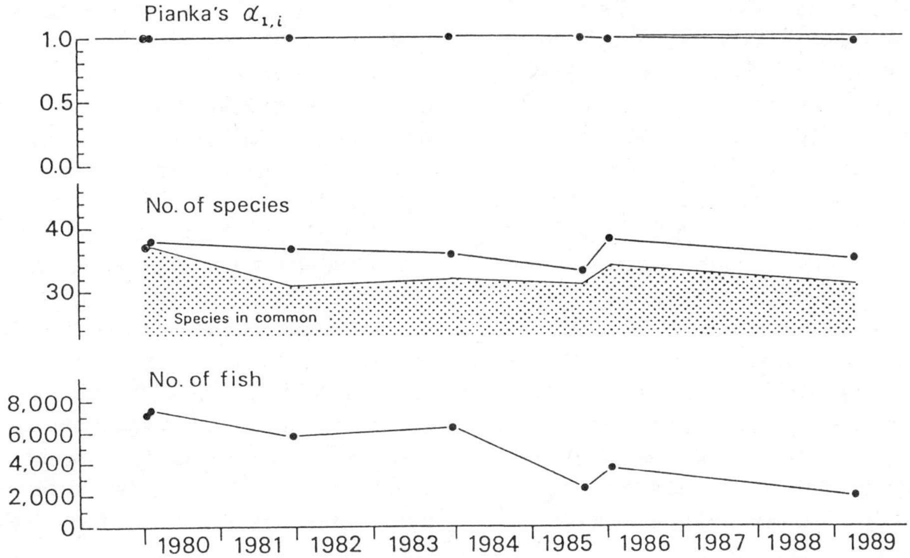

Figure 2. Changes of parameters of the rocky littoral fish community on the northern study site (Luhanga) through seven censuses during a 10-year period (Hori 1991). Pianka's index of overlap and number of species are calculated between the first and each of the other census times.

and has shown how they utilized the habitat differently from each other (Table 1). Those species can be categorized into three groups according to their mating system; temporarily haremic species, permanently haremic ones, and mono/bigamous ones. Successive removal experiments of a pair and its brood demonstrated that each species has a specific selectivity of breeding sites that corresponds with its reproductive characteristics, mating system, and parental care. Therefore, partitioning of spawning sites among species along diversity of microhabitats in rocky areas as well as different breeding strategies might be important for the species' coexistence.

Food Habit Groups and Interspecific Relationships

In contrast to the prominent tendency of differential space utilization among fish species in breeding, interspecific interactions through their feeding activities are

Table 1. Morphometric features of spawning and brooding sites of 10 allied species cohabiting on the northern study site (modified from Gashagaza 1991).

(Mating System) *Species*	*N*	*Spawning (S) or Brooding (B) site*	*Shape*	*Inclination a (°)*	*Inclination b (°)*	*ED (mm)*	*Ed (mm)*	*D (mm)*
(Temporarily Haremic)								
L. compressiceps	21	S+B	Hole	——	−14~87 (25)	42 ± 8	23 ± 6	49 ± 9
L. lemairii	16	S+B	Hole	——	21~88 (40,46)	98 ± 36	63 ± 22	119 ± 24
L. fasciatus	4	S+B	Pipe	——	−5~89 (48,86)	20 ± 0	20 ± 0	120 ± 4
L. profundicola	6	S	Rock surface	84~138 (118,122)	——	——	——	——
		B	Hole	——	18~87 (42,62)	33 ± 7	24 ± 8	115 ± 26
(Permanently Haremic)								
L. furcifer	8	S	Rock surface	95~148 (106,107)	——	——	——	——
		B	Pit	——	——	19 ± 4	13 ± 3	8 ± 2
L. mondabu	24	S+B	Tunnel	——	27~66 (36,38)	——	——	135 ± 14
(Mono/Bigamous)								
L. attenuatus	11	S	Stone surface	16~141 (84)	——	——	——	——
		B	Pit	——	——	46 ± 3	22 ± 4	32 ± 14
L. elongatus	14	S	Rock surface	68~174 (146,141)	——	——	——	——
		B	Crevice & hole	——	6~73 (34,39)	70 ± 20	47 ± 11	71 ± 17
L. toae	16	S	Stone surface	0~68 (24,32)	——	——	——	——
		B	Pit	——	——	27 ± 8	18 ± 5	15 ± 6
L. tretocephalus	10	S+B	Crevice & tunnel	——	22~78 (42,45)	——	——	145 ± 25

Inclinations of surfaces (inclination a*) range from 0° (horizontal) to 180° (upside-down horizontal) and inclinations of holes and hole-like structures (inclination* b*) from −90° (downward vertical) to 90° (upward vertical). The median is in parentheses, following the range. ED: maximum diameter of entrance, Ed: minimum diameter of entrance, D: depth.*

more complicated (Yamaoka 1982, 1991; Hori 1983, 1987, 1991; Takamura 1983, 1984; Kawanabe et al. 1985; Kawanabe 1986; Mbomba 1986). We first look at the feeding relationships of littoral fishes based on the food web among them and divide them into groups according to their food habits. Then we examine interactions separately between different food habit groups and within each group. In this examination, behavioral interactions are categorized as predation, aggressive behavior, and following behavior.

Figure 3 shows a food web among fishes in and around the quadrat of the northern study site. Subadult fish of piscivorous species, which mainly fed on benthic animals, are treated as if they were a different species from the adult fish in this figure. There were five piscivores that fed mainly on small fishes, and two scale eaters that attacked mainly large fish, including piscivores. Piscivores and scale eaters largely shared prey species. There were 17 herbivores, and most others ate benthic animals. Although some species were specialized for one item, such as unicellular algae, fish fry, and shrimp, most species fed on several items. Any one of these items was utilized by several fish species such as piscivores and scale eaters. Hori (1987) has reported another food web among fish in the central study site, which indicates a general pattern and also a locally specific variation of prey-predator interaction.

A cluster analysis of their dietary similarity in the northern study site showed that all the species were grouped into 12 food habit groups, each of which was composed of species sharing a 50% similarity in diet. The same analysis adopted for both the central and southern littoral fish communities revealed that both communities were also composed of the same 12 food-

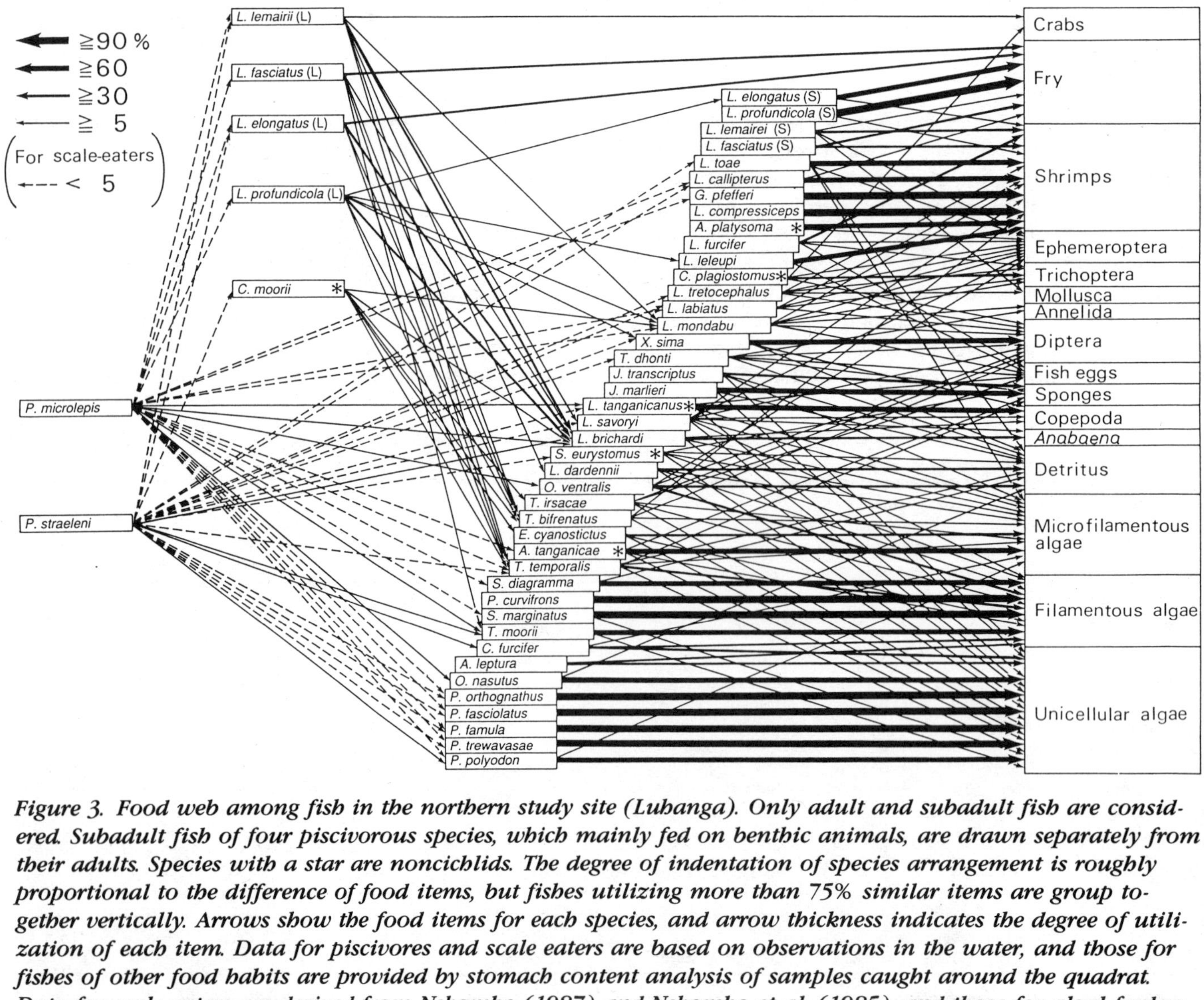

Figure 3. Food web among fish in the northern study site (Lubanga). Only adult and subadult fish are considered. Subadult fish of four piscivorous species, which mainly fed on benthic animals, are drawn separately from their adults. Species with a star are noncichlids. The degree of indentation of species arrangement is roughly proportional to the difference of food items, but fishes utilizing more than 75% similar items are group together vertically. Arrows show the food items for each species, and arrow thickness indicates the degree of utilization of each item. Data for piscivores and scale eaters are based on observations in the water, and those for fishes of other food habits are provided by stomach content analysis of samples caught around the quadrat. Data for scale eaters are derived from Nshombo (1987) and Nshombo et al. (1985), and those for algal feeders are from Takamura (1984).

habit groups as the northern site, but that the numbers of and/or composition of species in every group were not the same among the three communities (see also Hori 1991). We also note that morphologically similar species having similar foraging methods were distributed in different sites, suggesting that the behavioral difference is of primary importance for their coexistence. Thus, these food-habit groups seem to be composed of a limited number of feeding niches, each of which demands a distinct morphology and, accordingly, a foraging behavior diversified from each other. Divergent morphology and foraging behavior in each food habit group has been documented for benthos eaters (Hori 1983), algal feeders (Yamaoka 1982, 1983; Mbomba 1985, 1986; Yamaoka et al. 1986), scale eaters (Nshombo et al. 1985; Hori 1987), and piscivores (Hori 1983, 1987).

Hori (1987) studied behavioral interactions among species of different food-habit groups and those of fish in the same group in the central study site. Predatory interactions, including scale eating, were important among fish of different groups. Aggressive behaviors between groups were directed by many fish towards various potential predators—piscivores, scale eaters, fry eaters, and shrimp eaters that often preyed on guarded fry (Fig. 4).

However, following behavior was also prevalent in various combinations of species of different groups (Fig. 5). For example, some shrimp eaters followed a large insect eater to take shrimps flushed out by the insect eater, and some piscivorous fish follow other large innocuous algal feeders using the fish as cover for hunting. Such relations can be regarded as commensalism—more specifically, facultative commensalism—because the fish being followed (host) get no benefit from the following fishes (commensals). Takamura (1983) has reported another case of the commensal relationship between a filamentous algal feeder and a unicellular algal feeder. The combination of host and commensal species seemed to depend on what situation could be provided by the host species through their feeding activities (Kawanabe et al. 1985).

Among interactions between species of the same food-habit group, aggressive behavior was prominent (see Fig. 4). Moreover, the overlap of foraging sites between any two benthos eaters was inversely related to the overlap of their diet, and aggressiveness in general was high between fishes that showed high overlap of

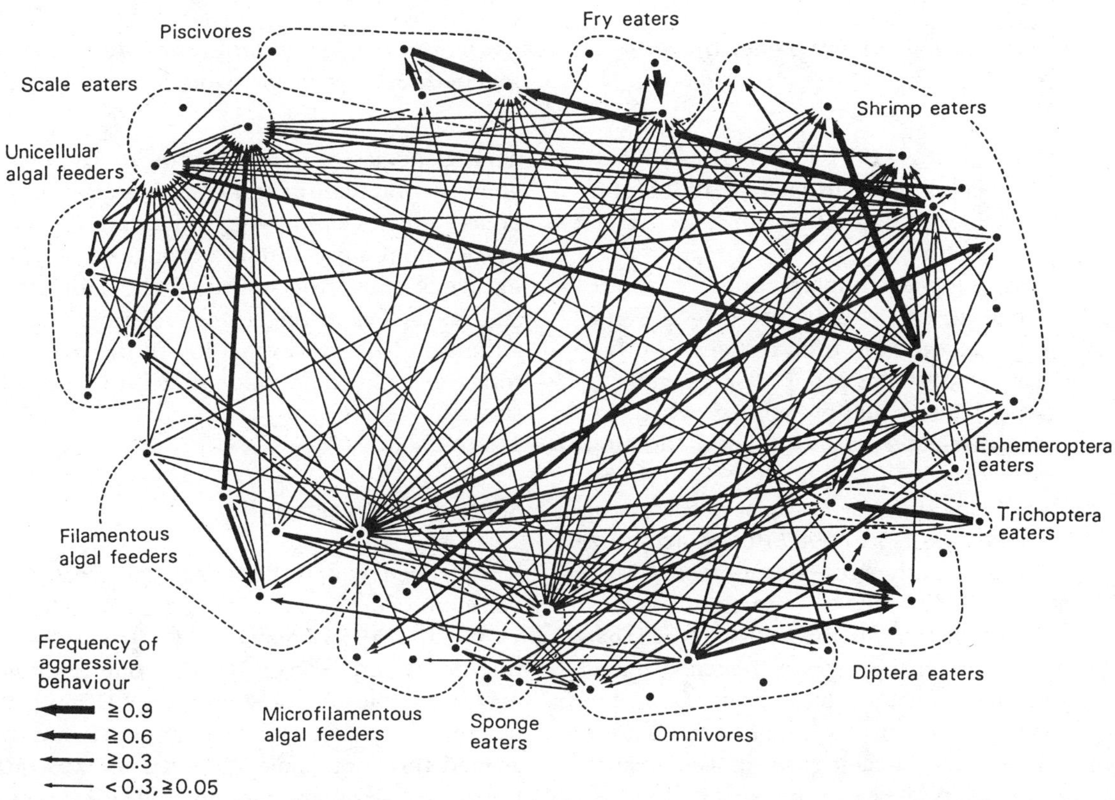

Figure 4. The direction and the frequency of aggressive behaviors among fish within and between food habit groups at the central study site. Circles define each of the 12 food habit groups, which are named after their common food item, and the dots within the circles represent each fish species belonging to the group. Each arrow shows the direction of aggressive behaviors from attacker to attackee. The frequency between two species is indicated by the arrow thickness, which is given by number of encounters resulting in aggressive behaviors per total number of encounters within a distance of three times either fish's size (TL).

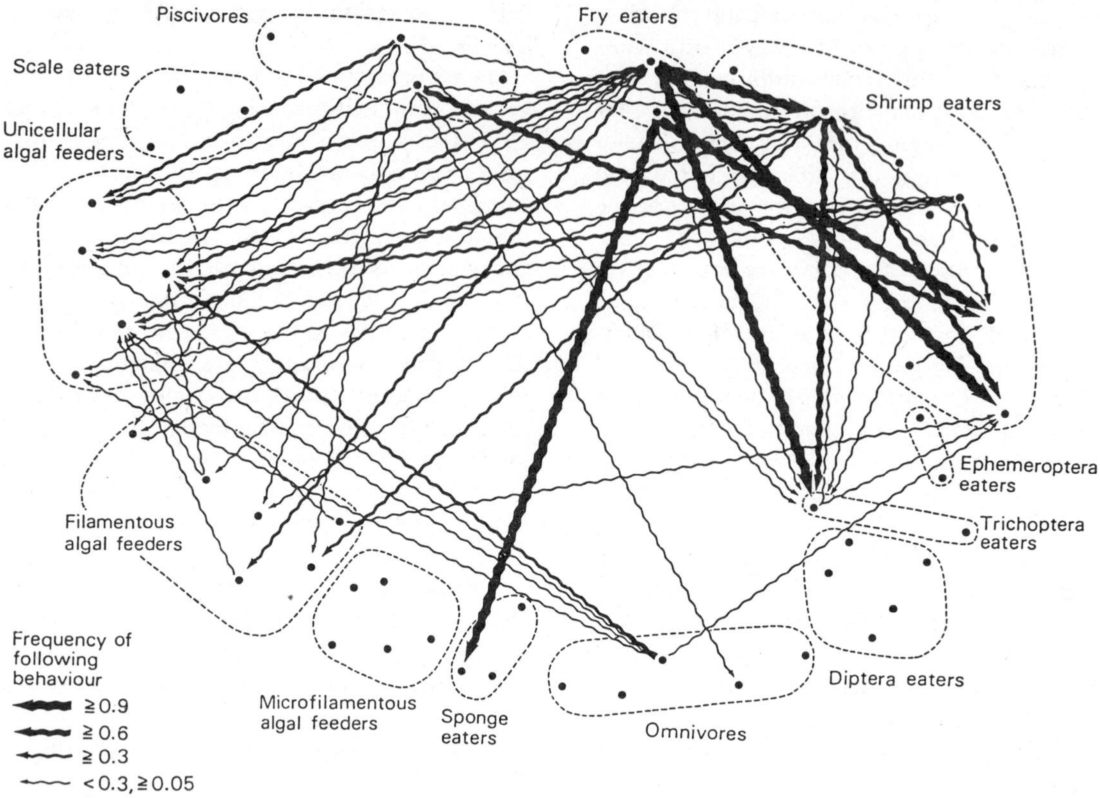

Figure 5. The direction and the frequency of following behaviors among fish within and between food habit groups at the central study site. The direction and the frequency are shown in the same manner as in Figure 4.

both diet and foraging sites (Hori 1991). These observations suggest that food is partitioned by differential foraging, which is achieved partly through direct interactions among competitors.

In spite of these competitive interactions, there are indirect interactions in which each species gets some benefit in foraging success from the presence of other species of the same food habit group. For example, a scale eater showed higher hunting success when another congeneric scale eater with a different hunting method was nearby (Fig. 6). This should be caused by the prey's caution being dispersed against predators of diverse hunting methods. Hori (1987) documented a similar relationship within the group of piscivores. Matsuda et al. (in preparation) called this situation "exploitative mutualism" and proved by a theoretical model that this mutualism is a factor that promotes the coexistence of species of similar food habit. An important point for community organization is that exploitative mutualism should operate within a food-habit group as a stabilizing mechanism for relevant populations, unlike interspecific competition and obligatory mutualism. The reason is that a population inferior in number will get higher feeding efficiency than a superior one by circumventing the prey's guard toward the abundant predator. Furthermore, behavioral divergence in every food-habit group probably develops the facultative commensalisms among species of different groups. Thus, it may be argued that mutualism and commensalism both act to increase species richness by operating on different parts of the community. But as a change in one group would influence other groups through the commensal relationship, as well as through other relationships such as predation, and overall evaluation of the effect of these interactions on community structure is open to future studies.

Aspects of Preservation

It had been widely believed that great diversity in a community itself brings about great stability (MacArthur 1955; Elton 1958). May (1972, 1973, 1981) questioned this belief and asserted that a complex community is more unstable than a simple one. During the last two decades, much controversial work has been done around this topic, and May's theory was in turn questioned from various standpoints (DeAngeles 1975; McNaughton 1977; Pimm 1979; Redfearn & Pimm 1987), though no positive support for the traditional belief has been established. At present, therefore, it is difficult to generalize about any clear relation between a community's diversity and its stability.

From our earlier argument, however, it is obvious that

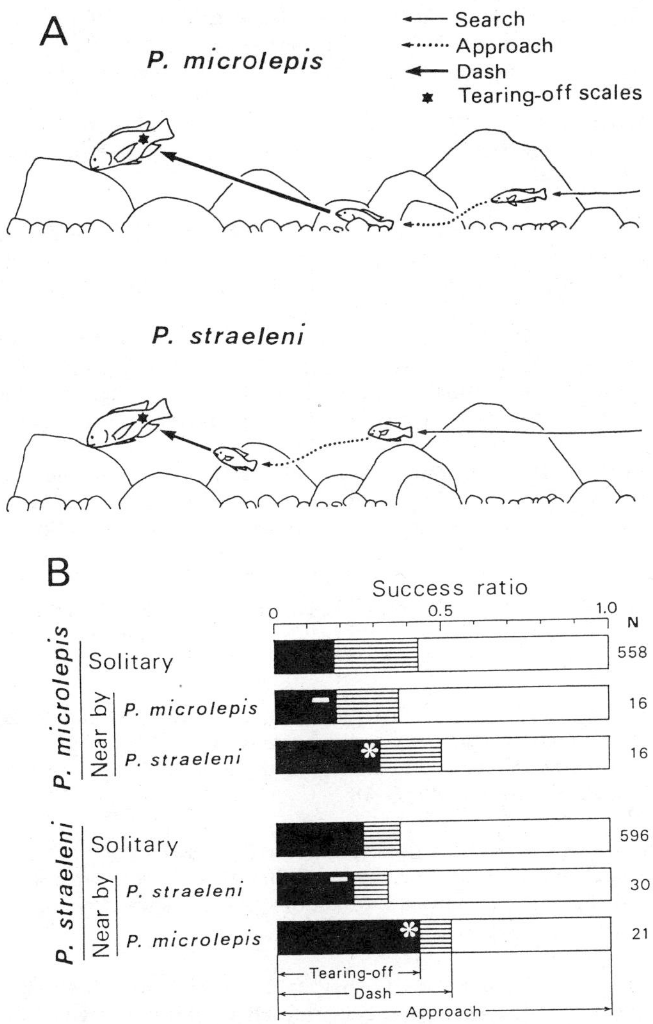

Figure 6. Hunting behavior and success ratio of two scale eaters, Perissodus microlepis *and* P. straeleni *(Hori 1987). (A) Sequence of hunting behavior of the species. (B) Hunting success ratio of the two species under different foraging situations. The success ratios of "Tearing-off" and "Dash" are shown as proportions of "Approach," which is set at 1.0. * indicates that the rates are significantly different from that in a solitary feeding situation at the 5% level. – indicates that the difference is nonsignificant. Significance level was tested with approximation to normal distribution. Both species had significantly higher hunting success when another species was present within 50 cm.*

the diversity of littoral fish communities in Lake Tanganyika is maintained through various kinds of direct and indirect interaction among species. This situation seems to be self-evident, because such a diverse community can come about only if the evolution of each species is closely connected and adjusted to the evolution of other species. This view suggests that maintenance of ecological networks among species, rather than preservation of rare or unique species, is most important.

Such communities generated through and supported by a web of interspecific interaction may stand against any change in density of some component species and against any equivalent perturbation on a local scale, as suggested in a removal experiment in this study. It may be valuable, however, to point out that even such stable communities are easily disrupted by any unusual perturbation to which they have never become adjusted during their evolution (Lowe-McConnell 1987). Any chemical pollution, including oil leaks and drastic civil engineering in littoral zones, represent such perturbations. In particular, any perturbation severing the web of interaction will seriously damage such communities (Elton 1958). This possibility was realized in Lake Victoria, where the introduction of an exotic predator (Nile perch) destroyed indigenous cichlid populations lakewide (Barel et al. 1991) and a large part of the unique fauna was probably lost forever. We must learn an important lesson from this event. Lake Tanganyika is an evolutionary reservoir for old lineages of East African cichlid fish (Nishida 1991), and its fauna is the most divergent among the African Great Lakes (Fryer & Iles 1972; Lowe-McConnell 1987). In order to preserve its irreplaceable diversity, any introduction of exotic species—unintended or deliberate—should be strictly controlled by the relevant authorities.

We must also consider the diversity among fish communities in different localities of the lake. Local communities in littoral areas were more or less isolated from each other; in other words communities specific to rocky areas were separated from those specific to sandy areas, and the genetic interchange among communities of each type was very low even between two adjacent rocky or sandy areas (Nishida, personal communication). This situation is in contrast to that of coral reefs, where the fish, unlike the cichlids, have pelagic larval stages. The species composition is much different among the communities in the three study sites, as shown earlier. Among communities of such different composition, patterns of interspecific interaction are also varied (Hori 1991). A poorly studied area, such as the southwest coast of the lake, may also contain a community much different from these three sites, as suggested by some faunal investigations (Poll 1953, 1956). Special care is needed, therefore, to preserve such lakewide diversity (Kawanabe & Gashagaza 1991). An integrated conservation of littoral areas in the form of national parks or reservations, each of which contain sets of major habitats, in various parts of the lake shore may be necessary. At least, some proper areas on the whole western coast, where no national park or reservation exists at present, should be preserved in an appropriate form through coordination among the countries concerned.

Acknowledgments

We wish to express our thanks to the Centre de Recherche en Sciences Naturelles de Zaire, the Tanzania National Scientific Research Council, and the Department of Fisheries of Zambia for their permission to conduct research in Lake Tanganyika. The director and the staff of the Uvira Station of Zaire, the Mahale Mountains Wildlife Research Center, and Lake Tanganyika Research Unit of Zambia kindly helped us in our field studies at each location. We are greatly indebted to members of the research teams for generous cooperation and stimulating discussions during the course of this work.

This work was partly supported by a Grant-in-Aid for Scientific Research on Priority Areas (#319), Project "Symbiotic Biosphere: An Ecological Interaction Network Promoting the Coexistence of Many Species" (#03269103 and 03269105) and for Overseas Scientific Survey (#404130, 504328, 56041032, 57343028, 58041043, 59043039, 60041036, 61043031, and 63041080) from the Ministry of Education, Science and Culture, Japan, and by the Joint Study Project on Ecological, Limnological and Fish-Biological Research in Lake Tanganyika and Its Environs for the Good and Stable Supply from the Japan International Cooperation Agency.

Literature Cited

Barel, C. D. N., W. Ligtvoet, T. Goldschmidt, F. Witte, and P. C. Goudswaard. 1991. The haplochromine cichlids in Lake Victoria: An assessment of biological and fisheries interests. Pages 258–278 in H. A. Keenleyside, editor. Cichlid fishes: Behaviour, ecology and evolution. Chapman & Hall, London, England.

Brichard, P. 1989. Pierre Brichard's book of cichlid and all the other fishes of Lake Tanganyika. T. F. H. Publications, Hong Kong, China.

Connell, J. H., and W. P. Sousa. 1983. On the evidence needed to judge ecological stability or persistence. American Naturalist **121(6)**:789–824.

DeAngeles, D. L. 1975. Stability and connectance in food web models. Ecology **56**:238–243.

Elton, C. 1958. The ecology of invasion by animals and plants. Methuen, London, England.

Fryer, G., & T. D. Iles. 1972. The cichlid fishes of the great lakes of Africa. Oliver & Boyd, Edinburgh, Scotland.

Gashagaza, M. M. 1991. Diversity of breeding habits in lamprologine cichlids in Lake Tanganyika. Physiological Ecology Japan **27**:29–65.

Greenwood, P. H. 1991. Speciation. Pages 86–102 in H. A. Keenleyside, editor. Cichlid fishes: Behaviour, ecology and evolution. Chapman & Hall, London, England.

Grossman, G. D. 1982. Dynamics and organization of a rocky intertidal fish assemblage: The persistence and resilience of taxocene structure. American Naturalist **119(5)**:611–637.

Harrison, G. W. 1979. Stability under environmental stress: Resilience, persistence, and variability. American Naturalist **113**:659–669.

Hori, M. 1983. Feeding ecology of thirteen species of *Lamprologus* (Teleostei; Cichlidae) coexisting at a rocky shore of Lake Tanganyika. Physiological Ecology Japan **20**:129–149.

Hori, M. 1987. Mutualism and commensalism in the fish community of Lake Tanganyika. Pages 219–239 in S. Kawano, J. H. Connell, and T. Hidaka, editors. Evolution and coadaptation in biotic communities. University of Tokyo Press, Tokyo, Japan.

Hori, M. 1991. Feeding relationships among cichlid fishes in Lake Tanganyika: Effects of intra- and interspecific variations of feeding behavior on their coexistence. Ecology International Bulletin **19**:89–101.

Hori, M., K. Yamaoka, and K. Takamura. 1983. Abundance and micro-distribution of cichlid fishes on a rocky shore of Lake Tanganyika. African Studies Monographs **3**:25–38.

Hynes, H. B. N. 1950. The food of fresh-water stickleback (*Gasterosteus aculeatus* and *Pygosteus pungitius*), with a review of methods used in studies of the food of fishes. Journal of Animal Ecology **19**:36–58.

Kawanabe, H. 1986. Cooperative study on the ecology of Lake Tanganyika between Japanese and African scientists, with special reference to mutual interactions among fishes. Physiological Ecology Japan **23**:119–128.

Kawanabe, H., and M. M. Gashagaza. 1991. Ecological diversity of Lake Tanganyika with special references to fishes. Page 69 in A. W. Cohen, editor. Report on the first international conference on the conservation and biodiversity of Lake Tanganyika University of Burundi, Burundi.

Kawanabe, H., M. Hori, and K. Takamura. 1985. Cooperative feeding among cichlid fishes in rocky shores of Lake Tanganyika. Verhandlungen der Internationale Vereinigung für theoretishe und angewandte Limnologie **22**:2661.

Kohda, M. 1991. Intra- and interspecific social organization among three herbivorous cichlid fishes in Lake Tanganyika. Japanese Journal of Ichthyology **39(2)**:147–163.

Kondo, T. 1986. Feeding habits of *Lamprologus savoryi* (Teleostei: Cichlidae) with reference to its social behaviour. Physiology and Ecology Japan **23**:1–15.

Kuwamura, T. 1986. Parental care and mating systems of cichlid fishes in Lake Tanganyika: A preliminary field survey. Journal of Ethology **4**:129–146.

Kuwamura, T. 1987. Distribution of fishes in relation to the depth and substrate at Myako, east-middle coast of Lake Tanganyika. African Studies Monographs **7**:1–14.

Kuwamura, T. 1992. Overlapping territories of *Pseudosimochromis curvifrons* males and other herbivorous cichlid fishes in Lake Tanganyika. Ecological Research **7**:43–53.

Lowe-McConnell, R. H. 1987. Ecological studies in tropical fish communities. Cambridge University Press, Cambridge, England.

MacArthur, R. H. 1955. Fluctuations of animal populations and a measure of community stability. Ecology **36**:533–536.

May, R. M. 1972. Will a large complex system be stable? Nature **238**:413–414.

May, R. M. 1973. Stability and complexity in model ecosystems, 2nd edition. Princeton University Press, Princeton, New Jersey.

May, R. M. 1981. Patterns in multispecies communities. Pages 197–227 in R. M. May, editor. Theoretical ecology: Principles and applications. Blackwell Scientific Publications, Oxford, England.

Mbomba, N. B. 1985. Comparative feeding ecology of 11 sympatric species of cichlid algal feeders in Lake Tanganyika. Verhandlungen der Internationale Vereinigung für theoretishe und angewandte Limnologie **22**:2662.

Kawanabe, H. 1986. Comparative feeding ecology of Aufwuchs eating cichlid fishes in Lake Tanganyika with reference to their developmental changes. Physiol. Ecol. Japan **23**:79–108.

McNaughton, S. J. 1977. Diversity and stability in ecological communities. American Naturalist **111**:515–523.

Nagoshi, M. 1983. Distribution, abundance and parental care of the genus *Lamprologus* (Cichlidae) in Lake Tanganyika. African Studies Monographs **3**:39–47.

Nishida, M. 1991. Lake Tanganyika as an evolutionary reservoir of old lineages of East African cichlid fishes: Inferences from allozyme data. Experientia **47**:974–979.

Nshombo, M. 1987. Scale eating behaviour of *Perissodus straeleni* (Cichlidae). Ecological Limnology of Tanganyika **4**:95–97.

Nshombo, M., Y. Yanagisawa, and M. Nagoshi. 1985. Scale-eating in *Perissodus microlepis* (Cichlidae) and change of its food habits with growth. Japanese Journal of Ichthyology **32(1)**:66–73.

Pianka, E. R. 1973. The structure of lizard communities. Annual Review of Ecological Systems **4**:53–74.

Pimm, S. L. 1979. The structure of food webs. Theoretical Population Biology **16**:144–158.

Poll, M. 1953. Poisson non Cichlidae. Résultats Scientifiques Exploration Hydrobiologique Lac Tanganika (1946–1947), **3(Fasc. 5a)**:1–251.

Poll, M. 1956. Poisson cichlidae. **3(Fasc. 5b)**:1–619.

Poll, M. 1986. Classification des Cichlidae du lac Tanganyida: Tribus, genres et espèces. Académie royale de Belgique, Memoires de la classe des sciences **8°(2)45(2)**:1–163.

Redfearn, A., and S. L. Pimm. 1987. Insect outbreaks and community structure. Pages 99–133 in P. Barbosa, and J. C. Schultz, editors. Insect outbreaks. Academic Press, New York.

Simpson, E. H. 1949. Measurement of diversity. Nature **688**:163.

Takamura, K. 1983. Interspecific relationship between two aufwuchs eaters *Petrochromis polyodon* and *Tropheus moorei* (Pisces: Cichlidae) of Lake Tanganyika, with a discussion on the evolution and functions of a symbiotic relationship. Physiology and Ecology Japan **20**:59–69.

Takamura, K. 1984. Interspecific relationships of Aufwuchs-eating fishes in Lake Tanganyika. Environmental Biology of Fishes **10**:225–241.

Tiercelin, J-J., and A. Mondeguer. 1991. The geology of the Tanganyika trough. Pages 8–48 in G. W. Coulter, editor. Lake Tanganyika and its life. Oxford University Press, London, England.

Yamaoka, K. 1982. Morphology and feeding behaviour of five species in genus *Petrochromis* (Cichlidae). Physiology and Ecology Japan **19**:57–75.

Yamaoka, K. 1983. Feeding behaviour and dental morphology of algae-scraping cichlids (Pisces: Cichlidae) in Lake Tanganyika. African Studies Monographs **4**:77–89.

Yamaoka, K. 1991. Feeding relationships. Pages 151–172 in H. A. Keenleyside, editor. Cichlid fishes behaviour, ecology and evolution. Chapman & Hall, London, England.

Yamaoka, K., M. Hori and S. Kuratani. 1986. Ecomorphology of feeding in "goby-like" cichlid fishes in Lake Tanganyika. Physiology and Ecology Japan **23**:17–29.

Yanagisawa, Y. 1987. Social organization of a polygynous cichlid *Lamprologus furcifer* in Lake Tanganyika. Japanese Journal of Ichthyology **34**:82–90.

Yuma, M. 1989. Community structure of benthos-feeding fishes in Lake Tanganyika. Ecological Limnology of Tanganyika **6**:39.

Contributed Papers

Conservation Lessons from Long-Term Studies of Checkerspot Butterflies

PAUL R. EHRLICH*

Center for Conservation Biology
Department of Biological Sciences
Stanford University
Stanford, California 94305

DENNIS D. MURPHY

Center for Conservation Biology
Department of Biological Sciences
Stanford University
Stanford, California 94305

Abstract: *Long-term studies of checkerspot butterflies (*Euphydryas *spp.) are used to draw conservation lessons with implications for the design and management of reserves for invertebrates. We discuss;*

1. *The importance of determining the structure of and patterns of mating within demographic units.*
2. *The lack of congruence between migration and gene flow.*
3. *The crucial role of habitat diversity in the protection of species which otherwise require only small areas of habitat.*
4. *The importance of identifying and preserving "reservoir" populations within metapopulations.*
5. *The necessity of identifying subtle habitat requirements.*
6. *The key role of environmental stochasticity in the extinction of small populations.*
7. *The difficulty of protecting metapopulations and of introducing populations into "empty" habitat patches.*

We conclude, among other things, that the size of reserves required to preserve herbivorous insects is often underestimated; that a "minimum viable metapopulation" approach to their conservation may be needed; that long-term studies of the biology of populations of sample invertebrate groups would greatly aid conservation biologists; and that more effort is needed to inform the public and decision makers about the need to conserve populations *as well as species.*

Resumen: *Se discuten las implicaciones que los estudios sobre las mariposas del género* Euphydryas *tienen para el diseño y manejo de reservas para invertebrados. Se desarrollan los siguientes temas:*

1. *La importancia que tiene el determinar la estructura y los patrones de apareamiento dentro de las unidades demográficas.*
2. *La falta de congruencia entre la migración y el intercambio de genes.*
3. *La importancia de la diversidad de hábitat en la protección de aquellas especies que requieren áreas pequeñas.*
4. *La importancia de identificar y proteger "reservas" de poblaciones dentro de metapoblaciones.*
5. *La necesidad de identificar aquellas características sutiles del hábitat que requiere la especie.*
6. *La importancia que tienen los eventos estocásticos del medio ambiente en la extinción de las poblaciones pequeñas.*
7. *La dificultad para proteger metapoblaciones y para introducir poblaciones en fragmentos de hábitat "vacíos" ("empty" habitat patches).*

Concluímos, entre otras cosas, que es común el subestimar el tamaño de las reservas necesario para la sobrevivencia de las poblaciones de insectos herbívoros; que se necesita enfocar la conservación de estos insectos desde un punto de vista de "metapoblación mínima viable"; que estudios a largo plazo de la biología de poblaciones de grupos de invertebrados selectos serán de mucho valor para los biólogos; y que es necesario informar al público y a las personas que toman decisiones en el campo de la conservación sobre la necesidad de conservar poblaciones *y especies.*

**Correspondence and requests for reprints should be addressed to this author.*

Paper submitted 1/5/87; revised manuscript accepted 4/7/87.

Introduction

Field research in population biology tends to be short in duration and narrow in geographic scope. This situation presents a major impediment to the application of basic research to problems confronting conservation biologists, since conservation biology is by definition concerned with the maintenance of populations and species over the long term. "Snapshots" of the histories of single populations over brief periods repeatedly have proven to give at best a partial, and at worst an erroneous, impression of the factors controlling the distribution and abundance of plants and animals.

One of the few long-term studies of a suite of invertebrate species over a substantial portion of their ranges has been carried out on checkerspot butterflies (genus *Euphydryas*). The ecology and population genetics of checkerspot butterflies in the western United States have been under intensive study for the past 27 years (e.g., Ehrlich 1961, 1965; Singer 1971; Ehrlich et al. 1975; Singer & Ehrlich 1979; Brown & Ehrlich 1980; Singer 1982; Murphy et al. 1983, 1984, 1986; Murphy 1984; Brussard et al. 1985). Here we summarize some of that work, highlighting findings of particular importance to conservation biology, including the design and management of nature reserves. Our findings should give particular insight into the conservation biology of herbivorous insects.

Population Biology

Physical Environment and Population Structure

1. *An important lesson for conservation biology from the earliest studies of* Euphydryas *is that the movement of individuals between habitat patches or reserves does not necessarily mean that gene flow is occurring. It also does not mean that, in case of extinction, recolonization of one patch by individuals from the other will occur readily.*

The most extensively studied colony of *Euphydryas* is that of *E. editha* on Stanford University's Jasper Ridge Biological Preserve. There the butterflies occupy an island of serpentine-based grassland approximately 9 hectares in extent, surrounded by chaparral and sandstone-based grassland. Despite essentially continuous habitat, uninterrupted by barriers to the butterfly's flight (Ehrlich 1961), the colony was found to consist of three separate demographic units—populations sufficiently isolated to have independent dynamics, though not necessarily enough to be independent genetic units. In each season, fewer than 5 percent of the individuals marked in one demographic unit were recaptured in another, so migration between extant units did not dramatically influence their dynamics. Nonetheless, this consistent finding suggests that *E. editha* exists as "metapopulations"; that is, populations subdivided into groups that occupy clusters of habitat patches and interact extensively.

Furthermore, because of the combination of late-season timing of migration and the reproductive strategy of the butterflies, individual movements are clearly much more frequent than the movement of genes. Males "plug" females after they inseminate them, effectively preventing another insemination for a week or more (Labine 1964). Most females are mated and plugged within hours of eclosion, so that after the first week of the female flight period an immigrant male has a comparatively small chance of being able to inseminate females. Later, when females become receptive to second inseminations, the probability of offspring survival becomes very small because of time constraints on larval development *(discussed below)*. These same constraints make it unlikely that any but the earliest migrant females will make a genetic contribution to the recipient population, or serve as effective founders.

2. *Understanding the scale and structure of populations is crucial to designing and managing reserves.*

The dynamic histories of the three demographic units on Jasper Ridge, in areas C, G, and H (Ehrlich 1965), have been documented since 1960 (Fig. 1). Note that Jasper Ridge area G went extinct in 1964, was re-established in 1966, and then went extinct again in 1974. If the mark–recapture studies had addressed the entire grassland as a single area, and the butterflies had been considered as a single demographic unit (solid line in Fig. 1), the extinction events would have been overlooked, and the safety of the Jasper Ridge population from stochastic extinctions of its component demographic units would have been overestimated.

3. *Understanding the patterns of mating within demographic units is also required if conservation efforts are to succeed, because those patterns may change with density.*

We have found population structure in some locations other than Jasper Ridge to be much looser (demographic units much less well defined), and this structure is reflected in the genetics of the populations. Spatial autocorrelation of allele frequencies in Great Basin populations, for example, suggests that discrete demographic units may occupy habitat areas of dozens of square kilometers and individuals may move much greater distances (Wright et al. 1987). Yet, even in more or less continuous habitat, special physical features may be crucial to population structure. In Colorado, where diffuse demographic units of *Euphydryas editha* occupy many square kilometers of montane sagebrush communities, hilltops appear to constitute important rendezvous points at which males aggregate when population densities are low and to which females travel to seek mates (Ehrlich & Wheye 1986).

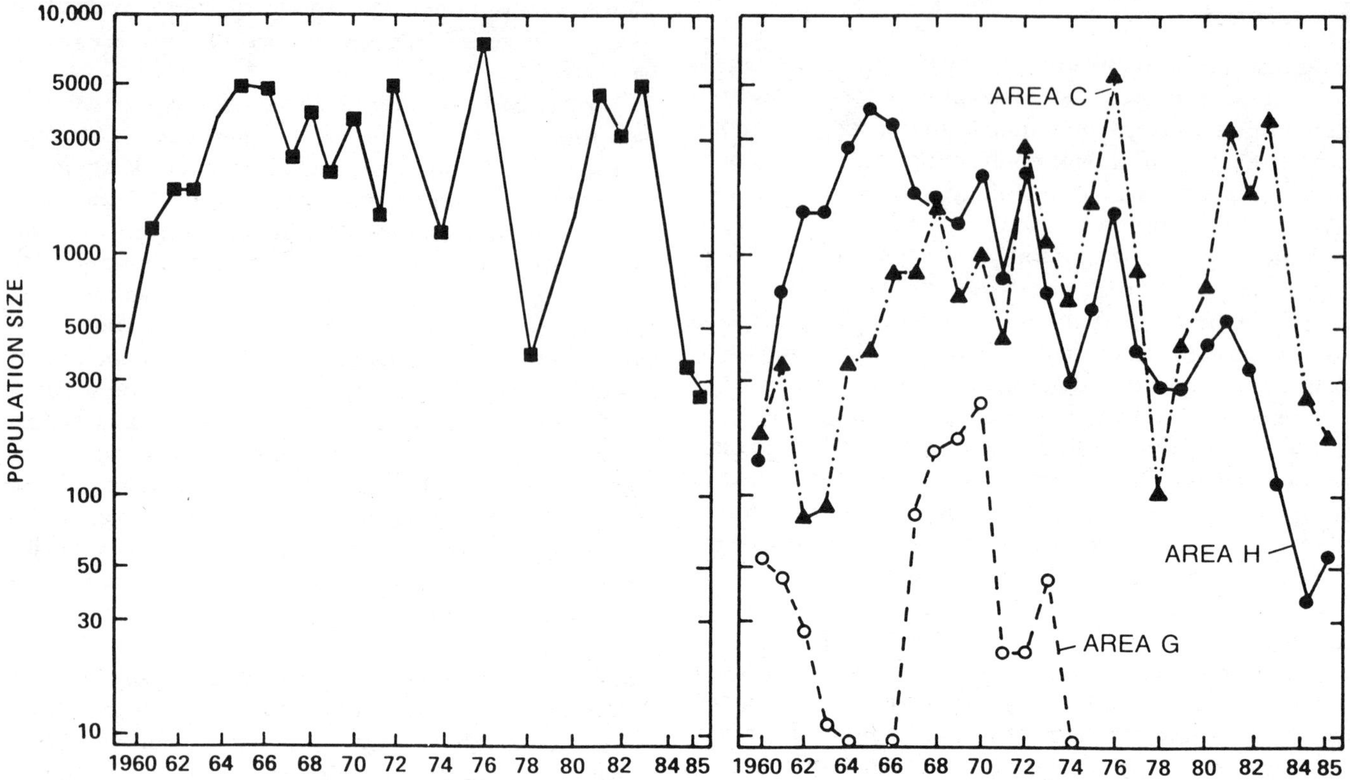

Figure 1. Jasper Ridge population size changes had the separate demographic units not been identified (left). Population size changes of the demographic units in Jasper Ridge areas C, G, and H (right). Note extinction and recolonization events in area G.

At low population densities, such mating aggregations may play a crucial role in keeping a population viable. At high densities, however, males that move to hilltops may have relatively small chances of mating (Ehrlich & Wheye 1986). Thus, while such an environmental feature may seem unimportant if the study of a demographic unit was done in a year of abundance, its inclusion may prove critical if a reserve is sliced out of continuous habitat.

4. *Topographic heterogeneity or other aspects of habitat diversity may be key determinants of the habitat suitability of an area for the maintenance of populations of many insect herbivores that otherwise require only small areas of habitat.*

Topography is a crucial consideration in the conservation biology of threatened populations of *Euphydryas editha bayensis* in the San Francisco Bay area of California, the ecotype represented at Jasper Ridge. For more than a century, the ecotype has been confined to patches of serpentine-based grassland. Earlier it may have been much more widespread, but lack of knowledge of the California grassland flora before the introduction of now virtually ubiquitous Eurasian weeds makes reconstruction of prior distributions impossible. For the last century, areas of suitable habitat have been declining in the face of development, and populations in those serpentine habitats that remain are often subject to extinction as a result of climatic events such as the California drought of the mid-1970s (Ehrlich et al. 1980) and the "El Nino" weather of 1982–1983 (Dobkin et al. 1987).

The sizes of populations of the Bay Area ecotype are controlled in a density-independent fashion. Eggs are laid in the spring, and upon hatching, larvae are immediately in a race to reach the size at which they can enter an obligatory diapause in which they pass the summer dry season of California's Mediterranean climate. The larvae lose this race if they cannot grow large enough before their annual food plants senesce. There is a complex phase relationship in the timing of the insect's developmental cycle and that of the plant (Singer 1972, Singer & Ehrlich 1979). This relationship is mediated by the macroclimate interacting with the microclimate, and the latter varies significantly with slope and exposure (Dobkin et al. 1987).

The remaining patches of serpentine habitat most suitable for the long-term maintenance of populations are the largest and most topographically diverse. A large area of habitat alone is clearly not sufficient to guarantee the persistence of checkerspots. Extinctions have been observed in large habitats that lack topographic heterogeneity. The Coyote Reservoir population, the most densely populated of known Bay Area colonies in 1971, was extinct by 1976. The Coyote Reservoir site consists almost entirely of an east-facing slope. Topographic heterogeneity allows some areas within a serpentine patch to have a favorable microclimate for larval development in any weather regime; north-facing slopes will be favored in most years, while south-facing slopes contribute to population increases in wet years. High survival rates on south-facing slopes in wet years appear to push population numbers high enough in topographically diverse locations to allow demographic units to survive substantial declines during inevitable drought years.

5. *If conservation options are limited, top priority should be given to identifying and preserving reservoir populations, if any.*

The overall pattern in the San Francisco Bay area appears to be one in which just one population south of the bay (Morgan Hill, Santa Clara County) is large enough (Fig. 2) and stable enough to act as a "reservoir population" of *E. editha bayensis* (Fig. 3). A reservoir population may be defined as a demographic unit which can be expected to persist through essentially all natural environmental perturbations. A reservoir population is one that can act as a relatively permanent source of colonists to habitats that cannot permanently maintain what might be called "satellite" populations. We suggest that since all demographic units will eventually become extinct, that the terms "reservoir population" and "satellite population" only be used where it is estimated that the former will persist at least 10 times as long as the latter.

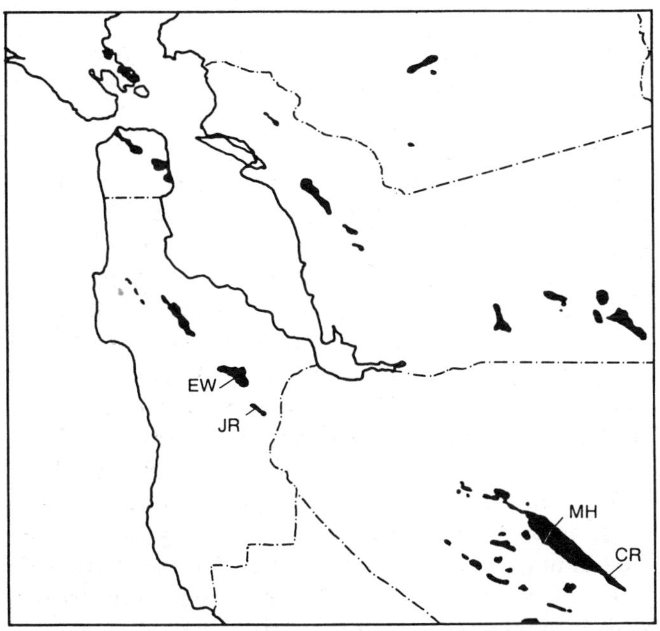

Figure 2. Distribution of San Francisco Bay Area serpentine soils, including grassland habitats of Euphydryas editha. *EW = Edgewood Park (part of the San Mateo-Woodside grassland system), JR = Jasper Ridge, CR = Coyote Reservoir, and MH = Morgan Hill.*

The extensive San Mateo-Woodside system of serpentine-based grasslands has been severely fragmented, and only at Edgewood Park is the remnant habitat greater than 10 ha in extent. The ability of the population there to function as a reservoir is doubtful, as it has undergone dramatic population fluctuations itself ($>10^5$ individuals in 1981 to $<10^2$ in 1985). All other sites, including Jasper Ridge Reserve, are too small for truly long-term maintenance of the butterfly's populations through extended drought. It seems likely that if the California drought had continued for another year, all three demographic units in the reserve might have become extinct. Indeed, weather conditions seem not to have been optimal since 1981, and neither the Jasper Ridge area C nor area H population has been able to rebuild to the high density levels of the early 1970s.

Resources and Population Dynamics

6. *Natural history investigations should be undertaken before a reserve is designed in an attempt to identify subtle habitat requirements that could potentially play key roles in the long-term survival of populations to be protected.*

Independent fluctuations in size of the demographic units on Jasper Ridge result, at least in part, from rather subtle differences in the means by which larvae survive to diapause each spring. In both populations, survival is greatest for eggs laid on north-facing slopes or other comparatively mesic areas by early emerging females from south-facing slopes. In the area H demographic unit, many survivors are larvae that wander from drying *Plantago* and locate the flowers of later-senescing *Orthocarpus densiflorus* on which the larvae feed to diapause. At nearby Edgewood Park, females often oviposit directly on *Orthocarpus* (Singer 1971). In contrast, many survivors to diapause in Jasper Ridge area C come from eggs laid on *Plantago* on soil tilled by gophers *(Thomomys),* where plants are large and exhibit delayed senescence—although *Orthocarpus* plays a role as a secondary food source in this area also.

While the primary resource determinants of habitat suitability clearly are oviposition and larval host plants, other resources may play crucial roles in population persistence and habitat suitability. The availability of nectar, for instance, is very important in maintaining egg production by females in the latter stages of reproduction. A paucity of middle- and late-season nectar has been implicated in the extinction of the Jasper Ridge area G population (Murphy et al. 1983). The presence of standing water has been shown to affect population structure

Figure 3. Newly emerged Euphydryas editha bayensis *female in the area H serpentine grassland of the Jasper Ridge Biological Preserve.*

in *Euphydryas* (Ehrlich & White 1980) and in *Chlosyne,* a genus closely related to *Euphydryas* (Schrier et al. 1976). Males travel relatively long distances to sip from moist earth around puddles, presumably to obtain sodium (Arms et al. 1974, Adler & Pearson 1982). The presence of both nectar and standing water, additionally, have been shown to be determinants of overall butterfly species numbers in Great Basin riparian communities (Murphy & Wilcox 1986).

7. *Under severe environmental conditions, the degree of oligophagy may be a crucial determinant of extinction proneness.*

We used mark–recapture studies and detailed statistical analysis in an attempt to identify adult population dynamic parameters that might be unique to particularly extinction-prone species. But, population studies of highly endangered *Euphydryas editha* and ubiquitous, unthreatened *Euphydryas chalcedona* on Jasper Ridge revealed few differences (Murphy et al. 1986). Indeed, both species appear to move freely within suitable habitat and exchange individuals at low rates with other colonies at distances of several kilometers or more.

Ease of recolonization well may be determined, however, primarily by patterns of resource use, rather than interspecific differences in vagility. The Bay Area ecotypes of *Euphydryas chalcedona* feed on perennial host plants and appear to accept a wide variety of hosts. *E. editha,* by contrast, feeds on annuals and exhibits narrower host tolerance. Furthermore, within the California range of *E. editha,* populations feeding on perennials (e.g., on *Pedicularis densiflora* in the inner coast ranges and on *Castilleja nana* in the alpine Sierra Nevada) have suffered fewer documented "natural" extinctions than those feeding on annuals (e.g., on *Plantago* and *Orthocarpus* in outer coast ranges and on *Collinsia tinctoria* in the Sierra foothills). These observations suggest that wide host tolerance and the incorporation of less ephemeral hosts into the larval diet may confer resistance to extinction.

Dispersal, Patch Dynamics, and Gene Flow

8. *Recolonization of empty patches may not occur in most years; many populations may be re-established in rare years of explosive dispersal.*

The need to monitor populations for many generations is underscored by our observation of populations of *Euphydryas editha* in southern California. Individuals in those populations exhibit rather sedentary behavior similar to Jasper Ridge butterflies. Dispersal in southern California populations increases, however, in particularly dry years when oviposition plants and adult nectar sources are sparse. Yet, in 1977, a year of substantial rainfall in southern California, the populations exploded in size, and adults dispersed en masse from population centers (Murphy & White 1984). These circumstances (not repeated since) allowed insight into the abiotic and biotic forces that can determine patterns of dispersal in the species.

Dry winters seem to result in increased vagility because larval host plants and adult nectar sources are sparse from the lack of precipitation and because postdiapause larvae defoliate a higher proportion of available host plants before adult females emerge to oviposit. When wet years precede dry years, defoliation is intensified in the dry years because larval populations are larger. Furthermore, consecutive years of adequate rainfall apparently result in butterfly population build-ups such that entire standing crops of host plants may be defoliated; massive postdiapause mortality of larvae then occurs and virtually all emerging adults disperse. Population size, dispersal, and the likelihood of local extinction, therefore, are determined by density-independent factors in most years, but by density-dependent factors in a very few years. These rare years, however, appear to be one of the keys to the natural recolonization of habitats unable to support populations over the long term. Thus not only may gene flow and recolonization potential be difficult to estimate from data on movement of individuals, but the amount of dispersal itself may vary dramatically from year to year.

Minimum Viable Population Size

9. *Environmental stochasticity, especially variation in host quantity, quality, and phenology, is the major cause of extinction, particularly in small populations.*

A general constancy has been found in allozyme frequencies and in average heterozygosity in California *Euphydryas* populations, despite wide fluctuations in population sizes and after local "bottlenecks" where N_e may have fallen well below 50. This suggests that deleterious genetic effects, such as inbreeding depression or loss of variation, may be relatively unimportant in the persistence of *Euphydryas* populations.

Although rates of dispersal from Jasper Ridge demographic units appear to be inversely correlated with population size (Gilbert & Singer 1973), recent observations in area H indicate that *E. editha* populations can persist for numerous generations at surprisingly small sizes. It appears that levels of dispersal in *Euphydryas* are complexly related both to population density itself, and to the abundance (or scarcity) of resources. The population in Jasper Ridge area H has been estimated to consist of less than 100 individuals for four consecutive years (in two of those years, N_e was certainly less than 10!). Thus, while the role of demographic stochasticity has not been determined and could be significant in certain circumstances, our present evidence indicates that small population size in and of itself will not ordinarily lead to population extinction. Small populations, of course, are particularly susceptible to extinction due to environmental events such as drought.

Even large populations are at risk, however, when drought reaches dramatic proportions (on the order of half "normal" rainfall), or if several less extreme drought years occur in succession. Such conditions are effectively "natural" catastrophes. Likewise, the major cause of anthropogenic extinction of *Euphydryas editha* populations has not been a gradual deterioration of local environments leading eventually to population extinctions, but rather a catastrophic loss of habitats. For example, in the San Francisco Bay area the paving over of patches of serpentine grassland has led to the endangerment of *E. editha bayensis.*

Management and Politics

10. *Habitat patches and metapopulations are difficult to protect.*

In the Outer Coast Ranges of central California, native grassland persists as remnant patches on serpentine and other highly weathered soils. Unique chemical characteristics, including high levels of magnesium, nickel, and chromium, and low levels of calcium, nitrogen, and phosphorus, appear to account for the resistance of serpentine soils to invasion by non-native grasses and forbs, and the resultant exclusion of many native plant species. The high degree of ecotypic differentiation and endemism in serpentine-based grassland communities make them an obvious target for conservation efforts, but they exist as dispersed small units, ranging from less than a hectare to a few thousand hectares in extent, with most less than 10 ha. If a continuous national park or other reserve of reasonable configuration was established that contained all the serpentine patches in the San Francisco Bay area, it would necessarily include the homes and businesses of millions of people.

Most of the native grassland remnants are privately owned, hence conservation efforts must follow one of two paths: acquisition of property, obviously an expensive proposition; or invocation of the Endangered Species Act. Unfortunately, the 1982 reauthorization of the Act largely removed protection from "take" for endangered and threatened plants. *Euphydryas editha bayensis* is, at present, the only animal restricted to the native grassland community that has been proposed for federal endangered species protection.

Two major problems exist in using this butterfly as a tool to protect serpentine-based grassland habitat. First, because some number of demographic units within the metapopulation are always extinct and their sites awaiting recolonization, the butterfly may not be present in many of the remnants with the highest plant diversity (Murphy & Ehrlich 1987). Since the Endangered Species Act focuses on protecting "species," temporarily vacant habitat patches have no legal protection. Second, invertebrates can be protected only if an entire described subspecies is threatened or endangered. An opponent to the listing of *Euphydryas editha bayensis* has challenged the validity of the traditional taxonomic treatment of the subspecies. As biologists well know, the subspecies is an arbitrary taxonomic category. This challenge, which promises to become a protracted battle, ultimately may compromise future attempts to protect infraspecific genetic diversity of invertebrates.

Finally, because small reserves in largely urban areas have neighbors, maintaining good relations with those neighbors may be critical to the integrity of the reserve. Jasper Ridge has an extremely active docent program, and docents lead carefully guided tours of the facility on which, among other things, the importance of *Euphydryas* work is explained. Tours were routed away from area H after it was discovered that, even with the most careful supervision, it was impossible to prevent trails from widening in response to increased foot traffic. The area around the preserve is suburban, and some homes share common boundaries with it. Educational programs, fencing, and patrolling at Jasper Ridge have generally been able to keep intrusion to a minimum in the past decade, but constant vigilance has been required, and in one case Stanford University has had to resort to extensive and expensive litigation to maintain the integrity of the preserve.

Simply finding the funds to establish small reserves to protect the elements of a metapopulation is only the first, and in many cases the easiest, step. Finding the additional funds and engaging the time of interested people to manage them are clearly major challenges. Furthermore, it has yet to be determined whether, even if the sites that can support the demographic units of a metapopulation (such as that of *E. e. bayensis*) can be preserved, the metapopulation itself can be protected. It may be, for example, that complete urbanization of intervening areas would so interfere with dispersal and recolonization that the entire system would collapse to extinction.

Thus, political problems often involve more than finding ways to protect reserves. Conservation biologists working with butterflies and other vagile organisms must always be concerned with land use patterns adjacent to reserves. They must work to make those areas more hospitable to organisms that disperse out of the protected area. Conservation biologists also must actively campaign for abatement threats such as acid precipitation and rapid anthropogenic climate change. No amount of careful management of reserves will prevent extinctions from such adverse conditions.

11. *The ease of maintaining a metapopulation or re-establishing extinct metapopulations by introducing insects into "empty" habitat is difficult to predict.*

It may be a long time before the success of a transplant experiment can be evaluated, even when a univoltine insect is involved. Re-establishment of demographic units known to have gone extinct and their establishment in "empty" habitats that seem suitable both appear to be possible management options for the conservation of *Euphydryas*. Two transplantation experiments have been carried out in connection with our studies of the allozyme genetics of *E. editha*. The failure of one transplant of 500 larvae to Agua Fria in the foothills of the California Sierra Nevada was due, at least in part, to human disturbance of a portion of the habitat. Another, to Arroyo Bayou in the Inner Coast Range of central California, again with 500 larval colonists, appears to have resulted in establishment of a demographic unit, although there are signs that natural immigration also played a part.

Our most interesting result with transplant experiments thus far have been achieved with another checkerspot species, *Euphydryas gillettii*. One of two attempts to transplant a colony for northwestern Wyoming to central Colorado was successful (Holdren & Ehrlich 1981). A population founded in 1977 with 83 egg masses (about 10^4 eggs) has persisted through the 1986 field season. Most important to conservation is that in nine generations the population has fluctuated in size, but has never exceeded 200 individuals, and has not spread from the immediate vicinity of the release, despite extensive adjacent and apparently suitable habitat (Ehrlich & Holdren, unpublished data). The population thus has never expanded to occupy a variety of sites and slopes and, therefore, may yet go extinct. Indeed, this underscores the difficulty of assessing habitat quality or suitability in general.

Unfortunately, population density constitutes the best bioassay of habitat quality. Furthermore, even if habitat is adequate to sustain a population for many ordinary years, the key to long-term habitat suitability is whether a habitat can support a population during extreme years (for *E. editha*, years of drought or deluge).

We are now embarking on a series of experiments to re-establish *E. editha bayensis* populations in isolated patches of habitat that we know previously supported demographic units, and in others that appear suitable but whose previous record of occupancy (or lack of it) is unknown. This work, which may involve as many as 70 patches, should greatly expand our understanding of the problems that managers will face in maintaining these and related organisms with similar "metapopulation dynamics." It is, however, already clear from work with

Euphydryas and with many other organisms (e.g., Ehrlich 1986, 1987) that simply introducing a group of colonists into entirely suitable habitat is no guarantee of establishment.

12. *Introduced mammalian herbivores may have complex impacts on endangered herbivorous insects.*

In the Bay Area, grassland habitat supporting high densities of *Euphydryas editha* host plants are early and intermediate successional stages leading to a stable climax community dominated by perennial bunchgrasses (Murphy & Ehrlich 1987). Native ruminant grazers (including the now extirpated tule elk and pronghorn antelope) undoubtedly played a role in creating successional habitat. In the absence of these species, domesticated livestock now affect the landscape physically and biologically and can alter profoundly the plant species composition, hence suitability of habitat for *Euphydryas editha.*

The physical presence of cattle directly affects the butterfly in its nonvagile stages. Mortality from being stepped on or buried under cow pats obviously can be significant, and is proportional to the density of cattle in a given area (White 1986). A more subtle effect of cattle is the physical alteration of the landscape from the action of hooves, including the terracing of steep slopes. Larvae are often found exclusively on cattle terraces on north-facing slopes in winter when the sun is low in the sky, sunlight is at a premium, and the terraces receive more isolation than the slopes.

The effect of cattle on the species composition of grasslands is more complicated. The physical action of hooves provides continual ground disturbance, which may or may not favor the butterfly's host plants. Selective grazing of introduced annual grasses reduces their ability to invade and outcompete the native flora, although native forbs and bunchgrasses are also consumed by cattle. Cattle droppings add concentrated nutrients that favor the establishment of introduced grasses in localized areas. The areas around salt licks and watering troughs can be almost completely dominated by introduced grasses. On both grazed and ungrazed sites, the butterfly host plants are interspersed with patches of introduced grasses. In ungrazed sites, invasion of introduced grasses appears to be facilitated by large gopher mounds. The dynamic balance of these processes is poorly understood.

13. *Endangered populations of insects may be "studied to death."*

In the course of the intensive study in 1981 in Jasper Ridge area H *(above),* it became apparent that roughly150–200 investigator hours spent collecting in an area of approximately 2.5 ha had a significant impact on the vegetation of the area. Although investigators took care always to wear tennis shoes (rather than cleated boots), some crushing of areas of herbage was obvious at the end of the experiment, although no long-term damage to the vegetation appears to have been done. We were concerned, however, that some pupae, egg masses, and larvae of the butterflies (which are close to the soil surface) may have been destroyed by trampling, which, in turn, could have serious consequences for the population. Sadly, research areas, especially those close to major centers, tend to be small and thus vulnerable to serious disturbance from research activities themselves and to interference between research projects. At the Jasper Ridge Preserve and the Rocky Mountain Biological Laboratory, the field stations we know best, both problems are causing growing concern among researchers and managers. Since it is questionable whether intensive mark–recapture experiments in such limited habitat areas can add to one's knowledge of the population (Murphy et al. 1986), it seems to us that managers should advocate instead the use of noncapture methods to estimate population sizes over the long-term.

Conclusions

What are the most important lessons for conservation biology to come out of the *Euphydryas* work? We think the first is that the size requirements of reserves designed to protect herbivorous insects may often be underestimated. Populations of such insects seem often to be subject primarily to density-independent population regulation, and thus their waiting time to extinction may be relatively short. A reserve that simply encompasses the location of a single demographic unit, even if it provides adequate marginal buffering and security from intrusion, may soon lose the organisms it was designed to protect. If such insects are to be conserved, in many if not most cases, a series of demographic units, selected for the diversity of their habitats, should be set aside. Where necessary, either corridors among the sites should be established or active management planned so that stochastic extinctions of demographic units can be followed by assured recolonizations.

In short, the "minimum viable population" sort of approach that many conservation biologists favor for vertebrate species, while risky enough for them, may not work at all for many invertebrates. Instead, concern there should be shifted toward "minimum viable metapopulations." This means that, in some cases, vertebrates or plants will prove inadequate as "umbrella species" for invertebrates. While the grizzly bear, because of its gigantic home range requirements, may protect many invertebrates, including checkerspot butterflies, in the Yellowstone ecosystem, preserving one patch of serpentine soil-based grassland for the last known population of the San Mateo Thornmint *(Acanthomintha obovata duttoni)* will do little to aid in the long-term maintenance of *Euphydryas editha bayensis.*

It is also clear that the level of effort that has gone into uncovering the details of *Euphydryas* biology cannot be expended on more than a minute fraction of potentially endangered invertebrate species—most of which, even in the temperate zones, have yet to be recognized as endangered. This leads to two conclusions. First, strategies for invertebrate conservation must be established on the basis of long-term studies of a few species or taxonomic groups chosen as representative. Second, in the absence of better information, conservative assumptions should be made about the long-term requirements of invertebrates. We probably should make a best guess about the habitat area required to preserve a species, and then at least double that area to reach a recommended minimum reserve size.

We cannot emphasize too strongly the importance of making studies of sample groups over the *long term*. We have, for example, drastically altered our original conclusions repeatedly after observing populations under different climatic regimes. Doing a long-term study also often allows time for observing the behavior of different ecotypes of the same species, which in the *Euphydryas* system led to drastic revisions of many early ideas about the biology of these organisms. Fortunately, the National Science Foundation is already aware of the need to conduct long-term studies (e.g., Callahan 1984), and conservation biologists in the United States should do all they can to encourage increased funding by Congress of the Foundation's program to support such research.

Finally, there is a great need for conservation biologists to emphasize the conservation of *populations* along with the conservation of species. We have seen here how the attrition of populations of *Euphydryas editha bayensis* has endangered the entire ecotype. A similar attrition, we believe, is gradually endangering many, if not most, other organisms. Focusing on the loss of species is focusing on the closing scene of what is ordinarily a long process. It leads to underestimates of the scale and severity of the current extinction crisis. When public statements are made on that crisis, the critical importance of the loss of populations should always be mentioned.

Literature Cited

Adler, P.E.; Pearson, D.L. Why do male butterflies visit mud puddles? *Can. J. Zool. 60*:322–325;1982.

Arms, K.; Feeney, P.; Lederhouse, R. Sodium: Stimulus for puddling behavior by tiger swallowtail butterflies *(Papilio glaucus). Science 185*:373–374;1974.

Brown, I.L.; Ehrlich, P.R. Population biology of the checkerspot butterfly *Euphydryas chalcedona*. Structure of the Jasper Ridge colony. *Oecologia 47*:239–251;1980.

Brussard, P.F; Ehrlich, P.R.; Murphy, D.D.; Wilcox, B.A.; Wright, J. Genetic distances and the taxonomy of checkerspot butterflies (Nymphalidae: Nymphalinae). *J. Kans. Ent. Soc. 58*:403–412;1985.

Callahan, J.T. Long term ecological research. *BioScience 34*:363–367;1984.

Dobkin, D.S.; Olivieri, I.; Ehrlich, P.R. Rainfall and the interaction of microclimate with larval resources in the population dynamics of checkerspot butterflies *(Euphydryas editha)* inhabiting serpentine grasslands. *Oecologia 71*:161–166;1987.

Ehrlich, P.R. The attributed of invaders and invasion processes: vertebrates. In: H.A. Mooney, J.A. Drake, eds. *Biological Invasions: A Global Perspective*.

Ehrlich, P.R. Which animal will invade? In: H.A. Mooney, J.A. Drake, eds. *Ecology of Biological Invasions of North America and Hawaii. Ecological Studies, Vol. 58*. New York: Springer-Verlag;1986:79–95.

Ehrlich, P.R. The population biology of the butterfly, *Euphydryas editha*. II. The structure of the Jasper Ridge colony. *Evolution 19*:327–336;1965.

Ehrlich, P.R.; Murphy, D.D. The population biology of checkerspot butterflies *(Euphydryas)* – a review. *Biol. Zentral. 100*:613–629;1981.

Ehrlich, P.R.; Murphy, D.D.; Singer, M.C.; Sherwood, C.B.; White, R.R.; Brown, I.L. Extinction, reduction, stability and increase: The responses of checkerspot butterfly populations to the California drought. *Oecologia 46*:101–105;1980.

Ehrlich, P.R.; White, R.R. Colorado checkerspot butterflies: Isolation, neutrality, and the biospecies. *Amer. Nat. 115*:328–341;1980.

Ehrlich, P.R.; White, R.R.; Singer, M.C.; McKechnie, S.W.; Gilbert, L.E. Checkerspot butterflies: A historical perspective. *Science 188*:221–228;1975.

Ehrlich, P.R.; Launer, A.E.; Murphy, D.D. Can sex ratio be determined? The case of a population of checkerspot butterflies. *Amer. Nat. 124*:527–539;1984.

Ehrlich, P.R.; Wheye, D. Non-adaptive "hilltopping" behavior in male checkerspot butterflies *(Euphydryas editha). Amer. Nat. 127*:447–483;1986.

Gilbert, L.E.; Singer, M.C. Patterns of dispersal and gene flow in a butterfly species. *Amer. Nat. 107*:58–72;1973.

Holdren, C.E.; Ehrlich, P.R. Long-range dispersal in checkerspot butterflies: Transplant experiments with *Euphydryas gillettii*. *Oecologia 50*:125–129;1981.

Holdren, C.E.; Ehrlich, P.R. Ecological determinants of food plant choice in the checkerspot butterfly *Euphydryas editha* in Colorado. *Oecologia 52*:417–423;1982.

Labine, P.A. Population biology of the butterfly, *Euphydryas editha*. I. Barriers to multiple inseminations. *Evolution 18*:335–336;1964.

McKechnie, S.W.; Ehrlich, P.R.; White, R.R. Population genetics of *Euphydryas* butterflies. I. Genetic variation and the neutrality hypothesis. *Genetics 81*:571–594;1975.

Mueller, L.E.; Wilcox, B.A.; Ehrlich, P.R.; Murphy, D.D. A direct assessment of the role of genetic drift in determining allele frequency variation in populations of *Euphydryas editha. Genetics 110*:495–511;1985.

Murphy, D.D. Nectar sources as constraints on the distribution of egg masses by the checkerspot butterfly, *Euphydryas chalcedona* (Lepidoptera: Nymphalidae). *Environ. Entomol. 12*:463–466;1983.

Murphy, D.D. Butterflies and their nectar plants: The role of the checkerspot butterfly *Euphydryas editha* as a pollen vector. *Oikos 43*:113–117;1984.

Murphy, D.D.; Launer, A.E.; Ehrlich, P.R. The role of nectar feeding in egg production and population dynamics of the checkerspot butterfly, *Euphydryas editha. Oecologia 56*:257–263;1983.

Murphy, D.D.; Menninger, M.S.; Ehrlich, P.R. Nectar source distribution as a determinant of oviposition host species in *Euphydryas chalcedona. Oecologia 62*:269–271;1984.

Murphy, D.D.; Menninger, M.S.; Ehrlich, P.R.; Wilcox, B.A. Local population dynamics of adult butterflies and the conservation status of two closely related species, San Francisco Bay, USA. *Biol. Conserv. 37*:201–233;1986.

Murphy, D.D.; White, R.R. Rainfall, resources, and dispersal in southern populations of *Euphydryas editha* (Lepidoptera: Nymphalidae). *Pan-Pac. Ent. 60*:350–354;1984.

Murphy, D.D.; Wilcox, B.A. Butterfly diversity in natural forest fragments: A test of the validity of vertebrate-based management. In: J. Verner et al., eds. *Wildlife 2000: Modeling Wildlife Habitat Relationships.* Madison, WS: Univ. Wisconsin Press;1986:287–292.

Murphy, D.D.; Ehrlich, P.R. The conservation biology of California's remnant native grasslands. In: Mooney, H.A., L.F. Huenneke, eds. *California Grasslands: Structure and Productivity;* (in press)1987.

Schrier, R.D.; Cullenward, M.J.; Ehrlich, P.R.; White, R.R. The structure and genetics of a montane population of the checkerspot butterfly, *Chlosyne palla. Oecologia 25*:279–289;1976.

Singer, M.C. Evolution of foodplant preferences in the butterfly *Euphydryas editha. Evolution 35*:383–389;1971.

Singer, M.C. Complex components of habitat suitability within a butterfly colony. *Science 173*:75–77;1972.

Singer, M.C. Quantification of host specificity by manipulation of oviposition behavior in the butterfly *Euphydryas editha. Oecologia 52*:224–229;1982.

Singer, M.C.; Ehrlich, P.R. Population dynamics of the checkerspot butterfly *Euphydryas editha. Fortrschr. Zool. 25*:53–60;1979.

Wheye, D.; Ehrlich, P.R. The use of fluorescent pigments to study insect behavior: Investigating mating patterns in a butterfly population. *Ecol. Entomol. 10*:231–234;1985.

White, R.R. Pupal mortality in the bay checkerspot butterfly. *J. Res. Lepid. 25*:52–62;1986.

White, R.R.; Singer, M.C. Geographical distribution of host plant choice in *Euphydryas editha* (Nymphalidae). *J. Lepid. Soc. 28:*103–107;1974.

Wilcox, B.A.; Murphy, D.D. Conservation strategy: The effects of fragmentation on extinction. *Amer. Nat. 125*:879–887;1985.

Wright, J.; Brussard, P.F.; Murphy D.D.; Ehrlich, P.R.; Wilcox, B.A. Genetic population structure in *Euphydryas editha:* The search for an appropriate spatial scale. *Evolution* (submitted for publication)

Geometrical Factors and Metapopulation Dynamics of the Bush Cricket, *Metrioptera bicolor* Philippi (Orthoptera: Tettigoniidae)

OSKAR KINDVALL

Department of Wildlife Ecology
The Swedish University of Agricultural Sciences
Box 7002, S-750 07 Uppsala, Sweden

INGEMAR AHLÉN

Department of Wildlife Ecology
The Swedish University of Agricultural Sciences
Box 7002, S-750 07 Uppsala, Sweden

Abstract: *This paper presents a metapopulation study of the bush cricket,* Metrioptera bicolor, *living in a recently fragmented landscape. The species inhabits grass and heathland patches of varying area and isolation. Analyses are made of how these geometrical factors affect local population size and density, distribution pattern, and the probability of local extinction and colonization. The proportion of available patches occupied varied between 72 and 79% during 1985–1990. Unoccupied patches were smaller and more isolated than those that were occupied. Patches where populations became extinct during this period were smaller than those with persisting populations. Since local population size was well correlated with patch area, it was clear that stochastic extinctions only occurred in small populations. Critical patch size for population extinction was approximately half a hectare. Colonized patches were less isolated than those that had not been colonized. Critical inter-patch distance for colonization was about 100 meters. The turnover was restricted to an identifiable share of the available patches. Only 33% of the patches were so small that extinction due to stochastic causes could be considered highly probable. This metapopulation will therefore most likely persist over a considerable period in its present spatial structure. There are apparent threats of further fragmentation, however, and*

Paper submitted March 28, 1991; revised manuscript accepted January 17, 1992.

Factores geométricos y dinámica metapoblacional del grillo *Metrioptera bicolor* Philippi (Orthoptera: Tettigoniidae)

Resumen: *Este trabajo presenta un estudio metapoblacional del grillo* Metrioptera bicolor *que vive en un paisaje recientemente fragmentado. Esta especie habita parches de pasto y matorrales cuyas superficies y grados de aislamiento varían. Se analiza como estos factores geométricos afectan el tamaño y densidad poblacional local, los patrones de distribución y la probabilidad de extinción local y colonización. La proporción de parches disponibles que estaban ocupados osciló entre un 72–79% durante el período 1985–1990. Los parches desocupados fueron mas pequeños y estaban mas aislados que aquellos que se encontraban ocupados. Los parches donde las poblaciones se extinguieron durante este período fueron más pequeños que aquellos con poblaciones que persistieron. Dado que el tamaño poblacional local estaba bién correlacionado con el area de los parches, fue evidente que extinciones estocásticas sólo ocurrieron en poblaciones pequeñas. El tamaño de parche crítico para la extinción poblacional fue de aproximadamente media hectarea. Los parches colonizados estaban menos aislados que aquellos que no fueron colonizados. La distancia crítica entre patches para la colonizacion fue de aproximadamente 100 metros. El recambio estuvo restringido a una porción identificable de los parches disponibles. Sólo un 33% de los parches fueron lo suficientemente pequeños* *nothing is known about the likelihood of large-scale extinctions resulting from extremely unfavorable weather conditions. Nevertheless, our results show that it is appropriate to include geometrical factors in metapopulation models.*

como para que la extinción debido a causas estocásticas pudiese ser considerada altamente probable. Por consiguiente, es muy probable que esta metapoblación persista con su estructura espacial actual durante unperíodo de tiempo considerable. Sin embargo, existen claras amenazas de fragmentación adicional y nada es sabido sobre la probabilidad de extinciones a gran escala, resultantes de condiciones climáticas extremadamente desfavorables. De cualquier modo, nuestros resultados indican que es apropiado incluir factores geométricos en modelos metapoblacionales.

Introduction

Many populations of endangered species have become divided into a number of local populations as a result of habitat fragmentation. Some of these sets of local populations may fit the metapopulations concept (Levins 1970; Hanski & Gilpin 1991), with both extinctions and (re)colonizations occurring over time. The long-term survival of a species, or metapopulation resistance, requires that the extinction rate be compensated by an equally high rate of colonization (Levins 1970; Gilpin 1987; Hanski 1991).

Levins's simple metapopulation model (1970) ignored all aspects of geometrical factors, such as patch area and inter-patch distance. A more realistic model, in which these factors are included, was discussed by Gilpin (1987); it assumes that the probability for a local population to become extinct is inversely related to the patch area and that the rate of immigration and (re)-colonization declines exponentially with distance from the source patch (see MacArthur & Wilson 1967).

There are both deterministic and stochastic causes of local extinction. Deterministic causes, such as habitat destruction, may strike populations independently of population size. Stochastic causes are expected to be more or less dependent on population size, however, making small populations more vulnerable than large ones (Shaffer 1981, 1987). Therefore, a negative relationship between extinction rate and patch area could be expected if population size is positively correlated with patch area.

We analyze the distribution pattern of the Swedish population of *Metrioptera bicolor,* which lives in an archipelago of habitat fragments of various size and isolation, in order to evaluate how well this system fits with the model proposed by Gilpin (1987). We focus on how the geometrical factors, patch area and inter-patch distance, affect distribution pattern, extinctions, colonizations, population size, and density.

Material and Methods

The Species

Metrioptera bicolor (Orthoptera: Tettigoniidae) is a medium-sized (12–19 mm), ground-living bush cricket or katydid without flight ability. However, a macropterous form with fully-developed wings has been observed occasionally (Ander 1947; Berglund 1974). These specimens can fly (by our own observation), but their relative importance as colonizers is not known. Nymphs emerge in May, having six nymphal instars, and become adults during July. The mating period lasts until late October. No specimens survive the winter.

The main distribution of *M. bicolor* is from the steppes of Mongolia and Siberia, and through eastern Europe, where it occurs south to Yugoslavia and north to Poland. It is also known in a few localities in northern Italy, Switzerland, Germany, and eastern France (Harz 1957, 1969). The Swedish population is fairly isolated: the nearest known localities are situated about 300 km away in northernmost Germany, to the south of Lübeck (Holst 1986). That population was also isolated, however, and is assumed to be extinct today (Bellmann 1985).

The Study Area and Its History

The Swedish metapopulation is restricted to parts of the Vomb valley in Skåne (Fig. 1A, 55°40'N 13°35'E). The area is sandy and was an almost treeless heathland in the Eighteenth Century. The temporary fields were extremely poor, and the area suffered from wind erosion. The vegetation was characterized by *Festuca ovina, Hieracium pilosella,* and *Scleranthus annuus* in most of the fields when Linnaeus travelled through the area in 1749. *Calluna vulgaris* and *Cladonia* lichens colonized fields that had been left undisturbed for several years (Linnaeus 1751).

The fragmentation process began about 170 years ago; the first successful pine plantation originates from about 1812–1820 and covers a small area east of the village of Vomb. However, large areas remained open until the Twentieth Century. By 1955, most of the area had become pine forest (Emanuelsson et al. 1985). Today, only a small fraction of the previously open landscape is left. Some of it is still used for agriculture, but considerable areas are now occupied by buildings, gardens, roads, and so forth. There are also several small patches of dry grassland, consisting of different types of grass communities and successions in the colonization process that started once agricultural activity had ceased. Such fragments are the ones of special interest in this study.

The area is limited by natural barriers, such as the stream Björkaån to the northeast, lake Vombsjön to the north, and the stream Klingavälsån to the west and south. The eastern boundary runs through the village of Sjöbo (Ander 1947; Berglund & Sandhall 1981). No observation of *M. bicolor* has ever been made outside this area, although its presence has been known since 1930, when it was discovered by Kemner and Ander (Ander 1933). The streams, therefore, being about ten meters wide, appear to be effective barriers.

The Field Work

The metapopulation was surveyed systematically three times. The first survey was made in September–October 1985, with some complementary work done in September 1986. The second was made in July–September 1989, and the third in the same period in 1990. All suitable habitat patches were visited to determine presence or absence of *M. bicolor.* It is easy to detect *M. bicolor* due to its characteristic stridulation, especially when an ultrasound detector is used (Ahlén 1981, 1982). The surveys were carried out during periods when males in reference areas were stridulating vigorously. In cases when no specimens were observed and no stridulation heard, the localities were revisited.

During the period from August 28 to September 12, 1989, estimates of local population size and density were made for all patches of suitable habitat, except for two where only presence or absence was checked. Patches were classified by type of habitat, and apparent threats were noted.

To minimize the influence of meteorological factors, all fieldwork was done during the day when the majority of males call constantly. During sunny weather when the temperature is about 15° C or more, almost every male stridulates. Seven patches were surveyed two or three times to check whether the estimates were independent of the day on which the survey was made. No detectable drop in population size occurred during the period.

The number of calling males was used as a relative measure of population size. Estimates of population size were achieved in two ways. In small or narrow patches all calling males were counted systematically by walking in straight lines spaced at four meters. This four-meter spacing was chosen because at that distance it is possible to distinguish how many males near each other are calling at the same time.

In larger patches (more than one hectare) another method was used. An iron rod with a rope attached was thrust into the ground at points chosen at random on a map. The rope, with knots spaced at four meters, made it possible to walk in circles with known areas. Calling males within a number of such circles were counted. The number of samples per patch was determined to give a coefficient of variance less than 30%. In a circle size of 0.18 hectare, this usually meant that four to seven samples were enough.

During 1985–1990, frequent visits were made to patches of suitable habitat outside the known area of the metapopulation, especially on the opposite sides of the streams Björkaån and Klingavälsån, where suitable habitats occur near the bridges. Surveys were also made in the large areas of dry grassland north of the lake Sövdesjön, five kilometres south of the stream Klingavälsån.

Suitable Habitats and Local Populations

When analyzing a regional distribution of a species as a metapopulation, one must answer two important questions. The first concerns which habitats are suitable and which should be considered more as barriers from an island biogeographical point of view. The second concerns the delimitation of local populations.

M. bicolor has high temperature requirements and inhabits only places exposed to the sun (Holst 1986). Accordingly, dense pine forest is a barrier, and no specimens have ever been observed there. Not all types of open habitats are suitable, however, since *M. bicolor* oviposits only on grass stems (Hartley & Warne 1972). The lack of oviposition sites in pastures, lawns, and tillage makes these habitats unsuitable for reproduction. *M. bicolor* is expected to avoid these habitats since both nymphs and adults feed mainly on grass pollen (by our own field observations), which is a scarce resource there. It is reasonable to assume that these habitats are much less effective barriers than are forests.

Among the natural habitats mentioned in Table 1, habitats six and seven are quite rare and of minor importance in this system. These habitats occur only within patches dominated by various grass communities (habitats 1–5, Table 1). To get some idea of the quality of habitats 1–5, an experiment was conducted in 1990. Recently hatched nymphs were caged in the field. Two cages (bottom area two square meters) with ten nymphs in each were placed in each habitat, and the

Table 1. Classification of open habitats in the *Metrioptera bicolor* area; grass and dwarf-shrub communities named by vegetation features, with dominating plant species mentioned.

1	Poor sandy grassland: *Corynephorus canescens, Festuca ovina, Helichrysum arenarium, Hieracium pilosella, Cladonia* spp.
2	Sparsely overgrown and unstable sand: *Carex arenaria, Jasione montana, Helichrysum arenarium*
3	Dense and low grassland: *Agrostis capillaris, A. gigantea, Festuca rubra, Campanula rotundifolia, Achillea millefolium, Galium verum*
4	Dense and tall grassland: *Arrhenatherum elatius, Festuca polesica, Dactylis glomerata*
5	Homogeneous *Deschampsia* grassland: *Deschampsia flexuosa*
6	Dwarf-shrub heath: *Calluna vulgaris*
7	Shrub and tall herb vegetation: *Rubus idaeus, Cirsium* spp., *Urtica dioica, Epilobium angustifolium*
8	Pastures
9	Lawns
10	Tillage

Nomenclature according to Krok and Almquist (1984).

number of nymphs that survived to adulthood was counted.

The results of the cage experiments (Table 4), together with the fact that no specimens have ever been found in patches of homogeneous *Deschampsia* grassland, led to only habitats 1–4 (Table 1) being classified as suitable. Consequently, all other habitats are assumed to be barriers of various effectiveness.

The narrowest barriers separating different habitat patches within the study area consisted of 10-meter-wide roads. Studies were also conducted to determine whether this barrier was sufficient to separate local populations by making capture and recapture surveys at three patches surrounded by forests and roads during July–September 1990. Both males and females were caught as young adults and individually marked before being released again at the same spot. Recapture surveys were made both within the patches and in surrounding areas. Movement distances were measured on one patch using the least square mapping technique described by Rohlf and Archie (1978).

Map Analyses

All patches of suitable habitat were digitized from an economic map (1:10 000) to a computer. The digitized coordinates were used to compute the patch area. The variable that describes the remoteness or insularity of the patches, inter-patch distance, was measured directly on the map as the distance between the patch and the nearest patch with *M. bicolor* present.

Statistical Analyses

Local population size and density are variables that consist of both accurate counts and approximations. The population size in large patches has been approximated from mean values. Separate correlation analyses were made with accurate counts to check whether approximations affected the results (Table 2).

The analyzed variables were more or less intercorrelated. Multiple regression analyses were used to minimize these effects; for this purpose some variables had to be log-transformed. Data were tested with the chi-square test for normal distribution. The dependent variables (log) population density and (log) population size did not deviate significantly from a normal distribution, $p = 0.23$ and $p = 0.076$, respectively. The independent variables patch area and inter-patch distance were also log-transformed prior to the regression analyses.

Results

Distribution Pattern

Altogether, 72–79% of the available patches of suitable habitat were occupied during 1985–1990 (Fig. 1). Unoccupied patches were significantly more remote and smaller than occupied ones (Table 3).

Adult Dispersal

Out of a total number of 315 marked adult specimens in three different populations, 80% were recaptured one or more times during the mating period in 1990. Only one male was observed to cross a 10-meter-wide road and reach a nearby patch of suitable habitat. Mean daily movements measured at one patch did not differ between sexes (males: 2.0 m, females: 2.6 m; $n = 90, p = 0.24$, t-test). Most individuals remained in the same part

Table 2. Spearman rank correlations of the geometrical variables and population size and density.

	Distance	*Area*	*Density*	*Population size*
Population size	**−0.358****	**0.871******	**0.387*****	——
Density	**−0.303***	**0.025**	——	0.154
Area	**−0.227***	——	−0.449***	0.769****
Distance	——	−0.023	−0.278*	−0.243

Correlation coefficients from the dataset where all observations are included are written in bold figures (n = 83). The other figures represent a data set where the approximations are excluded (n = 64). Both sets of data give similar results, but the latter generates equal or lower p *values, except in the case of the negative density-area correlation.*
* $p < 0.05$, ** $p < 0.01$, *** $p < 0.001$, **** $p < 0.0001$.

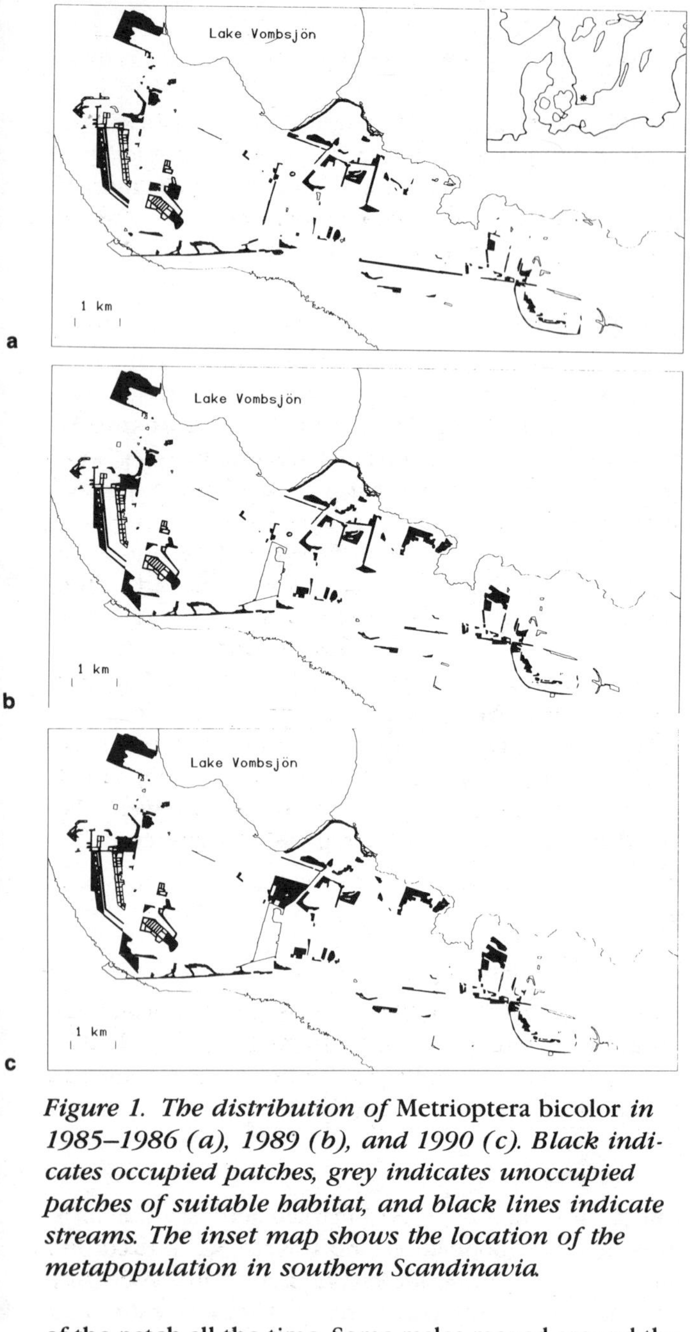

Figure 1. The distribution of Metrioptera bicolor *in 1985–1986 (a), 1989 (b), and 1990 (c). Black indicates occupied patches, grey indicates unoccupied patches of suitable habitat, and black lines indicate streams. The inset map shows the location of the metapopulation in southern Scandinavia.*

of the patch all the time. Some males moved around the patch, however, and one male was observed to move at least 292 m over 50 days.

Colonizations

Altogether, 18 patches were colonized or recolonized during the survey. These patches were less isolated than patches that were not colonized. Median inter-patch distances for colonized versus uncolonized patches were 40/245 metres ($n = 9/22$, $p < 0.001$, Mann-Whittney U-test) in 1989 and 50/265 m ($n = 9/22$, $p < 0.05$, Mann-Whittney U-test) in 1990. The same result was obtained when only observations of colonizations to patches created during the survey period were analyzed, as well as those that were empty for at least two years (median distance for colonized versus uncolonized patches: 40/265, $n = 10/22$, $p < 0.01$, Mann-Whittney U-test). The chance of colonization decreases dramatically when the inter-patch distance exceeds about 100 meters (Fig. 2). The longest inter-patch distance recorded for colonization was 250 m over tillage. Patch size did not differ significantly between colonized and uncolonized patches (median area for colonized versus uncolonized patches: 0.49/0.57 hectares, $n = 18/22$, $p = 0.19$, Mann-Whittney U-test).

There were 14 additional examples of colonization in former tillage and pasture that became suitable habitats during the study period. There were no barriers in those cases. There is no case of colonization across the streams that limit the range of the metapopulation (Fig. 1).

Extinctions

From 1986 to 1990 a total of 18 local populations became extinct. Six of these were deterministic extinctions caused by habitat destruction or alteration, such as house-building, grazing, or application of pesticides. The other twelve extinctions occurred on patches without any noticeable change in habitat.

The twelve patches with stochastic extinctions were significantly smaller than those with persisting populations. The median area for patches with extinct versus extant populations was 0.16/1.17 and 0.06/1.32 hectares in 1989 and 1990, respectively (1989: $n = 9/83$, $p < 0.001$, Mann-Whittney U-test; 1990: $n = 3/78$, $p < 0.05$, Mann-Whittney U-test). The maximum area of a patch with extinction was 0.73 hectares. Three of the patches in which stochastic extinctions occurred in 1989 were again occupied in 1990. The difference in patch size is still significant when these patches are excluded (median patch area for extinct/extant populations: 0.29/1.32 hectares, $n = 6/78$, $p < 0.01$, Mann-Whittney U-test). The risk of local extinction apparently increases when patch size is less than half a hectare (Fig. 3), which corresponds to a critical population size of about 12 males (median density = 25 males/hectare). In 1990, only 67% of the suitable patches were larger than half a hectare.

The extinct populations were not more isolated than extant ones (median distances for patches with extinct/extant populations: 40/20 m, $n = 12/78$, $p = 0.054$, Mann-Whittney U-test).

Population Density

There is no clear effect of geometrical factors on population density. A negative correlation is found between population density and inter-patch distance, but the r^2 value is very low (Table 5A).

Population density in the poor sandy grassland (num-

Table 3. Geometrical differences between unoccupied and occupied patches in 1985–1986, 1989 and 1990.

	*Occupied**		*Unoccupied***		*Differences*
Year	*Median*	*Range*	*Median*	*Range*	p-*Value*
Patch area (hectares)					
1985–1986	0.95	0.03–43.2	0.48	0.03–4.13	<0.05
1989	1.17	0.03–43.2	0.44	0.03–46.5	<0.0001
1990	1.09	0.03–43.2	0.48	0.05–46.5	<0.05
Inter-patch distance (meters)					
1985–1986	30	10–630	310	10–1380	<0.00001
1989	20	10–630	210	10–750	<0.0001
1990	20	10–630	230	10–750	<0.0001

Differences are tested with the Mann-Whitney U-test.
*n = *91, 85, 91;* **n = *28, 33, 24.*

ber 1 in Table 1) was distinctly different from the others. The density of *M. bicolor* in that habitat was half as high as in the other suitable habitats (Table 4). The density in each habitat corresponds to the survival of nymphs in the outdoor cages.

The density in former tillage that has become a suitable habitat during the survey period was half the density in habitat numbers 2–4 (median = 15 males/hectare, $n = 20$, $p < 0.005$, Mann-Whittney U-test).

Population Size

Local population size was positively correlated to patch area. Results from both simple and multiple regression are in accordance (Fig. 4 and Table 5B). There was no significant correlation between population size and inter-patch distance (Table 5B).

The numbers of females and males were counted at some localities in 1990. The sex ratio did not deviate significantly from 1:1 during the nymphal stage (x^2 = 0.96, d.f. = 3, $p > 0.05$) and the early mating period (x^2 = 4.86, d.f. = 2, p > 0.05). As a result of higher death rates among adult females, the sex ratio later became skewed towards males (x^2 = 136.7, d.f. = 7, $p < 0.01$).

A higher death rate among females was also found when the adult dispersal was studied. It was not possible to observe the females for more than half the corresponding time for the males (mean observation time for males versus females was 17.4/8.9 days, $n = 43/47$, $p < 0.05$, t-test).

Altogether, 16,707 calling males were counted in 1989 in the whole metapopulation, giving a total number of approximately 35,000 individuals. The largest local population consisted of 3714 males.

Environmental Changes

The most important environmental change is that more patches of suitable habitat have appeared, such as tillage, pastures and, in one case, a garden, where management ceased and the vegetation went through a rapid succession in the direction of habitat numbers 3 and 4 (Table 1). Two clear cut areas have also become suitable habitats (numbers 2 and 3). Most of the clear cut areas are unsuitable because the vegetation is usually dominated by the homogeneous *Deschampsia* grassland (habitat

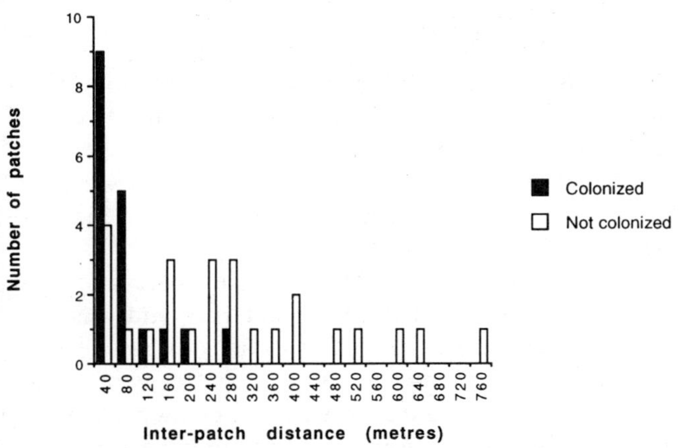

Figure 2. The frequency distribution of inter-patch distance of patches that have become colonized and those that have not during 1989 and 1990.

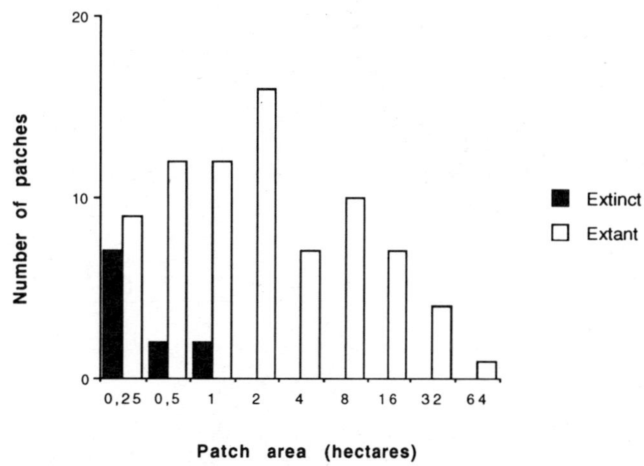

Figure 3. The frequency distribution of areas of patches with extant populations and patches where the population became extinct in 1989 and 1990.

Table 4. The medians of density (males/hectare) and nymphal survival probability in different types of habitats.

Habitat	*Density*	n	*Range*	*Survival*
1	11.0	5	4–15	0%*
2	35.5	4	18–138	20–30%
3	32.5	18	6–134	30–40%
4	30.5	36	5–126	0–10%
5	never observed			0%**

Habitat numbers refer to the list in Table 1. Densities of habitats 1 and 5 differ significantly from the others ($p < 0.05$, Mann-Whitney U-test). Estimates of survival rate are based on two replicates and are not testable.

** 70% survival after two weeks. One specimen reached the fifth nymphal instar.*

*** All nymphs died before the age of two weeks.*

number 5). A total of 158 hectares of new suitable habitats appeared during the study period.

There has been additional fragmentation and loss of habitat area due to plantation and natural colonization of pine and spruce. Three patches were destroyed by house-building activities. These patches were all very small, 0.05–0.17 hectares.

The fragmentation process has worked in two different ways. In twelve cases, a reduction in habitat area occurred and in three cases formerly large patches became divided into two smaller ones. The total loss of habitat due to fragmentation amounted to 28 hectares between 1985 and 1989. An additional loss of 2.68 hectares occurred due to house building in 1990.

The number of patches varied between years from 119 in 1985–1986 to 117 in 1989. Twenty new patches occurred and eight disappeared due to house building or grazing. The rest of the decrease in number of patches was caused by the joining of previously separate patches as former tillage and clear cut areas became suitable habitat. In one case, a former road became overgrown; this resulted in connection of three separate patches.

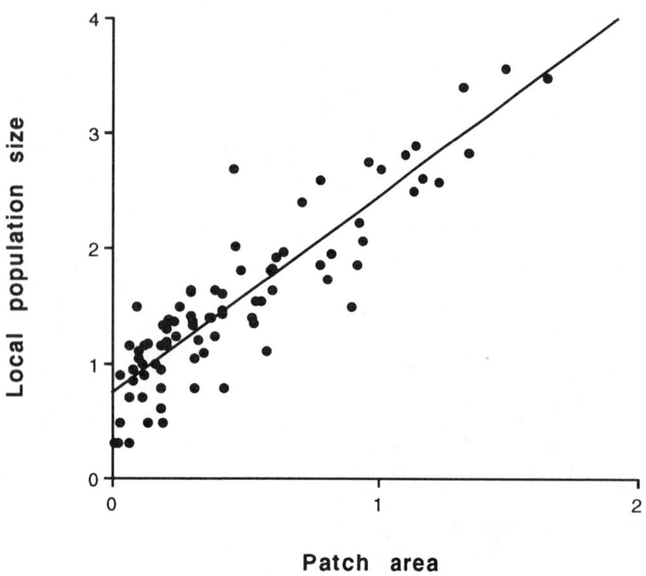

Figure 4. The relationship between local population size and patch area in the metapopulation of Metrioptera bicolor. *Log-transformed data. The regression line:* $y = 1.70x + 0.74$, $r^2 = 80.65\%$, $n = 83$, $p < 0.00001$.

Table 5. Multiple regression analysis of the dependent variables population density (A) and population size (B) and geometrical factors, patch area, and inter-patch distance.

Independent variables	*Partial coefficient*	*S.E.*	t-*value*	p-*value*
(A) (log) Population density				
(log) Area	0.082	0.101	0.805	0.42 NS
(log) Distance	−0.164	0.077	−2.137	<0.05
Constant	1.604	0.139	11.55	<0.00001
$r^2 = 5.43\%$ S.E. = 0.345 $n = 83$				
(B) (log) Population size				
(log) Area	1.647	0.095	17.40	<0.00001
(log) Distance	−0.142	0.072	−1.97	0.052 NS
Constant	0.968	0.130	7.46	<0.00001
$r^2 = 81.6\%$ S.E. = 0.322 $n = 83$				

Threats

Further fragmentation clearly threatens 39% of the 85 patches where *M. bicolor* occurred in 1989. The most obvious threat is the natural spread of pine, which has started in 20% of the patches and will reduce the area markedly in the next 10 years if allowed to continue. There are also five cases of recent forest planting in patches where *M. bicolor* occurs.

Discussion

Metapopulation Dynamics

The *p* value (Levins 1970), which is the fraction of occupied patches, was fairly constant during the five-year period. However, not the same patches were occupied every year. Both colonizations and extinctions were observed. The term metapopulation is therefore justified (see Hanski & Gilpin 1991).

There is an obvious risk that some observations of colonization and extinction are not real, and this also applies to the three recolonizations observed. Bush crickets pass the winter in the egg, and eggs of *M. bicolor* may hatch after one or two winters depending on photoperiodic conditions and temperature (Ingrisch 1986). To eliminate the possibility of pseudo-turnover (Nilsson & Nilsson 1983), it is necessary to exclude observations where a prolonged egg stage might explain the absence of individuals in one year. However, the results concerning geometrical factors are the same.

Effects of Area

Our results support the hypothesis that the probability of local extinctions should be inversely related to the patch area (Gilpin 1987; MacArthur & Wilson 1967). This has also been demonstrated in studies on fruit flies (Forney & Gilpin 1989), waterfleas (Bengtsson 1989), orb spiders (Toft & Schoener 1983), and on mammals such as pikas (Smith 1974).

Very few studies have been made of the relationship between population size and area. Inter-specific competition or predation may possibly cause a negative or discontinuous relationship between population size and area, if any at all. In such a case, the relationship between extinction rate and patch area will probably not be negative.

The population size of *Metrioptera brachyptera,* which occurs in sympatry with *M. bicolor* in the study area, does not correlate with patch area (Kindvall, unpublished data). There is no correlation between the population size of pool frogs, *Rana lessonae,* and pond area (Sjögren 1988). A pond may not be equivalent to a patch, because ponds are used mainly for reproduction. Large patches of suitable habitat were often underpopulated in terms of density by the lygaeid bug, *Lygaeus equestris,* although large patches were more likely to be colonized in that system (Solbreck & Sillén-Tullberg 1990).

The positive relationship found in *M. bicolor* is in accordance with orb spiders (Toft & Schoener 1983) and fruit bats (Wiles et al. 1989) on true islands, and with Pikas (Smith 1974) and bay checkerspot butterflies (Harrison et al. 1988) in habitat islands.

It would appear that only a well-defined proportion (33%) of the available patches of suitable habitat are affected by local extinctions in the metapopulation of *M. bicolor* (the small ones with the small population). This is in accordance with the work by Schoener and Spiller (1987) on Bahamian orb spider metapopulations, where extinctions were restricted to 10–40% of the smallest populations, depending on the species.

Local extinctions may be frequent and yet relatively unimportant for the long-term survival of a metapopulation with high variance in population size, since local extinctions will tend to strike the smallest populations with the least influence on the metapopulation (Schoener & Spiller 1987; Harrison et al. 1988; Harrison 1991). This is the case with the metapopulation of *M. bicolor* during the short term of the study. However, catastrophes and other less size-dependent extinctions may cause far different dynamics in the future. The metapopulation of the lygaeid bug, *Lygaeus equestris,* for example, experienced a large-scale extinction due to extremely bad weather conditions in 1987, after many years of moderate fluctuations and a high proportion of available patches occupied. This large-scale extinction resulted in a reduced probability of recolonization at a time when the need for propagules was greatest (Solbreck 1991).

As long as demographic stochasticity (Shaffer 1987) is considered to be the main cause of local extinction, and provided that further fragmentation does not change the spatial structure of available habitat patches, it is realistic to assume that the Swedish metapopulation of *M. bicolor* will persist for many years (see Schoener & Spiller 1987).

There are apparent threats of further fragmentation in the Vomb valley. In order to preserve *M. bicolor,* clearance work has to be done and further planting has to be avoided in several areas. Given limited financial resources, it would be better to save some of the large patches rather than several small ones of equal total area. This is the same priority suggested by laboratory experiments with fruit flies (Forney & Gilpin 1989) and in general for the preservation of high diversity (Diamond 1975; Frankel & Soulé 1981).

The poor sandy grassland seems to have half the value for *M. bicolor* that other suitable habitats do. In other words, patches of poor sandy grassland have to be larger than other patches to sustain a viable population of *M. bicolor.* However, it is possible that the poor sandy grassland might be more resistant to natural colonization of pine trees. In the long term, this may compensate for the quality difference.

Effects of Inter-Patch Distance

Inter-patch distance affects the possibility of recolonization on patches where populations have become extinct. A negative effect was found in this study, as in many others (such as Smith 1974; Toft & Schoener 1983; Sjögren 1988; Harrison 1989). The inter-patch distance may also affect local populations in a less striking way—the so-called "rescue effect" (Brown & Kodric-Brown 1977). The weak negative relationship between population density and inter-patch distance indicates that the immigration rate in remote patches is perhaps too low to compensate for fluctuations in reproductive success from year to year, which would certainly increase the chances of extinction (Gilpin 1987; Forney & Gilpin 1989). However, no such effect of isolation on extinction rate was found.

The fact that the p value (see Levins 1970) is larger than the fraction of patches large enough to sustain a viable population of *M. bicolor* is probably a result of a compensating effect of recolonizations on the extinction rate.

In terrestrial systems there are usually several types of inter-patch habitat with different degrees of resistance to migrants. Many animals are assumed to disperse only within the type of habitat that is suitable for reproduction and feeding (Stamps et al. 1987), which seems to be the case for *M. bicolor.* Thus the narrow corridors of grass vegetation along roads may be expected to be important for the migration of *M. bicolor* through the pine forest.

Movements of adult *M. bicolor* within patches correspond very well to the maximum inter-patch distance observed in colonization. However, the adult stage might not be the most important one for colonization. Introduction experiments with a morphologically similar species, *Metrioptera roeseli,* showed that most specimens stayed at their releasing point throughout their entire life, but two males were recaptured as adults 0.2 km and 1.0 km away (de Jong & Kindvall 1991). These males remained within a few square meters during the rest of their lives (4–6 weeks). That study also revealed a positive correlation between nymph size and mean daily dispersal distances, making late nymphal instars potential emigrants.

Nevertheless, results from the capture and recapture surveys of adult *M. bicolor* support the assumption that individuals occupying different patches are from different local populations (see Hanski & Gilpin 1991), even if the barrier is only a narrow road. Despite roads having been observed to be almost complete barriers for other species of bush crickets (Samways 1989), this finding is by no means general for this group of insects.

Conclusion

Conservation biologists often use the metapopulation concept when trying to understand how to manage endangered species living in fragmented habitats. However, most empirical evidence about metapopulation dynamics comes from nonfragmented systems, such as real islands (Toft & Schoener 1983; Hanski 1986; Schoener & Spiller 1987) or island-like habitats, that are not subject to a fragmentation process. Examples from the latter group are ponds (Sjögren 1988), rockpools (Pajunen 1986; Bengtsson 1991), and serpentine grassland (Harrison et al. 1988). However, the Swedish occurrence of *M. bicolor* is an example of a formerly continuous population that has recently become fragmented by afforestation of a formerly open heathland.

We conclude that the studied population of *M. bicolor* can be regarded as a real metapopulation with turnover. However, its dynamics do not fit Levins's model because the turnover is restricted to patches with special geometrical features—the smallest ones and ones that are not too isolated. Our results support the inclusion of geometrical factors in metapopulation models.

Acknowledgments

The research was supported by the Swedish Council for Forestry and Agricultural Research and WWF-Sweden. We thank Jan Bengtsson, Per Sjögren, Jeremy Robertson, Staffan Ulfstrand, Per Edenhamn, and Johnny de Jong for helpful comments.

Literature Cited

Ahlén, I. 1981. Ultraljud hos svenska vårtbitare. Entomologisk Tidskrift **102**:27–41.

Ahlén, I. 1982. Songs of Swedish bush crickets. Tape recording. Department of Wildlife Ecology, Uppsala, Sweden.

Ander, K. 1933. Två för Sveriges fauna nya Orthopterer. Entomologisk Tidskrift **54**:246–248.

Ander, K. 1947. Flygförmågan hos våra hopprätvingar. Fauna och Flora **42**:210–221.

Bellmann, H. 1985. Heuschrecken, beobachten, bestimmen. Verlag J. Neumann-Neudamm, Melsungen, Germany.

Bengtsson, J. 1989. Interspecific competition increases local extinction rate in a metapopulation system. Nature **340**:713–715.

Bengtsson, J. 1991. Interspecific competition in metapopulations. Pages 219–237 in M. E. Gilpin and I. Hanski, editors. Metapopulation dynamics: empirical and theoretical investigations. Linnean Society of London, Academic Press, London, England.

Berglund, B. 1974. Intressant vårtbitarfynd i Skåne (Vombs socken). Skånes Natur **61**:120–121.

Berglund, B., and Å. Sandhall. 1981. Några sydsvenska vårtbitare. Entomologisk tidskrift **102**:23–26.

Brown, J. H., and A. Kodric-Brown. 1977. Turnover rates in insular biogeography: effect of immigration on extinction. Ecology **58**:445–449.

de Jong, J., and O. Kindvall. 1991. Cikadavårtbitaren *Metrioptera roeseli*—nykomling eller hotad relikt? Fauna och Flora **86**:214–221.

Diamond, J. M. 1975. The island dilemma: lessons of modern biogeographic studies for the design of natural reserves. Biological Conservation **7**:129–146.

Emanuelsson, U., C. Bergendorff, C. Carlsson, N. Lewan, and O. Nordell. 1985. Det Skånska kulturlandskapet. Bokförlaget Sigma, Lund, Sweden.

Forney, K. A., and M. E. Gilpin. 1989. Spatial structure and population extinction: a study with Drosophila flies. Conservation Biology **3**:45–51.

Frankel, O. H., and M. E. Soulé. 1981. Conservation and evolution. Cambridge University Press, Cambridge, England.

Gilpin, M. E. 1987. Spatial structure and population vulnerability. Pages 125–139 in M. E. Soulé, editor. Viable populations for conservation. Cambridge University Press, Cambridge, England.

Hanski, I. 1986. Population dynamics of shrews on small islands accord with the equilibrium model. Biological Journal of the Linnean Society **28**:23–36.

Hanski, I. 1991. Single-species metapopulation dynamics: concepts, models and observations. Pages 17–38 in M. E. Gilpin and I. Hanski, editors. Metapopulation dynamics: empirical and theoretical investigations. Linnean Society of London, Academic Press, London, England.

Hanski, I., and M. E. Gilpin. 1991. Metapopulation dynamics: brief history and conceptual domain. Pages 3–16 in M. E. Gilpin and I. Hanski, editors. Metapopulation dynamics: empirical and theoretical investigations. Linnean Society of London, Academic Press, London, England.

Harrison, S. 1989. Long-distance dispersal and colonization in the bay checkerspot butterfly. Ecology **70**:1236–1243.

Harrison, S. 1991. Local extinction in a metapopulation context: an empirical evaluation. Pages 73–88 in M. E. Gilpin and I. Hanski, editors. Metapopulation dynamics: empirical and theoretical investigations. Linnean Society of London, Academic Press, London, England.

Harrison, S., D. D. Murphy, and P. R. Ehrlich. 1988. Distribution of the bay checkerspot butterfly, *Euphydryas editha bayensis:* evidence for a metapopulation model. The American Naturalist **132**:360–382.

Hartley, J. C., and A. C. Warne. 1972. The developmental biology of the egg stage of Western European Tettigoniidae (Orthoptera). Journal of Zoology **168**:267–298.

Harz, K. 1957. Die Geradflügler Mitteleuropas. Gustav Fischer Verlag, Jena, Germany.

Harz, K. 1969. Die orthopteren europas. Vol. 1. Vorlag Dr. W. Junk, The Hague, The Netherlands.

Holst, K. T. 1986. The saltatoria of Northern Europe. Entomologica Scanidinavica, Vol. 16. Brill/Scandinavian Science Press, Leiden, The Netherlands.

Ingrisch, S. 1986. The plurennial life cycles of European Tettigoniidae (Insecta: Orthoptera) 2. The effects of photoperiod on the induction of an initial diapause. Oecologia **70**:617–623.

Krok, T. O. B. N., and S. Almquist. 1984. Svensk flora. Esselte Herzogs, Uppsala, Sweden.

Levins, R. 1970. Extinction. Pages 77–107 in M. Gerstenhaber, editor. Some mathematical questions in biology, Vol. 2. American Mathematical Society, Providence, Rhode Island.

Linnaeus, C. 1751. Skånska resa. Stockholm, Sweden.

MacArthur R. H., and E. O. Wilson. 1967. The theory of island biogeography. Princeton University Press, Princeton, New Jersey.

Nilsson, S. G., and I. N. Nilsson. 1983. Are estimated species turnover rates on islands largely sampling errors? The American Naturalist **121**:595–597.

Pajunen, V. I. 1986. Distributional dynamics of *Daphnia* in a rockpool environment. Annales Zoologici Fennici **23**:131–140.

Rohlf, F. J., and J. W. Archie. 1978. Least-squares mapping using interpoint distances. Ecology **59**:126–132.

Samways, M. J. 1989. Insect conservation and landscape ecology: a case-history of bush crickets (Tettigoniidae) in southern France. Environmental Conservation **16**:217–226.

Schoener, T. W., and D. A. Spiller. 1987. High population persistence in a system with high turnover. Nature **330**:474–477.

Shaffer, M. L. 1981. Minimum population sizes for species conservation. Bioscience **31**:131–134.

Shaffer, M. L. 1987. Minimum viable population: coping with uncertainty. Pages 69–86 in M. E. Soulé, editor. Viable populations for conservation. Cambridge University Press, Cambridge, England.

Sjögren, P. 1988. Distribution and extinction patterns among local populations of pool frog (*Rana lessonae*) on the northern periphery of its range. Paper V in P. Sjögren. Metapopulation biology of *Rana lessonae* Camerano on the northern periphery of its range. Uppsala University, Uppsala, Sweden. Ph.D. Thesis.

Smith, A. T. 1974. The distribution and dispersal of pikas: consequences of insular population structure. Ecology **55**:1112–1119.

Solbreck, C. 1991. Unusual weather and insect population dynamics: *Lygaeus equestris* during an extinction and recovery period. Oikos **60**:343–350.

Solbreck, C., and B. Sillén-Tullberg. 1990. Population dynamics of a seed feeding bug, *Lygaeus equestris.* 1. Habitat patch structure and spatial dynamics. Oikos **58**:199–209.

Stamps, J. A., M. Buechner, and V. V. Krishnan. 1987. The effect of edge permeability and habitat geometry on emigration from patches of habitat. American Naturalist **129**:533–552.

Toft, C. A., and T. W. Schoener. 1983. Abundance and diversity of orb spiders on 106 Bahamian islands: biogeography at an intermediate trophic level. Oikos **41**:411–426.

Wiles, G. J., T. O. Lemke, and N. H. Payne. 1989. Population estimates of fruit bats (Pteropus marianus) in the Mariana Islands. Conservation Biology **3**:66–76.

Puerto Rican Parrots and Potential Limitations of the Metapopulation Approach to Species Conservation

MARCIA H. WILSON
CAMERON B. KEPLER*

U.S. Fish and Wildlife Service
Patuxent Wildlife Research Center
Laurel, MD 20708, U.S.A.

NOEL F. R. SNYDER

Parrot Programs, Wildlife Preservation Trust International
P.O. Box 426
Portal, AZ 85632, U.S.A.

SCOTT R. DERRICKSON

Ornithology, Conservation Research Center
National Zoological Park
Front Royal, VA 22630, U.S.A.

F. JOSH DEIN

U.S. and Wildlife Service
National Wildlife Health Research Center
6006 Schroeder Road
Madison, WI 53711, U.S.A.

JAMES W. WILEY

U.S. Fish and Wildlife Service
Grambling Cooperative Wildlife Project
Grambling State University
P.O. Box 815
Grambling, LA 71245, U.S.A.

JOSEPH M. WUNDERLE, JR.
ARIEL E. LUGO

Institute of Tropical Forestry
Southern Forest Experiment Station
USDA Forest Service
Rio Piedras, PR 00928

DAVID L. GRAHAM

Schubot Exotic Bird Health Center
Texas Veterinary Medical Center
Texas A & M University
College Station, TX 77843, U.S.A.

WILLIAM D. TOONE

San Diego Wild Animal Park
15500 San Pasqual Valley Road
Escondido, CA 92027, U.S.A.

Abstract: *Population viability analyses for a number of endangered species have incorporated a metapopulation approach. The risk assessments of these viability analyses have indicated that some extant populations should be subdivided into numerous subgroups with exchange of individuals among them in order to reduce the chance of catastrophic loss of the species. However, routine application of a policy of extensive subdivision may have detrimental consequences for certain endangered species. We examine the Puerto Rican Parrot as a case history in which this policy is ill-advised. In 1989, a population viability analysis was conducted for the*

* *Mailing address: Southwest Research Group, School of Forest Resources, University of Georgia, Athens, GA 30602, U.S.A.*
Paper submitted March 16, 1992; revised manuscript accepted January 19, 1993.

Las cotorras portorniqueñas y las limitaciones potenciales de la estrategia de metapoblaciones aplicada a la conservación de especies

Resumen: *Los análisis de viabilidad de poblaciones hechos para un gran número de especies en peligro de extinción ahora tienen un enfoque de metapoblación. Para reducir las probabilidades de pérdidas catastróficas de individuos de una especie, poblaciones existentes se subdividen en numerosos subgrupos; ésto se hace de acuerdo a los resultados de los análisis de riesgo que son parte integral del enfoque de análisis de viabilidad de poblaciones. Sin embargo, la subdivisión contíinua y rutinaria de poblaciones de especies en peligro de extinción puede ser detrimental para la sobrevivencia de la especie. Examinamos al caso de la Cotorra Puertorriqueña dónde este tipo de enfoque no es recomen-*

parrot. The document recommended subdivision of the existing small captive flock into three groups. One of these captive flocks would consist of individuals transferred to a multi-species facility in the continental United States. Subsequently, individuals from this facility would be exchanged with the insular captive population(s) and the relict wild flock. For two reasons, implementation of this recommendation might have led to serious repercussions. First, this parrot, like many endangered species, has gone through a genetic bottleneck and may have a heightened susceptibility to disease. Multi-species facilities are a high-risk environment favoring the transmission of pathogens, especially when the facilities are located outside the natural ranges of a particular species. Second, the parrot is a K-selected species for which mate selection is idiosyncratic. This type of species often proves difficult to breed in captivity in small groups. Part of the problem in mate selection may be reduced by a policy allowing frequent transfers of individuals among facilities, but such movements increase the chances of spreading disease in the metapopulation. Thus, population viability analyses need to acknowledge that proliferation of captive subgroups accompanied by exchanges of individuals can in themselves carry substantial risks that must be weighed against the presumed benefits of subdivision.

dable. En el 1989, un análisis de viabilidad poblacional se efectuó para la Cotorra Puertorriqueña. Como resultado del análisis se recomendó la subdivisión en tres grupos de la pequeña población cautiva existente. Se recomendó, además, transferir uno de los tres grupos a una facilidad con múltiples especies en los Estados Unidos de América. Subsiguientemente, individuos de esta facilidad serían intercambiados con las poblaciones cautivas en Puerto Rico y con la población de cotorras aún sobreviviendo bajo condiciones naturales. Hay dos razones por las cuales tales recomendaciones podrian haber repercutido seriamente sobre la cotorra. En primer lugar, esta especie, como muchas otras en peligro de extinción, ha atravesado por un cuello de botella genético y como resultado de esto su suceptibilidad a enfermedades ha incrementado. Facilidades con múltiples especies son ambientes de alto riesgo para especies como la cotorra pues en ellos se favorece la transmisión de patógenos, especialmente cuando los facilidades están localizadas lejos del ambiente natural de especies particulares. En segundo lugar, la cotorra es una especie tipo K para la cuál la selección de parejas es idiosincrática. Este tipo de especie es generalmente difícil de propagar en cautiverio particularmente cuando las poblaciones son pequeñas. Parte del problema de selección de parejas se puede aliviar intercambiando frecuentemente individuos entre facilidades, pero tal estrategia aumenta el riesgo de transmisión de enfermedades entre las metapoblaciones. Por lo tanto, el análisis de viabilidad de poblaciones debe considerar que la proliferación de subgrupos cautivos acompañados por el intercambio continuo de individuos en esos subgrupos, por su cuenta, conlleva riesgos sustanciales que se deben contraponer ante los presuntos beneficios de la subdivisión de las poblaciones.

Introduction

A primary objective of applied conservation biology is to minimize extinction rates of threatened species (Goodman 1987; Quinn & Hastings 1987). This orientation led to defining minimum viable populations and to incorporating the metapopulation concept as a frame of reference into species conservation (Shaffer 1985; Simberloff 1988).

Gilpin and Soulé (1986) introduced the process known as population viability (sometimes vulnerability) analysis (PVA) to estimate minimum viable populations and to plot conservation strategies for endangered species. The PVAs produced to date have involved both intrinsic population characteristics (morphology, physiology, reproduction, disease resistance, mobility, behavior, etc.) and environmental constraints (habitat quantity and quality, interacting species, disease vectors, etc.). These features may allow estimation of extinction probabilities and development of strategies to minimize these probabilities. As noted by Shaffer (1990), PVA is an emerging technique of risk analysis that does not as yet have a well-developed methodology or widely accepted standards. The PVA process has been applied to about 46 species, including a variety of threatened vertebrates, invertebrates, and plants (U. S. Seal, personal communication).

Recently, a metapopulation approach has been incorporated into many PVAs (for example, the 1989 Florida Panther Viability Analysis and Species Survival Plan and the 1989 Puerto Rican Parrot PVA). For wild populations, a metapopulation is a complex of interdependent subpopulations affected by recurrent extinctions and linked by recolonization from one or more large reservoir populations (Murphy et al. 1990). For many endangered species, the PVA process has been expanded to include wild and captive subpopulations (Lacy et al. 1989; Seal & Lacy 1989).

To reduce chances of catastrophic loss of all members of an endangered species and to counter genetic drift, Lacy (1987) recommended that managers subdivide a managed population into numerous breeding subgroups that exchange a few individuals per generation. For species with small total populations, the subdivisions can involve very small numbers of individuals in each subpopulation—for example, as few as 12 Puerto Rican Parrots (Lacy et al. 1989).

The costs of dividing wild populations into small subpopulations involve an increased probability of extinction due to demographic stochasticity and a possible loss of genetic variation. If members of each subpopulation reproduce, these costs can theoretically be mitigated in captivity by regularly exchanging individuals among facilities or sites (in the wild). However, blanket application of a metapopulation approach to species conservation can be detrimental for certain endangered species, especially for species with a high susceptibility to diseases or with difficulty of breeding in small captive groups. In this report, we address these two factors for optimal management of endangered species in captivity, with special reference to the endangered Puerto Rican Parrot (*Amazona vittata*).

Disease

Disease can play a major role in the conservation of endangered species, as evidenced by a canine distemper epizootic which led to the extirpation of the black-footed ferret (*Mustela nigripes*) in the wild (Thorne & Williams 1988). Endangered species may often have heightened susceptibility to disease because of reduced genetic variation in small populations (O'Brien et al. 1985; Watkins et al. 1990) and because many endangered species (especially insular species) occur in isolated populations without previous exposure to certain pathogens (van Riper et al. 1986; Cooper 1989; Jenkins et al. 1989). This heightened susceptibility to disease is often apparent when endangered species are compared with closely related but nonendangered species. For example, the endangered Whooping Cranes (*Grus americana*) at the Patuxent Wildlife Research Center have been much more susceptible than the Sandhill Cranes (*G. canadensis*) to coccidia and viral encephalitis (Dein et al. 1986).

Snyder et al. (1985) reported nestling mortalities from herpesvirus infections of the endangered Mauritius Pink Pigeon (*Columba mayeri*) fostered under domestic pigeons (*C. livia*) at the Rio Grande Zoo in Albuquerque. The foster pigeon flock apparently had been free of disease for several years. Four fertile Mauritius Pink Pigeon eggs were fostered to the surrogate birds. However, latently infected foster pigeons shed herpesvirus during the reproductive period, and each of the Mauritius Pink Pigeon hatchlings died suddenly at 5 to 11 days of age. No deaths among nestling domestic pigeons in the foster flock were due to viral infection. The Mauritius Pink Pigeons seemed to have had no previous experience with this virus.

The transfer of live animals or eggs to non-native areas always increases the risk of exposing immigrants to pathogens to which they are not adapted and the risk of spreading pathogens from one area to another (Ashton & Cooper 1989; Gaskin 1989). For example, O'Brien et al. (1985) described the spread of a coronavirus throughout a cheetah (*Acinonyx jubatus*) colony. These authors attributed the cause to a clinically healthy female transferred to Wildlife Safari. Despite intensive therapy, this female and 17 other cheetahs died. O'Brien et al. (1985) suggested that a dearth of genetic variation rendered this colony extremely susceptible.

Gaskin (1989) suggested that congregation of a diverse array of species from all around the world in quarantine facilities and large, multi-species holdings in zoos, private collections, and the pet trade is one reason captive psittacines are plagued with so many diseases. These multi-species congregations have provided enhanced opportunities for the transmission of infectious agents.

Quarantine, a common practice in the transfer of live animals to new locations, does not guarantee that an individual is free of diseases. Many disease agents are not necessarily fatal on contact but can remain latent in carriers and be shed intermittently, particularly under stress. Carriers can harbor certain diseases for their entire lives and may transmit infections to their offspring. The most disturbing fact is that carriers of some disease agents can be asymptomatic. Such is the case with a psittacine herpesvirus known as Pacheco's disease (Gaskin 1989), and with a number of other avian diseases.

Shima and Osborn (1989) identified a pair of newly acquired Red-Collared Lorikeets (*Trichoglossus haematodus rubritorquis*) as the "point source" of an epornitic of *Salmonella typhimurium* in the San Diego Zoo collection of lories and lorikeets. Seven birds, representing six species, died. The originally infected pair was imported from an Australian zoo, went through standard federal quarantine, and then a 30-day quarantine at a holding facility at the San Diego Wild Animal Park before being introduced into the collection at the zoo. Routine screening failed to identify the problem during quarantine.

The purpose of most zoos is to display animals, and many species are housed in close proximity for viewing by the public. To replace animals that die and to diversify collections for public viewing, zoos continually acquire new individuals and species from around the world. In part, because these animals often come from aggregations of species in quarantine or from other facilities, zoos may be high-risk environments that favor transmission of pathogens. Not surprisingly, many epizootics have been documented in zoological facilities (see Montali et al. 1975*a*; May 1988; Thorne & Williams 1988; Cooper 1989; Derrickson & Snyder 1992).

For threatened birds that might be highly susceptible to diseases and yet are in need of captive propagation,

Derrickson and Snyder (1992) recommended the following principles:

(1) Facilities should, whenever possible, be located in the species' natural range.
(2) The species should be housed in two or three geographically isolated, single-species facilities.
(3) The facilities should be staffed by individuals who are not simultaneously caring for other captive birds.
(4) As much as possible, facilities should be in areas free from arthropod vectors and feral populations of exotic birds.
(5) Established husbandry protocols should emphasize disease prevention.

These precautions do not preclude the involvement of zoological institutions in breeding of endangered species that are susceptible to diseases, but they emphasize proper locations and carefully controlled conditions. Likewise, Kennedy-Stoskopf (1989) pointed out that an appropriate environment and good husbandry practices are essential to limiting infectious diseases in captive situations.

Recovery programs for endangered species increasingly include augmenting or reestablishing wild populations with captive individuals or with wild, translocated conspecifics. If either type of individual harbors pathogenic organisms, relict wild populations can be severely jeopardized. For example, many reintroduced populations of Wild Turkeys (*Meleagris gallopavo*) in the U.S. are infected with *Plasmodium kempi,* a hematozoan parasite that appears to have originated from the translocation of infected birds (Castle & Christensen 1990). In a like manner, there is convincing evidence that a movement of rabies up the East Coast was initiated by the translocations of infected raccoons from endemic areas in Florida and Georgia (Nettles et al. 1979). Similarly, the upper respiratory disease in wild desert tortoises (*Xerobates agassizii*) is believed to stem from a probable release of infected pet (or captive-held) tortoises (Anonymous 1989; Jacobson & Gaskin 1990). This disease now threatens the species.

Infectious and parasitic diseases can cause premature death in wild populations (Scott 1988; Cooper 1989). Fatal diseases are notoriously difficult to detect in wild animal populations because sick and dead individuals commonly go unrecovered, in part because they are taken quickly by predators and scavengers (Balcomb 1986; Scott 1988; Linz et al. 1991). Even when a causative agent can be identified in a diseased wild population, treatment is often impractical, particularly if species are difficult to catch or easily stressed by handling or if practical means of treatment are unknown.

Furthermore, disease is most likely to affect a population without previous contact with the pathogen. For example, as noted by Cooper (1989), if the herpesvirus that afflicted the Pink Pigeons in the Rio Grande Zoo had been introduced into the wild Pink Pigeon population in Mauritius, results could have been disastrous. With good reason, Ritchie et al. (1989) stressed that, because of the widespread movement of birds in the pet trade, the risks of introducing highly virulent viruses into wild populations of endangered psittacine species are substantial.

Besides causing premature mortality, introduction of infected captive animals into the wild can also reduce reproduction of wild populations. For example, Hudson (1986) found that reproduction in Red Grouse (*Lagopus scotius*) can be reduced by the nematode (*Trichostrongylus tenuis*). Similarly, Saumier et al. (1986) found that captive American Kestrels (*Falco sparverius*) infected with the parasite *Trichinella pseudospiralis* had severely reduced reproductive success compared with control birds. Infected pairs initiated laying later, laid smaller clutches, and hatched fewer eggs than controls. Identifying an infection, rather than some other factor, as the actual cause of low reproductive success in a wild population is difficult.

Because some avian pathogens can be transmitted in eggs, one cannot avoid transport of pathogens by limiting movements of animals to the egg stage. The bacterium *Salmonella pullorum* can spread from a parent flock via the egg to chicks by true transovarian vertical transmission (Ashton & Cooper 1989). Other avian pathogens can also enter eggs as they cool following surface contamination of the shell (Ashton & Cooper 1989). Vertical transmission is also known to occur with avian leukosis virus, reticuloendotheliosis, adenoviruses, and *Mycoplasma* spp. (Reece 1989). If an infected embryo does not die, infection of newly hatched chicks may occur and spread throughout a captive flock.

Because novel pathogens or their vectors can be spread with increasing ease in the modern world, Cooper (1989) made an urgent plea for minimizing the spread of infectious diseases from one locality to another. As Ashton and Cooper (1989) pointed out, exclusion of an avian pathogen is much easier than its elimination.

Elimination of a disease requires that it is well defined; that it can be easily diagnosed or detectable by a simple, cheap, and rapidly applied test (which is often not available); and that it can be eradicated. But understanding of many diseases, especially avian diseases, is still far from perfect. Part of the difficulty stems from the sheer number of species from all over the world that are involved. Host ranges are unknown for some diseases such as psittacine beak and feather disease, and causative agents are still unknown for others such as psittacine proventricular dilatation syndrome (parrot wasting disease) and internal papillomatous disease. Of the many infectious agents, Gaskin (1989) reported that viruses are the most troublesome mainly because they can occur in epornitic proportions with little possibility of chemotherapeutic intervention.

Once certain pathogens become established in a facility, they can be extremely difficult to control or eliminate. One such disease that has repeatedly plagued zoological and private collections and the poultry industry is avian tuberculosis. This disease affects a wide variety of avian taxa, and it is readily spread by fecal contamination of substrates, feed, or water sources, and possibly also vertically through egg contamination (Montali et al. 1978). Genetic factors can also play a role in transmission (Cromie et al. 1991). Control or eradication of this disease is particularly difficult because the causative bacillus, *Mycobacterium avium,* can remain viable in the soil for years, and because reliable methods to diagnose and treat the disease in live animals remain unavailable. Traditional eradication includes the elimination of the entire collection and either the replacement of existing substrates, the destruction of the entire facility, or movement of the facility to a new site (Montali et al. 1975*b;* Beehler 1990). Such measures obviously entail severe financial, biological, and ethical costs (particularly when dealing with endangered and threatened species), and do not preclude accidental reinfection. Controlling the spread of avian tuberculosis among facilities or between captive and wild populations clearly will not be possible until efficient and accurate diagnostic techniques become available to detect the disease in live animals. Unfortunately, avian tuberculosis has been detected post-mortem in several endangered species, and this insidious disease represents a substantial threat to many ongoing breeding and reintroduction programs.

Polyomaviruses, which are significant pathogens of psittacines, are also resistant to inactivation. For example, Gough (1989) reported on an aviary in which an outbreak of a Budgerigar fledgling disease occurred. After the initial outbreak, breeding operations were curtailed for eight months, but young continued to die when breeding was resumed. The entire aviary was depopulated, fumigated with formaldehyde gas, and left vacant for seven months. Then new breeding stock was purchased from an aviary presumably free of the disease. Despite these massive efforts to eliminate the disease, it soon broke out again.

Likewise, Thorne and Williams (1988) reported that canine parvoviral enteritis, a very contagious infectious virus of canids, is resistant to environmental inactivation. Apparently carrier wolves may have been the source of this virus, which caused mortalities among maned wolves (*Chysocyon brachyurus*) both at the San Antonio Zoo (Fletcher et al. 1979) and at the National Zoological Park (Mann et al. 1980).

In summary, because of the slow-acting but nevertheless lethal or debilitating nature of many pathogens, because of the difficulties in detecting the presence of many pathogens in asymptomatic carriers, because of the widespread movement of animals carrying pathogens through captive-breeding collections around the world, because of the heightened susceptibilities of many endangered species to disease, and because of severe risks of contamination of wild populations of endangered species by infected individuals from captivity, it is questionable that extensive subdivision of captive populations of endangered species, especially in multispecies facilities, represents a risk-reduction strategy rather than a risk-enhancement strategy. This is especially true when movements between subpopulations are routinely required to meet genetic and demographic management objectives. Presumably, many of the risks of disease can be greatly reduced if subdivision of captive populations is limited to single-species facilities within the natural range of the endangered species, and if personnel serving the facilities practice rigorous disease prevention measures, including avoidance of contact with other animals that may be harboring disease agents.

Poor Reproduction of Small Populations Housed in Captivity

The productivity of many species in captivity is affected strongly by population size. In many colonial species, social facilitation is a necessary element for reproduction. For this reason, flamingos and many other colonial waterbirds such as storks, ibises, and spoonbills, usually do not breed in captivity in small groups (Kear & Palmes 1980; Luthin et al. 1985).

For many endangered species, obtaining adequate levels of reproduction in captivity poses difficulties. For species with idiosyncratic mate selection, subdivision of an already small population reduces the number of possible pairings at any one site and therefore the overall productivity of the species.

In many large psittacines, reproduction occurs in only a small fraction of a captive population, and success in captive breeding seems to depend on providing birds with an ample array of potential mates. Even with numerous trials with potential mates, many birds remain nonreproductive, presumably because of their failure to find compatible partners among the available choices. The psittacine literature is rife with examples of the contortions aviculturists have been through to obtain compatible pairs. Even in psittacines, such as Cockatiels (*Nymphicus hollandicus*), that breed relatively freely in captivity, forced pairing commonly results in much less successful reproduction than free mate choice (Yamamoto et al. 1989). The spontaneous choice of mates seems to be crucial for successful reproduction in many avian species, such as Canvasbacks (*Aythya valisineria*) (Bluhm 1985) and Whooping Cranes (Kepler 1978).

With species in which formation of compatible pairs is especially difficult, extensive subdivision of captive populations can pose significant impediments to securing sustained reproduction and meeting genetic management objectives. Both of these problems can be mitigated only by extensive and sometimes expensive transfers of individuals among facilities. Such movements, however, can greatly increase the chance of spreading disease among subpopulations and can increase the vulnerability of the entire metapopulation to disease.

Thus, the perceived reduction of risk from splitting a captive population to prevent loss of an entire genome to some site-specific catastrophe, such as fire, disease, tornados, or local wars, must be balanced against the potential increase in disease exposure and decrease in reproductive potential that may result from subdivision. Viable or optimal group size varies greatly from species to species.

A Case Study: The Puerto Rican Parrot

The Puerto Rican Parrot was abundant and widespread on the islands of Puerto Rico, Culebra, and Vieques. By the late 1960s, it was reduced to a tiny remnant population in the Luquillo Experimental Forest, also known as the Caribbean National Forest. This reduction was caused by a variety of factors, including habitat destruction, capture for pets, and killing for food and protection of crops (Snyder et al. 1987). The wild population reached a low point of 13 birds at the beginning of the 1975 breeding season. Since then it has recovered at a low to moderate rate, reaching a high of 45–47 individuals after the 1989 breeding season. This increase was a result of identification and reduction of several limiting factors affecting the wild population, and of fostering captive-reared birds into wild nests. Meanwhile, a captive population was started in the early 1970s largely through removal of eggs from wild nests, and reached a total of 58 individuals by January 1993 (A. Valido, personal communication).

Although successful captive breeding has been achieved in all years since 1979, the number of breeding pairs has remained low in both wild and captive flocks, averaging about four pairs in each flock. The captive flock is in a single facility within the natural range of the wild population in the Luquillo Experimental Forest. With the exception of captive Hispaniolan Parrots (*Amazona ventralis*) brought in for surrogate studies, the aviary has been essentially closed to importation of other avian species. No significant disease problems have occurred at the facility in the 20 years of its existence.

A second aviary for the species, about 100 km distant, has recently been completed. The site of this second aviary is in the Rio Abajo Commonwealth Forest, which has been acknowledged to be the best location to reestablish a second wild population of the species (U.S. Fish and Wildlife Service 1987).

In June of 1989, the U.S. Fish and Wildlife Service hosted the Captive Breeding Specialist Group of the International Union for the Conservation of Nature and Natural Resources in a PVA on the Puerto Rican Parrot. The PVA report made several recommendations for management of wild and captive flocks, but most notably it advocated the near-term splitting of the captive flock into three parts: the existing Luquillo Aviary in the Luquillo Experimental Forest, the second aviary in Rio Abajo, and a third aviary at some mainland zoo in the United States (Lacy et al. 1989). Ultimately, the PVA recommended the establishment of some 10 captive flocks in various locations, including zoos. These recommendations were based on the application of the metapopulation concept.

The principal justifications offered for sending birds to a mainland zoo were: (1) enhanced reduction of risks from catastrophe, and (2) enhancement of captive reproduction utilizing the resources and expertise at a mainland zoo. The 12 birds recommended for shipment to a mainland zoo were all birds that had not bred as yet in captivity.

The primary threat to both wild and captive populations of the Puerto Rican Parrot has long been perceived to be a major hurricane disaster. By 1989, the last wild habitat of the species had been spared any major hurricane for over 50 years. The effects of a major storm were difficult to predict but were perceived to be significant for both wild and captive flocks, and a consensus has long existed that establishment of a second captive flock and a second wild flock as soon as practical would be best.

Consensus on the advisability of starting a third captive flock, especially one in a mainland zoo location, has been more elusive. Debate on this issue occupied the greater part of 1990 and centered on the two major issues raised in the earlier part of this report: disease risks and projected effects on reproduction. The Puerto Rican Parrot is perhaps a classic example of a species for which high disease susceptibility is predicted. As a member of the Psittacidae, it is a potential host of more than 30 known diseases and potential disease syndromes. Moreover, it is a highly endangered, insular species that has been in a genetic bottleneck for nearly 25 years, a period during which the number of breeding pairs has been low, very likely resulting in a high degree of relatedness and reduced genetic diversity among surviving individuals.

Opponents of the move of birds to a mainland zoo pointed out that a variety of psittacine diseases are very difficult to detect, yet are widespread in mainland zoos and could potentially be spread to captive and wild flocks in Puerto Rico if interchanges of birds were carried out. Conversely, if interchanges were not carried out, it was difficult to see how a mainland zoo flock would have any conservation value for the existing wild population or for a prospective new wild population.

Of special concern, the mainland zoo recommended to host Puerto Rican Parrots has had recurrent cases of avian tuberculosis, a disease that is very difficult to detect in carriers and extremely difficult to eradicate from a facility. Furthermore, this facility maintained a diverse, multi-species psittacine collection and, as a result, the risk of Puerto Rican Parrots contacting other diseases in this facility appeared to be substantial. Opponents of the move suggested that it would be wiser to export the expertise of a mainland zoo to facilities in Puerto Rico rather than to move birds to such a mainland environment. This position was ultimately upheld by the U.S. Fish and Wildlife Service as a result of consultations with its own veterinary experts.

Of equal concern were the potential effects on reproduction from subdividing the captive flock three ways. The captive flock had never supported more than six successful breeding pairs, despite major efforts to establish compatible pairings and despite the fact that surrogate Hispaniolan Parrots had been bred much more successfully at the same facility, using the same techniques. Production of captive young Puerto Rican Parrots has rarely exceeded six individuals in a year, despite experimentation with a variety of husbandry techniques.

What would be the projected effects of splitting the existing captive flock three ways? Clearly, the choices for trying new pairings or allowing maturing birds to select mates would be greatly reduced for each facility from what is now possible. This could conceivably result in a failure to achieve any successful new pairings in any facility. This expectation is based on a track record of needing more than a dozen individuals to form each new productive pair. If, indeed, the major problem in captive reproduction of this species is achieving compatible pairs, a three-way split of the existing captive population could have severe consequences.

Other possible causes of low captive production of the Puerto Rican Parrot likely exist, but the potential importance of these other causes must always be viewed against the consistently successful captive production of Hispaniolan Parrots at the same facility using the same diet and techniques as with the Puerto Rican Parrot. In any event, a case has yet to be made that solutions to the low captive reproduction of Puerto Rican Parrots can be found any more easily in a mainland environment than within the natural range of the species.

Almost as if to test the strength of conservation strategies under debate, Puerto Rico was directly hit by a major storm, Hurricane Hugo, in September of 1989. This hurricane caused massive damage to the remnant habitat of the Puerto Rican Parrot. The parrot aviary, however, survived almost completely unscathed, and with no loss of birds. The wild population was reduced from 45–47 birds to about 25 birds. Interestingly, parrot breeding resumed in 1990. In 1991, a record six pairs of parrots bred in the wild, the highest total since the 1950s. The next year, these six pairs fledged a record of 11 young, and the post-breeding population estimate was between 34 and 37 parrots (A. Valido, personal communication). It presently appears that chances for a resumption of sustained recovery of the wild population are relatively good. Evidently, the amount of risk attributed to a major storm can be overestimated, to both captive and wild populations.

How many subpopulations of the Puerto Rican Parrots are needed at the present time? Clearly the answer to this question is crucially dependent on how many birds are available and on how many birds constitute a critical mass for successful reproduction. Splitting either the existing wild flock or the existing captive flock carries penalties to these flocks that must be weighed against the benefits. The existing two subpopulations have just come through a major environmental disaster in relatively good shape, which should perhaps temper enthusiasm for splitting them up. Had the wild population been only half as large at the time the storm hit, things could look much worse now than they do.

In our view, the wild flock has shown a surprisingly encouraging resilience and ability to increase over the past two decades (at a rate very similar to the recovery of the wild population of Whooping Cranes). Now it has also shown an encouraging survival through a major storm. We believe that preserving the integrity of this population represents the highest priority in future efforts to preserve the species.

In view of the long history of relatively poor performance of the species in captivity, conservation efforts should not be skewed toward large-scale proliferation of captive populations; instead, the fruits of captive production in the near term should be used largely to bolster the existing wild population. Once that population reaches perhaps 70 to 100 individuals, it may obtain a fairly comfortable status with respect to future hurricanes, and it will become important to begin the establishment of a second wild flock.

Meanwhile, we believe that establishment of a second captive flock in Rio Abajo represents a reasonable strategy, but it should not take place at the expense of the productivity of the existing captive or wild subpopulations. This second aviary should be established with current nonbreeders and juveniles from the existing captive flock. No attempts to establish a second wild flock should be made until the existing wild flock exceeds 70 birds, because such efforts would conflict with bolstering the extant wild flock as quickly as possible while it is still at a critically low level. We see no clear advantage in establishing a third captive flock at the present time.

Conclusions

The overriding goal of securing wild populations can sometimes be forgotten in the recent climate of enthusiasm for captive breeding. The primary rationale usually offered for establishing multiple captive populations is one of risk reduction from catastrophe. Unfortunately, it is not often acknowledged that proliferation of captive populations itself carries significant risks, especially with respect to disease vulnerability and reproductive penalties.

The recent case history of the critically endangered Puerto Rican Parrot indicates that even major catastrophes can have relatively limited effects on some populations. This experience underscores the need for incorporating more rigorous information on the species' ecology into genetic models and into concepts of metapopulations. Species and populations have resiliency mechanisms that may be as important as genetic make-up in their day-to-day survival. Therefore, decisions on their management must balance ecological and genetic arguments.

The recovery potential of some species may also depend on critical-mass effects that argue against oversplitting populations. When subdivision involves moving endangered species into non-native areas, it places them in close proximity to potential vectors of new diseases. The risks involved cannot be rigorously quantified in magnitude, but they appear to be well worth avoiding when alternatives exist. The consequences of mistakes can easily prove fatal to species.

We wish to make perfectly clear that we do not oppose an important role for zoological institutions in the conservation of endangered species, and in fact we strongly support such a role. Much remains to be improved, however, regarding the precautions taken in breeding endangered species in captivity, particularly in disease prevention and control. Attempts to house species that are arguably sensitive to disease in ex situ multi-species environments are generally unwise, and the should be replaced by a preference for breeding endangered species in situ in closed single-species facilities operating under rigorous disease-prevention protocols. While this approach is in some cases more expensive, such is not always the case, especially considering the costs of routine international transport in ex situ approaches. Furthermore, the benefits of in situ programs that involve further communities in the conservation of their native fauna can be as substantial as the benefits in reducing unnecessary risks of exposure to disease.

Acknowledgments

We thank M. Friend, W. Barrow, E. Rockwell, and two anonymous reviewers for comments on the manuscript.

Literature Cited

Anonymous. 1989. Emergency action taken to protect the desert tortoise. Endangered Species Technical Bulletin **14:**1, 5–6.

Ashton, W. L. G., and J. E. Cooper. 1989. Exclusion, elimination and control of avian pathogens. Pages 31–38 in J. E. Cooper, editor. Disease and threatened birds. Technical publication No. 10. Cambridge, England.

Balcomb, R. 1986. Songbird carcasses disappear rapidly from agricultural fields. Auk **103:**817–820.

Beehler, B. A. 1990. Management of *Mycobacterium avium* in a mixed species aviary. Pages 125–129 in Proceedings of the American Association of Zoo Veterinarians.

Bluhm, C. K. 1985. Mate preferences and mating patterns of Canvasbacks (*Aythya valisineria*). Ornithological Monographs **37:**45–56.

Castle, M. D., and B. M. Christensen. 1990. Hematozoa of Wild Turkeys from the midwestern United States: Translocation of wild turkeys and its potential role in the introduction of *Plasmodium kempi.* Journal of Wildlife Diseases **26:**180–185.

Cooper, J. E. 1989. The role of pathogens in threatened populations: An historical review. Pages 51–61 in J. E. Cooper, editor. Disease and threatened birds. Technical publication No. 10. , Cambridge, England.

Cromie, R. L., M. J. Brown, D. J. Price, and J. L. Stanford. 1991. Susceptibility of captive waterfowl to avian tuberculosis. Tubercle **72:**105–109.

Dein, F. J., J. W. Carpenter, G. G. Clark, R. J. Montali, C. L. Crabbs, T. F. Tsai, and D. E. Docherty. 1986. Mortality of captive Whooping Cranes caused by eastern equine encephalitis virus. Journal of the American Veterinary Medical Association **189:**1006–1010.

Derrickson, S. R., and N. F. R. Snyder. 1992. Potentials and limits of captive breeding in parrot conservation. Pages 133–163 in S. R. Beissinger and N. F. R. Snyder, editors. New World parrots in crisis: Solutions from conservation biology. Smithsonian Institution Press, Washington, D.C.

Fletcher, K. C., A. K. Eugster, R. E. Schmidt, and G. B. Hubbard. 1979. Parvovirus infection in maned wolves. Journal of the American Veterinary Medical Association **175:**897–900.

Gaskin, J. M. 1989. Psittacine viral diseases: A perspective. Journal of Zoo and Wildlife Medicine **20(3):**249–264.

Gilpin, M. E., and M. E. Soulé. 1986. Minimum viable populations: Processes of species extinction. Pages 19–34 in M. E. Soulé, editor. Conservation biology: The science of scarcity and diversity. Sinauer Associates, Sunderland, Massachusetts.

Goodman, D. 1987. How do any species persist? Lessons for conservation biology. Conservation Biology **1(1)**:59–62.

Gough, J. F. 1989. Outbreaks of budgerigar fledgling disease in three aviaries in Ontario. Canadian Veterinary Journal **30**:672–674.

Hudson, P. J. 1986. The effect of a parasitic nematode on the breeding production of Red Grouse. Journal of Animal Ecology **55**:85–92.

Jacobson, E. R., and J. M. Gaskin. 1990. Clinicopathologic investigations on an upper respiratory disease of free-ranging desert tortoises, *Xerobates agassizii.* U.S. Contract report. Bureau of Land Management, Department of the Interior, Riverside, California.

Jenkins, C. D., S. A. Temple, C. van Riper, and W. R. Hansen. 1989. Disease-related aspects of conserving the endangered Hawaiian Crow. Pages 77–87 in J. E. Cooper, editor. Disease and threatened birds. Technical publication No. 10. Cambridge, England.

Kear, J., and P. Palmes. 1980. Andean and James Flamingos. International Zoo Yearbook **20**:17–23.

Kennedy-Stoskopf, S. 1989. Infectious diseases: Cause or effects? Journal of Zoo and Wildlife Medicine **20(3)**:247–248.

Kepler, C. B. 1978. Captive propagation of Whooping Cranes: A behavioral approach. Pages 231–241 in S. A. Temple, editor. Endangered birds: Management techniques for preserving threatened species. University of Wisconsin Press, Madison, Wisconsin.

Lacy, R. C. 1987. Loss of genetic diversity from managed populations: Interacting effects of drift, mutation, immigration, selection, and population subdivision. Conservation Biology **1(2)**:143–158.

Lacy, R. C., N. R. Flesness, and U. S. Seal. 1989. Puerto Rican Parrot (*Amazona vittata*) population viability analysis and recommendations. Captive Breeding Specialist Group, Apple Valley, Minnesota.

Linz, G. M., J. E. Davis, Jr., R. M. Engeman, D. L. Otis, and M. L. Avery. 1991. Estimating survival of bird carcasses in cattail marshes. Wildlife Society Bulletin **19**:195–199.

Luthin, C. S., G. W. Archibald, L. Hartman, C. M. Mirande, and S. Swengel. 1985. Captive breeding of endangered cranes, storks, ibises, and spoonbills. International Zoo Yearbook **24/25**:25–39.

Mann, P. C., M. Bush, M. J. G. Appel, B. A. Beehler, and R. J. Montali, 1980. Canine parvovirus infection in South American canids. Journal of the American Veterinary Medical Association **177**:779–783.

May, R. M. 1988. Conservation and disease. Conservation Biology **2(1)**:28–30.

Montali, R. J., M. Bush, and G. A. Greenwell. 1976*a.* An epornitic of duck viral enteritis in a zoological park. Journal of the American Veterinary Medical Association **169(9)**:954–958.

Montali, R. J., M. Bush, C. O. Thoen, and E. Smith. 1976*b.* Tuberculosis in captive exotic birds. Journal of the American Veterinary Medical Association **169(9)**:920–927.

Montali, R. J. M. Bush, and E. E. Smith. 1978. Pathology of tuberculosis in captive exotic birds. Pages 209–215 in Proceedings: Mycobacterial infections of zoo animals symposium. Smithsonian Institution Press, Washington, D.C.

Murphy, D. D., K. E. Freas, and S. B. Weiss. 1990. An environment-metapopulation approach to population viability analysis for a threatened invertebrate. Conservation Biology **4(1)**:41–51.

Nettles, V. F., J. H. Shadduck, R. K. Sikes, and C. R. Reyes. 1979. Rabies in translocated raccoons. AJPH **69(6)**:601–602.

O'Brien, S. J., M. E. Roelke, L. Marker, A. Newman, C. A. Winkler, D. Meltzer, L. Colly, J. F. Evermann, M. Bush, and D. E. Wildt. 1985. Genetic basis for species vulnerability in the cheetah. Science **227**:1428–1434.

Quinn, J. F., and A. Hastings. 1987. Extinction in subdivided habitats. Conservation Biology **1(3)**:198–208.

Reece, R. L. 1989. Avian pathogens: Their biology and methods of spread. Pages 1–23 in J. E. Cooper, editor. Disease and threatened birds. Technical publication No. 10. , Cambridge, England.

Ritchie, B. W., F. D. Niagro, P. D. Lukert, K. S. Latimer, W. L. Steffens, and N. Pritchard. 1989. A review of psittacine beak and feather disease. Journal of the Association of Avian Veterinarians **3(3)**:143–148.

Saumier, M. D., M. E. Rau, and D. M. Bird. 1986. The effect of *Trichinella pseudospiralis* infection on the reproductive success of captive American Kestrels (*Falco sparverius*). Canadian Journal of Zoology **64**:2123–2125.

Scott, M. E. 1988. The impact of infection and disease on animal populations: Implications for conservation biology. Conservation Biology **2(1)**:40–56.

Seal, U. S., and R. C. Lacy. 1989. Florida panther (*Felis concolor coryi*) viability analysis and species survival plan. Captive Breeding Specialist Group, Apple Valley, Minnesota.

Shaffer, M. L. 1985. The metapopulation and species conservation: The special case of the Northern Spotted Owl. Pages 86–99 in R. J. Gutierrez and A. B. Carey, editors. Ecology and management of the Spotted Owl in the Pacific Northwest. General technical report PNW-185. U.S. Department of Agriculture–Forest Service,

Shaffer, M. L. 1990. Population viability analysis. Conservation Biology **4(1)**:39–40.

Shima, A. L., and K. G. Osborn. 1989. An epornitic of *Salmonella typhimurium* in a collection of lories and lorikeets. Journal of Zoo and Wildlife Medicine **20(3)**:373–376.

Simberloff, D. 1988. The contribution of population and community biology to conservation science. Annual Review of Ecology and Systematics **19**:473–511.

Snyder, B., J. Thilsted, B. Burgess, and M. Richard. 1985. Pigeon herpesvirus mortalities in foster reared Mauritius Pink Pigeons. Pages 69–70 in Proceedings of the American Association of Zoo Veterinarians.

Snyder, N. F. R., J. W. Wiley, and C. B. Kepler. 1987. The parrots of Luquillo: Natural history and conservation of the Puerto Rican Parrot. Western Foundation of Vertebrate Zoology, Los Angeles, California.

Thorne, E. T., and E. S. Williams. 1988. Disease and endangered species: The black-footed ferret as a recent example. Conservation Biology **2(1)**:66–74.

U.S. Fish and Wildlife Service. 1987. Recovery plan for the Puerto Rican Parrot, *Amazona vittata.* U.S. Fish and Wildlife Service, Atlanta, Georgia.

van Riper, C., S. G. van Riper, M. L. Goff, and M. Laird. 1986. The epizootiology and ecological significance of malaria in Hawaiian land birds. Ecological Monographs **56**:327–344.

Watkins, D. L., N. L. Letvin, A. L. Hughes, and T. F. Fedder. 1990. Molecular cloning of cDNAs that encode MHC class I molecules from a New World primate (*Saguinus oedipus*): Natural selection acts at positions that may affect peptide presentation to T cells. Journal of Immunology **144**:1136–1143.

Yamamoto, J. T., K. M. Shields, J. R. Millam, T. E. Roudybush, and C. R. Grau. 1989. Reproductive activity of force-paired Cockatiels (*Nymphicus hollandicus*). Auk **106**:86–93.

Delayed Sexual Maturity and Demographics of Blanding's Turtles (*Emydoidea blandingii*): Implications for Conservation and Management of Long-Lived Organisms

JUSTIN D. CONGDON
Savannah River Ecology Laboratory
Drawer E
Aiken, SC 29820, U.S.A.

ARTHUR E. DUNHAM
Department of Biology
University of Pennsylvania
Philadelphia, PA 19104, U.S.A.

R. C. VAN LOBEN SELS
Red Mountain High School
7301 East Brown Road
Mesa, AZ 85207, U.S.A.

Abstract: *A study of Blanding's turtles conducted during 27 of the last 37 years provided demographic data sufficient to examine how life-history characteristics may constrain population responses of long-lived organisms. Eight independent estimates of annual adult survivorship exceeded 93%. Nest survival was variable and ranged from 0.0 to 63% annually, with a mean of 44% from 1976 to 1984 and 3.3% from 1985 to 1991. Recruitment of juveniles and adults was sufficient to replace individuals estimated to have died during the study. A life table for the population resulted in a cohort generation time of 37 years and required a 72% annual survivorship of juveniles between 1 and 13 years of age to maintain a stable population. Population stability was most sensitive to changes in adult or juvenile survival and less sensitive to changes in age at sexual maturity, nest survival, or fecundity. The results from the present study indi-*

Paper submitted October 6, 1992; revised manuscript accepted March 23, 1993.

Maduración sexual retardada y demografía de las tortugas de Blanding (*Emyoidea blandingii*): Implicaciones para la conservación y manejo de organismos longevos

Resumen: *Un estudio de las tortugas de Blanding conducido durante 27 de los últimos 37 años aportó datos demográficos suficientes para examinar como las características de la historia de vida pueden limitar las respuestas poblacionales de organismos longevos. Ocho estimaciones independientes de la supervivencia anual de adultos excedió el 93%. La supervivencia en el nido fue variable y estuvo entre 0.0 y 63% anual, con una media de 44% para el período 1976–1984 y 3.3% para el período 1985–1991. El reclutamiento de juveniles y adultos fue suficiente para reemplazar individuos que se estimó murieron durante el estudio. Una tabla de vida para la población resultó en un período de regeneración de 37 años, y requirió un 72% de supervivencia anual de juveniles entre 1 y 13 años de edad para mantener una población estable. La estabilidad pobla-*

cate that life-history traits of long-lived organisms consist of co-evolved traits that result in severe constraints on the ability of populations to respond to chronic disturbances. Successful management and conservation programs for long-lived organisms will be those that recognize that protection of all life stages is necessary. Programs such as headstarting or protection only of nesting sites, in the absence of programs to reduce mortality of older juveniles and adults, appear to be less than adequate to save long-lived organisms such as sea turtles and some tortoises. The concept of sustainable harvest of already-reduced populations of long-lived organisms appears to be an oxymoron.

cional fue más sensible a cambios en supervivencia de adultos o juveniles y menos sensible a cambios en edad de maduracion sexual, supervivencia en nidos o fecundidad. Los resultados del estudio presente indican que las características de la historia de vida de organismos longevos consiste en caracteres coevolucionados que resultan en limitaciones severas sobre la habilidad de las poblaciones para responder a perturbaciones crónicas. Programas de manejo y conservación exitosos para organismos longevos serán aquellos que reconozcan que se necesita la protección de todos los estadíos de vida. Programas tales como "headstarting" o protección de los sitios de anidamiento solamente, en ausencia de programas para reducir la mortalidad de juveniles mayores y adultos, parecen ser poco adecuados para salvar organismos longevos tales como las tortugas marinas y algunas otras tortugas. El concepto de cosecha sostenible de poblaciones ya reducidas de organismos longevos parece ser una incongruencia.

Introduction

Development of conservation programs often must proceed without adequate data on life-history trait values of target species. Reasons for lack of data include technical and logistic problems related to obtaining life-history data on some species, and the difficulty of obtaining reliable data from populations that are already reduced or in decline. The problems that impede life-history studies in general are magnified when the species being studied are long lived (Tinkle 1979).

Studies of the variation of traits of organisms provide evidence that there are constraints on the breadth of functions that a single organism can perform. Constraints have been demonstrated at all levels of organization (genetic, historical, physiological, biomechanical, behavioral), including life histories. Examination of the limits of single traits and covariation among possible life-history trait values can suggest (1) possible selective factors that shape or constrain life histories; (2) ways that specific life-history traits may constrain the population responses of long-lived organisms; and (3) how life-history traits must be considered in the design and implementation of management and conservation programs.

A life-history feature that long-lived vertebrates have in common is delayed sexual maturity (Charlesworth 1980; Dunham et al. 1988; Charnov 1990). Benefits attributed to delaying the onset of reproduction include increased quality of young produced, increased number of young per reproductive bout, decreased costs associated with reproduction, and decreased risk of mortality as an adult (Gadgil & Bossert 1970; Tinkle et al. 1970; Wiley 1974; Bell 1977; Stearns & Koella 1986). Theoretically, such benefits combine to result in a higher lifetime reproductive success among individuals that delay sexual maturity than that attained by individuals maturing earlier. Costs of delaying sexual maturity include increased risk associated with death prior to first reproduction and lengthened generation times.

An often overlooked intergenerational life-history feature associated with delaying sexual maturity is the substantial increase in the annual survival of juveniles required to maintain a stable population (Dunham et al. 1989). Even with high annual fecundity, values of alpha approaching 20 years require annual juvenile survivorship greater than those reported for most extant vertebrates (Ricklefs 1973; Turner 1977; Dunham et al. 1988; Wilbur & Morin 1988) to maintain a stable population.

It is unknown how general conclusions drawn from the Dunham et al. (1989) model on dinosaur life histories may be, because almost all life-history attributes had to be estimated from little or no data. We first examine the generality of the results of the dinosaur model using relatively complete life-history data from Blanding's turtles (*Emydoidea blandingii*). We then examine how life-history traits may constrain potential responses to the disturbance of populations of long-lived organisms, and we discuss results from this study in relation to the conservation and management long-lived organisms in general, and of tortoises and freshwater and sea turtles in particular. Blanding's turtles are excellent models because they have life history and demographic characteristics similar to those of other freshwater turtles, sea turtles, and tortoises, and to those proposed for dinosaurs. In addition, Blanding's turtles and many other freshwater turtle species are of conservation concern in some parts of their range.

Materials and Methods

Blanding's turtles on the University of Michigan's E. S. George Reserve have been studied for 27 of the past 37 years (1953 to 1957 by O. Sexton, 1968 to 1973 by H.

Wilbur, 1975 to 1979 by D. Tinkle and J. Congdon, and 1980 to 1991 by J. Congdon). Estimates of long-term adult survival used in this paper are based on data from Blanding's turtles marked during all periods of study. Data on clutch size and reproductive frequency are taken from Congdon et al. (1983) and Congdon and van Loben Sels (1990), and data on age at maturity from Congdon and van Loben Sels (1993).

Each year from 1975 through 1986, and in 1991, intensive aquatic trapping was carried out from early May through early September, and drift fences were usually monitored from April through June and during September and October. From 1987 through 1990, the study was conducted only from early May to early July. A total of 711 marked individuals were recaptured 2968 times (see Congdon et al. [1983] and Congdon and van Loben Sels [1991] for descriptions of capture methods). Each turtle was individually marked by notching the margins of the carapace, weighed, measured (straight line plastron length and carapace length), and then released at the point of capture.

Age class zero was assigned to the period from egg laying to emergence from the nest, which occurred from late August to early October. Age class 1 began at emergence from the nest and covered the hatchlings' first fall, winter, and first full activity season. Turtles that were first captured with fewer than 19 growth rings and evidence of recent rapid body growth (a light-colored area near the medial line on the plastron) were assigned an age based on the assumptions that rapid growth indicated young individuals and that growth rings were laid down annually in juveniles and young adults. Recaptures of juveniles during the past 15 years support both assumptions. Unaged turtles of adult size that were first marked by Owen Sexton between 1954 and 1957 and recaptured during the present study were assigned a minimum age of 20 years at first capture. The minimum age assigned is conservative in that most turtles would be assigned an age less than their actual age.

Life tables were constructed from long-term means of all parameters except for annual survivorship (s_x) of juveniles between 1 and 13 years old; juvenile survivorship values were estimated to obtain values resulting in a stable population. For the life table, fecundity (m_x) was defined as the number of female eggs produced annually (average clutch size divided by 2 [to adjust for production of males by making an assumption of an equal primary sex ratio] and then multiplied by clutch frequency).

To examine the consequences of a reduction in survivorship of juveniles or adults, we performed two qualitatively distinct types of analyses. In both types, we set the population parameters at slightly more than long-term values to minimize the probability that the required juvenile survival necessary to maintain a stable population would be overestimated. For example, (1) all females were allowed to mature at the actual minimal alpha (14 years) for the population rather than at the average alpha of 17.5 years; (2) all females were allowed to produce a single average clutch of 10 eggs annually (annual fecundity = 5 female eggs) rather than adjusting for some females skipping reproduction each year; (3) maximum adult survival was allowed to reach 0.96 to adjust for emigration, which was not considered in adult survival values.

In the first type of analysis, we explored the impact of changes in life-history parameters on population stability. We examined which combinations of demographic variables would permit population persistence—$r \geqslant 0$, r is the solution to Euler's equation:

$$1 = \sum_{x=0}^{\infty} e^{-rx} l_x m_x$$

where l_x is the survival from birth to age x, m_x is the expected fecundity of a female of age x, and e is the base of natural logarithms. Three subsets of analyses were calculated with (1) fecundity fixed at five and nest survival at 0.26, combinations of average annual juvenile and adult survival for which $r = 0$ were calculated for values of alpha varying from 14 to 20 years; (2) alpha fixed at 14 years and nest survival at 0.26, combinations of average annual juvenile and adult survival for which $r = 0$ were calculated for annual fecundity values from 2 to 8; and (3) alpha fixed at 14 years and annual fecundity at five, combinations of average annual juvenile and adult survival for which $r = 0$ were calculated for nest survival values from 0.1 to 1.0.

In the second type of analysis, a single demographic variable was allowed to vary over a wide range of values while all others were held constant at the long-term average for the population. Life-table variables were calculated for each unique combination of demographic variables, and the intrinsic rate of increase (r) was calculated for each result. Four subsets of analyses were carried out allowing (1) age-specific annual fecundity, (2) nest survivorship, (3) juvenile survivorship, and (4) adult survivorship to vary in turn. Adult and juvenile survival rates varied from 0.005 to 0.99 in steps of 0.005. Age-specific fecundity varied from 2 to 8 in steps of 1, and nest survivorship varied from 0.1 to 1.0 in steps of 0.1.

Results

Data on reproduction used in construction of life tables were obtained from previously published literature on the E. S. George Reserve's Blanding's turtles (Table 1) and from analysis of the survival of nests and adults from the present study. Annual survivorship of observed nests

Table 1. Reproductive characteristics of Blanding's Turtles on the E. S. George Reserve.

Variable	*Value*	*Reference*
Mean Clutch Size	10.2	1
Clutch Frequency	0.80	1
Minimum Alpha	14.0 years	2, 3
Mean Alpha	17.5 years	3
Maximum Alpha	20.0 years	3

[1] Congdon et al. 1983.
[2] Congdon & van Loben Sels 1991.
[3] Congdon & van Loben Sels 1993.

over 16 years averaged 26.1% (min = 0%; max = 78%; 1SE = 6.6; n = 193). Nest survival of 119 observed nests constructed from 1976 to 1984 averaged 43.8% (min = 7%; max = 78%; 1SE = 7.2) was significantly higher than that observed for 75 nests constructed from 1985 to 1991 (mean = 3.3%; min = 0; max = 17%; 1SE = 2.4).

Independent measures of minimum survival (s_x) of adults over eight periods ranging in duration from 31 years (1955–1986) to six years (1980–1986) averaged 93.5% (min = 91.5%; max = 94.7%; 1SE = 0.5%).

Based on an annual mortality rate (q_x) of 0.04 and an estimated mean (111) and maximum (150) number of adult females in the population (Congdon et al. 1986), the estimated mean and maximum number of adult females that die each year was 4.4 and 6.0, respectively. The number of new juveniles and adult females marked each year on the E. S. George Reserve (mean over 13 years = 9.4) is slightly higher than the maximum estimate of deaths in the population.

Values of annual survival (s_x) for age 0 for the entire study period were used to calculate a life table. Variation in alpha was accounted for by increasing fecundity (m_x) values used in the life table between ages 14 and 17 years. Values of m_x used in the life table are the number of female offspring produced per female (using average clutch size, assuming an unbiased hatchling sex ratio, and adjusting for an 0.8 annual reproductive rate for each female). The long-term mean nest survival (0.26) requires annual juvenile survival (S_x) between ages 1 and 13 years to average 0.78 to maintain a stable population, and it results in a cohort generation time ($\Sigma x l_x m_x / R_0$; or, in a general sense, the average age of mothers of neonates in a population with a stable age distribution) of 37 years (Table 2). A second life table calculated using average nest survival of 0.44 for the period prior to 1985, holding all other parameters constant, resulted in r = 0.015, R = 1.67, and a reduction in the estimated average annual juvenile survival required to maintain a stable population by less than 4%.

The first set of simulations examined the relationships between juvenile and adult survival rates while allowing alpha (Fig. 1), fecundity (Fig. 2), and nest survivorship (Fig. 3) to vary. If the average alpha in Blanding's turtles

Table 2. A life tale for *Emydoidea blandingi* on the E. S. George Reserve. Data are long-term means for the population (see text).

Age	*s(x)*	*l(x)*	*m(x)*	*l(x)m(x)*
0	0.2610	1.00000	0	0.0000
1	0.7826	0.26100	0	0.0000
2	0.7826	0.20426	0	0.0000
3	0.7826	0.15985	0	0.0000
4	0.7826	0.12510	0	0.0000
5	0.7826	0.09790	0	0.0000
6	0.7826	0.07662	0	0.0000
7	0.7826	0.05996	0	0.0000
8	0.7826	0.04693	0	0.0000
9	0.7826	0.03672	0	0.0000
10	0.7826	0.02874	0	0.0000
11	0.7826	0.02249	0	0.0000
12	0.7826	0.01760	0	0.0000
13	0.7826	0.01378	0	0.0000
14	0.9600	0.01078	1	0.0108
15	0.9600	0.01035	2	0.0207
16	0.9600	0.00994	3	0.0298
17	0.9600	0.00954	4	0.0382
18	0.9600	0.00916	4	0.0366
19	0.9600	0.00879	4	0.0352
20	0.9600	0.00844	4	0.0338
21	0.9600	0.00810	4	0.0324
22	0.9600	0.00778	4	0.0311
23	0.9600	0.00747	4	0.0299
24	0.9600	0.00717	4	0.0287
25	0.9600	0.00688	4	0.0275
26	0.9600	0.00661	4	0.0264
27	0.9600	0.00634	4	0.0254
28	0.9600	0.00609	4	0.0244
29	0.9600	0.00584	4	0.0234
30	0.9600	0.00561	4	0.0224
31	0.9600	0.00539	4	0.0215
32	0.9600	0.00517	4	0.0207
33	0.9600	0.00496	4	0.0199
34	0.9600	0.00477	4	0.0191
35	0.9600	0.00457	4	0.0183
36	0.9600	0.00439	4	0.0176
37	0.9600	0.00422	4	0.0169
38	0.9600	0.00405	4	0.0162
39	0.9600	0.00389	4	0.0155
40	0.9600	0.00373	4	0.0149
45	0.9600	0.00304	4	0.0122
50	0.9600	0.00248	4	0.0099
55	0.9600	0.00202	4	0.0081
60	0.9600	0.00165	4	0.0066
65	0.9600	0.00134	4	0.0054
70	0.9600	0.00110	4	0.0044
75	0.9600	0.00089	4	0.0036
80	0.9600	0.00073	4	0.0029
85	0.9600	0.00059	4	0.0024
90	0.9600	0.00048	4	0.0019
95	0.9600	0.00040	4	0.0016
100	0.9600	0.00032	4	0.0013
105	0.9600	0.00026	4	0.0011
110	0.9600	0.00021	4	0.0009

Annual fecundity (4 eggs) is based on a mean clutch size of 10 eggs, a reproductive frequency of 0.85, and the assumption that half of all eggs produce daughters. s_x *is the annual probability of survival,* l_x *is the probability of survival from age x to age x + 1,* m_x *is the expected fecundity of a female aged x,* r = *the intrinsic rate of population increase or the implicit solution of* $1 = \Sigma l_x m_x e^{-rx}$, *and* T_c *is the cohort generation time for the population. Population parameters are* R_o = *0.995,* r = *0.0001, and* T_c = *37.5 years.*

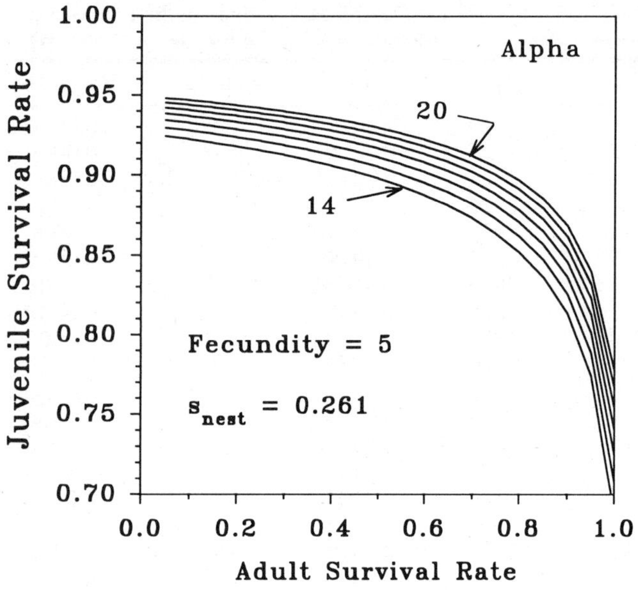

Figure 1. The relationship between adult survival rate and juvenile survival rate while average age at sexual maturity (alpha) is allowed to vary from 14 to 20 years. Values for fecundity and nest survival are fixed.

increased from 14 to 20 years, the average juvenile survivorship required to maintain a stable population would increase from 76% to approximately 85% (Fig. 1). To maintain a stable population, increases in fecundity from 2 to 8 female-producing eggs per adult female

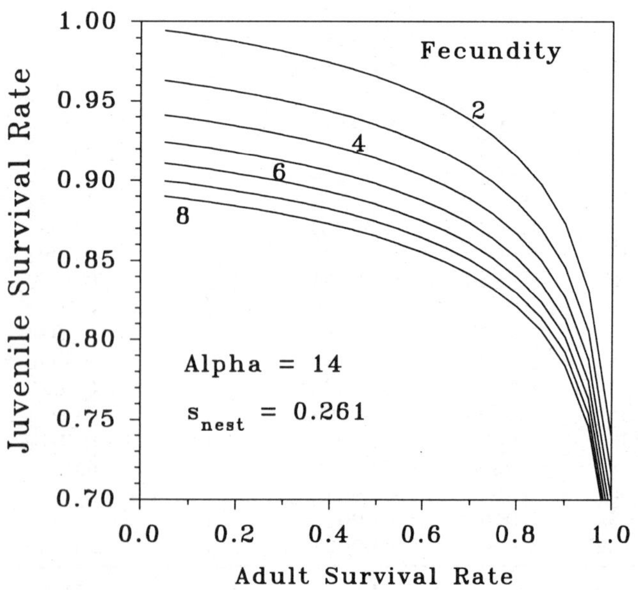

Figure 2. The relationship between adult survival rate and juvenile survival rate while annual average fecundity is allowed to vary from 2 to 8 female-producing eggs per adult female. Values for alpha and nest survival are fixed.

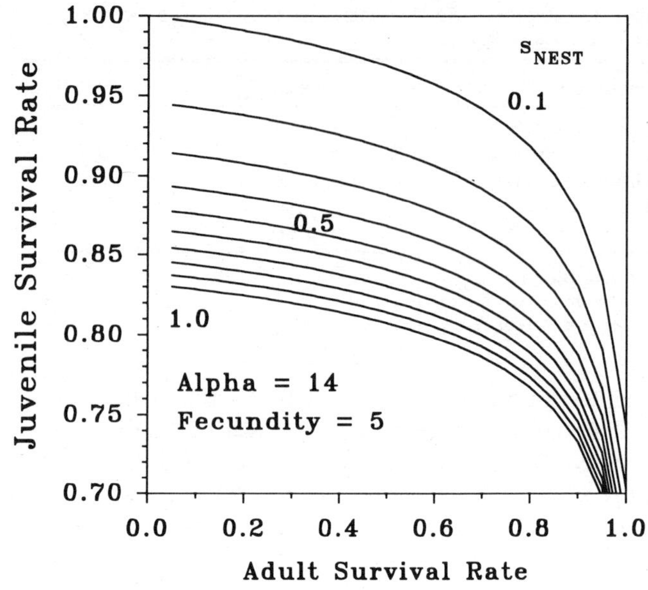

Figure 3. The relationship between adult survival rate and juvenile survival rate while nest survivorship (s_x) is allowed to vary from 0.1 to 1.0. Values for alpha and fecundity are fixed.

would allow a 10% reduction from the 84% average required juvenile survivorship (Fig. 2). When nest survival is reduced from 70% (approximately the highest value recorded in all years) to 10%, juvenile survivorship must increase by approximately 11% (Fig. 3) to maintain a stable population. The relationships between annual juvenile and adult survival for all simulations (Figs. 1, 2, and 3) have the characteristic of being much steeper as adult survivorship exceeds 85%. The rate of change in population increase or decrease (deviation from $r = 0$) was least rapid for variation in nest survivorship and fecundity (Fig. 4*a,b*) and was most rapid for variation in juvenile and adult survivorship (Fig. 4*c,d*).

Discussion

Blanding's turtles are among the longest-lived freshwater turtles for which data are available from the field (Brecke & Moriarty 1989). Some individuals marked on the E. S. George Reserve as adults in the mid-1950s remain in the population, are now approaching a minimum of 60 years of age, are still reproductive (Congdon & van Loben Sels 1991), and exhibit no signs of increased mortality or reproductive senescence (Congdon & van Loben Sels 1993). Annual survival rates of Blanding's turtles are among the highest reported for freshwater turtles (Gibbons & Semlitsch 1982; Mitchell 1988; Frazer & Gibbons 1990; Brooks et al. 1991; Frazer et al. 1991).

The decline in nest survival from 44% for the nine years prior to 1985 to 3% for the seven years after 1984

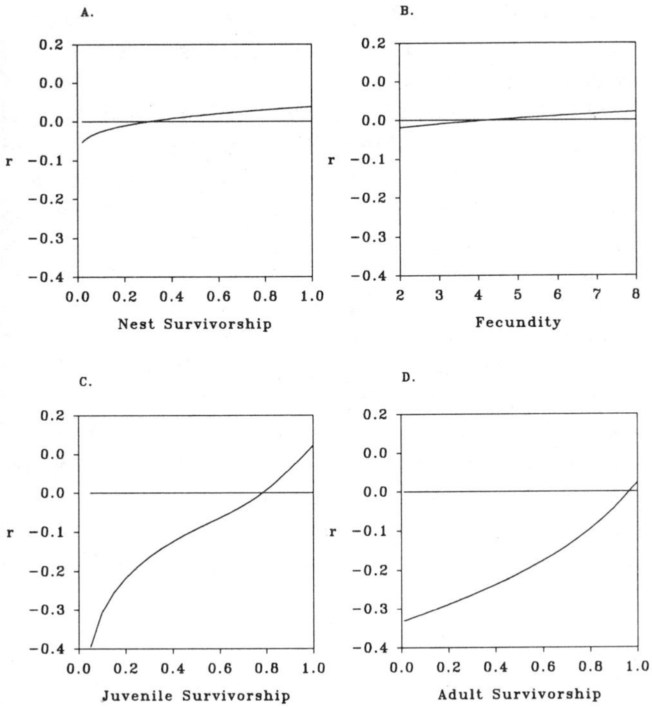

Figure 4. The relationships between (a) fecundity, and survivorships of (b) nests, (c) juveniles, and (d) adults to population stability (r = *0) set by long-term mean values for Blanding's turtles on the E. S. George Reserve.*

represents a substantial problem to the population if the reduction is chronic—if nest survival remains low for more than a single generation time. Using the reduced nest survival data and keeping all other variables the same as in Table 2, the average juvenile survival from 1 to 13 years of age that was necessary to maintain a stable population increased from 75% to greater than 85%. Although the cause of the decline in nest survival is unknown, it coincided with a collapse in the fur market in general and, most important for turtles, the demand for the pelts of major nest predators such as raccoons and foxes. If (1) the decline in nest success of Blanding's turtles is caused by increasing predator populations; (2) the populations of raccoons and foxes remain high due to the absence of large predators; and (3) the growing public opinion against trapping furbearers and wearing or using wild animal furs continues, predator population control other than fur-bearer trapping may have to be implemented to maintain some turtle populations.

Female Blanding's turtles mature between the ages of 14 and 20 years, with the slowest-growing individuals maturing at the greatest ages (Congdon & van Loben Sels 1993). The average alpha of Blanding's turtles (17.5 years) approaches the limit of about 20 years predicted by demographic constraints (Dunham et al. 1989). In contrast to expectations based on (1) extended delay in alpha; (2) the marking of over 300 hatchlings prior to 1985 (Congdon and van Loben Sels 1991); and (3) the high annual survival rates of juveniles (>70%) required to maintain a stable population, few recaptures of juveniles below 9 years of age have been made. A low number of captures of juveniles has also been reported for other populations of Blanding's turtles (Gibbons 1968; Graham & Doyle 1979; Petokas 1986). Suggested reasons for the low numbers of juveniles captured include the following facts: (1) juveniles occupy habitats not trapped or typically searched by investigators (but see Pappas & Brecke 1992); (2) juveniles are very secretive, are not detected during searches, and are not susceptible to trapping; and (3) juveniles are rare and represent actual recruitment into a population of long-lived adults. The present analyses indicate that a high annual survival of juveniles is required to maintain the stable population on the E. S. George Reserve, which in turn suggests that the juveniles are present on the Reserve but are eluding capture. Being cryptic and secretive may be one way that juveniles attain high survivorships at early ages.

Two long-lived chelonians of conservation concern are sea turtles as a group and desert tortoises. Sea turtles (National Research Council 1990), desert tortoises (Turner et al. 1987), and Blanding's turtles share many life-history traits, including delayed sexual maturity, iteroparity, and high adult survival rates. Desert tortoises reach sexual maturity at between 12 and 20 years of age (Turner et al. 1987; Kristin Berry, personal communication), which is very similar to Blanding's turtles. Although the model and data from this study indicate that it is unlikely, some sea turtles may delay alpha for more than 20 years (Colin Limpus, personal communication). Demographic constraints on delaying sexual maturity makes attainment of sexual maturity at 50 years of age (Davenport 1988) virtually impossible. In years that sea turtles reproduce, they produce more and larger clutches (Lenarz et al. 1981) than do freshwater or terrestrial turtles. Since their reproductive interval may be over three years, however, their average annual fecundity is not as high as it appears from their large egg output in reproductive years (Frazer 1984). For example, a loggerhead turtle that produces four clutches of 120 eggs every four years has an average annual fecundity of 60 female-producing eggs. Desert tortoises have clutch sizes that average 4–5 eggs, and some females are capable of producing more than one clutch; however, some adult females do not reproduce every year (Turner et al. 1984; Turner & Berry 1986). As a result, fecundity in desert tortoises is approximately three to four female eggs and is similar to Blanding's turtles. Therefore, sea turtles and desert tortoises have life histories similar enough to Blanding's turtles to share the demographic constraints demonstrated in this study.

People have historically harvested both eggs and adult female sea turtles at nesting beaches for food (National Research Council 1990). More recently, nesting beaches have been restricted due to housing and recreational development, and mortality of adults and juveniles has increased through capture and drowning in fishing and shrimping trawls operating in near-shore waters. Desert tortoises recently have also been subjected to higher mortality associated with habitat destruction, upper respiratory tract disease, increased raven predation on hatchlings and young juveniles, harvesting for pets, and poaching (Berry 1986).

The model and data from Blanding's turtles point out that, in long-lived organisms in general and sea turtles and desert tortoises in particular, chronic reduction in the survival of adults requires an increase in the already high level of survivorship of juveniles to maintain a stable population. The probability that juvenile survival can increase substantially from the present high levels through compensatory release from density dependence, to offset even a relatively small chronic decrease in the survival of adults in long-lived organisms, seems very low (Brooks et al. 1991). Even under protected conditions on the E. S. George Reserve, the required survivorship of juveniles necessary to maintain a stable population is substantially higher than that documented for any other vertebrate (Ricklefs 1973; Turner 1977; Dunham et al. 1988; Wilbur & Morin 1988).

A general conclusion of this study is that the suite of life-history traits that coevolve with longevity results in populations that are severely limited in their ability to respond to chronic increases in mortality of neonates and even less so to increased mortality of juveniles or adults. Headstarting hatchlings in many cases will have little effect without a concomitant reduction in the causes of mortality among older juveniles and adults. In addition, the relatively low fecundity, low nest survival, and high adult survival coupled with extremely high juvenile survival required to maintain stable populations argue strongly against applying the concept of sustained harvest to populations of long-lived organisms. Effective management and conservation programs (Frazer 1992) will recognize the integrated nature of life histories and the extreme limitation that the evolution of longevity has placed on the ability of populations of long-lived organisms to withstand and respond to increased mortality or reduced fecundity of any life-history stage. In addition, programs developed to aid in the recovery of depleted populations of long-lived organisms must recognize that there will be long delays before population responses can be detected.

Acknowledgments

We acknowledge the late Donald W. Tinkle for his participation in the early years of this study. We thank J. Whitfield Gibbons for his continued enthusiastic support, Joyce Klevering for assistance with X-radiography, and Owen Sexton and Henry Wilbur for providing their data on the E. S. George Reserve's Blanding's turtles. Special thanks go to Mr. Russell W. Reister, Director of Plant Operations at the University of Michigan, for his leadership in responding to the problems caused by the introduced plant (autumn olive) that was beginning to overrun the turtle nesting areas on the E. S. George Reserve. We recognize and appreciate the University of Michigan, the Museum of Zoology, and the maintenance crew for maintaining the E. S. George Reserve as a world-class research area. The following people made notable contributions to this study: Harold and Sue Avery, Margaret Burkman, Carl and Melvin Congdon, Ruth Estes, Rose Fama, Robert Fischer, Matthew Hinz, Mark Hutcheson, David Kling, Shawn Meager, Roy Nagle, Tal Novak, Patricia Orleans, John Stegmeir, Julie Wallin, and Bradley, John, and Dick Wiltse. Earlier drafts of this manuscript were improved by comments from H. Avery, W. Boarman, J. Lovich, G. Meffe, J. Pechman, R. Fischer, D. Schultz, and S. Weeks. Funding was provided by NSF Grants DEB–74–070631, DEB–79–06301, BSR–84–00861, and BSR–90–19771. Editing of computer files, data analysis, and manuscript preparation were aided by contract DE–AC09–76SROO–819 between the University of Georgia and the U.S. Department of Energy.

Literature Cited

Bell, G. 1977. The life of the smooth newt (*Triturus vulgaris*) after metamorphosis. Ecological Monographs **47**:279–299.

Berry, K. H. 1986. Desert tortoises (*Gopherus agassizii*) research in California, 1976–1985. Herpetologica **41**:62–67.

Brecke, B., and J. J. Moriarty. 1989. *Emydoidea blandingii* (Blanding's Turtle) longevity. Herpetology Review **20(2)**:53.

Brooks, R. J., G. P. Brown, and D. A. Galbraith. 1991. Effects of a sudden increase in natural mortality of adults on a population of the common snapping turtle (*Chelydra serpentina*). Canadian Journal of Zoology **69**:1314–1320.

Charlesworth, B. 1980. Evolution in age structured populations. Cambridge University Press, New York.

Charnov, E. L. 1990. On evolution of age at maturity and adult lifespan. Journal of Evolutionary Biology **3**:139–144.

Congdon, J. D., and R. C. van Loben Sels. 1991. Growth and body size variation in Blanding's turtles (*Emydoidea blandingi*): Relationships to reproduction. Canadian Journal of Zoology **69**:239–245.

Congdon, J. D., and R. C. van Loben Sels. 1993. Relationships of reproductive traits and body size with attainment of sexual maturity and age in Blanding's turtles (Emydoidea blandingii). Journal of Evolutionary Biology **6**:317–327.

Congdon, J. D., D. W. Tinkle, G. L. Breitenbach, and R. C. van Loben Sels. 1983. Nesting ecology and hatching success in the turtle *Emydoidea blandingi.* Herpetologica **39**:417–429.

Congdon, J. D., J. L. Greene, and J. W. Gibbons. 1986. Biomass of freshwater turtles: A geographic comparison. American Midland Naturalist **115**:165–173.

Davenport, J. 1988. The turtle industry of Bali. Bulletin of the British Herpetology Society **25**:16–24.

Dunham, A. E., D. M. Miles, and D. Reznick. 1988. Life history patterns in squamate reptiles. Pages 443–511 in C. Gans and R. Huey, editors. Biology of the Reptilia, 16*b.* Alan R. Liss, New York.

Dunham, A. E., K. L. Overall, W. P. Porter, and K. A. Forster. 1989. Implications of ecological energetics and biophysical and developmental constraints for life-history variation in dinosaurs. Special Paper 238. Geological Society of America. Boulder, Colorado.

Frazer, N. B. 1984. A model for assessing man age-specific fecundity in sea turtle populations. Herpetologica **40**:281–291.

Frazer, N. B. 1992. Sea turtle conservation and halfway technology. Conservation Biology **6**:179–184.

Frazer, N. B., and G. W. Gibbons. 1990. Life tables of a slider population. Pages 183–200 in J. W. Gibbons, editor. Life history and ecology of the slider turtle. Smithsonian Institution Press, Washington, D.C.

Frazer, N. B., J. W. Gibbons, and J. L. Greene. 1991. Life history and demography of the common mud turtle *Kinosternon subrubrum* in South Carolina, USA. Ecology **72**:2218–2231.

Gadgil, M., and W. H. Bossert. 1970. Life historical consequences of natural selection. American Naturalist **104**:1–24.

Gibbons, J. W. 1968. Observations on the ecology and population dynamics of the Blanding's turtle, *Emydoidea blandingi.* Canadian Journal of Zoology **46**:288–290.

Gibbons, J. W., and R. D. Semlitsch. 1982. Survivorship and longevity of long-lived vertebrate species: How long do turtles live? Journal of Animal Ecology **51**:523–527.

Graham, T. E., and T. S. Doyle. 1979. Dimorphism, courtship, eggs, and hatchlings of the Blanding's turtle, *Emydoidea blandingi* (Reptilia, Testudinies, Emydidae) in Massachusetts. Journal of Herpetology **13**:125–127.

Lenarz, M. S., N. B. Frazer, M. S. Ralston, and R. B. Mast. 1981. Seven nests recorded for a loggerhead turtle (*Caretta caretta*) in one season. Herpetological Review **12**:9.

Mitchell, J. C. 1988. Population ecology and life histories of the freshwater turtles *Chrysemys picta* and *Sternotherus odoratus* in an urban lake. Herpetological Monographs **2**:40–61.

National Research Council. 1990. Decline of sea turtles: Causes and prevention. National Academy Press, Washington, D.C.

Pappas, M. J., and B. J. Brecke. 1992. Habitat selection of juvenile Blanding's turtles, *Emydoidea blandingii.* Journal of Herpetology **26**:233–234.

Petokas, P. J. 1986. Patterns of reproduction and growth in the freshwater turtle *Emydoidea blandingii.* Ph.D. dissertation. University of New York, Binghamton, New York.

Ricklefs, R. E. 1973. Fecundity, mortality, and avian demography. Pages 366–434 in D. S. Farner, editor. Breeding biology of birds. National Science Foundation, Washington, D.C.

Stearns, S. C., and J. C. Koella. 1986. The evolution of phenotypic plasticity in life-history traits: Predictions of reaction norms for age and size at maturity. Evolution **40**:893–913.

Tinkle, D. W., H. M. Wilbur, and S. G. Tilley. 1970. Evolutionary strategies in lizard reproduction. Evolution **24**:55–74.

Tinkle, D. W. 1979. Long-term field studies. BioScience **29**:717.

Turner, F. B. 1977. The dynamics of populations of squamates, crocodilians, and rhynchocephalinians. Pages 157–264 in C. Gans and D. W. Tinkle, editors. Biology of the reptilia, vol. 7. Academic Press, New York.

Turner, F. B., and K. H. Berry. 1986. Population ecology of the desert tortoise at Goffs, California, in 1985. Report #12-1544 to the Southern California Edison Company. UCLA Laboratory of Biomedical and Environmental Sciences, University of California, Los Angeles, California.

Turner, F. B., P. A. Medica, and C. L. Lyons. 1984. Reproduction and survival of the desert tortoise (*Scaptochelys agassizii*) in Ivanpah Valley, California. Copeia **1984**:811–820.

Turner, F. B., P. A. Medica, and R. B. Bury. 1987. Age-size relationships of desert tortoises (*Gopherus agassizi*) in southern Nevada. Copeia **1987**:974–979.

Wilbur, H. M., and P. J. Morin. 1988. Life history evolution in turtles. Pages 396–447 in C. Gans and R. Huey, editors. Biology of the reptilia, vol. 16*b.* Alan R. Liss, New York.

Wiley, R. H. 1974. Evolution of social organization and life-history patterns among grouse. Quarterly Review of Biology **49**:201–227.

Island Extinction Rates from Regular Censuses

MICHAEL L. ROSENZWEIG

Department of Ecology and Evolutionary Biology
University of Arizona
Tucson, AZ 85721, U. S. A.

COLIN W. CLARK

Institute of Applied Mathematics
University of British Columbia
Vancouver, BC V6T 1Z2, Canada

Abstract: *Regular censuses conducted over a long time allow the calculation of both extinction and immigration rates. We present formulae for estimation of those rates. We use them on bird censuses of three British islands. These formulae improve on previous estimators of extinction but reaffirm that smaller populations have a higher probability of becoming extinct. On the other hand, they suggest no correlation between extinction rate and either body size or migratory status among birds.*

Tasas de extinción de islas a partir de censos regulares

Resumen: *Censos regulares efectuados a lo largo de períodos prolongados de tiempo permiten el cálculo tanto de tasas de extinción como de inmigración. Presentamos fórmulas para la estimación de dichas tasas. Las usamos en censos de aves de tres islas Británicas. Estas fórmulas representan mejoras de estimadores de extinción anteriores, pero reafirman que poblaciones pequeñas tienen una mayor probabilidad de extinción. Por otro lado, ellas sugieren una ausencia de correlación entre la tasa de extinción y el tamaño corporal o el estatus migratorio entre las aves.*

Island biogeographical data help us understand the causes of extinction. Such data can be analyzed according to the predictions of a substantial body of theory (see Preston 1962*a*; MacArthur & Wilson 1963; Richter-Dyn & Goel 1972; Leigh 1981; Goodman 1987). One prediction these data confirm is that population size and extinction rate are inversely correlated (Diamond 1984).

The relationship between population size and extinction rate is so strong and pervasive that it must be factored out of all data before any other influence on extinction rate can be examined (Pimm et al. 1988). Pimm et al. (1988) tried to do this by relying on a theoretical form for *T*, time to extinction. The attempt was ingenious and elegant but required pages of ad hoc technical comments on data treatment and many ad hoc rules for excluding data.

Tracy and George (1992) accepted the need to neutralize the effect of population size on extinction rate estimates. But they rejected the method suggested by Pimm et al. They also pointed out the value of Cox regressions on data such as finite sequences of censuses. We believe they are right about Cox regressions, but we also believe that the need to use them adds still another unfortunate statistical complication.

Both Pimm et al. (1988) and Tracy and George (1992) rely on a measure of extinction rate that invites statistical complications. This is *T*, the number of consecutive years a species appears in an island's censuses. The larger this number, the lower the extinction rate. Thus, their measure of extinction rate is $1/T$. Clearly, as Tracy and George have noted, the sequences of census data that yield $1/T$ are censored: the length of T for

Paper submitted January 4, 1993; revised manuscript accepted August 6, 1993.

species present at the first census is underestimated. So is T for species present at the last census. It is the censorship that calls for the Cox regressions. (As Pimm et al. note, the error of censorship deminishes rapidly as T increases).

An alternative measure of extinction rate, μ (Diamond & May 1977), causes none of these troubles. μ is the probability that a species becomes extinct (on an island, for example) in a year subsequent to its being found there.

The parameter μ may be estimated from census data such as exists for birds on several British islands. Such data contain four possible transitions. If a species is found one year, it may or may not be found the next. If it is not found one year, it may or may not be found the next. Symbolizing presence in the census with a P and absence with an A, we may therefore expect the transitions PP, PA, AA, or AP. The number of events of each type we symbolize by $N(ij)$.

The result AP implies an immigration event, whereas PA implies an extinction. But PP may also mean an extinction has occurred—one followed in the same interval by an immigration. For example, suppose one year Reed Buntings breed on the island of Skokholm, but the next year none of these adults nor any of their progeny survive. Reed Buntings have suffered an extinction. But if some Reed Buntings from another place move onto the island and breed during that second year, we shall not realize that the extinction has occurred. We will record a PP.

To calculate the proportion of PP's that hide an extinction, we need to define symbols for the immigration probability and for the disappearance probability—that is, the probability of extinction followed during the same census interval by failure to reimmigrate.

Call the immigation probability λ and the disappearance probability δ. Then we can use that part of the data consisting of transitions beginning with A to estimate λ:

$$\lambda \simeq N(AP)/[N(AP) + N(AA)]. \quad (1)$$

And we use the rest to estimate δ:

$$\delta \simeq N(PA)/[N(PA) + N(PP)]. \quad (2)$$

Now using Fig. 1, you will see that

$$\mu \simeq \delta/(1 - \lambda). \quad (3)$$

Elsewhere (Clark and Rosenzweig 1994), we develop arguments to support these formulae of estimation. They are actually the maximum likelihood estimates for the Markov chain shown in Fig. 1 (subject to a consistency constraint discussed below).

Of course, the true value of μ cannot exceed 1.00. But sometimes, owing to stochasticity, the estimate of μ (Equations 1, 2, 3) does exceed 1.00. As an example,

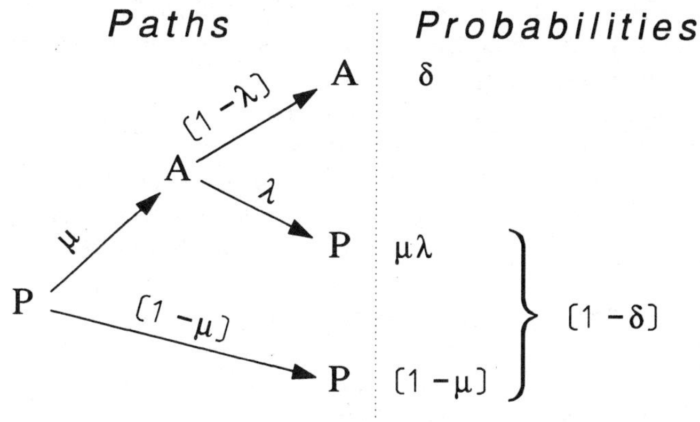

So: $\delta = \mu[1 - \lambda]$

$\mu = \delta/[1 - \lambda]$

Figure 1. The Markov pathways that lead to a second census from a Presence *during one census. The probability of immigration is λ; that of extinction is μ. The probability of a disappearance is δ. Repeated* Presence *may or may not hide an extinction. Both δ and λ can be estimated from the data. Together they estimate μ.*

consider a species with the following sequence of records:

Year	1	2	3	4	5	6	7	8	9	10
Presence	*P*	*A*	*A*	*P*	*P*	*P*	*A*	*A*	*P*	*A*

Thus $N(PP) = 2$
$N(PA) = 3$
$N(AP) = 2$
$N(AA) = 2$
And $\lambda \simeq 0.5$
$\delta \simeq 0.6$
$\mu \simeq .6/.5 = 1.2$

We know this cannot be the correct value of μ because by definition $0 \leq \mu \leq 1$ and $0 \leq \lambda \leq 1$, and (see Fig. 1) $\delta = \mu(1 - \lambda)$. Thus we have a constraint on δ:

$$0 \leq \delta \leq (1 - \lambda). \quad (4)$$

Equations 1, 2, and 3, applied to the data above, produced estimates that violate this constraint.

Clark and Rosenzweig (1994) also provide formulae for the most likely estimates of λ and μ when the constraint is violated:

$$\mu \simeq 1 \text{ and} \quad (5)$$
$$\lambda \simeq [N(PP) + N(AP)]/\Sigma N, \quad (6)$$

where ΣN is the number of all transitions.

Hence, the best estimates for these data are $\lambda \simeq 4/9 \simeq 0.444$ and $\mu \simeq 1.00$. In comparison, the value of T for these same data is $(3 + 1)/2 = 2$, and the estimate of extinction rate, $1/T$, is 0.5. (Or, if we include the initial P too, it is $(1 + 3 + 1)/3 = 1.67$ and $1/T = 0.6$)

The measure μ uses all the data. Interior censuses count twice: once as initial records in a transition and once as terminal records in the next transition.

We calculated μ for published records of three of the 16 British islands whose bird censuses were analyzed by Pimm et al. (1988): Bardsey (Diamond 1984), Inner Farne (Diamond & May 1977), and Skokholm (Williamson 1983). However, we excluded results based on very small samples. We decided in advance to exclude estimates that came from data with only one transition determining δ or λ. Also, many of the species' census records consist entirely of an unbroken string of presences; these by themselves tell nothing about extinction or immigration rates. They could result from very low extinction, very high immigration, or both. We did not use them either.

The values of $1/T$ (listed in Pimm et al 1988) closely resemble the values of μ (Fig. 2). The only noteworthy exceptions are a few species that $1/T$ estimates to have high extinction rates (about 1.00 in fact) but that actually have more moderate rates judging from their transition probabilities.

Because we did not use the most tenuous estimates of extinction, our results (Fig. 3) are much less noisy than those based on $1/T$ (compare with the similar figure in Pimm et al. 1988). As expected, the overall importance

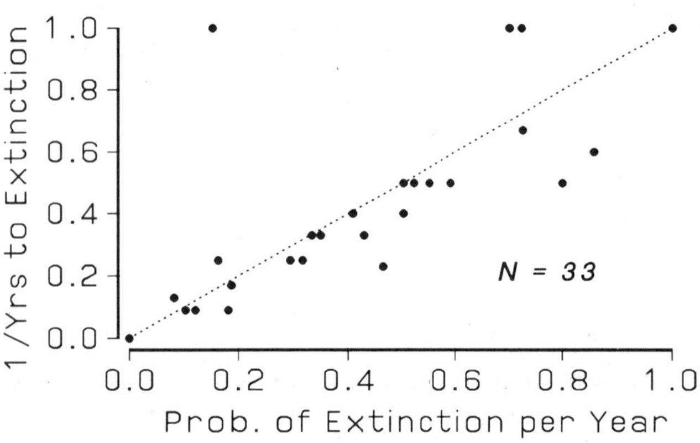

Figure 2. Comparison of two methods to estimate extinction rate. Estimating extinction from the length of sequences of Presence, 1/T, *is slightly less reliable than estmating it from the whole set of transition data. The line is* y = x, *not a regression. The statistic 1/*T *usually underestimates extinction. But the three species with the poorest agreement between methods are all overestimated. These are Blackbird, Swallow, and Whitethroat (all) of Skokholm Island. All had 1/*T *= 1.00 (Pimm et al. 1988). But all showed both low immigration (0.065, 0.077, and 0.048, respectively) and either two or four* PP *transitions. Together, these data make it unlikely that the species always become extinct in a single year. We estimate their extinction probabilities per year as 0.153, 0.722, and 0.700, respectively.*

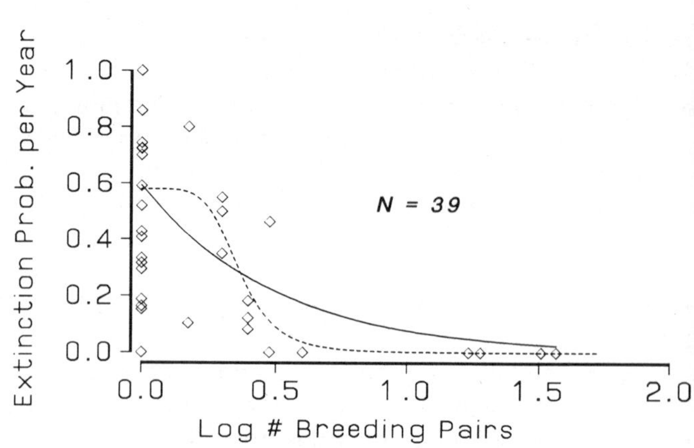

Figure 3. As have other studies, ours found that smaller populations have an increased risk of extinction. The two regression lines were fit by nonlinear regression and differ only slightly in their R^2 *value. The exponential decay equation:* $y = 0.59e^{(-2.03x)}$. *The inverse sigmoid equation:* $y = 0.58[0.0043/(0.0043 + x^{5.40})]$.

of small population size to high extinction rate remains clear.

Figure 3 still contains residual variance after being fit by sigmoidal ($R^2 = 0.379$) or by exponential ($R^2 = 0.358$) regression. It is this residual variance that both Pimm et al. and Tracy and George tried to explain. They could not agree, however, on the effect of body size on extinction rate, nor on the effect of the interaction of body size and population size on extinction rate. They also studied the influence of residency (versus migratoriness) on extinction.

We calculated the residuals of extinction probability from both regressions in Fig. 3. Then, using the same designations for each species as Pimm et al. (large versus small body; resident versus migrant), we studied the effects of body size and residency on extinction. We found none (Fig. 4)

Both Pimm et al. and Tracy and George used a larger data set (351 species-island combinations versus 39 species-island combinations in our study). Most of their data were not available to us, so it is possible that type II error caused the lack of any body size or migratory influences in our results. It is also possible that such effects emanate from only some of the islands (as suggested by Tracy and George); we had only three islands to work with. But it seems equally possible that we came up with no trends because no such trends exist. At least there is no hint of them in our analyses, a conclusion confirmed by the substantial overlaps in Fig. 4.

Working with extinction probability, μ, instead of time to extinction, $1/T$, has several advantages. It uses more of the data. It uses data in a simple way. It leads to straightforward regression of extinction against population size. And the residuals from such regression can be studied to elucidate secondary effects on extinction

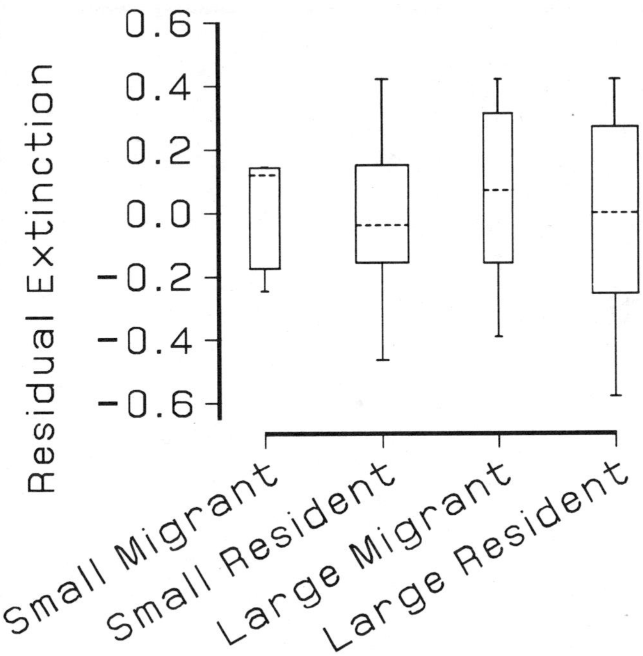

Figure 4. Residuals from the curves of Fig. 3 do not depend on whether a species is large or migratory. Boxes enclose half the data. Horizontal dashed lines (inside boxes) mark medians. Box width is proportional to the square root of sample size. These residuals come from the inverse sigmoid fit, but those from exponential decay tell the same story.

rate. The estimates given by Equations 1–6 are maximum likelihood estimates of μ and λ. Clark and Rosenzweig (1994) extend these estimators to include the possibility of irregular censuses.

Acknowledgments

Both Stuart Pimm and Richard Tracy offered advice and encouragement. Matt Daugherty helped reduce the data to a presence/absence matrix. M. L. Rosenzweig was supported by grant BSR-8905728 from the U. S. National Science Foundation. C. W. Clark was supported partly by grant 83990 from the Natural Sciences and Engineering Research Council of Canada.

Literature Cited

Clark, C. W., and M. L. Rosenzweig. 1994. Extinction and colonization processes: Parameter estimates from sporadic surveys. American Naturalist Pages **00**:000–000.

Diamond, J. M. 1984. "Normal" extinctions of isolated populations. Pages 191–246 in M. H. Nitecki, editor. Extinctions. University of Chicago, Chicago, Illinois.

Diamond, J. M., and R. M. May. 1977. Species turnover on islands: Dependence on census interval. Science **197**:266–270.

Goodman, D. 1987. The demography of chance extinction. Pages 11–34 in M. E. Soule, editor. Viable populations for conservation. Cambridge University Press, Cambridge, England.

Leigh, E. G. 1981. The average lifetime of a population in a varying environment. Journal of Theoretical Biology **90**:213–239.

MacArthur, R. H., and E. O. Wilson. 1963. An equilibrium theory of insular zoogeography. Evolution **17**:373–387.

Pimm, S. L., H. L. Jones, and J. Diamond. 1988. On the risk of extinction. American Naturalist **132**:757–785.

Preston, F. W. 1962. The canonical distribution of commonness and rarity. Ecology **43**:185–215.

Richter-Dyn, N., and N. S. Goel. 1972. On the extinction of a colonizing species. Theoretical Population Biology **3**:406–433.

Tracy, C. R., and T. L. George. 1992. On the determinants of extinction. American Naturalist **139**:102–122.

Williamson, M. 1983. The land-bird community of Skokholm: Ordination and turnover. Oikos **41**:378–384.

Conservation Genetics in the Management of Desert Fishes

GARY K. MEFFE*

University of Georgia's Savannah River Ecology Laboratory
Drawer E
Aiken, SC 29801, U.S.A.

ROBERT C. VRIJENHOEK

Center for Applied Genetics
Cook College
Rutgers University
New Brunswick, NJ 08903, U.S.A.

Abstract: *The status and security of fishes in North American deserts have steadily declined in this century due to man's activities in this naturally fragile region. We address genetic aspects of the population structure of desert fishes as applicable to conservation and recovery programs by developing two zoogeographic models of isolation and gene flow. In the Death Valley model, populations are isolated, with no chance of natural gene flow among them. Genetic diversity within populations tends to be low, but genetic divergence among populations within a species is high. In the Stream Hierarchy model, a complicated hierarchical genetic structure exists and is a function of geographic proximity and connectivity of habitats. Within-habitat genetic diversity tends to be higher, and among-habitat differentiation lower, than in the Death Valley model. These two systems must be recognized as distinct and managed differently. We also suggest three areas of experimentation needed to better understand and manage genetic stocks of desert fishes: relationships between heterozygosity and fitness, experimental mixing of similar stocks to examine effects of increased heterozygosity, and analysis of the relative roles of genetic adaptation and phenotypic plasticity in local differentiation.*

Resumen: *El estatus y la seguridad de los peces en los desiertos Norteamericanos ha empeorado a un ritmo sostenido durante este siglo debido a las actividades del hombre en esta región de naturaleza frágil. Discutimos aspectos genéticos de la estructura poblacional de peces del desierto con respecto a su aplicabilidad a programas de conservación y recuperación mediante el desarrollo de dos modelos zoogeográficos de aislamiento y flujo genético. En el modelo de Death Valley, las poblaciones están aisladas, sin ninguna oportunidad para el flujo de material genético entre ellas. La diversidad genética dentro de estas poblaciones tiende a ser baja, pero la divergencia genética entre poblaciones de una especie dada es alta. En el modelo de stream hierarchy, existe una complicada estructura genética jerarquizada y ésta esta dada en función de la proximidad geográfica y la conectividad de habitats. La diversidad genética al interior de un habitat tiende a ser más alta y la diferenciación entre habitats más baja que las del modelo de Death Valley. Debe reconocerse que estos dos sistemas son distintos y deben ser manejados diferentemente. También sugerimos tres areas de experimentación requeridas para comprender y manejar mejor los stocks genéticos de peces del desierto: relaciones entre heterocigoticidad y adecuación, mezcla experimental de stocks similares para examinar los efectos al aumento de heterocigoticidad, y el análisis de los roles relativos de la adaptación genética y la plasticidad fenotípica con respecto en la diferenciación local.*

* *Correspondence should be addressed to G. K. Meffe.*

Introduction

Fishes in the Sonoran, Mohave, Great Basin, and Chihuahuan deserts of North America are declining at an alarming rate. At least 20 taxa have gone extinct in the last several decades (Meffe, in press), and many more are in immediate danger of similar fates (Williams et al. 1985). This major conservation problem has prompted governmental agencies and academic scientists to conduct research on basic ecologies and life histories, develop propagation programs for species perpetuation and reintroduction, and restore degraded habitats. We address here a different aspect of conservation of this highly endangered fauna, the application of population genetics to management and long-term perpetuation of desert fishes.

Desert Aquatic Systems

North American deserts are extremely arid, with high geographic relief, heavily dissected drainages, and widely dispersed surface waters. Natural aquatic systems in these regions are fragmented on both broad geographic and local scales. In addition to being isolated, desert aquatic systems can experience enormous physico-chemical fluctuations, as in streams and rivers (Minckley & Meffe 1987), or constant but harsh conditions, as in some springs (Deacon & Minckley 1974). The extreme isolation and severe abiotic conditions have led to a high degree of phenotypic divergence of fishes, often with small populations living in small habitats (Miller 1958). Many taxa are relicts from better-watered times that have been trapped in isolated springs and streams during the last 10,000- to 12,000-year post-pluvial period. This has resulted in a high degree of endemism (Williams et al. 1985).

Several genetic and demographic consequences of such an isolated, fragmented distribution are likely, including (1) local divergence via natural selection or genetic drift; (2) little or no gene flow among isolated demes that might otherwise moderate losses of genetic variability after population crashes; and (3) little or no recolonization of isolated habitats after local extinction. Southwestern desert fishes are thus a vulnerable, "extinction-prone" group due to the limited geographic range and isolation of many species.

Added to this natural fragility are major anthropogenic disturbances of two types: habitat destruction (impoundment and diversion of rivers, groundwater pumping, and drying of surface springs) and the introduction of a huge diversity of exotic fishes, many of whom prey on, compete with, or hybridize with native species (Naiman & Soltz 1981). Management of such a fauna over the long term must therefore consider the genetic consequences both of natural fragmentation with small population sizes and of artificial disruption by man. General and philosophical aspects of genetic management of fishes have been outlined elsewhere (Meffe 1986, 1987; Ryman & Utter 1987). Here we consider more specific approaches of genetic studies and experimental manipulations of rare and endangered southwestern fishes.

Divergence, Migration, and Gene Flow

Two demographic factors, genetically effective population size and migration rate, affect the degree of divergence among local colonies of fishes. Colonies diverge from one another as a consequence of local selection pressures, mutation, and random genetic drift; drift alone can cause considerable divergence among small colonies. The rate of divergence due to drift in turn depends on the genetically effective size of the local populations (N_e). N_e represents the number of breeding adults in each generation, and is affected by sex ratio, pattern of mating, and variance in reproductive output among individuals (Crow & Kimura 1970).

Gene flow via migration (m, or the proportion of individuals exchanged between colonies per generation) maintains genetic variation within colonies and retards divergence among colonies. Divergence occurs as a product of N_e and m. If N_e is small, colonies tend to diverge rapidly, as a result of random processes. High rates of gene flow (m) are needed to maintain genetic homogeneity among them. As long as $N_e m > 1$, local colonies will tend not to diverge significantly with regard to the types of alleles present (Allendorf 1983). Thus, a pair of large populations with mean $N_e > 10{,}000$ that exchange individuals at a rate of $m = 0.001$ will not diverge significantly by chance alone, since $N_e m > 10$. However, a pair of small populations with $N_e < 1000$ and with a similar rate of gene exchange would diverge, since $N_e < 1$.

This scenario is based on the assumption that colonies are dispersed geographically as if they were islands and gene flow can occur equally among all the islands. As we discuss below, this scenario may be an appropriate description for some aquatic systems (the Death Valley model), but for most freshwater fishes, riverine habitats impose a more complex population structure (the Stream Hierarchy model).

Zoogeographic Models of Gene Flow

We envision two models of genetical population structure in desert fishes. The existence of these different zoogeographic models must be recognized in management programs aimed at habitat restoration and restocking of fish populations.

The Death Valley Model

Death Valley and similar intermontane basins of the Southwest contain a series of extremely isolated springheads and small streams that were connected in pluvial times when the basins were filled by inland lakes and seas (Miller 1948; Hubbs & Miller 1948). As a consequence of natural post-Pleistocene drying trends, these habitats now contain remnant fish populations that were formerly part of a larger interbreeding population. Presently, geographical isolation among colonies is complete (m = 0; Fig. 1), and divergence is accelerated by small colony sizes and possibly strong, local selection pressures.

Total genetic diversity of a species in this model (H_t) consists of diversity within colonies (H_c) plus differences among colonies within the total range (D_{ct}) (Nei 1975). If a species is endemic to a single locality, then the latter term drops out and total genetic diversity consists of heterozygosity within the single site. As colonies lose variation and become more homozygous via genetic drift, random differences among colonies also accumulate through time. Since we envision no migration (m = 0) to counteract these processes, both the decline of H_c and the growth of D_{ct} are simple functions of N_e within colonies.

Depending on its mode of action, natural selection will alter the effects of genetic drift. Directional selection may accelerate divergence for locally adaptive traits, but divergence will be slowed for characters that are under some form of stabilizing or balancing selection throughout the range of the species. This model is exemplified by pupfishes (genus *Cyprinodon*) and killifishes (genus *Empetrichthys*), which are naturally isolated in many remnant springs and streams in Death Valley and adjacent basins in California and Nevada (Miller 1948).

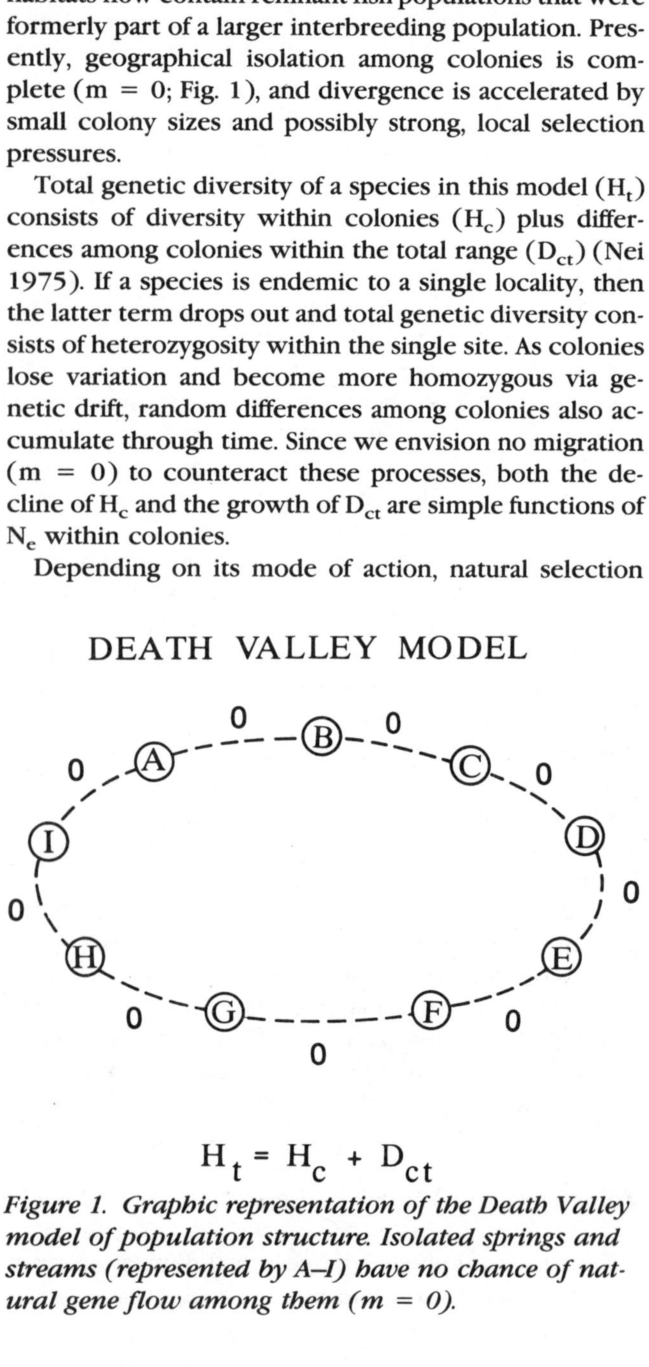

Figure 1. Graphic representation of the Death Valley model of population structure. Isolated springs and streams (represented by A–I) have no chance of natural gene flow among them (m = 0).

Considerable evidence exists for morphological, physiological, behavioral, and life historical differences among populations of desert pupfishes (Naiman & Soltz 1981), supporting this model. Turner (1974) was the first investigator to study genetic population structure in these fishes. His electrophoretic studies of pupfishes in and near Death Valley revealed little overall allelic variation. However, most variation occurred among, rather than within, these remnant populations. For example, *Cyprinodon nevadensis* contains three recognized subspecies that occur in isolated habitats of the Amargosa basin. Most of the thirty gene loci he examined were homozygous within populations, but four were polymorphic in one population. *Cyprinodon diabolis* and *C. salinas,* which apparently derived from *C. nevadensis* since the drying of Death Valley, presently occur in single isolated habitats, and each species appears to be homozygous at most loci for alleles that occur in *C. nevadensis.* However, each species also contains some unique alleles that may have arisen via mutation since isolation from the presumed ancestor. Unfortunately, one cannot compute genetic diversity statistics from the data presented in Turner's pioneering study, but in general the data are consistent with our model: very little polymorphism occurs within, and considerable divergence occurs among, populations.

In managing a genetic population structure represented by the Death Valley model, we need not be concerned with human interruption of gene flow. Such gene flow has historically not existed, at least not since ancestral lakes have desiccated. On the contrary, we should take precautions to avoid creating gene flow among populations that are naturally isolated and may therefore be in the process of adaptive radiation. Artificial gene flow would tend to reduce or eliminate genetic divergence among localities (D_{ct}), which may represent a substantial proportion of the genetic diversity in the species.

The major concern for fishes in the Death Valley model is maintenance of high N_e within localities. Crashes due to man's intervention have occurred, jeopardizing several species' existence. In August 1969, the Owens pupfish (*Cyprinodon radiosus*) nearly went extinct when its only known habitat completely dried (Miller & Pister 1971). Fortunately, some 800 fish were rescued, literally at the last minute, and held in captivity until spring flow resumed. Likewise, existence of the Devil's Hole pupfish (*Cyprinodon diabolis*) was jeopardized when the water level in its single habitat was being lowered by groundwater pumping. Only a Supreme Court decision to halt development in the region restored flow and maintained this habitat (Pister 1979, 1985). Thus, the goal for conservation efforts of fishes in this model is simply to maintain strong, isolated populations.

A more complicated version of the Death Valley model occurs with *Cyprinodon macularius* in the vicinity of the Salton Sea sink in southern California. Using Turner's (1983) electrophoretic data, Echelle, Echelle, and Edds (1987) were able to partition total genetic diversity into H_c and D_{ct} components. Polymorphism within lakeshore colonies accounted for about 70% of the total genetic variance, and differences among colonies accounted for about 30%, contrary to expectations of the Death Valley model. However, this is easily explained in the context of the hydrologic history of this region. These populations have been repeatedly subjected to isolation and possible reconnection during several rounds of drying and flooding in the lake basin. According to Turner (1983), Holocene Lake Cahuilla occupied this basin, followed by drying, followed by Wisconsinan Lake LeConte, followed by drying once again. From 1905 to 1907, the basin was flooded with water that broke out of the Colorado River irrigation system. Apparently, these cycles of reconnection have permitted the maintenance of significant variation within local populations. The recent flooding of the Sal-

STREAM HIERARCHY MODEL

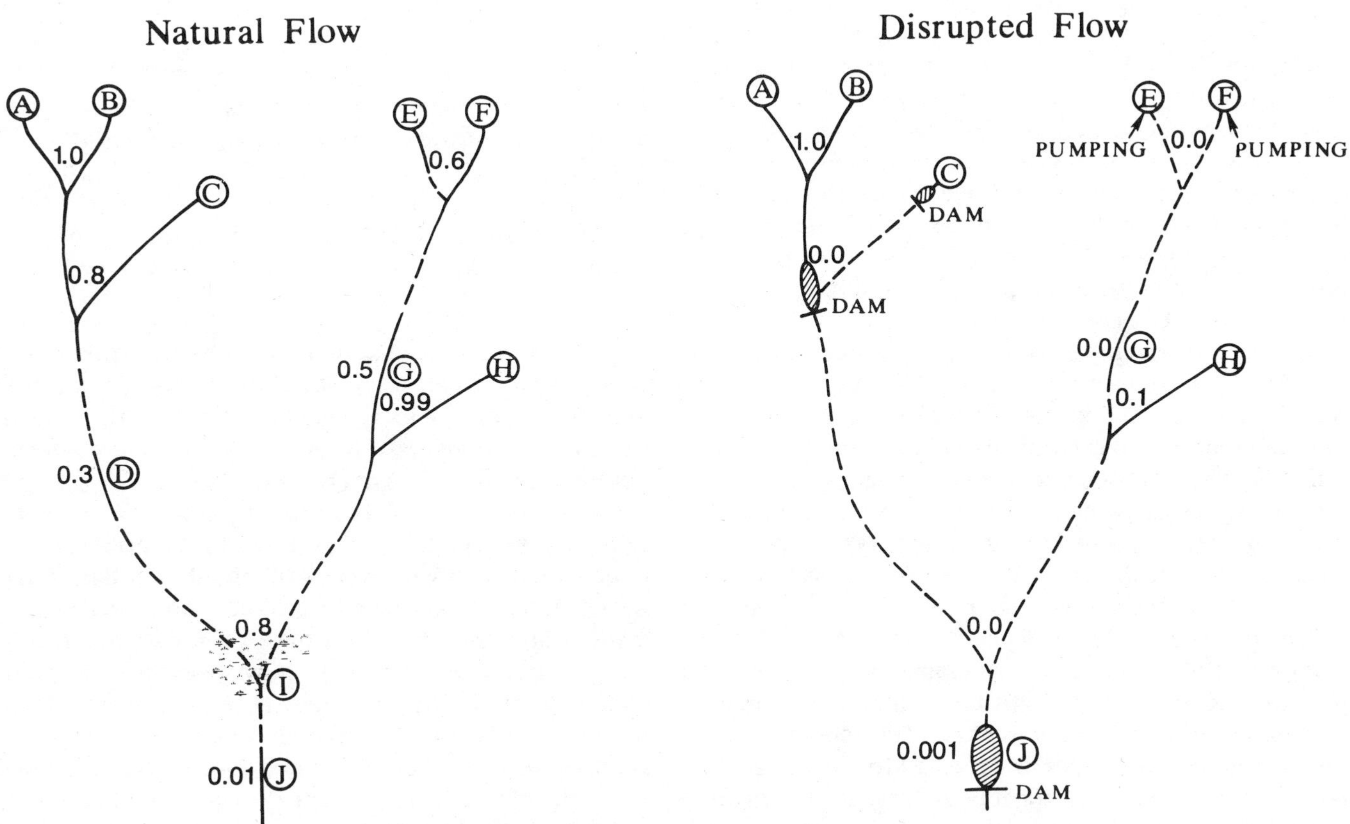

$$H_t = H_c + D_{cr} + D_{rs} + D_{st}$$

Figure 2. Graphic representation of the Stream Hierarchy model of population structure. (Left) Hypothetical natural situation. Sites A–J represent 10 natural populations separated by various probabilities of connectivity. Sites A, B, C, E, F, and H are springheads, I is a marsh, and the remainder are stream segments. Numbers refer to hypothetical probabilities of connection between that habitat and habitats further upstream over the long term, and should not be interpreted as estimates of gene flow (m). (Right) Hypothetical disruption of natural situation. In this case, dams and pumping of springheads have severely altered habitats and probabilities of gene flow. Two populations (D and I) are extinct, and most others have reduced connectivities.

ton Sea sink due to human intervention may also have prevented the pattern of local differentiation observed in Death Valley.

It is instructive that the most isolated population of *C. macularius,* in Quitobaquito Spring in the upper Sonoyta basin of Arizona, does exhibit a level of genetic differentiation that is roughly comparable to that observed among Death Valley pupfish populations (Turner 1983). This isolated population was recently described as a distinct subspecies, *C. macularius eremus* (Miller & Fuiman 1987).

The Stream Hierarchy Model

The second pattern of population structure occurs in desert fish that inhabit one or more dendritic river systems, with varying degrees of connectedness and gene flow (Fig. 2a). In this case, local demes are only partially isolated from other gene pools, with some probability of gene flow among them. Periodic or permanent connections allow some migration and gene flow among habitats, and local population divergence is less extreme than in the Death Valley model. A hierarchical genetic structure results, wherein higher frequencies of gene flow correspond to greater similarity among demes. This hierarchy of genetic divergence is a function of geographic connectedness of habitats: localities that are frequently connected and exchange population members will have greater genetic (and presumably phenotypic) similarities than localities that infrequently or never experience gene exchange.

In such a hierarchical system, total genetic diversity of a species (H_t) consists of within-colony diversity (H_c) plus differences among colonies within a river (D_{cr}) plus differences among rivers within major drainage systems (D_{rs}) plus divergence among drainages occupied by the species as a whole (D_{st}). More complex hierarchies may develop depending upon levels of connectivity and extent of the species' range. During wet years, patches connect and allow migration by fishes; dry years isolate habitats and stop gene exchange. Natural climatic variation thus connects and separates local demes in a hierarchical structure that is a function of geographic distance and river discharge. This model is typified by numerous species in the Gila River drainage of southern Arizona and probably represents fishes in many lotic systems in the southwest.

Interruption of the stream hierarchy by habitat alteration can change patterns of gene flow among localities and disrupt natural genetic structure in two ways. Movement of fishes among naturally isolated populations introduces gene flow that historically might not have occurred, thus decreasing D_{cr}, D_{rs}, or D_{st}. Conversely, impoundment of rivers, pumping of springheads, or other alteration of drainage patterns can fragment and isolate populations that would normally experience gene exchange (Fig. 2b). Such actions should cause a general decline in local genetic diversity (H_c) and a corresponding increase in divergence among colonies within a drainage system (D_{cr} and D_{rs}).

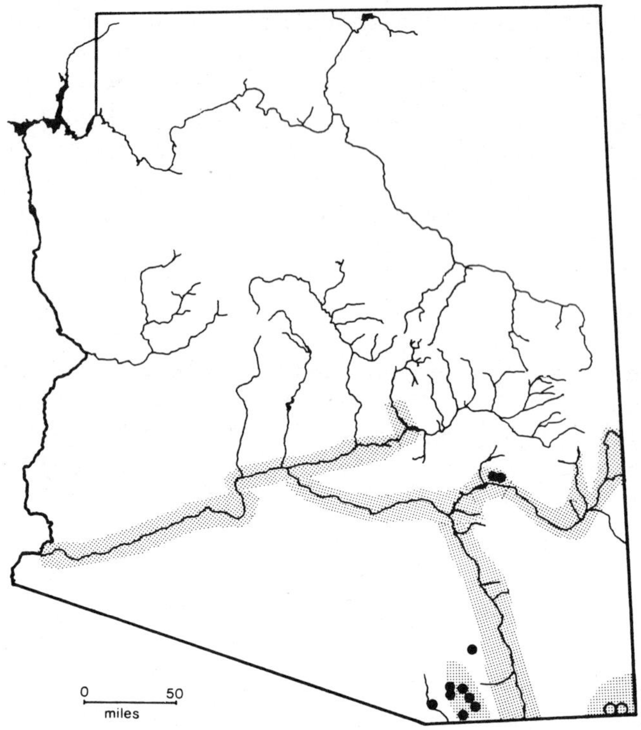

Figure 3. Distribution of Poeciliopsis occidentalis *in Arizona. The historical distribution, as of about 1940, is shaded; circles are distribution as of late 1987. Solid circles are the Gila subspecies (*P. o. occidentalis, *genetic group I) and open circles are the Yaqui subspecies (*P. o. sonoriensis, *genetic group II).*

Habitat destruction resulting in population segregation is presently the greatest threat to the preservation of genetic diversity in river-dwelling fishes of the American Southwest. Because of limited water availability, many southwestern streams have been dammed or diverted, completely changing flow characteristics and connectivity patterns (Minckley & Meffe 1987), and extensive groundwater pumping has depleted ancient aquifers and caused many isolated springs to cease flowing (Williams et al. 1985; Pister & Unkel, in press).

An example of hierarchical population structure in an endangered fish species occurs in the Sonoran topminnow, *Poeciliopsis occidentalis.* This livebearing fish is native to the lower Colorado River drainage of Arizona and New Mexico and to several major drainages in Sonora, Mexico. It was extremely abundant through the Gila River basin of southern Arizona, including numerous springs along watercourses, as recently as the 1940s (Hubbs & Miller 1941; Minckley & Deacon 1968; Fig.

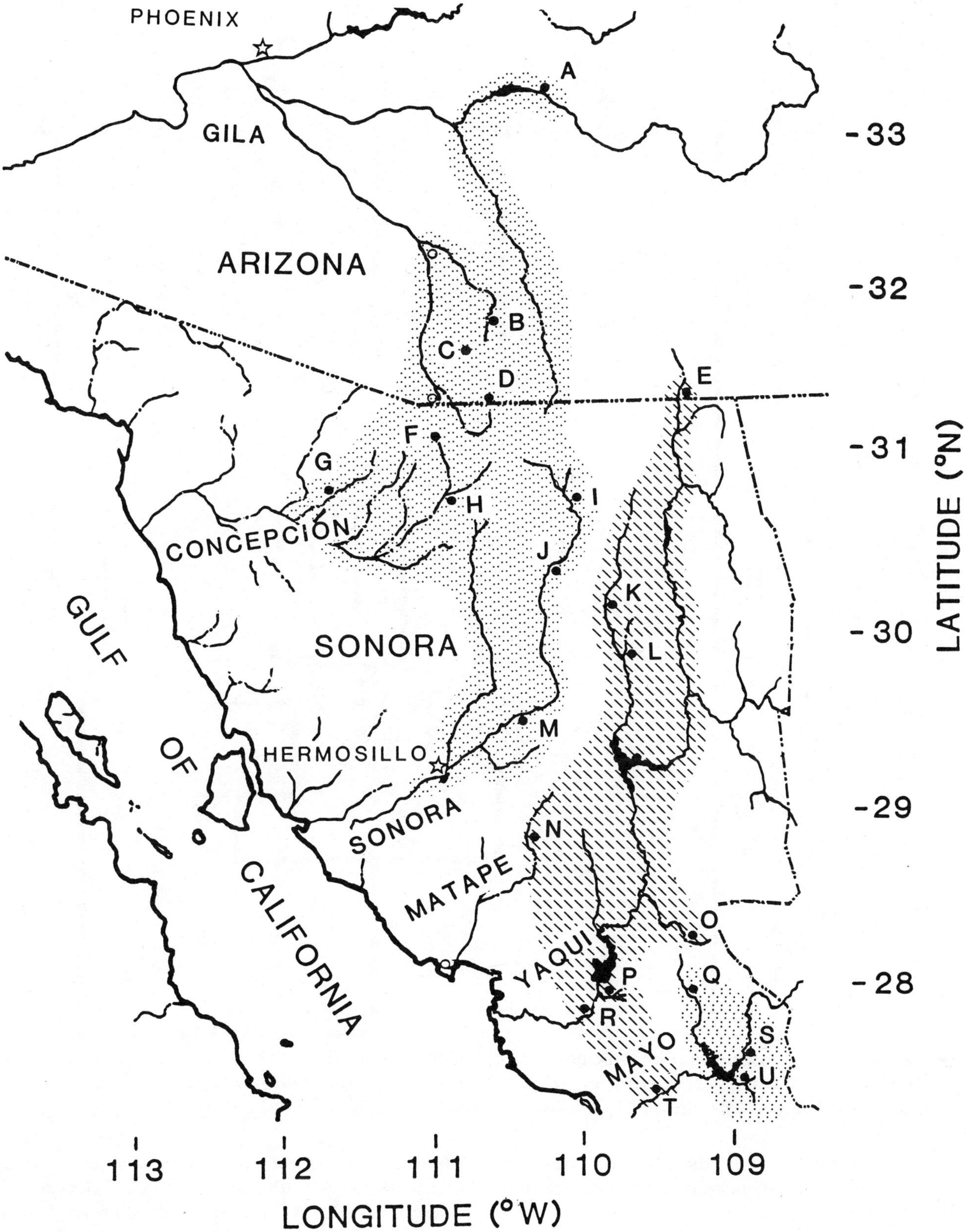
PHOENIX
GILA
ARIZONA
SONORA
CONCEPCION
HERMOSILLO
SONORA
MATAPE
YAQUI
MAYO
GULF OF CALIFORNIA
A
B
C
D
E
F
G
H
I
J
K
L
M
N
O
P
Q
R
S
T
U
-33
-32
-31
-30
-29
-28
LATITUDE (°N)
113
112
111
110
109
LONGITUDE (°W)

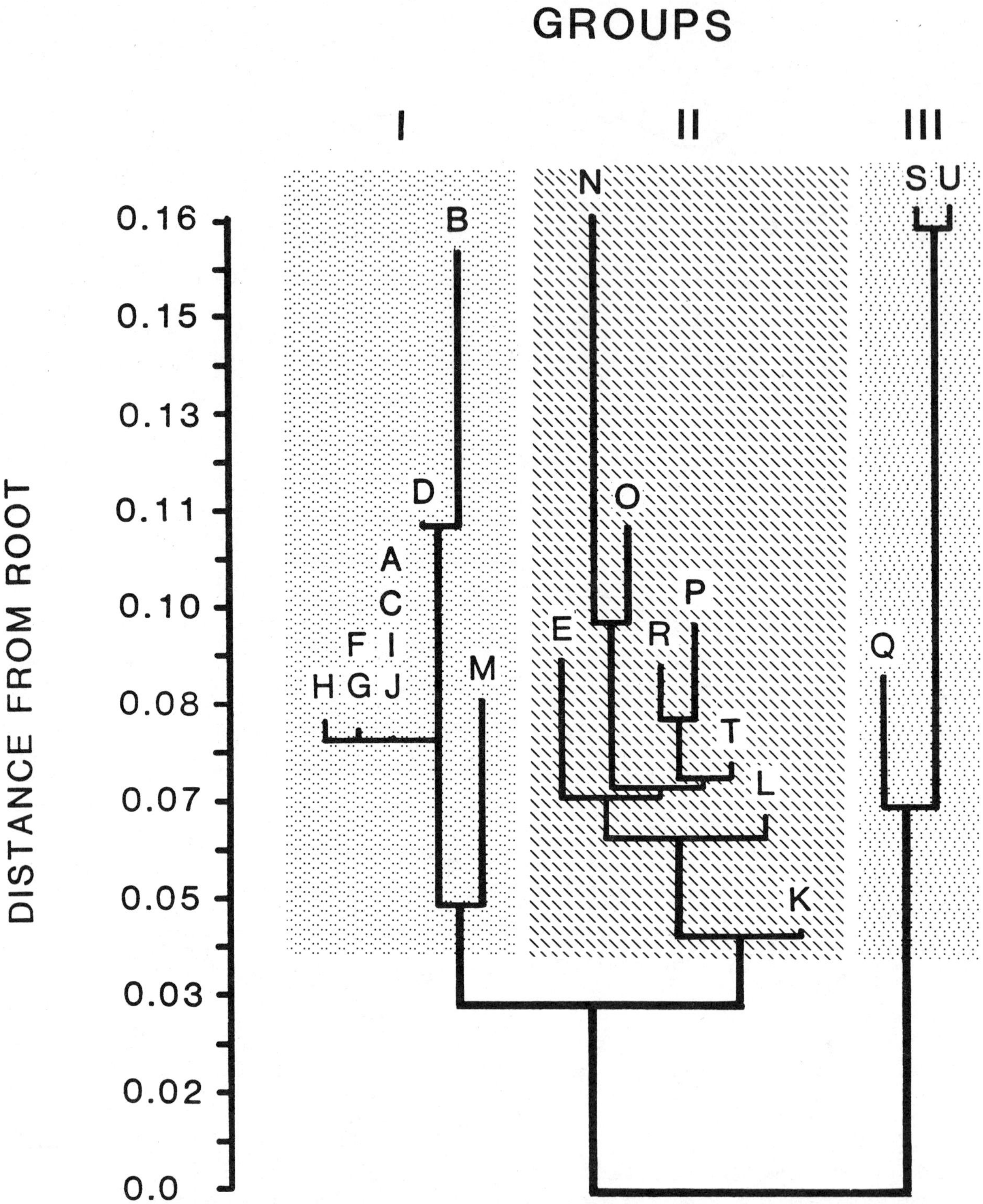

Figure 4. Genetic relationships among Poeciliopsis occidentalis *populations. The dendogram produced three clusters that correspond with geographical distributions on the map. Apparent nonlinearity of the dendogram scale is due to rounding. (Reprinted with permission from Vrijenhoek, Douglas, & Meffe 1985.)*

3). A rapid decline ensued over the last four decades to the point that the fish was included on the Federal Endangered Species List in 1973.

Decline of the topminnow is attributed to widespread habitat destruction and replacement by the introduced mosquitofish (*Gambusia affinis*) (Meffe et al. 1983; Meffe 1985). In 1983 the topminnow occurred in 15 natural habitats in the United States. Presently, the fish occurs in 12 natural localities in Arizona (Fig. 3), all of which are isolated headwater springs or streams, but it remains abundant in Mexico where mosquitofish have not been systematically introduced.

Vrijenhoek et al. (1985) studied genetic diversity in five remnant Arizona populations of Sonoran topminnows and 16 populations throughout its range in Mexico. In general, local diversity (H_c) declined at the northern periphery of the species' range and in upstream sections within rivers (Fig. 4, Appendix). Fluctuations in population size that lead to losses of genetic variance are more likely to occur in these peripherally isolated montane localities.

This broadly distributed riverine species exhibits the expected hierarchical population structure. Pairwise genetic distances among the 21 populations (Appendix) were clustered by the Distance-Wagner method, resulting in three genetically defined groups (Fig. 4). Group I consists of fishes in the Gila River of Arizona and riós de la Concepcion and Sonora, of Mexico. Fishes from the Rió Matape, Rió Yaqui, and the lower Rió Mayo constituted Group II, and fishes from the Upper Rió Mayo are in Group III. The first two groups correspond to morphologically defined subspecies previously recognized (Minckley 1973); the third group, in the partially isolated headwaters of the Rió Mayo, was the most genetically divergent.

We performed a hierarchical partitioning of genetic diversity using the differences among the three groups as the highest level of the hierarchy (D_{st}). The bulk of genetic diversity in this species is represented by divergence between the three geographical groups (52.8%; Fig. 4); 21.3% of the variance is contained within colonies (H_c), 25.5% is due to differences between colonies within rivers (D_{cr}), and only 0.4% is attributable to differences between rivers within the major subdivisions (D_{rs}). This hierarchical approach, which recognizes the degree of connectedness between habitat patches, indicates that conservation of genetic diversity in this species should distinguish geographic groupings and include plans for their continued separation. Combination of fishes from different groups could erode the largest proportion of genetic variance in the species; management and manipulation should thus be conducted only within each grouping.

Sonoran topminnows have been propagated at the Dexter National Fish Hatchery near Dexter, New Mexico, since 1976 for reintroduction throughout the natural range of the species in Arizona (Minckley & Jensen 1985). At least 100 field sites had been stocked through 1985 with fishes from Dexter (Brooks 1986; Simons 1987).

The initial stock for hatchery breeding originated in Monkey Spring, Santa Cruz County, Arizona (site C, Fig. 4). This is an isolated, thermally stable springhead in the Santa Cruz drainage that has probably not fluctuated in temperature more than 1 C for thousands of years (Constantz 1975). Topminnows in Monkey Spring have no detectable allelic diversity at the 25 loci examined in our study (Appendix). This genetically invariant group also has low fecundity (Schoenherr 1977; Constantz 1979) and is likely to be physiologically intolerant due to its constant environment. We recommended that the hatchery stock be replaced with fish from Sharp Spring (site D), a population from the same genetic subdivision (Vrijenhoek, Douglas, and Meffe 1985). The Sharp Spring population has higher fecundity (Meffe 1985) and growth rate (Vrijenhoek & Sadowski, unpublished data) than the Monkey Spring stock. This population is also a better candidate for broad-scale reintroduction because it naturally experiences wide environmental fluctuations (Meffe et al. 1983; Meffe 1984) and has detectable allelic variation (H = 3.7%; Appendix). Our recommendation has been followed, and as of 1986 Sharp Spring fish constitute the hatchery stock for all introductions and experimental work within the Group I region.

A special case of the Stream Hierarchy model is the situation in which a species occurs in only one river system. As in the Death Valley model, total genetic diversity can be partitioned into two components:

$$H_t = H_c + D_{ct}$$

Because connections can occur between neighboring stream sites during annual rainy seasons, or between more distant sites during periods of extreme flooding, we expect much less diversification among populations than was observed in Death Valley pupfishes.

Echelle and coworkers (1987) recently described such a population structure for four pupfish species inhabiting streams in the Chihuahuan desert of New Mexico and Texas. *Cyprinodon pecosensis* is the most broadly distributed of these four species, and only 7.7% of its total genetic diversity is attributable to differences among populations spread along a 200 km portion of the Pecos River in Texas. *Cyprinodon elegans* and *C. bovinus* are each restricted to single creeks within the Pecos River drainage. The differences among local colonies (D_{ct}) that are only a few kilometers apart accounted for 10.8% (*C. elegans*) and 1.4% (*C. bovinus*) of the total genetic diversity in each species. *Cyprinodon tularosa* is restricted to several partially isolated habitats that are spread over several hundred kilometers in the Salt Creek drainage of New Mexico, and D_{ct} accounts for 19.0% of its total genetic diversity.

Assuming that the values of D_{ct} obtained for these four species reflect an equilibrium between the divergent effects of genetic drift and the convergent effects of gene flow, we can roughly estimate the effective migration rate (N_em) among colonies from the equation $D_{ct}/H_t = 1/(4N_em + 1)$ (modified from Crow & Kimura 1970). For *C. tularosa*, N_em is approximately 1, which can be interpreted as the equivalent of one migrant between populations per generation. The effective migration rate is much higher along the 10 km portion of Salt Creek inhabited by *C. bovinus,* being roughly equivalent to an exchange of 17 to 18 individuals between colonies per generation. Values for *C. pecosensis* and *C. elegans,* respectively, are three and two individuals per generation.

These estimates, however, are sensitive to the underlying model of population structure (Slatkin 1985). Although accurate in a true island situation, they probably underestimate gene flow in a hierarchical situation, and more realistic models of population structure are needed. Even if such models were available, though, it is likely that each dendritic system would have idiosyncratic properties that would require simulation to more accurately estimate gene flow. Thus, no single analytical approach would pertain in every case. Nevertheless, a useful rule of thumb is that an exchange of one individual between populations per generation within a river system would maintain most of the allelic diversity, but would not swamp the opportunity for local adaptation (Allendorf 1983).

Management of species such as these *Cyprinodon* should be aimed at preserving the integrity of each species while maintaining its genetic variability. Movement of fish of the same species among colonies within a river (except, perhaps, for *C. tularosa,* which has the highest D_{ct}) would probably pose no specific problems since this appears to occur naturally. However, precautions should be taken to avoid mixing between species. If mixed, it is likely that all four of these *Cyprinodon* species could hybridize (see Turner & Liu 1977; Cokendolpher 1980); introduced *Cyprinodon variegatus* have already hybridized with native *C. pecosensis* (Echelle, Echelle, & Edds 1987).

Into the Future—How Do We Proceed with Management of Southwestern Fishes?

The coming decades will present tremendous challenges to the manager of dwindling aquatic resources in the Southwest. Continued population growth in the region will further drain limited natural resources, the spread of exotics will further influence local biota, and the status of native species will certainly continue to decline. However, careful management, based on sound genetic and ecological studies, can slow or reverse these negative trends. One of the more powerful tools in the effort to maintain healthy, diverse populations is the application of genetic principles to management.

We particularly emphasize the need to incorporate experimental studies of population genetics and fitness into management of endangered fishes to better predict the survival and long-term success of remnant populations. Such research would directly benefit the target species, and the principles could be applied to management of other taxa as well. Desert fishes provide opportunities that do not exist with most endangered species. Populations of whooping cranes, cheetahs, or condors cannot be manipulated to determine optimal genetic structure. However, such studies are feasible with many endangered fishes. We thus recommend pursuit of at least three areas of experimentation that would benefit management of threatened and endangered fishes in the Southwest.

1. *Test the relationship between heterozygosity and individual fitness.* Studies with a variety of organisms have resulted in a "heterozygosity concensus," the belief that overall fitness is enhanced by genetic diversity (Frankel & Soulé 1981; Mitton & Grant 1984). However, we know little of actual mechanisms of heterozygote superiority, and we do not know whether all species or populations are affected similarly by heterozygosity. Experimental studies of relationships between genetic diversity and growth rates, survivorship, fecundity, developmental stability, and competitive ability are needed to determine the importance of maintaining populations with high heterozygosity. Such studies are currently underway in our laboratories with poeciliid fishes, and early results indicate superior fitness of more heterozygous individuals in some cases; we encourage exploration of such questions in other groups.

2. *Experimentally mix genetic stocks.* Although we strongly discourage mixing stocks among major genetic subdivisions, we encourage stocking **experiments** that combine populations from within the same subdivision. Such experiments would permit evaluation of the effects of mild outcrossing and may prove to be valuable in restocking efforts. For example, in *P. occidentalis* from Group I in Arizona, Ciénega Creek and Monkey Spring fish are both homozygous for all loci examined (Appendix). Preliminary laboratory results indicate that Monkey Spring fish may grow slower than individuals from the more polymorphic Sharp Spring population (Vrijenhoek & Sadowski, unpublished data). Because they are fixed for alternative alleles at two loci, crosses of Ciénega Creek (CC) and Monkey Spring (MS) stocks should increase heterozygosity for these and many other loci. Tests of relative fitness of the CC/MS "hybrids," the parental stocks, and the Sharp Spring stock would indicate how outcrossing and concomitant increase in heterozygosity affects fitness. If such experimental crosses prove to be more robust than single-locality fish, their use in stocking new field sites should be considered. Existing natural localities should always remain genetically pure, however.

tionists," attempting to develop and implement a policy that would restore a system to an approximation of its former self. In a stream hierarchical system that has undergone radical changes to the point that some natural populations are extinct and others are artificially isolated, we favor restoration of gene flow and diversity to mimic the natural situation, rather than maintaining an unnatural and zoogeographically perverse "status quo." Estimates of effective migration rates from genetic structure statistics can give us a rough guideline for the extent of gene flow that should be considered.

3. *Test the relative roles of genetic adaptation and developmental plasticity in local phenotypic differentiation.* Desert fishes are notorious for phenotypic differentiation among localities (Miller 1948; Minckley 1973). This differentiation may be the result of genetic differences or plastic developmental processes influenced by the local environment. Identification of the causes of phenotypic differentiation is important to management of individual populations. If differentiation is genetic, it is critical to maintain the diversity represented by differences among colonies (D_{cr}). If differentiation is simply a plastic response to local conditions, then a particular phenotype may be induced from a broad range of genotypes by creating the proper environment. The latter may occur in a goodeid fish in Mexico (Turner et al. 1983), a cichlid in Cuatro Ciénegas, Mexico (Kornfield et al. 1982; Liem and Kaufman 1984), several African cichlids (Greenwood 1965), and Artic charr (Vrijenhoek, Marteinsdottir, and Schenck 1987). Thus, we need critical experiments that separate true genetic differentiation (local evolutionary change) from simple environmental induction of unique morphs.

A Final Question

We conclude by raising a question, one that we cannot definitively answer but one that is critical to all conservation programs. What are our objectives in conservation biology? Specifically, what should we conserve? Species? Subspecies? Individual populations? Unique alleles? This is a difficult question, and perhaps the most important philosophical issue facing us, for we cannot design effective conservation programs until we identify specific goals.

The answer may depend upon the organism of concern, and economic constraints. For a tiger or rhinoceros, we may have to settle for a representative type of the species, for it may be difficult to do otherwise, since natural and financial resources are limiting. It is also unrealistic to expect to conduct experimental work on the genetic structure of such taxa. However, we can do much more with desert fishes. It is far less costly to maintain regional, local, or even allelic diversity in these fishes—they can be easily manipulated, local habitats are available or can be made available, and fish can be kept in small refugia such as hatcheries or even aquaria.

In particular, the Sonoran topminnow is a model organism with which to develop a conservation program with a sound genetic basis. The topminnow is more abundant than many endangered species, and much is known about its ecology, physiology, and genetics (Constantz 1975; Minckley, Rinne, & Johnson 1977; Schoenherr 1977; Bulger & Schultz 1982; Meffe 1984, 1985; Vrijenhoek, Douglas, & Meffe 1985). It is easy to raise in captivity and hardy under field and experimental conditions. Concerted efforts in recovery of this species, currently underway by several groups, could benefit other endangered species programs by developing sound management schemes, including application of genetic principles and field and laboratory experimentation of fitness responses to management programs. With this species at the forefront, we have an excellent opportunity to restore an entire endangered fauna to some semblance of its historical condition; as conservation biologists, this should be our goal.

Acknowledgments

We thank M. E. Douglas for his efforts in fieldwork and analysis of original *P. occidentalis* samples and J. B. Coleman and L. Orebaugh for assistance with figures. Two anonymous reviewers raised important points for clarification. GKM was supported by contract DE-AC09-76SR00-819 between the U.S. Department of Energy and the University of Georgia. RCV was supported by National Science Foundation grant BSR86-00661.

Literature Cited

Allendorf, F. W. 1983. Isolation, gene flow, and genetic differentiation among populations. Pages 51–65 *in* C. M. Schonewald-Cox, S. M. Chambers, B. MacBryde, and L. Thomas, editors. Genetics and Conservation: A Reference for Managing Wild Animal and Plant Populations. Benjamin/Cummings, Menlo Park, California.

Brooks, J. E. 1986. Status of natural and introduced Sonoran topminnow (*Poeciliopsis o. occidentalis*) populations in Arizona through 1985. Report to Offices of Endangered Species and Fishery Resources, U.S. Fish & Wildlife Service, Albuquerque, New Mexico.

Bulger, A. J., and R. J. Schultz. 1982. Origin of thermal adaptations in northern versus southern populations of a unisexual hybrid fish. Evolution **36:**1041–1050.

Cokendolpher, J. C. 1980. Hybridization experiments with the genus *Cyprinodon* (Teleostei: Cyprinodontidae). Copeia **1980:**173–176.

Constantz, G. D. 1975. Behavioral ecology of mating in the male Gila topminnow, *Poeciliopsis occidentalis* (Cyprinodontiformes: Poeciliidae). Ecology **56:**966–973.

wald-Cox, S. M. Chambers, B. MacBryde, and L. Thomas, editors. Genetics and Conservation: A Reference for Managing Wild Animal and Plant Populations. Benjamin/Cummings, Menlo Park, California.

Brooks, J. E. 1986. Status of natural and introduced Sonoran topminnow (*Poeciliopsis o. occidentalis*) populations in Arizona through 1985. Report to Offices of Endangered Species and Fishery Resources, U.S. Fish & Wildlife Service, Albuquerque, New Mexico.

Bulger, A. J., and R. J. Schultz. 1982. Origin of thermal adaptations in northern versus southern populations of a unisexual hybrid fish. Evolution **36:**1041–1050.

Cokendolpher, J. C. 1980. Hybridization experiments with the genus *Cyprinodon* (Teleostei: Cyprinodontidae). Copeia **1980:**173–176.

Constantz, G. D. 1975. Behavioral ecology of mating in the male Gila topminnow, *Poeciliopsis occidentalis* (Cyprinodontiformes: Poeciliidae). Ecology **56:**966–973.

Constantz, G. D. 1979. Life history patterns of a livebearing fish in contrasting environments. Oecologia **40:**189–201.

Crow, J. F., and M. Kimura. 1970. An Introduction to Population Genetics Theory. Harper and Row, New York.

Deacon, J. E., and W. L. Minckley. 1974. Desert fishes. Pages 385–488 *in* G. W. Brown, Jr., editor. Desert Biology. Volume 2. Academic Press, New York.

Echelle, A. A., A. F. Echelle, and D. R. Edds. 1987. Population structure of four pupfish species (Cyprinodontidae: *Cyprinodon*) from the Chihuahuan Desert region of New Mexico and Texas: Allozymic variation. Copeia **1987:**668–681.

Frankel, O. H., and M. E. Soulé. 1981. Conservation and Evolution. Cambridge University Press, New York.

Greenwood, P. H. 1965. Environmental effects on the pharyngeal mill of a cichlid fish, *Astatorheochromis alluaudi,* and their taxonomic implications. Proceedings of the Linnaean Society of London **176:**1–10.

Hubbs, C. L., and R. R. Miller. 1941. Studies of the order Cyprinodontes, XVII. Genera and species of the Colorado River system. Occasional Papers of the Museum of Zoology, University of Michigan **433:**1–9.

Hubbs, C. L., and R. R. Miller. 1948. The zoological evidence: Correlation between fish distribution and hydrographic history in the desert basins of western United States. *In* The Great Basin, with Emphasis on Glacial and Postglacial Times. Bulletin of the University of Utah **38:**17–166.

Kornfield, I. L., D. C. Smith, P. S. Gagnon, and J. N. Taylor. 1982. The cichlid fish of Cuatro Ciénegas, Mexico: Direct evidence of conspecificity among distinct morphs. Evolution **36:**658–664.

Liem, K. F., and L. S. Kaufman. 1984. Intraspecific macroevolution: Functional biology of the polymorphic cichlid species *Cichlosoma minckleyi.* Pages 203–216 *in* A. A. Echelle and I. Kornfield, editors. Evolution of Fish Species Flocks. University of Maine Press, Orono.

Meffe, G. K. 1984. Effects of abiotic disturbance on coexistence of predator-prey fish species. Ecology **65:**1525–1534.

Meffe, G. K. 1985. Predation and species replacement in American southwestern fishes: A case study. Southwestern Naturalist **30:**173–187.

Meffe, G. K. 1986. Conservation genetics and the management of endangered fishes. Fisheries **11(1):**14–23.

Meffe, G. K. 1987. Conserving fish genomes: Philosophies and practices. Environmental Biology of Fishes **18:**3–9.

Meffe, G. K. In press. Fish utilization of springs and ciénegas in the arid Southwest. *In* R. R. Sharitz and J. W. Gibbons, editors. Freshwater Wetlands and Wildlife: Perspectives on Natural, Managed and Degraded Ecosystems. Office of Scientific and Technical Information, U.S. Department of Energy. Oak Ridge, Tennessee.

Meffe, G. K., D. A. Hendrickson, W. L. Minckley, and J. N. Rinne. 1983. Factors resulting in decline of the endangered Sonoran topminnow *Poeciliopsis occidentalis* (Atheriniformes: Poeciliidae) in the United States. Biological Conservation **25:**135–159.

Miller, R. R. 1948. The cyprinodont fishes of the Death Valley system of eastern California and southwestern Nevada. Miscellaneous Publications of the Museum of Zoology, University of Michigan **68:**1–155.

Miller, R. R. 1958. Origin and affinities of the freshwater fish fauna of western North America. Pages 187–222 *in* C. L. Hubbs, editor. Zoogeography. American Association for the Advancement of Science Publication 51.

Miller, R. R., and L. A. Fuiman. 1987. Description and conservation status of *Cyprinodon macularius eremus,* a new subspecies of pupfish from Organ Pipe Cactus National Monument, Arizona. Copeia **1987(3):**593–609.

Miller, R. R., and E. P. Pister. 1971. Management of the Owens pupfish, *Cyprinodon radiosus,* in Mono County, California. Transactions of the American Fisheries Society **100:**502–509.

Minckley, W. L. 1973. Fishes of Arizona. Arizona Game & Fish Department, Phoenix.

Minckley, W. L., and J. E. Deacon. 1968. Southwestern fishes and the enigma of 'endangered species.' Science **159:**1424–1432.

Minckley, W. L., and B. L. Jensen. 1985. Replacement of Sonoran topminnow by Pecos gambusia under hatchery conditions. Southwestern Naturalist **30:**465–466.

Minckley, W. L., and G. K. Meffe. 1987. Differential selection by flooding in stream-fish communities of the arid American Southwest. Pages 93–104 *in* W. J. Matthews and D. C. Heins, editors. Community and Evolutionary Ecology of North American Stream Fishes, University of Oklahoma Press, Norman.

Minckley, W. L., J. N. Rinne, and J. E. Johnson. 1977. Status of the Gila topminnow and its co-occurrence with mosquitofish. U.S.D.A. Forest Service Research Paper RM-**198:**1–8.

Mitton, J. B., and M. C. Grant. 1984. Associations among protein heterozygosity, growth rate, and developmental homeostasis. Annual Review of Ecology and Systematics **15:**479–499.

Naiman, R. J., and D. L. Soltz, editors. 1981. Fishes in North American Deserts. John Wiley, New York.

Nei, M. 1975. Molecular Population Genetics and Evolution. American Elsevier, New York.

Pister, E. P. 1979. Endangered species: Costs and benefits. Environmental Ethics **1:**341–352.

Pister, E. P. 1985. Desert pupfishes: Reflections on reality, desirability, and conscience. Environmental Biology of Fishes **12:**3–11.

Pister, E. P., and W. C. Unkel. In press. Conservation of western wetlands and their fishes. *In* R. R. Sharitz and J. W. Gibbons, editors. Freshwater Wetlands and Wildlife: Perspectives on Natural, Managed and Degraded Ecosystems. Office of Scientific and Technical Information, U.S. Department of Energy. Oak Ridge, Tennessee.

Ryman, N., and F. Utter. 1987. Population Genetics and Fisheries Management. University of Washington Press, Seattle.

Schoenherr, A. A. 1977. Density dependent and density independent regulation of reproduction in the Gila topminnow, *Poeciliopsis occidentalis* (Baird and Girard). Ecology **58:**438–444.

Simons, L. H. 1987. Status of the Gila topminnow (*Poeciliopsis occidentalis occidentalis*) in the United States. Special Report on Project E-1 Title VI of the Endangered Species Act of 1973, as amended. Arizona Game & Fish Department, Phoenix.

Slatkin, M. 1985. Gene flow in natural populations. Annual Review of Ecology and Systematics **16:**393–430.

Turner, B. J. 1974. Genetic divergance of Death Valley pupfish species: Biochemical versus morphological evidence. Evolution **28:**281–294.

Turner, B. J. 1983. Genic variation and differentiation of remnent natural populations of the desert pupfish, *C. macularius.* Evolution **37:**690–700.

Turner, B. J., and R. K. Liu. 1977. Extensive interspecific compatibility in the New World killifish genus *Cyprinodon.* Copeia **1977:**259–269.

Turner, B. J., T. A. Grudzien, K. P. Adkisson, and M. M. White. 1983. Evolutionary genetics of trophic differentation in goodeid fishes of the genus *Ilyodon.* Environmental Biology of Fishes **9:**159–172.

Vrijenhoek, R. C., M. E. Douglas., and G. K. Meffe. 1985. Conservation genetics of endangered fish populations in Arizona. Science **229:**400–402.

Vrijenhoek, R. C., G. Marteinsdottir, and R. A. Schenck. 1987. Genotypic and phenotypic aspects of niche diversification in fishes. Pages 245–250 *in* W. J. Matthews and D. C. Heins, editors. Evolutionary and Community Ecology of North American Stream Fishes. University of Oklahoma Press, Norman.

Williams, J. E., D. B. Bowman, J. E. Brooks, et al. 1985. Endangered aquatic ecosystems in North American deserts with a list of vanishing fishes of the region. Journal of the Arizona-Nevada Academy of Science **20:**1–62.

Appendix

Allelic frequencies in *Poeciliopsis occidentalis* (in total, 25 loci were examined; see Vrijenhoek etal. 1977). Sample size (N), percentage of loci that are polymorphic (P), and mean percentage of heterozygous loci per individual (H) are given for each site.

RIVER:		*Gila*				*Concepción*			*Sonora*			*Matape*	*Yaqui*						*Mayo*			
Site[a]:		*A*	*B*	*C*	*D*	*F*	*G*	*H*	*I*	*J*	*M*	*N*	*E*	*K*	*L*	*O*	*P*	*R*	*Q*	*S*	*T*	*U*
	N:	*24*	*22*	*23*	*62*	*30*	*68*	*30*	*30*	*30*	*60*	*60*	*24*	*30*	*44*	*30*	*30*	*110*	*30*	*30*	*30*	*30*
Ldh-1	a	1.00	1.00	1.00	1.00	1.00	1.00	1.00	1.00	1.00	1.00	.70	1.00	1.00	1.00	.98	.83	.76	1.00	1.00	.94	1.00
	c	—	—	—	—	—	—	—	—	—	—	.30	—	—	—	.02	.17	.24	—	—	.60	—
Ldh-2	b	1.00	1.00	1.00	1.00	1.00	1.00	1.00	1.00	1.00	1.00	1.00	1.00	1.00	1.00	1.00	.63	.96	1.00	1.00	1.00	1.00
	c	—	—	—	—	—	—	—	—	—	—	—	—	—	—	—	.37	.04	—	—	—	—
Pgd	f	—	1.00	—	.55	.02	—	—	—	—	—	—	—	—	—	.53	—	—	—	—	—	—
	a	1.00	—	1.00	.45	.98	1.00	1.00	1.00	1.00	1.00	.79	1.00	1.00	1.00	.47	1.00	1.00	1.00	1.00	1.00	1.00
	b	—	—	—	—	—	—	—	—	—	—	.21	—	—	—	—	—	—	—	—	—	—
Pgm	a	—	—	—	—	—	.02	—	—	—	—	—	—	—	—	—	—	—	—	—	—	—
	b	1.00	1.00	1.00	1.00	1.00	.98	1.00	1.00	1.00	.70	—	—	.31	.21	—	—	—	—	—	—	—
	c	—	—	—	—	—	—	—	—	—	.30	1.00	1.00	.69	.79	.95	1.00	1.00	1.00	1.00	1.00	1.00
	d	—	—	—	—	—	—	—	—	—	—	—	—	—	—	.05	—	—	—	—	—	
Gpi-1	a	—	—	—	—	—	—	—	—	—	.29	.27	—	—	—	—	.10	.10	—	—	—	.06
	b	1.00	1.00	1.00	1.00	1.00	1.00	1.00	1.00	1.00	.71	.73	1.00	1.00	1.00	1.00	.90	.90	1.00	1.00	1.00	.94
Gpi-2	a	1.00	1.00	1.00	1.00	1.00	1.00	1.00	1.00	1.00	1.00	.93	1.00	1.00	1.00	1.00	.88	.96	1.00	1.00	1.00	1.00
	c	—	—	—	—	—	—	—	—	—	—	.07	—	—	—	—	.12	.04	—	—	—	—
Aat-3	f	—	—	—	—	—	.01	—	—	—	—	—	—	—	—	—	—	—	—	—	—	—
	b	1.00	1.00	1.00	1.00	1.00	.99	1.00	1.00	1.00	1.00	1.00	1.00	1.00	1.00	1.00	1.00	1.00	1.00	1.00	1.00	1.00

Appendix—*Continued*

RIVER:		*Gila*				*Concepción*			*Sonora*			*Matape*	*Yaqui*						*Mayo*			
Site[a]:		*A*	*B*	*C*	*D*	*F*	*G*	*H*	*I*	*J*	*M*	*N*	*E*	*K*	*L*	*O*	*P*	*R*	*Q*	*S*	*T*	*U*
N:		*24*	*22*	*23*	*62*	*30*	*68*	*30*	*30*	*30*	*60*	*60*	*24*	*30*	*44*	*30*	*30*	*110*	*30*	*30*	*30*	*30*
Idh-1	a	—	—	—	—	—	—	—	—	—	—	—	—	—	—	—	—	.12	—	—	.01	—
	b	1.00	1.00	1.00	1.00	1.00	1.00	1.00	1.00	1.00	.61	.66	1.00	1.00	1.00	1.00	1.00	.79	1.00	1.00	.99	1.00
	d	—	—	—	—	—	—	—	—	—	.39	.34	—	—	—	—	—	.08	—	—	—	—
Mdh-1	f	—	—	—	—	—	—	—	—	—	—	—	—	—	—	—	.04	.03	—	—	—	—
	a	1.00	1.00	1.00	1.00	1.00	1.00	1.00	1.00	1.00	1.00	1.00	1.00	1.00	1.00	1.00	.96	.97	1.00	1.00	1.00	1.00
Mdh-2	a	1.00	1.00	1.00	1.00	1.00	1.00	1.00	1.00	1.00	1.00	1.00	1.00	1.00	1.00	1.00	1.00	1.00	—	—	.97	.01
	c	—	—	—	—	—	—	—	—	—	—	—	—	—	—	—	—	—	1.00	1.00	.03	.99
Es-4	a	1.00	—	1.00	.70	1.00	1.00	1.00	1.00	1.00	1.00	.38	.42	1.00	1.00	1.00	1.00	1.00	1.00	—	.99	—
	b	—	1.00	—	.30	—	—	—	—	—	—	.62	.58	—	—	—	—	—	—	1.00	.01	1.00
Es-5	á	—	—	—	—	—	—	—	—	—	—	—	—	—	—	—	—	—	—	1.00	—	1.00
	a	1.00	1.00	1.00	1.00	1.00	1.00	.92	1.00	1.00	.73	.40	1.00	1.00	.97	.32	1.00	.96	1.00	—	1.00	—
	b	—	—	—	—	—	—	.08	—	—	.27	.60	—	—	.03	.68	—	.04	—	—	—	—
Sod	a	1.00	1.00	1.00	1.00	1.00	1.00	1.00	1.00	1.00	.64	—	—	.29	—	—	—	—	—	—	—	—
(TO)	b	—	—	—	—	—	—	—	—	—	.36	1.00	1.00	.71	1.00	1.00	1.00	1.00	—	1.00	.98	.90
	c	—	—	—	—	—	—	—	—	—	—	—	—	—	—	—	—	—	1.00	—	.02	.10
	P:	0.0	0.0	0.0	8.0	4.0	8.0	4.0	0.0	0.0	20.0	32.0	4.0	12.0	16.0	12.0	20.0	28.0	0.0	0.0	20.0	12.0
	H:	0.0	0.0	0.0	3.7	2.0	0.3	0.7	0.0	0.0	8.7	11.6	1.5	3.2	1.1	4.2	4.4	4.0	0.0	0.0	1.2	1.4

[a] Sites: (A) Bylas Spring; (B) Ciénega Creek (C) Monkey Spring; (D) Sharp Spring; (E) San Bernardino Wildlife Refuge; (F) La Atascosa Ciénega; (G) Rió Altar; (H) Rió Alisos; (I) Rió Bocobachi; (J) Rió Arizpe; (K) Cumpas; (L) Rió Moctezuma; (M) Ures; (N) San Jose de Pima; (O) Rió Nuri; (P) Agua Caliente; (Q) Rió Cedros; (R) Est. Corral; (S) San Bernardo; (T) Navojoa; (U) El Tabelo.

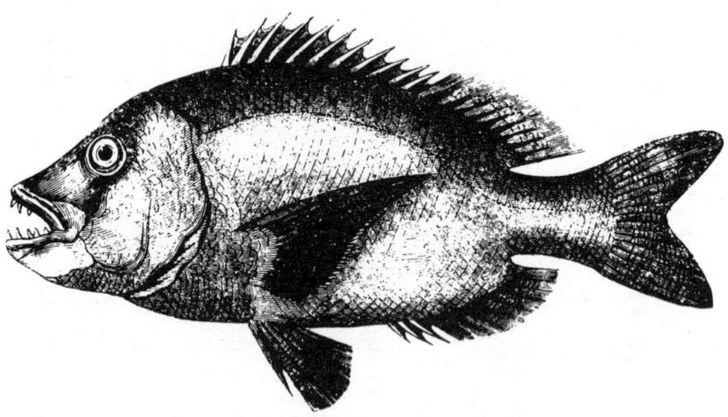

Conservation Genetics of Pacific Salmon I. Temporal Changes in Allele Frequency

ROBIN S. WAPLES

National Marine Fisheries Service
Northwest Fisheries Center
2725 Montlake Blvd. East
Seattle, WA 98112, U.S.A.

DAVID J. TEEL

National Marine Fisheries Service
Northwest Fisheries Center
2725 Montlake Blvd. East
Seattle, WA 98112, U.S.A.

Abstract: *Recent reductions in the abundance of all Pacific salmon species (*Oncorhynchus *spp.), coupled with large increases in artificial production, demand that careful attention be paid to genetic changes occurring in both wild and cultured populations. Analysis of electrophoretic data for chinook salmon (*O. tshawytscha*) from the Pacific coast of Oregon revealed substantial allele frequency changes over 2–4 years in hatchery, but not wild, populations. Unfortunately, our understanding of the causes of this result is hampered by a lack of theoretical models designed for organisms with life history features like those of Pacific salmon. We used computer simulations to provide a context for understanding genetic changes observed in the hatchery populations.*

Simulation results indicated that annual fluctuations in population allele frequencies due to genetic drift can typically be expected to be several percent, with the absolute magnitude determined primarily by the effective number of spawners each year rather than the age structure. Changes over 10- to 25-year periods were only slightly greater than short-term changes (1–5 years). The magnitude of allele frequency change over time was different for juvenile and adult samples. The probability of a significant test statistic comparing allele frequencies in temporally spaced samples increased with the ratio of sample size to effective number of breeders per year. This is an important consideration for conservation biologists, who typically are concerned with populations of small effective size.

Simulation results indicate that it is necessary to postulate

Paper submitted 1/25/89; revised manuscript accepted 6/2/89

Resumen: *Se han observado reducciones recientes en la abundancia de la especie de salmón del Pacifico (*Onoorhynchus *spp.). Estos cambios están relacionados con el incremento de la producción artificial, lo cual requiere que se preste atención especial a los cambios genéticos que están ocurriendo en las poblaciones silvestres y cultivadas de esta especie. Un análisis de los datos electroforéticos para el salmón chinook (*O. tshawytsha*) de la costa del Pacífico de Oregón reveló cambios substanciales en la frecuencia de alelos durante un período de 2–4 años en los criaderos, pero no en las poblaciones silvestres. Desafortunadamente, nuestro entendimiento de las causas de este resultado es limitado dado a la falta de modelos teóricos diseñados para organismos con ciclos de vida semejantes a las del salmón del pacifico. Nosotros utilizamos la simulación por computadoras para proporcionar un contexto para entender los cambios genéticos observados en las poblaciones del criadero.*

Los resultados de la simulación indicaron que las fluctuaciones anuales en la frecuencia de alelos de la población dado a la deriva genética, se espere que sean de varios porcentages. La magnitud absoluta es determinada principalmente por el número efectivo de hembras reproductivas cada año en vez de la estructura de edades. Cambios genéticos durante un período de 10 a 25 años fueron ligeramente mayores que cambios a corto plazo (1–5 años). La magnitud en los cambios en la frecuencia de alelos a través del tiempo fue diferente para las muestras de juveniles y de adultos. La probabilidad de un significativo test estadístico comparando frecuencias de alelos en muestras espaciadas en el tiempo aumentó con la relación de tamaño de la muestra a número efectivo de reproductores por año. Esta es una consideración importante para biólogos de conserva-

unrealistically large selection coefficients to explain the genetic changes in the hatchery populations by natural selection. The changes observed are consistent with a pure drift model provided that the effective number of breeders was as small as about 25–50 per year. Analysis of brood stock information for the hatcheries indicates that the effective population number may indeed have been this low, although the number of returning adults was often much larger. This conclusion underlines the importance of monitoring the genetic consequences of the large-scale artificial propagation programs involving Pacific salmon.

ción, a quienes típicamente les concierne el tamaño efectivo de poblaciones pequeñas.

Los resultados de la simulación indican que es necesario postular coeficientes de selección irrealisticamente grandes para explicar los cambios genéticos en la población reproductiva por selección natural. Los cambios observados son consistentes con un modelo puro de deriva provisto que el efectivo número de reproductores fue tan pequeño como 25 – 50 por año. El análisis de la información del linaje reproductive para los criaderos indica que el efectivo número poblacional en realidad pudo haber sido tan bajo, a pesar de que el número de adultos devueltos fue con frecuencia mucho más grande. Esta conclusión subraya la importancia de evaluar periodicamente las consecuencias genéticas de los programas de propagación artificial a larga escala que involucran al salmón del Pacífico.

Introduction

Anadromous Pacific salmon (*Oncorhynchus* spp.) spend one or more years at sea before returning to fresh water to spawn. The tenacity these fish demonstrate during the rigorous spawning migration — massive physiological changes associated with entry into fresh water, imposing natural obstacles such as rapids and waterfalls, predation by bears and other carnivores — has long been a source of wonderment to human observers. Unfortunately, Pacific salmon have not been as successful in overcoming obstacles created by man. Destruction of spawning habitat, blockage of migratory routes by dams, and overfishing have all made major inroads into spawning runs in most river systems. Current abundances of the five species in northwestern North America are substantially below historical levels (Fredin 1980; Fraidenburg & Lincoln 1985), and some populations have disappeared entirely or are on the verge of extinction (e.g., Williams 1989). To mitigate losses to wild stocks, considerable efforts have been made to boost artificial production in hatcheries. These efforts are carried out on a colossal scale: in 1987 in Washington state alone, over 330,000,000 Pacific salmon were released from over 130 facilities (Abrahamson 1988).

These two factors, depletion of wild stocks and increased hatchery production, can greatly accelerate the process of genetic change in Pacific salmon. Accelerated change may compromise the long-term fitness of the species by reducing overall levels of variability and eliminating adaptive gene complexes evolved over thousands of generations. Surviving wild populations are subject to selection in altered environments, immigration (from transplantation programs or strays from hatcheries), and genetic drift (through reduced population size). Rapid genetic change is an even greater concern in cultured populations: the opportunities for artificial selection are obvious, transfer of eggs or fish between hatcheries may introduce new genes or create novel genetic combinations through hybridization, and brood stock practices may cause the effective number of breeders to be much less than the number of returning adults (e.g., Simon et al. 1986).

Our interest in the problem of genetic changes in Pacific salmon populations grew out of the analysis of electrophoretic data for a series of populations of chinook salmon (*O. tshawytscha*) from the Pacific coast of Oregon. Nine wild and nine hatchery populations were sampled in two different years, allowing comparisons of allele frequency at an average of about 10 polymorphic loci per population (Fig. 1). In the wild populations, the percentage of single-locus contingency chi square tests showing statistically significant allele frequency change (8%) was only slightly higher than the nominal α level for the test (5%). In eight of the nine hatcheries, however, the genetic changes were much more substantial (significant results in 22%–63% of single-locus tests; overall hatchery average = 36%).

This striking result emphasizes the need for monitoring the genetic consequences of large-scale artificial propagation programs. Identifying potential problems, however, is only the beginning; efforts to reduce rates of genetic change in cultured populations are unlikely to succeed unless the causes of the change can be identified. Distinguishing between the most likely causes of short-term genetic change (selection, migration, genetic drift) is difficult in any event, and the problem is even more complex in Pacific salmon because of the lack of an adequate model for genetic change. Most of the seminal work of Wright, Fisher, and Haldane was based on a discrete generation model, but among the Pacific salmon species this assumption is reasonable only for

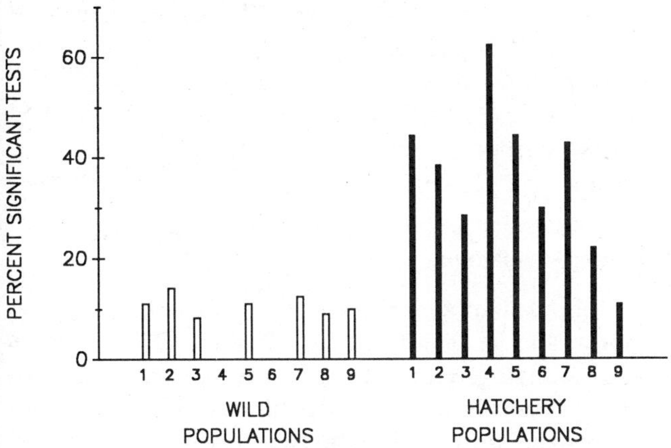

Figure 1. Contrasting patterns of genetic change over 2–4 years in nine wild and nine hatchery populations of chinook salmon from Oregon. For each population, allele frequencies in two temporally spaced samples (of approximately 100 juveniles each) were compared at an average of 10 polymorphic loci; the percentage of single-locus contingency chi square tests showing statistically significant allele frequency change is indicated by the height of the columns. All of the hatchery populations except #9 showed much higher rates of genetic change than did any of the wild populations.

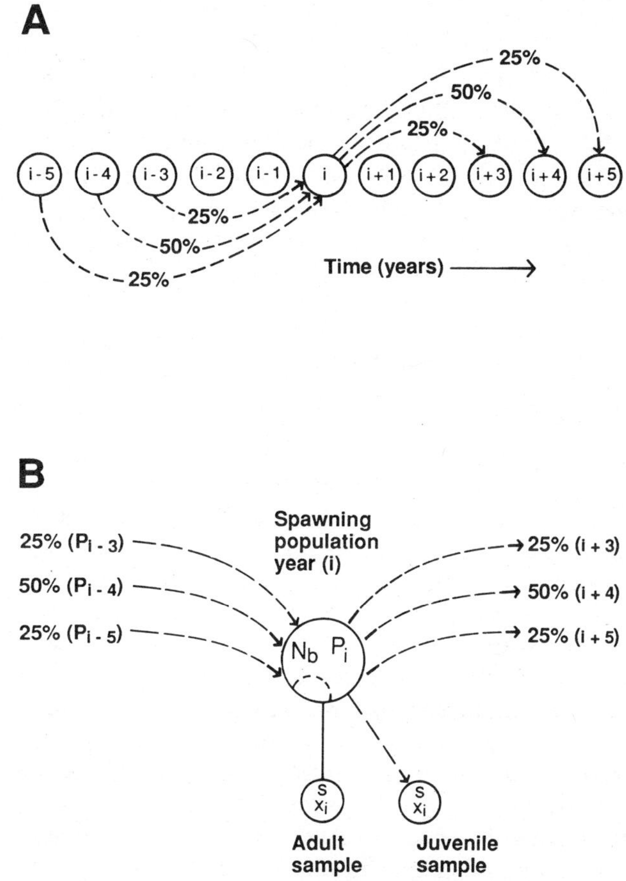

Figure 2. ***A.*** *Pattern of overlapping generations and migration between year classes used to model genetic change in Pacific salmon populations. Each circle represents a spawning population from a different brood year in the same stream. In the basic model, probability of spawning at age 4 was 0.5, so on average 50 percent of the spawners in year* i *were derived from offspring produced in brood year* i − 4*, with the remainder divided between years* i − 3 *and* i − 5*.* ***B.*** *Sampling plan. For each brood year* i*, samples of S individuals were drawn for genetic analysis.* P_i *is the population allele frequency for the* N_b *spawners in year class* i*;* X_i *is the allele frequency in the sample. Adults were sampled without replacement from the* N_b *spawners after reproduction (solid line); juveniles were sampled binomially from the pool of gametes produced by the spawning population (broken line).*

the pink salmon, *O. gorbuscha*, which has a rigid 2-year life cycle in the Pacific Northwest (Aspinwall 1974). More recently (e.g., Hill 1979), it has been shown that many of the results obtained under more restrictive assumptions also hold for organisms in which generations overlap. Pacific salmon, however, do not have generations that overlap in the same sense as, say, most mammals or birds (Fig. 2). In most overlapping generation models, individuals breeding in one time period may survive to reproduce again; in such a population, composition of the breeding population changes gradually over time through death or senility and the maturity of juveniles. In contrast, *Oncorhynchus* spp. die after spawning, so mating occurs only with others returning in the same year. There is, however, some variation in age at spawning in all species except the pink salmon. A particular year's spawning population of chinook salmon, for example, might include fish aged 2, 3, 4, and 5 (or more) years. Thus, although there is 100 percent turnover in the spawning population each year (in contrast to a gradual turnover in other organisms with overlapping generations), a Pacific salmon spawning population typically includes genetic information derived from two or more previous brood years. The result is a pattern of overlapping year classes distinct from that of any existing population genetics model.

These life history features of Pacific salmon make the problem of temporal genetic change difficult to approach analytically, but the system is amenable to study by simulation. Using the model described below, we address three major questions: (1) What magnitude of change in population allele frequency between years is expected for Pacific salmon in the absence of any directional forces? (2) How should the results of statistical tests of allele frequency change over time be interpreted? (3) Based on results for (1) and (2), can we identify the probable causes of the genetic changes observed in the Oregon hatchery populations? Answers to questions (1) and (2) are important because they provide a context for understanding changes in allele frequency observed in salmon populations. As we shall see, the answer to the third question provides important insight for those interested in conserving the genetic resources of cultured populations.

Materials and Methods

Electrophoresis

Samples of approximately 100 juveniles each were taken in 1981 (1980 brood year) from nine wild and nine hatchery populations along the Pacific coast of Oregon. Comparable samples were taken from the wild populations in 1983 (1982 brood) and from the hatcheries in 1985 (1984 brood). The hatchery and wild populations were not paired in a statistical sense; rather, they are part of a geographically more extensive program to sample the major chinook-producing populations in North America (cf. Utter et al. 1989). Sample preparation, starch gel electrophoresis, and interpretation of banding patterns followed Aebersold et al. (1987) and Utter et al. (1987). Loci found to be polymorphic in the hatchery samples are indicated in Table 1. For each locus in each population, a contingency chi square test was used to test the hypothesis that population allele frequencies did not change between the times of sampling. Isoloci (pairs of duplicated loci sharing alleles with identical electrophoretic mobility; see Allendorf and Thorgaard 1984; Waples 1988) were not used in this analysis, nor were they modeled in the simulations. Analysis of temporal changes at duplicated loci is somewhat more complicated (Waples, manuscript II, in press).

Computer Simulations

We adopted a model similar to that used by several authors (Nei & Tajima 1981; Pollak 1983; Waples 1989*a*) studying temporal variation in organisms with discrete generations, with certain modifications dictated by the unique life history features of Pacific salmon. Consider a population in which the effective number of breeders returning to spawn each year is N_b (N_b is analogous to effective populations size, N_e, except that it refers to the number per year rather than per generation). Change was monitored at a single gene locus with two alleles (*A*,*a*). In a given year's spawning population (e.g., brood year i), the breeders include individuals of different ages (ages $k = 3,4,5$) in the expected proportions f_k ($\Sigma f_k = 1$). The $2N_b$ genes representing the breeding population in year i were selected in a series of binomial draws from previous brood years (the selection being from brood year $i - k$ with probability f_k), and the probability of drawing an *A* allele was P_{i-k} = the frequency of allele *A* in brood year $i - k$. Offspring produced from the spawn also return to breed at various ages, thus continuing the pattern of overlapping year classes.

Samples for genetic analysis in a particular brood year

Table 1. Significance level of single-locus χ^2 tests of temporal stability of allele frequencies in chinook salmon from nine Oregon hatcheries. Sample sizes in 1980 and 1984 brood years, respectively, are given in parentheses below hatchery names. Samples from two different stocks (spring and fall run) were taken at two hatcheries. For loci showing significant temporal changes (n.s. = not significant), the direction of frequency change of the common allele is indicated in parentheses. A dash indicates no test was performed (locus monomorphic in both samples or no data from one sample).

Locus	*Cedar Creek (99,100)*	*Cole Rivers (spring) (11,350)*	*Cole Rivers (fall) (50,100)*	*Elk River (100,100)*	*Fall Creek (100,100)*	*Rock Creek (100,100)*	*Salmon River (99,100)*	*Trask River (fall) (100,100)*	*Trask River (spring) (100,100)*
Aat-3	n.s.	n.s.	n.s.	—	—	n.s.	n.s.	—	n.s.
Ada-1	—	—	—	0.01(+)	n.s.	—	n.s.	n.s.	—
Adh	—	n.s.	—	—	—	n.s.	—	—	n.s.
Ah-4	0.05(+)	n.s.	n.s.	n.s.	0.001(−)	0.01(+)	0.05(+)	n.s.	0.01(−)
Gl-1	n.s.	0.05(−)	n.s.	n.s.	n.s.	0.05(+)	0.01(+)	n.s.	n.s.
Gr	0.001(−)	—	0.05(+)	n.s.	—	n.s.	—	—	n.s.
Idh-2	—	—	0.01(+)	—	—	—	—	—	—
Ldh-4	—	—	0.01(+)	n.s.	—	—	—	—	—
Ldh-5	0.05(−)	n.s.	0.05(+)	—	—	—	—	—	n.s.
Lgg-1	n.s.	n.s.	n.s.	n.s.	0.05(−)	0.01(+)	0.001(+)	0.001(−)	0.05(+)
Ltp	—	—	n.s.	—	—	—	—	—	—
Mpi	n.s.	n.s.	n.s.	0.001(−)	n.s.	n.s.	0.05(+)	0.05(+)	0.05(−)
Pgdh	—	—	n.s.	—	—	—	—	—	—
Pgk-2	n.s.	n.s.	n.s.	n.s.	n.s.	0.01(−)	0.05(+)	n.s.	n.s.
Sod-1	0.001(−)	n.s.	0.05(−)	n.s.	n.s.	n.s.	n.s.	0.05(−)	n.s.
Combined χ^2 test over all loci:									
Significance level	0.001	n.s.	0.01	0.001	0.01	0.001	0.001	0.001	0.01

(e.g., year i) were chosen in two ways, corresponding to procedures commonly used in studies of Pacific salmon. Adults for genetic analysis are generally taken as post-spawn mortalities (in the wild) or as poststripping sacrifices (in hatcheries). These adults (and the genes they carry) represent a sample *without replacement* of the actual spawning population in brood year i. Juvenile samples are generally taken from the large pool of offspring resulting from the previous year's spawn. A juvenile sample can be considered to be a binomial sample (i.e., a sample with replacement) from gametes produced by the spawners of brood year i. Figure 2 illustrates these sampling schemes and the pattern of overlapping year classes typically found in chinook salmon.

Monte Carlo methods were used to simulate the pattern of genetic change in Pacific salmon. To initialize the simulations, allele frequencies in the N_b breeders in years $i = 1–5$ were set to 0.5. Beginning in year 6, population allele frequencies (P_i) were determined by sampling from brood years up to 5 years before, as described above. Sample allele frequencies (X_i) were also computed for adults and juvenile progeny taken each year. This process was repeated until year 50 to allow the system to reach a dynamic equilibrium. For each year from $i = 51$ to 100, the difference in allele frequencies ($|X_i - X_{i-k}|$) was computed for samples taken in the current year and $k = 1, 2, 3, 4, 5, 10$, and 25 years previously. Differences in population allele frequencies ($|P_i - P_{i-k}|$) were computed in a similar manner. After year 100, allele frequencies were reinitialized to avoid gene extinction, and the process was repeated. Two hundred replicates were performed for each set of initial parameters, with data taken over the last 50 years in each replicate. Thus, $50 \times 200 = 10{,}000$ data points ($|X_i - X_{i-k}|$ or $|P_i - P_{i-k}|$) were available for each value of k years separating two samples. Means and variances of these differences in allele frequency were computed over the 10,000 data points. A 2×2 contingency chi square test of equality of allele frequencies was performed for each comparison of juvenile or adult samples, and the percentage of statistically significant test results ($\alpha = 0.05$) over the 10,000 trials was recorded.

Considerable variation in age at spawning occurs both within and between species of Pacific salmon. In the basic model (Fig. 2), probability of spawning at age 4 (f_4) was 0.50, and probability of spawning at ages 3 and 5 was 0.25. This age structure is similar to that found in chinook salmon populations from many areas of the Pacific Northwest, including coastal areas of Oregon (Nicholas & Hankin 1988). In the simulations, an average of 50% of the genes were "immigrants" from brood years other than the one 4 years before, resulting in a "migration" rate (m) or rate of gene flow among year classes of 0.5. We also performed simulations in which m ranged from 0.1 (90% from year $i - 4$, 5% each from years $i - 3$ and $i - 5$) to 0.7 (30% from year $i - 4$, 35% each from years $i - 3$, and $i - 5$) to encompass some of the intra- and interspecific variation in age structure found in Pacific salmon.

Results

Magnitude of Allele Frequency Change

Figure 3A shows the mean difference in allele frequency $|P_i - P_{i-k}|$ in spawning populations k years apart. Even with 50% of the breeders "migrating" between year classes, P_i typically varied several percent from year to year. In the simulations shown in Figure 3A, initial allele frequency was set to 0.5. As expected, the magnitude of allele frequency change was somewhat smaller in simulations with initial frequency closer to 0 or 1 (unpublished data). Relatively small frequency differences (about 1–2%) were found if the number of breeders each year was very large ($N_b = 500$ to 1000), whereas

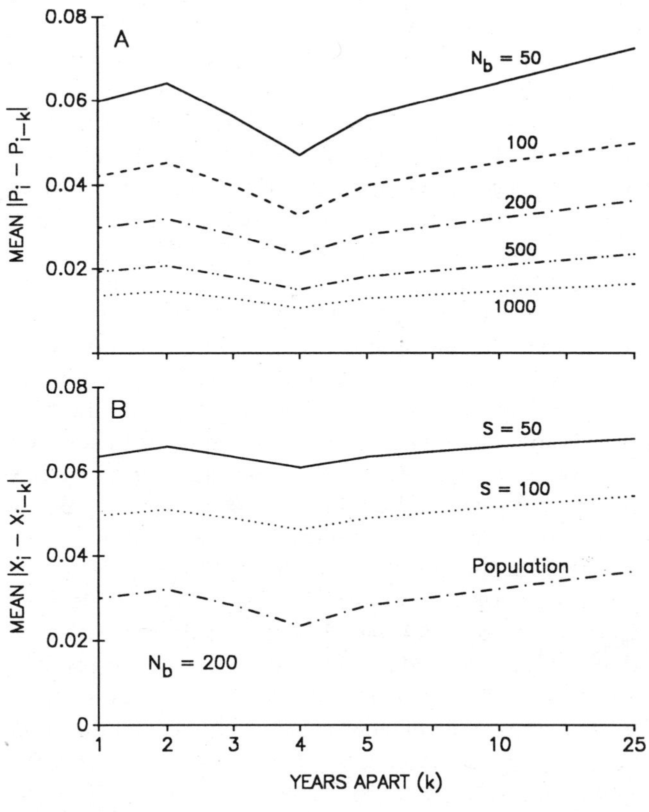

Figure 3. The magnitude of temporal changes in allele frequency observed in the Pacific salmon model. A. Mean change in frequency (P_i) in spawning populations over a period of k *years as a function of the effective number of spawners per year (N_b). **B.** Mean change in sample allele frequency (X_i) for juvenile samples of $S = 50$ and $S = 100$ taken* k *years apart from a population with $N_b = 200$ breeders per year. Results are from simulations with $P = 0.5$ and age structure as shown in Figure 2A.*

simulations using fewer spawners (N_b = 50 to 100) resulted in larger changes between years (about 4–6%). The inverse relationship with N_b was expected because the changes in population allele frequency are a consequence of drawing a finite number of breeding individuals each year. Because of the complete turnover each year in the spawning population, annual fluctuations of allele frequency were larger in the Pacific salmon model than are expected using standard models assuming overlapping or discrete generations (Waples, manuscript I, in press). Estimating N_b in spawning populations is difficult; however, given the wide range of run sizes found in most species of Pacific salmon, this number probably differs by several orders of magnitude among populations.

Another important result illustrated in Figure 3A is that the change in population allele frequencies over 10 or 25 years was not much larger than the change over much shorter periods (1–5 years). This result is due to the pattern of overlapping year classes found in Pacific salmon. Each spawning population incorporates genetic information from several brood years, thus dampening the effects of random drift for the system as a whole. Whereas micro variation (differences between consecutive brood years) is a function of the number of breeders each year, macro variation (over generations, or 3- to 5-year periods) is a function of the much larger number of spawning individuals that comprise an entire generation (about 4 years in the present model).

Normally, data are not available for the entire spawning population, and one must estimate allele frequencies from a sample. Two points can be made regarding temporal differences observed in samples drawn in the simulations (Fig. 3B). First, because sampling from the population introduces a source of random error in addition to genetic drift, these samples show greater frequency differences than are found in the entire breeding population. Second, the magnitude of this sampling error is inversely proportional to sample size, so larger absolute differences are found in smaller samples (e.g., larger differences for S = 50 than S = 100 in Figure 3B). The data in Figure 3B and subsequent figures are for juvenile samples. Allele frequency differences in adult samples were generally somewhat smaller, but results for adult sampling are more complicated because they also depend on the ratio of N_b to N, the number of actual spawners (see below).

Test for Equality of Allele Frequencies

The results depicted in Figure 3 indicate that tests of the equality of allele frequencies in temporally spaced samples must be interpreted with caution. This is because standard tests (such as the homogeneity χ^2 test) assume as a null hypothesis that both samples are binomially drawn from the same probability distribution. In the case of temporal variation, the population allele frequencies change over time, so this assumption is invalid. Temporally spaced samples will on average differ more than independent binomial samples drawn from the same population, resulting in a higher proportion of "significant" test results than the nominal α level of the test would suggest, even if no directive forces are involved. The probability of a significant test comparing temporally spaced samples can be obtained analytically in the discrete generation model (Waples 1989*b*), but the complicated life history features of Pacific salmon required a simulation approach.

Results of these simulations (Fig. 4) clearly show that for a given spawning population size (N_b), more tests were significant when a larger sample was taken. The explanation for this result is straightforward: power of a

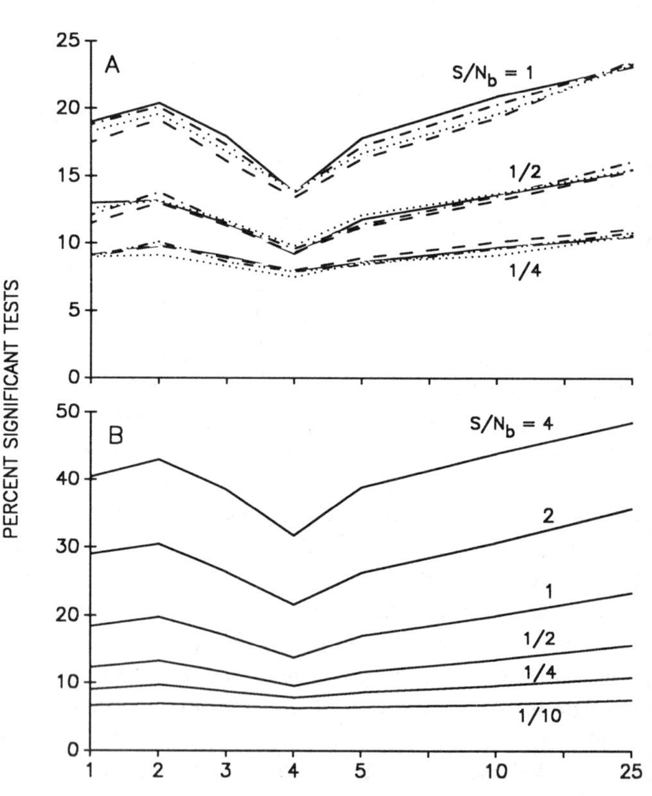

Figure 4. Percent significant χ^2 tests ($p < 0.05$) of homogeneity of allele frequencies in juvenile samples from the same population taken k *years apart.* ***A.*** *Results from simulations with N_b = 100 (solid line), 200 (dashed line), 500 (dashed-dotted line), and 1000 (dotted line) and sample sizes such that S/N_b = 0.25, 0.5, and 1. Probability of a significant test result depends not on absolute value of S or N_b, but rather their ratio.* ***B.*** *Means of all simulations in A for indicated ratios of sample size to effective number of breeders. In the simulations, initial allele frequency was 0.5, and age structure was as shown in Figure 2A.*

statistical test increases with sample size if the null hypothesis is false. In analyzing temporal variation, the null hypothesis for the standard chi square test (two samples taken from the same probability distribution) is always violated, and larger samples more accurately reflect the actual temporal differences that exist in the breeding populations. In fact, the probability of a significant test depends not on the actual values of S or N_b, but on the ratio S/N_b. Figure 4A depicts results of simulations with $N_b = 100, 200, 500$, and 1000 and juvenile samples with $S = 0.25N_b$, $S = 0.5N_b$, and $S = N_b$. It is clear that the ratio S/N_b explains almost all of the differences between simulations. Waples (1989*b*) reported a similar result in the discrete generation model. This means that for a sample of any size S, the probability of a significant test (assuming only drift and sampling error are involved) can be predicted as a function of N_b. Figure 4B, which can be used for this purpose, shows the mean percentage of significant tests for all simulations with values of S/N_b between 0.1 and 4.

Note that results in Figures 4, 5, and 7 are for χ^2 tests at a single locus with two alleles. In comparing temporally spaced samples, the probability of a significant test increases with both the number of alleles per locus and the number of loci used (if an overall test is used combining results for multiple loci). See Waples (1989*b*) for a discussion of these points for the case of discrete generations.

The Oregon Hatchery Data

The simulation results provide a context for evaluating possible causes for the relatively large allele frequency changes observed in the hatchery populations. One possibility — that the changes were larger in the hatchery samples because they were taken 4 years apart as opposed to 2 years for the wild samples — can be ruled out. In simulations with $m \leq 0.5$, the magnitude of genetic change (and the proportion of significant χ^2 tests) was lowest for samples taken exactly 4 years apart. This is a consequence of the assumption in the model that chinook salmon are more likely to return at age 4 than at any other age: samples taken 4 years apart can be expected to be the most similar if the majority of the spawners are 4 years old. If anything, therefore, we would expect the timing of the samples to produce slightly larger changes in the wild populations.

The simplest explanation for the temporal changes is that they are due to stochastic factors alone (genetic drift and sampling error). Examination of Figure 4B indicates that these factors can be expected to produce the proportion of significant tests observed in the Oregon hatcheries (36%) only if the ratio S/N_b was somewhat larger than 2. Because most of the hatchery samples were of 100 individuals (Table 1), it is necessary to postulate $N_b < 50$ to explain the observed changes by stochastic factors alone.

Before examining the possibility of restricted population size in more detail, it is necessary to consider several other factors that might be responsible for the changes:

1. Inaccurate or artifactual data for at least one year's data. This is possible, but it would not explain why only the hatchery samples were affected. Furthermore, the enzymes used in the analyses have been routinely surveyed for many years, and breeding experiments have verified the Mendelian basis of observed variation for most systems (James B. Shaklee, Washington Department of Fisheries, personal communication).

2. Nonrandom sampling of juveniles for genetic analysis. It is possible that allele frequencies in the hatchery populations as a whole were relatively stable, and that changes observed in the samples were due to one or both samples being nonrepresentative. This might occur, for example, if one of the samples were progeny of only a few adults (see Allendorf and Phelps 1981 for discussion of this point). Available data, however, indicate that eggs from different families were mixed shortly after spawning and long before our sampling (Edward Cummings, Oregon Department of Fish and Wildlife, personal communication). It thus seems reasonable to assume that the samples were approximately random with respect to the surviving progeny.

3. Natural or artificial selection. Although it is often assumed that the majority of electrophoretically detectable polymorphisms are selectively neutral, a number of studies suggest that natural selection is important in maintaining polymorphisms at some electrophoretically assayed gene loci in some organisms. Artificial selection was presumably absent in the wild populations, but a variety of hatchery practices might result in the disproportionate representation of certain genotypes in the spawners or offspring.

If strong selection involving these electrophoretically screened loci (or closely linked loci) was occurring in the hatcheries, it apparently was not uniform in direction in all hatcheries. Table 1 gives the direction of change in frequency of the common allele for all loci with significant χ^2 tests, and no clear pattern is evident. At most loci, the common allele increased in frequency in some hatcheries and decreased in others. Furthermore, significant test results were distributed over 11 different polymorphic loci. Nevertheless, selection cannot entirely be ruled out on this basis because it might act differently in different environments or genetic backgrounds.

To examine the more general possibility that selection was responsible for the observed changes, we considered a two-allele model with the fitness of the three genotypes (*AA, Aa, aa*) given by $w_1 = 1$, $w_2 = 1 - s$, $w_3 = 1 - 2s$, respectively, where s is the selection coefficient. If a population with allele frequency P is in Hardy-Weinberg equilibrium before selection, expected genotypic frequencies after one generation of selection are $P^2w_1/\overline{w}$, $2P(1 - P)w_2/\overline{w}$, and $(1 - P)^2w_3/\overline{w}$, respectively, where $\overline{w}$ (mean fitness) $= P^2w_1 + 2P(1 - P)w_2 + (1 - P)^2w_3$. This model is based on the assumption that the alleles are codominantly expressed (as in most electrophoretic phenotypes), and the result is strong directional selection against the *"a"* allele. This type of selection will not, however, cause substantial departures from Hardy-Weinberg equilibrium because genotypic frequencies after selection will be close to those expected based on the new allele frequencies (cf. Lewontin & Cockerham 1959). This is important because no substantial Hardy-Weinberg departures were observed in samples from the Oregon hatcheries (unpublished data).

We performed a series of simulations as described previously, with the following exception: an episode of natural selection was introduced between the formation of the spawning population in year *i* and the samples of 100 juveniles taken from their progeny. For years $i = 51$ to 100, an array of samples (corresponding to the array of selection coefficients $s = 0, 0.01, 0.05, 0.1, 0.2, 0.3, 0.5$) were binomially drawn from the spawning population. In each sample, 100 genotypes were chosen by random draw with the probabilities of each genotype as given above for the corresponding value of *s*. Allele frequencies (x_i) in these samples were compared with those in samples taken 4 years previously which were not subject to selection ($s = 0$). Sampling was performed in this way because we are interested in the effects of selection in a single generation, and selection was presumed to have occurred after the initial sample was drawn. Long-term selection should lead to similar results when considering comparisons over a single 4-year period, but such a model was not used to avoid the problem of gene extinction.

Results of these simulations are shown in Figure 5. Selection coefficients in the range 0.01–0.1 had little effect on the probability of a significant test. To explain the differences observed (36% significant tests) in terms of selection and drift, it is necessary to postulate selection coefficients of $s > 0.2$ over all the loci surveyed. Selection of this intensity causes more than a 40 percent reduction in fitness of the *aa* homozygote — a far greater loss of fitness than has been convincingly demonstrated for other electrophoretically detectable polymorphisms (see review by Eanes 1987). Other models of selection are possible, of course, but most do not result in such intense directional change in allele frequency. It seems unlikely, therefore, that natural or artificial selection was responsible for all, or even a substantial part, of the changes observed in the Oregon hatchery populations.

4. Mixing of gene pools. Our simulation model incorporated gene flow only between year classes, but a certain amount of straying is known to occur between different streams in all species of Pacific salmon (Ricker 1972; Lister et al. 1981). In addition, it is not uncommon for hatcheries with surplus production in a particular year to transfer progeny (usually eyed eggs) to other hatcheries that have been less successful. Such mixtures of gene pools can considerably change allele frequencies in the recipient population, and there is some indirect evidence that this may have happened to some of the Oregon hatchery populations.

Nei and Li (1973) pointed out that gametic disequilibrium (nonzero correlations between alleles at different gene loci) is expected in mixed populations containing two or more distinct gene pools. Waples and Smouse (1990) found statistically significant ($p < 0.05$) gametic disequilibrium in four of the nine Oregon hatchery samples from the 1980 brood and in five of the nine from 1984. In contrast, none of 18 samples from Oregon wild populations showed significant disequilibrium. The existence of substantial gametic disequilibrium is consistent with the hypothesis that fish from distinct gene pools made genetic contributions to the hatchery populations prior to 1985. However, the absence of a consistent heterozygote deficiency (Wahlund effect) in samples showing significant gametic disequilibria (Waples & Smouse 1990) argues against the hypothesis that the allele frequency differences in general were due to the transfer of eggs that produced the 1984 brood. Gametic disequilibrium without the Wahlund effect would be expected if an episode of random mating occurred after the admixture event (e.g., if parents of the 1984 brood were a mixture), but Waples and

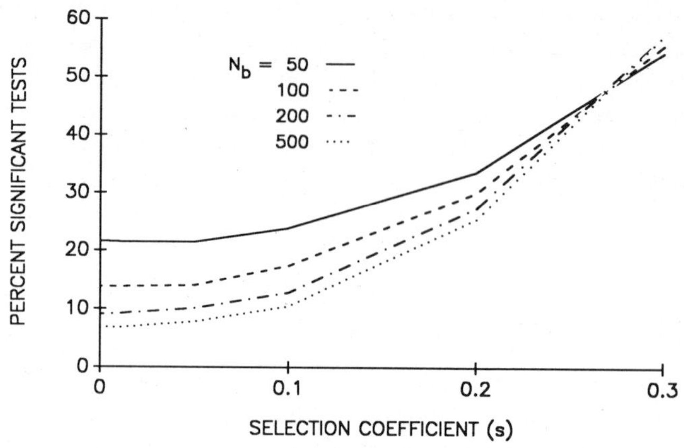

Figure 5. Percent significant χ^2 tests of homogeneity of allele frequencies as a function of selection coefficient (s) and effective number of breeders per year (N_b). Results are from simulations with two samples (100 juveniles each) taken 4 years apart, with one episode of selection before the second sample. Relative fitnesses of the three genotypes (AA, Aa, aa) were 1, 1 − s, and 1 − 2s, respectively. Initial allele frequency was 0.5, and age structure was as shown in Figure 2A.

Smouse (1990) showed that in that case the allele frequency differences between the contributing populations must have been quite substantial for the residual disequilibrium to be as large as observed. Furthermore, hatchery records indicate that in the period 1979–1984, Cedar Creek and Fall Creek were the only two hatcheries receiving imports that were released on site. Natural straying into the hatcheries might have contributed to the disequilibrium but, again, the immigrants would have to have been fairly numerous and genetically distinct. Mixtures of gene pools thus may have contributed to the temporal changes in allele frequency in at least some populations, but this phenomenon seems an unlikely cause of all the changes observed.

Interestingly, Waples and Smouse (1990) found that the multilocus test they used is also very sensitive to disequilibrium arising from genetic drift (all progeny produced by a finite number of parental genotypes). Simulation results demonstrated that it is necessary to postulate N_b of no more than about 50 per year to account for the observed levels of disequilibrium by drift—about the same population size necessary to account for the allele frequency changes by drift.

It is thus possible to explain both the allele frequency changes and the gametic disequilibrium by assuming N_b was no larger than about 50 per year in these hatcheries. In view of the fact that over 1000 adults a year return to some of these hatcheries (Nicholas & Hankin 1988), this number may seem unrealistically small. However, some hatcheries have many fewer returning adults, and even those hatcheries with a large run size may use only a small percentage of the total for spawning. In addition, at some hatcheries progeny (usually eggs) are exported to other hatcheries or are released off site. Exports often involve the entire production of some families, thus effectively eliminating them from the breeding pool responsible for the remaining progeny. High variance of reproductive success among families may further limit the effective number of breeders per year. Simon et al. (1986) showed that the above factors can cause a substantial reduction in effective population size in coho salmon.

The data necessary to compute N_b for the nine Oregon hatchery populations are not all available, but we obtained the following information for the period 1979–1984 (Edward Cummings, Oregon Department of Fish and Wildlife, personal communication; see Table 2). The number of females used for spawning ranged from a low of 2 in one year at Cedar Creek to over 1800 at Cole Rivers (spring run). In most cases, eggs from four to six females were mixed with milt from at least two males; hence the typical sex ratio of spawners was about 1:2 male:female. Exports (generally eggs) ranged from none (at several hatcheries) to over 90 percent of the total production in one year for Cole Rivers fall run. Using data for the number of females that produced the exported eggs, we have estimated the number of females (F^*) responsible for progeny remaining in the hatchery. For each population, F^* was computed each year, as well as the harmonic mean $\tilde{F}^*$ over the 6-year period.

The effective number of breeders can be computed using a variation of Wright's (1938) formula: $N_b = 4MF/(M + F)$, which, assuming the ratio of spawning males (M) to females (F) was 1:2, reduces to $N_b = 4F/3$. Under this assumption, N_b in the period 1979–1984 was less than 100 for five populations (Cedar Creek, Cole Rivers (fall run, Fall Creek, Rock Creek, Salmon River), about 100–200 for Trask River (spring and fall run), over 200 for Elk River, and over 1100 for Cole Rivers (spring run) (Table 2). These numbers are, in most cases, somewhat higher than the average of $N_b = 50$ or less required to explain the allele frequency changes by genetic drift alone. However, N_b computed using only the sex-ratio adjustment assumes that each male (or each female) has an equal opportunity to produce progeny that survive to

Table 2. Breeding information for nine chinook salmon stocks from Oregon hatcheries during the period 1979–1984.

Hatchery	*Number of females spawned*	*Percent exported*[a]	$\tilde{F}^{*b}$	*Maximum* N_b[c]	*Percent significant tests*[d]
Cedar Creek	2–25	0	8	11	44.4
Cole Rivers (spring)	610–1861	0–36	863	1151	11.1
Cole Rivers (fall)	22–210	0–93	26	35	38.4
Elk River	149–378	0–23	209	279	22.2
Fall Creek	19–69	0–12	41	55	28.6
Rock Creek	43–153	0–59	72	96	44.4
Salmon River	53–79	0–23	60	80	62.5
Trask River (spring)	94–160	0–35	104	139	30.0
Trask River (fall)	164–324	25–61	129	172	42.9

Based on information provided by Oregon Department of Fish and Wildlife.
[a] *Estimated percentage of females whose entire production was exported off-site.*
[b] *Harmonic mean (years 1979–1984) number of females responsible for progeny remaining in hatchery after adjusting for exports.*
[c] *Effective number of breeders computed using* $\tilde{F}^*$*, assuming 1:2 ratio of spawning males and females and a binomial distribution of progeny number per parent.*
[d] *Single locus* χ^2 *tests comparing allele frequencies in samples from 1980 and 1984 brood years (see Tables 1 and 2).*

reproduce the next generation. This calculation does not consider possible reduction in N_b due to high variance in reproductive success among individuals of the same sex, and it therefore must be considered a maximum estimate. Apart from the probability that survival to spawning age of individuals from different families is nonrandom, the practice of pooling sperm and eggs from multiple individuals for fertilization has been shown to result in disproportionate contributions by some individuals at the expense of others (Simon et al. 1986; Withler 1988), further reducing the effective size. In the absence of data on variance in reproductive success, it is impossible to calculate N_b more accurately for these Oregon hatcheries. However, it is instructive that in a coho salmon hatchery to which thousands of adults returned each year, Simon et al. (1986) found that (1) as few as 4 percent of the returning adults were used for spawning in a given year; (2) the variance in family size was typically several times the mean; and (3) in some years the effective number of breeders may have been as low as 16–22 of each sex. Therefore, it is clearly possible that actual N_b may have been 50 or less, at least in the five hatcheries with maximum $N_b < 100$, and perhaps in Trask River as well.

Additional support for the hypothesis that drift was a major factor contributing to the changes observed in the hatcheries is the clear, inverse relationship between the number of females used for spawning and the magnitude of temporal change. This relationship was apparent even using the crude measure of the percentage of significant single-locus χ^2 tests (Fig. 6). For example, the three hatcheries with the highest percentages of significant tests (Salmon River, Cedar Creek, Rock Creek) used relatively few females for spawning, and the two hatcheries using the most females (Elk River, Cole Rivers spring run) had the lowest proportions of significant tests.

Variations on the Basic Model

The simulation results described above were obtained with $m = 0.5$ and initial allele frequency set to $P = 0.5$. Certain simulations were repeated with different m and P values to evaluate the sensitivity of the analysis to these parameters. Because the rate of change in allele frequency per generation is a function of $P(1 - P)$, it is clear that the absolute magnitude of the changes shown in Figure 3A is specific for $P = 0.5$. The probability of a significant test statistic, however, in theory is independent of allele frequency (Waples 1989*b*). In practice, this independence will break down if P is too close to 0 or 1 or if S is too small. In simulations using values of P larger than 0.5, little effect of allele frequency was observed for $P \leq 0.9$ and $S \geq 50$. For smaller sample sizes or very extreme allele frequencies, the probability of a significant test is slightly less than shown in the figures.

The effects of different age structures were evaluated by varying m from 0.1 to 0.7. On average, each spawning population comprised $(m/2)N_b$ breeders derived from years $i - 3$ and $i - 5$ and $(1 - m)N_b$ breeders from year $i - 4$. Figure 7 shows results of simulations with $N_b = 200$ and juvenile samples of $S = 50$ fish. For values of m that are realistic for most chinook salmon populations ($m \approx 0.3$ to 0.7), migration had little effect on the probability of a significant test. Apparently, 30 percent migration from other brood years is sufficient to smooth out allele frequencies over time about as much as they can be equalized, given that each spawning pop-

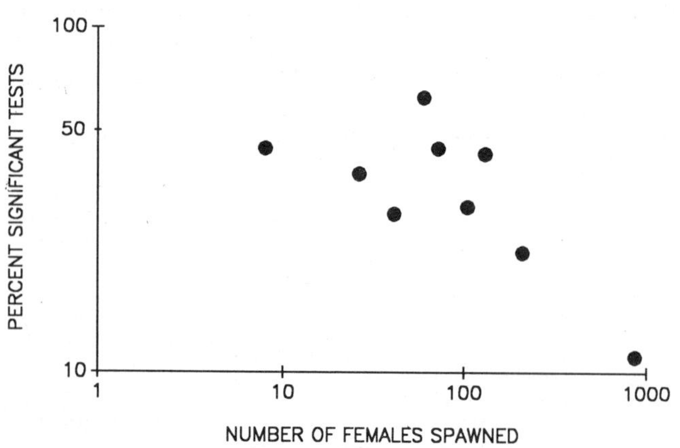

Figure 6. Relationship between temporal variability in allele frequencies in nine Oregon hatchery stocks of chinook salmon and the harmonic mean number of female spawners for the period 1979–1984. Temporal variability was measured as the percentage of statistically significant ($p < 0.05$) single-locus χ^2 tests of equality of allele frequencies in juvenile samples taken from 1980 and 1984 brood years.

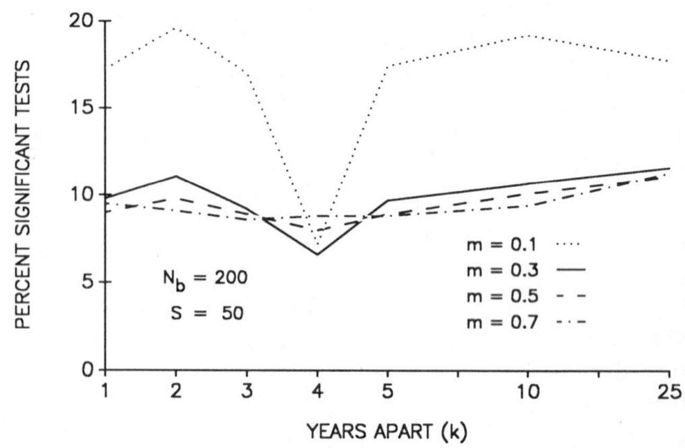

Figure 7. Chinook salmon: effects of gene flow or "migration" between year classes (m) on probability of a significant test of homogeneity of allele frequencies in juvenile samples from the same population taken k *years apart. Results are from simulations with effective number of breeders per year (N_b) set to 200, sample size = 50, and initial allele frequency = 0.5.*

ulation is of finite size and sampling error in choosing allele frequencies in the N_b breeders is inescapable. The similarity of the results for a broad range of migration parameters suggests that they are also qualitatively true for other anadromous Pacific salmonids (sockeye salmon [*O. nerka*], chum salmon [*O. keta*], and steelhead trout [*O. mykiss,* formerly *Salmo gaidnerii*]) with broad overlap in year classes. Quantitatively, the absolute magnitude of allele frequency change will also depend on the average age at reproduction (assumed to be 4 years in the present model). This topic will be addressed in a subsequent paper (Waples, manuscript I, in press).

A dramatic difference in the pattern of change over time was seen if the proportion of gene flow between year classes was as small as 10 percent (Fig. 7). With 90 percent of the spawners returning at age 4, the probability of a significant test for samples not taken exactly 4 years apart was about twice as high as it was for lower migration rates. Such a large preponderance of a single class is unusual in chinook salmon but is typical of coho salmon and some populations of chum salmon. In such cases, interpretation of genetic change requires particular care.

Adult Sampling

Two factors contribute to allele frequency differences in temporally spaced samples: changes in the population frequency and random error in drawing the samples. A juvenile sample of S progeny from brood year *i* is equivalent to a binomial sample of 2S gametes from the N_b spawners in year *i* (allele frequency P_i), so the variance of allele frequency with respect to P_i [$V(X_i|P_i)$] in juvenile samples is just the binomial variance:

$$\text{(juvenile sample) } V(X_i \mid P_i) = \frac{P_i(1 - P_i)}{2S}. \qquad (1)$$

If, as will typically be the case, the effective number of breeders is less than the actual number (N), a juvenile sample still estimates allele frequency (P_i) in the $2N_b$ genes making up the effective breeding population in brood year *i* (with variance still the binomial variance given above). An adult sample, however, provides an estimate of allele frequency (call it P_i') in all the spawning adults subject to being sampled (whether or not they contribute to N_b). The variance of X_i relative to P_i, the allele frequency that is relevant for the evolutionary history of the population, depends on the relationship between N and N_b. It can be shown that this quantity is approximately equal to

$$\text{(adult sample) } V(X_i \mid P_i) \approx \frac{P_i(1 - P_i)}{2N - 1}\left[\frac{N - S}{S} + \frac{N - N_b}{N_b}\right]. \qquad (2)$$

The sampling variance in equation (2) is less than that for juvenile sampling (equation 1) provided that the ratio N_b/N is larger than 0.5. Under these circumstances, adult sampling can be expected to result in smaller allele frequency differences between samples and fewer statistically significant test results. Because of the dependence on the ratio N_b/N, temporal changes in adult samples are not shown here but they can be modeled for any parameter values desired.

Discussion

The simulation results provide insight into the process of temporal change in allele frequency in Pacific salmon. These insights, in turn, allow an evaluation of the probable causes of the relatively large changes observed in samples from the Oregon hatcheries. Major points to emerge from this analysis include:

1. For populations with broad overlap of year classes, the magnitude of short-term allele frequency change is determined primarily by the effective number of breeders (N_b) per year, rather than the age structure of the spawners. However, in populations or species (e.g., coho salmon) with a large majority of adults spawning at the same age, the magnitude of genetic change is much more sensitive to the elapsed time between the brood years being compared.
2. The magnitude of sampling error in estimating population allele frequencies is different for juvenile and adult samples of Pacific salmon. Therefore, the life history stage sampled must be taken into consideration when interpreting allele frequency data. Interpretation of results for juvenile samples is more straightforward because sampling error is not a function of the ratio N_b/N, as it is for adult samples.
3. Because genetic drift causes annual fluctuations in allele frequency in Pacific salmon populations, comparisons of samples from different brood years have an inflated probability of being statistically significant. This is not an artifact of small sample size; in fact, it increases with the ratio S/N_b and is therefore more pronounced for larger samples. This has important implications for the field of conservation genetics. Interest typically focuses on populations with reduced effective size, in which case statistical tests of allele frequency change are more likely to be significant. Conclusions based on such tests can be seriously misleading if they do not take the confounding effects of genetic drift into consideration. In the context of the results presented here, it is possible to determine whether observed changes are large enough that it is necessary to consider explanations other than genetic drift and sampling error (selection, genetic admixture, etc.).

Identifying the cause (or causes) of the genetic changes observed in the hatchery populations of chinook salmon is the necessary first step before strategies to minimize rapid changes can be effective. Additional simulations, performed in this study to model natural selection and previously (Waples & Smouse 1990) to model genetic admixture, indicate that it is difficult to explain a substantial portion of the changes by either of these factors. In contrast, all available evidence is consistent with the conclusion that the effective number of breeders per year averaged 50 or less at most of the hatcheries. This suggests that more serious changes such as inbreeding depression and erosion of genetic variability may also be occurring. Although such changes do not necessarily characterize hatchery populations in general (cf. Utter et al. 1989), they have been documented in a number of cases (reviewed by Allendorf & Ryman 1987).

An important question thus becomes, "Can we expect serious problems from inbreeding in Pacific salmon populations in which N_b is as small as 25 or 50 per year?" In attempting to answer this question, we again face a problem posed by the unusual life history features of Pacific salmon. Most (if not all) discussions of the deleterious effects of inbreeding (e.g., Franklin 1980; Simberloff 1988) refer to effective population size per generation (N_e), whereas we have shown that short-term genetic change in Pacific salmon populations is a function of the effective number of breeders per year (N_b). Because it is not clear exactly how to relate these two effective numbers, it is difficult to adapt results from the extensive literature on conservation genetics for use with Pacific salmon.

The second paper in this series (Waples, manuscript I, in press) addresses this problem. It establishes the relationship between N_b and N_e, evaluates the rate of loss of genetic variability in Pacific salmon populations, and suggests guidelines for a minimum effective number of breeders per year. The minimum acceptable number varies somewhat depending on management goals and other considerations, but an N_b of 25–50 would appear to be too small for long-term genetic health under any circumstances. A variety of measures, apart from simply using more fish, can be adopted to increase the effective number of breeders in hatcheries (see also Simon et al. 1986; Allendorf & Ryman 1987). Differences among individuals in reproductive success can be minimized by using equal numbers of each sex in pairwise spawnings and fertilizing approximately the same number of eggs from each female. If a fixed number of eggs are needed for the next brood year, taking relatively few eggs from each of a large number of females increases the effective number of breeders. Choosing breeders from throughout the spawning run helps to preserve genetically based behavior patterns related to time of spawning. If eggs or progeny are exported, they should be a random sample from the entire brood stock rather than derived from a limited number of individuals.

Measures such as these can do a great deal to prevent the erosion of genetic variability in Pacific salmon hatcheries, but even if they are implemented there is no guarantee that N_b will be as large as desired. Because most mortality occurs after release from the hatchery, a major factor in reducing N_b below the census size (high variance in reproductive success among individuals) can only be evaluated through elaborate tagging programs. Therefore, indirect estimates of N_b can be very valuable in monitoring the success of efforts to minimize inbreeding. For the Oregon hatchery populations, the observed percentage of significant χ^2 tests was used to obtain a crude estimate of N_b. Further analysis has indicated that a more powerful approach is available, analogous to methods that have been used for organisms with discrete generations (Krimbas & Tsakas 1971; Nei & Tajima 1981, Waples 1989*a*). The third paper in this series (Waples, manuscript II, in press) deals with this topic.

Although this study explicitly modeled the unique life history pattern of Pacific salmon, many other organisms with overlapping generations present some of the same problems in understanding possible causes of genetic change. In particular, the ways in which the samples for genetic analysis and the breeding population at a given point in time are drawn from the population as a whole can profoundly affect the perceived rate of change. Computer simulations that incorporate realistic values for life history parameters can contribute substantially to our understanding of genetic changes (e.g., Harris & Allendorf's 1989 study modeling a grizzly bear population).

Finally, it should be stressed that regardless of the organism, results of standard statistical tests comparing temporally spaced samples should be interpreted with caution. Even if nothing but stochastic factors are involved, the probability of a significant test statistic is an increasing function of S/N_b. Conservation biologists, who typically are interested in small populations, can therefore expect to detect "significant" genetic changes fairly often. It is important to realize that such results are sensitive to sampling effort as well as population size and may not reflect any directional evolutionary forces.

Acknowledgments

We thank Jay Nicholas and Rod Kaiser for providing samples and Ed Cummings and Leslie Schaeffer for supplying brood stock information. Eric Loudenslager brought the allele frequency changes to our attention, and Donald Campton, John Emlen, Gary Meffe, and Fred Utter made helpful comments on earlier drafts of the manuscript.

Literature Cited

Abrahamson, P. 1988. A detailed listing of the liberations of salmon into the open waters of the state of Washington during 1987. Progress Report No. 267. Washington Department of Fisheries, Olympia, Washington. 447 pp.

Aebersold, P. B., G. A. Winans, D. J. Teel, G. B. Milner, and F. M. Utter. 1987. Manual for starch gel electrophoresis: a method for the detection of genetic variation. NOAA Technical Report **61**:1–19.

Allendorf, F. W., and S. R. Phelps. 1981. Use of allele frequencies to describe population structure. Canadian Journal of Fisheries and Aquatic Sciences **38**:1507–1514.

Allendorf, F. W., and N. Ryman. 1987. Genetic management of hatchery stocks. Pages 141–159 in N. Ryman and F. Utter, editors. Population genetics and fishery management. University of Washington Press, Seattle, Washington.

Allendorf, F. W., and G. H. Thorgaard. 1984. Tetraploidy and the evolution of salmonid fishes. Pages 1–53 in B. Turner, editor. Evolutionary genetics of fishes. Plenum Press, New York.

Aspinwall, N. 1974. Genetic analysis of North American populations of the pink salmon (*Oncorhynchus gorbuscha*): possible evidence for the neutral mutation–random drift hypothesis. Evolution **28**:295–305.

Eanes, W. F. 1987. Allozymes and fitness: evolution of a problem. Trends in Ecology and Evolution **2**:44–48.

Fraidenburg, M. E., and R. H. Lincoln. 1985. Wild chinook salmon management: an international conservation challenge. North American Journal of Fisheries Management **5**:311–329.

Franklin, I. R. 1980. Evolutionary change in small populations. Pages 135–150 in M. E. Soulé and B. A. Wilcox, editors. Conservation biology: an evolutionary-ecological perspective. Sinauer, Sunderland, Massachusetts.

Fredin, R. A. 1980. Trends in North Pacific salmon fisheries. Pages 59–119 in J. W. McNeil and D. C. Himsworth, editors. Salmonid ecosystems of the North Pacific. Oregon State University Press, Corvallis, Oregon.

Harris, R. B., and F. W. Allendorf. 1989. Genetically effective population size of large mammals: an assessment of estimators. Conservation Biology **3**:1–11.

Hill, W. G. 1979. A note on effective population size with overlapping generations. Genetics **92**:317–322.

Krimbas, C. B., and S. Tsakas. 1971. The genetics of *Dacus oleae*. V. Changes of esterase polymorphism in a natural population following insecticide control — selection or drift? Evolution **25**:454–460.

Lewontin, R. C., and C. C. Cockerham. 1959. The goodness-of-fit test for detecting natural selection in randomly mating populations. Evolution **13**:561–564.

Lister, D. B., D. G. Hickey, and I. Wallace. 1981. Review of the effects of enhancement strategies on the homing, straying, and survival of Pacific salmonids. Volume I. Department of Fisheries and Oceans, Vancouver, British Columbia.

Nei, M., and W-H. Li. 1973. Linkage disequilibrium in subdivided populations. Genetics **73**:213–219.

Nei, M., and F. Tajima. 1981. Genetic drift and estimation of effective population size. Genetics **98**:625–640.

Nicholas, J. W., and D. G. Hankin. 1988. Chinook salmon populations in Oregon coastal river basins: description of life histories and assessment of recent trends in run strengths. Oregon Department of Fish and Wildlife, Information Report **88-1**:1–359.

Pollak, E. 1983. A new method for estimating the effective population size from allele frequency changes. Genetics **104**:531–548.

Ricker, W. E. 1972. Hereditary and environmental factors affecting certain salmonid populations. Pages 27–160 in R. C. Simon and P. A. Larkin, editors. The stock concept of Pacific salmon. H. R. MacMillan Lectures in Fisheries. University of British Columbia, Vancouver.

Simberloff, D. 1988. The contribution of population and community biology to conservation science. Annual Review of Ecology and Systematics **19**:473–511.

Simon, R. C., J. D. McIntyre, and A. R. Hemingsen. 1986. Family size and effective population size in a hatchery stock of coho salmon (*Oncorhynchus kisutch*). Canadian Journal of Fisheries and Aquatic Sciences **43**:2434–2442.

Utter, F., P. Aebersold, and G. Winans. 1987. Interpreting genetic variation detected by electrophoresis. Pages 21–45 in N. Ryman and F. Utter, editors. Population Genetics and Fishery Management. University of Washington Press, Seattle, Washington.

Utter, R., G. Milner, G. Stahl, and D. Teel. 1989. Genetic population structure of chinook slamon in the Pacific Northwest. Fisheries Bulletin **87**:239–263.

Waples, R. S. 1988. Estimation of allele frequencies at isoloci. Genetics **118**:371–384.

Waples, R. S. 1989*a*. A generalized method for estimating population size from temporal changes in allele frequency. Genetics **121**:379–391.

Waples, R. S. 1989*b*. Temporal stability of allele frequencies: testing the right hypothesis. Evolution **43**:1236–1251.

Waples, R. S. Conservation genetics of Pacific salmon. II. Effective population size and the rate of loss of genetic variability. (Manuscript I, Journal of Heredity, in press.)

Waples, R. S. Conservation genetics of Pacific salmon. III. Estimating effective population size. (Manuscript II, Journal of Heredity, in press.)

Waples, R. S., and P. E. Smouse. 1990. Gametic disequilibrium analysis as a means of identifying mixtures of salmon populations. American Fisheries Society Symposium **7**:439–458, in press.

Williams, J. G. 1989. Snake River spring and summer chinook salmon: can they be saved? Regulated Rivers **4**:17–26.

Withler, R. E. 1988. Genetic consequences of fertilizing chinook salmon (*Oncorhynchus tshawytscha*) eggs with pooled milt. Aquaculture **68**:15–25.

Wright, S. 1938. Size of population and breeding structure in relation to evolution. Science **87**:430–431.

Conservation Genetics of the Black Rhinoceros (*Diceros bicornis*), I: Evidence from the Mitochondrial DNA of Three Populations

MARY V. ASHLEY*

Genetics Laboratory
Department of Anthropology
Columbia University
New York NY 10027 U.S.A.

DON J. MELNICK

Genetics Laboratory
Department of Anthropology
Columbia University
New York NY 10027 U.S.A.

DAVID WESTERN

Wildlife Conservation International
New York Zoological Society
Bronx, NY 10460, U.S.A.

Abstract: *A drastic decline in the number of black rhinoceroses (*Diceros bicornis*), primarily as a result of poaching, places this species in imminent danger of extinction. The remaining black rhinos are divided into small, isolated populations that are vulnerable to demographic extinction, disease epidemics, genetic drift, and inbreeding. Some conservationists have suggested minimizing these threats by moving as many animals as possible from different isolated populations to a few safe "rhino sanctuaries." To examine the possible long-term genetic consequences of such a strategy, we focused our efforts on determining the level of genetic differences among the remaining black rhino populations by examining restriction fragment length polymorphisms of the rapidly evolving mitochondrial DNA molecule. The 23 black rhinos in our survey, including animals from*

* *Requests for reprints should be addressed to Mary V. Ashley, Department of Biology, Lake Forest College, Lake Forest, IL 60045.*
Paper submitted 3/8/89; revised manuscript accepted 8/4/89.

Resumen: *Una disminución drástica en el número de rinocerontes negros (*Diceros bicornis*), principalmente debido a la craza ilegal, pone a dicha especie en peligro de extinción inminente. Los rinocerontes negros restantes estan divididos en poblaciones pequeñas y aisladas, que son vulnerables a la extinción demográfica, las epidemias, la deriva genética y la endogamia. Algunos conservacionistas han sugerido disminuir estas amenazas, trasladando tantos animales como sea posible, de diferentes poblaciones aisladas, a un par de "santuarios seguros" para rinocerentes. Para examinar las posibles consecuencias genéticas a largo plazo de dicha estratégia, enfocamos nuestros esfuerzos en determinar el nivel de diferencia genética entre las poblaciones restantes de rinocerontes negros. Examinamos polimorfismos de fragmentos de longitud restringida de la molécula ADN, de evolución mitocondrial rápida. Los 23 ejemplares de rinocerontes negros de nuestro estudio, que incluyen a animales provenientes de tres regiones geográficas y de dos subespecies descritas, denotaron muy poca diferenciación del ADN mitocondrial. Unicamente 4 de 18 enzimas re-*

three geographic regions and two named subspecies, showed very little mitochondrial DNA differentiation. Only 4 out of 18 restriction enzymes revealed any mtDNA polymorphisms, and the average estimated percent sequence divergence between the four mtDNA genotypes observed as 0.17%. Mitochondrial DNA divergence between the two named subspecies, D. b. minor *and* D. b. michaeli, *was estimated to be only 0.29%. These results indicate a very close genetic relationship among the black rhinos in our survey. Thus, the mitochondrial DNA data suggest that within national boundaries, the black rhino populations we sampled may be considered single populations for breeding purposes, which might increase the species' probability of survival.*

strictivas, denotaron poliformismo mitocondrial del ADN (ADNmt) y, el porcentaje promedio estimado de divergencia secuencial entre los cuatro genotipos ADNmt observados, fue de 0.17%. La divergencia de ADN mitocondrial entre las dos subespecies descritas, D. b. minor *y* D. b. michaeli, *se estima que fue solo de 0.29%. Estos resultados indican una relación genética muy cercana entre los rinocerontes negros de nuestro estudio. Por lo tanto, los datos de ADN mitocondrial sugieren que, dentro de las fronteras nacionales, las poblaciones de rinocerontes negros que estudiamos pueden considerarse una sola población para finalidades de reproducción, lo cual puede incrementar la probabilidad de supervivencia de dicha especie.*

Introduction

Despite its reputation as a powerful and invincible beast, the rhinoceros has suffered precipitous declines in number and is threatened with extinction. Fewer than 11,000 individuals of all five species survive in small scattered populations throughout Africa, India, and Southeast Asia. The black rhino (*Diceros bicornis*), the focus of this study, has suffered the most dramatic decline, disappearing faster than any other large mammal. The species once occupied most of sub-Saharan Africa and numbered in the hundreds of thousands (Fig. 1). Even by the turn of the century, large, nearly contiguous populations of black rhino were spread across much of central, eastern, and southern Africa. However, by 1970 their numbers had declined to 65,000 and over the past 18 years poaching has reduced this number by 95%. The remaining 3,800 animals are split into some 75 populations, only ten of which have more than 50 animals (Western & Vigne 1985; Du Toit et al. 1987; Wildlife Conservation International News 1988).

Although the loss of any species is tragic, the plight of the rhinoceros is particularly apalling because the overwhelming cause of their demise is not destruction of their preferred habitat but continued slaughter by poachers to supply two major markets. Horns are fashioned into ornamental dagger handles costing up to $30,000 in the Near East, particularly in North Yemen. In many parts of East Asia, rhino horn is valued at $8,000 to $15,000 per kilogram for various medicinal purposes (Martin 1983; Penny 1988). Increasing affluence in the Fat East and oil wealth in the Near East, coupled with declining availability, have drastically increased the value of rhino horn in recent years, creating devastating repercussions for rhinos in Africa. The market value of rhino horn is currently so high that all age and size classes are susceptible to poachers (Leader-Williams 1988).

Assuming for the moment that the governments of countries still harboring sizeable numbers of black rhino (e.g., Kenya, Tanzania, Zimbabwe, and South Africa) are successful in controlling poachers, another major problem exists for the species. The small, isolated populations of black rhino that remain are vulnerable to the effects of demographic fluctuations, local ecological perturbations and disease epidemics, and loss of genetic variability due to drift and inbreeding (e.g., Crow & Kimura 1970; Gilpin & Soulé 1986; Wilcove et al. 1986). Additionally, low population density reduces the probability that a male will find a female during the 1–2-day period in her estrous cycle when she is sexually

Figure 1. Map of Africa showing black rhino distributions during historical times (heavy black line), 1900 (hatched area), and 1987 (black area). Redrawn from Western & Vigne (1985), incorporating data from WCI for 1987.

receptive (Hitchins & Anderson 1983). Hence, what were once thought to be minimum viable populations on the basis of genetic parameters (i.e., inbreeding depression and loss of genetic heterogeneity) are now in many instances considered underestimates (Soulé 1987).

The shortage of manpower and resources within national conservation departments across Africa is a primary factor in the decline of the rhino (Leader-Williams & Albon 1988). Consolidating groups is an easier and less expensive way to reestablish former densities. The idea of creating more sanctuaries, which are effectively "species parks," has received widespread interest following the increase in Kenya's rhino populations in such sanctuaries. However, management strategies such as this are complicated by the fact that each remaining population has been assigned to one of several subspecies, based upon aspects of external morphology such as horn shape and body size. The genetic and evolutionary relationships of these morphologically defined subspecies are unknown. The most widely accepted classification (Groves 1967) recognizes seven subspecies of *Diceros bicornis,* one of which *D. b. ladoensis,* is probably extinct. Three other subspecies — *D. b. brucii,* found in Ethiopia and Somalia; *D. b. longipes,* which remains only in Cameroon and perhaps Chad; and *D. b. chobiensis,* found in Angola — are down to a few dozen animals, if they are not already extinct (Western & Vigne 1985). *D. b. bicornis,* if it can be considered a distinct taxon at all, is found in Namibia and probably numbers less than 100 (Hall-Martin 1985; Du Toit et al. 1987). The remaining two subspecies, *D. b. michaeli* and *D. b. minor,* or populations designed as such, will figure most importantly in the return of the black rhino, should this be accomplished. *D. b. michaeli,* found in Kenya and Tanzania, has declined drastically as well, but still numbers between 500 and 1,000 and has increasingly received more protection in Kenya. The vast majority of the approximately 150 black rhinos in North American and European zoos are of this subspecies (Du Toit et al. 1987) and so will be important for future captive breeding efforts. *D. b. minor* is the most common remaining race, ranging from Kenya to South Africa, and with numbers at about 2,500, it is relatively the most secure.

Should all the remaining black rhinos be considered as a single population for breeding purposes? This tactic might increase their chances of survival by increasing effective population sizes and thus forestalling stochastic demographic extinctions, inbreeding depression, and loss of the species' existing genetic variability (Soulé 1983; Gilpin & Soulé 1986; Ralls et al. 1986; Goodman 1987). Alternatively, do different populations (which may or may not coincide with subspecies designations) merit separate conservation as genetically and possibly ecologically distinct units? The latter strategy might prevent outbreeding depression or the production of animals with genetic makeups inappropriate for a given environment (Templeton 1986). These questions should, ideally, be tackled from both an ecological and genetic standpoint. Ecologically, it might be possible to distinguish locally adapted traits. Statistically significant differences in serum vitamin E levels, for example, have been found between Kenyan and southern African samples, which may reflect substantial differences in diet (Dierenfield, personal communication). Ecological differences also distinguish the desert rhinos of Namibia from the highland forest rhinos of Kenya (Du Toit 1987).

In the absence of any clear morphometric differences, IUCN's African Elephant and Rhino Specialist Group has placed a high priority on genetic studies of black rhinos to resolve whether discrete populations could be identified (Du Toit et al. 1987). As a first step in applying molecular genetic techniques to questions of black rhino conservation, we have examined the mitochondrial DNA (mtDNA) of 23 black rhinos representing two morphologically defined subspecies and three geographic populations. We chose mtDNA because its rapid evolutionary rate has shown it to be a useful molecule for determining intraspecific relationships of many animals (e.g., Wilson et al. 1985; Avise & Lansman 1983). If the rhino populations surveyed here have had separate evolutionary histories for a considerable length of time, it should be reflected in the divergence of mtDNA's from animals in different populations.

The mitochondrial genome consists of a closed circular DNA molecule which codes for 13 proteins and a complete set of transfer RNAs. It is extremely conserved in size (about 16,000 base pairs in all mammals that have been examined) and gene arrangement (Brown 1983). It lacks the complicating features of repetitive DNA or introns; therefore, a relatively simple restriction enzyme analysis of the molecule can be undertaken to yield good estimates of genetic relationships among fairly large numbers of individuals. It is maternally inherited without recombination and thus represents an unambiguous marker of maternal phylogeny. Because it evolves 5–10 times more rapidly than single-copy nuclear DNA (Brown et al. 1979) and intraspecific mtDNA variability has been widely demonstrated, this approach seemed the most likely to uncover genetic differentiation among black rhinos, should it exist.

Materials and Methods

With the cooperation of field biologists, wildlife managers, and zoo personnel in both Africa and the United States, we were able to obtain whole blood from both captive and wild-caught black rhinos (Table 1). Our sample included 11 *D. b. michaeli* of Kenyan origin now kept in U.S. zoos, 11 *D. b. minor* taken from wild populations in Zimbabwe, and one captive (U.S.) *D. b. minor* of South African origin. While blood was separated into plasma, red blood cells, platelets, and white blood cells or buffy coats, the latter two components being our primary source of DNA. Total DNA was extracted from white blood cells or buffy coats by standard procedures. We also obtained frozen organ tissue from three animals that died during the period of our study (Table 1). This frozen tissue served as a source of purified mtDNA, which was isolated by the method of differential centrifugation (Lansman et al. 1981).

We have used restriction enzymes to survey the black rhinos for mtDNA polymorphisms. Restriction enzymes recognize specific oligonucleotide sequences, usually 4 to 6 base pairs in length, and cleave double-stranded DNA wherever these sequences occur. By surveying mtDNAs with a set of restriction enzymes, we can obtain an accurate estimation of similarity by determining the proportion of restriction fragments and/or restriction sites they share.

Samples of total DNA were digested with 14 restriction enzymes (Bethesda Research Laboratories) having 5 or 6 base pair (b.p.) recognition sites, according to manufacturer's instructions. These enzymes typically cleave mtDNA into 1–7 fragments. The DNA fragments once obtained were separated electrophoretically in 1% agarose gels along with a radioactively labeled (α-32-P) one-kilobase ladder (Bethesda Research Laboratories), then transferred to GeneScreen-plus membranes (New England Nuclear) by an alkaline blotting procedure (Southern 1975; Reed & Mann 1985). Purified mtDNA obtained from tissue was then nick-translated with α-32-P labeled nucleotides and was used to "probe" the southern blots. Membranes were then washed under high-stringency conditions and exposed to Kodak XAR film.

Additionally, to increase our resolution, we digested the three purified mtDNA samples (one from each population) with four enzymes having 4 b.p. recognition sites. These enzymes cleave the mtDNA into 20–30 fragments and thus have a greater likelihood of revealing differences between individuals. Because each sample contained only purified mtDNA, the resulting restriction fragments could be directly labeled with α-32-P (Brown 1980) before being separated electrophoretically on 3.5% polyacrylamide gels. Again, an appropriate radioactively labeled molecular weight/size standard was included in the gel. Gels were subsequently dried under vacuum and exposed to Kodak XAR film.

The proportion of shared restriction fragments was calculated between the observed mtDNA genotypes. The percent sequence divergence between the mitochondrial genotypes was estimated using equation 6b of Upholt (1977). Calculations for restriction enzymes having 6, 5, and 4 b.p. restriction sites were calculated separately, then weighted according to the total number of base pairs recognized by each type of restriction enzyme. This procedure allowed an overall estimate or weighted average of nucleotide sequence divergence, based on the differences revealed by all the restriction enzymes used, between the mtDNA of different individuals.

Results

Each restriction fragment pattern produced by a given enzyme was arbitrarily assigned a letter. The results of the enzymes having 5 or 6 base-pair recognition sites for all 23 animals are listed in Table 2. The results of enzymes for a smaller set of three animals, including en-

Table 1. Black rhino samples.

Sample	*Origin**	*Subspecies***	*Tissue*	*Source*
1–8	Zimbabwe	*D. b. minor*	Buffy coat	Department of Parks and Wildlife Management Zimbabwe
9	Zimbabwe	*D. b. minor*	Frozen liver	Department of Parks and Wildlife Management Zimbabwe
10, 11	Zimbabwe	*D. b. minor*	W.B.C.***	Los Angeles Zoo
12	South Africa	*D. b. minor*	Frozen brain	Calvin Bentsen Ranch Brownsville, Texas
13–16	Kenya	*D. b. michaeli*	W.B.C.	Denver Zoo
17–19	Kenya	*D. b. michaeli*	W.B.C.	St. Louis Zoo
20	Kenya	*D. b. michaeli*	W.B.C.	Zoo Atlanta
21	Kenya	*D. b. michaeli*	W.B.C.	Busch Gardens
22	Kenya	*D. b. michaeli*	Frozen liver	Kansas City Zoo
23	Kenya	*D. b. michaeli*	W.B.C.	Detroit Zoo

* *For animals born in captivity, the origin of the individual's mother is given.*
** *Subspecific designation according to range distributions given by Groves, 1967.*
*** *White blood cells.*

Table 2. mtDNA Patterns for enzymes with 5 and 6 b.p. sites.

		D. b. minor		*D. b. michaeli*
		South Africa	*Zimbabwe*	*Kenya*
Enzyme	*# Sites**	*n = 1*	*n = 11*	*n = 11*
AvaI	2	A	A	A
BamHI	2	A	A	A
BglII	1	A	A	A
ClaI	1	A	A	A
DraI	5	A	A	A
EcoRI	2(1)	A	A(B)	A
EcoRV	2	A	A	A
HaeII	4	A	A	A
HindIII	4	A	A	A
ScaI	7	A	A	A
XbaI	4	A	A	A

* *The number in parentheses represents the number of restriction sites for haplotype B.*

Table 3. mtDNA Patterns for additional enzymes.

		D. b. minor		*D. b. michaeli*
		South Africa	*Zimbabwe*	*Kenya*
Enzyme	*# Sites**	*n = 1*	*n = 1*	*n = 1*
AvaII	4	A	A	A
BclI	6(5)	A	A	B
HincII	7	A	A	A
HinfI	30(29)	A	A	B
HpaII	16	A	A	A
MboI	23	A	A	A
TaqI	24(25)	A	A	B

* *The number in parentheses represents the number of restriction sites for haplotype B.*

zymes having 4 b.p. recognition sites (i.e., HinfI, HpaII, MboI, TaqI), are presented in Table 3. The restriction enzymes used in our survey yielded an average of 140 restriction sites per mitochondrial genome. This corresponds to a recognized total of over 630 b.p., or 3.9% of the mitochondrial genome. For 14 out of a total of 18 restriction enzymes, absolutely no mtDNA variability was observed. That is, all rhinos surveyed had the identical restriction fragment pattern (designated as "A" in Tables 2 and 3) for any one of these 14 restriction enzymes. One enzyme, EcoRI, was found to be polymorphic among the Zimbabwe rhino, with 3 of 11 animals possessing only one EcoRI restriction site instead of the two sites found in the other 8 members of this population. Three enzymes, BclI, HinfI, and TaqI, revealed a difference between the Kenyan population and the Zimbabwe and South African populations. In each case, the result could be interpreted as a single loss or gain of a restriction site.

In total, then, for our sample of 23 animals, only three mtDNA haplotypes could be distinguished: (1) the Kenyan haplotype with fragment pattern "B" for BclI, HinfI, and TaqI; (2) the Zimbabwe haplotype with fragment pattern "B" for EcoRI; and (3) the Zimbabwe and South African haplotype with fragment pattern "A" for all 18 restriction enzymes. These three mtDNA haplotypes are extremely similar to one another (Table 4), with an average estimated percent sequence difference between any pair of haplotypes and/or populations of 0.17%. The average difference between subspecies was only slightly higher, 0.29%.

Discussion

The results of the mtDNA analysis strongly suggest a very close genetic relationship among all the black rhinos in our survey. Because of the generally rapid rate of mtDNA evolution in mammals, differences observed among rhino populations appear to indicate a very recent common ancestry. If mtDNA evolves at a rate of 2% per million years as suggested (Brown et al. 1979; Wilson et al. 1985), this common ancestry probably dates back no farther than 100,000 years. Indeed, the level of differentiation between the so-called subspecies is well within the range (0–4%) observed among members of other mammalian species (e.g., Avise & Lansman 1983), and even within the range (0–2%) that has been observed among members of the same local population (Ashley & Wills 1987). Thus, there is no evidence from these data that the black rhinos we sampled represent "evolutionarily distinct units."

These findings for the black rhino stand in sharp contrast to the level and distribution of mtDNA differences reported for the white rhino. The southern white rhino, *Ceratotherium simum simum,* has recovered quite well from a population bottleneck that occurred at the turn of the century, and now more than 3,000 members of this population can be found in South Africa and other African countries (Penny 1988). The status of the northern white rhino, *Ceratotherium simum cottoni,* is much bleaker, with only 22 individuals known to exist in the wild (Western 1987). Management plans originally proposed supplementing the northern race with members from the southern population. Based on a survey of one individual from each of the two races, however, George et al. (1983) reported a very high level of mtDNA divergence (approximately 4.0%). Partially as a result of this study, managers decided against interbreeding the two races. Unlike the black rhino, the white rhino subspecies have existed in nonoverlapping

Table 4. Estimated percent sequence divergence between mtDNA types, based on the proportion of shared restriction fragments (Upholt, 1977).

	Zimbabwe 1	*Zimbabwe 2*	*S. Africa*
Kenya	0.24	0.39	0.24
Zimbabwe 1	—	0.08	0.00
Zimbabwe 2		—	0.08
S. Africa			—

ranges, at least during historical times. George et al. suggest, on the basis of their molecular data, that the two white rhino subspecies have been isolated from each other for at least two million years. However, recent research on variability of nuclear-coded allozymes found little differentiation between the northern and southern subspecies (Merenlender et al. 1989), suggesting a more recent isolation.

We chose mtDNA analysis because we thought it would be most likely to uncover genetic differences between the black rhino populations, should they exist. It seems unlikely from our results that significant barriers to successful interbreeding would exist, given what we estimate to be a brief history of separation between the populations in question. However, more information should be obtained before final management decisions are made. The mitochondrial genome represents only a tiny fraction of an organism's genetic makeup, and problems that might arise from interbreeding might not necessarily be reflected in mtDNA differentiation. For this reason, we are conducting an allozyme survey in our laboratory to determine if the findings regarding the mtDNA hold for nuclear-coded genes as well. The allozyme survey will also be more informative for determining if the black rhinos suffer from reduced levels of genetic variability, as has been reported for some species that have passed through recent population bottlenecks (Bonnell & Selander 1974; O'Brien et al. 1983). Thus far, we have found no allozyme polymorphisms within or between populations, despite the fact that (1) we have included in our analysis animals from Kenya, Zimbabwe, and three different populations in South Africa (Etosha, Ado, and Zululand) and (2) our initial surveys have included all three allozyme loci identified by Merenlender et al. (1989) as polymorphic in African rhinos (Amato & Melnick, unpublished data).

Karyotype analysis is also recommended, because chromosomal differences reducing the fertility of hybrids could conceivably exist in the absence of either allozyme or mtDNA differentiation. There have been no known crosses of black rhinos from different subspecies in captivity, which might indicate reduced viability or fertility.

The application of genetics to conservation issues is a practical endeavor and should yield concrete recommendations for management strategies. Black rhinos from the populations included in our study will probably be the ancestors of all future black rhinos, as their successful breeding is the only hope for the survival of the species. Our results provide strong evidence for a very close genetic relationship among these populations. At the national level, the level at which management decisions are currently made, the pooling of black rhinos carries with it little risk of mixing distinct genetic adaptations worthy of separate conservation efforts. This finding should allow managers to aggregate individuals to create larger local populations or demes. Preserving the black rhino in relatively large local populations would have several beneficial effects. These include retarding the rate of loss of genetic variability, buffering each aggregate against the possibility of demographic extinction, restoring previous population densities, and allowing the wildlife managers with limited resources to provide better production against poachers. Taken together, these effects should, in the long run, increase the probability of survival of this critically endangered species.

Acknowledgments

This research grew out of the participation by two of us (DJM and DW) in the October 1986 African Rhino Workshop organized by the AAZPA SSP, the Cincinnati Zoo, and the King's Island Wild Animals Habitat. It represents one part of a larger effort to understand the genetic structure of the black rhinoceros and help preserve this highly endangered species. We are grateful to many institutions and individuals who helped put together the research materials and provided financial support for the laboratory analysis. In particular we wish to thank David Cumming, Raoul Du Toit, and the Wildlife Department of Zimbabwe, and Perez Olindo and the Department of National Parks and Wildlife Management of Kenya, for supplying us with blood and organ tissue samples critical to this study. We would also like to thank the administration and veterinary personnel of Busch Gardens, Tampa, Florida; the Calvin Bentsen Ranch, Brownsville, Texas; Denver Zoo; Detroit Zoo; Kansas City Zoo; Los Angeles Zoo; St. Louis Zoo; and Zoo Atlanta for supplying samples from captive black rhino of Kenyan and South African origin. We appreciate the comments of two anonymous reviewers and Robert Vrijenhoek on an earlier version of this article. Finally, we would like to single out Mary Pearl, Tom Foose, Ed Maruska, and Eric Miller for their logistical support and encouragement throughout this project. Financial support for this research was supplied by Wildlife Conservation International, a division of the New York Zoological Society, the Cincinnati Zoo, the AAZPA, and those other U.S. zoos that contributed to the Black Rhino SSP's fund-raising effort.

Literature Cited

Ashley, M., and C. Wills. 1987. Analysis of mitochondrial DNA polymorphisms among Channel Island deer mice. Evolution 41:854–863.

Avise, J. C., and R. A. Lansman. 1983. Polymorphism of mitochondrial DNA in populations of higher animals. Pages 147–164 in M. Nei and R. K. Koehn, editors. Evolution of genes and proteins. Sinauer, Sunderland, Massachusetts.

Bonnell, M. L., and R. K. Selander. 1974. Elephant seals: genetic variation and near extinction. Science **184**:908–909.

Brown, W. M. 1980. Polymorphism in mitochondrial DNA of humans as revealed by restriction endonuclease analysis. Proceedings of the National Academy of Sciences of the United States of America **77**:3605–3609.

Brown, W. M. 1983. Evolution of animal mitochondrial DNA. Pages 62–88 in M. Nei and R. K. Koehn, editors. Evolution of genes and proteins. Sinauer, Sunderland, Massachusetts.

Brown, W. M., M. George, Jr., and A. C. Wilson. 1979. Rapid evolution of animal mitochondrial DNA. Proceedings of the National Academy of Sciences of the United States of America **76**:1967–1971.

Crow, J. F., and M. Kimura. 1970. An introduction to population genetics theory. Harper & Row, New York.

Du Toit, R. 1987. Conservation biology of the black rhino. Zimbabwe Science News **21**:63–67.

Du Toit, R., T. J. Foose, and D. H. M. Cumming. 1987. Proceedings of the African Rhino Workshop, Cincinnati, October, 1986. Pachyderm Special Issue, No. 9.

George, M., Jr., L. A. Puentes, and O. A. Ryder. 1983. Genetic differences between subspecies of white rhinoceroses. International studbook of African rhinos. Berlin Zoological Garden. Berlin, East Germany.

Gilpin, M. E., and M. E. Soulé. 1986. Minimum viable populations: the process of species extinction. Pages 13–34 in M. E. Soulé, editor. Conservation: the science of scarcity and diversity. Sinauer, Sunderland, Massachusetts.

Goodman, D., 1987. The demography of chance extinction. Pages 11–34 in M. E. Soulé, editor. Viable populations for conservation. Cambridge University Press, Cambridge, England.

Groves, C. P. 1967. Geographic variation in the black rhinoceros *Diceros bicornis* (L., 1758). Z Saugetierkunde **32**:267–276.

Hall-Martin, A. 1985. The nabab of Aukoerebis. African Wildlife **39**:245–247.

Hitchins, P. M., and J. L. Anderson. 1983. Reproduction, population characteristics and management of the black rhinoceros *Diceros bicornis minor* in the Hhuhluwe/Corridor/Umfolozi game reserve complex. South African Journal of Wildlife Research **13**:78–85.

Lansman, R. A., R. O. Shade, J. F. Shapira, and J. C. Avise. 1981. The use of restriction endonucleases to measure mitochondrial DNA sequence relatedness in natural populations. III. Techniques and potential applications. Journal of Molecular Evolution **17**:214–226.

Leader-Williams, N. 1988. Patterns of depletion in a black rhinoceros population in Luangwa, Zambia. African Journal of Ecology **26**:181–187.

Leader-Williams, N., and S. D. Albon. 1988. Allocation of resources for conservation. Nature **336**:533–535.

Martin, E. B. 1983. Rhino exploitation. World Wildlife Fund, Hong Kong.

Merenlender, A. M., D. S. Woodruff, O. A. Ryder, R. Kock, and J. Váhala. 1989. Allozyme variation in African and Indian rhinoceroses. Journal of Heredity **80**:377–382.

O'Brien, S. J., D. E. Wildt, D. Goldman, D. R. Merril, and M. Bush. 1983. The cheetah is depauperate in genetic variation. Science **221**:459–462.

Penny, M. 1988. Rhinos: endangered species. Facts on File, New York.

Ralls, K., P. H. Harvey, and A. M. Lyles. 1986. Inbreeding in natural populations of birds and mammals. Pages 35–56 in M. E. Soulé, editor. Conservation biology: the science of scarcity and diversity. Sinauer, Sunderland, Massachusetts.

Reed, K. C., and D. A. Mann. 1985. Rapid transfer of DNA from agarose gels to nylon membranes. Nucleic Acid Research **13**:7207–7221.

Soulé, M. E. 1983. What do we really know about extinction? Pages 111–124 in C. M. Schonewald-Cox, S. M. Chambers, B. MacBryde, and L. Thomas, editors. Genetics and conservation. Benjamin/Cummings, London, England.

Soulé, M. E. 1987. Introduction. Pages 1–10 in M. E. Soulé, editor. Viable populations for conservation. Cambridge University Press, Cambridge, England.

Southern, E. M. 1975. Detection of specific sequences among DNA fragments separated by gel electrophoresis. Journal of Molecular Biology **98**:503–517.

Templeton, A. R. 1986. Coadaptation and outbreeding depression. Pages 105–116 in M. E. Soulé, editor. Conservation genetics: the science of scarcity and diversity. Sinauer, Sunderland, Massachusetts.

Upholt, W. B. 1977. Estimation of DNA sequence divergence from comparison of restriction endonuclease digests. Nucleic Acids Research **4**:1257–1265.

Western, D. 1987. Africa's elephants and rhinos: flagships in crisis. Trends in Ecology and Evolution **2**:343–345.

Western, D., and L. Vigne. 1985. The deteriorating status of African rhinos. International Union for Conservation of Nature/Species Survival Commission Report **1985**:1–15.

Wilcove, D. S., C. H. McLellan, and A. P. Dobson. 1986. Habitat fragmentation in the temperate zone. Pages 237–256 in M. E. Soulé, editor. Conservation biology: the science of scarcity and diversity. Sinauer, Sunderland, Massachusetts.

Wildlife Conservation International News. 1988. Rhino rescue fund to implement emergency action plan. Summer issue.

Wilson, A. C., R. L. Cann, S. M. Carr, M. George, U. B. Gyllensten, K. M. Helm-Bychowski, R. G. Higuchi, S. R. Palumbi, E. M. Prager, R. D. Sage, and M. Stoneking. 1985. Mitochondrial DNA and two perspectives on evolutionary genetics. Biological Journal of the Linnear Society **26**:375–400.

Restoration Considerations for Wiregrass (*Aristida stricta*): Allozymic Diversity of Populations

TERRENCE W. WALTERS

Fairchild Tropical Garden
11935 Old Cutler Road
Miami, FL 33156, U.S.A.
and
Department of Biological Sciences
Florida International University
Miami, FL 33199, U.S.A.

DEENA S. DECKER-WALTERS

Fairchild Tropical Garden
11935 Old Cutler Road
Miami, FL 33156, U.S.A.

DORIA R. GORDON

The Nature Conservancy
Florida Museum of Natural History
University of Florida
Gainesville, FL 32611, U.S.A.

Introduction

Millar & Libby (1989) discuss the consequences restorationists face in using genetically inappropriate stock for restoration and re-introduction projects. They point out that using nonlocal populations to restore sites near or within areas of native populations can cause serious genetic contamination. The introduced stock, if it survives, can interbreed with natives, passing on nonnative genes or gene complexes to the offspring of native individuals. This, in turn, may affect the dynamics of the entire community and the course of evolution of the community as well as the taxon. Another problem with using propagules from nonlocal sites is that these introductions might not tolerate the edaphic and climatic conditions of the new site. Use of such germplasm in restoration efforts may result in high mortality and inefficient projects over the long term (Bacchus 1992).

These concerns are relevant for wiregrass (*Aristida stricta* Michx.; Poaceae), a perennial bunchgrass that is an understory pineland dominant across much of the southeastern coastal plain. The original range extended from southeastern North Carolina to the Florida Everglades and west to Mississippi. Wiregrass grows in deep, sandy soils in sandhills, pine flatwoods, open bogs, and savannas (see Clewell 1989). Areas presently dominated by wiregrass support another 191 known species of rare plants (Stolzenburg 1991).

Restoration of wiregrass and the fire it carries may now be possible over large areas from which the species has been extirpated because of recent success in prop-

Paper submitted August 5, 1992; revised manuscript accepted September 23, 1993.

agating this species (Seamon et al. 1989; Seamon & Myers 1992). Both the public and private sectors have an interest in converting improved pasture and agricultural and silvicultural sites back to pinelands, the native community for wiregrass. Extensive wiregrass population augmentation and restoration efforts across the southeast are believed to be imminent.

Although wiregrass is not rare, its present distribution covers only a fraction of its original area. Destruction and degradation of habitat through agriculture, urbanization, silviculture, grazing, and fire suppression have significantly decreased both the extent and density of wiregrass populations. Populations of wiregrass have been predicted to become rare within a few decades (Clewell 1989).

The decline of this species becomes more critical when its role in the community is examined. Because of foliage, growth, and density characteristics, wiregrass is highly flammable and supplies much of the fine fuel that supports fire in the pyrogenic communities in which it is found (Clewell 1989). Loss of wiregrass from these systems can accelerate loss of species richness and overall degradation.

Several characteristics of this species suggest that genetic differentiation among populations is likely. Wiregrass is distributed over a wide geographical area that differs regionally and locally in edaphic and climatic conditions. Individuals are long-lived (Clewell 1989), seed viability and dispersal distance are low, and seedling establishment in the field appears rare (Parrott 1967). Furthermore, flowering is fire- and seasonally-dependent across regions. For example, in North Carolina, plants flower after a fire occurring between November and August; in Florida, only fires between April and August stimulate flowering (Parrott 1967; Seamon & Myers 1992). Thus, evidence for phenological differentiation exists.

As a supplement to continued morphological and life-history studies of Florida populations of wiregrass by the third author, we analyzed allozyme variation within and among wiregrass populations from three distinct regions within the species' geographical range. A better understanding of the distribution of genetic variation in wiregrass should assist workers in choosing the most genetically appropriate stock for restoration projects (Maunder 1992).

Materials and Methods

Plant Material

Thirteen wiregrass populations from Florida, Alabama, Mississippi, and North Carolina were sampled (Fig. 1, Table 1). The Florida populations were distributed over roughly 440 km. Distances between FL1 and FL2, FL4 and FL5, and FL8 and FL9 were 8, 1.6, and 0.4 km, respectively. Distances between AL10 and MS11 and between MS11 and MS12 were 2–3 km each.

We analyzed allozymes in the leaf blades of nine to 17 individuals per population (Table 1). For each of the seven Florida populations, seeds from 15 plants were collected. These seeds were sown, and the second-generation plants were grown to maturity under uniform conditions at the University of Florida. Various second-generation plants, representing as many different wild parents as possible, were tested electrophoretically. Electrophoretic material for populations outside of Florida (Table 1) was collected directly from individual wild plants.

Extraction

Approximately 0.030 grams of mature blade tissue was homogenized in two drops of extraction buffer (see Walters & Decker-Walters 1991) with a mortar and glass rod. The resulting supernatant was absorbed onto 18 mm by 3 mm Whatman #3 filter paper wicks.

Electrophoresis

Using the procedures and gel/electrode buffer systems described in Walters & Decker-Walters (1993), we assayed eight enzyme systems: aspartate aminotransferase (AAT), glucose-6-phosphate isomerase (GPI), isocitrate dehydrogenase (IDH), leucine aminopeptidase (LAP), malate dehydrogenase (MDH), phosphoglucomutase (PGM), phosphogluconate dehydrogenase (PGD), and shikimate dehydrogenase (SKD).

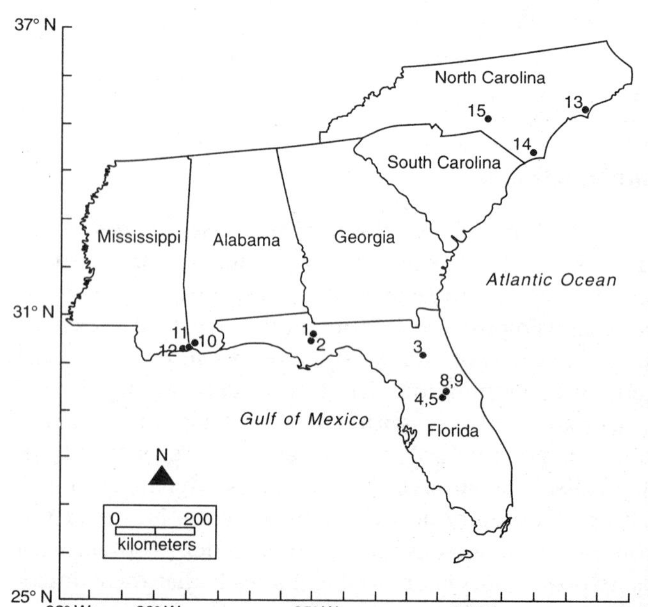

Figure 1. Locations of the 13 Aristida stricta *populations sampled in the southeastern United States.*

Table 1. Location data and number of individuals sampled for 13 populations of *Aristida stricta*.

Population Designation	*State*	*County*	*Locality*	*Plant Community*[1]	*Number of Individuals*
FL1	Florida	Liberty	Torreya State Park	S	11
FL2	Florida	Liberty	Appalachicola Bluffs and Ravines Preserve	S	10
FL3	Florida	Putnam	Katharine Ordway-Swisher Memorial Preserve	S	13
FL4	Florida	Orange	Wekiwa Springs State Park—east	S	17
FL5	Florida	Orange	Wekiwa Springs State Park—west	S	14
FL8	Florida	Orange	Rock Spring Run State Park—north	F	12
FL9	Florida	Orange	Rock Spring Run State Park—south	F	14
AL10	Alabama	Mobile	Grand Bay	P	10
MS11	Mississippi	Jackson	Pecan—north	P	10
MS12	Mississippi	Jackson	Pecan—south	P	10
NC13	North Carolina	Carerett	Croatan National Forest	P	10
NC14	North Carolina	Brunswick	Green Swamp Nature Reserve	P	9
NC15	North Carolina	Hoke	Holland Lake, Fort Bragg	S	10

[1] *S = sandhill, F = pine flatwood, P = pine savanna.*

Data Interpretation

Isozyme loci were identified numerically according to the relative migration of their enzymatic products on the gel; loci producing the most anodal variants received the lowest numerical designation. Alleles were identified with an alphabetic symbol reflecting the position of the associated enzyme product on the gel; alleles producing more anodal enzymes were assigned lower alphabetic characters.

The BIOSYS-1 computer package of Swofford & Selander (1989) was used to generate allelic frequencies, mean number of alleles across all loci (A_L), mean number of alleles across polymorphic loci (A_P), the percentage of loci polymorphic (PLP), genetic identities (Nei 1972) among populations, and a cluster analysis (UPGMA) of the genetic identities.

Results and Discussion

Approximately 25 enzyme loci were discovered for the eight enzyme systems. Generally, the phenotypes and population segregation patterns were as expected for a diploid organism. However, the high number of loci apparent in a few enzyme systems (for example, four for PGM and four for IDH) suggests polyploid and/or aneuploid changes in this species' past.

Only 15 putative loci were consistently interpretable and scorable. *Idh-3* was monomorphic. Allelic frequencies at the 14 polymorphic loci for each population are available from the first author.

For the wiregrass populations, the mean number of alleles per locus (A_L) ranged from 1.8 to 2.1, with an overall average of 1.9. The mean number of alleles for polymorphic loci (A_P) ranged from 2.2 to 2.7, with an overall average of 2.3. The percentage of loci that were polymorphic (PLP) ranged from 60 to 87%, with an overall average of 72%. Species means of A_L, A_P, and PLP were 3.6, 3.8, and 93%, respectively.

Populations of widespread angiosperm species have an average A_P value of 2.67 and a PLP value of 43%; species means, on average, are 3.2 for A_P and 59% for PLP (Hamrick et al. 1991). Although the values calculated here for wiregrass are high relative to those of other widespread, long-lived, herbaceous perennials, high outcrossing levels probably account for the increased genetic variation within wiregrass populations (see Loveless & Hamrick 1984; Hamrick & Godt 1989; Hamrick et al. 1991).

Nei's (1972) genetic identities among the wiregrass populations ranged from 0.83 to 0.99; these values are typical for conspecific populations (Crawford 1983). Cluster analysis of the identities produced three major groupings corresponding to three geographic regions: Florida, western Alabama/eastern Mississippi, and North Carolina (Fig. 2). These groupings resulted largely from alleles at the five loci listed in Table 2. For exam-

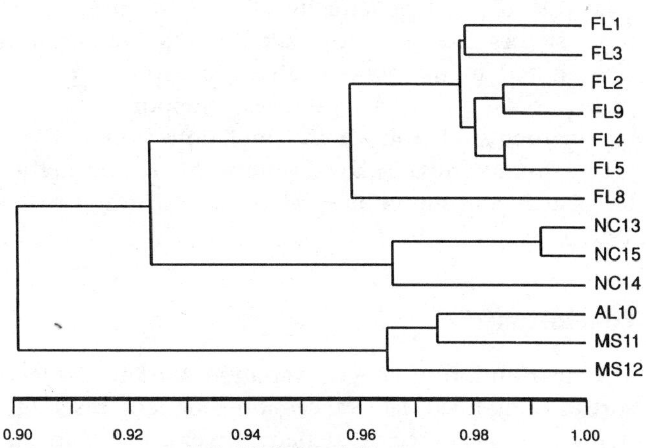

Figure 2. Phenogram resulting from cluster analysis (UPGMA) of Nei's (1972) genetic identities among 13 populations of Aristida stricta *from Florida (FL), Alabama (AL), Mississippi (MS), and North Carolina (NC). See Table 1 for population descriptions.*

Table 2. Genetic profiles at five enzyme loci for 13 populations of *Aristida stricta*.[1]

Population	*Aat-1*	*Gpi-3*	*Lap-2*	*Pgm-1*	*Skd-2*
FL1	mk	usw	ded^n	hi	omn
FL2	mkn	usw	def	hi	oml
FL3	mkj	usw	de	hi	oml
FL4	mk	us	de	hi	omlnr
FL5	m	us	def	hi	omlnr
FL8	m	us	de	hi	moln
FL9	mn	su	de	hi	oml
AL10	mj	uxs	ed	hi	roq
MS11	mkj	xu	ed	h	roqm
MS12	mkj	xu	ed	hi	rom
NC13	m	uxs	ed	ih	oml
NC14	m	uxs	ed	ih	noml
NC15	mo	uxs	ed	ih	omln

[1] *These loci account for the three major geographic subdivisions in Figure 1. Alleles at a locus are listed according to their relative frequencies within a population. See Table 1 for population descriptions.*

ple, Florida populations lacked *Gpi-3x,* an allele present in the other states and dominant in the two populations from Mississippi. At *Lap-2, d* was the dominant allele in Florida populations whereas *e* was more common in the remaining populations. *Aat-1j* was common in Alabama and Mississippi, rare in Florida, and not detected at all in North Carolina. *Aat-1k* was also missing in North Carolina populations. These populations were further characterized by a higher frequency of allele *i* at *Pgm-1.* Finally, Alabama and Mississippi populations displayed unique variation at *Skd-2,* which included allele *f,* the dominance of *e,* and the lack of *b* and *d.*

Within Florida, geographical distance was not always correlated with genetic distance (Fig. 2). Two sandhill populations, FL4 and FL5, were within 6.5 km of two flatwood populations, FL8 and FL9. However, FL8 was relatively distinct genetically (Fig. 2). Consequently, other factors such as ecological differentiation may have contributed to the genetic divergence of some wiregrass populations. This hypothesis is supported by common garden studies in which populations from different plant communities displayed genetic differences in their life-history and morphological characteristics (Gordon 1992).

Conclusions

The distribution of genetic variation within this widespread, long-lived, late-succession monocot resembles that of similar species: genetic diversity within and among wiregrass populations is high. Although the breeding system of wiregrass has not been documented, the allozyme variation supports the hypothesis that plants of this species are outcrossing and wind-pollinated. Genetic divergence at geographical extremes has probably resulted from long periods of relative reproductive isolation. Other factors affecting gene-flow opportunities, such as ecologically based phenological differentiation, may have contributed to genetic divergence of more closely spaced populations.

Ex situ collections of wiregrass have been and are being developed to provide germplasm for damaged habitat. The development of these collections should take into account the findings of this study. For example, the allelic differences among the three major geographic regions of this study suggest the occurrence of locally adapted genotypes. Recent reciprocal transplant experiments among sandhill and flatwood communities support these results (Kindell 1993). Consequently, the most appropriate stock for regenerating a particular wiregrass population may come from nearby, ecologically similar populations. While local movement of wiregrass propagules across similar sites appears appropriate, we recommend that restorationists consider the development and maintenance of regional *ex situ* collections.

For a widespread and genetically diverse species such as wiregrass, a variety of *ex situ* collections should reflect the distinct allelic diversity represented in geographically disjunct native stands, as well as cover the extensive range of the species (see Millar & Libby 1991). These separately maintained collections would provide appropriate genetic stock for reintroduction into different geographical areas. Also, given the large amount of allozyme variation we detected within wiregrass populations, stock collecting should include propagules from as many parent plants within a population as possible, as opposed to collecting many propagules from a minimum number of plants.

Acknowledgments

The authors thank Richard Broadwell, Margit Bucher, Ann Burruss, Jan Coyne, Richard LeBlond, Chris Oberholzer, Greg Seaman, Paula Seaman, and Bruce Sorrie for their assistance. This research was supported by Fairchild Tropical Garden and The Nature Conservancy.

Addendum

Prior to resubmission of the revised mansucription, Peet (1993) published a taxonomic study of *Aristida stricta.* He proposed that grasses traditionally assigned to this species include two morphologically and geographically separate species: *A. stricta* is confined to North Carolina and the northern counties of South Carolina, and *A. beyrichiana* Trin. & Rupr. occurs from southern South Carolina to the Gulf of Mexico.

Based on the genetic identity values and genotypes derived from the populations examined in this study and on a comparison of our genetic identity values to those of other angiosperm species, the allozyme data suggest that populations in Florida, Alabama, Mississippi, and North Carolina belong to a single species. We will withhold judgement on this issue, however, until allozymes can be sampled from a larger number of populations throughout the range of wiregrass.

Literature Cited

Bacchus, S. T. 1992. The creation and restoration of wetlands. Part 3. Variations due to age, localized conditions, and successional factors within species. The Palmetto **12**:14–16.

Clewell, A. F. 1989. Natural history of wiregrass (*Aristida stricta* Michx., Gramineae). Natural Areas Journal **23**:223–233.

Crawford, D. J. 1983. Phylogenetic and systematic inferences from electrophoretic studies. Pages 257–287 in S. D. Tanksley and T. J. Orton, editors. Isozymes in plant genetics and breeding, part A. Elsevier Science Publishers, Amsterdam, The Netherlands.

Gordon, D. R. 1992. Population variation in wiregrass (*Aristida stricta*) grown in a common garden experiment. Bulletin of the Ecological Society of America **73**:189.

Hamrick, J. L., and M. J. W. Godt. 1989. Allozyme diversity in plants species. Pages 43–63 in A. H. D. Brown, M. T. Clegg, A. L. Kahler, and B. S. Weir, editors. Plant population genetics, breeding and genetic resources. Sinauer Associates, Sunderland, Massachusetts.

Hamrick, J. L., M. J. Godt, D. A. Murawski, and M. D. Loveless. 1991. Correlations between species traits and allozyme diversity: Implications for conservation biology. Pages 75–86 in D. A. Falk and K. E. Holsinger, editors. Genetics and conservation of rare plants. Oxford University Press, New York.

Kindell, C. E. 1993. Population differentiation in *Aristida stricta*: Differences between sandhill and flatwood populations in Florida. M. S. thesis. Florida State University, Gainesville, Florida.

Loveless, M. D., and J. L. Hamrick. 1984. Ecological determinants of genetic structure in plant populations. Annual Review of Ecology and Systematics **15**:65–95.

Maunder, M. 1992. Plant reintroduction: An overview. Biodiversity and Conservation **1**:51–61.

Millar, C. I., and W. J. Libby. 1989. Disneyland or native ecosystem: Genetics and the restorationist. Restoration and Management Notes **7**:18–24.

Millar, C. I., and W. J. Libby. 1991. Strategies for conserving clincal, ecotypic, and disjunct population diversity in widespread species. Pages 149–170 in D. A. Falk and K. E. Holsinger, editors. Genetics and conservation of rare plants. Oxford University Press, New York.

Nei, M. 1972. Genetic distance between populations. American Naturalist **106**:891–900.

Parrott, R. T. 1967. A study of wiregrass (*Aristida stricta* Michx.) with particular reference to fire. M. S. thesis. Duke University, Durham, North Carolina.

Peet, R. K. 1993. A taxonomic study of *Aristida stricta* and *A. beyrichiana*. Rhodora **95**:25–37.

Seamon, P. A., and R. L. Myers. 1992. Propagating wiregrass from seed. The Palmetto **12**:6–7.

Seamon, P. A., R. L. Myers, L. Robbins, and G. S. Seamon. 1989. Wiregrass reproduction and community restoration. Natural Areas Journal **9**:264.

Stolzenburg, W. 1991. The wiregrass mystery. Nature Conservancy **41**:28–29.

Swofford, D. L., and R. B. Selander. 1989. BIOSYS: A computer program for the analysis of allelic variation in genetics. University of Illinois, Urbana, Illinois.

Walters, T. W., and D. S. Decker-Walters. 1991. Patterns of allozyme diversity in the West Indies cycad *Zamia pumila* (Zamiaceae). American Journal of Botany **78**:436–445.

Walters, T. W., and D. S. Decker-Walters. 1993. Systematics of the endangered Okeechobee Gourd (*Cucurbita okeechobeensis*: Cucurbitaceae). Systematic Botany **18**:175–187.

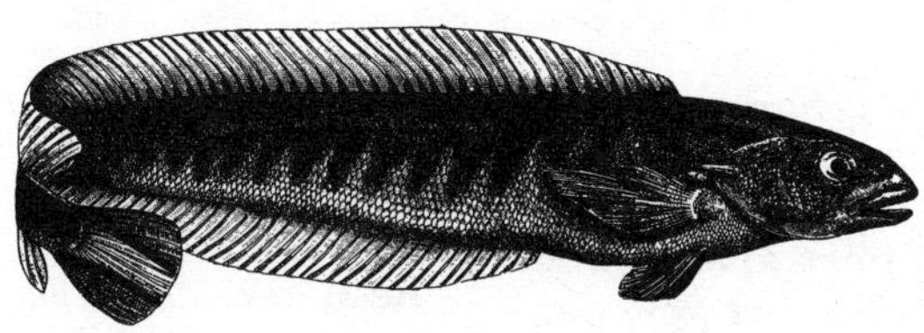

Population Structure of Loggerhead Turtles (*Caretta caretta*) in the Northwestern Atlantic Ocean and Mediterranean Sea

BRIAN BOWEN*
JOHN C. AVISE
Department of Genetics
University of Georgia
Athens, GA 30602, U.S.A.

JAMES I. RICHARDSON
Institute of Ecology
University of Georgia
Athens, GA 30602, U.S.A.

ANNE B. MEYLAN
Florida Marine Research Institute
100 Eighth Avenue, S.E.
St. Petersburg, FL 33701, U.S.A.

DIMITRIS MARGARITOULIS
Sea Turtle Protection Society of Greece
Solomou 35, GR-106 82
Athens, Greece

SALLY R. HOPKINS-MURPHY
South Carolina Wildlife and Marine Resources Department
P.O. Box 12559
Charleston, SC 29422, U.S.A.

Abstract: *To assess population genetic structure and evolutionary relationships among nesting populations of loggerhead turtles (*Caretta caretta*), we analyzed mitochondrial (mt) DNA variation in 113 samples from four nesting beaches in the northwestern Atlantic Ocean and from one*

* *Current address: BEECS Genetic Analysis Core, P. O. Box 110699, University of Florida, Gainesville, FL 32611, U.S.A.*
Paper submitted November 5, 1992; revised manuscript accepted July 7, 1993.

Estructura poblacional de la tortuga caguama (*Caretta caretta*) en el océano Atlántico noroeste y el Mar Mediterraneo

Resumen: *Con el propósito de evaluar la estructura genética de las poblaciones y las relaciones evolutivas entre poblaciones anidadoras de la tortuga caguama (*Caretta caretta*), analizamos la variación en el ADN mitocondrial (ADNmt) de 113 muestras provenientes de cuatro playas de anidación en el Océano Atlántico noroccidental y de una en el Mar Mediterráneo. Las diferencias significativas en la frecuencia de haplotipos entre poblaciones anidadoras en Flo-*

nesting beach in the Mediterranean Sea. Significant differences in haplotype frequency between nesting populations in Florida and in Georgia/South Carolina, and between both of these assemblages and the Mediterranean nesting colony, indicate substantial restrictions on contemporary gene flow between regional populations, and therefore a strong tendency for natal homing by females. Nonetheless, this regional genetic structure appears shallow, indicating recent evolutionary connections among rookeries. Data from tag returns and mtDNA, as well as geological considerations, suggest that over short evolutionary time scales (perhaps a few thousand years), dispersal by female loggerheads is sufficient to allow colonization of appropriate habitat in proximity to established rookeries but is too low to significantly affect the population dynamics of rookeries on a contemporary time scale. These data indicate that nesting populations of the loggerhead turtle must be managed as demographically independent units. The population subdivisions based on mtDNA analyses are concordant with previously reported distinctions between Florida and Georgia/South Carolina nesting populations based on environmental markers, tag recaptures, and morphology.

rida y en Georgia/Carolina del Sur, y entre estos dos grupos y la colonia anidadora en el Mediterráneo, indican limitaciones substanciales en el flujo genético contemporáneo entre poblaciones regionales y, por lo tanto. una tendencia de las hembras a anidar en sus playas de nacimiento ("natal homing"). A pesar de lo anterior, dado que las distancias genéticas son pequeñas se deducen conecciones evolutivas recientes entre colonias anidadoras. Los resultados de marcaje y del ADNmt, además de consideraciones geológicas, sugieren que a lo largo de escalas evolutivas pequeñas (quizá unos cuantos miles de años), la dispersión de caguamas hembras es suficiente para permitir la colonización de hábitats apropiados en la proximidad de playas de anida ción establecidas pero es demasiado leve para afectar significativamente la dinámica de poblaciones de las colonias anidadoras en una escala de tiempo contemporánea. Nuestra información indica que las poblaciones anidadoras de la tortuga caguama deben ser manejadas como unidades demográficas independientes. La subdivisión poblacional basada en los análisis del ADNmt concuerdan con las distinciones ya reportadas entre las poblaciones anidadoras en Florida y Georgia/Carolina del Sur basadas en marcadores ambientales. recapturas de tortugas marcadas y morfología.

Introduction

The loggerhead turtle, *Caretta caretta*, is distributed widely in warm temperate and subtropical oceans. At intervals averaging two to three years, adult loggerheads depart from the foraging grounds on reproductive migrations that range in distance from a few to thousands of kilometers (Meylan 1982; Margaritoulis 1988*a*). Tagging data indicate that most females return faithfully to the same nesting beach, and both sexes return to the same foraging areas between reproductive migrations (Limpus et al. 1992). Loggerhead hatchlings leave the nesting beach to occupy oceanic current systems (such as the North Atlantic gyre), where they drift passively for five years or more before recruiting to coastal neritic zones (Carr 1986). Subadults may occupy coastal feeding grounds for a decade or more before their first reproductive migration (Carr 1987). Estimates of the age at maturity for western Atlantic loggerheads range from 12 to 30 years (Frazer & Ehrhart 1985; Zug et al. 1986; Klinger & Musick 1992).

In recent reports, mtDNA analyses have proven useful for defining demographic and evolutionary units among the marine turtles *Chelonia mydas* (green turtle) and *Lepidochelys* spp. (ridley turtles) (Bowen et al. 1991, 1992). The maternal inheritance of mtDNA lends this approach a special significance in evaluating aspects of marine turtle life history, including the possibility of natal homing by females (Meylan et al. 1990). Tag-recapture experiments indicate that female loggerheads typically return to the same nesting area (Bjorndal et al. 1983), but it is unclear whether this site fidelity is a product of natal homing behavior. In principle, a variety of mechanisms could explain female nest-site fidelity, including imprinting of hatchlings, genetic programming, or social interaction (see Owens et al. 1982). Notably, social interaction could account for nest-site fidelity without invoking natal homing; first-time nesters may follow experienced females from feeding grounds to suitable nesting locations, and then focus on that location for subsequent nesting efforts. Under this "social facilitation" scenario, recruitment to non-natal rookeries would provide an avenue of female-mediated gene flow between rookery populations that have overlapping feeding grounds. Alternatively, under a "natal homing" scenario, female-mediated gene flow between rookeries would be restricted or absent. Thus, the contrasting models of "social facilitation" versus "natal homing" generate distinct predictions about the distribution of mtDNA lineages among nesting assemblages.

Based on significant differences in the concentrations of heavy metals in eggs (Stoneburner et al. 1980) and in the composition of epibiota on adult females (Caine 1986), loggerhead nesting populations in the southeastern U.S. may be divided into at least two demographic units corresponding to Florida and Georgia/South Carolina. However, metal concentrations and epibiota are acquired through prolonged contact with a particular environment, and therefore are a product of the habitats utilized by females during non-nesting intervals. Thus, these "acquired" markers reflect differences in feeding areas or migration pathways but leave open the question of whether Georgia/South Carolina and Florida rookeries are genetically distinct. Here we seek to add an evolutionary perspective to the understanding of loggerhead population structure in the southeastern U.S. through analyses of innate, genetically inherited markers.

A second concern addressed in this study is the population structure of Mediterranean loggerhead turtles. It has been suggested that feeding grounds in the western Mediterranean contain more juveniles than could be generated by Mediterranean rookeries alone (Argano & Baldari 1983; Laurent 1990*a*, 1990*b*). Several researchers have speculated that North Atlantic turtles may enter the Mediterranean system via currents of the North Atlantic gyre (Carr 1987; Laurent 1990*b*), become trapped in the Mediterranean Basin by strong currents at the Straits of Gibraltar (Carr 1987), and eventually recruit to Mediterranean nesting colonies (Groombridge 1988). If Atlantic loggerheads recruit at high frequency to Mediterranean rookeries, this fact should be reflected in the genetic composition of western Atlantic and Mediterranean nesting populations.

Thus, this study was designed to assess the genetic and demographic integrity of nesting populations both within one region (the northwestern Atlantic), and among rookeries on a geographic scale consistent with known loggerhead migrations (the North Atlantic Ocean and Mediterranean Sea). Given the magnitude of loggerhead movements, both micro- and macrogeographic scales are relevant to loggerhead conservation. As noted by Limpus et al. (1992), "no one country controls the fate of a given turtle population."

Materials and Methods

Between 1987 and 1991, 113 nests were sampled from: (1) Kiparissia Bay, western Peloponnesus, Greece (n = 21); (2) Cape Romain, South Carolina, U.S.A. (n = 19); (3) Cumberland Island and Little Cumberland Island, Georgia, U.S.A. (n = 44); (4) Broward County and St. Lucie County, eastern Florida, U.S.A. (n = 15); and (5) Key Island (Collier County), western Florida, U.S.A. (n = 14).

Nests were sampled for two eggs or for one hatchling during one typical internesting interval (12–14 days) to assure that the same female was not resampled. Two eggs were taken to offset mortality during transportation, because loggerhead eggs are very sensitive to motion during the first few weeks of development (Limpus et al. 1979). Eggs were incubated for six to eight weeks prior to processing, and hatchlings were sacrificed at appropriate lab facilities. Because siblings are expected to be identical with respect to mtDNA genotype, sample sizes refer to the number of nests sampled.

Closed-circular mtDNA was isolated from soft tissues (hatchlings) or whole embryos (eggs) by CsCl-ethidium bromide density gradient centrifugation (Lansman et al. 1981). Purified mtDNAs were digested with the 17 informative four-, five- and six-base cutting restriction enzymes listed in Table 1. In addition, representative samples were digested with *Bam*HI, *Cla*I, *Eco*RI, *Kpn*I, *Nsi*I, *Sac*I, *Sal*I, and *Sma*I, but these enzymes proved to be phylogenetically uninformative, producing either one or no cuts in loggerhead samples. Digestion fragments were end-labelled with ^{35}S nucleotides and separated on 1.0–1.7% agarose gels. Restriction fragments were visualized by autoradiography and assigned molecular weights on the basis of comparison to a 1-kb ladder standard.

Estimates of nucleotide sequence divergence (p values) were calculated by the "site" approach of Nei and Li (1979); haplotype and nucleotide diversities were estimated by the methods described by Nei and Tajima (1981) and Nei and Li (1979), respectively. Restriction fragment profiles were described using composite letter codes and were joined into a parsimony network that interrelates observed restriction fragment patterns.

Because we are interested in genetic relationships among turtles from particular pairs of nesting beaches (such as those that may occupy the same feeding grounds), some of the analyses described below include pairwise rookery comparisons (although results of multiple pairwise comparisons are not statistically independent). Pairs of rookeries were tested for significant differences in haplotype frequency by the G test, with Yates's correction for small sample size (Sokal & Rohlf 1981). Pairwise estimates of inter-rookery gene flow (Nm) were also calculated from G_{st} values ($Nm = 1/2a\,(1/G_{st} - 1)$), where $a = (L/L - 1)^2$ and L is the number

Table 1. Description and distribution of mtDNA genotypes observed in loggerhead turtles.*

Composite Code	*mtDNA Genotype*	*Rookery Location*	*Number of Nests*
A	*DCCCCCCCCACCCCCBC*	Georgia, U.S.A.	2
B	*DCCCCCCCCBCCCCCBC*	South Carolina, U.S.A.	19
		Georgia, U.S.A.	41
		East Florida, U.S.A.	4
		West Florida, U.S.A.	5
C	*DCCCCBCCCBCCCCCBC*	East Florida, U.S.A.	1
D	*ACCCCDCCCBBCCCCCC*	East Florida, U.S.A.	10
		West Florida, U.S.A.	9
		Kiparissia, Greece	21
E	*ACCCCDCBCBBCCCCCC*	Georgia, U.S.A.	1

* *Italicized letters refer to mtDNA restriction-fragment profiles produced by (from left to right):* AvaII, BclI, BglI, BglII, BstEII, BstNI, DraII, EcoRV, HindII, HindIII, MspI, NdeI, PvuII, SpeI, SstII, StuI, *and* XbaI. *For each enzyme, adjacent letters in the alphabet indicate that fragment profiles differ by a single restriction-site gain or loss; nonadjacent letters differ by at least two sites.*

of demes (Takahata & Palumbi 1985; Nei 1987). Calculations were conducted under assumptions of large L (hence $a \approx 1$) rather than small L (hence $a = 4$), providing a more conservative evaluation of population structure in the sense that the resulting *Nm* values are four-fold higher. Finally, an estimate of mean migration rate (*Nm*) among rookeries was calculated by the private-allele method (Slatkin 1985), using the equation in Slatkin and Barton (1989).

Results

Five genotypes were observed among the 113 loggerhead nests sampled (Table 1), with a mean of 82 restriction sites scored per individual. Digestion profiles are available from the senior author upon request. All restriction site changes could be explained by specifiable gains or losses of particular restriction sites.

Turtle species tend to exhibit relatively low levels of genetic variation and differentiation (Avise et al. 1992; Karl et al. 1992; but see Scribner et al. 1986), and previous studies indicate that loggerhead turtles have comparatively low genetic diversity at protein electrophoretic loci (Smith et al. 1978; Gyuris & Limpus 1988). The current study extends this qualitative conclusion to loggerhead mtDNA: overall haplotypic and nucleotide diversities in surveyed loggerheads were 0.505 and 0.0018, respectively (Table 2). These estimates, and the levels of sequence divergence among all observed haplotypes (Table 3), are at the low end of the spectrum of such values reported for comparisons of conspecific vertebrates (Avise et al. 1987, 1989).

The size of the loggerhead mtDNA molecule was estimated to be 16.6 kilobases (kb). Variation in mtDNA genome size, however, was revealed by many of the restriction enzyme fragment profiles. Discrete size classes differed by roughly 80–100 bases and spanned a total range of about 0.4 kb in a single region of the mtDNA molecule. Because the evolutionary dynamics of these size polymorphisms are uncertain, and because size polymorphisms were distributed more or less continuously across our assayed locations, these variants were excluded from population analyses. All subsequent discussion will concern restriction-site changes only.

Table 2. Loggerhead turtle haplotype and nucleotide diversities (Nei & Tajima 1981; Nei & Li 1979).

	Haplotype Diversity (h)	*Nucleotide Diversity* (π)
Greece	0.000	0.0000
South Carolina	0.000	0.0000
Georgia	0.132	0.0002
East Florida	0.514	0.0018
West Florida	0.495	0.0018
Overall	0.505	0.0018

Table 3. Estimates of sequence divergence between the mtDNA clones described in Table 1, based on the site approach of Nei and Li (1979).

mtDNA clone	A	B	C	D	E
A	—	0.0010	0.0023	0.0089	0.0100
B		—	0.0012	0.0078	0.0089
C			—	0.0092	0.0103
D				—	0.0010
E					—

Two distinct groups of haplotypes were observed in the mtDNA phylogeny (*A, B, C* versus *D, E* haplotypes in Figure 1), and these groups differ at a mean level of sequence divergence $p = 0.008$. This magnitude of divergence is similar to the deepest fork observed in a global phylogeny for green turtle mtDNA ($p = 0.007$), which partitioned Atlantic-Mediterranean from Indian-Pacific phylads (Bowen et al. 1992). In contrast, both major groupings in the loggerhead mtDNA phylogeny are observed within the Atlantic Ocean and co-occur in the Florida and Georgia nesting populations.

To some extent, the low level of genetic variation observed in this study impairs our ability to resolve regional population issues. Rookery samples are dominated by two mtDNA genotypes (*B* and *D* in Table 1 and Figure 2). Nevertheless, the distribution of these genotypes has a strong geographic component. Genotype *D*,

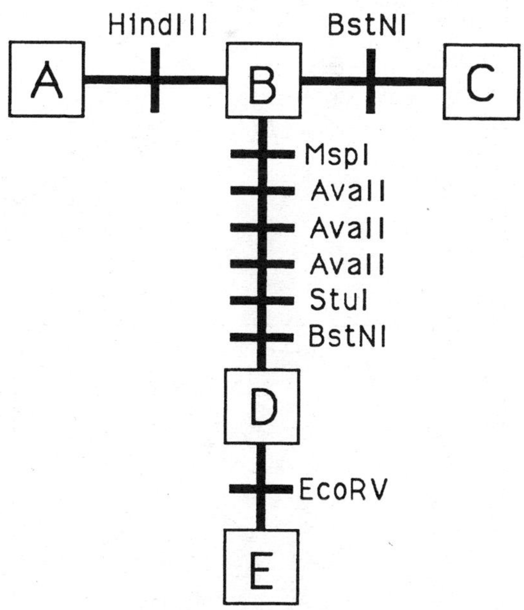

Figure 1. Parsimony network summarizing the interrelationships among mtDNA haplotypes observed in the survey of northwestern Atlantic and Mediterranean loggerhead rookeries. Letters refer to the composite genotypes described in Table 1. Each dash corresponds to a single restriction-site gain or loss, with the corresponding enzyme indicated above or to the left of the dash.

observed in 67% of eastern Florida samples (n = 15) and 64% of western Florida samples (n = 14), is completely absent from the Georgia (n = 44) and South Carolina (n = 19) collections (Fig. 2). Genotype frequencies differed significantly in eight of ten pairwise rookery comparisons (Table 4). Notably, the cases where genotype frequencies were not significantly different involved proximal nesting beaches: (1) eastern Florida and western Florida; and (2) Georgia and South Carolina. We conclude that loggerhead rookeries of the southeastern U.S. comprise at least two genetic populations, between which contemporary gene flow is low (Table 4).

The Mediterranean sample was fixed for one mtDNA genotype (haplotype *D* in Table 1) that was observed at 64–67% frequency in Florida samples and absent in

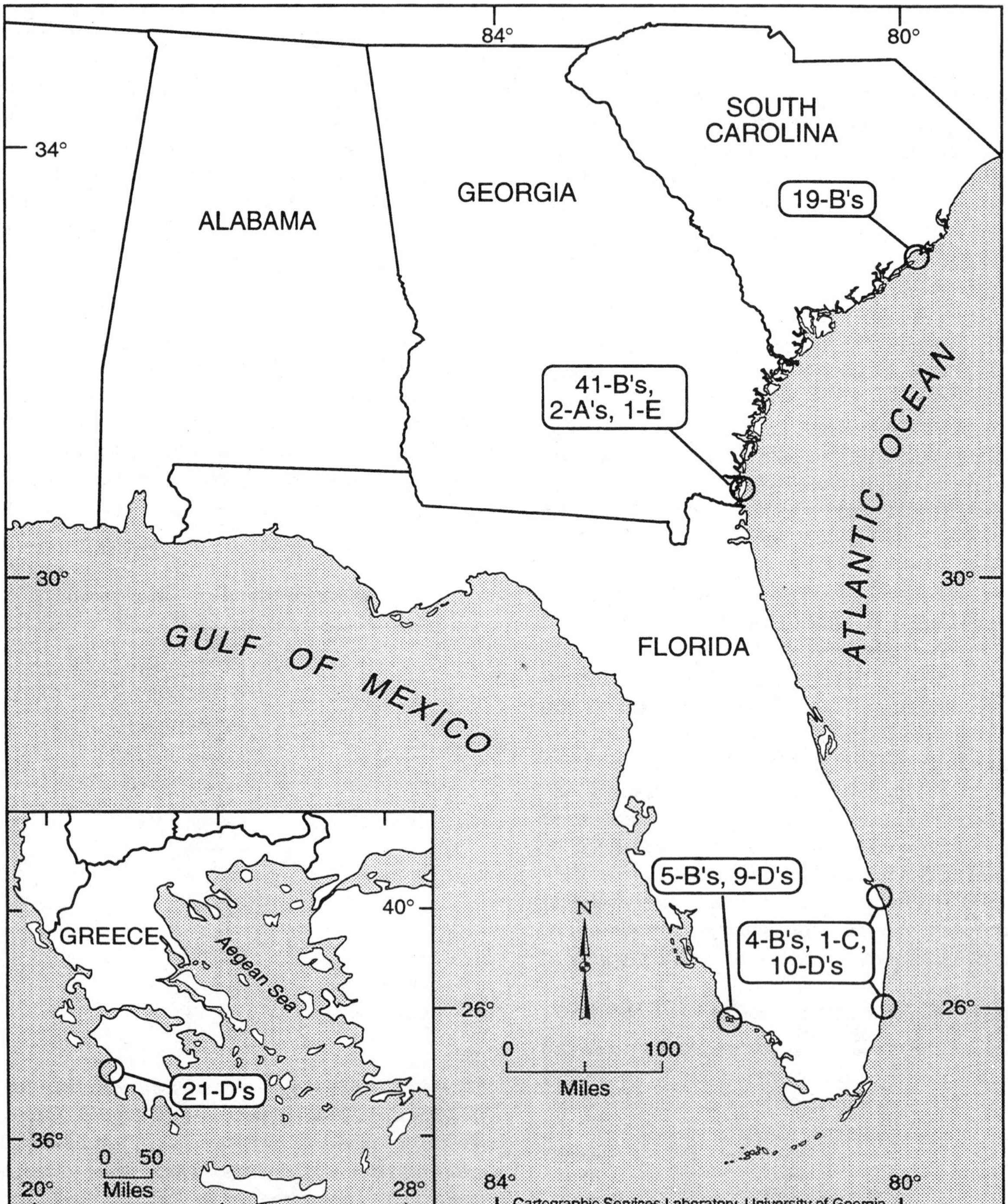

Figure 2. Collection sites and genotypes observed at each of five Atlantic and Mediterranean rookeries of the loggerhead turtle. Each letter refers to the composite mtDNA genotype of an individual turtle (Table 1).

Table 4. Pairwise comparisons of haplotype frequencies among nesting beaches with a *G*-test of independence* (above diagonal) and *Nm* values based on the G_{st} estimates (below diagonal).**

	I	*II*	*III*	*IV*	*V*
I. South Carolina	—	0.3 N.S.	19.9 $P < 0.001$	15.3 $P < 0.001$	46.0 $P < 0.001$
II. Georgia	1.1	—	33.0 $P < 0.001$	25.8 $P < 0.001$	71.6 $P < 0.001$
III. East Florida	0.8	3.5	—	0.1 N.S.	6.0 $P < 0.02$
IV. West Florida	0.8	3.5	high	—	6.4 $P < 0.02$
V. Greece	0.0	0.4	1.3	1.2	—

* *Using Yates's correction (Sokal & Rohlf 1981).*
** *Takahata & Palumbi 1985.*

Georgia and South Carolina. Pairwise comparisons of the Mediterranean versus West Atlantic rookeries yielded significant *G* values and low estimates of inter-rookery migration (ranging from $Nm = 0.0$ to $Nm = 1.3$ migrants per generation [Table 4]). Thus, the Mediterranean rookery differs significantly in genetic composition from nesting beaches in the northwestern Atlantic.

Based on the private allele method (Slatkin 1985), a mean value of $Nm \approx 2.0$ is estimated for the assayed loggerhead nesting populations. This value is qualitatively similar to values derived from the G_{st} approach (Table 4). In general, values of *Nm* greater than about one to four indicate that gene flow is sufficient to maintain a relatively homogeneous gene pool, whereas lower values indicate that gene flow is not sufficient to prevent divergence of isolated gene pools by genetic drift (Birky et al. 1983; Slatkin 1987). Several caveats concerning *Nm* estimates merit consideration, however. First, the estimates generated here are from a single gene genealogy. More accurate estimates of gene flow would be expected from an analysis of multiple independent loci (although they would not then apply strictly to female lineages that are of special interest here). Second, some of the rookeries surveyed may be the product of recent colonization events (see below), such that assumptions of population equilibrium are not met (although Slatkin and Barton [1989] suggest that the approaches used here to estimate *Nm* are relatively insensitive to this type of bias). Finally, the theoretical basis for these estimates is still under development, and empirical calibrations are currently unavailable. For these reasons, *Nm* estimates should be interpreted as *general* indicators of the magnitude of genetic exchange.

Discussion

The nesting beaches of the southeastern United States, taken together, comprise the second largest reproductive aggregate of loggerhead turtles in the world, with approximately 35,000 nesting females (Murphy & Hopkins 1984). Within this region, however, nesting habitat is not continuous. The nesting beaches in Florida and Georgia/South Carolina are separated by several hundred kilometers of beach in which loggerhead nesting density is low (Murphy & Hopkins-Murphy 1989). In the Mediterranean, prominent nesting aggregates are reported from Zakynthos and Peloponnesus in Greece and along the adjacent coastline of Turkey (Groombridge 1988). Notably, the loggerheads that nest at these Mediterranean locations are significantly smaller than those that nest in the western Atlantic (Margaritoulis 1982, 1988*b*).

In terms of the distribution of mtDNA lineages, the various Atlantic and Mediterranean nesting populations assayed in this study are significantly different, a conclusion consistent with results of a protein electrophoretic analysis of loggerhead rookeries in Queensland, Australia (Gyuris & Limpus 1988). But we observed more sharing of mtDNA genotypes among loggerhead rookeries than was observed in green turtles, a fact reflected in somewhat higher estimates of mean migration rate based on Slatkin's (1985) private allele method: $Nm \approx 2.0$ for Atlantic loggerheads, compared with $Nm \approx 0.3$ for Atlantic green turtles (Bowen et al. 1992). It is possible that these migration estimates reflect a higher level of dispersal in female loggerhead turtles relative to green turtles; movement between spatially distinct green turtle rookeries is known to be extremely rare (Meylan 1982). On the other hand, the higher migration estimate for loggerheads may be an artifact of sampling design, because four of the five sampled nesting beaches are in one geographic province (the southeastern United States). Additional data will be required to determine whether nesting loggerhead turtles are less site-specific than are nesting green turtles.

Evolutionary History of the Western Atlantic and Mediterranean Rookeries

Climatic processes have undoubtedly influenced the contemporary distribution and population structure of loggerhead nesting assemblages. Loggerhead eggs require a minimum of 60 days above 25°C to incubate successfully, such that cold temperate conditions in the Mediterranean (Buckley et al. 1982; McCoy 1980) may have precluded nesting here during the most recent glacial period (18,000 to 12,000 years before the present). In the western Atlantic, loggerheads possibly nested in southern Florida during glacial intervals, but almost certainly not at present-day rookery locations in Georgia, South Carolina, and North Carolina (see Hedgpeth 1954). Thus, the contemporary distribution of nesting beaches in the southeastern United States may be the product of colonization events over the last

12,000 years, as loggerhead females extended the northern boundary of the nesting range. One consequence of this colonization process could be a progressive loss of mtDNA diversity in more northerly rookeries, as maternal lineages are filtered through a series of colonization bottlenecks. Estimates of haplotypic and genotypic diversity (Table 2) are consistent with this scenario.

In the southeastern U.S., this colonization process has apparently been sufficient to extend the northern limits of nesting by 1000 km within the last 12,000 years (fewer than 600 loggerhead generations). In the Mediterranean, habitats that were too cold to support nesting and feeding 12,000 years ago are now utilized extensively by loggerhead turtles. Furthermore, the presence of the *D* genotype at 100% frequency in samples from Greece indicates that this colony shared a recent common ancestor with western Atlantic populations. Thus, conclusions drawn from climatic history, mtDNA data, and tagging studies are concordant in indicating that loggerheads are active colonizers that can occupy newly opened habitat over relatively short evolutionary time scales. In other words, contact between loggerhead rookeries in the West Atlantic and Mediterranean has been sufficient to prevent pronounced evolutionary divergence, despite a restriction of contemporary gene flow between nesting assemblages (Table 4).

Regional Population Structure—Southeastern United States

MtDNA data indicate that nesting loggerhead turtles in the southeastern United States are divided into Georgia/South Carolina and Florida cohorts, a conclusion also supported by subtle differences in morphology (Stoneburner 1980). Many coastal marine organisms of the southeastern United States show a phylogeographic discontinuity in this area (Bowen & Avise 1990; Avise 1992), including an estuarine terrapin (Lamb & Avise 1992). But the apparent separation in loggerheads is difficult to reconcile with some aspects of tag-recapture data. While most nesting females return to the same beach in successive nesting seasons, a small fraction of tagged turtles has been observed nesting at alternative sites (Dodd 1988). Bjorndal et al. (1983) reviewed 25 cases of nesting-beach relocations in the southeast region, and LeBuff (1974) reported that a female tagged on a western Florida nesting beach was observed nesting on the east coast of Florida (550 km distant) four years later. More directly relevant to this discussion, 11 tagged loggerheads were observed to nest at both Georgia and eastern Florida sample locations during the period of 1978–1985 (J. I. Richardson, unpublished data).

The significant difference in observed mtDNA haplotype frequencies between Florida and Georgia/South Carolina rookeries appears to be inconsistent with this propensity for nesting relocations, because the exchange of even one migrant per generation is sufficient in theory to maintain alleles in similar frequencies in populations at equilibrium (Slatkin 1987; but see Allendorf 1983). One possible explanation is that Georgia/South Carolina and Florida nesting populations have not reached an equilibrium condition. Support for this possibility stems (1) from climatic data: the Georgia and South Carolina rookeries were probably unsuitable for nesting 12,000 years (or about 600 loggerhead generations) before the present, and (2) from contemporary demographic data: nesting populations have been reduced in the last several decades by mortality associated with incidental capture in the shrimp fishery (Henwood & Stuntz 1987; National Research Council 1990). Either of these factors could be sufficient to abrogate assumptions of population equilibrium. A related hypothesis is that these nesting relocations are a relatively recent phenomenon, perhaps induced by human encroachment on nesting grounds and adjacent internesting habitat. Inferential support for this scenario is provided by field observations: turtles disturbed during nesting are more likely than undisturbed turtles to relocate to an adjacent beach, and they may use the new beach for subsequent nesting efforts (T. M. Murphy, personal communication).

The heavy metal concentration of loggerhead egg shells differs significantly between Florida and Georgia/South Carolina rookeries (Stoneburner et al. 1980), a distinction that is further supported by differences in the epibiota assemblages of nesting turtles (Caine 1986). Since both epibiota and heavy metals accumulate during non-nesting intervals, these environmental markers indicate some level of segregation on feeding grounds or migratory routes. Notably, tag recoveries support a similar partition: Florida nesting turtles are recovered along the southeastern U.S., in the Bahamas, and in the Caribbean (Meylan 1982; Meylan et al. 1983), while Georgia nesting turtles have been recovered almost exclusively on the east coast of the U.S. (Bell & Richardson 1978; Richardson 1982, unpublished data). Thus, both environmentally-acquired markers and human-applied tags indicate some segregation on feeding grounds.

Acquired markers (heavy metals, epibiota, and tags) and innate genetic markers (mtDNA) support a concordant partitioning of nesting populations into Florida versus Georgia/South Carolina units. But these two classes of markers elucidate very different aspects of loggerhead natural history. The genetic markers indicate that loggerhead females tend to nest in the vicinity of their natal rookery, whereas the acquired markers indicate that Florida and Georgia/South Carolina nesting populations tend to segregate on feeding grounds as well. Taken together, these distinct classes of information allow a more complete picture of loggerhead migratory behavior to emerge.

Mediterranean Recruitment and Natal Homing

The mtDNA data indicate that regional loggerhead nesting populations are genetically distinct, but a complete test of natal homing requires information on feeding ground composition as well. If nesting populations maintain separate feeding grounds, then genetic expectations under the natal homing and social facilitation models converge because turtles are not confronted with the option of following experienced breeders to a non-natal rookery. Therefore, the optimal genetic tests of these competing hypotheses involve nesting populations that share feeding areas.

The supposition that juvenile loggerheads from North Atlantic nesting beaches occur on Mediterranean feeding grounds is based on three lines of evidence. First, Carr (1987) noted that more juvenile loggerheads occupy Mediterranean feeding grounds than could be generated by the Mediterranean rookeries alone (see also Argano & Baldari 1983). Second, a loggerhead tagged in the Azores subsequently was recovered in the Mediterranean (Bolten et al. 1992). Third, the North Atlantic current system (believed to passively transport juvenile loggerheads) branches into the Mediterranean (Estrada et al. 1985). Groombridge (1988) speculated that surface currents and oceanic topology may trap pelagic-stage (juvenile) loggerheads in the Mediterranean Basin, and that some of these strays may remain there to breed.

If one accepts the premise that juveniles from North Atlantic nesting beaches occupy Mediterranean feeding grounds, then mtDNA data provide a critical test of natal homing for these turtles. Under a social facilitation model, considerable recruitment of North Atlantic juveniles onto Mediterranean nesting beaches is expected, resulting in a sharing of common mtDNA lineages. Contrary to these expectations, samples from one of the largest Mediterranean rookeries (Kiparissia Bay, Greece) contained only genotype *D* of the two genotypes (*B* and *D*) that dominated northwestern Atlantic samples. Thus, the mtDNA data indicate that female-mediated gene flow between the northwestern Atlantic and Mediterranean rookeries is limited (Table 4). These data are inconsistent with a major role for social facilitation. If juveniles are trapped in the Mediterranean, as Groombridge (1988) suggests, they apparently are not recruiting to Mediterranean rookeries at levels sufficient to homogenize haplotype frequencies or to affect contemporary demographics.

Conservation Concerns

Mitochondrial lineages are useful for addressing the female component of population structure, but they leave parallel questions about male dispersal unresolved. In terms of conservation genetics and management of marine turtles, the population structure of females is of special interest because nesting females ultimately govern the reproductive success of these species. However, other aspects of marine turtle conservation may benefit from an understanding of the male component of population genetic structure. For example, do males mate with females from other rookeries, facilitating nuclear gene flow between nesting populations that are isolated with respect to female lineages? If regional nesting populations are linked by male-mediated gene flow, then these assemblages may be buffered to some extent from the reduced genetic diversity associated with population bottlenecks. While a paternal analog to mtDNA is unavailable, many of these issues may be addressed with analysis of nuclear (n) DNA loci. In green turtles, restriction fragment analysis of nDNA loci demonstrate that nesting populations are less structured with respect to these biparentally-inherited markers than is the case for maternally-inherited lineages (Karl et al. 1992). The implications of these data are that male green turtles provide an avenue of gene flow between nesting populations. It remains to be seen whether the same conclusion applies to loggerhead turtles. Nonetheless, these complementary approaches (in this case, nDNA and mtDNA analyses) are necessary to resolve the complex genetic architecture of marine turtle populations.

The mtDNA data reported here indicate a significant population genetic structure for loggerhead turtle rookeries in the North Atlantic Ocean and Mediterranean Sea. Contemporary female-mediated gene flow between regions is negligible (Table 4), yet all loggerhead populations are related very closely in an evolutionary sense. What do these results imply for the management of threatened and endangered populations? Over short evolutionary time scales (perhaps a few thousand years), female dispersal apparently is sufficient to allow colonization of appropriate habitats in proximity to established rookeries. Both tag returns and mtDNA data indicate, however, that female gene flow is too low to have a significant impact on population dynamics on a contemporary scale. Therefore, if nesting females are depleted or extirpated at one rookery, regional dispersal will not be sufficient to replenish this resource over a time scale that is meaningful to wildlife management agencies. This conclusion holds regardless of the magnitude of inter-rookery gene flow mediated by males. Accordingly, nesting populations must be considered to be demographically independent.

Acknowledgments

For field assistance we gratefully recognize M. Camhi, P. Castaneda, C. Coogan, L. Ehrhart, G. Garris, L. Fisher, S. Karl, R. Klinger, L. Letson, R. Mezich, N. Richardson, B.

Schroeder, C. Sears, G. Smith, W. Weber, M. Zacks, and J. Zurita. For logistic support we are indebted to G. Acres, R. Carthy, R. Ferl, M. Harris, J. A. Huff, S. Jacobsen, J. Musick, W. Nelson, E. Possardt, T. Richardson, D. Walker, J. Woody, the Broward County Environmental Quality Control Board, the Caribbean Conservation Corporation, The Conservancy, the Florida Department of Natural Resources, the Georgia Department of Natural Resources, the National Marine Fisheries Service (U.S.), the Sea Turtle Protection Society of Greece, the South Carolina Wildlife and Marine Resources Department, and the U.S. Fish and Wildlife Service. K. Bjorndal, A. Bolten, L. Laurent, C. Limpus, and N. Mrosovsky provided insightful reviews of an earlier version of this report. We thank A. Abreu for translation and constructive comments. The map was drafted by J. Ingram at the Cartographic Lab, Geography Department, University of Georgia. This research was supported by the National Geographic Society, the National Science Foundation, the Interdisciplinary Center for Biotechnology Research (University of Florida), and by a Genetics Training grant to B. W. Bowen from the National Institutes of Health. Funding for the South Carolina rookery samples was provided by the National Oceanographic and Atmospheric Administration (NOAA). The views expressed herein are those of the authors and do not reflect the views of NOAA or any of its subagencies.

Literature Cited

Allendorf, F. W. 1983. Isolation, gene flow, and genetic differentiation among populations. Pages 51–65 in C. M. Schonewald-Cox, S. M. Chambers, B. MacBryde, and L. Thomas, editors. Genetics and Conservation. Benjamin/Cummings, London, England.

Argano, R., and F. Baldari. 1983. Status of Western Mediterranean sea turtles. Rapports de Commission Internationale de la Mer Mediterranee **28(5)**:233–235.

Avise, J. C. 1992. Molecular population structure and the biogeographic history of a regional fauna: A case history with lessons for conservation biology. Oikos **63**:62–76.

Avise, J. C., J. Arnold, R. M. Ball, E. Bermingham, T. Lamb, J. E. Neigel, C. A. Reeb, and N. C. Saunders. 1987. Intraspecific phylogeography: The mitochondrial DNA bridge between population genetics and systematics. Annual Review of Ecology and Systematics **18**:489–522.

Avise, J. C., B. W. Bowen, and T. Lamb. 1989. DNA fingerprints from hypervariable mitochondrial genotypes. Molecular Biology and Evolution **6**:258–269.

Avise, J. C., B. W. Bowen, T. Lamb, A. B. Meylan, and E. Bermingham. 1992. Mitochondrial DNA evolution at a turtle's pace: Evidence for low genetic variability and reduced microevolutionary rate in the Testudines. Molecular Biology and Evolution **9**:457–473.

Bell, R., and J. I. Richardson. 1978. An analysis of tag recoveries from loggerhead sea turtles (*Caretta caretta*) nesting on Little Cumberland Island, Georgia. Florida Marine Research Publications **33**:1–66.

Birky, C. W., Jr., T. Maruyama, and P. Fuerst. 1983. An approach to population and evolutionary genetic theory for genes in mitochondria and chloroplasts, and some results. Genetics **103**:513–527.

Bjorndal, K. A., A. B. Meylan, and B. J. Turner. 1983. Sea turtle nesting at Melbourne Beach, Florida, I. Size, growth, and reproductive biology. Biological Conservation **26**:65–77.

Bolten, A. B., H. R. Martins, K. A. Bjorndal, M. Cocco, and G. Gerosa. 1992. *Caretta caretta* (loggerhead) pelagic movement and growth. Herpetological Review **23**:116.

Bowen, B. W., and J. C. Avise. 1990. Genetic structure of Atlantic and Gulf of Mexico populations of sea bass, menhaden, and sturgeon: Influence of zoogeographic factors and life-history patterns. Marine Biology **107**:371–381.

Bowen, B. W., A. B. Meylan, and J. C. Avise. 1991. Evolutionary distinctiveness of the endangered Kemp's ridley sea turtle. Nature **352**:709–711.

Bowen, B. W., A. B. Meylan, J. P. Ross, C. J. Limpus, G. H. Balazs, and J. C. Avise. 1992. Global population structure and natural history of the green turtle (*Chelonia mydas*) in terms of matriarchal phylogeny. Evolution **46(4)**:865–881.

Buckley, H. A., L. R. Johnson, N. J. Shackleton, and R. A. Blow. 1982. Late glacial to recent cores from the eastern Mediterranean. Deep Sea Research **29**:739–766.

Caine, E. A. 1986. Carapace epibionts of nesting loggerhead sea turtles: Atlantic coast of U.S.A. Journal of Experimental Marine Biology and Ecology **95**:15–26.

Carr, A. 1986. Rips, FADS, and little loggerheads. Bioscience **36(2)**:92–100.

Carr, A. 1987. New perspectives on the pelagic stage of sea turtle development. Conservation Biology **1**:103–121.

Dodd, C. K., Jr. 1988. Synopsis of the biological data on the loggerhead sea turtle *Caretta caretta* (Linnaeus 1758). U.S. Fish and Wildlife Service Biological Report **88(14)**:1–110.

Estrada, M., F. Vives, and M. Alcaraz. 1985. Life and the productivity of the open sea. Pages 148–197 in M. Margalef, editor. Western Mediterranean. Key Environments Series. Pergamon Press Ltd., London, England.

Frazer, N. B., and L. M. Ehrhart. 1985. Preliminary growth models for green, *Chelonia mydas*, and loggerhead, *Caretta caretta*, turtles in the wild. Copeia **1985**:73–79.

Groombridge, B. 1988. Marine turtles in the Mediterranean: Distribution, population status, conservation. Report T-PVS (88) 42. Council of Europe, Environment Conservation and Management Division, Strasbourg, France.

Gyuris, E., and C. J. Limpus. 1988. The loggerhead turtle *Caretta caretta* in Queensland: Population breeding structure. Australia Wildlife Research **15**:197–209.

Hedgpeth, J. W. 1954. An introduction to the zoogeography of the northern Gulf of Mexico with reference to the invertebrate fauna. Pages 111–211 in Publication 3. Institute of Marine Science, University of Texas, Port Aransas, Texas.

Henwood, T. A., and W. E. Stuntz. 1987. Analysis of sea turtle captures and mortality during commercial shrimp trawling. Fisheries Bulletin **85:**813–817.

Karl, S. A., B. W. Bowen, and J. C. Avise. 1992. Global population structure and male-mediated gene flow in the green turtle (*Chelonia mydas*): RFLP analyses of anonymous nuclear DNA regions. Genetics **131:**163–173.

Klinger, R. C., and J. A. Musick. 1992. Annular growth layers in juvenile loggerhead turtles (*Caretta caretta*). Bulletin of Marine Science **51:**224–230.

Lamb, T., and J. C. Avise. 1992. Molecular and population genetic aspects of mitochondrial DNA variability in the diamondback terrapin, *Malaclemys terrapin*. Journal of Heredity **83:**262–269.

Lansman, R. A., R. O. Shade, J. F. Shapira, and J. C. Avise. 1981. The use of restriction endonucleases to measure mitochondrial DNA sequence relatedness in natural populations. III. Techniques and potential applications. Journal of Molecular Evolution **17:**214–226.

Laurent, L. 1990*a*. Les tortues marines en Algerie et au Maroc (Mediterranee). Bulletin de la Societe Herpetologique de France **55:**1–23.

Laurent, L. 1990*b*. L'origine des tortues caouannes, *Caretta caretta* (Linnaeus 1758) de Mediterranee Occidentale. Rapports de Commission Internationale de la Mer Mediterranee **32:**240.

LeBuff, C. R., Jr. 1974. Unusual nesting relocation in the loggerhead turtle, *Caretta caretta*. Herpetologica **30:**29–31.

Limpus, C. J., V. Baker, and J. D. Miller. 1979. Movement induced mortality of loggerhead eggs. Herpetologica **35:**335–338.

Limpus, C. J., J. D. Miller, C. J. Parmenter, D. Reimer, N. McLachland, and R. Webb. 1992. Migration of green (*Chelonia mydas*) and loggerhead (*Caretta caretta*) turtles to and from eastern Australian rookeries. Wildlife Research **19:**347–358.

Margaritoulis, D. 1982. Observations on loggerhead sea turtle, *Caretta caretta*, activity during three nesting seasons (1977–1979) in Zakynthos, Greece. Biological Conservation **24:**193–204.

Margaritoulis, D. 1988*a*. Post-nesting movements of loggerhead sea turtles tagged in Greece. Rapports de Commission Internationale de la Mer Mediterranee **31(2):**284.

Margaritoulis, D. 1988*b*. Nesting of the loggerhead sea turtle, *Caretta caretta*, on the shores of Kiparissia Bay, Greece. Mesogee **48:**59–65.

McCoy, F. W. 1980. Climate changes in the eastern Mediterranean area during the past 240,000 years. Pages 79–100 in C. Doumas, editor. Thera and the Aegean World II: Papers and Proceedings of the Second International Scientific Congress, Santorini, Greece, August 1978. Thera and the Aegean World, London, England.

Meylan, A. B. 1982. Sea turtle migration—evidence from tag returns. Pages 91–100 in K. A. Bjorndal, editor. Biology and conservation of sea turtles. Smithsonian Institution Press, Washington, D. C.

Meylan, A. B., K. A. Bjorndal, and B. J. Turner. 1983. Sea turtle nesting at Melbourne Beach, Florida, II. Post-nesting movement of *Caretta caretta*. Biological Conservation **26:**79–90.

Meylan, A. B., B. W. Bowen, and J. C. Avise. 1990. A genetic test of the natal homing versus social facilitation models for green turtle migration. Science **248:**724–727.

Murphy, T. M., and S. R. Hopkins. 1984. Aerial and ground surveys of marine turtle nesting beaches in the southeast region, U.S. Report no. NA83-GA-C-00021. National Marine Fisheries Service, Southeast Fisheries Center Miami, Florida.

Murphy, T. M., and S. R. Hopkins-Murphy. 1989. Sea turtle and shrimp fishing interactions: A summary and critique of relevant information. Center for Marine Conservation, Washington, D.C.

National Research Council. 1990. Decline of the sea turtles: Causes and prevention. National Academy Press. Washington, D.C.

Nei, M. 1987. Molecular evolutionary genetics. Columbia University Press, New York.

Nei, M., and W.-H. Li. 1979. Mathematical models for studying genetic variation in terms of restriction endonucleases. Proceedings of the National Academy of Science U.S.A. **76:**5269–5273.

Nei, M., and F. Tajima. 1981. DNA polymorphism detectable by restriction endonucleases. Genetics **97:**145–163.

Owens, D. W., M. A. Grassman, and J. R. Hendrickson. 1982. The imprinting hypothesis and sea turtle reproduction. Herpetologica **38:**124–135.

Richardson, J. I. 1982. A population model for adult female loggerhead sea turtles (*Caretta caretta*) nesting in Georgia. Ph.D. Dissertation. University of Georgia, Athens, Georgia.

Scribner, K. T., J. E. Evans, S. J. Morreale, M. H. Smith, and J. W. Gibbons. 1986. Genetic divergence among populations of yellow-bellied slider turtle (*Pseudemys scripta*) separated by aquatic and terrestrial habitats. Copeia **1986(3):**691–700.

Slatkin, M. 1985. Rare alleles as indicators of gene flow. Evolution **39:**53–65.

Slatkin, M. 1987. Gene flow and the geographic structure of natural populations. Science **236:**787–792.

Slatkin, M., and N. H. Barton. 1989. A comparison of three indirect methods for estimating average levels of gene flow. Evolution **43:**1349–1368.

Smith, M. H., H. O. Hillstad, M. N. Manlove, D. O. Straney, and J. M. Dean. 1978. Management implications of genetic variability in loggerhead and green sea turtles. Pages 302–312 in Proceedings of the 13th International Congress of Game Biologists. Wildlife Society, Washington, D.C.

Sokal, R. R., and F. J. Rohlf. 1981. Biometry. 2nd edition. W. H. Freeman, San Francisco, California.

Stoneburner, D. L. 1980. Body depth: An indicator of morphological variation among nesting groups of adult loggerhead sea turtles (*Caretta caretta*). Journal of Herpetology **14(2)**:205–206.

Stoneburner, D. L., M. N. Nicora, and E. R. Blood. 1980. Heavy metals in loggerhead sea turtle eggs (*Caretta caretta*): Evidence to support the hypothesis that demes exist in the western Atlantic population. Journal of Herpetology **14(2)**:171–175.

Takahata, N., and S. R. Palumbi. 1985. Extranuclear differentiation and gene flow in the finite island model. Genetics **109**:441–457.

Zug, G. R., A. H. Wynn, and C. Ruckdeschel. 1986. Age determination of loggerhead sea turtles, *Caretta caretta*, by incremental growth marks in the skeleton. Smithsonian Contributions to Zoology **427**:1–34.

Mitochondrial DNA Variability of the Gray Wolf: Genetic Consequences of Population Decline and Habitat Fragmentation

ROBERT K. WAYNE
NILES LEHMAN*

Department of Biology
University of California at Los Angeles
Los Angeles, CA 90024, U.S.A.

MARC W. ALLARD

Department of Zoology
University of Florida
Gainesville, FL 32611, U.S.A.

RODNEY L. HONEYCUTT

Department of Wildlife and Fisheries Sciences
Texas A & M University
College Station, TX 77843, U.S.A.

Abstract: *The gray wolf is a large, highly mobile predator whose original geographic range included most of the Northern Hemisphere. High rates of genetic exchange probably characterized even distantly-separated populations in the past, but recent population declines and habitat fragmentation have isolated previously contiguous populations, especially in the Old World. We examine mitochondrial DNA (mtDNA) variability among twenty-six populations of wolves from throughout their geographic range. We find eighteen mtDNA genotypes in gray wolves: seven of these are derived from hybridization with coyotes, four are confined to the New World, six are confined to the Old World, and one is shared by both areas. Genetic differentiation among wolf populations is significant but small in magnitude. In the*

Paper submitted April 10, 1991; revised manuscript accepted October 29, 1991.

** Present address: Department of Molecular Biology, The Scripps Research Institute, 10666 N. Torrey Pines Road, La Jolla, CA 92037, U.S.A.*
Address all correspondence to R. K. Wayne.

Variabilidad del ADN mitocondrial del lobo gris: concecuencias genéticas de la declinación poblacional y fragmentación del hábitat.

Resumen: *El lobo gris es un predador grande y altamente móbil cuyo distribución geográfica original abarcaba la mayor parte del Hemisferio Norte. Es muy probable que en el pasado, altas tasas de intercambio genético caracterizaran, incluso, poblaciones muy distantes. Sin embargo, recientes declinaciones poblacionales y fragmentación del habitat han aislado poblaciones que antiguamente se escontraban contiguas, especialmente en el Viejo Mundo. Examinamos la variabilidad en el ADN mitocondrial (mtDNA) de 26 poblaciones de lobos existentes a lo largo de toda su area de distribucion geográfica. Encontramos 18 genotipos de mtDNA en lobos grises; de estos 7 son derivados de hibridación con coyote, 4 están confinados al Nuevo Mundo, 6 están confinados al Viejo Mundo y uno está presente en ambas areas. La diferenciación genética entre poblaciones de lobos es significativa pero de pequeña magnitud. En el Viejo Mundo, la*

Old World, most localities have a single unique genotype, whereas in the New World several genotypes occur at most localities and three of the five genotypes are nearly ubiquitous. The pattern of genetic differentiation in the gray wolf contrasts with that of another large, highly vagile canid, the coyote, in which genetic differentiation among populations is not significant even among widely separated localities. We suggest that the difference between these two species reflects the rapid, recent increase in coyote numbers and expansion of their geographic range, and the coincident decline in gray wolf populations. Apparent genetic differences among extant wolf populations may be a recent phenomenon reflecting population declines and habitat fragmentation rather than a long history of genetic isolation.

mayoría de las localidades tienen un único genotipo específico para cada loalidad, mientras que en el Nuevo Mundo, la mayoria de las localidades poseen varios genotipos, y tres de los cinco son aproximadamente cosmopolitas. El patrón de diferenciación genética en el lobo gris contrasta conaquel de otro cánido grande y altamente móvil, el coyote, en el cual la diferenciación genética entre poblaciones no es significativa incluso en el casode localidades muy distantes. Sugerimos que la diferencia entre estas dos especies refleja el rápido y reciente incremento en el número de coyotes, la expansión en su distribución geográfica y la declinación en las poblaciones de lobos grises que ocurrió coincidentemente. La diferenciación genética entre las poblaciones de lobos existentes en la actualidad, puede ser un fenómeno reciente que refleja, antes que una larga historia de aislamiento genético, la declinación poblacional y fragmentación del habitat.

Introduction

The Gray Wolf, *Canis lupus*, is the most widely-distributed noncommensal terrestrial vertebrate. Its geographic range includes Mexico, the Holarctic, Europe, Asia, and the Middle East (Mech 1974). Gray wolves vary considerably in size, morphology, and pelage, and approximately twenty-four North American and eight Old World subspecies are currently recognized on these criteria (Pocock 1935; Goldman 1944; Nowak 1979; Hall 1981). Given the apparent morphologic differentiation among subspecies and the large geographic distances between populations, molecular measures of divergence might be expected to reveal differences among many populations. However, gray wolves are highly mobile carnivores and frequently move over long distances (Van Camp & Glukie 1979; Fritts 1983; Mech 1987). Individual gray wolves have been recorded to travel several thousand kilometers, and typical dispersal distances may exceed 100 kilometers (Mech 1987). While high rates of mobility do not necessarily result in high rates of gene flow (Shields 1987), long-range dispersal and colonization of new areas by individuals are so typical of wolves that high rates of genetic exchange are likely (Mech 1987); consequently, genetic divergence among widely-separated populations may be small. In fact, a genetic analysis of the North American coyote, *C. latrans*, a close relative of the gray wolf, demonstrates that high rates of genetic exchange have limited differentiation even among distant populations (Lehman & Wayne 1991).

Although wolves and coyotes are closely related, they have different ecological requirements, and the advance of civilization has had opposite effects on their abundances. Coyotes can persist in agricultural areas near humans and have recently increased dramatically throughout their range (Nowak 1979). In contrast, gray wolves have suffered large population declines over the last few hundred years (Young 1944; Nowak 1979). The historic range of the gray wolf included all of the United States, large parts of Mexico, and the Old World (Fig. 1). As towns and agriculture developed, gray wolves were systematically eliminated in many areas because of their perceived threat to livestock, game, and humans. Except in northern Russia and China, gray wolves in the Old World now exist predominantly as small populations of less than a few hundred animals that are well separated and highly isolated from other populations (Ginsberg & Macdonald 1990). In the New World, abundant wolf populations are now common only in northern Canada and Alaska. The southern populations are small and live in more disturbed and fragmented habitats (Mech 1970; Kolenosky 1983; Carbyn 1987). In southeastern Canada and Minnesota, coyotes have invaded areas where wolves were once abundant and have interbred with them (Kolenosky & Stanfield 1975; Lehman et al. 1991). As a consequence of being divided into small, highly isolated populations, the gray wolf may in fact be more genetically structured than its smaller but more abundant relative, the coyote.

Restriction-site analysis of mitochondrial DNA has been an effective tool for deducing population structure (Avise et al. 1987). Populations within species often have unique mtDNA genotypes that are related in a phylogenetic network which reflects the distance between populations or the presence of geographic barriers. This relationship has been designated "phylogeographic partitioning" by Avise and colleagues (1987), who predict that partitioning may not be evident in species in which gene flow among populations is high and average dispersal distances are large. In such species, different populations may possess many of the same genotypes, and genetic distance among genotypes in different populations may not increase with the geographic distance between them. This population structure has been observed in North American coyotes (Lehman & Wayne 1991). Using mtDNA restriction-site data, we evaluate partitioning among populations of gray wolves worldwide. We assess the significance of among-population variation and the phylogeographic structure suggested by maximum parsimony and distance analyses of restriction-site data. Our results demonstrate the probable consequences of recent population declines and habitat fragmentation for the genetic variability of the gray wolf.

Methods

Samples of organs or whole blood were collected from 350 gray wolves from twenty-six localities (Table 1, Fig. 1). These samples represent gray wolf populations from throughout their geographic range, including most of North America (United States, Mexico, and Canada), Europe (Sweden, Finland, Estonia, Italy, and Portugal), China, and the Middle East (Israel and Iran) (see Table 1). Genomic DNA was isolated from these samples and analyzed for mitochondrial DNA restriction-site polymorphisms as previously described (Wayne et al. 1990; Lehman et al. 1991). Twenty-one restriction enzymes, *Acc* I, *Apa* I, *Bam* HI, *Bcl* I, *Bgl* I, *Bgl* II, *Bst* EII, *Bst* UI, *Cla* I, *Dra* I, *Eco* RI, *Eco* RV, *Hha* I, *Hinc* II, *Hind* III, *Nco* I, *Sca* I, *Sst* I, *Stu* I, *Xba* I, and *Xmn* I, were used separately to digest samples of approximately 10 μg of DNA from each individual. The digested DNA was separated electrophoretically on a 1% agarose gel, transferred to nylon membranes, and hybridized with the entire mtDNA genome cloned from a domestic dog. Mitochondrial DNA fragments were visualized by autoradiography, and their identity was determined by comigration of fragments.

We also analyzed samples from nine domestic dogs, each representing the nine distinct mtDNA genotypes found in our initial survey of thirty-three dog breeds. The genotype designations and breeds of the nine domestic dogs are as follows: D1, Chow; D2, Beagle; D3, Doberman; D4, Eskimo Dog; D5, Maremma; D6, New Guinea Singing Dog; D7, Doberman; D8, Bulldog; and

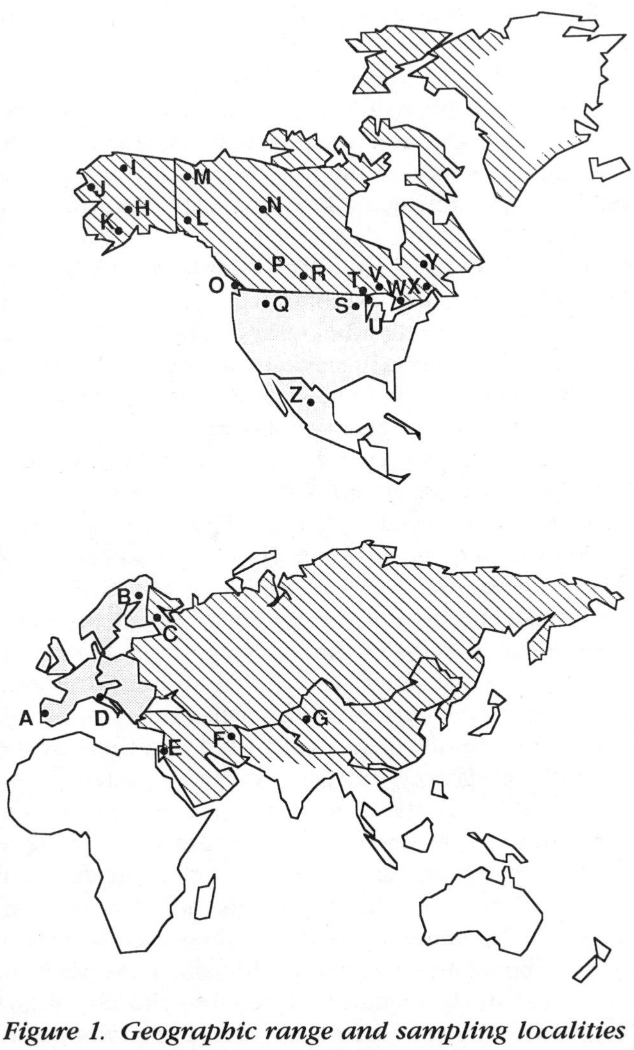

Figure 1. Geographic range and sampling localities of gray wolves (see Table 1). The diagonal lines indicate areas where gray wolf populations may still be relatively large and contiguous, and the stippled background indicates areas where gray wolf populations have gone extinct in recent times and where some remnant populations still exist.

Table 1. Map location (Fig. 1), locality, sample size (number of individuals), source of wolf specimens, and sample sets used in the mtDNA analysis.

Map Location	*Locality*	*Sample Size*	*Source*
A	Portugal—Central	3	F. Fonseca
B	Sweden—North and Finland	9	L. Laikre
C	USSR—Estonia	3	L. Lairke
D	Italy—North	14	E. Randi
E	Israel	2	M. Ben-David
F	Iran	2	V. O'Toole
G	China[1]	2	J. Shima
H	Alaska—Denali	22	L. D. Mech
I	Alaska—Anaktuvik	14	L. Adams
J	Alaska—Nome	8	W. Ballard
K	Alaska—Kenai Peninsula	6	T. Bailey
L	Yukon Territory	10	P. Merchant
M	Northwest Territories—Inuvik	53	P. Clarkson
N	Northwest Territories—Yellowknife	8	F. Jackson
O	British Columbia—Vancouver Island	15	L. Fritts
P	Alberta	4	P. Paquet
Q	Montana	5	M. Fairchild
R	Manitoba	8	H. D. Cluff
S	Minnesota—Northeast	66	L. D. Mech
T	Minnesota—North	20	P. Gogan
U	Michigan—Isle Royale	7	R. Peterson
V	Western Ontario	23	R. Peterson
W	Central Ontario	25	R. Peterson
X	Eastern Ontario	3	G. Forbes
Y	Quebec	14	F. Potvin
Z	Mexico[2]	4	J. Patton
Total		350	

SET I: all 26 populations, A through Z
SET II: 12 populations, (A) (B, C) (D) (E) (F) (G) (H–K) (L–N) (O) (P–R) (S–Y) (Z)
SET III: all 19 North American populations, H through Z
[1] *Wild-caught wolves from an unknown locality in China.*
[2] *Samples represent three distinct matrilines that originated in female wolves likely captured somewhere in the Sierra Madre Mountains in Mexico. Fewer than ten Mexican wolves are thought to exist in the wild.*

D9, Anatolian. We analyzed these samples using the same approach described above for the gray wolf samples.

Mitochondrial DNA genotypes were constructed from composite restriction-fragment patterns on all twenty-one enzymes (Lehman et al. 1991). Presence or absence of restriction sites was deduced from the pattern of fragment loss or gain among genotypes and the size distribution of restriction fragments (Lansman et al. 1983; Wayne et al. 1990; Lehman et al. 1991). In every comparison, we assumed that the minimum number of restriction site changes had occurred (Lehman et al. 1991). A phylogenetic tree of all genotypes was constructed by maximum parsimony analysis of a presence-absence matrice of restriction sites using the branch-and-bound algorithm of PAUP, version 3.0 for the Apple Macintosh (Swofford 1990). A representative coyote genotype, C1, was used to root the tree of gray wolf genotypes (Lehman et al. 1991). The percent of nucleotide sequence divergence between genotypes was calculated by the restriction-site method of Nei and Li (1979), with modifications by Nei and Tajima (1983), and corrected for the use of restriction enzymes with multiple recognition sites. A UPGMA distance tree of genotypes was constructed based on the sequence divergence data using the NTSYS-pc computer program, version 1.4 (Applied Biostatistics Inc., Setauket, New York).

To describe gene flow and genetic partitioning at different hierarchical levels, three sets of geographic groupings were made (Table 1, see Lehman & Wayne 1991). Set I includes all localities; Set II arranges the localities into twelve groupings representing large geographic subregions; Set III includes only the North American localities. Genetic partitioning of populations based on gray wolf mtDNA genotype frequencies was quantified using a variation on classical F-statistics for mtDNA proposed by Nei and Tajima (1983) in which maximum-likelihood estimates of within- and among-population nucleotide diversity are generated by iteration. These calculations were performed on an IBM PC compatible using the Quick-BASIC program HAPLO written by Lynch and Crease (1990).

Results

We observed eighteen genotypes in gray wolves using twenty-one distinct restriction enzymes (Table 2). Ninety-five restriction sites were detected representing approximately 500 bp of mtDNA sequence or 3% of the 16,800 bp genome found in canids (Wayne et al. 1990). Seven of the genotypes are derived from hybridization with coyotes (Table 2, Lehman et al. 1991) and are not included in the subsequent analyses because differences among them do not reflect differentiation among gray wolf populations. The remaining eleven gray wolf genotypes are divided among Old and New World localities (Table 3), with six found only in the Old World, four found only in the New World, and one found in both areas. Each of the Old World genotypes is found at a single locality, except for genotype W3, which exists at two neighboring localities as well as in the New World. Considering the small sample sizes at some localities, however, additional sampling might reveal more genotypes.

By contrast, in North America, genotypes W1, W3, and W4 are nearly ubiquitous, being found at eight, ten, and six localities, respectively (Table 3). Genotype W4 is found at localities as distant as northeastern Minnesota and Inuvik, Northwest Territories (3100 kilometers), and genotype W3 is found at localities as distant as Nome, Alaska and northern Montana (3600 kilometers). The two remaining North American gray wolf genotypes are found at only one locality. Genotype W2 is found in Manitoba wolves and W14 in Mexican wolves. Fourteen of the nineteen North American localities reveal more than one genotype, with the greatest number occurring in Minnesota, Ontario, and Quebec. The diversity at these localities reflects the presence of coyote-derived genotypes in individuals phenotypically identified as gray wolves. If coyote-derived genotypes are excluded, the most diverse localities are Montana and Inuvik, Northwest Territories (Table 3, Q and M). In North America, the correlation between number of genotypes and sample size is not significantly greater than zero ($p < 0.05$), indicating that our sample sizes are in general sufficient to estimate genotype diversity.

All gray wolf genotypes are genetically very similar, differing by at most eight restriction sites out of a total of ninety-six surveyed (Table 4). Thus, the estimated sequence divergence among gray wolf genotypes is small, ranging from 0.10% to 0.83% (Table 4). The average sequence divergence, without respect to genotype frequency, is 0.16% among New World genotypes and 0.49% among genotypes specific to the Old World. These within-species values are small relative to the average sequence divergence calculated between coyote and gray wolf genotypes, which is approximately 3.1% (Table 4). The relationships among genotypes can be represented in phylogenetic trees constructed through maximum-parsimony analysis of restriction-site data and distance analysis of sequence divergence data (Figs. 2 and 3). The maximum-parsimony tree suggests several relationships: Estonian, Italian, Iranian, and Mexican genotypes cluster together; Portuguese and Israeli genotypes cluster; and Chinese wolves are the most primitive gray wolf genotype (Fig. 2). However, these relationships are weakly defined due to the small number of informative characters uniting genotypes and may be strongly influenced by reversals or parallelisms. In fact, except for the clustering of Iranian and Italian geno-

Table 2. Restriction-fragment patterns in gray wolves and a sample of nine domestic dogs.

	Enzymes																				
Genotype	*a*	*b*	*c*	*d*	*e*	*f*	*g*	*h*	*i*	*j*	*k*	*l*	*m*	*n*	*o*	*p*	*q*	*r*	*s*	*t*	*u*
Gray Wolves																					
W1	A	C	F	A	A	E	F	A	A	B	J	D	F	B	C	C	B	D	C	A	C
W2	A	C	F	A	A	E	F	A	A	D	J	D	F	C	C	C	B	D	C	A	C
W3	A	C	F	A	A	E	F	A	A	B	J	D	F	C	C	C	B	D	C	A	C
W4	A	A	F	A	A	E	F	A	A	B	J	D	F	C	C	C	B	D	C	A	C
W5	A	C	F	A	A	E	F	A	A	B	K	D	F	C	C	C	B	D	C	A	C
W6	A	C	G	A	A	E	F	A	A	B	L	D	F	C	C	C	B	D	C	A	C
W7*	C	C	H	C	C	C	C	C	C	C	A	C	C	C	C	C	C	F	C	C	C
W8*	C	C	H	C	C	C	D	C	C	C	A	C	C	C	C	C	C	F	C	C	C
W9*	C	C	B	C	A	D	E	C	C	C	E	C	C	C	C	C	C	E	C	C	C
W10*	C	C	C	C	C	C	C	C	C	C	D	C	C	C	C	C	C	C	C	C	C
W11*	C	C	B	C	A	C	C	C	C	C	E	C	C	C	C	C	C	E	C	C	C
W12*	C	C	C	C	C	C	C	C	C	C	G	C	C	C	C	C	C	C	A	C	C
W13*	C	C	C	C	C	C	C	C	C	C	A	C	C	C	C	C	C	C	C	C	A
W14	B	C	F	A	A	E	F	A	A	B	K	D	F	C	C	C	B	D	C	A	C
W15	A	C	F	A	A	E	F	A	A	B	K	D	F	C	C	C	B	H	C	A	C
W16	A	C	F	F	A	E	F	A	A	B	N	D	F	C	C	C	B	D	C	A	C
W17	A	C	F	D	A	E	C	A	A	B	J	D	F	C	C	C	B	D	C	A	C
W18	A	C	I	A	A	E	F	A	A	B	K	D	F	C	C	C	F	D	C	A	C
Domestic Dogs																					
D1, D3, D4	A	C	F	A	A	E	F	A	A	A	K	D	F	A	C	C	B	Y	C	A	X
D2	A	C	F	A	A	E	F	A	A	A	K	D	F	A	X	C	B	D	C	A	D
D5	A	C	F	A	A	E	F	A	A	X	K	D	F	A	C	C	B	D	B	A	D
D6	A	C	F	A	X	E	F	A	A	X	K	D	F	A	C	C	B	Y	C	A	D
D7	A	C	F	A	A	E	F	A	A	X	K	D	F	C	C	C	B	X	C	A	D
D8	A	C	F	A	A	E	F	A	A	X	J	D	F	C	C	C	B	X	C	A	D
D9	A	C	X	A	A	E	F	A	A	X	J	D	F	C	C	C	B	X	C	A	D

Dog samples indicated by upper case letters. The most common pattern in coyotes is given a designation C (Lehman et al., 1991). Domestic dog specific patterns are indicated by an X or Y and underlined. The restriction enzymes used are as follows: a) Bgl *II, b)* Bam *HI, c)* Bcl *I, d)* Bgl *I, e)* Bst *EII, f)* Bst *UI, g)* Stu *I, h)* Cla *I, i)* Dra *I, j)* Eco *RV, k)* Hinc *II, l)* Sca *I, m)* Hha *I, n)* Nco *I, o)* Eco *RI, p)* Hind *III, q)* Xba *I, r)* Acc *I, s)* Apa *I, t)* Sst *I, u)* Xmn *I. Asterisks denote coyote-type genotypes found in wolves.*

types and the primitive position of the Chinese wolf genotype, the distance tree does not corroborate these relationships and instead indicates a grouping of North American genotypes (Fig. 3).

In North America, genotype frequency differences exist among the localities that have more than one genotype (Table 3). Genotype W1 tends to have a more eastern distribution and is common in the Northwest Territories, Minnesota, and southeast Canada but is not found in Alaska. Genotype W3 is common in western localities such as Alaska, Alberta, British Columbia, and Montana, is less common in Northwest territories, and is not found in Minnesota or southeast Canada. This genotype is also found in northern Europe. Genotype W4 is widely distributed but rare from Alaska to southeast Canada. Such differences in genotype frequency among localities and the presence of unique genotypes at several New and Old World localities suggests geographic partitioning. This notion can be tested by estimating N_{ST}, an analog of F_{ST}, which provides a value for the proportion of nucleotide-sequence variation due to among-population differences. This value is significant for the three groupings of our data (Table 5). The largest N_{ST} value is for the aggregated data set (Table 5, Set II) with a value of 0.86, and the smallest is for the North American localities (Set III) with a value of 0.76. Thus, approximately 75% to 85% of the restriction-site variation is due to among-population differences.

The domestic dog is a recent derivative of the gray wolf and is thought to hybridize with them in some areas such as Italy, where feral dogs are common and gray wolves are rare (Mech 1970; Bibikov 1982; Boitani 1982; Olsen 1985). Nine dog genotypes were found in our initial restriction-site survey of thirty-three dog breeds. A different panel of restriction enzymes was used in the initial survey, however, and only seven distinct dog genotypes were found with the panel of restriction enzymes described in Table 2. Each dog genotype differed from that in gray wolves by only one or two restriction sites, or by approximately 0.2% sequence divergence (Table 2). These data show that dogs are a very close relative of gray wolves and that hybridization between gray wolves and domestic dogs is not apparent in the populations that we surveyed, as none of the gray wolves had a mtDNA genotype common in domestic dogs. Moreover, because our results suggest the existence of restriction-fragment polymorphisms unique to domestic dogs, it may now be possible to determine the species of the female parent in individuals suspected to be dog–gray wolf hybrids.

Table 3. Frequencies and sample sizes (in parenthesis) of gray wolf genotypes from localities in the Old and New Worlds (see Fig. 1 and Table 1).

Old World Gray Wolf Genotypes	*W3*	*W5*	*W6*	*W15*	*W16*	*W17*	*W18*
A. Portugal	—	—	—	—	—	1.00 (3)	—
B. Sweden	1.00 (9)	—	—	—	—	—	—
C. Estonia	0.67 (2)	—	—	—	—	—	0.33 (1)
D. Italy	—	—	—	1.00 (14)	—	—	—
E. Israel	—	—	—	—	1.00 (2)	—	—
F. Iran	—	1.00 (2)	—	—	—	—	—
G. China	—	—	1.00 (2)	—	—	—	—
New World Gray Wolf Genotypes	*W1*	*W2*	*W3*	*W4*	*W14*		
H. Alaska—Denali	—	—	0.73 (16)	0.27 (6)	—		
I. Alaska—Anatuvik	—	—	1.00 (14)	—	—		
J. Alaska—Nome	—	—	1.00 (8)	—	—		
K. Alaska—Kenai	—	—	0.17 (1)	0.83 (5)	—		
L. Yukon Territory	—	—	1.00 (10)	—	—		
M. Northwest Territories—Inuvik	0.75 (40)	—	0.23 (12)	0.02 (1)	—		
N. Northwest Territories—Yellowknife	0.88 (7)	—	0.12 (1)	—	—		
O. British Columbia—Vancouver Island	—	—	1.00 (15)	—	—		
P. Alberta	0.25 (1)	—	0.75 (3)	—	—		
Q. Montana	0.20 (1)	—	0.60 (3)	0.20 (1)	—		
R. Manitoba	—	1.00 (4)	—	—	—		
S. Minnesota—Northeast	0.95 (19)	—	—	0.05 (1)	—		
T. Minnesota—North	1.00 (11)	—	—	—	—		
V. Western Ontario	0.35 (8)	—	—	—	—		
W. Central Ontario	0.67 (8)	—	—	0.33 (4)	—		
Z. Mexico	—	—	—	—	1.00 (4)		
Coyote-Derived Gray Wolf Genotypes	*W7**	*W8**	*W9**	*W10**	*W11**	*W12**	*W13**
R. Manitoba	1.00 (4)	—	—	—	—	—	—
S. Minnesota—Northeast	0.52 (24)	—	0.48 (22)	—	—	—	—
T. Minnesota—North	0.22 (2)	—	0.78 (7)	—	—	—	—
U. Michigan—Isle Royale	—	1.00 (7)	—	—	—	—	—
V. Western Ontario	0.53 (8)	—	0.47 (7)	—	—	—	—
W. Central Ontario	0.39 (5)	0.08 (1)	0.46 (6)	—	—	—	0.08 (1)
X. Eastern Ontario	—	—	—	0.33 (1)	0.33 (1)	0.33 (1)	—
Y. Quebec	—	—	—	0.07 (1)	0.07 (1)	0.43 (6)	0.43 (1)

Designations of genotypes are the same as in Lehman et al. (1991).

Discussion

In the New World, most well-sampled localities have more than one mtDNA genotype and frequently share the same genotypes. However, the two frequent New World genotypes W1 and W3 have different geographic distributions and the genotypes W2 and W14 are locality specific. Together these two conditions result in the existence of significant genetic differentiation among North American gray wolf populations. In the Old World, most localities have a distinct genotype not shared with other localities, suggesting greater genetic subdivision.

Table 4. Percent sequence divergence (above diagonal) and number of different restriction sites (below diagonal) between gray wolf (W1–W18) and coyote (C1) genotypes.

	W1	*W2*	*W3*	*W4*	*W5*	*W6*	*W14*	*W15*	*W16*	*W17*	*W18*	*C1*
W1	—	0.205	0.101	0.205	0.205	0.507	0.202	0.310	0.306	0.519	0.513	3.101
W2	1	—	0.101	0.205	0.205	0.507	0.202	0.310	0.306	0.519	0.513	2.851
W3	2	1	—	0.101	0.101	0.402	0.100	0.205	0.202	0.412	0.407	2.955
W4	2	1	2	—	0.205	0.507	0.202	0.310	0.306	0.519	0.513	3.101
W5	2	1	2	2	—	0.507	0.202	0.103	0.306	0.519	0.306	3.099
W6	5	4	5	5	5	—	0.501	0.614	0.608	0.824	0.814	2.918
W14	3	2	3	3	1	6	—	0.306	0.303	0.513	0.507	3.059
W15	4	3	4	4	2	7	3	—	0.412	0.629	0.411	2.994
W16	3	2	3	3	3	6	4	5	—	0.412	0.614	3.208
W17	5	4	5	5	5	8	6	7	4	—	0.833	2.996
W18	5	4	5	5	3	8	4	5	6	8	—	3.457
C1	27	26	25	27	27	26	26	27	28	26	30	—

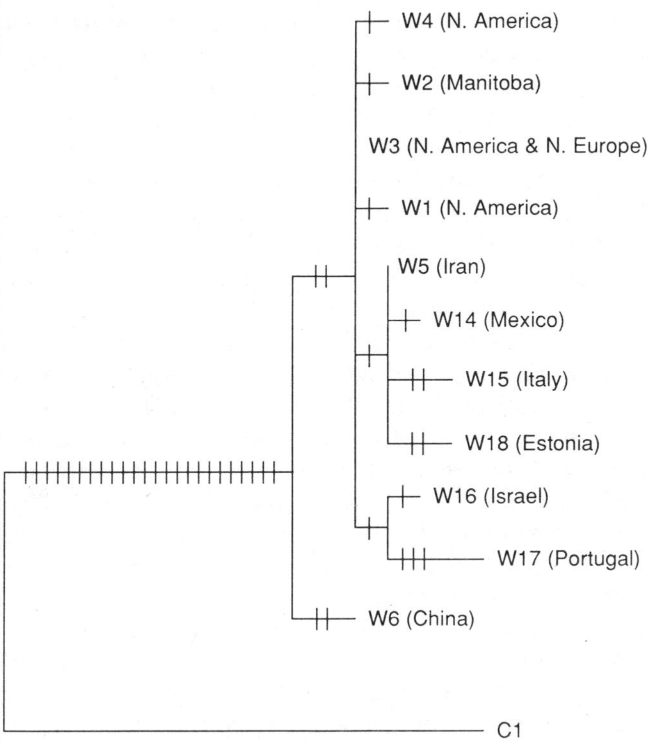

Figure 2. The single most parsimonious phylogenetic tree of gray wolf genotypes based on analysis of restriction-site data using twenty-one distinct restriction enzymes (see Table 1 for locality information). The tree was generated using the branch-and-bound option of the PAUP program (Swofford 1990) and rooted using the coyote genotype, C1 (Lehman et al 1991). Tree length = 42, overall consistency index = 0.88. The tree is identical to the bootstrap consensus tree produced by one thousand iterative subsamplings of our data. In the bootstrap tree, none of the ingroup nodes were significant according to the 95% criterion (Felsenstein 1985). The number of unique restriction sites shared by the group defined at each phylogenetic node is indicated by a small vertical line.

These results contrast dramatically with those from research on a similar North American canid, the coyote (Lehman & Wayne 1991). In the coyote mtDNA survey, 327 individuals were analyzed and thirty-two distinct genotypes were found (more than six times that in our similar-sized sample of North American wolves). Many coyote genotypes were widespread and most localities had a greater number of genotypes (mean was four genotypes per locality) than in the wolf survey. The N_{ST} values for three groupings of coyote populations ranged from 0.20 to 0.30 and were not significantly different from zero. Thus, gene flow appears to have stymied genetic differentiation among North American coyote populations.

The number of mtDNA genotypes in a stable population reflects the effective number of females (Birky et al. 1983, 1989; Avise et al. 1988; Lehman & Wayne 1991). For populations with little geographic partitioning, the effective number of females may be estimated as $N_f = (a - 1)/2\mu$, where a is the estimated total number of mtDNA genotypes and μ is the substitution rate (Lehman & Wayne 1991; for alternative method see Avise et al. 1988). We estimate empirically the total number of gray wolf genotypes in North America as 5.5 and the substitution rate as 1.53×10^{-5} restriction-site substitutions per genome per generation as in Lehman and Wayne (1991). We assume that in the past the North American gray wolf populations were more like those of present-day coyotes and approached panmixia. Using these parameters, the effective number of females in the recent past is $(5.5 - 1)/[(2)(1.53 \times 10^{-5})] \approx 150{,}000$ breeding females. Given that 1) approximately 50% of the population are adults, 2) 60% of the adult females breed, and 3) the sex ratio is 1:1 (Mech 1970; Packard et al. 1983), the past population of North American wolves would need to be about 1.0 million to sustain 5.5 mtDNA genotypes at equilibrium. In contrast, the coyote population size, based on its estimated genotype diversity, is predicted to be an order of magnitude larger, 1.75 million breeding females or about 11.5 million individuals (Lehman & Wayne 1991). Presently,

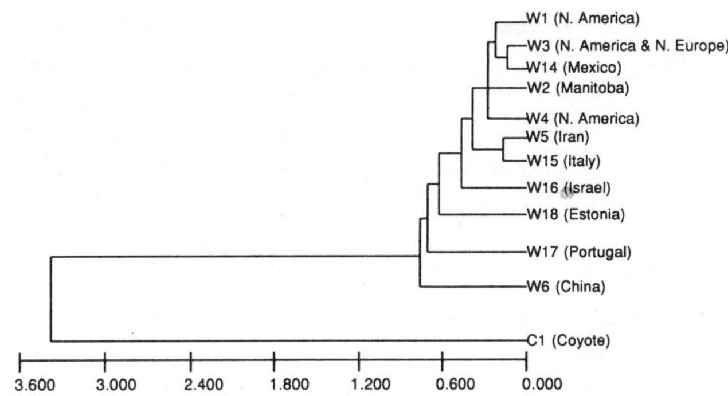

Figure 3. UPGMA tree based on sequence divergence values among gray wolf genotypes (Table 4). Tree was generated using the NTSYS-pc computer program, version 1.4 (Applied Biostatistics Inc., Setauket, New York). See Table 1 for locality information.

Table 5. Number of populations, N_{ST}, standard error (SE), and D-statistic and significance levels for analysis of among-population variation (Lynch & Crease, 1990; See Table 1 for population set information).

	Set I	*Set II*	*Set III*
Number of Populations	26	12	16
N_{ST}	0.79	0.86	0.76
SE	0.14	0.08	0.11
D-Statistic	30.4	118.2	46.8
Significance	$p < 0.01$	$p < 0.01$	$p < 0.01$

60,000 individuals is a generous estimate based on census data of the gray wolf population in North America (Carbyn 1987). Thus, our result indicates that a much larger wolf population existed in the recent past. In fact, Seton (1925) estimated the prehistoric wolf population to have been two million individuals, with a decrease to 200,000 by 1908. Our estimate of the effective number of females is consistent with a larger wolf population existing in the recent past.

The differences in population size and genetic variability of gray wolves and coyotes may reflect their contrasting recent histories. Previously abundant only in the south central region of the United States, coyotes have expanded their geographic range several-fold in the last few hundred years to include nearly all of North America (Nowak 1979; Voigt & Berg 1987). This rapid increase was coincident with a precipitous decrease in the abundance and geographic range of gray wolves due to habitat destruction, loss of large game, and direct exploitation by humans (Young 1944; Mech 1970; Nowak 1979; Carbyn 1987). Ironically, the contrast in the recent history of the two species reflects their differing response to human-induced habitat changes and predator control efforts. The coyotes' smaller size and flexible social behavior allow it to persist in smaller habitat patches and on smaller prey that may exist there (Bekoff & Wells 1980, 1986; Bowen 1981). The long-term persistence of gray wolves generally requires prey greater than several kilograms and large uninterrupted forested habitats (Mech 1970; Carbyn 1987; Voigt & Berg 1987). Moreover, because predator control programs are much more effective in controlling gray wolves than coyotes (Bekoff & Wells 1986; Carbyn 1987; Voigt & Berg 1987), any competitive exclusion of coyotes from an area by gray wolves that might prevent their colonization is mitigated (see Berg & Chesness 1978; Fuller & Keith 1981; Carbyn 1982; Berg & Kuehn 1986). Thus, the actions of humans have favored both the numerical increase of coyotes and the decrease of gray wolves, which have consequently had dramatic effects on the relative variability of the two species.

In comparison to North America, the Old World populations included in our survey are not contiguous but are small and separated by large tracts of unoccupied habitat. Populations in Sweden, Finland, Italy, Portugal, and Israel number from fewer than a dozen to as many as 250 individuals (Ginsberg & Macdonald 1990). Only in the northern areas of Russia and China do wolf populations potentially exist over a large region similar to Alaska and Northern Canada. Unfortunately, we have few samples from Russia or China. The large number of genotypes (seven) we find in the limited area of the Old World spanned by our localities suggest that past population sizes of wolves were considerably larger. As recently as the early 1800s gray wolves were abundant over much of the European continent but were subsequently eliminated in large areas through vigorous extermination efforts. By 1900, gray wolves had disappeared from most of the continent and now occupy only a few percent of their former geographic range (Zimen 1978; Zimen & Boitani 1979; Ginsberg & Macdonald 1990). The situation is thus more perturbed than in the northern reaches of the New World. A possible scenario for Europe and the Middle East is that as wolf numbers declined, genetic variation was partitioned among small, isolated populations, and a single genotype may have been rapidly fixed in each population through drift. The time to fixation of a single genotype in any one population is approximately $2N_f$ (Hartl & Clark 1989); in a population of 250 wolves, N_f should be near 40, and the time to fixation about 80 generations or 160 years (assuming a generation time of two years, Mech 1970). Thus, small isolated populations in the Old World may be expected to show a single genotype given isolation even in historic times.

All gray wolf genotypes are genetically very similar, differing by less than about 0.8% sequence divergence. This contrasts with a maximum intraspecific value of 2.3% among coyote genotypes (Lehman et al. 1991). Assuming a constant rate of 2–4% sequence divergence per million years (Shields & Wilson 1987), the last common ancestor of the present gray wolf genotypes likely existed 200,000 to 400,000 years ago. This date is less than that predicted by the fossil record, which indicates a first appearance of gray wolves about 600,000 to 800,000 years ago (Kurtén & Anderson 1980). The gray wolf is thought to have originated in the Old World and soon thereafter immigrated to the New World across the Bering Land Bridge (Kurtén 1968; Nowak 1979; Kurtén & Anderson 1980). This is supported by the presence of genotype W3 in North America and northern Europe.

Unfortunately, further details of the phylogenetic history of gray wolf populations are difficult to reconstruct from the mtDNA data because few restriction-site polymorphisms are phylogenetically informative. The grouping of the Mexican wolf genotype with those from the Old World (Fig. 2) was unexpected on geographic grounds but is supported by the existence of only one shared restriction site, which may represent a homoplasy. This association is not supported by the distance tree, which instead suggests a grouping of North American genotypes and an early divergence of Old World genotypes. Thus, the distance tree is more consistent with fossil evidence, which indicates that gray wolves originated in the Old World and migrated into North American (Kurtén & Anderson 1980). However, immigration events may have occurred multiple times throughout the Ice Ages and, considering that many gray wolf genotypes are likely to have vanished with the extinction of wolf populations, many pieces are probably missing from the phylogenetic puzzle. Moreover, mtDNA may record only a small and biased fraction of the phylogenetic history of a species, so analysis of nuclear loci may also be desirable.

Given the small amount of genetic differentiation among New and Old World gray wolf populations, the large number of apparent subspecies defined by morphologic criteria and pelage characteristics is paradoxical. Some of these subspecies have been defined based on analysis of very small samples sizes, and many taxonomists believe the number of subspecies should be reduced (Young 1944; Jolicoeur 1959; Skeel & Carbyn 1977; Pederson 1982; Nowak 1983). Moreover, the cranial and dental proportions used to define gray wolf subspecies may change rapidly in small populations, especially if selection on certain characters is intense (Berry et al. 1978; Fitch & Atchley 1985; Wayne 1986; Wayne & O'Brien 1986). Thus, morphologic differences need not be indicative of a long history of genetic isolation. Genetic differentiation among historic populations of wolves was probably similar to that among extant coyotes: gene flow among localities was high and little genetic differentiation was apparent even among widely spaced populations. As wolf numbers declined due to depredation programs and habitat fragmentation, gene flow among populations was reduced. This was particularly true in Europe and the Middle East, where wolf populations are small and highly isolated. As mentioned above, under such conditions genotype diversity was rapidly lost, and a single genotype fixed in each population through drift. Thus, the genetic partitioning we observe today among wolf populations may reflect the recent influence of agriculture and human expansion and may not have been a common pattern in the past.

Despite small population sizes, hybridization with dogs does not seem to have been an important influence on mtDNA diversity in gray wolves. In contrast, hybridization between wolves and coyotes accounts for seven of the twelve genotypes found in North American gray wolves, and coyote-derived genotypes are found in wolves over a wide area from Manitoba eastward to Quebec. Thus, under disturbed conditions, hybridization between wolves and coyotes is more common and extensive than between wolves and their much closer relative, the domestic dog. Subspecific hybridization can also be detected through the use of mtDNA restriction-site polymorphisms. For example, we have used the presence of a unique *Bgl* II restriction-site polymorphism found only in Mexican wolves (Table 2) to determine the purity of matrilines in the captive population of Mexican wolves. While interspecific hybridization may be undesirable, however, differences among gray wolf populations appear small and may be an artifact of human-induced habitat fragmentation. Thus, conservation programs might consider behavioral or demographic factors to be more important than maintenance of the genetic purity of putative wolf subspecies when making decisions about captive breeding or reintroduction programs (O'Brien & Mayr 1991).

Acknowledgments

We thank the many collaborators on Table 1 who provided gray wolf samples, and we thank I. L. Brisbin, R. Coppinger, A. Eisenhawer, T. Fuller, P. Gogan, K. Hansen, L. Lairke, T. Meier, D. Wurster-Hill, and P. Vrana. We also appreciate comments on the manuscript made by D. Girman, T. Fuller, L. Lairke, L. D. Mech, R. Peterson, and B. Van Valkenburgh. This work was supported in part by a USPHS National Service Award GM-07104 to N. L., a U.S. National Park Service agreement with R. K. W. and a National Science Foundation grant to R. K. W. (BSR 9020282).

Literature Cited

Avise, J. C., J. Arnold, R. M., Ball, E. Bermingham, T. Lamb, J. E. Neigel, C. A. Reeb, and N. C. Saunders. 1987. Intraspecific phylogeography: the mitochondrial DNA bridge between population genetics and systematics. Annual Review of Ecology and Systematics **18**:489–522.

Avise, J. C., R. M. Ball, and J. Arnold. 1988. Current versus historical population sizes in vertebrate species with high gene flow: a comparison based on mitochondrial DNA lineages and inbreeding theory for neutral mutations. Molecular Biology and Evolution **5**:331–344.

Bekoff, M., and M. C. Wells. 1980. The social ecology of coyotes. Scientific American **242**:130–148.

Bekoff, M., and M. C. Wells. 1986. Social ecology and behavior of coyotes. Advances in the Study of Behavior **16**:251–338.

Berg, W. E., and R. A. Chesness. 1978. Ecology of coyotes in northern Minnesota. Pages 229–247 in M. Bekoff, editor. Coyotes: biology, behavior and management. Academic Press, New York.

Berg, W. E., and D. W. Kuehn. 1986. Elusive coyote. Minnesota Volunteer **49**:9–13.

Berry, R. J., M. E. Jackobson, and J. Peters. 1978. The house mice of Faroe Islands: a study in microdifferentiation. Journal of Zoology **185**:73–92.

Bibikov, D. I. 1982. Wolf ecology and management in the USSR. Pages 120–133 in F. H. Harrington and P. C. Paquet, editors. Wolves of the world. Noyes Publishers, Park Ridge, New Jersey.

Birky, C. W., Jr., T. Maruyama, and P. Fuerst. 1983. An approach to population and evolutionary genetic theory for genes in mitochondria and chloroplasts, and some results. Genetics **103**:513–527.

Birky, C. W., Jr., P. Fuerst, and T. Maruyama. 1989. Organelle gene diversity under migration, mutation, and drift: equilibrium expectations, approach to equilibrium, effects of heteroplasmic cells, and comparison to nuclear genes. Genetics **121**:613–627.

Boitani, L. 1982. Wolf management in intensively used areas of Italy. Pages 158–172 in F. H. Harrington and P. C. Paquet, editors. Wolves of the world. Noyes Publishers, Park Ridge, New Jersey.

Bowen, W. D. 1981. Variation in coyote social organization: the influence of prey size. Canadian Journal of Zoology **59**:639–652.

Carbyn, L. N. 1982. Coyote population fluctuations and spatial distribution in relation to wolf territories in Riding Mountain National Park, Manitoba. Canadian Field Naturalist **96**:339–341.

Carbyn, L. N. 1987. Gray wolf and red wolf. Pages 358–377 in M. Novak, J. A. Baker, M. E. Obbard, and B. Malloch, editors. Wild furbearer management and conservation in North America. Ministry of Natural Resources, Ontario, Canada.

Felsenstein, J. 1985. Confidence limits on phylogenies: an approach using the bootstrap. Evolution **39**:783–791.

Fitch, W. M., and W. R. Atchley. 1985. Evolution in inbred strains of mice appears rapid. Science **228**:1169–1175.

Fritts, S. H. 1983. Record dispersal by a wolf from Minnesota. Journal of Mammalogy **64**:166–167.

Fuller, T. K., and L. B. Keith. 1981. Non-overlapping ranges of coyotes and wolves in northeastern Alberta. Journal of Mammalogy **62**:403–405.

Ginsberg, J. R., and D. W. Macdonald. 1990. Foxes, wolves, jackals and dogs. An action plan for the conservation of canids. International Union for Conservation of Nature and Natural Resources, Gland, Switzerland.

Goldman, E. A. 1944. Classification of wolves, Part 2. Pages 389–639 in S. P. Young and E. A. Goldman. The wolves of North America. American Wildlife Institute, Washington, D.C.

Hall, E. R. 1981. The mammals of North America. John Wiley and Sons, New York.

Hartl, D. L., and A. G. Clark. 1989. Principles of population genetics. Sinauer Press, Sunderland, Massachusetts.

Jolicoeur, P. 1959. Multivariate geographic variation in the wolf, *Canis lupus* L. Evolution **13**:283–299.

Kolenosky, G. B. 1983. Status and management of wolves in Ontario. Pages 35–40 in L. N. Carbyn, editor. Wolves in Alaska and Canada: their status, biology and management. Canadian Wildlife Service Report Series Number 45, Ontario, Canada.

Kolenosky, G. B., and R. O. Standfield. 1975. Morphological and ecological variation among gray wolves (*Canis lupus*) of Ontario, Canada. Pages 62–72 in M. W. Fox, editor. The wild canids. Their systematics, behavioral ecology and evolution. Van Nostrand Reinhold Company, New York.

Kurtén, B. 1968. Pleistocene mammals of Europe. Aldine Publishing Company, Chicago.

Kurtén, B., and E. Anderson. 1980. Pleistocene mammals of North America. Columbia University Press, New York.

Lansman, R. A., J. C. Avise, C. F. Aquadro, J. F. Shipira, S. W. Daniel. 1983. Extensive genetic variation in mitochondrial DNA's amony geographic populations of the deer mouse, *Peromyscus maniculatus*. Evolution **37**:1–16.

Lehman, N., and R. K. Wayne. 1991. Analysis of coyote mitochondrial DNA genotype frequencies: estimation of the effective number of alleles. Genetics **128**:405–416.

Lehman, N., A. Eisenhawer, K. Hansen, D. L. Mech, R. O. Peterson, and R. K. Wayne. 1991. Introgression of coyote mitochondrial DNA into sympatric North American gray wolf populations. Evolution **45**:104–119.

Lynch, M., and T. J. Crease. 1990. The analysis of population survey data based on DNA sequence variation. Molecular Biology and Evolution **7**:377–394.

Mech, L. D. 1970. The wolf: the ecology and behavior of an endangered species. University of Minnesota Press, Minneapolis, Minnesota.

Mech, L. D. 1974. *Canis lupus*. Mammalian Species **37**:1–6.

Mech, L. D. 1987. Age, season, distance, direction, and social aspects of wolf dispersal from a Minnesota pack. Pages 55–74 in B. D. Chepko-Sade and Z. Tang Halpin, editors. Mammalian dispersal patterns. University of Chicago Press, Chicago.

Nei, M., and W.-H. Li. 1979. Mathematical model for studying genetic variation in terms of restriction endonucleases. Proceedings of the National Academy of Science of the U.S.A. **76**:5269–5273.

Nei, M., and F. Tajima. 1983. Maximum likelihood estimation of the number of nucleotide substitutions from restriction sites data. Genetics **105**:207–217.

Nowak, R. M. 1979. North American Quaternary *Canis*. Museum of Natural History, University of Kansas, Lawrence.

Nowak, R. M. 1983. A perspective on taxonomy of wolves in North America. Pages 10–19 in L. N. Carbyn, editor. Wolves in Canada and Alaska: their status, biology and management. Canadian Wildlife Service Report Series Number 45, Ontario, Canada.

O'Brien, S. J., and E. Mayr. 1991. Bureaucratic mischief: recognizing endangered species and subspecies. Science **251**:1187–1188.

Olsen, S. J. 1985. Origins of the domestic dog: the fossil record. The University of Arizona Press, Tucson.

Packard, J. M., L. D. Mech, and U. S. Seal. 1983. Social influences on reproduction in wolves. Pages 78–85 in L. N. Carbyn, editor. Wolves in Alaska and Canada: their status, biology and management. Canadian Wildlife Service Report Series Number 45, Ontario, Canada.

Pederson, S. 1982. Geographic variation in Alaskan wolves. Pages 345–361 in L. N. Carbyn, editor. Wolves in Alaska and Canada: their status, biology and management. Canadian Wildlife Service Report Number Series 45, Ontario, Canada.

Pocock, R. I. 1935. The races of *Canis lupus*. Proceedings of the Zoological Society of London **43**:617–686.

Seton, E. T. 1925. Lives of game animals, Vol. 1, Part 1. Charles T. Branford, Boston.

Shields, L. 1987. Dispersal and mating systems: investigating the causal connections. Pages 3–27 in B. D. Chepko-Sade and Z. Tang Halpin, editors. Mammalian dispersal patterns. University of Chicago Press, Chicago.

Shields, G. F., and A. C. Wilson. 1987. Calibration of mitochondrial DNA evolution in geese. Journal of Molecular Evolution **24**:212–217.

Skeel, M. A., and L. N. Carbyn. 1977. The morphological relationship of gray wolves (*Canis lupus*) in national parks of central Canada. Canadian Journal of Zoology **55**:737–747.

Swofford, D. L. 1990. PAUP: Phylogenetic Analysis Using Parsimony (version 3.0). Illinois Natural History Society, Champaign.

Van Camp, J., and R. Glukie. 1979. A record long distance move by a wolf (*Canis lupus*). Journal of Mammalogy **60**:236–237.

Voigt, D. R., and W. E. Berg. 1987. Coyote. Pages 345–356 in M. Novak, J. A. Baker, M. E. Obbard, and B. Malloch, editors. Wild Furbearer Management and Conservation in North America. Ministry of Natural Resources, Ontario, Canada.

Wayne, R. K. 1986. Cranial morphology of domestic and wild canids: the influence of development on morphologic change. Evolution **40**:243–261.

Wayne, R. K., and S. J. O'Brien. 1986. Empirical demonstration that structural gene and morphometric variation of mandible traits are uncoupled between mouse strains. Journal of Mammalogy **67**:441–449.

Wayne, R. K., A. Meyer, N. Lehman, B. Van Valkenburgh, P. W. Kat, T. K. Fuller, D. Girman, and S. J. O'Brien. 1990. Large sequence divergence among mitochondrial DNA genotypes within populations of East African black-backed jackals. Proceedings of the National Academy of Science of the U.S.A. **87**:1772–1776.

Young, P. S. 1944. The wolves of North America. Part I. Dover Publications, New York.

Zimen, E. 1978. The wolf: his place in the natural world. Souvenir Press, London.

Zimen, E., and L. Boitani. 1979. Status of the wolf in Europe and the possibilities of conservation and reintroduction. Pages 43–83 in E. Klinghammer, editor. The behavior and ecology of wolves. Garland STPM Press, New York.

Use of Genetic Characters in Conservation Biology

The recent paper by Wayne and Jenks (1991), in addition to adding fuel to controversies concerning the specific and conservation status of the red wolf (Gittleman & Pimm 1991; O'Brien & Mayr 1991, Rennie 1991), has important implications for the application of genetic characters in conservation biology. We believe it has led to confusion on two separate but interrelated issues: (1) the use of genetic characters in identifying species and hybridization, and (2) the role of hybridization in evolution. Both issues pertain to treatment of hybrids and species concepts within the framework provided by the United States's Endangered Species Act (ESA).

While molecular techniques have produced impressive discoveries in evolutionary biology, it is important to realize their limitations. Molecular methods provide additional character sets for testing hypotheses of divergence and phylogenetic relationships; molecular data do not supersede other kinds of information. As repeatedly discussed before (Buth 1984; Hillis 1987), failure to find diagnostic molecular characteristics among morphologically distinctive forms does not imply conspecificity, but rather indicates that the available genetic data (typically representing only a small fraction of the genome) do not allow rejection of the null hypothesis (i.e., no divergence). Speciation could conceivably result from very few genetic changes (or even one); therefore, levels of genetic divergence are not valid measures of specific status.

The red wolf study is a case in point. Mitochondrial DNA (mtDNA) data were used to argue that red wolves may have originated through hybridization between grey wolves and coyotes. The data leave no doubt that coyote and red wolf have hybridized, but implication of grey wolf in a hybrid origin is unsupported. If one ignores introgression from coyotes, many systematists would conclude from available data (morphology, proteins, mtDNA) that the red wolf is a closely related, sister taxon to the grey wolf. Despite assertions to the contrary, failure to detect divergence between red and grey wolf mtDNAs would not, by itself, be used as evidence of either hybridization or conspecific status, but would suggest instead recent common ancestry. In addition, the authors themselves recognize difficulties with the proposed hybrid origin. They noted that hybridization between grey wolves and coyotes in the northern United States and Canada does not produce red wolves, and further indicated that "unique environmental and genetic conditions are required for the genesis of a red wolf phenotype." Therefore, red wolves are historical products of speciation, regardless of their mode of origin. Available information suggests that red wolves represent a morphologically distinct, close relative of the grey wolf that has hybridized with coyotes.

Hybridization is often assumed to be maladaptive and a threat to the continued existence of the affected species. This view is largely anthropocentric and based in part on aesthetic ideals of purity. It is also important to recognize that hybridization has played an important role in evolution. The role of hybridization in polyploid speciation of plants is renowned. In addition, introgressive hybridization has been suggested as a means of rapidly introducing genetic variation, allowing species to adapt to changing environments (i.e., Lewontin & Birch 1966; Grant 1985; DeMarais et al. 1992). This is not to say that hybridization is always an important evolutionary force or that it is never detrimental, but it does indicate that all circumstances must be carefully considered when evaluating the nature and impact of introgression.

Once again, the red wolf provides an example (albeit weak) for discussion. Analysis of mtDNA from six pelts collected between 1905 and 1930 indicated that red wolves and coyotes hybridized even then. Have coyotes and red wolves always hybridized or is introgression a result of human habitat modification? The answer to this question could have serious implications for management strategies for these species. Unfortunately, specimens analyzed are far too few and collected too recently to allow evaluation of human impact.

Treatment of hybrids and species concepts must be considered when evaluating taxa relative to the ESA. One must realize that the legislation was designed to protect distinctive forms (not just species) in danger of extinction, irrespective of mode of origin, making the species concept one chooses to apply irrelevant. We also assert that endangered taxa affected by introgression should not be discriminated against because of their lack of genetic "purity." Genetic markers, which are often selectively neutral and decoupled from distinctive phenotypic features (i.e., the target of selection), may be transmitted readily among taxa and provide little or no information on the impact of introgression.

With increasingly sophisticated methods of analysis, instances of hybridization and introgression are likely to be recognized more often. In most cases, these processes do not destroy the integrity of recognized forms. Considerable introgression of neutral markers has no effect on the fitness or specific status of the affected form; therefore, to deny protection due to lack of "purity" defeats the purpose of our efforts.

The issues addressed here are critical to conservation biology. In science, misunderstandings or false acceptance of hypotheses may lead to confusion and retard progress; however, further research typically corrects the errors. In the case of conservation biology, such errors have disastrous and irreversible impacts. Condemnation of distinct historical entities due solely to lack of "purity" or apparent lack of genetic divergence is unacceptable, as it leads to extinction of nonrenewable resources.

Thomas E. Dowling
Bruce D. DeMarais
W. L. Minckley
Michael E. Douglas
Department of Zoology
Arizona State University
Tempe, AZ 85287-1501, U.S.A.

Paul C. Marsh
Center for Environmental Studies
Arizona State University
Tempe, AZ 85287-3211, U.S.A.

Literature Cited

Buth, 1984. In Evolutionary genetics of fishes. Plenum Press, New York, New York.

DeMarais, B. D., et al. 1992. Origin of *Gila seminuda* (Teleostei: Cyprinidae) implications for evolution and conservation. Proceedings of the National Academy of Sciences USA **89.**

Gittleman, and Pimm 1991. Nature **351**:524–525.

Grant 1985. The evolutionary process: a critical review of evolutionary theory. Columbia University Press, New York, New York.

Hillis 1987. Annual review of ecology and systematics. **18**:23–42.

Lewontin, and Birch 1966. Evolution **20**:315–336.

O'Brien, and Mayr 1991. Science **251**:1187–1188.

Rennie 1991. Scientific American **265**:18–20.

Wayne, and Jenks 1991. Nature **351**:565–568.

On the Use of Morphologic and Molecular Genetic Characters to Investigate Species Status

ROBERT K. WAYNE

Institute of Zoology
Zoological Society of London
Regent's Park
London NW1 4RY, U.K.

I disagree with some of the conclusions of a letter by Dowling et al. (1992) regarding the use of "genetic" characters to delineate species. The authors of the letter maintain that: (1) "genetic" data cannot be used to disprove the species status of closely related taxa especially if morphologic data provides evidence of distinction; and (2) interspecific hybridization must not be regarded necessarily as an undesirable event between endangered species. Hybrids should not be uniformly condemned. The authors use our mitochondrial DNA results on the red wolf (Wayne & Jenks 1991) to provide illustrations of their points.

Regarding the use of genetic data to determine species status, Dowling et al. make no mention of the intense debate surrounding species concepts (Baum 1992). The species concept that is adopted largely determines the appropriate use of genetic and morphologic characters to define species. Current philosophical extremes are represented by the phylogenetic species concept which defines species as clusters of organisms possessing uniquely shared characters (McKitrick & Zink 1988; Baum 1992) as opposed to reproductively isolated groups of interbreeding organisms, often termed the biological species concept (BSC) (O'Brien & Mayr 1991). Although other canid species we have examined can be defined under the phylogenetic species concept by mtDNA analysis, the red wolf as a separate species cannot because we find no unique mtDNA restriction site or nucleotide substitutions that unite all red wolves in a single clade (Wayne & Jenks 1991). Similarly, conclusions based on past morphologic analyses are invalid under the phylogenetic species concept because they utilize phenetic similarity rather than character state data (Nowak 1979). It remains for the morphologists to use modern phylogenetic techniques (Wiley 1981; Mayr & Ashlock 1991) to demonstrate that the red wolf can be defined by discrete character states uniquely shared by all red wolves.

Under the biological species concept, evidence for reproductive isolation is often inferential and thus reproductive isolation is difficult to test. Indeed this is a common criticism of the BSC. Dowling et al. are correct in stating that only a few genetic changes may lead to speciation under this definition of species and therefore measures of "genetic" divergence are not valid measures of specific status. However, in this respect, measures of morphologic divergence are not sufficient either as they may not correspond with reproductive isolation. A critical point is that both morphological and molecular techniques intend to uncover heritable differences among organisms. Dowling et al. make an unnecessary distinction in this respect between "genetic" techniques which I assume they mean to involve direct assessment of DNA sequence divergence and "morphologic" techniques which more indirectly assess variation at the DNA level.

In fact, if we accept the BSC, species status cannot be assessed directly because there are no general methods for directly assessing genetic variation in genes that govern interfertility. However, in the absence of gene flow, populations might be expected to diverge through drift in both morphology and DNA sequence. Thus, the presence of significant morphologic or molecular divergence between closely related *sympatric* forms can be used to infer that a long history of reproductive isolation has existed between them. Selection complicates this conclusion because it may affect the extent of morphologic and molecular differentiation.

Is there evidence for the absence of gene flow and the presence of genetic divergences between red wolves and other sympatric canids? As Dowling et al. acknowledge, the presence of mtDNA genotypes in red wolves that are identical to those in coyotes clearly indicates that widespread interspecific gene flow (hybridization) has occurred between the two species. However, the additional presence of genotypes in red wolves, which are indistinguishable from those in gray wolves, they interpret as consistent with the red wolf being a closely related form to gray wolves. If this is true, red wolves and gray wolves are very closely related indeed as the amount of sequence divergence between them is less than that among genotypes from single gray wolf populations (Wayne et al. 1992). Moreover, two gray wolf related genotypes are found in red wolves, one otherwise found only in Mexican wolves and the other common in northern gray wolves (Wayne & Jenks 1991). Such heterogeneity likely reflects past gene flow between northern and southern gray wolf populations and red wolves. Given the high rates of gene flow which characterize populations of wolf like canids today (Lehman et al. 1991; Wayne et al., 1992) and the absence of ecologic or topographic barriers separating red wolves from nearby gray wolf populations prior to European colonization, to conceive of the red wolf persisting as a distinct subspecies of gray wolf is difficult. Furthermore, recent discussions of subspecies require that they be defined by uniquely shared traits (Avise & Ball 1990; Dizon et al. 1992). The mix of gray wolf genotypes found in red wolves indicates they are not a gray wolf subspecies under this definition.

Does the morphologic data provide a more definitive test of speciation? Contrary to the assertion by Dowling et al., the most recent and comprehensive morphologic analysis of red wolves does not suggest they are a subspecies or a very close relative of gray wolves (Nowak 1979). In fact, the red wolf is intermediate in morphologic dimensions between gray wolves and coyotes.

> In nearly all measurements and other features in which *C. rufus* [the red wolf] differs from *C. lupus* [the gray wolf], the former approaches *C. latrans* [the coyote]. Indeed, available specimens of the red wolf almost bridge the gap between the proximal extremes of the other two species. Hybrid origin for *C. rufus* thus seems to be one possibility, but there are other solutions to the problem. The most reasonable explanation is that *C. rufus* represents a primitive line of wolves that has undergone less change than *C. lupus*, and has retained more characters found in the ancestral stock from which both wolves and coyotes arose. (Nowak 1979, pg. 87)

Nowak maintains that the red wolf is a descendent of an early Pleistocene form (about 1.5 million years ago) existing before the appearance of modern gray wolves and thus we might expect morphologic and molecular genetic techniques to define it as a separate monophyletic group. This is not the case of the mtDNA results which show an absence of phylogenetically distinct red wolf genotypes and only the presence of genotypes very similar or identical to gray wolves and coyotes. The morphologic data indicate red wolves are distinct but this is based on phenetic similarity rather than character state data which, as the above quote implies, is circular with regard to species status. Hybrids between distinct forms are *expected* to be intermediate in morphology and thus the morphologic position of the red wolf is consistent with both a hybrid origin and a separate taxon status. For example, Nowak's multivariate morphologic positions of supposed gray wolf-coyote hybrids overlap with the position of the red wolves (Fig. 1). *Because of the possibility of a hybrid origin for the red wolf, discrete character state differences uniquely shared by red wolves are needed in order to define them as a separate taxon.*

Dowling et al. remark that the apparent absence of red wolf-like individuals from the more recent hybrid zone defined between gray wolves and coyotes in Eastern Canada, supports the separate species hypothesis. This is not entirely true. For many years, intermediate sized canids have been described from eastern Canada but not compared directly to red wolves (Kolenosky & Standfield 1975; Hilton 1978; Nowak 1979; Schmitz & Kolenosky 1985; Thurber & Peterson 1991). A similar approach to that used by Nowak has shown these canids

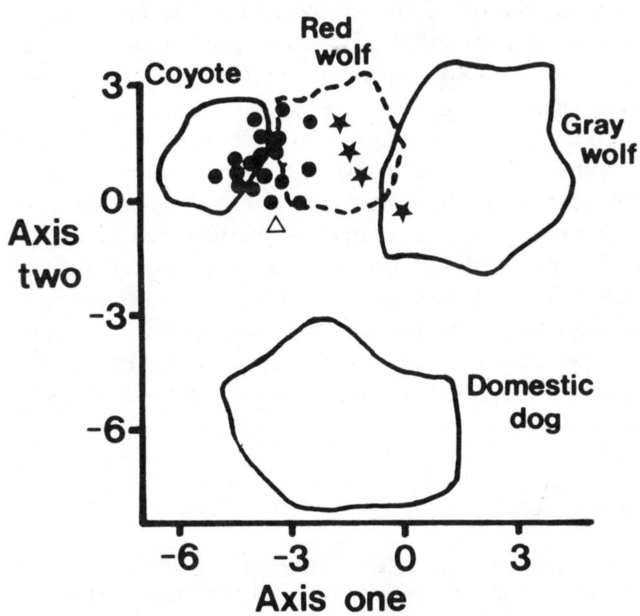

Figure 1. Multivariate analysis of 15 cranial measurements of wolf-like canids (Nowak 1979). Figures 10 and 15 from Nowak have been superimposed in this figure. The positions of suspected gray wolf-coyote hybrids from southeastern Canada (stars), a specimen from the Royal Ontario Museum, thought to be a gray wolf-coyote hybrid (triangle), and New England canids (dots) are indicated.

to be distinct and intermediate in form to gray wolves and coyotes, but these researchers, rather than assuming they represent a separate species, suggested their intermediate form reflected hybridization between coyotes and gray wolves or a response to changes in prey size and abundance.

Should hybrids receive protection similar to that given distinct endangered species? Dowling et al. argue that because hybridization may play an important role in evolution and because the Endangered Species Act recognizes any distinct population as potentially qualifying for protection, red wolves should be protected. They should not be excluded because of an absence of "genetic purity". Indeed interspecific hybridization is common in vertebrates and may be a very important source of evolutionary novelty in plants, birds and lower vertebrates. However, hybrid zones are generally small and restricted in mammals (Barton & Hewitt 1989) and do not appear to have the same evolutionary importance as in other groups. Moreover, a "species" as defined by the ESA is vague in its definition, perhaps in part to allow flexibility in the application of changing taxonomic concepts. Systematists are irresponsible if they let the vagueness of the ESA prevent them from applying their knowledge to the ordering of taxa according to phylogenetic distinctiveness (May 1991). We should not treat hybrids between very closely related forms with the same priority as distinct species in monospecific genera. Given limited resources we cannot hope to preserve all taxonomic units and thus we must order them according to a taxonomic hierarchy in which hybrids have lower priority than distinct species.

If the red wolf is confirmed as a hybrid with additional morphologic and molecular data, does it follow that we argue for its preservation? If so, do we also argue to preserve coyote-wolf hybrids in eastern Canada, gray wolf-domestic dog hybrids in Italy and domestic dog-dingo hybrids in Australia? When such preservation is at the expense of the genetic integrity of distinct endangered species and when hybridization likely reflects human-induced habitat changes, we need to be cautious in our reasoning.

Literature Cited

Avise, J. C., and R. M. Ball. 1990. Principles of genealogical concordance in species concepts and biological taxonomy. Pages 45–67. In D. Futuyma and J. Antonovics, editors. Oxford Surveys in Evolutionary Biology. Volume 7. Oxford Press, Oxford, England.

Baum, D. 1992. Phylogenetic species concepts. Trends in Ecology and Evolution **7**:1–2.

Barton, N. H., and G. M. Hewitt. 1989. Adaptation, speciation and hybrid zones. Nature **141**:497–503.

Dizon, A. E., C. Lockyer, W. F. Perrin, D. P. Demaster, and J. Sisson. 1992. Rethinking the stock concept: a phylogenetic approach. Conservation Biology **6**:24–36.

Dowling, T. E., B. D. Demarais, W. L. Minckley, M. E. Douglas, and P. C. Marsh. 1992. Use of genetic characters in Conservation Biology. Conservation Biology **6**:7–8.

Hilton, H. H. 1978. Systematics and ecology of the eastern coyote. Pages 209–228 in M. Bekoff, ed. Coyotes: biology, behavior, and management. Academic Press, New York.

Kolenosky, G. B., and R. O. Standfield. 1975. Morphological and ecological variation among gray wolves (*Canis lupus*) of Ontario, Canada. Pages 62–72 in M. W. Fox, ed. The wild canids. Van Nostrand Reinhold, New York.

Lehman, N., A. Eisenhawer, K. Hansen, D. L. Mech, R. O. Peterson, P. J. P. Gogan, and R. K. Wayne. 1991. Introgression of coyote mitochondrial DNA into sympatric North American gray wolf populations. Evolution **45**:104–119.

Lehman, N., and R. K. Wayne. 1991. Analysis of coyote mitochondrial DNA genotype frequencies: estimation of effective number of alleles. Genetics **128**:405–416.

May, R. 1990. Taxonomy as destiny. Nature **347**:129–130.

Mayr, E., and P. D. Ashlock. 1991. Principles of systematic zoology. Mcgraw-Hill, New York.

McKitrick, M. C., and R. M. Zink. 1988. Species concepts in ornithology. The Condor **90**:1–14.

Nowak, R. M. 1979. North American Quaternary *Canis.* Mongrel Museum of Natural History University, Kansas, Lawrence, Kansas.

O'Brien, S. J., and E. Mayr. 1991. Bureaucratic mischief: recognizing endangered species and subspecies. Science **251**:1187–1188.

Schmitz, O. J., and G. B. Kolenosky. 1985. Wolves and coyotes in Ontario: morphological relationships and origins. Canadian Journal of Zoology **63**:1130–1137.

Thurber, J. M. and R. O. Peterson. 1991. Changes in body size associated with the range expansion in the coyote (*Canis latrans*). Journal of Mammology **72**:750–755.

Wayne, R. K., and S. M. Jenks. 1991. Mitochondrial DNA analysis supports extensive hybridization of the endangered red wolf (*Canis rufus*). Nature **351**:565–568.

Wayne, R. K., N. Lehman, M. W. Allard, and R. L. Honeycutt. 1992. Mitochondrial DNA variability of the gray wolf: genetic consequences of population decline and habitat fragmentation. Conservation Biology **6**:559–569.

Wiley, E. O. 1981. Phylogenetics: The theory and practice of phylogenetic systematics. Wiley Interscience, New York.

The Red Wolf is Not a Hybrid

RONALD M. NOWAK

Office of Scientific Authority
Mail Stop 725, Arlington Square
U.S. Fish and Wildlife Service
Washington, DC 20240, U.S.A.

The red wolf (*Canis rufus*), an endangered species that was extirpated in its original range in southeastern North America by 1970, was recently reintroduced successfully in North Carolina (Henry 1992; Parker & Phillips 1991; Phillips & Parker 1988). Simultaneously, a longstanding interest in its systematic status intensified as Wayne and Jenks (1991), on the basis of an analysis of mitochondrial DNA, proposed that at some time in the past the red wolf had originated as a hybrid between the gray wolf (*Canis lupus*) and the coyote (*Canis latrans*). They also held that the mtDNA component of the current red wolf population is predominantly coyote. Their conclusions, while questioned even from the standpoint of genetics (Dowling et al. 1992), have received widespread attention (Begley et al. 1991; DeBlieu 1992; O'Brien & Mayr 1991).

That *C. rufus* and *C. latrans* had indeed hybridized was suspected long ago (Goldman 1944) and was generally considered one of the major factors in the near disappearance of the red wolf (McCarley 1962; Nowak 1979). However, it also was thought that this process was the result of modern human environmental disruption and that hybridization did not engulf the entire red wolf population prior to conservation efforts. The critical new suggestions made by Wayne and Jenks (1991) are: (1) that the red wolf actually originated as a hybrid and had never been a valid species or subspecies; and (2) that the existing captive and reintroduced red wolf population is substantially coyote in ancestry.

The study by Wayne and Jenks follows closely on reports of another "hybrid zone" in the Great Lakes region, said to involve the spread of coyote mtDNA to 62 percent of gray wolves in Minnesota and to 100 percent of those on Isle Royale (Lehman et al. 1991). Observations by field personnel in those areas, however, indicate no change in the morphological, behavioral, or ecological characters of the wolves (L. D. Mech & R. O. Peterson, personal communication). Therefore, even if the mtDNA situation has been reported accurately for both the gray wolf and the current red wolf population, the latter still could retain all other characters of the original species, and thus not represent a hybrid zone in the usual sense. Indeed, Wayne and Jenks support continued protection of the red wolf.

Unfortunately, a likely typical reaction is that of Gittleman and Pimm (1991), who combined Wayne and Jenks' study with a background of their own to criticize efforts being made to reestablish the red wolf. They implied that this animal's "unusually modern discovery" in 1851 somehow supports a view that it is not distinctive. Actually, 1851 only marks the first publication of a valid scientific name for the animal. Its presence had been recognized long before, published under another name in 1791, based on observations in 1774 (Harper 1942). The coyote, the distinctiveness of which is unquestioned, was not named until 1823, based on observations in 1819 (Young & Jackson 1951). Contrary to what is stated by Gittleman and Pimm (1972), the red wolf was not "resurrected as a species in 1972, by Paradiso and Nowak." Most authors have accepted specific status since the studies of Goldman (1937, 1944). Such workers included Ferrell et al. (1980), who were incorrectly cited by Gittleman and Pimm as not having found the red wolf to be distinctive. There was a minority view that the red wolf is a subspecies of the gray (Lawrence & Bossert 1967, 1975), but until 1991 there was only one published suggestion of hybrid origin (Mech 1970) and it was unsupported by study.

Gittleman and Pimm suggested that if the red wolf is not a full species it may not be worth saving. However, the Endangered Species Act of 1973 also covers subspecies and certain vertebrate populations. Of the 49 native United States terrestrial mammals now on the List of Endangered and Threatened Wildlife, 38 are subspecies or populations; they include such entities as the Key deer, the Mount Graham red squirrel, and the conterminous U.S. populations of the gray wolf and grizzly bear. The Endangered Species Act does not address hybrids, but recent interpretations indicate that a species may continue to be covered, even if it has experienced limited genetic introgression from another species (Henry 1992; Refsnider 1990).

Wayne and Jenks considered the plausible view that the red wolf is a subspecies of the gray, but rejected it in favor of hybrid origin. The latter notion is unsupported by a large series of modern museum specimens (Nowak 1979). Had the red wolf always represented nothing more than a hybrid zone between gray wolf and coyote, we would expect a complete blend of all three named species throughout the designated range of the red wolf. However, specimens from approximately 1890–1930 suggest that hybridization was then just getting underway in the region where the western part of the red wolf's range met the southeastern part of the coyote's (Nowak 1979). In most of the region the two species were then sympatric or in close proximity, with hybrid individuals having appeared at a few restricted localities. Only in central Texas was there a substantial hybrid zone, where *C. latrans* and *C. rufus* (but not the sympatric *C. lupus*) blended into one another.

It could be argued, in an attempt to resolve the matter, that long ago the gray wolf and coyote did interbreed in southeastern North America, and that a resulting hybrid population subsequently stabilized and became reproductively isolated from its parental species. *C. rufus* would then be considered a valid species but one of hybrid origin. Such taxa are relatively common among plants and have been discovered among fishes (DeMarais et al. 1992), but have not been reported for mammals. Fossil history does not support this origin for *C. rufus*. The red wolf, in much the same form as now, was present in North America through the Irvingtonian and Rancholabrean ages (Nowak 1979). It seems to represent an intermediate, surviving stage in the course of wolf evolution from a small coyote-like ancestor to the modern gray wolf. As already noted, hybridization with the coyote did occur, but only within the last 100 years or so.

Material collected after 1930 shows the spread of hybridization between *C. rufus* and *C. latrans*, and the steady reduction of unmodified red wolf populations (Nowak 1979). By the 1960s, such groups had largely disappeared, though available skulls from 1963–70 in extreme southeastern Texas indicate that the population there was still predominantly like original *C. rufus*. A number of animals subsequently were live-captured in that area and adjacent southwestern Louisiana, and used to form a breeding pool for eventual reintroduction.

I recently carried out a canonical discriminant analysis of measurements of relevant skulls, using the Statistical Analysis System (SAS Institute 1987). Three known groups were used: 50 coyotes of the subspecies *C. latrans lestes* from before 1925 in Idaho, 38 gray wolves of the subspecies *C. lupus pambasileus* and *C. l. tundrarum* from Alaska, and 28 gray wolves of the subspecies *C. l. lycaon* from Minnesota and Isle Royale in the 1960s and 1970s. In addition, 22 red wolves from before 1940 in Florida, Alabama, Mississippi, Louisiana, and the Big Thicket area of southeastern Texas were compared as individuals to the known groups. Also compared as individuals were 9 of the founders of the red wolf captive population and 5 of their descendants from the ongoing North Carolina reintroduction project. All series are composed only of complete skulls thought to represent full-grown males, and each series comprises every such specimen available from the indicated time and/or location. The pre-1940 red wolf sample is from a geographic and temporal range in the southeastern United States that previous work (Nowak 1979) suggests was not affected by hybridization with the coyote; no other appropriate specimens of wild *Canis* of any kind are available from this range. The 10 cranial measurements numbered, 1, 2, 4, 5, 6, 8, 11, 12, 14, and 15 in my previous work (Nowak 1979) were entered into the analysis.

Graphical results (Fig. 1) show the three species—*C. latrans, C. rufus,* and *C. lupus*—to be distinct from one another. Minnesota and Isle Royale gray wolf populations have affinity to those of Alaska and are not indicative of hybridization with the coyote. The red wolf falls roughly between the coyote and gray wolf, but does not merge with either. The captive and reintroduced red

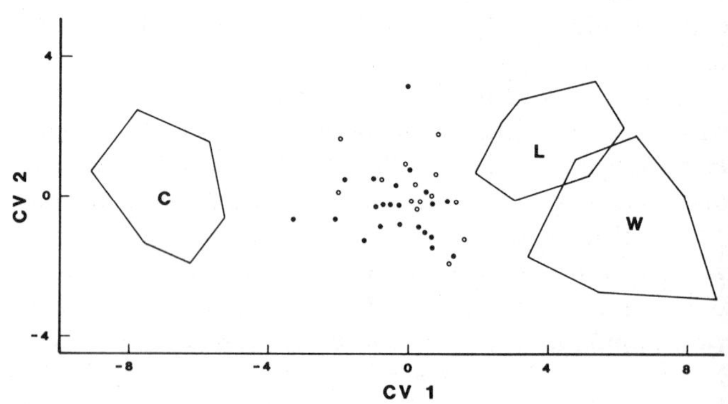

Figure 1. Plot of first and second canonical variables (CV 1 and CV 2) of a canonical discriminant analysis of skulls of Canis. *Letters indicate mean positions of known groups: coyote (C), Minnesota and Isle Royale gray wolf (L) Alaska gray wolf (W). Solid lines show boundaries of distributions of known groups. Dots indicate positions of 22 specimens of red wolf collected before 1940. Open circles indicate positions of 9 individuals used to found existing captive and reintroduced population of red wolf, and 5 individuals that had been released in North Carolina.*

wolf stock has a statistical distribution close to that of the original *C. rufus*.

Acknowledgments

I am grateful to Lois Loges, David Waddington, and Jay Sheppard for assistance with computer work.

Literature Cited

Begley, S., L. Wilson, M. Hager, and P. Annin. 1991. Return of the wolf. Newsweek **118(7)**:44–50.

DeBlieu, J. 1992. Could the red wolf be a mutt? New York Times Magazine, 14 June:30–46.

DeMarais, B. D., T. E. Dowling, M. E. Douglas, W. L. Minckley, and P. C. Marsh. 1992. Origin of *Gila seminuda* (Teleostei: Cyprinidae) through introgressive hybridization: implications for evolution and conservation. Proceedings of the National Academy of Science **89**:2747–2751.

Dowling, T. E., B. D. DeMarais, W. L. Minckley, and M. E. Douglas. 1992. Use of genetic characters in conservation biology. Conservation Biology 6:7–8.

Ferrell, R. E., D. C. Morizot, J. Horn, and C. J. Carley. 1980. Biochemical markers in a species endangered by introgression: the red wolf. Biochemical Genetics **18**:39–47.

Gittleman, J. L., and S. L. Pimm. 1991. Crying wolf in North America. Nature **351**:524–525.

Goldman, E. A. 1937. The wolves of North America. Journal of Mammalogy **18**:37–45.

Goldman, E. A. 1944. Classification of wolves. Pages 389–636 in S. P. Young and E. A. Goldman, The wolves of North America. American Wildlife Institute, Washington, D.C.

Harper, F. 1942. The name of the Florida wolf. Journal of Mammalogy **23**:339.

Henry, V. G. 1992. Finding on a petition to delist the red wolf (*Canis rufus*). Federal Register **57**:1246–1250.

Lawrence, B., and W. H. Bossert. 1967. Multiple character analysis of *Canis lupus, latrans,* and *familiaris,* with a discussion of the relationships of *Canis niger.* American Zoologist 7:223–232.

Lawrence, B., and W. H. Bossert. 1975. Relationships of North American *Canis* shown by a multiple character analysis of selected populations. Pages 73–86 in M. W. Fox, editor, The wild canids, their systematics, behavior, ecology and evolution. Van Nostrand Reinhold, New York.

Lehman, N., A. Eisenhawer, K. Hansen, L. D. Mech, R. O. Peterson, P. J. P. Gogan, and R. K. Wayne. 1991. Introgression of coyote mitochondrial DNA into sympatric North American gray wolf populations. Evolution **45**:104–119.

McCarley, H. 1962. The taxonomic status of wild *Canis* (Canidae) in the south central United States. Southwestern Naturalist **7**:227–235.

Mech, L. D. 1970. The wolf: the ecology and behavior of an endangered species. Natural History Press, Garden City, New York.

Nowak, R. M. 1979. North American Quaternary *Canis.* Monograph No. 6. Museum of Natural History, University of Kansas, Lawrence, Kansas.

O'Brien, S. J., and E. Mayr. 1991. Bureaucratic mischief: recognizing endangered species and subspecies. Science **251**:1187–1188.

Paradiso, J. L., and R. M. Nowak. 1972. *Canis rufus.* Mammalian Species. No. 22.

Parker, W. T., and M. K. Phillips. 1991. Application of the experimental population designation to recovery of endangered red wolves. Wildlife Society Bulletin **19**:73–79.

Phillips, M. K., and W. T. Parker. 1988. Red wolf recovery: a progress report. Conservation Biology **2**:139–141.

Refsnider, R. L. 1990. Notice of finding on a petition to delist the gray wolf (*Canis lupus*). Federal Register **55**:48656–49659.

SAS Institute. 1987. SAS/STAT guide for personal computers. Version 6 edition. SAS Institute Inc., Cary, North Carolina.

Wayne, R. K., and S. M. Jenks. 1991. Mitochondrial DNA analysis implying extensive hybridization of the endangered red wolf, *Canis rufus.* Nature **351**:565–568.

Young, S. P., and H. H. T. Jackson. 1951. The clever coyote. Wildlife Management Institute, Washington, D.C.

Comments On Red Wolf Taxonomy

MICHAEL K. PHILLIPS

U.S. Fish and Wildlife Service
P.O. Box 1969
Manteo, NC 27954, U.S.A.

V. GARY HENRY

U.S. Fish and Wildlife Service
330 Ridgefield Court
Asheville, NC 28801, U.S.A.

Introduction

Dowling et al. (1992) point out that results presented by Wayne and Jenks (1991) added to the controversy concerning the specific and conservation status of the red wolf (*Canis rufus*). Their assessment was correct as the results were used to challenge the integrity of the red wolf recovery program (Gittleman & Pimm 1991), and as rationale by the American Sheep Industry to petition the Secretary of Interior to remove the red wolf from the List of Endangered and Threatened Wildlife. The U.S. Fish and Wildlife Service determined that the petition did not present substantial information to warrant delisting (Henry 1992).

The debate over the origin of the red wolf is not new but rather has been brewing since the animal was classified as nearly extinct about 30 years ago (McCarley 1962). Some authorities consider the red wolf a full species, while others think it is a subspecies of the gray wolf (*Canis lupus*) or a hybrid resulting from interbreedings of gray wolves and coyotes (*Canis latrans*).

Dowling et al. (1992) point out potential problems with the interpretations presented by Wayne and Jenks (1991) and emphasize that no single data set should receive priority consideration when addressing taxonomic issues. We agree with Dowling et al. (1992) and below present data on red wolf behavior, ecology, and morphology that we feel should be considered when debating the origin of the species.

* *Send correspondence to Michael K. Phillips.*

Behavioral, Ecological, and Morphological Studies

During the 1970s, four studies documented behavioral and ecological differences between red wolves and coyotes (Riley & McBride 1972; Carley 1975; Shaw 1975; McCarley 1977). All these researchers concluded that *C. rufus* was a valid taxon.

In 1986 the Red Wolf Recovery Program initiated a reintroduction project in the Alligator River National Wildlife Refuge in northeastern North Carolina (Phillips & Parker 1988; Phillips 1990; Parker & Phillips 1991). From September 1987 through August 1992, 42 captive-born wolves were released in the refuge and a minimum of 23 wolves were born in the wild. Intensive telemetric monitoring (over 7000 relocations) of these animals led to additional insight into red wolf behavior and ecology.

Reintroduced and wild-born red wolves were very social and most belonged to packs (i.e. extended families) that occupied territories. It was not unusual for yearling and 2-year-old wolves to associate with their parents, assist with pup-rearing, and restrict movements to natal home ranges. Red wolves were relatively intolerant to strange conspecifics, and intraspecific aggression led to the death of 7 wolves. A similar social structure was observed for red wolves in Texas and Louisiana (Riley & McBride 1972). Thus, in terms of sociality red wolves are similar to gray wolves (*Canis lupus*) (Mech 1970:38–146). In contrast, coyotes (*Canis latrans*) are often more asocial: animals belong to breeding pairs or small family groups, home ranges of groups often evince considerable overlap, pups often disperse before their second summer, and intraspecific aggression is not thought to be an important source of mortality (Gier 1975; Andrews & Boggess 1978; Danner & Smith 1980; Althoff & Gipson 1981; Bekoff & Wells 1982; Roy & Dorrance 1985; Windberg et al. 1985; Harrison 1986: 1–55; Gese et al. 1989; Person & Hirth 1991).

The ecological role of the red wolf is largely defined by its food habits. Analysis of 2000 scats indicated that white-tailed deer (*Odocoileus virginianus*) and raccoons (*Procyon lotor*) were the primary year-round food items for wolves in the refuge. Although some of the deer probably were eaten as carrion, we documented wolf predation of apparently healthy adult deer. Most raccoons probably were taken as live prey. Gray wolves also rely heavily on ungulates and sometimes make extensive use of medium-sized mammals (Mech 1970:168–192; Peterson 1977:49–66). In contrast, deer and raccoons are of tertiary or lesser importance to coyotes in southern States, use of deer tends to be seasonal (greatest during the fawning period and hunting season), and adult deer are usually eaten as carrion (Korschgen 1957; Fooks 1961; Wilson 1967; Gipson 1974; Meinzer et al. 1975; Michaelson & Goertz 1977; Smith & Kennedy 1983; Wooding et al. 1984; Lee 1986; Leopold & Krausman 1986; Blanton & Hill 1989; Windberg & Mitchell 1991).

Fourteen individuals comprised the founding stock for the red wolf captive breeding program. Every animal generated from the founding stock has conformed to the morphological standards for the species (Paradiso & Nowak 1972; Carley 1975). The red wolf is intermediate to the coyote and gray wolf in most physical characteristics (Bekoff 1977; Mech 1974; Paradiso & Nowak 1972). The disproportionately long legs and large ears of the red wolf are the two most obvious external features separating the species from both the coyote and gray wolf. Experienced observers usually can distinguish adults of the three species on the basis of these external features (Riley & McBride 1972). In addition to unique external features, the morphology of the red wolf brain and skull differ from other canids (McCarley 1962; Paradiso 1968; Atkins & Dillon 1971; Paradiso & Nowak 1971, 1973; Gipson et al. 1974; Freeman 1976; Elder & Hayden 1977; Atkins 1978; Nowak 1979)

Conclusions

In terms of behavior, ecology, and morphology red wolves have never exhibited the variability one would expect if the species originated from interbreedings of gray wolves and coyotes. Indeed, Wayne and Jenks (1991) noted that hybridization between coyotes and gray wolves in the northern United States and Canada does not produce red wolves. Data concerning red wolf form and function support that contention that red wolves possess wolf-derived nuclear DNA that produces a wolf-like rather than a hybrid-like organism. Thus, it seems that *Canis rufus* is a valid taxon or, at the very least, a subspecies of *Canis lupus*.

We realize that the taxonomic identity of the red wolf cannot be unequivocally defined from available data. Accordingly, the Red Wolf Recovery Program has sponsored an analysis of the species' nuclear DNA and further investigations of wolf and coyote skull morphology.

Because young *Canis* show distinct and significant differences in behavioral ontogeny (Bekoff & Jamieson 1975; Bekoff 1978), the Recovery Program is also interested in sponsoring a study to analyze the development of agonistic and play behavior in red wolves. Linear discriminant function analysis would permit direct comparison to other studies of canid behavioral development and skull morphology. Bekoff (1978) and Bekoff et al. (1975) demonstrated the utility of using behavioral phenotypes to assess taxonomic relationships.

Additionally, the Recovery Program is considering sponsoring captive hybridization studies involving coyotes, red wolves, and gray wolves as this line of investigation has not been adequately addressed (Kolenosky 1971; Schmitz & Kolenosky 1985).

Literature Cited

Althoff, D. P., and P. S. Gipson. 1981. Coyote family spatial relationships with reference to poultry losses. Journal Wildlife Management **45**:641–649.

Atkins, D. L. 1978. Evolution and morphology of the coyote brain. Pages 17–35 in M. Bekoff, editor. Coyotes: biology, behavior, and management. Academic Press, Inc. New York.

Atkins, D. L., and L. S. Dillon. 1971. Evolution of the cerebellum in the genus *Canis*. Journal of Mammalogy **52**:96–107.

Andrews, R. D., and E. K. Boggess. 1978. Ecology of coyotes in Iowa. Pages 249–265 in M. Bekoff, editor. Coyotes: biology, behavior, and management. Academic Press, Inc., New York.

Bekoff, M. 1977. *Canis latrans*. Mammalian Species Account No. 79. American Society of Mammalogists.

Bekoff, M. 1978. Behavioral development in eastern coyotes. Pages 97–126 in M. Bekoff, editor. Coyotes: biology, behavior, and management. Academic Press, Inc. New York.

Bekoff, M., and R. Jamieson. 1975. Physical development in coyotes (*Canis latrans*) with a comparison to other canids. Journal of Mammalogy **56**:685–692.

Beckoff, M., and M. C. Wells. 1982. Behavioral ecology of coyotes: social organization, rearing patterns, space use, and resource defense. Z. Tierpsychologie **60**:281–305.

Bekoff, M., H. L. Hill, and J. B. Mitton. 1975. Behavioral taxonomy in canids by discriminant function analysis. Science **190**:1223–1225.

Blanton, K. M., and E. P. Hill. 1989. Coyote use of white-tailed deer fawns in relation to deer density. Proceedings Annual Conference Southeastern Association Fish Wildlife Agencies **43**:470–478.

Carley, C. J. 1975. Activities and findings of the red wolf recovery program from late 1973 to July 1, 1975. U.S. Fish Wildlife Service, Albuquerque, New Mexico.

Danner, D. A., and N. S. Smith. 1980. Coyote home range, movement, and relative abundance near a cattle feedyard. Journal Wildlife Management **44**:484–487.

Dowling, T. E., B. D. DeMarias, W. L. Minckley, and M. E. Douglas. 1992. Use of genetic characters in conservation biology. Conservation Biology **6**:7–8.

Elder, W. H., and C. M. Hayden. 1977. Use of discriminant function in taxonomic determination of canids from Missouri. Journal of Mammalogy **58**:17–24.

Fooks, L. G. 1961. Food habits of indigenous Canidae and Felidae in Arkansas based on complete and sample analysis of stomach contents. University of Arkansas, Fayetteville, Arkansas. M. S. Thesis.

Freeman, R. C. 1976. Coyote x dog hybridization and red wolf influence in the wild *Canis* of Oklahoma. M. S. Thesis, Northeastern Oklahoma State University, Tahlequah, Oklahoma.

Gese, E. M., O. J. Rongstad, and W. R. Mytton. 1989. Population dynamics of coyotes in southeastern Colorado. Journal of Wildlife Management **53**:174–180.

Gier, H. T. 1975. Ecology and social behavior of the coyote. Pages 247–262 in M. W. Fox, editor. The wild canids: their systematics, behavioral ecology, and evolution. R. E. Krieger Publishing Co., Inc., Malabar, Florida.

Gipson, P. S. 1974. Food habits of coyotes in Arkansas. Journal of Wildlife Management. **38**:848–853.

Gipson, P. S., J. A. Sealander, and J. E. Dunn. 1974. The taxonomic status of wild *Canis* in Arkansas. Systematic Zoology **23**:1–11.

Gittleman, J. L., and S. L. Pimm. 1991. Crying wolf in North America. Nature **351**:524–525.

Harrison, D. J. 1986. Coyote dispersal, mortality, and spatial relationships with red foxes in Maine. University of Maine, Orono, Maine. Ph.D. Dissertation.

Henry, V. G. 1992. Finding on a petition to delist the red wolf (*Canis rufus*). Federal Register **57**:1246–1250.

Kolenosky, G. B. 1971. Hybridization between wolf and coyote. Journal of Mammalogy **52**:446–449.

Korschgen, L. J. 1957. Food habits of the coyote in Missouri. Journal of Wildlife Management **21**:424–434.

Lee, R. M. 1986. Food habits of the coyote (*Canis latrans*) in Tennessee. Memphis State University, Memphis, Tennessee. M.S. Thesis.

Leopold, B. D., and P. R. Krausman. 1986. Diets of 3 predators in Big Bend National Park. Journal of Wildlife Management **50**:290–295.

McCarley, H. 1962. The taxonomic status of wild Canis (Canidae) in the southcentral United States. Southwestern Naturalist **7**:227–235.

McCarley, H. 1977. Vocalizations of red wolves (*Canis rufus*). Journal of Mammalogy **59**:27–35.

Mech, L. D. 1970. The wolf: the ecology and behavior of an endangered species. Natural History Press, Garden City, New York.

McCarley, H. 1974. *Canis lupus*. Mammalian Species Account No. 37. American Society of Mammalogists.

Meinzer, W. P., D. N. Ueckert, and J. T. Flinders. 1975. Foodniche of coyotes in the rolling plains of Texas. Journal Range Management **28**:22–27.

Michaelson, K. A., and J. W. Goertz. 1977. Food habits of coyotes in northwest Louisiana. Louisiana Academy of Science **40**:77–81.

Nowak, R. M. 1979. North American quaternary *Canis*. Monograph No. 6, Museum of Natural History, University of Kansas, Lawrence, Kansas.

Paradiso, J. L. 1968. Canids recently collected in east Texas, with comments on the taxonomy of the red wolf. American Midland Naturalist **80**:529–534.

Paradiso, J. L., and R. M. Nowak. 1971. A report on the taxonomic status and distribution of the red wolf. U.S. Fish and Wildlife Service, Special Wildlife Report No. 145. U.S. Fish and Wildlife Service, Washington, D.C.

Paradiso, J. L. 1972. *Canis rufus*. Mammalian Species Account No. 22. American Society of Mammalogists.

Paradiso, J. L. 1973. New data on the red wolf in Alabama. Journal of Mammalogy **54**:506–509.

Parker, W. T., and M. K. Phillips. 1991. Application of the experimental population designation to recovery of endangered red wolves. Wildlife Society Bulletin **19**:73–79.

Person, D. K., and D. H. Hirth. 1991. Home range and habitat use of coyotes in a farm region of Vermont. Journal of Wildlife Management **53**:361–376.

Peterson, R. O. 1977. Wolf ecology and prey relationships on Isle Royale. National Park Service Scientific Monograph Series No. 11. U.S. National Park Service, Washington, D.C.

Phillips, M. K. 1990. Measures of the value and success of a reintroduction project: red wolf reintroduction in Alligator River National Wildlife Refuge. Endangered Species Update **8**:24–26.

Phillips, M. K., and W. T. Parker. 1988. Red wolf recovery: a progress report. Conservation Biology **2**:139–141.

Riley, G. A., and R. T. McBride. 1972. A survey of the red wolf (*Canis rufus*). Scientific Wildlife Report No. 162. U.S. Fish and Wildlife Service, Washington, D.C.

Roy, L. D., and M. J. Dorrance. 1985. Coyote movements, habitat use, and vulnerability in central Alberta. Journal of Wildlife Management **49**:307–312.

Schmitz, O. J., and G. B. Kolenosky. 1985. Wolves and coyotes in Ontario: morphological relationships and origins. Canadian Journal of Zoology **63**:1130–1137.

Shaw, J. H. 1975. Ecology, behavior, and systematics of the red wolf (*Canis rufus*). Yale University, New Haven, Connecticut. Ph.D. Disseration.

Smith, R. A., and M. L. Kennedy. 1983. Food habits of the coyote (*Canis latrans*) in western Tennessee. Journal of Tennessee Academy of Science **58**:27–28.

Wayne, R. K., and S. M. Jenks. 1991. Mitochondrial DNA analysis implying extensive hybridization of the endangered red wolf, *Canis rufus*. Nature **351**:565–568.

Wilson, W. C. 1967. Food habits of the coyote, *Canis latrans*, in Louisiana. Louisiana State Univesity, Baton Rouge, Louisiana. M. S. Thesis.

Windberg, L. A., H. L. Anderson, and R. M. Engeman. 1985. Survival of coyotes in southern Texas. Journal of Wildlife Management **49**:301–306.

Windberg, L. A., and C. D. Mitchell. 1991. Winter diets of coyotes in relation to prey abundance in southern Texas. Journal of Mammalogy **71**:439–447.

Woodling, J. B., E. P. Hill, and P. W. Sumner. 1984. Coyote food habits in Mississippi and Alabama. Proceedings Annual Conference Southeastern Association Fish Wildlife Agencies **38**:182–188.

Response to Wayne, Nowak, and Phillips and Henry: Use of Molecular Characters in Conservation Biology

THOMAS E. DOWLING
W. L. MINCKLEY
MICHAEL E. DOUGLAS
Department of Zoology
Arizona State University
Tempe, AZ 85287-1501, USA

PAUL C. MARSH
Center for Environmental Studies
Arizona State University
Tempe, AZ 85287-3211, USA

BRUCE D. DEMARAIS
Department of Zoology
Arizona State University
Tempe, AZ 85287-1501, USA

In our recent letter (Dowling et al. 1992), we stated our concerns over two interrelated issues: (1) the use of genetic (i.e., molecular) characters in identifying species and their hybrids, and (2) the role of hybridization in evolution. The letter was stimulated by Wayne and Jenks' (1991) analysis of mitochondrial DNA (mtDNA) variation in the red wolf, *Canis rufus,* and subsequent interpretations presented in both the professional (e.g., Gittleman & Pimm 1991) and popular literature (e.g., Rennie 1991). Although consequences for conservation of the red wolf were addressed in those papers, our letter was focused more on the broader implications for conservation of endangered taxa.

In this issue, Nowak (1992) and Phillips and Henry (1992) have presented views generally in accord with ours, providing additional behavioral, ecological, morphological, and paleontological evidence supporting distinctiveness of the red wolf and its continued protection under the United States' Endangered Species Act. Nowak concludes that the red wolf is (1) neither the product of ongoing or recent hybridization between coyote (*C. latrans*) and gray wolf (*C. lupus*) nor, (2) a taxon of hybrid origin, but (3) represents a descendant from an intermediate stage in wolf evolution. We concur with his first conclusion but await further data to evaluate the second and third. The only current evidence inconsistent with the last alternative is the lack of distinctiveness among mtDNA sequences from red and gray wolves. Wayne and Jenks (1991) presumably expected differences as predicted from the "standard" rate of mtDNA evolution based on the calibration from primates (Hillis & Moritz 1990). Explanation of such a discordance requires only a reduction in rate of mtDNA evolution in wolves relative to primates, and such rate variation has been documented for a variety of organisms (e.g., Avise et al. 1992; Martin et al. 1992). The most appropriate test of an hybrid origin will be provided by analysis of codominant, single locus, nuclear gene markers alluded to by Phillips and Henry (1992).

Also in this issue, Wayne (1992) takes exception to

our views and presents further interpretation of the red wolf situation, misconstruing some of our ideas and attributing to us viewpoints we neither presented nor endorse. His discussions and criticisms focus on three aspects of our letter: (1) limitations of available molecular genetic data in evaluating hybrid origin and specific status of the red wolf, (2) protection of hybrids under the U.S. Endangered Species Act, and (3) failure to consider the debate over species concepts. Given the apparent misinterpretation and the importance of these issues to conservation biology, we are compelled to respond and discuss our thoughts in further detail.

Firstly, Wayne indicates that we believe molecular data cannot be used to refute the specific status of a taxon. This is not the case. In our original letter, we simply expressed concern over a general perception that molecular characters are "better" (i.e., Gittleman & Pimm 1991) than other types of data for identifying species, and argued that all available information (rather than a subset) should be applied to the question at hand (e.g., Buth 1984; Hillis 1987; Lande 1991). To our knowledge, this analytical approach has not yet been applied in taxonomic evaluation of the red wolf.

Consider just two sets of data for the red wolf discussed by Wayne (1992): (1) cranial measurements, and, (2) mtDNA restriction site/sequence data. Multivariate analysis of cranial measures identified the red wolf as an entity distinct from both coyote and gray wolf (Figure 1 in Nowak 1992 and Wayne 1992). Wayne chooses to ignore this morphological information (from Nowak 1979) because it has not been analyzed cladistically, and instead bases his evaluation solely on information from mtDNA. MtDNAs from a sample of red wolves were virtually indistinguishable from those in samples of coyotes and gray wolves. Coyote mtDNA is believed to have been introduced by recent hybridization (reviewed in Nowak 1992 and Phillips & Henry 1992). Again (see above), the point of dispute is the limited divergence between red and gray wolf mtDNAs.

Following a perplexing treatment of concepts relative to species, subspecies, and populations, Wayne then seems to conclude that the red wolf does not deserve taxonomic recognition: "Although other canid species we have examined can be defined under the phylogenetic species concept by mtDNA analysis, the red wolf as a separate species cannot because we find no unique mtDNA restriction site or nucleotide substitutions that unite all red wolves in a single clade (Wayne & Jenks 1991)" and later "Furthermore, recent discussions of subspecies require that they be defined by uniquely shared traits (Avise & Ball 1990; Dizon et al. 1992). The mix of gray wolf genotypes found in red wolves indicates they are not a gray wolf subspecies under this definition."

This treatment is confusing and deeply flawed. Firstly, the phylogeny of a single gene (or linked group of genes such as found in mtDNA) may differ from that of the species in which the gene occurs due to random sorting of inherited polymorphisms among evolving lineages (Pamilo & Nei 1988). This phenomenon is known to confound phylogenetic analysis of mtDNA data (Neigel & Avise 1986). By ignoring the morphological data, Wayne thus bases his decision solely on characters representing a small fraction of the genome. Given the lack of diagnostic mtDNA characters, it is essential to evaluate variation in additional characters to resolve the issue. Secondly, many morphologically distinctive taxa exhibit little or no allozymic or mtDNA divergence (e.g., pupfishes [genus *Cyprinodon*], Turner 1974; Echelle & Dowling 1992), while others are morphologically indistinguishable but highly divergent for molecular characters (e.g., plethodontid salamanders, reviewed in Larson 1989). Pupfish species are just as valid as those of plethodontids, but the two groups have diverged at different sets of loci.

We neither rejected nor accepted the hypothesis of hybrid origin for red wolf, but submitted, and do so again, that the available data are insufficient to distinguish between competing hypotheses. Morphological intermediacy of a taxon is neither necessary nor sufficient to demonstrate hybrid origin. In multivariate analyses of more than two taxa, some will, by necessity, be morphologically intermediate to others (Wilson 1992). In addition, hybrid taxa are not always morphologically intermediate to their progenitors (e.g., DeMarais et al. 1992), and even F_1 hybrids are not always intermediate to their parental forms (e.g., Neff & Smith 1978; Dowling et al. 1989; Meagher & Dowling 1991). Inference of hybrid origin from mtDNA data alone also is suspect. Neigel and Avise (1986) specifically cautioned against using mtDNA for inferring hybridization among closely related taxa because random extinction of mtDNA lineages can cause discordance between organismal and mtDNA phylogenies. A true test of hybrid origin for red wolf requires joint consideration of additional characters, a view also presented by O'Brien and Mayr (1991*a*).

Secondly, Wayne (1992) disagrees with our views on treatment of hybrids under the U.S. Endangered Species Act. He is understandably vague and indecisive in his discussion of the origin of red wolves (as were we, since more data are required for an objective decision). Unfortunately, he has similar difficulty discussing our views of the potential evolutionary significance of hybridization and the advisability for protection of imperiled hybrid taxa. His hesitation appears to stem from confusion over or apprehension to address the difference between hybrid taxa and hybrids (see also Wayne & Jenks 1991).

Taxa of hybrid origin are self-perpetuating groups of individuals of mixed ancestry following an evolutionary trajectory independent from that of their progenitors. The U.S. Endangered Species Act applies to recognized entities or populations deemed important enough in some way to preserve, and does not address their origin (Dowling et al. 1992). Genetically distinct taxa of hybrid origin must not be denied protection due to mixed ancestry. If the red wolf proves to represent an historically stable entity generated by long past (maybe even ancient) hybridization between gray wolf and coyote, then it is a taxon of hybrid origin that clearly should be protected.

We define a hybrid as an individual of mixed ancestry tracing its origin to the mating of a single male and a single female of different taxa. Hybrid individuals may or may not give rise to a hybrid taxon, and the point at which such occurs is unknown. Thus, if a red wolf phenotype originated and was perpetuated through hybridization between gray wolf and coyote following colonization by Europeans, it would not be a hybrid taxon nor merit protection until it became self-perpetuating and achieved an independent evolutionary track.

Wayne's discussion further includes the disturbing inference that we might advocate preservation of domestic dog-wild canid hybrid individuals at the "expense" of endangered species. We do not do so now, nor did we originally suggest such action. We stated: "This is not to say that hybridization is always an important evolutionary force or that it is never detrimental, but does indicate that all circumstances must be carefully considered when evaluating the nature and impact of introgression." Like many others (e.g., Anderson 1949; Lewontin & Birch 1966; Short 1972; Grant 1981), we recognize that hybridization can play an important role in evolution of plants and animals. Our approach is conservative. Each instance of hybridization *must* be evaluated separately.

Furthermore, when endangered taxa (including those of hybrid origin) have hybridized with related taxa due to human activity, conservation strategies must take into account the consequences of management actions as well as introgression (Allendorf & Leary 1988, Dowling & Childs 1992). When sufficient numbers persist to preserve genetic integrity, hybrids and conditions promoting hybridization may be eliminated to prevent further introgression. If, however, a taxon is rare, it may be necessary to protect all individuals, even if introgressed, to preserve remaining genetic diversity (Echelle 1991). Whether or not the red wolf is a taxon originating through hybridization or descended from a coyote-like ancestor (Nowak 1992), we will in fact argue strongly for preservation of hybrids of any kind if no other means exists to prevent extinction of a lineage. For another example, introduction of South American cougars to Florida has resulted in introgression of their mtDNA into the endangered Florida panther (*Felis concolor coryi*). We concur with O'Brien and Mayr (1991*b*) that the rarity and distinctiveness of Florida panther requires continued protection be afforded these individuals, despite introgression.

Lastly, despite Wayne's assertions to the contrary, we addressed the debate over species concepts in our first letter, which applies equally well here (brackets ours): "One must realize that the legislation [referring to the U.S. Endangered Species Act] was designed to protect distinctive forms (not just species) in danger of extinction, irrespective of mode of origin, making the species concept one choses to apply irrelevant." Wayne indicates that systematists would be irresponsible to allow the ESA to influence taxonomic decisions. We agree, but submit the converse as also valid. Conservation biologists must not allow long-standing disagreements over species definitions to interfere with decisions to conserve evolutionary entities. Conservation biologists should indeed be concerned with species concepts, but arguments over their application should not confound or otherwise deter conservation efforts. A goal of conservation biology is to preserve the diversity of life, requiring consideration of evolutionary potential as well as taxonomy. The red wolf is a taxon morphologically distinct from coyote and gray wolf, requiring recognition. We leave it to mammalian systematists to decide its taxonomic rank. As conservation biologists, however, we must continue to protect entities like the red wolf to preserve the evolutionary potential of each species.

Literature Cited

Allendorf, F. W., and R. F. Leary. 1988. Conservation and distribution of genetic variation in a polytypic species, the cutthroat trout. Conservation Biology **2**:170–184.

Anderson, E. 1949. Introgressive Hybridization. Wiley and Sons, New York.

Avise, J. C., and R. M. Ball, Jr. 1990. Principles of genealogical concordance in species concepts and biological taxonomy. Pages 45–67 in D. Futuyma and J. Antonovics, editors. Oxford Surveys in Evolutionary Biology Vol. 7. Oxford University Press, Oxford, U.K.

Avise, J. C., B. W. Bowen, T. Lamb, A. B. Meylan, and E. Bermingham. 1992. Mitochondrial DNA evolution at a turtle's pace: evidence for low genetic variability and reduced microevolutionary rate in Testudines. Molecular Biology and Evolution **9**:457–473.

Buth, D. G. 1984. The application of electrophoretic data in systematic studies. Annual Review of Ecology and Systematics **15**:501–522.

DeMarais, B. D., T. E. Dowling, M. E. Douglas, W. L. Minckley, and P. C. Marsh. 1992. Origin of *Gila seminuda* (Teleostei: Cyprinidae) through introgressive hybridization: Implications for evolution and conservation. Proceedings of the National Academy of Sciences, USA **89**:2747–2751.

Dizon, A. E., C. Lockyer, W. F. Perrin, D. P. DeMaster, and J. Sisson. 1992. Rethinking the stock concept: a phylogeographic approach. Conservation Biology **6**:24–36.

Dowling, T. E., and M. R. Childs. 1992. Impact of hybridization on a threatened trout of the southwestern United States. Conservation Biology **6**:355–364.

Dowling, T. E., B. D. DeMarais, W. L. Minckley, M. E. Douglas, and P. C. Marsh. 1992. Use of genetic characters in conservation biology. Conservation Biology **6**:7–8.

Dowling, T. E., G. R. Smith, and W. M. Brown. 1989. Reproductive isolation and introgression between *Notropis cornutus* and *Notropis chrysocephalus* (family Cyprinidae): comparison of morphology, allozymes, and mitochondrial DNA. Evolution **43**:620–634.

Echelle, A. A. 1991. Conservation genetics and genic diversity in freshwater fishes of western North America. Pages 141–153 in W. L. Minckley and J. E. Deacon, editors. Battle Against Extinction: Native Fish Management in the American West. University of Arizona Press, Tucson, Arizona.

Echelle, A. A., and T. E. Dowling. 1992. Mitochondrial DNA variation and evolution of the Death Valley pupfishes (*Cyprinodon,* Cyprinodontidae). Evolution **46**:193–206.

Gittleman, J. L., and S. L. Pimm. 1991. Crying wolf in North America. Nature **351**:524–525.

Grant, V. 1981. Plant Speciation, second edition. Columbia University Press, New York.

Hillis, D. M. 1987. Molecular versus morphological approaches to systematics. Annual Review of Ecology and Systematics **18**:23–42.

Hillis, D. M., and C. Moritz. 1990. An overview of applications of molecular systematics. Pages 502–515 in D. M. Hillis and C. Moritz, editors. Molecular Systematics. Sinauer Associates, Sunderland, Massachusetts.

Larson, A. 1989. The relationship between speciation and morphological evolution. Pages 579–598 in D. Otte and J. A. Endler, editors. Speciation and Its Consequences. Sinauer Associates, Sunderland, Massachusetts.

Lande, R. 1991. Applications of genetics to management and conservation of cetaceans. Pages 301–310 in A. R. Hoelzel, editor. Genetic Ecology of Whales and Dolphins. International Whaling Commision, special issue 13. The Red House, Cambridge, U.K.

Lewontin, R. C., and L. C. Birch. 1966. Hybridization as a source of variation for adaptation to new environments. Evolution **20**:315–336.

Martin, A. P., G. J. P. Naylor, and S. R. Palumbi. 1992. Rates of mitochondrial DNA evolution in sharks are slow compared to mammals. Nature **357**:153–155.

Meagher, S., and T. E. Dowling. 1991. Hybridization between the cyprinid fishes *Luxilus albeolus, L. cornutus,* and *L. cerasinus,* with comments on the hybrid origin of *L. albeolus.* Copeia 1991:979–991.

Neff, N. A., and G. R. Smith. 1978. Multivariate analysis of hybrid fishes. Systematic Zoology **28**:176–196.

Neigel, J. E., and J. C. Avise. 1986. Phylogenetic relationships of mitochondrial DNA under various demographic models of speciation. Pages 515–534 in E. Nevo and S. Karlin, editors. Evolutionary Processes and Theory. Academic Press, New York, New York.

Nowak, R. M. 1979. North American Quaternary *Canis.* Monograph of the Museum of Natural History, number 6. University of Kansas, Lawrence, Kansas.

Nowak, R. M. 1992. The red wolf is not a hybrid. Conservation Biology **6**:593–595.

O'Brien, S. J., and E. Mayr. 1991a. Species hybridization and protection of endangered animals. Science **253**:251–252.

O'Brien, S. J., and E. Mayr. 1991b. Bureaucratic mischief: recognizing endangered species and subspecies. Science **251**:1187–1188.

Pamilo, P., and M. Nei. 1988. Relationships between gene trees and species trees. Molecular Biology and Evolution **5**:568–583.

Phillips, M. K., and V. G. Henry. 1992. Comments on red wolf taxonomy. Conservation Biology **6**:596–599.

Rennie, J. 1991. Howls of dismay. Scientific American **265**:18–20.

Short, L. L. 1972. Hybridization, taxonomy, and avian evolution. Annals of the Missouri Botanical Garden **59**:447–453.

Turner, B. J. 1974. Genetic divergence of Death Valley pupfish species: Biochemical versus morphological evidence. Evolution **28**:281–294.

Wayne, R. K. 1992. On the use of morphologic and molecular genetic characters to investigate species status. Conservation Biology **6**:590–592.

Wayne, R. K., and S. M. Jenks. 1991. Mitochondrial DNA analysis supports extensive hybridization of the endangered red wolf (*Canis rufus*). Nature **351**:565–568.

Wilson, P. 1992. On inferring hybridity from morphological intermediacy. Taxon **41**:11–23.

Estimates of Lethal Equivalents and the Cost of Inbreeding in Mammals

KATHERINE RALLS

Department of Zoological Research
National Zoological Park
Smithsonian Institution
Washington, D.C. 20008, U.S.A.

JONATHAN D. BALLOU

Department of Zoological Research
National Zoological Park
Smithsonian Institution
Washington, D.C. 20008, U.S.A.

ALAN TEMPLETON

Department of Biology
Washington University
St. Louis, Missouri 63130, U.S.A.

Abstract: *The costs of inbreeding in natural populations of mammals are unknown despite their theoretical importance in genetic and sociobiological models and practical applications in conservation biology. A major cost of inbreeding is the reduced survival of inbred young. We estimate this cost from the regression of juvenile survival on the inbreeding coefficient using pedigrees of 40 captive mammalian populations belonging to 38 species.*

The number of lethal equivalents ranged from −1.4 to 30.3, with a mean of 4.6 and a median of 3.1. There was no significant difference between populations founded with wild-caught individuals, a mixture of wild-caught and captive-born individuals, and individuals of unknown origin. The average cost of a parent-offspring or full sibling mating was 0.33, that is, mortality was 33% higher in offspring of such matings than in offspring of unrelated parents. This is likely to be an underestimate.

Resumen: *Los costos de procreación en consanguinidad en poblaciones naturales de mamíferos son desconocidos a pesar de su importancia teórica en los modelos genéticos y sociobiológicos y en sus aplicaciones prácticas para la biología de la conservación. Uno de los costos mayores de la procreación en consanguinidad es la disminución en la sobrevivencia de las crías consanguíneas. Estimamos este costo por medio de la regresión de la sobrevivencia juvenil en el coeficiente de procreación en consanguinidad utilizando pedigrís de 40 poblaciones de mamíferos en cautiverio pertenecientes a 38 especies.*

El número de equivalentes letales varió de −1.4 a 30.3, con una media de 4.6 y una mediana de 3.1. No hubo diferencia significativa entre poblaciones formadas a partir de individuos silvestres capturados, a partir de una mezcla de individuos silvestres capturados, y a partir de individuos de origen desconocido. El costo promedio del apareamiento de padre-cría o hermanos completamente consanguíneos fue de 0.33, es decir, la mortalidad fue 33% más alta en las crías de tales apareamientos que en las crías de especies no relacionadas. Es probable que este cálculo sea una subestimación.

Paper submitted 8/5/87; revised manuscript accepted 11/11/87.

Introduction

Many studies of laboratory, domestic, and zoo animals have documented reduced survival and fecundity of inbred young (Wright 1977; Ralls & Ballou 1983; Sausman 1984; Templeton & Read 1984). Inbreeding depression is thus a major concern in the management of small populations, and estimates of the cost of inbreeding are of considerable importance to conservation biology.

However, inbreeding can increase an individual's inclusive fitness by producing young that share more of its genome. Thus, when inbreeding has little or no genetic cost, there should be strong selective advantage for inbreeding as well as recognition and cooperation among kin (Wilson 1976; May 1979). The cost of inbreeding is therefore of theoretical importance as well.

Calculations of the total cost of inbreeding in natural populations would involve considering the effects of inbreeding on several components of fitness. However, the "cost of inbreeding" that appears in a variety of theoretical models (Dawkins 1976; Bengtsson 1978; Parker 1979; Smith 1979; Feldman & Christiansen 1984) is defined solely in terms of the survival of inbred young relative to non-inbred young. There are almost no estimates of this quantity in natural populations of mammals (Packer 1979).

We estimate this cost from pedigrees of 40 captive mammalian populations belonging to 38 species.

Methods

Morton, Crow, & Muller (1955) developed a log model for estimating the cost of inbreeding from the rate at which juvenile survival decreases with increasing amounts of inbreeding. Specifically,

$$S = e^{-(A + BF)} \quad (1)$$

where S is the proportion of individuals surviving to some age, F is the inbreeding coefficient, A is considered a measure of death due to environmental causes and the genetic damage expressed in a randomly mating population, and B is a measure of the rate at which survival decreases with increasing inbreeding.

Makov & Bittles (1986) evaluated the use of this and several other equations to estimate effects of inbreeding in humans. They found that many different models could adequately detect significant inbreeding effects; however, different models resulted in different values of A and B. Because of the limited range of inbreeding levels in available data from human populations (F = 0–0.125), they were unable to determine which equation most adequately modeled data on inbreeding effects in humans. They suggested that different equations could more effectively be evaluated in animal populations with wider ranges of inbreeding levels.

We evaluated the log transformed equation (1) and two other equations, using several of our largest data sets with relatively wide ranges of inbreeding levels (F = 0–0.5). The two additional equations were

$$S = A + B(F) \quad (2)$$

$$\arcsin\sqrt{S} = A + B(F) \quad (3)$$

where S, A, B, and F are the same values as in equation (1). Model 2 was used because it represents the simplest linear relationship between the variables. Model 3 (angular transformation) was used since it is often recommended for estimating proportions (Sokal & Rohlf 1969). Weighted least squares regression, with a small sample size correction (Templeton & Read 1984), was used to estimate the parameters for each of the models. The total percentage of variation explained by the equation (R^2) was used to evaluate which model best fitted the data.

When analyzing pedigrees of zoo animals, care must be taken to distinguish inbreeding depression from hybridity effects or "outbreeding depression" (Templeton & Read 1984; Templeton et al. 1986). We therefore carried out the analysis developed for this purpose by Templeton & Read (1984) on those pedigrees with adequate sample sizes but found no evidence of outbreeding depression (Templeton & Read 1984; unpublished data).

Inbreeding coefficients (F) were calculated for each animal in each pedigree, relative to the founders of the population. Methods for calculating F from pedigree data are given by Ballou (1983). F is the probability that the two alleles present at a given locus are "identical by descent"—that is, are derived by replication of a single allele from a common ancestor. F ranges from 0 in a non-inbred individual to 1.0 in a completely inbred (homozygous) individual (Crow & Kimura 1970). The effect of inbreeding is often less severe in individuals with inbred ancestors (Bowman & Falconer 1960; Lorenc 1980; Templeton & Read 1984), but we were unable to exclude them from the analysis because this eliminated all levels of inbreeding except F = 0.25 in many pedigrees.

Levels of inbreeding varied among pedigrees (Table 1). For each level of inbreeding represented in a particular pedigree, we calculated the proportion of animals that survived to a criterion age. This was 180 days for the larger species and one-half the age at sexual maturity for the smaller ones (Table 2). Ideally, studies of the relationship between inbreeding and juvenile mortality should be based upon the total mortality before reaching reproductive age (Cavalli-Sforza & Bodmer 1971), but we were unable to follow many individuals for this period because zoo animals are often transferred to other institutions before reaching reproductive age. Considering survival to a criterion age less than repro-

Table 1. Comparison of models used for estimating cost of inbreeding.

		Comparison of R^2 Values MODEL		
SPECIES[a]	*Maximum Inbreeding level*	*Log (1)*	*Linear (2)*	*Arcsin (3)*
Short bare-tailed opossum	.328	.80	.79	.77
Elephant shrew	.125	.05	.06	.07
Golden lion tamarin	.375	.35	.26	.26
Greater galago	.250	.17	.14	.13
Maned wolf	.312	.77	.83	.83
Bush dog	.500	.02	.00	.00
Pygmy hippopotamus	.375	.45	.55	.55
Dorcas gazelle	.375	.64	.66	.63

[a] *Scientific names listed in Table 2.*

ductive maturity tends to underestimate the cost of inbreeding, as inbred mortality increases more rapidly than non-inbred mortality with increasing age in some species (Ralls, Brugger, & Glick 1980; unpublished data).

Results

Table 1 shows the results of the three models applied to eight of the largest data sets. R^2 values were highest for the Linear model (2) in 2 populations, highest for the Arcsin model (3) in 2 populations, and highest for the log model (1) in 4 populations. As Makov and Bittles (1986) concluded, no one model was clearly better than the others; R^2 values ranged over only a few percentage points across the models.

The log transformed model (1) has been used extensively in the literature to estimate number of lethal equivalents and is the theoretically expected model, if it is assumed that genetic and environmental influences are independent of each other with respect to survival (Morton, Crow & Muller 1955). Use of this model also facilitates comparisons with A and B values already published in the literature. We therefore selected it for all subsequent analyses.

Estimates for A and B are shown in Table 2. Values of A ranged from 0.03 to 1.11 with a mean of 0.33 and a median of 0.32. Values for B ranged from −0.68 to +15.16, with a mean of +2.33 and a median of +1.57 (Fig. 1). Of the 40 populations, 36 had positive slopes, which clearly indicates an overall trend towards higher levels of juvenile mortality with increasing inbreeding coefficients (Sign test, $P < .001$). This relationship was statistically significant—that is, the slope of the line was significantly greater than zero—in only 9 (23%) of the populations. However, most of our sample sizes were small and distributed over only a few levels of inbreeding. The statistical power to detect slopes significantly greater than zero was therefore limited. Considering only those populations in which the relationship between inbreeding and survival is significant would be likely to greatly overestimate the average cost of inbreeding in mammals. Limiting the analysis to only those species with relatively large data sets increases the power of the statistical comparisons but reduces the number of species that can be analyzed. Only 10 species had more than five levels of inbreeding and total sample sizes over 100. Six of these 10 had slopes significantly different from zero; the average B value was 1.98, with a median of 1.64. These B values did not differ significantly from those in the overall data set (Mann-Whitney U test, $P > 0.05$).

The distributions of B by order are shown in Figure 2. Median values were between one and two except for the Carnivora. There were no statistically significant differences between average B values in populations founded with wild-caught individuals ($\bar{x} = 2.57$, $n = 18$), a mixture of wild-caught and captive individuals ($\bar{x} = 2.42$, $n = 11$), and individuals of unknown origin ($\bar{x} = 1.95$, $n = 10$) (Kruskal-Wallis Test, $P = 0.88$).

The number of lethal equivalents per gamete lies between B and A but is usually very close to B (Cavalli-Sforza & Bodmer 1971; Crow & Kimura 1970). The number per zygote or individual is twice the number per gamete, thus our estimates of the average number of lethal equivalents per individual are twice the values of B in Table 2, with a mean of 4.6 and a median of 3.1. We estimated the cost of inbreeding for matings between first-degree relatives (parents and their offspring or full siblings) by solving equation (1) for each species using F = 0 and F = 0.25 to obtain the predicted survivorship at these levels of inbreeding. The cost of inbreeding (i) at F = 0.25 is then equal to

$$i = 1 - \left[\frac{\text{Survivorship at F} = 0.25\text{: } e^{-(A+.25B)}}{\text{Survivorship at F} = 0\text{: } e^{-(A)}}\right]$$

$$= 1 - e^{-.25B.} \qquad (4)$$

The average cost of inbreeding between first degree relatives, calculated by averaging the costs across all populations, was 0.33 (Table 2). Solving equation (4)

Table 2. The cost of inbreeding in 40 mammalian populations.

TAXON	Survival to Age (Days)	N	Founder[a] Type	No. of Inbred Levels	Model Estimates A	Model Estimates B	Model R^2	Cost of Inbreeding[c] at F = 0.25	Data Source
MARSUPIALIA									
Short bare-tailed opossum (*Monodelphis domestica*)	75	251	W	6	0.03	0.43[b]	0.80	.10	National Zoo
Parma wallaby (*Macropus parma*)	180	17	W	5	0.32	1.69	0.47	.34	National Zoo
INSECTIVORA									
Elephant shrew (*Elephantulus rufescens*)	21	218	W	7	0.28	2.12	0.05	.41	National Zoo
PRIMATES									
Black spider monkey (*Ateles fusciceps robustrus*)	180	23	W	3	0.23	2.22	0.88	.43	National Zoo
Saddle-backed tamarin (*Saguinus fuscicollis*)	180	233	U	2	1.11	1.86	—	.37	Monell Chemical Senses Center
Illiger's saddle-backed tamarin (*Saguinus f. illigeri*)	180	406	U	4	0.40	7.92	0.40	.82	Rush-Presbyterian St. Luke's Medical Center
Golden lion tamarin (*Leontopithecus r. rosalia*)	180	974	W	18	0.54	2.15[b]	0.35	.42	1984 Studbook
Ring-tail lemur (*Lemur catta*)	180	53	M	4	0.34	0.13	0.01	.03	Oregon Primate Research Center
Black lemur (*Lemur macaco*)	180	43	W	3	0.52	2.78	0.87	.50	Oregon Primate Research Center
Brown lemur (*Lemur fulvus*)	180	136	M	6	0.32	9.17[b]	0.94	.90	Oregon Primate Research Center
Greater galago (*Galago c. crassicaudatus*)	180	251	M	29	0.45	1.69[b]	0.17	.34	Oregon Primate Research Center
Melanotic galago (*Galago c. argentatus*)	180	54	M	4	0.36	0.48	0.19	.11	Oregon Primate Research Center
Crab-eating macaque (*Macaca fascicularis*)	180	237	U	3	0.37	0.29	0.56	.07	New England Primate Research Center
Celebes black ape (*Macaca nigra*)	180	86	U	3	0.38	2.84	0.70	.51	Oregon Primate Research Center
Chimpanzee (*Pan troglodytes*)	180	247	U	4	0.35	1.05	0.67	.23	Yerkes Primate Center
RODENTIA									
Climbing rat (*Tylomys nudicaudus*)	45	49	U	5	0.23	−0.14	0.02	−.04	National Zoo
Wied's red-nosed rat (*Wiedomys pyrrhorhinos*)	30	23	W	2	0.05	15.16	—	.98	National Zoo
Rock cavy (*Kerodon rupestris*)	90	132	U	3	0.12	0.77	0.87	.18	National Zoo
Salt-desert cavy (*Dolichotis salinicola*)	90	17	W	2	0.08	7.21	—	.34	National Zoo
Acouchi (*Myoprocta pratti*)	135	36	U	5	0.30	2.20	0.17	.42	National Zoo
Boris (*Octodontomys gliroides*)	75	53	U	6	0.26	1.15	0.33	.25	National Zoo
Punare (*Cercomys cunicularus*)	60	161	W	4	0.10	0.94[b]	0.91	.21	National Zoo
CARNIVORA									
Maned wolf (*Chrysocyon brachyurus*)	180	338	M	4	0.52	−0.68	0.77	−.19	1983 Studbook
Bush dog (*Speothos venaticus*)	180	176	W	9	0.54	0.24	0.02	.06	1983 Studbook
Sumatran tiger (*Panthera tigris sumatrae*)	180	427	M	12	0.49	0.01	0.00	.003	1983 Studbook
PERISSODACTYLA									
Zebra (*Equus burchelli*)	180	50	U	2	0.30	1.56	—	.32	National Zoo
ARTIODACTYLA									
Pygmy hippopotamas (*Choeropsis liberiensis*)	180	419	W	12	0.33	1.59[b]	0.45	.33	1982 Studbook

Table 2. Continued

TAXON	Survival to Age (Days)	N	Founder[a] Type	No. of Inbred Levels	Model Estimates A	Model Estimates B	Model R^2	Cost of Inbreeding[c] at F = 0.25	Data Source
Reeves muntjac (*Muntiacus reevesi*)	180	75	M	9	0.19	1.20	0.37	.26	National Zoo
Eld's Deer (*Cervus eldi thamin*)	180	24	M	2	0.31	7.57	—	.85	National Zoo
Pere David's Deer (*Elaphurus davidianus*)	180	39	C	7	0.17	0.63[b]	0.74	.15	National Zoo
Reindeer (*Rangifer tarandus*)	180	50	W	4	0.32	4.20	0.71	.65	National Zoo
Giraffe (*Giraffa camelopardalis*)	180	19	W	2	0.29	2.24	—	.43	National Zoo
Kudu (*Tragelaphus strepsiceros*)	180	25	W	2	0.37	−0.03	—	−.01	National Zoo
Bongo (*Tragelaphus eurycerus*)	180	74	W	3	0.23	−0.55	0.74	−.15	1984 Studbook
Gaur (*Bos gaurus*)	180	182	W	6	0.18	0.51	0.36	.12	Hinz & Foose, 1982
Scimitar-horned oryx (*Oryx dammah*)	180	81	M	2	0.09	4.63	—	.69	National Zoo
Wildebeest (*Connochaetes taurinus*)	180	42	W	11	0.33	0.28	0.02	.07	National Zoo
Dik-dik (*Madoqua kirki*)	180	20	M	3	0.80	0.59	0.12	.14	National Zoo
Dorcas gazelle (*Gazella dorcas*)	180	143	M	15	0.34	1.85[b]	0.64	.37	National Zoo
Spekes gazelle (*Gazella spekei*)	30	64	W	5	0.22	3.08[b]	0.92	.54	Templeton & Read, 1983
		Mean:			0.33	2.33		0.33	
		Median:			0.32	1.57		0.33	
		Lower Quartile:			0.23	0.45		0.09	
		Upper Quartile:			0.39	2.81		0.47	

[a] Founder Type: W = All founders wild-caught.
C = Founders captive-born.
M = Founders were a mix of wild-caught and captive-born.
U = Source of founders unknown.

[b] B (slope) significantly different than zero at the 0.05 level.

[c] Cost of inbreeding for F = 0.25: $= 1 - \left[\dfrac{\text{Predicted inbred survival: } e^{-(A+.25B)}}{\text{(Predicted non = inbred survival: } e^{-(A)}}\right] = 1 - e^{-.25B}$.

using the average B value (2.3) results in a cost of inbreeding of 0.44. However, the statistic of interest here is the estimate of the expected value of the cost of inbreeding rather than the cost of inbreeding calculated from the expected value of B. We therefore base our discussion on an average cost of inbreeding of 0.33. The distribution of the cost of inbreeding between first degree relatives is shown in Figure 3.

Discussion

The costs of inbreeding varied widely among captive populations. This is not surprising since one would expect populations to differ in their level of susceptibility to inbreeding. However, in many cases, the models fit the data very poorly and only a small proportion of the variance was explained. These variable results probably reflect the heterogeneous data used for the analysis. The available data for the populations surveyed differed in sample size and the range and number of levels of inbreeding. Nevertheless, these results do provide data on the costs of inbreeding and number of lethal equivalents in a wide variety of captive populations and allow analyses of general trends and patterns.

The median number of estimated lethal equivalents for the captive mammalian populations we examined was 3.1. This figure is similar to estimates for other animal populations. Humans (May 1979), *Drosophila* (Dobzhansky 1970), and the great tit, *Parus major* (Bulmer 1973), are thought to have about two lethal equivalents per individual, and the Japanese quail, *Coturnix coturnix japonica*, is thought to have about 3.4 (Sittmann, Abplanalp & Fraser 1986). Our estimates for captive carnivores, although based on only three populations, were quite low. More carnivore populations should be studied to determine if this is characteristic of the order or unique to the data sets we examined.

May (1979), assuming the number of lethal equivalents in humans was 2.2, estimated the cost of breeding in humans at F = 0.25 to be .42. However, his equation for calculating the inbreeding cost contained an error. The correct cost, based on formula (4), is .24. This es-

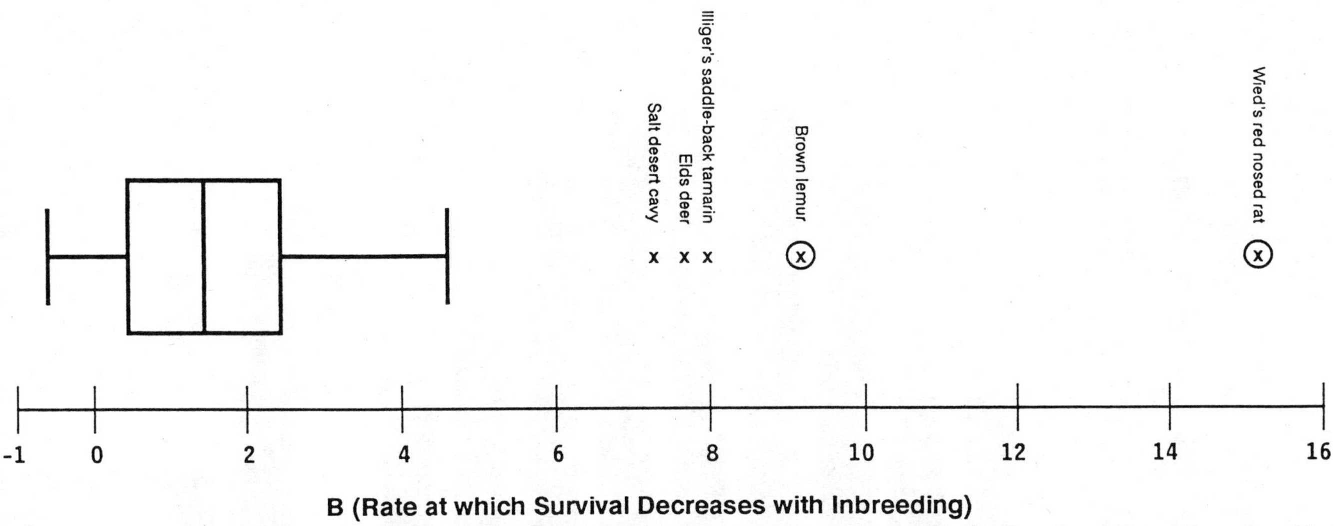

Figure 1. Box plots of B, a measure of the rate at which survival decreases with increasing inbreeding, for 40 mammalian populations. The median (middle vertical line in box), upper and lower quartiles (left and right ends of box), upper and lower inner fences (vertical lines), outlying values (x), and values beyond the outer fences (⊗) are shown (Hoaglin, Mosteller, & Tukey 1983).

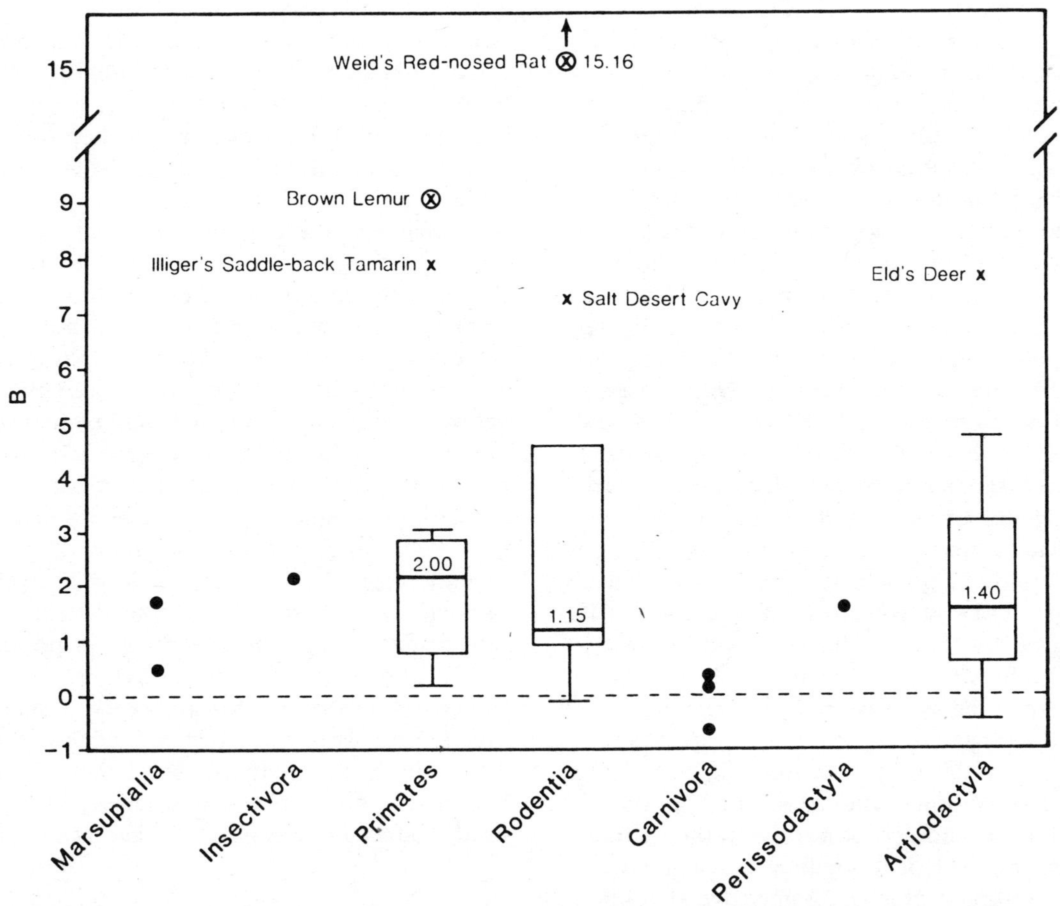

Figure 2. Box plots of B across 40 mammalian populations by order. Median effects (middle horizontal line in box), upper and lower quartiles (upper and lower ends of boxes), upper and lower inner fences (horizontal lines), outlying values (x), and values beyond the outer fences (⊗) are shown for the distribution of B in primates, rodents, and artiodactyls. Results for individual populations in other orders are shown by solid dots.

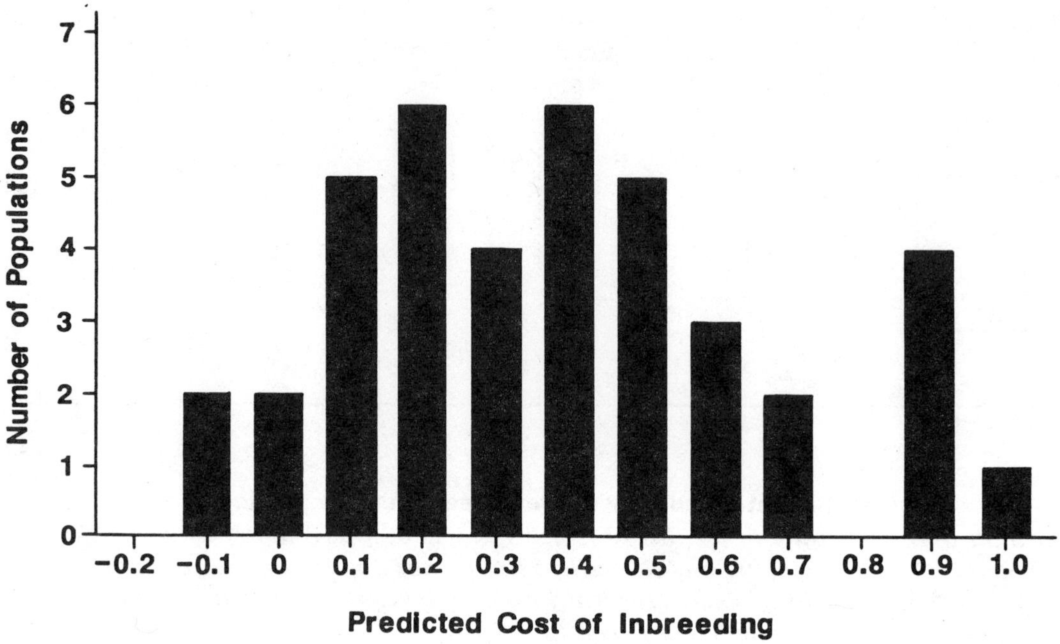

Figure 3. Distribution of the predicted cost of inbreeding in matings resulting in young with an inbreeding coefficient of 0.25 (i.e., matings between parents and offspring or full siblings) for 40 mammal populations.

timate is slightly lower than the average .33 cost of inbreeding found in our mammal populations.

The total costs of inbreeding in natural populations are probably considerably higher than our estimates. First, our estimate of the cost based on only one component of fitness (survival of young) is probably low. We were unable to count early embryonic deaths, exclude individuals with inbred ancestors, and follow individuals until the age of reproductive maturity. Furthermore, mortality rates of inbred young may be higher in natural populations, because many weak young that might die in the wild survive in captivity with the assistance of veterinary care. Second, there are likely to be additional costs of inbreeding in other components of fitness, such as litter size in species that normally bear multiple young and a reduction in fecundity of the inbred young that do survive to reproductive age (Wright 1977). (The reported higher recruitment rate of inbred young in the great tit (van Noordwijk & Scharloo 1981) is not supported by the data (Greenwood & Harvey 1982).) Third, inbred individuals with low levels of heterozygosity may be highly susceptible to viral epidemics (O'Brien et al. 1985).

Considering only the cost of inbreeding relative to the gain in inclusive fitness due to inbreeding, theory suggests that females should not mate with their fathers or sons unless the cost of inbreeding is less than .33 (Smith 1979). Although this is a highly oversimplified model, our data suggest that the cost of inbreeding in mammals is usually high enough (mean = .33) that females should not mate with their closest relatives. The limited data on the frequency of such matings in natural populations of mammals agree with this prediction. Estimates based on observations of identifiable individuals during long-term field studies range from zero to 2% in 9 of 14 well-studied mammalian populations, and the highest documented frequency is 5.5% (Ralls, Harvey & Lyles 1986).

Estimates of the cost of inbreeding also have important applications to conservation biology. The effects of the accelerated rate of inbreeding in small populations, in both captivity and the wild, can potentially drive a population towards extinction (Gilpin & Soulé 1986). The susceptibility of most small populations of conservation interest to elevated levels of inbreeding is unknown, and predicting the degree to which mortality may be increased as a result of inbreeding is impossible. The results presented here provide estimates of the general relationship between the rates of inbreeding and juvenile mortality in a large variety of captive mammal populations and will be useful in developing conservation management programs for small populations (Ballou, in press). Unfortunately, however, our estimates of the cost of inbreeding for individual populations varied greatly and were not clustered near the mean value. Thus, the severity of inbreeding effects in any unstudied mammalian population is quite likely to differ from that predicted by models based on average values.

Acknowledgments

We thank the Friends of the National Zoo for financial support, Karl Kranz, Steve Osofsky, and Becky Rorabaugh for data entry, Judith Block for assisting with the National Zoo records, the institutions that contributed pedigree data, and James Crow for helpful comments on the manuscript.

Literature Cited

Ballou, J. 1983. Calculating inbreeding coefficients from pedigrees. Pages 509–520 *in* C. M. Schonewald-Cox, S. M. Chambers, B. MacBryde, and L. Thomas, editors. Genetics and Conservation: A Reference for Managing Wild Animal and Plant Populations. Benjamin/Cummings, Menlo Park, California.

Ballou, J. Considerations of inbreeding and outbreeding depression in the captive propagation of black-footed ferrets (*Mustela nigripes*). *In* U. S. Seal, editor. Reproductive Biology of the Black-footed Ferret and Small Population Biology. Yale University Press, New Haven, Connecticut, in press.

Bengtsson, B. O. 1978. Avoid inbreeding: At what cost? Journal of Theoretical Biology **73**:439–444.

Bowman, J. C., and D. S. Falconer. 1960. Inbreeding depression and heterosis of litter size in mice. Genetical Research **1**:262–274.

Bulmer, M. G. 1973. Inbreeding in the great tit. Heredity **30**:313–325.

Cavalli-Sforza, L. L., and W. F. Bodmer. 1971. The Genetics of Human Populations. W. H. Freeman, San Francisco.

Crow, J. K., and M. Kimura. 1970. An Introduction to Population Genetic Theory. Harper and Row, New York.

Dawkins, R. 1976. The Selfish Gene. Oxford University Press, Oxford, U.K.

Dobzhansky, T. 1970. Genetics of the Evolutionary Process. Columbia University Press, New York.

Feldman, M. W., and F. B. Christiansen. 1984. Population genetic theory of the cost of inbreeding. American Naturalist **123**:642–653.

Gilpin, M. E., and M. E. Soulé. 1986. Minimum viable populations: Processes of species extinction. Pages 19–34 *in* M. Soulé, editor. Conservation Biology: The Science of Scarcity and Diversity. Sinauer, Sunderland, Massachusetts.

Greenwood, P. J., and P. Harvey. 1982. The natal and breeding dispersal of birds. Annual Review of Ecology and Systematics **13**:1–21.

Hinz, R. L., and T. J. Foose. 1982. Inbreeding, mortality and sex ratio in gaur (*Bos gaurus*) under captivity. Journal of Heredity **73**:297.

Hoaglin, D. C., F. Mosteller, and J. W. Tukey. 1983. Understanding Robust and Exploratory Data Analysis. John Wiley and Sons, New York.

Lorenc, E. 1980. Analysis of fertility in inbred lines of mice derived from populations differing in their genetic load. Zeierzeta Laboratoryjne **17**:3–16.

Makov, E., and A. H. Bittles. 1986. On the choice of mathematical models for the estimation of lethal equivalents in man. Heredity **57**:377–380.

May, R. M. 1979. When to be incestuous. Nature **279**:192–194.

Morton, N. E., J. F. Crow, and H. J. Muller. 1955. An estimate of the mutational damage in man from data on consanguineous marriages. Proceedings of the National Academy of Sciences of the U.S.A. **42**:855–863.

O'Brien, S. J., M. E. Roelke, L. Marker, et al. 1985. Genetic evidence for species vulnerability in the cheetah. Science **227**:1428–1434.

Packer, C. 1979. Inter-troop transfer and inbreeding avoidance in *Papio anubis*. Animal Behavior **27**:1–36.

Parker, G. A. 1979. Sexual selection and sexual conflicts. Pages 123–166 *in* M. S. Blum and N. A. Blum, editors. Sexual Selection and Reproductive Competition in Insects. Academic Press, New York.

Ralls, K., and J. Ballou. 1983. Extinction: Lessons from zoos. Pages 164–184 *in* C. M. Schonewald-Cox, S. M. Chambers, B. MacBryde, and L. Thomas, editors. Genetics and Conservation: A Reference for Managing Wild Animal and Plant Populations. Benjamin/Cummings, Menlo Park, California.

Ralls, K., K. Brugger, and A. Glick. 1980. Deleterious effects of inbreeding in a herd of captive Dorcas gazelle. International Zoo Yearbook **20**:137–146.

Ralls, K., P. H. Harvey, and A. M. Lyles. 1986. Inbreeding in natural populations of birds and mammals. Pages 35–56 *in* M. Soulé, editor. Conservation Biology: The Science of Scarcity and Diversity. Sinauer, Sunderland, Massachusetts.

Sausman, K. A. 1984. Survival of captive-born *Ovis canadensis* in North American zoos. Zoo Biology **3**:111–121.

Sittmann, K., B. Abplanalp, and R. A. Fraser. 1986. Inbreeding depression in the Japanese quail. Genetics **54**:371–379.

Smith, R. H. 1979. On selection for inbreeding in polygynous animals. Heredity **43**:205–211.

Sokal, R. R., and F. J. Rohlf. 1969. Biometry. W. H. Freeman, San Francisco.

Templeton, A. R., and B. Read. 1983. The elimination of inbreeding depression in a captive herd of Speke's gazelle. Pages 241–261 *in* C. M. Schonewald-Cox, S. M. Chambers, B. MacBryde, and L. Thomas, editors. Genetics and Conservation: A

Reference for Managing Wild Animal and Plant Populations. Benjamin/Cummings, Menlo Park, California.

Templeton, A. R., and B. Read. 1984. Factors eliminating inbreeding depression in a captive herd of Speke's gazelle. Zoo Biology **3**:177–199.

Templeton, A. R., H. Hemmer, G. Mace, U. S. Seal, W. M. Shields, and D. S. Woodruff. 1986. Local adaptation, coadaptation, and population boundaries. Zoo Biology **5**:115–126.

van Noordwijk, A. J., and W. Scharloo. 1981. Inbreeding in an island population of the great tit. Evolution **35**:674–688.

Wilson, E. O. 1976. The central problems of sociobiology. Pages 205–217 *in* Theoretical Ecology: Principles and Application. R. M. May, ed., Blackwells, Oxford.

Wright, S. 1977. Evolution and the Genetics of Populations. Vol. 3. University of Chicago Press, Chicago.

Hereditary Blindness in a Captive Wolf (*Canis lupus*) Population: Frequency Reduction of a Deleterious Allele in Relation to Gene Conservation

LINDA LAIKRE
NILS RYMAN
Division of Population Genetics
Stockholm University
S-106 91 Stockholm, Sweden

ELIZABETH A. THOMPSON
Department of Statistics
University of Washington
Seattle, WA 98195, U.S.A.

Abstract: *Numerous cases of hereditary diseases and disorders have been reported in wild animals bred in captivity, but little attention has been paid to the particular genetic management problems that arise when such defects occur. These problems include the obstacle of eliminating the deleterious allele(s) without contemporary loss of genetic variability. In this paper we use the statistical methods of pedigree analysis to address questions regarding a previously presumed hereditary form of blindness observed in a captive wolf population bred for conservation purposes in Scandinavian zoos. The most likely mode of inheritance coincides with an autosomal recessive allele with either a full penetrance or a reduced penetrance of 0.6 (depending on the reliability of studbook records). Using these two models of inheritance, we calculate the probability of carrying the blindness allele for each living animal. Analysis of the effect of removing high-probability carriers on founder allele survival and level of inbreeding demonstrates that the frequency of the deleterious allele can be significantly reduced without seriously affecting founder allele survival or current degree of inbreeding in the wolf population.*

Paper submitted February 14, 1992; revised manuscript accepted February 24, 1993.

Ceguera hereditaria en una población cautiva de lobos (*Canis lupus*): Reducción en la frecuencia de un alelo deletereo en relación con la conservación de genes

Resumen: *Numerosos casos de enfermedades y desórdenes hereditarios han sido reportados en animales salvajes criados en cautiverio, sin embargo, poca atención se ha prestado a los particulares problemas de manejo genético que se presentan cuando tales defectos ocurren. Estos problemas incluyen el obstáculo que se presenta para eliminar lo(s) alelo(s) deletereos sin la pérdida simultánea de variabilidad genética. En este trabajo nosotros utilizamos los métodos estadísticos de análisis de pedigre para considerar preguntas concernientes a una forma de ceguera, previamente considerada como hereditaria, que se observó en una población cautiva de lobos criada con fines concervacionistas en zoológicos Escandinavos. El modo de transmisión hereditaria coincide con un alelo recesivo autosómico con penetrancia total o con una penetrancia reducida de 0.6 (dependiendo de la confiabilidad de los registros cría). Usando estos dos modelos de transmisión hereditaria calculamos la probabilidad de portar el alelo de ceguera para cada animal. El análisis del efecto de remover los individuos con alto riesgo de ser portadores sobre la supervivencia del alelo fundador y el nivel de endocria demuestra que la frecuencia de los alelos deletereos puede ser significativamente reducida sin afectar seriamente la supervivencia del alelo fundador o el nivel actual de endocria de la población de lobos.*

Introduction

An increasing number of wild animal species rely on captive breeding for their survival (Seal 1986; International Union for the Conservation of Nature and Natural Resources 1987; Foose & Ballou 1988), and there is a growing awareness of the problems associated with the genetic management of small, captive populations (see Flesness 1977; Chesser et al. 1980; Senner 1980; Ralls & Ballou 1986). One of the main objectives of careful genetic management is avoidance of the negative effects of inbreeding (inbreeding depression; Wright 1977:44–96), which are frequently expressed as a general decrease of viability and fecundity (Falconer 1981:224–238) or as the occurrence of hereditary defects and diseases (Strong 1978).

Numerous cases of diseases and disorders that are presumed or proven to be hereditary have been reported for wild animals bred in captivity (see Guillery & Kaas 1973; Blampied & Allchurch 1979; Leipold 1980; Bovée et al. 1981; Cooper et al. 1982; Karesh 1983; Snyder & Richard 1984; Keller & Kramer 1985; Burton & Ramsay 1986; Cambre & Fowler 1988). Little attention, however, appears to have been paid to the particular difficulties of genetic management that arise when such hereditary defects occur.

The first major obstacle usually involves establishing the mode of inheritance of the defect. Frequently, this matter must be addressed on the basis of information that can be derived from existing data on the incidence of the defect among interrelated individuals, whereas the results of planned test crosses are typically not available. The next step, removing or reducing the frequency of the deleterious allele(s), includes determining, as accurately as possible, which animals of the potential breeders are likely to carry the allele(s). Such genotype probability computations are often complicated, particularly if the genealogy incorporates several generations of inbreeding, which is typical of pedigrees of wild animals bred in captivity. The final planning of the breeding includes combining the goal of removing the deleterious allele(s) with the general aim of preserving genetic variability (minimizing the loss of founder alleles). The elimination of single alleles by means of selective breeding without contemporary loss of others is an intriguing task that requires extensive statistical analysis of pedigree data (see Thompson 1986).

In a previous paper we reported a presumably hereditary form of blindness and other indications of inbreeding depression in a captive wolf (*Canis lupus*) population bred in Scandinavian zoos (Laikre & Ryman 1991). Here we use the methods of pedigree analysis to address questions regarding the mode of inheritance of the eye defect and the future breeding of the wolf population. There are several intuitively based contentions among managers regarding the blindness: they include the notions that the allele(s) causing blindness has already been eliminated from the population since none of the presently live wolves are blind, that it has not been eliminated and that it would be difficult or impossible to reduce its frequency, and that attempts in that direction would inevitably result in an extensive increase in the level of inbreeding. Our analysis shows that none of these intuitive assumptions necessarily is correct.

The Wolf Population

The wolf conservation project in Scandinavia zoos aims at preserving the genetic characteristics of the Fennoscandic wolf. It is one of the most extensive captive breeding programs conducted in Scandinavia.

The captive population descends from four founders of Fennoscandic origin, two Finnish and two Swedish ones (Fig. 1). In addition to the Fennoscandic founders, two Russian animals have contributed genes to the population; two Russian female litter mates were imported from a zoo in Russia in 1980. The Russian animals were introduced into the program to reduce inbreeding, but little or no attention was paid to the effects of this step on the basic goal of preserving the Fennoscandic genes. The introduction has resulted in the spread of genes of non-Fennoscandic origin, but as yet the probability of survival of Fennoscandic alleles in live animals is not affected by the introgression of Russian alleles, since the Fennoscandic parents of all "hybrids" are still alive (Fig. 1).

Studbook records have been kept since the breeding program started, and a previous analysis revealed extensive inbreeding resulting in inbreeding depression expressed as reductions in juvenile weight, reproduction, and longevity (Laikre & Ryman 1991). Blind animals have also been recorded in the population. On average, 3.5 of the initial eight presumably unique (not identical by descent) Fennoscandic founder alleles per locus have been lost.

The first blind animal was detected in the Finnish branch in 1967 (Fig. 1). Initially, the blindness was not assumed to be hereditary (Hans-Ove Larsson, present studbook keeper, personal communication), and siblings of the blind animals were not discarded when wolves were selected for breeding. Since then a total of 24 blind animals (10 females and 14 males) distributed over seven families have been recorded. The affected animals exhibit a variety of eye abnormalities, such as cataract, retinal dysplasia, and abnormal blood-vessels in the retina and vitreous, causing complete blindness or severe sight reduction (Kristina Narfström, veterinarian, personal communication). The blind animals also appear to have reduced viability (Laikre & Ryman 1991).

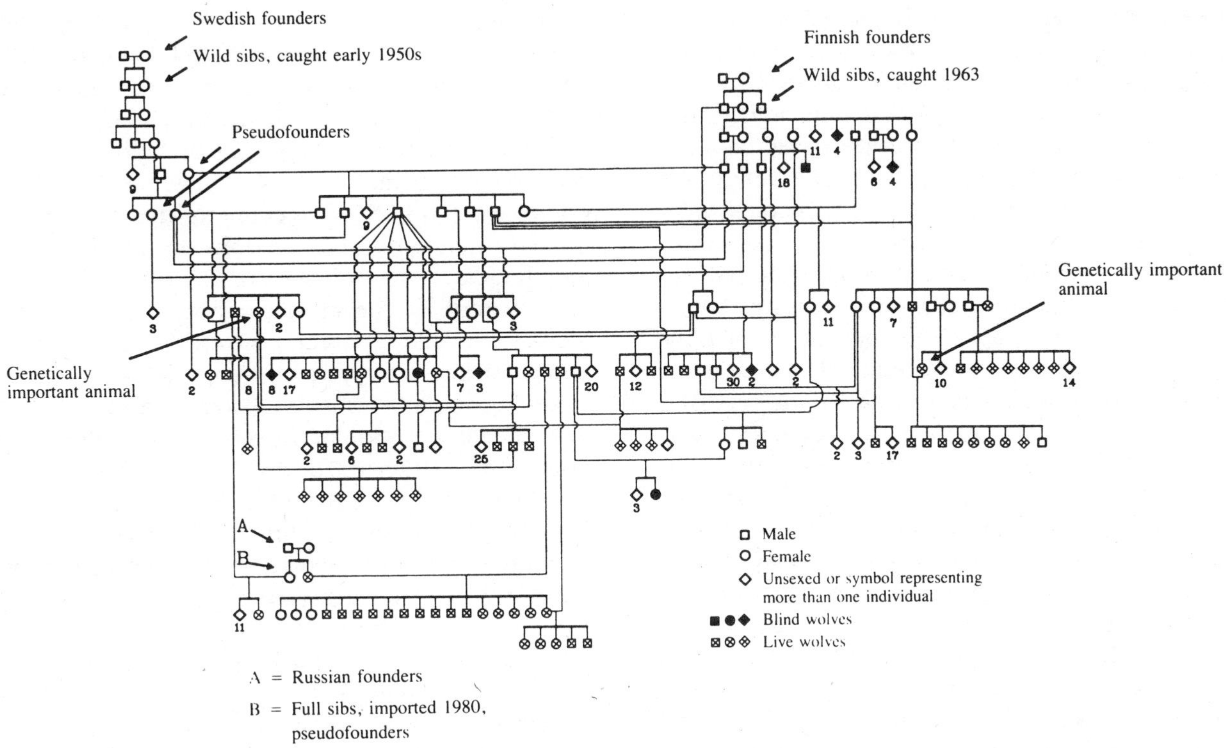

Figure 1. ***Pedigree of the total captive wolf population bred in Scandinavian zoos (January 1988). Siblings (both sexes) dying without reproducing are represented by a single diamond symbol, and figures indicate the number of such animals within a particular family. The pedigree depicts a total of 442 individuals. The animals denoted as genetically important have low probabilities of carrying the blindness allele while retaining high proportions of Swedish and Finnish founder alleles, respectively.***

The defect may be difficult to detect without a close eye examination because blind animals learn to find their way in the enclosure and appear to behave normally.

Our analysis refers to the population that existed in January 1988. At that time the full wolf pedigree comprised 442 individuals; there were a total of 77 live animals (37 males, 22 females, and 18 unsexed), 55 of which were of "pure" Fennoscandic origin (Fig. 1). These 55 animals will subsequently be referred to as the "present population."

Mode of Inheritance

A genetic basis for the eye defect is indicated by the segregation pattern (Fig. 1) (Laikre & Ryman 1991); the occurrence of blindness is family dependent, and affected animals are found only in families that are inbred for genes from the Finnish founders. All blind wolves have nonblind litter mates, which appears to preclude an explanation based on environmental factors common to entire litters. A hereditary etiology is also supported by observations of domestic dogs, which have a genomic organization similar to that of wolves (Iljin 1941), and in various breeds of which comparable defects are known to be hereditary (Barnett 1969; Aguirre 1973; Narfström 1981).

As pointed out previously (Laikre & Ryman 1991), the most straightforward explanation for the segregation pattern of the blindness is an autosomal recessive allele carried by at least one of the Finnish founders. Obviously, for a single allele with complete penetrance (penetrance = probability of phenotype given genotype; Thompson 1986), sex-linked or autosomal dominant inheritance is impossible since none of the parents producing blind offspring exhibit the trait (Fig. 1). A sex-linked recessive with reduced penetrance (hemi- or homozygosity for the allele results in blindness with a probability less than unity) necessitates the presence of the blindness allele in the Swedish branch as well, a scenario which appears unlikely as it would imply that the blindness allele is common in wild wolf populations. Moreover, sex-linked inheritance is contradicted by the even sex ratio among affected animals.

Although an explanation based on an autosomal dominant allele with reduced penetrance cannot be readily excluded, several factors support an autosomal recessive mode of inheritance. First, with an autosomal dominant it is expected that an equal proportion of animals will express the trait among parents and offspring in families where blind animals have occurred; the probability of the observed case is fairly low (0/14 versus 24/115; $p = 0.08$ one-tailed). Second, only animals that are inbred for Finnish genes exhibit the defect; none of the animals that carry Finnish genes exclusively in the heterozygous form are blind (24/321 versus 0/68; $p = 0.02$). Third, similar eye abnormalities in domestic dogs are caused by autosomal recessive alleles (Rubin 1968; Barnett 1969; Barnett et al. 1970; Aguirre 1973).

There is, however, a significant deficiency of blind wolves in comparison to the predicted 3:1 ratio expected for a single autosomal, fully penetrant, recessive allele (Laikre & Ryman 1991). This deficiency could imply incomplete penetrance (two copies of the allele may not necessarily result in blindness), or that more than one locus is involved. In contrast, the deficiency may be due entirely to difficulties in detecting all blind animals, or failure in recording observed cases in the studbook. The latter explanation is supported by the fact that only two of the seven families in which blind individuals have occurred show a pronounced lack of blindness, whereas in the other families approximately 25 percent of the offspring exhibit the defect (in accordance with 3:1 segregation).

The conditions described above indicate that the blindness in the wolf population is most easily explained by a single autosomal locus segregating for a recessive allele. Looking for the least complicated model to explain the observed data, we applied a likelihood analysis to the question of penetrance at that locus. Penetrance may be described by two parameters, α and β (Thompson 1990), where α = the probability of blindness given two copies of the allele and β = the probability of blindness given one copy of the allele. In the likelihood analysis we compared the likelihoods of the observed pedigree for various combinations of α and β, initially assuming a fully recessive allele ($\beta = 0$). Penetrance is subsequently used synonymously with α.

Methods of Computation

Like many pedigrees of zoo animals, the wolf genealogy is quite complex, with complicated relationships and numerous inbreeding loops (Fig. 1). In the present population all the animals are more or less interrelated; this implies that phenotypic observations of any particular animal in any generation contribute information about most other individuals of the pedigree with respect to their probability of having a particular genotype. Obviously, tasks such as determining the most likely mode of inheritance or calculating the probability that a particular animal has a specific genotype (for example, whether it carries the blindness allele in the heterozygous state) are not trivial ones—the probabilities of all possible genotype combinations of all other animals in the pedigree have to be considered. Such computations are virtually impossible to carry out by hand in a pedigree such as the present one; they require specially designed computer algorithms to account for the large number of possible genotype combinations and their associated likelihoods.

To determine the most likely values of α and β, as well as the corresponding probabilities of living wolves being carriers, we applied the so-called peeling method (Cannings et al. 1976, 1978; Thompson et al. 1978). For a detailed description of the method of peeling see Cannings et al. (1976, 1978) and Thompson et al. (1978). Briefly, peeling is a procedure for computation of likelihoods on complex pedigrees. The principle of the method is to work recursively through the pedigree, successively removing ("peeling") individuals while combining the information they represent into probability functions defined on some subset of the remaining individuals (Cannings et al. 1978). The likelihoods obtained reflect the probability of the observed phenotype combinations given the mode of segregation and the penetrance parameters. In this case Mendelian inheritance was assumed and the penetrance of hetero- and homozygous carriers was gradually altered to find the most likely α and β parameters based on the observed pedigree (Fig. 1).

The complexity of the wolf pedigree made "trimming" necessary: individuals and matings not contributing to solving a particular problem were removed to make computations feasible (Geyer et al. 1989). Accordingly, the Swedish branch and the two Russian founders were cut off and replaced with five "pseudo-founders" (Fig. 1) when carrier probabilities and inheritance model likelihoods were calculated.

The effects of attempting to reduce the frequency of the blindness allele were examined by successively removing animals with decreasing probabilities of being carriers and by calculating the effect on the average number of surviving founder alleles per locus and the average level of inbreeding among the remaining individuals. The degree of inbreeding was measured by three coefficients (F_T, F_F, and F_S) corresponding to the probabilities of having the two alleles at a locus identical by descent with the alleles originating from any founder (total inbreeding), Finnish founders, and Swedish founders, respectively ($F_T = F_F + F_S$; Laikre & Ryman 1991).

Because the goal of the captive breeding program is to conserve Fennoscandic wolf genes, the Russian founder alleles have not been included in the gene survival analysis. The Russian descendants were considered in the calculations only when information on their phenotype affected genotype probabilities of Fennoscandic animals.

The likelihood and gene survival computations were performed using the software package Pedpack (Thomas 1987). The present analysis includes the calculation of a large number of likelihoods; for brevity only a sample of them are presented. Because a likelihood is the probability of the observation under a particular hypothesis, a specific likelihood contains little information by itself—it must be related to the likelihoods of the same observation under other hypotheses. For example, the likelihood of the observed wolf pedigree assuming $\alpha = 0.6$ and $\beta = 0$ is not very informative unless it is compared to the likelihoods of the observed pedigree assuming other values of α and β. We present the majority of likelihoods in terms of relative values, that is, each likelihood has been divided by the largest likelihood obtained from a series of computations appropriate to a particular question (in other words, the most likely hypothesis has a relative likelihood of unity).

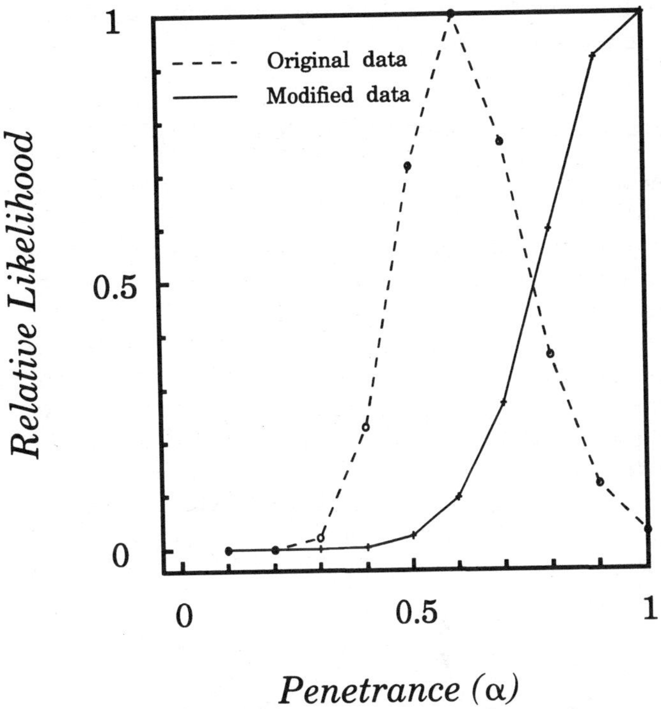

Figure 2. Relative likelihoods of the observed wolf pedigree under various assumptions regarding the penetrance (α) of the recessive blindness allele. In the modified data set the number of blind animals in two families with deficiency of blind offspring (compared to the expected 3:1 ratio) have been artificially increased to 25%.

Penetrance

The likelihood computations for a hereditary model assuming a fully recessive allele ($\beta = 0$) yielded the largest likelihood for $\alpha = 0.6$ (Fig. 2). To test for any tendency of dominance, β values in the range of $0 \leqslant \beta \leqslant 1$ were combined with values of α in the same range. There were no indications of dominance; no value of $\beta > 0$ yielded a likelihood larger than the one obtained for $\beta = 0$ and $\alpha = 0.6$. Thus, the single-locus model of inheritance best fitting the data agrees with an autosomal, fully recessive allele with a reduced penetrance of 0.6.

To validate the genetic basis for the blindness, the genetic model ($\alpha = 0.6$ and $\beta = 0$) was compared with a nongenetic one. The nongenetic model assumed the same prior probability of blindness for all individuals, and that probability was estimated from the observed frequency of blind animals. The likelihood ratio indicates that the genetic model is over 8000 times more likely than the nongenetic one.

As mentioned above, only two out of seven families exhibit a deficiency of blind animals, compared with the 3:1 ratio expected under complete penetrance ($\alpha = 1$). To examine whether the apparently reduced penetrance of the homozygote ($\alpha = 0.6$) is explained entirely by the lack of blindness observed in those two families, the most likely penetrance was estimated after the proportion of blindness in these two families had been artificially increased to 25 percent. In a similar test these two families were removed, leaving only the reproducing individuals, in order not to interrupt the pedigree. These computations (keeping $\beta = 0$) yielded most likely α values of 1.0 and 0.9, respectively. The two modified data sets were also evaluated for evidence of dominance, but no such tendency was observed in either case. The observations indicate that there is no strong evidence of reduced penetrance if the two zoos keeping these two families have indeed failed to observe and record all cases of blindness.

The explanation implying incomplete recording appears quite plausible considering the above-mentioned difficulties in identifying affected animals and the typically bad publicity for the zoo associated with producing blind wolves. The fact that significantly fewer animals have been sexed in the two families with a deficiency of blindness, as compared to the remaining five families with blind individuals, also supports this contention (12/58 versus 4/81 individuals without sex recording; $p = 0.006$, two-tailed).

In spite of the quite strong indications of a fully recessive allele with complete penetrance, further analyses have been carried out assuming both complete and reduced penetrance ($\alpha = 1$ and $\alpha = 0.6$, respectively). The case of reduced penetrance was included to evaluate the effects of a potentially inaccurate conclusion regarding penetrance on the estimation of carrier probabilities and the associated prospect of reducing the frequency of the blindness allele.

Frequency Reduction and Gene Conservation

Reducing the frequency of the blindness allele in the population requires estimates for each living wolf of the probability that it carries one or two copies of the allele (carrier probabilities). For the estimation of carrier probabilities, it is necessary to specify either the frequency of the defective allele in the wild population from which the founders were taken or the genotypes of the founders. We have no idea of the allele frequency in the wild; further, the founders of this particular pedigree originate from different geographic regions (Finland and Sweden) and may represent distinct wild populations with different frequencies of the blindness allele. Therefore, we chose the approach of specifying the genotypes of the founders.

Assuming Full Penetrance (α = 1; Case 1)

The likelihood of possible genotype combinations of the Fennoscandic founders was computed, and the results support the contention that the blindness allele originates from the Finnish branch. For α = 1.0, the most likely founder genotype combination is the one where one of the Finnish founders was a heterozygous carrier. The question of whether one or both of the Finnish founders carried the blindness allele, however, is irrelevant for the computation of carrier probabilities in the case of full penetrance: both of the reproducing Finnish wild caught sibs must have been heterozygotes (Fig. 1).

The distribution of carrier probabilities among the live, "pure" Fennoscandic wolves is shown in Figure 3.

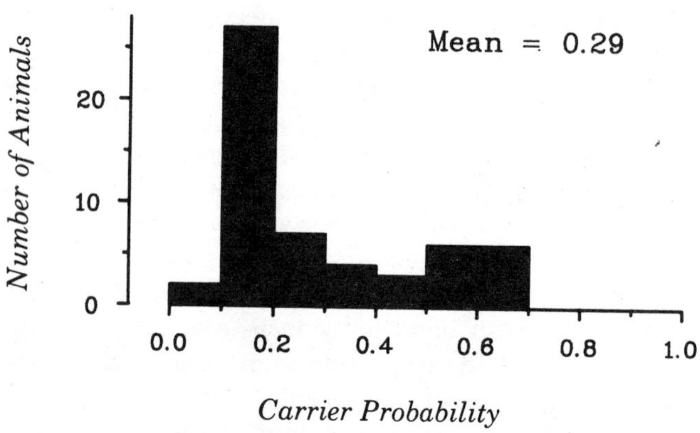

Figure 3. The distribution of probabilities of carrying the blindness allele in the present Scandinavian wolf population (n = 55). The carrier probabilities are based on a model of inheritance assuming an autosomal, fully recessive (β = 0) allele with complete penetrance (α = 1). The Russian descendants are not included.

Notably, all live animals have a probability greater than zero of being a carrier, the carrier probabilities ranging from 0.06 to 0.67 (mean = 0.29; SD = 0.19).

Individuals with high probabilities of carrying the blindness allele can be removed without any major effect on gene survival—in other words, on the survival of remaining unique founder alleles (Fig. 4). Over 90% of the still surviving founder alleles (on average 4.5 alleles per locus; Laikre & Ryman 1991) can be maintained even if all animals with a carrier probability greater than 0.3 are removed. If the carrier probability for removal is reduced further, eliminating all animals with a carrier probability exceeding 0.2, 87% of the founder alleles would still be preserved.

The reduction in the number of remaining founder alleles when high probability carriers are removed is slightly more pronounced for the Finnish than for the Swedish founder alleles (Fig. 4, Case 1). While only 7% of the surviving Swedish founder alleles are lost if a carrier probability of 0.2 is used as the limit for removal, it would result in extinction of 16% of the remaining Finnish founder alleles.

The elimination of high-probability carriers does not have a critical impact on the current level of inbreeding in the population (Table 1). Removing the 26 animals with a carrier probability exceeding 0.2 increases the average F_T by 0.04. The increase is due entirely to a higher degree of homozygosity for the Finnish founder genes. The removal would result in an increase of 0.03 in the average level of kinship (degree of inbreeding in the offspring) among all live animals.

Assuming Reduced Penetrance (α = 0.6; Cases 2–4)

The question of the Finnish founder genotype combination is more relevant when α = 0.6 is assumed. Under this assumption both of the reproducing Finnish wild-caught sibs were not necessarily heterozygous carriers; either or both of them could have been homozygous carriers, which potentially affects the genotypic array of the present population.

The most likely Finnish founder genotype combinations include *Bb* × *Bb, BB* × *Bb, and Bb* × *bb,* where *B* denotes the "normal" and *b* the blindness allele. These three combinations have relative likelihoods of 1, 0.76, and 0.45, respectively, and they will subsequently be referred to as Cases 2–4. It should be noted that these likelihood computations did not consider gene or genotype frequencies in the wild populations of the founders. Obviously, Cases 2 and 4 may appear somewhat unlikely considering that the founders were from a natural population and that the allele is deleterious. Nevertheless, those combinations may have occurred if the founders were closely related.

The carrier probabilities among live animals are quite similar among all the models of reduced penetrance

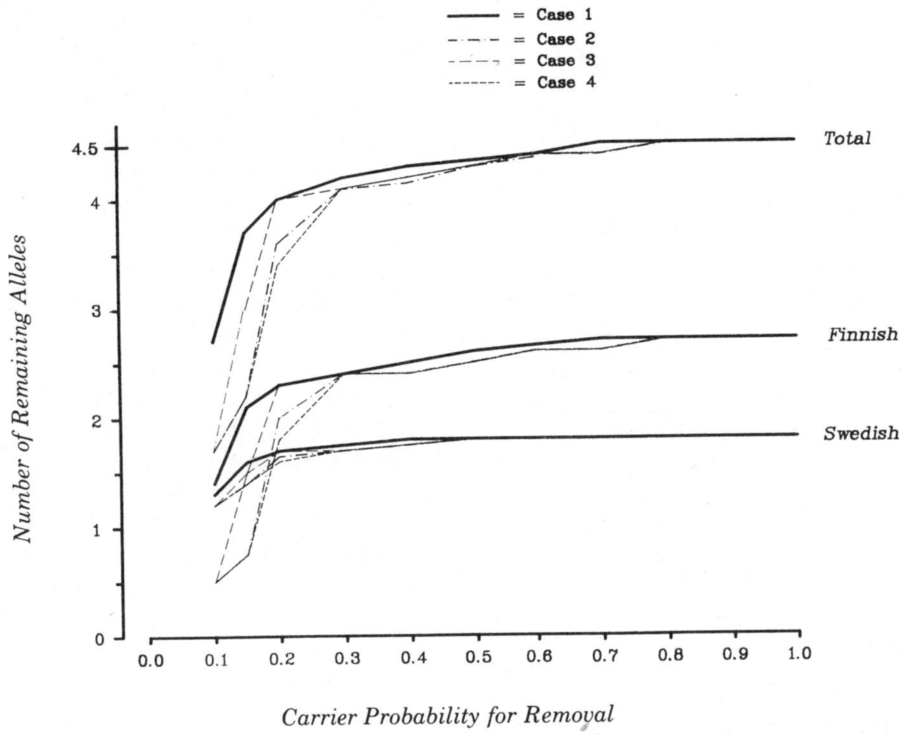

Figure 4. Average number of founder alleles per locus that will survive in the present population when animals with different probabilities of being carriers of the blindness allele are removed. The carrier probabilities were obtained assuming an autosomal, fully recessive allele with either full penetrance (α = 1; Case 1) or reduced penetrance (α = 0.6; Cases 2–4). The three cases under the reduced penetrance model represent different genotype combinations of the two Finnish founders (the blindness allele occurring in one, two, or three copies among those founders). The Russian descendants are not included.

(Cases 2–4), and they are also very close to those obtained assuming α = 1 (Fig. 5). Animals classified as high-probability carriers are essentially the same individuals in all four cases; the same is true for low-probability carriers, although the variation is somewhat larger for this group of animals. The pairwise coefficients of correlation (r) between individual carrier probabilities for the different cases are all high (six coefficients, all 0.97 or larger, $df = 53$). Thus, for the present pedigree, the assumption regarding the penetrance does not appear to have any major effect on carrier probabilities among live animals. Therefore, the effects on loss of founder alleles and increase of inbreeding associated with attempts to reduce the frequency of the blindness allele should be essentially the same regardless of the above assumptions concerning the penetrance (Fig. 4).

Table 1. Effect of removing animals with different probabilities of being carriers of the blindness allele on population size and current levels of total (F_T), Finnish (F_F), and Swedish (F_S) inbreeding.

Minimum Carrier Probability for Removal	*Number of Animals Remaining*	F_T	F_F	F_S
1.0 (no removals)	55	0.34	0.17	0.17
0.6	49	0.35	0.17	0.18
0.5	43	0.36	0.18	0.18
0.4	40	0.36	0.18	0.18
0.3	36	0.37	0.19	0.18
0.2	29	0.38	0.22	0.16
0.1	2	0.34	0.16	0.18

Discussion

The problems associated with hereditary defects in captive wild animals will probably increase with the growing practice of captive breeding (Benirschke 1977; Fagan 1980). The present study further demonstrates the applicability and potential of statistical analysis of pedigree data for genetic management of endangered species bred in captivity (see Geyer & Thompson 1988; Geyer et al. 1989).

Our analysis demonstrates the opportunity for the simultaneous conservation of unique founder alleles and significant frequency reduction of a segregating deleterious allele in the captive wolf population bred in Scandinavian zoos. This opportunity is only marginally affected by potentially erroneous assumptions regarding penetrance, and to avoid unnecessary details we restrict

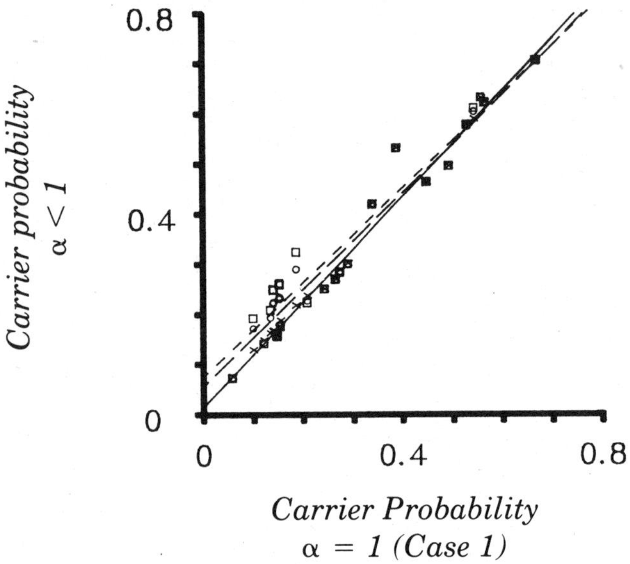

Figure 5. Linear relationships between the carrier probabilities derived for live animals under the assumption of full penetrance (Case 1) and those obtained under the assumption of reduced penetrance for different genotype combinations of the two Finnish founders (Cases 2–4). The regression coefficients are 0.98, 1.05, and 0.95 for Cases 2 (– – – ○), 3 (—— ×), and 4 (- - - -□), respectively ($p \ll 0.001$ in all cases).

the following discussion to the case of an autosomal and fully penetrant recessive allele (Case 1).

The likelihood analyses indicate a frequency of the blindness allele in the present population (55 animals) of 0.15. The fact that all live animals have a carrier probability exceeding zero implies that elimination of the deleterious allele must be a stepwise procedure. By removing animals with a carrier probability of 0.2 or larger, a 50% reduction of the allele frequency is possible, which reduces the allele frequency to 0.07.

Reducing the frequency of the blindness allele further is difficult at the first step because few animals exhibit carrier probabilities lower than 0.2. If, for instance, the carrier probability for removal is set to 0.1, only two females would be left. It is interesting, however, that these two animals alone retain as much as 60% (2.7 out of 4.5 alleles per locus) of the remaining Fennoscandic founder alleles. Obviously, these two wolves (referred to as genetically important in Fig. 1) combine low probabilities of carrying the deleterious allele with high probabilities of carrying relatively large numbers of unique founder alleles.

It should be noted that controlled test crossing is the most reliable method of establishing the mode of inheritance of a phenotype. Statistical analysis constitutes an important complement when test breeding data are not available or are difficult to obtain. For zoo populations, lack of available space may obstruct the performance of test crossing concurrently with the conservation breeding program. Also, "forced" matings may be difficult from an ethological point of view. In the case of the present wolf population, no test breedings to establish the inheritance of the blindness have been conducted.

Among the Russian descendants (not included in Fig. 3), all of the live animals except for the imported female (Fig. 1) also have carrier probabilities exceeding zero. The probabilities range from 0.06 to 0.30 (mean = 0.23; SD = 0.05).

Our results contradict some contentions among managers involved with the breeding program. First, the idea that the blindness allele has already been eliminated is not supported by our findings. Second, the contrasting notion that the allele is spread in the population to the extent that little can be done to reduce its frequency is evidently also erroneous. Third, the Russian descendants are frequently believed to be free from the blindness allele; this belief is based on the assumption that guaranteed noncarriers have been chosen as mates for the descendants of the Russian founders. As is obvious from the pedigree, this is not true, since a sibling of two of the chosen mates has sired a blind offspring. Rather, our analysis shows that those two males have high probabilities of being carriers (0.45), whereas the third male has a fairly low carrier probability (0.12). Fourth, the two females characterized by relatively high proportions of remaining unique founder alleles while having low probabilities of carrying the blindness allele had not been identified as genetically important prior to our analysis.

It has been suggested that the wolf conservation program should be terminated because of the currently high level of inbreeding, the rapid loss of founder alleles, and the occurrence of blindness in the population. We disagree with that opinion. The present analysis indicates that there is a good chance of considerably reducing the frequency of the blindness allele without significant further loss of remaining founder alleles. Indeed, the level of inbreeding is conspicuously high, and several traits are affected by inbreeding depression (Laikre & Ryman 1991). There are, however, no indications that the inbreeding depression constitutes an immediate threat to the continued survival of the population. Therefore, it appears quite feasible to conserve the genetic characteristics retained in the present population provided that future decisions regarding breeding are based on scientifically accepted principles of sound genetic management for conservation.

Conclusions

This study demonstrates the applicability of advanced pedigree analysis for addressing management problems related to the occurrence of hereditary defects in small, captive animal populations bred for conservation. Results regarding a presumed hereditary form of blindness observed in a captive wolf population bred in Scandinavian zoos are as follows:

(1) The blindness is most likely caused by an autosomal recessive allele, possibly with reduced penetrance.
(2) All live wolves have a probability greater than zero of carrying the blindness allele.
(3) The possibility is demonstrated of considerably reducing the frequency of the deleterious allele in the population without seriously affecting founder allele survival or degree of inbreeding.
(4) The allele appears to be spread in a group of animals previously regarded as definite noncarriers.
(5) Two animals were identified as genetically important in the sense of combining a low probability of carrying the blindness allele with retention of a large proportion of remaining founder alleles.

These results contradict some current contentions regarding the genetic status of the population.

Acknowledgments

We thank Hans-Ove Larsson for providing the studbook data, Charles Geyer and Alun Thomas for valuable suggestions and aid with the analytical procedures, and Per Erik Jorde for comments on the manuscript. The facilities made available to L. Laikre during a visit to the Department of Statistics, University of Washington, Seattle, are gratefully acknowledged. The study was supported by grants to E. Thompson from the National Science Foundation (grant BSR-8921839) and to N. Ryman from the Swedish Natural Science Research Council. Financial contributions from Hierta-Retzius' Stipendiefond and Wallenbergsstiftelsens Jubileumsfond to L. Laikre are greatly appreciated.

Literature Cited

Aguirre, G. 1973. Hereditary retinal diseases in small animals. Veterinary Clinics of North America **3**:515–528.

Barnett, K. C. 1969. Genetic anomalies of the posterior segment of the canine eye. Journal of Small Animal Practice **10**:451–455.

Barnett, K. C., G. R. Bjorck, and E. Kock. 1970. Hereditary retinal dysplasia in the Labrador Retriever in England and Sweden. Journal of Small Animal Practice **10**:755–759.

Benirschke, K. 1977. Genetic management. International Zoo Yearbook **17**:50–60.

Blampied, N. L. Q., and A. F. Allchurch. 1979. Veterinary report. The Dodo, Journal of the Jersey Wildlife Preservation Trust **16**:99–102.

Bovée, K. C., M. Bush, J. Dietz, P. Jezyk, and S. Segal. 1981. Cystinuria in the maned wolf of South America. Science **212**:919–920.

Burton, M., and E. Ramsay. 1986. Cryptorchidism in maned wolves. Journal of Zoo Animal Medicine **17**:133–135.

Cambre, R. C., and M. E. Fowler. 1988. Congenital anomalies in the llama (Lama glama): An update. Annual Proceedings of the Joint Conference of the American Association of Zoo Veterinarians and the American Association of Wildlife Veterinarians **1988**:175–176.

Cannings, C., E. A. Thompson, and M. H. Skolnick. 1976. The recursive derivation of likelihoods on complex pedigrees. Advances in Applied Probability **8**:622–625.

Cannings, C., E. A. Thompson, and M. H. Skolnick. 1978. Probability functions on complex pedigrees. Advances in Applied Probability **10**:26–61.

Chesser, R. K., M. H. Smith, and I. L. Brisbin. 1980. Management and maintenance of genetic variability in endangered species. International Zoo Yearbook **20**:146–154.

Cooper, J. E., L. Arnold, and G. M. Henderson. 1982. A developmental abnormality in the Round Island skink *Leiolopisma telfairii*. The Dodo, Journal of the Jersey Wildlife Preservation Trust **19**:78–81.

Fagan, D. A. 1980. The pathogenesis of dental disease in carnivores. Annual Proceedings of the American Association of Zoo Veterinarians **1980**:128–131.

Falconer, D. S. 1981. Introduction to quantitative genetics. 2nd edition. Longman, London, England.

Flesness, N. R. 1977. Gene pool conservation and computer analysis. International Zoo Yearbook **17**:77–81.

Foose, T. J., and J. D. Ballou. 1988. Management of small populations. International Zoo Yearbook **27**:26–41.

Geyer, C. J., and E. A. Thompson. 1988. Gene survival in the Asian wild horse (*Equus przewalskii*). I. Dependence of gene survival in the Calgary breeding group pedigree. Zoo Biology **7**:313–327.

Geyer, C. J., E. A. Thompson, and O. A. Ryder. 1989. Gene survival in the Asian wild horse (*Equus przewalskii*): II. Gene survival in the whole population, in subgroups, and through history. Zoo Biology **8**:313–329.

Guillery, R. W., and J. H. Kaas. 1973. Genetic abnormality of the visual pathways in a "white" tiger. Science **180**:1287–1288.

Iljin, N. A. 1941. Wolf-dog genetics. Journal of Genetics **42**:359–421.

International Union for the Conservation of Nature and Natural Resources (IUCN) (The World Conservation Union). 1987.

The IUCN policy statement on captive breeding, September 4, 1987. IUCN Council, Gland, Switzerland.

Karesh, W. B. 1983. Urolithiasis in Asian small-clawed otters (*Amblonyx cinerea*). Annual Proceedings of the American Association of Zoo Veterinarians **1983**:42–44.

Keller, G. L., and L. Kramer. 1985. Bilateral patellar dislocation in a gelada baboon. Journal of Zoo Animal Medicine **16**:55–58.

Laikre, L., and N. Ryman. 1991. Inbreeding depression in a captive wolf (*Canis lupus*) population. Conservation Biology **5**:33–40.

Leipold, H. W. 1980. Congenital defects of zoo and wild mammals: A review. Pages 457–470 in R. J. Montali and G. Migaki, editors. The comparative pathology of zoo animals. Smithsonian Institution Press, Washington, D.C.

Narfström, K. 1981. Cataract in the West Highland White Terrier. Journal of Small Animal Practice **22**:467–471.

Ralls, K., and J. D. Ballou, 1986. Preface to the proceedings of the workshop on genetic management of captive populations. Zoo Biology **5**:81–86.

Rubin, L. F. 1968. Hereditary retinal dysplasia in Bedlington Terriers. Journal of the American Veterinary Medical Association **152**:260–262.

Seal, U. S. 1986. Goals of captive propagation programmes for the conservation of endangered species. International Zoo Yearbook **24/25**:174–179.

Senner, J. W. 1980. Inbreeding depression and the survival of zoo populations. Pages 209–224 in M. E. Soulé and B. A. Wilcox, editors. Conservation biology. An evolutionary-ecological perspective. Sinauer Associates, Sunderland, Massachusetts.

Snyder, S. B., and M. J.. Richard. 1984. Apparent familial hypothyroidism in jaguars (*Panthera onca*). Annual Proceedings of the American Association of Zoo Veterinarians **1984**:132–133.

Strong, L. C. 1978. Inbred mice in science. Pages 45–67 in H. C. Morse III, editor. Origins of inbred mice. New York Academic Press, New York, New York.

Thomas, A. 1987. Pedpack: User's manual. Technical Report 99. Department of Statistics, University of Washington, Seattle, Washington.

Thompson, E. A. 1986. Pedigree analysis in human genetics. The Johns Hopkins University Press, Baltimore, Maryland.

Thompson, E. A. 1990. Inference of a gene and its paths of descent: The Newfoundland example. American Journal of Human Biology **2**:291–301.

Thompson, E. A., C. Cannings, and M. H. Skolnick. 1978. Ancestral inference. I. The problem and the method. Annals of Human Genetics **42**:95–108.

Wright, S. 1977. Evolution and the genetics of populations. A treatise in four volumes. Vol. 3. Experimental results and evolutionary deductions. University of Chicago Press, Chicago, Illinois.

Allocating Scarce Resources for Conservation of Endangered Subspecies: Partitioning Zoo Space for Tigers

LYNN A. MAGUIRE*

School of Forestry and Environmental Studies
Duke University
Durham, NC 27706, U.S.A.

ROBERT C. LACY

Department of Conservation Biology
Chicago Zoological Park
Brookfield, IL 60513, U.S.A.

Abstract: *All remaining subspecies of tigers (*Panthera tigris*) are endangered in the wild and space for captive breeding in western zoos is limited. How should this space be allocated to enhance survival and retention of genetic variation in the face of uncertainty about genetic relationships among tiger subspecies and about the future of tigers in the wild? We solicited expert judgments on subspecies population trends and on the ratio of effective to census size in captivity and analyzed different scenarios for dividing the 1000 captive spaces among four tiger subspecies. To maximize survival and retention of gene diversity in individual subspecies, it is best to divide the space equally among the four subspecies. To maximize retention of gene diversity for the species as whole, it is better to allocate more space to the subspecies that appears most variable on the basis of limited electrophoretic data (*P. t. tigris*). Allocating half of the captive space to *tigris *and dividing the remainder equally among the other three subspecies is a satisfactory compromise between species welfare and subspecies welfare that ensures survival of all four subspecies (at least in captivity) and retains about 80 percent of existing gene diversity within subspecies and about 93 percent of diversity for the species as a whole. Sensitivity analyses showed that our recommendations were robust to uncertainty about the demographic and genetic status of tiger subspecies. For other species with several endangered subspecies, such as rhinos, resource limitations*

* *Address correspondence to this author.*

Paper submitted 7/19/88; revised manuscript accepted 12/19/89.

Resumen: *Todas las subespecies que quedan del tigre (*Panthera tigris*) se encuentran en peligro de extinción en su estado salvaje y en el espacio para la crianza en zoológicos occidentales es limitado. ¿Cómo debería ser asignado este espacio para asegurar la sobrevivencia y la retención de la variación genética de la especie? El estado actual de las subespecies de tigres es incierto en lo que se refiere a las relaciones genéticas y al futuro de los tigres en estado salvaje. Nosotros solicitamos el juicio de experto 1) en la tendencia de poblaciones de subespecies, 2) en la relación entre el tamaño efectivo y 3) de los censos en cautiverio. También análizamos el escenario de dividir los 1000 espacios en cautiverio entre cuatro subespecies de tigres. Para maximar la sobrevivencia y retención de la diversidad genética en subespecies individuales, es preferible el dividir el espacio de igual manera entre las cuatro subespecies. Para maximar la retención de la diversidad genética para la especie como un todo, es mejor el asignar más espacio a las subespecies que aparecen más variables sobre las bases de datos electroforéticos limitados (*P. t. tigris*). Asignando la mitad del espacio en cautiverio a *tigris *y dividiendo el restante igualmente entre las otras tres subespecies es un compromiso satisfactorio entre el bienestar de la especie y de las subespecies que asegura la sobrevivencia de todas las cuatro subespecies (por lo menos en cautiverio) y retiene cerca del 80% de la diversidad genética existente dentro de las subespecies y cerca del 93% de diversidad para la especie como un todo. Análisis de sensitividad mostraron que nuestras recomendaciones fueron sólidas a la incertidumbre acerca del estatus demográfico y genético de las subespecies de tigres. Para otras* *may be too severe to permit an easy compromise between species and subspecies welfare and acquiring better genetic data becomes critical.*

especies, con varias subespecies en peligro, tales como rinocerontes, las limitaciones de recursos pueden ser demasiado severas para permitir un compromiso sencillo entre el bienestar de especies y subespecies y el adquirir mejores datos genéticos comienza a ser critico.

Introduction

Many endangered species exhibit geographic variation formally recognized by subspecies designation. Modern taxonomists reserve subspecific names for geographically restricted populations that differ consistently over a number of characters (e.g., size, pelage, skeletal measurements). Many previously named subspecies, however, differ in only one or a few characteristics; these may not correspond to geographic distribution and may show little concordance with other variable traits (Wilson & Brown 1953). For other species, geographic variation with adaptive significance for local populations has no formal taxonomic designation. Whether recognized by taxonomy or not, the genetic basis and adaptive significance of differences among subspecies and geographic races are often unclear. Yet, conservation of intraspecific variation is an important part of efforts to preserve diversity (Chambers & Bayless 1983).

Resources for conserving diversity are limited (Soulé et al. 1986) and scarce resources must be allocated among a subset of the species and subspecies needing help. Because information on the population and genetic status of endangered species and subspecies is limited and the outcomes of management decisions uncertain, it is rarely obvious how to do this. The distribution of space for captive breeding of tiger (*Panthera tigris*) subspecies in North American and European zoos is a case in point. We will show how a combination of decision analysis, expert judgment, and sensitivity analysis can be used to extract the most from limited demographic and genetic data to guide managers of tiger subspecies in captivity.

Current Status of Tigers

Tigers were once distributed throughout Asia. Eight or more subspecies were recognized on the basis of size, pelage, and skeletal dimensions (Herrington 1987). Habitat destruction and conflict with humans have reduced the wild population to five subspecies, all endangered. The map in Figure 1 shows the approximate distributions and current estimates of the numbers of each subspecies in the wild and in captivity (Seal et al. 1987). Because there are no short-term prospects for conserving *amoyensis* outside of China, we excluded this subspecies from our analysis.

The genetic structure of the surviving tiger subspecies is known only from limited electrophoretic analysis of blood proteins from captive tigers (Newman et al. 1985; O'Brien et al. 1987; S. O'Brien, unpublished data; Appendix). Genetic diversity in tigers is typical for mammals; the total expected heterozygosity, 0.036, is the same as the mean H from a survey of 46 mammalian species (Nevo 1978). F_{ST}, a measure of between-subspecies genetic divergence (Crow & Kimura 1970), is 0.073. The low F_{ST} value means that tiger subspecies differ very little genetically, less than one-third the divergence seen among human racial groups and less than half the separation among lion populations in Africa (O'Brien et al. 1987).

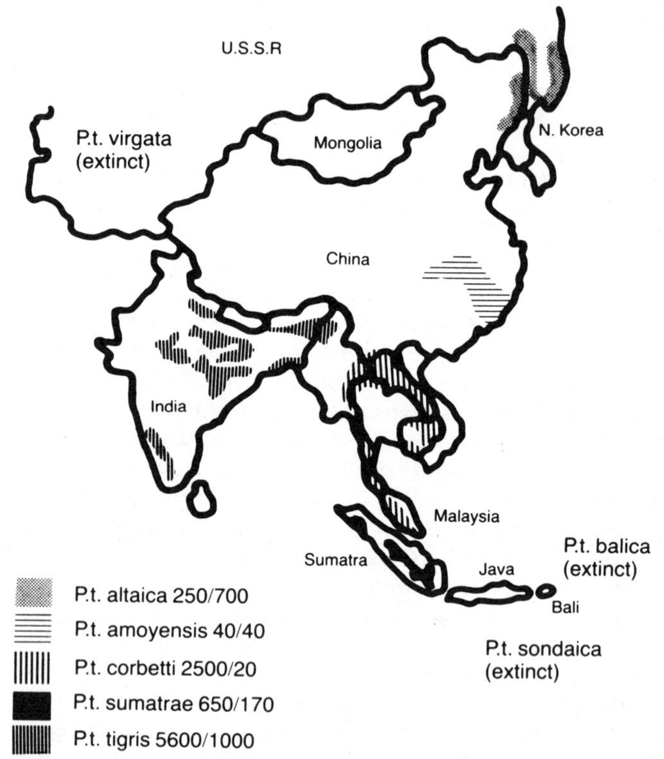

Figure 1. Current geographic distribution and population status of Panthera tigris *subspecies. Estimated wild population/estimated captive population in zoos worldwide after Seal et al. (1987). (Map modified from R. L. Tilson and U. S. Seal, editors. 1987.* Tigers of the World, *with permission of Noyes Press, Park Ridge, New Jersey.)*

Captive Management of Tigers

The tiger Species Survival Plan (SSP) is a coordinated captive management program intended to maximize the contribution that western zoos can make to conserving tigers worldwide. A committee of experts recommends allocations of available space among tiger subspecies and prescribes schedules of breeding exchanges and demographic management to meet captive conservation objectives.

The representation of the four subspecies in the world captive population is far from equitable (Fig. 1). About 700 of the 1000 or so SSP spaces for tigers are occupied by *altaica* and most of the rest by *tigris* or *tigris x altaica* crosses. This situation is partly the result of a conscious effort to use captive resources to bolster a subspecies severely imperiled in the wild, and partly the accidental outcome of the success of the captive breeding program for *altaica*. Recently, concern about threats to *sumatrae* in the wild prompted zoo managers to consider converting some spaces currently occupied by *altaica* to *sumatrae*. This suggestion accords with the conventional wisdom that captive programs should place highest priority on taxa that are least likely to survive in the wild.

But is this the best way to use zoo space for tiger conservation? All tiger subspecies are endangered in the wild, captive space is limited, and the taxonomic validity of and genetic relationships among subspecies are unclear. What is the best allocation of captive space among subspecies to support the overall goal of preserving tigers?

Objectives and Criteria for Tiger Conservation

To optimally allocate captive space among tiger subspecies, we must identify specific conservation goals. We focused on three: (1) maximizing the number of subspecies surviving, whether in the wild or in captivity; (2) maximizing genetic diversity at the species level; and (3) maximizing genetic diversity at the subspecies level. We chose a 200-year time frame (about 32 tiger generations) to match the recommendation of Soulé et al. (1986) for long-term conservation plans.

The first objective assigns value to individual subspecies without regard for the taxonomic or genetic validity of subspecies designations. Maintaining individual subspecies focuses attention on geographic isolates, which helps to maintain the geographic distribution of tigers in the wild. Recognizing the uniqueness of tiger populations in their countries of origin motivates sociopolitical support for conservation programs that benefit not only tigers, but other species and habitats as well. The success of Project Tiger in securing reserves and controlling poaching, especially in India (Panwar 1987), is a good example. The second objective ensures that the species as whole retains enough variation to maintain fitness in captive and wild environments (Ralls & Ballou 1983; Allendorf & Leary 1986). The third objective combines elements of the first two. It preserves variability in all four subspecies, but allows greater variability in one to be exchanged for lower variability in another.

Several metrics for assessing genetic variation have been proposed (e.g., total heterozygosity, allelic diversity, inbreeding coefficients), all of which have some drawbacks in calculation and interpretation (Hedrick et al. 1986; Fuerst & Maruyama 1986; Lacy 1987). We used the panmictic index (P_t), which measures the fraction of initial heterozygosity retained at time t, to measure retention of gene diversity for individual subspecies and for the species as a whole (Crow & Kimura 1970; see Appendix for calculations).

Allocating Captive Space

The problem then is to allocate the 1000 captive spaces among the four tiger subspecies in a way that best meets the objectives outlined above. Success in meeting the objective of subspecies survival requires that either the wild or the captive population (or both) remain large enough over the 200-year planning period to avoid extinction. Success in meeting the objectives for retention of genetic variation at the species and subspecies levels requires that the combined captive and wild populations of each subspecies have a genetically effective size as large as possible.

Effective population sizes in the wild and in captivity are influenced by both biological and sociopolitical events over which wildlife managers have only partial control. In the wild, effective population size depends on preservation of suitable habitat, mortality from natural and human causes, movements of tigers among isolated subpopulations, and other stochastic events. In captivity, effective population size depends on the allocation of captive space among subspecies, and on the ratio of effective to census size (N_e/N) achieved through management of captive reproduction and mortality (Seal 1986). N_e/N in captivity depends partly on biological factors, such as the distribution of births and natural deaths, but also on institutional and sociopolitical factors, such as cooperation in breeding exchanges and public and professional reaction to culling captive tigers.

The decision tree in Figure 2 summarizes the factors affecting how well alternative allocations of captive space achieve tiger conservation goals. The decision node (square) represents the choice of captive popula-

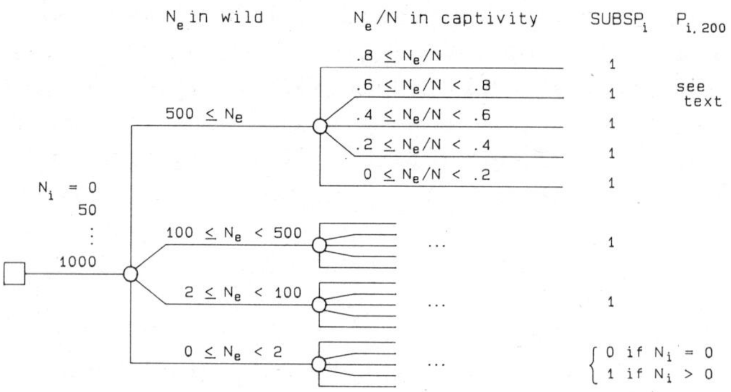

Figure 2. Decision tree for selection of subspecies captive population sizes (N_i) under uncertainty about subspecies effective population size in the wild (N_e) and about the ratio of effective to census population size in captivity (N_e/N). $SUBSP_i$ = number of subspecies surviving; $P_{i,200}$ = panmictic index, the proportional gene diversity of subspecies i after 200 years of captive management.

tion size for each subspecies. The random nodes (circles) represent uncertainty about effective population sizes in the wild and in captivity. Values to the right of the terminating branches of the tree show the impact of each combination of captive and wild effective population size on subspecies survival and genetic variability.

Since these effective population sizes are uncertain, the best we can do to anticipate their impacts on subspecies survival and genetic variation is to estimate the probability of each scenario and use these probabilities to weight the possible outcomes. We elicited probabilities from a group of experts on tiger biology and management using procedures similar to those described by Behn and Vaupel (1982). These probabilities reflect the experts' knowledge of current conditions affecting tigers in the wild and in captivity and their best judgments about natural and cultural factors that will influence tiger populations over the next 200 years. The Bengal tiger (*tigris*) has greater numbers in the wild, better habitat protection, and better representation in western and Indian zoos than the other subspecies. The status of the Indochinese tiger (*corbetti*) is very uncertain due partly to political instability; only about twenty are in captivity, none in western zoos. The Sumatran tiger (*sumatrae*) is in great jeopardy; several hundred remain in habitat increasingly threatened by resource exploitation and human population growth. Moderate numbers are scattered among western zoos. The Siberian tiger (*altaica*) lingers in forest reserves in northeastern Russia and China, where conflict with human activities outside the reserves is increasing. Breeding of this subspecies in western zoos has been very successful, with the captive population several times as large as the wild. The experts' "optimistic," "realistic," and "pessimistic" interpretations of these events are reflected in the probabilities in Table 1; the "pessimistic" consensus was that all subspecies would be extinct in the wild within 200 years.

We elicited probabilities for the N_e/N ratio in captive populations from a group of experts on captive breeding (Table 2). Recent estimates of the N_e/N ratio in captive tiger populations range from 0.2 to 0.4 (Flesness & Cronquist-Jones 1987; Ballou & Seidensticker 1987). We are uncertain how difficult it will be to raise N_e/N above this level through more careful captive management.

Both effective population size in the wild and the ratio of effective to census size in captivity will fluctuate over the next 200 years. The long-term effective size of a fluctuating population may be approximated by the harmonic mean of the population sizes over time, so the sizes listed in Tables 1 and 2 may be viewed as harmonic means of the fluctuating populations.

Calculating Subspecies Survival and Genetic Variability

For a given subspecies, each combination of captive population size, effective population size in the wild, and ratio of effective to census size in captivity represents a scenario for which we can calculate subspecies survival and retention of genetic variability at the subspecies level (see Appendix). Expected values for subspecies survival and genetic variability were obtained by weighting each scenario by its probability. Not surprisingly, larger captive population sizes improve both survival and retention of variability. If there were no constraints on total captive population size, the best strategy would be to have 1000 (or more) of each sub-

Table 1. Probability that the harmonic mean of subspecies effective population sizes (N_e) in the wild over the next 200 years falls within certain ranges. Probabilities derived from consensus of expert opinion using methods of Behn and Vaupel (1982). T = *tigris*; A = *altaica*; S = *sumatrae*; C = *corbetti*.

	Probability			
	Subspecies			
N_e in Wild	*T*	*A*	*S*	*C*
	Optimistic			
$N_e > 500$	0.1	0.0	0.0	0.0
$100 < N_e < 500$	0.4	0.0	0.0	0.2
$2 \leq N_e < 100$	0.4	0.5	0.3	0.5
$0 \leq N_e < 2$	0.1	0.5	0.7	0.3
	Realistic			
$N_e > 500$	0.0	0.0	0.0	0.0
$100 < N_e < 500$	0.05	0.0	0.0	0.0
$2 \leq N_e < 100$	0.1	0.0	0.05	0.05
$0 \leq N_e < 2$	0.85	1.0	0.95	0.95
	Pessimistic			
	All subspecies extinct in wild.			

Table 2. Probability that the ratio of effective to census population size (N_e/N) in captivity falls within a certain range.

Captive N_e/N	*Probability*
$N_e/N > .8$	.05
$.6 < N_e/N \leq .8$	.15
$.4 < N_e/N \leq .6$	.5
$.2 < N_e/N \leq .4$*	.2
$0 \leq N_e/N \leq .2$	.1

** = current level (Ballou & Seidensticker 1987; Flesness & Cronquist-Jones 1987). Probabilities derived from consensus of expert opinion. Assumed to be identical for all subspecies.*

species, ensuring survival of all subspecies and retention of more than 96% of the genetic variation in each. Unfortunately, zoo space is limited, and the 1000 spaces must be divided among the four subspecies.

To evaluate what combinations of captive population sizes totaling 1000 best meet the conservation objectives, we combined the expected values for subspecies survival and gene diversity over the four subspecies for several captive management strategies to obtain the number of surviving subspecies (SUBSPP); the average gene diversity retained within subspecies (P_W); and the gene diversity retained by the species as a whole (P_T) (Tables 3 and 4; see Appendix for calculations).

The values of the subspecies-level criteria (P_W and SUBSPP) are independent of genetic structure, but retention of variation at the species level under different allocations of captive space (P_T) depends on the distribution of variation within and between tiger subspecies. We can see the possible impact of genetic structure by first looking at two extremes: (1) no alleles shared by the four subspecies, and (2) all subspecies genetically identical. The first case, where the variation in each subspecies is unique, is adequately covered by criterion P_W, which assigns equal weight to the variability expressed in each subspecies, regardless of how variable each subspecies is or how many unique alleles each carries. In the second case, where all subspecies are genetically identical, allocation of space among subspecies is irrelevant from a genetic perspective and the nongenetic considerations are adequately covered by criterion SUBSPP, which assigns equal value to the survival of each subspecies. Neither of the two extremes is at all likely, although the low F_{ST} value from electrophoretic studies suggests that the genetic structure of tigers is closer to the second extreme than to the first.

We can use these electrophoretic results to make a best guess at retention of genetic variability by the species as a whole (P_T), based on subspecies population sizes (in the wild and in captivity) and on the initial levels of heterozygosity within each subspecies. Subspecies with larger effective population sizes and with higher initial levels of heterozygosity (e.g., *tigris* with $H_0 = 0.041$) have a greater impact on retention of variation in the species as a whole because they have more variation to lose than do subspecies with lower initial levels of heterozygosity (e.g., *altaica* or *sumatrae* with $H_0 = 0.015$ and 0.014, respectively).

Evaluating Allocation Strategies

We used expected values of the three criteria (number of surviving subspecies, SUBSPP; average gene diversity within subspecies, P_W; and species gene diversity, P_T) to rank different strategies for allocating the 1000 captive spaces among the four tiger subspecies (Table 3). Allocation strategies with the largest representation of *tigris* in the captive population are best for maximizing genetic variation in the species as a whole (P_T), but they are poor for maximizing variability and survival of individual subspecies (P_W and SUBSPP) (S1, S2 vs. S3, Table 3). If fewer than four subspecies can be accommodated in the captive program, including the most variable subspecies, *tigris* and *corbetti*, it is most important for maximizing P_T (S1 vs. S4, S5 vs. S6, Table 3). Since these subspecies are the ones most likely to survive in the wild (Table 1), this recommendation contradicts the conventional wisdom that captive programs should emphasize species or subspecies least likely to survive in the wild. This contradiction arises because none of the tiger subspecies are likely to survive in large numbers in the wild; the primary hope for retaining diversity lies in the captive populations. Maintaining the most variable

Table 3. Combined expected values of the three decision criteria for "realistic" probabilities of subspecies survival in the wild. P_T = gene diversity retained by the whole species after 200 years; P_W = average gene diversity retained by individual subspecies after 200 years; SUBSPP = number of subspecies surviving after 200 years. N_i = captive population size of subspecies *i*.

	N_i				*Expected Values*		
Strategy	*T*	*A*	*S*	*C*	*P_T*	*P_W*	*SUBSPP*
S1	1000	0	0	0	1.08	0.26	1.1
S2	850	50	50	50	1.06	0.59	4.0
S3	250	250	250	250	0.75	0.84	4.0
S4	0	0	1000	0	0.42	0.28	1.2
S5	500	0	0	500	0.95	0.47	2.05
S6	0	500	500	0	0.46	0.50	2.2
S7	550	150	150	150	0.93	0.80	4.0
S8	700	100	100	100	1.00	0.73	4.0

Table 4. Combined expected values of the three decision criteria for "optimistic" probabilities of subspecies survival in the wild. Headings as in Table 3.

	N_i				Expected Values		
Strategy	*T*	*A*	*S*	*C*	P_T	P_W	*SUBSPP*
S5	500	0	0	500	0.97	0.62	2.8
S9	333	333	0	333	0.88	0.74	3.3
S6	0	500	500	0	0.67	0.79	3.6
S4	0	0	1000	0	0.66	0.66	3.1

subspecies in captivity preserves greater diversity for the species as a whole than allocating space to the less diverse subspecies.

In contrast, strategies that include all four subspecies in equal numbers (S3, Table 3) maximize survival and variation for individual subspecies (P_W and SUBSPP), but do not maximize variation for the species as a whole (P_T). If it is necessary to omit one or more subspecies from the captive program, it is best for P_W and SUBSPP to omit those with the best chances for survival in the wild, i.e., *tigris* (S6 vs. S5, S4 vs. S1, Table 3). This contradicts the strategy for maximizing P_T, where it is best to omit those subspecies with lowest variability (which are also those least likely to survive in the wild).

In compromising between maximizing retention of variability at the species level and at the subspecies level, we placed a high priority on survival of all four subspecies and restricted our attention to strategies which include all four in the captive program. Deviating from approximately equal numbers of the four subspecies in the direction of including more *tigris* increases P_T substantially, with only modest drops in P_W (S7, S8, S2, Table 3). Perhaps the best compromise is strategy S7, which allocates a little more than half the captive space to *tigris,* with the remainder divided equally among the other three subspecies. This strategy retains about 93% of the genetic variation currently represented in the species, an average of about 80% of the variation in each subspecies, and assures the survival of all four subspecies, at least in captivity.

Discussion

Since our results are based on limited and uncertain information on tiger genetics and population dynamics, it seems prudent to ask under what circumstances our recommendations could prove misleading. The first source of error we will consider is uncertainty about the fate of wild tiger populations over the next 200 years, as represented by the range of probabilities in Table 1. The results reported in Table 3 are based on the "realistic" probability estimates for effective population size in the wild. In general, results from the "optimistic" probabilities give higher values for all three criteria (Table 4), and results from the pessimistic scenario, where all subspecies become extinct in the wild, give slightly lower values. The patterns of response and recommendations for allocation do not change. One interesting exception to this generalization is that, under the "optimistic" probabilities, some strategies allocating space to larger captive populations of fewer subspecies are better for both subspecies survival and subspecies gene diversity than strategies which include more subspecies at smaller population sizes. For example, 1000 *sumatrae* (S4, Table 4) gives higher P_W and SUBSPP than does 500 *tigris* plus 500 *corbetti* (S5, Table 4); and 500 each of *altaica* and *sumatrae* (S6, Table 4) gives higher P_W and SUBSPP than 333 each of *tigris, altaica,* and *corbetti* (S9, Table 4). In these cases, limiting captive programs to the subspecies least likely to survive in the wild (*altaica* and *sumatrae*) is a better strategy than also including *tigris* and *corbetti.* It is the combination of subspecies survival probabilities and subspecies gene diversities that produces these patterns.

Another possible source of error is our probability distribution for N_e/N under careful captive management. To explore the impact of different N_e/N ratios on the three evaluation criteria, we calculated the three measures under fixed values of N_e/N ranging from 0.2 to 0.5. Within this range, there were few differences in qualitative conclusions about which strategies are best for P_T versus P_W and what the best compromises might be.

The last sources of error we will discuss stem from limitations of the genetic data used to estimate the current distribution of variability among subspecies. With only 26 tigers sampled, it is quite likely that failure to identify variation unique to particular subspecies is an artifact and that variation between subspecies is higher than estimated. If so, allocations of captive space should be shifted toward those strategies that maximize subspecies survival and retention of subspecies variation, i.e., allocations that maximize our first two criteria.

Sample sizes of some subspecies may be small enough (e.g., *sumatrae,* with n = 3) that some of the apparent differences in variability among subspecies may be due to sampling error. We estimated the likelihood that the more limited variability in the samples from *altaica* and *sumatrae* could have occurred by chance alone when sampling from the same pool of variation exhibited by *tigris*. This probability is less than 0.001 for *altaica* and about 0.08 for *sumatrae* (G-tests of significant differences in allelic frequencies), indicating that lower variability in *altaica* and *sumatrae* relative to *tigris* cannot be attributed to small sample size.

It is possible that the electrophoretic data do not reflect the distribution among subspecies of genetic traits important to fitness. If the level of variation important to fitness were disproportionately underestimated for subspecies showing lower levels of electrophoretic variation, allocations of captive space should again shift toward those suggested by the subspecies survival and subspecies variability criteria.

The only circumstance in which the electrophoretic data might be truly misleading would be if the less variable subspecies electrophoretically were actually the most variable for traits important to fitness. Such a reversal seems unlikely because the patterns of electrophoretic variation are consistent with those expected from the biogeography and population histories of the different subspecies. The most variable subspecies (*tigris*) is centrally located within the historic range (Fig. 1) and has maintained larger wild populations than the other subspecies. The subspecies with lowest levels of electrophoretic variation (*altaica* and *sumatrae*) are at the extremes of the current distribution. *Sumatrae* is the last island subspecies (the Bali and Javan subspecies having disappeared in recent years) and presumably has been isolated for some time. *Altaica* went through a bottleneck, perhaps as low as 20–30 animals, about 50 years ago. Although our genetic data are admittedly limited, there is little reason to believe that recommendations based on our results would misdirect tiger conservation plans.

Implications for the Tiger Species Survival Plan

An early version of the analysis presented here showed that excluding *tigris* and *corbetti* from the tiger SSP could undermine retention of genetic variability for tigers as a species and prompted a plan to allocate approximately equal space to each of the four subspecies (U.S. Seal, personal communication). The current analysis suggests that it may be better to tip the balance in favor of *tigris*, despite its relatively well-protected status. Over the next 200 years, even *tigris* is not assured of large enough populations to avert genetic or demographic disaster in the wild. The other three subspecies may well have unique genetic adaptations that should be conserved, but from the standpoint of maximizing genetic diversity for the species as a whole, including *tigris* should be a priority. Opportunities to construct a captive *tigris* population that represents current levels of variation in this subspecies will not last forever; waiting until *tigris* is as imperiled as some of the other subspecies can only jeopardize survival of the species as a whole. Because of biological and institutional inertia in reallocating captive space among taxa, there will be plenty of opportunity to reevaluate the allocation of space (perhaps as a consequence of more complete genetic data), provided that all four taxa are included in the SSP now, and provided that they are managed to maximize retention of variability in whatever subspecies population sizes are initially chosen.

Implications for Management of Other Species

Factors Influencing Tiger Management Strategies

The optimal strategy for allocating captive space among tiger subspecies depends on the goals of tiger conservation and on the details of tiger genetics, taxonomy, and population ecology. Among the factors determining the optimal strategy are: (1) the structure of genetic variation among tiger subspecies populations; (2) the relative probabilities of survival of different subspecies in the wild; (3) the absolute probabilities of subspecies survival in the wild; (4) the size of the captive population relative to that of the wild population; and, (5) interactions between genetic variation within subspecies and probability of survival in the wild. We will not explore the implications of all combinations of these factors here, but we will discuss how our results for tigers can and cannot be applied to conservation of other taxa.

For tigers, there appears to be a tradeoff between maximizing retention of genetic variation for the species as a whole and maximizing retention of genetic variation by individual subspecies. This tradeoff arises because (1) the most variable subspecies, *tigris*, contains all the variation found in the less variable subspecies (at least according to available data); and (2) the most variable subspecies is the most abundant and best protected in the wild and, therefore, least likely to become extinct. The circumstance where all the measured variability for a species is expressed in a single subspecies, and where that subspecies is most abundant and least likely to become extinct, may be common for large, mobile species that once were widely distributed but have been fragmented recently by habitat destruction and exploitation. The apparent conflict between species welfare and subspecies welfare that arises in tiger management may occur for other endangered taxa as well.

For tigers, resources for both captive and wild conservation are relatively abundant. Tigers are popular exhibit animals, and they breed easily in captivity. The amount of zoo space for tigers, although limited, is sufficient to maintain all four subspecies with negligible risk of extinction of each captive population. Wild tigers have enormous public appeal, evidenced by the millions of dollars contributed to Project Tiger to design and protect reserves in India and Southeast Asia. Because of these resources, the plight of tigers is not nearly so critical as that of many other large mammals. Even the compromise strategies would maintain as much as 80% of the variation for each subspecies and more than 90% of total species variation over 200 years.

Impact of Resource Limitations

For many other taxa, resources are more limited relative to the size of the conservation task. For example, there are four genera of rhino worldwide, all threatened or endangered. The African black rhino alone has seven designated subspecies, several of which are probably already extinct. There are about 800 zoo spaces for rhino of all genera worldwide (Cumming 1987). Some rhino species are difficult to breed in captivity, perhaps because they require large social groups (Lindemann 1982). In situations such as these, finding a satisfactory compromise between the welfare of the species as a whole and the welfare of individual subspecies is more difficult. Managers must set clear priorities among evaluation criteria in order to decide which to emphasize. It may be necessary to omit some subspecies from the captive program, or to merge subspecies populations. In these cases, information about the genetic relationships among subspecies is critical.

Limitations of Genetic Information

Even the meager data on tiger genetics is more than is available for many endangered species. The very circumstances that make a species endangered — small, fragmented populations in the wild; absence of stable captive populations — make it difficult to obtain the genetic data needed for conservation decisions. Given the limitations on number of animals and number of loci that can be sampled, it may not be possible to demonstrate that a subspecies lacks unique genetic variation. We cannot examine every locus, nor even all loci that promote local adaptation. Indeed, the relationship between the types of variation measured by existing genetic techniques and fitness is so poorly understood that the application of genetic management to conservation is more a matter of principle than of well-tested practice.

Given these pessimistic views on the utility of available genetic information for conservation planning, one direction for future research is to analyze the allocation of resources among subspecies in the face of uncertainty about their genetic relationships. We intend to repeat our allocation analysis for several alternative genetic structures (e.g., some subspecies less variable, but with unique variation; no measurable genetic differences among subspecies; all subspecies with some unique variation). We also intend to vary other factors that influence the optimal allocation of resources to see which strategies, if any, are robust to changes in genetic and demographic circumstances. Such robust strategies can be used in situations where genetic information is unavailable. In addition, knowing the circumstances in which these robust strategies perform poorly will focus research on those cases where genetic information is most needed to resolve planning dilemmas.

Although this analysis has dealt with the allocation of resources for captive management among endangered subspecies, the issues raised here, and the methods used to address them, are equally applicable to conservation of higher taxa and to management of endangered populations in the wild. Allocating limited resources, coping with uncertainty in the biological and sociopolitical environment, resolving tradeoffs among conflicting objectives, and planning in the absence of complete information are common to most conservation problems.

Appendix

Electrophoretic Data for Tigers

Stephen J. O'Brien and colleagues electrophoretically analyzed blood proteins at 50 loci from a small sample of captive tigers (Newman et al. 1985; O'Brien et al. 1987; S. O'Brien, unpublished data). Five loci varied within or among subspecies. Sample sizes and expected heterozygosity for each subspecies (H_i, expressed as an average across all sampled loci) are as follows: *tigris* ($n = 9$), 0.041; *altaica* ($n = 14$), 0.015; and *sumatrae* ($n = 3$), 0.014. Assuming that the numbers of each subspecies are as shown in Figure 1, and that allele frequencies in *corbetti* equal the averages of the three subspecies that have been studied, the total expected heterozygosity (H_T), the weighted mean within-subspecies expected heterozygosity (H_W), and the between-subspecies expected heterozygosity ($H_B = H_T - H_W$) are 0.036, 0.033, and 0.0026, respectively.

The most variable subspecies, *tigris*, contains all the allelic variation found in the species at the sampled loci. The two less variable subspecies studied, *altaica* and *sumatrae*, contain subsets of the variation found in *tigris*. Given the current geographic separation among subspecies, each may have unique genetic variants at some unstudied loci. Electrophoresis of proteins from 50 loci (out of perhaps 100,000 protein-coding and regulatory loci in the mammalian genome) provides a good statistical sample for estimating average genetic divergence, but very limited assessment of the genetic variants in each population.

Because no samples were obtained for *corbetti*, we assigned this subspecies the average of the allele frequencies in the other subspecies sampled, then calculated its heterozygosity based on these average allele frequencies ($H_i = 0.028$). *P. t. corbetti* occupies a geographic range in between *tigris, sumatrae,* and *altaica* (Fig. 1). It is thought to be fairly widely distributed in the wild, numbering fewer than *tigris* but more than *altaica* or *sumatrae* (Seal et al. 1987). It seems plausible, therefore, that *corbetti* is intermediate in heterozygosity and allele frequencies between *tigris* and the two less variable subspecies.

Calculation of Evaluation Criteria for Individual Subspecies

We imposed a minimum size of 50 for subspecies populations in captivity so that the risk of extinction of the captive subpopulations due to demographic stochasticity would be negligible. A subspecies survives if it is represented in the captive population or if the effective size of the wild population is at least two. The panmictic index at 200 years (32 generations) for each subspecies ($P_{i,200}$) is calculated according to the formula:

$$P_{200} = (1 - 1/2N_e)^{32} \qquad \text{(Eq. 1)}$$

(Crow & Kimura 1970), where N_e is the effective population size in the wild plus the product of captive population size and N_e/N in captivity. (We used the values 0, 50, 250, and 1000 to represent the ranges of effective population sizes in the wild in Table 1; and the values 0.1, 0.3, 0.5, 0.7, and 0.9 to represent the ranges of N_e/N in Table 2.) We used the probability distributions listed in Tables 1 and 2 to calculate statistical expectations for the subspecies panmictic index ($P_{i,200}$) and subspecies survival ($SUBSP_i$) by multiplying the values at each terminating branch of the decision tree by the probability of that branch. (See Behn & Vaupel 1982 and Maguire 1986 for decision tree calculations.)

Combining Criteria over the Four Subspecies

To form an overall expected value for number of subspecies surviving, we summed the four $E(SUBSP_i)$ values. Each of these can range from 0 to 1, so the possible range for the combined expected value (designated SUBSPP in Tables 3 and 4) is 0 to 4. To form a combined expected value for gene diversity retained by all subspecies (panmictic index: P_W), we averaged $P_{i,200}$ over the four subspecies. P_W can range from 0 to 1. This measure of within-subspecies gene diversity assigns equal weight to each of the four subspecies regardless of their abundances or initial levels of heterozygosity.

Gene diversity within each subspecies declines at the rate described by Equation 1 above. With s extant subspecies (equal to or less than four, depending on the choice of captive population sizes), a fraction $(s - 1)/s$ of the within-subspecies gene diversity that is lost each generation is converted to between-subspecies gene diversity as the subspecies diverge genetically ($H_{B,t} = H_{B,0} + [s - 1][H_{W,0} - H_{W,t}]/s$). The total gene diversity ($H_T = H_W + H_B$) decays according to the equation $H_{T,t} = H_{W,t} + H_{B,t} = H_{T,0} - (H_{W,0} - H_{W,t})/s$. This calculation ignores the partial separation of the captive from the wild population of each subspecies, which is appropriate if there is occasional migration between the captive and wild populations (Lacy 1987). The loss of diversity will be more rapid if no such interchange takes place. We calculated $H_{T,t}$ as the expected heterozygosity at time t for the species as a whole, as if the subspecies populations were mixed and immediately attained Castle-Hardy-Weinberg equilibrium at that point.

After calculating $H_{T,200}$ for each combination of wild and captive population sizes, we obtained an expected value for $H_{T,200}$ from the joint probabilities of effective population size in the wild (Table 1) and N_e/N ratio in captivity (Table 2). Finally, we expressed the expected value for $H_{T,200}$ as a proportion of $H_{T,0}$ to obtain P_T, the gene diversity retained by the species as a whole, as reported in Tables 3 and 4.

Because the captive tiger population is a substantial fraction of the total world population (about 20% currently, and likely to increase as wild populations decline), the relative representation in the captive population of tiger subspecies having different levels of heterozygosity can have a marked impact on H_T for the species. Allocation strategies that increase or decrease the relative representation of the more variable subspecies from current levels produce an apparent increase or decrease in H_T, independent of the loss of variation due to drift and inbreeding that will occur over the 200-year period of concern. That is why P_T values for allocations with large proportions of *tigris,* the most variable subspecies, may exceed 1 (Table 3).

Acknowledgments

We would like to thank those who helped with the initial formulation of this analysis: U. S. Seal, T. Foose, R. Tilson, D. Wharton, J. Ballou, N. Flesness, and M. Don-Carlos. We thank J. Ballou for doing some of the initial calculations. We especially thank S. O'Brien for generously sharing published and unpublished data on tiger genetic structure. Thanks to Michael E. Soulé, anonymous reviewers, and the editor for their guidance on revisions of an earlier draft.

Literature Cited

Allendorf, F. W., and R. F. Leary. 1986. Heterozygosity and fitness in natural populations of animals. Pages 57–76 in M. E. Soulé, editor. Conservation biology. Sinauer Associates, Sunderland, Massachusetts.

Ballou, J. D., and J. Seidensticker. 1987. The genetic and demographic characteristics of the 1983 captive population of Sumatran tigers (*Panthera tigris sumatrae*). Pages 329–347 in

R. L. Tilson and U. S. Seal, editors. Tigers of the world: the biology, biopolitics, management and conservation of an endangered species. Noyes Publications, Park Ridge, New Jersey.

Behn, R. D., and J. W. Vaupel. 1982. Quick analysis for busy decision makers. Basic Books, New York.

Chambers, S. M., and J. W. Bayless. 1983. Systematics, conservation and the measurement of genetic diversity. Pages 349–363 in C. M. Schonewald-Cox, S. M. Chambers, B. MacBryde, and L. Thomas, editors. Genetics and conservation: a reference for managing wild animal and plant populations. Benjamin/Cummings, Menlo Park, California.

Crow, J. F., and M. Kimura. 1970. An introduction to population genetics theory. Harper and Row, New York.

Cumming, D. 1987. Small population management of black rhinos. Pages 12–13 in R. duToit and T. Foose, editors. Proceedings of the 1986 Cincinnati African Rhino Workshop. Pachyderm No. 9, December 1987.

Flesness, N. R., and K. G. Cronquist-Jones. 1987. Possible selection in captive *Panthera tigris altaica.* Pages 363–370 in R. L. Tilson and U. S. Seal, editors. Tigers of the world: the biology, biopolitics, management and conservation of an endangered species. Noyes Publications, Park Ridge, New Jersey.

Fuerst, P. A., and T. Maruyama. 1986. Considerations on the conservation of alleles and of genic heterozygosity in small managed populations. Zoo Biology **5**:171–179.

Hedrick, P. W., P. F. Brussard, F. W. Allendorf, J. A. Beardmore, and S. Orzack. 1986. Protein variation, fitness, and captive propagation. Zoo Biology **5**:91–100.

Herrington, S. J. 1987. Subspecies and the conservation of *Panthera tigris:* preserving genetic heterogeneity. Pages 51–62 in R. L. Tilson and U. S. Seal, editors. Tigers of the world: the biology, biopolitics, management and conservation of an endangered species. Noyes Publications, Park Ridge, New Jersey.

Lacy, R. C. 1987. Loss of genetic diversity from managed populations: interacting effects of drift, mutation, immigration, selection, and population subdivision. Conservation Biology **1**:143–158.

Lindemann, H. 1982. African rhinoceroses in captivity. Ph.D. dissertation. University of Copenhagen, Copenhagen, Denmark.

Maguire, L. A. 1986. Using decision analysis to manage endangered species populations. Journal of Environmental Management **22**:345–360.

Nevo, E. 1978. Genetic variation in natural populations: patterns and theory. Theoretical Population Biology **13**:121–177.

Newman, A., M. Bush, D. E. Wildt, D. Van Dam, M. Th. Frankenhuis, L. Simmons, L. Phillips, and S. J. O'Brien. 1985. Biochemical genetic variation in eight endangered or threatened felid species. Journal of Mammalogy **66**:256–267.

O'Brien, S. J., G. E. Collier, R. E. Bienveniste et al. 1987. Setting the molecular clock in Felidae: the great cats, *Panthera.* Pages 10–27 in R. L. Tilson and U. S. Seal, editors. Tigers of the world: the biology, biopolitics, management and conservation of an endangered species. Noyes Publications, Park Ridge, New Jersey.

Panwar, H. S. 1987. Project Tiger: the reserves, the tigers and their future. Pages 110–117 in R. L. Tilson and U. S. Seal, editors. Tigers of the world: the biology, biopolitics, management and conservation of an endangered species. Noyes Publications, Park Ridge, New Jersey.

Ralls, K., and J. Ballou. 1983. Extinction: lessons from zoos. Pages 163–184 in C. M. Schonewald-Cox, S. M. Chambers, B. MacBryde, and L. Thomas, editors. Genetics and conservation: a reference for managing wild animal and plant populations. Benjamin/Cummings, Menlo Park, California.

Seal, U. S. 1986. Goals of captive propagation programmes for the conservation of endangered species. International Zoo Yearbook **24/25**:174–179.

Seal, U. S., P. Jackson, and R. L. Tilson. 1987. A global tiger conservation plan. Pages 487–498 in R. L. Tilson and U. S. Seal, editors. Tigers of the world: the biology, biopolitics, management and conservation of an endangered species. Noyes Publications, Park Ridge, New Jersey.

Soulé, M. E., M. Gilpin, W. Conway, and T. Foose. 1986. The millenium ark: how long a voyage, how many staterooms, how many passengers? Zoo Biology **5**:101–113.

Wilson, E. O., and W. L. Brown, Jr. 1953. The subspecies concept and its taxonomic application. Systematic Zoology **2**:97–111.

Breed Structures of Rare Pigs: Implications for Conservation of the Berkshire, Tamworth, Middle White, Large Black, Gloucester Old Spot, British Saddleback, and British Lop

S. J. G. HALL

Research Group in Mammalian Ecology and Reproduction
Physiological Laboratory
Downing Street
Cambridge, CB2 3EG England

Abstract: *Registration details of seven rare British breeds of pig were studied over the period 1978–1986 inclusive. For four breeds, numbers of pigs registered have shown a slight upward trend. About 70% of males and 40% of females that breed, do so in herds other than the natal herd. In the breed for which most pedigree data were analysed (the British Lop), 95–100% of recent pig crops were inbred. In the rare breeds generally, mean inbreeding at around 6% signifies an inbreeding rate of about 1% per generation, higher than has been found in commercial breeds. Breeds did not differ greatly in the ways they were structured.*

About a third of herds supplied boars to other herds, a relatively high proportion, indicating that the breeds have a structure that is not completely hierarchical, and this is favorable for genetic conservation. Pigs whose parents were from different herds were significantly less inbred than those with both parents from the same herd. In some breeds it was clear that pigs sharing the same bloodline name were more closely related than pigs within the same herd but with different bloodline names.

Conservation procedures applied to these pigs have been designed to conform with the customs and procedures of the British pedigree livestock industry. The most important single such procedure is the registration, with a central authority, of breeding stock.

Resumen: *Detalles de registro de siete variedades escazas de cerdos Britanicos fueron estudiados por el periodo de 1978–1986 inclusive. Para cuatro variedades los números de cerdos registrados han manifestado una tenue tendencia a aumentar. Acerca de 70% de los machos y 40% de las hembras que procrean lo hacen en manadas que no son sus originales. En la variedad de la cual la mayoría de datos de pedigrí fueron analizados ("British Lop"), el 95–100% de la parición reciente de cerdos fueron consanguíneos. En variedades raras generalmente la media de consanguinidad es mas o menos 6% lo cual significa un índice de consanguinidad de 1% por generación, más alto que lo encontrado en variedades comerciales. La estructura de los grupos de reprodución no difieren mayormente.*

Mas o menos 1/3 de las manadas suministraron verracos a las otras manadas, lo cual da una proporción relativamente alta, indicando que los grupos tienen una estructura que no es completamente jerárquica, y esto es favorable a la conservación genética. Los cerdos cuyos padres provenían de diferentes manadas eran significativamente menos homogeneas que aquéllos donde los dos padres venían de la misma manada. En algunas razas era claro que los cerdos que compartían el mismo nombre y linea de sangre estaban relacionados mas cercamente que a aquellos cerdos dentro de la misma manada con diferentes nombres y lineas de sangre.

Los procedimientos conservacionistas aplicados a estos cerdos han sido designados para ajustarse con las costumbres y procedimientos de la industria de ganado de pedigrí Britanica. El procedimiento singular mas importante es el registro con una autoridad central de crías de ganado.

Paper submitted April 12, 1988; revised manuscript accepted May 3, 1988.

Introduction

The need for conservation of the genetic resources represented by traditional breeds of farm animals is appreciated by many organizations, including the Food and Agricultural Organization of the United Nations. Worldwide there are nine species of pig (Suidae) in five genera (Macdonald 1985). In Southeast Asia pigs are particularly abundant and diverse, with wild and feral domestic pig *Sus scrofa* and five other species (*Sus barbatus, S. celebensis, S. verrucosus, S. salvanius,* and *Babyrousa babyrussa*), of which the last three are not secure in numbers. Many of these Asian species interbreed, and there are also local races, which presumably reflect local adaptation, that ought to be conserved (National Research Council 1983; Macdonald 1985).

In the Western world, conservation of livestock has been attempted in many countries using different approaches (see, for example, Brooke 1984; Crawford 1984; Bodo 1985; Henson 1986). In Britain the *Rare Breeds Survival Trust* (RBST) is at the forefront of this work (Jewell 1985). The work of the RBST and of its members in relation to certain breeds of sheep has been described by Hall (1986).

The RBST applies a large proportion of its resources to the conservation of rare pigs. Modern pig farming (Pond 1983) depends on the heterotic and additive genetic effects resulting from crossbreeding and the breeds favoured for the production of commercial crossbred stock have shown very great changes over recent years in British farming, where the Landrace and the Large White breeds are most important (the latter breed is known in the United States as the Yorkshire).

Population structures have been studied in many breeds of livestock (Vu Tien Khang 1983) because an understanding of them can help in the dissemination of genetic improvements. The conservation of rare breeds can also be advanced by this understanding, particularly where there is a policy of fostering private ownership of livestock (Hall 1986). Pigs, however, have not been studied extensively in this respect. It is remarkable that in spite of their economic importance and the availability of pedigree data there have been very few studies of their breed structures (Lush & Anderson 1939; Bekedam 1959; McPhee 1965; Jonsson 1971; Hanset 1973; Smith et al. 1978). It is an urgent priority for genetic conservation that such a study should be carried out in this species.

In Britain, the Berkshire, Tamworth, Middle White, Large Black, Gloucester Old Spot, British Saddleback, and British Lop (cover illustration, this issue) are defined as rare, that is, with fewer than 150 breeding females (Anon. 1986, 1987*a*, 1988), though the Gloucester Old Spot is probably above this number by now (Anon. 1988). The histories of the breeds and of their breed societies are summarized by Davidson (1966). Interest in the pedigree breeding of pigs is long established; the National Pig Breeders' Association (NPBA) and the British Berkshire Society were founded in 1884. Of the seven breeds discussed here, the British Lop has the youngest breed society, founded in 1921, which continues to be independent, though the other six rare breeds are all now catered for by the NPBA. Registration of the NPBA breeds is encouraged by the RBST, whose members are able to register their pigs through a RBST-subsidized scheme. The RBST arranges for semen to be stored from boars considered of particular conservation significance and for incentives to be paid for the support of boars at stud ("incentive boars") that are considered to represent rare lines within the breeds (Anon. 1988). The RBST Show and Sale, held annually in September, is a major marketplace (Anon. 1987*b*); 55 females and 16 males of these seven breeds were sold at this event in 1987.

In British pedigree pigs, the "bloodline" system of breed subdivision is practiced. At registration a pig receives an official number and a binomial consisting of (1) the designation of the herd in which it was bred, and (2) a name, which in the case of males is that borne by the sire, and in the case of females by the dam. In most cases a numeral is appended to differentiate siblings or part-siblings from each other.

Bloodlines were founded in the more-or-less distant past by particularly notable foundation animals, and these patronymics or matronymics have thus been transmitted down through the generations. As Lush (1943) pointed out, discussion of bloodlines is "rare ... in textbooks on animal breeding and still rarer in textbooks on genetics." However, the concept is very widely used by practical pedigree breeders and an analysis of breed structures would be incomplete without some mention of it. This paper considers some genetic relationships within and among bloodlines.

The present study adopts a comparative approach as there may be implications for conservation management of any differences among the seven breeds under discussion. For example, the Gloucester Old Spot is the closest in type to a modern, intensively reared breed; the Large Black and the British Saddleback are particularly associated with outdoor pig farming. The Berkshire, Large Black, and British Saddleback are black-skinned breeds while the others are white-skinned, and this can influence their value as meat animals. In addition, the Berkshire and Middle White are extreme pork breeds, suited for early slaughter at light weights. Such differences in agricultural function may influence population processes, so generation intervals, seasonality, and the production of registered piglets were investigated.

This paper outlines the principles, practice, and genetic consequences of the conservation techniques applied to rare British pigs. It may be possible to modify these techniques to suit other pig populations. Unlike those of cattle and sheep, embryos of pigs cannot at present be stored cryogenically and the maintenance of breeding populations would be a necessary feature of a complete conservation program. This paper shows how such populations might be structured.

Materials and Methods

In the late 1970s the NPBA pig registration system was transferred to a computer, and while some parentage records prior to 1978 are on the computer record it is convenient to take 1978 registrations as a starting point for the six breeds registered by the NPBA. Pedigree data in the form of conventional herd books go back much further — to 1884 in the case of the Berkshire — but 1978 is sufficiently far back for pedigrees of at least four generations to be deduced for early- to mid-1980s registrations. The NPBA registration details were transcribed onto files on the Cambridge University IBM 3081 computer.

British Lop registration details are in a conveniently accessible form dating back to 1970 and these were entered manually into the computer.

Registration details were sorted by the computer in various ways so animals were classified by different combinations of their attributes of sex, date of birth, sire, dam, and herd of registration. Statistical tests were performed as appropriate (Helwig & Council 1979; Sokal & Rohlf 1981). The computer program described by Falco (1975) was used to compute inbreeding coefficients, following pedigrees back for up to six generations.

Inbreeding coefficients were calculated for all animals born in the period 1982–1986 inclusive. These data were used to investigate mean inbreeding of male and female piglet crops in each year. Further investigations were made of the inbreeding of a sample of animals classified according to bloodline and to herd in which they were bred. For the NPBA breeds this sample comprised the animals listed in the 1986 NPBA registration printout, published at the end of 1986 but including many animals of 1985 birth and some from previous years. For the British Lop the sample comprised the animals born in 1985 and 1986.

The coefficient of kinship (Jacquard 1974), which for a pair of animals equals the inbreeding of their progeny (see Bennett 1987 for references) was calculated for selected pairs of animals. Coefficients of kinship were calculated among present-day piglets to find out whether animals bred in the same herd, of different bloodlines, were more or less closely related than animals bred in different herds. Another measure of relatedness, the coefficient of relationship of Wright (1922), could also have been used. It takes into account the inbreeding of the animals whose relationship is to be found, but is rather more difficult to calculate.

Results

Population Trends

Details of the numbers of pigs born each year, and the numbers of herds and of sires and dams, are summarized in Table 1 and Fig. 1. Full details are available from the author. Data from 1978 and 1986 were excluded from much of the analysis because registration details of animals born in those years were not complete in the pedigree records. Altogether, registration details of 8,960 pigs were available.

The Tamworth and British Saddleback have shown a slight downward trend in numbers born, paralleled by a decrease in numbers of herds registering. The British Lop has declined slightly in numbers, but has shown an increase in number of herds registering.

Registration sex ratios were similar among the breeds

Table 1. Summary of population statistics 1979–1985; number of registrations and of herds, and numbers and origins of sires and of dams.

Breed	*Mean annual registrations*		*Mean number of herds registering*	*Mean number of registrations per herd*	*Mean number of sires employed*	*Mean number of dams employed*	*Mean percentage of herds using more than one sire*
	M	*F*					
Berkshire	25	81	19.7	5.4	19.6	38.9	32.6
Tamworth	27	76	26.6	3.9	28.3	47.1	24.7
Middle White	18	55	13.4	5.4	16.6	27.0	35.1
Large Black	26	79	17.1	6.1	23.6	45.1	42.5
Gloucester Old Spot	46	253	35.1	8.5	47.6	119.1	27.2
British Saddleback	49	150	27.3	7.3	47.9	99.9	51.3
British Lop	28	105	17.7	7.5	21.0	46.6	35.5

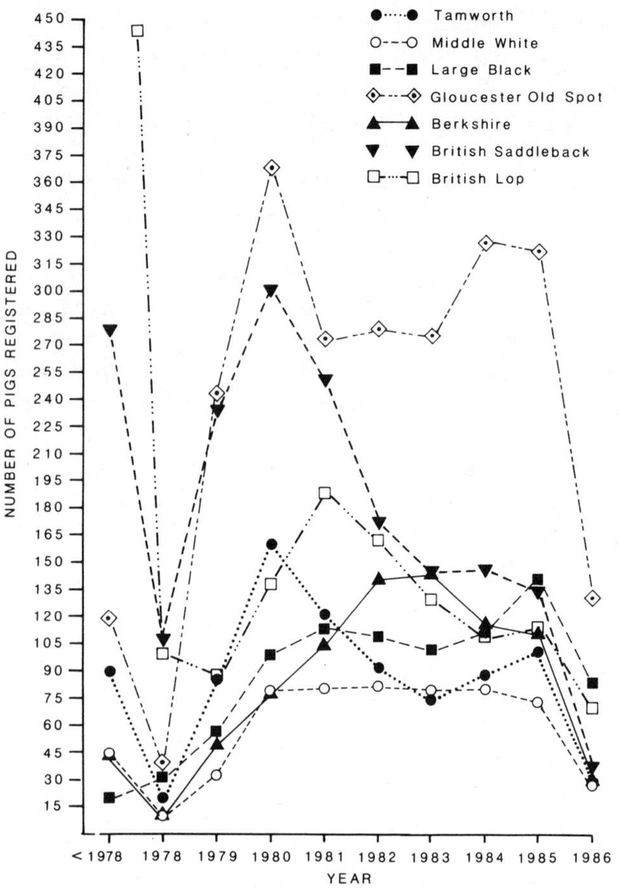

Figure 1. Numbers of registered pigs (sexes combined) born in each year since 1978. Numbers before 1978 are not comparable because registration procedures differed among breeds, and data for 1986 are incomplete.

at around 30% (males/females) except for the Gloucester Old Spot, which with a mean ratio of 18.91% was the main contributor ($G = 46.81$, d.f. $= 1$) to the heterogeneity indicated by a G-test ($G = 74.93$, d.f. $= 6$, $p < 0.01$).

The numbers of sires and dams used follow the same trend for each breed as number of registrations; the number of dams used is a minimum estimate for the number of breeding females. Excepting the Berkshire and the Large Black, there has been a reduction in mean registrations per herd. There is a general increase in the percentage of dams bred in a different herd from that in which they were used but sires did not show this so clearly.

Use of Homebred and Bought-in Males and Females

Around half of all males registered, and about a third of females, produced registered offspring subsequently (Table 2). In all breeds, some males were used to sire piglets in other herds as well as in the natal herd.

"Bought-in" boars (these sired offspring in some herd other than the natal herd) had each, on average, rather more registered offspring than homebred boars (Table 3), except in the Gloucester Old Spot. In all breeds the registration sex ratio was more biased toward females in the piglet crops sired by bought-in boars, although statistical significance was only achieved in the Tamworth and the Middle White.

Each bought-in female had on average about the same number of registered offspring as each homebred female (Table 4). Sex ratio was slightly lower, indicating that, as in the case of boars, a bias existed toward registering the female progeny of bought-in animals.

Generation Intervals

The mean ages at which boars were used for production of registered progeny (Table 5) ranged from 1 year 5 months (Berkshire) to 1 year 9 months (Tamworth). The mean ages at which boars were first used were not calculated. For females the Berkshire had the lowest mean age at breeding (1 year 10 months) and the Large Black the greatest (2 years 5 months). Maximum ages at which boars and sows were mated were 7 years 6

Table 2. Breeding careers of pigs.

	Number registered 1978–1984		*Number breeding (total)*		*Percentage breeding*		*Of those breeding: Percentage breeding in herd of registration only*		*Percentage breeding in other herd only*		*Percentage breeding in herd of registration and in other herd*	
	Males	*Females*	*Males*	*Females*	*Males*	*Females*	*Males*	*Females*	*Males*	*Females*	*Males*	*Females*
Berkshire	154	485	71	151	46	31	18	52	68	44	14	4
Tamworth	175	465	80	151	46	32	19	47	70	52	13	1
Middle White	109	336	61	103	56	31	21	46	64	50	15	4
Large Black	159	465	84	174	53	37	31	66	54	33	15	1
Gloucester Old Spot	270	1,537	133	378	49	25	25	51	54	44	21	5
British Saddleback	318	1,042	142	278	45	27	55	79	35	20	12	1
British Lop	281	1,081	145	319	52	30	32	65	61	31	7	4
All breeds	1,466	5,411	716	1,544	49	29	31	60	56	37	14	3

Table 3. Sex ratios of progeny of homebred and bought-in boars.

	Progeny of homebred boars			*Progeny of bought-in boars*		
Breed	*Total boars breeding in herd of registration*	*Mean offspring per boar*	*Ratio males/total*	*Total boars breeding in other herd(s)*	*Mean offspring per boar*	*Ratio males/total*
Berkshire	23	9.21	26.9%	58	9.79	22.9%
Tamworth	24	6.88	33.9%	66	9.18	23.3%
Middle White	22	7.05	31.0%	48	8.19	21.1%
Large Black	39	5.97	27.5%	58	10.65	24.8%
Gloucester Old Spot	61	15.51	17.7%	100	13.14	13.5%
British Saddleback	93	8.60	24.6%	66	11.06	22.2%
British Lop	56	5.57	26.3%	99	12.49	19.4%
All breeds	318	8.88	23.8%	495	11.04	19.9%

months (Tamworth) and 8 years 11 months (Large Black) respectively. The mean dam-offspring interval for the black-skinned breeds (Berkshire, Large Black, British Saddleback) was 913 days; for the others, 794 days.

Seasonality

Breeding was not seasonal but there were peaks in January and July. Of the 3,991 registered litters (i.e., litters including registered piglets) born to all seven breeds, 36% were born in these months. The Gloucester Old Spot had the least pronounced seasonality, with a total of 28% of its litters being born in January and July. At the other extreme was the Large Black (52%).

A proportion of sows produced two registered litters a year — this ranged from 4% (Berkshire 1984) to 28% (Gloucester Old Spot 1984). The number of such litters produced by each breed over the period 1978–1986 (for the British Lop, since 1970) was expressed as a proportion of the maximum potential number that could have been produced by the sows known to be reproductively active. The black-skinned breeds produced 56% of the registered litters potentially possible while for the other breeds the value was 58%; this difference falls short of statistical significance (chi^2 = 2.99, d.f. = 1; $p > 0.05$).

Relationships Among Herds

A high proportion (overall, more than a third; Table 6) of the herds that registered pigs supplied boars to other herds that sired registered stock in their new herds. A further analysis (available from the author) was made showing the two or three most important herds that supplied boars. Herd importance was taken as being indicated by the number of herds to which boars were supplied. It should be noted that "supplied" in this context can mean a boar was sold, or made available at stud, or semen was supplied. The fourteen principal herds (two from each breed) constituted only 3% of all the herds, but they supplied 45% of the 716 boars that bred. These boars sired 28% of the total (n = 8,960) registrations.

Inbreeding of Recent Piglet Crops

The mean inbreeding coefficients of piglet crops in 1986 were between 2.2% and 7.6% (Table 7). There was no consistent difference between the sexes. In most NPBA breeds, a majority of the piglets were inbred. In the British Lop, all were measurably inbred, and this was because of the more extensive pedigree records used for the analysis of this breed. Generally, the mean inbreeding did not unequivocally show an increase in the

Table 4. Sex ratios of progeny of homebred and bought-in females.

	Progeny of homebred females			*Progeny of bought-in females*		
Breed	*Total females breeding in herd of registration*	*Mean offspring per female*	*Ratio females/total*	*Total females breeding in other herd(s)*	*Mean offspring per female*	*Ratio females/total*
Berkshire	85	4.89	24.3%	72	5.06	23.6%
Tamworth	73	5.21	31.3%	80	4.89	20.0%
Middle White	51	4.63	·27.1%	56	5.71	23.4%
Large Black	116	5.01	26.7%	60	4.50	23.0%
Gloucester Old Spot	211	5.82	16.5%	186	5.55	13.8%
British Saddleback	222	5.64	23.1%	58	4.78	25.3%
British Lop	219	4.48	22.2%	113	5.03	18.3%
All breeds	977	5.19	22.6%	625	5.15	19.1%

Table 5. Generation intervals of pigs, in days.

		Days from date of birth of sire to date of birth of progeny		*Days from date of birth of dam to date of birth of progeny*	
	n	mean	S.D.	mean	S.D.
Berkshire	551	631.98	286.44	777.18	392.16
Tamworth	432	766.39	475.42	870.28	444.31
Middle White	381	640.36	286.75	789.49	362.93
Large Black	665	719.06	372.59	1,006.55	602.16
Gloucester Old Spot	1,348	702.75	312.99	776.29	399.08
British Saddleback	701	675.86	286.55	930.11	545.01
British Lop	1,230	647.91	292.80	788.54	417.64

period 1982–1986, but the proportion of pigs measurably inbred did so.

Inbreeding Within Herds

In the Large Black the mean inbreeding of the group of pigs (n = 35) whose parents were bred in the same herd was 9.7%. That of the pigs whose parents were bred in different herds (n = 169) was 2.5%. This difference is statistically significant (chi^2 approximation of Kruskal-Wallis test: chi^2 = 26.4, d.f. = 1, $p < 0.001$). All other breeds showed the same pattern with a similar level of statistical significance.

Relationships Within Bloodlines

The coefficients of kinship between animals bearing the same bloodline name are rather greater than are those between animals of different bloodlines but bred in the same herd, which are in turn greater than those between animals of different bloodlines and bred in different herds (Table 8). This is most clearly seen in the Berkshire, Large Black, and British Saddleback; sample sizes are rather small in the other breeds on account of the multiplicity of bloodlines and breeders, or, in the case of the Middle White, because the breed is not numerous. In the Gloucester Old Spot the picture is confused because of the coexistence of two bloodline naming systems. Differences among the Large Black samples were significant (chi^2 approximation of Kruskal-Wallis test: chi^2 = 63.8, d.f. = 2, $p < 0.001$).

Table 6. Number of herds named in the pedigree records, and number and percentage of them that have supplied boars that bred in other herds.

Breed	*Herds in records*	*Herds that supplied boars*	%
Berkshire	57	23	40.4
Tamworth	87	31	35.6
Middle White	42	16	38.1
Large Black	47	21	44.7
Gloucester Old Spot	140	29	20.7
British Saddleback	66	34	51.5
British Lop	60	21	35.0
Totals	499	175	35.0

Discussion

Numerical Trends and the Nature of Breed Structures

Rare breeds of pig are not showing the same consistently strong, upward trends in numbers of herds and of registration as have been recorded in rare sheep (Hall 1986). They continue to be very rare. When it is noted that there are about 828,000 commercial breeding sows in Britain (MLC 1985), the numerical status of the rare pigs is put into perspective. Only in the Gloucester Old Spot have more than a hundred sows been cited annually as producing registered litters, in recent years. In this breed there was a peak in female registrations in 1980 that has not been repeated or seen in other breeds. The increasing popularity of the Gloucester Old Spot underlies its markedly different registration sex ratio where males are registered at only 19% of the rate of females, compared with 31% in the other breeds.

Since 1979, however, four other breeds have also shown increases in numbers of herds registering pigs. The exceptions are the Tamworth and the British Saddleback (Table 1). These two, with the British Lop, are also the breeds showing a decline in number of registrations. The other trends summarized in Table 1 have less bearing on the continued existence of the breeds in the short term but are likely to affect the degree of inbreeding; it is advantageous that a large number of sires and dams should be employed and if they are bought-in rather than homebred, they are less likely to produce inbred offspring. In the Large Black, for instance, pigs whose parents were bred in the same herd had coefficients of inbreeding four times the magnitude of those whose parents came from different herds.

The breeding careers of pigs (Table 2) show strong similarities among breeds. Indeed so far as male pigs are concerned the only proportions that are markedly different from the overall picture relate to the British Saddleback, where a high proportion of males breed in the herd of registration only and a rather low proportion are used purely as "bought-in" sires. Grouping the seven breeds, it is seen that 44.4% of sires were homebred. In some other farming contexts the use of homebred sires, coupled with selection, has been seen as a means to genetic improvement; Smith et al. (1978) recorded that in nucleus herds of Large White, British Landrace, and Welsh pigs, 64% of sires were homebred. The popularity among breeders of rare pigs, of bought-in sires, would therefore tend to favor genetic conservation.

The British Saddleback also stands out as using homebred females to a greater extent than other breeds. The

Table 7. Percentages of 1986 pig crops that were inbred, with mean coefficients of inbreeding, the sexes being considered separately.

Breed	Percentage of crop that was inbred	Males			Females		
		n	Mean inbreeding %	SD	n	Mean inbreeding %	SD
Berkshire	56.7	8	4.1	8.6	22	5.1	7.8
Tamworth	38.0	5	5.2	11.1	25	2.4	7.0
Middle White	79.3	6	4.9	4.9	23	4.6	5.2
Large Black	66.3	27	2.7	3.7	57	2.4	3.4
Gloucester Old Spot	48.1	22	2.8	5.0	109	2.7	5.1
British Saddleback	60.0	11	2.2	4.8	24	6.8	6.6
British Lop	100.0	19	7.6	6.1	53	5.0	2.6

Gloucester Old Spot showed a tendency to use boars in both the herd of registration and in other herds. The possible origins of these anomalies in the ways the breeds are structured are discussed below.

Overall, 49% of registered boars sired registered progeny, 69% of them in herds other than the natal herd. Of registered females, only 29% produced litters from which pigs were registered, 37% of these females being used only in herds other than the natal herd. In rare sheep (Table 2 of Hall 1986), 36% of registered ram lambs subsequently breed, indicating that from the market point of view too few ram lambs are castrated and that breeders of rare pigs make a rather more accurate assessment of how many entire males the market will take. Only a small proportion (3% overall) of sows bred in other herds as well as the one in which they were born. This is in sharp contrast to the 14% noted in rare sheep (Hall 1986), where female sheep are frequently bought and sold in the course of their breeding career. The female pigs that are sold from one herd to another are an important contributor to gene flow.

In the rare pigs, bought-in boars do not differ greatly from homebred boars in the sex ratio of their registered offspring, but bought-in boars do tend to sire more registered piglets on average. It must be noted that this is not necessarily an indication of higher litter size, simply that more piglets were deemed to be worth registering (Table 3). This difference is not apparent with homebred compared with bought-in females, which are very similar in terms of mean numbers of offspring registered, and of registration sex ratio (Table 4).

Generation Interval, Seasonality, and Breed Differences

There do not seem to be any published figures for generation intervals in British pigs, other than estimates of 1.8 years for Large White and 1.7 years for British Landrace and Welsh (Smith et al. 1978). Table 5 shows how generation intervals differ among breeds, being rather longer in the black-skinned breeds, and the interval from date of birth of a sire to that of his registered sons is longer than that relating to the dam.

Seasonality of breeding is least marked in the Gloucester Old Spot. This might be connected with this breed being the only rare breed to have been selected strongly for indoor husbandry (Foster 1983), but it is perhaps more likely (G. Cloke, personal communication) that the pronounced seasonality of the other breeds could occur because owners schedule breeding so as to have stock ready for the big summer and Christmastime agricultural shows. The lack of seasonality of the Gloucester Old Spot may be a reflection of its more "commercial" status, breeding being scheduled according to the market for pig meat.

The black-skinned rare breeds would perhaps have been expected to be used extensively for crossbreeding, producing white-skinned pigs for sale, to avoid butchers' prejudice against black pigs. Although the proportion of litters from which piglets are registered is indeed slightly lower than for the white-skinned breeds, it is not significantly so. Evidently the two groups of breeds do not differ in the extent to which they are crossed for commercial purposes with other breeds. In general, dif-

Table 8. Dependence of coefficients of kinship of pairs of animals upon whether they bore the same bloodline name, and upon whether they were bred in the same herd. Mean coefficients of kinship are in parentheses.

Breed	Number of males born in 1986 (n)	Number of pairwise relationships calculable (n (n − 1)/2)	Number of relationships calculable between pairs of animals of:		
			Same bloodline, regardless of herd of birth	Different bloodlines, same herd of birth	Different bloodlines, different herds of birth
Berkshire	8	28	11 (9.0%)	2 (4.7%)	15 (2.5%)
Large Black[a]	27	351	69 (12.2%)	50 (4.7%)	232 (3.9%)
British Saddleback	11	55	7 (8.7%)	0	48 (4.7%)

[a] *Differences among the Large Black samples were significant (chi² approximation of Kruskal-Wallis test: $chi^2 = 63.78$, $d.f. = 2$, $p < 0.001$).*

ferences among breeds in the ways they are structured are not marked; had there been pronounced differences, a conservation priority might have been indicated.

Relationships Among Herds

In rare sheep about a third of flocks supply males to other flocks (Hall 1986) and the proportion is similar in rare pigs (Table 6). It is lowest in the Gloucester Old Spot, and this reflects the dominance of the Ribbesford herd (Foster 1983). Only in the Gloucester Old Spot has a major herd supplied boars to more than half of the other herds. There is a strong tendency toward using homebred boars in the British Saddleback (Table 3), where individual major herds have not supplied boars to large numbers of other herds. This breed was formed only recently (1967) by the merger of the Essex and Wessex Saddlebacks, two breeds with widely separated centers of distribution and probably a rather local trade in boars as a result.

Inbreeding

The greater extent of the British Lop pedigree data is clear from Table 7, where the computer program, which traces ancestors to the sixth generation (Falco 1975), has found common ancestors for virtually all recent registrations of this breed. With a mean inbreeding coefficient of around 6% the rate of inbreeding is therefore about 1% per generation. This is higher than has been found in other, commercial breeds (Poland-China in the United States: 0.61% per generation, Lush & Anderson 1939; Large White in Australia: 0.59%, McPhee 1965; Danish Landrace: 0.50%, Jonsson 1971; Large White, Welsh, British Landrace: 0.52%, Smith et al. 1978). Pedigrees in the NPBA breeds do not go quite so far back, as far as the data used in this study are concerned; they probably represent four or five generations of ancestry and as such inbreeding is similar in rate to the 1% per generation found in the British Lop.

There is only space here to present results from the Large Black, but it is clear that outcrossing within the breed has the general effect of curbing inbreeding. In all breeds, animals whose parents were both bred in the same herd were more inbred than those whose parents had different origins. The breeders of British Lop pigs are perhaps particularly aware of this: of 187 pigs born in 1985 and 1986, only seven were by parents both bred in the same herd.

There was some evidence (S. J. G. Hall, unpublished data) that different herds have incurred different degrees of inbreeding. In the Large Black (S. J. G. Hall, unpublished results), particular bloodlines were probably not particularly inbred, and although the bloodline name by definition indicates a line of descent to the founder of the bloodline it only represents a concentration of genes from that ancestor if the latter is cited as a common ancestor in the pedigree.

This possible attribute of bloodlines does not seem to have been considered in the literature and it could mean that the bloodline concept has a biological reality that animal breeding scientists have overlooked. Practical pig breeders use the concept for at least part of the process of planning matings, and this is evident from Table 8, where the coefficients of kinship of animals with the same bloodline name were higher than those of animals in the same herd but of different bloodline name.

The planning of conservation measures should take this into account but the coefficients of kinship between "incentive boars" (see Introduction) and foundation boars show how (S. J. G. Hall, unpublished results) if the aim is to preserve or propagate the genes of an ancestral boar by the medium of semen from a present-day boar, assessments of kinship must be carried out. Indeed the genes of a particular ancestral male can sometimes be concentrated in a present-day female.

Responsibility for conservation of rare breeds in Britain rests firmly upon individual breeders, who usually apply other criteria besides those of genetic conservation to the planning of their breeding programs. The population parameters given in this paper summarize the genetic consequences of this. In a centrally run, planned program, the parameters would be expected to take on different values, probably more favorable to genetic conservation. However, public sympathy and support for genetic conservation may in some societies and cultures be more readily enlisted when the conservation is practised by individual breeders and this may represent an economic benefit to be set against the genetic costs.

In Britain, support for the conservation efforts of individual breeders has come in large measure from the RBST, and this support has been of a kind that conforms to the customs and procedures of the British pedigree livestock business. Registration of stock is particularly encouraged by the RBST because, for British pedigree breeders, without this documentation an animal is not accepted as being of any particular breed, however close it may be in appearance, although it will be acceptable as such for many commercial farmers. Registration is the first essential for the perpetuation of husbanded breeds. The RBST also supports the sale of breeding pigs, mainly through the medium of its annual Show and Sale, and this contributes to gene flow and to the limitation of inbreeding.

Storage of semen, the granting of incentives, surveys of the numerical status of breeds, and the periodic evaluation of the genetic consequences of these programs, also contribute to breed conservation.

This paper has presented evidence that these conservation efforts have been successful, though no prediction is made as to how far into the future the breeds can be maintained under the system described here. The single most important element in this kind of conservation program is the registration of stock of each breed with its breed society. This provides the documentation that underlies breed integrity, and permits the kind of assessment of which this paper is an example.

Acknowledgments

This study was suggested by Mr. M. M. Rosenberg, CBE, and supported by him through the Rare Breeds Survival Trust. The National Pig Breeders' Association, and Mr. Peter Cooper of the Royal Agricultural Society of England, together with the British Lop Pig Society, provided pedigree data. Dr. M. Wallace made available the program to calculate inbreeding. I am grateful to Dr. G. Wiener (AFRC Institute of Animal Physiology and Genetics Research), Mr. G. Cloke (Chairman of the National Pig Breeders' Association, and Vice President of the Rare Breeds Survival Trust), Mr. A. J. Dymond (chief executive, Rare Breeds Survival Trust), and two anonymous reviewers for their helpful comments on the manuscript. Mrs. D. Hughes drew the figures.

Literature Cited

Anonymous. 1986. Rare pig bloodline survey. Ark **13**:415–420.

Anonymous. 1987*a*. Rare breeds acceptance procedure. Ark **14**:12.

Anonymous. 1987*b*. Show and Sale report. Ark **14**:339–349.

Anonymous. 1988. Rare pig bloodline survey. Ark **15**:46–53.

Bekedam, I. M. 1959. Inbreeding in the Dutch swine-breeds. Zeitschrift fur Tierzuchtung und Zuchtungsbiologie **73**:218–220.

Bennett, B. 1987. Measures of relatedness. Ethology **74**:219–236.

Bodo, I. 1985. Hungarian activities in the conservation of domestic animal genetic resources. Animal Genetic Resources Information **4**:16–22.

Brooke, C. 1984. Organizational activities for the conservation of farm animals in the United States. Animal Genetic Resources Information **3**:31–36.

Crawford, R. D. 1984. Assessment and conservation of animal genetic resources in Canada. Canadian Journal of Animal Science **64**:235–251.

Davidson, H.R. 1966. The production and marketing of pigs. London: Longmans.

Falco, M. J. 1975. The calculation of inbreeding coefficients. Biometrics **31**:993.

Foster, J. 1983. Gloucestor Old Spots going strong. Ark **10**:83–85.

Hall, S. J. G. 1986. Genetic conservation of rare British sheep: the Portland, Manx Loghtan and Hebridean breeds. Journal of Agricultural Science, Cambridge **107**:133–144.

Hanset, R. 1973. Consanguinité et parenté chez le porc de Piétrain. Annales de Génétique et Sélection Animale **5**:177–188.

Helwig, J. T., and K. A. Council. 1979. SAS user's guide, 1979 ed. Cary, North Carolina: SAS Institute.

Henson, E. 1986. American update. Ark **13**:300–303.

Jacquard, A. 1974. The genetic structure of populations. New York: Springer-Verlag.

Jewell, P. A. 1985. Rare breeds of domestic livestock as a gene bank. Ark **12**:158–168.

Jonsson, P. 1971. Population parameter estimates of the Danish Landrace pig. Acta Agriculturae Scandinavica **21**:11–16.

Lush, J. L. (1943). Animal breeding plans. 2nd ed. Ames, Iowa: Iowa State College Press.

Lush, J. L., and A. L. Anderson. 1939. A genetic history of Poland-China swine. Journal of Heredity **30**:149–156.

Macdonald, D. W. 1985. The encyclopaedia of mammals. Vol. 2. London: George Allen & Unwin.

McPhee, C. P. 1965. Inbreeding, migration and structure of the pedigree Large White pig population in Australia. Australian Journal of Experimental Agriculture and Animal Husbandry **5**:270–278.

MLC. 1985. Pig yearbook 1985. Bletchley, England: Meat and Livestock Commission.

National Research Council 1983. Little-known Asian animals with a promising economic future. Washington, DC: National Academy Press.

Pond, W. G. 1983. Modern pork production. Scientific American **248**:78–87.

Smith, C., C. H. C. Jordan, D. E. Steane, and M. B. Sweeney. 1978. A note on inbreeding and genetic relationships among British tested pigs. Animal Production **27**:125–128.

Sokal, R. R., and F. J. Rohlf. 1981. Biometry. 2nd ed. San Francisco: W. H. Freeman.

Vu Tien Khang, J. 1983. Méthodes d'analyse des données démographiques et généalogiques dans les populations d'animaux domestiques. Génétique Sélection et Evolution **15**:263–298.

Wright, S. 1922. Coefficients of inbreeding and relationship. The American Naturalist **56**:330–338.

Diversity

Minor Breeds and Major Genetic Losses

During 1987, the American Minor Breeds Conservancy (Pittsboro, North Carolina) conducted a survey of feral livestock populations present in the United States. The survey incorporated herds on public lands, including properties dedicated to the preservation of native flora and fauna, such as the U.S. National Parks. According to a letter sent by Elizabeth Henson in March of 1987: "Feral populations of cattle, pigs, sheep, goats, horses, and burros may provide a valuable source of genetic variation not present in modern breeds of livestock.... Some traits may be of present or future commercial, scientific, historical or aesthetic value. They may also have novel or rare characteristics or adaptations. For example, these traits may include adaptations to extreme environmental conditions such as cold, drought or high parasite load or other characters of potential commercial or scientific value." The American Minor Breeds Conservancy is thus an advocate for the maintenance of feral populations as a potential genetic resource.

The conservation of these feral populations can, unfortunately, present a direct threat to the preservation of native species populations and gene pools. It is well known that feral goats and hogs have severely reduced or eliminated many species endemic to oceanic islands. Feral animals may also disturb continental ecosystems, thus causing intensive grazing, soil erosion, reduction of favored forage plants, and elimination of native fauna and their habitats. The problem is more than mere "overpopulation" of feral species. Some types of indigenous ecosystems may have low resistance to grazers, and very small populations of feral animals may have adverse impacts. Feral species may compete with native ungulates for limited resources, including food and water, and may spread diseases and parasites into native populations, even when the population of the feral species is well below its environmental "carrying capacity." Desert big horn sheep, for example, are easily displaced by feral burros. A small population of feral hogs, roaming unchecked, can eliminate a majority of the nests in a sea turtle rookery.

The conservation of minor breeds is certainly legitimate and has potential social benefits. The question becomes: are natural areas, such as national parks, appropriate places to accomplish this, and to what extent should feral animals be allowed to reside in habitats that otherwise have relatively low levels of anthropogenic disturbance? We can also ask: how much natural diversity dare we lose to support a feral species?

The situation is complicated by the goals of maintaining the genetic resources present in feral populations. If the feral animals are removed to a new site or returned to an agricultural habitat, the environmental conditions influencing the feral population will change, and new selection pressures may eventually affect the gene pool. To preserve the genetic resource through multiple generations, conservationists must leave the feral population in the habitat in which it developed or it must be moved to a similar location. Further, since many feral animal populations are relatively small (sometimes fewer than 200 individuals), preserving a few animals in a pen may eliminate "genetic resources." Feral animal populations are not usually prone to the problems associated with inbreeding, but one or two dozen animals (an expensive herd to support on hay) may not be enough to perpetuate a reproductively viable population for any length of time.

An irony of the situation is that some of the oldest and most biologicaly interesting feral animal populations are on islands or in remote habitats. These populations are also among the most damaging to indigenous species. The goats and wild pigs on Isla Mona, off the coast of Puerto Rico, have probably been there for centuries and have certainly experienced little recent outbreeding. These animals may be slowly destroying endemic plant populations and the pigs are certainly feeding on limited "genetic resources," such as hawksbill turtle nests (Anastasia Kontos, personal communication). Feral horses and ponies have existed on the coastal barrier islands of the eastern United States since colonial times, and some contemporary populations have distinctive morphological, and perhaps, genetic characteristics. Grazing by livestock on these islands has modified salt marsh structure, reduced forbs in interdune grasslands, and, in some areas, almost eliminated maritime forest.

Currently, the managerial questions concerning feral animal populations tend to be decided site by site. Even the U.S. National Park Service is not consistent. On the coastal barrier islands, the agency is maintaining some feral equid populations, while removing feral sheep, goats, cattle and hogs (the latter of which may be among the oldest emigrants to the islands). Research scientists and resources managers dedicated to the conservation of biological diversity need to establish some principles for resolving the potential conflict between maintaining both minor breeds and native species and ecosystems. To this end, I would like to suggest some policies for nature reserves, national parks, public lands and other similar sites:

(1) Feral animals should be removed from a natural area if there is any evidence that they are a threat to an endemic or endangered native species. If the feral population is genetically distinct, very ancient, or of some scientific or aesthetic value, individuals should be transported to another suitable site (not a natural area) or genetic materials should be otherwise preserved. A feral animal population can be replaced, at least at the specific level, but an extinct native species has no substitute.

(2) When a feral species is found in large numbers (several thousand animals) over a wide range, populations should, if possible, be removed from parks, natural areas and nature reserves.

(3) When a feral population is found over a very limited range (a single island), or has a small population, retention of a representative population in a natural area may be considered if:

   a. The feral population can be shown to be genetically unique through appropriate scientific technique, such as electrophoresis, analysis of mitochondrial DNA, or detailed morphological comparisons. The standards used should be similar to those for determining a local race or subspecies of a native animal.
   b. The feral population has been on the site for an extended period of time, at least a century and preferably longer, and has not been subject to recent outbreeding with populations from other localities. Folklore, such as tales of Spanish shipwrecks releasing horses, is not acceptable as verification of a venerable lineage, but genetic study and appropriate historic accounts (particularly written evidence) are.
   c. The feral population has some potential future value (historic, scientific or aesthetic).
   d. The feral population is not threatening any endemic or endangered species or any native species genetic resources (including subspecies and unique local populations) and is not threatening or disturbing any unique native communities or ecosystems.
   e. Any native plant or animal communities disturbed by the feral population are well represented in other parks or nature reserves.
   f. The feral population can be kept at a low level and no major damage to native ecosystems will occur, and the feral population is not spreading diseases or parasites to native species.

(4) The burden of proof should rest on the advocates for feral species not on the advocates for native species. Advocates for feral species must demonstrate that the animals are not damaging the populations of any endangered species present and are not causing major environmental deterioration. They must also prove the local feral animal population is genetically unique.

(5) Fencing may be considered in some cases where feral species can be successfully confined to a small portion of the habitat, and disturbance can be eliminated in key native ecosystems. When fencing is used, the feral animals should always be fenced into a resistant area, rather than fenced out of endangered species habitats or other sensitive areas. The majority of the natural area should be managed for native flora and fauna, not for feral species. Species should not be fenced if they can not be securely kept in enclosures (i.e., hogs and goats can escape and reestablish populations in sensitive areas).

On the public lands, a great deal of time and effort has already been expended proving feral species damage indigenous ecosystems and threaten native species. Although some of this work is valuable and should continue, an extensive "case" against a nonnative species should not be necessary before the species is removed from a national park or a nature reserve. Research efforts should center on more cost-effective and more humane population control strategies. If a park or nature reserve is considering maintaining a feral population, preliminary research should include genetic and historic study of the feral species, and very thorough surveys of rare and endangered native species within the feral species' range.

Under the policies suggested above, the management of feral animals would differ from area to area, but the direction of the programs would depend on the value and integrity of the genetic resources involved. For example, in the case of feral animal populations on southeastern United States seashores, Cumberland Island would remove feral horses because the herd is thought to be of recent origin and was outbred as recently as the 1950s. If the horse herd on Shackleford Bank (Cape Lookout) is found to be genetically distinct (a study is underway at the University of Georgia), the herd might be maintained if no native species are threatened by its presence and measures can be taken to protect native plant communities, particularly maritime forest, from grazing and trampling impacts. Continuing to maintain the herd of "Banker ponies" in a pen on Ocracoke Island would be appropriate, although the management strategy could be modified. Currently, a population of 15 to 20 individuals is maintained in a fenced enclosure. This is not, by itself, a reproductively viable group, and is expensive to keep because the ponies must be fed hay part of the year. This effort should either have a formal outbreeding program with other "Banker" herds, or should be reduced to a smaller group for interpretation to the public, and replacement animals could be obtained by purchase, as necessary. In the case of feral hogs, the Cumberland herd should be eliminated unless it can be shown to be genetically distinct from mainland razor backs. This is unlikely because the hogs have been outbred in recent decades. If the hogs are found to be genetically unique, the large amount of hog-initiated environmental disturbance on the island, and the potential threat to sea turtles and forest understory plants, would favor removing a representative population to some other site.

In managing "minor breeds," one feral animal population is not equal to another. There is very little literature to indicate which populations really are genetically distinct or valuable. Some populations are much younger than others, and some have been continually outbred with domestic livestock and escapes. It remains to be seen how commercially and scientifically useful some of these populations are. In 15 years of work with feral species, I have never reviewed a request for animals for breeding experiments or commercial stock improvement. I have encountered requests for wild hogs for stocking as game animals. It shold be noted that in this case, the preference is for obtaining European wild boar (or wild boar–feral hog hybrids) rather than obtaining feral hogs.

In conclusion, feral animals have such heavy impacts on so many native species, that managing feral animals in natural habitats must be conducted with great caution. If extinctions and declines in the number of native species continue to occur in important natural areas because of feral animal presence, the only viable policy for public lands may be no feral animals in any native ecosystem, period.

Susan P. Bratton
U.S. National Park Service
Cooperative Studies Unit, Institute of Ecology
University of Georgia
Athens, Georgia 30602, U.S.A.

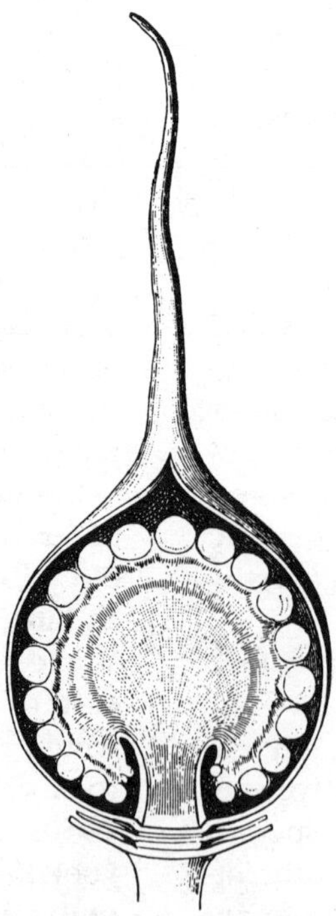

Genetic Conservation in Feral Populations of Livestock

DIRK VAN VUREN*

Department of Systematics and Ecology
University of Kansas
Lawrence, KS 66045, U.S.A.

PHILIP W. HEDRICK

Department of Biology
Pennsylvania State University
University Park, PA 16802, U.S.A.

Abstract: *Feral livestock may be sources of genetic variation with potential commercial, scientific, historical, or aesthetic value. Important variants may include primitive traits absent in modern breeds and novel or rare adaptations; the presence of these variants may be suggested directly by morphological markers, quantitative traits, fitness characters, or rare or unique alleles, and indirectly by the extent and duration of isolation, founder number, ancestry, or environmental conditions. Both direct and indirect evidence indicated potentially valuable traits in a population of feral sheep on Santa Cruz Island, California, and recently 18 lambs were captured to found a mainland population. We discuss several genetic goals important in maintaining a feral population, but caution that such goals must not conflict with the preservation of natural diversity threatened by feral livestock.*

Resumen: *El ganado asilvestrado puede ser fuente de variación genética con un potencial valor comercial, científico, histórico o estético. Variaciones importantes pueden incluír caracteristicas primitivas ausentes en razas modernas y en adaptaciones nuevas o raras. La presencia de estas variaciones puede ser señalada directamente por marcadores morfológicos, rasgos cuantitativos, características adaptativas, o por alelos raros o únicos e, indirectamente, por la amplitud y duración de aislamiento, tamaño de la población ancestral, linaje o condiciones ambientales. Evidencias directas e indirectas indican características potencialmente valiosas en una población de ovejas asilvestradas en la Isla de Santa Cruz, California. Recientemente, 18 ovejas fueron capturadas para iniciar una población en el continente. Discutimos varios objetivos genéticos que son importantes, para mantener una población asilvestrada, pero advertimos que tales objetivos no deben ser conflictivos con la preservación de la diversidad natural amenazada por ganado asilvestrado.*

Introduction

Genetic conservation usually focuses on maintaining genetic variation in rare, endangered, or captive species. Genetic conservation is also practiced by some agricultural scientists who for the past several decades have tried to maintain the genetic variation in domesticated plants and animals. In plant species, these conservation efforts include the maintenance of land races, collections of samples from wild relatives, and sometimes collections from naturalized populations.

For domesticated livestock, however, nearly all the close wild relatives have been exterminated. Feral livestock (domesticated animals that have reverted to the wild state), such as cattle (*Bos taurus*), pigs (*Sus scrofa*), sheep (*Ovis aries*), goats (*Capra hircus*), horses (*Equus caballus*), and donkeys (*Equus asinus*), offer potential sources of genetic variation that may not

* *Correspondence should be sent to this author.*
Paper submitted 9/14/88; revised manuscript accepted 2/22/89.

be present in modern breeds of livestock. The first coordinated efforts to identify and preserve valuable traits of feral animals began in New Zealand in the mid 1970s (Whitaker & Rudge 1976). Subsequent research discovered possible genetic differences between feral animals and their domestic conspecifics (Heath 1983; Rudge 1983; Orwin & Whitaker 1984) and led to the preservation of some remnant feral populations in New Zealand (Rudge 1986*a*). These efforts have begun to generate interest elsewhere (e.g., Munton et al. 1984), but at present few populations have been surveyed for their possible genetic value.

We begin with an outline of a logical genetic basis for conservation of populations of feral livestock and discuss factors that may help identify particularly important feral populations. Second, we describe a recent effort to conserve feral sheep from Santa Cruz Island, California, U.S.A., and the genetic basis for it. Finally, we summarize the genetic goals of maintaining a feral population and discuss the constraints imposed by the overriding importance of preserving natural diversity threatened by feral livestock.

Genetic Basis for Conservation

Populations of feral livestock may have two primary categories of attributes that could be of significance for genetic conservation (Table 1). First, feral populations may have relict characteristics or genetic variants that either are absent in modern breeds or exist only in rare or minor breeds of livestock. Some of these traits may be of present or future commercial or scientific value. In addition, feral populations may have traits of historical or aesthetic value not found in modern domestic livestock. Examples are the Soay sheep on the island of Hirta, Scotland, which display the characteristics of unimproved and primitive sheep (Doney et al. 1974) possibly introduced by Vikings (Morton Boyd & Jewell 1974), and feral cattle on the island of Swona, Scotland, which may be an unusual relict breed of triple-purpose (beef, milk, draught) cattle (Hall & Moore 1986).

Table 1. Categories of attributes that may be present in feral populations, and their possible indicators.

Attributes	Relict characteristics or genetic variants
	Novel or rare adaptations
Indicators	
Direct	Morphological markers
	Quantitative traits
	Fitness characters
	Rare or unique alleles
Indirect	Generations since isolation
	Extent of isolation
	Founder size
	Effective population size
	Ancestral breeds
	Physical and biotic environment

Second, feral populations may have novel or rare characteristics or adaptations. These traits may include adaptations to extreme environmental conditions such as temperature stress, drought, high parasite load, or other characters of potential commercial or scientific value. The presence of these traits in feral populations may be the result of extreme selection pressures increasing the frequency of very rare genetic types; of mutation within the population followed by subsequent selection genetic drift, or both; or of introgression from domestic or other feral populations followed by recombination, selection, genetic drift, or some combination of these. Additionally, feral populations might contain individuals that have both relict and novel attributes.

An example of an adaptation of commercial significance has been found in the Louisiana native sheep derived from the white-faced scrub sheep found along the Gulf Coast of the United States until the late 1800s. Comparison of the native sheep with Suffolk sheep showed that the native sheep had significantly lower parasite populations throughout most of the year (Fig. 1; Lemarie et al. 1987). Because gastrointestinal nematode infection is considered the main sheep production problem in Louisiana, the nematode resistance found in native sheep has potential economic significance.

An example of a character useful in scientific research is found in the feral pigs of Ossabaw Island, a coastal barrier island off of Georgia, U.S.A. It appears that this population has been isolated for over 400 years and that few additional animals have been introduced throughout its history (Mayer & Brisbin 1989). The Ossabaw Island pigs are genetically distinct from other feral and domestic populations (Smith et al. 1980) and have a unique lipid physiology that makes them useful in medical and nutritional research (e.g., Wangsness et al. 1977; Buhlinger et al. 1978; Stribling et al. 1984).

How can we determine directly or indirectly if a feral

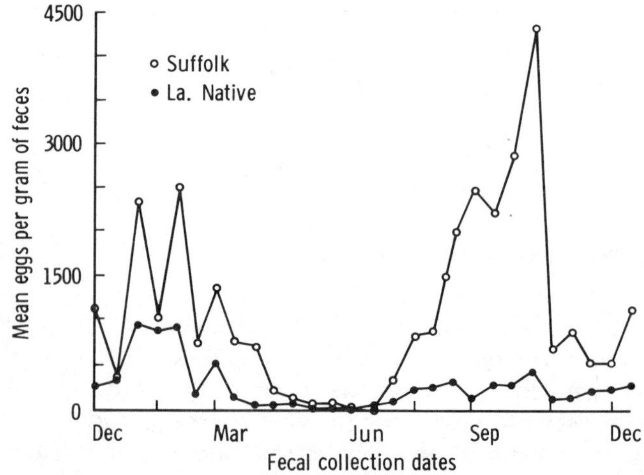

Figure 1. Mean number of gastrointestinal nematode eggs per gram of feces in Louisiana native and Suffolk sheep. Redrawn from Lemarie et al. (1987).

population might have important genetic characteristics? Direct indicators are best (Table 1). First, morphological markers such as genetic variants affecting hair color, coat length, or presence of horns may indicate ancestry or adaptation. On Pitt and Arapawa islands, New Zealand, feral sheep have pigmented wool that is self-shedding; their domestic ancestors were nonshedding white sheep (Rudge 1983; Orwin & Whitaker 1984).

Second, quantitative or fitness characters such as body size, degree of sexual dimorphism, litter size, or breeding season may indicate significant differences in ancestry or selective pressure. Feral sheep from four populations in different parts of the world weighed considerably less than domestic sheep, but weights were remarkably consistent among the four populations (Orwin & Whitaker 1984). Other traits, such as resistance to particular parasites or pathogens, could also indicate important adaptations. When evaluating these characters one should be careful to account for environmental differences. For example, low parasite load in feral animals may result from lack of intermediate hosts, and small body size might result from poor nutrition. Only by comparison with other breeds in a common environment (as in Fig. 1) could differences in parasite resistance or body size be determined.

Finally, if the population is found to have rare or unique alleles when blood group, electrophoretic, or other genetic markers are examined, then it may have a different ancestry from extant breeds, or possibly be a mutation that has subsequently increased by selection or genetic drift. On the other hand, the absence of unique genetic markers does not necessarily indicate the absence of important or unique adaptive characteristics because it is usually assumed that biochemical variants are not adaptive.

If direct information about the population is not immediately available, then indirect factors may be useful as indicators of possible genetic difference. First, the number of generations since the feral population was isolated from domestic populations may indicate the likelihood of novel adaptations evolving (as in the Ossabaw swine). Second, if isolation has been nearly complete, many of the characteristics of the founders should be retained, undiluted. Note, however, that interbreeding between feral and domestic populations need not result in a change in the feral gene pool if the hybrids do not become part of the mating feral population. Feral and domestic sheep coexisted on Pitt Island for many years, but interbreeding was rare, in part because of behavioral differences (Rudge 1983).

Third, the size of the founder population and the effective population size since the establishment of the feral population may be significant. For example, a small initial population, low effective numbers in subsequent generations, or both may result in genetic drift, which in turn may result in chance fixation of unusual characteristics. Feral goats on one island of Aldabra Atoll had just two coat color phenotypes, both probably associated with recessive alleles at a single locus, whereas on an adjacent island 10 different phenotypes were identified (Van Vuren & Coblentz 1988). Fourth, if a feral population was known to have originated from a now-extinct breed, it may contain important relict variants. An example is the Soay sheep of St. Kilda.

Finally, the type of environment in which the feral population lives may indicate potential significance. Extreme climatic conditions such as temperature stress or drought, interspecific interactions with predators, parasites, or competitors (either wild or domestic), or intense intrapopulation competition for food or mates may suggest that adaptations to these factors may be present. Rudge (1986*b*) proposed that population changes among feral sheep on Campbell Island, New Zealand, resulted in part from genetic adaptation to heat stress. In the same vein, past or present management practices such as hunting may be important selective pressures. Hunters of feral sheep on Santa Cruz Island preferentially shot horned males but avoided shooting sheep with black or brown wool (R. Bello, personal communication).

Santa Cruz Island Sheep

Recently the American Minor Breeds Conservancy (AMBC), with the help of The Nature Conservancy (TNC), has established a mainland population of feral sheep removed from Santa Cruz Island (SCI). Santa Cruz Island, located 40 km south of Santa Barbara, California, has supported a large population of sheep that have been feral for at least 70 years (Van Vuren & Coblentz 1987). Because of the ecological damage these sheep have caused, numbers have been reduced on the western 88 percent of the island from an estimated 21,240 in 1980 (Van Vuren & Coblentz 1989) to fewer than 40 by March 1987 (Schuyler 1989). The eventual goal of TNC, owner of the western 88 percent of the island, is the complete eradication of the sheep (Schuyler 1989).

In January 1988 a team of seven people from AMBC and TNC was able to catch 12 lambs (7 female and 5 male) and remove them successfully from SCI to form the basis of a mainland population. A subsequent attempt by TNC in November 1988 resulted in the capture of six more lambs (2 female and 4 male). All 18 lambs were taken from an area near the eastern end of the island called No Man's Land, a region that still contains a large sheep population. With a proper breeding scheme that attempts to avoid inbreeding and equalizes founder contribution, and by rapidly expanding population size, these sheep should be representative of the attributes of the feral SCI sheep.

What are the attributes of these sheep? From the 1988 capture efforts and from Van Vuren & Coblentz (1984, 1987, 1989), several characters may help indicate ancestry. For example, the wool of SCI sheep is very fine, 56 percent of adult males are horned, and only about 1 percent of the sheep have colored wool (ranging in extent from a few small, rusty spots to entirely black); these attributes are shared by some contemporary breeds. On the other hand, in several characters these sheep may differ from many others. First, many SCI sheep have only short wool on the legs, face, or belly. More dramatic, the "rat tail" (tail with very short wool) is relatively common in these sheep (Table 2) but apparently is absent in all common domestic breeds. It is possible that this is an adaptive, derived character that confers resistance to flystrike and related infections. Furthermore, the sheep are much smaller than domestic sheep, they lamb almost exclusively in December and January, and they consume a substantial amount of browse.

What is known historically about these sheep or can be concluded about their environment? The sheep were on SCI by 1850, when a sheep ranch had been firmly established. Most of the 50,000 sheep in 1890 were categorized as Merino (Towne & Wentworth 1945; see also Brumbaugh 1980). An introduction of 10,000 domestic sheep in the 1930s is of unknown breed composition (C. Stanton, personal communication), but may have been Churro (G. Spurlock, personal communication). In any case, there appears to have been a substantial founder number, predominantly Merino.

The environment of SCI appears to have been quite harsh. Although the island is large and vegetation is diverse, strongly seasonal precipitation means that there is virtually no growth of herbaceous vegetation from May through October. Moreover, the density of sheep in 1980 averaged as high as 2.1/ha, over twice the normal stocking rate for domestic sheep in mainland California (Van Vuren & Coblentz 1989). The sheep had denuded much of the landscape so that there was much less forage than in managed mainland pastures (Van Vuren & Coblentz 1987). In addition, they had been hunted for many years by a hunt club and by the previous owner of the island, Santa Cruz Island Company.

Overall, the SCI sheep may have some adaptive characteristics that could be important for commercial use. In particular, their exceptional foraging ability and high lambing rate on poor forage may be advantageous on the mainland. Although tails on most domestic sheep are docked, having rat-tailed sheep may be a cost-effective alternative to docking. Research on wool, milk, and other characteristics of SCI sheep is now being planned (M. E. Stanley, personal communication).

Table 2. Frequency of the "rat tail" among feral sheep from three geographically distinct localities on Santa Cruz Island, California.

Locality	n	*Frequency*
Central Valley	76	0.13
Lagunitas Secas	60	0.55
No Man's Land	25	0.48

Methods and Goals of Genetic Conservation

To maintain a feral population and its genetic variation, there are four general alternatives:

(1) maintain in situ as a free-living population
(2) maintain in situ but in a restricted area
(3) move the population to less sensitive land but allow it to be either free living or restricted
(4) maintain a captive population

An example of alternative (1) is the ca. 1100–1500 Soay sheep that live unrestricted on the 637-ha island of Hirta (Morton Boyd 1974). An example of alternative (2) is on 6500-ha Pitt Island, which until recently supported about 3000 feral sheep. The population was surveyed for color, sex ratio, and horn frequency, a founding group of 305 representative animals was fenced inside a 200-ha scientific reserve, and the remainder of the sheep were eradicated (Rudge 1983). Additionally, some feral horse populations are being maintained in situ on the coastal barrier islands of the eastern United States (Bratton 1988), and Berger (1986) has proposed several in situ reserves for feral horses in the western United States.

Alternatives (1) and (2) are the most desirable from the standpoint of retaining the genetic variation in the population; however, for most feral populations, including virtually all those living on islands, in situ preservation is not an acceptable option because of the severe impacts that feral livestock have on native ecosystems. These impacts are well documented (e.g., Bratton 1975; Spatz & Mueller-Dombois 1975; Coblentz 1978; Scowcroft & Giffin 1983; Fowler de Neira & Roe 1984; Singer et al. 1984; Scowcroft & Hobdy 1987; Van Vuren & Coblentz 1987) and are particularly devastating on island ecosystems, which typically are rich in endemic taxa that may lack defenses against large mammalian herbivores. The risks posed by in situ preservation are serious; moreover, these risks probably cannot be assessed adequately without first removing all feral livestock. For example, an insular endemic plant species (*Arabis hoffmannii*) thought to be extinct on SCI reappeared after eradication of feral sheep (R. C. Hansen, personal communication), and a previously undescribed plant species (*Canavalia kauensis*) was discovered on Hawaii after exclosure of feral goats (Baker & Reeser 1972).

We agree with Bratton's (1988) cogent review of the dangers of maintaining in situ feral populations, and we strongly support her suggested management policies. Preservation of natural diversity threatened by feral livestock is of paramount importance.

Alternative (3) might be a workable option for some feral populations. For example, long-range plans for SCI sheep include establishing a population in a large pasture with climate and vegetation generally similar to SCI. The last alternative, maintaining the feral population in close confinement, is almost a contradiction in terms (Rudge 1986*a*). In most cases, though, this is the only immediate possibility, as in the case of the SCI sheep.

Several genetic conservation goals are important in designing a program for a captive feral population:

(1) retain genetic variation by maximizing effective population size (N_e), avoiding inbreeding, and equalizing founder contribution
(2) retain unusual characteristics of the feral population
(3) allow continued selection for feral traits
(4) avoid selection for nonferal traits

The maintenance of the important characters in feral populations initially should be based on the same principles as genetic conservation in other populations (e.g., Hedrick 1989). Genetic variation is best retained by maximizing N_e (Lande & Barrowclough 1987), minimizing inbreeding, or equalizing founder contribution. Of course all of these approaches are means to reduce the extent of genetic drift, and in many instances they are similar.

In addition, other goals similar to those in breeding captive endangered species must be kept in mind (see Frankham et al. 1986). For example, one may wish to directly preserve particular traits, such as the rat tail in SCI sheep, to allow continued feral selection, and to avoid selection for some domestic traits. Obviously these goals may sometimes conflict if, for instance, retaining unusual characteristics results in inbreeding. As a result, a conservation-breeding-management program must be carefully constructed to balance these goals.

Conclusion

The evidence we have reviewed suggests that some feral populations of livestock may constitute potentially valuable genetic resources. Organizations such as AMBC have shown an interest in conserving these resources, and several efforts to do so have resulted. These efforts, of course, must not conflict with the preservation of natural diversity threatened by feral livestock.

Interest in conserving feral populations appears to be most developed in New Zealand, where the extent of feral populations is well documented, some feral animals have been used for commercial purposes, and several populations have been preserved (Rudge 1986*a*). In North America, a recently completed livestock census (Henson 1986) mentions some feral populations, and Mayer & Brisbin (1989) have completed a thorough study of feral pigs. The AMBC has initiated an extensive inventory of feral populations and is now in the process of cross-checking their findings with those of an earlier survey by McKnight (1964). This inventory might serve as a starting point for identifying potentially valuable feral populations.

Acknowledgments

We thank I. L. Brisbin, Jr., B. E. Coblentz, J. L. Koprowski, and M. R. Rudge for helpful criticisms of an earlier draft of the manuscript.

Literature Cited

Baker, J. K., and D. W. Reeser. 1972. Goat management problems in Hawaii Volcanoes National Park. Natural Resources Report No. 2, National Park Service, U.S. Department of the Interior, Washington, D.C.

Berger, J. 1986. Wild horses of the Great Basin. University of Chicago Press, Chicago, Illinois.

Bratton, S. P. 1975. The effect of the European wild boar, *Sus scrofa*, on gray beech forest in the Great Smoky Mountains. Ecology **56:**1356–1366.

Bratton, S. P. 1988. Minor breeds and major genetic losses. Conservation Biology **2:**297–299.

Brumbaugh, R. W. 1980. Recent geomorphic and vegetal dynamics on Santa Cruz Island, California. Pages 139–158 in D. M. Power, editor. The California Islands: proceedings of a multidisciplinary symposium. Santa Barbara Museum of Natural History, Santa Barbara, California.

Buhlinger C. A., P. J. Wangsness, R. J. Martin, and J. H. Ziegler. 1978. Body composition, *in vitro* metabolism and skeletal muscle characteristics in fast-growing lean and in slow-growing obese pigs at equal age and weight. Growth **42:**225–236.

Coblentz, B. E. 1978. The effects of feral goats (*Capra hircus*) on island ecosystems. Biological Conservation **13:**279–286

Doney, J. M., M. L. Ryder, R. G. Gunn, and P. Grubb. 1974. Colour, confirmation, affinities, fleece and patterns of inheritance in the Soay sheep. Pages 89–125 in P. A. Jewell, C. Milner, and J. Morton Boyd, editors. Island survivors: the ecology of the Soay sheep of St. Kilda. Athlone Press, London, England.

Fowler de Neira, L. E., and J. H. Roe. 1984. Emergence success of tortoise nests and the effect of feral burros on nest success on Volcan Alcedo, Galapagos. Copeia **1984:**702–707.

Frankham, R., H. Hemmer, O. A. Ryder, E. G. Cothran, M. E. Soulé, N. P. Murray, and M. Synder. 1986. Selection in captive populations. Zoo Biology **5:**127–138.

Hall, S. J. G., and G. F. Moore. 1986. Feral cattle of Swona, Orkney Islands. Mammal Review **16:**89–96.

Heath, A. C. G. 1983. Parasites of the feral sheep of Pitt Island, Chatham group. New Zealand Journal of Zoology **10:**365–370.

Hedrick, P. W. 1989. Genetic conservation in captive populations and endangered species. In S. Jain and L. Botsford, editors. Applied population biology. Martinus Nijhoff/Dr. W. Junk, Dordrecht, Netherlands. (In press.)

Henson, E. 1986. North American livestock census. American Minor Breeds Conservancy, Pittsboro, North Carolina.

Lande, R., and G. F. Barrowclough. 1987. Effective population size, genetic variation, and their use in population management. Pages 87–123 in M. Soulé, editor. Viable populations for conservation. Cambridge University Press, Cambridge, England.

Lemarie, S. L., F. G. Hembry, J. E. Miller, T. Gray, and L. C. Staggs. 1987. Sheep susceptibility to internal parasites and efficacy of some anthelmintics. Louisiana Agriculture **30:**12–21.

Mayer, J. J., and I. L. Brisbin, Jr. 1989. Wild pigs in the United States. University of Georgia Press, Athens, Georgia. (In press.)

McKnight, T. 1964. Feral livestock in Anglo-America. University of California Publications in Geography **16:**1–78.

Morton Boyd, J. 1974. Introduction. Pages 1–7 in P. A. Jewell, C. Milner, and J. Morton Boyd, editors. Island survivors: the ecology of the Soay sheep of St. Kilda. Athlone Press, London, England.

Morton Boyd, J., and P. A. Jewell. 1974. The Soay sheep and their environment: a synthesis. Pages 360–373 in P. A. Jewell, C. Milner, and J. Morton Boyd, editors. Island survivors: the ecology of the Soay sheep of St. Kilda. Athlone Press, London, England.

Munton, P. N., J. Clutton-Brock, and M. R. Rudge, editors. 1984. Feral mammals — problems and potential. IUCN, Gland, Switzerland.

Orwin, D. F. G., and A. H. Whitaker. 1984. Feral sheep (*Ovis aries* L.) of Arapawa Island, Marlborough Sounds, and a comparison of their wool characteristics with those of four other feral flocks in New Zealand. New Zealand Journal of Zoology **11:**201–224.

Rudge, M. R. 1983. A reserve for feral sheep on Pitt Island, Chatham group, New Zealand. New Zealand Journal of Zoology **10:**349–364.

Rudge, M. R. 1986*a.* A role for feral mammals in conserving the genetic diversity of livestock. Pages 7–23 in Animal genetic resources information. United Nations Environment Programme, FAO, Rome.

Rudge, M. R. 1986*b.* The decline and increase of feral sheep (*Ovis aries* L.) on Campbell Island. New Zealand Journal of Ecology **9:**89–100.

Schuyler, P. 1989. Control of feral sheep on Santa Cruz Island. In F. G. Hochberg, editor. Recent advances in California Islands research: proceedings of the third California Islands symposium. Santa Barbara Museum of Natural History, Santa Barbara, California. (In press.)

Scowcroft, P. G., and J. G. Giffin. 1983. Feral herbivores suppress mamane and other browse species on Mauna Kea, Hawaii. Journal of Range Management **36:**638–645.

Scowcroft, P. G., and R. Hobdy. 1987. Recovery of goat-damaged vegetation in an insular tropical montane forest. Biotropica **19:**208–215.

Singer, F. J., W. T. Swank, and E. E. C. Clebsch. 1984. Effects of wild pig rooting in a deciduous forest. Journal of Wildlife Management **48:**464–473.

Smith, M. W., M. H. Smith, and I. L. Brisbin, Jr. 1980. Genetic variability and domestication in swine. Journal of Mammalogy **61:**39–45.

Spatz, G., and D. Mueller-Dombois. 1975. Succession patterns after pig digging in grassland communities of Mauna Loa, Hawaii. Phytocoenologia **3:**346–373.

Stribling, H. L., I. L. Brisbin, Jr., J. R. Sweeney, and L. A. Stribling. 1984. Body fat reserves and their prediction in two populations of feral swine. Journal of Wildlife Management **48:**635–639.

Towne, C. W., and E. N. Wentworth. 1945. Shepherd's empire. University of Oklahoma Press, Norman, Oklahoma.

Van Vuren, D., and B. E. Coblentz. 1984. Impacts and adaptations of feral sheep on Santa Cruz Island, California, USA. Pages 43–53 in P. N. Munton, J. Clutton-Brock, and M. R. Rudge, editors. Feral mammals — problems and potential. IUCN, Gland, Switzerland.

Van Vuren, D., and B. E. Coblentz. 1987. Some ecological effects of feral sheep on Santa Cruz Island, California, USA. Biological Conservation **41:**253–268.

Van Vuren, D., and B. E. Coblentz. 1988. Dental anomalies of feral goats (*Capra hircus*) on Aldabra Atoll. Journal of Zoology, London **216:**503–505.

Van Vuren D., and B. E. Coblentz. 1989. Population characteristics of feral sheep on Santa Cruz Island. Journal of Wildlife Management **53:**306–313.

Wangsness, P. J., R. J. Martin, and J. H. Gahagan. 1977. Insulin and growth hormone in lean and obese pigs. American Journal of Physiology **233:**E104–E108.

Whitaker, A. H., and M. R. Rudge, editors. 1976. The value of feral farm mammals in New Zealand. Information Series, vol. 1, New Zealand Department of Lands and Survey, Wellington.

Radish as a Model System for the Study of Engineered Gene Escape Rates Via Crop-Weed Mating

TERRIE KLINGER

Department of Botany and Plant Sciences and Program in Genetics
University of California
Riverside, California 92521-0124, U.S.A.

DIANE R. ELAM

Department of Botany and Plant Sciences and Program in Genetics
University of California
Riverside, California 92521-0124, U.S.A.

NORMAN C. ELLSTRAND*

Department of Botany and Plant Sciences and Program in Genetics
University of California
Riverside, California 92521-0124, U.S.A.

Abstract: *Engineered genes in transgenic crops may escape into the ambient environment via crop-weed hybridization. However, natural crop-weed mating rates (gene flow) are largely unknown. We measured mating between wild and cultivated radishes in an experiment that simulated natural stands around seed multiplication plots. We used a genetic marker to identify crop-weed mating events. Although weeds at the cultivar plot margin (1 m distance) received much more gene flow than distant plants, detectable gene flow occurred at our most distant site (1000 m). For insect-pollinated, outcrossing crops like radish, strategies other than distance must be employed to ensure complete isolation.*

Resumen: *Los genes fabricados en los cultivos transgénicos pueden escapar al medio ambiente a través de la hibridización entre los cultivos y las malezas. Sin embargo, casi se desconocen las tasas de fertilización natural (flujo de genes) entre los cultivos y las malezas. Medimos la fertilización entre rábanos silvestres y cultivados en un experimento que simulaba los plantíos que naturalmente rodean los parcelas de multiplicación de semillas. Usamos un marcador genético para identificar la incidencia de fertilización entre malezas y cultivos. Aunque las malezas que estaban en los márgenes del cultivo (a 1 m de distancia) recibieron un flujo genético mucho mayor que las plantas más lejanas, hubo flujo genético detectable en nuestro sitio más distante (1000 m). Para los cultivos polinizados por insectos en cultivos con caracteristicas como las del rábano, hay que usar estrategias distintas a la de la distancia para asegurar un aislamiento completo.*

Molecular biologists, ecologists, and evolutionary geneticists have suggested that the primary risk of releasing genetically engineered crops is the sexual transfer of engineered genes to related wild plants via natural cross-pollination (e.g., Colwell et al. 1985; Ellstrand 1988; Ellstrand & Hoffman 1990; Goodman & Newell 1985; Regal 1988). Traits targeted for transfer to crops by biotechnologists, such as salinity tolerance and herbicide resistance (Gasser & Fraley 1989), are traits that could increase aggressiveness in natural species (Colwell et al. 1985; Ellstrand 1988; Ellstrand & Hoffman

* *Correspondence should be addressed to this author.*
Paper submitted August 14, 1990; revised manuscript accepted June 25, 1991.

1990). If crops mate with wild relatives (e.g., cultivated sunflowers with wild sunflowers), the resulting aggressive hybrids could present an economic threat as weeds or an environmental threat as competitors in natural assemblages. Dozens of genetically engineered crops are either being field tested or are awaiting field tests (Gasser & Fraley 1989). Data on rates of crop-weed mating are critical for the safe release of such crops (Ellstrand & Hoffman 1990; Tiedje et al. 1989). While sexually compatible crops and related wild species co-occur frequently in agroecosystems (Ellstrand 1988; Ellstrand & Hoffman 1990; National Research Council 1989), little is known of natural hybridization rates between them (Ellstrand & Hoffman 1990; but see Langevin et al. 1990; Wilson 1990). Furthermore, virtually nothing is known of the ecological factors that modify such rates (Ellstrand 1988; Ellstrand & Hoffman 1990). Therefore, we created a model experimental system employing natural, innocuous genetic markers to measure the rates of mating between crop plants and their wild relatives under agricultural conditions.

We chose cultivated and wild radish (*Raphanus sativus* L.; Brassicaceae) for our model species because radish is outcrossing an insect-pollinated (Panetsos & Baker 1968; Stanton 1987). This mating system is common to many varied cultivated species, including grains (e.g., rye, buckwheat), vegetables (e.g., cabbage, carrot, cucumber, pepper), oil crops (e.g., sunflower, canola), fruits (e.g., apple, melon), forage crops (e.g., alfalfa), herbs (e.g., sage), and ornamentals (e.g., zinnia) (Fryxell 1957; Frankel & Galun 1977). The ecological factor we studied, the distance separating the crop from the wild species, was chosen because physical isolation is the most frequently cited method suggested to prevent transgenic crops from mating with their sexually compatible relatives (see Kareiva & Morris 1991; Manasse & Kareiva 1990).

Materials and Methods

Experimental stands of a radish cultivar (Round White, W. A. Burpee Co.) were planted in January 1989 at the University of California's South Coast Field Station in Irvine and its Moreno Valley Field Station in Moreno Valley. The field stations lie about 100 km apart and differ in annual rainfall and in seasonal and daily temperatures. At each site, the cultivar was planted and maintained to simulate breeder's blocks for seed multiplication plantings (George 1985). Both locations were well isolated from commercial plantations (>>10 km). Replicated experimental populations of nine wild radishes grown from seed or seedlings collected from natural populations were planted in 3 × 3 arrays with 1 m spacing, at varying distances from the crop: 200, 400, 600, and 1000 m, and at the crop margin (1 m). All naturally occurring wild radishes at both stations were eradicated.

Our marker was a codominant allozyme locus (leucine amino-peptidase: *Lap*). This gene is expressed in both parents and progeny (Ellstrand & Devlin 1989). The cultivar is homozygous for *Lap6*, an allele apparently absent in natural populations surrounding each field station (T. Klinger, unpublished data). Prior to planting in the field, the experimental wild plants were first prescreened to be homozygous for *Lap2*, homozygous for *Lap3*, or heterozygous for these alleles. By June 1989, fruits had matured on the wild plants. Thirty-six seeds were harvested from each surviving wild plant and genotyped. Seeds heterozygous for *Lap6* allele represented the escape of cultivar genes.

Data were analyzed as follows:

Incidence of gene flow as a function of distance was estimated by the presence or absence of the crop marker among seed progeny of target plants on a per-plant basis at each of the two field stations. These data were then subject to the chi-square test for homogeneity of variance (Snedecor & Cochran 1967).

Rate of gene flow as a function of distance was estimated as the proportion of hybrid (crop × weed) seeds per target plant. These data were also subject to the chi-square test. Finally, to account for our fixed number of target weeds arranged along concentric circles of varying circumference, we multiplied the frequency measurements by πd (where d is the diameter). This correction gives a more accurate estimation of relative pollen dispersal across a two-dimensional (planar) surface containing other plants that share pollinators with the radish crop and target weeds (Grosberg 1991; Kareiva & Morris 1991).

Results and Discussion

Crop-weed hybridization rates varied greatly among plants and plots (Figs. 1a and 1b). Hybridization rates for immediately adjacent weeds (1 m distance) ranged from 14% to 100%. According to our one-dimensional analysis, rates declined sharply at greater distances; however, the greatest distance (1000 m) did not give complete isolation. Both the incidence and rate of crop-weed hybridization varied significantly with distance (Table 1). Average frequencies of the crop marker among seed progeny are presented graphically in Figures 1a and 1b; the results of one-way analysis of variance following arc-sin square-root transformation (Zar 1984) are reported in the figure legend. These results are consistent with the results of the chi-square test;

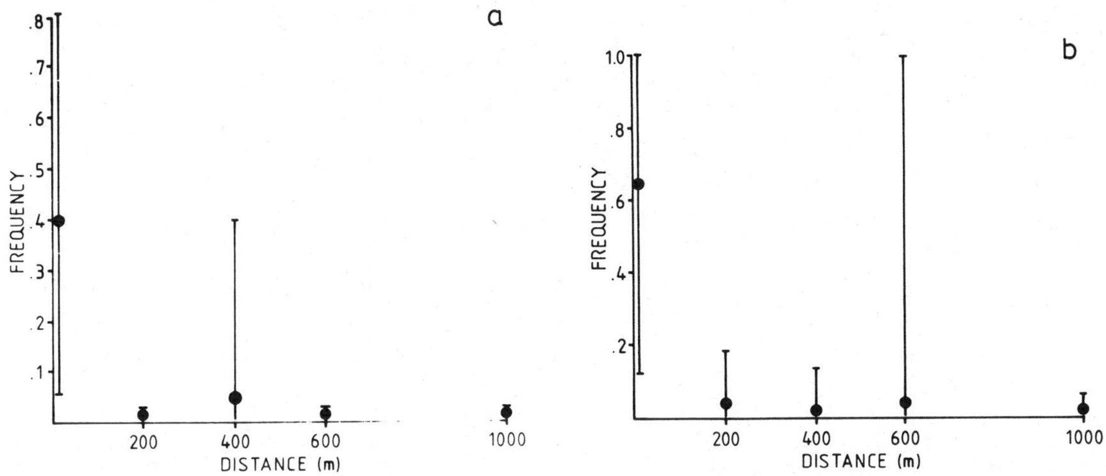

Figure 1. Frequency of the radish crop marker allele as measured among seed progeny of target weeds. Values given represent the mean ± 1 S.D. for all replicates at each distance. a) at South Coast Field Station. b) at Moreno Valley Field Station. Results of ANOVA for frequency vs. distance were significant: South Coast Field Station, P = .0001, D.F. = 4; Moreno Valley Field Station, P = .0001, D.F. = 4.

however, the chi-square test requires fewer assumptions regarding normality of the data.

In contrast to our one-dimensional analysis, estimates of relative pollen dispersal across the two-dimensional landscape were greater at distance (200 to 1000 m) than along the crop margin (1 m) (Figs. 2a and 2b). These values do not represent measured gene flow; rather, they indicate the relative amount of pollen necessary to effect the rates of gene flow that we observed at each distance.

Ecological and genetic variables may account for the observed variability in gene flow incidence and rates. Variation in the number of plants per plot, the number of flowers produced per plant or per plot, the position of the weeds relative to other crop species, and pollinator behavior could all influence gene flow (Handel 1983). For example, if pollinators spend more time in plots with more plants, gene flow is expected to decrease. However, in this experiment, gene flow per plant did not vary with the number of surviving plants per plot. Also, different levels of compatibility between the crop and individual weeds could explain gene flow variation from plant to plant within a plot. However, infrequent stochastic events and our limited ability to detect them could also explain this pattern. Therefore, more experimental work will be necessary to identify sources of gene flow variation. These factors might be exploited to reduce the opportunities for the escape of engineered genes (Ellstrand 1988; Ellstrand & Hoffman 1990).

Table 1. Results of the chi-square test for homogeneity of variance.

	Parameter	
Site	*Incidence*	*Rate*
South Coast Field Station	$\chi^2 = 63.80$ $p < 0.005$ D.F. = 4	$\chi^2 = 1046.0$ $p < 0.005$ D.F. = 4
Moreno Valley Field Station	$\chi^2 = 49.13$ $p < 0.005$ D.F. = 4	$\chi^2 = 864.5$ $p < 0.005$ D.F. = 4

Little is known of gene exchange patterns for other crop-weed pairs. Two prior studies are relevant. Gene flow between cultivated *Cucurbita pepo* and wild *C. texana* (Kirkpatrick & Wilson 1989) can occur at long distances (exceeding 1300 m) and can vary greatly among recipients. Gene flow among rice cultivars and a locally sympatric conspecific weed varied from 1.0% to 52% in mixed stands (Langevin et al. 1990). Neither study specifically addressed the role of distance as a factor in gene flow variation.

These data demonstrate that gene flow can occur between crop plants and their wild relatives in agricultural situations. Such gene flow can act as an avenue for the escape of engineered genes. For radish, spatial separation of crops and weeds by 1000 m will not prevent gene escape. The variation in gene flow at distances of 200 to 1000 m demonstrates that the distance conferring complete isolation may be difficult to predict. Our data suggest distances from engineered crops to compatible weeds must be at least several kilometers for spatial isolation to prevent the escape of engineered genes. Alternative strategies (Ellstrand 1988; Ellstrand & Hoffman 1990) must be sought and employed to ensure complete isolation.

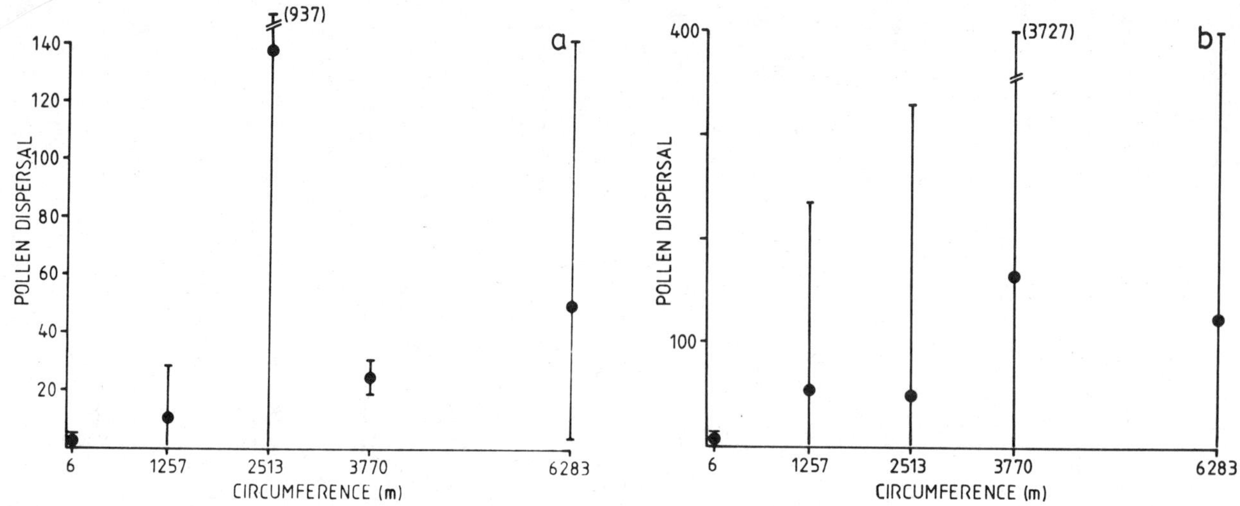

Figure 2. Estimated pollen dispersal across the two-dimensional landscape (see text). Values represent the mean ± 1 S.D. a) at South Coast Field Station. b) at Moreno Valley Field Station.

Acknowledgments

We are grateful for both the electrophoretic expertise of J. M. Clegg and for her comments on an earlier draft of this manuscript. Additional comments from R. K. Colwell and P. Kareiva were especially helpful. We thank M. Bartholomew and B. Lacerra for their help with field work and electrophoresis. Finally, the personnel of UC South Coast and Moreno Valley Field Stations made valuable contributions to our experimental plans with their horticultural advice. Supported by grants from the University of California Biotechnology Research and Education Program and the National Science Foundation (BSR-8505982 to N.C.E.).

Literature Cited

Colwell, R. K., E. A. Norse, D. Pimentel, F. E. Sharples, and D. Simberloff. 1985. Genetic engineering in agriculture. Science **229:**111–112.

Ellstrand, N. C. 1988. Pollen as a vehicle for the escape of engineered genes? Pages S30–S32 in J. Hodgson and A. M. Sugden, editors. Planned release of genetically engineered organisms. Trends in Biotechnology Special Publication. Elsevier, Cambridge, England.

Ellstrand, N. C., and B. Devlin. 1989. Transmission genetics of isozyme loci in *Raphanus sativus* (Brassicaceae): stress-dependent non-Mendelian segregation. American Journal of Botany **76:**40–46.

Ellstrand, N. C., and C. A. Hoffman. 1990. Hybridization as an avenue for escape of engineered genes. BioScience **40:**438–442.

Frankel, R., and E. Galun. 1977. Pollination mechanisms, reproduction, and plant breeding. Springer-Verlag, Berlin, Germany.

Fryxell, P. A. 1957. Mode of reproduction in higher plants. Botanical Review **23:**135–233.

Gasser, C. S., and R. T. Fraley. 1989. Genetically engineered plants for crop improvement. Science **244:**1293–1299.

George, R. A. T. 1985. Vegetable seed production. Longman, New York.

Goodman, R. M., and N. Newell. 1985. Genetic engineering for plants for herbicide resistance: status and prospects. Pages 47–53 in H. O. Halvorson, D. Pramer, and M. Rogul, editors. Engineered organisms in the environment: scientific issues. American Society of Microbiology, Washington, D.C.

Grosberg, R. K. 1991. Sperm-mediated gene flow and the genetic structure of a population of the colonial ascidian *Botryllus schlosseri.* Evolution **45:**130–142.

Handel, S. N. 1983. Pollination ecology, plant population structure, and gene flow. Pages 163–211 in L. Real, editor. Pollination biology. Academic Press, Orlando, Florida.

Kareiva, P., and W. Morris. 1991. Using models to integrate data from field trials and estimate risks of gene escape and gene spread. D. MacKenzie, editor. Biosafety and risk assessment in biotechnology. U.S.D.A. Technical Publication. In press.

Kirkpatrick, K. J., and H. D. Wilson. 1989. Interspecific gene flow in *Cucurbita: C. texensis* vs. *C. pepo.* American Journal of Botany **75:**519.

Langevin, S. A., K. Clay, and J. B. Grace. 1990. The incidence and effects of hybridization between cultivated rice and its related weed, red rice (*Oryza sativa* L.). Evolution **44:**1000–1008.

Manasse, R., and P. Kareiva. 1990. Quantifying the spread of recombinant genes and organisms. Pages 215–231 in L. Ginzberg, editor. Assessing ecological risks of biotechnology. Butterworth, New York.

National Research Council. 1989. Field testing genetically modified organisms: framework for decisions. National Academy Press, Washington, D.C.

Panetsos, C. P., and H. G. Baker. 1968. The origin of variation in "wild" *Raphanus sativus* (Cruciferae) in California. Genetica **38**:243–274.

Regal, P. J. 1988. The adaptive potential of genetically engineered organisms in nature. Pages S36–S38 in A. Hodgson and A. M. Sugden, editors. Planned release of genetically engineered organisms. Trends in Biotechnology Special Publication. Elsevier, Cambridge, England.

Snedecor, G. W., and W. G. Cochran. 1967. Statistical methods. Iowa University Press, Iowa City, Iowa.

Stanton, M. L. 1987. Reproductive biology of petal color variants in wild populations of *Raphanus sativus.* I. Pollinator response to color morphs. American Journal of Botany **74**:178–187.

Tiedje, J. M., R. K. Colwell, Y. L. Grossman, et al. 1989. The planned introduction of genetically engineered organisms: ecological consideration and recommendations. Ecology **70**:298–315.

Wilson, H. D. 1990. Gene flow in squash species. BioScience **40**:449–455.

Zar, J. H. 1984. Biostatistical analysis. Prentice-Hall, Englewood Cliffs, New Jersey.

References

Ashley, M. V., D. J. Melnick, and D. Western. 1990. Conservation genetics of the black rhinoceros (*Diceros bicornis*), I: evidence from the mitochondrial DNA of three populations. Conservation Biology **4(1)**:71–77.

Bowen, B., J. C. Avise, J. I. Richardson, A. B. Meylan, C. Margaritoulis, and S. R. Hopkins-Murphy. 1993. Population structure of loggerhead turtles (*Caretta caretta*) in the northwestern Atlantic Ocean and Mediterranean Sea. Conservation Biology **7(4)**:834–844.

Braithwaite, R. W. and W. M. Lonsdale. 1987. The rarity of *Sminthopsis virginiae* (Marsupialia: Dasyuridae) in relation to natural and unnatural habitats. Conservation Biology **1(4)**:341–343.

Congdon, J. D., A. E. Dunham, and R. C. van Loben Sels. 1993. Delayed sexual maturity and demographics of Blanding's turtles (*Ermydoidea blandingii*): implications for conservation and management of long-lived organisms. Conservation Biology **7(4)**:826–833.

Dowling, T. E., B. D. DeMarais, W. L. Minckley, M. E. Douglas, and P. C. Marsh. 1992. Use of genetic characters in conservation biology. Conservation Biology **6(1)**:7–8.

Dowling, T. E., W. L. Minckley, M. E. Douglas, P. C. Marsh, and B. D. Demarais. 1992. Response to Wayne, Nowak, and Phillips and Henry: use of molecular characters in conservation biology. Conservation Biology **6(4)**:600–603.

Ehrlick, P. R. and D. D. Murphy. 1987. Conservation lesson from long-term studies of checkerspot butterflies. Conservation Biology **1(2)**:122–131.

Harris, R. B. and F. W. Allendorf. 1989. Genetically effective population size of large mammals: an assessment of estimators. Conservation Biology **3(2)**:181–191.

Hori, M., M. M. Gashagaza, M. Nshombo, and J. Kawanabe. 1993. Littoral fish communities in Lake Tanganyika: irreplaceable diversity supported by intricate interactions among species. Conservation Biology **7(3)**:657–666.

Kindvall, O. and I. Ahlén. 1992. Geometrical factors and metapopulation dynamics of the bush cricket, *Metrioptera bicolor* Philippi (Orthoptera: Tettigoniidae). Conservation Biology **6(4)**:520–529.

Klinger, T., D. R. Elam, and N. C. Ellstrand. 1991. Radish as a model system for the study of engineered gene escape rates via crop-weed mating. Conservation Biology **5(4)**:531–535.

Laikre, L., N. Ryman, and E. A. Thompson. 1993. Hereditary blindness in a captive wolf (*Canis lupus*) population: frequency reduction of a deleterious allele in relation to gene conservation. Conservation Biology **7(3)**:592–601.

Lamberson, R. H., B. R. Noon, C. Voss, and K. McKelvey. 1994. Reserve design for territorial species: the effects of patch size and spacing on the viability of the Northern Spotted Owl. Conservation Biology **8(1)**:185–195.

Lesica, P. and F. W. Allendorf. 1992. Are small populations of plants worth preserving? Conservation Biology **6(1)**:135–139.

Mace, G. M. and R. Lande. 1991. Assessing extinction threats: toward a reevaluation on IUCN threatened species categories. Conservation Biology **5(2)**:148–157.

Maguire, L. A. and R. C. Lacy. 1990. Allocating scarce resources for conservation of endangered subspecies: partitioning zoo space for tigers. Conservation Biology **4(2)**:157–166.

Manges, E. S. 1991. Seed germination percentage increases with population size in a fragmented prairie species. Conservation Biology **5(2)**:158–164.

Meffe, G K. and R. C. Vrijenhoek. 1988. Conservation genetics in the management of desert fishes. Conservation Biology **2(2)**:157–169.

Nowak, R. M. 1992. The red wolf is not a hybrid. Conservation Biology **6(4)**:593–595.

Oliver, I. and A. J. Beattie. 1993. A possible method for the rapid assessment of biodiversity. Conservation Biology **7(3)**:562–568.

Phillips, M. K. and V. G. Henry. 1992. Comments on red wolf taxonomy. Conservation Biology **6(4)**:596–599.

Ralls, K., J. D. Ballou, and A. Templeton. 1988. Estimates of lethal equivalents and the cost on inbreeding in mammals. Conservation Biology **2(2)**:185–193.

Rojas, M. 1992. The species problem and conservation: what are we protecting? Conservation Biology **6(2)**:170–178.

Rosenzweig, M. L. and C. W. Clark. 1994. Island extinction rates from regular censuses. Conservation Biology **8(2)**:491–494.

Van Vuren, D. and P. W. Hedrick. 1989. Genetic conservation in feral populations of livestock. Conservation Biology **3(3)**:312–317.

Walters, T. W., D. S. Decker-Walters, and D. R. Gordon. 1994. Restoration considerations for wiregrass (*Aristida stricta*): allozymic diversity of populations. Conservation Biology **8(2)**:581–585.

Waples, R. S. and D. J. Teel. 1990. Conservation genetics of the Pacific salmon I. Temporal changes in allele frequency. Conservation Biology **4(2)**:144–156.

Wayne, R. K. 1992. On the use of morphologic and molecular genetic characters to investigate species status. Conservation Biology **6(4)**:590–592.

Wayne, R. K., N. Lehman, M. W. Allard, and R. L. Honeycutt. 1992. Mitochondrial DNA variability of the gray wolf: genetic consequences of population decline and habitat fragmentation. Conservation Biology **6(4)**:559–569.

Wilson, M. H., C. B. Kepler, N. F. R. Snyder, S. R. Derrickson, F. J. Dein, J. W. Wiley, J. M. Wunderle, Jr., A. E. Lugo, D. L. Graham, and W. D. Toone. 1994. Puerto Rican parrots and potential limitations of the metapopulation approach to species conservation. Conservation Biology **8(1)**:114–123.

DRUG ABUSE

Proceedings of the

International Conference

THE WHITE HOUSE

WASHINGTON

October 27, 1970

Those who take part in the International Conference on Drug Abuse are responding to a challenge that is reaching deplorable proportions not only in our country but throughout the world.

As most of you know I have directed a major Federal involvement in the area of drug education. But the final success of any such effort depends on the fullest possible participation of civic groups and private citizens, strengthening and complementing the work of Government.

I commend your positive approach to turn the tide in an area that poses such an unparalleled threat to our society. May your deliberations be successful, and may all mankind benefit from the work you do.

Richard Nixon

DRUG ABUSE

Proceedings of the International Conference

Edited by
Chris J. D. Zarafonetis, M.D.
Director, The Thomas Henry Simpson Memorial Institute for Medical Research, The University of Michigan, Ann Arbor, Michigan

59 Contributors

Lea & Febiger • PHILADELPHIA • 1972

ISBN: 0-8121-0366-1

Published in Great Britain by Henry Kimpton Publishers, London

Library of Congress Catalog Card Number 72-154243

Printed in the United States of America

PREFACE

This volume consists of the proceedings of the International Conference on Drug Abuse which was held at Ann Arbor, Michigan, November 9-13, 1970. The symposium was designed to help the practicing physician deal more effectively with the problems of drug abuse—not merely in the context of the individual physician-patient relationship, but also as a participant in community action programs with school officials, the courts, and representatives of social agencies. Over 1,150 registrants attended the Conference with representation from many parts of the world.

All of the presentations and relevant discussions are included in the *Proceedings*. It was deemed appropriate, however, to omit various informal remarks, irrelevant commentary, and the biographical introduction of speakers. In a few instances authors added new relevant information, but for the most part, the principal editorial changes are those which occur normally in adapting manuscripts prepared for oral presentation to publication form. This aspect of the editorial process was carried out by the authors of the respective papers.

A program of this magnitude could not have been held without much assistance and support on the part of many individuals and organizations. Grateful acknowledgment is made for the generous financial support contributed by the following: Abbott Laboratories, North Chicago, Illinois; The Allstate Foundation, Northbrook, Illinois; Bristol Laboratories, New York, N.Y.; CIBA Pharmaceutical Company, Summit, New Jersey; Herbert H. and Grace A. Dow Foundation, Midland, Michigan; Ford Motor Company Fund, Dearborn, Michigan; Glenwood Laboratories, Inc., Tenafly, New Jersey; Hoechst Pharmaceutical Company, Cincinnati, Ohio; Eli Lilly and Company Foundation, Indianapolis, Indiana; McNeil Laboratories, Inc., Fort Washington, Pennsylvania; Merck Sharp & Dohme Postgraduate Program, West Point, Pennsylvania; Metropolitan Life Insurance Company, New York, New York; Miles Laboratories, Inc., Elkhart, Indiana; Parke, Davis & Company, Detroit, Michigan; Pfizer Pharmaceuticals, Pfizer Laboratories Division, J. B. Roerig Division, New York, New York; The Proprietary Association, Washington, D.C.; Roche Laboratories, Division of Hoffmann-La Roche, Inc., Nutley, New Jersey; William H. Rorer, Inc., Fort Washington, Pennsylvania; Sandoz Pharmaceuticals, Hanover, New Jersey; Schering Corporation, Bloomfield, New Jersey; E. R. Squibb & Sons, Inc., New York, New York; The University of Michigan Highway Safety Research Institute, Ann Arbor, Michigan; The Upjohn Company, Kalamazoo, Michigan; Warner-Lambert Charitable Foundation, Morris Plains, New Jersey; Wyeth Laboratories, Philadelphia, Pennsylvania.

Several faculty members of The University of Michigan Medical School gave valuable assistance and guidance in the planning and preparation for this conference. A special debt of gratitude is owed Dr. Maurice H. Seevers, not only for his fine scientific contribution but also for his many helpful suggestions to the Planning Committee. The Planning Committee included Dr. Edward A. Carr, Jr., Mr. Louis Graff, Mr. Donald C. Kraushaar, Dr. William J. Oliver, Dr. John C. Pollard, Mr. Robert K. Richards, Dr. Melvin L. Selzer, Dr. Harry A. Towsley, Dr. Neal A. Vanselow, and myself as Chairman. I wish to acknowledge and commend the logistical support provided by the Department of Postgraduate Medicine through Dr. Towsley, Dr. Vanselow, Mr. Richards, Mrs. Jan Johnson, Mr. Ralph Lewis, and Mr. William L. Schmalgemeier. I am especially grateful and appreciative for the administrative and secretarial assistance of Miss Barbara Stoner of the Simpson Memorial Institute staff. Miss Stoner carried a major share of this effort from its inception to its conclusion with the publication of this volume.

Finally, I am most grateful to all of the essayists and moderators who participated in this conference. They deserve credit for any success that the symposium may have achieved in bringing into perspective the problems of drug abuse in man. It is my hope that their efforts will provide a basis of understanding upon which can be constructed rational and effective programs to cope with this serious cultural phenomenon.

Chris J. D. Zarafonetis, M.D.
Ann Arbor, Michigan

CONTRIBUTORS

BURTON M. ANGRIST, M.D., Research Psychiatrist, Department of Psychiatry and Neurology, New York University Medical Center, New York, N.Y.

ELGIE L. BELLIZIO, A.B., Executive Director, Salinas Area Youth Information/Crisis Center, Teacher, Salinas Union High School District, Salinas, Calif.

HERBERT BENSON, M.D., Assistant Professor of Medicine, Harvard Medical School, Thorndike Memorial Laboratory, Boston City Hospital, Boston, Mass.

THOMAS A. BEWLEY, M.A., M.D., F.R.C.P.I., D.P.M., Consultant Psychiatrist, Drug Dependence Treatment Clinics, St. George's, St. Thomas' & Tooting Bec Hospitals, London, England

PHILIP BOYD, M.D., St. Lukes-Woodside Hospital, The Middlesex Hospital Medical School, London, England

JOSEPH H. BRENNER, M.D., Medical Department, Massachusetts Institute of Technology, Cambridge, Mass.

HENRY BRILL, M.D., Director, Pilgrim State Hospital, West Brentwood, L.I., N.Y.

DALE C. CAMERON, M.D., M.P.H., Chief, Drug Dependence, World Health Organization, Geneva, Switzerland

SIDNEY COHEN, M.D., Consultant, National Institute of Mental Health, Los Angeles, Calif.

DONALD F. COOKE, B.S., Department of Medicine, Boston City Hospital, Boston, Mass.

ERIC C. DAHL, B.A., Department of Medicine, Boston City Hospital, Boston, Mass.

JUDIANNE DENSON-GERBER, J.D., M.D., Odyssey House, New York, N.Y.

EDWARD F. DOMINO, M.D., Professor of Pharmacology, Director, Neuropsychopharmacology Research Program, Department of Pharmacology, The University of Michigan, Ann Arbor, Mich.

HENRY W. ELLIOTT, M.D., PH.D., Professor and Chairman, Department of Medical Pharmacology and Therapeutics, University of California, Irvine, Orange, Calif.

CARL F. ESSIG, M.D., Chief, Research Sections, National Institute of Mental Health, Lexington, Ky.

MAX FINK, M.D., Professor of Psychiatry, Department of Psychiatry, New York Medical College, New York, N.Y.

HAVELOCK F. FRASER, M.D., Consultant, Medical Research Division, The Lilly Research Laboratories, Indianapolis, Ind.

A. J. FRIEDHOFF, M.D., Department of Psychiatry, New York University Medical Center, New York, N.Y.

John H. Frykman, B.D., Community Counselor, Carmel Unified School District, Carmel, Calif.

B. J. George, Jr., J.D., Professor of Law, Director, Center for Administration of Justice, Wayne State University, Detroit, Mich.

Samuel Gershon, M.D., Department of Psychiatry and Neurology, New York University Medical Center, New York, N.Y.

Leonard Goldberg, M.D., Professor of Research on Alcohol and Analgesics, Karolinska Institut, Stockholm, Sweden

John A. Gronvall, M.D., Dean of the Medical School and Director of the Medical Center, The University of Michigan, Ann Arbor, Mich.

Leo E. Hollister, M.D., Medical Investigator, Department of Medicine, Veterans Administration Hospital and Stanford University School of Medicine, Palo Alto, Calif.

Bo Holmstedt, M.D., Professor, Department of Toxicology, Swedish Medical Research Council, Karolinska Institut, Stockholm, Sweden

Carl-Magnus Ideström, M.D., Chief, Laboratory of Experimental Psychiatry, Department of Psychiatry, Karolinska Institut, Stockholm, Sweden

John E. Ingersoll, A.B., Director, Bureau of Narcotics and Dangerous Drugs, U. S. Department of Justice, Washington, D.C.

Harris Isbell, M.D., Clinical Pharmacologist, Department of Internal Medicine, University of Kentucky Medical Center, Lexington, Ky.

Jerome H. Jaffe, M.D., Special Consultant to the President for Narcotics and Drug Abuse, Director, Special Action Office, Drug Abuse Prevention, Executive Office of the President, Washington, D.C.

Ian Pierce James, M.A., B.A., M.B., B. Chir., D.P.M., M.A.N.Z.C.P., Consultant in Forensic Psychiatry, Bristol Clinical Area, London, England

Abraham Kaplan, Ph.D., Professor of Philosophy, Department of Philosophy, The University of Michigan, Ann Arbor, Mich.

Masaaki Kato, M.D., Chief, Division of Adult Mental Health, National Institute of Mental Health, Chiba Ken, Japan and Department of Psychiatry, Tokyo Medical College, Tokyo, Japan

Paul Kielholz, M.D., Director, Psychiatric Clinic, University of Basel, Basel, Switzerland

Alfred J. R. Koumans, M.D., Staff Psychiatrist, Medical Department, Massachusetts Institute of Technology, Boston, Mass.

John C. Kramer, M.D., Chief of Medical Research, Department of Medical Pharmacology and Therapeutics, University of California, Irvine, Orange, Calif.

Allan Linnarson, D.Sc., Department of Toxicology, Karolinska Institut, Stockholm, Sweden

Donald B. Louria, M.D., Professor and Chairman, Department of Public Health and Preventive Medicine, New Jersey College of Medicine and Dentistry, Newark, N.J.

Thomas Mahon, B.S., M.S., Assistant Research Director, Law Enforcement Planning Council, Gainesville, Fla.

William R. Martin, M.D., Chief, National Institute of Mental Health Addiction Research Center, Lexington, Ky.

E. W. Maynert, M.D., Professor of Pharmacology, Department of Pharmacology, University of Illinois College of Medicine, Chicago, Ill.

Frederick H. Meyers, M.D., Professor of Pharmacology, Department of Pharmacology, University of California Medical Center, San Francisco, Calif.

James P. Murphy, Odyssey House, New York, N.Y.

Jean Nepote, Secretary-General, INTERPOL, St. Cloud, France

Warren W. Nichols, M.D., Ph.D., Head, Department of Cytogenetics, Institute for Medical Research, Camden, N.J.

John W. Oliver, L.L.B., District Judge, U.S. District Court, Western District of Missouri, Kansas City, Mo.

Kenneth D. Parker, M.Crim., Toxicologist, The Hines Laboratories, Inc., San Francisco, Calif.

*John C. Pollard, M.D., Professor of Psychiatry, Department of Psychiatry, The University of Michigan, Ann Arbor, Mich.

Charles C. Rohrs, M.D., Assistant Director, Odyssey House Programs, Director, Medical Services, Odyssey House, New York, N.Y.

James H. Scheuer, L.L.D., Representative, 21st District of New York, House of Representatives, Congress of the United States, Washington, D.C.

Stanley H. Schuman, M.D., Dr.P.H., Associate Professor of Epidemiology, Department of Epidemiology, School of Public Health, The University of Michigan, Ann Arbor, Mich.

J. W. Schweitzer, Ph.D., Assistant Professor of Experimental Psychiatry, Department of Psychiatry, New York University Medical Center, New York, N.Y.

Maurice H. Seevers, Ph.D., M.D., Professor and Chairman, Department of Pharmacology, The University of Michigan, Ann Arbor, Mich.

David E. Smith, M.D., M.S., Founder & Director, Haight-Ashbury Clinic, Assistant Clinical Professor, Department of Pharmacology, University of California, San Francisco, Calif.

Roger C. Smith, D.Crim., Executive Director, Marin Open House, Tiburon, Calif.

Harry A. Towsley, M.D., Director, Department of Postgraduate Medicine, The University of Michigan, Ann Arbor, Mich.

J. Thomas Ungerleider, M.D., Associate Professor-in-Residence, The Neuropsychiatric Institute, U.C.L.A. Center for the Health Sciences, Los Angeles, Calif.

R. Keith Wallace, Ph.D., Department of Medicine, Boston City Hospital, Boston, Mass.

Norman Weiner, M.D., Professor and Chairman, Department of Pharmacology, University of Colorado, Denver, Colo.

Chris J. D. Zarafonetis, M.D., Professor of Internal Medicine and Director, Simpson Memorial Institute, The University of Michigan, Ann Arbor, Mich.

*Deceased August 15, 1971.

CONTENTS

Morning Session
Monday, November 9, 1970

Presiding: Chris J. D. Zarafonetis, M.D.

WELCOMING REMARKS

JOHN A. GRONVALL, M.D.

It is a real pleasure to welcome you to the University of Michigan and to this symposium on drug abuse. This symposium deals with a worldwide trans-cultural problem of great magnitude. It is a problem not only of the current time, but is one of great historical importance. The problem of drug abuse in many ways represents the dark face of a coin, the bright face representing the success that medicine has had in identifying and formulating potent pharmacologic agents. While these agents are becoming increasingly effective in dealing with medical problems, they also create increasing problems as their use is abused.

I recollect that some two and one-half years ago in several conversations with Dr. Zarafonetis the subject of drugs and drug abuse came up and Dr. Zarafonetis commented that one of our problems is that we, as physicians, do not have the kind of broad information that will allow us to deal effectively with the drug abuse problem. This symposium, then, is an outgrowth of that kind of interest expressed by Dr. Zarafonetis and other faculty colleagues here and elsewhere. The symposium brings together an impressive group of international experts who can, we hope, give us information that will help us deal with this problem.

Drugs have been of importance to physicians throughout history, but knowledge about drugs has always been inadequate. Voltaire called attention to this in the eighteenth century when he said, "Doctors pour in drugs about which they know little, to treat diseases of which they know less, in human beings of whom they know nothing." From that state of ignorance we have advanced considerably, but our current ignorance is still considerable. In a recent drug abuse symposium held at Rutgers, Dr. Daniel Freedman, Chairman of Psychiatry at the University of Chicago, said, "The point is that those who seek information about drugs must be prepared to search through complex social, psychological and medical facts; to *not* hear ready answers; to *not* be satisfied *either* with hastily publicized scientific findings *or* with hastily legislated 'rehabilitative' programs—both of which are substitutes for a sober analysis of what is possible and the ways things are. A part of the contemporary scene taps our deep sense of injustice. We impatiently demand instant application in the name of medical and scientific findings which, in fact, may not exist and which—in the absence of serious and sustained support of basic inquiry—may never exist."

This symposium, I think, points out admirably the role of the university in dealing with complex medical and social problems. For the University of Michigan this interest in drugs is a natural one. Under Dr. Seevers' direction, in our Department of Pharmacology for 25 or more years there have been extensive studies on drug addiction. More recently, the establishment of the Upjohn Center for Clinical Pharmacology points out the serious involvement of this university in the study of drugs. But, of course, it would be fractious for any one university to attempt to deal with such a complicated problem, and in this symposium the University of Michigan then serves as a coordinating point to bring together a large group of people for a week-long meeting to define the current state of knowledge in the multiple fields dealing with this drug problem. I am particularly pleased that this symposium will be published so that this work will be available, not only to you who are here today, but to a worldwide audience. Once again, let me welcome you on behalf of the University of Michigan, and I know that this symposium will be a worthwhile and a beneficial effort as we in the medical community seek continually to deal with this very difficult problem. Thank you.

WELCOMING REMARKS

HARRY A. TOWSLEY, M.D.

On behalf of the Department of Postgraduate Medicine of the University of Michigan Medical School, it is my very great pleasure to welcome each of you to this International Symposium on Drug Abuse. I know that you share with us the very great and rapidly growing concerns for the mental and physical well-being, not only of those people who are living today, but also for the many unborn who will follow in a world in which the indiscriminate use of drugs of all descriptions is more widespread than it has ever been before.

Dr. Zarafonetis, our course chairman for this program, and his committee have brought together an outstanding faculty of experts from this country and abroad in an effort to help each of us gain a clearer understanding of the magnitude of the problems that we face and, hopefully, to guide us in finding acceptable solutions for their prevention and management in the daily practice of medicine, in the community, in the nation, and in the world.

We are deeply indebted to the corporations, foundations, and associations which gave material and moral support in order to make this symposium a reality. Without their help it would have been impossible to bring together such a distinguished faculty from the four corners of the earth.

It is the fervent hope of the program committee for this symposium, the members of the Department of Postgraduate Medicine, and the faculty and administration of our Medical Center that each of you will be able to take home from this conference new knowledge and new ideas that will be useful to you and to the community from which you come, in combating the widespread, indiscriminate use of drugs, most of which were developed to *alleviate* man's suffering, not to destroy him. To this end, we are indeed happy to have you with us.

INTRODUCTION

CHRIS J. D. ZARAFONETIS, M.D.

Some 30 months ago, it became apparent to me that many physicians, including myself, were not sufficiently informed as to the nature, extent, and impact that the drug abuse phenomenon was having on our society. It was also evident that the problem extended beyond the professional boundaries of medicine with implications for, and involvement of, the home, school, church, and industrial and governmental elements of our society. Experience has shown that a major difficulty with problems which cross social, political, and professional boundaries lies in the definition of roles and the fixing of authority and responsibility of those who deal with such problems. In the case of drug abuse, the situation is further complicated by certain international factors which involve the foreign services of governments and various international agencies. Obviously, it is necessary to delineate clearly the roles of all who are involved in activities to reduce the abuse of drugs, not only to avoid overlap and duplication of effort, but also to avoid deficiencies in programs undertaken to cope with the problem.

What then are the responsibilities of the physician—your responsibilities and mine—in dealing with the drug abuse problem?

First, there is our traditional professional role which attends the patient-doctor relationship. Unfortunately, medical education and postgraduate training in recent years have emphasized disease and system-specialization which, while meeting important clinical objectives, has left many ill-prepared to deal with drug abuse as a medical entity.

It is for this reason that considerable time at this conference will be devoted to a medical definition of the problem, and to a consideration of some fundamental questions including epidemiology and potential genetic implications of drug abuse. Basic and clinical pharmacology are supplied for each major category of abused drugs. There is a systematic, up-dated review of methods of detection and principles of diagnosis, as well as the acute and chronic medical-psychiatric aspects of patient management.

The second and less well-defined responsibility of the physician lies in his role as a participant and adviser in community action programs along with school officials, social agencies, and the courts. Much of the program before us is directed toward a better understanding of how the physician may relate to these efforts and contribute more effectively in this role of social responsibility.

I referred earlier to the difficulty of dealing with problems which in addition to a medical component also include social and governmental agencies. A key question that must be asked in these circumstances is: "Who is in charge?" In our system of government, the ultimate responsibility lies in the office of our elected political leadership, whether it be local, state, or national. Unfortunately, the respective political offices are not structured to deal with such problems at the working level.

This limitation in governmental structure and capabilities is, in the case of the drug abuse problem, further compounded by wide differences in the national and individual state laws pertaining to drugs and drug traffic. In regard to the latter, we may look forward to some helpful guidelines when we hear from our experts on legislation, law, justice, criminology, and law enforcement.

Finally, the Program Committee is most grateful to the distinguished scientists who have come from near and far to serve as our faculty in this conference. A conscious effort was made to bring together workers of wide and differing experiences, and we should all benefit from the results of literally hundreds of man-years of study that they have collectively devoted to the problem.

We trust that attitudes have not been fixed in advance, and, that while consensus may prevail at the end of this conference, an open and receptive posture will be retained throughout in keeping with medical tradition.

DRUG DEPENDENCE VIS-A-VIS DRUG ABUSE

MAURICE H. SEEVERS, Ph.D., M.D.

The complexities of drug dependence and drug abuse as scientific, medical, and social problems are so great that the pharmacologic and psychologic principles common to all forms of drug dependence and drug abuse are easily obscured in the mass of detail and in semantic arguments about definitions.

One of the most common errors, one which stands as an almost absolute barrier to defining the issues in a logical manner as they apply to psychoactive drugs, is to use "drug dependence" and "drug abuse" as if they were synonymous. Note that they are not synonyms.

"Drug dependence" involves predictable and reproducible *individual-drug* interactions which can be described precisely in pharmacologic, psychologic and medical terms. In contrast, the term "drug abuse" represents only a value judgment of "society."

As commonly employed, the term "drug abuse" may refer to any type of drug or chemical without regard to its pharmacologic actions. This includes the whole gamut of over-the-counter and prescription drugs, even those which are not psychoactive. The term is used to imply either individual injury or social harm without regard to the nature of the injury or the organ systems involved. It is also commonly used to differentiate legal from illegal use or to distinguish self-administration from medical treatment. Drug abuse may encompass all forms of acute and chronic intoxication. It embraces all of the broad connotations of harm which have come to be implicit in the use of the terms "habit-forming" and "addiction."

The term "drug abuse" has only one uniform connotation, the disapprobation of "society."

Unfortunately, society uses any or all of this wide variety of meanings as a basis for establishing control measures or legal sanctions. It is not surprising, therefore, that great confusion exists even among those who are most knowledgeable about this multifaceted problem. Nor is it surprising that one "society" may ignore completely while another may deplore and impose irrational sanctions against the same chemical substance.

In an effort to sift out and expose the major components of this perplexing problem, it seems logical to define *psychoactive drug dependence* strictly in terms of science and medicine and to arbitrarily define *drug abuse* exclusively in sociologic terms. This concept permits a relatively clear-cut separation, at least as a basis for analysis, between the effects on the individual which result from individual-drug interactions and the direct adverse effects on society which result from the antisocial behavior of the drugged individual.

It is essential to recognize that this concept is based upon the arbitrary assumption that drug damage to the individual as a person is of no concern to society as long as his behavior does not constitute a direct threat to the well-being of others.

Clearly, this is a false assumption. Injury to the individual resulting from psychoactive drugs is obviously a concern of society. In spite of this reservation, let us pursue the concept further.

Many of the perplexing problems which relate to psychoactive drugs can be rationalized and resolved if primary emphasis is placed on the direct hazard to society posed by the antisocial behavior of the drugged person.

When drug abuse is defined thus, the abuse potential of any drug, old or new, can be assessed accurately in terms of its capacity to induce certain types of psychotoxicity which lead to disorders of behavior. Furthermore, measures designed to control abuse can be aimed primarily at protecting society from the drugged individual. In situations where individual injury does occur the primary thrust of control measures can be directed toward protecting the individual from himself.

Possibly the most compelling reason for defining drug abuse in this manner is that, with most psychoactive drugs, little if any evidence of somatic damage in the physical or pathologic sense is associated even with extreme degrees of psychotoxicity. Some of the most socially dangerous types of psychotoxicity—acute amphetamine and cocaine psychoses, chronic heroin dependence, acute alcohol and barbiturate poisoning—exhibit no specific neurologic or other organ pathology. Of all the potent psychoactive agents, only chronic alcohol and barbiturate dependence have a well-established and specific pathology.

The evidence is clear that psychologic and psychiatric stigmata may persist as permanent or semi-permanent functional residues following some forms of drug abuse, even of short duration. It is an open question whether an individual ever returns to his pre-drug state either physiologically or psychologically during a remission from, or following a "cure" of, drug dependence.

Since the total impact of drug abuse on society is a sum of the effects on its individual members, drug-induced pathology or stigmata in the drug abuser cannot logically be considered to be of importance only to the individual. Therefore, an arbitrary line must be drawn in order to eliminate individual injury as a component of any definition of drug abuse.

But direct injury to the individual from psychoactive drugs is so ephemeral—so inconsistent in appearance, so difficult to substantiate by retrospective analysis, and shows so little correlation with drug-induced functional disorders which are inimical to society—that to designate its presence or absence as a significant factor in the assessment of drug abuse potential of psychoactive drugs is neither scientific nor rational.

Let us now consider drug dependence. Dependence on drugs in the broadest sense involves a vast majority of the world's population. It is not necessarily harmful and in most instances carries no social stigma.

It is important to recognize that all drug therapy, regardless of the pharmacologic nature of the drug or whether the substance is self-administered or physician-prescribed, has a psychologic component. This is true even with substances which have no pharmacologic effect whatsoever. Were it not so, advertising to the public would lose much of its force and physicians would lose a valuable aid in the treatment of the sick, or those who believe they are sick.

With pharmacologically inert substances, dependence is based solely on the so-called placebo effect. It is strictly psychogenic and a reasonable measure of the existing psychologic state of the user. Many other substances which have no direct effect on the central nervous system may, because of a strong peripheral action, reinforce psychologically and create dependence by drug-induced sensory input or by abolishing unpleasant sensations which arise from the periphery; for example, the burning sensation from counterirritants, intense catharsis, relief of heartburn by antacids, and the like.

Dependence on non-psychoactive drugs or placebos is introduced here only to dismiss it as a minuscule problem which should not be confused, as it has been, with the principal issues of drug dependence and abuse of psychoactive drugs.

As we now consider psychoactive drugs, it may appear fatuous to say that an individual who has had no experience with these substances will never become drug-dependent.

In effect, however, this is the only means by which drug dependence could be eliminated in human society. Having once experienced such drug effects, a large majority of the world population will inevitably become drug-dependent. Some drugs are so weak that little harm results. But many are such powerful reinforcers that if all the population were to be given a drug trial with every major psychoactive drug by intravenous administration and then permitted free access to the drug of their choice, the disaster for society would be enormous.

Lesser animals, such as monkeys or even rats, having once experienced the effects of a drug such as cocaine will, without exception, if given the opportunity, self-administer the drug until they die. It should be clearly understood by those who decry the depravity of the drug addict that susceptibility is only relative, and if conditions are optimal, almost any individual can be made to be drug-dependent in the most serious sense of the term, even against his will.

The only *common* denominator of all drug dependence is psychologic reinforcement resulting from reward associated with the first individual-drug interaction and the subsequent increasing desire for repeated reinforcement.

Whereas a large number of people are psychologically dependent in varying degrees upon some psychoactive drug or chemical substance, only a limited number of these individuals are drug abusers by any reasonable definition of the term.

To establish a rational relationship between dependence and abuse, and to aid in defining drug abuse in precise terms, it seems desirable to group dependence-producing drugs into three main categories. This grouping is based on the nature of the dependence the drugs produce and their capacity to induce psychotoxicity and antisocial behavior.

Group 1 consists of substances with such weak psychoactive properties that they are measurable largely in subjective rather than objective terms. The principal members of this group are tobacco, coffee, tea, betel, and the analgesic-antipyretics. They do not induce physical dependence or result in psychotoxicity even with prolonged or excessive use. Heavy and prolonged chronic intoxication may damage organ systems and result in injury to the individual; at least one of these substances, tobacco, is such a powerful reinforcer psychologically that the most strenuous deconditioning methods often fail to break a chronic habit. But none of these substances leads to antisocial behavior, a factor that distinguishes them from all other dependence-producing drugs.

Characteristics of Weak Psychoactive Drugs (Group 1—Tobacco, Coffee, Tea, Betel, Non-narcotic Analgesics [Aspirin, Phenacetin, etc.])

Very weak psychoactivity even with large quantities
Strong psychologic dependence
No psychotoxicity
No antisocial behavior

Group 2 includes the so-called psychomotor stimulants such as the amphetamines and the hallucinogens (lysergic acid diethylamide [LSD], marihuana, etc.) whose fundamental actions are also excitatory. A common property of the members of this broad group is their inducement of psychotoxicity, distortion of perception, hallucinations, illusions, and disorders of behavior.

The hallucinogens, although less potent reinforcers than the central nervous system (CNS) stimulants, are no less capable of initiating antisocial behavior, because they induce psychotoxicity with any effective dose which involves perceptual distortion. In contrast to the psychomotor stimulants, many of them are utilized only for "spree" use. When large quantities are used chronically, some may induce sedation rather than stimulation. The latter is clearly evident with derivatives of cannabis. The weak dosage forms, marihuana as commonly used in the United States and bhang as used in India, are essentially excitatory as with low doses of alcohol. With the potent hashish, charas, and toxic amounts of any cannabis preparation, this merges into sedation or even anesthesia. Two compelling reasons for classifying cannabis preparations as stimulants are: (1) Like the stimulants and unlike the depressants, they are hallucinogenic. In general, hallucinatory behavior, whether of endogenous or toxic origin, is evident only in association with other manifestations of excitation. With the depressants, hallucinations become manifest only during the rebound excitation of withdrawal. (2) Unlike the depressants and like the stimulants, cannabis preparations do not induce physical dependence.

An important characteristic of the stimulants, distinguishing them from the depressants, is that antisocial behavior is induced as a direct action of the drug and usually stems from false bravery, paranoid delusions, and other extremes of perceptual distortion without impairment of, or with even greater, capacity to perform physically. In contrast to the abuser of depressants, who will be physically incapacitated by an overdose, the abuser of stimulants lacks mental control over the fully active musculature and is likely to be more dangerous to society.

Physical dependence does not develop to any substance in this group, demonstrating that only strong psychologic reinforcement is necessary to perpetuate the most enslaving type of drug dependence. While it is true that the person psychologically dependent on the stimulants may have a compulsion for more drugs, this is not related to any physical need but rather to the desire for reward.

The occurrence of aversive symptoms and signs following the termination of a prolonged bout with massive doses of amphetamines or other psychomotor stimulants does not signify the development of physical dependence. Such reactions are manifestations of extreme exhaustion and stress resulting from absolute insomnia, fatigue, hunger and malnutrition, and disturbances in acid-base and water balance.

Whereas tolerance develops with the use of some of the drugs in this group—notably the amphetamines and LSD—it does not develop with the use of cocaine, the most powerful reinforcer of all, indicating that tolerance is not necessary for the development of strong psychologic dependence.

Characteristics of Stimulant Psychoactive Drugs (Group 2—Cocaine, Amphetamines, Pipradrol, Methylphenidate, Phenmetrazine, etc.; LSD, Mescaline, Marihuana [Hashish], etc.)

- Strong psychoactive effects, predominantly stimulant, which may merge into depression with large dosage
- Strong psychologic dependence
- No physical dependence
- Withdrawal symptoms and signs due to stress of insomnia, fatigue, disturbances in metabolism, and psychologic factors.
- Both acute and chronic psychotoxicity
- Severe and often violent antisocial behavior during drug administration and sometimes during withdrawal
- No established permanent brain damage

Group 3 is comprised of depressants capable of inducing physical dependence of two types. One type of dependence, the morphine type, extends to all morphine-like drugs. The other, the alcohol-barbiturate type, includes many sedatives, hypnotics, and hydrocarbons in addition to alcohol and the barbiturates.

Psychotoxicity following the "spree" use of excessive quantities of alcohol, many hydrocarbon solvents, anesthetics, the barbiturates, and other sedative-hypnotic drugs may result in personal injury or disorders of behavior. As with the stimulants, this results from the direct pharmacologic action of the drugs on the central nervous system. Unlike the stimulants, however, the depressants can cause loss of consciousness and motor function, thereby exerting a controlling influence on continued excessive drug intake.

The chronic abuse of the depressants is associated with two related but mutually independent phenomena, tolerance and physical dependence. Tolerance development modifies drug actions by reducing their effects on certain vital mechanisms, such as the cardiovascular and respiratory systems. Because all parts of the central nervous system do not become tolerant to these drugs at the same rate, the brain as a whole is exposed to larger and larger drug concentrations as tolerance develops in those systems mentioned above which are necessary for survival.

Continuous uninterrupted exposure of the nervous system to large concentrations of depressant drugs induces physical dependence and a state of latent hyperexcitability which manifests itself as a clearly defined syndrome only when administration of the drug is discontinued.

Physical dependence is a fundamental biologic phenomenon demonstrable in many mammals, involving depressant drug-cell interactions in the CNS. A unique, characteristic, and reproducible abstinence syndrome follows drug withdrawal. But physical dependence is *not* the only parameter which establishes dependence on this broad class of substances. Primary dependence relates solely to the reinforcing capacity of these drugs and is based therefore on pschologic dependence as with all other psychoactive drugs. But when physical dependence *does* become established following chronic use, it serves as a powerful secondary reinforcer of a negative nature to continued drug use because of the anticipated or actual aversive symptoms of withdrawal.

Great scientific and medical effort has been expended in studying physical dependence. Its parameters are well understood in the laboratory and in the clinic. It can be produced in animals and man without fail and it may be treated successfully. But we do not yet understand the cellular mechanisms that determine its existence. This concentration on physical dependence has resulted in its being designated one of the most important, and, in the minds of many, the most important factor in conditioning and maintaining drug-seeking behavior. In fact, physical dependence was for many years the principal factor which distinguished addiction from habituation, two terms that were assigned scientific prestige and definition by Tatum and Seevers nearly forty years ago, before anything was known about the synthetic morphine-like substances or their antagonists, the short-acting barbiturates, the minor tranquilizers, the amphetamines or related stimulants, or any of the synthetic hallucinogens. Interdisciplinary studies on drug-induced behavior by pharmacologists, psychologists, and psychiatrists were not even visualized. So little was known at that time that cocaine, recognized even then not to induce physical dependence, was classified as an addicting drug, although such classification violated the definition.

The terms "habituation" and "addiction" plagued scientists for years. Addiction, for example, implied dependence upon any substance, e.g., tobacco, which was considered to be inimical to the individual in the public health sense rather than in its capacity to induce disorders of behavior. Fortunately, better perspective has been brought to this complex field by the abandonment of the terms "habituation" and "addiction" as without meaningful scientific value.

The complex of symptoms and signs following drug withdrawal is known as the abstinence syndrome and exists in two distinct types.

The morphine-abstinence syndrome produced by morphine and all morphine-like drugs is a unique clinical syndrome associated with hyperactivity of all responsive tissues and characterized by changes in behavior and a marked increase in all somatic and autonomic activity. All morphine-like drugs may be used interchangeably to suppress this hyperactivity and may, therefore, be substituted one for another. Other depressants, such as alcohol, barbiturates, and phenothiazine, do not substitute specifically for morphine-like drugs and only suppress this hyperactivity if given in quantities which will generally suppress most nervous function.

The development of morphine-like physical dependence follows the classical dose-response curve. There is a no-effect level with small infrequent doses. Physical dependence develops and becomes more intense with increase in daily intake and frequency of administration until a peak dosage is reached above which further increments in dosage do not increase the intensity of the dependence. The morphine-abstinence syndrome, although distressing subjectively, is rarely life-threatening, and withdrawal can usually be accomplished without much difficulty.

The alcohol-barbiturate abstinence syndrome differs from the morphine-type in several important respects. The rate of development of physical dependence does not follow the typical dose-response curve. It is not produced with even moderate doses and especially with intermittent administration of these drugs. Prolonged and continuous administration of large and partially incapacitating doses is necessary over months or years when these drugs are taken orally in the ordinary manner. This is, however, only a matter of total daily intake since the completed syndrome with alcohol or the barbiturates can be induced in monkeys in a few weeks with continuous intravenous administration of toxic doses.

This type of abstinence syndrome is characterized by delirium, hallucinations, tremor, hyperthermia, and grand mal seizures. It is so intense as to be life-threatening. Abrupt withdrawal is a carefully calculated risk, the safe procedure being gradual withdrawal over several weeks. This syndrome may be induced by alcohol and other hydrocarbons, the barbiturates, and many of the so-called minor tranquilizers, but not by the phenothiazines. Any of these drugs except the phenothiazines may be used interchangeably by substitution to suppress the syndrome induced by the withdrawal of other agents in this category, suggesting a common mechanism of action. However, morphine-like drugs, although they may partially mask this syndrome, do not suppress it.

Characteristics of Depressant Psychoactive Drugs (Group 3—Morphine and Alternates; Barbiturates and Sedative Hypnotics; Alcohols and Hydrocarbons)

- Strong psychoactive effects of depressant type
- Strong psychologic dependence
- Strong physical dependence of two types, each with a unique and characteristic abstinence syndrome
- Acute and chronic psychotoxicity
- Antisocial behavior associated with acute and chronic intoxication during administration and during withdrawal
- Established brain damage with alcohol and barbiturates but not with morphine-like drugs

In briefly summarizing drug dependence it is evident that:

1. It involves most, if not all, of the world population.
2. It exists in innumerable patterns and in all degrees of intensity depending upon the nature of the drug, the route of administration, the dose and frequency of administration, and many other pharmacologic variables; it also varies according to the state of the individual and his environment.

3. Although it is not necessarily harmful to either the individual or to society, at the other extreme it may lead to serious injury or even death of the individual and to psychotoxicity and disorders of behavior with inimical effects on society.
4. The primary and common base of dependence for all drugs is psychologic reinforcement based on reward. When physical dependence is also present following chronic drug administration, the threat of the aversive effects of withdrawal serves as a powerful secondary reinforcer often leading to compulsive abuse.

With this background we can now relate drug dependence to drug abuse and define the latter in rather precise terms which may serve as a basis for the determination of abuse potential of psychoactive drugs.

For the moment we will pass over a discussion of the weak psychoactive agents of Group 1 and turn to the powerful stimulants, hallucinogens and depressants described in Groups 2 and 3.

From the known pharmacologic effects of these drugs, three general types of psychotoxicity which may lead to antisocial behavior can be described. Based upon the concepts hitherto outlined, I would define psychoactive drug abuse as follows:

1. Distortion of perception and other psychotic manifestations which can lead to violence. Most common in Group 2 drugs, and may persist beyond the drug action itself. Also may occur during withdrawal from drugs in Group 3, especially those of alcohol-barbiturate type.
2. Mental and physical incapacity to perform perceptual-motor tasks. Occurs most commonly with Group 3 drugs. Because it may occur during first trial with drugs such as alcohol, barbiturates, and volatile hydrocarbons, this type of abuse does not necessarily involve drug dependence. Dependence on this group of drugs perpetuates the problem by making it a chronic state.
3. Mental and/or physical lethargy leading to inability or refusal to assume individual or social responsibility. Can occur with any drug in Groups 2 or 3, but most typically with the opiates, barbiturates, alcohol, and hashish.

When defined in this manner, the objectives of measures designed to control or minimize the problem of drug abuse are more clearly discernible. In simple terms, the principal objective is to protect the large majority, who are not drug abusers, even though they may be drug-dependent, from the direct threat that the drug-abusing minority poses to their orderly existence as individuals and to the well-being of the organized society of which they are a part.

One truth is so easily discernible that it stands out like a beacon.

History tells us, and scientific facts and common sense support the conclusion, that no highly organized society could maintain a high position of achievement or establish superior goals for the future in any sophisticated field of endeavor without adopting measures designed to control or minimize effects of drug abuse as herein defined.

Let us turn for the moment to the problem of individual injury.

Humanistic considerations demand that society should also adopt measures to protect the individual against himself. The nature and extent of such measures are not easily definable because in many instances such measures to be effective must impinge on the rights of a large majority in order to protect a very small minority.

The whole problem of individual injury is such a murky area, even with drugs which are capable of inducing psychotoxicity leading to drug abuse, that I believe this facet of the problem should be dealt with separately and considered under the general heading of acute and chronic toxicity as with any other drug.

The situation becomes even more complex when one deals with the weak psychoactive agents. Even though these substances are incapable of inducing psychotoxicity in any dose, they may create strong drug dependence in a few emotionally disturbed persons. Following long-continued chronic exposure to massive doses, injury to certain organ systems may occur.

The case with tobacco is too well known to require further amplification. Furthermore, justification for use of tobacco is much more difficult to establish, at least in the minds of the non-users who represent a substantial fraction of the population.

The case with the analgesic-antipyretics (phenacetin, aspirin, and the like) is a better example, especially since use is almost universal and the total consumption of these substances is so large. For a vast majority of the population these substances may be used safely and effectively when taken according to the recommendations printed on the labels. A very, very small minority of emotionally unstable persons, mostly women, develop strong dependence on phenacetin to the point where they may consume 30 to 100 tablets daily for many years. This may lead to severe kidney injury and even death.

Claims are being made that the same situation may exist for aspirin and there are physicians who recommend that aspirin be placed on prescription. Let us view this problem as a hypothetical case.

Clearly, we have here dependence on a drug which is almost undetectable subjectively in recommended dosage but which may become more psychoactive in excessive dose. We also have specific organ pathology which may be fatal. As far as these few individuals are concerned these are very serious problems.

Do we call this psychoactive drug abuse? I think not. Chronic toxicity? Yes! But drug abuse? No. Do we propose unreasonable control measures, such as placing these drugs on prescription, thus penalizing the whole population in order to protect a few and creating logistic problems for the physician who is already swamped with paper and circumvented by rules? I hope not.

At this point in history, when drug abuse has surfaced as a major world problem, everyone is hell-bent to get into the act either of protecting the individual against himself or protecting society against the depredations of the drug abuser. It may be well to remind ourselves that drug dependence and drug abuse are not new. They only have new faces on old bodies.

Under such pressures the inevitable tendency is to over-react, either to press for abolition of all control and permit the situation to seek its own level, or, on the other hand, to toss all kinds of drugs into a hamper labeled "drug abuse" and propose controls and sanctions designed to protect the individual and society against every conceivable injury.

Without a careful analysis of history and studied efforts to maintain perspective, we Americans can again make the same blunders that are responsible in part for our present dilemma. These blunders can be attributed to the very simple fact that control measures were not consonant with the elementary scientific facts about drug dependence and drug abuse.

THE MANY FACES OF DEVIANT DRUG USE

DALE C. CAMERON, M.D., M.P.H.

Before taking up the main theme of this discussion, may I take this opportunity to pay public tribute to Dr. Maurice Seevers—long the distinguished Chairman of the Department of Pharmacology here at the University of Michigan Medical School—for his outstanding contributions to the field of drug dependence. Not only has he contributed greatly to clarity of thought in this field, but also he has done much to interest and train others and to identify those dependence-producing substances that constitute a significant threat to public health because of their liability to abuse.[1,2] Under certain international instruments, such as the Single Convention on Narcotic Drugs, 1961, the World Health Organization is responsible for identifying drugs having such properties in order that designated controls may be imposed on their availability. That task has been greatly facilitated by the steady flow of data from Professor Seevers' laboratory concerning the dependence liability of candidate drugs. Not willing to rest on past achievements in perfecting techniques for determining the physical dependence capacity of drugs, his laboratory has, for the past several years, been engaged in developing needed methods to determine the degree to which a drug is liable to induce psychic dependence.[3] Dr. Seevers, your colleagues in the fields of drug dependence and medical education are deeply indebted to you.

Turning now to the topic of deviant use of dependence-producing drugs, let me say at the outset that the word "deviant" is used in the sense of "different" or "apart from" the usual or normal, and *not* in the sense of social unacceptability. There is no intended implication that the deviation is necessarily harmful or beneficial, nor right or wrong. The choice of the phrase "deviant use" was made in a perhaps not altogether successful effort to get away from terms such as "abuse" and "addiction" which, for many persons, have strong moral overtones. Of course, evaluations must necessarily continue to be made relative to the consequences, and therefore the appropriateness, of the self-administration of dependence-producing drugs. It would be useful if such evaluations could be described in relatively neutral terms rather than to employ, from the outset, words suggesting that certain value judgements have already been made. "Drug dependence," as defined by the WHO Expert Committee on Drug Dependence, describes only a state characterized by certain behaviour and implies nothing more.[3] It is, therefore, a relatively neutral term. On the other hand, the word "addiction" is usually understood to indicate not only a particular pattern of behaviour but also one that is so harmful to society that certain controls should be placed on the availability of the dependence-producing drug involved. Again, as has been pointed out by Dr. Seevers, "defining drug-taking as abuse represents only the value judgement of an organized 'society,' nothing more. As commonly employed, the term 'abuse' when applied to drugs is all things to all men and a dumping ground for any adverse effect."[2] Use of the phrase "drug abuse" may often lead to confusion rather than clarification since it may carry connotations of inappropriate medical or other use, individual and social harm, or illegal action. Conversely, the phrase "deviant use of drugs" is intended to carry only the idea that such use is somehow unusual and may thus bear looking into.

This is not to say that only the so-called deviant use of dependence-producing drugs is interesting. Indeed, study of the non-deviant use of alcohol may help us to learn more about the development of patterns involving its deviant use by alcoholics. Most people regard tobacco as a dependence-producing substance, but its use is certainly not deviant in most cultures. Cigarettes are, nevertheless, of great health interest. Further, one might well endeavour to learn why the repeated use of tobacco is so strongly reinforcing in the absence of major psychotropic effects comparable to those due to dependence-producing drugs of the types under discussion at this symposium.

In discussing the deviant use of drugs I shall touch upon the following topics: (1) broad patterns of use in different regions of the world, (2) current beliefs and assumptions about drug use, (3) some factors apparently associated with drug-taking behaviour, and (4) some areas of needed research, with especial reference to possible approaches to the study of the prevalence and incidence of drug dependence and the associated human and environmental factors.

PATTERNS OF USE

Man has long sought ways of easing his discomforts and enhancing his pleasure; the use of drugs to these ends is an ancient practice. The availability of a drug at a given time and place is, necessarily, a precondition to its use. In earlier times, the substances used as chemical comforts were primarily local in origin. The choice among available psychoactive substances was not large and one of them usually became the local drug of choice, for example, alcoholic beverages in some countries, opium in others, and cannabis preparations in still others.

While there has always been some multiple drug use, especially the substitution of one drug for another when a preferred drug was unavailable, there has been a rather remarkable increase in recent years of the successive or simultaneous use of several drugs by a given user, apparently in order that he may experience a variety of effects. But even this is not an invention of today's youth, as those of you will know who can recall the combined use of opiates and cocaine of an earlier period, or who are currently called upon to treat middle-aged housewives suffering from the heavy combined use of alcohol and other drugs, particularly those of the barbiturate type.

Problems of multiple drug use, plus the achievements of modern chemistry and the advent of rapid and extensive means of transportation and communication, make increasingly difficult the already complex task of determining the incidence and prevalence of the deviant use of drugs. An additional and perhaps even greater difficulty arises from the fact that deviant drug use is usually associated with illegal or, at least, socially unacceptable behaviour and therefore is often hidden from view. And again, what is obviously deviant behaviour in one country may not be such in another.

For the reasons stated, there are few reliable, and even fewer comparable, data about the extent of deviant drug use and drug dependence of various types in different parts of the world. Despite this well-known situation, I have been asked to include in this presentation sufficient information on that subject to make it evident that the problems involved are not confined to a few countries or regions. This certainly is not a difficult task. First, several other essayists at this meeting are to discuss the situation in at least five separate countries or regions, evidence that this is not a local problem. Let me take up in alphabetical order the several types of dependence-producing drugs. This order is chosen to bring alcohol to the head of the list where I suspect it belongs.

Alcohol Type

The moderate use of alcohol in its various beverage forms is not deviant behaviour in most countries of the world. In certain countries in North Africa, the Middle East, and the Indian-Pakistani subcontinent, any use may well be deviant, but, even in some of these countries, the prevalence of moderate drinking is apparently increasing. However, the immoderate use of alcohol, measured in terms of manner and frequency, is deviant behaviour in essentially all countries. It would appear that, of those who drink at all, a certain percentage become dependent on alcohol. The prevalence of alcoholism varies considerably in different countries, and may depend both on the prevalence of those who drink and the prevalence of alcoholism among drinkers. Estimates made between 1945 and 1963 on the rate of alcoholism per 100,000 persons in the population aged 20 or over in selected countries include the following: England and Wales—865 to 1100, Switzerland—2100 to 2700, Chile—3610 to 4150, various states of the United States—1580 to 7090, and France—5200 to 7300.[4] The highest rate is about 8.5 times that of the lowest in the particular set of figures utilized to illustrate variability in the prevalence of alcoholism.

2

Amphetamine Type

Drugs of the amphetamine type are used in deviant ways in a number of places in the world. One pattern involves the oral self-administration of relatively stable amounts ranging from one or two to several therapeutic doses daily or intermittently. Persons of all ages, especially those of middle-age, appear to be involved. A second pattern involves a rapid escalation over periods of one or two weeks to the point where massive oral and often intravenous doses are taken. Such deviant use is to be found primarily among late adolescents and young adults. One or both of these patterns of use have been identified in Japan,[5] Sweden,[6] the United Kingdom, and the United States of America[7] among others. Such use in Japan and the United Kingdom has markedly decreased and the trend is said to be downward in Sweden. Deviant use of amphetamine-type drugs is beginning to appear in certain African,[8] European, South American, and other countries.[9]

Barbiturate Type

Deviant use of barbiturate-type drugs has much the same geographic distribution as those of the amphetamine type. However, their use by young persons in Sweden is relatively small, while their intravenous use is said to be gaining in the United Kingdom. In Australia and the United States of America, the oral use of barbiturate-type drugs is prominent among middle-aged housewives.[10]

Cannabis Type

When I said that I "suspected" that alcohol belongs at the "head of the list," I had in mind the possibility that cannabis might well belong there, at least in terms of the number of users. Its use in one or another of its several forms—bhang, ganja, charas, marihuana, maconha, kif, and hashish, to mention but a few—is particularly widespread in those areas where alcohol is least used. It is also relatively prevalent in some countries where alcohol is widely used. In some parts of India, where alcohol is not widely used, the use of bhang is neither illegal nor socially unacceptable, while the availability of ganja is controlled and the use of charas is proscribed. Ganja is usually more potent than bhang and marihuana; charas is a resinous material, comparable to hashish, which is more potent than other commonly used preparations. Various preparations of cannabis are widely used, not only in the Indian-Pakistani subcontinent but also in certain Middle Eastern, African,[8,11] and South American countries and their use is gaining in North America and Europe.[9,12]

Cocaine Type

The chewing of coca leaves in the Andean regions of South America is only a marginally deviant practice and perhaps not even that in some localities. Zapata-Ortiz estimates that 6,000,000 persons are regular users in Bolivia and Peru, the principal countries involved.[13] This would mean that 1 of every 2.5 persons of all ages uses coca. Goddard et al. say that "it is regarded as quite 'natural' that a young man when he begins work also begins chewing coca. Yet he is not required to do so for membership in the group of men to which he has just been admitted, nor is he particularly expected to do so."[14]

Hallucinogen Type

Naturally occurring hallucinogens are used in religious ceremonies by certain groups of native people. Such use is non-deviant in the particular group. Another pattern involves the recreational or so-called "mind-expanding" use of predominantly synthetic hallucinogenic materials by deviant subcultures, some members of which appear to pursue their use with an almost religious zeal. The latter pattern is to be found primarily in developed countries of the West.

Morphine Type

Drugs of the morphine type are most widely used in certain countries of the Middle East and in Southeast Asia. Prior to the cessation of poppy culture in Iran in the mid 1950's it was estimated that 1,500,000 persons in a population of under 25,000,000 (1 person in 16) used opium. Following the ban, the number was reduced, but those who remained or subsequently became dependent have variously been estimated to number 100,000 to 300,000 and possibly more (1 in 83 if there were 300,000). Heroin began to be used instead of opium, especially by younger persons who were disinclined to emulate the opium-smoking of their elders. Iran has recently rescinded its ban on poppy culture and is making opium available to certain of its narcotic-dependent persons. It is hoped that epidemiologic data can be obtained during the course of this important natural experiment. Other Middle Eastern and some North African countries are thought to be affected but to a lesser extent.

There is believed to be very substantial use of opium in the hill areas where Thailand, Burma, Laos, and China are in close proximity. Opium and morphine are used in Singapore while heroin is the drug of choice in Hong Kong and the flat-land areas of Thailand. Very rough estimates of users in this general region of the world would be measured in the hundreds of thousands and the rates in some countries could run as high as 1 person in 300 to 1 in 45.

Ingersoll has suggested that there are 120,000 to 180,000 heroin-users in the United States of America (1 in 1,100 to 1,700 persons).[15] In the United Kingdom the number of heroin- and other opioid-users coming to the attention of the Home Office was reported by Spear to be in excess of 2,240 during 1968, or about 1 person in 24,500.[16] During 1969, about the same number were known to the Home Office, but there was a marked reduction in heroin and a compensatory increase in methadone use. The use of opioids in Europe is on a small scale despite a recent sharp rise in some countries. There is thought to be relatively little deviant use of morphine-type drugs in South America and Africa, although there are reports of sporadic seizures in those areas.[9]

Looking at the problem another way, it has been estimated that substantially more than 1200 tons of opium flow into the illicit market each year.[17] Taken as opium, this would sustain about 800,000 to 1,600,000 opium-dependent persons, assuming that the usual daily consumption of an opium-smoker is approximately 2 to 4 gm per day. Taken as heroin, this would sustain about 2,200,000 to 6,500,000 heroin-dependent persons.* Of course, only a relatively small, but unknown, proportion of illicit opium is converted to heroin. Regardless of the uncertainty involved, it is clear that, if the assumptions made relative to amounts available and average consumption per day approach reality, a substantial number of persons could be involved in the use of drugs of the morphine type.

CURRENT BELIEFS AND ASSUMPTIONS

The deviant use of drugs and drug dependence involve a complex interaction between the characteristics of the drug-taker, the nature of his several environments, and the properties of the drugs taken. One of the many facets of these complex interactions has to do with the ways in which the phenomena are seen or understood by various groups of observers; that is, by members of different drug-taking and non-drug-taking cultures and subcultures in various parts of the world. Not surprisingly, the deviant use of drugs presents differing faces to different observers. The "eye of the beholder" as well as the nature of the phenomenon influence perception, particularly where certain beliefs and values are strongly held.

There are two sets of beliefs that influence a given observer's perceptions in this field. The *first* has to do with the character of the prevailing conscious and unconscious beliefs and assumptions about the causes and effects of drug-taking behaviour. Without going into the "effects" side of the question, a review of some of the beliefs and assumptions relating to cause will indicate the wide divergence of views. The WHO Expert Committee on Drug Dependence, meeting in August 1970, noted that "no single 'cause' has been demonstrated." It then listed the following as some of the many etiological hypotheses currently put forward in relation to designated types of drug dependence:

> (1) that such drug dependence may be a manifestation of an underlying character disorder in which immediate gratification is sought in spite of the possibility of long-term adverse consequences and at the price of immediate surrender of adult responsibilities;
> (2) that it may be a manifestation of delinquent-deviant behaviour in which there is pursuit of personal pleasure in disregard of social convention, so that to some this is primarily a moral problem;
> (3) that it may be an attempt at self-treatment by persons suffering from (a) psychic distress either of the normal variety seen, for instance, in adolescence or as a reaction to social and/or economic stress, frustration or blocked opportunity; or the more persistent problem of depressive illness, chronic anxiety, or other psychiatric disorders; (b) physical distress—hunger, chronic fatigue, or disease; (c) a belief that

$$^*\frac{1{,}200{,}000{,}000 \text{ gm opium}}{365 \text{ days} \times 2 \text{ gm per user}} = 1{,}643{,}835 \text{ persons}$$

Assuming that the usual daily consumption of a heroin-user is approximately 50-150 mg, the average morphine content of opium is about 10%, and the average yield of heroin per kg of morphine is roughly 1 kg; then

$$\frac{120{,}000{,}000 \text{ gm heroin}}{365 \text{ days} \times 0.050 \text{ gm per user}} = \begin{matrix}6{,}575{,}342\\ \text{persons}\end{matrix} \text{ or } \frac{120{,}000{,}000 \text{ gm}}{356 \times 0.15 \text{ gm}} = \begin{matrix}2{,}191{,}781\\ \text{persons}\end{matrix}$$

the drug has special powers to prevent disease or to increase sexual capacity;
(4) that it may provide a means of achieving social acceptance in a social subculture, particularly for the socially inadequate;
(5) that it may be a manifestation of a permanent or reversible metabolic lesion brought about by the repeated use of high doses of drugs;
(6) that it may be part of a rebellion against conventional social values relating to pleasure, tradition, success, and status;
(7) that even in the absence of pre-existing psychopathology, it may result from the acquisition of a complex set of instrumental and classically conditioned responses and may therefore be a form of learned behaviour;
(8) that even in the absence of underlying psychopathology, it may result from socio-cultural pressures leading to heavy use of a drug, for example, alcohol;
(9) that any or all of these factors may play a role in the causation of drug dependence in a given individual.[18]

There is a *second,* broader and all-pervasive set of beliefs and values that influences one's perception of deviant use of drugs. It also colours one's choice of etiological hypotheses. As I have noted elsewhere:

> This broad system of values has to do with such matters as (1) the relative worth of material possessions and spiritual, cultural and traditional beliefs and experiences, (2) the respective importance of individual rights, prerogatives and responsibilities as opposed to those of society, (3) the nature of practices which are considered to be 'good' or 'evil' and (4) the meaning and value of life itself.[19]

FACTORS ASSOCIATED WITH DRUG USE

The capacity to induce psychic dependence is the only characteristic shared by all dependence-producing drugs. It has too long and too widely been assumed that the most important factor in the perpetuation of drug dependence of the morphine type was the wish to avoid the physical and other discomfort associated with withdrawal. This assumption has been held despite the fact that former users of drugs of this type often resume their drug-taking after having completed withdrawal. Further, many persons become dependent on drugs that produce little or no physical dependence.

In judging the presence or absence of psychic dependence in an individual, it is important to ascertain to what extent the use of drugs "(1) appears to be an important life-organizing factor and (2) to take precedence over the use of other coping mechanisms."[20] Drug-taking is an important life-organizing factor if substantial amounts of time and energy are devoted to obtaining, thinking about, using, or discussing the drug or drugs in question. Drug-taking has assumed a precedence over other coping mechanisms when the persons concerned tend to respond to differing life situations and personal moods by taking drugs in preference to any of the other possible responses in his repertoire.

Among the more important psychic "motives" that often appear to be associated with the deviant use of drugs are: (1) to escape from something, (2) to have a new, pleasurable or thrilling experience, (3) to achieve improved "understanding" or "insight," (4) to achieve a sense of belonging, and (5) to express independence and sometimes hostility.[19]

Turning from personal "motives" to sociocultural influences, it may be noted that, in settings where taking a particular drug is socially acceptable, moderate use tends to be widespread. Where large numbers of persons use a drug, and particularly where attitudes towards intoxication are lenient or ambivalent, a significant proportion of users tends to become dependent. Where the use of a drug is not culturally accepted, such use as does occur tends to be individually excessive, though generally not widespread in the population. The deviant use of drugs often appears to be associated with rapid socioeconomic change. If so, it may be related to a weakening of cultural controls as old patterns of living are replaced by new. Leaving a familiar locale to live in different circumstances with different persons may contribute, at least temporarily, to the diminution of cultural controls.[20]

Deviant drug use is found among all socio-economic classes. Though the desire to escape momentarily from the impact of grinding poverty has often been mentioned as a "cause" of drug use, it must be noted that leisure made possible by relative affluence also appears to be associated with recourse to drugs.

The extent and rapidity of mass communication and transportation now enable many persons in one part of the world to learn quickly of the doings of others in distant places. The sensational manner in which news of drug use and users is sometimes presented and the ease with which rumors and misinformation about alleged "benefits" of such use are disseminated may stimulate some persons to seek out and experiment with drugs. As I look at some of the so-called news reports on drugs, I am sometimes reminded of a perhaps apocryphal entrepreneur, active in the United States of America during the alcohol prohibition era, who marketed, at least temporarily, a packet containing sugar and a fruit flavour. The label read somewhat as follows: "Instructions: Use only for the preparation of syrup! Warning: Do *not* place in an earthenware container with yeast and cover with two gallons of water. To do so might result in the production of an alcoholic beverage, which is unlawful."

RESEARCH NEEDS

Four areas are identified as requiring markedly intensified research effort.[19] The first has to do with the *effects on man* of taking drugs over long periods of time and at various dose levels. Studies in this area should endeavour to make a clear distinction between pharmacodynamic effects *per se* and those associated with society's reaction to a person who indulges in non-acceptable behaviour. The existence in different parts of the world of varying socio-cultural attitudes towards the use of a given drug will facilitate the design of projects addressed to these questions.

A second important area of inquiry has to do with the *natural history* of drug-taking. More information is needed on the factors associated with the initiation, perpetuation, and discontinuation of deviant drug use. In connection with the latter, Scharze observed that of 40 neophyte heroin-users studied, 22 discontinued drug use because of such external events as the arrest of the supplier.[21] No decision had been taken prior to the "event" to discontinue drugs. On the other hand, 18 persons discontinued after making a conscious decision to do so. This decision was often triggered by an external event such as the conviction or death of a friend. Nine persons quit because they experienced physical dependence. Not only is it important to learn why certain persons start or discontinue drug use but also why others, possessing essentially similar personality characteristics and subjected to the same socio-cultural environment, *do not* adopt drug-taking as a way of life. Information on these points should be of value in developing increasingly effective preventive and rehabilitation activities.

A third area deserving markedly increased research effort has to do with *evaluating the effectiveness of preventive, treatment, and rehabilitation services*. Suitable criteria must be established against which to measure the results of various services. Such criteria should include not only the presence or absence of drug-taking behaviour but also its extent if it still continues, occupational performance, and the character of personal and social relationships. Beyond the question of criterion measures is the matter of making meaningful comparisons between different treatment methods or techniques, including comparisons between treatment and no treatment or placebo.

The final research area to be mentioned has to do with the *identification of dependence-producing substances*. Since the deviant use of certain dependence-producing substances *may* result in individual or public health problems, it is important that substances with such potential be identified as early as possible. The need for control of such a drug and the level of that control should be determined on the basis of "(a) the degree of risk to public health and (b) the usefulness of the drug in medical therapy."[3]

WHO POLICY AND PROGRAMME

Beyond WHO's activities related to the identification of dependence-producing drugs and the eventual international control of some of them, the provision of consultative services on request of governments and the development of technical reports, the Organization has convened nine WHO Expert Committees and one Scientific Group that have concerned themselves with measures designed both to minimize the chances of potential users becoming interested in drugs and to foster the development of effective treatment, including rehabilitation programmes. In short, WHO has long been concerned with both the demand as well as the supply side of the "supply and demand" balance.

The Twenty-third World Health Assembly, in a resolution on drug dependence adopted in May 1970, noted "with great concern the extensive and serious public health problems resulting from the self-administration of dependence-producing drugs, in particular among the younger age-group" and expressed the conviction "that there is an urgent need at local, national and international levels to implement improved preventive, treatment and rehabilitation measures." The Assembly recommended (1) "that the World Health Organization should encourage and assist the development of improved preventive, treatment and rehabilitation programmes and the pursuit of needed knowledge in the field of drug dependence;" (2) urged Member States to promote these same ends; and (3) requested the Director-General "to develop means for the international collection and exchange of data on the prevalence and incidence of drug dependence, and on the human and environmental factors associated therewith."[22]

In compliance with the resolution, WHO will continue to (1) utilize the advice of experts from all over the world, (2) make available technical reports and publications, (3) furnish, on request of governments, consultation on preventive, treatment, and rehabilitation services, and (4) cooperate with international and other organizations having common interests and goals.

Beyond these continuing activities, a further undertaking is called for by the Assembly resolution: the development "of means for the international collection and exchange of data on the prevalence and incidence of drug dependence, and on the human and environmental factors associated therewith." As was noted earlier in this presentation, it is difficult to collect data on the prevalence and incidence of activities considered socially unacceptable or unlawful, or both. Especially is this so when an official governmental agency undertakes this task with respect to previously undetected behaviour. Data gathered directly by such agencies clearly must be augmented by information from other sources. Some of the means now being considered are:

1. The establishment of *reference facilities* in selected regions of the world to provide relatively brief, factual, and interpretive reports on a regular basis on drug dependence in their geographic areas, these reports to contain (a) published objective data or informed opinion on the prevalence and incidence of drug dependence of various types, (b) morbidity and mortality associated with drug-taking, (c) the nature and extent of treatment, rehabilitation, and preventive services provided, and (d) the nature of policy changes relative to drug-taking and the manner of implementing these changes.

2. The convening of *consultation groups*

of *selected investigators* who have entrée to high drug-use groups to discuss (a) their experiences in obtaining data about the patterns of drug use in their areas of study, (b) the human and environmental factors associated with the initiation, perpetuation, and discontinuation of self-administration of dependence-producing drugs, and (c) the feasibility of collecting data along reasonably comparable parameters.

3. The carrying out of *cross-sectional* and *longitudinal studies* in the *general population* and in *selected high-risk population groups* relative to (a) the prevalence and incidence of use of dependence-producing drugs of various types, (b) patterns and methods of use, (c) means of transmission, and (d) the social and personality factors associated with *non-use* as well as *use* in a given socio-cultural setting.

4. The establishment of *epidemiologic intelligence teams* composed of persons knowledgeable about problems of deviant use of drugs and representing such disciplines as anthropology, biostatistics, clinical psychopharmacology, clinical psychology, epidemiology, internal medicine, public health administration, psychiatry, and sociology. The members would be available as a team to study selected populations in which there was reason to believe that an outbreak of drug dependence or change of drug-abuse pattern might be occurring. The members would meet at regular intervals to plan their methods of approach and to develop research instruments.

These several possible approaches, as well as others, will require much further study as to their technical feasibility, anticipated fruitfulness, and cost before decisions can be taken relative to implementation.

CONCLUSION

In this presentation, we have looked briefly at some of the many faces of the deviant use of drugs. In the main, the nature of their broad features is quite readily apparent. However, I have no doubt that during this exercise some persons may have perceived at least a portion of these images in somewhat differing ways. Nevertheless, I am sure one would find quite general agreement among those present here, on the proposition that, where important features of deviant use of drugs are not now clearly seen, increasing clarity is more likely to be achieved by the application of additional light through research than by adding greater warmth through heated repetition of firmly held views.

REFERENCES

1. Eddy, N., Halbach, H., Isbell, H., and Seevers, M.: Drug dependence: its significance and characteristics. *Bull WHO 32*:723-733, 1965.
2. Seevers, M. H.: Drug dependence and drug abuse: a world problem. *Pharmacologist 12* (2):172-181, 1970.
3. WHO Expert Committee on Drug Dependence: *WHO Techn Rep Ser* No. 407, 6-18, 1969.
4. WHO Expert Committee on Mental Health: *WHO Techn Rep Ser* No. 363, p. 43, 1967.
5. Kato, M.: An epidemiological analysis of the fluctuation of drug dependence in Japan. *Int J Addict 4*(4):595, 1969.
6. Goldberg, L.: Drug abuse in Sweden. *Bull Narcot XX* (1):3, 1968.
7. Bewley, T.: Drug dependence in the USA. *Bull Narcot XXI* (2):14, 1969.
8. Lambo, T.: Medical and social problems of drug addiction in West Africa. *Bull Narcot XVII* (1):1-13, 1965.
9. United Nations Commission on Narcotic Drugs: (Document E/4455) *Economic and Social Council: Official Records*, pp. 16, 22-23, 1968.
10. Rankin, J.: *The Size and Nature of the Use and Misuse of Alcohol and Drugs in Australia.* (In press.)
11. Tanner, R.: Drug addiction in East Africa. *Int J Addict 1* (1):11, 1966.
12. United Nations Commission on Narcotic Drugs: (Document E/4606/Rev. 1) *Economic and Social Council: Official Records*, pp. 42-47, 1969.
13. Zapata-Ortiz, V.: The chewing of coca leaves in Peru. *Int J Addict 5* (2):288, 1970.
14. Goddard, D., Goddard, S., and Whitehead, P.: Social factors associated with coca use in the Andean Region. *Int J Addict 4* (4):579, 1969.

15. Ingersoll, J.: Statement before the UN Commission on Narcotic Drugs. September 28, 1970.
16. Spear, H.: The growth of heroin addiction in the United Kingdom. *Brit J Addict 64*:247, 1969.
17. United Nations International Narcotics Control Board: *Report to the Economic and Social Council on the Work of the Board* (Document E/INCB/5) *Economic and Social Council: Official Records,* p. 21, 1969.
18. WHO Expert Committee on Drug Dependence: *WHO Techn Rep Ser* No. 460, pp. 11-12, 1970.
19. Cameron, D.: Drug dependence: some research issues. *Bull WHO 43*:590-591, 1970.
20. Cameron, D.: Abuse of alcohol and drugs: concepts and planning. *Chron WHO, 25*: 10, 12, 1971.
21. Scharze, R.: Cessation patterns among neophyte heroin-users. *Int J Addict 1* (2):23-32, 1966.
22. Twenty-third World Health Assembly: *Off Rec WHO, 184*:22 (Resolution WHA 23.42), 1970.

EPIDEMIOLOGY OF DRUG ABUSE IN SWEDEN

LEONARD GOLDBERG

Use and abuse of drugs, whether as medicaments to combat disease by helping the individual to find relief from hunger, pain, anxiety, and discomfort, or as a means to induce a different state of mind, as part of religious rites in many cultures, in others as a celebration of feats of bravery after wars or marking great events in the life of the individual or the tribe, are widespread. In certain individuals the use of drugs may form part of their daily life, and play a substantial role by infringing on the work and normal occupation of the drug taker. In some individuals the indulgence in and occupation with drugs may eventually interfere with their health, with their physical, psychic, and social well-being, and with the normal functioning of the society in which they live.

The border line between use and abuse, between health and disease, whether for an individual or a society, as part of ordinary life or a symptom of an underlying disease, as a commended occupation or a misuse of compounds for purposes outside the recommended use, may be difficult to assess. The proper place varies with the type of drug used, with the type of individual indulging in the use, with the acute and long-term medical, social, and economic consequences of the intake, and with the environment where the drug is taken, its mores and habits, legislation and social norms.

THE EPIDEMIOLOGICAL OR ECOLOGICAL MODEL

One method of attempting to analyze the phenomenon of drug use and abuse within a society is to employ an epidemiological or ecological approach, originally used for the study of epidemics in the classical sense —influenza, malaria, tuberculosis, where this approach encompasses the *agent* or causative microorganism, the *host* or individual engaged, and the *environment*. The actual phenomenon will be the result of the interaction or interplay between these three groups of factors. With regard to drug use and abuse the agent represents the drug, and its properties, the host represents the individual, his psychic and somatic make-up and his response to the action of the drug, and the environment will include all factors coming from outside, and interacting with drug and man. In this interaction also phenomena such as resistance, tolerance, interaction, and interplay are included.

Since the thirties an epidemiological or ecological approach has been used for the analysis of a great number of phenomena in society, not limited to transferrable or communicable diseases. This mode of analysis has been used to study such widely differing entities as traffic accidents, mental diseases, or diabetes and lately the incidence and prevalence of alcohol use and abuse and its consequences for the individual and for society.

Employing this mode of approach on a study of drug use and abuse would denote:

I. *Agent,* i.e., the drug and its properties; some pertinent factors are listed in Table 1.

Table 1. Factors of Importance for the Emergence of Drug Abuse and Drug Dependence

Agent

1. Chemical structure
2. Metabolic fate
3. Physiological effects
4. Therapeutic effects
5. Toxic effects
6. Tolerance phenomena
7. Emergence of dependence
8. Abuse liability
9. Risk with prolonged use—therapeutic and/or non-medical
10. Dose
11. Mode of administration
12. Type of drug abused
13. Intervals between doses
14. Frequency and duration of administration
15. Increase in dose
16. Psychotoxicity
17. Interaction with other functions and other drugs

Individual

1. Genetic and acquired properties
2. Metabolism
3. Physiology
4. Psychic make-up
5. Personality structure—neuroticism, deviant personality
6. Influence of age, occupation
7. Tolerance phenomena
8. Dependence liability
9. Psychotoxicity

Environment

1. Infancy
2. Home-family relations
3. Social structure
4. Subcultures
5. Availability of drug—legally, illegally
6. Use—abuse
7. Spread of habit—"contamination"
8. Legislation
9. Other social phenomena—alcohol, tobacco, crime
10. Risk to public health

II. *Individual,* the individual using the agent, and his physiological, psychological, normal, and abnormal responses to the acute and/or chronic administration of the drug. See Table 1.

III. *Environment,* the factors outside the agent and individual, modifying the agent—individual interplay, its characteristics ranging from the influence of age, sex, family and group relations and habits and practices, to emergence of subcultures and the effects of legislation and of other social and economic conventions and mores (Table 1).

The *aims* of an epidemiological approach, for the purpose of the present study, can be summarized into:

1. *Fact finding surveys,* presenting prevalence and incidence, distribution of types of drugs used, age, sex, and marital status, distribution in various strata of the population, medical and social consequences, dynamics of development and changes with time.

2. *Elucidation and analysis* of elements involved and isolation of possible factors of importance and of underlying interaction patterns, of mechanisms, and of possible causal relationships.

3. Stating possible theoretical and practical *consequences* and *implications* of the findings obtained.

4. *Counter-measures* to cope with an undesired situation are important ultimate *aims* of an epidemiological approach. They fall outside the objectives of the present study, but range from possible measures to be taken, aimed at the agent, the individual, and the environment, elucidated from different points of view, ranging from prevention, treatment and aftercare to social, educational, and legislative measures. Important items include attempts to reduce availability of agent, to reduce contamination by actual pathogenic agents (bacteria, hepatitis virus), and to interfere with the transfer of certain undesirable habits and practices.

The *aim* of the present study is to analyze the situation in Sweden with regard to drug use and abuse from an epidemiological or ecological point of view, attempting to analyze the agents involved, the individuals and their characteristics and some factors in the environment, and to attempt to understand some of the elements involved in the interplay between the three groups of factors involved, with special reference to:

1. description of the situation in various strata of the population with regard to prevalence, incidence, and intensity of drug use;
2. elucidation of the possible differences between groups with regard to drug preference, and medical and social consequences;
3. assessment of possible ill-effects on the individual and on society of prolonged drug use, and pointing to possible differences between various types of drugs in this regard;
4. evaluation of some of the factors and possible underlying causes that characterize the emergence of widespread drug use and abuse, from the role of age, marital status, social and economic status to legislation and other possible regulatory forces;
5. characterization of some of the changes that have occurred within the last two decades with regard to choice of agents, types of individuals involved and the build-up of the environment, of legislation, and of society.

DEFINITIONS

Narcotic Drugs

From a biological or purely medical point of view a "narcotic" drug means an agent inducing "narcosis," i.e., a depression of cell activity, e.g., reduction of cell metabolism, reactivity to stimulation, growth, or reproduction. This definition comprises, for example, drugs inducing general anesthesia, analgesia, sleep, sedation, or tranquilization.

From a legal point of view, however, narcotic drugs are defined as those encompassed by international or national conventions and having certain elements in common, independent of whether they have depressant, stimulant, hallucinogenic, or other mind-disturbing action.

The international conventions cover three main groups of narcotic drugs: opiates—opium, morphine derivatives and synthetic drugs with morphine-like action; coca leaves and cocaine; and cannabis and its preparations.

Sweden has added two more groups of substances to its National Narcotics Drug List, stimulant drugs and hallucinogens, because these drugs are being abused in Sweden.

The definitions used in the present survey for "narcotic drugs" follow those of the Swedish legislation, and comprise:

1. *Opiates, coca leaves* and *cocaine, cannabis* and preparations thereof; these drugs being included in international conventions to which Sweden is a party.
2. *Stimulant* drugs acting on the central nervous system, added to the National Narcotics Drug List and handled as narcotics in Sweden owing to their being abused. These drugs comprise: amphetamines (1944) amphetamine, metamphetamine, dexamphetamine, phenmetrazine, e.g., Preludin (1959), methylphenidate, e.g., Ritalina (1959), and pipradrol (1959).
3. *Hallucinogens,* mainly mescaline, psilocybin, and lysergide (lysergic acid diethylamide—LSD-25), added to the National Narcotics Drug List in 1966.

Addiction, Habituation, and Dependence

The World Health Organization through its Expert Committee on Drug Dependence has defined various states and types of drugs: addiction and habituation and addiction-producing and habit-forming drugs in 1950–1952 and 1956–1957; and dependence of various types in 1963–1964.

When addiction and habituation are mentioned in this survey they refer to the definitions laid down by WHO in 1956–1957;

when dependence is discussed, the definitions of 1963–1964 apply.

Drug Abuse

Drug abuse means essentially the use of drugs for non-medical purposes or in doses, usually excessive amounts, not founded on ethical medical grounds.

In Sweden drug abuse may mean several things, in some cases that the abuse has been of a rather short duration; in others the term may mean "leading to medical and/or social ill-effects." In many cases the term *drug abuse* is used in the meaning of "addiction" or "habituation" according to the old terminology (WHO, 1950–1957) or "drug dependence of a specified type" according to the WHO 1963–1964 definition.

Because many of the studies reported in this survey have not gone into details regarding the types of abuse observed, it is impossible at present to elucidate the term further than was done in the original studies.

When speaking of "drug" in this connection, the term usually comprises (1) the narcotic drugs according to the Swedish definition, and (2) hypnotics and sedatives (tranquilizers).

In a number of studies also (3) volatile solvents (trichlorethylene, thinner) and glue used by inhalation are included.

"Knark"

The majority of the agents abused in Sweden are colloquially referred to by the Swedish word *"knark."* "Knark" means an agent that (1) can bring about "queer" psychic symptoms, i.e., a psychoactive agent in modern terms; (2) can be "abused," i.e., used for "self-medication," and not according to prescription or medical advice on strict indications; (3) can be, and often is, taken in excessive doses; and (4) usually is, or can be, obtained illegally.

The term, originating in the early sixties, includes (1) the narcotic drugs according to the Swedish definition; and (2) hypnotics and tranquilizers; usually neither the volatile agents (solvents), alcoholic beverages, nor tobacco are included in the term.

INTRODUCTION

Although abuse of alcoholic beverages in Sweden has been a social problem for many hundreds of years, its medical consequences being recognized by Linnaeus in the eighteenth century and by Huss in 1849 who introduced the disease concept and coined the term "Alcoholismus chronicus," the abuse of drugs has not been recognized as a public health problem until in the last few decades.

Abuse of barbiturates and organic bromide derivatives was described by Rylander in 1933; intake of large doses, 20 to 60 tablets daily, emergence after prolonged use of tolerance and in many cases chronic ill-effects, mainly affecting the central nervous system, the spinal cord, and peripheral nerves. Mental changes, ataxia, and peripheral neuritis, resembling *dementia paralytica,* characterized many of the cases described. Chemical diagnosis of high barbiturate or bromide concentrations in plasma and spinal fluid, and negative results of Wasserman and Kahn tests (for syphilis) helped to secure the proper diagnosis and treatment; slow withdrawal of the drugs abused.

Cases of abuse of opiates by oral intake or by injection of soluble opioids were known but occurred only sporadically.

An enquiry in 1940 to all state and municipally engaged physicians disclosed a total of 70 cases, mainly of opiates.

A beginning abuse of amphetamines and a more widespread use of other drugs led the Board of Health and Welfare (Medicinalstyrelsen) to estimate the number of abusers of narcotic drugs. Lindgren and Krook (1956) concluded, by means of questionnaires, that during 1954 a total of 146 patients abusing narcotic drugs had been treated in psychiatric clinics and mental hospitals. Males dominated, 63 per cent, against

37 per cent of females. A total of 78 per cent of patients abused opiates. Of these, morphine was abused in 43 per cent, morphine derivatives in 21 per cent, and synthetic morphine-like drugs in 14 per cent. Less than 1 per cent (two persons) abused heroin.

The rest, 22 per cent, abused stimulant drugs, 21 per cent amphetamines, or a total of 35 patients, and less than 1 per cent cocaine (two cases).

The mean age was 44 years; only 17 per cent were under 30, and only 1.4 per cent were under 21 years of age.

About half of the persons were former patients, i.e., had acquired the habit as a consequence of prolonged medication for various diseases, owing to the physician not being aware of the inherent risk of the drugs prescribed, hence they were "iatrogenic" or medically induced.

About one fourth of the people belonged to the medical (physicians) or paramedical (nurses) professions, showing the increased "occupational" risk of these professions.

Few individuals used drugs in a group setting (together with other persons), and abuse of a combination of different drugs, e.g., morphine and stimulants such as amphetamines, was extremely rare.

A later survey, carried out by a Mental Health Committee, showed that in 1960 a total of 190 individuals had been treated in hospitals for drug addiction or drug abuse. About 65 per cent used opiates and 35 per cent amphetamines. The mean age of this group was now lower than in 1954, 36.5 years, somewhat higher for those taking opiates than for those taking stimulants; 7 per cent of the group were under 20 years.

The finding that *amphetamines* ("pep pills") had stimulating properties and also might lead to appetite-suppression, together with widespread and intensive communication of their beneficial effects by mass media—newspapers, weekly journals, and radio—led to a rapid increase in their use. The preparations, introduced in Sweden in 1938, were already put on prescription in 1939, curbing the spread for about one year. The increased use then progressed rapidly, especially among students, housewives, and authors. A survey in 1942–1943 showed that 70 to 80 per cent of students had used these drugs at least occasionally, i.e., "spree" use, as stimulant drugs in preparation for examination (often with disastrous effects!).

During this time only a small number of severe cases of abuse were known: an estimated total of some 200 excessive users, and some 20 to 40 cases of severe abuse with somatic and psychic complications, from vasomotor consequences to overt psychoses, sometimes displaying a schizophrenia-like syndrome, disappearing after withdrawal.

The small number of severe cases, 20 to 40 out of an estimated total of some 200,000 occasional users, corresponds to an incidence of 100 to 200 per million users, or 1:5000 to 1:10000, or 3 to 6 per million inhabitants (Goldberg 1944, 1968 a, b).

The sale of drugs increased eightyfold in a five-year period, 1939–1943 (Goldberg, 1944). Warnings sent to the medical profession by The Board of Health and Welfare and the placing of the amphetamines on the Swedish National Narcotics List in 1944 reduced the sale to less than half.

During the rest of the forties and the first half of the fifties there were still only sporadic cases of amphetamine abuse; 32 people treated in hospitals in 1954 and 67 people in 1960. A slowly rising phenmetrazine (Preludin) abuse and a beginning abuse of methylphenidate (Ritalina) in 1958–1959 led to the placing of these stimulant drugs on the Swedish Narcotics List (1959).

The following development is characterized by a slow shift from a widespread use of small doses with few cases of abuse of large doses and somatic and psychic complications to a curb in the use and a significant rise in the number of abusers on high doses, injecting the drug, and suffering from mental and somatic complications.

The similarity between the pattern described for amphetamines and the picture characterizing the first years of introduction of cannabis, in a society where cannabis had been unknown, is obvious (Goldberg 1944, 1968 a, b). The use of cannabis is now characterized by an "epidemic"-like, exponential spreading, the majority using small doses only occasionally, most out of curiosity, following group mores; in the beginning only a small percentage used the drug daily in high doses, leading to complications in a selected number of cases.

Starting in the late fifties but accelerating in the sixties, the picture changes. Younger age groups are involved, individuals indulge in drug-taking in groups ("drug-joints"), the motivation for drug-taking changes, criminal acts become involved more and more, and new motifs, often based on new views, on a different outlook on life, a new philosophy, become new elements in the drug abuse pattern.

This general picture will be elaborated and supported by a number of studies aimed at throwing light on the present situation and its background, the interactions between agent, individual and environment, and some of the consequences in society and to public health.*

SURVEYS

Hospitalized Patients

Several studies on the prevalence of drug abuse among hospitalized patients have been carried out in the last few years. Two point-prevalence studies have been made on two occasions, October 5, 1966, and April 3, 1968.

Questionnaires were sent to all Swedish hospitals, including psychiatric clinics and mental hospitals, in order to obtain information on all patients with the diagnosis drug abuse treated on one and the same day.

The results of the 1966 survey, comprising 556 patients, have been reported (SOU 1967:25; Goldberg 1968 a). Some of the main results of the 1968 study, comprising

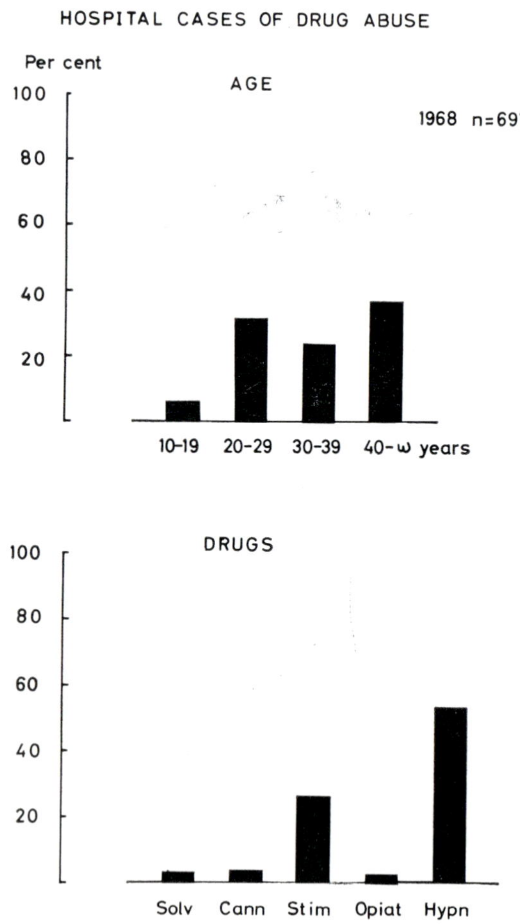

FIG. 1. Point-prevalence study. Hospital cases of drug abuse, seen in Swedish hospitals on 3 April 1968 (n = 697). Age distribution and types of drugs abused. Only cases with abuse of one type of drug mainly are given in the graph; in addition 9.3 per cent have combined different types of drugs. Adapted from data by Håkanson and Hoffman (1969 a).

* Detailed accounts of the development up to 1967 are given by the Swedish Commission on Treatment of Drug Addiction (Narkomanvårdskommitén) in two monographs in Swedish (SOU 1967:25 and 1967:41); the development up to 1969 is elucidated by the Commission in two additional monographs in Swedish (SOU 1969:52 and 1969:53). Detailed accounts in English and critical reviews are given by Goldberg 1968 a,b, 1970) and by Inghe (1969). The present communication attempts to analyze the development in a wider context and to make the results of these and other studies available to readers outside of Scandinavia.

a total of 697 patients (64.8 per cent men, 35.2 per cent women), will be given here, mainly based on the data given by Håkanson and Hoffmann (1969 a).

The age distribution is shown in Figure 1. The mean age is 34.3 years, ranging from under 20 to over 60 years. The distribution shows two peaks, one with a maximum at 25 to 29 years, and one with a maximum around 40 to 49 years and older, inferring a heterogeneous material, most probably composed of two different samples, one younger with a lower mean age, about 25 years, and one older with a mean age over 40 years.

When looking at the types of drugs abused (Fig. 1) it is noted that the majority (53.8 per cent) abused hypnotics—both barbiturate and non-barbiturate derivatives—and tranquilizing and sedative drugs. One fourth abused central stimulants (27.1 per cent); only a minority inhaled volatile solvents (2.3 per cent), smoked cannabis (3.4 per cent), or used opiates (2.3 per cent). Somewhat less than one tenth (9.4 per cent) have combined different types of drugs. The majority of these cases combined stimulants with hypnotics and opiates; the younger users also added cannabis.

With regard to modes of administration, oral intake predominates (69 per cent). One fourth injected the drug (25 per cent); inhalation was noted in 6 per cent.

When analyzing the marital status, a clear difference from what is normal in the age groups studied is noted (Fig. 2). The majority of the patients were single (54 percent), against 12 per cent in the same age groups in the total population. A minor proportion was married, 25 per cent against 83 per cent in the whole population. There was a higher proportion of divorced people, 19 per cent against only 3.5 per cent in the normal population, hence the distribution differed completely from that of the general population.

When analyzing the *age* distribution of those abusing each single type of drug, something very interesting appears. Each type of

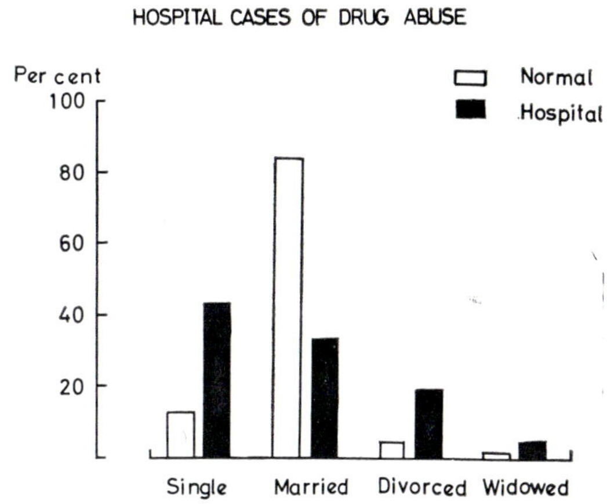

FIG. 2. Hospital cases. Marital status. "Normal," refers to general population with same age distribution as hospital material. For details, see text and Figure 1.

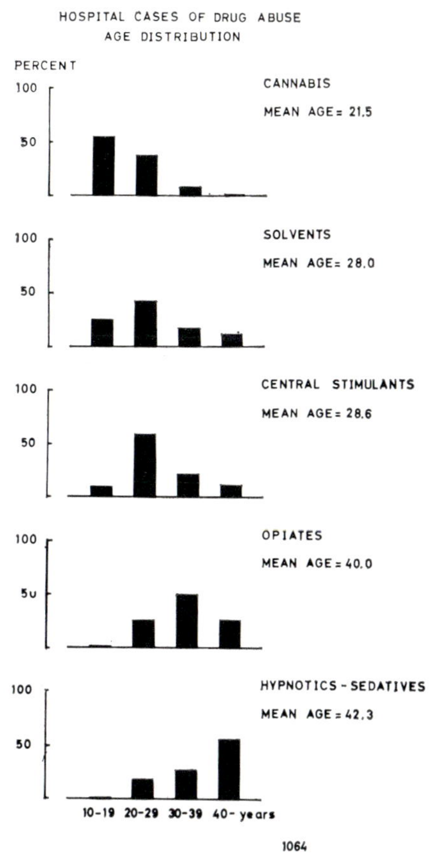

FIG. 3. Hospital cases. Distribution of age groups on different types of drugs. For details, see text and Figure 1.

drug is abused by a characteristic range of ages (Fig. 3).

Cannabis was mostly smoked by adolescents under 20 (54 per cent of all smokers) and a few over 20; 38 per cent in the age group 20 to 29 years, and only 8 per cent in the age group 30 to 39 years, none over 40. The mean age was 21.5 years.

Volatile solvents, as was cannabis, were to a large extent inhaled ("sniffed") by juveniles (25 per cent), but "sniffers" were also found in older age groups, 44 per cent in the 20 to 29 year group, and a total of 31 per cent among those over 30, hence not so definitely confined to younger ages as cannabis smoking. The mean age was 28.0 years.

Abuse of central stimulants had its peak at 21 to 25 years; about 10 per cent were under 20 years and one third (32 per cent) were over 30 years. The mean age was 28.6 years.

The peak age of opiate abusers was between 30 and 39 years, none were under 20. The mean age was 40 years.

In the abuse of hypnotics and sedatives, tranquilizers were used most frequently by those over 40 (56 per cent); one fourth (26 per cent) were between 30 and 39 years, 18 per cent between 20 and 29, and only 0.5 per cent under 20. Thus a completely different distribution from that for cannabis or stimulants was apparent. The mean age was 42.3 years.

Those combining different types of drugs, about 10 per cent of all, had the peak proportion, 40 per cent, at 20 to 29 years; only 11 per cent were below 20, and 25 per cent over 40. The mean age was 32.2 years; the distribution reflected the combined picture of hypnotics, stimulants, and cannabis in the younger age groups.

The total number of abusers of hallucinogens, mainly LSD, was so small, about 5 per cent of all, that a proper age distribution was not obtained. The abuse of LSD, however, seems to predominate in younger age groups and to be rare in older individuals.

The age of the individual is thus one important factor with respect to the type(s) of drug(s) abused and the mode of administration. This was stressed already in the analysis of the survey made in 1966, when the types of drugs abused were referred to the age distribution of those abusing drugs (Goldberg 1968).

When analyzing the distribution of types of *drugs* abused within each age group (Fig. 4), again characteristic distributions are found.

Smoking of cannabis (28 per cent), inhalation of volatile solvents (9 per cent), and intravenous injection of central stimulants (41 per cent) predominated in the adolescent age group and not less than 15 per cent combined different types of drugs.

In the age groups between 20 and 30 years, the central stimulants, taken by in-

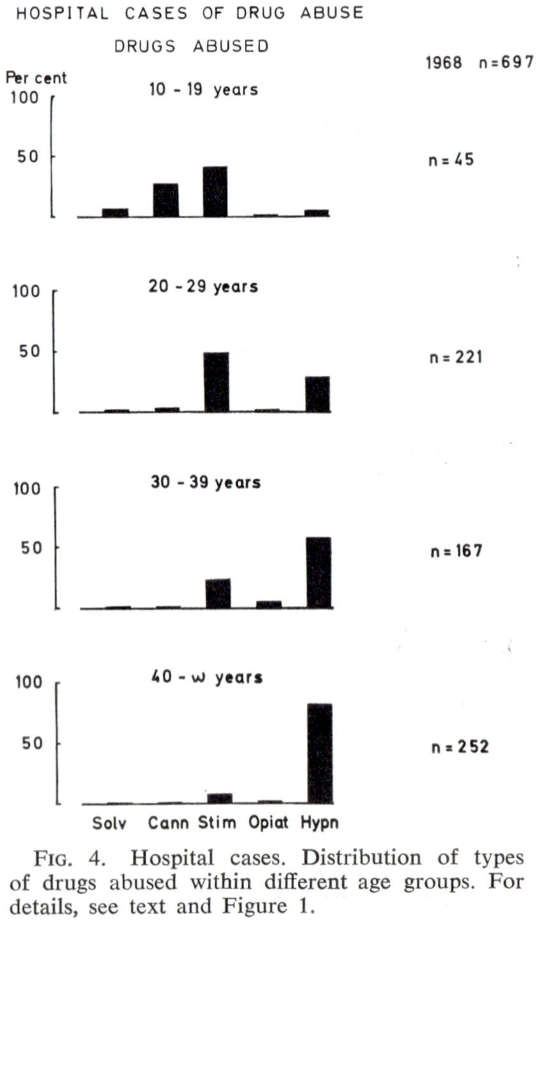

FIG. 4. Hospital cases. Distribution of types of drugs abused within different age groups. For details, see text and Figure 1.

jection, predominated (49 per cent), often combined with injection of opiates. Hypnotics and sedatives were taken by 30 per cent. Inhalation of solvents (3 per cent) and cannabis smoking (4 per cent) were noted only in a minority. Somewhat more than one tenth (11.6 per cent) combined different types of drugs.

In older age groups, especially over 40 years, oral intake of hypnotics and sedatives predominated, 57 per cent in the 30 to 39 year group and 81 per cent in those over 40. Injection of central stimulants was less frequent. It occurred in 24 per cent in the 30 to 39 year group, and in 8 per cent of those over 40. The same was true with regard to the other types of drugs. Opiate intake was rare (2 to 5 per cent), and cannabis smoking or inhalation of solvents was very unusual in these age groups (0 to 2 per cent).

It is interesting to see whether any differences are noted between the 1966 findings (SOU 1967:26, analyzed further by Goldberg 1968 a) and those in 1968. The difference in total number, 566 in 1966 and 697 in 1968, may be season-dependent (October 1966 versus April 1968), and does not necessarily have to be interpreted as an increase in prevalence.

Other data, however, indicate that a true increase in incidence may have occurred. The total number of cases of drug abuse treated in hospitals rose from 1614 in 1965 to 5223 in 1967; the total number of individuals treated for drug abuse rose during the same three-year period from 605 to 955 (Hoffmann and Håkanson, 1969).

There are also changes seen in the pattern of drugs abused in 1966 and in 1968. A higher proportion abused stimulant drugs, including cannabis, in 1968, as compared to 1966. Thirty-one per cent used stimulants in 1968 against 18 per cent in 1966, and 3 per cent used cannabis in 1968 against only 1 per cent in 1966.

A corresponding decrease was seen in the abuse of hypnotics, 61 per cent in 1968 against 69 per cent in 1966.

This change is also reflected in the age distribution. The proportion of juvenile users under 20 is about the same in 1968 as in 1966, 7 per cent, but the group 20 to 29 years has increased to 32 per cent, from 28 per cent in 1966, and the group over 40 has decreased to 37 per cent from 41 per cent in 1966. The mean age was correspondingly lower in 1968, 34.3 years against 37.1 years in 1966.

Oral intake of hypnotics and sedatives is thus one characteristic of the hospital drug abuse population, being the background of the often heard saying among hospital personnel: "Abuse of hypnotics and sedatives is our largest abuse problem."

The analysis of the material clearly shows this to be a characteristic of older age groups (Figs. 3, 4), whereas younger age groups tend to abuse central stimulants by injection, and the youngest age groups mainly inhale volatile solvents or smoke cannabis (Figs. 3, 4). Hence a substantial and increasing proportion of hospital patients abuse other types of drugs.

Patients with different types of drug abuse are not a random sample of a hospital patient population. They tend to show two age maxima, representing two age groups with different types of drugs abused (Fig. 1), cannabis, solvents and central stimulants dominating among the younger age groups, and hypnotics, sedatives (and opiates) among the older groups. They have a distribution of marital status (Fig. 2) that differs from that of the general population with the same age distribution, with an overrepresentation of single and divorced individuals, and an underrepresentation of married persons.

With regard to background data a high proportion of alcohol abuse, about 35 per cent, was present among the drug abusers. Studies in progress show that in a non-selected, male hospital population about 20 per cent of all hospital patients would be expected to have chronic alcohol problems (Goldberg, in prep.).

Very few, about 3 per cent of the drug

abusers in the hospital study, have a criminal background. Additional data concerning the 1966 material were analyzed by Goldberg (1968); a detailed analysis with regard to social grouping, employment, diagnoses, and the like of the 1968 material is made by Håkanson and Hoffmann (1969 a).

Individuals in Contact with Police

A totally different group of individuals with drug abuse is seen by the police. Two surveys were made of the individuals coming in contact with the police (the Swedish police were nationalized in 1965), one covering the period 1–30 June 1966 and one the same period in 1968.

The 1966 survey, comprising 510 individuals, has been reported and analyzed (SOU 1967:25, Goldberg 1968 a); the 1968 study will be presented here, based mainly on figures given by Håkanson and Hoffmann (1969 b).

The 1968 study, based on questionnaires sent out to the police, comprises all cases seen by the police during one month, June 1968, comprising a total of 658 individuals.

The age distribution is given in Figure 5. The mean age was 24.9 years. Not less than 40 per cent of all were under 20; 36 per cent were in the 20 to 29 year group, 10 per cent in the 30 to 39 year group, and 12 per cent were over 40, hence a completely different distribution from that seen among the hospital patients (Figs. 1, 5).

The distribution of marital status shows an overrepresentation of single individuals, 82 per cent against 22 per cent in the normal population with the same age distribution, an underrepresentation of married, 8 per cent against 76 per cent, and an overrepresentation of divorced, 10 per cent against 3 per cent in the general population.

The shift in marital status, as compared to the normal distribution, is even more accentuated than that found among the hospital drug abuse population (Fig. 2). The proportion of single individuals is larger, and

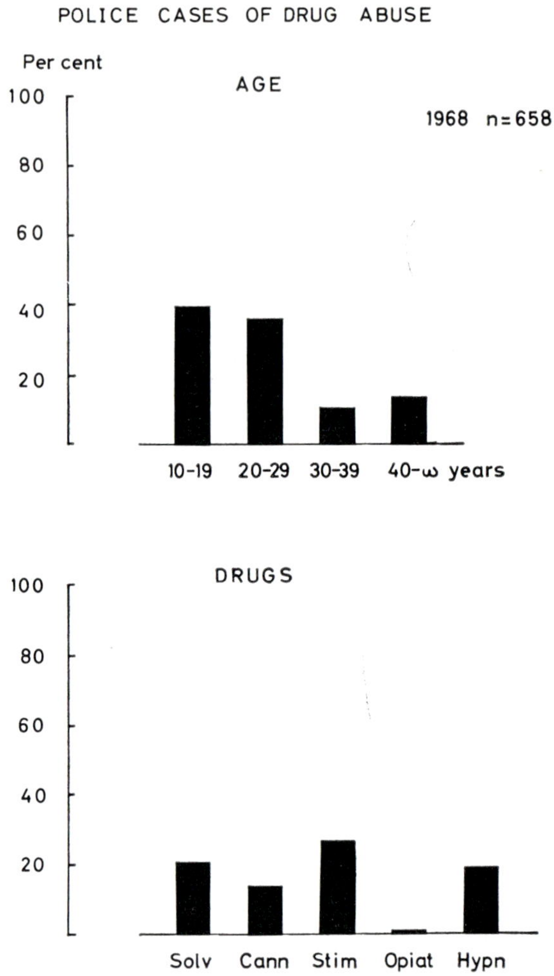

FIG. 5. Prevalence study. Cases of drug abuse seen by the Swedish Police 1 to 30 June 1968 (n = 658). Age distribution and types of drugs abused. Only cases with abuse of one type of drug mainly are given in the graph; in addition, 18.6 per cent have combined different types of drugs. Adapted from data by Håkanson and Hoffmann (1969 b).

the proportion of married is less, even when the difference in age distribution is taken into account and comparisons are made within each age group.

The types of drugs abused are seen in Figure 5. Abuse of central stimulants dominates (26 per cent of all). Inhalation of volatile solvents was reported in 21 per cent, intake of hypnotics and sedatives in 19 per cent, and cannabis smoking in 14 per cent.

A total of 19 per cent combined different types of drugs, mainly cannabis and stimulants; some of them also took opiates by injection.

A high proportion injected the drugs abused, 84 per cent of those taking stimulants, 100 per cent of those taking opiates, and 60 per cent of those combining different types of drugs; the average was 35 per cent.

When analyzing the age distribution of those abusing different types of drugs, a typical and characteristic age distribution was seen (Fig. 6). The mean ages were 16.7 years among solvent "sniffers," 19.7 years among the cannabis smokers, 28.0 years in those injecting central stimulants, 31.7 years in the opiate abusers, and 37.2 years in the abusers of hypnotic and sedative drugs. The mean age among those combining different types of drugs was 24.8 years, pointing to the combination of cannabis, stimulants, and opiates.

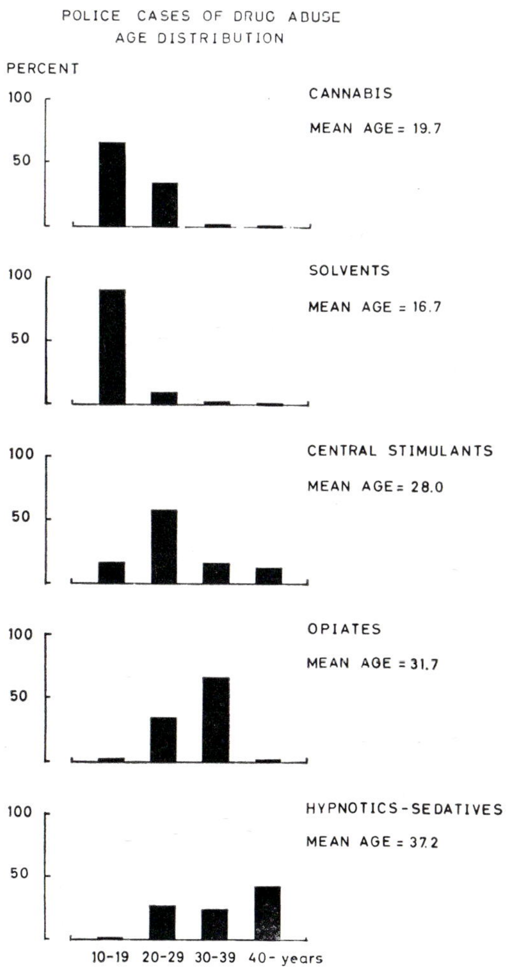

FIG. 6. Police cases. Distribution of age groups on different types of drugs. For details, see text and Figure 5.

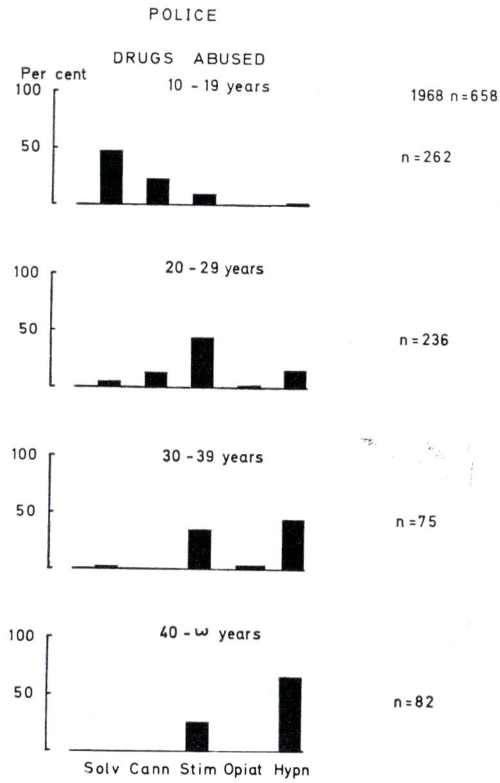

FIG. 7. Police cases. Distribution of types of drugs abused within different age groups. For details, see text and Figure 5.

The age distribution of users of each type of drug abused was essentially the same as that found in the hospital patients (Figs. 3, 6), with three notable differences. The solvent "sniffers," the opiate abusers, and those combining drugs were younger than corresponding groups in the hospital patients; the other groups—cannabis, stimulants, and hypnotics—agreed extremely well (Figs. 3, 6).

When analyzing each separate age group

(Fig. 7) the same characteristic drugs were found in each group as were found in the hospital patients. Cannabis smoking and solvent inhalation were found mainly among juveniles, central stimulants taken by injection among the 20- to 40-year-old group, and opiates and hypnotics, sedatives, among those 30 years and older. Combination of drugs was observed in all age groups.

With regard to criminal background about 25 per cent of those seen by the police had a criminal background (were arrested for criminal activity, etc.); 75 per cent had contacted, or were contacted by, the police for other reasons.

With regard to the changes from 1966 to 1968 the following differences are noted. The proportion of individuals under 20 had increased from 19 per cent in 1966 to 41 per cent in 1968, corresponding to a lower mean age in 1968, 24.9 years against 29.1 years in 1966.

Paralleling this decrease in age of users in 1968 a higher proportion of cannabis smokers were found in 1968, 14 per cent against 2 per cent in 1966.

About 5 per cent were abusing hallucinogens, mainly LSD; no proper age analysis of distribution can be made on this small sample.

The proportion of "sniffers" had also increased to 21 per cent in 1968 against 11 per cent in 1966, whereas the proportion of abusers of stimulants had decreased, from 36 per cent in 1966 to 26 per cent in 1968. The proportion of abusers of hypnotics and sedatives, partly taken by injection, had not changed substantially, being 19 per cent in 1968 against 21 per cent in 1966. The same was true of opiates, 1 per cent in 1966, 0.5 per cent in 1968.

The large differences between the hospital and the police material have already been stressed: the difference in age distribution (Figs. 1, 5), in distribution of drugs abused (Figs. 1, 5), and in the proportion of those injecting drugs. When comparing the age distributions of users for each type of drug, a striking similarity is seen (Figs. 3, 6), which also to a certain extent holds for the distribution of drugs within each age group (Figs. 4, 7).

Case-finding Survey

A case-finding survey was carried out in the Greater Stockholm region (total population 1.3 million), during the period July 1 to December 31, 1967. A detailed analysis has been made by Frej (1969 a); only some typical traits will be given here.

All state and municipal agencies, hospitals, doctors, social workers, police, and others that could be expected to have professional contacts with drug abusers were approached and asked to report cases of drug abuse to a scientific body, the Department of Social Medicine of the Karolinska Institutet Medical School, with the provisions that no data were disseminated further and the information obtained would be kept confidential and used for statistical purposes only. A total of 5628 agencies and persons were approached, and a total of 4392 cases were reported, corresponding to 3027 individuals after checking for duplicate reporting.

The figure, 3027 individuals, corresponds to about 0.3 per cent of the total number of people in the area, or about 0.5 per cent of the population over 15 years of age.

The age distribution shows a two-peak distribution, one peak between 21 and 25 years of age, and one about 31 to 40 years. The mean age is low, 25.6 years.

With regard to the drug spectrum the low mean age, with 37 per cent of individuals below the age of 20 years, would indicate central stimulants, taken by injection, to dominate the picture. This was also found to be the case when analyzing the distribution of drugs abused with regard to main type of drug abused. Central stimulants dominated the picture (60 per cent); cannabis smoking (14 per cent) and oral intake of hypnotics and sedatives (8 per cent) came next in order. Opiates were abused by about 6 per cent, and LSD by about 5 per cent; solvent inhalation was dominant in 3 per cent. In many of the abusers several different types of drugs were abused ("mixed abuse").

When studying the age distribution of each type of drug abused (Fig. 8), cannabis was found to have its peak among the juveniles, the mean age being 17.7 years; solvent inhalation coming next, the mean age being 18.7 years. Central stimulants dominated in the 20 to 29 year range, mean 26.0 years; these drugs, however, were also used both by younger and by older age groups, even if to a lesser extent than among the dominating group.

The abusers of opiates had a mean age of 27.8 years, and hypnotic and sedative abusers, 38.7 years.

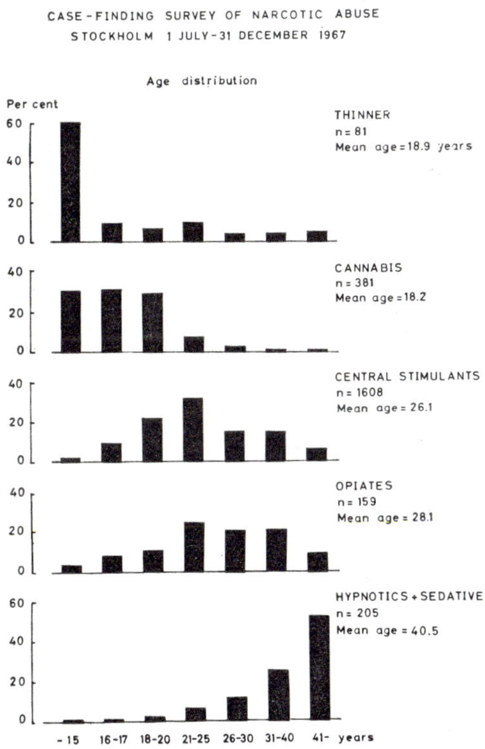

FIG. 8. Case-finding survey in Greater Stockholm 1 July to 31 December 1967 (n = 3027). Distribution of age groups on different types of drugs; only cases abusing one type of drug mainly are included (n = 2659). Adapted from data by Frej (1969 a).

The results with regard to age distribution of users within each type of drug abused are in complete agreement with both the hospital (Fig. 3) and the police material (Fig. 6), pointing to the important role of age for the choice of type of drug abused, provided that a number of other conditions are fulfilled, e.g., availability of drug, opportunity to use it and to obtain the necessary means (money), and an environment favorable for the emergence of drug use.

Random Selections of Whole Population

Only few and limited studies have been carried out so far on randomly selected samples from the general population (SOU 1969:52).

Two studies, on 282 and 311 persons respectively, were carried out on two occasions in 1965, in connection with a TV series on narcotics by The Swedish National Radio and Television Corporation. A total of 2.5 and 1.9 per cent respectively of those interviewed admitted having taken drugs on one or several occasions.

In a later study, in 1969, carried out by the same body on 1000 randomly selected individuals aged 15 to 70 years, 3 per cent had taken drugs. The highest proportions were found in men (5 per cent), in age group 15 to 24 years (8 per cent), and in students (10 per cent).

Random Selection of Younger Age Groups

A group of 1288 juveniles, aged 16 to 25 years, randomly selected from the vital statistics records of Greater Stockholm, were interviewed in 1966 in a study by N. Gustafsson and co-workers (1969).

A total of 11.9 per cent of the males (n = 656), and 3.5 per cent of the females (n = 632) had taken drugs, or 7.8 per cent of the whole group. The highest frequency was seen in the 18 to 20 year age group.

About half (3.7 per cent), 5.5 per cent of the males and 1.9 per cent of the females, had used drugs only once, and about 1.6 per cent, 2.9 per cent of the males and 0.3 per cent of the females, had used drugs more than 10 times. About 1 per cent of all males had taken drugs by intravenous injection.

Not less than 25 per cent of all males and 10 per cent of the females had been offered drugs.

In a later nationwide study in 1968 carried out by the Swedish Institute for Opinion Research (SIFO) on 2120 persons aged 12 to 24 years stratified according to SIFO standards, a total of 6 per cent of the males and 4 per cent of the females were found to have taken drugs; in Stockholm the proportion having taken drugs was 11 per cent.

Pupils in Grade 9

A nationwide study in 1967, carried out by the National Board of Schools (reported by Österlind and reviewed by Goldberg 1968) on selected groups of pupils in elementary and higher schools in Sweden, showed the highest frequency of drug taking to exist in grade 9, the last class in the compulsory school system, the mean age being 16 years.

It was therefore thought to be of interest to assess the drug taking pattern in more detail in the group at risk, grade (class) 9, with a mean age of 16 years. Two studies with this aim have been carried out.

In the *first* study, Herulf, engaged by the Children's Welfare Board in Stockholm, carried out a point-prevalence study by means of questionnaires on all children in class 9 in the Greater Stockholm area on 27 April 1967. The study comprised 8585 children in 72 classes, only 45 (0.5 per cent) not having answered.* On one or several occasions, 22.7 per cent of the boys and 17.2 per cent of the girls had taken drugs, or 19.6 per cent of all.

A substantial proportion had taken drugs during the last month, 30.8 per cent of all. Of those having taken drugs one to ten times in all, one fifth (20.0 per cent) had taken drugs during the preceding month, against three fourths (76.5 per cent) of those having taken drugs more than ten times.

Cannabis was used by 68 per cent, stimulant drugs by 17 per cent, opiates by 5 per cent, and other drugs, mainly hypnotics and LSD, by 10 per cent (Fig. 9). There was a clear difference in drug choice among those taking drugs only occasionally, one to ten times, and those having taken drugs more frequently, more than ten times. In the former group, 80 per cent had smoked cannabis and 9 per cent took stimulants, against 50 per cent smoking cannabis and 30 per cent using stimulants among those with more frequent use. An increase in frequency of drug use seems to parallel a switch in type of drug taken.

The same difference is seen with regard to mode of intake (owing to combined intake the sum of percentages is more than 100). In those who had taken drugs only occasionally, one to ten times, 91 per cent smoked cannabis, 11 per cent took drugs by mouth, and 3 per cent injected, against 85 per cent smoking cannabis, 38 per cent taking drugs orally, and 19 per cent injecting drugs among those who took drugs frequently, more than ten times. Hence there was a three- to sixfold increase in the proportion of those taking more dangerous drugs with the increase in frequency of drug taking.

When studying the reasons for variations in incidence of drug taking in different schools, or classes within one and the same school, the incidence ranging from a minimum of 9 per cent to a maximum of 40 per cent, it was found that a high incidence of drug abuse paralleled a higher number of children coming from broken or incomplete homes,

*The data were presented by Herulf in a mimeographed manuscript in 1967, and studied in detail by Goldberg (1968 a); new and partially more extended presentations were given by Herulf (1969 a, b). The percentages given may differ within 1 per cent between the various presentations, depending on the underlying number of cases varying between 7793 and 8345, and on the type of question studied.

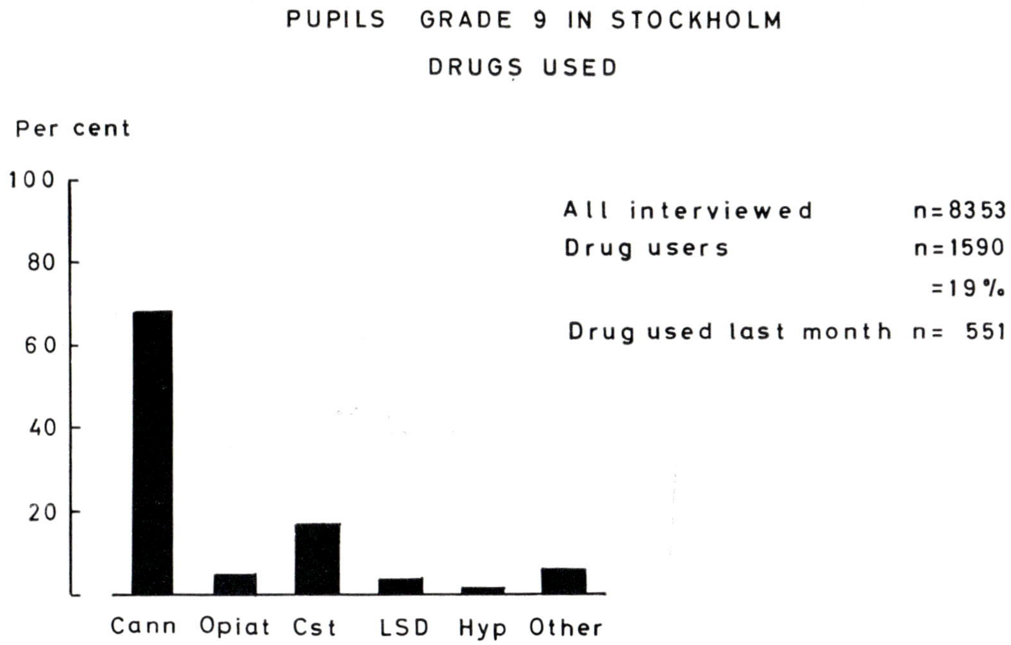

FIG. 9. Prevalence study. Types of drugs abused among all pupils in grade 9, Stockholm (n = 8353), last grade of compulsory school. Mean age, 16 years. Adapted from data by Herulf (1969 a, b).

a higher proportion of children choosing technical versus scholastic courses, and a higher frequency of individuals with heavy alcohol intake.

A high proportion of children had been "offered" drugs. More than half of the children that did not take drugs but had been offered drugs had been approached by a "stranger," and only rarely by classmate or friends.

When studying the possible importance of other factors for the drug abuse pattern a parallelity was found between intensity of use and type of drug used, and a number of environmental and other factors. A higher frequency of drug taking and a switch to stimulant drugs paralleled a higher proportion of children having been in contact with police and/or youth welfare boards, who had been treated for nervousness, who came from incomplete homes, and who showed a higher intensity of drinking and/or smoking practices (Figs. 10 to 12).

It is seen that with increasing use of narcotics the proportion of those smoking tobacco and smoking heavily rises steeply. The distribution is shown in Figure 10. In Figure 12 the rise in frequent smoking, more than 10 cigarettes per day, is accompanied by an increase in frequency of drug taking. The same applies to both boys and girls (Fig. 12).

The same trend is also seen with regard to drinking practices. The proportion of those drinking at all increases with increase in frequency of drug use (Fig. 11); the same tendency again was noted for both boys and girls (Fig. 12).

The distribution on different intensities of drinking is shown in Figure 11; the increase in frequency of more intense drinking, more than 5 times per month, is given in Figure 12.

In the *second* study, on all pupils in grade 9 in Göteborg carried out in 1968 by Jonsson and Svedugård (1969), a total of 3408 questionnaires could be evaluated out of 4414 pupils, or 77 per cent. A total of 13.5

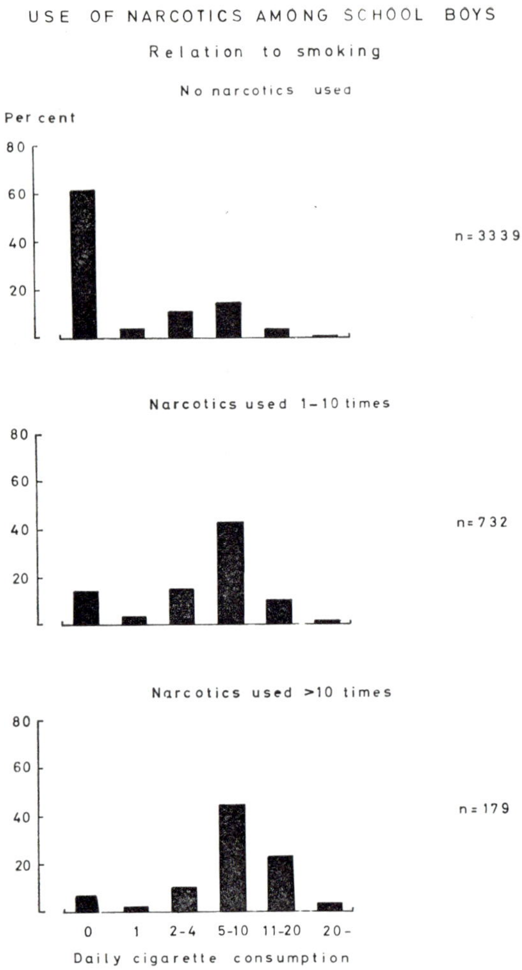

FIG. 10. Relation between intensity of smoking of tobacco and abuse of drugs in male pupils in grade 9 (last grade in compulsory school) in Stockholm (n = 4250). Mean age, 16 years. Adapted from data by Herulf (1969 b).

per cent were found to have taken drugs on one or several occasions.

With regard to types of drug used, a fair proportion combined several types of drugs. When attempting to evaluate the main type of drug used, 64 per cent listed cannabis as their main drug of use, 17.5 per cent listed stimulants, 4.5 per cent opiates, 7.4 per cent LSD, and 6.8 per cent other drugs, mainly hypnotics; 5 per cent injected drugs.

As to intensity of drug use, almost half (41 per cent) stated they used drugs one or several times a week up to daily, 41 per cent used drugs one to several times a month, and 18 per cent used drugs only occasionally, a few times per year. Girls seemed to use drugs somewhat more often than boys, 51 per cent stated weekly or daily use against 35 per cent of the boys. The same is seen when looking at drug taking during the week preceding the questionnaire; 63 per cent of the girls had used drugs and 50 per cent of the boys.

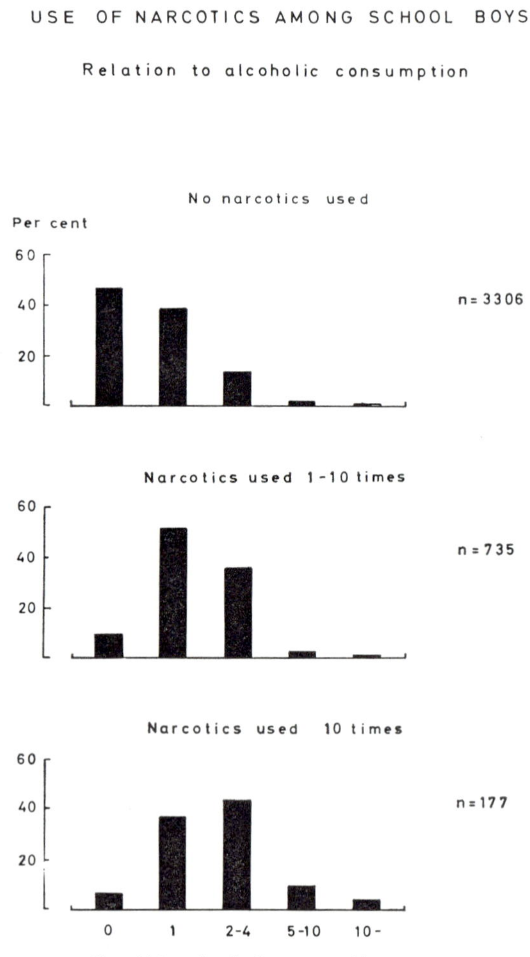

FIG. 11. Relation between intensity of drinking and abuse of drugs in male pupils in grade 9 (last grade of compulsory school) in Stockholm (n = 4218). Mean age, 16 years. Adapted from data by Herulf (1969 b).

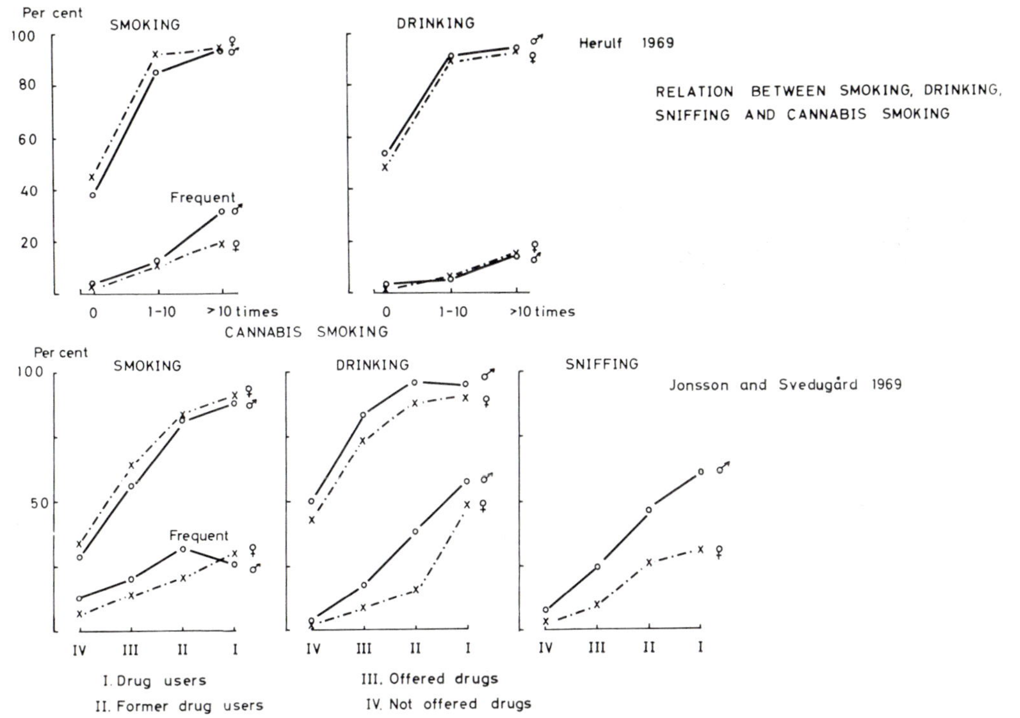

FIG. 12. Relation between intensity of tobacco smoking, drinking and inhalation of volatile solvents ("sniffing") and abuse of drugs in pupils in grade 9 (last grade in compulsory school). Mean age, 16 years. Adapted from data by Herulf (1969 b; n = 5173) for Stockholm and by Jonsson and Svedugård (1969; n = 3408) for Göteborg.

Data actually refer to abuse of all types of drugs; the majority (84 per cent) smoked cannabis.

Per cent = proportion of all pupils.

Figures on abscissa in upper graph refer to frequency of drug use.

Lower set of curves in each graph refers to more intense or more frequent use. Groups in lower set of graphs refer to four categories of drug use: I, Present use; II, Former use, now stopped; III, No drug use, but offered drugs; IV, No drug use, not offered. The graphs show how intensity of smoking, drinking, or sniffing increases parallel to the increase in frequency and intensity of drug abuse.

With regard to the development of the drug use, it is characteristic that a majority started with cannabis (82 per cent), the rest started on stimulants, opiates, or hypnotics. Those using cannabis at the time of the questionnaire had to a large extent also started with cannabis, whereas a higher proportion of those on stimulants had started on stimulants. Further, it is noted that a fair proportion add other drugs to the initial choice, i.e., start on several drugs. Stimulants were also used by 15 per cent of the initial cannabis smokers, in a minority of cases also opiates, LSD, or hypnotics.

For an analysis of possible factors influencing drug use the material was divided into four groups, I to IV. Of the 13.5 per cent using drugs, 3.8 per cent were still taking drugs (group I), and 9.7 per cent had stopped (group II). Group III comprised those having been offered drugs but not taking drugs (24.8 per cent), and group IV comprised those not having been offered drugs (58.0 per cent).

There was a parallelity between degree of drug experience, ranging from no contact (group IV) to actual and maintained use (group I), and a large number of items.

When looking into a possible relation between intensity of drug use and tobacco smoking, it was found (Fig. 12) that the proportion of smokers increased from 31 per cent in those not using drugs (group IV) to 89 per cent in group I (drug users). Intensive smoking (more than ten cigarettes a day) was found to increase from 10 per cent in group IV to 20 per cent in group I.

With regard to alcohol intake, it was found (Fig. 12) that the proportion of those taking alcohol increased from 47 per cent in group IV to 93 per cent in group I. More intense drinking, at least once per week, increased from 2 per cent in group IV to 31 per cent in group I, and the proportion of those often being intoxicated rose from 3 per cent in group IV to 54 per cent in group I.

Sniffing was also studied. The proportion of those sniffing increased from 6 per cent in group IV to 54 per cent in group I (Fig. 12).

Other items also showed a parallelity. Not being at ease in school ranged from 5 per cent in group IV to 20 per cent in group I, and being away from school without leave at least one hour per week ranged from 8 per cent in group IV to 40 per cent in group I.

Impaired contact with parents showed the same trend. The proportion of those telling parents about what they were doing in their free time, informing parents about school progress, and being able to discuss problems with parents ranged from 50, 43, and 54 per cent respectively in group IV to only 18, 24, and 38 per cent in group I, i.e., only half as many.

Military Conscripts (Eighteen Years Old)

In 1967 a pilot study was carried out by Agrell, Netz, and Wolff at the Military Psychological Institute (MPI) on individuals examined for acceptance for military service (conscription) in Stockholm and Göteborg.

A second, extended study was carried out in 1968/69 by Agrell, Netz, and Wolff (1969, 1970). As part of the medical and psychological examination a questionnaire had to be answered anonymously; individual results, which could not be identified, were kept confidential. Eight regions were studied, five comprising the three largest cities, Stockholm, Göteberg, and Malmö, and three were representative of provinces, Bohuslän, Skåne, and Bergslagen. A total of 23,305 questionnaires were collected; of these only 816, or 3.5 per cent, had to be discarded. Of the population studied 49.5 per cent came from the three largest cities, 25.4 per cent from smaller cities, and 24.1 per cent from rural areas; 0.7 per cent had lived abroad. The mean age was 18.3 years, varying from 16 to 20 years, 97.1 per cent were in the 18 to 19 year range.

It was found that 18.1 per cent of all respondents had on one or several occasions used drugs of the narcotic or hypnotic-sedative group; the proportion was 23.3 per cent in the large cities and 9.4 per cent in the provincial areas (Fig. 13).

The distribution on frequency of taking is also given in Figure 13. It is seen that the proportion of those taking a drug more than once is higher in cities (upper two graphs) than in provinces (lower graphs).

The type of drug mainly used by drug takers (Fig. 14) was cannabis, taken by 88 per cent. Seven per cent used central stimulants, orally or by injection, 1 per cent opiates, 0.7 per cent LSD, and 3 per cent hypnotics-sedatives. A total of 3.1 per cent of the users were injecting drugs, mainly stimulants. There was no essential difference between urban and rural areas in this respect.

A special analysis showed that a total of 10.1 per cent of the drug users had on one or several occasions injected drugs. This corresponds to a total of 1.7 per cent of all individuals studied in the age group 18 to 20 years.

When studying the number of times a drug had been taken it was noted that one fourth (24 per cent) of all users had only taken the drug once against one fifth (18 per cent) more than 50 times. This pattern was different in large cities and in other areas. In

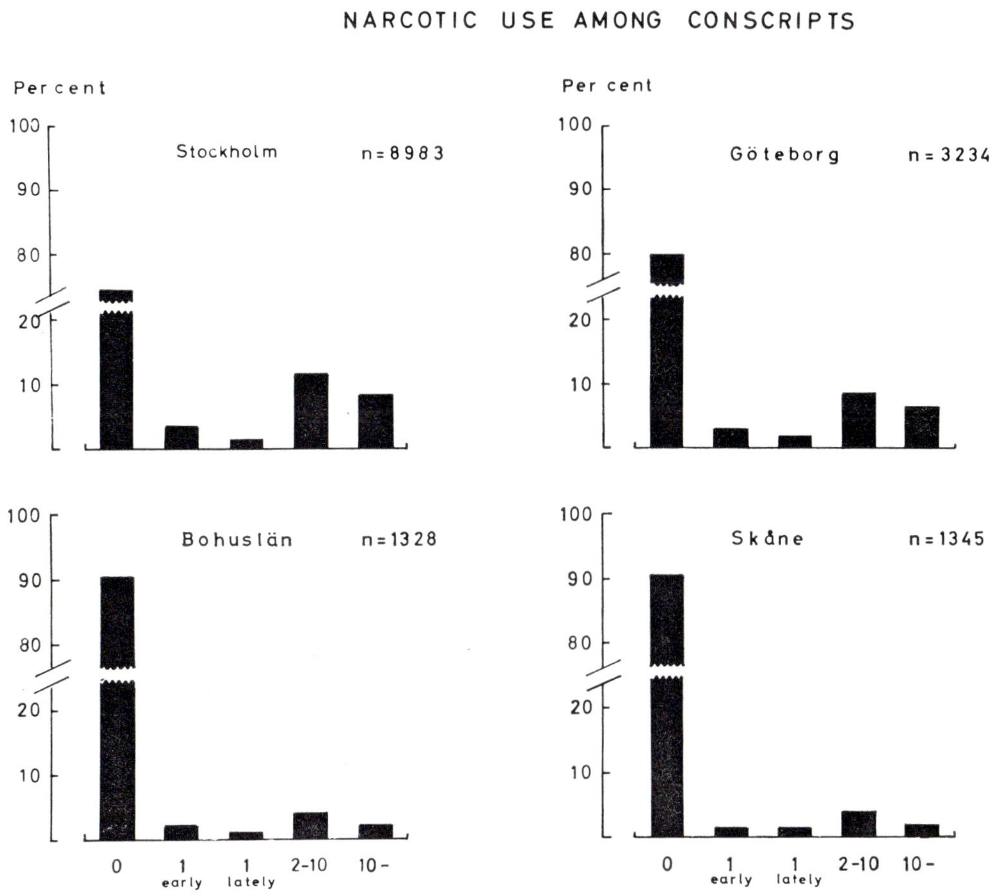

FIG. 13. Abuse of drugs among Swedish military conscripts 1968/1969 (n = 14890). Males; mean age, 18 years. Distribution of frequency of drugs taken, from none to more than 10 times, among all individuals. Adapted from data by Agrell, Netz, and Wolff (1969, 1970).

the cities 23 per cent had taken a drug only once, against 32 per cent outside cities. Nineteen per cent had taken drugs more than 50 times in cities against only 12 per cent in provinces.

When evaluating frequency of intake for each type of drug, great differences between different types of drugs were noted. Those smoking cannabis had the lowest frequency, only 15 per cent had taken the drug more than 50 times. The highest frequency was noted among those injecting drugs; 71 per cent had taken drugs more than 50 times. The other drugs fell between these limits. Fifteen per cent of those taking stimulant drugs orally, 32 per cent of LSD users, and 40 per cent of opiate users had used these drugs more than 50 times.

A covariation between certain forms of drug taking and the tendency to injection is also seen. The proportion of those combining smoking or oral intake with injecting drugs increases from 1.5 per cent among those smoking cannabis one to ten times, to 8.5 per cent when taking cannabis 11 to 50 times, to 23.4 per cent when taking cannabis more than 50 times. The same is seen among those taking stimulants orally; 7.8 per cent among those taking stimulants orally less than ten times also injected, against 40 per

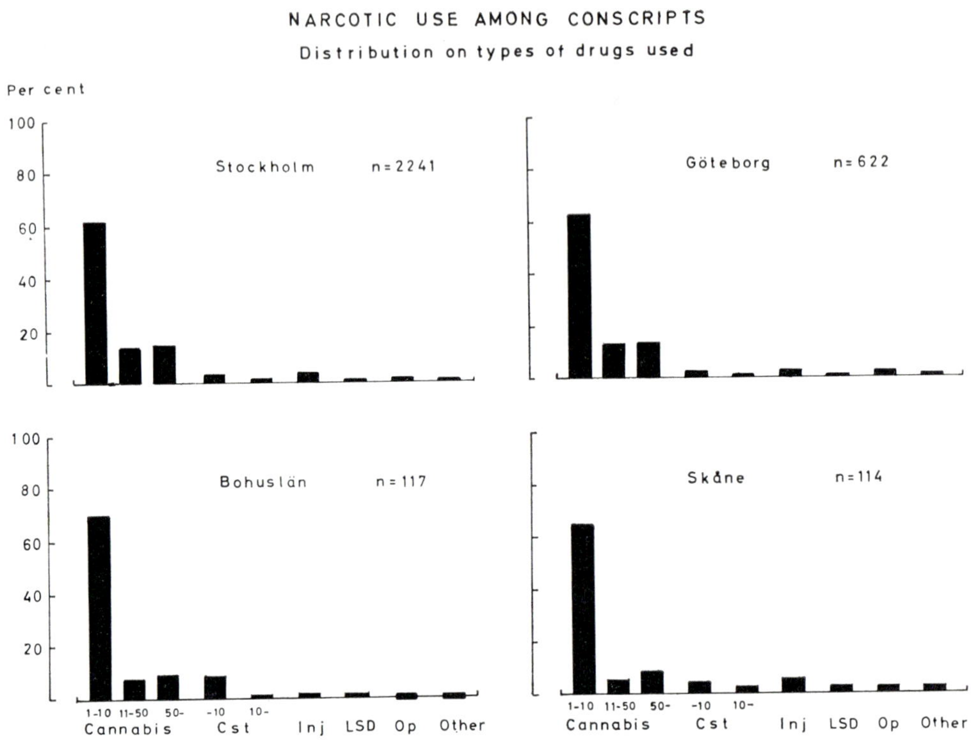

FIG. 14. Drug abuse among military conscripts 1968/1969. Distribution of types of drugs abused among drug abusers only (n = 3094). For details, see text and Figure 13.

cent among those taking stimulants orally more than ten times. Opiate users (37.8 per cent) and LSD users (57 per cent) also injected drugs. Hence injection also followed increase in frequency, as well as the type of drug taken.

When evaluating the type of drug that had been the initial one used, it was found that the majority of users of any type of drug had started on the *same* type of drug. Of the cannabis users, 96 per cent had started on cannabis, 71 per cent of the stimulant users on stimulants, 58 per cent of LSD users on LSD, and 50 per cent of opiate users on opiates.

It is noteworthy, however, that in all categories a substantial number had also started on cannabis, ranging from 28 per cent in the stimulant and LSD groups to 43 per cent in the opiate group.

It can be inferred from these results that of these 18-year-olds starting on cannabis, 97 per cent continued with cannabis.

Of those starting on other drugs, 51 to 65 per cent continued with the same type, the rest, 37 to 49 per cent, mainly switched to cannabis.

With regard to motives for taking drugs curiosity stands out as one major reason. With regard to those who had taken the drug on a few occasions only and then stopped, curiosity as the main motive for drug taking was noted in 81 to 87 per cent. When the intensity of drug taking increased, other motives appeared. Curiosity was mentioned in 65 per cent, but also influence from friends (14 per cent) and "other reasons" in 21 per cent.

An attempt was also made to study whether drug taking induced a "need" to continue taking the drug. The categories to be answered ranged from feeling a "very great" and "rather great" need to a "rather small," "very small," or "no" need to continue.

The results are illustrated in Figure 15. Two main features are seen.

1. The need to continue drug taking increased with the frequency of intake, e.g., in cannabis smokers from 6.7 per cent in those who had used cannabis one to ten times, to 51 per cent in those having smoked more than 50 times. The same is also seen in those taking stimulants orally. The proportion increased from 5.7 in those taking the drug less than ten times, to 44 per cent in those taking the drug more than ten times.

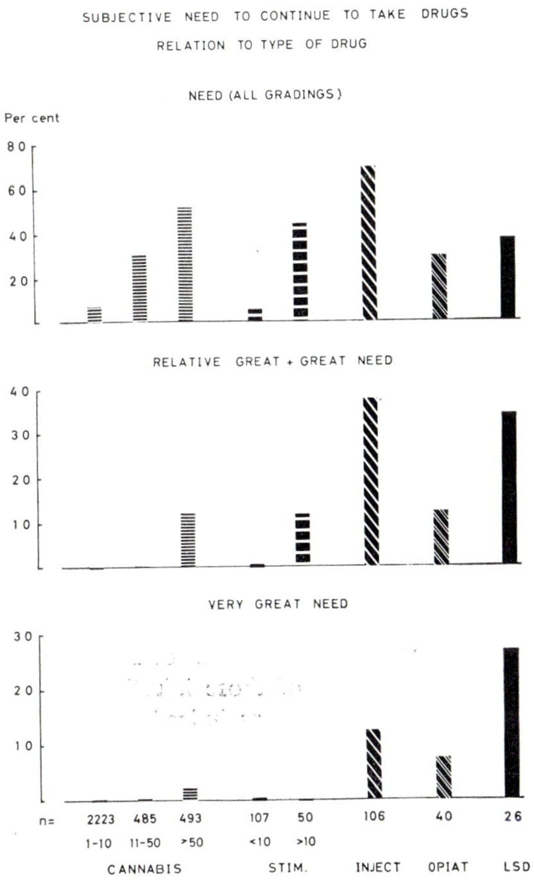

FIG. 15. Relation between various degrees of "need" to continue drug taking and intensity of drug taking and types of drugs abused in male male military conscripts (n = 3530). Adapted from data by Agrell, Netz, and Wolff (1969).

Stim. refers to stimulants taken orally; inject. refers to individuals injecting drugs, mainly stimulants.

The graph illustrates, among other things, the difference in emergence of need to continue between stimulants taken orally and injected, and also shows that a need also exists in cannabis smokers who have smoked more than 50 times.

The same picture is seen when looking at frequency only, independent of type of drug. Need to continue ranged from 1.7 per cent in those taking the drug once, to 5.8 per cent when taking the drug two to four times, 15.9 per cent when taking five to ten times, 32 per cent in those taking 11 to 50 times, and 56 per cent in those having taken drugs more than 50 times.

If only referring to more intense need, the proportion ranged from 1.2 per cent in those taking the drug five to ten times, to 2.4 per cent when taking 11 to 50 times, and 19.1 per cent in those taking the drug more than 50 times.

2. The proportion of those stating they felt a need to continue was different for the various types of drugs (Fig. 15), ranging from 17.1 per cent in those smoking cannabis, to 17.2 per cent in consumers of stimulants orally, to 30.0 per cent of opiate users and 38.5 per cent of LSD users, to 68.9 per cent of those injecting drugs, mainly stimulants.

When evaluating only more intense degrees of need to continue, the proportions found were 2.3 per cent for cannabis smokers, 3.8 per cent for those taking stimulants orally, 12.5 per cent for those taking opiates, 34.6 per cent for LSD users, and 37.8 per cent for those injecting drugs, mainly stimulants.

These figures illustrate two additional facts, viz. (1) the importance of type of drug for the emergence of need, illustrated by the increase in intensity of need to continue when switching from cannabis to other types of drug, and (2) the importance of the mode of administration, as seen when switching from smoking and/or taking drugs orally to injection.

When analyzing the individuals in the

study in more detail, it was found that not only drug users could be grouped according to intensity of drug taking mode of administration and type of drug taken, but also non-users. Three groups can be discerned, depending among other things on attitude to drug taking: those who were abstaining completely and were opposed to drug taking (83 per cent), those who had not taken a definite stand (10 per cent), and "potential" drug takers (7 per cent) who felt inclined to take drugs.

Whereas the first two categories in all other respects are alike and differ from drug takers, the third category, the "potential" drug takers, in many respects showed some of the same characteristics as those smoking cannabis or taking other drugs on single occasions only (less than 10 times), but deviated less from the non-users than those taking drugs more frequently or taking drugs such as LSD or opiates or injecting drugs.

When attempting to go into more detail concerning possible factors that would shed light on the tendency to abuse drugs and to continue taking drugs, a number of conditions were evaluated in detail.

With regard to social background three factors could be isolated and further analyzed: the proportion of individuals from "broken homes," the place of living (urban or rural), and the level of education.

1. The proportion of individuals having lived in incomplete homes (one parent only, foster home) increased from 9.6 per cent among non-users to 27.8 per cent in drug takers with heavy abuse, injecting drugs.

2. A higher proportion of heavy abusers were living in large cities, 82 per cent against 52 per cent of non-users, and only a small number of heavy abusers were living in rural areas, 8 per cent against 23 per cent of non-users.

3. With regard to level of education two tendencies are clear. Those non-users who were inclined to test drugs, but had not actually taken drugs, had a more advanced school background in 60.6 per cent of the cases, against 43.9 per cent of non-users not inclined to drug taking. However, individuals with heavy abuse (62.2 per cent) had a lower educational background than potential users (37.3 per cent), or non-users (55.1 per cent).

When attempting to evaluate social adaptation, Agrell et al. found two items to correlate with type and frequency of drug taking:

1. The tendency to be away from school without leave followed closely the drug taking pattern. When identifying those who had been away without leave often, many hours per week, only 4.3 per cent of the non-users were found in this category. The proportion was much higher among drug users, ranging from 13.4 to 32.4 per cent among cannabis users with different intensity of intake, and from 17.2 to 60.0 per cent in those taking stimulants orally. The proportion was 33.3 per cent in opiate users, 40.7 per cent in LSD users, and 56.5 per cent in those injecting drugs, mainly stimulants.

2. A similar tendency was seen with regard to contacts with police or youth welfare boards. Of non-users 2.5 per cent had had several contacts with police or youth welfare boards against 7.6 to 19.8 per cent of cannabis users, 18.8 to 39.2 per cent of users of stimulant drugs (oral use), 22.2 per cent of opiate users, 14.8 per cent of LSD users, and 54.3 per cent of those injecting drugs.

Emotional stability, as judged from questions about liking to be at home, thinking of leaving home (escaping), contact with parents, and feeling nervous, also correlated with intensity and type of drug use. Among non-users 73.5 per cent liked their home very much and 53.3 per cent had good contact with their parent(s); only 2.5 per cent had thought many times of escaping from home, and 5.1 per cent had often felt nervous.

The trend was quite different among those taking drugs. The greatest change was noted among those injecting drugs; only 46.9 per cent felt well at home and 27.1 per cent had good contact with their parent(s); 26.8 per cent had thought of escaping many times, and 39.7 per cent often felt nervous.

Also a relation between intensity of drug taking and smoking and drinking practices was noted. Both the proportion of those drinking and/or smoking and the frequency of intake increased from non-drug users to drug users, and increased in proportion to intensity and type of drug, the highest values seen in those injecting drugs.

The percentage of those taking alcohol increased from 85.6 per cent in non-drug users to 92.6 per cent in heavy drug users. Drinking more than ten times per month increased from 7.8 per cent among non-users to 34.7 per cent among heavy drug users.

The proportion of those smoking tobacco increased from 55 per cent in non-drug users to 90.4 per cent in heavy drug users, and heavy smoking (more than 20 cigarettes a day) increased from 2.2 per cent in non-users to 26.3 per cent in heavy drug users.

Hence these results show a parallel increase in intensity of drinking and smoking with that of drug use, contrary to the belief often forwarded that drug taking and drinking would exclude each other. The agreement between these findings and those in the 16 year old is obvious (Figs. 10–12).

Finally the analysis shows not only a clear difference with regard to the factors studied between different types of drugs used, as well as between those using a certain drug occasionally only and those using it very frequently, but also between those smoking or eating drugs as opposed to injecting.

Studies on Soldiers in Various Stages of Training

In order to see whether there were any differences between the 18-year-old group and soldiers of various ages, two studies have been carried out on limited samples of soldiers at two different stages of their training (Agrell, Netz, and Wolff, 1970).

1. In a sample of 2146 soldiers, aged 20 to 21 years, making their first military service in 1968, hence two years after examination for conscription, a total of 11 per cent were found to have used drugs.

The proportion was thus only half of that seen among the 18-year-olds (no correction made for differences in background and urban-rural distribution).

Cannabis dominated among the types of drugs abused, 81.4 per cent of all drug users, 8.3 per cent used stimulants, 3.7 per cent LSD, 1.8 per cent opiates, and 4.5 per cent other drugs, mainly hypnotics and/or sedatives.

More than one fourth (28.7 per cent) had only used drugs once, a total of 79.5 per cent less than ten times, and 9.7 per cent more than 50 times.

2. A second sample comprised 1961 soldiers with a mean age of 28 years, ranging from 20 to 47 years, in a later stage of military training. Only 2.8 per cent of drug abusers were found in this group; the proportion was somewhat higher among those under 30 years.

Among the drug users 52 per cent smoked cannabis, 26 per cent used stimulants, and 4 per cent opiates. In 18 per cent other drugs (mainly hypnotics) were used, hence there was a different distribution from that seen among the two younger age samples, those examined for conscription and those making their first military service.

The total figure, 2.8 per cent, agrees with that found in the general population.

These results may be interpreted in several ways:

1. A majority of drug takers, especially cannabis smokers, stop drug taking altogether, only a minority continues to take drugs.

2. Drug taking is dominant especially among younger age groups and has not yet reached the older age groups.

3. Juveniles with severe consequences of drug taking will not be admitted for military service and will drop out.

All three inferences seem to be borne out by a number of facts. A high percentage of cannabis smokers actually use the drug for a few times only ("out of curiosity") and then stop, whereas those abusing stimulants are the ones that predominantly continue their drug use. These two trends are seen in the shift in proportion between users smoking cannabis or taking stimulants.

It also seems likely that drug taking out of curiosity, limited to a few occasions only, then the use is stopped, is characteristic of a high proportion of juveniles and is less often seen in adults. Adults, especially of older age, tend more to take hypnotics and sedatives, which also is disclosed in the distribution found and adds to the shift in drug pattern.

Finally it can be added that the high incidence of drug taking among 18-year-olds, almost every fifth, as well as the occurrence of severe cases of drug taking in this category, with the risk of continuing to take drugs and to acquire chronic damage (psychotoxicity) on one hand and the considerably lower incidence among soldiers in service brings up two questions: (1) whether juveniles with overt drug abuse as diagnosed at the medical examination, e.g., by examining possible injection marks, scars, infection, should be free from military service and (2) whether juveniles with overt drug abuse should be taken into military service and be under thorough medical supervision, using the environment of the military service in a country such as Sweden as part of a therapeutic approach. The results of both procedures should be thoroughly tested and evaluated. Tentative results from the second approach, speaking of several individuals succeeding in stopping the drug use, encourage this type of therapeutic approach to be continued.

University Students

A survey of possible use of narcotic drugs among university students in Stockholm in 1968 was carried out by Ahrens, Kihlbom, and Näs (1969). A random sample of students of Scandinavian origin between the ages of 19 and 35 years obtained questionnaires; a total of 491 male students (out of 583, or 84.2 per cent) and 247 female students (out of 296, or 83.4 per cent) responded. The highest percentage of those not responding were found in the oldest age group, over 28 years.

A total of 20.8 per cent of the male and 15.2 per cent of the female students had used drugs (cannabis) on one or several occasions. Of the males, 5.1 per cent had used cannabis more than ten times, and 1.4 per cent more than 50 times.

Of the males, 9.7 per cent, and of the females, 8.3 per cent, had used stimulant drugs orally, about one third of them only once. None of the respondents had injected drugs. LSD had been used by 1.4 per cent of the males.

Of the drug users about one third had combined the use of cannabis and stimulants, about two thirds used cannabis only, and one eighth stimulants only.

The majority of users took the drugs only occasionally, about half to one third had taken drugs during the two months preceding the survey. After one year, one in three seems to continue using drugs.

With regard to exposure to the drug, 37 per cent of all the males interviewed and 30 per cent of the females had been offered drugs, more among the younger students, e.g., 46 per cent of 19- to 21-year-old males, against 30 per cent of 26- to 35-year-olds.

Two thirds knew friends or others who used cannabis, and as many as 14 per cent knew others who injected drug.

One interesting feature is the difference between students not willing to use drugs and those who might consider taking drugs. Up to 93 per cent of those willing to take drug had had several offers to obtain cannabis, against 68 per cent in the group opposed to drug taking.

There were differences noted between nondrug users and drug users with regard to a number of background data. When comparing non-users to users with the highest intensity, 87 per cent of non-users were brought up to age 18 in homes with both parents, against 64 per cent of the drug users. Similar differences were also noted with regard to a number of other factors, ranging from family relations, from conflicts with parents (16 per cent versus 56 per cent), being critical of parents' norms (29 per cent versus 69 per cent), impossibility to discuss personal problems with parents (25 per cent versus 46 per cent), not liking school (11 per cent versus 41 per cent), tendency to skip lessons and to leave school without leave (9 per cent versus 38 per cent) and to finding himself different from other individuals (20 per cent versus 59 per cent).

Another example is the presence of signs of nervousness, and feelings of being depressed, suicidal ideas and suicidal attempts, and having sexual problems. The results are illustrated in Figure 16, showing the difference between non-users and users on one hand, and the increase in proportion of individuals with these signs with increase in intensity of drug taking on the other.

The proportion of individuals (males) with nervous disturbances ranged from 17 per cent in non-users to 64 per cent in those using drugs five times or more (Fig. 16). The proportion of those depressed ranged from 39 to 74 per cent, of having contemplated suicide from 9 to 33 per cent, and of those attempting suicide from 1 to 8 per cent; the proportion of those with sexual problems ranged from 16 to 54 per cent (Fig. 16).

Also in this group of respondents, as well as in pupils in grade 9 and in military conscripts, there was a clear correlation between

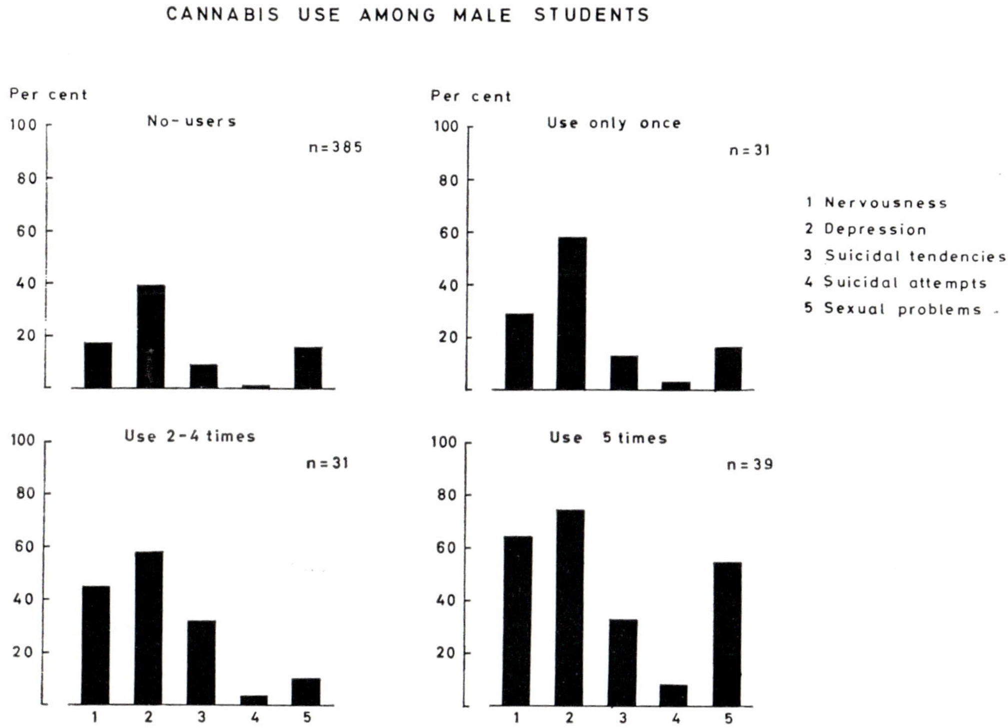

FIG. 16. Relation between proportion of individuals displaying different types of neurotic signs and intensity of cannabis use in male students in Stockholm (n = 486), aged 22 to 35 years. Adapted from data by Ahrens, Kihlbom, and Näs (1969).

use of drugs and intensity of drug use and intensity of smoking and drinking practices. A higher proportion smoked tobacco and used alcoholic beverages among drug users, and the intensity rose with intensity of drug use.

Similar, but less marked differences, were also noted among the non-users, not willing to take drugs. Out of those not willing to take drugs 56 per cent smoked tobacco and 18 per cent consumed wines at least once per month, against 78 per cent smoking and 29 per cent drinking among non-users willing to take drugs.

The same tendency is seen with regard to earlier use of medicaments, the drug users also showing a greater proportion of individuals having obtained medicaments, 10 per cent on an average in non-users against 22 per cent in drug users with the most intense use.

THE CHANGING PATTERN OF DRUG ABUSE

Based on the surveys made on drug taking in Sweden and other facts an attempt will be made to characterize the present situation and to stress the elements that are implicated in the development of drug taking and drug abuse within the framework of the epidemiological model chosen: the interplay between agent, individual, and environment.

The Agent

New Groups of Drugs Being Abused. Opiates dominated the spectrum of abused narcotic drugs up to the nineteen fifties among the few drug abusers. Among the non-narcotic drugs the hypnotics dominated, mainly used by elderly persons (Figs. 3, 4, 6–8). A host of new groups of drugs are now entering the drug scene, bringing about a change in the drug spectrum.

The number of opiate users has increased somewhat, but abuse of opiates has now become a very small part of the drug spectrum (Figs. 1, 2, 4, 5, 7, 13, 14).

The stimulant drugs (amphetamines, phenmetrazine and methylphenidates) appearing in the forties and fifties occurred among one third of the drug abusers at the end of the fifties, and dominated the picture in the first half of the sixties (Figs. 3, 4, 6–8).

The volatile solvents administered by inhalation ("sniffing"), became the fashion, especially in juveniles, in the fifties and early sixties; their popularity seems to have declined in the last few years (Figs. 3, 4, 6–8, 12).

Cannabis has now taken the scene, mainly among the juveniles (Figs. 3, 4, 6–9, 13, 14). Starting in the sixties, the number of users is rapidly increasing.

The hallucinogens, mainly LSD, have come in during the last few years, and are still seen in a limited proportion only (Figs. 1, 5, 15).

These changes are also disclosed in the types of drugs involved in trials for narcotic offenses (Fig. 18).

Switch from One Type of Drug to Another. Previously the drug abuser usually stayed with his preferred type of drug, mainly opiates or occasionally stimulant drugs; hypnotics were usually used by the elderly. When changing drugs, the switch was usually between members of the same group of drugs, e.g. from one type of opiate to another, or within the same pharmacological group.

Now drug users not only switch from one drug to another within the same type of drug, from amphetamines to phenmetrazine, but also from one type of drug to another, from opiates to stimulants, or from cannabis to stimulants to opiates, or from stimulants to hallucinogens.

Abuse of Combination of Different Types of Drugs. The picture of the drug abuser only using one type of drug (opiates or stimulants or hypnotics) is no longer the typical pattern of drug abuse. The use of different types of drugs is more and more common. Some examples are the combination of stimulant drugs with opiates and hypnotics, or of cannabis with stimulants and hypnotics, or of hallucinogens with stimulants and occasionally opiates, or of cannabis with inhalation of volatile solvents ("sniffing") in juveniles. In the surveys made, the proportion of individuals combining different types of drugs varies from 10 to 50 per cent, and this tendency is increasing.

Hence this combined intake, "polytoxicomania," is a characteristic trait of drug use today.

New Modes of Administration. Whereas oral intake of hypnotics and stimulants and occasionally injection of opiates was the characteristic mode of intake before the present era, new modes of administration have now appeared.

Inhalation ("sniffing") of volatile solvents, injection of stimulant drugs, opiates, and in poly-drug use with injections also of other types of drugs (hypnotics, sedatives and LSD), and smoking of cannabis preparations, sometimes together with an alleged (?) admixture of opiates or hypnotics, are the three new modes of administration, playing an increasing role in the present drug situation (Figs. 1, 5, 13–15).

Higher Doses. A typical feature of the present pattern is also the intake of high, often excessive, doses of some of the drugs abused. Rylander (1966, 1969) has found many patients in a forensic-psychiatric ward, abusing high doses of phenmetrazine, who showed drug-induced psychoses, that a solution of, for example, 20 to 30 phenmetrazine tablets (25 mg each), crushed and laked out, actually containing 500 to 750 mg substance, yielded 200 to 300 mg phenmetrazine; 60 tablets (1500 mg) gave 547 to 561 mg phenmetrazine, and 100 tablets (2500 mg) 1009 mg! Thus some of these individuals have injected up to several thousand milligrams of phenmetrazine daily; this should be compared to the therapeutic dose (e.g. used as a weight-suppressing agent) of 75 mg orally.

The character of today's drug scene, the drugs being furnished illicitly, often makes it impossible for the abuser to know the actual potency of the compound he is injecting, whether it is a pure substance or mixed with an innocuous substance to dilute the main compound or adulterated in various ways, sometimes being something quite different.

Spree Use of Drugs. Another characteristic trait in the present drug-taking pattern is "spree" use, i.e., use of high doses of an agent, commonly of a nature unknown to the user, taken in order to experience some new sensations. Often high doses and dangerous modes of administration are employed. Injections of phencyclidine, wine, or crushed and laked influenza-tablets are some examples.

Preference for Different Types of Drugs by Age Groups. The difference in types of drugs used and abused in different age groups has been stressed (Goldberg 1968 a, b) and is borne out by the surveys made: sniffing and cannabis-smoking in the young, injection of stimulants in those between 20 and 30, and oral intake of hypnotics and sedatives in those over 40; additional drugs being used by these age groups in a varying proportion (Figs. 3, 4, 6–9, 13–15).

The background for this characteristic choice must be sought on one hand in the effects of the drug on the individual and the preferences for one or another of the main effects of the drug on the individual, as related to his psychic (and metabolic?) make-up and to CNS structural and biochemical findings, and on the other the needs of the individual and his relations to his environment. Balter (personal communication) has suggested, based on the results of a study of prescriptions in the United States, that younger people ask for drugs with stimulant effects, whereas older people ask for drugs with depressant and sedative actions to obtain sedation and relief from anxiety and pain, hence tendencies that agree with the choice of drug abusers in different age groups.

The Individual

Younger Age Groups. The change in the last two decades to younger age groups engaged in drug taking is clearly seen in the hospital studies when repeated at intervals, and still more dominant when other groups at risk are seen.

In the hospital survey from 1954 (Lindgren and Krook 1956), the mean age was found to be 44 years; only 17 per cent were under 30 years of age, and only 1.4 per cent were under 21 years of age.

The change came around 1960. In a second hospital study from that year (Mental Health Commission 1964), the mean age had decreased to 36.5 years; the proportion of those under 20 years of age had increased but was still rather small, only 7 per cent.

During the sixties this picture did not change substantially. The mean ages in the 1966 and 1968 hospital surveys were 37.4 and 34.3 years, and the proportion of abusers under 20 years was 7 and 9 per cent (Fig. 1).

The great change was seen in other groups of users. In the 1966 and 1968 police studies, the mean ages were low, 30 and 24.9 years respectively, and the proportion of individuals under 20 years of age was high, 18.7 per cent in 1966 and 40.8 per cent in 1968 were under 20 years of age (Fig. 5).

In the 1967 case-finding survey (Frej 1969 a), the mean age was 25.6 years, not less than 37 per cent were under 20.

In a study of 200 cases of infectious, drug-induced hepatitis, the mean age was 24.8 years, a total of 30 per cent were under 20 years.

The same tendency was seen in a study of out-patients treated for drug abuse (Ljungberg et al. 1969), the mean age was 26.3 years, and 16.3 per cent were under 20.

As already stated this change in age also leads to a change in the types of drugs abused (Figs. 1–8), from the opiates and hypnotics as seen earlier, and still manifest in the hospital material (Fig. 1), to cannabis, solvents, and stimulants (Figs. 5, 13, 14).

It should also be noted that there is a difference in the age distribution with regard to drug taken between males and females, the females usually taking drugs at an earlier age. In Herulf and Sunesson's study (1967) of 169 young drug addicts (over 80 per cent abusing stimulants), the mean age among the boys was 18.6 years, and among the girls 16.9 years. In a study of drug-induced hepatitis, the mean age of males was 25.8 years, ranging from 15 to over 50 years of age, against a mean age of 22.2 years in females, ranging from 10 to 44 years.

Change in Motivation to Take Drugs. In earlier days the majority of the group that abused drugs were patients who had obtained the drug by prescription for relief of pain of some illness (opiates), and as weight-reducing agents (the amphetamines).

The emergence of dependence was a consequence of prolonged medication, the physician often not being aware of the dependence-producing properties of the medication. Continuation or perpetuation of drug taking mainly occurred to avoid the emergence of unpleasant abstinence or withdrawal phenomena; this category of patients could be looked on as a random sample of the patient population. This is still true today of the majority of older patients (over 40 years of age) indulging in oral intake of hypnotics and sedatives; also many housewives are found in this category.

Today the picture is different. A large proportion of the drug takers actively seek drugs as a means to obtain sensations or experiences of new and different kinds, from the intensive sensations termed "a kick" to "mood elevation" and "mood expansion" and other sensations that are experienced as more or less "pleasurable." Some take drugs to solve their problems, to cope with the stress of the present world, to conform with the mores of the group, to obtain status, to get a new identification. Others take drugs to fill the "need to continue," which must be looked on as part of a development of tolerance. The continuation of drug taking in order to create new sensations, often felt as a drive or an urge to take drugs, is quite different from the continuation of taking a drug in order to avoid discomfort.

The effects also may be different, more intensive, more complex and more dangerous, often leading to complications, from psychic disturbances to overdoses and toxic phenomena of various kinds, in a higher proportion than in yesterday's drug-taking generation.

Changes in Personality Structure. Among the earlier generation of drug takers, in whom a majority were patients who had become dependent by "accident" and who continued taking drugs to prevent the emergence of withdrawal symptoms, or who were elderly and took hypnotics, sedatives, or minor analgesics to alleviate pain and anxiety, the proportion of individuals with deviant personalities was small.

Today, with a large proportion of drug takers actively seeking drug sensations, a high proportion of individuals with deviant personalities is found among many groups of drug users; in younger age groups, in those attending out-patient clinics (Ljungberg et al. 1969), and in groups in which a high proportion of individuals with deviant personalities is found even without the syndrome of drug abuse, e.g., among criminals and severe alcoholics.

All of the studies referred to in detail, by Herulf (1969 a, b) in 16-year-old pupils in grade 9, by Agrell et al. (1969, 1970) in 18-year-old military conscripts, and by Ahrens et al. in students in whom this problem has been elucidated, have shown that the proportion of individuals displaying indications of emotional instability such as signs of nervousness, sexual problems, depression and suicidal tendencies, increases from a minimum in non-drug users to a maximum in those drug users who show the highest frequency of drug use (Fig. 16).

The histories elucidated make it likely that the symptoms noted to a large extent already existed *before* the drug use, paralleling the phenomenon of maladjustment.

Another indication is the difference in marital status observed between drug takers and non-drug takers, exemplified in Fig. 2: a high proportion of single and divorced individuals, and a low proportion of married. Also this change from the general population may be looked on as a result of different personality structure.

Increased Risk of Complications. The higher incidence of injection as the dominating mode of administration is one reason for the increase in frequency and severity of complications. The complications range from infection at the site of injection, abscesses, thrombophlebitis, and scar formation to systemic infections, including septicemia.

One common complication is inoculation serum hepatitis, mainly transferred by the joint use of infected cannulas and syringes. The incidence has increased rapidly from 53 cases in 1962 to over 929 drug-induced cases in 1968. The proportion rose from 6 per cent of all cases of hepatitis in 1962 to 38 per cent of all cases in 1968. In 1970, 85 per cent of all cases of inoculation hepatitis were drug-induced.

The highest proportion of hepatitis is seen among abusers of stimulants, from 36 per cent in the 1967 Stockholm case finding study (Frej 1969 a) to 62 per cent among the drug takers under 20 at Maria Hospital in Stockholm. Also, among opiate abusers who also injected amphetamines, a high proportion had had inoculation hepatitis.

The age distribution among hepatitis cases tends to be younger and younger. Whereas in 1964 only 13 per cent of all cases were between 13 and 15 years of age, every second was in that age range (49 per cent) in 1967 (Ekman and Ström 1967, 1969).

The disease may not only be dangerous to the patient himself but also to others. Hospital personnel have become infected by handling blood samples from these patients. When blood from drug abusers with latent serum hepatitis is being transfused, hepatitis virus may be transferred to other patients, causing serious complications; for a period of time even some types of operations needing large amounts of blood (among others, transplantation) had to be postponed.

Other severe complications may occur also, e.g. shock, respiratory failure, or death, resulting from an overdose. Because of unknown composition and strength, the preparation may contain higher concentrations of active material than expected. When tolerant individuals who have lost their tolerance (e.g., after a prolonged withdrawal or abstinence period) take the high, formerly tolerated dose, it now is toxic and may be fatal. A third mechanism is seen in novices who are given the high dose tolerated by a tolerant, dependent individual; in these people also this high dose may prove to be fatal.

Combinations of different types of drugs may also result in increased toxicity, e.g., the combination of opiates and hypnotics or alcohol and hypnotics.

Mental disturbances of different kinds are noted in increasing proportion in drug abusers, especially after the introduction of central stimulants, cannabis, and hallucinogens. Toxic psychic manifestations are seen, both acute after large doses and chronic after prolonged administration. They range from mental disturbances of various kinds to psychoses and other severe mental changes, perseveration, depersonalization, depression with suicidal tendencies, paranoia with violence against alleged persecutors, delusions, hallucinations, and other derangements.

Excessive drug taking, either alone or combined with alcohol, may have other consequences when taken by drivers. Of all cases of drunken driving in Sweden, 1 to 2 per cent involve intake of excessive doses of drugs, mainly hypnotics and sedatives (81 per cent), amphetamines or other stimulant drugs (phenmetrazine or methylphenidate) in 14 per cent, and morphine (1 per cent), or intake of volatile agents (4 per cent).

In the Stockholm area, the proportion was different, stimulants being noted in 57 per cent of the 42 "drug-driving" cases in 1966 against a proportion of 9 per cent in the rest of the country.

Finally, drug abusers show a higher mortality rate, amounting to 10 to 20 times that seen in the normal population. The increase is due to an increased rate of accidents with a fatal outcome, to a high rate of suicides, and to fatal complications after intake of an overdose. Often a lowered resistance to infections and other environmental stress factors occurs, combined with a deterioration in nutritional state, in the general hygienic level, and in living conditions.

Relation of Smoking, Drinking, and Inhalation of Solvents ("Sniffing"). All studies, in which an attempt has been made to study the possible correlation between intensity of drug use on one hand and presence and intensity of smoking tobacco, drinking, and sniffing on the other, have shown a clear relationship (Agrell et al. 1969, 1970; Ahrens et al. 1969; Herulf 1969 a, b; Jonsson and Svedugård 1969). Examples have been given and are illustrated in Figures 10 and 12 showing non-users to be at one end of the spectrum in frequency of using tobacco, drinking, or sniffing, and users to be at the other, showing an increase in proportion of those using tobacco, drinking, and sniffing, with increasing severity of drug use.

One inference seems to be that some of the same elements (emotional instability, insecurity and maladjustment in the family and in society) that make the individual more inclined to use drugs, also make him more prone to indulge in smoking, drinking, and sniffing.

One finding in the same direction is that non-drug-using individuals, who have been offered drugs (Jonsson and Svedugård 1969) and would be willing to test drugs, are more like drug users in certain respects than non-drug users who oppose the idea of taking drugs; similar observations were made by Ahrens et al. (1969).

Emergence of Need to Continue Taking the Drug. One of the characteristics of drug dependence is "psychic dependence," the term coined by WHO in 1963 (Wld Hlth Org. techn. Rep. Ser: 1964:273 and 1964: 287) denoting a compulsive drive to continue taking a drug. In this definition also elements as liking, and drug seeking behavior will be included.

In one survey (Agrell et al. 1969, 1970) on 18-year-old military respondents, an attempt was made to evaluate whether any sensations of this type were observed by the respondents. The questions was: "How great is your need to take narcotics?" The answers comprised five grades, from "very great" and "rather great" to "fairly small," "very small," and "none at all."

As already discussed and illustrated in Figure 15, this item correlated with type of drug taken; opiate users, LSD users, and those injecting drugs showed definite need to continue, and among cannabis users and those taking stimulants orally only those with the highest intensity of drug use felt a definite need (Fig. 15).

The results are thus in line with previously stated findings that both type of drug and intensity of drug use have a bearing on the emergence of symptoms that may be an expression of underlying dependence or of the changes that constitute the basis for later emergence of dependence; the terminology used was not so strict, possibly with the exception of those displaying "great need," as to tell exactly whether a dependence was less developed in a certain individual than in another.

The results are interesting, however, and should be further analyzed, e.g., cross tabulated against factors such as emotional instability or degree of social adaptation.

Environment

Family Relations. Several studies have attempted to assess the role played by changes in family relations and other conditions within this sphere (Agrell et al. 1969, 1970; Frej 1969 a; Herulf 1969 a, b; Jonsson and Svedugård 1969). All these studies have shown that a relation exists between variations in family relations and the emergence of drug use or no drug use on one hand, and between degree of deterioration in family relations and intensity of drug use on the other. The factors studied have ranged from complete to incomplete homes to foster homes, contact or no contact with parents, desire to stay at home or to escape from home, and degree of contact with parents. The non-user is at one end of the scale with well-adjusted family relations, and the intensive drug user at the other end, showing the highest degree of disruption.

Social Adjustment. The same is true of degree of social adjustment, as judged from contact with school, liking or disliking school or having escaped from school without leave, feeling peculiar or different from others, and the like. Again the non-user is at one end of the scale with a higher proportion of well-adjusted individuals and only a few less well-adjusted individuals, and the user at the other end of the scale with the highest intensity of intake showing the highest proportion of individuals with various forms of social maladjustment.

Emergence of Subcultures (Contracultures). One typical trait in the present society is the emergence of groups of individuals with common habits and ideals, with certain common behavioral patterns and interests, often marked by the way in which the individuals in the group behave and live —clothing, hairdress, outlook on life. In certain connections, these groups have been termed subcultures, or contracultures in others. Opposition to the existing society, to other age groups, to the "ruling class" ("the Establishment"), sometimes the trespassing of traditional social boundaries and the creation of new ways of living are some of the characteristic features which can become the breeding ground for various forms of deviant behavior. Sometimes these and other traits are looked on as integral parts of a new philosophy ("escapism"), which is thought to favor the use of means to "widen the horizon" and help the individual to cope with the existing society. Drug taking may be looked on by the members of the group as one of these means.

Special environments ("milieus") are also an important factor in the emergence of drug taking. Drug takers often gather in one place or another, depending on the atmosphere and the opportunity to obtain drugs, ranging from certain discotheques and youth clubs to homes.

One characteristic trait of the drug abuse picture in Stockholm is the occurrence of special places where groups of abusers gather to smoke and/or inject drugs; such a place is called a "drug-joint" or "drug-quarter" ("geggarkvart"). Usually this place is located in a slum area, often in houses left by their former inhabitants because they are to be torn down. The "quarter" usually belongs to an individual, the host, who serves as the center of the group and procures the drug(s) and often at the same time teaches the novice how to use the drug, e.g., how to inject. Sometimes the host himself is not a drug taker. The number of participants varies and is usually anywhere between 10 and 20. The individual members are not constant but are always coming or going. The furniture is the simplest possible, some mattresses, a table, some kitchen utensils. Filth and poverty characterize the situation. Some people are discussing, others look absent-mindedly around far away in their dreams, others are excited, constantly moving or repeating the same thing in a stereotypic way ("pounding"), some may be in deep stupor.

Active field work in 1969 by a "field searching group" within the Stockholm City Social Board disclosed a total of 492 drug-quarters located in Stockholm (compare also the studies by Stålberg et al. [1969] in Uppsala). Some 100 to 200 were in use simultaneously, hence corresponding to a total of some 1500 to 3000 abusers in this type of locality, and representing all grades from the recently initiated user to the inveterate, chronic abuser (Hogstedt 1969, SOU 1969:52, 53).

When comparing the location of the drug-quarters with the distribution of housing facilities, it was noted that the drug-quarters clustered in the same areas where the highest proportion of inadequate housing facilities existed, i.e., in slum areas. The number of drug-quarters varied between 23 and 65 per city region in six of the most threatened areas; in areas with adequate housing, the number of quarters was small, from zero to 16 per city region in twelve regions of this type.

Special environments play a role for emergence of drug taking and drug abuse; the drug-quarter is one typical example. This is a place where people know they find drugs and other individuals with some common way of thinking, and where a habit can be "transferred" from one individual to another, where the novice can be introduced to the use of drugs and to different modes of administration. The common use of syringes and cannulas is one explanation of the high incidence of serum hepatitis among drug users injecting drugs.

However, it should be added that intake of drugs within a group setting, even if characteristic of today's drug use, especially among the younger generation, is not the only pattern of intake. Some studies (Frej 1969 b) indicate that, with increase in the severity of drug abuse, the drug taker tends to take the drug alone and no longer as part of a group. A similar tendency is seen with chronic alcoholism; after a period of using large amounts of alcohol as part of social drinking together with others, the individual begins to drink alone, and may even deny in public that he drinks.

Use and Availability of Medicaments. The sale of medicaments with psychoactive effects, excluding the narcotic drugs (opioids, cocaine and coca leaves, cannabis, stimulants and hallucinogens) has traditionally been on prescription only, sold through pharmacies under the supervision of the

Board of Health and Welfare (Socialstyrelsen, formerly called Medicinalstyrelsen, the Medical Board). The Board controls the handling of drugs in pharmacies, prices, and the like. The pharmacies have been state-owned since January 1971.

The use, advertising, and sale of medicaments are strictly controlled by a regulation, essentially going back to 1934; the latest additions are from the sixties. Only between 3000 and 4000 medicaments are allowed on the market in Sweden; indications for use and efficacy are strictly controlled; the ratio between efficacy and risk of untoward actions is one important factor in the consideration of acceptance or rejection of a medicament for possible introduction on the market.

The sale of narcotic drugs is strictly surveyed in accordance with international conventions and national narcotics laws; the sale of other psychoactive drugs follows in certain respects.

The total turnover of medicaments is increasing constantly, by about 15 per cent per year, corresponding to an almost threefold increase in a ten-year period, from 354 million Swedish crowns in 1958 to 1011 million in 1968.

The number of prescriptions has risen in the same ten-year period from 23.8 to 38.1 million prescriptions, corresponding to about 4.7 per cent per year, or 60 per cent in ten years.

As part of an over-all program to reduce drug abuse, a number of measures have been taken during the years by the Board of Health, aimed at informing the medical profession and restricting the availability and the sale of narcotic drugs.

Essentially the following steps were taken:

1. The placing of amphetamines (1944) and of phenmetrazine and methylphenidate (1959) on the Swedish Narcotics List.

2. Information issued in 1960 by the Board of Health to the medical profession, about the risks involved in medication with analgesic, stimulant, and hypnotic-sedative drugs.

3. Information issued by the Board in 1962 to the medical profession about the risks in prescribing psychoactive drugs, together with restrictions on the right of physicians to prescribe narcotics by telephone, and information to physicians and pharmacists with regard to the necessity to control the identity of an alleged patient and the risks of theft of medicaments and of forgery of prescriptions.

4. The termination of the sale of phenmetrazine in December 1965.

5. Restriction of the sale of the whole group of stimulants to a few indications only, mainly narcolepsy, and only on license, issued by the Board for specified patients and specified indications, other indications not being accepted as a rule, on 1 July 1968.

These measures substantially brought down the number of prescriptions on narcotic drugs and reduced drastically licit trade in narcotics. The number of prescriptions was reduced by 89 per cent in nine years, from 762,913 in 1959 to 86,179 in 1968. The sale of narcotic drugs also fell drastically, stimulant drugs by 97 per cent (from 33 million doses in 1959 to less than 0.5 million in 1968) and analgesic narcotics by 80 per cent (from 19 million doses in 1959 to 3.4 million doses in 1968).

For comparison it can be stated that the total sale of hypnotic and sedative drugs (combinations not included) amounted to 350 million tablets in 1968, and it seems not to have changed substantially in the last few years.

The sale of all drugs of this type, combinations included, amounts to 800 to 900 million a year.

There is at present no diversion of narcotic drugs from the licit to the illicit market; thus in Sweden illegal narcotics, including stimulants and hallucinogens, are furnished by an illegal market.

The size of the illicit market is difficult to estimate. Some estimations, based on one hand on the amount of seized material, knowing that Police and Customs only seize a small fraction of what is coming in, or on illegally manufactured material, and on the other on an estimation of the number of abusers and their average consumption would put the illicit market at a daily influx of 50,000 to 200,000 tablets, or the equivalent in substance.

The illicit trade is outside the scope of this review; suffice it to state that drugs are being brought into the country by all possible routes: in suitcases carried by travelers, in cars, trucks, and boats, and dropped by parachute in the countryside. According to available accounts part of this "trade" is highly organized on an international as well as a national basis, including organized prison escapes of those serving sentences for narcotic offenses.

The need of international cooperation in the vigil against the illicit trade has been stressed repeatedly and is one of the most urgent needs in this field. Possible measures range from extension of already existing cooperation between countries via Interpol to synchronization and equalization of legislation and of actual measures of enforcement. The realization that the measures of any one country in this field, e.g., with regard to intensity of search and enforcement, severity of sentences, and consequences for illicit trade, have a bearing on all the other countries involved strengthens the necessity to act.

One example of the connection between illicit trade in narcotics and other criminal activity can be seen in the results obtained when the activities of the police, the prosecutors, and the courts were intensified in the spring of 1969. Besides making drugs more difficult to obtain and probably reducing the number of new cases of drug abuse, the action also led to a reduction in crime in general along with a substantial reduction in check forgery. The inference seems adequate that money obtained from forged checks, to a certain extent, is being used for obtaining and trading in illicit narcotics.

Reduction in the availability of narcotic drugs is one important prerequisite for reducing the number of new cases of drug abuse and for attempting to reduce the perpetuation of drug abuse in people already using narcotic drugs.

Economical Viewpoints. Among the many factors involved in the emergence of drug abuse, economical viewpoints also play a role. Methods to procure the money needed to obtain the quantity of the drugs necessary to maintain and satisfy the inveterate and dependent drug user range from using salary or income to living on sick insurance money or sick pension; from male and female prostitution, pilfering, theft, check forgery, and illicit trade in narcotic drugs (buying small amounts, using half and peddling the rest) to the highly organized large-scale operations.

A detailed and penetrating study of the relationships between availability of narcotic drugs and of psychoactive medicaments, income, and other means of obtaining money, and the labor situation, wages, price index, and other economic indices, as has been carried out in the study of alcoholism, is necessary in order to increase our knowledge concerning the possible relationships between these various factors and to give a background for possible countermeasures aimed at those links in the chain where a counterattack has the greatest chance to succeed.

With regard to *alcohol,* it is demonstrated that changes in price of beverage, in the wage level, and in legislative and other measures have a bearing on the sale of alcohol, causing definite changes in the quantities of alcohol beverages bought: an increase in price, within certain limits, is accompanied by a decrease in consumption. But other effects also can be noted.

A strike in the Swedish Alcohol Monopoly distributing system in 1963, lasting for 8 weeks, brought down the sale of spirits within a few weeks to 10 per cent of normal. No change occurred in the sale of wines, and an increase by several hundred per cent was noted in the sale of beers. This decrease in the availability of spirits was accompanied by a decrease in drunkenness, in drunken driving, in delirium tremens, in accidents, in male suicides, and in accidents in industry. The figures went back to the level existing before the distribution conflict within weeks after the termination of the conflict and the return of the sales to the preconflict level.

Detailed studies month for month should disclose whether similar relationships obtain within the drug field, e.g., the exact effects of the concentrated police efforts in Sweden with regard to narcotic offenses in the spring of 1969; the bearing of possible findings for the evaluation of the effects of various countermeasures is obvious.

Number of Trials for Narcotic Offenses and Types of Drugs Involved. A review of sentences for narcotic offenses, principles for dismissal of trials, for giving a suspended sentence, for probation, and the like has been published (SOU 1967:25 and 41, SOU 1969:52 and 53, Goldberg 1968 a, b).

In Figure 17 the development for the period from 1954 to 1965 is illustrated; the number of sentences has risen from 6 in 1954 to over 300 in 1965. The rise follows an exponential course and that is to be expected assuming the mechanism behind drug offenses to be a spreading of drug abuse by epidemiological mechanisms, e.g., by transfer from individual to individual.

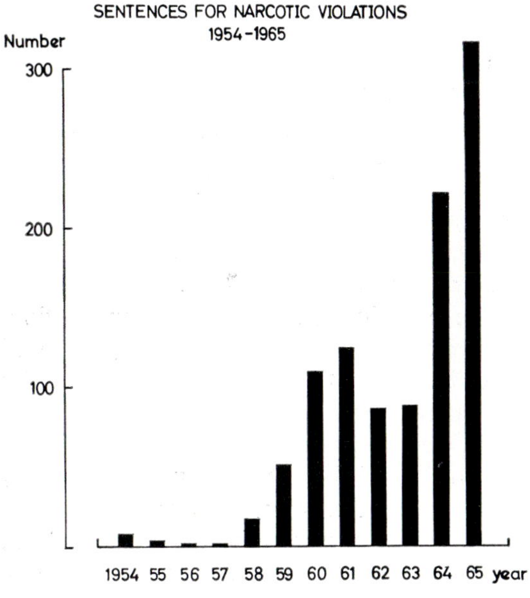

FIG. 17. Number of trials for narcotic offenses (mainly illicit trade) during the period 1954–1965. Adapted from SOU 1967:25.

In the present work, another part of the problem will be touched on, namely, the development in the five-year period from 1966 to 1970 and the distribution of these sentences, mainly comprising cases of illicit "trade" in narcotics—pushing, peddling, illegal manufacture, and smuggling—and on different types of drugs (Friberg, 1971).

Figure 18 shows the total number of trials increasing sharply from 279 in 1966 to 3154 in 1969 and falling to 2254 in 1970.* In the graph on the left of Figure 18, the absolute figures are given; in the middle graph the figures are in a logarithmic scale, i.e., indicating the rates of change; in the right graph, the number of sentences for each separate drug are expressed in per cent of the total number of sentences, hence indicating proportions.

When looking at each separate drug it is seen that in 1966 sentences for violation of narcotic legislation involving stimulant drugs (amphetamines and the like) dominated the picture. Of the trials in 1966, 79 per cent concerned stimulants. It is then seen how the proportion of sentences involving stimulants decreases to 32.3 per cent in 1970, in spite of the rise in the total number of trials, from 220 in 1966 to 1479 in 1969, then decreasing to 696 in 1970.

Opiates play a minor role in this connection. The number of trials for this type of drug rose from 14 in 1966 to 94 in 1969 and 60 in 1970, corresponding to 0.5 per

*During the first few months of 1969 an extra-severe law enforcement campaign against drug abuse was instituted by a manifold increase in personnel from police, prosecutors, and courts, making a concentrated effort possible. By the second half of 1969, the forces put to work on narcotic cases were reduced to approximately the same size as before the campaign. The concentrated effects of the campaign are noticeable in the graphs in Figure 18.

TRIALS FOR NARCOTIC OFFENCES

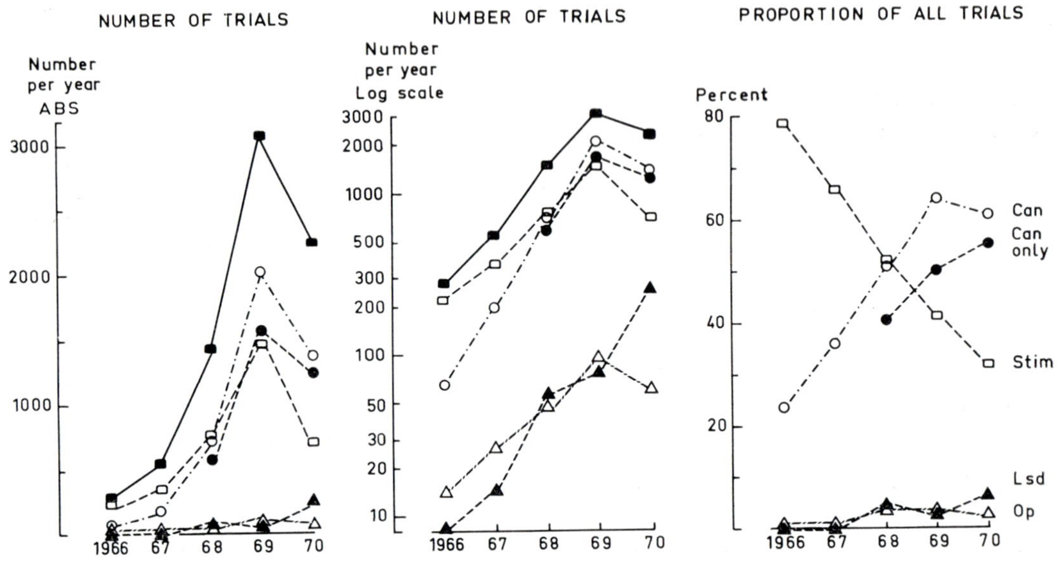

FIG. 18. Number of trials for narcotic offenses (mainly illicit trade) during the period 1966–1970. Distribution on different types of drugs: Can = cannabis; Can only = cannabis only; Stim = stimulant drugs; Op = Opiates.

The development is shown in three scales: *left,* absolute number; *middle,* rate of change (semi logarithmic scale); *right,* per cent, indicating proportion of all narcotic offenses. Note rise in 1969; in that year an intensified police campaign was carried out.

cent in 1966 to 2.7 per cent of all cases in 1970; the maximum was 3.3 per cent in 1968.

Cannabis is now taking the lead as the most-smuggled drug, mainly in the form of hashish, from 64 trials in 1966 to 2,017 in 1969 and 1,387 in 1970. These figures correspond to 23.5 per cent of all trials in 1966, to a three-fold increase, 64.0 per cent, in 1969, to 61.7 per cent in 1970. In most cases these trials concerned cannabis only, the proportion of these cases ranging from estimated 18 per cent in 1966 to 55 per cent in 1970.

Finally hallucinogens, mainly LSD, are coming into the picture, from one trial for LSD in 1966 to 251 trials in 1970. The numbers correspond to 0.4 per cent of all trials in 1966, rising to 11.1 per cent in 1970.

This change in the pattern on the illicit drug market reflects the change in use and abuse by the consumer, with an increase in use of all types of narcotic drugs, and a shift over to cannabis and LSD; the stimulants decreasing somewhat in proportion to the other drugs, and opiates still playing a minor role only.

Increase in Mass Communication. One factor of importance for the spread of drug abuse is the increase in mass communication. The wide, and immediate, dissemination of information in the field of drug abuse, by newspapers, journals, broadcasting, TV, records, and other routes, means that a large proportion of those at risk will immediately be informed about use, alleged effects, and, to a certain extent, also about side effects. This knowledge, depending on the type of information given and the personality of the exposed individual, will stop some and look inviting and tempting to others.

These facts, together with the increased means of traveling from one country to another, making it possible to experience the effects on the spot together with others and to partake in various activities and to procure the drug where it is cheapest and most easily available, explain to a certain extent not only the widespread use and abuse of a number of drugs, but also why some of these habits spread so rapidly and penetrate a population so thoroughly in such a short time.

No studies exist that have attempted to evaluate the role of mass media in this process. This is an important task, because studies of this kind may provide data on how mass-media should be employed as an active and integral part of prevention and information measures.

COURSE OF DRUG ABUSE IN THE INDIVIDUAL

Results of a Follow-up Study

It is important to have knowledge about the actual course of the drug abuse state in individual cases. A follow-up study by Frej (1969 b) has attempted to answer some of the questions that might be posed.

As a starting point, the patients with drug abuse, treated in a hospital in 1964 for drug-induced hepatitis, were chosen. A total of 74 patients, 62 men and 12 women, were found and interviewed and closely examined in 1967, after three to three-and-one-half years.

It was found that 77 per cent had mainly used stimulants (61 per cent mainly phenmetrazine and 16 per cent mainly methylphenidate); 18 per cent had used opiates mainly; and 5 per cent cannabis or hypnotics. Half of the patients had used narcotic drugs for more than five years, i.e., more than two years before the hepatitis occurred.

The mean age was 27.5 years, the range being 20 to 43 years in the males, from 20 to 29 years in the women. Thus the women were younger. Fifty-nine per cent were single, 19 per cent married, and 20 per cent divorced, thus a distribution on marital status that was clearly different from the normal distribution in the general population; the drug abuse population showed more single, less married, and considerably more divorced individuals (about 3 to 3.5 per cent divorced in the normal population, in corresponding age groups); the drug abuse distribution was comparable to the distributions found in the hospital or in the police studies.

The patients had started their drug abuse at a mean age of 19.6 years, the women at a younger age than the males. Stimulants had been the starting drug in 65 per cent of the cases. (Cannabis smoking was only rarely seen in Sweden in 1960 to 1962.) The debut drug was taken orally in 69 per cent of the cases and injected in 20 per cent. About one third of those starting on oral intake went over to intravenous injection within one year, and more than half within two years.

Injection was the preferred route in 87 per cent of the cases, oral intake in 11 per cent, and smoking in 2 per cent.

An attempt was also made to estimate frequency of intake. Fifteen per cent had taken the drug less than 100 times, 36 per cent less than 1000 times, 51 per cent less than 2000 times, and 49 per cent more than 2000 times; the median was around 2000 times. Those taking opiates (over 90 per cent) had taken more than 2000 injections, the median in those injecting stimulants was between 1500 and 2000.

Between half and two thirds had also taken other drugs beside the preferred one, mainly other stimulants, further opiates, cannabis, or sedatives, in 22 per cent also LSD.

Based on these findings the severity of the drug abuse could be ranked operationally into one of three categories: (I) limited abuse = duration of less than four years and not more than 1000 injections, 14 patients (19 per cent); (II) considerable abuse = four years or more and maximally 1000 injections, or less than four years and more than 1000 injections, 15 patients (20 per cent); (III) severe abuse = more than four years' duration and more than 1000 injections, 45 patients (61 per cent).

The costs to procure the narcotics were estimated to be less than 200 crowns ($40 or £16) per month in half of the cases, the median being around 120 crowns; in the rest, up to 3000 crowns and more were given as the upper limit.

With regard to employment and work conditions the situation had clearly deteriorated during the period of drug abuse. Before taking drugs 57 per cent had been fully employed, and 15 per cent had been without a job. At the time of the interview, three-and-one-half years later, only 28 per cent were fully employed, thus less than half, and as many as 43 per cent had no job.

Depending on the length of time without a job (from less than half a year to more than three years) drug abuse was said to be the main factor causing unemployment, ranging from 27 to 75 per cent of these cases, or 42 per cent of the total.

Thirty-nine per cent had no income at the time of the interview and in 43 per cent the income had decreased during the period of drug abuse; this was indirectly due to long periods in the hospital, 23 per cent, or more directly due to the narcotic abuse in 56 per cent.

When looking at family, friends, and other contacts a high incidence of alcoholism (67 per cent), narcotic abuse (33 per cent), psychic diseases (23 per cent), and deviant behavior was found.

The incidence of psychic disturbances, treated on an out- or in-patient basis, rose sharply during the observation period, from a proportion of 5 per cent before the narcotic abuse, to 9 per cent at the time of acquiring hepatitis, to 42 per cent at the time of the interview.

When analyzing criminal background, a total of 81 per cent of these patients had been sentenced for criminal offenses at the time of the interview, against 19 per cent at the time before the drug abuse, thus showing a severe deterioration during the period of narcotic abuse.

The same tendency is seen with regard to alcoholism offenses (drunkenness, drunken driving). Twenty-four per cent were found in registers before their drug debut, against 65 per cent after; when considering all available records, 76 per cent were found to have been recorded. This applied mainly to abusers of stimulant drugs; among the opiate abusers only 30 per cent had alcoholic offenses.

With regard to periods of sickness, as judged from the records of the national sick insurance, 97 per cent had obtained sick insurance for longer or shorter periods of time; the median was four to five months, 16 per cent had been sick for more than one year. A total of 91 per cent were incapacitated for somatic diseases for shorter or longer periods, and 59 per cent for psychic diseases. *In no case has narcotic abuse been mentioned as the diagnosis to the national insurance authorities!* Forty-six per cent had been sick the year before their narcotic debut, the median time being less than one month, against 76 per cent the year before the interview, the median time being one to two months. Fifteen per cent of all these patients had a psychic disease before the narcotic debut, against 74 per cent after. These findings illustrate the deterioration in state of health brought about by the narcotic abuse.

For analytical reasons the material was then grouped according to outcome into three groups: (1) those having *stopped* taking drugs for at least six months (31 per cent); (2) those *having used* narcotics the preceding month (34 per cent); (3) those *doubtful* as to having stopped, having only taken drugs occasionally during the last six months and intending to stop (31 per cent).

The group that had stopped was younger, had a lower proportion of divorced, and a higher proportion of married; the proportion of women was higher.

Further, a higher proportion were employed, had not been absent from work during the preceding month, had had shorter periods of unemployment during the preceding three years, had higher incomes, and there were fewer in the group who needed help from social agencies.

The contrary was seen in the groups not having stopped, or doubtful.

When analyzing the situation with regard to severity, it was found that all of the patients with limited abuse had stopped, 40 per cent of those with considerable abuse, and 13 per cent only of those with severe abuse; or 35 per cent on an average after three to three-and-one-half years.

When looking at the drugs abused, 44 per cent of those abusing stimulants had stopped, none of the opiate abusers, and one of the two abusers of hypnotics.

With regard to criminal offenses, there were 70 per cent sentenced among those having stopped taking drugs, against 88 per cent among those continuing or who were doubtful as to stopping.

In order to attempt to check the results obtained and at the same time to follow the further development, a second interview, in 1969, was carried out with 66 of the 74 patients.

It was found that during these additional two years, corresponding to a total observation period of four-and-one-half to five years, an additional 12 per cent, or a total of 47 per cent, had stopped taking drugs; 36 per cent continued, and only 17 per cent were now doubtful as to stopping. Hence half of those in the original "doubtful" group had now stopped.

Seventy per cent of those with a limited abuse had stopped, 60 per cent of those with considerable abuse, and 27 per cent of those with severe abuse.

With regard to type of drug, 47 per cent of abusers of stimulant drugs (54 per cent, if only known cases are considered), and 15 per cent of opiate abusers had stopped.

These findings have a bearing not only on the prognosis of drug abuse, but also on a large number of other factors.

On one hand, the parallelity is noted between the development of abuse and medical, social, economic, and criminal deterioration, and, on the other, the fact that patients with severe forms of drug abuse (long duration, high frequency of intake and somatic, psychic, and social consequences) can stop taking the drug for longer or shorter periods of time and improve their condition. This improvement seems not to be confined to the time immediately following release from the hospital, but actually continues for many years, one third (35 per cent) having stopped after an observation time of three to three-and-one-half years, and 47 per cent after four-and-one-half to five years.

REFERENCES

SOU = Statens offentliga utredningar (Governmental Commission Reports).

Agrell, J., Netz, B., and Wolff, B.: Drug use in military conscripts 1968 (Swedish). SOU 53: 68, 1969.

Agrell, J., Netz, B., and Wolff, B.: Narcotic Abuse in Military Conscripts 1968/1969 (Swedish). Military Psychological Institute (MPI). A-Report No. 8, June 1970, pp. 1-71.

Ahrens, U., Kihlbom, M., Näs, N. O.: Students and narcotics (1968). SOU 53:109, 1969.

Ekman, A., and Ström, J.: Narcotic drug abuse: Somatic complications (Swedish). Läkartidn. *64*:4083, 1967.

Ekman, A., and Ström, J.: in SOU 52, 1969.

Frej, G.: Case-finding survey in Stockholm 1 July-31 December 1967 (Swedish). SOU 53: 194, 1969 a.

Frej, G.: Follow-up of narcotic abusers treated for hepatitis in Stockholm in 1964 (Swedish). SOU 53:281, 1969 b.

Friberg, H.: Trials for narcotic offenses 1966–1970. Manuscr. 1971.

Goldberg, L.: Habituation and abuse after the use of amphetamine preparations (Swedish). Svenska Läkartidn. *41*:561, 1944.

Goldberg, L.: Drug Abuse in Sweden (I). Bull. Narc. 1968 a, XX, No. 1, pp. 1–31.

Goldberg, L.: Drug Abuse in Sweden (II). Bull. Narc. 1968 b, XX, No. 2, pp. 9–36.

Goldberg, L.: Epidemiology of Drug Abuse and Drug Dependence. Proc. Internat. Inst. Prev. Treatm. Drug Dependence. Editors: E. and A. Tongue. Internat. Council on Alcohol and Addictions, Lausanne 1970, pp. 217–240.

Gustafsson, N.: in SOU 52:19, 1969.

Herulf, B.: Narcotic consumption among school children in Stockholm. In: Abuse of Central Stimulants. Editors: F. Sjöqvist and M.

Tottie. Stockholm, Almqvist and Wiksell, 1969 a, p. 215.
Herulf, B.: Narcotic consumption in Stockholm schools, grade 9, April 1967 (Swedish). SOU 53:13, 1969 b.
Herulf, B., and Sunesson, T.: Young narcotic drug abusers in Stockholm (Swedish). Läkartidn. *64*:1936, 1967.
Hoffmann, O., and Håkanson, K.: Analysis of cases of abusus alii found in the National Medical Statistics Register (Swedish). SOU 53:218, 1969.
Hogstedt, C.: "Narcotic drug joints" (Swedish). SOU 53:404, 1969.
Huss, M.: Alcoholismus chronicus. Stockholm, Johan Beckman, 1849.
Håkanson, K., and Hoffmann, O.: Point prevalence study on drug abuse in hospitals 3 April 1968 (Swedish). SOU 53:229, 1969 a.
Håkanson, K., and Hoffmann, O.: Drug abusers seen by police. June 1968 (Swedish). SOU 53:247, 1969 b.
Inghe, G.: The present state of abuse and addiction to stimulant drugs in Sweden. In: Abuse of Central Stimulants. Editors: F. Sjöqvist and M. Tottie. Stockholm, Almqvist and Wiksell, 1969, pp. 187–214.
Jonsson, E., and Svedugård, L.: Use of narcotics among school children (Swedish). SOU 53: 43, 1969.
Kvarby, G.: Drunken drivers with drug offenses 1966. Mimeographed manuscript, Swedish Narcotics Drug Commission, 1967.
Lindgren, G., and Krook, G.: Number of drug addicts in Sweden 1954 (Swedish). Nord. Med. *55*:929, 1956, and Svenska Läkartidn. *53*:361, 1956.
Ljungberg, L., Bjerkenstedt, L., Sarap, E., Björn, M., and Fridén, A.: Out-patient treatment of drug abusers in Stockholm. 26 January 1966–31 January 1969 (Swedish). SOU 53:459, 1969.
Mental Health Legislation (Swedish). SOU 40, 1964.
The Narcotic Problem (Narkotikaproblemet). Parts I–IV. Official Reports of the Swedish Commission on Treatment of Drug Addiction (Narkomanvårdskommittén). SOU 25, 1967 (I), 41, 1967 (II), 52, 1969 (III) and 53, 1969 (IV).
Rylander, G.: Barbiturism—a new addiction (Swedish). Nord. Med. *5*:647, 1933.
Rylander, G.: Preludin addicts from clinical and medical-criminological view-points (Swedish). Läkartidn. *68*:4973, 1966.
Rylander, G.: Clinical and medico-criminological aspects of addiction to central stimulating drugs. In: Abuse of Central Stimulants. Editors: F. Sjöqvist and M. Tottie. Stockholm, Almqvist and Wiksell, 1969, pp. 251–274.
Stålberg, E.: Field studies and social intervention in Uppsala. In: Abuse of Central Stimulants. Editors: F. Sjöqvist and M. Tottie. Stockholm, Almqvist and Wiksell, 1969, pp. 291–296.
World Health Organisation. Techn. Rep. Ser., No. 57 (1952), 116 (1957), 273 (1964), 287 (1964), 312 (1965), 343 (1966), 407 (1969), 437 (1970).
Österlind, O.: Smoking, drinking and drug abuse among Swedish school pupils. Manuscript, 7 July 1967, Swedish National Board of Schools.

EPIDEMIOLOGY OF DRUG DEPENDENCE IN JAPAN

MASAAKI KATO, M.D.

In prewar Japan, licensed pharmacologists were able to deal in almost all ordinary drugs, not only vitamins and antibiotics but also analgesics and some hypnotics, without physicians' prescriptions.

In spite of this universal availability of dependency-producing drugs, there were very few instances of drug dependence before the end of World War II. Drug abusers were mainly narcotic addicts in minority groups, the medical profession, and prostitutes. Some neurotic individuals were dependent on bromisovalerylurea and barbiturates.

METHAMPHETAMINE ABUSE

The methamphetamine epidemic was, therefore, the first experience of epidemic drug abuse and dependence among Japanese. This was also the first epidemic of methamphetamine abuse and dependence in the world.

In 1940, methamphetamine came into the Japanese market under the trade name of "Benzedrine," and was used only for research purposes. Later it was employed officially in the Air Force and other military forces to promote the soldiers' fighting activities; it was called "Senryoku Zokyo Zai" (drug to inspire the fighting spirits).

After the war, the contract for a large amount of methamphetamine, which had been stored by Japanese pharmaceutical companies for military use, was suddenly cancelled. These companies tried to sell their large stocks of methamphetamine, advertising the drug as one that would inspire the fighting spirits in daily life, without knowing that it had terrible dependency-producing effects. The advertisements of pharmaceutical companies were very effective in spreading methamphetamine abuse among those who wanted to recover their subjective feeling of self-confidence, in particular among Bohemians, novelists, entertainers, racial minorities, and juvenile delinquents. The spread of the drug's abuse was furthered by veterans who had experienced its effects in wartime. The lack of self-confidence among Japanese at that time was caused by the demoralized, anomic state that existed after the war.

The use of methamphetamine, largely in ampule form, spread from urban to rural areas throughout Japan. In July 1948, methamphetamine was listed as a dangerous drug; in October 1949, a law was passed to control this "awakening drug" (Kakuseizal) and the Ministry of Health and Welfare warned the principal pharmaceutical companies to suspend its manufacture completely. However, the abuse of methamphetamine, made and sold in the black market, began to increase, and finally the Awakening Drug Control Law was passed in June 1951.

Notwithstanding these intensified measures, the number of methamphetamine abusers increased rapidly, until the Law was revised in 1954, and again in 1955, in order to tighten its penal provisions and to control the raw materials used in its manufacture. Smugglers, carriers, and manufacturers were placed under penalty of ten years' imprisonment or the fine of 500,000 yen at the maximum. The law also provided for compulsory admission of addicts to mental hospitals.

The peak of the epidemic was reached in 1954, by which time it was estimated that approximately 550,000 persons might be called abusers and 200,000 showed some psychotic symptoms as a result of methamphetamine intake, out of a population of 90 million at that time.

Due to revision of the Law, compulsory hospital admission, and the antiamphetamine campaign provided by the Cabinet's decision, the number of abusers declined rapidly after 1957.

Although there probably were several factors leading to the reduction of the epidemic of methamphetamine use at that time, it should be noted that the suicide rate in Japan, particularly among young people, has rapidly decreased since 1959, after reaching a peak in 1955 and in 1958. However, the incidence of violent crimes among youth has increased since 1958. Thus, approximately ten years after the end of the war, the incidence of methamphetamine abuse and of suicide in Japan reached a peak and began to decline. It might be presumed that these two phenomena, stimulant abuse and suicide, were caused mainly by the universally pessimistic mood among Japanese, in particular among the youth, at that time. Thus, not only rigid drug control and community campaign but also recovery from a depressive, pessimistic mood might have been important factors in the reduction of the incidence of stimulant abuse and suicide. However, violent and aggressive behavior and the number of violent crimes among youth has increased following the reduction in stimulant abuse and suicide incidence.

NARCOTICS ABUSE

Although heroin abuse began during the end of 1940's and continued into the mid-1960's, its peak was reached in 1961 and 1962; since 1965, following the amendment of the Narcotic Control Law in 1963, the incidence of heroin abuse has declined. At the peak of heroin abuse, an estimated 40,000 to 50,000 persons were implicated—approximately one tenth of the peak number of methamphetamine abusers.

Although the decline of narcotics abuse followed the establishment (in 1955) of the counsellor system for narcotics addicts, the amendment of the Narcotic Control Law, and an anti-narcotic campaign in the mass-communication media, one of the main factors accounting for the reduction was the breakup of gang-groups in the metropolitan areas brought about by strong pressure of community and government. Besides, in Japan, with a population of 100 million in islands covering an area smaller than that of the state of California, the apprehension of smugglers from abroad is much easier than it is in large continents such as America and Europe.

THE FLUCTUATION OF DRUG ABUSE AMONG YOUTH

It was notable that, during the last two decades, approximately 75% of heroin addicts were in their twenties and thirties, and the dependency-producing drug abused by those under the age of twenty years fluctuated from methamphetamine to hypnotics, then to analgesics and organic solvents.

After 1960, a non-barbiturate hypnotic called methaqualone (2-methyl-3-*o*-tolyl quinazone) began to be abused by groups of juvenile delinquents. In 1961 and 1962, 73.7% of methaqualone abusers were from fourteen to seventeen years old. In 1964, the Ministry of Health and Welfare designated 46 sedatives as powerful drugs that were not to be sold to or delivered to anyone suspected of incompetence in their normal handling. Following this, the number of methaqualone abusers decreased and the number of analgesics abusers began to increase. Almost all of the analgesics were mixtures of allopyrabital, phenacetin, and caffein. Among these, allopyrabital, a molecular compound of allobarbital and aminopyrine, was the dependency-producing drug. Some of the analgesics abusers had delirium during their withdrawal from drugs.

Since 1967, analgesics abusers have increased in number, exceeding that of methaqualone abusers, particularly in urban areas. In addition, the abuse of organic solvent (glue) has increased since 1968; some 20,000 to 30,000 youths are arrested every year because of antisocial behavior associated with glue-sniffing. It was reported that 100 to 150 glue-sniffers died of suffocation each year. In the past two years, glue-sniffing has largely replaced abuse of methaqualone and analgesics.

It is well-known that the effects of organic solvent include dream-like, fantastic visual hallucinations something like those produced by the hallucinogens. Although marihuana and LSD abuse had never given rise to any social problem in Japan, a Cannabis Control Law was enacted in 1948 and an LSD Control Law in 1970.

G. M. Carstairs in Edinburgh explained why marihuana was practically unknown to the people of China and Japan: "Chinese people had little inclination to despise material pleasure of this world and the Japanese would be the last to renounce the active life."[1] However, the possibility of marihuana abuse in Japan is watched with keen interest because the traditional value systems are changing rapidly, particularly among the young.

EPIDEMIOLOGY OF DRUG DEPENDENCE IN THE GENERAL POPULATION

From the epidemiological standpoint, the diagnostic criteria and case-finding measures must be analyzed in order to compare Japanese experience with international and interdisciplinary standards of drug abuse and dependence. To clarify the diagnostic differences, evaluation of the causes of drug abuse and dependence in Japan were made by 213 psychiatrists. It was determined that the terms "drug abuse," "drug dependence," "drug addiction," and "intoxication" should be more strictly defined. For case-finding of drug abuse and dependence, a survey of 853 mental hospitals was carried out from 1963 to 1966. Of the total of 411 cases of drug abuse and dependence, in 42.8%, methaqualone was implicated; in 27.2%, meprobamate; in 11.4%, bromisovalerylurea; and in 5%, Librium was responsible. Although such surveys do not reflect the actual prevalence of drug abuse and dependence, but rather the extent of utilization of psychiatric resources, they do seem to reflect the attitudes of the abusers' relatives toward them. Psychiatric help for methaqualone abusers was sought by relatives of the patients mainly because of the patients' antisocial behavior. In contrast, withdrawal symptoms were responsible for the admission of meprobamate abusers to mental hospitals.

To investigate the prevalence of drug abuse and dependence in the general population, a survey was carried out in 1969 in four areas near Tokyo, one each in a metropolitan, industrial, commercial and agricultural area. The study was accomplished through Parent-Teacher's Associations of primary and secondary schools in these areas. The results showed that the prevalence of drug abuse and dependence among those using dependency-producing drugs, such as analgesics, minor tranquilizers, and hypnotics, daily or two or three times a week averaged 0.6%, with 0.8% in the metropolitan area, 0.7% in the industrial area, 0.6% in the commercial area, and 0.3% in the agricultural area. Dependency-producing drugs were kept as household medicine in 12.7% of the families in the metropolitan area, in 7.4% of those in the industrial area, in 12% of those in the commercial area, and 9.3% of those in the agricultural area. Thus, the metropolitan area ranked highest and the agricultural area lowest in both analyses. Among the dependency-producing drugs kept in households, analgesics were the most prominent (8.9%), minor tranquilizers were next (1.2%), and hypnotics the least (0.3%). Almost all of these users did not know the actual danger of drug-dependence, although some of them showed EEG changes upon withdrawal.

In conclusion, I wish to inquire as to the experiences of others at this conference regarding the similarities and dissimilarities of the epidemiology of drug dependence in various countries of the world. It would be of special interest and importance to learn the socio-cultural background of host, the environment, and the drug or drugs being misused in each country

REFERENCE

1. Carstairs, G. M.: *Bhang and Alcohol: Cultural Factors in the Choice of Intoxicants.* The Marihuana Papers. Signet Books. North American Library, 1968, pp. 103-120.

Afternoon Session
Monday, November 9, 1970

Presiding: Chris J. D. Zarafonetis, M.D.

EVALUATION OF THE EFFECTIVENESS OF PRESCRIBING CLINICS FOR NARCOTIC ADDICTS IN THE UNITED KINGDOM (1968–1970)

THOMAS H. BEWLEY, M.A., M.D., F.R.C.P.I., D.P.M.,
IAN PIERCE JAMES, M.A., B.A., M.B., B.Chir., D.P.M., M.A.N.Z.C.P.,
AND THOMAS MAHON, B.S., M.S.

CHANGES IN INCIDENCE AND PATTERN OF DRUG DEPENDENCE

In the past two decades there have been striking changes in the incidence of drug misuse in the United Kingdom. Twenty years ago any type was uncommon except for those few people who had become addicted following medical treatment or because of their professions (doctors, nurses, or pharmacists). There was virtually no drug addiction among people who had obtained drugs illicitly. Now, cannabis, amphetamines, combinations of amphetamines and barbiturates, lysergic acid diethylamide, and opiates are all increasingly misused by people who have obtained these drugs illegally in the first instance. The most striking recent change has been the increase in the number of known intravenous misusers of heroin (at one time combined with cocaine or methamphetamine), and of methadone and barbiturates (also taken intravenously). There has also been an increase in the consumption of amphetamines, barbiturates, and alcohol, all obtained legally. Although there has been a decrease in the total consumption of tobacco, the tendency for young people to start smoking early continues. The history of recent changes in the pattern of drug abuse in the United Kingdom has been charted in a number of sources including studies involving non-dependent, occasional drug users.[1-4] The best overview of the epidemiology of drug dependence in the United Kingdom and on which much of this introduction is based, is by David Hawks.[5]

Opiate Dependence

Since 1950 there has been a marked change in the incidence of dependence on opiates and in the characteristics of those known to be dependent.[6] Before 1950 the number known to be dependent on narcotic drugs was relatively stable, varying between 400 and 600. Those involved tended to be middle-aged and to be disproportionately from the medical and paramedical professions. In most cases, addiction had a therapeutic origin and the drug involved was usually morphine. Almost coincidental with publication of the first report of the Interdepartmental Committee on Drug Addiction (the "Brain Committee"), there originated a series of changes which have continued in the same direction until the present day. There appeared both an increase in the number of people known to be dependent on narcotic drugs and an increasing tendency for their addiction to have non-therapeutic origins. The persons concerned also tended to be younger and predominantly male. There was evidence that the drug of preference was heroin rather than morphine and indications that other drugs were being increasingly used. Whereas the misuse of illicitly obtained drugs was originally confined to London, the problem spread to other parts of the country.

It needs to be emphasized that the figures on which these observations are based have never until recently been comprehensive. Prior to the 1968 legislation it was not obligatory for doctors treating dependent patients to report their dependence and the figures published by the Home Office were largely derived from the routine examination of pharmacists' records. The annual statistics published by the Home Office until 1968 showed only those addicts known to have been taking drugs during the previous year. They excluded those who were not known to be currently taking drugs, for example, addicts who were in prison or abroad or who obtained their drugs entirely from illicit sources.

While the total number of addicts to opiates in the United Kingdom is still small by international standards—estimated to be of the order of four to eight per 100,000[6]—the rate at which new cases have been reported leaves no cause for complacency. Ten times as many new cases were being recorded in 1968 as in the ten years previously. The most striking difference between these addicts and the earlier ones was their youth, the majority being in their early twenties.[7] This led to the recalling of an Inter-departmental Committee, which made several recommendations for control, including restrictions on the prescribing of heroin. *Cocaine* was for several years used in combination with heroin. The majority of heroin addicts injected both drugs together intravenously ('main lining'), with very few using cocaine alone. There has been a decrease in this in the past two years, and also in the use of intravenous methamphetamine.

Significant Events Affecting Incidence

In 1969, Spear attempted to analyze the events of the 1950's which heralded the present epidemic.[8] He observed that the first significant change in the situation occurred after the Second World War when there was an increased number of seizures of cannabis made by H. M. Customs and Excise. The first post-war annual report (1946) acknowledged that there was a small traffic in cannabis, largely confined to two Negro groups with colored seamen being the principal importers. A further "considerable increase" was noted in the following year and the same sources implicated. By 1950 it was clear that the traffic in cannabis was of greater importance than the traffic in opium and the total number of persons prosecuted for offenses involving cannabis has shown a steady rise ever since. While the use of this drug continued to be linked mostly with the colored population, in 1950 it became apparent that cannabis was also being used by the indigenous population.

A second major case of drug trafficking occurred when approximately 1,200 grains of morphine, mostly in tablet form and 14 ounces of cocaine were stolen from a firm of wholesale chemists. Although the drugs had been stolen by one of the firm's employees, it transpired that the instigator was a person having contacts with drug addicts in the West End of London. However, the persons to whom he supplied drugs were already known addicts and there was no evidence that others were recruited to the habit by his activities. A further major case of trafficking occurred in 1951 when the dispensary of a hospital on the borders of London was broken into and considerable quantities of morphine, heroin, and cocaine were stolen. The person involved turned out to be a former employee of the hospital. Although the full extent of his activities cannot be charted, it appears that a majority of those supplied drugs by him had no previous experience of heroin addiction. All of those whom it could be ascertained had received their initial supplies from this source subsequently came to the notice of the Home Office as heroin addicts.

While Spear did not attempt a causal analysis of the change in the pattern of drug abuse in the United Kingdom, he observed that the appearance of a pusher in the West End and the scarcity of cannabis following increased police activity appeared to have played a significant part.

THE "BRITISH SYSTEM"

For many years it was believed that there was a "British System" whereby opiate addicts could have opiates prescribed for them and that this led to the very small number of people dependent on drugs. In fact, there was no system, but as there was very little in the way of misuse of drugs this did not matter.[9] Original guidelines for prescribing of opiates to opiate addicts were laid down in a Departmental Committee Report.[10] Addicts for whom morphine or heroin could be prescribed, included "Persons for whom, after every effort has been made for the cure of addiction, the drug cannot be completely withdrawn, either because:

> "(1) complete withdrawal produces serious symptoms which cannot be satisfactorily treated under the ordinary condition of private (general) practice; or
>
> "(2) the patient, while capable of leading a useful and fairly normal life so long as he takes a certain non-progressive quantity, usually small, of the drug of addiction, ceases to be able to do so when the regular allowance is withdrawn."

It was also recommended that this course of treatment not be embarked on without the opinion of a second doctor.

In practice, from the early 1950's, a certain number of doctors were prepared to prescribe heroin for heroin addicts. It was found that when more than was required by an individual was prescribed for him, he might part with some of this to another person who could thus become addicted. Because of this an Interdepartmental (Brain) Committee was recalled and in their second report it was recommended that there should be limitations on the prescribing of heroin and cocaine to persons dependent on these drugs.[11] Following this, special legislation was passed in the Dangerous Drugs Acts of 1967 to enable the Home Office to license certain doctors to prescribe heroin and cocaine for addicts, and at the same time the Department of Health and Social Security set up special treatment centers, the majority of these being in the London area.[12]

Thus the situation came about where special clinics had been set up with the aim of prescribing opiates for addicts, although at this time there was no evidence that this was necessarily sound treatment.

Heroin Maintenance ("Ambulatory Clinics")

Whether it is possible to supply heroin to a patient who has become addicted without causing more harm to himself and society is still widely debated. In the United States it is considered that the maintenance of stable dosage levels is 'generally inadequate and medically unsound.' Heroin has been banned and the prescription of morphine to support an addict would not be considered ethical.[13] In Britain, maintenance with heroin had been considered (in theory) an accepted method of treatment only when other methods had failed. In practice, up to the present, heroin (and other opiates, cocaine and methamphetamine) had been prescribed fairly liberally to addicts. Because of the spread of addiction this has been changed and now only certain licensed doctors are entitled to prescribe heroin for addicts. Surprisingly, despite the fact that this is a controversial subject on which much has been written, no one has yet evaluated exactly what happens when heroin is prescribed for addicts; e.g., whether their work records get worse or get better, whether the patients take the drug themselves or sell or give it to their friends, whether the high death rate can be substantially lowered by training heroin addicts to inject themselves using aseptic techniques, and in general whether more harm than good is done if heroin is carefully prescribed in this way. At present, in Britain new clinics have been established as recommended by the Brain Committee and these clinics are prescribing heroin for ambulatory maintenance. Some of the assumptions underlying the British approach have been reviewed by Griffith Edwards.[14]

Working of Treatment Clinics

After the implementation in April 1968 of the 1967 Dangerous Drugs Acts a large number of patients appeared at the new clinics. The majority of doctors working there were swamped and at that time all that could be done was to find out what had been previously prescribed for the individual and continue to prescribe this. Gradually histories were obtained, a better assessment was made of individual patients, and generally the amounts prescribed were reduced. Later it was possible to act in a more rational manner and the new patients who attended could be assessed as to whether they were using drugs or not, and whether they were physically dependent. If this latter were the case the question arose of whether and how much to prescribe for them, if they did not wish to "come off" all drugs. At this time, through regular meetings of the doctors working in various clinics, a general system of control was introduced whereby all patients who were having drugs prescribed for them attended at least once weekly and were given prescriptions which could be collected daily. The prescriptions were sent to individual chemists and they were made out in such a way that the patient could only get one day's supply of drugs at a time on Mondays through Fridays and two days' supply on Saturdays. This has remained the system up to the present.

At some of the treatment clinics, routine urine testing for the presence of drugs is done weekly on all patients, and it has been found that although the majority of patients use the drugs that are prescribed for them, a good number use other drugs as well. Occasionally one finds that a person is having a particular drug prescribed which is not found in his urine. This can lead to difficulties for the doctor who does not know whether this is a labeling fault or laboratory error, or whether the person has not used the drug and is possibly selling it on the black market.

Problems in Management

When treating a patient dependent on drugs the aims of the doctor may not be those of the patient. A person who is using drugs or who wishes to use drugs may want to use them, not to alleviate illness or suffering, but for their pleasurable effect. If this is the sole reason he wishes to use the drug (for example, taking amphetamines intravenously at a party) it is unlikely that a doctor would accede to the request to provide drugs for this purpose. Most doctors would not consider it their function to provide pleasure at low cost to the individual through the National Health Service. In general, as doctors, our aim is to deal with such states as pathological depression. Doctors may aim to make unhappy people "normal," but not to make normal people euphoric or give them a "buzz" or a "high."

This prescribing role has both advantages and disadvantages. Doctors working in treatment clinics are dealing with captive patients who must attend weekly to obtain their drugs—an unusual doctor/patient relationship. In other situations a patient who is dissatisfied with his doctor can go to another doctor, but if a patient attending one of the treatment clinics is dissatisfied and goes to another clinic, that clinic will check with the Home Office and contact the first clinic. The second clinic may accept the patient or advise him to remain at the first clinic since the treatment first prescribed would be continued no matter what clinic he attended. In practice the patient who is attending a treatment clinic has little choice.

Type of Dependence and Treatment

The type of drug dependence is generally considered to be determined by:

1. Whether or not the patient is continuing to take the drug.
2. Whether or not the patient is functioning efficiently irrespective of whether he is using drugs or not.
3. Whether the patient himself is satisfied with the treatment.
4. Whether the treatment limits or further spreads the addiction.

The first point to be made about any treatment is that the amount of "noise" made by proponents of the different types of treatment is inversely proportional to the amount of scientific evaluation that has been made of their effectiveness. Secondly, the one statement that appears to be true is that all types of treatment have had limited effectiveness for a limited number of patients. Thirdly, since our knowledge of the natural history of different types of dependence on drugs is much poorer than it should be, it is difficult to know whether various treatment methods have improved or worsened the situation. Finally, it is very important to make the point that all methods of treatment may be of very little value at all except when viewed as experiments for finding better methods of treatment.

Efficacy of Treatment

Very few papers have been published in which the efficacy of different programs of treatment have been compared or the effect of treatment has been assessed. Chapple and Gray have reported on one year's work at a center for the treatment of addicts which operates three major treatment

Table 1 Nationality of 491 addicts, by sex

Nationality	Male N	Male %	Female N	Female %	Total N	Total %
British	334	80.87	62	79.49	396	80.65
Irish	34	8.24	5	6.42	39	7.94
"Scottish"	12	2.91	2	2.56	14	2.85
"Welsh"	1	.24	2	2.56	3	.61
Canadian	15	3.63	1	1.28	16	3.26
American	6	1.45	1	1.28	7	1.43
Other (outside U.K.)	5	1.21	4	5.13	9	1.83
None stated	6	1.45	1	1.28	7	1.43
Total	413		78		491	100.00

programs: (1) a general supportive program which includes the prescribing of heroin, (2) a methadone maintenance program, and (3) an "off narcotics" program.[15] The fact that the allocation of patients to programs was discretionary, however, precludes comparison of the programs.

Review of the Working of Three Prescribing Clinics

The Population Treated. The three clinics reviewed are the out-patient clinic at St. George's Hospital, the out-patient clinic at St. Thomas' Hospital, and the in-patient unit at Tooting Bec Hospital. The characteristics of the patients attending these clinics appear to be typical of those of the addict population of London as a whole. The case records of 491 addicts who came under my care in these three clinics were reviewed and reported by T. A. Mahon.[16] Mahon recorded the nationality, place of birth, racial characteristics, home address, social class, marital status, and age of the 491 addicts by sex (Tables 1 through 7). The case notes of

Table 2 Place of birth of 491 addicts, by sex

	Male		Female		Total	
Place of birth	N	%	N	%	N	%
London	108	26.15	19	24.36	127	25.87
United Kingdom (other than London)	167	40.44	36	46.15	203	41.34
Eire	37	8.96	4	5.13	41	8.35
Canada	21	5.08	2	2.56	23	4.68
United States	8	1.94	2	2.56	10	2.04
Other (outside the U.K.)	20	4.84	5	6.41	25	5.09
None stated	52	22.59	10	12.82	62	12.63
Total	413		78		491	100.00

Table 3 Racial characteristics of 491 addicts, by sex

	Male		Female		Total	
Race	N	%	N	%	N	%
White	402	97.35	77	98.72	479	97.56
Negro	6	1.45	....	....	6	1.21
Mixed*	2	.48	....	....	6	.42
Other†	2	.48	....	....	2	.41
None stated	1	.24	1	1.28	2	.41
Total	413		78		491	100.00

* This is the heading under which two patients who were Eurasians appeared in the case histories.
† These two patients were Orientals.

Table 4 Home address of 491 addicts, by sex

	Male		Female		Total	
Home address	N	%	N	%	N	%
London	268	64.89	63	80.77	331	67.41
Other	84	20.34	11	14.10	95	19.35
No fixed abode	56	13.56	3	3.85	59	12.01
Not stated	5	1.21	1	1.28	6	1.22
Total	413		78		491	100.00

Table 5 Social class distribution of the 491 addicts, by sex

	Male		Female		Total	
Social class	N	%	N	%	N	%
Class I (professional)	4		1		5	1.02
Class II (intermediate)	58		4		62	12.63
Class III (skilled)	115		26		141	28.72
Class IV (partly skilled)	36		3		39	7.94
Class V (unskilled)	88		4		92	18.73
None stated	47		19		66	13.44
Not in labor force*	65		21		86	17.52
Total	413	84.31	78	15.69	491	100.00
*Not in labor force						
Student	14		4		18	20.93
Housewife	....		8		8	9.30
Unemployed	51		9		60	69.77
Total	65		21		86	100.00

Table 6 Marital status of the 491 addicts, by sex

	Single		Married		Divorced		Separated		Not stated		Total	
Sex	N	%	N	%	N	%	N	%	N	%	N	%
Male	314	76.07	51	12.35	8	1.94	34	8.23	6	1.45	413	84.31
Female	51	65.38	16	20.51	3	3.85	6	7.69	2	2.57	78	15.69
Total	365	74.34	67	13.65	11	2.24	40	8.14	8	1.63	491	100.00

every third patient were scrutinized closely and this group of patients, which will be referred to as the intensive sample, were studied more thoroughly than the others. Tables 8, 9 and 10 give information about the marital status of the intensive-sample group and Tables 11 and 12 outline the degree of satisfaction in sexual relationships in this group. Table 13 shows that 78% of the intensive-sample group were unemployed at the time they came first for treatment; Table 14 confirms the impression that most of these patients associated extensively with other addicts and were in fact part of a drug subculture. Tables 15, 16, and 17 show that the majority of the group had criminal rec-

Table 7 Age of 482 drug-dependent patients upon first admission to hospital or clinic, by sex

	Male		Female		Total	
Age (years)	N	%	N	%	N	%
Under 15	1	.25	—	—	1	.21
15–19	102	25.06	25	33.33	127	26.35
20–24	188	46.18	27	37.00	215	44.61
25–29	52	12.77	14	18.67	66	13.69
30–34	38	9.34	2	2.67	40	8.29
35–39	19	4.67	4	5.33	23	4.77
40–44	3	.74	—	—	3	.62
45–49	3	.74	2	2.67	5	1.04
50–54	—	—	—	—	—	—
55 and over	1	.25	1	1.33	2	.42
Total	407		75		482	100.00

Table 8 Age at marriage in comparison with age at first admitted use of heroin, by sex (intensive sample)*

	Male		Female		Total	
	N	%	N	%	N	%
Married before first heroin use	15	60.00	5	71.42	20	62.50
Married in the same year as first heroin use	1	4.00	1	14.29	2	6.25
Married after first heroin use	9	36.00	1	14.29	10	31.25
Total	25		7		32	

* In the case histories of four patients who had been married and had used heroin the comparison could not be made because of missing information. Two other patients who had been married claimed to have never used heroin.

Table 9 Marital status at time of entry into hospital or clinic, by sex (intensive sample)

Sex	Single		Married		Divorced		Separated		Widowed		Total	
	N	%	N	%	N	%	N	%	N	%	N	%
Male	106	77.37	7	5.11	5	3.65	17	12.51	2	1.42	137	
Female	15	68.18	4	18.18	1	4.55	2	9.09	....	....	22	
Total	121	76.10	11	6.92	6	3.77	19	11.95	2	1.26	159	

Table 10A Number of children of married addicts in intensive sample, by sex

	Male		Female		Total	
Number of children	N	%	N	%	N	%
1	22	75.86	2	40.00	24	70.59
2	4	13.79	1	20.00	5	14.71
3	1	3.45	2	40.00	3	8.82
4	2	6.90	....	....	2	5.88
Total	29		5		34	

Table 10B Location of children of addicts in intensive sample, by sex

	Male		Female		Total	
Location	N	%	N	%	N	%
Living with patient	5	17.24	1	20.00	6	17.65
Living with others	22	75.86	3	60.00	25	73.53
Not stated	2	6.90	1	20.00	3	8.82
Total	29		5		34	

Table 11 Opinion concerning level of satisfaction in sexual relations, by sex (intensive sample)*

	Male		Female		Total	
Level of satisfaction	N	%	N	%	N	%
Very good	1	1.67	—	—	1	1.45
Satisfactory	15	25.00	2	22.22	17	24.64
Good	1	1.67	—	—	1	1.45
Poor	43	71.66	7	77.88	50	72.46
Total	60		9		69	100.00

* This information was not contained in the case histories of 90 of the patients in the intensive sample.

Table 12 Change in interest in sexual relations since addiction, by sex (intensive sample)*

	Male		Female		Total	
Stated change	N	%	N	%	N	%
No change	13	22.41	2	22.22	15	22.39
Decreased interest	26	44.83	5	55.56	31	46.27
Impotence	19	32.76	2	22.22	21	31.34
Total	58		9		67	100.00

* This information was not contained in the case histories of 92 of the patients in the intensive sample.

Table 13 Employment status at time of entry into hospital or clinic, by sex (intensive sample)

	Male		Female		Total	
Employment status	N	%	N	%	N	%
Employed (full time)	23	16.79	3	13.64	26	16.35
Employed (part time)	2	1.46	2	9.09	4	2.52
Unemployed	109	79.56	16	72.73	125	78.62
Not in labor force*	3	2.19	1	4.54	4	2.51
Total	137		22		159	100.00
*Not in labor force						
Housewife	—		1		1	
Student	3		—		3	
Total	3		1		4	

Table 14 Extent of contact with other addicts, by sex (intensive sample)*

	Male		Female		Total	
Extent of contact	N	%	N	%	N	%
All associates addicts	24	32.00	2	18.18	26	30.23
Most associates addicts	39	52.00	7	63.64	46	53.49
Few associates addicts	10	13.33	2	18.18	12	13.95
No associates addicts	2	2.67	....	....	2	2.33
Total	75		11		86	100.00

* This information was not contained in the case histories of 73 of the patients in the intensive sample.

Table 15 Criminal records, by sex (intensive sample)

	Male		Female		Total	
Record	N	%	N	%	N	%
Has criminal record	104	75.91	9	40.91	113	71.07
No criminal record*	23	16.79	11	50.00	34	21.38
Not stated	10	7.30	2	9.09	12	7.55
Total	137		22		159	100.00

* Eight of these (one female, seven males) stated that they had been searched but never arrested.

Table 16 Number of arrests of addicts, by sex (intensive sample)

Number of arrests	Male N	Male %	Female N	Female %	Total N	Total %
None	33	24.09	13	59.08	46	28.93
One	24	17.52	1	4.55	25	15.72
Two	21	15.33	2	9.08	23	14.47
Three	16	11.68	2	9.08	18	11.32
Four	10	7.30	1	4.55	11	6.92
Five	15	10.94	1	4.55	16	10.06
Six	10	7.30	1	4.55	11	6.92
Seven	5	3.65	....	....	5	3.14
Eight	....	....	....	....	....	....
Nine	3	2.19	1	4.55	4	2.52
Total	137		22		159	100.00
Mean number of arrests:		2.43	For males: 2.56 For females: 1.59			

Table 17 Age at first arrest in comparison to age at first use of heroin, by sex (intensive sample)*

	Male N	Male %	Female N	Female %	Total N	Total %
First arrest before first heroin use	59	69.41	1	25.00	60	67.42
First arrest in the same year as first heroin use	11	12.94	2	50.00	13	14.61
First arrest after first heroin use	15	17.65	1	25.00	16	17.97
Total	85		4		89	100.00

* The records of twenty-four patients did not contain enough data to make this comparison possible.

Table 18 Admitted involvement in black market, by sex (intensive sample)*

Extent of involvement	Male N	Male %	Female N	Female %	Total N	Total %
Illicit buying	43	72.88	6	66.67	49	72.06
Illicit selling	1	1.69	1	11.11	2	2.94
Illicit buying and selling	15	25.43	2	22.22	17	25.00
Total	59		9		68	100.00

* This information was not contained in the case histories of 91 patients in the intensive sample.

Table 19 Stated first source of heroin for eighty-two addicts, by sex (intensive sample)*

	Male		Female		Total	
Source	N	%	N	%	N	%
Friend	34	50.00	7	46.66	41	50,00
"Black market"	17	25.00	3	20.00	20	24.39
Another addict	10	14.71	4	26.24	14	17.07
"Person," "chap"	6	8.82	....	....	8	7.32
Physician	1	1.47	....	....	1	1.22
Total	68		15		82	100.00

* This information was not contained in the case histories of 77 of the patients in the intensive sample.

ords, less than 30% had never been arrested, and that the first arrest for 70% of those having criminal records had been made before first heroin use. All had been involved at some time in the black market. Table 18 shows that three-quarters of a group of 68 patients admitted to buying and one-quarter to buying and selling narcotics. Table 19 shows that three-quarters of a group of 82 first obtained heroin either from a friend or from the black market.

Follow-up. In order to review the effects of the three clinics a follow-up study of all the patients who first attended the clinics in April 1968 to April 1969 was made. The records of 397 patients were reviewed at that time and again one year later (April 1970). Table 20 shows the follow-up status of these patients. The majority were continuing to have opiates prescribed for them in the clinics; 40% were currently working. Less than 20% had discontinued the use of all drugs. Patients were continuing to use drugs not prescribed for them. A four-week survey of urine tests made in 1969 and 1970 showed that there had been a slight decrease in the number of people using drugs not prescribed for them (Table 21). Table 22 shows the admitted misuse of sedatives.

Table 20 Follow-up status

	April 1969	April 1970	No change between April 1969 and April 1970
Attending St. Thomas (Lambeth) Clinic or St. George's Clinic	148	123	93
Attending other clinics	75	83	49
In-patient	42	21	3
Penal institution	28	27	10
Died	3	9	—
One attendance only (nil prescribed)	32	22	22
Other	36	38*	30
Not known to be using drugs	33	74†	26
Total	397	397	233

*Deported, 2; abroad, 7; in care of general practitioner, 13; using amphetamine only, 15; other, 1.

†Since prison, 29; since probation/court appearance, 8; since hospital, 14; since attending as outpatient, 14; other, 5.

Over-all Effects of the Prescribing Clinics. When special prescribing clinics were established in England in 1968 one of the hopes was that they would play a role in preventing the spread of narcotic addiction, particularly among young people. One way in which the effectiveness of the clinics can be measured is to determine whether there is an increase or decrease in the number of new cases that come to light each year, and also whether there are any changes in the rates at which new cases appear. From this point of view it would seem, as shown in Table 23, that

Table 21 Four-week survey of urine tests

	1969		1970	
	Persons	Times	Persons	Times
Prescribed drugs only (or none prescribed)	29		41	
Drugs not prescribed but present				
Amphetamines	39	61	15	16
Morphine	14	30	12	19
Methadone	3	4	2	4
Barbiturates	43	64	28	64
Drugs prescribed but not found				
Morphine (heroin prescribed)	6	10	4	4
Methadone (prescribed)	10	15	2	3
Total attending	97		93	

Table 22 Admitted misuse of sedatives

St. Thomas (Lambeth) Clinic	May 1969	May 1970
Not using sedatives	36	21
Using sedatives (one or more)	36	35
Route of administration		
Oral use only	16	15
Intravenous use	11	10
Both oral and intravenous use	9	10
Source		
Obtained from legal source	8	18
Obtained illicitly	28	23
Types		
Barbiturates	27	25
Methaqualone*	20	23
Other	4	15
Total	72	56

* With diphenhydramine hydrochloride (Mandrax).

the clinics have had some success in that the number of new cases, which had been increasing by 50% annually, increased by only 3% in 1969, with the over-all number of addicts known to the Home Office increasing slightly in that year (Home Office reports). Table 23 also shows that there have been marked changes in prescribing narcotics since this had become the responsibility of the clinics, with much more methadone and much less heroin now being given. In 1969, 1201 more patients took methadone and 823 less took heroin. In other words the clinics were responsible for changing the administration of heroin to that of methadone in more than 800 patients. In September 1968, 2.855 kg of heroin were prescribed each month; by December 1969 the amount was 1.638 kg. In fact, many more people are being supported on methadone than are being supported on heroin. Currently about 1 kg of methadone is prescribed by the clinics each month.[17] Table 24 shows the change in age of addicts under 20, and Table 25 details all drug offenses in the United Kingdom from 1945 through 1969.

Another way one can judge the efficacy of the treatment clinics is whether they have had any other over-all effects, for example, on the number of addicts being arrested and going to prison. In comparing the three-month period prior to attending a clinic with a three-month period after a year's attendance, Dr. C. LeFevre found that for a small group of 68 patients studied, the arrest rates were lower in the second period.[18] In a survey made in Brixton Prison, which is the remand prison for all adults in the London area, Dr. James showed that the number of addicted prisoners declined steadily over a two-year period[19] (Fig. 1). Gardner and Connell studied the findings in 107 opioid users attending a drug-dependence clinic in London, between March 1968 and

Table 23 United Kingdom statistics of drug addiction and criminal offenses involving drugs

												1969*	
Year	1958	1959	1960	1961	1962	1963	1964	1965	1966	1967	1968	Year total	As at Dec. 31, 1969
Total number of drug addicts	442	454	437	470	532	635	753	927	1349	1729	2782	2881	1466
Drugs†													
No. taking heroin	62	68	94	132	175	237	342	521	899	1299	2240	1417	499
" " methadone	12	60	68	59	54	55	61	72	156	243	486	1687	1011
" " cocaine	25	30	52	84	112	171	211	311	443	462	564	311	81
" " morphine	205	204	177	168	157	172	162	160	178	158	198	345	111
" " pethidine	117	116	98	105	112	107	128	102	131	112	120	128	83
Origin													
No. of therapeutic origin	349	344	309	263	312	355	368	344	351	313	306	289	247
" " non-therapeutic origin	68	98	122	159	212	270	372	580	982	1385	2420	2533	1196
No. of unknown origin	25	12	6	18	8	10	13	3	16	31	56	59	23
Ages													
Under 20 years	—	—	1	2	3	17	40	145	329	395	764	637	224
Taking heroin‡	—	—	1	2	3	17	40	134	317	381	709	598	221
20–34 years	—	50	62	94	132	184	257	347	558	906	1530	1789	897
Taking heroin‡	—	35	52	87	126	162	219	319	479	827	1390	1709	872

35–49 years	—	92	91	95	107	128	138	134	162	142	146	174	116
Taking heroin‡	—	7	14	19	24	38	61	52	83	66	78	101	69
50 years and older	—	278	267	272	274	298	311	291	286	279	260	241	204
Taking heroin‡	—	26	27	24	22	20	22	16	20	24	20	46	40
Age unknown	—	34	16	7	16	8	7	10	14	7	82	40	25
Taking heroin‡		—	—	—	—	—	—	—	—	1	43	26	13
Sex													
No. of male addicts	197	196	195	223	262	339	409	558	886	1262	2161	2295	1067
No. of female addicts	245	258	242	247	270	296	344	369	463	467	621	586	399
Professional classes (medical or allied)													
Total number	74	68	63	61	57	56	58	45	54	56	43	43	26

* It will be seen that the statistical data for 1969 is presented differently from that of the preceding years. Previously these statistics have been based on the total number of addicts coming to the notice of the Home Office during the course of the year. New recording procedures have made it possible to give details of those addicts known to have been receiving supplies of drugs at the end of the year, as well as the total number of cases coming to notice during the year.

† These figures refer to drugs used alone or in combination with other drugs. Thus an addict using both heroin and cocaine will be included under both drugs, and it must be pointed out that all but a handful of the cocaine addicts shown are also using heroin.

‡ For 1969 this figure is for addicts to *heroin and/or methadone*. The reason for this is that, as a result of a deliberate policy adopted by hospital clinics in the treatment of heroin addiction of weaning patients from heroin onto methadone; methadone has supplanted heroin as the drug most commonly used by addicts.

Produced by: Drugs Branch, Home Office, Romney House, Marsham Street, London SW1.

February 1969, in relation to clinical groups, management policy, dosage of drugs prescribed, diagnosis of dependence, and the use of biochemical tests to determine the presence of drugs in urine.[20] Establishing the diagnosis and determining the maintenance doses were especially difficult problems. Heroin or methadone linctus was prescribed for some patients even though the urine showed no evidence of the presence of the drug, and some of these patients became addicted during their attendance at the clinic. Gardner and Connell considered that heroin maintenance therapy should be re-evaluated, and used in young people with great caution if at all. Testing the urine for the presence of drugs was a vital part of diagnosis, treatment, and a rigorous management regimen.

Gardner reported the deaths of opioid users occurring between 1965 and 1969 in the United Kingdom.[21]

Stimson and Ogborne, in a survey of heroin addicts attending clinics, reported that 84% used drugs other than those prescribed for them by the clinic in the month before the interview; 34% stated that they had engaged in criminal activities, other than those covered by the Drugs Acts, during the three months before interview.[22]

Table 24 Ages of "under 20" addicts known to Home Office

	14	15	16	17	18	19	Total
1960	—	—	—	—	—	1	1
1961	—	—	—	—	1	1	2
1962	—	—	1	—	—	2	3
1963	—	—	2	2	2	11	17
1964	1	—	1	8	11	19	40
1965	—	8	5	19	42	71	145
1966	1	17	26	68	111	106	329
1967	—	3	38	82	100	172	395
1968	—	10	40	141	274	299	764
1969 total	—	—	24	83	218	312	637
Dec. 31, 1969	—	—	6	33	73	112	224

SUMMARY

There have been marked changes in the pattern of drug dependence in England in the last twenty years and particularly since 1960. The most striking increase has been in the number of known opiate addicts. The majority of these were dependent on heroin

Number of Physically Dependent Narcotic Addicts Received at Brixten Prison. Each Month (Averaged for Three Month Period) From January 1968 to August 1970.

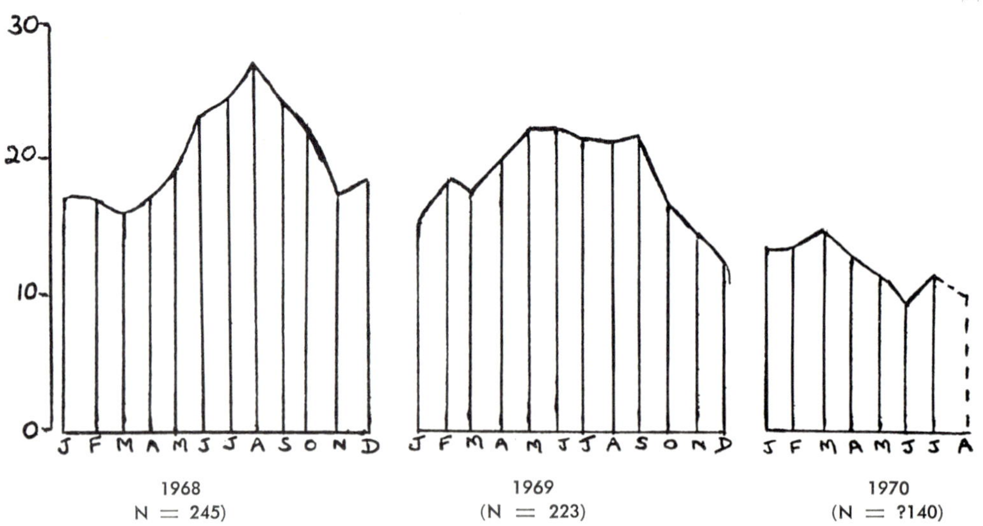

SOURCE: DR. I. P. JAMES.

Fig. 1

Table 25 Offenses involving drugs

	D.D.A.—1965			Drugs (prevention of misuse) Act 1964
	Cannabis	Opium	Manufactured Drugs	
1945	4	206	20	
1946	11	65	27	
1947	46	76	65	
1948	51	78	48	
1949	61	52	56	
1950	86	41	42	
1951	132	64	47	
1952	98	62	48	
1953	88	47	44	
1954	144	28	47	
1955	115	17	37	
1956	103	12	37	
1957	51	9	30	
1958	99	8	41	
1959	185	18	26	
1960	235	15	28	
1961	288	15	61	
1962	588	16	71	
1963	663	20	63	
1964	544	14	101	
1965	626	13	128	958*
1966	1119	36	242	1216
1967	2393	58	573	2486
1968	3071	73	1099	2957
1969	4683	53	1359	3762

From 1945 to 1953 inclusive the figures relate to prosecutions.

From 1954 onwards, the figures relate to convictions.

*This figure covers the period Dec. 10, 1964 to Dec. 12, 1965.

(originally either combined with cocaine or methamphetamine) taken intravenously. All the new cases were found in non-therapeutic addicts, chiefly in the 18- to 25-year group. The ratio of men to women was four to one. In 1965 a Governmental Committee recommended certain changes to deal with this problem. Legislation in the Dangerous Drugs Act 1967 was implemented in April 1968. Notification of addiction to drugs controlled under the Dangerous Drugs Act became compulsory and the prescribing of heroin or cocaine to an addict could henceforth be carried out only by a doctor officially licensed. At the same time the Department of Health set up a number of special clinics (the majority of which were in the London area) to treat heroin addicts. In general, only doctors working at these clinics were licensed to prescribe the controlled drugs. Addiction to heroin was still considered an illness and it continued to be official policy to supply drugs to addicts to maintain their addiction.

The experience during the first two years of operation of the treatment clinics and the effects of the Act have been reviewed. Initially, a large number of patients were registered in the clinics during the first few months after the Act came into force, but there was a gradual diminution in the number of new cases. Eventually the total number of patients attending all the clinics ceased to increase. The aims and working methods of the clinics are described and a follow-up is presented of all the patients who attended two of the out-patient clinics in the first year. During the two years after the clinics were started there was a decrease in the total amount of heroin and an increase in the amount of methadone prescribed. Virtually no central nervous system stimulants were prescribed after the first few months. Further changes in the patterns of misuse of drugs were noted in this period with a marked increase in the use of intravenous methamphetamine associated with very heavy prescribing by two general practitioners; this trend was checked following withdrawal of intravenous methamphetamine by the manufacturers, who now only supply hospital pharmacies and not retail chemists. Following this there was some misuse of barbiturates taken intravenously.

There is some evidence that the clinics have had a limited success combating the problem. The price of heroin and methadone on the black market increased significantly (heroin from £1 to £6 per 60 mg) and addiction became both more expensive and less attractive. The number of cases of dependence on opiates in 1969 was only 3.5% greater than that in 1968, whereas previously there had been a 50% increase each year. Also the total number of new cases seen in 1969 was lower than in 1968. There was also evidence of a decrease in the number of opiate addicts who were admitted to prison and this decrease continued during the two-year period. It would appear that at the time of admission to the program in a prescribing clinic some 20% of addicts were working while currently some 40% are doing so. Thus the evidence presented points to an improved work record and less criminality in the addicts attending clinics, along with a decrease in both black market activities and the number of new cases.

REFERENCES

1. Linken, A.: The psycho-social aspects of student drug taking. *Proceedings of Conference on Psycho-Social Aspects of Drug Taking.* Long Island City, Pergamon Press, Inc., 1965.
2. Weiner, R. S. P.: *Drugs and Schoolchildren.* London, Longmans, 1970.
3. Wright, J. D.: A survey of the knowledge and attitudes about drug misuse among fourth-year pupils at three Wolverhampton schools. Unpublished report: Health Department, Wolverhampton, 1968.
4. Murphy, F. W., Brady, J., and Paul, S.: Drug abuse in a London suburb. Unpublished report, 1969.
5. Hawks, D. V.: The epidemiology of drug dependence. *Bull Narcot 22* (3):15–24, 1970.
6. Bewley, T. H.: Recent changes in the incidence in all types of drug dependence in Great Britain. *Proc Roy Soc Med 62*:175–177, 1968.
7. Bewley, T. H.: Heroin addiction in the United Kingdom (1954-1964). *Brit Med J 2*:1284–1286, 1965.
8. Spear, H. B.: The growth of heroin addiction in the United Kingdom. *Brit J Addict 64*: 245–255, 1969.
9. Brill, H.: Misapprehensions about drug addiction. Some origins and repercussions. *Compr Psychiat. 4* (3): 150-159, 1963.
10. Ministry of Health: Report of the Departmental Committee on Morphine and Heroin Addiction (Rolleston report). London, H.M.S.O., 1926.
11. Ministry of Health and Department of Health for Scotland: Drug Addiction. Report of the Interdepartmental Committee, London 1961. Drug Addiction. Second Report of the Interdepartmental Committee, London, 1965.
12. Dangerous Drugs Act, 1967. London, H.M.S.O.
13. JAMA Narcotics and Medical Practice. AMA Report. 186–976, 1963.
14. Edwards, G.: The British approach to the treatment of heroin addiction. *Lancet 1*: 768–772, 1969.
15. Chapple, P. A. L., and Gray, G.: One year's work at a centre for the treatment of addicted patients. *Lancet 1*:908–911, 1968.
16. Mahon, T. A.: An exploratory study of hospitalized narcotic addicts in Great Britain. Doctoral Thesis, Florida State University, 1970.
17. Cahal, D.: Personal communication.
18. Le Fevre, C.: Personal communication.
19. James, I: P.: Personal communication.
20. Gardner, R., and Connell, P. H.: One year's experience in a drug dependence clinic *Lancet, 2*:455–458, 1970.
21. Gardner, R.: Death in United Kingdom opioid users 1965-1969. *Lancet 2*:650–653, 1970.
22. Stimson, G. V., and Ogborne, A. C.: Survey of addicts prescribed heroin at London clinics. *Lancet 1*:1163–1166, 1970.

ACKNOWLEDGMENTS

We are very much indebted to the Drugs Branch of the Home Office, and particularly Mr. C. G. Jeffery and Mr. H. B. Spear, without whose help we would not have been able to follow-up our patients.

GENETIC HAZARDS OF DRUGS OF ABUSE

WARREN W. NICHOLS, M.D., Ph.D.

Drugs of abuse are in no way unique in the potential dangers they present to genetic material. Rather, they share this hazard with a variety of medicinal compounds and environmental chemicals and agents. It was, however, the hallucinogen LSD that focused attention on this problem. The risks of drugs and chemicals in the environment have been recognized by geneticists for a number of years. Levan in 1951[1] pointed out this hazard, as did Barthelmess in 1956,[2] and Ehrenberg and Gustafsson in 1959.[3] Actually, it was predicted that chemicals would have the capability of producing genetic damage shortly after Muller's historic demonstration, in 1927, of x-ray induced damage to the genetic material.[4]

The first proof that chemicals were capable of producing genetic damage came during World War II with the use of nitrogen mustards. Mustards and similar compounds had little public exposure since they were used mainly as agents in warfare and cancer chemotherapy. Widespread concern by physicians and the public was not really aroused until it was demonstrated that LSD as well as pesticides and some food additives and food substitutes, such as the cyclamates, had similar potential. The hazards to genetic material from environmental agents may well be one of the major problems faced by civilization at the present time and in the future. In reviewing this problem we will discuss the types of genetic damage that occur, the significance of the induced genetic changes, ways to detect and test for the potential of an agent to produce genetic change, and finally, a consideration of specific drugs of abuse in relation to their genetic potential.

TYPES OF GENETIC DAMAGE

The types of genetic damage that can occur all fit into the broad definition of mutation. This term was first used by DeVries in 1901[5] to describe an abrupt heritable change in the evening primrose. Today a usual textbook definition is: "An abrupt hereditary change not due to the recombination of genes."[6] The definition is imprecise and includes a wide range of genetic change. It includes changes in chromosome number and changes in chromosome structure in the general category of mutations. If the change is limited to the approximate 1,000 base pairs that comprise the average gene, it is spoken of as a "gene mutation." If a single base within a gene is altered, it is termed a "point mutation."

SIGNIFICANCE OF INDUCED GENETIC CHANGES

The significance of damage or change to the genetic material varies with the type of damage produced. The most severe types of damage, including chromosome breakage and unstable chromosomal rearrangements, usually lead to cell death. Cell death is of greatest importance when it occurs during the organogenesis period of embryogenesis. During this time it can result in fetal wastage or a teratogenic event. Changes in chromosome number at an early stage of embryogenesis can also lead to an abnormal or dead fetus.

The significance of changes resulting in gene mutations depends on whether they occur in a somatic or germ cell. Gene mutations in somatic cells are a potential hazard in carcinogenesis and aging, while those in germ cells may increase the genetic load of the population and initiate clinical hereditary disease.

TESTING FOR GENETIC DAMAGE

When discussing methods of detecting or testing for genetic damage by drugs of abuse or any environmental agent, several factors must be evaluated. These include the type of damage that is being tested; how relevant the test system is for the human; the sensitivity of the method; the reproducibility of the method; and the technical and economical factors involved in carrying out the procedures. Some of the tests available are listed in Table 1. Each has various attributes and shortcomings. The details of these tests are beyond the scope of this presentation but can be found in other articles by interested readers.[7–10]

If I may be permitted to make some generalizations, it can be said that the ancillary methods provide excellent means of testing gene and point mutations, and in some cases chromosome mutations as well. But these methods are lacking in relevance to the human. All the mammalian methods are more relevant to the human, but the specific locus test, at the present, lacks the sensitivity required for testing of drugs and chemicals, although it has been successfully used to assess radiation damage. This leaves the cytogenetic methods, the host-mediated assay, and the dominant lethal test, as currently best meeting the required standards. These will be briefly outlined.

Table 1 Testing genetic damage

I. Mammalian methods
   A. Cytogenetic
      1. Chromosomal
      2. Biochemical
   B. Host-mediated assay
   C. Dominant lethal test
   D. Specific locus test
II. Ancillary methods
   A. Microbial
      1. Bacterial
      2. Neurospora
      3. Phage and transformation
   B. Plant
   C. Insect
III. Population monitoring

Cytogenetic Methods

The cytogenetic methods can be divided into chromosomal and biochemical types. The chromosomal types can be carried out either in vivo or in vitro, and on metaphase or anaphase; they have the distinct advantage of being able to be performed on humans or human cells in culture. Basically these methods seek chromosome breaks, or resultant chromosome rearrangements, or changes in chromosome number. Some of these abnormalities are illustrated in Figures 1, 2, and 3. These methods are probably the best available for the large category of chromosomal mutations. However, whether chromosome breakage is an adequate indicator of, or test system for, gene mutations is less clear. There are excellent correlations between gene mutations and the chromosome breakage induced by irradiation and chemical agents. The first of these correlations is found in the similarity of the dose response of the single-hit type of chromosome aberration and of gene mutations induced by irradiation. Both of these follow a straight-line relationship.[11]

The second correlation is found in the data collected by Kihlman in his book, *Actions of Chemicals on Dividing Cells.*[12] At the time he wrote the book, Kihlman studied all of the chemical substances that he could find that had been tested for both gene mutations and chromosome breakage. Chromosome breakage was observed in both plant and animal systems, and gene mutations were studied in microbial, insect, or plant materials. These data are shown in Table 2. Note that although there was a very good correlation between chromosome breakage induced in plant cells and gene mutations, it was not perfect. However, the correlations between chromosome breakage

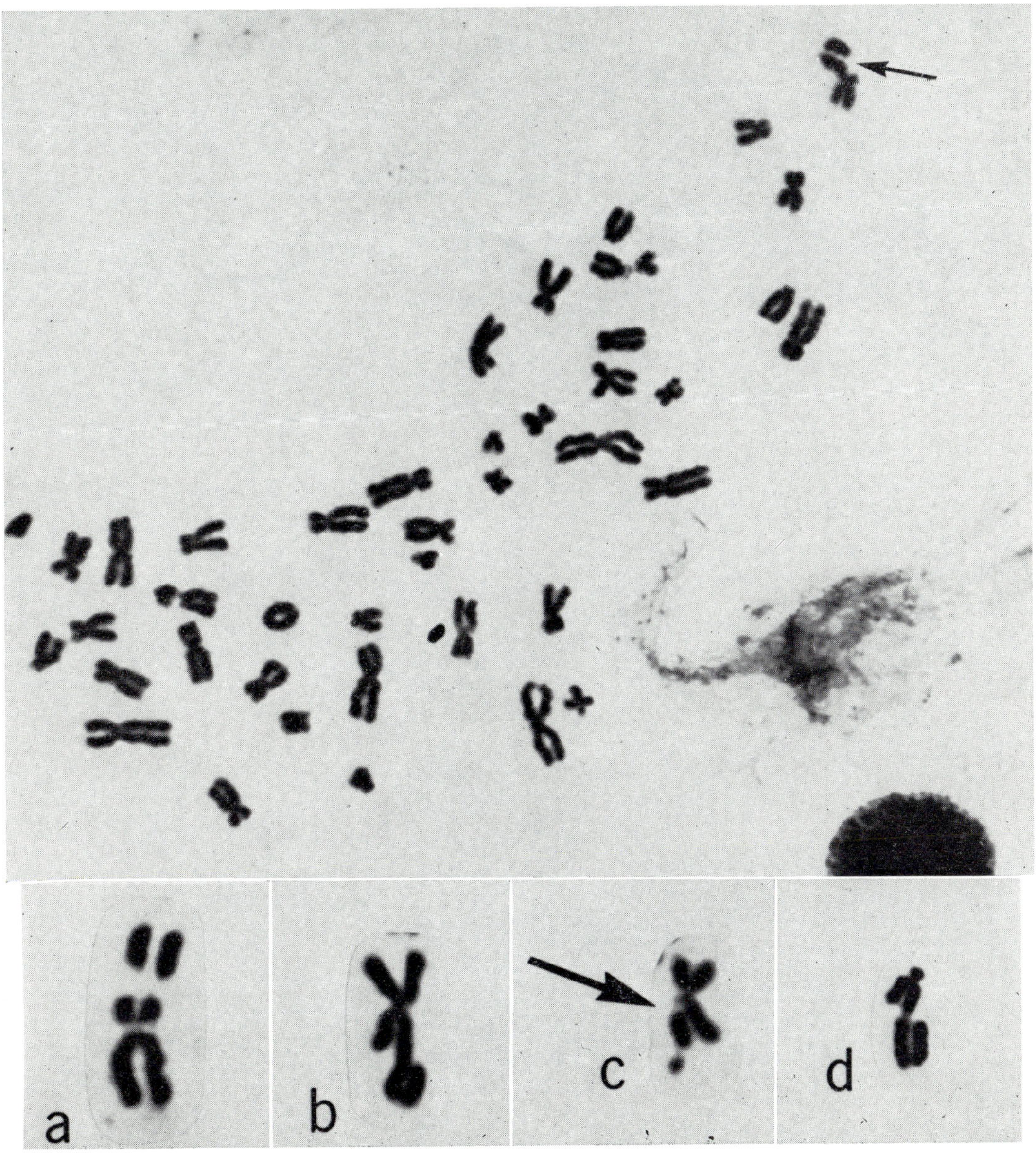

Figure 1 Examples of chromosome breakage. *Above,* Intact cell in metaphase with break indicated at arrow. *Below,* Selected types of chromosome breaks: a, chromosome break; b, chromatid break; c, chromatid gap; d, delayed isolocus break.

induced in mammalian cell systems and gene mutations were 100%. Whether these correlations are consistent enough for chromosome breaks to be considered an adequate test system for gene mutations is not currently known. The biochemical cytogenetic test systems described by Puck and Kao[13] and Chu and Malling[14] should answer this question over the next several years. These studies utilize cells in culture and selective methods to obtain specific biochemical mutants, allowing the study of gene mutations

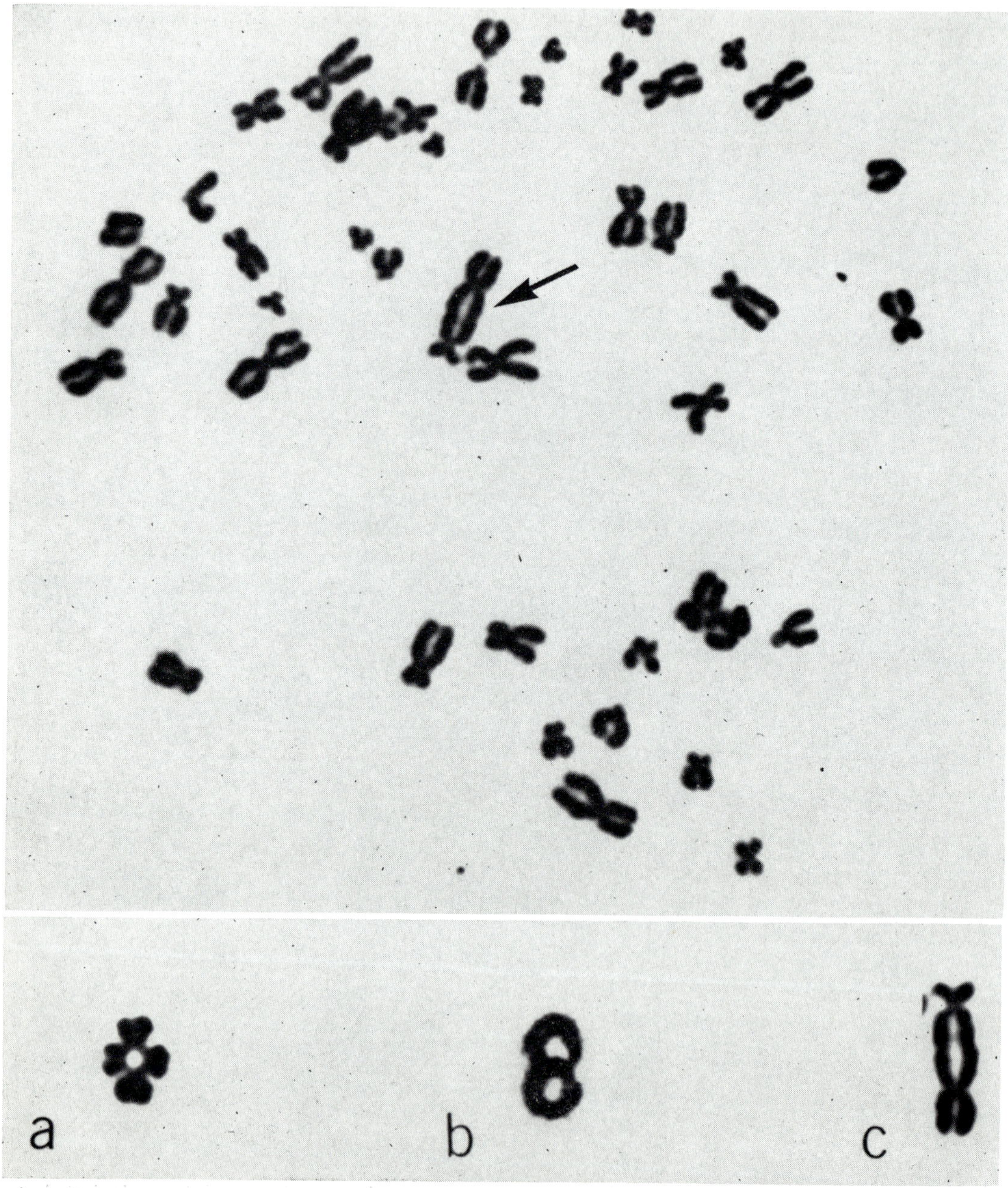

Figure 2 Examples of chromosome rearrangements. *Above,* Intact cell in metaphase with dicentric at arrow. *Below,* Selected chromosomal interchange: a, quadriradial; b, interlocking rings; c, dicentric.

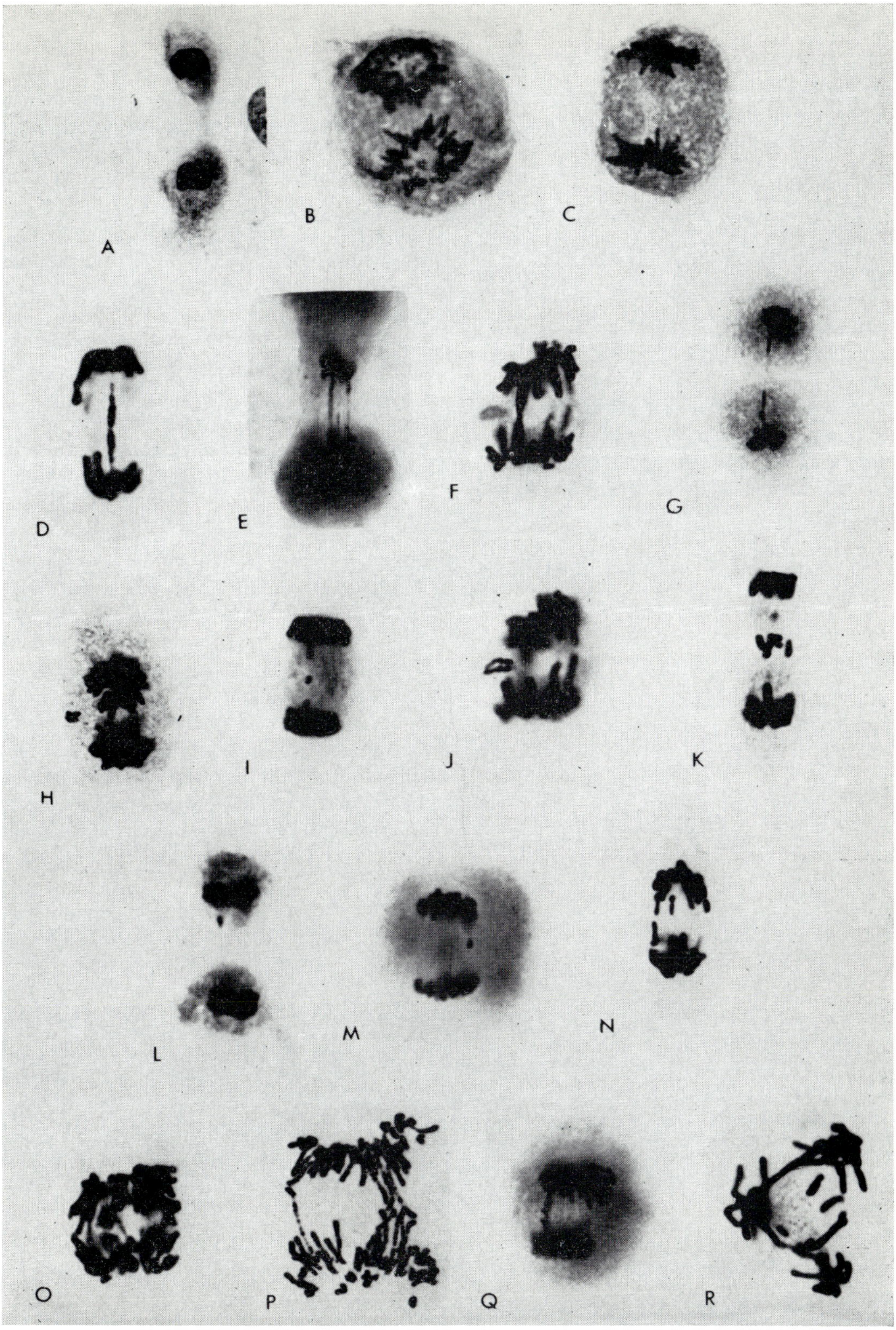

Figure 3 A variety of normal and abnormal anaphase figures: a-c, normal anaphase; d-g, a variety of anaphase bridges; h-k, a variety of acentric fragments; l-n, attached fragments; o-q, pseudochiasmata; r, multipolar spindle with a chromosome bridge, attached fragments, and pseudochiasmata.

and chromosomal mutations in the same cell system at the same time.

Host-Mediated Assay

The host-mediated assay is a test designed to utilize the sensitivity of microbial systems for the detection of point mutations in a setting that is more relevant to mammals and humans. This is accomplished by injecting a microorganism into the peritoneal cavity of a mammal. The drug or agent under test is administered to the mammal. At suitable time periods the microorganism is removed and mutants sought by growing the microorganism on selective media. In this way the microorganism is subjected to the original agent to be tested, but it is in the milieu of mammalian detoxification processes and metabolic products.

Table 2 Comparison between chromosome-breaking and mutagenic effects of chemicals in plant and animal materials

	Chromosomal aberrations		
Compound	Plant root-tips	Mammalian cells in tissue culture	Mutagenic effect
Adenine	+	+	+
2,6 Diaminopurine	—	+	+
Caffeine	+	±	+
8-Ethoxycaffeine	+	±	±
Purine riboside	—	+	+
Deoxyodenosine	+	+	No data
5-Fluorodeoxyuridine	+	+	No data
5-Bromodeoxyuridine	—	+	+
Cytosine arabinoside	—	+	No data
Maleic hydrazide	+	—	—
Azaserine	+	+	+
Streptonigrin	+	+	+
Mitomycin C	+	+	+
Hydroxylamine	±	+	+
Nitrogen mustard	+	+	+
Triethylenemelomine	+	+	+
Diepoxybutane	+	+	+

+ = marked effect; — = no effect; ± = effect very low, although just about significant.

From Kihlman, B. A.: *Actions of Chemicals on Dividing Cells*. Englewood Cliffs, New Jersey, Prentice-Hall, Inc., 1966, p. 198.

Dominant Lethal Test

The dominant lethal test employs male rodents that are given the drug or agent to be tested. They are then mated with a series of females during the entire period of spermatogenesis. The females are sacrificed and autopsied at specified intervals following the matings, and corpora lutea and the total number of living and dead implants are enumerated. This method tests for chromosomal damage and rearrangements by determining their lethality to the developing embryo. It is not known if the method is capable of detecting gene mutations, and its sensitivity would seem to be somewhat less than that of the cytogenetic method.

Population Monitoring

Population monitoring is not a test method but rather a means of detecting mutagenic agents that may not have been uncovered by testing. This system utilizes the incidence or rate of selective hereditary defects to detect an increased mutation rate. The defects sought should be dominant so that they are expressed in the first generation after their occurrence. They also should be recognizable at birth to minimize the delay in reporting. Should an increased mutation rate be found by population monitoring, a great deal of work would still remain to be done to pinpoint the causal agent.

ASSESSING DRUGS OF ABUSE

Although the above methods have been used to test some agents, genetic tests on drugs of abuse have employed predominantly chromosome cytogenetic methods. The drug of abuse that has received most attention is LSD. There is a paucity of information concerning other drugs of abuse, but some of the tranquilizers such as chlorpromazine (Thorazine)[15-18] and perphenazine (Etrafon, Triavil, Trilafon)[16] have yielded contradictory or preliminary positive results, and two negative studies on marihuana have been reported.[19,20]

Concerning LSD, considerable confusion followed the several reports on observed chromosome effects. Although several factors probably contribute to the contradictory nature of the results and the subsequent confusion, predominant would seem to be the populations studied. The initial study by Cohen[15] was of a "hippie" population of drug users. There were six subsequent reports of drug users, and the majority of these studies showed that LSD did affect chromosomes.[21–26] It is highly likely that these observations are correct, but they are quite difficult to interpret because of the large number of variables encountered. Drug users often are malnourished, may have a high incidence of hepatitis and virus infection, may be on multiple drugs, and probably obtain the LSD from illicit sources, making dosage and purity uncertain. In contrast to these positive studies, we can examine the two best studies of LSD made under medical supervision. The first of these, by Hungerford et al.,[27] showed an increase in chromosome abnormalities compared with controls, but this was much smaller than that observed in drug users. The study by Tjio et al.,[28] of the largest series of cases reported, showed no increase over control levels. Problems of this type resulting from experimental design can best be resolved by employing expert committees to standardize test procedures and aid in the interpretation of results. It should also be pointed out that several drugs utilized in cancer chemotherapy and as immunosuppressants yield marked chromosome effects with no ambiguity or confusion in the results. These include some of the precursors of DNA, analogues of deoxyribosides, and antifolic acid drugs that inhibit cellular DNA synthesis. Also included are such antibiotics as Azaserine, Mitomycin C, and streptonigrin, and alkylating agents such as the nitrogen mustard (Chapter 10 of reference 12). These results are clear and unambiguous because of both the consistency of the results and the large amount of breakage they produce. Some of these drugs are also known to produce abortion or teratogenic events in humans.

SUMMARY

Genetic damage induced by drugs and all environmental agents is of great import to the community. Depending on the type of damage produced, there is possible significance in teratogenesis, carcinogenesis, aging, and increase in the genetic load. While at the present time it cannot be determined with certainty which of the several test methods will best answer all requirements, it is imperative that a group of tests that offer most promise be selected and utilized.

REFERENCES

1. Levan, A.: Chemically induced chromosome reactions in Allium cepa and Vicia faba. *Cold Spring Harbor Symposium on Quantitative Biology 16*:233, 1951.
2. Barthelmess, A.: Mutagene arzneimittel. *Arzneimittelforschung 6*:157, 1956.
3. Ehrenberg, L., and Gustafsson, Å.: Chemical mutagens: Their uses and hazards in medicine and technology. *A Report of February 1959 to the National Board of Health.* Lund, Sweden, 1970.
4. Muller, H. J.: Artificial transmutation of the gene. *Science LXVI* (#1699):84, 1927.
5. DeVries, H.: *Die Mutationstheorie Bd. 1. Die Enstehung der Arten durch Mutation.* Leipzig, Veit & Comp. 1901.
6. Muntzing, A.: *Genetics: Basic and Applied.* Stockholm, Sweden, Lts. Forlag, 1961, p. 442.
7. *Report of the Secretary's Commission on Pesticides and Their Relationship to Environmental Health*: U.S. Dept. Health, Education and Welfare, Washington, D.C., 1969.
8. Epstein, S., and Legator, M. (Eds.): *Mutagenicity of Pesticides.* Cambridge, Mass., M.I.T. University Press. 1971.
9. Hollaender, A. (Ed.): *Environmental Chemical Mutagens.* New York, Plenum Pub. Corp., 1971.
10. Nichols, W. W.: The relationship of chromosome aberrations to drugs. *Drug-Induced Diseases,* Vol. IV. Amsterdam, Excerpta Medica. (In press.)
11. Sparrow, A. H.: Types of ionizing radiation and their cytogenetic effects. *Mutation and Plant Breeding,* publication #891, Nat. Acad. of Sci. Nat. Research Council, Washington, D.C., 1961, p. 55.

12. Kihlman, B. A.: *Actions of Chemicals on Dividing Cells.* Englewood Cliffs, New Jersey, Prentice-Hall, Inc., 1966.
13. Puck, T. T., and Kao, F.: Genetics of somatic cells. V. Treatment with 5-bromodeoxyuridine and visible light for isolation of nutritionally deficient mutants. *Proc Nat Acad Sci USA 58*:1227, 1967.
14. Chu, E. H. Y., and Malling, H. V.: Mammalian cell genetics II. Chemical induction of specific locus mutations in Chinese hamster cells in vitro. *Proc Nat Acad Sci USA 61*:1306, 1968.
15. Cohen, M. M., Hirschorn, J., and Frosch, W.: In vivo and in vitro chromosomal damage induced by LSD-25. *New Eng J Med 277* (20):1043, 1967.
16. Cohen, M. M., Hirschorn, J., and Frosch, W. A.: Cytogenetic effects of tranquilizing drugs in vivo and in vitro. *JAMA 207*: 2425, 1969.
17. Nielsen, J., Friedrich, U., and Tsuboi, T.: Chromoscme abnormalities in patients treated with chlorpromazine, perphenazine, and lysergide. *Brit Med J 3*:634, 1969.
18. Abdullah, S., and Miller, O. J.: Effect of drugs on nucleic acid synthesis and cell division in vitro. *Dis Nerv Sys 29*:829, 1968.
19. Martin, P. A.: Cannabis and chromosomes. *Lancet 1*:370, 1969.
20. Dorrance, D., Janiger, O., and Teplitz, R. L.: In vivo effects of illicit hallucinogens on human lymphocyte chromosomes. *JAMA 212* (9):1488, 1970.
21. Irwin, S., and Egozcue, J.: Chromosomal abnormalities in leukocytes from LSD-25 users. *Science 157*:313, 1967.
22. Loughman, W. B., Sargent, T. W., and Israelstam, D. M.: Leukocytes of human exposed to LSD: Lack of chromosomal damage. *Science 158*:508, 1967.
23. Egozcue, J., Irwin, S., and Maruffo, C. A.: Chromosomal damage in LSD users. *JAMA 24*:122, 1968.
24. Sparkes, R. S., Melnyk, J., and Bozzetti, L. P.: Chromosomal effect in vivo of exposure to lysergic acid diethylamide. *Science 160*: 1343, 1968.
25. Valenti, C.: *Neurobiological Aspects of Psychopathology* (Edited by J. Zubin and C. Shagass). New York, Grune & Stratton, 1969, p. 275.
26. Judd, L. L., Brandkamp, W. W., and McGlothin, W. H.: Comparison of chromosomal patterns obtained from groups of continual users, former users and non-users of LSD-25. *Amer J Psychiat 126*:626, 1969.
27. Hungerford, D. A., Tayler, K. M., Shagass, C., La Badie, G. U., Balban, G. B., and Paton, G. R.: Cytogenetic effects of LSD-25 therapy in man. *JAMA 206*:2287, 1968.
28. Tjio, J. H., Phanke, W. N., and Kurland, A. A.: LSD and chromosomes. A controlled experiment. *JAMA 210*:849, 1969.

PANEL DISCUSSION—INTRODUCTORY SESSION

Moderator: DR. ZARAFONETIS

Panel Members: DRS. BEWLEY, CAMERON, GOLDBERG, KATO, NICHOLS, and SEEVERS

Dr. Zarafonetis: For the next hour our speakers of the day will field questions from the audience; I hope that each of you will feel free to enter the discussion and provoke the panel in any way you believe appropriate. There are microphones in the aisles; the questions and responses will be recorded and edited for inclusion in the proceedings. For this reason, I would appreciate it very much if each questioner would first identify himself and his home.

Dr. Harris Isbell (Lexington, Ky.): I have questions for Dr. Bewley. First, what is the source of the heroin used by the new addicts who come into your clinic? Is this by diversion from the clinic supply, smuggling, or how?

Dr. Bewley: They obtain it from two sources. Firstly, some of the heroin that is prescribed leaks into the black market. We know that a certain percentage of what we prescribe is sold or given to friends. The second source of heroin is a small amount that is smuggled in from Hong Kong. This is heroin that would be smoked in Hong Kong but is used intravenously in London by heroin addicts. This is usually referred to as "Chinese heroin." There is a small amount of that, not much.

Dr. Isbell: The second question is, what accounts for the marked increase in new methadone addicts in the last year or two?

Dr. Bewley: Many heroin addicts have switched to methadone. In 1969, roughly 1200 who were attending the clinics were switched to methadone, so this is really no change in a sense. They are still opiate addicts. Three years ago it was heroin, one year ago this was changed to methadone.

Dr. Isbell: Were the addicts changed by the clinics or by private physicians?

Dr. Bewley: Primarily by the clinics. Since the clinics were set up, private physicians almost never prescribe for addicts.

Dr. Isbell: My next question is a technical one concerning different drugs being taken. One of your charts showed tests for detection of both heroin and morphine in urine. Is that correct?

Dr. Bewley: Only morphine in urine. We test for morphine in the urine of those who are having heroin prescribed. I think the table may have misled you. It said "heroin prescribed" but "morphine in urine" or "morphine not in urine."

Dr. Isbell: I was thinking that it was probably impossible to differentiate between the two in the urine.

Dr. Bewley: We don't try. We just test for morphine in urine.

Dr. Isbell: What were the leading causes of death?

Dr. Bewley: The most common cause of death was overdose (often of a sedative—for example, an overdose of barbiturates) and also suicide and septic complications. These are the three major causes of death.

Dr. Isbell: Do you have any idea of the death rate in this young population as compared with the death rate in the non-drug using population?

Dr. Bewley: Yes, we did a survey about three years ago. Very roughly, the death rate in the drug-using group was 28 times that of the expected death rate in a similar population of young people.

Dr. Isbell: Do you have any data on social adjustment, productivity, sexual adjustment, and so on, in addicts maintained on heroin?

Dr. Bewley: To answer very briefly, when we did some surveys of people coming into the clinics, we found that 20% of them were working when they first came. Currently about 40% are working, so there has been some slight improvement in work adjustment. As far as sexuality is concerned, there is a generalized decrease in libido and sexual activity of those who are using opiates.

Dr. Isbell: What do you do with the non-narcotic drug-dependent patient? Obviously he's not a candidate for maintenance on heroin.

Dr. Bewley: Sometimes a person is a candidate because he is a poly-drug misuser. If a patient is using amphetamines, barbiturates, methadone, heroin, LSD, cannabis, and whatever else he can lay his hands on, we may continue to prescribe methadone for him. Most of the people, for example, who use barbiturates intravenously take other drugs intravenously as well, often including the opiates. We treat people who take amphetamines in the form of pills in the usual way—with exhortation and good advice. Users who take amphetamines intravenously are nearly always using opiates as well, and we may then consider prescribing opiates for them.

Dr. Isbell: What about the hallucinogen users? LSD?

Dr. Bewley: This is a small problem in England as compared with that in the United States. There was an increase in use, and then there was a slight decrease, and now there is a slight increase again, but it still remains a small problem. Not many of these addicts come to the clinic. They come to the clinic mainly because they are in trouble with the law or something like that. We see an occasional person who has had a bad trip. We don't see very many people on hallucinogens apart from those who have other problems, or those who are using opiates intravenously in addition to the LSD. We don't do very much about the LSD use in such a case.

Dr. Seevers: I would like to ask Dr. Bewley whether he still prescribes cocaine?

Dr. Bewley: Very little cocaine is prescribed at all. I can give my personal experience. Out of 400 patients, I prescribe cocaine for two. One of these is a retired major-general who is deaf, bedridden, and has been having cocaine prescribed for him for the last 30 years. Initially he was given cocaine when he was a patient in the ear, nose, and throat department and he had been having the same amount prescribed for many years. I am continuing to prescribe cocaine for him. For people who were getting heroin and cocaine, we stopped prescribing cocaine; we also stopped prescribing intravenous methamphetamine which was widely given at the time the clinics were started.

Dr. Eric Chivian (Washington, D. C.): Dr. Seevers, in your classification of drugs, I understood that you included marihuana and hashish as stimulants. I wondered if you could explain this, since my understanding is that in most pharmacologic classifications they are considered as sedative-hypnotic drugs.

Dr. Seevers: Well, they're a little like alcohol. Small doses produce the stimulant effect which is desired—the marihuana or small-dose situation. With hashish and large doses of purified resins, one talks about depression. The stimulant effects may involve the same release mechanisms as those experienced with alcohol. It's hard to make an absolute classification. As you recall, I mentioned in the classification that some of these drugs have sedative properties when taken in large doses.

Dr. Chivian: I thought you classified alcohol as a depressant.

Dr. Seevers: I do primarily, but it's also a stimulant. Either alcohol or cannabis can be put into either category depending on the dose. I didn't find it necessary to put them in both categories, and prefer to classify cannabis with the stimulants since it is a hallucinogen like other stimulants and because it does not produce physical dependence.

Dr. Leonard Sigell (Cincinnati, O.): Drs. Goldberg, Kato and Bewley have all described the special dependence problems of Preludin and methaqualone in their country. I recall reading that, in Sweden, restrictions were placed on the use of Preludin, and, I believe, on its availability. I would be interested to know why Preludin and methaqualone were selected for use out of categories in which they fall, and whether the restrictions placed on their availability had any impact on the incidence of the use of other drugs in the same categories.

Dr. Goldberg: The reaction of the authorities to the abuse of amphetamines—in the beginning comprising amphetamine (racemic form), dexamphetamine and methamphetamine—led first to the amphetamines being sold on prescription only (1939), and then to placing them on the Swedish Narcotics List (1944). This, together with warnings sent to the medical profession led to a sharp decrease in the legal sale, lasting for a few years.

When, in the fifties, a new wave of abuse of stimulant drugs started, now also comprising phenmetrazine (Preludin) and methylphenidate (Ritalin), these two substances were also placed on the Swedish Narcotics List (1959).

The next step was to forbid the sale of phenmetrazine completely (Dec. 1965), this drug having become the drug of choice for abuse, especially for injection, its prolonged abuse leading to a host of severe complications, including overt psychoses such as paranoia. It is among the stimulants leading to the most intense feelings of dependence and urge to continue.

The illegalization of phenmetrazine was followed by a switch to other stimulant drugs, including methylphenidate, by juveniles and other abusers, as well as to a decrease in the number of seizures of illicit phenmetrazine—coming in by smuggling or produced in clandestine laboratories—and an initial increase in seizures of methylphenidate.

With a whole set of studies showing the abuse of stimulant drugs to be the dominant trait and the therapeutic value to be very limited, the sale of these drugs finally (1968) became restricted. They could be prescribed, by special license only, for the treatment of patients with certain rare diseases, such as narcolepsy. The prescriber was required to give the name of the patient, the diagnosis and reason for the use of stimulants. These measures led to a decrease in the number of prescriptions of stimulants drugs from several hundred thousand to a few hundred per year, and a corresponding decrease in turnover.

At present the illicit market in stimulants is dominated by substances in capsules for intravenous injection, amphetamines and phenmetrazine being the most frequently employed. The incidence of use of other drugs is increasing—cannabis mainly as hashish, and the hallucinogens, LSD in particular. Any direct connection between restriction in sale of stimulants and increase in the abuse of other drugs is not obvious. It is probably an expression of the increased pressure by forces on the illicit market to sell illicit drugs and promote drug-taking.

Dr. Kato: As to methaqualone use and abuse in Japan, it was listed as one of the 46 dangerous drugs that are allowed to be sold only to doctors. Pharmacists who sold it to young people are considered drug abusers and are punished.

Dr. Edward Pierce (Ann Arbor, Mich.): I would like to ask Dr. Bewley, or anyone, what is methaqualone? Is there any other name for it?

Answer from the audience: Quaalude, Mandrax in England.

Dr. Pierce: Thank you. Dr. Bewley, did I understand you to say that none of the deaths among the addict population were due to overdoses of the heroin itself?

Dr. Bewley: Oh, no. Some were due to overdoses of heroin itself but some resulted from overdoses of other drugs.

Dr. Pierce: I ask because addicts in this country feel that overdose occurs in those taking large doses who suddenly get a purified form of the drug. You may overcome that problem because your patients are getting a standardized product.

Dr. Bewley: Occasionally someone, who has been in a hospital or prison, who has been off all drugs, and has lost his tolerance, takes the large amount he had previously been accustomed to taking; this may be fatal.

Dr. Pierce: That's basically an error on the user's part. I think I know something of the properties of methadone versus heroin, but I don't understand why you are using methadone intravenously rather than orally.

Dr. Bewley: The reason we use it intravenously is that if we give the drug to the patient and he picks it up himself, he then decides how he will take it. If we give a person tablets or linctus or another form of oral methadone, unless he takes it under our supervision, he may take it intravenously anyway. If he is going to do this, we think it better to prescribe it in injectable form. One of my patients, for example, for whom I prescribed tablets, crushed them up and injected them into an artery; he lost half his hand, a very undesirable result.

Dr. Pierce: Would you ever change to oral methadone?

Dr. Bewley: It is possible. The essence of the "British system" is that we are playing the whole thing by ear. It may well be that we will switch over at some stage to dispensing methadone for oral consumption on the clinic premises. If necessary, we can do it. At the present, we don't. And since, at present, the patients are given their drugs to take themselves they are probably going to take them intravenously and therefore we prescribe them for intravenous use.

Dr. Pierce: Is there a "high" associated with intravenous methadone that is similar to the heroin "high?"

Dr. Bewley: I think the answer is "yes." But this varies from person to person and from time to time; if a person is on a regular dose of methadone, taking the same amount all the time, he doesn't get very much of a high when he takes the drug intravenously.

Dr. Pierce: You say that you are playing it by ear and that if the addict is used to injecting the drug he probably will continue in this practice. Are you suggesting that these addicts go from heroin to methadone? Or is that at their suggestion?

Dr. Bewley: Heroin is a shorter-acting drug. We may have a patient, for example, who takes all his heroin in the morning and comes in to the clinic saying, "Dr. Bewley, you're not giving me enough; I use it all up in the morning. I have to take barbiturates in the afternoon when I start getting withdrawal symptoms." At this stage I would say, "The best thing for me to do is to switch you over to methadone. It's a long-acting drug." So that's one reason. The other reason is that methadone for injection comes in ampules, whereas heroin comes in tablet form—a much less sterile preparation. You can see, from the medical point of view, the advantages of the person changing to methadone.

Dr. Pierce: On one of the tables that you showed I wasn't able to add the figures as you went along, but it looked to me that you had a considerable number of addicts who had entered the program but left it after a length of time. I got the feeling from you that there is at least a percentage who are leaving it altogether. Would you elaborate on that point?

Dr. Bewley: Very roughly we would expect somewhere between 10 and 20% of the people who come into the program to be off all drugs. Some others, of course, are in prison, some have died, some are currently inpatients, but follow-ups show that the majority of the people remain in the program although they may have moved to a clinic in another part of London or in another part of the country.

Dr. Pierce: One of the programs that will be discussed tomorrow by Dr. Jaffe is the long-term maintenance of people on methadone, which is similar to what you are doing. One of our problems is what to do with someone who says that he has been addicted to straight heroin for, say, four months. We have asked ourselves, how does one insure that one is not dealing with a temporary addict who could become a methadone addict? In other words, do you have a criterion for acceptability? Suppose someone comes to you and says, "I have been using straight heroin for the past four months; I am addicted because when I tried to get off I had withdrawal symptoms and now I would like to get heroin legally from you rather than from the street." Would you admit that kind of person into the program, or does he have to be addicted for a longer period of time before you admit him?

Dr. Bewley: In the early days of the clinics we admitted some people who, in fact, I think weren't addicted, and who became addicted for the first time to what we were prescribing. This we did because we were admitting a large number of people and the situation was very confused. I don't think we are doing that at present. If somebody comes along now, we make sure that he is physically dependent on the drug before we will prescribe. If necessary, we say, "We won't prescribe for you unless you come into the hospital for evaluation." Very often we will say to the person, "If you do come into this program, we will only prescribe methadone for you and only a limited amount of methadone at that," and with this understanding he may come into the program. We are in a better position now, I think, and we can be sure from urine tests that the person is using the drug. We can, if necessary, admit the user into the hospital, observe him to see whether he shows any withdrawal symptoms, and then estimate how much of the drug he has been using.

Dr. Pierce: May I ask this question once more, because I think it is very important. If a person tells you, and you have reason to believe him, that he has been addicted for six months or less, what program do you put him on?

Dr. Bewley: If he is dependent on the drug, we would be prepared to take him into the program on a methadone-prescribing regimen. It would vary a bit. If we thought he hadn't become dependent, then we wouldn't prescribe. We might say to the person, "Look, we think at this stage you would do better to stay off all drugs." If he had definitely become physically dependent in that six months, then we would be prepared to admit him to the program.

Dr. Pierce: For an indefinite period of time?

Dr. Bewley: For an indefinite period. Again we would try to evaluate what we were doing in terms of how satisfied he was, whether he was working or not, whether he was keeping out of jail—we would use these criteria for evaluating what we were doing for him.

Dr. Joseph S. Hupert, (St. Louis, Mo.): I would like to clarify something with Dr. Nichols. You mentioned two studies on LSD, and I'd like to throw in a third one, by Savage, Pahnke, and the people at Spring Grove, reported in the Journal of the Americal Medical Association (Pahnke, W. N., *et al.*: JAMA *212*:1856-1863, 1970). Now, is it my understanding that, with the best assay methods we now have, there is no real hard evidence that LSD can lead to the genetic transmission of birth defects?

Dr. Nichols: There are several studies that have been done on LSD under medical supervision. I mentioned those reported by Tjio, *et al.,* and by Hungerford, *et al.,* because they are examples of the most complete and the best. I think it can be said that the best evidence at this time is that there are few or no genetic defects that have been shown to be produced by LSD when it is taken in pure form by healthy individuals. But I do think that the observations made of the drug-user populations are very likely also valid. We may have a coincident factor due to the many variables involved or a synergistic or additive effect along with malnutrition, virus infections, multiple drug use, and so on.

Dr. Hupert: I understand that some investigators have been studying not only the chromosomes but the bone marrow and the sexual reproductive cells in monkeys and have not found associated bone marrow changes that they find, for instance, in association with radiation damage. Would you elaborate on this?

Dr. Nichols: I didn't bring in any of the animal studies. There are a great many animal studies, and these are just as contradictory as the human studies. Some people are finding defects, others are not. I don't know of any instances in which the bone marrow results have differed from peripheral leukocyte results in the same study. I don't think that really is an important point. It is important if there are positive-negative studies, but this goes across the board. The only ones that have been consistently positive have been *in vitro* studies in which LSD is added to culture systems.

Dr. Hupert: Then pure LSD in human beings has not been shown to be positive in terms of chromosome damage.

Dr. Nichols: For the healthy individual, that is essentially correct.

Dr. William R. Edmundson, (East Orange, N.J.): I have two questions or comments; the first is directed to Dr. Cameron. Did I understand you to say that the opiate addicts in the United States number somewhere in the neighborhood of 125,000?

Dr. Cameron: I said that Mr. Ingersoll suggested that the number was somewhere between 120,000 and 180,000. It's an order of magnitude estimate.

Dr. Edmundson: I think that this is an underestimate, because in New York City alone last year there were 900 deaths directly or indirectly due to intravenous or non-intravenous narcotism. Those of us who work in the field in the New York area estimate conservatively that there must be in the range of 300-plus thousand addicts in the New York City area alone. The death rate in New York this year is keeping up with that of last year.

Dr. Cameron: Of course, there are no precise data available on the number of heroin users in the United States. We are aware of the death rate that you speak of, and I also understand that, among those who have died, only some 20 or 25% were known to the registers. Therefore, some have assumed that there would be four or five times as many users as were on the register. Such a projection relies on the assumption that those who die as a consequence of overdose are typical users. I am not sure that such is the case. Such a projection, therefore, may be helpful but not altogether conclusive.

Dr. Edmundson: It is true that, in the 15- to 35-year age group, opiate addiction or drug abuse in general is a leading cause of death.

Dr. Bewley, when you prescribe opiates in your clinic, are the addicts permitted to take the drug home with them or do they take it at the clinic under your direct supervision?

Dr. Bewley: The system is a rather more complex one. We write a prescription which we mail to a particular retail pharmacist. The patient has to go to that particular retail pharmacist and pick up his drugs there. He then can use them that day. We only prescribe on a daily basis. We send a week's prescription, if you like, by writing a prescription for a week to be picked up daily, and on a Saturday the addict picks up two day's supply because retail pharmacies are not open on a Sunday.

Dr. Edmundson: How do you evaluate the addict's opinion of his own tolerance? When he tells you he needs 50 pills, whereas in effect he probably may be able to get by with just 30, do you trust him?

Dr. Bewley: Well, I suppose the answer is partly yes, partly no. What we have found is that apart from the fact that people overestimate the amount that they are using, most of the information we are given by people coming into the clinics is surprisingly accurate, such things as their past criminal records, for example. I think that we would be rather cautious in accepting at face value what somebody said that he was using. In fact, when in doubt, we will act on what we think the person has been using. This is one of the things the doctor decides; the amount that he is going to have is not something that the addict himself determines. This is ultimately the doctor's decision and again sometimes what we are doing in effect is making a guess. Of course, with the passage of time we make changes. We may start off prescribing too much and we later reduce the amount. One of the slides indicated that the total amount of heroin that had been prescribed had been halved in a period of two or three years and I think this is closer to what the patients are actually using now.

Dr. Marshall M. Johnson, Jr. (Pittsburgh, Pa.): I am associated with a methadone maintenance program. Dr. Seevers, do you believe that there is a lifelong physiologic change from heroin addiction, or from any opiate addiction?

Dr. Seevers: Do you mean are there any pathologic changes?

Dr. Johnson: Is there a lifelong physiologic-pathologic change within the cellular structure such that even if a person abstains, he still maintains a hunger for heroin?

Dr. Seevers: This a moot question. No one has been able to find a satisfactory assay method for studying functional changes, although there were studies made at Lexington years ago indicating that some physiologic changes persisted for a considerable period of time, possibly months. To answer the question concerning a permanent pathologic injury, we don't have information to prove that one way or the other. Damage to the central nervous system was described even in the literature of 100 years ago and there are sporadic pathologist's reports showing diffuse injury in the central nervous system. But these have been criticized by pathologists for the same reason that Dr. Nichols mentioned in relation to genetic changes in these individuals. Most street addicts buy drugs in preference to food; many have nutritional problems, intercurrent disease, and environmental situations capable of producing diffuse pathologic conditions. We have attempted to study this in the monkey. Animals that were on morphine for several years were examined by a neuropathologist. Some diffuse changes pretty well distributed throughout the whole central nervous system were found. These lesions did not appear to be of anoxic origin and were not distributed in any particular area. Here again we have the problem of the significance of this type of diffuse change. As in man, these animals don't always remain in the best nutritional state. All I can say at this stage of the game is that no one really has a solid answer to your question.

Dr. Arthur Bieganowski, (Bronx, N. Y.): I first would like to say that it is a little distressing to find that at this international symposium being held in the United States, the only epidemiologic data we have comes from Japan and Sweden. I would like to direct my question to Dr. Bewley, whose words seem most practical today. Dr. Bewley, what do you do with the psyche of your patients? Have you in British medicine succumbed to the illusion that heroin users are like pancreatectomized dogs who need only physiologic supplementation?

Dr. Bewley: I didn't quite get the question.

Dr. Bieganowski: I want to know what you do with the "person" of your patients. I mean, do you just give the patients heroin and write them off as a chronic problem, or do you try to develop their personalities and their characteristics away from the problems which led them into drug addiction? Do you try to return them to society as useful people both to themselves and to your country?

Dr. Bewley: Yes, we do. We don't put much emphasis on psychotherapy because we don't think that it's likely to work and it's very time-consuming. But most of our effort, in fact, is on social manipulation, encouraging people to go back to work, to get into training programs, to lead lives that are more satisfactory to themselves. We have had a small limited success in this; for example, the number of people working has risen from 20% to 40%, which I think is a step in the right direction. Because we don't feel we are just prescribing drugs, we try to do what we do with other patients—help them to lead lives that are more satisfactory to themselves and to work and function socially as well as possible.

Dr. John G. Cowl, (Towson, Md.): Dr. Seevers, in discussing characteristics of drugs, you mentioned antisocial behavior secondary to acute and chronic intoxication during administration and withdrawal of the drug. I wonder if you could describe some of the antisocial behavior that you have observed or experienced in the administration or withdrawal of agents such as marihuana, hashish, the hallucinogens, and even some of the harder narcotics.

Dr. Seevers: It's pretty well established that theft and other crimes involving acquisition of drugs or money to buy them occur during withdrawal. In other words, the dependent person may turn to crime in order to relieve the signs and symptoms.

Dr. Cowl: Well, it's my understanding that no withdrawal symptoms have ever been attributed to the hallucinogens, including marihuana and a number of other stimulants that you described.

Dr. Seevers: That doesn't mean that an individual will not try to get the drug if he doesn't have it. Withdrawal signs are psychologic as well as physical. A dependent individual may have a strong psychologic drive to get to these by theft or any other mechanism, regardless of whether there are specific identifiable withdrawal symptoms or signs.

Dr. Cowl: In the case, then, of the psychologic dependence that we so frequently hear about, specifically what kind of antisocial behavior, aside from getting the drug, is seen during withdrawal from marihuana or its administration?

Dr. Seevers: We are talking about hashish, not marihuana.

Dr. Cowl: Or hashish, then?

Dr. Seevers: Hashish is a horse of a different color than marihuana. Why is it proscribed, for instance, in India and in Egypt, even though bhang, which is a very weak preparation, is permitted? Hashish produces types of antisocial behavior that even the Indians don't want.

Dr. Cowl: Well, I assume, then, that since you have not described specifically any antisocial behavior attached to withdrawal from marihuana that you don't know of any personally. I'd like to ask two questions of Dr. Cameron. First of all, is marihuana still considered a narcotic by the World Health Organization?

Dr. Cameron: I think the World Health Organization has never considered it to be a narcotic in the pharmacologic sense. It is true that it is controlled under the Single Convention on Narcotic Drugs but I might say that this also controls cocaine. I doubt that anyone ever suggested that marihuana was a narcotic in the pharmacologic sense even though it is under international control as a consequence of being included in a convention on narcotic drugs.

Dr. Cowl: Well, then, sir, can you please explain why those three particular classes of drugs are described under the same convention?

Dr. Cameron: I believe this is perhaps a result of the historical development of international controls. The first such controls of dependence-producing drugs involved the production of opium and coca-substances of plant origin. The major concern was with opium. Synthetic narcotics and cannabis were brought under international control at a later time. By the mid-1950's there were a number of international instruments providing a variety of controls for these substances. The Single Convention on Narcotic Drugs, 1961, was adopted in an attempt to codify all the existing control instruments. Consideration is now being given to the international control of certain substances producing dependence of the amphetamine, barbiturate, and hallucinogen type. It is generally agreed that a new protocol will be required to achieve this end if international controls are to be imposed.

Dr. Cowl: Dr. Cameron, you mentioned that in regard to the deviant use of drugs, a country or society determines whether the use of the chemical is deviant or not, and what is deviant for one society may not be for another. This holds true for societies. It would seem that this is not a medical problem but a sociological one, and therefore I wonder why we spend so much time dealing with it when it is not really within the province of medicine.

Dr. Cameron: I think few would suggest that dependence on these drugs is purely either a medical or a social problem; rather there are both medical and social aspects involved. However, we can speak of only one facet at a time. In relation to some drugs, cannabis and alcohol, for instance, it is not unusual to use them in some countries, but very unusual to use them in others.

Dr. Burton M. Angrist, (New York, N.Y.): Dr. Bewley, if I understood one slide correctly, it showed a striking decrease in compulsive methamphetamine users—down to 0, in fact. Now, when these stimulant substances have been controlled in Sweden and in the United States, we have seen the rapid emergence of illicit labs. Can it be that the British are less resourceful? Why is it that you are not seeing continued use? I am curious about your description of the current "speed" scene in England.

Dr. Bewley: If I could very briefly say something about the problem of methamphetamine, which in one way was hardly a drug problem at all. We had two physicians who prescribed irresponsibly; one of them prescribed 24,000 ampules of 30 mg methamphetamine to 100 people in the course of a month. We obviously have a problem here, but this is essentially a physician problem rather than a drug problem. In fact, by dealing with this particular physician and also by restricting the supply of methamphetamine ampules to hospitals, the problem was controlled. This was a voluntary restriction on the part of the medical profession and the manufacturers. What happened when methamphetamine was withdrawn in this way was that a certain number of people moved on to other amphetamines and methylphenidate, or Ritalin in tablets which they broke up and injected. There has also been a trend among this group of people to use barbiturates intravenously. Intravenous barbiturate use tends to be self-limiting, because it is painful. It leads to thrombosis in the veins and sterile abscesses, and although we had a wave of people using barbiturates intravenously, this seems to have settled down. There is rather less of this particular kind of misuse. At the moment we have to say, "Wait and see," because we don't know what is going to happen next. I think one has to be careful that when he deals with a particular misuse of drugs, he isn't merely replacing it with something else.

Mr. Robert Winfield, (Ann Arbor, Mich.): Dr. Goldberg, in your data you presented a number of prevalence rates which indicated that the older population was using depressant drugs while the younger population was using another set of drugs, including marihuana. The implication, though it was never stated, was that this may have something to do with the longitudinal development of drug habits. You didn't present any data on where the drug user starts in your culture, what drugs he progresses to, and what the outcome is. I wonder if you could say something about that.

Dr. Goldberg: Time did not permit me to elaborate on the problems of choice of the initial drug of abuse and of the outcome of continued drug-taking. Studies having a bearing on these problems have been made, however.

A number of prevalence studies covering various strata of the population, interviews regarding drug of choice and initial drug choice, a longitudinal study now in progress on juveniles in Uppsala carried out by Dr. Stålberg, and several follow-up studies have shown that initial drug choice, possible progression, and ultimate outcome depend, among other things, upon age, availability of drug, and the types of drugs taken initially and those continuously abused.

The majority of juveniles in Sweden who continue smoking cannabis, mainly hashish, after 1965 have started with smoking cannabis, whereas most of those injecting stimulant drugs—the amphetamines such as amphetamine (racemic form), dexamphetamine and methamphetamine, phenmetrazine (Preludin) and methylphenidate (Ritalin)—have also started on stimulant drugs. A fair proportion of those on cannabis discontinue its use altogether, some continue with smoking, whereas others progress to stimulants, taken orally or by injection, or to injection of opiates.

Before 1965, drug-taking among juveniles often started by inhalation of volatile solvents ("sniffing")—trichloroethylene or thinner; in this age group today, cannabis smoking seems to have been substituted for "sniffing."

The members of the older generation still seem to start with the oral intake of hypnotics and sedatives, and only rarely progress to injection of stimulant drugs. It has been observed, however, that in wards where younger and older patients are brought together, young drug abusers have made older patients begin injecting drugs, necessitating a separation of younger and older drug abusers under certain circumstances.

With regard to outcome, spontaneous discontinuation, even by persons abusing drug use severely, has been observed. In those continuing the abuse, especially those injecting stimulants, a high incidence of complications, ranging from sepsis and infectious hepatitis to severe psychoses, and a high mortality rate, due to suicides, overdose, accidents, and intercurrent diseases, have been noted.

Details will be presented in the final version of my report.

Dr. Hugh D. Kirkpatrick, (Charlotte, N.C.): Dr. Bewley, I would like to know your personal opinion of certain American programs such as that functioning at Odyssey House, where maintenance, whether on heroin or methadone, is considered to be the wrong approach to detoxification and rehabilitation. In the words of a junkie I worked with this summer, "Do you think that methadone maintenance or heroin maintenance is not society's cop-out?" I am not sure you're familiar with that term; it's a street term which means "passing-the-buck," or "throwing-the-pound," perhaps.

Dr. Bewley: If I understand your question correctly, I think the answer probably is "yes."

Dr. Stephen Wilson, (Ann Arbor, Mich.): Dr. Bewley, what are your criteria, if any, for short-term withdrawal therapy instead of maintenance therapy for addicts? And secondly, you showed a beautiful slide of boxes upon boxes of disposable syringes. How do you control the use and re-use of those syringes in your patients?

Dr. Bewley: To control re-use, we supply a specific number of syringes; if a person is using heroin and taking it six times a day, we would give him 42 syringes a week. Of course, he may not ask for more for a month's time, so he would obviously have been re-using them. But we do endeavor to get people to use disposable syringes.

Dr. Wilson: What I am saying in essence is that these are the type of syringes that get into the black market of illegal drug abuse also.

Dr. Bewley: Syringes can be bought over the counter, and people can also make them at home. So we don't really feel that by cutting down on disposable syringes we are going to affect this situation. Also, we feel that it is reasonable to provide syringes in this way if people are going to inject themselves anyway.

Dr. Wilson: How about question number one? Do you have criteria for short-term withdrawal therapy instead of maintenance therapy?

Dr. Bewley: It very much depends on what the patient himself wants to do. We encourage people to discontinue the use of all drugs because we think that this is a healthier and more sensible way of living their lives. If they feel that they can do it, we will encourage them to, usually as an outpatient. We will gradually reduce the amount of drug that they are taking, and then change the drug to oral methadone. We gradually reduce the methadone dosage, depending on the person's total situation, how he is living, how he is functioning, whether he is feeling comfortable in his lot—this kind of thing. We encourage the individual to "come off" all drugs but we haven't got any hard and fast rules about when he does it.

Dr. Beny J. Primm, (Brooklyn, N.Y.): I would like to question Dr. Seevers and also Dr. Cameron. The New York area has been subscribing to the theory that there is a metabolic need for an opiate once an individual becomes chronically addicted. I would like some comment concerning whether that theory is true because the greater New York area has now begun methadone maintenance programs in 22 hospitals.

Dr. Seevers: I don't know of any evidence to support this hypothesis in any way. In the first place, what is meant by a metabolic need?

Dr. Primm: Some of the literature from the Dole-Nyswander school of thought subscribes to the fact that once you become chronically addicted to heroin, there is a metabolic need, and since that is so, methadone maintenance *ad infinitum* is the treatment of choice for this individual.

Dr. Seevers: It means nothing to me. I don't know of any evidence to support it. I know Dole very well, and I know his hypothesis, but a metabolic need implies that methadone is a natural product of body metabolism. I am certain heroin isn't and neither is methadone. One comment that might be made is that there is absolutely no basic difference between methadone and heroin except in duration of action. Fundamentally, there is no other difference as far as one can determine.

Dr. Primm: Would you comment on that, too, Dr. Cameron? The problem is that in the greater New York area we are subscribing to one kind of theoretical knowledge that has been provided by research. If you people feel differently, why hasn't something been written to refute that kind of knowledge that is printed in the literature, to give the practitioner some guidelines to work with when directing a methadone program based on the theory that people have a metabolic need for opiates once they have become chronically addicted?

Dr. Cameron: All I can do is echo what Dr. Seevers has already said. I know of no data which would firmly support such a hypothesis. Further, one could point to some data suggesting that the hypothesis might not be valid. For example, there are a number of people who "come off" drugs and "stay off" drugs for the rest of their lives. This would suggest that at least they did not have a metabolic defect. But the question of whether there is or is not a metabolic defect is perhaps not really fundamental and crucial to the use of methadone in the management of certain patients.

Dr. Henry W. Elliott (Irvine, Cal.): Since Dr. Dole isn't here, I'd like to say that I don't think he actually means a real metabolic need in the biochemical sense of the term. He talks more of a "drug hunger" and this is not to him a real biochemical thing.

Dr. Cameron: I think one further point needs to be made. The questioner noted that methadone was being used in New York City and may have implied that this was the only approach in that city. Of course, this is not the only approach because there are many other programs that use narcotic antagonists while others use individual psychotherapy and group programs.

Dr. James N. Oss (Montclair, N.J.): Dr. Seevers, one of your slides listed three general types of psychotoxicity that created antisocial behavior. The third one showed that there was mental or physical lethargy, or both, which led to inability or refusal to assume individual or social responsibility. Would you please expand your reasoning concerning that kind of antisocial behavior?

Dr. Seevers: Well, it's similar to the reason we don't want the chronic alcoholic, although we have about ten million of them. They become a serious social problem which the rest of the population must support. In places where other drugs such as hashish are involved, the same thing is true. Three percent of the Egyptian and 8% of the Afghanistan population are heavy hashish users. Even these governments don't want a dead load that they have to support, and that is about all I am saying.

Dr. Oss: Can I just pose what might be a rhetorical question: Why does American society accept a certain kind of social disability while discriminating against other kinds of social disabilities, refusing in fact to support them? Is there a logical or valid reason for this kind of separation?

Dr. Seevers: Logic or not, it is society's decision. We had prohibition against alcohol, but society decided that they were willing to accept the bad with the good. Whether this was a good decision or a bad decision is not the point. I believe we do not want another dead load on this society even if we could afford it. We are probably the only country in the world that could afford it, but I don't believe we want to do so. No other nation in the world has ever wanted such a load.

Dr. Alexander Katz, (Columbus, O.): I would like to address two questions to Dr. Bewley. In one of the charts that you presented, the figures showed an increase in opiate addiction that seems to have leveled off when the clinics opened. Would you expand on that?

Dr. Bewley: I would just say that hopefully maybe the clinic system, a kind of institutionalized way of prescribing drugs, is working better than the free-for-all that we had previously. But I think it is too soon to do more than echo the Home Office which suggested that these figures be interpreted with "cautious optimism."

Dr. Katz: Could you also expand on your ending comments about why you don't think that the "English system" can be implemented in this country, the AMA notwithstanding?

Dr. Bewley: What I meant was this: what we have been doing in England depends partly on the structure and the way that the medical services are organized. You might look at what we are doing, determine the principles behind this particular approach, and consider introducing those principles. What I am saying is that it is not because we are in a position to prescribe heroin that it works this way with us. If you could prescribe heroin it would not work exactly the same way with you. The point I was trying to make in my presentation was that it wasn't only a question of whether or not we prescribed heroin. The "English system" depends more on the way we organize our services, and on our philosophy of treatment.

Dr. Bejan B. Markovich, (Chicago, Ill.): I would like to ask Dr. Bewley if there are any noticeable differences in the rate and magnitude of social readjustment in the clinic population being maintained on heroin versus that on methadone.

Dr. Bewley: I can only give you a rather vague answer for this one, because I haven't the exact figures and I think they might be meaningless. If a patient is doing badly on heroin we change to methadone so those who are left behind on heroin are doing better. This doesn't mean that I think prescribing heroin is more effective than prescribing methadone. I don't think I can give you a meaningful answer to that particular question.

Dr. Zarafonetis: The hour for our panel has passed. We will have additional opportunities to raise questions throughout the rest of the week, more specifically tomorrow and the following day on the pharmacologic actions, and to get to the more complete management of patients and so forth. Therefore, please save your questions.

Again, I wish to express my appreciation to the panelists for their excellent contributions to the day's proceedings. All will be present for the next several days and I'm sure we will be hearing more from them. Thank you very much, gentlemen.

DRUGS AND SOCIETY

JOSEPH H. BRENNER, M.D.

John Kaplan is a professor of law at Stanford University. In 1966, the California Legislature decided on a thorough examination and revision of the state's criminal laws, last revised in 1872. Kaplan was one of a task force of six reporters, all of them law professors in California, engaged by the legislature to draft initial reports and proposals. By chance, Kaplan was assigned those aspects of the law which have to do with drug abuse and controls. A preliminary tentative draft on marihuana was issued in 1969, signed by Kaplan and the other five reporters. Shortly afterwards, all six reporters were summarily fired by the California Legislature and replaced with a prosecutor from the California Attorney General's Office; it is said that some attempt was made to suppress that report. The California Legislature might have responded in a number of other ways, and we should be glad that in this country these do not include sending Kaplan off to manage a power station in a distant mountain range, or declaring him insane and putting him out of action in that way. However, it is clear that the report on marihuana was so antithetical to the beliefs and attitudes of the legislators that perhaps they had no choice but to fire the reporters and by so doing declare their report incompetent.

This strange episode in California points up the enormous contradictions and complexities that attend any attempt at resolving problems associated with drugs and the drug scene. It also illustrates some of the complexities and contradictions within the minds of legislators, and perhaps even in the mind of Dr. Kaplan. In any event, Dr. Kaplan was sufficiently fascinated by what he had learned that he decided to explore in greater depth the issues raised by marihuana, its properties, its history, its usage, and the mass of social and legal restraints that surround it. What he found he has described and analyzed with remarkable clarity and precision in his book, *Marijuana, the New Prohibition,* published in 1970. It is to sources such as that book that we must turn if we are to attempt to make any sense at all of the drug scene; and of course it is best to listen to those who document well what they say.

I believe that the handling of the drug situation is too serious and too difficult to be entrusted to politicians; what we need now are people who are more directly concerned with individual welfare—physicians, for example—to take it upon themselves to become interested and knowledgeable in this field, for the use and abuse of all drugs have more to do with medicine than with politics. Despite many assertions to the contrary, a vast amount is known about the various drugs that are illicitly used, and especially about marihuana. To suggest that the gaps in our knowledge of the effects of marihuana are so great as to paralyze our ability to formulate reasonable policies governing prescription and/or proscription of its use, is to allow free rein to all those who advocate the attachment of severe penalties to the use of marihuana.

The literature on marihuana is enormous; the earliest reference to cannabis appears in a work of pharmacy written by the Chinese emperor Shen-Neng in 2734 B.C., approximately 2000 years before the plant reached India. But it is in India that marihuana has been most carefully cultivated, highly valued, and widely used. Indian literature is replete with references to the cannabis plant, the different uses of which have been tightly interwoven with the fabric of Indian life. For at least 2000 years, the drug has been richly used for medicinal, religious, social, and recreational purposes. Marihuana and hashish have been seen as protectors from evil influence, support in times of famine, stimulants to labor, a source of self-confidence, the key to divine revelation, and, by Indians of the 19th century, as the "poor man's heaven," "soother of grief," and "heavenly guide." Indeed, the wide use of marihuana is not surprising in India—a highly civilized country with a complex culture, diverse religions, and exquisite art forms, where appreciation of the esthetic and high regard for individual feelings are by tradition highly valued, and where sensuousness has not withered before waves of self hate.

Even leaving aside the earlier history of marihuana, for much of it could be dismissed as "too unscientific," a reasonable body of up-to-date knowledge has accumulated on marihuana within a scientific framework. While it is true that psychopharmacology is in its infancy, in empirical and practical terms we already know enough about marihuana to suggest major reforms in laws and attitudes that would benefit all of us. It is evident, then, that to further our understanding of the complex interaction of drugs and society, we must examine information and issues that are usually considered outside the purview of medicine. I have found it both arbitrary and convenient to divide any such study into four components. The first component I have already referred to: we should be aware of and be prepared to look at the history of the use of drugs throughout the world in different countries, where, in differing degrees, there is involvement with sociology, politics, religion and so on just as in this country. Although countries, like individuals, are usually reluctant to learn from one another, we cannot afford to ignore information from other countries about drugs, their use and abuse, their sanction, and their control.

The second component we must look at is the particular historical unfolding of attitudes toward drugs in the United States. We have to understand how it is that particular drugs in this country came to be regarded as they presently are by legislators, by the law, by the police, by parents, by schools, and so on. Here we will find contradictions and paradoxes, wrong assumptions, and laws and practices based on flimsy evidence; contradictions that leave us wondering with incredulity why more has not been done to reform the laws on drugs, particularly as they refer to marihuana.

The third component is the drugs themselves: their origin, their chemistries, their pharmacology and psychopharmacology; how they act, at what sites in the body, what psychologic and religious meanings they might have for the individual and for society. This leads to the fourth component: the motivations of those people who take drugs. We must try to understand why it is that some people turn to drugs and not others. What is learned is that most people who turn to drugs do so not for the pursuit of sheer pleasure, but for relief from pain and inner anguish. We realize, for example, that hippies and other street people do not exactly enjoy continuous days and nights of delight. They are, in fact, where they are because they are lonely and frightened, seeking and searching, and often when they turn to drugs they do so more to escape loneliness and pain than for any other reason. I will now make a few random observations on each of these four components.

FIRST COMPONENT

I would like to begin with three anecdotal observations of a cultural/historical nature. Early in the 1950's, I read a report in an English newspaper that the then Prime Minister of Burma, U Nu, temporarily stepped down from his office so that he could go to the mountains to contemplate. He handed over his duties to his Deputy Prime Minister; his action brought about no crisis in government, nor, it seems, did his colleagues think he was losing his mind. He said that he would return within a certain time, took no aides with him, and then journeyed to the mountains to think and reflect, free from the distractions of office and social life. After several weeks, he returned and resumed office. Most would agree that such an event could happen only in a culture where values or qualities such as reflectiveness, contemplation, self-examination, and other-worldliness are held in high esteem. While we know that one culture can be very different from another, we should strive to understand why it is that the same drugs are adjudged differently in different parts of the world. Alcohol, for example, is not looked upon too happily in Eastern nations while drugs like marihuana are looked upon, if not with favor, at least with little or no disfavor. There is, clearly, an ethic, a set of values in the Far East and the Middle East that is very different from our own. In the West, generally speaking, we value aggressiveness, getting ahead, something called progress. We desire to master the outside world, to harness nature, to control the physical environment. Alcohol feeds this desire, making control and power seem more important, while decreasing concern for others and for the environment. In those Eastern countries where reflectiveness, passivity, and the ability to look inwards are valued, we find that drugs that can help a person to such a frame of mind are not usually frowned upon, provided they do no obvious damage to the body or mind or to the social order. So there appear to be deeply rooted cultural reasons why alcohol has been looked upon less happily in the East while being accepted and even embraced by the West.

The next anecdotal comment takes us to India, where during the halcyon days of the British Raj, ladies from England would often take their afternoon tea with marihuana tea cakes. In fact, one of the cookbooks written by Mrs. Beeton, published, I believe, in 1924, contains recipes for the use of marihuana in cooking and baking—and a more respectable writer of cookbooks than Mrs. Beeton there never was. It would be difficult to determine whether the tea-time habits of those days in India helped sustain the British empire for so long or whether they helped in its demise. But there it is, and there is no record of a steady train of English ladies invalided back to England with cannabis psychoses.

For the third anecdote, we move to home territory and Margaret Mead. A few months ago Margaret Mead asserted that marihuana was less toxic than tobacco and milder than alcohol and suggested that an adult with a cocktail in one hand and a cigarette in the other who tells an adolescent not to smoke pot is not likely to be listened to. Governor Kirk of Florida immediately issued an angry rejoinder: "In our schools we teach our children patriotism and respect for the laws and then along comes this dirty old lady on television." Apart from the style, this is not too different from the reaction of the California Legislature to Professor Kaplan.

SECOND COMPONENT

Let us consider the historical unfolding of the drug scene in this country. Before the Harrison Narcotic Act of 1914, there were no effective controls over the use of narcotics in the United States. Indeed, through the 19th century and up until 1914, narcotics had been used in great quantities; a large number of patent medicines containing opiates were freely advertised and could be bought openly and cheaply. It has been estimated that before narcotics were brought under federal control by the Harrison Act, one in four hundred of the population of the United States was addicted to opium or its derivatives. Among those addicted there was probably a preponderance of white women who were poor and had no easy access to physicians. The simplest remedy to ease physical pain was to buy a cheap narcotic preparation, and frequent use often led to addiction. At that time there was no black market in narcotics and no known correlation between their use and the crimes of the urban poor. After the Harrison Act went into effect, the rate of addiction is said to have fallen to one in four thousand. These figures cannot be assumed to be accurate, or even nearly accurate, because of the enormous difficulties in obtaining such information, but unquestionably the Harrison Act greatly reduced the number of people addicted to narcotics in the United States. Every effort was made to help those people who had been addicted to opiates to get off the drugs, and there is no record of terrible and wide-reaching consequences when these drugs were taken out of patent medicines and put on prescription. As an aside, I happen to believe that it is not that difficult to stop taking heroin, provided that the reason the heroin was started in the first place was not related to a sense of complete hopelessness and despair. The poor white housewife from a country town in Michigan could make the transition away from opiates in 1914 more readily than some of the present "junkie" inhabitants of our ghettos. (Testifying at a recent trial in New York City, a witness told of becoming addicted to heroin at the age of thirteen: "Heroin helped my nose not to smell the urine-soaked hallways, I didn't feel the garbage underfoot, I didn't hear the sound of police sirens tearing through the black jungle.")

Marihuana, of course, when viewed with even minimal scientific objectivity cannot be lumped with the opiates and called a narcotic, for it has very different pharmacologic properties and clinical effects. How then did it come to be classified as a narcotic? Until the late 1920's, marihuana was in common use only in the southern and southwestern regions of the United States and most marihuana users were Mexican-Americans. But gradually, during the 1920's and 1930's, the use of marihuana began to spread among the poor, predominantly in rural areas where the hemp plant, which is the source of marihuana, was readily available. Smoking the crushed leaves of the female plant, and to a lesser extent those of the male plant, produced in the user a pleasant sense of euphoria and relaxation, often without a hangover or other unpleasant side-effects. Smoking marihuana was cheaper than smoking tobacco, and was, in fact, absolutely free to most of its users. It soon came to be used by many for its calming and tranquilizing effects, no small thing for fearful men and women. Eventually, the use of marihuana spread from the rural areas to the cities, but, again, predominantly among the poor. At the same time smaller special groups, among them artists and jazz musicians, began smoking marihuana not only for the relaxation and euphoria that it produced, but also for particular effects on their aesthetic sensibilities, or so they claimed.

One can only speculate on the reasons for the massive attack on marihuana, spearheaded by the Federal Bureau of Narcotics and supported by Congress, that took place in the 1930's. At the time that this attack was mounting, marihuana was not smoked by students or middle- and upper-class whites, but in the main by poor blacks and poor whites whose roots were in rural America. The Depression, which brought mass migration from the country to the cities, accounted partly for the upsurge in marihuana use in the cities as well as a number of other social and political changes. The country was beaten down; massive unemployment, hunger, malnutrition, and despair followed the collapse of the economy. An anxious, nervous Congress was told that marihuana was eating its way into the very soul of the nation, ravaging bodies and minds—a clear and present danger to society. And if the Depression was not so easy to abolish, the use of marihuana certainly could be forbidden. The campaign to outlaw marihuana was led by the Federal Bureau of Narcotics, a government agency created in 1930. The new Bureau, headed by Harry J. Anslinger, subjected the press and Congress to a steady stream of alarming statements concerning marihuana—statements that were frightening, lurid, and sometimes tantalizing. Again and again the accusations were made: marihuana drove people to violence, crime, sexual excess, and even madness; it sapped the will, destroyed the moral fiber, and frequently led the user to opium and heroin addiction. Marihuana was a terrible curse visited on the nation and its use had to be eradicated.

In 1937, the Marihuana Tax Act was passed, attaching severe penalties to the possession of even the smallest amount of the herb. The spate of legislation that quickly followed, state by state, thoroughly criminalized marihuana. In recent years, with the rapid increase and widening use of marihuana, the police have found themselves in impossible situations where they must do things that are inconsistent. On the one hand they are asked to enforce laws against a crime that is committed in private and where there is no victim, and on the other hand they are continually reminded to take careful note of constitutional guarantees that exist to protect the privacy of the individual. The confusion of the police is matched by the dilemma of the courts, where the censures and punishments that can be imposed on the user or dealer of marihuana range so widely that the power of the courts ceases to be discretionary and becomes arbitrary. For example, in some states certain marihuana offenses might—at least as the law is written —make the offender liable to the death penalty. What actually happens in those states and others, if the decision is made to prosecute, is that the range of punishment extends from probation to a long term in prison for seemingly identical offenses against the marihuana laws.

There are other curious matters in the history of marihuana in this country; one of them is tied to differing views on the right of the individual to pleasure. The pursuit of happiness, it would appear, is a right that is not altogether inalienable in the United States, for attitudes that embrace hedonism conflict with the dominant Protestant ethic which avers that all rewards come from hard work. Note the reasoning of the Massachusetts Superior Court in a test case brought three years ago to test the constitutionality of the state laws on marihuana. In this case, the two accused were apprehended at the Boston Airport with several pounds of marihuana. This particular action was brought to test the constitutionality of the law without reference to the guilt or innocence of the accused. Counsel on both sides invited experts from all over the country to testify on the nature of marihuana and its effects; as far as I know, until that time no other court had given marihuana, the real defendant, such a long and detailed hearing. The judge concluded, thoughtfully but unconvincingly, that on the basis of all the evidence before him it seemed likely that marihuana was a dangerous drug and that the Massachusetts' laws were, indeed, not unconstitutional. In the discussion leading up to his ruling, the judge concluded that an individual does not have the right to pursue pleasure in any way he chooses, even privately and without affecting others.

That kind of statement tells us that, despite the separation of church and state in the United States, there are commonly shared notions of morality which borrow more from some religions than others; this morality determines the nature of some of our laws or at the very least modifies the ways in which some laws are applied. Lawyers and judges say that the law is not the same as morality; yet if we allow that courts and the laws are beholden more to an ill-defined ethic than to morality, it becomes easy to see how prosecutors and courts can, apparently with no felt contradiction, sentence two priests, the brothers Berrigan, to long prison terms for the destruction of draft files. True, they were guilty of militantly symbolic acts, but the response of the courts denies with violence the right of the Berrigans to hold a different and more rigorous standard *within the same general ethic*. All this may appear to be a long way from drugs; perhaps so. However, if we are to understand the ways in which the laws proscribing marihuana operate in this country, we must have some idea of the ways in which our society works. And of course, it works rather well, for some.

THIRD COMPONENT

Now let me move for a moment to the third component that needs to be studied, the drugs themselves. Usually I am much more comfortable talking about the drugs themselves for I am a physician, not a sociologist. My experience then, as a physician, suggests that in our consideration we should long since have separated marihuana from the other drugs of abuse—from the opiates, from hallucinogens like LSD, and from amphetamines and barbiturates. Nor should we confuse marihuana and hashish, the resin of the cannabis plant; marihuana is marihuana and hashish is hashish. A glass of beer is one thing and a bottle of scotch another. And I believe that we now have enough evidence to say that marihuana is a relatively innocuous drug.

Hallucinatory drugs like LSD and mescaline require a comment of a different order. It is possible to say—and many do—that those people who tell us that they experience a profound mystical experience or derive insight from the use of LSD are deluding themselves. Viewed in this way, the acid user is judged to have arrived at less rather than more insight and may even be considered psychotic. However, just as when a patient comes to us complaining of a headache and, upon examination, we find no sign or cause for the headache we cannot tell him that he does not have a headache, thus we must grant that if someone says he is having a particular kind of experience of the mind that is filled with rich meaning and significance for him, then he may indeed be having such an experience. And the experience may indeed be endowed with all the qualities and consequences that the individual describes even though we cannot perceive, let alone measure them. At the very least we should allow the possibility that what he claims has happened might well have happened, even while we remain unconvinced. The ineffable is, after all, uncommunicable. Upon reading Aldous Huxley's eloquent account of experiences with mescaline in the *Doors of Perception; Heaven and Hell* one is left strangely unsatisfied; the book is fascinating but provides no contagion of ecstasy to the reader. One is left with the impression that a trip, on LSD or mescaline, is the only route to discovery of what Huxley and others have found. It is worth noting that several leaders in the fields of philosophy, mysticism, and theology share Huxley's view that a mystical experience artificially induced by a drug is no less a mystical experience for being induced in this way.

In 1962, on Good Friday, Dr. Walter Pahnke carried out a remarkable double-blind experiment with the cooperation of twenty Christian theological students. Ten received Psilocybin and ten nicotinic acid as a non-hallucinogenic control (the nicotinic acid caused various physical symptoms to make the subject feel that he had taken a hallucinogen). Extensive data were collected during the experiment and for several months afterwards. Nine out of the ten who had taken Psilocybin reported a deeply felt religious experience; only one of the others had similar feelings. Just as remarkable was the fact that during the several months of follow-up, those who reported the initial religious experience felt that their lives had been significantly changed with new reaches of self-understanding and deeper religious insight.

It is clear that we must pay a great deal of attention to drugs that can alter the state of the mind not only because of those drugs that already exist, but because chemists all over the world are hard at work on a wide range of psychoactive drugs. The number of new substances that can be synthesized for potential use by human beings is limitless. What, then, is to happen if drugs are developed which induce the pleasurable symptoms, the profound insights, and other desirable effects claimed for the acid trip without any of the unpleasant ones and without dangerous consequences? Such drugs are bound to present us with serious moral dilemmas—dilemmas surely worth pondering now.

Before continuing, I want to read from a letter to the *London Times* written by Christopher Mayhew, a member of the British House of Commons, describing an early "psychedelic experience" in the middle 1950's when he had taken "an LSD-type hallucinogenic drug to test a theory:" "I experienced the beatific vision, eternal life, heaven. It was all there as the saints had described it—ecstasy, timelessness, illumination and unity, or if you prefer it, depersonalization, time disturbances, light hallucinations and the disintegration of the ego." An amusing commentary that pokes fun at 'objectivity' while emphasizing the power and the subtlety of language.

FOURTH COMPONENT

The fourth component of the drug scene that we must try to understand is what goes on in the minds and lives of those who take drugs, what motivates them. A good way to do this is to listen carefully to what drug users have to say, while we guard against interpreting too rigidly what we hear within one or another theoretical framework. A surer guide to understanding is a willingness to refer what one hears to one's own life experiences. For example, consider these excerpts from two tape-recorded interviews with adolescents. The first adolescent is a young resident of a northern ghetto, a fourteen-year-old black youth:

> I've been smoking pot so long I couldn't give you a figure on how long—five years, I'd say, more or less. I don't keep track. I don't count the times, the years. My cousin and his friends, they started us. They gave us the stuff. They taught us how. My cousin, he said to me: "George, you've got to know how to leave, man, you really do." I didn't know what he was meaning. I told him he was talking crazy. He said I could find out, if I wanted, if he was crazy or not. He gave me the cigarette and I took it in, the pot, and I liked it. I saw what he was saying. I got a little high, I must have. That was a long time ago. I don't remember that far back; I don't remember everything. All I know is this: you use the stuff, and you get real loose and you're off on a cloud a little. You're gone. You've left. You see?
>
> I like pot. It's good. You don't go *too* high. You're cool. It's good to use when you want to cool yourself down. My friends, all of them, use it sometimes. Some are fooling around with stronger stuff, but not me. I see too many people around here that are hooked. They're ruined They're done, all washed up. I've taken other things. I've experimented, you know. But I don't go for those big "highs," not like you hear people describe them . . . The people who drive big cars, they don't use a lot of drug stuff. They sell to others. You got to be smart. My old man knows the ropes around here. He says to let other people want to buy, and you be ready to sell what they want, and that's how to get money. He's right. . .
>
> I've never used a needle; I wouldn't let a needle get near any vein of mine. I'd kill someone who tried to get me hooked. I like pot, but that's no needle game. Who the hell thinks pot will lead to the needle? That's up to you. If you're heading for the needle, you are. If you're not, they couldn't make you touch a needle even if *they* pay you—and there are pushers who will. They'll tell you: look, the

first few times we'll give you the stuff free to try, and we'll give you some money, so you don't have to work, and you can relax and enjoy yourself. You know what I say? They can try someone else. I let them know. They know anyway, though. They know who's ready and who isn't. One pusher I know better than the others (he's my brother's friend) told me it takes him half a minute to figure who's going to get hooked and who isn't. He can read their minds and see what their minds say by the look in their eyes, he says.

What does smoking pot have to do with the needle? Nothing. . . Unless you've got the hunger in you and you're fooling around. . . . you're looking for something to take for the hunger, you'll never be happy until you're hooked. Then you stop smoking pot. If you're on the needle, you don't want pot. If you take pot and you're not satisfied, you go to the needle. If you take pot and it's what you want, you stay right there. I'm right here. I'm staying right here. No one's moving me. No one's going to push me. No one's moving my friends. No one's pushing them. We harvest the grass and burn it up and that's all we're after—except money, you know. The trouble is, there's not a lot of money around here—more grass and more needles than dollar bills, I believe that.

Some ten miles away—ten very long miles—a fifteen-year-old youth also talks about marihuana. Although the two speakers are about the same age, both Americans, and almost neighbors, this youth is white, middle-class and well-to-do.

I've tried pot. I don't use it all the time, just certain times. That's a lot of crazy talk. We go to someone's house and we turn on. What's so bad about that? We like to do it. You feel good. You're free. You don't lose your mind. You don't have a hangover. You just feel good—good and free. Is there anything wrong with feeling free? Anything bad? Why shouldn't a guy take in a few deep breaths, inhale some grass, then lean back and feel good? I don't see anything, no images or pictures or anything. I don't hear anything—nothing. All that happens is that I feel myself loosening up. My muscles relax; I feel good, that's all I can say. I feel free.

Sometimes I feel better than other times. Maybe it's the pot, how good it is. Maybe it's just me. I wouldn't take pot before an exam, or if I were going for an interview at a college. If I did, I might start asking myself *why*—why I memorize all the junk I do in school, and why everyone climbs all over everyone else to get ahead, always to get ahead. When you start thinking like that, you're in danger of dropping out. I don't want to drop out, but I have the right to stop sometimes and go slow. That's what smoking grass does for me—that's *all* it does, but that's a lot. I mean it helps me cool down and look, really look; it slows the world down for you, so you can ask yourself and your friends what you're doing and what you're after. I told my father he ought to ask himself questions like that. You know what he said. It's predictable, it really is: "I haven't got the time. You won't either when you're my age, and you've got a family to support and all the tuition bills of your kids."

I asked him why he wanted us all to go to fancy schools and colleges, and he said we're spoiled, asking questions like that. We should be glad we're there. I am. I don't argue with him on that score. He's right. But you can't be blind all the time—and work, work like him, my dad. I get good grades. I'm not on top, but I'm not on the bottom. I play a good game of squash. I'm on the baseball team. I make out; I make out pretty good. Most of my friends use pot; they smoke grass and so do I—not a lot, but sometimes, once a week or twice a month, it depends. If my father knew that, he'd think I was a criminal or something. I could tell him I'm doing fine in school and was on the baseball team, but he wouldn't care. He says if you start with drugs, you end up in jail and you're finished for life. I hope I never get like that. I hope I never get so old I'm completely out of it. I hope I keep my cool on things.

You can't talk to older people about drugs. In school they give you this story that "one thing leads to another"; that's all they tell you. And if they're not saying that, they hit you with your 'problems'—your problems, your problems; that's all they talk about ("You're going through a difficult time," and all that). If that doesn't work, if you tell them there's nothing wrong with you, nothing troubling you, and you're not going to take heroin—tomorrow or any other day—they'll tell you that last thing: you can't violate the law. The law is the law—I've heard that about one million times, I'd say, conservatively speaking. They don't know how funny they sound when they get up and start preaching at you for turning on—once a week at the most. And if you tell them about liquor and cigarettes, the millions of alcoholics and the deaths caused by driving under the influence, and cancer of the lungs, they stiffen up—the teachers, the school doctor, and the psychologists they bring in here—and they say "two wrongs don't make a right," or "I'm not talking about alcohol or cigarettes, but drugs."

> No one has dared tell them to their face that turning on with grass on Friday night with your friends is not becoming a *drug addict.* God, they love to call you that, a drug addict. They seem to think they can scare you or something. They're scared; or they just don't know how it is. They've never smoked grass, and they talk as if we're going to end up criminals, you know, or in a mental hospital or something.

We have learned from statements like these something of the ordinary lives of the young and of their struggle to grow up in this country. Ordinary because war and racial strife and many other ugly things have been with us for so long and so constantly that they appear to be thoroughly incorporated into the American way of life; they are, as someone has said, "as American as apple pie." The inability, so far, of our society to deal intelligently with the drug situation is just one example of the contradictions that confront our young. And just as our wars produce a quota of dead and wounded Americans each week, so inexorably do our drug laws destroy a quota of Americans each week, by incarceration in prisons, by overdose of drugs taken in secret and in fear, and by other more subtle ways.

RECOMMENDATIONS FOR REDUCTION OF MARIHUANA PROBLEM

I will end with some tentative thoughts on how we might go about reducing the marihuana problem to more sensible—and therefore manageable—proportions. Marihuana needs to be decriminalized. This may take time and it might have to come about by degrees, for we all know that the rehabilitation of a criminal is usually long and arduous. But we could, I suppose, borrow from the rhetoric of the time. As with our war in Southeast Asia, we could say that our policies in the drug field seem to be slowly headed towards victory, however imperceptible this may seem to those of us outside government circles. Because of this trend toward victory, there is no longer a need for all-out war; marihuana can then be slowly decriminalized and the "troops" pulled out. Legislators, police, and courts could all back away from drug problems, handing them over to physicians, social scientists, and community representatives who might together go about the business of healing and restructuring not only the wounds caused by drugs themselves, but also the particular piece of chaos in our society that results in policies that do little more than mete out draconic penalties for drug offenses.

To regulate the production, supply, and usage of marihuana, a central federal control board could be constituted—structured perhaps along the lines of the Federal Commission on Civil Rights—which would be strong enough to resist the kind of political pressures that might neutralize its usefulness. The members of the board could be both professional and non-professional, with the boundaries blurred so that "professionalism" and "expertise" could not drown out the voices of some of the members. At the same time, it would probably be sensible to appoint a second group, a group of ombudsmen to monitor the functioning of the board, because otherwise the federal board might, with the passage of time, have a tendency to lean towards serving those who appointed them, neglecting those they were appointed to serve.

The details of any such plan to decriminalize marihuana and perhaps other drugs would have to be worked out over time, and no doubt changed and modified as circumstances dictate. One thing that might be considered would be a heavy tax on marihuana, the revenue from which might be used for the care and treatment of those in trouble with drugs.

A few of the other positive consequences of a move toward decriminalization of drugs would include the release of a great number of law enforcement personnel for other duties and a sharply diminished caseload in the courts, as thousands upon thousands of our young cease to be *de facto* criminals. It is, of course, possible to list negative as well as positive consequences of any changes in our approach to drug problems along the lines that I have sketched. However, I believe that if a new approach is kept fluid enough, it can be worked on and refined, worried about, and tested so that eventually a policy would emerge that is acceptable both to the hawks and the doves of the drug scene.

DISCUSSION

Dr. Gusztav A. Batizy (Brecksville, O.): I worry about this country's youth; I am particularly concerned with Dr. Brenner's statement about a 15-year-old girl who used marihuana and had wonderful experiences from it. I am wondering, Dr. Brenner, at what age do you recommend that people begin participating in this wonderful experience? I also wish to comment that I do not think the drug would be the solution for any minority ethnic group or for any one who feels economically deprived.

Dr. Brenner: I do not think that drugs can be used or should be used to solve social problems. Clearly, drugs are more likely to be taken as a consequence of problems. However, I did make the point that I worry about the notion that people cannot pursue their pleasures, provided they are not damaging themselves or society. As to the age question, I feel it is reasonable to draw the line at 18, by which time, hopefully, much of the turmoil of adolescence has subsided. Today, in most states of India, marihuana is legally available but sales are restricted to those 18 and over. As far as I am aware, and I have talked to many Indians and reviewed the literature, there is no evidence that would stand up to critical examination which suggests that the use of this drug for thousands of years in India has given rise to serious moral, political, religious, or social problems.

What I have not mentioned, and would like to touch upon briefly, are some of the possible therapeutic uses of marihuana. For example, marihuana can be very useful during the second stage of labor and has been used extensively in the Middle East for just this purpose. Marihuana eases labor by allowing the muscles used in childbirth to relax when contraction serves no purpose, so that damage to pelvic and perineal muscles is minimized. It is particularly helpful for women who deliver their children in villages remote from hospital facilities. I also think marihuana could be very useful for those people, particularly the aged, who have a variety of physical disabilities which are chronic and irremediable and who might or might not be psychologically depressed. For such patients, marihuana could offer a degree of calm and relaxation and make sleep easier. These two examples should indicate that the possible therapeutic uses for marihuana have not been completely explored.

Dr. James Jackson (Muskegon, Mich.): As a black physician, I am really discouraged this first day; I hope the remaining days will dispel some of my misgivings. Here we are talking about drug problems and drug abuse, which we have seen the white community throw on to the black community in years past. Now, as Malcolm X used to say, "the chicken comes home to roost." Yet nowhere on this program do I see a discussion of racism in relation to our drug problem. Many black physicians are experts on the relationship of drugs to sociological problems. Those who put together this international symposium seemed to ignore the black situation and racism in relation to drug abuse, as well as to other problems. Unless these things are considered in relationship to racism in the United States, it becomes a joke to even talk about what we are going to do in regard to the drug problem.

Dr. Brenner: Anyone involved with drug problems in the United States knows only too well that racism has determined the particular pattern of enforced poverty and depression, despair and hopelessness that in turn has determined the high incidence of addiction to drugs in the ghettos. It is worth noting that in the first few months of 1951, no less than 260 boys and girls, their ages ranging from 14 to 20, were admitted to Bellevue Hospital in New York City because of addiction to heroin. Other children were killed by the drug and never made it to the hospital. All of those children were black, all were from Harlem, but as the problems of heroin had not yet spilled across racial, class and town boundaries, their deaths made no headlines in the newspapers. The concerns touched on by Dr. Jackson are horrifyingly well-documented by facts.

Dr. Beatrice K. Rose (Portland, Oregon): In view of the fact that drug dependence is the only illness that we treat with abstinence, and in view of the fact that you speak so highly of the need for decriminalization of the marihuana user, I am curious to know why you have not suggested that drug dependence be made a medical problem rather than a crime.

Dr. Brenner: Drug dependence is a problem that is medical as well as legal and social. Physicians as a group have been no more eager than others to tangle with the difficult, painful and generally unrewarding problems of non-medical, usually illicit, use of drugs; in this sense I believe that the medical profession has been delinquent. Until recently, physicians have given little help to the courts, to the police and to the Federal Bureau of Narcotics in problems relating to drugs. The guidance that has been offered has often been biased and polemical. However, the fact that more and more individual doctors, and groups of doctors, some with the support of the federal government, are now willing to move into this very demanding area, is, I hope, indicative of the willingness of society as well as of the medical profession, to shift some of the burden of drug dependence problems away from the police and the courts towards physicians, social workers and the like.

Mr. Joseph S. Hupert (St. Louis, Mo.): I am not a doctor, but in defense of the medical association I would like to point out that when the law was passed in 1937, a Dr. Woodward, I believe, testified for the AMA against its passage. He said the law was ill-conceived and not based on scientific information.

I also wish to mention the economic issue, detailed so very well by Mr. Kaplan in his book. In 1967, there were 51,000 arrests for marihuana in California. This cost approximately 72 million dollars, exclusive of police investigation costs. In 1968, it was estimated that the cost to the state of California came to 100 million dollars. The question which comes to mind is, Can we afford this kind of escalation? Can we really afford it, or whether we wish to or not, might it not be better to tax marihuana and to divert these resources to something else? I would also like to congratulate Dr. Brenner on a moderate and courageous talk which I appreciated very much.

Dr. Robert G. B. Bjornson (St. Paul, Minn.): Concerning the adulteration question, we know, of course, that the street drugs are grossly adulterated. In eleven consecutive samples of what was sold on the street in Minneapolis as mescaline, for example, there was no mescaline found in any of them. There was acid in one degree or another, every one with various adulterants. However, my prime reason for coming to the microphone is to say that I think in a symposium like this, it is essential that we not "fiddle while Rome burns." The issue to which Dr. Brenner has addressed himself today is, it seems to me, central. I rather suspect it is the one which brought virtually all of us here. We all, I am sure, remember Dr. James Goddard and what happened to him when he expressed himself mildly on the subject of marihuana. I think his job as director of the

Food and Drug Administration lasted something like ten days. But this is really our job as physicians. It is our job to be involved and I want to say that I hope in some of the panels that follow we can pursue some of the thoughts which Dr. Brenner has expressed, whether we agree with them or not. This is certainly a good place to try to straighten ourselves out on these questions.

Morning Session
Tuesday, November 10, 1970

Presiding: Henry W. Elliott, M.D., Ph.D.

DRUG ABUSE DETERMINED BY URINALYSIS—AVAILABILITY AND APPLICABILITY OF TESTING SERVICES, DETECTABILITY AND RELIABILITY OF INTERPRETATION OF RESULTS

KENNETH D. PARKER, M.Crim.

The Hine Laboratories, Inc., located in San Francisco, California, has been engaged in the detection of drug abuse by chemical means for approximately 10 years. During this period we have gained considerable experience and have increased our capabilities in several ways:

1. The number of drugs which can be detected has increased.
2. The cost per unit test has decreased.
3. The sensitivity of the tests has increased in terms of the quantity detectable.
4. The interpretations of the concentrations of the drug or metabolites found in blood and urine samples have been facilitated.

Studies have been conducted by our group with subjects receiving known doses of narcotics, amphetamines, and barbiturates by different routes of administration.[1–21] Excretion of the unchanged drugs or metabolites and blood levels achieved have been correlated with drug effects.

Analyses of biological fluids for determination of drug abuse are of three essential types, requiring different degrees of refinement of technique applicable to handling various numbers of samples. They are:

1. Mass screening tests on urine, utilizing thin-layer chromatography.
2. Qualitative and quantitative tests on urine and blood, using gas-liquid chromatography to detect a number of drugs.
3. Special tests requiring the application of microtechniques, such as the determination of morphine or amphetamine concentrations in blood utilizing gas-liquid chromatography.

Mass screening tests require 2 ounces of urine. The 21 compounds most frequently detected by these tests, including eight narcotics of the opiate or opioid type referred to as Group A, six barbiturates referred to as Group B, and seven amphetamines and phenethylamines referred to as Group C, are listed in Table 1.

These screening tests reliably detect the presence of these compounds or their metabolites in the urine in concentrations of 1 to 3 mcg/ml of urine. They are qualitative tests only. They are used principally to check for drug abuse and to ascertain that

Table 1 Detectability of drugs in urine specimens by means of mass screening techniques with thin-layer chromatography

Compound	Concentration in urine (mcg/ml)
Opiates and opioids (Group A)	
Morphine*	1
Codeine	1
Dihydrohydroxycodeinone (oxycodone)†	1
Dihydromorphinone (Dilaudid)	1
Dihydrocodeinone (Dicodid)	1
Methadone (Dolophin)	1
Meperidine (Demerol)	1
d-Propoxyphene (Darvon)	3
Barbiturates (Group B)	
Barbital (Veronal)	3
Phenobarbital (Luminal)	3
Butabarbital (Butisol)	3
Amobarbital (Amytal)	3
Secobarbital (Seconal)	3
Pentobarbital (Nembutal)	3
Amphetamines and phenethylamines (Group C)	
Amphetamine (Benzedrine, Dexedrine)	3
2,5-Dimethoxy-4-methylamphetamine (STP, DOM)	3
Methamphetamine (Methedrine, Desoxyn)	3
Mephentermine (Wyamine)	3
Mescaline (Peyote)	3
Methylene dioxyamphetamine (MDA)	3
Phenmetrazine (Preludin)	3

* Including morphine derived from heroin use.
† Ingredient present in Percodan.

persons on the methadone program are taking that drug.

The cost of these tests depends on number of samples submitted at one time. Individually submitted samples cost approximately $5.00.

Presently, The Hine Laboratories perform 24,000 tests of this type per month or about 1,000 urine samples per working day. Approximately 10% of the samples test positive for one or more of the drugs. The positive samples are retested using a different thin-layer chromatography solvent system for confirmation. The tests are usually performed within one day of the receipt of the samples. Results are reported within three days. We receive specimens from 90 agencies located throughout the United States.

Industrial medical departments use these services to assess the prevalence of drug abuse by employees. Management finds that employees with a history of frequent absences, poor production, serious accidents, or adjustment problems with fellow workers or superiors may be suspected of possible drug use. Many companies are developing educational programs when drug use by employees is confirmed by screening techniques.

The qualitative and quantitative tests on urine and blood using gas-liquid chromatography consist of two procedures. The first, requiring 10 ml of blood, tests for agents of the hypnotic-sedative class, including volatile organics and the acid or neutral organics. Table 2 lists the 28 componds or agents most frequently detected. When the sample is submitted on an individual basis, the cost of the test is $20 if the reaction is negative and $30 if positive. Positive samples cost $10 more because they require additional confirmation and quantitative determination. With information obtained from these tests, an opinion may be expressed concerning the probability that the person was under the influence of the drug, or exhibited diminished capacity, altered behavior, or altered ability to form specific intent. In some cases the cause of unconsciousness or death is specified. This test determines the extent of use and effects, and furnishes information as to the probable behavior of the subject.

The second part of the test analyzes urine for the presence of narcotics and other agents of abuse grouped as amphoteric and basic organics. Table 3 lists a total of 51 compounds of the amphoteric and basic organics class: 12 narcotics, 14 amphetamines and phenethylamines, 16 antihistamines, 6 phenothiazines and 3 phenothiazine isoteres. The cost of this test is $15 or $25, the latter amount for those samples which are positive for one or more of the drugs

Table 2 Abuse drugs or agents of the hypnotic-sedative class which are detectable in the blood

Volatile organics	Acid and neutral organics
Ethyl alcohol (alcoholic beverage)	Barbiturates*
Paraldehyde (Parol)	Carbamates†
Chloral hydrate (Kessodrate, Noctec)	Piperidinediones‡
Ethchlorvynol (Placidyl)	Hydantoins§
Benzene	Benzodiazepines‖
Toluene (glue solvent)	Methaqualone (Quaalude)
Gasoline	

* *Barbiturates*
barbital (Veronal)
phenobarbital (Luminal)
butabarbital (Butisol)
amobarbital (Amytal)
secobarbital (Seconal)
pentobarbital (Nembutal)
amobarbital + secobarbital (Tuinal)

† *Carbamates*
meprobamate (Miltown, Equanil)
ethinamate (Valmid)
tybamate (Solacen, Tybatran)
mebutamate (Capla, Cadlaril)
hydroxyphenimate (Listica)
carisoprodal (Soma)

‡ *Piperidinediones*
glutethimide (Doriden)
methyprylon (Noludar)
piperidione (Sedulon)

§ Hydantoins
diphenylhydantoin (Dilantin)
mesantoin (Mephenytoin)

‖ *Benzodiazepines*
chlordiazepoxide (Librium)
diazepam (Valium)
oxazepam (Serax)

Table 3 Narcotics and other agents of abuse detected in the urine (amphoteric and basic organics)

Narcotics
morphine
heroin (as morphine)
codeine
dihydrohydroxycodeinone (oxycodone in Percodan)
dihydromorphinone (Dilaudid)
dihydrocodeinone (Dicodid)
levorphanol (Levorphan)
methadone (Dolophin)
meperidine (Demerol)
d-propoxyphene (Darvon)
alphaprodine (Nisentil)
cocaine

Amphetamines and Phenethylamines
amphetamine (Benzedrine, Dexedrine)
benzphatamine (Didrex)
chlorphentermine (Lucophene)
diethylpropion (Tenuate)
methamphetamine (Methedrine, Desoxyn)
tranylcypramine (Parnate)
mephentermine (Wyamine)
methyl phenidate (Ritalin)
phenmetrazine (Preludin)
phentermine (Ionamin)
methylene dioxyamphetamine (MDA)
phencyclidine (PCP, Serenyl)
2,5-dimethoxy-4-methylamphetamine (STP, DOM)
mescaline (Peyote)

Antihistamines
antazoline (Antitine)
chlorcyclizine (Perazil)
chlorothen (Tagathen)
chlorpheniramine (Chlortrimeton)
dimenhydrinate (Dramamine)
diphenylpraline (Diafen, Hispril)
doxylamine (Decapryn)
methaphenilene (Diatrin)
methapryline (Histadyl)
orphenadrine (Mephenamin)
phenindamine (Thephorin)
pheniramine (Trimeton)
promethazine (Hiberna)
pyrilamine (Neoantegran)
thonzylamine (Neohetramine)
tripellenamine (Pyribenzamine)

Phenothiazines
chlorpromazine (Thorazine)
prochlorperazine (Compazine)
promethazine (Phenergan)
promazine (Sparine)
thioridazine (Mellaril)
trifluoperazine (Stelazine)

Phenothiazine Isoteres
amitriptyline (Elavil)
imipramine (Tofranil)
nortriptyline (Aventyl)

Table 4 Information pertinent to the determination of common abuse drugs following a single administration

Substance	Group	Usual route of administration	Abuse dose	Duration of action	Body substance for analysis	Period of detectability	Abuse levels
Amphetamine	Central stimulant	Oral	10 mg	12 hours	Urine	24–36 hours	10 ppm
Barbiturates	Hypnotic-sedative	Oral	200 mg	3–8 hours	Blood & urine	24 hours	4 ppm in blood
Codeine	Narcotic	Oral & I.V.	100 mg	2–4 hours	Urine	10 hours	10 ppm
Diethyl tryptamine	Hallucinogen	Smoked	10 mg	2 hours	Urine	?	?
2,5-Dimethoxy-4-methyl amphetamine (STP)	Hallucinogen	Oral	10 mg	10–20 hours	Urine	24 hours	5 ppm
Ethyl alcohol	Hypnotic-sedative	Oral	More than 50 grams (4 drinks) in a short time)	1 hour for each 12 grams	Blood, urine, breath	1 hour for each drink containing 12 grams	1000 ppm (0.1%) in blood
Glutethimide (Doriden)	Hypnotic-sedative	Oral	1.0 gram	3–5 hours	Blood	16 hours	10 ppm
Heroin	Narcotic	I.V.	5 gm	1–3 hours	Urine	10 hours	1 ppm as morphine
Lysergic acid diethylamide (LSD)	Hallucinogen	Oral	150 mcg	6–10 hours	Blood	2 hours (special tests required; no detection by screen)	10 ppb
Marihuana	Hallucinogen	Inhalation	200 mg plant leaves containing 2 mg tetrahydrocannabinol	5 hours	Sputum or washing from hands	Not detected in urine or blood with present methods	?

Meprobamate	Hypnotic-sedative	Oral	600 mg	2–6 hours	Blood	16 hours	20 ppm
Mescaline (Peyote)	Hallucinogen	Oral	300 mg	2–4 hours	Urine	24 hours	10 ppm
Methamphetamine	Central stimulant	I.V.	20–100 mg	12 hours	Urine	24–36 hours	50 ppm
Methylenedioxy-amphetamine (MDA)	Central stimulant	Oral	150 mg	6–12 hours	Urine	12 hours	20 ppm
Methyprylon (Noludar)	Hypnotic-sedative	Oral	500 mg	2–6 hours	Blood	8 hours	10 ppm
Morphine	Narcotic	I.V.	15–100 mg	1–6 hours	Urine	6–18 hours	5 ppm
Myristicin (in nutmeg)	Toxicant	Oral	500 mg	24 hours	?	?	?
Oxycodone (in Percodan)	Narcotic	Oral	8 mg	4 hours	Urine	8 hours	1 ppm
Phencyclidine (PCP, Serenyl)	Hallucinogen	Oral	10–50 mg	4–8 hours	Urine	24 hours	5–15 ppm
Psilocybin	Hallucinogen	Oral	1 mg	2 hours	Urine	?	?
Solvents (glue)	Hypnotic-sedative	Inhalation	5–10 cc	2–30 minutes	Breath (special technique required for collections; call for information), and blood	4 hours	1–50 ppm in breath

and require further confirmatory quantitative tests.

Special testing procedures are used in selected cases for the determination of heroin, morphine, or the amphetamines after intravenous administration or when the drug is taken in other than the usual therapeutic doses. Special microtechniques are utilized on blood samples. These tests require a minimum of 5 ml of whole blood and involve special methods of sample preparation and derivative formation for determination by two or more gas-liquid chromatographic systems.

We have become interested in deaths which occur after rapid heroin injection. We are presently determining morphine levels in the blood and relating them to history of narcotic abuse in order to learn whether a specific blood level is related to death from rapid administration. After the administration of 10 mg of morphine intramuscularly four times a day for five days, we find blood levels in the range of 0.05 to 0.3 mcg per milliliter in volunteer subjects. In contrast, in cases of death following recent intravenous injection of heroin or morphine, we find levels of total (free plus conjugated) morphine in the amount of 1 to 10 mcg/ml. We have data on approximately 60 such deaths to date, representing approximately one year's experience in the San Francisco Bay Area.

Table 4 lists our experience regarding the abuse dose, duration of action, and period of detectability of 21 of the commonly encountered agents. We wish to emphasize that these data reflect minimal values only and detectability by the mass screening tests utilizing thin-layer chromatography. In many instances, the period of detectability would be increased in the event of repeated heavy drug usage. Special tests would offer greater sensitivity. As can be seen, there is no test presently available which will satisfactorily detect lysergic acid diethylamide (LSD) or marihuana use. Tests have been successfully performed on the sputum or washings from the hands of a recent smoker of marihuana and have been found useful for inference of use and knowledge.

In conclusion, chemical tests on urine and blood samples are being used to a considerable extent. They are being requested and utilized by physicians engaged in community action programs which interact with parents, pupils, patients, law enforcement officers, trial attorneys, courts, treatment and detoxification centers, probation and parole systems, and representatives of social agencies and industry. Data developed by such tests help the physician by identifying the drugs involved and determining the extent of use so that appropriate courses of action may be taken.

REFERENCES

1. Parker, K. D.: *Some Effects of Alcohol and Meprobamate, in Combination, on Vision and Psychomotor Performance.* Thesis, University of California, Berkeley, California, 1960.
2. Parker, K. D., and Kirk, P. L.: Separation and identification of barbiturates by gas chromatography. *Anal Chem 33*:1378, 1961.
3. Parker, K. D., Fontan, C. R., and Kirk, P. L.: Separation and identification of tranquilizers by gas chromatography. *Anal Chem 34*:757, 1962.
4. Parker, K. D., Fontan, C. R., Yee, J. L., and Kirk, P. L.: Gas chromatographic determination of ethyl alcohol in blood for medicolegal purposes; separation of other volatiles from blood or aqueous solution. *Anal Chem 34*:1234, 1962.
5. Parker, K. D., Fontan, C. R., and Kirk, P. L.: Separation and identification of some sympathomimetic amines by gas chromatography. *Anal Chem 34*:1345, 1962.
6. Parker, K. D., Fontan, C. R., and Kirk, P. L.: A rapid gas chromatographic screening method utilizing a solid injector in the examination of toxicological extracts for alkaloids, barbiturates, sympathomimetic amines, and tranquilizers. *Anal Chem 35*: 356, 1963.
7. Parker, K. D., Fontan, C. R., and Kirk, P. L.: Improvements in gas chromatography of barbiturates. *Anal Chem 35*:418, 1963.
8. Parker, K. D., Elliott, H. W., Nomof, N., Dewey, M. L., and Way, E. L.: Comparison of the nalorphine test and urinary analysis in the detection of narcotic use. *Clin Pharmacol Ther 5*:405–413, 1964.

9. Parker, K. D., Hine, C. H., Nomof, N., and Elliott, H. W.: Urine screening techniques employed in the detection of users of narcotics and their correlation with the nalorphine tests. *J Forensic Sci 11*:152–166, 1966.
10. Parker, K. D., and Hine, C. H.: Manual for the determination of narcotics and dangerous drugs in the urine. *Psychopharmacol Bull 3*:18–42, 1966.
11. Elliott, H. W., Nomof, N., and Parker, K. D.: Correlation of the nalorphine test with concentration of metabolites of codeine in the urine after single doses and continuous administration of codeine. *Clin Pharmacol Ther 8*:78–85, 1967.
12. Parker, K. D., and Hine, C. H.: Manual for the determination of narcotics and dangerous drugs in the urine. *Bull Narcotics 19*: 51–57, 1967.
13. Parker, K. D., Wright, J. A., and Hine, C. H.: A system for identification of compounds present in the microgram range utilizing gas-liquid chromatography in combination with simultaneous fraction collection and separation by thin-layer chromatography. *J Forensic Sci Soc 7*:162–170, 1967.
14. Parker, K. D., Wright, J. A., Halpern, A. F., and Hine, C. H.: The determination of barbiturates present in blood and urine by gas-liquid chromatography with a peak shift for qualitative confirmation. *J Forensic Sci Soc 8*:125–140, 1968.
15. Parker, K. D., Wright, J. A., Halpern, A. F., and Hine, C. H.: Preliminary report on the detection and quantification of opiates and certain other drugs of abuse as trimethylsilyl derivatives by gas-liquid chromatography. *J Forensic Sci Soc 10*:(1) 17–22, 1970.
16. Parker, K. D., Wright, J. A., Halpern, A. F., and Hine, C. H.: Preliminary report on the separation and quantification of cannabis constituents present in plant material and when added to urine by thin-layer and gas chromatography. *Bull Narcotics 20*:9–14, 1968.
17. Nomof, N., Elliott, H. W., and Parker, K. D.: The local effect of morphine, nalorphine and codeine on the diameter of the pupil of the eye. *Clin Pharmacol Ther 9*:358–365, 1968.
18. Elliott, H. W., Nomof, N., Parker, K. D., and Turgeon, G. R.: Comparison of the nalorphine (pupil) test with chemical tests for the detection of narcotic use. *Calif Med 109*:121–125, 1968.
19. Parker, K. D., Elliott, H. W., Wright, J. A., Nomof, N., and Hine, C. H.: Blood and urine concentrations of subjects receiving barbiturates, meprobamate, gluthethimide or diphenylhydantoin. *Clin Toxicol 3*:(1) 131–145, 1970.
20. Hine, C. H.: The role of the industrial nurse in the detection and prevention of drug abuse. *Occup Health Nursing 17*:15–17, April 1969.
21. Hine, C. H., Wright, J. A.: A program for control of drug abuse in industry. *Occup Health Nursing 18*:(4) 15–18, April 1970.

PHARMACOLOGY OF NARCOTICS AND ANTAGONISTS AS RELATED TO DRUG ABUSE

HENRY W. ELLIOTT, M.D., Ph.D.

For the purposes of this symposium on drug abuse I will limit my discussion of the pharmacology of the morphine-like drugs and their antagonists to actions which are pertinent to the problem under discussion.

We know very little about the mechanism of action of these drugs at the molecular or cellular level, but in the intact organism they are powerful analgesics which can relieve certain types of pain without seriously clouding consciousness. They do this by elevation of the pain threshold, an action which can be measured objectively, and by altering the patient's reaction to pain, an action difficult to measure objectively, but which can be observed and may be described by the patient as "I feel the pain but it doesn't bother me anymore." In sufficient dosage the drugs induce lethargy and sleep. They are tranquilizers in the sense that they make a patient oblivious to his surroundings and yet leave him responsive to direct stimulation. The latter two actions of these drugs, plus the fact that they may produce an extremely pleasurable sensation, especially when injected intravenously, are probably responsible for their high abuse potential. In the absence of pain there can obviously be no alteration of reaction to pain; instead many people experience a distinct euphoria which may persist for several hours. The euphoria may be pure pleasure or the pleasure associated with relief from an unpleasant state. The tranquilization alluded to above may also be pleasurable.

There are, in addition, several other actions of these drugs on the central nervous system which are pertinent to the present discussion. For instance, they depress the respiratory center and death may ensue if a drug user unsuspectingly injects an overdose of a narcotic. They also produce pupillary constriction or miosis by a centrally mediated mechanism; and this can be used to recognize an individual who is on narcotics and is the basis for the pupil test for narcotic abuse.[1] These drugs also stimulate the vomiting center and some drug users recognize an adequate dose by its ability to produce nausea and vomiting ("good sick"). In fact, all narcotics and antagonists are central nervous system stimulants in high doses even though in man they usually kill by respiratory depression. Under certain conditions, convulsions have been reported following administration of narcotics or antagonists and it is always possible that a new drug will be developed which will be less depressant to the respiratory center, and thus unmask the stimulant effect. The stimulation that some people experience following a dose of narcotic is not related to this action. Instead, it is probably disinhibition like that produced by alcohol and similar depressants. Other actions of these drugs which are pertinent to our discussion include the ability of most of them to release histamine. In man, this is usually manifested by itching of the nose or more generalized tingling ("pins and needles") enabling some drug users to identify the drug which they have just injected as a narcotic.[2] Finally, the ability of these drugs to decrease the activity of the gastrointestinal tract leads to the problem of constipation which always complicates chronic use of narcotics. The development of tolerance to the action of these drugs leads to the use of increasing doses by some people. This, of course, increases the cost of the habit but it is noteworthy that some users, especially those with ready access to narcotics, may maintain themselves for long periods on a constant daily dose. The phenomenon of development of physical dependence to narcotics may or may not be related to the development of tolerance, but, since it is manifested by an uncomfortable withdrawal syndrome if narcotics suddenly become unavailable, dependence has in the past been considered a principal factor responsible for the abuse of narcotics. Doubtless the discomfort which accompanies abstinence contributes to the desire to look for another fix, but present evidence indicates that the pleasurable effects of the drug are more important in drug abuse than the fear of unpleasant withdrawal symptoms.

Two narcotic antagonists are available on the market—nalorphine and levallorphan. The pharmacologic effects of these compounds differ according to their use. If they are given to an individual who has no narcotic in his tissues they have many of the effects of narcotics. They produce analgesia, respiratory depression, miosis, and possibly euphoria in some individuals, but usually they produce dysphoria and are rarely abused because of this. If these drugs are given to an individual who is depressed by an overdose of a narcotic they act as narcotic antagonists, stimulating respiration, counteracting lethargy, and, in short, constitute specific therapy for narcotic poisoning. If given to an individual who has developed tolerance to a narcotic they induce an abrupt, short-lived withdrawal syndrome which is similar to the syndrome seen when narcotics are suddenly withheld from an addict. The severity of the reaction is dose-dependent and low doses will produce mydriasis and not much else. This led to the development of the pupil test for narcotic abuse, and experimentally it has been shown that a narcotic antagonist will produce mydriasis after a single injection of a narcotic.

Examination of the planar formulas of narcotics and antagonists suggests that they belong to different chemical types. This is, of course, true since morphine is a phenanthrene derivative, meperidine is a phenylpiperidine derivative, and methadone is derived from the aliphatic chain compound heptane. However, these three apparently different chemical types differ only quantitatively in their pharmacologic effect and the available narcotic antagonists which are derivatives of the phenanthrene nucleus actually antagonize all types of narcotics. This apparently parodoxical situation is explained by the fact that when the molecules of analgesics and antagonists are examined in three-dimensional form they appear to be so similar in shape that they could fit into a common receptor.[3] This concept helps explain why the various narcotics have similar pharmacologic effects and why antagonists may act like narcotics if given alone, but antagonize the narcotics if given to individuals with a sufficient amount of narcotic in their tissues.

Looking at each of the groups of narcotics classified by chemical type from the standpoint of drug abuse, we find that morphine and its homologues have high abuse potential. Heroin, which is diacetyl morphine, is the most frequently abused drug of this group possibly because it passes the blood-brain barrier more rapidly than do the other compounds. However, in a controlled study a group of postaddicts expressed no preference between morphine and heroin.[2] Codeine is less potent than morphine and does not have as high an abuse potential, but it should not be equated with other more potent codeine-like drugs such as dihydrohydroxycodeinone (constituent of Percodan) which is much more potent. Meperidine and related compounds such as anileridine and piminodine, alphaprodine and fentanyl have pharmacologic actions similar to those of morphine, but may differ in potency or onset and duration of effect. Meperidine has the somewhat dubious distinction of being the favorite drug of abuse by members of medical and allied professions. Methadone and congeners are also pharmacologically similar in action to morphine, but methadone has a longer duration of action which is the basis for its use in the methadone maintenance treatment of narcotic addiction. The only other methadone derivative readily available in this country is dextropropoxyphene, which is ordinarily taken orally and does not have a high addiction liability by this route. Methadone derivatives such as dextromoramide are used in other parts of the world and it should be pointed out that potent methadone derivatives all have high abuse potential.

Not all of the important morphine-like compounds can be discussed here, but several derivatives of morphine should be mentioned for one reason or another. Levorphanol is a synthetic phenanthrene-type compound that is a useful substitute for morphine and is the parent compound of the narcotic antagonist levallorphan. Oxymorphone is another potent morphine homologue; it is the parent compound of the very interesting narcotic antagonist naloxone, which apparently has almost no morphine-like properties when given alone, but is a very effective antagonist of the depressant and euphorogenic actions of narcotics and other narcotic antagonists. Phenazocine is of interest because it represents one of the first compounds used which lacked one of the rings of the phenanthrene nucleus and yet retained morphine-like activity. The basic nucleus of this compound is found in the narcotic antagonist analgesic pentazocine and in narcotic antagonists such as cyclazocine. Finally, we should mention the very potent morphine-like compounds recently used for tranquilizing large animals.[4] These compounds are of the tetrahydrooripavine series. They are interesting from a chemical standpoint because they show that chemical manipulation of the C ring of the phenanthrene nucleus can markedly increase potency. A few milligrams of one of these compounds, etorphine, are sufficient to tame a wild rhinoceros, and if an overdose of the compound is administered, treatment with one of the narcotic antagonists can be given.

Research in the area of narcotics is presently aimed at producing compounds that are more specific analgesics. The finding that nalorphine was an effective analgesic, even though it produced dysphoria in a large percentage of patients who received it, was largely responsible for this trend in research. Thus pentazocine is a weak narcotic antagonist with analgesic properties and probably a relatively low addiction potential, although cases of pentazocine dependence have already been reported. In the treatment of narcotic addiction, some feel that if the effects of a narcotic are blocked by maintaining a constant concentration of a narcotic-like compound that has no euphorogenic effect the addict under treatment will not use his energy to acquire narcotics, but instead will revert to a more normal and socially acceptable life pattern. Experimental treatment programs are under way in which the antagonists cyclazocine and naloxone are used to block the effect of intravenously injected narcotics. Methadone is also being used in the treatment of narcotic addiction. Because of its long duration of action it is an ideal substitute for a compound such as heroin in addicts who have developed physical dependence to a narcotic. Rapid withdrawal of methadone then produces only a minimal abstinence syndrome. Furthermore, methadone has also been used in treatment programs to block the action of other narcotics and to prevent "drug hunger" which some feel is an essential component of drug addiction.

Below is a list of generic and trade names of drugs mentioned in this discussion:

Nalorphine—Nalline
Levallorphan—Lorfan
Meperidine—Demerol, Dolantin
Anileridine—Leritine
Piminodine—Alvodine
Alphaprodine—Nisentil
Fentanyl—Sublimaze
Methadone—Dolophine
Dextropropoxyphene—Darvon
Dextromoramide—Palfium
Levorphanol—Levo-Dromoran
Oxymorphone—Numorphan
Phenazocine—Prinadol
Pentazocine—Talwin

SUMMARY

The pharmacology of narcotics and narcotic antagonists pertinent to drug abuse is reviewed. Euphoria and tranquilization are probably the most important pleasurable central nervous effects of the narcotics; development of tolerance to and physical dependence on them contribute to their abuse potential. By virtue of its long duration of action, methadone is useful as a substitute for other narcotics to ameliorate the withdrawal syndrome and for maintenance of addicts in such a condition that intravenous injection of narcotics is not pleasurable. Narcotic antagonists when given alone have morphine-like effects, but dysphoria is more frequent than euphoria so their abuse potential is low. They are specific agents for treatment of an overdose of narcotic but if given to someone physically dependent on a narcotic they induce a withdrawal syndrome.

REFERENCES

1. Elliott, H. W., et al.: Detection of narcotic use. *Calif Med 109*:121–125, 1968.
2. Martin, W. R., and Fraser, H. F.: A comparative study of physiological and subjective effects of heroin and morphine administered intravenously in postaddicts. *J Pharmacol Exp Ther 133*:388–399, 1961.
3. Beckett, A. H., and Casy, A. F.: Synthetic analgesics; stereochemical considerations. *J Pharm Pharmacol 6*:986–999, 1954.
4. Harthoorn, A. M.: Comparative pharmacological reactions of certain wild and domestic mammals to thebaine derivatives in the M-series of compounds. *Fed Proc 26*:1251–1265, 1967.

PATTERNS OF ABUSE OF NARCOTICS: AN HISTORICAL VIEW

HAVELOCK F. FRASER, M.D.

DEFINITION OF ADDICTION AND ABUSE AND THE SCOPE OF THIS REVIEW

In 1964, the World Health Organization Expert Committee on Addiction-Producing Drugs recommended using the term "drug dependence" instead of "drug addiction" or "drug habituation."[1] Drug dependence is defined as a state arising from the repeated administration of a drug on a periodic or continuous basis and connotes "psychic" as well as "physical" dependence. Its characteristics vary with the agent involved, but drug dependence is a general term selected for its applicability to all types of drug abuse; it carries no connotation regarding the degree of risk to the public or need for a particular type of control.

In order to evaluate the concept of drug dependence in terms of its degree of risk to the public, we define abuse for the purposes of this review, as follows: Abuse of a drug exists if its use so harmfully affects the individual and/or society as to require its control. The degree of risk to the public is usually reflected in the type and extent of the controls promulgated.

Dr. Seevers has aptly described abuse as follows: "Drug abuse has only one uniform connotation, the disapprobation of 'society.' But there are hundreds of 'societies,' and the meaning of the term 'drug abuse' is therefore conditioned by the existing mores of the society which makes the judgment."[2] In this review the criteria for designating a narcotic drug as an abuse item are those specified by the World Health Organization, although it is recognized that such a judgment represents a compromise.

This paper will review the historically changing pattern of drug abuse, starting with the alkaloids of opium, and advancing to synthetic narcotic analgesics, including heroin. We will discuss the recent tendency to abuse narcotics concurrently with diverse pharmacologic agents such as cocaine and barbiturates. We will note the varying abuse potential of certain narcotics and the preference for various routes of administration, as well as the influence of age, sex, race, availability, and local law on the incidence of narcotic abuse. We will also discuss the sociologic characteristics of the narcotic addict population.

ABUSE OF OPIUM ALKALOIDS

The pleasant effects of opium may have been known to the ancient Sumerians (4,000 B.C.), whose description of the poppy was *hul* ("joy") plus *gil* ("plants"). However, the first definite mention of "poppy juice" is found in the writings of Theophrastus in the 3rd century B.C. Physicians in Arabia were familiar with the uses of opium, and Arabian traders introduced it to the Orient and China, where it was used principally for the control of dysentery. By the middle of the 16th century, the clinical use of opium in European medicine was much the same as it is today. It was Paracelsus who prepared and named Laudanum, a tincture of opium still in use today.

In the 18th century, opium smoking became popular and socially acceptable in the Orient. In Europe, the easy availability of opium led to some degree of abuse.[3]

In 1803, a young German pharmacist, Sertürner, isolated and described an opium alkaloid that he named morphine after Morpheus, the Greek god of dreams. Robiquet, in 1832, discovered another important alkaloid of opium, and by the middle of the 19th century the therapeutic use of morphine and codeine had progressively replaced the use of the crude preparations of opium.[3]

The abuse of morphine was accelerated by the invention of the hypodermic needle and the parenteral use of morphine. The intense subjective effect that followed its injection tended to produce more compulsive use of the drug. In the United States, widespread morphine abuse resulted from the extensive and prolonged treatment of wounded Civil War soldiers. The use of opium was further augmented by an influx of opium-smoking Chinese laborers and by the fact that opium could be freely purchased in grocery stores and pharmacies.[3] This ready availability of opium persisted in the United States until 1914, when the Harrison Narcotic Act was enacted. Prior to this law, the incidence of abuse of opium and related alkaloids was several times greater than today's.

Those interested in the early history of opium and its alkaloids and the problems of addiction are referred to the studies of Macht (1915)[4] and Terry and Pellens (1928).[5]

ABUSE OF SYNTHETIC NARCOTIC DRUGS

Diacetylmorphine, or heroin, should be classified as a semisynthetic alkaloid, although the ease of its synthesis would place it almost in the category of a natural product. Discovered in 1874 by C. R. Wright, it was not until 1900 that heroin was promoted as a drug to replace morphine and to prevent narcotic addiction.[6]

The relative use of heroin, morphine, paregoric, Dilaudid, meperidine, codeine, methadone, and other synthetic narcotics is illustrated by the diagnosis of addiction among those during the calendar year 1962 from the U.S. Public Health Service Hospital in Lexington, Kentucky. Table 1 indicates that heroin was the most abused of the narcotic drugs, although it is noteworthy that codeine, which many physicians consider relatively nonaddicting, had quite a high incidence of abuse, 114 diagnoses as compared with 1,731 for heroin.

Table 1 Drug addiction diagnoses of patients discharged from Lexington during the calendar year 1962
(*U.S. Public Health Service Hospital Statistics*)

Drug	Frequency of diagnosis
Heroin	1731
Barbiturates	684
Morphine	278
Paregoric	270
Dilaudid	195
Meperidine	115
Codeine	114
Methadone	58
Pantopon	22
Cocaine	5
Meprobamate	4
Anileridine	4
Dihydrohydroxycodeinone	4
Levo-dromoran	2
Opium	2
Tincture of opium	2
Marihuana	2
Glutethimide (Doriden)	2
Dihydrocodeinone	2
Alphaprodine (Nisentil)	1
Dextroamphetamine (Dexedrine)	1

The drugs listed in Table 1 include only the active isomer judged by the capacity of the drug to relieve pain.[7,8] The inactive isomers are not included, but a limited number of cases of dextromethorphan abuse have been reported even though the pattern of its subjective effects differs from that of morphine.

PREFERENCE OF ADDICTS FOR CERTAIN NARCOTIC DRUGS AND ROUTES OF ADMINISTRATION

Fraser and co-workers[9,10] demonstrated that veteran narcotic addicts in an experimental setting were able to differentiate subjectively between a placebo and active drug and between active drugs of varying potencies. The report of these investigators stated, however, that in several instances there was considerable discrepancy between responses recorded by narcotic addicts (subjective evaluation) and responses of observers (behavioral evaluation). In this study of eight subjects, the drugs used and the initial and final daily doses attained during an intoxicating interval of 18 to 20 days were as follows: morphine, subcutaneously, 32 and 240 mg; morphine, orally, 32 and 240 mg; codeine, orally, 200 and 1500 mg; phenazocine, 4.8 and 12 to 36 mg; levophenacyl morphan, 4 to 4.8 and 24 to 36 mg; diphenoxylate, 40 to 64 and 160 to 480 mg; d-3-methoxy-N-phenethyl-morphinan, 180 to 1200 mg; oral placebo. The chronic dosage questionnaire was completed daily by patients and aides. The extent to which patients and observers identified the various agents as "dope" and the extent to which these drugs were classified in the categories "would like to take daily," "don't care" and "no" were recorded for both the patient (P) and the observers (A), as tabulated in Figure 1.

It is noteworthy, particularly since the study was double-blind, that although each patient had approximately 80 opportunities to err in identifying the placebo as an active drug, the patients made no errors on this point, and the observers made very few. Patients and observers were able to identify the potent drugs as "dope" in a parallel manner. However, in the case of less potent agents or those given orally, there was considerable discrepancy between patients and observers regarding identification as dope. For the parameter "would like to take daily," there was a radical difference between the judgment of patients and of observers (Fig. 1). It is evident from these responses that patients discriminated among drugs to a high degree, while the observers rated many of them similarly. For example, morphine taken subcutaneously and codeine taken orally received equivalent ratings from the aides, while patients apparently had "private" and probably different criteria which were not available to the observer. Observers were much more impressed than patients by effects such as miosis, hypnosis, and respiratory depression. The responses analyzed were "identification as dope," "strength" (not herein described), and "would like to take daily." On the basis of the patients' responses to these questions, the seven drugs and a placebo were differentiated and a significant correlation of the rank order of the various agents was found for three subjective measures. In addition, these three subjective measures were significantly correlated with an objective measure, intensity of abstinence, obtained at a later time. Furthermore, the patients' preference ratings of the drugs (morphine by three routes, heroin by two, and codeine by one) correspond well with their actual abuse rates as determined by both the incidence of abuse according to the drug used at the time of admission to the hospital at Lexington and the incidence reported by the Bureau of Narcotics.

Codeine further illustrates the importance of subjective effects in determining the preference of addicts for specific drugs. The moderate "flush," a reaction due primarily to histamine release, is enjoyed by the addict

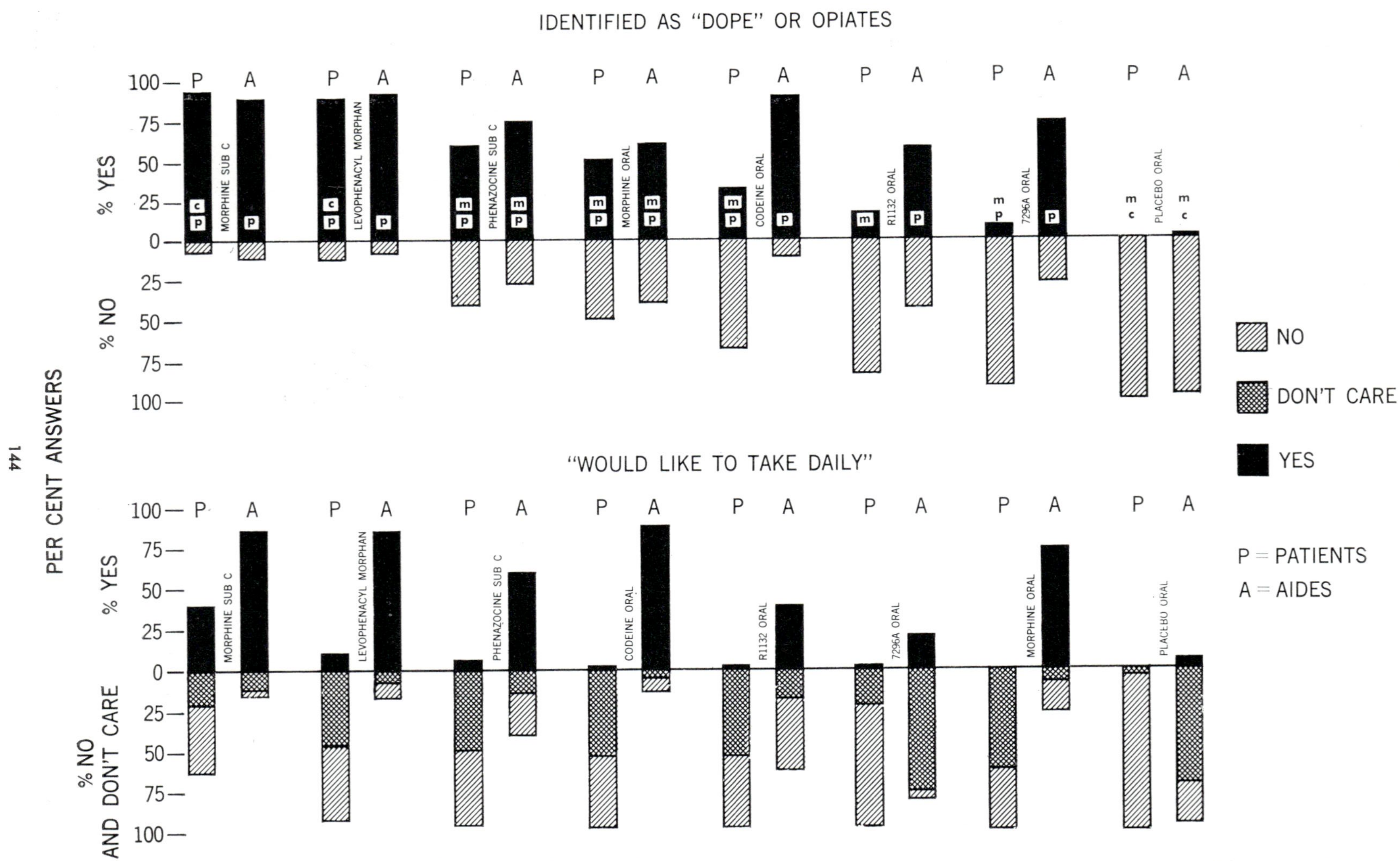

Figure 1 Comparison of average ratings of patients and aides (observers) using the chronic dosage questionnaire for opiates in a "short" addiction test of 18 to 20 days involving seven drugs and a placebo.
The variables measured included: "identified as 'dope' or opiates" and "would like to take the drug daily." (Reproduced from *The Addictive States,* Wikler, A. [Ed.]. Association for Research in Nervous and Mental Disease, Vol. 46. Baltimore, Williams & Wilkins Co., 1968, p. 176.)

when a drug is given intravenously; apparently morphine and heroin produce this effect in the desired amount. On the other hand, the more intense "flush" of intravenous codeine, caused by a larger release of histamine, is avoided by the addict. Furthermore, codeine when rapidly injected intravenously causes severe hypotension, an effect which is disturbing to the addict.

There are other factors which influence the preference of addicts for certain narcotic drugs, particularly when such drugs are administered intravenously. For example, certain drugs may be irritating to tissue, progressively occluding veins so that the addict is incapable of chronic abuse by the intravenous route. An important psychic factor in abuse relates to the addict's desire for immediate subjective effects in maximal intensity. A good example of a drug which does not exhibit these qualities is diphenoxylate.[11] When diphenoxylate was given intravenously to non-tolerant morphine addicts, the onset of effects was very gradual producing no pronounced subjective opium-like effects until 30 minutes after injection. The most probable reason for this delay in onset of effects is that diphenoxylate is insoluble in water and must be dissolved in pure propylene glycol. If even one drop of water is present in the syringe, diphenoxylate precipitates out completely. Evidently when injected into the vein, diphenoxylate precipitates promptly and produces its pharmacologic action very slowly. The potential abuser of diphenoxylate would need to carry a solvent like propylene glycol, a significant interference with his chronic abuse of this drug.[11]

It is thus seen that experienced narcotic addicts have individual preferences among drugs and definite preferences concerning route of administration, factors which undoubtedly influence their pattern of drug abuse. (Incidentally, observations of preference in man have been confirmed by studies of self-administered drug use by monkeys.[12])

CONCURRENT ABUSE OF NARCOTICS WITH OTHER DRUGS

The barbiturate-sedative class of drugs is most commonly abused concurrently with narcotics. Addicts frequently become dependent on both narcotics and barbiturates, and on admission to hospitals must be withdrawn from both agents at a gradual rate. At the USPHS Hospital in Lexington, a total of 684 patients were diagnosed as addicted to barbiturates compared with 1,731 addicted to heroin for the calendar year 1962 (Table 1). The point was also emphasized in a report by Hamburger[13] on the abuse of barbiturates by narcotic addicts and more recently by Chambers and Moldestad.[14] Table 2 gives the incidence of barbiturate-sedative abuse and addiction among narcotic abusers at Lexington for the years 1944, 1948, 1957, and 1966.

Glutethimide is frequently abused in conjunction with narcotics, and a particular cult of individuals abuse codeine and glutethimide (Doriden) in a ritual which involves the oral administration of glutethimide followed 15 minutes later by 4 ounces of codeine cough mixture.[15] The "high," which develops about one-half hour after the codeine is taken, is characterized as a depressive or "down" effect, to contrast it with the "high" produced by such drugs as the amphetamines.

Although at least 85% of the narcotic addicts admitted to the hospital at Lexing-

Table 2 The incidence of concurrent opiate and sedative abuse*

Year of admission	Sedative abusers No.	Sedative abusers %	% Increase
1944 (N = 100)	8	8.0	—
1948 (N = 100)	17	17.0	9.0
1957 (N = 100)	39	39.0	22.0
1966 (N = 100)	54	54.0	15.0

* A representative sample of 100 admissions to the U.S. Public Health Service Hospital was selected for 1966. Insofar as feasible, this sample was matched with patients for the preceding years.

ton, Kentucky, state that marihuana-smoking ("weed" or "pot") was their initial exposure to drugs, Table 1 suggests that very few of them persist in smoking marihuana after they have become narcotic addicts.

The frequency of abuse of other drugs in conjunction with heroin was surveyed by Perkins and Block in New York City.[16] They investigated the drugs abused by 521 patients admitted to their methadone maintenance program and 712 applicants not accepted. Forty-seven percent used one to four drugs in addition to heroin; 27%, other opiates; 8%, barbiturates; 8%, amphetamines; 4%, psychedelics. Sixty percent consumed alcohol, although alcohol was considered a problem in only 10% of these patients. There is relatively little concurrent abuse of narcotics and the major and minor tranquilizers. Concurrent abuse of heroin and cocaine, known as a "speed-ball," formerly had a high preference rating among narcotic addicts. At present, due probably to restricted supplies of cocaine, this combination is abused less.

Table 3 Major drugs used as reported by NARA admissions, July 21, 1967, through December 31, 1969*

Drug	Males	Females	Total
Opium	3	—	3
Paregoric	40	12	52
Morphine	100	26	126
Codeine	17	7	24
Terpin hydrate with codeine	5	2	7
Other codeine compounds	19	4	23
Heroin	1818	467	2285
Percodan	3	2	5
Numorphan	38	16	54
Normorphine	3	—	3
Dilaudid	72	18	90
Meperidine	54	27	81
Methadone	17	6	23
Leritine	2	—	2
Unspecified barbiturates	1	1	2
Pentobarbital	1	1	2
Seconal	1	—	1
Tuinal	2	—	2
Cocaine	9	2	11
Unspecified amphetamines	2	2	4
Methamphetamine	8	2	10
LSD	2	1	3
Marihuana	2	1	3
Glue	1	—	1
Carbrital	—	1	1
Desbutal	—	1	1
Darvon	—	2	2
Talwin	—	4	4
None	2	1	3
Total	2222	606	2828

* Admissions to the Clinical Research Center, Lexington, Ky., for treatment under the provisions of the Narcotic Addict Rehabilitation Act of 1966 (Dr. Conrad).

Conrad's observation of admissions to Lexington, reported in Table 3, indicates that if the drug of abuse is limited to the one which the addict used during the two weeks prior to admission, then concurrent abuse of barbiturates with narcotics is less prevalent than suggested by earlier statistics from this institution shown in Table 1.[17]

Dr. Way has described "chasing the dragon," a special technique for concurrent use of heroin and barbital used in the Far East. Small granules of heroin are mixed with several parts of barbital powder on a rectangle of foil which has been folded longitudinally. The mixture is heated gently with a lighted match or taper; as the mixture melts, the molten mass is rolled slowly back

Table 4 Characteristics of patients admitted to U.S. Public Health Hospital, Lexington, Ky.

Characteristics	1936 Total 1,036 (percent)	1965 Total 1,097 (percent)	1969 Total 2,081 (percent)
Sex			
Male	100	79	77.22
Female	0	21	22.78
Age			
Under 20	1	5	6.10
20-29	18	45	56.18
30-39	38	29	23.74
40 or more	43	21	13.98
Race			
White	88	60	47.57
Non-white	12	40	52.43

and forth on the foil while the addict inhales the fumes.

It is thus seen that there is considerable concurrent abuse of narcotics and other drugs, particularly barbiturates, amphetamines, psychedelics, and glutethimide. In this observer's opinion, part of this combined use may be explained by the fact that continuous supplies of heroin are uncertain and the addict resorts to the drugs which are more readily available.

INFLUENCE OF AGE, SEX, RACE, AVAILABILITY, NARCOTIC LAWS, AND SOCIOLOGICAL CHANGES ON THE PATTERNS OF NARCOTIC ABUSE

The influence of age, sex, and race on the pattern of narcotic abuse for the years 1936, 1965, and 1969 as revealed by statistics from the Clinical Research Center, Lexington, Kentucky, is shown in Table 4, prepared by Rasor.[18] In respect to the number of females reported addicted in 1936, the figures in the table are misleadingly low, since females were not eligible for admission to this institution at that time but were committed to Alderson Penitentiary in West Virginia. As compared with the rest of the U.S. population, the proportion of addicts under 20 years of age is also probably quite low, particularly for the year 1969, since first offenders—those just beginning the use of narcotics—are seldom institutionalized at Lexington. Nevertheless, the outstanding trends in the characteristics of the narcotic addict population are illustrated in this table, which shows distinct increase in the number of addicts under 30 years of age and a rapid shift from white to non-white.

The astounding increase in the number of persons under 18 years of age who are associated with drug abuse is evident from the statistics of drug arrests of juveniles, from 1960 to 1969, reported by the State of California Law Enforcement Agency, and illustrated in Table 5. One is astounded first by the increase in the number of juveniles arrested on charges of Dangerous Drugs, which include barbiturates, amphetamines, LSD, cocaine, and others. There is also a great increase in the number of arrests for possession or sale of marihuana. From 1960 to 1969 there was a threefold increase in the number of narcotic addicts or users arrested, but this is a much smaller increase than that observed for Dangerous Drugs and marihuana. A change in the pattern of drug abuse in general can be noted, along

Table 5 Drug arrests of juveniles (under 18 years) reported by California law enforcement agencies 1960 through 1969

	Specific offense by year of arrest						
Specific offense	1960	1962	1964	1966	1968*	1969*	% Change 1969 over 1960
Total	1,616	1,336	2,022	5,355	32,360	35,965	2,125.6
Marihuana	910	310	1,237	4,034	19,055	16,180	1,678.0
Heroin and other narcotics	38	25	38	61	205	255	—
Narcotic addict or user	122	58	66	57	345	495	305.7
Dangerous drugs	515	906	639	1,007	10,805	17,595	3,316.5
Other offenses	31	37	42	196	1,950	1,440	—

* Computed from expanded 20% sample.
Note: Percentages not computed on bases smaller than 50.

with a considerable reduction in the rate of increase of narcotic addiction. It is thought that the figures for California are applicable to other cities and states, although the number of arrests for Dangerous Drugs would not be as high elsewhere, since California is the only state to have a specific Dangerous Drug Act.

The Canadian pattern is outlined in Table 6 by statistics relating to student abuse of drugs in Toronto, Montreal, and Halifax. The proportion abusing opiates is quite low, 1.5 to 1.9%. The authors observed significant abuse of all these drugs, particularly tobacco and alcohol, in the seventh and eighth grades and a progressive increase of all types of abuse through the fourteenth grade, patterns in the three cities being very similar.[19]

The pattern of abuse of opiate drugs in Great Britain is shown in Table 7.[20] A progressive increase is noted in the number abusing opiates, particularly those classified in the "non-therapeutic origin" group; abusers under 20 years of age show the same remarkable increase noted in other areas.

Availability of drugs is a significant factor in determining which of several drugs will be abused and the extent of such abuse. Good examples of this are morphine and Dilaudid, two drugs whose subjective effects are similar to heroin's but whose incidence of abuse, as shown in Table 1, is much lower. Undoubtedly the disparity of abuse in this instance is due to the easier availability of heroin through illicit channels, in contrast to morphine and Dilaudid, which must be obtained from pharmacies, either by prescription or burglary.

Another illustration of the influence of availability is the passage of the Harrison Narcotic Act in 1914, which outlawed over-the-counter sales of opium and placed narcotics on prescription.

The country of Iran stopped the growing of opium in 1955 and instituted a rehabilitation program for narcotic addicts. As a result, the incidence of narcotic abuse declined from 1,500,000 addicts to an estimated 700,000 at present.[21]

A classic example of the influence of availability, social attitudes, laws, and the explosive epidemiologic aspects of addiction is seen in the history of methamphetamine abuse in Japan, described earlier by Seevers.[2] The Japanese, like all belligerents in World War II, were unaware of the abuse potential of amphetamines and used them to reduce air-pilot and industrial-worker fatigue. Civilian use began in 1945 and, since the drugs

Table 6 Percentage of students in Toronto, Montreal and Halifax using drugs, and the ranking of various drugs in terms of percentage of users

	Toronto		Montreal		Halifax	
Type of drug	% Users	Ranking	% Users	Ranking	% Users	Ranking
Tobacco	37.6	(2)	45.9	(2)	47.3	(1)
Alcohol	46.3	(1)	48.9	(1)	40.0	(2)
Marihuana	6.7	(5)	8.5	(3)	6.6	(3)
Stimulants	7.3	(4)	5.8	(5)	6.4	(4)
Tranquilizers	9.5	(3)	7.8	(4)	6.0	(5)
Glue	5.7	(6)	1.9	(9)	3.1	(6)
Barbiturates	3.3	(7)	2.7	(7)	3.0	(7)
LSD	2.5	(8)	3.2	(6)	2.4	(8)
Other hallucinogens	2.0	(9)	2.5	(8)	2.2	(9)
Opiates	1.9	(10)	1.5	(10)	1.6	(10)

Table 7 Known opiate addicts in Great Britain

Drug addicts	1958	1959	1960	1961	1962	1963	1964	1965	1966	1967
Total number	442	454	437	470	532	635	753	927	1349	1729
Origin										
Therapeutic origin	349	344	309	293	312	355	368	344	351	313
Non-therapeutic origin	68	98	122	159	212	270	372	580	982	1385
Unknown origin	25	12	6	18	8	10	13	3	16	31
Ages										
Under 20	—	—	1	2	3	17	40	145	329	395
20-34	—	50	62	94	132	184	257	347	558	906
35-49	—	92	91	95	107	128	138	134	162	142
50+	—	278	272	272	274	298	311	291	286	279
Age unknown	—	34	16	7	16	8	7	10	14	7

were freely available, it was estimated in 1954 that two million Japanese used amphetamines intravenously, with 600,000 estimated to be chronic users. The government tried initially to combat this extensive use by placing methamphetamine on prescription, but this measure was completely inadequate; subsequently, more stringent laws were passed and an all-out educational campaign was started which effectively curtailed the epidemic. The number of arrests declined from 55,664 in 1954 to 271 in 1958, and there has been no recurrence of epidemic methamphetamine abuse in the last 15 years. A lesser heroin epidemic from 1955 to 1962 was controlled by similar measures in Japan.

The epidemic abuse of drugs is a critical phenomenon throughout the world. In the United States, although the number of narcotic addicts has undoubtedly increased in recent years, nevertheless the proportion of persons abusing narcotics as compared with other drugs has unquestionably decreased. This may be explained in part by the differences between the current drug-abusing population and that of 10 or 15 years ago. Formerly drug abuse was confined primarily to the so-called ghetto areas of large cities. More recently, however, there has been a progressive extension of drug abuse to suburbia. This extension has been associated with abuse by all socioeconomic classes and, particularly, with a remarkable increase in abuse by teenagers. In New York City, for example, the Department of Health states that the leading cause of death among teenagers is heroin as such or complications associated with its intravenous abuse.

THE THERAPEUTIC APPROACH TO DRUG ABUSE

There is no simple explanation for the explosive increase in drug abuse in the United States and throughout the world. We are undoubtedly dealing with a major pathology of individuals and/or society in which at least 8 to 10% of our population are addiction-prone (an estimate which includes alcoholics but not tobacco-smokers). In view of the extent of abuse of narcotics and other drugs and the complexities of the problem, we should not anticipate miracles from the passage of any given set of laws. In the long run, we will probably profit most from patience and from an understanding that individuals abusing drugs are sick rather than criminal. Our therapeutic approach should be in accord with such a concept. The recent drug abuse law passed in the United States is a move in this direction, since penalties for use have been greatly reduced. The changing social patterns reflected in our laws may ultimately change the drug abuse pattern for narcotics as well as other drugs.[22]

REFERENCES

1. World Health Organization Expert Committee on Addiction Producing Drugs, *WHO* Tech. Rep., Series *273*:3–20, 1964.
2. Seevers, M. H.: Drug dependence and drug abuse: A world problem. *The Pharmacologist 12*:172–181, 1970.
3. Jaffe, J. H.: Narcotic Analgesics. Chapter 15 in Goodman, L. S., and Gilman, A. (Eds.): *A Pharmacological Basis of Therapeutics,* 4th ed. New York, Macmillan, 1970, pp. 237–275.
4. Macht, D. I.: The history of opium and some of its preparations and alkaloids. *JAMA 64*:477–481, 1915.
5. Terry, C. E., and Pellens, M.: *Opium Problem.* Committee on Drug Addictions, U.S. Bk. Cat. Supplement to 1928.
6. Editorial: The surprising extinction of the Charas traffic. *Bull Narcotics 5*:1–7, 1953.
7. Isbell, H., and Fraser, H. F.: Actions and addiction liabilities of Dromoran derivatives in man. *J Pharmacol Exp Ther 107*: 524–530, 1953.
8. Fraser, H. F., and Isbell, H.: Human pharmacology and addictiveness of certain dextroisomers of synthetic analgesics. *Bull Narcotics 14*:25–35, 1962.
9. Fraser, H. F., Van Horn, G. D., Martin, W. R., Wolbach, A. B., and Isbell, H.: Methods for evaluating addiction liability. *J Pharmacol Exp Ther 133*:371, 1961.
10. Fraser, H. F.: Methods for assessing the addiction liability of opioids and opioid antagonists in man. Chapter 13 in Wikler, A. (Ed.): *The Addictive States.* Association for Research in Nervous and Mental Dis-

ease, Vol. 46. Baltimore, Williams & Wilkins Co., 1968, p. 176.
11. Fraser, H. F., and Isbell, H.: Human pharmacology and addictiveness of ethyl 1-(3-cyano-3, 3-phenylpropyl)-4-phenyl-4-piperidine carboxylate hydrochloride (R-1132, diphenoxylate). *Bull Narcotics 13*:29–43, 1961.
12. Yanagita, T., Deneau, G. A., and Seevers, M. H.: *Proceedings of the International Union of Physiological Sciences, 23rd Int. Congress, Tokyo, Japan 4*:453, 1965.
13. Hamburger, E.: Barbiturate use in narcotic addicts. *JAMA 189*:366, 1964.
14. Chambers, C. D., and Moldestad, M.: *The Epidemiology of Opiate Addiction in the United States.* Springfield, Ill., Charles C Thomas, 1970, Chapter 8.
15. DiGiacomo, J. N., and King, C. L.: Codeine and glutethimide addiction. *Int J Addiction 5*:279, 1970.
16. Perkins, M. E., and Bloch, H. I.: Survey of a maintenance program on methadone. *Amer J Psychiat 126*:1389, 1970.
17. Conrad, H. T.: Personal communication.
18. Rasor, R.: Personal communication.
19. Smart, R. G., Laforest, L., and Whitehead, P. C.: *Report to the Committee on Drug Dependence, 1970.*
20. Zacune, J., Mitcheson, M., and Malone, S.: Heroin use in a provincial town: One year later. *Int J Addiction 4*:557, 1969.
21. Report of the International Control Board. *Bull Narcotics 22*:39, 1970.
22. McMorris, S. C.: Can we punish the acts of addiction? *Bull Narcotics 22*:43, 1970.

PATHOPHYSIOLOGY OF NARCOTIC ADDICTION: POSSIBLE ROLES OF PROTRACTED ABSTINENCE IN RELAPSE

WILLIAM R. MARTIN, M.D.

This presentation will summarize experiments done by me and my collaborators, testing the hypothesis that chronic use of narcotic analgesics may induce long-lasting physiologic changes. I will also present evidence that these long-lasting physiologic changes are associated with an increased responsivity to stress and will suggest that this increased responsivity is responsible in part either directly or indirectly for relapse to narcotic use and other forms of deviant behavior.

A study, initially designed to investigate the qualitative characteristics and duration of abstinence in the rat, revealed that the abstinence syndrome had two phases, called *primary* and *secondary abstinence* at first, but subsequently more descriptively named *early* and *protracted* abstinence.[1] The early abstinence syndrome in the rat emerged within 24 hours after the last dose of morphine and persisted for about 2 days; it was characterized by the emergence of "wet dog" shakes, loss of appetite and weight, decreased consumption of fluids, increased urination and defecation, a fall in body temperature and metabolic rate to subnormal levels, sleeplessness, hostility, irritability on handling, and enhanced norepinephrine and epinephrine excretion. Between the third and the fifth day of abstinence, the syndrome changed; water and food intake, body weight, temperature, and the number of "wet dog" shakes increased above control values, as illustrated in Figure 1. Catecholamine excretion during protracted abstinence in the rat was not significantly different from that of appropriate control animals.[2,3] The adrenal glands of the addicted animals 143 days after withdrawal were larger than those of appropriate control animals, and they contained somewhat more epinephrine and norepinephrine. The brains of the morphine-addicted animals contained more norepinephrine than those of appropriate controls; however, this difference was not statistically significant.[2] The changes observed in protracted abstinence persisted for three to more than six months following withdrawal.

Studies by Wikler and Pescor, which were conducted to test the hypothesis that abstinence signs (and presumably symptoms and drug need) could be conditioned, demonstrated that post-addict rats consumed more etonitazene (a potent orally effective narcotic) than did appropriate control animals.[4] As can be seen from Figure 2, both rats formerly addicted to morphine trained

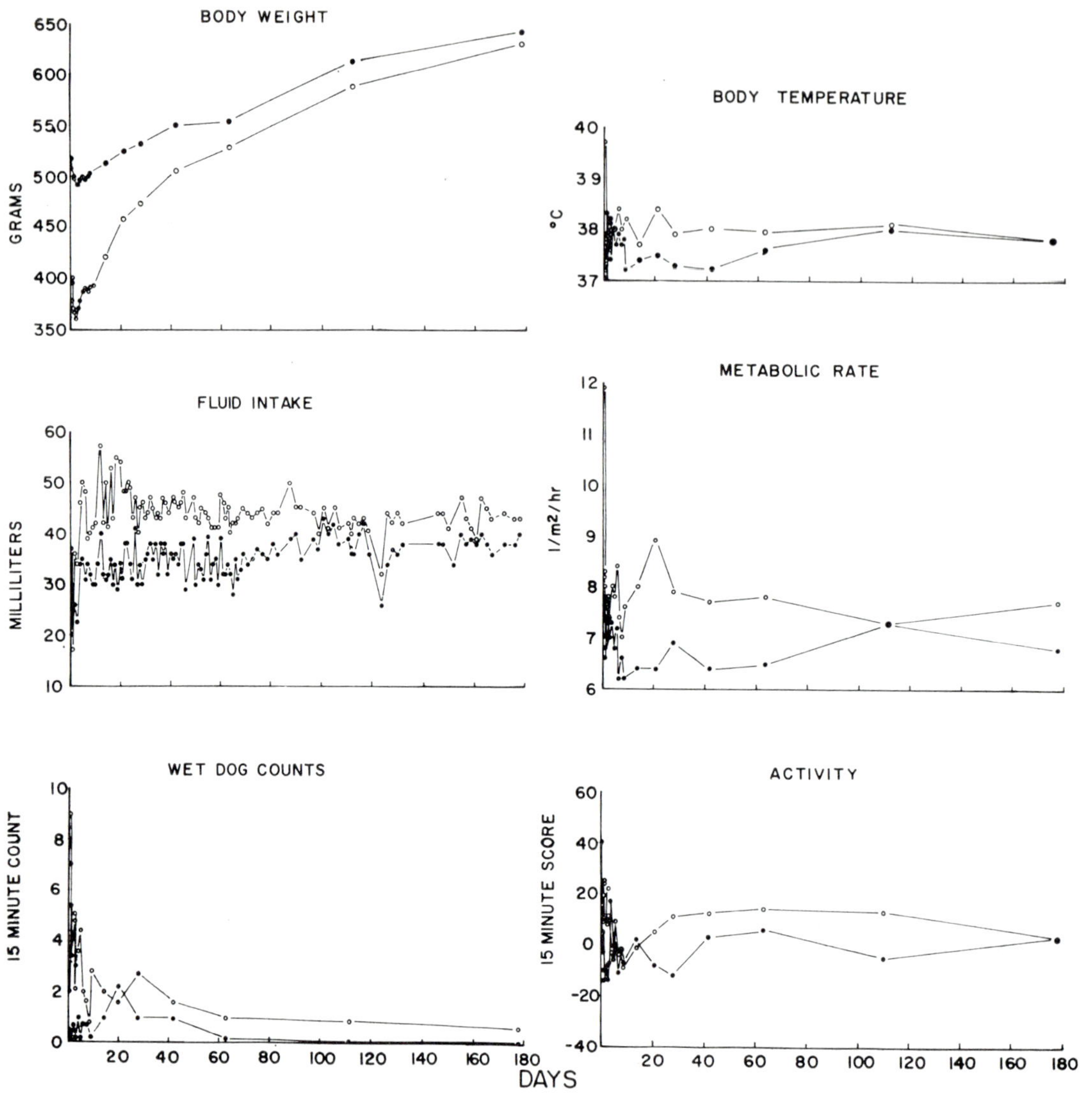

Figure 1 Time course of the secondary abstinence signs. Open circles (O) indicate mean value for abstinent rats; closed circles(●), mean values for saline controls. (Reproduced from a paper by Martin *et al.*[1] with the permission of Springer-Verlag.)

to obtain etonitazene in a linear maze and untrained rats (placed in a linear maze with etonitazene available at one end and water at the other) consumed more etonitazene than either trained or untrained rats that had never been addicted. This preference for etonitazene persisted for four to six months.

These studies suggested that chronic administration of morphine and subsequent withdrawal are associated with long-persisting physiologic changes, alterations in the metabolism of epinephrine and norepinephrine by the adrenal gland and perhaps the brain, and a continuing need for narcotics in the rat. Studies were then undertaken to test the relevance of these findings in man. Seven patients who were serving sentences

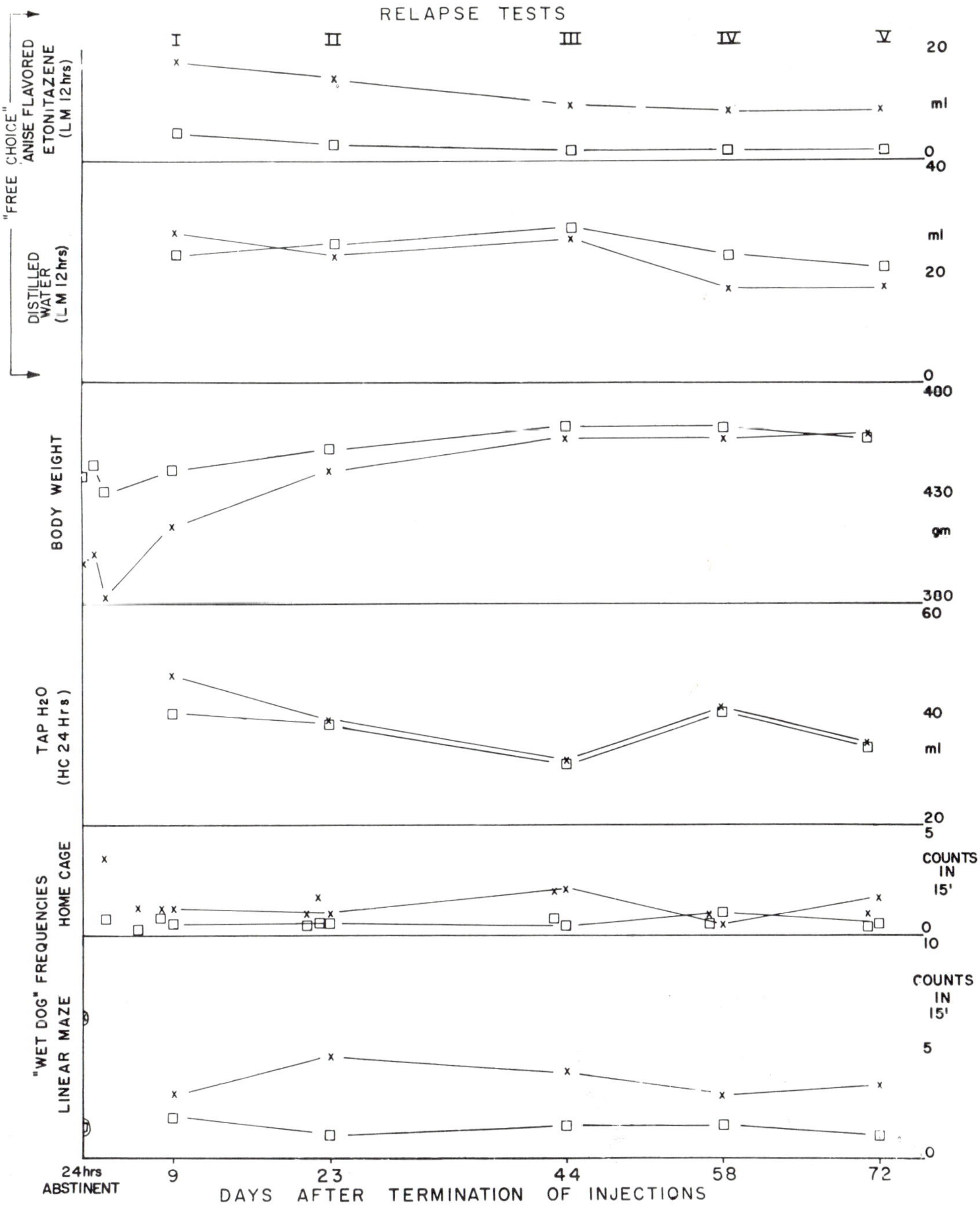

Figure 2 Behavior of previously morphine-injected rats (x) and previously saline-injected rats (□) after termination of injections. In this study, free choice was between water in the preferred tube and etonitazene solution in the non-preferred tube in all five relapse tests. (Reproduced from a paper by Wikler and Pescor[4] with the permission of the authors and Springer-Verlag.)

for violation of federal narcotic laws served as subjects and resided in the research ward of the Addiction Research Center throughout the study. Following an acclimation period, routine control observations and control respiratory studies were conducted over a seven-week period. The subjects then received morphine in increasing dose levels for five weeks, were stabilized on morphine (240 mg per day) for 34 weeks, and were then gradually withdrawn over a three-week period. Observations were continued for an additional 30 weeks.

Figure 3 illustrates some of the changes observed during the stabilization period. Blood pressure, pulse rate, and body temperature were elevated, pupils were constricted, and respiratory rate was depressed.[5] The responsivity of the respiratory center was depressed throughout the stabilization period; however, large doses of morphine (60 and 120 mg) produced only modest increases in depression.[6] During the stabilization phase, both clinical impression and responses to the MMPI and the Addiction Research Center Inventory indicated that

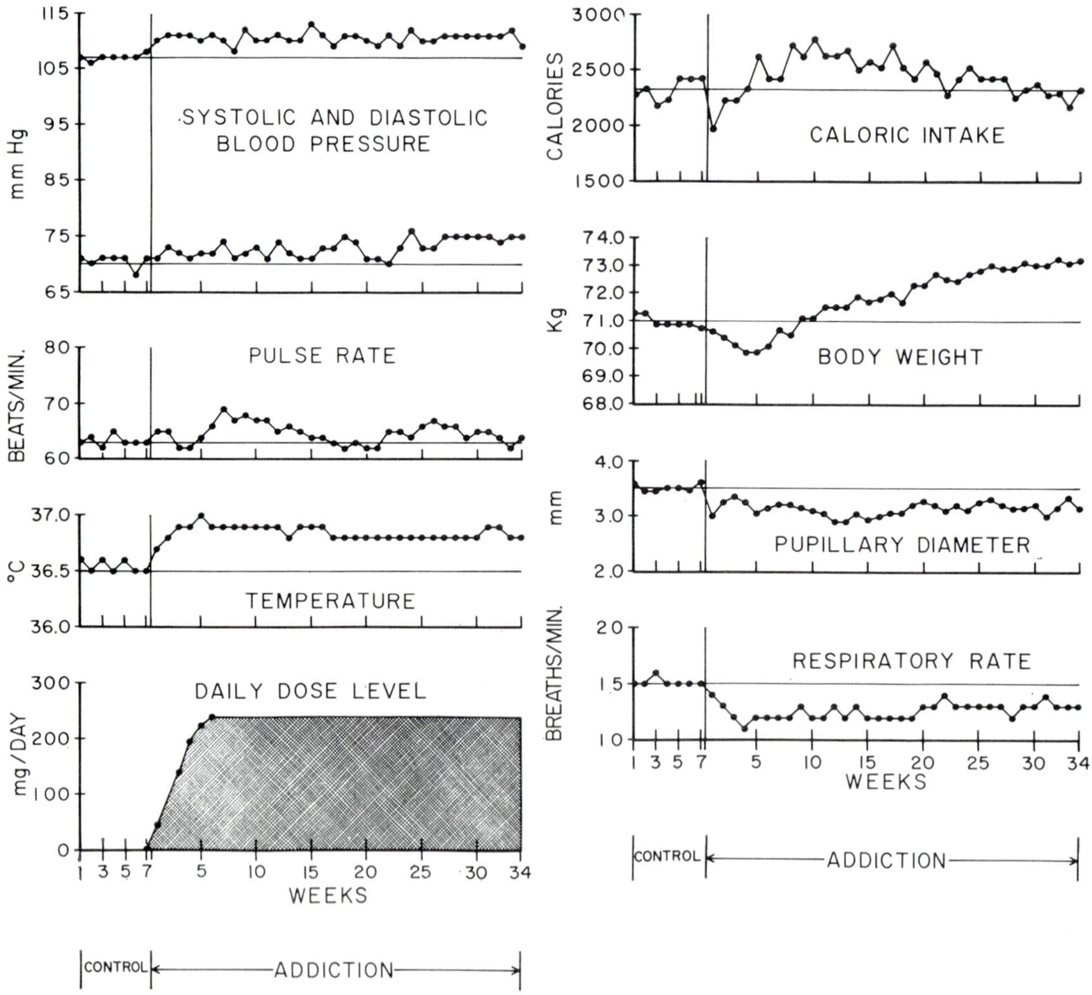

Figure 3 The effects of chronic administration of morphine on clinical variables. Each point represents the weekly A.M. observations for seven subjects. The horizontal line represents the mean of control determinations for seven subjects. (Reproduced from a paper by Martin and Jasinski[5] with the permission of Pergamon Press, Ltd.)

patients had become hypochondriacal, generally less motivated, bored with other people, withdrawn, and had fitful sleep.[7] Subsequent sleep studies have shown that although partial tolerance develops to the action of morphine in reducing rapid eye movement sleep (REMS), REMS continues to be depressed when patients are stabilized on morphine.[8] Early in the addiction cycle, epinephrine and norepinephrine excretion was increased, and corticosteroid excretion slightly depressed. During stabilization, excretion of epinephrine and corticosteroids was not different from that in the control period; however, norepinephrine excretion was somewhat elevated. These data on catecholamine excretion in man are similar to findings in the rat.[2,9]

When morphine was slowly withdrawn over a three-week period, blood pressure, pulse rate, and body temperature were elevated; there was a decrease in caloric intake and body weight; pupillary diameter increased to above control level; and respiratory rate, as well as the responsivity of the respiratory center to CO_2, was well above the preaddiction level (Figure 4).[5,6] The observations made in this study on blood pressure, body temperature, caloric intake, and respiratory rate, insofar as they are

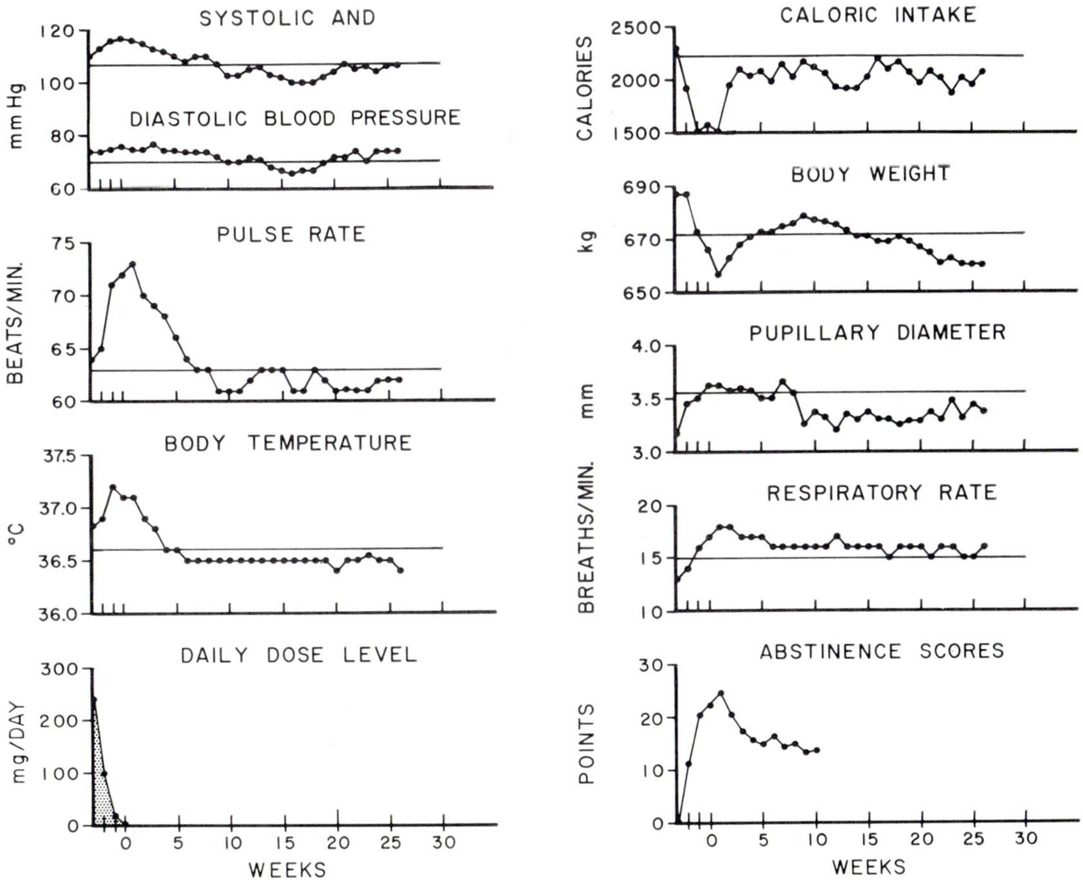

Figure 4 Changes seen during early and protracted abstinence. Each point represents the mean weekly A.M. values for six subjects. The first point of each curve represents the mean value for the last seven weeks of addiction. The horizontal line represents the mean control value for the six subjects. One subject was withdrawn from the study near the end of the chronic intoxication phase because of episodes of acute cholecystitis. (Reproduced from a paper by Martin and Jasinski[6] with the permission of Pergamon Press, Ltd.)

comparable, confirm the observations of Himmelsbach.[10] In addition, 17-OH-corticosteroid excretion was enhanced; surprisingly, however, there was no increase in the excretion of catecholamines when the early abstinence syndrome was maximal. Haertzen and Hooks have characterized the subjective state of patients during early abstinence, assessed by scale scores on the MMPI and ARCI, as being weak, tired, unmotivated, physically sick, tense, dysphoric, restless, impatient, uncooperative, irritable, and inefficient physically, socially, and intellectually.[7]

The early abstinence syndrome in man, as characterized above, persisted for six to nine weeks at which time the protracted abstinence syndrome emerged. The protracted abstinence syndrome was characterized by a small decrease in blood pressure, pulse rate, body temperature, pupillary diameter (Fig. 4), responsivity of the respiratory center to CO_2, and skin conductance below preaddiction levels.[6,11] In addition, patients excreted more epinephrine,[9] and the responsivity of skin conductance to hypercapnia was increased during the seventh and seventeenth weeks after withdrawal.[11] The other signs of protracted abstinence persisted through the twenty-sixth week of withdrawal.

The clinical significance of these findings is not clearly understood. In both rats and men who have been chronically intoxicated with morphine and withdrawn, an abstinence syndrome emerges which has two distinguishable phases: *early* or *primary abstinence* and *protracted* or *secondary abstinence*. The protracted abstinence syndrome has been shown to persist for more than four months following withdrawal in both species. In the rat, protracted abstinence has been associated with increased consumption of an opioid analgesic, etonitazene.[4] It is not known whether protracted abstinence is associated with an increased need for narcotics in man and contributes to relapse; there are, however, reasons why this may be a tenable hypothesis. Himmelsbach showed that postaddict patients exhibited increased responsivity to the cold pressor test for at least six months following withdrawal.[12] In the experiment herein described, subjects exhibited a somewhat increased responsivity of skin conductance to hypercapnia and increased epinephrine excretion two to four months after withdrawal. Although these findings do not prove that subjects are hyperresponsive to stress during protracted abstinence, they are consistent with the hypothesis. Further, the protracted abstinence syndrome may create a need for morphine-like narcotics because of their capacity to diminish responsivity to stress.[11]

At the time the concept of protracted abstinence was being formulated it was felt that it, as well as Wikler's concept of the role of conditioned abstinence and conditioned drug-seeking behavior in relapse, could be critically tested using narcotic antagonists in a clinical setting. However, the complexity of a clinical setting in which the role of many variables can neither be assessed nor controlled may well preclude the testing of hypotheses concerning the role of protracted abstinence and conditioning in relapse. In order to critically test the relevance of these theories to relapse, it may be necessary to develop new methodologies which will have the following characteristics:

1. Patients must be studied in an environment in which physical, biologic, and behavioral influences can be controlled for long periods of time.
2. Behavioral stimuli that are significant determinants of behavior must be identified and then simulated in the laboratory setting.
3. Sensitive techniques must be developed to assess the impact of behavioral stimuli on physiologic and psychologic functioning.

REFERENCES

1. Martin, W. R., Wikler, A., Eades, C. G., and Pescor, F. T.: Tolerance to and physical dependence on morphine in rats. Psychopharmacologia *4*:247-260, 1963.

2. Sloan, J. W., and Eisenman, A. J.: Long persisting changes in catecholamine metabolism following addiction to and withdrawal from morphine. In *The Addictive States,* Wikler, A. (Ed.). Association for Research in Nervous and Mental Disease, Vol. 46. Baltimore, Williams & Wilkins Co., 1968, pp. 96-105.
3. Martin, W. R., and Sloan, J. W.: The pathophysiology of morphine dependence and its treatment with opioid antagonists. *Pharmakopsychiatrie Neuro-Psychopharmakologie 1*:260-270, 1968.
4. Wikler, A., and Pescor, F. T.: Classical conditioning of a morphine abstinence phenomenon, reinforcement of opioid-drinking behavior and "relapse" in morphine-addicted rats. *Psychopharmacologia 10*:255-284, 1967.
5. Martin, W. R., and Jasinski, D. R.: Physiological parameters of morphine dependence in man—tolerance, early abstinence, protracted abstinence. *J Psychiat Res 7*:9-17, 1969.
6. Martin, W. R., Jasinski, D. R., Sapira, J. D., Flanary, H. G., Kelly, O. A., Thompson, A. K., and Logan, C. R.: The respiratory effects of morphine during a cycle of dependence. *J Pharmacol Exp Ther 62*:182-189, 1968.
7. Haertzen, C. A., and Hooks, N. T.: Changes in personality and subjective experience associated with the chronic administration and withdrawal of opiates. *J Nerv Ment Dis 148*:606-614, 1969.
8. Kay, D. C., and Kelly, O. A.: Human sleep during chronic intoxication with morphine. *Fed Proc 29*:384, 1970.
9. Eisenman, A. J., Sloan, J. W., Martin, W. R., Jasinski, D. R., and Brooks, J. W.: Catecholamine and 17-hydroxycorticosteroid excretion during a cycle of morphine dependence in man. *Psychiat Res 7*:19-28, 1969.
10. Himmelsbach, C. K.: Clinical studies of drug addiction. Physical dependence, withdrawal and recovery. *Arch Intern Med 69*:766-772, 1942.
11. Jones, B. E., Martin, W. R., and Jasinski, D. R.: Skin conductance responses to hypercapnia in man during a cycle of addiction to morphine. *Psychopharmacologia 14*:394-403, 1969.
12. Himmelsbach, C. K.: Studies on the relation of drug addiction to the autonomic nervous system: Results of the cold pressor tests. *J Pharmacol Exp Ther 73*:91-98, 1941.

THE MAINTENANCE APPROACH TO THE MANAGEMENT OF OPIOID DEPENDENCE

JEROME H. JAFFE, M.D.*

Lawrence Kolb (1962) concluded the first chapter of his book on drug addiction by listing the names and accomplishments of eight renowned Englishmen, all of whom were opiate addicts, all of whom contributed to their society's betterment, and six of whom lived to be more than 70 years of age.

We have known for years that, under certain conditions, the regular use of opiates is not incompatible with social productivity or general health. In spite of this knowledge, the acceptability of providing opiate users with regular supplies of opiates has varied greatly from country to country and from time to time within the same country. For example, in the United States at the turn of the century, the regular use of opiates was not viewed as a major social or medical problem, and there were a few individuals who advocated the administration of opiates as a technique for the treatment of alcoholism. Within a score of years attitudes had changed so radically that physicians who prescribed small amounts of opiates to confirmed opiate users were subjected to criminal prosecution.

* At the time this paper was originally presented, Dr. Jaffe was Director, State of Illinois Drug Abuse Program and Associate Professor, Department of Psychiatry, University of Chicago.

The history of these vicissitudes in attitude is too complex to detail here but is well described by Terry and Pellens (1928), Kolb (1962), and the ABA–AMA report (1962).

At present, we seem to be moving toward a consensus about certain aspects of the maintenance concept. There are now few responsible observers who would contest the following basic proposition: that there exists a group of compulsive narcotics users who, provided with appropriately supervised supplies of opioid drugs, can remain socially productive, but who, when deprived of such supplies through legitimate channels, eventually seek out illicit supplies to the detriment of themselves and the society in which they live.

The issues on which consensus has not yet been reached include the following:

1. What are the fundamental goals of the maintenance approaches?
2. What are the criteria for determining whether a given individual should be maintained on an opioid or offered (or compelled to accept) some form of treatment that does not involve maintenance?
3. What opioids and what routes of administration are best suited for use in maintenance programs, and what are

the relative advantages and disadvantages of each?

4. For any given drug, what are optimum dosages for achieving various goals for different individuals?
5. How can a health care delivery system best balance the need to make treatment readily available against the problems that might arise from illicit diversion of opioids?
6. What services (if any), other than the availability of opioids, should be considered essential, and what contribution do they make to the overall treatment?
7. What social objectives can a society reasonably expect to reach by utilizing the maintenance approach?

Any consideration of any treatment of narcotics users, including the maintenance concept, must recognize that one of the central issues is the tendency to relapse following withdrawal from the drug. There are a number of distinct explanatory hypotheses that attempt to account for this behavior. They range from the sociological to the biochemical (see Wikler, 1968). This paper will not attempt to review these explanatory models. However, it is worth pointing out that they are not necessarily mutually exclusive. For example, the self-administration of a narcotic can be viewed as behavior that is initiated out of curiosity and peer pressure, is maintained in its early stages by the positive reinforcing effects of the drug, and eventually becomes a highly reinforced pattern of behavior as a result of both operant and classical conditioning. The classical conditioning results in withdrawal phenomena conditioned to the environment in which the user lives. This, in turn, leads to situations in which, long after withdrawal, the environment evokes conditioned withdrawal symptoms that lead the individual to use drugs to alleviate the evoked distress. All of these factors theoretically could be superimposed on any permanent or long-lasting biochemical change that is induced by repeated use of narcotics.

Thus relapse may be due to a number of factors acting in concert. When considering the concept of maintenance it is useful to note that there are, in addition, at least two phenomenologically distinct varieties of subjective experience that many former narcotics users report as motivating them to return to narcotics use. One is an urge to reexperience one or more of the initial narcotic effects—for example, the "rush" ("flash") that follows intravenous use, or the euphoria ("high") or perhaps the relief of anxiety, pain, or depression.

Another distinct variety is a prolonged or persistent "emptiness" or sense of abnormality that causes many former opiate users to state that they feel "normal" only when using an opiatelike drug. In this instance, the urge to use—"narcotics hunger"—is not an urge to experience the "rush" or euphoria but rather to reachieve a state of normality. Dole and Nyswander (1968) have proposed that such a form of narcotics hunger could be due to a "metabolic" lesion arising as a result of the repeated use of narcotics.

Given at least two phenomenologically distinct varieties of "craving" or narcotics hunger—the urge to reexperience one or more opiate-induced effects and the desire to reexperience a remembered state of normality—it is possible to conceive of maintenance substances that would satisfy one variety of craving without affecting the other.

The possible goals, problems, and possibilities of any maintenance approach are virtually predetermined once it is decided whether one or the other or both of these varieties of drug hunger will be addressed.

The material summarized here is not intended to be a detailed review of the literature on maintenance but rather my own general impression of the progress and problems of the field organized around the seven issues outlined above. General references for the interested reader are supplied at the end of this paper.

GOALS

In general, all programs utilizing *oral* maintenance substances aim at permitting the compulsive narcotics user to become a law-abiding and productive citizen who does not use illicit narcotics. The programs do not attempt to provide a means whereby the patient can reexperience the "rush" or the narcotic "high." Indeed, the original program used high doses to induce cross-tolerance in order to prevent the patients from experiencing these effects from illicit drugs, and it originally excluded from treatment applicants who seemed overly interested in continuing to experience these effects. In those programs where the repetitive relapse to illicit narcotics is viewed as a manifestation of a persistent metabolic disturbance, the long-term goal is to stabilize the patient's dosage so that the regular use of oral medication causes minimal inconvenience. Most patients are viewed as being somewhat analogous to diabetics in that they will require methadone indefinitely. When patients in these programs are withdrawn from methadone most of them reexperience a sense of narcotics hunger and most of these eventually relapse to heroin use or are readmitted to maintenance programs.

Other investigators working with methadone maintenance reject the "metabolic defect" hypothesis and view the procedure as a form of "benign maintenance." While the short-term goals are the same, those espousing the benign-maintenance model hold out the possibility that most patients maintained on methadone can be entirely withdrawn from all narcotic drugs. A number of these investigators report that a few patients have been successfully withdrawn although relapse or readmission is statistically a more likely outcome of an attempt at withdrawal.

The goals of programs prescribing drugs for intravenous self-administration are somewhat different from those programs utilizing oral maintenance substances. Although it is difficult to make valid comparisons between the "British approach," in which heroin is prescribed, and the oral methadone programs in North America, both approaches share the goal of reducing the patient's incentive for seeking illicit supplies of narcotics. Each approach arose in its own peculiar historic and cultural context.

Until 1967, every physician in the United Kingdom was permitted to prescribe any narcotic to any narcotics addict according to his clinical judgment. Between 1964 and 1967 the number of known heroin and cocaine users increased more than threefold. It became evident that even a few physicians prescribing excessive amounts of heroin could create a major problem. Therefore, in 1967 the authorization to prescribe heroin to narcotics addicts was markedly restricted. The prescribing of heroin is now limited to a number of specialized clinics attached to major teaching hospitals. Patients registered at these clinics are given prescriptions for heroin which they can fill each day at certain chemists throughout the city. Most patients elect to use the heroin intravenously. The prescription of heroin for intravenous use is associated with the problems of infections, the transmission of disease due to shared needles, overdosage, the risk of illicit distribution, and the fluctuating level of drug action that must be expected when the effective drug action is short. However, unlike oral methadone, heroin does satisfy both types of hunger—the desire to get a narcotic effect as well as the desire to avoid the feeling of abnormality. In the sense that these programs do not insist that patients learn to forgo the subjective effects of the narcotic drugs, their goals for the patients are somewhat more modest than those of the oral methadone programs (see WHO Report, 1970).

In contrast to the British situation, the use of oral methadone arose in a situation in which heroin was illegal and where a physician prescribing any narcotic to narcotics addicts was subject to criminal prosecution. Thus, the idea of prescribing drugs for intravenous use, especially heroin, would have met with insuperable obstacles even if it had been adjudged to be medically useful. Fortunately, oral methadone maintenance avoids many of the problems associated with the medical distribution of heroin. The patient on oral methadone perceives little opiate effect and, if maintained on high dosage of methadone, is generally so tolerant to opiate actions that he perceives little opiate effect even when he uses illicit heroin intravenously. Because methadone is given orally exclusively and because its action usually lasts for 24 hours, it is possible to give every dose under direct supervision. Under these circumstances, the problems of infection, overdose, and illicit redistribution can be entirely eliminated. In practice, however, most patients are eventually given supplies of methadone to use at home, and the problems of misuse, redistribution, and accidental ingestion by non-tolerant individuals must be dealt with.

The only theoretical advantage of a medical distributive model prescribing heroin is that it would probably be universally acceptable to all heroin users. The percentage of heroin users who will accept oral methadone is currently a subject of lively debate. Yet it is already clear that the proportion of heroin users interested in maintenance approaches far exceeds the proportion who sought non-maintenance approaches. And it is the general acceptability of the maintenance approach as well as its capacity to move to clearly defined goals that distinguishes it from other treatment approaches to compulsive narcotics users.

In summary, most oral methadone programs attempt to move the patient toward socially productive behavior by supplying an orally effective narcotic substance that prevents withdrawal symptoms and allays one postulated form of narcotics hunger (a felt sense of abnormality) while (via cross-tolerance) making it even more difficult for the patient to gratify the other form of hunger (the desire to reexperience the "rush" and the "high").

The heroin maintenance system (the medical distributive model) permits the patient to obtain both effects and is thereby theoretically universally acceptable to the heroin-using population. It makes criminal behavior to get heroin unnecessary and puts an upper limit on the price of any heroin that is sold in illegal channels. However, the route of administration and mechanics of the distribution system leave open the possibility of illicit redistribution, overdose, and various infections.

The degree to which various oral maintenance programs achieve these goals varies considerably from program to program. Indeed, even a cursory analysis indicates that some of the goals could be at cross purposes. For example, some methadone programs place maximum emphasis on retaining patients in treatment with the hope that no matter how little progress the patient may seem to be showing if he remains in treatment there is at least some possibility of modifying his behavior. Such programs may be relatively tolerant of continued heroin use, alcoholism, or failure to seek legitimate sources of support; they may also retain most patients in treatment. Supporters of other programs may believe that patients should not be continued in treatment if they show persistent evidence of heroin use, abuse of other drugs, or antisocial behavior. Such programs may have a relatively high attrition rate, but very few patients who show evidence of heroin use and many patients who are legitimately employed.

The range of outcome possibilities is described in the many excellent papers presented at the Second and Third National Methadone Maintenance Conferences.

CRITERIA FOR TREATMENT

In the United States there is a growing consensus that any chronic heroin user who has not responded to other forms of treatment should be treated with methadone if he wishes to be so treated. Some of the factors that once served to disqualify an individual from maintenance treatment (such as concomitant psychosis, alcoholism, abuse of other drugs, and the like) are gradually being discarded by most programs. It is acknowledged that individuals with multiple problems are less likely to do well, but it is also clear that some movement toward socially acceptable goals is better than none. There is still considerable controversy about what constitutes chronicity, what is an adequate trial of non-maintenance approaches, and whether there should be any absolute lower age limit for eligibility. Many workers in the field are concerned that without clear criteria many youthful experimenters will be admitted to maintenance programs by well-meaning but inexperienced practitioners. However, a rigid age limit would be of little help since there are some communities in the United States in which chronic compulsive use of heroin of several years' duration has been observed in 14 to 16 year olds. Part of the difficulty arises from the metabolic lesion hypothesis. If, in fact, prolonged use of opiates leads to a permanent or semipermanent state of abnormality, then treatment with high doses of methadone would presumably cause such a state and the initiation of such treatment would be viewed in that light. However, if as many investigators are finding, low doses of methadone are often sufficient to achieve the goals of treatment, and many individuals can be eventually withdrawn from methadone without later relapse, then the questions of criteria will become considerably less important.

Which Specific Drugs? Some of the differences between oral methadone and intravenous heroin have already been discussed. In the United States, there is general agreement that given the several disadvantages of intravenous heroin maintenance, it should be considered as an approach only when a community has done everything possible to make oral maintenance procedures maximally effective and readily available.

It is now clear that methadone is only one of a number of opioids suitable for use in oral maintenance programs. Work with methadyl acetate (acetyl methadol) suggests that it may be as effective as methadone in facilitating social rehabilitation, while its longer duration of action offers certain additional advantages. Among these advantages is the possibility of directly supervising the ingestion of each dose by having the patient come to the dispensing site three times a week. This not only obviates any possibility of illicit diversion or accidental ingestion by non-tolerant individuals, but it also makes it possible for patients to report to clinics less often than six times per week even at the start of treatment.

It is not yet certain whether methadyl acetate will cause any side effects if used at doses higher than those explored in the limited clinical studies undertaken to date. It is clear, however, that its use could eventually lead to some problems that are not particularly acute in programs now using methadone. For example, assuming that at some point clinicians permit patients to take methadyl acetate home, two problems may arise. First, if there should be an accidental overdose with severe respiratory depression, treatment with antagonists would have to be continued for more than 24 hours lest the patient slip back into respiratory depression. Since many clinicians might at first be unaware of the extremely long duration of action of methadyl acetate, a period of education for clinicians would have to accompany any policy that permitted patients to take the drug from the clinic. The second problem arises from this substance's relatively slow onset of action. Some patients could conceivably respond to this slow onset by assuming that the medication was weak or had been diluted. If they then take additional medication, they will have ingested (or injected) a potentially lethal overdose (see Jaffe *et al.,* 1970; Jaffe and Senay, 1971).

DOSAGE

Most workers now agree that there is little to be gained clinically by increasing the daily dose of methadone beyond 160 mg. At such doses some patients still complain of feeling withdrawal symptoms between daily doses, but it is unlikely that an increase in the dosage will decrease the complaining.

Based on retrospective analyses of dosage records, several workers have observed that patients maintained on doses of methadone above 80 mg. tend to drop out of treatment less readily than those on doses below 80 mg. However, several double-blind prospective studies of the role of dosage suggest that within certain limits most patients obtain most of the benefits of the maintenance program even when maintained at doses of 30 to 50 mg. of methadone per day. What marginal differences do exist between high and low doses tend to favor the higher doses, particularly in the early phases of treatment. After the third or fourth month, the differences between groups become quite small. To a degree, these double-blind studies of high and low doses of methadone tend to support the idea that suppression of "narcotics hunger" may be a more significant factor in modifying heroin-seeking behavior than is cross-tolerance-induced "blockade." In general, as compared to lower doses of methadone (approximately 50 mg./day), high doses (approximately 100 mg./day) tend to be associated with more rapid and consistent suppression of heroin-using behavior, fewer complaints of withdrawal, more side effects (e.g., sweating and constipation), and a somewhat lower drop-out rate (see Perkins and Bloch, 1970; Goldstein, 1971; Jaffe, 1971, as well as other papers presented at the Second and Third National Methadone Conferences).

It is clear, however, that many patients can get along quite comfortably on less than the amount prescribed in most programs using high dosage. Once these patients have reached the stage when they are permitted to take medication home, high doses open the possibility of "skimming"—in which the patient gradually reduces his dosage by ingesting less than the full daily dose provided and saving or selling the amount remaining. Just how the various advantages and disadvantages of different dosage levels can be balanced best is not yet clear.

SOME SOCIAL ISSUES GENERATED BY THE MAINTENANCE APPROACH

Providing narcotics to narcotics users will not be as easy to integrate into the health care system as the provision of insulin to diabetics. There are several fundamental differences. First, the maintenance substances have a serious abuse potential. Given the well-established relationship between availability and addiction to narcotics, society has a responsibility to devise a system that will minimize illicit diversion of maintenance substances while making treatment available to those who need it. The problem of illicit diversion can arise in two ways—from patients who seek treatment from more than one program at a given time and from patients who use less than the amount prescribed to them. Possible means of handling these issues are described below.

Another difference from the diabetic model stems from the life style of certain subgroups of narcotics users whose general behavior is such as to make their presence somewhat objectionable to those who use general medical and psychiatric facilities. In addition, most narcotics users are loath to view themselves as needing traditional psychiatric treatment. Whether we should attempt to change these attitudes of the drug users as well as the non-drug users who will come into contact with them or develop specialized facilities for the treatment of addicts is another of the decisions that must be made in the near future.

Related to questions of health care delivery is the issue of just how much additional service should be part of a maintenance treatment program. There is some general agreement that there is a small group of narcotics users who require only medication to achieve maximum social benefits, but most of the others require varying degrees of social, psychological, vocational, and medical support. If there is a question of limited human and financial resources to be allocated among a number of social priorities then the costs of such components in maintenance programs must be examined in relation to the additional benefits obtained.

The efficacy of programs that should be helping to solve major social problems cannot be judged merely on their costs in treating an individual patient, without some consideration of the entire population that requires service. It is all too easy to accept for treatment those narcotics users who require little more than regular doses of methadone to become law-abiding, productive, and nonusers of illicit drugs. This would, indeed, result in a highly efficient program in terms of cost per patient achieving a given set of goals. The dispensing of medication is the least costly aspect of the treatment. However, from a social viewpoint, such a program might have an overall low level of effectiveness when we ask about its impact on the entire population of narcotics users, including those who require considerable amounts of ancillary services (e.g., alcoholics, psychotics, and the like).

The question of how treatment in the United States should be financed is, at present, unanswerable, since the financial support for the entire health care system is being reconsidered. But, in any event, it is becoming obvious that the cost of providing maintenance medication is only a fraction of the cost that the individual drug user and society must pay when such medication is unavailable. Making such treatment available should be given a high priority.

The very availability of treatment will bring about some very interesting changes in our attitude toward narcotics users who commit crimes. At present, society is somewhat ambivalent. We know that some criminals use narcotics; we know, too, that some individuals who were not criminals prior to using narcotics engage in criminal activity to obtain money to buy drugs. Generally, it is impossible to distinguish between these groups. Both usually claim that they committed crimes primarily to get drugs, hoping that their antisocial behavior will not be viewed harshly. Usually, it is not. It is difficult to hold an individual entirely responsible for his behavior if there are no reasonable alternatives to that behavior. Once maintenance treatment becomes available in all communities without any waiting period, no person will be able to say that he committed a crime to get drugs because he could not get treatment. The ready availability of treatment may permit society to identify the drug-using criminal with greater certainty and could result in more appropriate treatment of such individuals.

A GLIMPSE AT THE FUTURE

Most reasonable observers now feel that maintenance with methadone or similar oral drugs is neither the panacea that its ardent enthusiasts would have us believe nor the ineffectual and potentially dangerous fraud that its rabid detractors had envisioned. After more than six years of experience, it is apparent that there are no long-term effects that could not have been predicted from our observations of patients over the first three or four years. There are some side effects, but for the individual under treatment the risks of treatment with oral methadone are minimal compared to the health risks that arise from continuous use of illicit, intravenously administered heroin. This balance between risks and benefits would be shifted only slightly if heroin were supplied through medical channels as it is in Great Britain. Several studies there clearly demonstrate that the intravenous use of heroin, even when the drug is legitimately prescribed, commonly leads to a number of medical complications (see WHO Report, 1970). Similarly, the efficacy of the oral methadone maintenance procedure in reducing illicit heroin use is firmly established. A number of independent investigators, basing their conclusions on regular testing of urine specimens taken from methadone-using patients, have found that heroin use is reduced substantially in most patients and is eliminated in many. This is not to say that in every methadone treatment program all compulsive heroin users give up heroin entirely, but few treatment regimens in medicine claim 100 per cent success.

These reductions in illicit narcotics use seem to be one reasonable measure of treatment efficacy. Other parameters, however, are more difficult to assess. It is true that patients in most methadone programs show significant increases in legitimate employment and some decreases in criminal activity. But, it is perhaps unreasonable to infer that methadone by itself produces a change in life style. Changes in criminal activity and gainful employment are more fairly viewed as measures of the efficacy of the rehabilitative system rather than of the oral maintenance procedure per se. For some general estimates of the degree to which heroin use is reduced, criminal activity is reduced, and employment is increased in various programs, see the proceedings of the Second and Third Methadone Conferences.

It was estimated that in November of 1970 more than 6,000 patients were in some form of methadone maintenance program and by July of 1971 this figure had more than doubled. Using traditional criteria of safety and efficacy, it is becoming increasingly difficult to consider the use of oral methadone in maintenance programs as an investigatory procedure. However, there are problems related to changing its investigatory status that are still unresolved. For example, the hazards to the community if maintenance drugs are accidentally ingested or illicitly diverted have already been mentioned. Freed from the constraints of its investigational status and used on a large scale by individual physicians, these problems of diversion would multiply. The task for the near future will be the evolution of new and perhaps unique regulatory mechanisms that will permit society to make the procedure available to those who can benefit from it, while preventing illicit redistribution due to overprescribing or multiple registration (i.e., patients obtaining narcotics from several clinics).

As programs in the large metropolitan areas of the United States grow in size, it will become more and more difficult to rely on regional registration systems that depend on voluntary participation of program directors and that utilize the kinds of patient identifiers such as name, age, and social security number that are easily falsified. In the long run we will probably require a national registry (similar to that used in Great Britain), in which participation is a requirement for operating a maintenance program.

In all likelihood, there will be a considerable period of negotiation before the precise regulatory mechanisms evolve. Private practitioners will not cede their newly gained prerogative to treat narcotics users without protest. The health regulatory groups will want to play a major role, if not the exclusive regulatory role, while the Bureau of Narcotics and Dangerous Drugs (BNDD) will insist that it has a responsibility to make certain that maintenance programs do not become sources of illicit diversion.

Yet, some regulations must evolve. The experience in Great Britain demonstrated that even a handful of physicians prescribing narcotics in unlimited quantities can create the nidus for a major narcotics epidemic. The need to develop structures and regulations to prevent illicit diversion from maintenance programs will further complicate the problem of moving the treatment of the narcotics users into the mainstream of medical care. For example, it will not be as easy for a methadone-maintained individual to move from one treatment program to another as it now is for a diabetic maintained on insulin.

What proportion of the estimated 250,000 heroin users in the United States will seek treatment with methadone is not clear at present. The estimates range from 50 to 75 per cent of the total; this leads to a projection of a treatment network with a capacity of more than 100,000 people. In any event, it is likely that many of these individuals who start on an oral maintenance program will attempt to withdraw from the maintenance drug within a few years after starting. We do not know, at present, what techniques are best suited to help individuals sustain the social gains achieved after they are withdrawn. Perhaps the use of narcotic antagonists will be helpful in the post-maintenance situation. Certainly, clinical trials of these drugs can be expected.

In the United States, there are two divergent trends related to maintenance, the movements of which cannot be easily predicted at this time: the pressure from some to coerce heroin users into oral methadone maintenance treatment with some form of civil commitment procedure, and the recommendation of others to provide heroin users with heroin for intravenous self-administration in a system similar to that used in England.

It is apparent that, in the United States, the increase in the numbers of individuals who are using illicit heroin has moved the problem to a level of national priority. Bringing it under control will require a number of diverse approaches, and maintenance techniques will be prominent among them.

REFERENCES

In addition to the references specifically cited, I have made use of information from numerous sources, particularly papers presented at the Second and Third National Conferences on Methadone. Selected bibliographies on methadone maintenance and the British approach are included in the Eighteenth Report of the World Health Organization Expert Committee on Drug Dependence, and more than 100 references on methadone maintenance are included in the bibliography by Langrod.

American Bar Association–American Medical Association Joint Committee on Narcotic Drugs. Drug Addiction: Crime or Disease? Interim and final reports. Bloomington, Ind., Indiana University Press, 1961.

Dole, V. P., and Nyswander, M.: Methadone maintenance and its implications for theories of narcotic addiction. In: A. Wikler, Ed., *The Addictive States.* Baltimore, Williams & Wilkins, 1968, pp. 359–366.

Goldstein, A.: Blind Controlled Dosage Comparisons with Methadone in 200 Patients. Proceedings, Third National Conference on Methadone Treatment, Public Health Service Publication No. 2172. Washington, D.C., U.S. Government Printing Office, 1971.

Jaffe, J. H.: Methadone Maintenance: Variation in Outcome Criteria as a Function of Dose. Proceedings, Third National Conference on Methadone Treatment, Public Health Service Publication No. 2172. Washington, D.C., U.S. Government Printing Office, 1971.

Jaffe, J. H., and Senay, E.: Methadone and l-methadyl acetate. J.A.M.A., *216*:1303–1305, 1971.

Jaffe, J. H., Schuster, C., Smith, B., and Blachly, P.: Comparison of acetyl methadol and methadone in the treatment of long-term heroin users: a pilot study. J.A.M.A., *211*:1834–1836, 1970.

Kolb, L.: *Drug Addiction, a Medical Problem.* Springfield, Ill., Charles C Thomas, 1962.

Langrod, J.: A bibliography of the methadone maintenance treatment of heroin addiction. International J. Addictions, *5*:581–591, 1971.

Perkins, M., and Bloch, H.: Survey of methadone treatment programs. Amer. J. Psychiat., *126*: 1389–1396, 1970.

Proceedings, Third National Conference on Methadone Treatment. Public Health Service Publication No. 2172. Washington, D.C., U.S. Government Printing Office, 1971.

Second National Methadone Maintenance Conference. International J. Addictions, *5*:341-591, 1970.

Terry, C. E., and Pellens, M.: The Opium Problem. Committee on Drug Addictions, Bureau of Social Hygiene, Inc. New York, 1928.

Wikler, A., Ed.: *The Addictive States.* Baltimore, Williams & Wilkins, 1968.

W.H.O. Expert Committee on Drug Dependence, Eighteenth Report (World Health Org. Techn. Rep. Ser., No. 460), 1970.

A RATIONAL THERAPY OF OPIATE DEPENDENCE: NARCOTIC ANTAGONISTS*

MAX FINK, M.D.†

Proposed therapies of opiate dependence are chiefly either remedial and rehabilitative, attempting to redress injury, or substitutive, providing legal delivery of opiates to sustain the addicted individual and to relieve him of the necessity to engage in criminal activity to support his habit. The conditioning theory of opiate dependence provides a different model for treatment and prevention of opiate addiction based on the use of opiate antagonists.[1]

In this view, drug dependence proceeds in several phases. A subject first obtains relief from daily tensions or discomforts by the use of drugs, including opiates. With each relief by drug use, the habit is reinforced. But, opiates reduce tension or discomfort less and less efficiently as tolerance develops. Dependence is reinforced when the addict fails to maintain an adequate drug intake and periods of tension and discomfort increase in frequency and duration.

Sooner or later each addict is withdrawn, for the supply of opiates becomes inadequate. He is admitted to a hospital for relief of his dependence by a gradual reduction in doses of opiates, or he is arrested. This post-drug period, termed "conditioned abstinence," is characterized by severe physiologic disequilibrium, with marked changes in clinical responses resulting from chronic drug use. These states persist for many months, even when detoxification is accomplished slowly.

The conditioning theory suggests that extinction of drug-seeking behavior would provide a logical reversal of the pathophysiology of the addiction. If the relief afforded by narcotics during this period of conditioned abstinence were blocked, would extinction of physical dependence and the drug-seeking behavior occur? To accomplish extinction requires the development of a substance without narcotic effects, tolerance, or dependence, which can block administered opiates immediately and completely.

Narcotic antagonists are such substances. Substitution of the methyl group of the basic nitrogen of narcotic analgesics by an allyl group yields compounds capable of antagonizing the respiratory effects of narcotics. N-allylnorcodeine was described in 1915 as preventing the respiratory depression of morphine and heroin, and n-allylnormorphine (nalorphine) was extensively studied in

* Aided, in part, by MH-13358 and based upon research conducted under a contract with the New York State Narcotic Addiction Control Commission.

† I am indebted to my co-workers—Alfred M. Freedman, M.D., Richard Resnick, M.D., Arthur Zaks, M.D.—for the work done in these laboratories since 1966 on which this summary is based.

1950. Nalorphine precipitates an acute abstinence syndrome in post-addicts receiving morphine or methadone for short periods; produces dysphoria and anxiety in non-opiate dependent subjects; and exhibits analgesic activity. Similar chemical substitution yields the clinical antagonist levallorphan, and the experimental compounds cyclazocine and naloxone.

Nalorphine was first suggested for a clinical trial of blockade for extinction of dependence, but its short duration of action and high incidence of hallucinogenic effects made its use unsuccessful. In 1963, cyclazocine was shown to have a longer duration of narcotic antagonism and fewer side effects and it was introduced for clinical trial.[2,3] Cyclazocine use may also be associated with some of the unpleasant effects of opiates, it must be administered daily, and restrictions in supply have limited clinical programs. The development of a pure opiate antagonist, naloxone, rekindled interest in clinical trials.[4]

Since 1965, we have undertaken clinical trials of cyclazocine and naloxone to determine the feasibility of extinguishing heroin dependence. The subjects of our studies are male opiate addicts who volunteer for admission to an in-patient treatment center in a municipal hospital in the East Harlem section of New York City. The population of this community is predominantly Negro and Puerto Rican, with low income and educational levels; many receive extensive welfare assistance. They range in age from 17 to 54 with a mean of 26 years, and their duration of opiate addiction and substance abuse is from 2 to 30 years.

On admission to our study ward, the narcotic usage of each patient is estimated by inquiry. Based on their report and the severity of the early withdrawal symptoms, we prescribe a starting methadone dose, usually 20 to 50 mg. Methadone is given twice daily in reducing amounts for detoxification. Within four to seven days, the subject is usually drug-free. A period of observation without drugs allows medical and laboratory examinations before induction with an antagonist.

CYCLAZOCINE

Cyclazocine is an n-substituted benzomorphan derivative, chemically similar to nalorphine hydrochloride. It is an active analgesic and causes some subjects to experience dysphoria, vivid imagery, and anxiety with its use. In the initial clinical trials of cyclazocine, 4.0 mg/day was reported to be an effective dose for clinical antagonism to opiates.[2,3]

Induction

We first attempted to develop an induction schedule that would minimize dysphoria. In our initial study, 60 patients completed detoxification and began a cyclazocine induction schedule, and 58 completed the study period.[4] Induction was defined as achieving a daily therapeutic dose of 4.0 mg. In our first subjects, the dosage of cyclazocine was gradually increased in 40 days. To reduce this time span, groups of four patients each were tested in induction periods of 10, 15, 20, and 30 days. A daily increment of 0.4 mg (10-day schedule) brought complaints of somnolence, headache, irritability, "fuzzy thinking," illusions, and, in two instances, of visual hallucinatory experiences. These symptoms were not reported on the 15-day schedule with dosages divided in two increments, except that somnolence, weakness, and irritability appeared during the first four days. The 20- and 30-day induction periods were well-tolerated without secondary symptoms.

Since 1968, we have routinely recommended a 15-day cyclazocine induction. During the past year, however, we have been able to reduce the induction schedule to four days. Reports that naloxone antagonized the opiate-like effects of cyclazocine led us to induct patients with increments of 1.0 mg cyclazocine a day until a 4.0-mg single dose was reached in 4 days. Initially, we gave each patient 0.5 to 1.0 gm oral naloxone only when they complained of unpleasant symptoms, and found that naloxone has been necessary in only a few patients. Of 40 patients inducted on this rapid schedule, naloxone has been requested in less than one-third the subjects, usually from the third to the sixth day.[5]

Heroin Challenges

A principal feature of antagonist use is the extinction of the conditioned stimulus. Extinction is best achieved by repeated exposures to a conditioned stimulus without a physiologic or psychologic reward. Extinction depends on the complete absence of any response, particularly the euphoria to the conditioned stimulus (heroin). Because heroin is the drug of choice of New York opiate dependent subjects, we have measured the duration and extent of antagonism by intravenous heroin "challenges."[6]

During the induction phase, cyclazocine as an effective narcotic blocking agent is discussed in individual and group therapy sessions with all the resident addicts. They are told that the injection of a narcotic in their usual amounts would have neither euphoric nor systemic effects. To reinforce this suggestion, they are offered an opportunity for an intravenous injection of heroin in the laboratory when they are receiving a full daily dose of cyclazocine.[4,6,7] We have determined that 25 mg heroin/2 cc saline/2 minutes elicits both the behavioral effects and EEG changes in patients seven to ten days after a final dose of methadone during the detoxification schedule, when they are "clean."

In one study, 18 patients were challenged 24 hours after single daily doses of 4.0 mg cyclazocine. Twelve experienced no clinical effects and six became euphoric. The dose for one patient was raised to 5.5 mg and there was no response to a second heroin challenge 24 hours after a dose of cyclazocine. In five patients, cyclazocine was increased to 5.0 mg and in subsequent challenges these patients responded with a slight and short-lasting euphoria. From these experiences, it is probable that the effective duration of anti-narcotic activity of a single 4.0-mg dose on chronic administration is at least 20 hours, waning rapidly after 24 hours with a peak six to eight hours after a dose of cyclazocine.[8]

After Care

After discharge, patients are asked to return three times weekly. At each visit they are given a daily amount of cyclazocine and a supply for the intervening day or days. They are seen by a therapist at weekly intervals, and call upon a social worker, teacher, or rehabilitation counselor as the need arises.

Patients are encouraged to experiment with opiates when they are taking their cyclazocine, and, on occasion, have been admitted for heroin challenges—both to encourage the extinction process and to assure us that antagonism persists. Tolerance to the anti-narcotic activity of cyclazocine has not been observed. Patients receiving a heroin challenge up to 12 months after their induction to cyclazocine show as much blockade to heroin as when the challenge was given immediately after the induction period.

Patients report the ability to interrupt their cyclazocine schedule without symptoms of withdrawal. On the day of a missed cyclazocine dose, they are able to experience the euphoric and sedative effects of heroin. By re-instituting cyclazocine at the full dosage, they do not become re-addicted, and are able to continue their daily activities.

Patients receiving cyclazocine report various changes in behavior. Social activities such as going to parties and dances and leisure time spent with non-addicts, increase. Sexual activity is increased and more satisfactory. Anxiety is experienced less frequently and drug-seeking behavior decreases.

Criminality decreases and an interest in vocational activities is enhanced. But the use of marihuana, amphetamines, barbiturates, and alcohol has not been affected and their use continues or may even increase. In individuals with poor training and poor skills, rehabilitation is still difficult. In those with psychological, family, and social handicaps, these persist and require as intensive care as if the patient were not additionally handicapped by drug dependence.

In a 1968 follow-up of patients discharged on cyclazocine a year earlier, 74 patients were traced. Fifteen continued in rehabilitation from 5 to 16 months and were found to be in school or working. Eleven discontinued cyclazocine and entered a methadone maintenance program, and 48 returned to heroin use.

In a 1970 review of patients in our aftercare facility, 12 have continued on cyclazocine for more than two years; seven for 12 to 24 months; eight for 6 to 12 months; and 13 less than six months. In a conference of clinical investigators in June 1970, seven observers summarized their experiences with cyclazocine; in 400 subjects, they reported successful induction and cyclazocine maintenance in 40% of subjects.[9]

Present Program

We have examined the social characteristics of the patients who have been successfully maintained on cyclazocine and found that these are patients with a sustained heterosexual relationship who report on a questionnaire that they do not need heroin to function or to reduce anxiety (use it for social reasons).[5] At present, we are giving cyclazocine to all patients who meet these two criteria and who volunteer for this experimental treatment.

NALOXONE

Naloxone (n-allylnoroxymorphone) is an experimental opiate derivative that is five to eight times as effective as nalorphine in antagonizing opiate effects in animals.[10-13] Tolerance does not develop nor does withdrawal occur on sudden cessation after two weeks of repeated administration. It does not exhibit opiate-like effects, even at high dosage.[4] However, its duration of action is short—less than four hours.

We have reported the effects of naloxone given in acute single intravenous doses, and in chronic oral doses.[4] Using heroin challenges as our index, we observed complete blockade of 25 mg heroin by intravenous administration of 1.0 mg of naloxone. In one study we observed that divided daily 200-mg oral doses provided effective blockade to 20 mg heroin for ten hours. Our supply was exhausted and, after a year, we undertook a second study. By increasing the daily single oral dosages to 3.0 gm, we achieved blockade to 50 mg heroin for 24 hours.[13]

Most outstanding in these naloxone trials has been the absence of any toxic effects, even on extensive and detailed tests of clotting mechanisms. These studies also were discontinued because of a failure of supplies in December 1969.

DISCUSSION

It has been suggested that narcotic addiction is analogous to a conditioned response—the addict responding to stressful stimuli in his environment with drug-seeking behavior.[1] In this theory, the repeated use of heroin without the anticipated subjective effects should lead to extinction of a learned drug-seeking behavior.

This theory was first tested by the use of high doses of methadone as maintenance treatment for addicts.[14] Because methadone is an addicting substance with euphoriant properties, these trials were not an adequate test of the hypothesis. Cyclazocine, which has a length of antagonistic action greater than 24 hours, has been successfully used in the maintenance treatment of opiate-dependent individuals.[3,4,7,8] Our limited experience with naloxone suggests that it also establishes clinically effective narcotic antagonism.

These trials are a more adequate test of the hypothesis. We have observed drug-seeking behavior to continue in these subjects. Use of opiates is much reduced, however, and the patients are able to maintain their social and work commitments without re-addiction. We have found the antagonists to be less effective in reducing the conditioned drug-seeking response than in providing engagement to a therapy program and preventing the development of tolerance and withdrawal. Engagement makes possible continued job counselling, re-education, and social services; prevention of tolerance and withdrawal allows the subject to negotiate our social system without recourse to crime.

These trials indicate that antagonists provide a basis for treating opiate dependence. However, their duration of action is short, requiring daily administration. Since they are neither addicting nor euphorigenic, the principal incentive for the patient to continue treatment is verbal persuasion.

One solution to the problem of drug delivery is the development of a long-acting antagonist, preferably effective for periods of weeks or months, thus providing a basis for extinction of the conditioned dependence without daily cooperation of the subject. Such a formulation is possibie for the antagonist may be invested in a biodegradable plastic; this method is now in animal trial.[8] A limitation of cyclazocine may be the expression of secondary agonistic effects. These effects have not been observed with naloxone, however, and this antagonist uniquely satisfies the criteria: potent opiate antagonism without agonistic or toxic effects.[4,13] Another pure antagonist, M-5050, eight to ten times as effective as naloxone in animal trials, is also under study.[15]

It is of special interest that such a long-acting formulation may serve a prophylactic as well as a therapeutic use. Opiate dependence is so rampant in some urban centers that it is progressively affecting younger and younger candidates, with deaths in teenagers, once a rarity, now commonplace. Mass inoculation with a biodegradable plastic implant of a narcotic antagonist could, in susceptible populations, severely reduce both the death and addiction rates, and provide a basis for an active prophylaxis.

Efforts at developing a long-acting antagonist are imperative until a more logical means of preventing opiate addiction is developed. The present enthusiasm for the legal distribution of methadone or heroin substitutes a legal addiction for an illegal one, reducing neither the risks of addiction, nor of death. Maintenance schemes provide for an increase in the number of addicted persons, and an increase in the number of delivery centers. When thousands of addicts inhabit the cities of the nation, there will be an increased amount of licit, as well as illicit, drugs in the community. Most discouraging, however, is the failure to consider the means of withdrawal or treating these opiate-dependent individuals. It is an act of faith that means for withdrawal will be found, particularly since little effort is focused on this problem.

The narcotic antagonists provide a logical therapeutic and prophylactic possibility. In view of the need, further experimentation should be encouraged.

SUMMARY

Based on a conditioning theory of opiate dependence, various substances have been proposed to block the euphoric effects of opiates and reduce dependence by a process of negative reinforcement. Clinical experiences with cyclazocine and naloxone are described, indicating that both are effective therapeutic agents. The limitation of the present delivery systems and the characteristics of an ideal antagonist are described.

REFERENCES

1. Wikler, A.: Conditioning factors in opiate addiction and relapse. In *Narcotics,* Wilner, D. M., and Kassebaum, G. G. (Eds.). New York, McGraw-Hill Book Co., 1965.
2. Martin, W. R., Gorodetzky, C. W., and McLane, T. K.: An experimental study in the treatment of narcotic addicts with cyclazocine. *Clin Pharmacol Ther 7*:455–465, 1966.
3. Jaffe, J. H., and Brill, L.: Cyclazocine, a long-acting narcotic antagonist: Its voluntary acceptance as a treatment modality by narcotic abusers. *Int J Addict 1*:99–123, 1966.
4. Freedman, A., Fink, M., Sharoff, R., and Zaks, A.: Clinical studies of cyclazocine in the treatment of narcotic addiction. *Amer J Psychiat 124*:1499–1504, 1968.
5. Resnick, R., Fink, M., and Freedman, A. M.: A cyclazocine typology in opiate dependence. *Amer J Psychiat 126*:1256–1260, 1970.
6. Zaks, A., Bruner, A., Fink, M., and Freedman, A. M.: Intravenous diacetylmorphine (heroin) in studies of opiate dependence. *Dis Nerv Syst (Suppl) 30*:89–92, 1969.
7. Freedman, A., Fink, M., Sharoff, R., and Zaks, A.: Cyclazocine and methadone in narcotic addiction. *JAMA 202*:191–194, 1967.
8. Fink, M., and Freedman, A. M.: Antagonists in the treatment of opiate dependence. In *Modern Trends in Combatting Drug Dependence and Alcoholism,* Phillipson, R. V. (Ed.). London, Butterworth, 1970, pp. 49–59.
9. Fink, M.: Narcotic antagonists in opiate dependence. *Science 169*:1005–1006, 1970.
10. Sadovc, M. S., Balagot, R. C., Hatano, S., and Jobgen, E. A.: Study of a narcotic antagonist N-allylnoroxymorphone. *JAMA 183*:666–668, 1963.
11. Foldes, F. F., Lunn, J., Moore, J., and Brown, I.: N-allylnoroxymorphone: A new potent narcotic antagonist. *Amer J Med Sci 245*: 23–30, 1963.
12. Jasinski, D. R., Martin, W. R., and Haertzen, C. A.: The human pharmacology and abuse potential of N-allylnoroxymorphone (naloxone). *J Pharmacol Exp Ther 157*:420–426, 1967.
13. Zaks, A., Jones, T., Fink, M., and Freedman, A.: Treatment of opiate dependence with high dose oral naloxone. *JAMA* (in press).
14. Dole, V. P., and Nyswander, M.: A medical treatment for diacetylmorphine (heroin) addiction. *JAMA 193*:646–650, 1965.
15. Reckitt and Colman in Hull, England.

PANEL DISCUSSION—NARCOTICS

Moderator: DR. ELLIOTT

Panel Members: DRS. FINK, FRASER, JAFFE, MARTIN, and MR. PARKER

Dr. Carol C. Waterman (Glen Ellyn, Ill.): Dr. Fink, how did you select your cases and what were the side effects from the antagonists?

Dr. Fink: First, the patients are voluntary admissions who come to our institution after being told they may get a variety of treatments. Those who insist upon methadone maintenance or any specific treatment are not accepted. Secondly, if they have a long history of living without a female member in a family, whether it be a mate, a mother, a sister, or a lover, we tend to reject them. We have found it advantageous to have a female to whom we can give the medication, and who can supervise the patient's intake on the intervening days between clinic visits. This maneuver has been most successful. Among the patients without such a female companion, we had a high drop-out rate.

Dr. Arthur Bieganowski (Bronx, N. Y.): If I may be so bold, I would like to ask all my colleagues to take a short time at this meeting to discuss the pros and cons of substitution methods used in drug addiction because we have a terrible responsibility. This is an international symposium. Things that go down here will be read throughout the world and since we have so many distinguished guests our words will carry a lot of weight. It is up to us all to search our minds and our hearts, as well as all that we know of the literature concerning drug addiction, in order to come to some conclusion about the applicability of the substitution program. I do have a few questions, one of which I should like to direct to Dr. Fink. Do you think that methadone masks underlying personality structure in the interest of returning heroin misusers to a more acceptable functional state?

Dr. Fink: I regret I don't understand the concept of a personality "structure" which is masked. The theory we have is that methadone prevents the subject from experiencing the direct effects of illicitly prescribed opiates by cross-tolerance. Presumably, such blockade prevents him from going into the community to obtain opiates because he no longer gets an effect. Dr. Jaffe, would you comment on that question?

Dr. Jaffe: If Dr. Wikler were here, he would ask that you operationally define the term "mask." Certainly some kinds of behaviors that we see in the addict not maintained on methadone (i.e., attempting to maintain abstinence) are just not seen in the drug user who is maintained on methadone. Now, whether you want to call these behaviors the "underlying personality" structure or merely the responses of an individual who has grown up in a particular ambience which presents a social-cultural support for such behaviors as he tries to cope with a sense of abnormality, I don't know. Behavior exhibited by patients on methadone and those attempting to maintain abstinence are quite different. I am not willing to say that methadone "masks" anything. I will point out that I have seen people who for 20 years could not function socially, who, when maintained on methadone, function very, very responsibly and in most situations quite ethically.

Dr. Bieganowski: I just want to ask one more question. Would each of the panel answer this question in turn? If your son or daughter were a heroin user, would you put him or her on a methadone maintenance program?

Dr. Elliott: I don't think we have to answer that one at the moment. I would like to point out that as the week progresses, more and more methods of treatment are going to be discussed; certainly no one here believes there is only one treatment for addiction. I particularly like Teresa Graves' variation on a theme which we heard last year—unfortunately as you know, she is not on *Laugh-In* this year—"different strokes for different folk." So, let's look at all the methods as they are presented to us.

Dr. James M. Gracey (Denver, Colo.): I would like to address a question to Mr. Kenneth Parker. Since the reliability and accuracy of the thin-layer chromatography method are so important to the successful functioning of rehabilitation and methadone programs, please explain in more detail the degree of reliability of this test. I have had hundreds of consultations with addicts concerning the results of this test. On many occasions I have been convinced that the test results were wrong. On other occasions when no positive results were obtained, I have been equally convinced that an addict was using drugs. I might also add that I have had numerous experiences where I have been very satisfied with the results. Please comment on the reliability.

Mr. Parker: Our urinalysis screening test services are being used with a high degree of satisfaction by a number of agencies administering methadone as a means of treatment. Since large concentrations of methadone and its metabolites are encountered in the urine of these patients, these are a potential source of test interferences which could contribute to unreliability with some testing methods employing the thin-layer chromatographic technique. However, the test employed by our group was designed to circumvent this potential problem and does reliably detect narcotics, amphetamines, and barbiturates listed at the 1, 2, and 3 microgram per milliliter concentration, respectively, when present in the urine. Urines which test positive are retested using a different solvent system for confirmation before reporting the results. A negative test result would be expected in those instances where the extent of use was not great or where the period of time following the last use was such that sufficient quantities of the drug are no longer present in the urine sample provided. In some instances metabolites are detected which are used as secondary means of confirmation that certain drugs or combinations were used.

Dr. John G. Cowl (Towson, Md.): I should like to direct my questions to Dr. Jaffe. The first few will be short numerical answers and the others more of a theoretical nature. In your own project, Dr. Jaffe, do you actually accept all comers or not?

Dr. Jaffe: All comers who have used heroin for at least two years.

Dr. Cowl: I understand that at various times somewhat more than 25% of the people in your program have been charged with felonies after entering your program. Can you corroborate or negate that?

Dr. Jaffe: Could you clarify the question?

Dr. Cowl: How many people in the methadone program still have felony charges on them or have been removed from the program as a result of felony charges that were incurred after starting methadone?

Dr. Jaffe: The answer to that is rather complex. First, because of technical problems, we have been unable to gain access to the actual "rap" sheets in the state of Illinois. We have not had significant cooperation from the FBI or other agencies. We are, therefore, limited to the patient's own statement of the number of felonies and convictions and arrests preceding and after treatment. Now we do have some data with a certain amount of reliability, and we have gained access to some reliability checks that are being made. However, we have found that less than 10% of people entering the methadone aspect of our overall program did not have an arrest or conviction in the preceding two years. Thus, arrest-free patients are quite rare. We were dealing with a group of chronic heroin users with a mean age of 36 and average duration of narcotics use of 14 years on the south side of Chicago. Following treatment we looked at arrest rates and I can supply more precise data. The mean arrest rate for free-man week as reported by the patient is .03 arrests per free-man week. That calculation is made on the assumption that you cannot look only at raw arrest rates because the man who has been on the street for only six months for the two years immediately preceding treatment may have been arrested five times in six months, as opposed to a man who is continually non-incarcerated, and who with the same number of arrests has had fewer arrests per free-man week. Secondly, arrests reported by the patients following admission to a methadone program has been approximated at .01 arrests per free-man week, a decrease of one-third of pre-treatment rates. That includes arrests for any cause. There are occasions where patients have been harassed and arrested merely for loitering. That arrest is included in our data. We are not using conviction rates because our general experience has been that when jails and dockets are overcrowded, judges tend to give continuances, so that the conviction rate for those in a treatment program may become an erroneous index of decreased criminal activity. In other words, the tendency to convict as opposed to giving a continuance or something of this sort changes as a result of entry into a treatment program. Nevertheless, with all these biases, we have a significant drop. We think it is a minimum reduction of one-third of pre-treatment arrest levels. Lastly, we might point out that there is another error in most of this. When we cannot find someone who drops out of treatment, we do not report their arrests. Thus, we are only talking about those people who stay in; the others become fudge factors. These are the best data I can give you at this time.

Dr. Cowl: Earlier in your comments you stated that there is no substantial hope for characterological change in heroin addicts; this seems to set your goals for a humane, mature existence rather low in that particular patient population. We have also heard this morning from Dr. Martin and I heard you on the program, *The Advocates,* when you failed to answer the following question posed to you by Dr. Carl Deisler, the Medical Director of Synanon. What you and Dr. Martin have corroborated this morning is that, using the usual physical diagnostic techniques, one cannot tell the difference between a controlled abstinent patient and a methadone-maintained patient. I wonder how you explain, in light of both of those statements, the lack of hope for characterological change, and the lack of really measurable difference in a macroscopic way between methadone-maintained people and abstinent people. How can you support the program when you know that there are such places as Synanon where there are literally thousands of people living absolutely free of drugs, who are transformed characterologically, being directors and junior executives of a multi-million dollar corporation entirely within legal limits, serving their local neighborhoods, and so forth? Before you answer that I would just like to comment that I find it shamefully unfortunate that this symposium has excluded from its roster of experts a representative of Synanon, a group which has led the way in residential treatment for characterological disorders and drug addiction at no cost to the public. I underline *no cost* to the public and the cost to themselves of $3.00 a day, which is significantly lower than any Federal program that has been initiated, and that money they raise through their own initiative, work and through private donations.

Dr. Jaffe: I see that since you sat down it was a rhetorical question. You expected no answer and I will not attempt, therefore, to answer.

I don't think you heard me carefully when I made my statement. It was well qualified. I tried to point out that most of us have observed that fundamental characterological change is rare; one can expect that, when treated pharmacologically, those people most deeply involved in antisocial activity will show less change than those with less criminal involvement in terms of life-style under these circumstances. Therefore, the program as it is currently designed, emphasizing pharmacological rather than behavioral modification technique, cannot realistically be expected to produce those profound changes. I will not enter the lists with you under these circumstances with respect to whether Synanon produces such characterological change because we will get into questions of definition and of measurement; how such characterological change would manifest itself if we could define it ever—I doubt if we would ever agree on definitions—and whether it would be manifest on the street as opposed to the Synanon environment. We could also argue whether, in fact, the treatment is as effective as you think it is. First of all, I believe that you may not have been listening carefully to the qualifications of the methadone approach, and second, you ask something about, "Can you tell the difference between somebody on methadone and somebody who is not?" I tried to make it very clear that with the average techniques of clinical interview, you cannot. I hope I made it clear that with sophisticated instrumentation you can.

Dr. Cowl: My statement was a reiteration precisely about that, and therefore, with the phrase "drug hunger" over the Conference, I wonder why people insist on talking in terms of giving drugs.

Dr. Jaffe: I suppose it is also refuted from another point of view from the existence of thousands of people who have spent three or four or five or ten weeks at Synanon and are back on the street, some alive, some dead of an overdose. The issue to which we are addressing ourselves here as professionals interested in the broad topic of rehabilitation is not whether one method is superior to another, but how to develop a spectrum of services that will provide some kind of social rehabilitation for a wide variety of individuals. Now, there is enough data on Synanon-like organizations to indicate that they are effective in a rather select group. If you have data to indicate that the general population will respond to Synanon-like organizations, I wish that you or the people in Synanon would step forward and present those data. All reputable people who have looked into this have come to the conclusion that therapeutic communities may be extremely useful for certain groups and, indeed, some people who enter those communities do seem to exhibit marked characterological changes, at least changes in behavior. They also find that of the number who enter, a relatively small percentage stay long enough to exhibit those changes, but there are virtually no data to indicate that those changes are sustained once people leave that environment.

Dr. Elliott: I repeat that we are going to hear more about Synanon and all other types of treatment during the week; so please, in the words of the modern generation, "Cool it."

Dr. David G. Logan (Cleveland, O.): I would like either Dr. Fink or Dr. Jaffe to respond. It seems that there are two ways of nullifying the effects of intravenous heroin in the *established* heroin addict, that of cross-tolerance by treatment with methadone and related agents, and as opposed to that, the broad category of antagonists. I am curious, aside from the effects when these agents are withdrawn, what advantages are sought through searching for more antagonist agents?

Dr. Fink: There are a number of problems with the methadone model, and there are a number of potential advantages to the antagonist model. If one is willing to accept the Wikler and Martin experiments as a basis for a treatment model, you could conclude that one way to help a person stop his opiate dependence might be a deconditioning process. Possible advantages of a narcotic antagonist are its safety, lack of tolerance and dependence, ease of withdrawal, and non-salability. At the present, however, methadone is the treatment of choice because it is practical, inexpensive, operates in the majority of our volunteers, and is effective. But we are seeking other routes because of the potential and perhaps real limitations of methadone.

Dr. Logan: Is there any evidence that people withdrawn from cyclazocine remain abstinent for longer periods than people who have been maintained on methadone?

Dr. Fink: There is little evidence because we have been unsuccessful in withdrawing any patients from methadone. We have been successful in withdrawing a few patients from cyclazocine, but the numbers are much too small to be useful.

Mr. Leonard Williams (Flint, Mich.): Dr. Jaffe, you indicated that there is no euphoria from methadone. I am wondering then why you get a situation where methadone is being sold illegally on the white market?

Dr. Jaffe: I think it is a legitimate question. As a matter of fact, what I tried to say was that methadone maintenance was criticized as being a technique which substituted one euphoria agent for another. I then said that such criticism would be a moralistic objection even if it were true. My other comment was that it turns out that once somebody becomes tolerant to methadone, as is the case in a methadone maintenance program where someone is on the same dosage for many months, the amount of euphoria, if any, that they experience is indeed minimal; they cannot really tell whether they have methadone or placebo, and clinicians cannot distinguish them from other people in the same room. This does not contradict what Dr. Elliott said in his introductory remarks—that methadone was a narcotic substance capable of inducing euphoria. Indeed, in England it is used intravenously and produces a very nice high. It is a euphoriant agent, and does have abuse potential. Given to the non-tolerant individual, it produces a very pleasant "buzz." It is sold illicitly for two reasons and, as you heard, I really emphasized that one of the major limitations on methadone is our need to control illicit diversion. There are two reasons for this. First of all, as with any other drug, it can give the non-tolerant individual euphoria. Secondly, in most situations we have not developed sufficient treatment resources to provide methadone maintenance for all people who want it. There are a number of people who do not want to be on heroin, who would like to be on methadone and have very few compunctions about trying to buy some to create their own program while they are waiting for treatment. Now, what the situation is in your community—as to why it is being sold—I cannot say. I would say it is probably a combination of both, a few people who would like to get high and a few people who would like to treat themselves while they are waiting for the treating agencies to make resources available.

Mr. Williams: Dr. Jaffe, I am sure you are quite familiar with all the programs going on in Chicago. There are a number of abstinence houses; there is the drug antagonist program as well as the methadone program. Can you tell us whether the arrest rate went up or down? Is there a great deal of difference in the behavior of the people you know to be graduates of the various programs in terms of arrest, of employment, of going back to school, and so on?

Dr. Jaffe: I would like to start by describing the situation in its appropriate context. The groups, as it turns out, are not comparable. There are some graduates of Gateway House, for example, who are really doing remarkably well. There are some people who were on cyclazocine who are doing extremely well. But the characteristics of these groups turn out to be slightly different. The numbers in each group, the percentages of those who enter and finally complete treatment are different. We also have people who have relapsed, but the presentation of comparable data from different kinds of programs is far beyond what I think is appropriate in a panel discussion. I think that kind of discussion deserves considerably more time than I can possibly devote to it, since any short answer might be misleading.

Dr. Elliott: There is one other aspect of the problem that we are not going to discuss because there are no hard data on it, and that is the spontaneous drop-out rate from the use of heroin.

Mr. Williams: One more quick question. Dr. Fink, if a patient misses his dose of an antagonist, will he feel the effects of heroin if he is still shooting heroin? And if he cannot feel that dosage, is there a greater possibility of an overdose and death?

Dr. Fink: In theory and in practice, if a subject is taking an antagonist with the usual dosages of opiates available in the streets—I emphasize the word "usual"—the chances of an overdose are significantly reduced. In our laboratory we customarily give our subjects on an antagonist about 25 mg heroin as a challenge dose. This is between 10 and 15 times the usual street dose. From this large dose, they have neither the euphoric effects nor the physiological (EEG, pupillary) effects of heroin.

Mr. Williams: You mentioned your patients on an outpatient status. How do you determine whether they are using drugs?

Dr. Fink: We ask them. We encourage our patients to use opiates occasionally when they are protected by an antagonist. In theory, a subject taking an antagonist will feel no effect from an opiate for which he is paying dollars. The more opportunities he has to learn that opiates have no effect, the more likely he is to go through the process of deconditioning and lose the need for opiates. Recently, we have also begun spot-testing urine.

Dr. Alfred R. Heasty (Lafayette, Ind.): Maybe I came to the wrong conference. I have two patients in my private practice who have to take two or three shots of Demerol a week for headaches. Is this addiction, or am I talking to the wrong people?

Dr. Fraser: You would not get physical dependence on this dose. In other words, you need to take narcotics continuously around the clock, so to speak, for at least a week to acquire some degree of physical dependence. Yours is taken for acute symptoms and given by the doctor, not by the patient. In other words, if the patient does not have this available to take himself, I don't think it should be classified as addiction.

Dr. Marshall M. Johnson, Jr. (Pittsburgh, Pa.): I wish you would elaborate more on when we can expect acetylmethadone. What company makes it, and what are the dosage levels? My second question is, what are the inroads of Numorphan in the community at the present time?

Dr. Jaffe: Well, I can elaborate more on acetylmethadone and the dosage. Our experience with dl-acetylmethadone indicated there may be some side effects associated with that isomeric mixture. We recently did a smaller study, again double-blind, with just l-acetylmethadol. We found that a milligram for milligram substitution of l-acetylmethadol for methadone carried people very comfortably for 48 hours. Neither patients nor their physician could tell. It did take an increase to carry people for 72 hours. The ratio was about 1.3 mg of acetylmethadol to whatever the dose of methadone that the individual was on. Again, as far as we could tell, the clinician who was regulating the dosage could not tell the difference between the methadone-maintained patients and the acetylmethadol-maintained patients. It was a small group, but we did it reasonably well with double blinds, and I think it has promise. We also, as I said, have done another study since the one we published which seemed to support our original findings. As far as its availability is concerned, this is a very difficult problem. A few kilograms were synthesized for NIMH and are being used in further studies. I might say that this was made available through the courtesy of Merck, Sharp and Dohme who also made it available to us. In our discussions with the company, there did not seem to be any great enthusiasm for synthesizing large amounts, even when that was paid for. I don't understand the lack of enthusiasm but I might say in all fairness to them that Dr. Fink found a similar lack of enthusiasm on the part of a large company in synthesizing Naloxone. Is that correct? This does not have a great deal of commercial appeal. How soon it will become available, I cannot say. I know that when our supply runs out we are out of business. We hope by that time we will have demonstrated that it is a reasonable substitute and may have certain advantages.

Dr. Fink: We, also, have tested l-alpha-acetylmethadol, and found in our first six patients that 30 to 45 mg on Monday, Wednesday, and Friday maintained subjects without withdrawal symptoms and without drop-out for two months. If this is verified, such a long-acting synthetic will materially reduce the incidence of one of the principal objections to methadone, that of illegal spillage.

Dr. Johnson: How about the Numorphan? It is becoming more prevalent an addiction. I have heard complaints that when the heroin is too weak to get a good high, they switch to Numorphan.

Dr. Jaffe: I confess that if this is happening, it is not happening as yet in Chicago and I have no data on it at all.

Dr. Elliott: The pattern is, of course, that various drugs are used as they are available in different communities. Any one of these can be substituted for heroin if that becomes weak or unavailable.

Dr. Walter D. Meester (Grand Rapids, Mich.): I would like to direct my question to Dr. Jaffe. Dr. Pierce has been seeing patients and also has contemplated putting them on methadone. He directed a question to Dr. Bewley which was not completely answered, not because Dr. Bewley did not want to answer the question, but there were about five or six questions he had to answer in a row. This one escaped his attention. One of the problems we have in Grand Rapids and, I am sure, in the some 170 other programs throughout the country where they have methadone-maintenance programs is, what do you do in the interim period? There is a fair amount of pressure from referring physicians who want their patients to go on our methadone program, and who ask, "Why don't you take this patient right away? Why do you have to take time to process these people?" I understand that in your program in Chicago, which is, of course, much larger than ours and larger than most of the programs elsewhere in the country, you have 200 to 300 people on the waiting list at all times. I wonder if you could give some comments and advice to the people who are offering methadone maintenance, especially small programs like ours. What should we do in the interim period?

Dr. Jaffe: There are several possibilities. I do not think you should consider yourself unique in medicine. That is, there are people who wait for all kinds of services all over the world; there is always great pressure upon people in medicine to take somebody right away. This is true of elective gallbladder surgery or anything else, and there is a need for expanding clinical services to provide them when people need them. It seems quite desperate in the area of heroin addiction for obvious reasons. One can respond in one of several ways. Obviously you will eventually have to expand the services offered. Secondly, you can provide a kind of a holding operation which some cities have described as second-rate treatment, which does nothing more than provide them with medication as one option. The other option is to work out relationships with the medical community in general to do something in the interim. I think that the AMA Council and the National Academy of Science-National Research Council, as early as 1967, described it as ethical for physicians in the community to provide under carefully controlled conditions some medication for patients who have a verified appointment to enter into a treatment program. So there are options available; we should not use a pat formula, but the approach should be based on the resources available and the foreseeable potential for your making long-term treatment available. I think this is something that has to be discussed on a program-by-program basis with the realization that you are not going to solve in this area of medicine something that we have been unable to solve in every other medical situation. Our capacity to provide services has, in general, not kept pace with the demand for them, and I think that we will be under this pressure for some time.

Dr. Hugh DeArmond Kirkpatrick (Charlotte, N.C.): I direct my first question to Dr. Jaffe. Hypothetically, if you had a 16-year-old girl whose addiction to heroin or to narcotics was less than a year in duration, what would you do? What provisions are made? I understood you to say that a person could not enter your program unless he had been addicted for two years or more. What is done in Chicago for people of this nature?

Dr. Jaffe: When I described our program, the question at that time concerned the methadone-maintenance component, but the state of Illinois has built a multimodality program that would offer this person at least three options which do not include methadone maintenance. There would be treatment available, as a matter of fact, more readily available to people who are willing to enter a therapeutic community, for example, rather than wait for methadone. The waiting list is six months for methadone maintenance. There may be no waiting list at all for people who are willing to enter a residential therapeutic community which we also support, and which presently has excess capacity. It can take people who want to come in. We would strongly recommend to that girl that she seek admission to one of these facilities. If she refused, obviously we would have a dilemma. She would be dictating to us the treatment she wants at a time when we are not prepared to offer all of the treatments to her.

I think this gets us into a long philosophical discourse on the rights of people to decide their own form of treatment; if possible, I would like to withhold this for a more informal kind of discussion period.

Dr. Kirkpatrick: When you set a two-year addiction as one of your primary criteria for entering methadone maintenance, are you assuming that a person with an addiction like that cannot enter a therapeutic community where instead of substituting methadone or an antagonist for the drug, you utilize daily therapy sessions, hard physical labor, and reaction?

Dr. Jaffe: I make no assumptions. It is an arbitrary decision. What goes into such an arbitrary cut-off is very often as much political as it is rational. We had no data, and nobody can really say that people who enter a therapeutic community with the history that you describe will do better there or on methadone or on anything else. There are times when, like it or not, we do make arbitrary distinctions. Obviously we will not take an 8-year-old who shoots heroin once, if you want to take an extreme position. But we get into gray areas, and at some point it falls upon somebody to make policy. Over the last two years, it has fallen on me. We do have a flexible cut-off point. As data comes in, we may very well change that particular criterion. At this point it has not changed, but I want to make it clear that it is arbitrary. There are no data to indicate what would be best for that girl.

Dr. Eric Chivian (Landover, Md.): I should like to make a comment to Dr. Fraser and ask a question of Dr. Martin. Dr. Fraser, I would like to caution you, sir, that the data from people who show up at Lexington, Kentucky or from those arrested in California, in terms of drugs use may not be extrapolated to the national drug scene. From those data, you concluded that you did not feel the rate of narcotic abuse was rising as much as with some of the other drugs. I think that most of us who work in this field find that the rate of narcotics use is increasing tremendously and moving into areas where it has never been before. I wonder if you could comment on that.

Dr. Fraser: Well, I am sure part of the difficulty is that we do not have reliable statistics on the incidence of narcotic abuse and probably such statistics will not be forthcoming in the near future. It is true that the incidence of narcotic abuse has increased significantly; for example, in California there was a three-fold increase over 10 years. But, on the other hand, I think you will agree that the proportion abusing marihuana has increased enormously.

Dr. Chivian: Dr. Martin, I would like to ask about the experiment using addicted rats. You mentioned that the rate and use of narcotics were greater when compared to control rats not previously addicted. I wonder whether the intensity and duration of their use of narcotics paralleled the prolonged abstinence symptoms that they had. In other words, you drew a correlation between them, and I am wondering if in time and intensity they also were correlated?

Dr. Martin: I wish I could answer that question definitively. However, I can only answer it in broad terms. Several of the protracted abstinence signs have been observed in rats addicted to lower levels of morphine. We have studied this phenomenon only in patients who have been dependent on morphine until quite recently. We are currently studying protracted abstinence in patients who have been dependent on methadone.

Dr. Chivian: Did the time sequence correlate with the duration of the symptoms?

Dr. Martin: Yes, by and large, there is relatively good correlation between signs of protracted abstinence and drug-seeking behavior. Certain signs, however, return to normal levels earlier than others.

Dr. Chivian: Thank you.

Dr. Alfred J. Koumans (Cambridge, Mass.): I would like to ask some clarification of Dr. Fraser. In two of your slides you showed people who were admitted to the Lexington institution because of addiction to marihuana. Would you clarify what was meant by that?

Dr. Fraser: Actually, the regulations of the government under both acts require that you have to be addicted to a "narcotic" drug, specified under the law, to be admitted to Lexington. This does not mean that concurrently these patients are not addicted to other drugs. For example, approximately one-third of them were also addicted to barbiturates and had to be withdrawn from them. Actually, there are now very few patients recorded in Lexington as addicted to marihuana, even though for 85% their first experience may have been smoking marihuana. However, after they became heroin addicts, or other narcotic addicts, overall they discontinued the use of marihuana.

Dr. S. William Berg (Bloomington, Ind.): First, I can give you another use for methadone, at least by report. The pushers in Indianapolis are supposedly using it to decrease their heroin users to a level of heroin that is more economically viable for them. At that point they start to bring them back up again.

Secondly, Dr. Martin, I was impressed by your work with the physiological measures of your addicts, but the changes are so small, I am wondering if they have anything to do with addiction. First, are these changes statistically significant? Secondly, what were the addicts like in terms of their MMPI scores and their interviews and attitudes, both in the control period and during what you termed "the prolonged abstinence"? In other words, did these addicts seem sick in the control period, and also during the prolonged abstinence?

Dr. Martin: I will field these several questions as best I can. Most of the signs of protracted abstinence were statistically significant. With regard to the relationship of these findings to addiction and relapse, I hope that I modestly presented them and indicated that this is a working hypothesis. I tried, however, to spell out the reasons we think they may be of importance. We administered the MMPI and the ARCI in this particular experiment prior to addiction, during addiction, and during early abstinence. Most of our patients showed a marked elevation on the psychopathic deviate scale, and this was not changed by chronic administration of morphine. Chronic methadone administration does not alter psychopathic deviate scale scores. Chronic morphine and chronic methadone administration produce an increase in hypochondriasis, as well as decreased motivation and increased lethargy.

Dr. Berg: And these indicators paralleled your physiological changes?

Dr. Martin: No, the relationship is very complicated. In morphine addiction there is a complex of signs which I would like to call sympathetic activation. As you may have remembered, when patients were chronically intoxicated on morphine, their blood pressure, temperature, and pulse rate were elevated. In addition, patients sleep less; that is, they had less rapid eye movement (REM) sleep and less slow-wave sleep. Dr. Seevers would probably call these stimulatory changes. These types of changes are seen both in man and rat, and to a degree in the monkey. At the same time, the sedative actions of the morphine which Dr. Elliott has enumerated, such as "going on the nod" and "coasting," are present and are called depressant or narcotic effects. At first glance, the mood changes tend to be on the depressive side; however, curiously, they also have a dysphoric component.

Dr. Berg: Did the mood changes approach normal, then, on prolonged abstinence, or were you able to tell?

Dr. Martin: No. In prolonged intoxication, they do not approach normal behavior.

Dr. Elliott: A complex drug acting on a complex organism.

Dr. Gustave A. Batizy (Brecksville, O.): I have a brief question. Eight or nine years ago, there was a report of the use of antidepressants in morphine addiction. Since morphine is a sympathomimetic substance, and in view of the new catecholamine theory of affective disorder, would antidepressants be useful? In private practice, I have found antidepressants useful for addicted patients. I also believe antidepressants are extensively used for this purpose in Iran and Persia.

Dr. Fink: I think there is no hard evidence that the use of antidepressants is effective, but on the other hand, there is no evidence that they are not. It just has not been rigorously investigated. Perhaps it should be. The reports from Iran are quite inadequate, using monoamine oxidase (MAO) inhibitors in mixed addictions without controls.

Dr. Batizy: The second question would be for anyone of the panel. Could you give me some information about reported overdoses of methadone and methadone death?

Dr. Elliott: Do you mean death of patients on the methadone programs?

Dr. Batizy: There was a patient in October in Philadelphia, I believe, who died following self-administration of an overdose of methadone.

Dr. Elliott: Methadone is a standard narcotic, and in the non-tolerant person it kills by respiratory depression. Anything further anyone would like to add to that?

Dr. Kenneth Schooff (Detroit, Mich.): I have found methadone very useful at less than blocking levels in the young adult, as a means of getting patients into treatment. I wonder if the agonistic effect of cyclazocine might be useful, perhaps more useful, than something like Naloxone from this view. And also, do your cyclazocine patients come once a day for cyclazocine? How difficult is this, and how much do you need to use family members?

Dr. Fink: The difficulty in getting our patients to take cyclazocine is really no greater than with our methadone patients. They come, on the average, three times a week. Some cyclazocine patients are now coming in once or twice a week, and we have had a few patients on methadone also coming in once a week. There is little problem in getting their cooperation. This is a self-selected group of patients who volunteer for a treatment, being told they will receive either treatment A or B or C. The treatments are not popularly established; they are primarily experimental.

As far as using low doses of methadone in young persons, we have tried in a few cases to keep patients in the community on low doses, and we have failed each time. These were not young people, but adults from the professions in whom we felt that motivation was high. And it did not work. I don't know what would happen in a young person.

Regarding the agonistic effects of cyclazocine, it seems that when we have carried a few patients on low doses the agonistic effects are present, but tolerance develops very rapidly, and by the third or fourth day we do not see agonistic effects. This makes it possible for us to develop a rapid four-day treatment. An interesting question is why we were able to do it in 1969 and not in 1966. I think a large part of it has to do with the fact that in 1966, at the start of our program, with a new compound in a new setting, the anxiety of the staff, including myself was very high, and we were very careful. Given the support of an antagonist, such as Naloxone, we have been able to do things we could not do before. It has been a surprise to me as well as to my staff that we can now do it. By the fourth or sixth day, the agonistic effects of cyclazocine are gone. The patients tolerate cyclazocine very well and are willing to come back as often as three times a week. Of course, we give them something else when they come back, for this is not a drug delivery system alone. We make every effort to try to understand the environmental factors in their addiction, and help them where appropriate.

Dr. Elliott: Time passes, and everything must come to an end. At this time I would like to thank our panel for their stimulating presentations and discussion and our audience for their interest and participation.

Afternoon Session
Tuesday, November 10, 1970

Presiding: Jerome H. Jaffe, M.D.

THE EPISTEMOLOGY OF DRUG ABUSE

ABRAHAM KAPLAN, Ph.D.

"The world of the happy," Ludwig Wittgenstein once remarked, "is quite another than the world of the unhappy." What then is the world of the unhappy victims of drug abuse? What is their *Weltanschauung?* How do they see their world, themselves, and their lives? I raise the question of the relationship between the drug culture and society at large because in whatever ways the drug culture differs from the encompassing culture, it must be continuous with the latter, sustained by it, and perhaps to a significant degree engendered by it.

I shall present some philosophical speculations which demand, and hopefully lead to, scientific hypotheses which can be put to empirical test. I am not presuming to report any empirical findings. I hope only that sharing the perspective of the world of drug abuse will lead to the formulation of specific problems which lend themselves to empirical validation.

I find four components in this perspective. I shall be speaking, first, of the *neo-romanticism* which I believe characterizes the drug culture; second, the *cult of feeling* in it; third, the demand for *immediacy;* and finally, the *subjectivist madness.* All of these, I shall be urging, are not uniquely distinctive of those who use and abuse drugs, but can be identified as significant elements in the larger world.

The romantic movement from the end of the eighteenth century onward, in Rousseau and his successors, viewed society essentially as the locus of conformity, control, and repression of the individual. It did not occur to the romantic that society not only sets limits and imposes restrictions on the behavior of the individual, but also provides the resources with which the individual can live within those limits. Society gives us, as it were, not only the rules of grammar, but also the vocabulary, idiom, and even content of the language we employ. The romantic, however, experiences society only as repressive. He looks for what Plato had already described as a "divine release of the soul from the yoke of custom and convention." On the one side is the repressive society; on the other, the person whose individuality is experienced as being denied, indeed, whose very humanity is experienced as being denied.

Several generations after Rousseau, in our own country, Henry David Thoreau declared that if a man is out of step with his fellows it may be that he hears a different drummer; "let him step to the music which he hears." Thoreau is probably more widely influential today than he was in his own time. Particularly is he influential among those who experience themselves as alienated from society —alienated either because, as in the case of the young, they have not yet been *able* to feel themselves at home in this world, or alienated because, as in the case of the blacks and others on the fringes of our society, they have not been *allowed* to make themselves at home in this world.

I suggest that the distinctive language of the drug culture is not merely a manifestation of the criminal cant commonly developed among those who engage in unsocial or antisocial acts; I suggest that not only do members of the drug culture use a special language because they abuse drugs, but also that they abuse drugs because they use a special language—because, that is to say, the channels of communication in general are felt by them to be blocked. They feel that they can communicate only with a very special subgroup under very special circumstances, a subgroup with whom they share the romantic condemnation of a hostile, depersonalizing, dehumanizing social order.

If, as I believe, there is substance to this description of the perspective of the drug culture, it follows that today's emphasis on "permissiveness" as a causative agent in drug abuse is mistaken, perhaps dangerously so. Such an emphasis reinforces precisely this romantic image of society as essentially repressive. The image is reinforced not only by the doctrine that somehow it is freedom which is responsible for license, but also by actions undertaken by those who attack what they call "permissiveness." It is reinforced by the savagery—I know no more appropriate word—of the punishments which are inflicted, or urged to be inflicted, even for what could scarcely be objectively characterized as an *abuse* of some of the drugs.

What I find particularly distressing about this myth of "permissiveness" is both its failure to recognize the basic values of freedom in our society of individuality, and also the underlying assumption of an innate depravity in the young, as though if we give them permission to live as they choose, they will inevitably choose to be damned. I think that one could very reasonably say that young people today are not ready to exercise freedom; but I believe also that one could reasonably say, neither is anybody else. None of us is wholly ready either for freedom or for responsibility.

The social problem, in my opinion, does not stem from our failure to be more repressive. It lies, rather, in the absoluteness, and I would say also in the emptiness and the dishonesty, of so many of the norms that society imposes on the individual. I have elsewhere had occasion to characterize American morality as, from many standpoints, a tyranny tempered by hypocrisy. What many people perceive as a repudiation of morality is instead a repudiation of the hypocrisy.

I believe it is important to recognize that because our norms have been empty in a substantive human way, attempts have been made to fill that void with dangerous delusions. The young people of today are, whatever one might think of Marshall Macluhan, a TV generation; they have been exposed to TV from earliest childhood. Through TV, and other media and dimensions of experience, they have absorbed a widespread belief characteristic of our society, a belief in the *magic of ingestion*. The belief is that if only we take in the right things—the right pills, the right brands of cigarettes and beer, the right lotions, the right foods, the right soaps, and I don't know what all else—if we take in the appropriate substances all will be well with us. This belief in the magic of ingestion combines with the romantic image of a repressive society to generate the conclusion that there are certain things one could ingest and one should ingest even if a hostile society denies them to us.

When we have broadened or weakened the limits to be imposed on behavior, most often we have done so, I think, not for principled reasons, but to avoid unpleasantness. This observation points to the next theme I want to turn to—the *cult of feeling*. There is a widespread notion in our society that it is really feelings that are most significant in human life, and that the assessment of the moral or any other value which attaches to any proposed action must be carried out in terms of the specific quality of feeling that the action would engender.

The cult of feeling first of all sets itself against *thought* as a significant element in human experience. For this cult, thought, whatever its content, is in its very nature irrelevant to the most important concerns of life. In any case, thought belongs to the Establishment. It is one of the repressive institutions by which society maintains its control over the individual. At best, the greatest achievements of thought, in science and technology, are precisely the ones that confront our time with its greatest problems.

This is the perspective of the drug user. But it is also a perspective that is widespread among the anti-rationalists of our time. For instance, in the insightful play by Friedrich Duerrenmatt, *The Physicists,* which is set significantly in a madhouse, one of the characters, a physicist who has gone mad, describes himself and his colleagues: "Our knowledge has become a frightening burden. Our researches are perilous, our discoveries are lethal. . . . We have to take back our knowledge, and I have taken it back. There is no other way out, and that goes for you as well." This repudiation of thought we can also trace back to romanticism. "Gray is all theory and green the golden tree of life," wrote the young Goethe, and countless second- and third-rate romantic poets after him.

The contemporary neo-romantic reverses that great dictum of Socrates, "The unexamined life is not worth living." This declaration was the ground on which Socrates lived his life and faced his death. In our days we turn it around and say, "The unlived life is not worth examining." And we measure life only in terms of the richness and intensity of feeling that are to be found in it. To think too much is not worthwhile. It is even downright dangerous, and indeed for some, like Dostoevsky in his *Notes from the Underground,* "To be too conscious is a kind of illness." In this attack on reason the abuse of drugs is rationalized as a flight into health, away from the sickness of the man of thought.

The cult of feeling also denies that man can find in *action* any basis of value or meaning in his life. Action is futile, empty. Means defeat ends, and ends trivialize the means that are employed for their achievement. We must recognize, I believe, that for most of mankind in our time work is far from meaningful; it is instead a deadening routine of empty operations carried on for ends that are at best trivial and at worst destructive of everything decent and dignified in human life.

Thus, the claims of both thought and action are denied, in favor of the supposition that life can be based on sheer feeling. This supposition of drug abuse reflects and extends, I think, the widespread belief in our society that the norm of comfort, satisfaction, pleasure—in a word, happiness—suffices as a foundation of value. There is no realization that happiness itself is altogether rootless. It has no objective ground, but is a matter only of how a particular subjectivity responds in a particular moment. Many of you in the audience know how commonly it can happen that a man can die happy, a smile on his face—but die nevertheless—as in some types of anoxia, for instance, in the so-called "rapture of the deep." The victim may feel rapture, but only a madman can say that all is well with him. We make a mistake, I believe, when we try to counter the values of the victims of drug abuse by telling them that their ecstasy is purchased at the price of an eventual agony. We make a mistake because we are thereby tacitly accepting their principle of assessing the significance of action in terms of immediate feeling.

If we were to say that a man cannot achieve any lasting happiness in these terms, what reply can we make if we are confronted with the counter-question, "Very well, then, where *would* you go to achieve such happiness?" Freud says somewhere that it is the aim of psychoanalysis to replace neurotic misery with the realistic and healthy unhappiness which is the common lot of mankind. There is no need to construe this statement in a cynical way, but I *do* believe that a realistic awareness of the potential for meaning and value in human life discloses that we cannot measure the meaning and value of anything by asking simply how it might feel to us at a particular time.

At a particular time—for the drug abuser that always means at *this* particular time. This is the third of the elements that I want to discuss, the demand for *immediacy*. Such a demand is also, I think, a component of neo-romanticism. The romantics promulgated the myth of an Eternal Moment in which love and life are fulfilled, one instant in which time somehow is replaced by eternity, all truths are unfolded, and all aspirations are brought to achievement. I am convinced that this is a myth and a pernicious one. Quite the other way around, it seems to me that the foundation of every realistic perspective on human life and aspiration is a recognition of the reality of time.

It is perfectly true, to my mind, and of enormous importance, that only the present can serve as a locus of meaning and value. But the present comes to have meaning and value only because of what is imported into the present from both past and future. We are what we are now in full dimensionality, but our present experience possesses depth and content only because of all that we have been and of all that we might yet be. Ours is by no means the first generation to repudiate history, but we repudiate it today in a rather special way—not because history is bunk, as an earlier generation argued, but because it is irrelevant. We repudiate the future, not because it is uncertain and in such large measure uncontrollable, but because we take it to be unreal. And all that counts, baby, is right now and right here.

It seems to me that society perpetuates and reinforces this myth of the Eternal Moment in all the ways in which we are so exaggeratedly child-oriented. Let me not be misunderstood. I trust that I need not yield to anyone in the measureless value that I attach to all human life, and most particularly to the young, with all their unrealized potentialities. In being critical of a child-oriented society, I do not mean to derogate in the slightest degree the values, actual and potential, in the child; on the contrary, I am critical because what our child-oriented behavior amounts to is not a genuine caring for the child. It is not something that works for the welfare of the child, but is rather an indulgence that alleviates our own guilts and lessens our own anxieties.

It does both of these things at the cost of perpetuating precisely the childishness of the child. For childishness, in this connection, I see as essentially this: a refusal to postpone gratification; an incapability, indeed, of relating present effort to future outcome; an insistence that each moment and each situation be assessed as though it has no relationship with any other, does not reach out in any direction, and has neither conditions nor consequences. I venture to say, without any direct empirical evidence but based only on impressions garnered from my own limited experiences, that young people do not have to nearly the degree that used to be true a self-image of a career line, of a whole life-span—a cognitive map, as it were, of their world, where they are in it, and what roads open out before them into the emerging future. With a future as uncertain and uncontrollable as ours, such short-sightedness is understandable. Yet I think it also results in part from our reinforcing the impossible metaphysics: that only what is happening right now matters, only right now is real, nothing that has brought us to this point and nothing we will undergo as we leave this point need be taken into account in our present choices.

This leads me to the last of the mythologies of the drug culture, perhaps one which the other three already embrace—*subjectivist madness,* a phrase I take from Bertrand Russell, who with characteristic prescience coined it in his *History of Western Philosophy,* published about a quarter of a century ago. Subjectivism expresses itself very familiarly in *the principle of direct experience,* the principle that for a man to undergo a certain experience is enough for him to know everything important about that kind of experience, and if he has not himself undergone the experience, no other source of knowledge can serve. In short, the view is that what a person himself undergoes is both sufficient and necessary to an understanding of that kind of experience. One hears this view in contexts far removed from drug abuse; for example, in decisions as to staffing black studies programs, or of urban problems and such like. It is really extraordinary how widespread the notion is that if someone has lived in the ghetto he has therefore grasped the essentials of the situation, while academic standards of knowledge and understanding only serve the repressive society. And if a man has not himself had the experience, if he is not black (or whatever may apply to the case), then he simply cannot understand what is involved.

There are certain corollaries that I think would make this principle objectionable to many people if they considered its logical implications. For instance, the young, I think, would be less inclined to accept the principle of direct experience if they realized that one of its immediate logical consequences is that only the older generation is in a position to speak about power, responsibility, parenthood, and much else, because only they have directly experienced them. Another consequence is that no male can ever be a good obstetrician, and that only a madman can be a psychiatrist (though maybe here my argument breaks down).

There is also a more far-reaching implication, indeed as far-reaching an implication, I think, as any such proposition about human affairs can have. It is that the principle of direct experience is a denial of human culture as such. It is a denial of the possibility of learning from somebody else's experience, and I take culture to be nothing other than that. A group has a culture just insofar as it becomes possible, from the standpoint of the group, not to make the same mistake too often; that is, insofar as it becomes possible for those who have made the mistakes somehow to transmit what they have learned to others who have not yet made them. But in the perspectives of subjectivism what counts is not what we can learn as members of a culture, but only what we ourselves have gone through. To be sure, everything depends on what it counts *for.*

Plainly, nothing that anybody can say to me about nutrition will take the place of my own experience of eating the food, so far as satisfying my hunger is concerned—but not so far as concerns my learning whether it is safe for me to continue with that diet, and whether I might do better to provide myself with a more nourishing diet, regardless of how this one tastes to me.

Widespread in the subjectivist madness of our time is a confusion of the *relativity* of knowledge, to which we have become so sensitized in the twentieth century, with a presumed *subjectivity* of all knowledge. We know that what we observe depends to a significant degree on our observational standpoint; in the most advanced of our sciences we have learned how to specify precisely the exact form which that dependence takes. In doing so we transcend the relativity to arrive at results which are intersubjectively valid. This table before me presents a different shape to every person in this room. It must do so, in accord with the simple laws of geometrical optics. The conclusion does not follow that the table really has no shape at all, or that shape is only subjective. There is an objective shape, of which we have, to be sure, only approximate and probable knowledge, yet knowledge which does not depend on any individual's subjectivity.

Several generations back, Luigi Pirandello wrote a succession of effective plays with the theme which he took as the title for one of them, *Right You Are If You Think You*

Are. This is precisely what defines subjectivity—that thinking so makes it so. We have come to the point today where we imagine that we are being sophisticated or that we are standing up for the rights of the individual if, whenever anyone makes any statement whatsoever, we counter with "That's what *you* say." Well, of course, when I say something, that *is* what I say; I am not a ventriloquist. If I am told, "That's what *you* think," I hope so. I would like, when I declare an opinion, for it to be an honest expression of what *I* think and not of what someone else thinks and has put into my mouth. But the fact that it is what *I* say and what *I* think does not in itself make it false, and does not in itself make it true. It is not a counter to subjectivist madness to replace one subjectivity by another. I am afraid that this is what many people do, trying to reinforce authority by laying claim to some kind of absolute. The only counter to subjectivism is to insist on the reality of an objective world, a world in which our thinking so simply does not make it so.

There is much talk about the credibility gap in politics today. I am speaking here about what might be called an infinite credibility gap, an infinite gap because we imagine that there is nothing on the other side, that there just isn't any kind of truth to be established. All a man can do is to take his stand. You "pays your money and you takes your choice" and that's it; one choice is subjectively as good as another. The reality is constituted by what is within. "The mind is its own place and in itself can make a heaven of hell, a hell of heaven." This is a beautiful line which Milton very properly put into the mouth of Satan.

Coordinate with this epistemic reduction there is also a value reduction. The pluralism of values that is so important a feature of contemporary life is mistaken for a nihilism with regard to value. In the house of the Lord there are many mansions; somehow, those conditioned to absolutes suppose that in that case we must be altogether homeless. There is no place at all to go—I hear that over and over as the pathos of the unhappy victims of drug abuse. There come to mind the lines of the Gospel: "The birds of the air have their nests and the foxes have their holes, but the son of man has not where to lay his head." We are adrift in the void, so it is felt, because we recognize no other reality than our own inner experience.

In Peter Weiss's *Marat Sade,* another play which is set in a madhouse, the Marquis declares, "The only truths we can point to are the ever-changing truths of our own experience. What should you care about the world outside? For me the only reality is imagination, the world inside myself." I can very well imagine those words spoken by the victims of the abuse of drugs. There is a very good answer to his question in the very same discourse of that miserable madman. Because, as he himself says, "I do not know if I am hangman or victim." Only by looking to the world outside can we begin to discern the difference between hangman and victim, and possibly, hopefully, be able to so reconstitute the world outside and ourselves so as to make it unnecessary to be *either* hangman *or* victim.

I have been trying to make clear that the world of drug abuse is not a world of fulfillment and gratification. It is a world of anguish, of a piece with the anguish which is so deeply a part of this larger world of ours. The lives led by the victims of drug abuse are rightly described as lives of quiet desperation; but this is a description that applies as well to the lives of most of us. It is essentially because the drugs do not bring us any closer to the roots of the desperation that they constitute an abuse.

I conclude with a remark made by a master of Hassidism, a Jewish mystical sect which expressed its religious commitment in music and dance, and also in an occasional sip of brandy. An opponent, or at any rate a skeptic, of this movement's aspirations once asked the Rabbi, "Why do you Hassidim allow yourselves to become drunk on brandy?" "Oh no," he said, "you don't understand; we don't become drunk on brandy. The brandy only brings us joy; it is the joy that is intoxicating." I think the fundamental problem of drug abuse, in the largest social terms, is the problem of how to dispose of whatever resources we can yet command so as to make it possible still for men to experience joy.

DISCUSSION

Dr. Jaffe: Dr. Kaplan has graciously arranged his presentation so that there would be time for us to ask questions. My own feeling is that to do so would be somewhat like asking questions after a recital of poetry. I am sure he did not mean it that way, and yet that is how most of us experienced his superb presentation. Since this is a scientific symposium, perhaps we can put aside our perception of the poetic aspects of the presentation and ask if there might be some who wish to address questions to Professor Kaplan.

Mr. Robert Winfield (Ann Arbor, Mich.): Dr. Kaplan, you have delineated the dynamics behind the anguish of these individuals. How do you think we might approach people with this anguish—this problem—you have talked about?

Dr. Kaplan: I certainly don't want to commit myself to a specific program or platform of social action, but I would like to say something about the implications of the perspectives I have discussed. For instance, not much is gained by intensifying the severity of punishments. That does not address itself to underlying causes, which are, at bottom, a belief in another kind of magic. The history of a variety of devices of social control shows the utter futility of the kind of savagery manifested in the punishments for theft and counterfeiting, for example. In fact, there is reason to think that the repressiveness itself may contribute to the crime.

I would say also that we have to become much more discriminating than we are now concerning the various types of drugs, and the various kinds of behavior that are to be regarded as abuses. In my personal opinion, for what it is worth as a layman, marihuana is radically different from the heavy drugs, and to deal with them together as though the problem of their abuse is one and the same seems to me to be monstrous.

Finally, I think that there are implications here for dealing with the problem in an indirect way—much as we have come to recognize other illnesses—with a whole new perspective on the problem. In the history of medicine, as seen, at any rate, from the standpoint of the philosophy of science, we moved from a period in which disease was thought to be localized in particular tissues or organs to a period in which we recognized the illness of the whole organism, then beyond that to the notion—a very proper notion, in my opinion—that certain disturbances require family therapy. I am now saying that awareness of the health or disease of a whole *community* is what is here called for. The proper question is no longer, "What's wrong with that particular person?" or "What's wrong with me that I let him go wrong?" but rather, "What is wrong in this community, in which such and such segments of it try to cope with their problems in this particular way?" In these perspectives it seems to me that many specific answers would be forthcoming on which I would not presume to take a stand now.

Mr. Winfield: May I expand, then, on what you have just said, because it gets to the question of how we should work in the community to change the way the community approaches life. What ways do you see of changing the hypocrisy of our society, of changing the narrowmindedness of the individuals who grow up in our society? This is the issue which most concerns me. Do you have any thoughts on that?

Dr. Kaplan: I can only say I am very glad your question comes a week after the election instead of a week before. I would insist on the importance of creating and maintaining maximum opportunity for *everyone* to deal with just that question. That is why I said what I did about the dangerous myth, as I see it, of "permissiveness." We think that the way to cope with the problem is by imposing restrictions on freedom. At the same time we must do all we can to contribute to the power of knowledge and rationality to lead us in a realistic and responsible use of our freedom. I cannot speak more specifically here—for anybody to lay claim to wisdom would be downright foolish.

Unidentified lady member of audience: I am an American citizen, originally from Turkey, and was educated in a French lycée. So I want to say first that I am very much impressed with the presentation of the professor. I think it is very valuable in a rather dry and didactic meeting like this, where we look at things from a research point of view, to take a broad view of such problems.

Secondly, I want to present some of my ideas and humbly ask for some suggestions. You stated that perhaps feelings are emphasized in our society, subjective feelings, and naturally instead action, reason and reality should be brought to value, with which I agree. But there is a problem that I wish to refer to. Time and again you will hear from people who have a European background that in this wonderful American society there is something missing, and the very word "emptiness" that you mentioned is pertinent. When I work with my young patients, drug addicts, it is not so much that they have experienced anguish as they have not really experienced anything yet. They don't know what it is all about, but they talk of "boredom," and wonder where they'll get their next kick. "Joy" versus "pleasure" does not even have meaning. We try to explain what joy is, that there are certain joys in achieving, and so on. It is all so strange to them. What suggestions do you have, with your background and experience, to make the brandy unnecessary to the Hassidic master? How can they experience joy without the need of brandy, namely, the drugs?

Dr. Kaplan: I am not acquainted with a better analysis of joy, in psychologic terms, than the one which perceives joy as the experienced outcome of work and love, both of which are creative, and both of which reach out from the self to the other and to the world around us; in neither of these is there a preoccupation with our own inner psychic state, constantly asking, "How and what am I feeling now?" If you were now to ask what ideas I have on how to give people creative work, and on how to teach people to love, I can only reply, "Let me invite you to a symposium of philosophers, and then *you* can tell *us*."

PHARMACOLOGY OF SEDATIVE-HYPNOTIC DRUGS

E. W. MAYNERT

The sedative-hypnotic drugs are more widely used and abused than all the other psychotropic agents under current discussion. This group of chemicals (Table 1) includes ethyl alcohol, the barbiturates, some older remedies (chloral hydrate, paraldehyde, and the bromoureides), and an array of compounds more recently introduced into medicine. Although meprobamate is sometimes classified as a minor tranquilizer, it is an effective sedative and hypnotic, and its abuse differs little from that of the other drugs.

STRUCTURE-ACTIVITY RELATIONSHIPS

A striking feature of the sedative-hypnotic agents is the broad range of chemical structures that they embrace. Attempts to identify essential chemical groups or stereochemical configurations have failed, a result quite different from that obtained with the narcotics, central stimulants, and hallucinogens. Indeed, hypnosis and anesthesia can be produced by a substance as devoid of structure as the noble gas, xenon, which exists as unpaired atoms with a minimal propensity for forming bonds of any kind. Most scholars concerned with this group of drugs attribute pharmacologic activity to a physicochemical property such as lipid solubility or the ability to alter the structure of cellular water, particularly in the nervous system.[1,2] The fact that these properties are shared by many chemicals may be relevant to the finding that sedation is a common side effect of many kinds of drugs. The antihistaminics offer a familiar example.

Table 1 Some widely-used sedative-hypnotic drugs

Ethyl alcohol	Glutethimide Methyprylon
Barbiturates	Ethchlorvynol Ethinamate
Chloral hydrate Paraldehyde	Methaqualone
Carbromal Bromisovalum	Meprobamate

The large number of sedative-hypnotic drugs currently on the market raises an insufficiently recognized problem for the physician and the toxicologist. Inasmuch as the central depressant effects of these agents are additive, it is now possible to produce serious acute poisoning without exceeding the therapeutic dose of any single drug. Although the prospect is a little more remote, by successive additions of new agents to his dose regimen, a person could also become severely addicted to these drugs without ever exceeding therapeutic doses.

CENTRAL NERVOUS SYSTEM EFFECTS

Except in doses that pose a threat to life, the sedative-hypnotic drugs produce effects restricted almost entirely to the central nervous system. Under appropriate circumstances, increasing doses advance depression through a continuous spectrum ranging from mild sedation to deep anesthesia or coma. Drug-induced sleep usually differs from normal slumber in being deficient in rapid-eye-movement episodes and dreams, although recent research suggests that all hypnotic drugs may not act alike in this respect. For example, it is claimed that a new drug, Flurazepam, does not depress REM sleep or dream content.[3] In a stimulating environment, most of the sedative-hypnotic drugs may produce overt excitement resembling that seen in acute alcoholic intoxication. For many years, pharmacologists have attributed such excitement to the depression of inhibitory nervous connections rather than to direct stimulation. Electrophysiologic research has not contradicted this simple idea.

The sedative-hypnotic drugs contrast with the narcotics, central stimulants, and hallucinogens in being relatively non-selective in their actions. However, they do not depress all neural connections to the same extent. In recent years, interest has centered on the action of hypnotics on the ascending reticular formation of the brain stem. This system is viewed as providing a background of wakefulness by means of maintained activity within multineuronal circuits, and, accordingly, sedation and sleep are explained by an interference with this activity.[4] The results of many studies leave little doubt that the ascending reticular formation is readily depressed by small doses of sedative-hypnotic agents that have no effect on the classical afferent pathways. However, other parts of the nervous system including certain portions of the cerebral cortex, the limbic system, and the hypothalamus are equally or more sensitive.[5] Among the more resistant connections are some involved in the control of respiration. The loss of these reflexes after high doses is the usual cause of death from the hypnotic drugs.

Little is known about the mechanisms by which hypnotic drugs interfere with synaptic transmission in the central nervous system. The major action appears to be postsynaptic and to involve a reduction of the effects of the various transmitters. Both excitatory and inhibitory postsynaptic potentials are diminished. The drugs also impair the release of the transmitters at nerve endings. A vast amount of research has failed to relate these effects to inhibition of the oxidation of glucose or the formation or utilization of high-energy phosphates.

ABSORPTION, DISTRIBUTION, METABOLISM, AND EXCRETION

Ordinarily, addicts take sedative-hypnotic drugs by mouth. All the available compounds are rather efficiently absorbed from the gastrointestinal tract. The barbiturates available as soluble salts are sometimes injected intravenously. This route yields much more intense effects from pentobarbital or secobarbital, because these drugs rapidly reach high concentrations in the nervous system. In contrast, drugs such as barbital and phenobarbital enter the brain very slowly, and the difference between intravenous and oral doses is rather small.

After absorption all the sedative-hypnotic drugs are distributed throughout total body water. The placental barrier presents no obstacle to their widespread diffusion. The individual drugs differ greatly in their binding to plasma and tissue proteins, a fact which may be of some practical importance in connection with the efficiency of hemodialysis in the treatment of poisoning.

With the exception of barbital, which is excreted almost completely unchanged, and phenobarbital, which is eliminated unchanged to the extent of about 20%, all the sedative-hypnotic drugs are extensively metabolized. The rate of biotransformation appears to be the predominant factor in determining the duration of action of these compounds after oral administration. It is important to recognize that the rate of metabolism of a drug usually decreases steadily as the concentration in body water diminishes. Thus, it is impossible to determine precisely when the body has completely disposed of the drug. The disappearance of depressant effects is likewise not abrupt. Although pentobarbital and secobarbital are regarded as hypnotics with a short duration of effect, sensitive psychomotor tests have revealed deterioration of performance as long as 15 hours after the ingestion of the usual therapeutic doses.[6]

The final products of metabolism of the sedative-hypnotic drugs differ widely. Ethyl alcohol is largely converted to carbon dioxide and water. The other drugs are much less extensively degraded. Chloral hydrate is reduced to trichloroethanol, which is then oxidized to trichloroacetic acid or conjugated with glucuronic acid. Most of the other drugs appear to be metabolized primarily by oxidative mechanisms. In all cases, the pharmacologically inert products are more soluble in water than are the drugs. Thus, they resist reabsorption in the renal tubules and are eliminated in the urine.

ACQUIRED TOLERANCE

In contrast to the narcotics, the amount of tolerance developed to the sedative-hypnotic drugs is rather limited. Although tolerance is usually considered only in connection with continued drug administration, clearly it plays a role in the recovery from the effects of single doses. Acute tolerance can be readily demonstrated by measuring the plasma concentrations of a sedative-hypnotic drug after administration of a large dose by mouth. The common finding is that the concentration at which the subject first became ataxic is considerably lower than that at which ataxia later disappeared. Inasmuch as some of these drugs, particularly alcohol, equilibrate very rapidly between blood and brain, this result suggests some adjustment in the central nervous system that permits it to function properly in the presence of concentrations which earlier had deleterious effects.[7] Other experiments have revealed that after a large dose a person recovers with a higher plasma concentration of drug than after a small dose. A detailed study of acute tolerance in dogs demonstrated that (1) it is a limited phenomenon amounting to approximately a doubling of the concentrations corresponding to a very mild degree of ataxia, (2) it is dose-related and reaches a maximum at plasma concentrations corresponding to light anesthesia, (3) it may be fully developed in as short a time as 20 minutes, (4) it is related to the degree of neurologic derangement in that more tolerance develops to ataxia than to inability to stand and walk.[8]

The relationship of acute tolerance to chronic tolerance is still obscure. Dogs treated daily with alcohol for one year showed only a slightly greater tolerance than could be developed from a single large dose.[9] Investigations involving the repeated administration of therapeutic or somewhat larger doses of pentobarbital, secobarbital, or phenobarbital have demonstrated that tolerance to the sedative and hypnotic effects is practically complete within a few weeks. Most of the change appears within the first few days. Data obtained in human subjects taking phenobarbital provide a view of the quantitative significance of acquired tolerance. Persons experiencing severe drowsiness at plasma concentrations of 5 μg/ml on the first day were unaware of any effects of the drug at plasma concentrations of 25 μg/ml on the twelfth day.[10] However, chronic tolerance clearly has its limits. For example, a man tolerant to 1.2 gm of pentobarbital per day may show little evidence of intoxication. However, the addition of 0.1 gm per day to his schedule can produce serious poisoning.[11]

If tolerance is assessed in terms of dose rather than plasma concentrations, changes in the rates of metabolism of the drugs must be considered. The administration of certain sedative-hypnotic agents, particularly some barbiturates, chloral hydrate, glutethimide, methyprylon, or meprobamate, may increase the activity of mixed-function oxidases engaged in the metabolism of this series of drugs as well as a variety of other pharmaceuticals. This process appears to involve the synthesis of enzymes in the endoplasmic reticulum of the liver. Its contribution to chronic tolerance in man has not yet been quantified, but studies in rats suggest that it plays a major role.[12] To a large extent, tolerance developed to one hypnotic can be transferred to any of the others. Acquired tolerance disappears almost completely during one or two weeks of abstinence. Laboratory animals made tolerant to the soporific action of the barbiturates have a normal sensitivity to lethal doses. It is generally believed that man does not differ from other animals in this respect.

PHYSICAL DEPENDENCE

With the possible exception of the bromoureides, all the sedative-hypnotic drugs listed in Table 1 have proved capable of producing physical dependence in man. The abstinence syndromes involve delirium and grand mal seizures, manifestations of cerebral dysfunction rarely encountered during the withdrawal of narcotics. The differences between the hypnotic and narcotic abstinence syndromes can be accommodated by the homeostatic hypothesis of tolerance and physical dependence. The premise is that the body makes adjustments to correct the physiological distortions produced by drugs (tolerance) but that these alterations may be inimical in the absence of drugs (physical dependence). The hypnotics decrease seizure susceptibility and their withdrawal increases it. The narcotics have rather little effect on seizure thresholds.

The classical investigations of Isbell and his associates on pentobarbital and secobarbital demonstrated clearly the importance of the daily dose in producing physical dependence.[13,14] They found that when doses ranging from 0.6 to 2.2 gm were administered daily for weeks or months the incidence of convulsions and hallucinations during abstinence increased as a function of the dose. If patients were kept deeply intoxicated with amobarbital for 16 to 20 hours each day, 10 or 12 days of drug administration were sufficient to lead to a severe withdrawal syndrome.[15] Comparable data on dose and duration of abuse are not available for the other sedative-hypnotic drugs.

Recent laboratory studies in man have revealed an increase in the duration of rapid-eye-movement (REM) sleep on the night following the termination of ordinary hypnotic doses of the barbiturates. The profusion of eye movements during each period of REM was also increased. The fact that profusion of eye movements is positively correlated with the "activity" of dream content harmonizes these observations with clinical data that nightmares are more frequent and sleep duration is shorter when the use of hypnotics is stopped.[16] Although these findings may not be relevant to physical dependence, they do suggest that physiologic as well as psychologic factors may be involved in habituation to these drugs.

REFERENCES

1. Butler, T. C.: Theories of general anesthesia. *Pharmacol Rev 2*:121–160, 1950.
2. Pauling, L.: The hydrate microcrystal theory of general anesthesia. *Anesth Analg Curr Res 43*:1–10, 1964.
3. Kales, A.: Psychophysiological and biochemical changes following use and withdrawal of hypnotics. In *Sleep: Physiology and Pathology,* Kales, A. (Ed.). Philadelphia, J. B. Lippincott Co., 1969.
4. Magoun, H. W.: Brain mechanisms for wakefulness. *Brit J Anaesth 33*:183–193, 1961.
5. Domino, E. F.: Sites of action of some central nervous system depressants. *Ann Rev Pharmacol 2*:215–250, 1962.
6. Kornetsky, C., Vates, T. S., and Kissler, E. K.: A comparison of hypnotic and residual psychological effects of single doses of chlorpromazine and secobarbital in man. *J Pharmacol Exp Ther 127*:51–54, 1959.

7. Mirsky, I. A., Piker, P., Rosenbaum, M., and Lederer, H.: Adaptation of the central nervous system to varying concentrations of alcohol in the blood. *Quart J Stud Alcohol 2*:35–45, 1941.
8. Maynert, E. W., and Klingman, G. I.: Acute tolerance to intravenous anesthetics in dogs. *J Pharmacol Exp Ther 128*:192–200, 1960.
9. Newman, H. W., and Lehman, A. J.: Nature of acquired tolerance to alcohol. *J Pharmacol Exp Ther 62*:301–306, 1938.
10. Butler, T. C., Mahaffee, C., and Waddell, W. J.: Phenobarbital: Studies of elimination, tolerance and dosage schedules. *J Pharmacol Exp Ther 111*:425–435, 1954.
11. Isbell, H., and White, W. M.: Clinical characteristics of addictions. *Amer J Med 14*:558–565, 1953.
12. Remmer, H.: Drug tolerance. In *Enzymes and Drug Action,* Mongar, J. L., and de Reuck, A. V. S. (Eds.). Boston, Little, Brown & Co., 1962.
13. Isbell, H., and Fraser, H. F.: Addiction to analgesics and barbiturates. *Pharmacol Rev 2*:355–397, 1950.
14. Fraser, H. F.: Tolerance to and physical dependence on opiates, barbiturates, and alcohol. *Ann Rev Med 8*:427–440, 1957.
15. Alexander, E. J.: Withdrawal effects of sodium amytal. *Dis Nerv Syst 12*:77–82, 1951.
16. Oswald, I.: Drugs and sleep. *Pharmacol Rev 20*:273–303, 1968.

CHRONIC ABUSE OF SEDATIVE-HYPNOTIC DRUGS

CARL F. ESSIG, M.D.

In this presentation we define abuse as the self-administration of certain sedative-hypnotic drugs in excess of, or beyond, medical need.[1] Only those sedative-hypnotic (SH) drugs that are considered to have abuse characteristics like the barbiturates will be discussed. Some of these drugs have been called minor tranquilizers, central relaxants, neuroleptics, psychotropic agents, etc. Drugs that have been so categorized but which have abuse characteristics like the barbiturates will be referred to here as SH drugs. The phenothiazine and reserpine group of drugs (major tranquilizers) are not considered to have barbiturate-like abuse potential and will not be discussed in that context.

The general topic of SH drug abuse has been reviewed elsewhere chiefly for clinical purposes.[2,3] Another review is presented here both for that reason and to add some more recent information.

There is no intent to disparage the rational use of SH drugs in medical practice or to imply that excessive numbers of patients are prone to abuse such agents.

OCCURRENCE OF SH DRUG ABUSE

Bakewell and Wikler studied the prevalence of physical dependence on SH drugs in psychiatric patients.[4] They gave test doses of pentobarbital to determine which patients were physically dependent on SH drugs and reported that 6.8% of the 132 patients admitted to the psychiatric ward during a 14-month period had been abusing one or more SH drugs. Of the SH abusers, 66% had abused glutethimide or mixtures of it and either meprobamate or chlordiazepoxide.[4]

A chart study of admissions to the psychiatric service of another university hospital indicated that 7.6% of 1,686 patients admitted during a three-year period had abused SH drugs.[5]

Two surveys of admissions to the U. S. Public Health Service Hospital (now NIMH Clinical Research Center), Lexington, Kentucky, indicate that an increasing proportion of narcotic addicts also abuse SH drugs. The chief criterion in both studies was the need to undergo withdrawal treatment for SH drug dependence soon after admission to the hospital. In 1957 only 18.0% of the narcotic (opiates, opioids) addicts required treatment for withdrawal from SH drugs.[6] Another study done in the early 1960's indicated that 22.8% of narcotic addicts were physically dependent on SH drugs at the time of admission.[7] By 1966 the proportion of such patients who required treatment for SH withdrawal had increased to 35.0%.[6]

It is evident that data derived from psychiatric patients and narcotic addicts regarding SH drug abuse cannot be extrapolated to the general population. Unfortunately, various surveys of psychoactive drug use in the United States do not define terms such as "tranquilizer." Others fail to specify major or minor categories thereof or do not specify the drugs included under the collective term "tranquilizers."

It is doubtful that there is any accurate information concerning the incidence or prevalence of SH drug abuse in the general population of the United States.

BACKGROUND OF CONCEPTS CONCERNING SEDATIVE-HYPNOTIC DRUG ABUSE

During and soon after World War II an increasing number of patients entering the U. S. Public Health Service Hospital at Lexington, Kentucky, gave histories of barbiturate abuse. Clinical observations indicated that some were physically dependent on such drugs. For these reasons, controlled clinical experiments on the abuse characteristics of barbiturates were carried out by Isbell and his co-workers.[8] In these studies the clinical aspects of barbiturate intoxication and withdrawal were described and it was confirmed that the barbiturates can induce physical dependence. These data were published in 1950 or almost one-half century after the barbiturates were introduced into the clinic.[8] Since 1950 a series of new SH drugs have become available to the clinician. At present there is no requirement that each new SH drug be tested to determine if it can induce physical dependence in human subjects. Fortunately, the clear paradigm of barbiturate intoxication and abstinence established by Isbell and his co-workers facilitates the recognition of similar characteristics in clinical reports of the abuse of various new SH drugs. The use of such clinical data is justified because the tardy recognition of the abuse potential of the barbiturates was due largely to the failure of clinicians to heed the case reports of their colleagues during nearly a half century.[9,10]

SOME ETIOLOGIC FACTORS RELATED TO SEDATIVE-HYPNOTIC DRUG ABUSE

People with psychoneurotic or sociopathic personality traits are said to be more prone to abuse drugs than are individuals not having these traits. Alcoholics might be especially inclined to abuse SH drugs. More than 55% of psychiatric patients who had been abusing one or more SH drugs had also abused alcohol at some time.[4]

Abuse of SH drugs is not limited to people mentioned above or to those from the lower socioeconomic strata. Perhaps, in part because of the drugs' ready availability, some physicians, dentists, pharmacists, and nurses are known to abuse SH drugs.

It is difficult to determine what etiologic role, if any, is played by the introduction of increasing numbers of SH drugs into medical practice. In some patients, initial contact with and availability of an SH drug depends largely on the physician and pharmacist. The AMA Committee on Alcoholism and Addiction has published advisory cautions for the use of physicians in prescribing barbiturates and other SH drugs.[11]

STATES OF INTOXICATION ASSOCIATED WITH THE ABUSE OF SH DRUGS

Any of the following SH drugs, in excess dosage, can induce states of intoxication that would be very difficult to differentiate from those due to alcohol or barbiturate abuse: *meprobamate* (Miltown, Equanil, Cirponyl, Aneural, Ansietan, Ansil, Atraxin), *glutethimide* (Doriden), *ethchlorvynol* (Placidyl, Normoson, Serensil, Arvynol), *methyprylon* (Noludar), *ethinamate* (Valmid, Valamine, Valmidate, Valamin, Valaminettae), *chlordiazepoxide* (Librium, Metaminodiazepossido, Reliberan, Seren Vita, Sonia), *diazepam* (Valium), and *oxazepam* (Serax). Drowsiness, a staggering gait, and slurred speech are the common signs of such intoxication. The motor incoordination associated with abuse of these drugs can result in injury to the user or others.[2,3,8] Other manifestations of SH drug abuse are impairment of memory, confused thinking, and disorientation. Terms such as "acute brain syndrome" and "psychomotor retardation" have been used to describe this mental state. In some cases the patient appears agitated or demonstrates dissociative or rage reactions. In others, aggressive or destructive behavior occurs when an SH drug and alcohol are combined.[2,3]

The above signs of intoxication disappear after SH drug abuse is discontinued, but, if physical dependence has developed, abstinence manifestations influence the user to continue taking them. Prolongation of chronic abuse is likely to increase the chances that the user eventually will harm himself or others. Sedative-hypnotic drug intoxication is considered to be potentially more dangerous than that associated with opiate abuse because of the marked impairment of mental and motor performance that can be induced by SH drugs.[8] It is also thought that aggressive or violent behavior is more likely to be associated with SH than with opiate drug abuse.[8]

SITUATIONS THAT CAN PRECIPITATE ABSTINENCE

Abrupt discontinuation of SH drugs following chronic abuse may occur under several conditions. Some individuals decide to discontinue abruptly or markedly decrease their self-administration of such drugs. Occasionally a person who is physically dependent on one or more SH drugs is hospitalized for an unrelated illness and the physician who attends him does not learn that the patient has been abusing such drugs. This situation may result in an abstinence syndrome a day or more following admission. Other circumstances that precipitate drug withdrawal include incarceration, lack of funds, and failure of the only source of supply. Patients who are suspected of SH drug abuse should not have such medication withheld to the extent that a complete abstinence syndrome develops. This method of diagnosing the condition is considered too hazardous.

THE SEDATIVE-HYPNOTIC ABSTINENCE SYNDROME

Clinical descriptions of the withdrawal syndromes that follow discontinuation of the various SH drugs indicate that they are very similar to each other and to barbiturate abstinence. Therefore, a composite description of the abstinence syndrome will be given.

Minor abstinence manifestations appearing after the use of any of these drugs is discontinued can include one or more of the following: anxiety, apprehension, agitation, anorexia, nausea, vomiting, excessive sweating, tachycardia, syncopal episodes, hyperactive reflexes, insomnia, polyuria, tremulousness, and muscle twitches.[2,3]

Major abstinence phenomena are: generalized convulsions, a delirium or psychotic behavior, and sometimes death. The seizures resemble those seen in idiopathic grand mal epilepsy; they can occur as early as 12 to 16 hours after withdrawal or as late as the twelfth day thereafter. All patients do not develop abstinence seizures; it is probable that their occurrence is directly related, as in the case of the barbiturates, to the daily dose levels of SH drugs that had been used. Some patients only have a single abstinence convulsion, but others develop status epilepticus.[2,3,8] If a delirium occurs it might include tremulousness, agitation, disorientation, confusion, delusions, and auditory or visual hallucinations. Such states can also be associated with tachycardia and a progressive temperature elevation. If the body temperature exceeds 101°F or repeated convulsions occur, early and vigorous therapy is indicated.

There is little doubt that the SH abstinence syndrome is potentially more dangerous than that which follows opiate or opioid withdrawal.[8] If convulsions or a delirium or both occur during opiate withdrawal, the patient probably has also been abusing SH drugs.

SOME FEATURES OF THE ABSTINENCE SYNDROME AS RELATED TO VARIOUS SH DRUGS

Barbiturates. Controlled experiments with two barbiturates indicate that there is a safe daily dose level of these drugs which allows discontinuance without precipitating significant abstinence manifestations. Eighteen men received either secobarbital or pentobarbital in daily doses of 400 mg for 90 days. One man had some insomnia but none of them developed either convulsions or a delirium when use of these drugs was discontinued.[12] It is very likely that there is also a safe minimal daily dosage level for the other SH drugs discussed in this report.

Work with the barbiturates in man indicates that, as the daily dose is increased beyond the minimal safe level, the probability of developing major abstinence signs is also increased.[12] It would be surprising if the same did not hold true for other SH drugs.

Meprobamate. Animal experiments, case reports, and a controlled clinical study indicate that abuse of this drug can induce states of intoxication and physical dependence similar to those due to barbiturates.[2,13,14] Both minor and major abstinence manifestations developed in human subjects when the use of meprobamate was discontinued after it had been given for 40 days in dose levels of 3.2 gm daily.[14]

Glutethimide. Experiments in dogs and multiple clinical reports indicate that excessive use of glutethimide can simulate the intoxication and physical dependence produced by barbiturates.[2,15]

Based on clinical observations rather than controlled studies, Johnson and Van Buren estimated that discontinuation of 2.5 gm daily of glutethimide is likely to precipitate major abstinence manifestations.[16] There is insufficient data to indicate the daily dose of this drug that may be given without inducing a significant degree of physical dependence. Glutethimide has become one of the most often abused SH drugs among narcotic addicts from certain parts of the United States.

Methyprylon. A clinical report indicates that abuse of methyprylon to the extent of 2.4 gm daily can result in major abstinence signs if such a dose is abruptly discontinued.[17] Hallucinations, schizophreniform reactions, convulsions, and death have all been attributed to methyprylon withdrawal.[18]

Ethinamate. Minor abstinence signs, convulsions, delusions, and hallucinations have been observed following withdrawal of ethinamate when 13 gm have been taken daily.[19] It is not known whether smaller daily doses would induce a significant degree of physical dependence. An attempt to use a phenothiazine (promazine) in treating a patient with the ethinamate abstinence syndrome resulted in such severe hypotension that use of the promazine had to be discontinued.[19]

Ethchlorvynol. It is reported that a woman who had taken 1.5 gm of ethchlorvynol daily for "months" developed hallucinations, convulsions, and a delirium following discontinuation of use of this drug.[20] Similar abstinence manifestations have been described in patients who had been taking either 2 to 3 or 4 to 5 gm of ethchlorvynol daily.[21]

Chlordiazepoxide. Hollister and his coworkers demonstrated experimentally that chlordiazepoxide in doses of 300 mg/day or 600 mg/day could induce physical dependence in man. One of five patients who received 300 mg/day had two convulsions on the twelfth day after the use of chlordiazepoxide was discontinued. Six patients had been given 600 mg/day; one of these had a seizure on the seventh and another on the eighth day of withdrawal.[22]

Diazepam. A patient who had increased the daily dose of diazepam to 100 and 150 mg during each of two periods of its abuse had convulsions both times that use of the drug was discontinued.[23] In a study of 13 hospitalized schizophrenic patients, diazepam medication was gradually increased to daily dosages of 120 mg prior to the abrupt discontinuation of its use. Six of these patients developed "signs of withdrawal" and one had a convulsion on the eighth day of abstinence.[24]

A 34-year-old woman with muscle spasms (stiff-man syndrome) was treated with diazepam. She took the drug, as prescribed, in daily doses of 60 mg for 40 months. Because of a delay in ordering a new supply the use of the drug was discontinued. On the seventh day of abstinence, status epilepticus developed. The repeated seizures were terminated by administering pentobarbital intravenously. After recovery the patient was again given diazepam for the muscle spasms without further untoward effects.[25]

Oxazepam. Following an overdose of oxazepam, a child exhibited lethargy, ataxia of gait, and depression of reflexes.[26] Two cases of overdoses in adults resulted in lethargy and drowsiness.[26] It is estimated that doses of 8 to 12 mg/kg of body weight can induce signs of intoxication.[26]

It would be highly unusual if oxazepam were capable of inducing drowsiness and ataxia without also being likely to cause physical dependence if chronically abused.

A case has been reported as a "possible oxazepam abstinence syndrome." A woman who had been using 420 mg of oxazepam daily discontinued use of the drug and developed symptoms of a psychosis a week later. She talked irrationally, was disoriented as to time, had visual hallucinations, and became overactive. Desipramine and thioridazine were prescribed when oxazepam was discontinued. Despite the administration of these drugs, the author reporting this case believed that the temporary psychosis was due to oxazepam withdrawal.[27]

Methaqualone (Quaalude). Kato reported a survey (covering the years 1963 to 1966) of the drug abusers and addicts who were in 853 of the mental hospitals in Japan. Four hundred and eleven addicts were discovered; 176 (42.8%) of these were methaqualone abusers. The chief reason for hospital admission of the 176 patients was violent antisocial behavior associated with methaqualone abuse. Convulsions and the symptoms of delirium occurred in 7% and 9% respectively of those who discontinued the use of methaqualone.[28] Daily dosage levels and duration of drug abuse prior to drug withdrawal were not specified. Kato noted that in comparison to meprobamate and chlordiazepoxide "withdrawal symptoms were the severest for methaqualone abusers."[28]

A case of possible methaqualone abstinence has been described in which a patient had been taking as much as 9.0 gm daily during the weeks prior to withdrawal. The patient had been increasing his daily dose of methaqualone over a four-year period. Approximately a day after discontinuing use of the drug he became restless, confused, and complained of having visual hallucinations. Other manifestations attributed to methaqualone withdrawal in this patient were anxiousness, tremor, brisk reflexes, and an elevation of body temperature to 100°F. The patient was treated with a mixture of drugs and he began a rapid recovery on the fourth day after use of methaqualone was discontinued.[29]

DEATH ATTRIBUTED TO ABSTINENCE FROM SEDATIVE-HYPNOTIC DRUGS

Barbiturates. Fraser and co-workers reported a death attributed to secobarbital withdrawal in a physician who had abused the drug for eight months, taking 5.0 gm daily during the four months preceding discontinuing its use.[30] The patient developed tachycardia, tachypnea, and an axillary temperature of 107°F. He died six days after hospital admission. Impairment of circulatory function was postulated as cause of death.

Fraser and associates also reported a death attributed to withdrawal of cyclobarbital (Phanodorn). Use of the drug had been abused more than two years. Prior to withdrawal the daily dose had been 4 to 5 gm. The patient's body temperature rose to 107.6° F. and death occurred on the fifth day of abstinence. Autopsy revealed evidence of "circulatory death" with dilatation of the right heart and congestion of the lungs.[30]

Meprobamate. The death of a 38-year-old man was attributed to a sudden reduction of meprobamate intake. The patient had gradually increased his daily dose of the drug from 1.6 to 10.0 gm during an unknown period of time. When he abruptly reduced his daily dose to 1.6 gm, tremulousness, sweating, insomnolence, hyperthermia, and repeated grand mal convulsions occurred. Despite the use of cooling measures and other indicated treatment, the patient's temperature rose to 107.8°F., his blood pressure fell, and he expired about 80 hours after he began to reduce the dosage. Death was attributed to cardiovascular collapse precipitated by meprobamate withdrawal following chronic addiction.[31]

Methyprylon. A 24-year-old nurse died when she discontinued the use of methyprylon after abusing it for 18 months. Final daily dose levels ranged from 7.5 to 12.0 gm. The patient refused withdrawal treatment in the hospital and insisted on treating herself with a mixture that contained minute amounts of prochlorperazine (Compazine) and promazine (Sparine). She developed jerking movements, hallucinations, and generalized hyperreflexia. On the fifth day of abstinence she was found to be unconscious and "breathing noisily." She died before arriving at the hospital. At autopsy, all chambers of the heart were found to be dilated and the lungs were congested. Fatty degeneration of the myocardium was noted. The patient was said to have died in congestive failure. The phenothiazine "treatment" could not be excluded as a contributing factor in her death.[18]

Diazepam. Death followed diazepam withdrawal in a 20-year-old man in whom the diagnosis of dystonia musculorum deformans had been made. Diazepam was given in doses of 40 mg daily for six days to control the muscular spasms. The dose was increased to 60 mg daily during the following five days. The treatment was discontinued on the eleventh day because it was considered to be ineffective. The next day the patient developed an exacerbation of muscle spasms and tremors, profuse diaphoresis, mydriasis, a tachycardia of 190, and progressive temperature elevation to a maximum of 106.6°F. No evidence of infection was found clinically or at autopsy. Histopathologic examination indicated that the brain was affected by an "unclassifiable neuronal lipid storage disease" that was most severe in the caudate nucleus and putamen. It was speculated that the neurologic disorder played some role in this death and it was recommended that diazepam be withdrawn slowly in patients with diseases of the basal ganglia.[32]

DIAGNOSIS OF CHRONIC SEDATIVE-HYPNOTIC DRUG ABUSE

In a medical context, the aim of diagnosing chronic SH drug abuse is to prevent severe abstinence reactions. Because many patients will deny the abuse of SH or other drugs, the physician may use several guides to pinpoint history-taking for SH drug abuse. One of the more obvious may be noted in patients who appear to be "drunk" but lack evidence of alcohol abuse. Another appears in the patient in early withdrawal who is anxious, apprehensive, agitated, or tremulous. Some patients will have one or more drugs with them and not all will conceal such supplies. Occasionally patients will have injected the oral form of the barbiturates subcutaneously; they often have signs of ulceration or abscess formation. Such individuals might also be using other SH drugs. Sclerosed veins and needle marks are other presumptive but not diagnostic cutaneous signs that the patient might be taking one or more kinds of drugs.

Patients who have taken acute overdoses of SH drugs and are comatose should also be suspected of chronically abusing such agents. These individuals sometimes exceed their tolerance levels without intending to do so. Patients who recover from SH drug-induced coma should be followed for the development of abstinence signs so that proper treatment can be instituted.[33] After recovery from coma and other signs of intoxication the administration of test doses of pentobarbital should help in making the diagnosis of chronic SH drug abuse. (Test doses of pentobarbital are described under "treatment.") Clinical evidence indicates that cross-tolerance between the barbiturates and other SH drugs develops in individuals who chronically abuse such drugs. An increased tolerance to pentobarbital is possible when the patient is physically dependent on a barbiturate or other SH drug. The use of test doses of pentobarbital in establishing the degree of physical dependence is in itself a valuable diagnostic tool.[8]

An electroencephalogram sometimes supplies presumptive evidence of SH drug use. In the EEG of such patients, varying amounts of rhythmic fast (20 to 30 cycles/second) or mixed fast and slow activity appear if the record is made during wakefulness.[34] Failure to record such an EEG pattern does not rule out the use of SH drugs.

A thin layer chromatographic (TLC) method of identifying barbiturates and other SH drugs in the serum has been described by Fletcher and Basco (personal communication, 1969). These men used TLC to detect barbiturates, chlordiazepoxide, diazepam, ethchlorvynol, ethinamate, glutethimide, meprobamate, methaqualone, and methyprylon. By their method these drugs are detected in concentrations of less than 0.5 mg/100 ml of serum. Detecting SH drugs in the serum of a patient on more than one occasion may indicate the possibility of chronic use. However, such a patient is more likely to need test doses of pentobarbital in order to determine whether gradual withdrawal treatment should be carried out.

TREATMENT OF PHYSICAL DEPENDENCE ON SH DRUGS

Serious abstinence reactions can be prevented by using the basic principles developed by Isbell and his co-workers.[8,33] One of the most important of these is that the daily dose level of SH drug which the patient had been using be reduced by no more than the equivalent of 100 mg of pentobarbital daily. Thus, the patient's original degree of intoxication or physical dependence must be estimated so that stabilization can be attained before withdrawal is begun. This is more important than the choice of SH drug used. Clinical evidence indicates that cross-dependence develops among the various SH drugs and between any of them and the barbiturates. Since more experience has been gained with pentobarbital as a drug of substitution and withdrawal, its use will be referred to in the following discussion.

The gradual withdrawal procedure should be carried out in the hospital.

Many SH drug abusers either cannot or will not give an accurate history of the names or amounts of such agents that they had been taking. In these cases it is recommended that if the patients are intoxicated they be allowed to "sober up" before test doses of pentobarbital are given. Test doses of pentobarbital are used to estimate the daily amount that will keep the patient mildly intoxicated rather than showing signs of abstinence. The drug is given orally if possible in 200-mg doses at two- to six-hour intervals, depending on the effects of the drug. The patient is examined 60 to 90 minutes after each test dose to determine if signs of mild intoxication or abstinence are present. The aim is to avoid signs of abstinence (anxiety, tremor, pulse rise exceeding 20/minute on standing), and to induce a state of mild intoxication. In most cases the development of sustained nystagmus is a reliable indication of intoxication. Patients who do not become intoxicated but, instead, continue to have abstinence signs may need to receive 200 mg of pentobarbital every two hours. Those who remain sober without showing abstinence signs may need only 200 mg of pentobarbital at six-hour intervals. Thus, depending on the results of repeated examinations during "test dosing," the dosage and frequency of pentobarbital administration are regulated to maintain a mild state of intoxication throughout the day. Such a "stabilization" regimen is maintained for two or three days before a gradual withdrawal program is started. Pentobarbital reduction should not exceed 0.1 gm daily. Occasionally it is well to suspend the reduction a day if signs of abstinence develop.

It is emphasized that the procedure just outlined is one of estimation. Nurses can learn to detect signs of intoxication and withdrawal so that orders can be written to permit reductions or increases of doses as needed.

Determining the response to test doses is both diagnostic and therapeutic.[8] Moreover, it is a relatively safe procedure. Most patients who become intoxicated after the first test dose are not physically dependent. Those who require more than 400 mg of pentobarbital daily are considered to be physically dependent. Such patients need to be stabilized and the drug gradually withdrawn. The majority of physically dependent patients can be stabilized on daily dose levels between 800 and 1200 mg of pentobarbital.[4] However, it must not be forgotten that some individuals become physically dependent on huge daily doses of SH drugs so that 2.0 gm or more daily of pentobarbital may be required for stabilization in these patients.

If the more serious abstinence signs have developed before treatment has been initiated, it is necessary to give pentobarbital as active treatment. Pentobarbital is given at an accelerated rate if the patient has already had convulsions, an abstinence delirium, or psychosis. Patients who are agitated, have a fast pulse, and a fever of 101°F or above also require prompt treatment. It is recommended that 0.1 to 0.2 gm of pentobarbital be given (parenterally if necessary) hourly until a light sleep ensues. This treatment usually controls convulsions but psychotic manifestations may not disappear completely for several days or longer. In such cases the pentobarbital serves to sedate the agitated or tremulous patient who might otherwise become exhausted. Patients who have a fever higher than 104°F should be exposed to cooling procedures less severe than cold packs.[33] Fluid balance should be maintained and general medical care given as needed.

After the patient who required treatment for severe abstinence signs has awakened from the light sleep, it is best to let him "sober up" before the stabilization regimen is begun. The daily stabilization dose can be estimated by dividing the total amount of pentobarbital that had to be given to induce light sleep into separate daily dosages. This regimen is adjusted if necessary and maintained for three days before dosage reduction is begun. The rate of pentobarbital withdrawal should not exceed 100 mg per day.

It is important to recall that some patients who have taken an acute overdose of SH drugs might also be physically dependent on such agents. These patients might recover from coma and then gradually develop an abstinence syndrome. It is well to monitor those recovering from acute SH drug poisoning in order to institute a gradual reduction regimen if necessary.

Patients who are being stabilized sometimes become too ataxic after receiving extra test doses. They should not attempt to walk, bathe, or go to the bathroom unattended.

The pharmacotherapy of SH drug withdrawal does not preclude nondepressant medication for other medical conditions. Moreover, opiate withdrawal can be carried out simultaneously unless the patient has another serious illness.

The use of SH drugs other than pentobarbital during withdrawal treatment is not necessary. Reliance should not be placed on diphenyl hydantoin (Dilantin) as a partial or complete replacement for pentobarbital.[35,36] Diphenyl hydantoin is not indicated as an adjunctive anticonvulsant because it is not thought to be effective against barbiturate abstinence seizures in man or the dog.[35,36]

Phenothiazine derivatives are contraindicated as anticonvulsants or as adjuncts to treatment during SH withdrawal. These drugs have convulsant properties in their own right and do not prevent barbiturate abstinence convulsions in dogs.[37,38] It is doubtful that phenothiazines counteract autonomic disturbances associated with SH abstinence and their use may instead aggravate the hypotensive aspects of the SH abstinence syndrome.[19] At present the use of major tranquilizers during SH drug withdrawal is not recommended.

Discussion

Our understanding of the basic processes underlying physical dependence on depressant drugs and the abstinence phenomena related to such dependence is woefully inadequate. Much of our information thus far is merely descriptive.

It is known that mice, rats, cats, dogs, and monkeys can be made physically dependent on barbiturates.[39–41] Any of these species can develop various abstinence phenomena, including convulsions. Dogs and monkeys have been observed to develop hallucinatory-like behavior during barbiturate abstinence.[41,42]

It is likely that the basic mechanisms underlying physical dependence on the barbiturates are essentially the same as in the case of many other depressant drugs. The list of such drugs is growing, and with this the dissimilarity of their chemical structures is increasing. Freund has reported recently that mice develop abstinence convulsions if the administration of potassium bromide is discontinued after the drug has been ingested four to six days in high dosages (1,930 to 2,710 mg/kg of body weight).[40] This investigator suggests that an increased rate of bromide elimination might enhance the likelihood that abstinence signs will develop. If his thesis is correct, a too-rapid removal of bromide by diuresis or dialysis might result in convulsions or other abstinence signs in patients who have chronically abused the drug.[40] Freund also points out that although the bromide ion is not metabolized it is related to withdrawal seizures. He speculates that the important factor in the SH abstinence syndrome is relatively sudden release from central nervous system depression.[40]

Other experimental and clinical observations related to possible mechanisms underlying SH abstinence seizures are reviewed elsewhere.[34,39]

Death has been attributed to discontinuation of barbiturates, meprobamate, methyprylon, and diazepam.[18,30–32] Autonomic disturbances may well account at least in part for the deaths that occur under these conditions. Some of the abstinence signs related to SH drug withdrawal are autonomic in nature; dilated pupils, sweating, gastrointestinal complaints, postural hypotension, tachycardia, and temperature elevations have been observed as part of the barbiturate and other SH abstinence syndromes.[2,8]

Hyperthermia was observed clinically and evidence of some circulatory failure was found at autopsy in several of the patients whose deaths were reported above. If these phenomena play a role in causing death following SH drug withdrawal, is the threshold for these autonomic changes lower in patients who have certain diseases of the central nervous system than in those without

CNS disorders? Death following abrupt discontinuation of diazepam, after it had been given for less than two weeks in moderate dosage to a patient with dystonia musculorum deformans, is suggestive of this possibility.[32]

There are many other practical and theoretical questions related to physical dependence on SH drugs. Some questions related to withdrawing SH drugs have not been answered on the basis of controlled experiments in man because such studies cannot be done for humane or other reasons. Specific clinical problems arise in relation to acute overdosages (suicidal attempts) or the chronic abuse of each new SH drug. The rate of accumulation of such problems probably exceeds the rate at which solutions—if any—can be found for them. In view of such complexities it is relevant to consider whether the introduction of more drugs with barbiturate-like clinical actions serves any useful medical or public health need.

REFERENCES

1. Eddy, N. B.: Analgesic and dependence producing properties of drugs. In *The Addictive States,* Wikler, A. (Ed.). Baltimore, Williams & Wilkins Co., 1968, pp. 1–12.
2. Essig, C. F.: Addiction to nonbarbiturate sedative and tranquilizing drugs. *Clin Pharmacol Ther 5*:334–343, 1964.
3. Essig, C. F.: Symposium: nonnarcotic addiction. Newer sedative drugs that can cause states of intoxication and dependence of barbiturate type. *JAMA 196*:714–717, 1966.
4. Bakewell, W. E., Jr., and Wikler, A.: Symposium: nonnarcotic addiction. Incidence in a university hospital psychiatric ward. *JAMA 196*:710–713, 1966.
5. Ewing, J. A., and Bakewell, W. E., Jr.: Diagnosis and management of depressent drug dependence. *Amer J Psychiat 123*:909–917, 1967.
6. Chambers, C. D.: Barbiturate-sedative abuse: A study of prevalence among narcotic abusers. *Int J Addict 4*:45–57, 1969.
7. Hamburger, E.: Barbiturate use in narcotic addicts. *JAMA 189*:366–388, 1964.
8. Isbell, H., Altschul, S., Kornetsky, C. H., Eisenman, A. J., Flanary, H. G., and Fraser, H. F.: Chronic barbiturate intoxication. An experimental study. *Arch Neurol (Chicago) 64*:1–28, 1950.
9. Pohlisch, K.: Uber psychische Reaktionsformen bei Arzneimittelvergiftungen. *Mschr Psychiat Neurol 69*:200–367, 1928.
10. Wulff, M. H.: The barbiturate withdrawal syndrome. A clinical and electroencephalographic study. *Electroenceph Clin Neurophysiol* (Suppl.) *14*:173, 1959.
11. American Medical Association, Committee on Alcoholism and Addiction and Council on Mental Health: Dependence on barbiturate and other sedative drugs. *JAMA 193*:673–677, 1965.
12. Fraser, H. F., Wikler, A., Essig, C. F., and Isbell, H.: Degree of physical dependence induced by secobarbital or pentobarbital. *JAMA 166*:126–129, 1958.
13. Essig, C. F.: Withdrawal convulsions in dogs following chronic meprobamate intoxication. *Arch Neurol (Chicago) 80*:414–417, 1958.
14. Haizlip, T. M., and Ewing, J. A.: Meprobamate habituation. A controlled clinical study. *New Eng J Med 258*:1181–1186, 1958.
15. Essig, C. F.: Addictive and possible toxic properties of glutethimide. *Amer J Psychiat 191*:993, 1963.
16. Johnson, F. A., and Van Buren, H. C.: Abstinence syndrome following glutethimide intoxication. *JAMA 180*:1024–1027, 1962.
17. Jensen, G. R.: Addiction to Noludar. A report of two cases. *New Zeal Med J 59*:431–432, 1960.
18. Berger, H.: Addiction to methyprylon. *JAMA 177*:63–65, 1961.
19. Ellinwood, E. H., Ewing, J. A., and Hoaken, P. C. S.: Habituation to ethinamate. *New Eng J Med 266*:185–186, 1962.
20. Hudson H. S., and Walker, H. I.: Withdrawal symptoms following ethchlorvynol (Placidyl) dependence. *Amer J Psychiat 118*: 361, 1961.
21. Cahn, C. H.: Intoxication by ethchlorvynol (Placidyl). Report of four cases. *Canad Med Assoc J 81*:733–734, 1959.
22. Hollister, L. E., Motzenbecker, F. P., and Degan, R. O.: Withdrawal reactions from chlordiazepoxide (Librium). *Psychopharmacologia (Berlin) 2*:63–68, 1961.
23. Hollister, L. E., et al.: Diazepam in newly admitted schizophrenics. *Dis Nerv Syst 24*: 746–750, (Dec) 1963.
24. Aivazian, G. H.: Clinical evaluation of diazepam. *Dis Nerv Syst 25*:491–496, 1964.
25. Cohen, L.: Stiff-man syndrome (two patients treated with diazepam). *JAMA 195*:222–224, 1966.
26. Shimkin, P. M.: Oxazepam poisoning in a child. *JAMA 196*:662–663, 1966.
27. Selig, J. W.: A possible oxazepam abstinence syndrome. *JAMA 198*:951–952, 1966.

28. Kato, M.: An epidemiological analysis of the fluctuation of drug dependence in Japan. *Int J Addict 4*:591–621, 1969.
29. Ewart, R. B. L., and Priest, R. G.: Methaqualone addiction and delirium tremens. *Brit Med J 3*:92–93, 1967.
30. Fraser, H. F., Shaver, M. R., Maxwell, E. S., and Isbell, H.: Death due to withdrawal of barbiturates. *Ann Int Med 38*:1319–1325, 1963.
31. Swanson, L. A., and Okada, T.: Death after withdrawal of meprobamate. *JAMA* 184: 780–781, 1963.
32. Relkin, R.: Death following withdrawal of diazepam. *New York J Med 66*:1770–1772, 1966.
33. Fraser, H. F., and Grider, J. A.: Treatment of drug addiction. *Amer J Med 14*:571–577, 1953.
34. Wikler, A., and Essig, C. F.: Withdrawal seizures following chronic intoxication with barbiturates and other sedative drugs. In *Epilepsy Mod. Probl. Pharmacopsychiat.*, Vol. 4, Niedermyer, E. (Ed.). Basel/New York, Karger, 1970, pp. 170–184.
35. Cobb, S., and Miles, H. W.: Symptoms due to withdrawal of barbiturates. *Amer Pract Digest Treat 2*:768–771, 1951.
36. Essig, C. F., and Carter, W. W.: Failure of diphenylhydantoin in preventing barbiturate withdrawal convulsions in the dog. *Neurology 12*:481–484, 1962.
37. Fazekas, J. F., Shea, J. G., Ehrmantraut, W. R., and Alman, R. W.: Convulsant action of phenothiazine derivatives. *JAMA 165*:1241–1245, 1957.
38. Essig, C. F., and Fraser, H. F.: Failure of chlorpromazine to prevent barbiturate withdrawal convulsions. *Clin Pharmacol Ther 7*:466–469, 1966.
39. Essig, C. F.: Clinical and experimental aspects of barbiturate withdrawal convulsions. *Epilepsia (Amst) 8*:21–30, 1967.
40. Freund, G.: Alcohol, barbiturate and bromide withdrawal syndrome in mice (in press).
41. Woods, J. H., and Schuster, C. R.: Regulation of drug self-administration. In *Drug Dependence,* Harris, R. T., McIsaac, W. M., and Schuster, C. R. (Eds.). Austin and and London, University of Texas Press, 1970.
42. Fraser, H. F., and Isbell, H.: Abstinence syndrome in dogs after chronic barbiturate medication. *J Pharmacol Exp Ther 112*: 261–267, 1954.

SEDATIVE-HYPNOTIC DRUG ABUSE AMONG ADULTS

CARL-MAGNUS IDESTRÖM, M.D.

The classic dependence-producing drug in Scandinavia is, of course, alcohol. Opiates have never been a problem. The first dramatic and serious epidemic of drug abuse appeared in the late fifties when young people began injecting amphetamine and phenmetrazine. Cannabis-smoking has also spread during the last 10 to 15 years.

However, since World War II there has been an increase in the use of sedative and hypnotic drugs, primarily barbiturates, meprobamate, and benzodiazepines. In 1954 the consumption of barbiturates in Sweden was approximately 20 tons, equivalent to 29 tablets of 100 mg per person.[1] When meprobamate and later the benzodiazepines came on the market, the use of barbiturates decreased somewhat. For several years diazepam has been the leading product among sedatives, although barbiturates are still the most widely used hypnotics.

It is generally accepted that the benzodiazepines are less harmful than barbiturates and meprobamate with regard to acute toxicity and dependence liability. However, since these drugs have been on the market only about 10 years it is obviously too early to fully judge the effect of chronic administration.

In order to illustrate the abuse of sedative and hypnotic drugs in Scandinavia I will present two investigations, one from Norway and one from Sweden.

Retterstöl and Ropstad studied all 565 patients admitted to the psychiatric clinic of the University of Oslo during 1968.[2] The authors were especially interested in whether the pattern of abuse of benzodiazepines differed from that of barbiturates and meprobamate. They interviewed every patient, 279 men and 286 women, about their use of psychotropic drugs during a period of three years prior to admission. Twelve percent of all the patients had not taken any drugs during the three preceding years, 68% used drugs as prescribed, and 12% had taken drugs in excess. The authors defined drug

Table 1 Psychotropic drugs given during the last three years

	No. of patients	Percentage (of 565)
Chlordiazepoxide/diazepam	382	68
Barbiturates	43	8
Meprobamate	56	10
Major tranquilizers	92	16
Major antidepressives	97	17
Others	17	3
No drugs	69	12
Total	756	

Table 2 Duration of the use of drugs during three years prior to hospitalization

	1 month	1–6 months	½–1 year	1–2 years	More than 2 years
Chlordiazepoxide/diazepam	10	110	55	73	128
Barbiturates	0	6	8	3	26
Meprobamate	3	12	11	5	24
Major tranquilizers	5	39	19	11	17
Major antidepressives	8	52	24	7	5
Others	0	4	0	0	11

abusers as those who, on their own, increased the dose and consumed the drug over a longer period of time than their doctor had prescribed. All those described as abusers were mentally dependent on the drug, and some were also physically dependent.

Although the total number of patients was 565, many patients had been on several drugs so the total of drug exposures during the three years was 756. Ten percent of the patients had used two drugs; 23%, three or more drugs; 68% used chlordiazepoxide or diazepam (Table 1).

Most patients on major tranquilizers and antidepressives had taken them for one to six months, in contrast to the dependence-producing drugs which were used for longer periods of time, often more than two years (Table 2).

In the group of 68 abusers, 38 men and 30 women, 31 patients abused chlordiazepoxide or diazepam and 16 barbiturates or meprobamate (Table 3).

Table 3 Dependence to drugs

	No. of patients
Chlordiazepoxide/diazepam	31
Barbiturates	11
Meprobamate	5
Opiates	7
Amphetamine	3
Hallucinogens	1
Others	10
	68

The authors also compared the 68 abusers with the 555 non-abusers admitted to the clinic in the same year. They found no significant differences in sex distribution, educational level, or profession. The premorbid personality was more unstable among the abusers than among the other patients, at a level that was statistically significant. Those abusing drugs also had a higher incidence of alcohol problems. Previous attempts at suicide were more common among the abusers of benzodiazepine and barbiturate-meprobamate than among patients taking other drugs and this was also statistically significant. The authors recommended caution in prescribing sedative-hypnotic drugs to former drug- or alcohol-abusers and to chronic neurotics. They concluded by saying that they found no certain difference between abusers of benzodiazepines and barbiturate-meprobamate.

The second study is part of a large investigation of the drug problem initiated by the National Swedish Board of Health and Welfare and published 1969.[3] Sociologists made an inquiry by questionnaire to all psychiatric and general hospitals concerning the drug abusers hospitalized on April 3, 1968. The investigation was not designed to elucidate psychiatric or pharmacologic aspects of drug use, but nevertheless it yielded some important information about abuse and abusers. The study group comprised 697 patients, 452 men and 245 women.

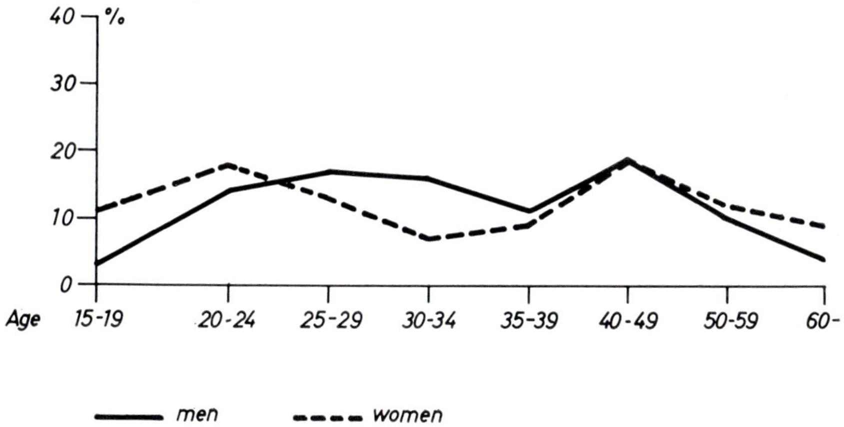

Figure 1 Abuse cases—age and sex distribution. Total nnmber of patients, 697—452 (64.8%) were men and 242 (35.2%) were women.

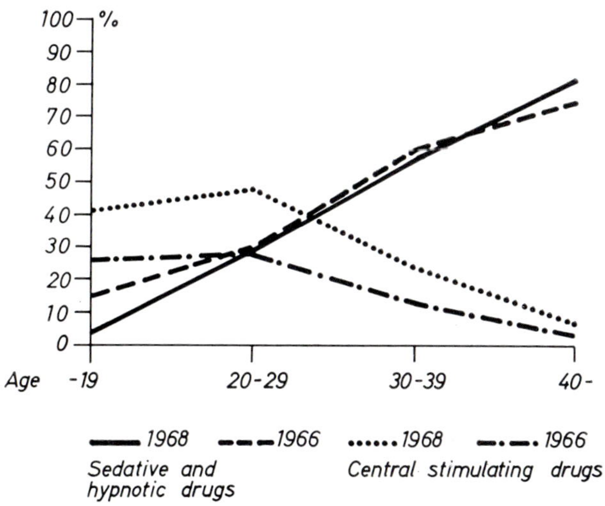

Figure 2 Distribution of misuse of sedative and hypnotic drugs versus central stimulating drugs, 1966 and 1968.

At the time of this study, in April 1968, sedative-hypnotic drugs and central stimulating drugs were the ones most commonly abused. Young people up to 30 years of age preferred amphetamine and phenmetrazine. The abuse of sedative and hypnotic drugs increased with age, becoming the most common drug used by all persons included in the study.

In 77.8% of those studied, the duration of the abuse was longer than one year, in 16.9% less than one year. Continuing alcohol abuse was reported in 44.8% of the subjects.

The Figures 3 and 4 show other differences between drug abusers and the total population. Among those abusing drugs, 53.5% were unmarried, 25.4% married, 18.5% divorced, and 2.6% widows or widowers.

A comparison with the normal population is not easy because the marital status of the

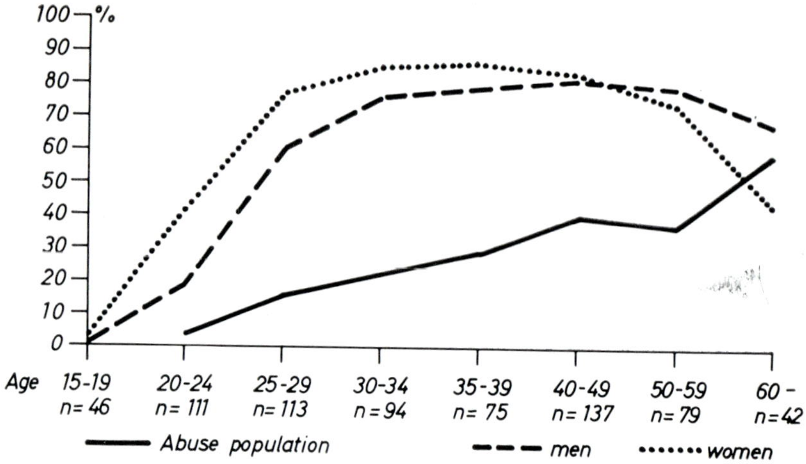

Figure 3 Marital status in different age groups in percent compared with total population of Sweden married.

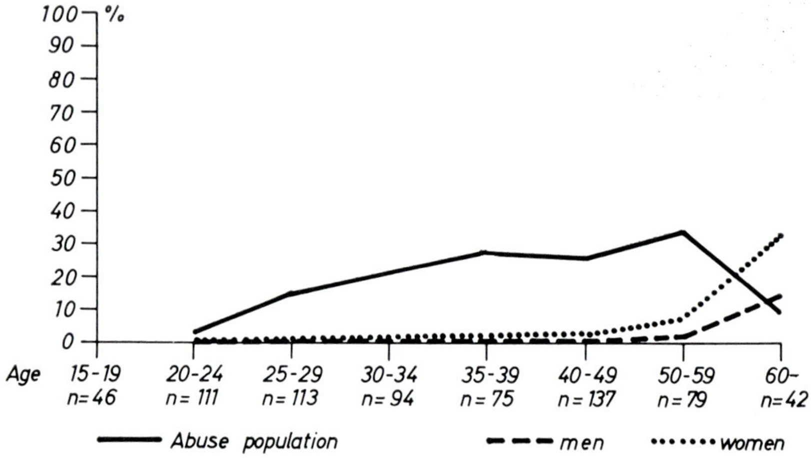

Figure 4 Marital status in different age groups in percent compared with the total population of Sweden divorced.

total population is calculated on the total number of individuals in the country. Another source of error is the larger proportion of women in the total population than in the abuse population. Thus, real differences between drug abusers and normal population may be less than shown in the figures, although the tendency seems correct.

The authors also studied the distribution of social groups among the drug-abusing population. A comparison with the total population is difficult, but there seems to be no significant difference.

The psychiatric classification in this investigation was heterogenous and superficial because there were many doctors involved. Roughly 50% of the patients were diagnosed as neurotic, psychotic, and sociopathic. Only

21.8% had no previous psychiatric treatment, most of the patients having had one or more courses in hospitals or institutions.

DISCUSSION

The sedative and hypnotic drugs all have a primarily tranquilizing effect, although in some cases a euphorizing and stimulating effect may be experienced early in use. It is thought that former alcohol addicts obtain the euphorizing effect more easily and more quickly than others and therefore run the risk of a new habit. I would like to remind you of the cross-tolerance phenomenon between alcohol and barbiturates experimentally shown by Isbell and associates[4] who found that the abstinence syndrome following abrupt and complete withdrawal of barbiturates in chronic abuse could be controlled by substitution with alcohol and vice versa. Later Sereny and Kalant reported cross-tolerance between alcohol and meprobamate and chlordiazepoxide.[5]

The reported Scandinavian investigations emphasize the point that the abusers of benzodiazepines are certainly as often neurotic and socially handicapped as are those who take barbiturates-meprobamate in excess. They are predominantly middle-aged and older, with a higher proportion of unmarried and divorced patients than in the normal population. They have used the drugs for a long time, often together with alcohol. These are some of the reasons why the use of sedatives can develop into a serious habit, even contributing to therapeutic difficulties.

It is said that 200 to 300 million people use benzodiazepines more or less regularly. It is reasonable to believe that even more people will use these drugs in the future. And it is probably true that more complications will appear with longer exposure to these drugs.

Even single doses of sedatives can sometimes give undesired side-effects, such as prolonged reaction time and higher threshold for visual stimuli.[6] Judgment can be impaired and emotional balance is often affected. It has also been emphasized by Essig that sedative and hypnotic drugs may contribute to accidents in work and traffic.[7] The future will show if decades of daily use of sedative and hypnotic drugs with or without alcohol causes serious mental and somatic complications.

Of course, the etiology of the development of the enormous use of sedative and hypnotic drugs is complex, and requires increasingly complex methods of handling the problem. One small part is certainly more personal communication between doctor and patient and less drug administration.

REFERENCES

1. Ideström, C.-M.: Flicker-fusion in chronic barbiturate usage. *Acta Psychiat Neurol Scand* suppl. 91, 1954.
2. Retterstöl, N., and Ropstad, O.: Misbruk av lette ataractica. *Nord Psyckiat T 23*:477, 1969.
3. Narkotikaproblemet. Statens offentliga utredningar 1969:53. Esselte AB, Stockholm, Sweden, 1969.
4. Isbell, H., Altschul, S., Kornetsky, C., Eisenman, A. J., Flanary, H. G., and Fraser, H. F.: Chronic barbiturate intoxication. An experimental study. *Arch Neurol Psychiat 64*:1, 1950.
5. Sereny, G., and Kalant, H.: Comparative clinical evaluation of chlordiazepoxide and promazine in treatment of alcohol-withdrawal syndrome. *Brit Med J 1*:92, 1965.
6. Ideström, C.-M., and Cadenius, B.: Chlordiazepoxide, dipiperon and amobarbital. Dose effect studies on human beings. *Psychopharmacologia 4*:235, 1963.
7. Essig, C. F.: Addiction to nonbarbiturate sedative and tranquilizing drugs. *Clin Pharmacol Ther 5*:334, 1964.

Editor's Note: Dr. Isbell kindly read the paper for Dr. Ideström, who was unable to be present because of illness. At the conclusion of the formal presentation, Dr. Isbell made the following remarks:

"I will take a minute or two to give you some impressions I have concerning barbiturate use in the United States. I think that an adequate survey in this country would show that we have a major problem in the misuse of barbiturates and similar type drugs by young people. There is an active illicit market in these things and the young people drop them for 'kicks' or use them to reinforce alcohol. There is also another group, primarily alcoholics, who take the drugs to reinforce or substitute for the alcohol. This group cuts across various ages and may extend from young people to older age groups.

"However, the people that we now see most frequently in my new environment are neither of these. They are primarily neurotic patients and Dr. Maynert also made the point that I am about to make. These patients, being neurotic, tend to go from doctor to doctor, and they give each physician all their nervous symptoms. Each physician pats them on the hand, writes another prescription for 'nerve pills,' and they take these. The patients are frequently compulsive and take their prescriptions as directed. Then they go to another physician with all kinds of complaints and get referred to a variety of subspecialists. Each gives them another prescription and this process goes on and on until we now have a person who has four or five bottles of different kinds of sedative-hypnotic drugs, all of which he or she carefully takes as directed. My particular record-holder brought in *17* bottles. I'm not kidding, 17 bottles. So these people become dependent, as Dr. Maynert said, on these drugs even though they are taking them in the prescribed dosage as directed. It appears that doctors forget to ask them, 'What other drugs are you taking?' and they forget to say, 'Please, if you take this, don't take any other medications.' This is the kind of patient one sees now and this is the kind of complication that has been noted in about 7% of patients receiving psychiatric care at the University of Kentucky Medical Center."

PANEL DISCUSSION—DEPRESSANTS

Moderator: DR. ZARAFONETIS

Panel Members: DRS. GOLDBERG, ISBELL, JAFFE, AND MAYNERT

Dr. Jaffe: I will try to respond for Dr. Essig's paper. If we can persuade Dr. Zarafonetis to preside and to recognize people from the floor, I will take off the chairman's hat.

Dr. Zarafonetis: Thank you, Dr. Jaffe. We will try to be more expeditious at the mikes, and I would suggest in fairness to all of the audience that each questioner limit himself to one question the first time around. I recognize Dr. Angrist at the microphone for the first question.

Dr. Burton Angrist (New York, N.Y.): The first question is addressed to any member of the panel. One unpleasant thought that I should like to share is this: In view of the increasing reports of sedative hypnotic dependence in opiate addicts, and the concomitant reports of tetanus in that population, I wonder if some points could be given for differential diagnosis between early tetanus and early sedative hypnotic withdrawal. This could be a distressing clinical problem.

Dr. Jaffe: Burt, I have never seen this problem, but you obviously have, and it may be that you could answer the question better than anybody on the panel. Now, I don't know much about tetanus, but I wonder whether a test dose of a short-acting barbiturate could be expected to suppress some of the hyper-reactivity of tetanus. Would that possibly be a way of differentiating the two? My own experience has been that it is very hard to tell if you only partially suppress barbiturate abstinence. By giving an inadequate suppressant dose, you are still left with the issue. For example, if you have somebody who is showing hyperreflexia and you cannot get a good history for whatever reason, you may give 200 mg of phenobarbital. If the individual remains relatively hyperreflexic, you cannot conclude that he does not have tetanus. Maybe you should have given a 300 mg test dose. Let us ask the internists who might know more than we psychiatrists do about how to differentiate the positive signs of tetanus from those of barbiturate withdrawal. Are there any?

Dr. Isbell: I really don't think there is a diagnostic problem unless both happen to coexist. People with tetanus ordinarily come in complaining of stiff jaws, and they have stiff jaws. Their history usually includes a puncture of some kind, so their diagnosis usually presents no difficulty. The barbiturate user on the other hand usually dose not complain of a stiff jaw. He comes in complaining that he is extremely nervous, about to come out of his skin, and he has the "shakes." He has been taking pills, maybe he has had a convulsion, and he is not complaining at all about a stiff jaw. So unless they happen to coexist, I don't think there would be any great reason for confusion. In the old days we used barbiturates in massive doses to treat people with tetanus in order to reduce the amount of external stimulation. Today we are more likely to use diazepam, or Valium, which would have the same effect; so really I don't see that it is any problem.

Dr. Walter D. Meester (Grand Rapids, Mich.): I should like to direct my question to Dr. Maynert since he covered the pharmacology of sedatives and hypnotics. It seems to me that very few of the youngsters coming to our drug rehabilitation program in Grand Rapids are abusing just one drug. Certainly there are very few that are abusing just one type of barbiturate or one type of sedative or hypnotic. They often end up in the emergency room because of a reaction. With this multiple use of sedatives and hypnotics, we who are in the business of treating these people are sometimes puzzled as to how we should treat them. We are also puzzled as to the interactions of drugs, and my question is whether some of the drugs that are most commonly known to potentiate, for instance barbiturates and alcohol, directly influence the development of tolerance; that is, would the combination of these drugs influence or have an effect on the severity of the abstinence syndrome?

Dr. Maynert: I am sorry that I cannot answer your questions. They are very fundamental, and precise answers are not yet available. It is still argued whether the effects of combinations of the sedative and hypnotic drugs involve more than simple addition. As you know, there has long been a view that barbiturates potentiate alcohol, and that the effects amount to something more than simple addition. A problem like this is very difficult to solve. I don't know of any evidence that persuades me that the pharmacology of these compounds involves more than simple addition. The problem becomes more complicated when we recognize that there are metabolic aspects to tolerance as well as adjustments in the central nervous system. This afternoon I mentioned that the continued use of these compounds may lead to the increased synthesis of drug-metabolizing enzymes, but other compounds are known that can inhibit them. In the presence of a poison of this kind, one may well observe potentiation or synergism, because a great deal of metabolic degradation may take place before the drugs enter the brain. I am unaware of any quantitative data relating to chronic tolerance or to abstinence syndromes resulting from combinations of the sedative-hypnotic drugs.

Dr. Zarafonetis: Dr. Goldberg, do you want to add to this?

Dr. Goldberg: In order to approach problems of drug interaction in man we have conducted a large series of studies in experimental subjects on the interplay between high doses of vitamins, and therapeutic doses of sedative and hypnotic drugs, as compared to the effect of placebo under double-blind conditions, on one hand, and small doses of alcohol, 0.7–0.8 gm per kilogram of body weight corresponding to 4 to 5 ounces of whisky, on the other, following the effects by a number of quantitative tests in relation to blood drug and alcohol levels.

We found high doses of vitamins to counteract part of the effects of alcohol. Among the benzodiazepines, 10 and 20 mg of chlordiazepoxide (Librium) also partially counteracted the alcohol effects, thus acting antagonistically, whereas the structurally similar diazepam (Valium) acted differently: 5 mg had very little effect, while 10 mg increased some of the alcohol effects, thus acting synergistically; the same was seen with 800 mg of meprobamate. Hypnotics, e.g., 200 mg methyprylon (Noludar) or 200 mg phenobarbital, increased the effects of alcohol, thus acting synergistically. Phenobarbital probably even potentiated the effects; this small therapeutic dose in combination with alcohol in some cases even caused a fall in blood pressure and unconsciousness.

There was thus a scale or spectrum of effects, from antagonism to synergism and potentiation (=supra-additive effects), most likely localized in the central nervous system, as no changes in blood alcohol or drug levels were observed.

In experiments in cats with chronically implanted electrodes, we have shown these pharmacodynamic differences in action, caused by small changes in chemical structure, to depend principally on different neurophysiological mechanisms of action, the various drugs differentially stimulating, respective inhibiting, selected structures of the central nervous system.

Some of the human experiments on drug-alcohol interactions were conducted in cooperation with Prof. P. Kielholz on drivers as field tests on a driving range; a film depicting the results will be shown by Prof. Kielholz on Thursday, Nov. 12, 1970.

Dr. Jaffe: Before you leave that topic, I would like to clarify something. You found this potentiation with only one of these several drugs. Is that correct? Only with phenobarbital? I think it is important that we not generalize. It may turn out, as it did with the anticonvulsant aspects of phenobarbital, that one cannot even generalize to all the barbiturates. So while it apparently has been demonstrated there is potentiation with a single barbiturate at this point with alcohol, your other data seem to indicate that it is not true of all sedative-hypnotic minor tranquilizer-type drugs. We have one demonstrated case of more than simple addition but the jury is not yet in on the full range of sedative hypnotics. Is that a correct statement of what you said?

Dr. Goldberg: Yes.

Dr. Maynert: While this afternoon I found it expeditious to stress the similarities among these drugs, in my view probably every one of these central nervous system depressants differs slightly from the others. This explains why, in constructing my table and seeking generalizations, I chose not to include chlordiazepoxide or diazepam. Clearly, these drugs show quite a few differences from alcohol and the barbiturates.

Dr. Goldberg: In support of Dr. Maynert's comments may I just stress that with regard to the combination of barbiturates and alcohol you can expect to find severe complications, even after relatively small doses of the two types of agents: primary and secondary shock with critical fall in blood pressure, and unconsciousness. Of course you know that quite a few deaths occurring among drunken people are due to the combination of alcohol and hypnotic drugs.

Dr. Maynert: I have been interested in this as a very difficult theoretical problem. Among the physiological factors that must be considered are blood pressure and temperature. When a drug lowers body temperature, metabolism slows down and potentiation can result.

Dr. Zarafonetis: Yes, Dr. Goldberg.

Dr. Goldberg: As to the theoretical point of view, our neurophysiological studies with chronically implanted electrodes in cats have given us some hints as to the underlying sites and mechanisms of action of the combined effects of CNS-active drugs and alcohol. The explanation for the actual spectrum of effects, from antagonism to synergism and potentiation, is not only neurophysiological, involving selected and specific changes in inhibitory and facilitatory pathways, acting on higher centers, as well as on basal functions such as maintenance of circulation and respiration. It involves also localized biochemical metabolic changes, affecting transmittor systems, receptor mechanisms, membrane function or enzymatic solute or particulate intracellular processes. With more knowledge accumulating, we will ultimately be able to formulate theories, and predict the outcome of the interplay between various types of CNS-active agents.

Dr. Zarafonetis: Thank you, gentlemen. Now to the next questioner.

Dr. Robert A. Braun (Detroit, Mich.): Since Dr. Jaffe was the chairman of this session, I would like to address this to him. This symposium for physicians, I notice, includes a lot of discussion of basic science, the findings of which are certainly very important to the physician. However, there has been so far very little, and, from what I have seen of the program, there will be very little of practical interest to the physician who deals with patients rather than one who deals with mice. The afternoon's topic of sedative-hypnotic drug abuse seems to be the most important one to the average physician who deals with patients who come to him for all kinds of problems for which he prescribes various drugs. He may prescribe one drug for the patient's lack of sleep, another for the patient's nervousness, another for the complaint of tightness of muscles in the back of the neck; of course, most of these drugs have pretty much the same effect. Also, as Dr. Isbell pointed out so aptly, the patient takes exactly what is prescribed; that is, all but 7% or 12% do, and they go from one physician to the other and take all the medications that are prescribed by each physician—in addition to the ethyl alcohol which is not prescribed but which many of them also take. This is why drug abuse among patients in general practice is quite a problem and why we have so many cases. The patients to which you have been referring, gentlemen, are not living in a vacuum. They have been to doctors and they have been given prescriptions. The most important question, I think, is how can the physician avoid making a drug abuser out of his patient? The comment that Dr. Isbell made earlier focused on one of the most important problems that the physician has these days. I am amazed that there is not more on this particular topic for this particular group. I should think physicians would be very interested in how they can prevent the patient from becoming a drug abuser. When the patient comes to the physician with symptoms of chronic drug abuse such as are pointed out to us this afternoon—restlessness, sleeplessness, or anxiety—in most cases he will be given drugs again for these symptoms which means adding more drugs to the ones which he is already taking. Instead the physician should be determining the cause of the patient's various symptoms. I would appreciate your giving us some discussion of this problem which, it seems to me, has not been discussed.

Dr. Jaffe: Obviously the time you have taken on this confirms my original feeling that the physicians here would much rather hear practical things than things applicable to mice. Dr. Maynert and I spent considerable time discussing this and then, as you heard, he spent some time emphasizing in the very beginning of his presentation that some people will take doses of each of these various drugs, some for relaxation of the neck, some for relaxing the GI tract, others for anxiety, never thinking they are taking too much of any one, but clearly taking too much in the aggregate. Dr. Isbell gave some specific cases and you have clearly underscored the problem again. It is regrettable that we do not have medical students here because the net effect of what we are all saying is that the basic skills of the physician really have not changed much. You might try, for a start, taking a complete history. It should include what drugs the patient has taken lately, from all doctors not just from you. Now, once you have done that and you find they are taking 11 other drugs, the first thing you might say is "I would like you to stop taking all of them. Bring them in, or throw them away." The latter is unlikely, since they are too expensive for most patients to do that, but certainly you should be aware of what the patient is taking. Secondly, if it sounds as though the patient has a lot of muscle tension in various parts of the body, it would be useful to consider the possibility that perhaps the answer is not simply another drug. Perhaps you should consider a response that is less pharmacologically oriented. I think it inappropriate to talk extensively about how to treat generalized tension in a patient who comes to a physician at this symposium, but obviously the physician should be aware that the drugs generally used for tension can lead to dependence. Physicians should take histories, and they should get patients to reduce dosage where possible. It may not be possible to get them to discontinue entirely because it may already be too late. In some cases, you may have a patient who, if he or she were to discontinue, will come and complain two nights later of a convulsion (or the family will). My only observation is that we are left with the situation the physician has always faced: what do you do with the tense patient, somebody in some kind of psychic distress. It is much too easy to say, "Well, talk to them." We all know how busy most physicians are. But I have no quick and easy solutions. Obviously, the one quick and easy solution—giving another drug that does the same thing—is inappropriate. I will let the rest of the panel respond to this question also. I think it is an important one.

Dr. Isbell: I think Dr. Jaffe has put his finger on it. Every history should include the record of the drugs that the patient is taking and has taken in the past, and if possible, these should be brought in and identified if he is a new patient to you. However, this is another one of those things that take time. Time is a commodity which most physicians have the least of, and actually this is a job that could be done by an office aide. It really does not take an M.D. degree to ask, "Are you taking any medicines? Do you know their names? If you don't know their names, what pharmacy did you get them from?" And you call the pharmacy. Someone else can do it for you. Actually, in our institution it is done for us by pharmacy interns.

Dr. Zarafonetis: Dr. Maynert or Dr. Goldberg, do either of you wish to add anything?

Dr. Maynert: I couldn't add anything to that. I agree perfectly.

Dr. Goldberg: I can summarize the discussion up to this point in stating that most doctors, at least the older generation, have not been trained to understand that interaction between different drugs, including alcohol, can and does occur. But also that drugs may act differently depending on the state of the reacting organism. One example, which we learned the hard way, is the thalidomide catastrophe, the interaction between the drug and the growing organism, the fertilized egg. Therefore, let us always assume that an interaction may occur and ask of what type.

Dr. Zarafonetis: Thank you, gentlemen. We will go to the next question.

Dr. Leonard Sigell (Cincinnati, O.): It seems to me that there is an interesting parallel between the question of the illegal use of marihuana and the legal and readily available prescription and non-prescription sedative-hypnotic agents. For example, the survey from California recently reported in the New England Journal of Medicine showed that in one year there were three hospital admissions associated with marihuana whereas there were more than 120 with acute barbiturate overdose and 40 due to over-the-counter drugs such as Sominex and Nytol. We have not heard much today about the over-the-counter sedative-hypnotics which are readily available, widely promoted on television, and frequently misused to the point of serious consequences. There was another study from Dr. Cluff in Florida showing that in the two months prior to the clinic visit the average patient admitted to taking 3.5 drugs, most of which were over-the-counter drugs, not drugs prescribed by the physician. So in view of our terrible experiences in handling the certainly less innocuous marihuana, how would you suggest that we attack the problem of over-the-counter drug abuse?

Dr. Jaffe: Well, I looked at the group that has generally supported this symposium and I wonder whether it would be an appropriate response of a guest to discuss that issue in any great detail. I can tell you that, in general, I think there may be too many over-the-counter drugs. I don't have any pat solutions to this problem. I am not sure that one need even link it with marihuana. If there had been no marihuana, I am sure your question would be equally relevant. Do we really need so many things that are sold for every minor distress? This is not a new question. Whether it is appropriately discussed under the rubric of sedative-hypnotics, however, is a relevant pharmacologic question. My own tendency would be to say that I think you may be generalizing when you say that it should be. One of the major criteria for making a drug "over-the-counter" is that it does not have the capacity to induce physical dependence of the barbiturate-sedative-hypnotic type. So we are dealing with a pharmacologically different class. Generally, we are dealing with a group of antihistamines that has sedative properties. Whether they can be "abused" in significant ways as opposed to "over-used" is certainly an issue and worthy of discussion. An even more significant question is whether or not the availability of over-the-counter drugs and their promotion by mass media has not significantly changed the ambience of our entire culture. But I am not sure that the question is as much pharmacological as it is sociological. The easy availability of over-the-counter preparations has its main impact by reducing the barriers to self-medication in general. It is a question, I think, of profound significance; I don't want to minimize it. However, I am not sure that you are correct in assuming that we should have discussed it here under the subject of the sedative-hypnotics. I would like other members of the panel to give their views on this question.

Dr. Isbell: I think that one might include the things that are sold over-the-counter to aid sleep, things like Sominex and so on, that contain essentially an antihistamine and a little bit of scopolamine. I do not believe there is really any significant problem of abusive use of these compounds. A lot of them are sold, and there are a few cases of suicidal attempts. I think we get an average of four or five a year at the University of Kentucky of patients who had taken this type of preparation. But I do not see that this particular group of mild sleeping medicines is really a big problem. The broad issue of over-the-counter drugs is too big to handle, and it is the Food and Drug Administration's problem to try to sort it out.

Dr. Zarafonetis: We will go on to another question.

Dr. Emil F. Pascarelli (New York, N.Y.): Just one quick comment on the question of tetanus and stiffness of the jaw. I have seen two instances of the presenting complaint of trismus in people who had been taking phenothiazines and interestingly enough, this syndrome responds to diphenhydramine administration or to barbiturates. It is something that we should keep in mind. I wondered if anyone else had seen that.

Dr. Zarafonetis: Thank you, I don't know if anyone else has. We will go to the next question.

Dr. Edward J. Lynn (East Lansing, Mich.): I think that over the years, doctors and their patients have engaged in a mutual conspiracy that has created a myth that anxiety or depression cannot be tolerated for more than five seconds. And although passing comments were made on the neurotic patient who shops around for pills, I wonder if the panel would care to make some comments on the physicians who over-prescribe, the pharmaceutical companies that over-produce, and the medical journals that are overloaded with advertisments for sedative-hypnotic agents.

Dr. Jaffe: I think your answer is self-evident. We must begin to get some perspective on whether we can continue to view the promotion of drug consumption by the public as something that lies entirely within the area of the profit-incentive system. The question requires that we take a much more ethical and self-examining view. If we do not eventually become more concerned with this issue, people outside of medicine will become more concerned with it and regulation will be imposed from the outside. Again, these are major sociologic questions. I think that many, many physicians would agree that we do tend to over-prescribe, and not only in the area of drugs of abuse. I think we have to point out that we have dealt with this problem in many situations. The pressure to prescribe does not come from the physician or advertising alone. If someone comes in with a cold, they may expect instantaneous cure. And woe to the physician who refuses to provide the most recent antibiotic available. There are people who look at you strangely if you refuse to prescribe for a runny nose; they think that there must be some magic pill on the market that will cure the common cold. I am viewed as nihilistic and antitherapeutic if I say that a runny nose can be survived. So I think it is a much more general question; if we can discuss questions of drug abuse now, we might then move to the broader questions of how our attitude toward drugs affects abuse patterns at a more general session. That would be my view. I think to move into this more general question would be to move away from the specific topic of sedative-hypnotics at this point in time.

Dr. Zarafonetis: If no other panelists wish to comment, we will go to the next question.

Dr. Carol C. Waterman (Glen Ellyn, Ill.): In the case of the chronically ill patient who needs continued medication for relief of pain and in whom you suspect that maybe you are encouraging drug dependency, is there any advantage in switching from one drug to another in an effort to avoid dependency or would methadone be applicable in this intsance?

Dr. Isbell: If you are talking about situations where you have to use the centrally active analgesics of the morphine type, either natural or synthetic narcotics, there is no point in switching. They all show a considerable amount of cross-tolerance, so you do little good by switching from one to the other. You would do just as well to settle down with one and stick with that and, if you have to, increase the dose. In a patient with an incurable disease, who cares how high you have to go to relieve his pain?

Dr. Zarafonetis: Thank you. Next question.

Dr. S. William Berg (Bloomington, Ind.): I have two comments I should like to make. First of all, we have seen three cases of dystonia and trismus after phenothiazines, as the gentleman from New York pointed out. The clue to these was that this was the first time they had taken phenothiazine and they developed the symptoms within a few hours afterwards. Secondly, I am not so inclined to dismiss over-the-counter products for relaxation and sleep, such as scopolamine, quite so quickly. It is true that scopolamine products are not abused as sedative-hypnotics very much, but I would point out that bromide products such as Miles' Nervine can be abused very much and these kinds of products have to be considered definitely hypnotics.

Dr. Zarafonetis: Thank you. Any comment from the panel? We shall go to the next question.

Unidentified member of the audience: In one of the slides in the last presentation, I noticed the major tranquilizers listed among one group of patients who were admitted. I should like the panel in general to comment on the habituating liability of both the major tranquilizers and the tricyclic antidepressants. I will phrase quickly the second question, if I may. In general practice, I often find it necessary, in spite of my realization that psychotherapy is indicated, to give the patient some relief from his environment by means of medication. In this instance, what medication does the panel feel is the most judicious choice?

Dr. Isbell: I believe part of the question was addressed to me for one of Dr. Idestrӧm's slides. This is the one from a Norwegian study in which they asked all patients admitted about any kind of psychotropic drug. These included the major tranquilizers of the phenothiazine, haloperidol classes, and the major antidepressives which I would imagine would be imipramine and Elavil. These are usually not regarded as drugs of dependence. There seems to be very little problem with them. The drugs of dependence listed on the slide were chlordiazepoxide, barbiturates, meprobamate and others which turned out to be amphetamine and opiates. Did that answer it for you?

Same questioner: Yes. The second question is, if you were in general practice and felt a real need for a patient who would not or could not obtain psychotherapy, the anxious neurotic patient who comes in complaining of insomnia, diarrhea, epigastric distress, and has no demonstrable pathology, what is the most judicious drug for the management of this patient, in your opinion? We have all said a great deal about what not to use, but what is judicious to use, realizing that occasionally we have to use something.

Dr. Isbell: Very often the most judicious drug is a drug with extremely little effect either by virtue of dose or by virtue of what it contains. Sometimes we give people tablets that contain digestive enzymes, a pretty yellow tablet, and it may work very well. If I think I have to give something, I am very inclined to use really placebo doses, chlordiazepoxide, 5 mg or phenobarbital, 15 mg, with reinforcement to be very careful about this stuff; "Don't take too much, these are powerful drugs," and so on. This would seem to work out very well. Actually, I think manipulation of a placebo effect is perfectly good medicine and among the things we lack today are some of the good old placebos our forefathers had and in which they had confidence. Back in the days they used nasty green-colored stuff—tincture of iron, quinine, strychnine, tincture of nux vomica—these did people worlds of good and very little harm. Unfortunately, today we are quite restricted in the choices we have of these essentially placebo medications that taste bad, smell bad, and generally work.

Questioner: Does the rest of the panel agree on that point?

Dr. Goldberg: May I add something to it? When I was in practice, about 25 years ago, I had a nurse patient about 40 years old. I gave her a sedative which in those days was realy phenobarbital, a dose of about 15 mg—"plus." And what was "plus"? "Plus" was some of these ill-smelling things given to stimulate the gastric juices. Being a pharmacologist, I knew what the pharmacist adds to preserve the solution so it should not mold. So I added .1% and a chemical name. The patient came back in a week and said, "You know, every single doctor I have seen has given me exactly the same but what really worked in your medicine was—" and then she named the preservative. Believe me, this is a true example. It was not intended to be a placebo effect, but I think this is quite clear. You must give something with confidence in it, and as Dr. Isbell said, it is very important to go back to treating patients, having contact with them, and to get away from some of these very potent drugs we have today. I think the point is confidence in small doses.

Dr. Zarafonetis: Was there another question?

Dr. Walter D. Meester (Grand Rapids, Mich.): Mr. Chairman, I would like to comment on that last question. I don't quite agree with the panel on the use of placebos; however, I do feel that the most judicious drug to be used in this case is the drug that you as a physician have the most experience with, and I think this is a very important principle to remember in the therapy of patients.

Dr. Zarafonetis: I think we have come to the placebo effect. I want to take this opportunity to thank Dr. Jaffe, Dr. Isbell, Dr. Maynert, and Dr. Goldberg for their participation on the panel. I also would like to express our appreciation to Dr. Iderström and Dr. Essig for their work in advance of this conference and wish them speedy recovery from illness which prevented their personal appearance today.

Morning Session
Wednesday, November 11, 1970

Presiding: Norman Weiner, M.D.

THE CHEMISTRY AND DETECTION OF AMPHETAMINES

J. W. SCHWEITZER, Ph.D., AND A. J. FRIEDHOFF, M.D.

The history of drug abuse of compounds of the amphetamine family can be traced back about 5000 years. The Chinese were aware of properties of extracts of the plant group collectively known as Ma-Huang. The active amine in these plants, ephedrine, or N-methyl-β-hydroxy amphetamine, was isolated in 1887,[1] but its therapeutic possibilities were ignored until 1924.[2] Amphetamine, although first synthesized in 1919,[3] was found to be a strong central stimulant by Alles in 1927.[4] By 1932 it was being tested as a decongestant. Aware of its central stimulant effects, Prinzmetal and Bloomberg, in 1935, found amphetamine to be effective in the treatment of narcolepsy.[5] Only three years later, Young and Scoville observed that the abuse of this drug by narcoleptics induced symptoms of paranoid psychosis.[6] Several reports over the next 10 years confirmed this.[7,8] As a result, the Benzedrine Inhaler, which contained 250 mg of amphetamine sulfate, was withdrawn in 1949. Some drug companies introduced an inhaler containing mephentermine, or N,α-dimethyl amphetamine. This drug also has strong central stimulatory effects. In 1966, Greenberg and Lustig described four cases of toxic psychosis in individuals who had ingested the contents of inhalers containing mephentermine.[9] This was confirmed in our laboratory by Angrist and co-workers, who, in addition, found large amounts of this drug and its metabolite, phentermine, in the urine of such patients.[10]

Reports on amphetamine psychosis continued. Herman and Nagler in 1954,[11] Connell in 1958,[12] and Bell and co-workers in 1965[13] have commented on the similarity between amphetamine and non-drug induced paranoid states. Most recently, Griffith and co-workers[14] and Angrist and co-workers[15] have induced psychotic manifestations by administering amphetamine to non-psychotic volunteers who were experienced amphetamine (ab)users.

With the introduction of amphetamine for use as an appetite suppressant, its misuse has become even more widespread. Methamphetamine, otherwise known as "speed," is another current favorite in enclaves of drug abuse such as New York's East Village and San Francisco's Haight-Ashbury district. Moreover, a number of new psychotomimetic amphetamines have been synthesized recently and at least one of these, "STP" or 2,5-dimethoxy-4-ethylamphetamine, is now well known. It appears likely that psychiatric admissions for psychoses induced by amphetamine or amphetamine-related compounds will not decrease in the near future. We would, therefore, suggest that all psychiatric patients be screened by testing their urine for the presence of one or more amphetamines.

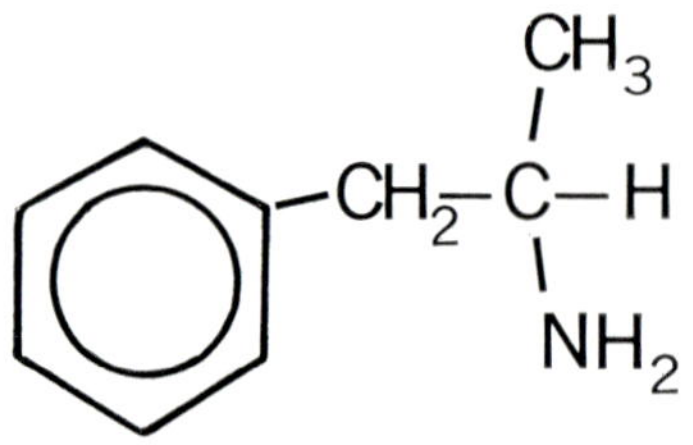

AMPHETAMINE

Figure 1

CHEMISTRY AND METABOLISM

The structure of amphetamine is shown in Figure 1. It can be described as α-methylphenethylamine. The methyl group alters the chemical behavior of the fundamental phenethylamine structure in at least two ways. It prevents amphetamine from undergoing oxidative deamination by monamine oxidase. This enzyme is located in mitochondria and is partly responsible for the

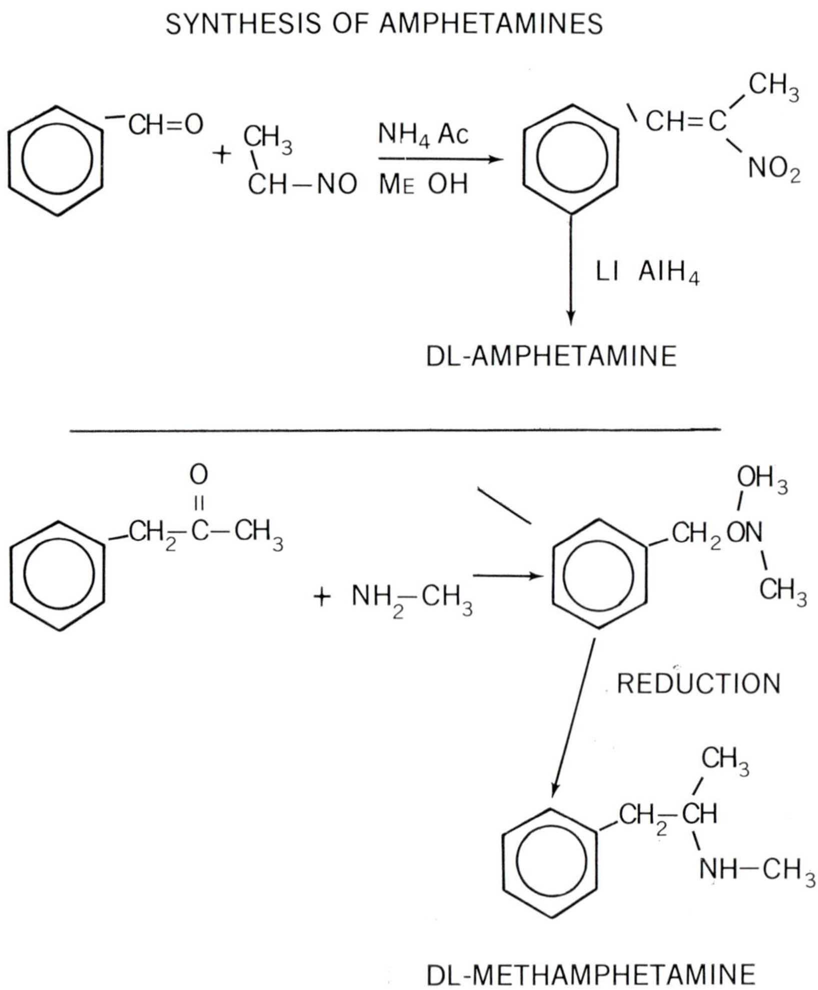

Figure 2

breakdown of catecholamines within neurons. It is also believed that amphetamine interferes with biochemical systems in the synaptic area. The nature of this interaction is under intensive study by groups throughout the world. The α-methyl group also introduces optical asymmetry into the molecule. Generally, the dextro or d-form is more biologically active than the levo form. It is only dextroamphetamine that is β-hydroxylated *in vivo* and is the more potent stimulant of the two isomers.

Unfortunately, the structure of amphetamine is extremely simple. It can be synthesized in two steps from commercially available starting materials such as benzaldehyde and nitroethane (Fig. 2). Methamphetamine can be synthesized by condensing phenylacetone with methylamine, both commercially available, and reducing the product. This synthesis was actually carried out by a 9-year-old child in an illicit "bathtub synthesis" operation.[16]

Among the earliest studies of the metabolism of amphetamines, Axelrod demonstrated, in 1954, that an enzyme in the microsomal fraction of liver can oxidatively deaminate amphetamine.[17] This is a relatively non-specific drug detoxifying enzyme, not to be confused with monamine oxidase. The product, phenylacetone, is further metabolized to benzoic acid, which is excreted in part as hippuric acid. In man, this is the dominant pathway and accounts for about 30% of an administered dose of amphetamine. Two minor, but possibly significant, pathways are also known (Fig. 3). These include p-hydroxylation and β-hydroxylation. The first yields a potent stimulant which is degraded in a manner similar to the degradation of amphetamine. In man, about 0.1 to 2.5% of an administered dose of amphetamine is excreted as p-hydroxyamphetamine. Smith and Dring have observed that there are marked individual variations in the excretion of this metabolite.[18] In a study with four subjects, one consistently excreted 4 to 30 times as much p-hydroxyamphetamine as did the other three. A typical excretion pattern is shown in Table 1. Both amphetamine and p-hydroxyamphetamine can be β-hydroxylated. Both products are pharmacologically active. There are conflicting reports on the detection of these β-hydroxyamines in the urines of subjects receiving amphetamine, but these compounds have been detected in studies with amphetamine in lower animals. It has been suggested by Kopin that p-hydroxynorephedrine may compete with the catecholamines for neuronal storage sites, which may account for differences between *in vivo* and *in vitro* observations on the action of amphetamine.[19]

N-Alkylamphetamines, such as methamphetamine and mephentermine, are N-dealkylated and the resulting primary amine then undergoes further metabolism as outlined above (Table 2). An investigation into the possibility that amphetamine might be

Table 1 Urinary metabolites of dl-C^{14}-amphetamine in man

Metabolite	% Excreted in 24 hours*
Amphetamine	30.0
Total benzoic acid	21.0
Phenylacetone	3.4
p-Hydroxyamphetamine	2.8
Total C^{14} in urine	66.0

* From Dring et al.[20]

Table 2 Metabolism of N-alkylamphetamines*

Drug	Dose (mg)	% Excreted unchanged	% Excreted as amphetamine
d-Amphetamine	5	36.5	—
d-Methamphetamine	5	43.5	6.5
d-Ethamphetamine	10	27.0	10.5
d-Isopropylamphetamine	10	2.0	24.5

* From Vree and Van Rossum.[21]

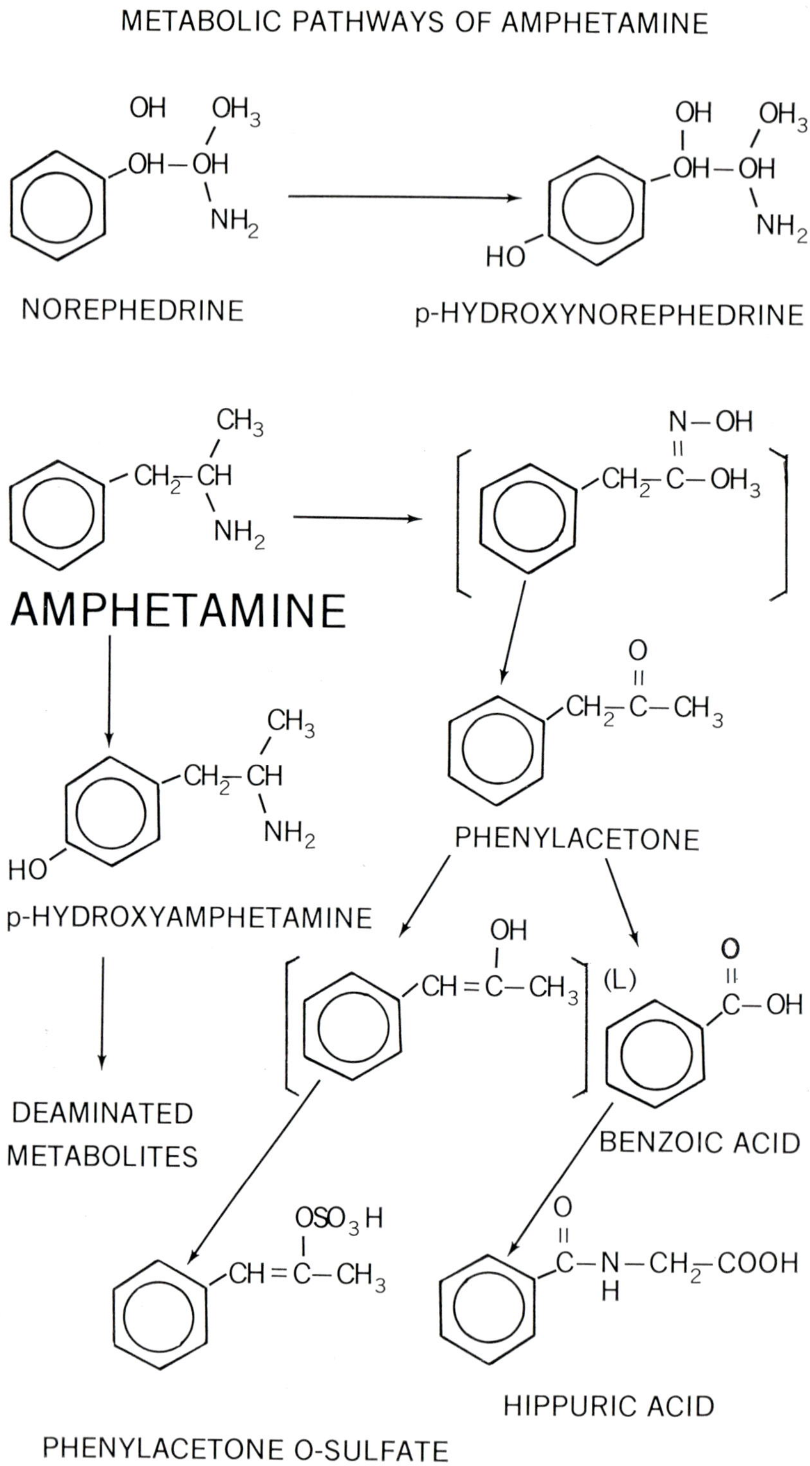
METABOLIC PATHWAYS OF AMPHETAMINE
NOREPHEDRINE
p-HYDROXYNOREPHEDRINE
AMPHETAMINE
p-HYDROXYAMPHETAMINE
PHENYLACETONE
(L)
DEAMINATED METABOLITES
BENZOIC ACID
HIPPURIC ACID
PHENYLACETONE O-SULFATE

Figure 3

converted metabolically to a known psychotomimetic was recently undertaken in our laboratories. It has been proposed that p-methoxyamphetamine might be formed from amphetamine, *in vivo*.[22] We administered amphetamine in doses of 5 to 50 mg/hour to five non-schizophrenic volunteers. An average total dose of 629 mg was administered over a mean of 51 hours.[23] Three of the subjects experienced a psychotic reaction. Urines were collected when behavioral effects were maximal and were analyzed for amphetamine and p-methoxyamphetamine. The limit of detection of the latter amine with our method is 70 to 80 mcg per 24-hour urine sample. None was found. In later studies, some of these subjects and some new ones were given p-methoxyamphetamine hydrochloride in doses of 10 to 65 mg.[24] A mean of 6.7% was excreted unchanged with a range of 0.3 to 15% (Table 3). By combining analytical sensitivity and data on the excretion of administered p-methoxyamphetamine, as much as 1.2 mg of this amine could have been formed without having been detected. Since p-methoxyamphetamine is very likely O-demethylated and deaminated, its metabolites would be identical with those of amphetamine. We are currently attempting to increase the sensitivity of our analytical procedure. In addition, by pretreating subjects with drugs that block O-demethylation, it may be possible to detect this aberrant amine if it is being formed from amphetamine.

Table 3 Excretion of p-methoxyamphetamine administered to man

Subject	Dose* (mg)	% Excreted unchanged
A	10	3.15/24 hours
A	65	0.26/36 hours
B	10	14.8/36 hours
C	20	4.25/24 hours
D	30	4.90/24 hours
E	57	11.8/15 hours

* As the hydrochloride.

The extent to which amphetamines are metabolized is related to the speed with which they are excreted. Beckett and coworkers have observed that subjects voiding urine at low pH, achieved by the prior administration of ammonium chloride, excrete about twice as much unchanged amphetamine as subjects whose urine is not pH-controlled.[25–27] Moreover, at pH values of 8 and above (by the prior administration of sodium bicarbonate) only a small fraction of administered amphetamine is excreted unchanged (Table 4). All other amphetamines investigated follow approximately the same pattern. The wide variations often found in

Table 4 Effect of urinary pH on the excretion of amphetamines*

Drug	Urinary pH	% Excreted unchanged
Dextroamphetamine	5	57.0
Dextroamphetamine	8	2.6
Dextromethamphetamine	5	63.0
Dextromethamphetamine	8	1.5

* From Wilkinson and Beckett.[27]

Table 5 Excretion of amphetamines in man*†

Amine	% Excreted unchanged	% Excreted as amphetamine
d-Amphetamine	60–70	—
d-Methamphetamine	40–50	5–10
l-Methamphetamine	60–80	2–5
d-Ethamphetamine	20–30	15–20
l-Ethamphetamine	65–80	5–10
dl-Ethamphetamine	40–50	10–15
d-Fenfluramine	15	—
l-Fenfluramine	20–30	—
Phentermine	80	—
p-Chlorophentermine	35	—
l-Ephedrine	95	—

* Urine rendered acidic by pretreatment with ammonium chloride.

† From Beckett and Brookes.[28]

Table 6 **Procedures employing gas-liquid chromatography for the analysis of amphetamines**

Compounds analyzed	Reference compound	Nature of Derivative	Column	Ref.
Amphetamine	None	Underivatized	Apiezon L grease	31
Amphetamine Methamphetamine	N,N-diMethylaniline	Underivatized	Carbowax 1500	32
Amphetamine	Methoxyphenamine	Acetone Schiff base	Polyethyleneglycol-KOH	28
Amphetamine Methamphetamine	N,N-diMethylaniline	Underivatized	Carbowax-KOH	33
Amphetamine	None	Trichloracetyl	JXR methyl silicone	34
Amphetamine Methamphetamine Fenfluramine	Chlorphentermine	Hepta-fluoro-butyroyl	OV-1	35
Chlorphentermine	Amphetamine	Hepta-fluoro-butyroyl	OV-1	35
Amphetamine Methamphetamine	N-propylamphetamine	Acetyl	SE 30	36
Amphetamine	N,N-diMethylaniline	Underivatized	Carbowax 6000 KOH	37
Amphetamine Methamphetamine Mephentermine Phentermine p-Methoxyamphetamine	Phenethylamine	Acetyl	SE 30-EGA-SE 30	38
d- and l-p-OH Amphetamine	None required	O-Methyl then trifluoroacetyl-l-prolyl	SE 30	39
Amphetamine	Triphenylamine	Propionyl	SE 30	40

amphetamine excretion studies are therefore related to diet since urinary pH is governed to a great extent by the kinds of foods eaten. Generally, N- or ring-substituted amphetamines are metabolized to a greater extent than is amphetamine. β-Hydroxyamphetamines, on the contrary, are hardly metabolized (Table 5).

DETECTION OF AMPHETAMINES

There are a number of methods available for the detection and estimation of amphetamines. Only a few are specific. Among non-specific methods are spot tests and spectrophotometric procedures that depend on the formation of a chromophore. In one of the latter, methyl orange is allowed to complex with amphetamines extracted from urine. Blank values in this procedure vary from 0.65 to 2 mcg as amphetamine per milliliter of urine.[17,28] Moreover, this procedure does not differentiate among amphetamines, and possibly may yield a number of false negatives. An interesting colorimetric test, so far applied exclusively for the analysis of amphetamine in capsules, involves the conversion of amphetamine to a pyrrole by reaction with 2,5-dimethoxytetrahydrofuran, and subsequent color development with Ehrlich's reagent. The sensitivity is poor, about 50 mg being required, and the test fails with methamphetamine.[29] Another recently discovered test employs cerium sulfate for the oxidation of amphetamines.[30] These tests are mentioned because they are new and perhaps should be further investigated.

By far the most specific and sensitive procedures for the analysis of amphetamines are the separation techniques of gas-liquid chromatography. Quantitation is assured if a reference compound, chemically and structurally related to the amphetamines, is added to a portion of the urine prior to analysis. Following extraction of the amines from urine and derivatization (to further increase specificity), the mixture is subjected to gas-liquid chromatographic separation. Not only are the substances separated from each other, but the precise relative amounts of each are also displayed. Subsequent introduction of a mixture of equal amounts of the amphetamines and reference compound in derivatized form yields the true instrument response of each amphetamine to the reference compound. With our procedure, about 20 urine samples can be analyzed during a working day, or about 10 samples in duplicate. Since a large part of the time is spent waiting for the sample to pass through the column, other routine work can be carried out as well. Thus, aside from the approximately two hours required to prepare the 20 urine samples for gas chromatographic analysis, little additional time is consumed.

A partial list of published procedures is outlined in Table 6. It can be seen that in those methods in which the amphetamines are analyzed in underivatized form, the column is relatively non-polar. After derivatization, which destroys the basicity of the amino group, a more polar column is desirable. In our system a polar polyethyleneglycol-adipate column is used and results in the separation of several amphetamines as shown in Table 7.

We have used the method for the screening of urines from patients admitted with a

Table 7 Relative retention times for amphetamines

Amine	Relative retention time*
Mephentermine	0.42
Phentermine	0.48
Methamphetamine	0.64
Amphetamine	0.72
Phenethylamine	1.00
p-Methoxyamphetamine	2.5

* For a 185 cm U-tube packed with 8% EGA on chromosorb W to within 30 cm of both ends. The remainder of each end was packed with 8% SE 30 on chromosorb W. Oven temperature was 205°C.

tentative diagnosis of amphetamine psychosis, and for the assay of amphetamines in subsequent urine samples from such patients in order to relate clearance with psychiatric state.[41]

Amphetamine is a simple molecule and can be synthesized quite easily, as noted earlier, even in the most makeshift of laboratories. Because of this and the recent development of new hallucinogenic amphetamines, such as "STP," psychiatric admissions following the use of such drugs can be expected to continue.

REFERENCES

1. Hamerslag, F. E.: *The Technology and Chemistry of Alkaloids.* New York, D. Van Nostrand Co., 1950, p. 7.
2. Chen, K. K., and Schmidt, C. F.: The action of ephedrine, an alkaloid from Ma Huang. *Proc Soc Exp Biol Med 21*:351, 1924.
3. Ogata, A.: α- and β-aminoalkyl (aryl) benzenes and their derivatives. *J Pharm Soc Japan 445*:193–216, 1919.
4. Leake, C. D.: *The Amphetamines, Their Actions and Uses.* Springfield, Ill., Charles C Thomas, 1958, p. 4.
5. Prinzmetal, M., and Bloomberg, W.: Use of benzedrine for treatment of narcolepsy. *JAMA 105*:2051–2054, 1935.
6. Young, D., and Scoville, W. B.: Paranoid psychosis in narcolepsy and the possible dangers of benzedrine treatment. *Med Clin N Amer 22*:637, 1938.
7. Monroe, R. R., and Drell, H. J.: Oral use of stimulants obtained from inhalers. *JAMA 135*:908, 1947.
8. Brown, C. T.: Benzedrine habituation. *Milit Surg 104*:365–379, 1949.
9. Greenberg, J. R., and Lustig, N.: Misuse of Dristan inhaler. *New York J Med 66*:613–617, 1966.
10. Angrist, B. M., Schweitzer, J. W., Gershon, S., and Friedhoff, A. J.: Mephentermine psychosis: misuse of the Wyamine inhaler. *Amer J Psychiat 126*:149–151, 1970.
11. Herman, M., and Nagler, S. H.: Psychoses due to amphetamine. *J Nerv Ment Dis 120*: 268–272, 1954.
12. Connell, P. H.: *Amphetamine Psychosis.* Maudsley Monographs No. 5. London, Oxford University Press, 1958.
13. Bell, D. S.: A comparison of amphetamine psychosis and schizophrenia. *Brit J Psychiat 3*:701–706, 1965.
14. Griffith, J. J., Oates, J., and Cavanaugh, J.: Paranoid episodes induced by drug. *JAMA 205*:39, 1968.
15. Angrist, B. M., and Gershon, S.: The phenomenology of experimentally-induced amphetamine psychosis—preliminary observations. *Biol Psychiat 2*:95–107, 1970.
16. *Chem Engin News,* No. 3, 1969, p. 11.
17. Axelrod, J.: Studies on sympathomimetic amines. II: The biotransformation and physiological disposition of d-amphetamine, d-p-hydroxyamphetamine and d-methamphetamine. *J Pharmacol Exp Ther 110*:315–326, 1954.
18. Smith, R. L., and Dring, L. G.: Patterns of metabolism of β-Phenylisopropylamines in man and other species. In *Amphetamines and Related Compounds,* Costa, E., and Garattini, S. (Eds.). New York, Raven Press, 1970.
19. Kopin, I. J., Fisher, J. E., Musacchio, J. M., Horst, W. D., and Weise, V. K.: "False neurochemical transmitters" and the mechanism of sympathetic blockade by monamine oxidase inhibitors. *J Pharmacol Exp Ther 147*:186, 1965.
20. Dring, L. G., Smith, R. L., and Williams, R. T.: The metabolic fate of amphetamine in man and other species. *Bio Chem J 116*: 425–435, 1970.
21. Vree, T. B., and Van Rossum, J. M.: Kinetics of metabolism and excretion of amphetamines in man. In *International Symposium on Amphetamines and Related Compounds,* Costa, E., and Garattini, S. (Eds.). New York, Raven Press, 1970.
22. Smythies, J. R., Johnston, V. S., Bradley, R. J., Benington, F., Morin, R. D., and Clark, L. C., Jr.: Some new behavior-disrupting amphetamines and their significance. *Nature 216*:128–129, 1967.
23. Angrist, B. M., Schweitzer, J. W., Friedhoff, A. J., and Gershon, S.: Investigation of p-Methoxyamphetamine excretion in amphetamine-induced psychosis. *Nature 225*: 651–652, 1970.
24. Schweitzer, J. W., Friedhoff, A. J., Angrist, B. M., and Gershon, S.: The excretion of p-methoxyamphetamine administered to humans. *Nature* (accepted for publication).
25. Beckett, A. H., and Rowland, M.: Urinary excretion kinetics of amphetamine in man. *J Pharm Pharmacol 17*:628–639, 1965.
26. Beckett, A. H., and Brookes, L. G.: The effect of chain and ring substitution on the metabolism, distribution and biological action of amphetamines. In *International Symposium on Amphetamines and Related Compounds,* Costa, E., and Garattini, S. (Eds.). New York, Raven Press, 1970.
27. Wilkinson, G. R., and Beckett, A. H.: Absorption, metabolism and excretion of the ephedrine in man: I. The influence of urinary pH and urine volume output. *J Pharmacol Exp Ther 162*:139–147, 1968.

28. Beckett, A. H., and Rowland, M.: A specific method for the determination of amphetamine in urine. *J Pharm Pharmacol* (Suppl.) *16*:27T–31T, 1964.
29. Fontari, F., and Morandini, F.: Colorimetric determination of amphetamine salts in dosage forms. *J Pharm Pharmacol 22*:411–413, 1970.
30. Wallace, J. E., Biggs, J. D., and Ladd, S. L.: Determination of amphetamine by ultraviolet spectrophotometry. *Anal Chem 40*: 2207–2210, 1968.
31. Thompson, G. F., and Smith, K.: Application of on-column acid-base reactions to the gas chromatographic separation of free fatty acids and amines. *Anal Chem 37*: 1591, 1965.
32. Cartoni, G. P., and De Stefano, F.: Determination of amphetamines by gas chromatography: urinary excretion of amphetamines in rat and man. *Ital J Biochem 12*: 296–309, 1963.
33. Beckett, A. H., and Rowland, M.: Determination and identification of amphetamines in urine. *J Pharm Pharmacol 17*:59–60, 1965.
34. Noonan, J. S., Murdick, P. W., and Ray, R. S.: A method for detecting amphetamine using gas chromatography of a halogenated derivative. *J Pharmacol Exp Ther 168*:205–209, 1969.
35. Bruce, R. B., and Maynard, W. R., Jr.: Determination of amphetamine and related amines in blood by gas chromatography. *Anal Chem 41*:977–979, 1969.
36. Lebish, P., Finkle, B. S., and Brackett, J. M., Jr.: Determination of amphetamine, methamphetamine, and related amines by gas chromatograph with hydrogen-flame ionization detector. *Clin Chem 16*:195–200, 1970.
37. Campbell, D. B.: A method for the measurement of therapeutic levels of (+)-amphetamine in human plasma. *J Pharm Pharmacol 21*:129–131, 1969.
38. Schweitzer, J. W., and Friedhoff, A. J.: Amphetamines in human urine: rapid estimation by gas-liquid chromatography. *Clin Chem 16*:786–788, 1970.
39. Gunne, L-M, and Galland, L.: Stereo-selective metabolism of amphetamine. *Biochem Pharmacol 16*:1374–1377, 1967.
40. Gunne, L-M.: The urinary output of d- and l-amphetamine in man. *Biochem Pharmacol 16*:863–869, 1967.
41. Angrist, B. M., Schweitzer, J. W., Friedhoff, A. J., Gershon, S., Kekimian, L. J., and Floyd, A.: The clinical symptomatology of amphetamine psychosis and its relationship to amphetamine levels in urine. *Int Pharmacopsychiat 2*:125–139, 1969.

PHARMACOLOGY OF CENTRAL NERVOUS SYSTEM STIMULANTS

NORMAN WEINER, M.D.

The most important group of central nervous system stimulants may be classified pharmacologically as sympathomimetic amines or substances with actions which mimic those of norepinephrine, the transmitter substance of the sympathetic nervous system. Of these, the prototype drug is amphetamine, an agent with multiple actions on the central nervous system of man and lower animals. d-Amphetamine modifies behavior in a number of fashions and in most instances its behavioral effects may be described as stimulant. However, both in man and other animals, amphetamine may reduce the behavior in situations in which a high output of behavior normally occurs.[1–3] Amphetamine most typically enhances psychomotor activity, produces restlessness and tremors, reduces fatigue, elevates mood, stimulates respiration, produces stereotyped behavior, and reduces food intake.[4]

Amphetamine may be considered as a chemical analogue of both norepinephrine, the adrenergic neurotransmitter substance, and 3,4-dihydroxyphenylethylamine (dopamine) (Fig. 1). The chemical features which distinguish this compound from norepinephrine endow this drug with physical-chemical properties required for maximal central nervous system activity. Because ampheta-

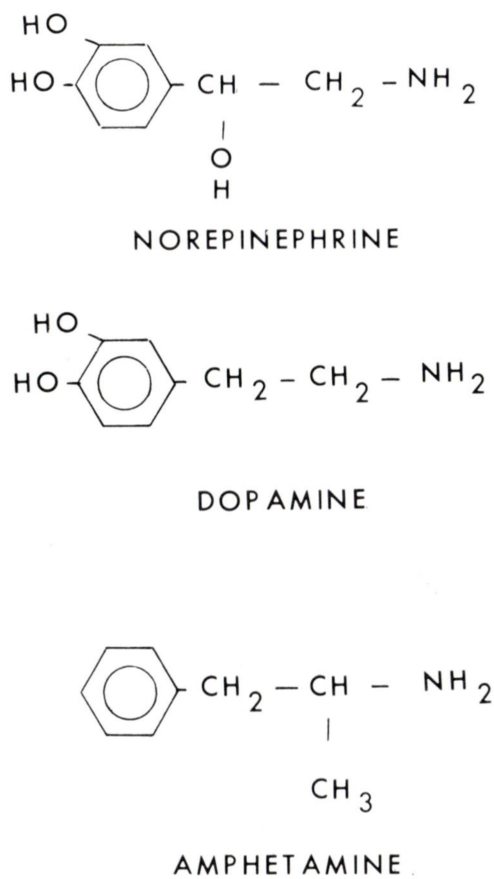

Figure 1 Structures of norepinephrine, dopamine, and amphetamine.

mine lacks the polar hydroxyl groups on the ring and side chain, it is much more lipid-soluble than norepinephrine and therefore is much better able to penetrate the lipid barriers surrounding the central nervous system.[5] By virtue of the α-methyl group on the side chain, amphetamine is able to resist oxidative deamination by the rather ubiquitous mitochondrial enzyme, monoamine oxidase, and therefore is much more stable and has a much longer duration of action than compounds lacking this α-methyl group.[6,7]

However, these important chemical differences between amphetamine and the catecholamines render amphetamine and its analogues incapable of directly stimulating the adrenergic receptor. With regard to its effects on the peripheral sympathetic nervous system, amphetamine is considered to be an indirectly acting sympathomimetic amine, i.e., it acts by releasing norepinephrine from peripheral adrenergic stores.[8,9] In the central nervous system, the action of amphetamine on adrenergic neuronal systems seems more complex. After reserpine treatment, when catecholamine stores are severely depleted, the central actions of amphetamine either are unaffected or are enhanced,[2,10] suggesting that amphetamine does not act solely by releasing endogenous norepinephrine. However, if the stores of norepinephrine and dopamine are depleted in the brain by the use of α-methyl-p-tyrosine, an inhibitor of the synthesis of norepinephrine, the central nervous system actions of amphetamine are markedly reduced.[11,12] Considerable evidence has accumulated which suggests that norepinephrine is stored in functionally distinctive compartments in adrenergic neurons and that newly synthesized norepinephrine may be extremely critical for adrenergic function. The newly synthesized amine appears to be most susceptible to release both by nerve stimulation[13,14] and by amphetamine and other indirect acting amines.[15] In contrast, reserpine depletes primarily that norepinephrine which is localized in the amine-containing storage vesicles.[9] It is possible that reserpine fails to deplete the quantitatively minor pool most susceptible to displacement by indirectly acting sympathomimetic amines and that this store may be a primary binding or storage site for newly synthesized norepinephrine.[16–18] Thus, the bulk of current evidence favors the view that amphetamine produces its central adrenergic effects by facilitating the interaction of endogenous norepinephrine with its receptor, rather than directly stimulating the adrenergic receptor system.

From many biochemical and pharmacologic studies, it is known that amphetamine has several actions on catecholamine metabolism, each of which may contribute to the central nervous effects of this compound. At low concentrations, amphetamine displaces norepinephrine and dopamine from intraneuronal storage sites and inhibits the uptake of catecholamines by adrenergic and dopaminergic nerve terminals in the region of the synaptic cleft (Fig. 2; Table 1).[19–24] These actions of amphetamine are believed to be primarily responsible for the pharmacologic effects observed after administration of usual therapeutic doses of the drug. The relative importance of these two effects in bringing about the actions of amphetamine has not been ascertained, but they will clearly act in concert to produce greater effects than might be predicted from either action alone. These biochemical actions of amphetamine are demonstrable either when the metabolism of dopamine in the caudate region is examined or when the metabolism of norepinephrine in cortex or brain stem is studied.[7,15,25,26] Taylor and Snyder have compared the effects of d- and l-amphetamine on *in vitro* uptake and retention both of labeled dopamine into the corpus striatum and of labeled norepinephrine into the brain stem, thalamus-hypothalamus-midbrain, and cerebellum.[27] They observed that d-amphetamine is much more potent than l-amphetamine in inhibiting the retention of ^{3}H-norepinephrine in various regions of the brain. The two isomers are approximately equally effective

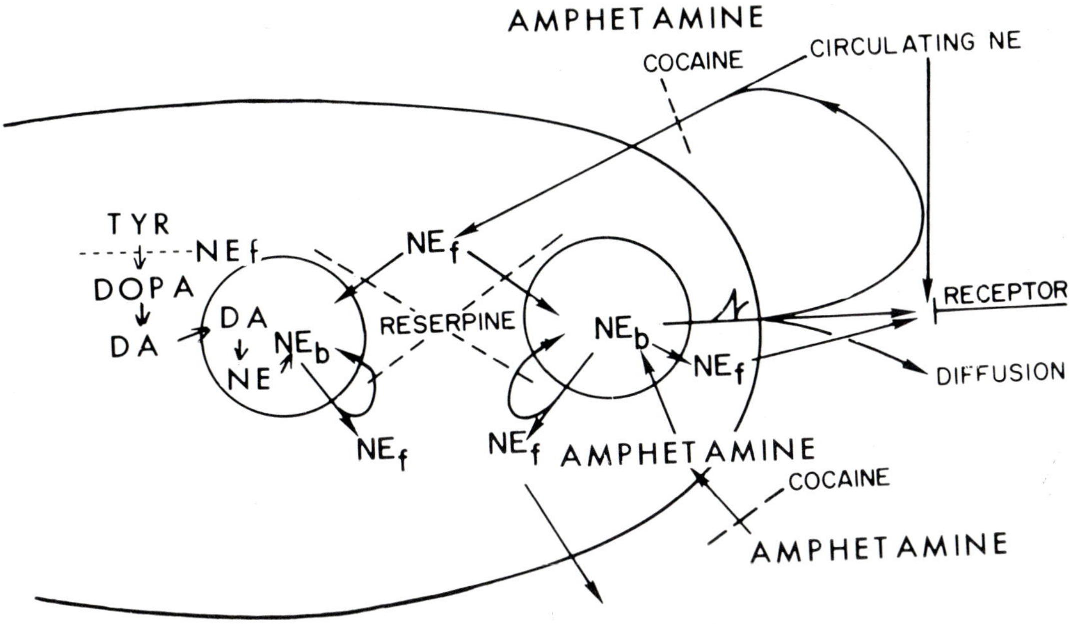

Figure 2 Schematic representation of norepinephrine metabolism in the adrenergic neuron and the effects of amphetamine, cocaine, and reserpine on norepinephrine metabolism. NE_b = bound norepinephrine; NE_f = free norepinephrine; TYR = tyrosine; DOPA = 3,4-dihydroxyphenylalanine; DA = dopamine; ---- = inhibitory effect on the process depicted: NE_f inhibits conversion of tyrosine to DOPA; cocaine and amphetamine block catecholamine uptake across the axonal membrane; reserpine blocks uptake of catecholamines into the storage particle. The ability of amphetamine to release bound norepinephrine also is shown in schematic fashion.

Table 1 Biochemical effects of amphetamine on amine-containing neurons

Norepinephrine-containing neurons
Enhanced release by displacement
Blockade of uptake mechanism
Inhibition of monoamine oxidase
Replacement of norepinephrine in storage vesicles by amphetamine metabolite
Dopamine-containing neurons
Enhanced release by displacement
Blockade of uptake mechanism
Inhibition of monoamine oxidase
5-Hydroxytryptamine-containing neurons
?Activation of 5-HT receptors
Inhibition of monoamine oxidase
Enhanced release by displacement (in high doses)

in blocking retention of ^{3}H-dopamine in the corpus striatum. These biochemical findings correlate well with behavioral studies wherein d-amphetamine is ten times more potent than l-amphetamine in enhancing locomotor activity, but is only twice as potent in producing the compulsive gnawing syndrome observed in rats. The latter action of amphetamine is presumed to be related to effects of the drug on striatal dopamine-containing neurons.

In much higher concentrations, d-amphetamine can, to a slight extent, accelerate the release of ^{3}H-5-hydroxytryptamine from brain tissue.[15,26] This action of amphetamine on tryptamine metabolism does not appear to be significant when usual pharmacologic doses of amphetamine are employed, but may be extremely important in the mediation of some of the toxic effects which are associated with high-dose amphetamine usage.

Amphetamine is able to competitively inhibit the oxidative deamination of norepinephrine and dopamine and as a consequence may prolong the duration of catecholamine effects in the central nervous system.[6,7] This action of amphetamine is probably relatively unimportant when usual pharmacologic doses of the drug are administered,[24] but may become increasingly important as tolerance develops to amphetamine and as higher and higher doses are administered.

It might be well to recall that the catecholamines are not the only endogenous amines which are subject to oxidative deamination by monoamine oxidase. 5-Hydroxytryptamine is also an excellent substrate for this enzyme and, in contrast to the catecholamines in which O-methylation can also inactivate these powerful agents, oxidative deamination is the only important route in the metabolism of 5-hydroxytryptamine.[28–30] Thus some of the actions of amphetamine, particularly after the administration of high doses, may be related to decreased breakdown of endogenous 5-hydroxytryptamine.

Vane demonstrated that amphetamine is able to exert a direct action on tryptamine receptors present in the rat stomach strip.[31] The contractions of the rat stomach strip observed with both amphetamine and 5-hydroxytryptamine are blocked by 2-bromolysergic acid diethylamide in a selective fashion which strongly suggests that these substances are acting on a common receptor. Vane speculated that analogous tryptamine receptors may exist in the brain and some of the effects associated with administration of amphetamine may be related to actions on 5-hydroxytryptamine receptors. Since similar psychotic manifestations are observed in users of high doses of amphetamine and in users of tryptamine derivatives, this interesting suggestion must be entertained seriously. Snyder and co-workers, in a series of provocative theoretical papers, presented evidence which suggested that amphetamine and tryptamine derivatives possess similar physical and chemical properties and indicated that these common features may provide the basis for interactions of these compounds with a common receptor. The activation (or inhibition) of this receptor may initiate effects in the brain which lead to distortion of perception and hallucinations.[32,33]

With this background on the biochemical effects of amphetamine on brain amines, some preliminary conclusions concerning the biochemical basis for the behavioral effects of amphetamine in animals and man may be derived (Table 2). It is now generally accepted that the psychomotor stimulant effects of amphetamine, the ability of amphetamine to suppress appetite, and the stereotyped behavior observed both in rats and in some amphetamine abusers are related to enhanced activity in catecholamine-containing neural systems. Although by no means conclusive, present evidence indicates that the anorectic effect of amphetamine is related to actions of the drug on norepinephrine-containing neurons in the hypothalamus, and psychic stimulant effects are related to actions on norepinephrine-containing neurons in the brain stem reticular formation. Psychomotor stimulation appears to be related at least in part to effects on noradrenergic neurons, although altered function of dopamine-containing neurons also may be involved. Stereotyped behavior and perhaps the compulsive repetitive behavior observed in high-dose amphetamine use may be related to actions on dopamine-containing neurons in the striatum.[34,35] Psychotic symptoms which are apparent after the chronic use of high doses of ampheta-

Table 2 Postulated amine neurotransmitters in neuronal pathways involved in actions of amphetamine

Behavioral effect	Neurotransmitter
Psychic stimulation	Norepinephrine
Reduced food intake	Norepinephrine
Enhanced psychomotor activity	Norepinephrine; ?dopamine
Stereotyped or repetitive behavior	Dopamine
Psychotic behavior, delusions	? 5-Hydroxytryptamine

mine are at the present time least understood. They do not appear to be related entirely to actions on catecholamine-containing neurons and these effects may be related either to actions of amphetamine on 5-hydroxytryptamine metabolism or to direct effects on tryptamine receptors in the central nervous system.

Amphetamines are metabolized in man and in other animals by two major pathways (Fig. 3).[36] One involves oxidative deamination through the microsomal enzyme system of the liver. This system requires NADPH and oxygen for activity. It should be emphasized that this oxidative deamination reaction does not involve the classical monoamine oxidase enzyme which is a major pathway in the deamination of catecholamines and 5-hydroxytryptamine. The second reaction in the metabolism of amphetamine involves aromatic hydroxylation *para* to the side chain of the molecule. To a smaller extent this phenolic derivative may be further hydroxylated on the β-carbon of the side chain to form p-hydroxynorephedrine.

One of the more fascinating manifestations of chronic amphetamine use is the development of tolerance to these agents. The exact mechanism by which tolerance develops to amphetamine is unknown, but several attractive theories have been proposed (Table 3). An intriguing possibility is predicated on the false neurotransmitter hypothesis developed by Kopin and co-workers.[37] According to this theory, β-hydroxylated phenylethylamines may be taken up and retained in the adrenergic storage vesicles, thus displacing norepinephrine. With nerve stimulation, this less active amine would be released instead of norepinephrine, and a smaller than usual expected response would result. In rats, p-hydroxynorephedrine tends

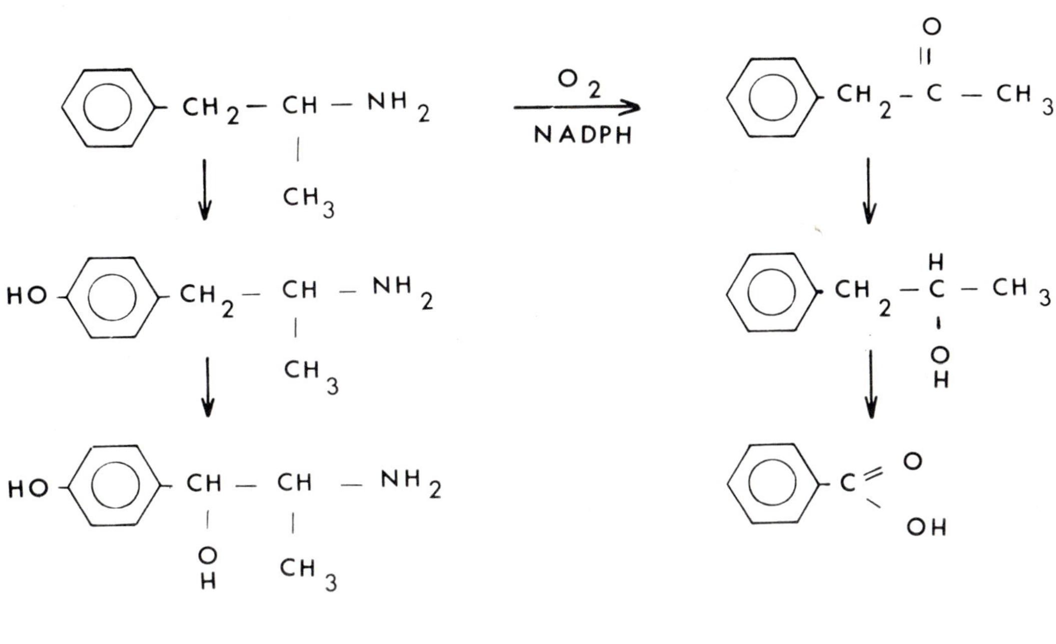

Figure 3 Metabolism of amphetamine. On the left are depicted the hydroxylation reactions which may ultimately lead to the formation of p-hydroxynorephedrine. The phenolic hydroxyl group may be subsequently conjugated with glucuronic acid or sulfate. On the right are shown the metabolic transformations subsequent to oxidative deamination of amphetamine. Ultimately, benzoic acid may be produced, which may be further transformed to hippuric acid by conjugation to glycine.

Table 3 Mechanisms of development of tolerance to amphetamine

1. Depletion of norepinephrine stores
   a. Release and block of uptake
   b. Inhibition of synthesis—indirect
2. Replacement of norepinephrine by p-hydroxynorephedrine ("false transmitter")
3. Gradual enhancement of activity of compensatory neurophysiologic systems
4. Induced synthesis of enzymes, enzyme systems, or receptor systems which are ordinarily inhibited by amphetamine

to accumulate in brain during chronic administration of amphetamine.[38,39] This agent accumulates in the storage granules of adrenergic neurons, presumably displacing the normal neurotransmitter, norepinephrine, and in effect functioning as a false transmitter. On nerve stimulation, p-hydroxynorephedrine may be released instead of norepinephrine, but this substance would be expected to exhibit a much smaller action on the adrenergic receptor and therefore reduced responses to nerve stimulation would result.[37] It has been speculated that tolerance may develop as a consequence of gradual replacement of norepinephrine by p-hydroxynorephedrine in the adrenergic nerve terminals.

A second possible mechanism for amphetamine tolerance involves direct displacement of norepinephrine from storage sites whose complement of norepinephrine is critical for the indirect sympathomimetic actions of amphetamine. If norepinephrine is released from storage sites into the axoplasm, the amine very likely will be more rapidly metabolized and neurotransmitter depletion will ensue (Fig. 2). Ordinarily, depletion of catecholamines is associated with enhanced synthesis of catecholamines in an attempt to maintain norepinephrine levels in the tissues. However, it has been shown that indirect sympathomimetic amines also inhibit norepinephrine synthesis.[18,40,41] The mechanism by which this inhibition of norepinephrine synthesis occurs presumably involves the release of norepinephrine intraneuronally and the inhibition of tyrosine hydroxylase by end-product feedback inhibition of this initial enzymatic reaction in the biosynthesis of catecholamines.[42] Thus, by two possible mechanisms, direct displacement of catecholamines from binding sites and inhibition of norepinephrine synthesis, depletion of norepinephrine may occur, particularly with excessive and chronic use of amphetamine. These effects could contribute to the tolerance or tachyphylaxis which develops with the use of these indirect sympathomimetic amines.

A third possible mechanism for the development of tolerance involves the more general theory that chronic stimulation of certain neuronal pathways ultimately leads to enhanced activity of compensatory neuronal pathways in an attempt to restore central nervous system homeostasis. Presumably these neurophysiologic compensatory mechanisms gradually become more effective and higher and higher doses of drug are required to overcome the developing compensations. When use of the drug is abruptly terminated, the compensatory mechanisms are not rapidly turned off, and these effects continue in an unrestrained fashion for varying lengths of time, accounting for the withdrawal or abstinence syndrome observed.[43]

A fourth mechanism that has been proposed to explain the development of tolerance relates to the induction of enzymes or receptors which have either been inhibited by the drug or whose activity is required in an increased fashion to overcome the actions of the drug. Because of either relatively or absolutely reduced activity of the particular enzyme or receptor system, the normal feedback inhibition of the synthesis of that enzyme or receptor is suppressed and it is synthesized at a more rapid rate. This concept has been more commonly applied to explain morphine tolerance, particularly with regard to alterations in 5-hydroxytryptamine synthesis and turnover, although the evidence that morphine tolerance occurs by this mechanism is inconclusive and controversial.[44–46] No definite evidence relating this mechanism to the development of amphetamine tolerance is available.

There is no evidence to support theories of altered metabolism or altered brain uptake or distribution of amphetamine as possible mechanisms for the development of tolerance to amphetamine.

From the foregoing remarks one might conclude that the complex effects of amphetamine on behavior are well matched by its equally complex biochemical effects on amine-containing neurons. Interesting correlations are beginning to emerge between these biochemical and behavioral phenomena, however, and only further investigation will allow us to determine whether these associated events are incidentally or causally related.

One of the most complex and controversial issues that has been raised as a result of the epidemic misuse of amphetamines and other agents which distort perception relates to the possible permanent alteration in behavior that may develop from prolonged use of these agents. Repeated self-administration of these agents may result in the evolution of new patterns of behavior that are intended to be compensatory in nature. Unfortunately, the drugs are not highly specific in their actions and, consequently, the compensations may not be specific and may not always be normalizing. Thus, untoward behavior patterns related to these compensations may develop. These new behavior patterns may include the delayed effects associated with chronic use of amphetamine and other drugs, such as reduced attention span and delusions. Perhaps more significant, these altered neurophysiologic and biochemical patterns, which arise as compensations for chronic stresses produced by the perception-distorting drugs, may themselves be responsible for long-term or permanent modifications of behavior. Whether the long-suspected but not definitively proven "mental deterioration" associated with chronic use of LSD, amphetamines, hashish, and other psychoactive agents is due to this type of neurophysiologic compensation remains a matter for conjecture, but, in my view, it seems a highly plausible speculation.

SUMMARY

Amphetamine and related centrally active sympathomimetic amines produce a variety of behavioral effects in man and animals. The patterns of responses are markedly influenced by the mode of administration, the dose, and the duration of administration of the drug. Many of the behavioral effects produced by amphetamine are associated with altered biochemical processes in specific amine-containing neural pathways. Decreased food intake, psychomotor stimulation, and enhanced alertness associated with administration of amphetamine appear to be dependent upon the release of norepinephrine from, and block of norepinephrine uptake into, norepinephrine-containing neurons. Compulsive gnawing in rats and analogous stereotyped, repetitive behavior in man which are associated with the administration of large doses of amphetamine appear to be correlated with similar biochemical effects of amphetamine in dopaminergic neurons. The psychotic symptomatology associated with amphetamine misuse is least completely understood but may be related to stimulant actions of amphetamine on 5-hydroxytryptamine receptors and inhibition of 5-hydroxytryptamine metabolism. During amphetamine administration, p-hydroxynorephedrine, a metabolite of amphetamine, appears to be taken up into and stored in norepinephrine-containing neurons in place of the natural transmitter substance. This displacement phenomenon, coupled with the ability of amphetamine itself to release norepinephrine from neurons, may contribute to the tolerance to amphetamine which develops after repeated administration of high doses of the drug.

REFERENCES

1. Dews, P. B.: Studies on behavior. IV. Stimulant actions of methamphetamine. *J Pharmacol Exp Ther 122*:137–147, 1958.
2. Smith, C. B.: Effects of d-amphetamine upon brain amine content and locomotor activity of mice. *J Pharmacol Exp Ther 147*:96–102, 1965.
3. Kelleher, R. T., and Morse, W. H.: Specificity of behavioral effects of drugs. *Ergebn Physiol 60*:1–56, 1968.
4. Innes, I. R., and Nickerson, M.: Drugs acting on postganglionic adrenergic nerve endings and structures innervated by them (sympathomimetic drugs). In Goodman, L. S., and Gilman, A.: *The Pharmacological Basis of Therapeutics,* 4th ed. New York, The Macmillan Co., 1970, pp. 478–523.
5. Vree, T. B., Muskens, A. Th. J. M., and van Rossum, J. M.: Some physicochemical properties of amphetamine and related drugs. *J Pharm Pharmacol 21*:774–775, 1969.
6. Blaschko, H., Richter, D., and Schlossman, H.: The oxidation of adrenaline and other amines. *Biochem J 31*:2187–2196, 1937.
7. Glowinski, J., Axelrod, J., and Iversen, L.: Regional studies of catecholamines in the rat brain. IV. Effects of drugs on the disposition and metabolism of H^3-norepinephrine and H^3-dopamine. *J Pharmacol Exp Ther 153*:30–41, 1966.
8. Fleckenstein, A., and Burn, J. H.: The effect of denervation on the action of sympathomimetic amines on the nictitating membrane. *Brit J Pharmacol 8*:69–78, 1953.
9. Weiner, N.: The catecholamines; biosynthesis, storage and release, metabolism, and metabolic effects. In *The Hormones,* Vol. 4, Pincus, G, Thimann, K. V., and Astwood, E. B. (Eds.). New York, Academic Press, 1964, pp. 403–479.
10. Smith, C. B.: Enhancement by reserpine and α-methyl-DOPA of the effects of d-amphetamine upon the locomotor activity of mice. *J Pharmacol Exp Ther 142*:343–350, 1963.
11. Weissman, A., Koe, B. K., and Tenen, S. S.: Antiamphetamine effects following inhibition of tyrosine hydroxylase. *J Pharmacol Exp Ther 151*:339–352, 1965.
12. Randrup, A., and Munkvad, I.: Role of catecholamines in the amphetamine excitatory response. *Nature 211*:540, 1966.
13. Sedvall, G. C., Weise, V. K., and Kopin, I. J.: The rate of norepinephrine synthesis measured *in vivo* during short intervals; influence of adrenergic nerve impulse activity. *J Pharmacol Exp Ther 159*:274–282, 1968.
14. Kopin, I. J., Breese, G. R., Krauss, K. R., and Weise, V. K.: Selective release of newly synthesized norepinephrine from cat spleen during sympathetic nerve stimulation. *J Pharmacol Exp Ther 161*:271–278, 1968.
15. Glowinski, J.: Effects of amphetamine on various aspects of catecholamine metabolism in the central nervous system of the rat. In *Amphetamines and Related Compounds; Proceedings of the Mario Negri Institute for Pharmacological Research, Milan, Italy,* Costa, E. and Garattini, S. (Eds.). New York, Raven Press, 1970, pp. 301–316.
16. Carlsson, A., Fuxe, K., Hamberger, B., and Lindqvist, M.: Biochemical and histochemical studies on the effects of imipramine-like drugs and (+)-amphetamine on central and peripheral catecholamine neurons. *Acta Physiol Scand 67*:481–492, 1966.
17. Hanson, L. C. F.: Evidence that the central action of amphetamine is mediated via catecholamines. *Psychopharmacologia 9*: 78–80, 1966.
18. Weiner, N: Regulation of norepinephrine synthesis. *Ann Rev Pharmacol 10*:273–290, 1970.
19. Stein, L.: Self-stimulation of the brain and the central stimulant action of amphetamine. *Fed Proc 23*:836–850, 1964.
20. Glowinski, J., and Axelrod, J.: Effect of drugs on the uptake, release, and metabolism of H^3-norepinephrine in rat brain. *J Pharmacol Exp Ther 149*:43–49, 1965.
21. Stein, L., and Wise, C. D.: Release of norepinephrine from hypothalamus and amygdala by rewarding medial forebrain bundle stimulation and amphetamine. *J Comp Physiol Psychol 67*:189–198, 1969.
22. Carr, L. A., and Moore, K. E.: Norepinephrine release from brain by d-amphetamine *in vivo. Science 164*:322–323, 1969.
23. Cook, J. D., and Schanberg, S. M.: The effects of methamphetamine on behavior and on the uptake, release and metabolism of norepinephrine. *Biochem Pharmacol 19*:1165–1179, 1970.
24. Rutledge, C. O.: The mechanisms by which amphetamine inhibits oxidative deamination of norepinephrine in brain. *J Pharmacol Exp Ther 171*:188–195, 1970.
25. McKenzie, G. M., and Szerb, J. C.: The effect of dihydroxyphenylalanine, pheniprazine, and dextroamphetamine on the *in vivo* release of dopamine from the caudate nucleus. *J Pharmacol Exp Ther 162*:302–308, 1968.
26. Rutledge, C. O.: Personal communication.
27. Taylor, K. M., and Snyder, S. H.: Amphetamine: differentiation by d and l isomers of behavior involving brain norepinephrine or dopamine. *Science 168*:1487–1489, 1970.
28. Axelrod, J.: Metabolism of epinephrine and of the sympathomimetic amines. *Physiol Rev 39*:751–776, 1959.
29. Hagen, P., and Weiner, N.: Enzymic oxidation of pharmacologically active amines. *Fed Proc 18*:1005–1012, 1959.

30. Weiner, N.: Substrate specificity of brain amine oxidase of several mammals. *Arch Biochem 91*:182–188, 1960.
31. Vane, J. R.: The actions of sympathomimetic amines on tryptamine receptors. In *Adrenergic Mechanisms,* Vane, J. P., Wolstenholme, G. E. W., and O'Connor, M. (Eds.). London, Churchill, 1960, pp. 356–372.
32. Snyder, S. H., and Merril, C. R.: A relationship between the hallucinogenic activity of drugs and their electronic configuration. *Proc Nat Acad Sci 54*:258–266, 1965.
33. Snyder, S. H., Richelson, E., Weingartner, H., and Faillace, L. A.: Psychotropic methoxyamphetamines: structures and activity in man. In *Amphetamines and Related Compounds; Proceedings of the Mario Negri Institute of Pharmacological Research, Milan, Italy,* Costa, E. and Garattini, S. (Eds.). New York, Raven Press, 1970, pp. 905–928.
34. Carlsson, A.: Amphetamine and brain catecholamines. In *Amphetamines and Related Compounds; Proceedings of the Mario Negri Institute of Pharmacological Research, Milan, Italy,* Costa, E. and Garattini, S. (Eds.). New York, Raven Press, 1970, pp. 289–300.
35. Andén, N.-E.: Effects of amphetamine and some other drugs on central catecholamine mechanisms. In *Amphetamines and Related Compounds; Proceedings of the Mario Negri Institute of Pharmacological Research, Milan, Italy,* Costa, E. and Garattini, S. (Eds.). New York, Raven Press, 1970, pp. 447–462.
36. Smith, R. L., and Dring, L. G.: Patterns of metabolism of β-phenylisopropylamines in man and other species. In *Amphetamines and Related Compounds; Proceedings of the Mario Negri Institute of Pharmacological Research, Milan, Italy,* Costa, E. and Garattini, S. (Eds.). New York, Raven Press, 1970, pp. 121–139.
37. Kopin, I. J.: False neurotransmitters. *Ann Rev Pharmacol 8*:377–394, 1968.
38. Brodie, B. B., Cho, A. K., and Gessa, G. L.: Possible role of p-hydroxynorephedrine in the depletion of norepinephrine induced by d-amphetamine and in tolerance to this drug. In *Amphetamines and Related Compounds; Proceedings of the Mario Negri Institute of Pharmacological Research, Milan, Italy,* Costa, E. and Garattini, S. (Eds.). New York, Raven Press, 1970, pp. 217–230.
39. Costa, E., and Groppetti, A.: Biosynthesis and storage of catecholamines in tissues of rats injected with various doses of d-amphetamine. In *Amphetamines and Related Compounds; Proceedings of the Mario Negri Institute of Pharmacological Research, Milan, Italy,* Costa E. and Garattini, S. (Eds.). New York, Raven Press, 1970, pp. 231–255.
40. Weiner, N., and Selvaratnam, I.: The effect of tyramine on the synthesis of norepinephrine. *J Pharmacol Exp Ther 161*:21–33, 1968.
41. Kopin, I. J., Weise, V. K., and Sedvall, G. C.: Effect of false transmitters on norepinephrine synthesis. *J Pharmacol Exp Ther 170*: 246–252, 1969.
42. Weiner, N., and Rabadjija, M.: The effect of nerve stimulation on the synthesis and metabolism of norepinephrine in the isolated guinea-pig-hypogastric nerve-vas deferens preparation. *J Pharmacol Exp Ther 160*:61–70, 1968.
43. Martin, W. R.: Pharmacological redundancy as an adaptive mechanism in the central nervous system. *Fed Proc 29*:13–18, 1970.
44. Way, E. L., Loh, H. H., and Shen, F. H.: Morphine tolerance, physical dependence and synthesis of brain 5-hydroxytryptamine. *Science 162*:1290–1292, 1968.
45. Way, E. L., Loh, H. H., and Shen, F. H.: Simultaneous quantitative assessment of morphine tolerance and physical dependence. *J Pharmacol Exp Therap 167*:1–8, 1969.
46. Cochin, J.: Possible mechanisms in development of tolerance. *Fed Proc 29*:19–27, 1970.

SOME OBSERVATIONS ON AND A REVIEW OF THE EFFECTS OF HIGH-DOSE USE OF AMPHETAMINES

JOHN C. KRAMER, M.D.

Stimulants have been used for centuries if not millennia. Xanthine-bearing plants are used by a majority of mankind, and produce a dependency which does not usually interfere with, and may, in fact, enhance the efficiency of the users.

In a separate category, the juices of coca leaves have provided a sense of well-being and endurance to Andean Indians since before the Conquistadors. Less well known, the fresh leaves of the khat plant have been chewed in East Africa and the Near East for centuries.[1] If the followers of Al Hasan used any drug to enhance their appetite for assassination and battle, one might speculate that they used khat rather than cannabis. When taken in the leafy form, both coca and khat may produce a dependency clearly detrimental to the user, although they often do not. Extracted from the plant, cocaine is much more likely to create a damaging dependency than are the phenethylamines (amphetamines in this report), the synthetic equivalents of khat.

Among the synthetic stimulants, amphetamine itself was first prepared in 1887 by Edelsano and methamphetamine in 1919 by Ogata,[2] but it was not until 1927 that the psychopharmacologic effects of amphetamine were described by Alles. From the 1930's through the 1950's medical use of amphetamines became extensive and it was looked upon as a useful and relatively safe agent although some toxic effects and some tendencies to produce dependence were described.[3] As recently as 1963 the AMA Council on Drugs, while recognizing the great potential for abuse of amphetamines, stated that "at this time compulsive abuse of the amphetamines (constitutes) a small problem (in the U.S.)."[4] By 1966 concern was greater. In that year the AMA Committee on Alcoholism and Addiction took note of the information that sufficient amphetamine products were available in the United States to supply 25 to 50 doses to every man, woman, and child in the country.[5] In 1966 Griffith[6] and Lemere[7] further warned of the extent and dangers of amphetamine use, and in 1967 Kramer[8] described the pattern and effects of a high dose administered intravenously.

PATTERNS OF USE

Though regular oral use of amphetamines may cause difficulties, including paranoid psychosis and a disabling dependence, these drugs are more likely to lead to such effects when used intravenously. Colonies of intravenous stimulant users have gathered in such areas as San Francisco, New York, Los Angeles, and elsewhere. At present the developmental history of the intravenous amphetamine user is typical enough to warrant a general description.

This individual has tried amphetamines orally; he may have liked them or not. He has used other drugs perhaps moderately, perhaps extensively, and he has been moving either in marihuana-psychedelic drug-using circles or in heroin-addict circles. His first intravenous use of amphetamine is an ecstatic experience and his first thought is "where has this been all my life." The effects somehow differ from those of oral amphetamines, not only quantitatively but also qualitatively. Early, his use of the drug is intermittent; doses probably equivalent to 20 to 40 mg per injection may be taken once or a very few times over a day or two. Days or weeks may intervene between sprees. Gradually the sprees become longer and the intervening periods shorter; doses become higher and injections more frequent.

After a period of several months the final pattern is reached in which the user (now called a "speed-freak") injects his drug many times a day, each dose in the hundreds of milligrams, and remains awake continuously for three to six days getting gradually more tense, tremulous, and paranoid as the "run" progresses. The runs are interrupted by bouts of profound sleep (called "crashing") which last a day or two. Soon after awakening from crashing, the drug is again injected and a new run starts. The periods of continuous wakefulness may be prolonged to weeks if the user is able to sleep even as little as an hour a day.

Although amphetamine used intravenously is a powerful reinforcer, there are individuals who have tried it once or several times and have chosen not to continue its use. Nevertheless, like heroin or cocaine, it is a form of drug use which may overwhelm even a casual dabbler.

It is interesting to note that in experiments in which rats had the opportunity for self-injection of amphetamines intravenously, the drug proved to be strongly reinforcing and the self-administration was characterized by periods of intake and abstinence entirely analogous to that seen in human amphetamine users.[9,10]

EFFECTS

The intravenous use of amphetamines produces a syndrome with a variety of behavioral and physical effects some of which, particularly the insomnia and the anorexia, may themselves produce symptoms or alter the effects specifically attributable to the drug. The high-dose user thus is not merely responding to a drug but also to altered sleep patterns, to undernourishment, to malnourishment, and often to infection. And amphetamines are seldom, if ever, used exclusively: "downers"—opiates, phenothiazines, and sedatives—are regularly used, as are cannabis and occasionally psychedelics. A surprising miscellany of other substances may be used experimentally and the "speed" itself is seldom pure, probably containing by-products not extracted during illicit manufacture as well as those added afterward to "cut" the product.

The Flash and the Euphoria

A few seconds following injection the user experiences a sudden, intense generalized sensation which has both physiologic and psychologic characteristics. It is ineffable and ecstatic, yet may differ in intensity and in quality from time to time. Because different batches of the drug produce a different intensity and quality of flash and because the pure, commercially produced products do not give a good flash, it is possible that the flash may depend upon substances other than the methamphetamine which have not been extracted in illicit manufacture, or have perhaps been added afterward.

The euphoria can be considered to have both primary and secondary characteristics. Part of the sense of well-being seems purely internal, and part stems from the feelings of ability and of invulnerability which are produced. Suddenly, magically, volubility and gregariousness appear and boredom departs.

These desired effects are extremely vulnerable to the impingement of tolerance. It takes ever more drug to re-create this chemical nirvana. It is the desire to re-experience the flash and the euphoria, and to avoid the fatigue and the depression of the "coming down" which drives the users to persist and necessarily to increase their dose and frequency of injection. And it is this persistence of use and those large doses which bring on all the other effects of these drugs.

Anorexia

One of the medical uses of amphetamines is to induce anorexia to aid in weight-reduction. In doses ordinarily prescribed, 5 to 30 mg per day, the anorexia produced is moderate, and some have questioned whether a placebo effect, rather than a drug effect, is responsible. With the large doses taken during abusive use there is no question that anorexia is produced. Users uniformly lose weight during periods of abuse. Appetite suppression may be so profound that users may find the very act of swallowing difficult. Some users diligently force themselves to take small amounts of highly nutritious foods or inject themselves with vitamins and other dietary supplements.

Upon arising from the profound sleep which follows a run, either immediately or perhaps a few hours later, the user becomes voraciously hungry. Though he has eaten little or nothing for several days bulimia like this seems to be related to release from the drug effect in large measure, because, in cases of non-drug-assisted starvation, appetite is diminished after several days of starvation.

Undernutrition and malnutrition result, and undoubtedly complicate all the other effects of high-dose amphetamine use.

In an unpublished study, Seevers, Gantz, and Deneau found that, in monkeys given high doses chronically, when the dose was raised past 32 mg/kg of body weight a day the animals became polyphagic rather than anorexic, but continued to lose weight although they ate more than three times their daily ration.[12]

Insomnia

Even early in this pattern of drug abuse, users remain awake for a day or two at a time. These periods gradually become longer so that the runs tend to last three to six days. Although longer runs have been reported they are generally isolated events. Some users will force themselves to lie down, close their eyes, and drift into a half-sleep for perhaps an hour or two. In this way, the user may be able to persist in a run for several weeks before crashing.

There is no question that sleep deprivation (or perhaps dream deprivation) alone can produce deterioration in performance, misperceptions, hallucinations, and other phenomena. All these occurred in one sleep deprivation experiment, but it was the impression of Pasnau and co-workers that withholding all stimulants during the eight and one-half days of wakefulness helped their subjects to cope with these effects.[11]

It is likely that both the insomnia and the drug contribute to the syndrome, and, as with other aspects of this phenomenon, are inextricably intertwined. The observation that many of the physical and psychologic symptoms are largely dissipated after sleeping for a day or two suggests that the insomnia alone is a major contributor to the syndrome. The fact that some symptoms persist after weeks or months of abstinence indicates that sleep deprivation is not alone responsible.

Considering that the usual pattern seen during well-established high-dose abuse is three to six days of wakefulness followed by one to two days of sleep, then users spend about one-fourth of their time in sleep—about the same proportion as non-users but distributed differently.

Tolerance develops to many of the effects of amphetamines, including that of wakefulness. When drug use is well developed, very large doses will be necessary to keep the user awake. At times when tremulousness develops after several days of wakefulness, users may take a moderate dose of their drug to calm them sufficiently so that they can relax and sleep. A "moderate" dose in this instance may be as much as several score milligrams.

Paranoia

A paranoid psychosis can be precipitated by either a single large dose or by chronic moderate doses of amphetamines. Two surveys of patients entering psychiatric units have suggested that amphetamines may be a causal or at least a precipitating event in the psychiatric hospitalization of patients not otherwise identified as users of amphetamines.[13,14] The presenting symptoms are those of paranoid psychosis.

High-dose intravenous users of amphetamines generally understand that they will sooner or later experience paranoia. Aware of this they are usually able to discount it. Moderate persecutory ideas and visual illusions will seldom be acted on because of the user's intellectual awareness of their nature and origin. However, when drug use has become very intense, or toward the end of a long run, even a well-practiced intellectual awareness may fail and the user may respond to his delusional system.

Leake has suggested in the past,[15] and Ellinwood more recently,[16,17] that the effect of amphetamine is to release underlying psychotic trends. Griffith, Cavanaugh, and Oates, however, precipitated a paranoid psychosis in all four subjects given d-amphetamine (120 to 220 mg a day for 24 to 120 hours) who had previously self-administered large amounts of the drug without experiencing psychosis.[18] In all of them a moderate personality disorder had been diagnosed previously.

Although there may be individual differences in sensitivity to the psychotogenic effects of amphetamines, it appears that anyone given a large enough dose for a long enough time will become psychotic. Although this hypothesis may be untestable, the experiences of a large number of high-dose amphetamine users suggest that it is more likely than the view that psychosis is precipitated only in those already so inclined.

As mentioned earlier, the paranoia does not usually start during the first few months of high-dose intravenous use. When it does finally begin it is mild, easily controlled, and is largely dissipated upon waking after crashing; it usually doesn't start again until after two or three days on a new run. As time goes on, the paranoia may start earlier in a run and may persist to some extent even after crashing. In some instances the first injection after a period of sleep will bring about a return of the paranoia. Once an individual has experienced amphetamine paranoia it will readily return even after a prolonged period of abstinence.

Violence

Public concern over use of psychoactive drugs often centers on the assumption that one effect of a specific drug is its tendency to induce unwarranted violence. Clearly, opiates do not possess this characteristic pharmacologically. Although an opiate user could, for instance, commit an act of violence during a robbery, there is nothing in the drug effect which would so incline him. If anything, opiates are more likely to inhibit than to further any tendency toward violent behavior.

From all evidence, amphetamine does tend to set up conditions in which violent behavior is more likely to occur than would be the case had the individual not used it. Suspiciousness and hyperactivity may combine to induce precipitous and unwarranted assaultive behavior. Under the influence of amphetamines, lability of mood is common, the user abruptly shifting from warmly congenial to furiously hostile moods for the most trivial of reasons.

Most high-dose amphetamine users describe involvement, either as aggressor or victim, in episodes in which murder or mayhem was avoided by the slimmest of margins. There are, of course, instances in which violence actually occurred. From descriptions of a number of these events it is clear that they would not have taken place had it not been for the use of amphetamines.

The role of barbiturates in violent behavior is difficult to assess. It has been regularly reported by many drug users that when barbiturate use is not followed by sleep they often experience considerable irritability (although opiates, cannabis, and the phenothiazines are calming). Users of amphetamines often use barbiturates for sedation and may possibly add to, rather than diminish, the tendency toward angry or perhaps violent behavior.

Compulsivity

Perhaps the most curious effect of amphetamines is its capacity to induce behavior which is persisted in or repeated for prolonged periods. If the user is not too disorganized the activity may, on the surface at least, be useful. Dwellings may be cleaned, automobiles polished, or items arranged to an inhuman degree of perfection. Sometimes these activities may be only partially completed when another compulsively pursued task intervenes. The behavior may be bizarre, as in the elaborate but nonfunctional reconstruction of mechanical or electrical devices, or it may take on a destructive character, as in skin-picking which may produce extensive ulcerations.

Analogous to this compulsive behavior in man is what has been termed stereotypy in animals. Rats, mice, guinea pigs, cats, and squirrel monkeys almost without exception performed repetitive acts, which although not unique for the species were characteristic for each animal.[19]

"Over-amping"

The term "over-amping" was probably coined when commercially produced ampules of methamphetamine were widely used. Users prefer this word to "overdose" which carries the connotation of an overdose of heroin, a condition which produces an entirely different set of symptoms. A variety of events may occur when the dose of amphetamine taken far exceeds the tolerance of the user. Descriptions have been too few in number for a clear pattern to emerge. One or several symptoms may occur, including chest pain which lasts minutes or hours. Unconsciousness, again lasting minutes or hours, may occur, the user waking and occasionally finding himself aphasic or paralyzed for hours or days in a pattern like that seen following a cerebrovascular accident. More frequent is the situation in which the user remains conscious, his mind racing with a myriad of thoughts, often in an ecstatic mood but unable, or perhaps unwilling, to move.

Under these circumstances the user's friends attempt to nurse him and may use opiates or sedatives in an endeavor to counteract the effects of the over-amping.

Death

The motto "Speed Kills" is cute, short enough to fit on a button, and carries a message of concern. It is not altogether accurate, since very few deaths have been recorded in which overdose of amphetamines has been the cause.

Although viral hepatitis and other infections are common and persistent among intravenous amphetamine users, only a few deaths related to infection have been recorded. Death by violence might add a few more names. Still, investigation of records at the offices of the Los Angeles County Coroner and of the San Francisco County Coroner revealed that only one or two deaths per year in each county could be attributed to overdose and a like number to other events which might be related to amphetamine use.

The rarity of death due to amphetamines may be due to the tolerance produced by these drugs and the relatively high ratio of effective dose to fatal dose. Dr. David Smith of the Haight-Ashbury Clinic and Amphetamine Research Project indicates that the two deaths he has noted occurred in individuals who were relative novices in amphetamine use.

Chlorpromazine (and probably other phenothiazines) has been shown to be efficacious in suppressing amphetamine effects both clinically[20] and in animals,[21] while barbiturates, although not without effect, are less valuable and introduce the hazard of barbiturate toxicity. These findings have been confirmed by acute, intravenous lethality studies in our laboratory. We have found that intravenous injection of methamphetamine, d-amphetamine, and d,l-amphetamine each produced rapid death in toxic doses (usually within two to three minutes, although ten minutes were allotted for experimental purposes). Those animals that did not succumb promptly usually survived for at least six hours. Because of the rapidity of death, such conditions as grouping, room temperature, or activity did not enter into consideration. Several findings of interest emerged. Unlike the results found with intraperitoneal injection, the LD_{50} curves were sharp and consistent; in addition, the lethality of each of the three substances was approximately the same. Chlorpromazine effectively raised the LD_{50} to about 55 mg/kg (with 5 mg/kg of CPZ) and about 65 mg/kg (with 10 mg/kg of CPZ), from the usual LD_{50} of about 35 mg/kg.

Besides confirming the usefulness of chlorpromazine in antagonizing the effects of phenethylamines, these experiments also suggest that the antagonism depends not only on the tendency of CPZ to diminish hyperactivity (though this may be one mechanism of action), but also on the altering of biochemical events within the tissues.

Pathology. Although human deaths due to amphetamines are not common, some clinical and pathologic descriptions have been published. In a recent report, Cravey and Baselt describe a young man who swallowed two packets of methamphetamine when confronted by police.[22] Within a half-hour he showed both delusional and hallucinatory behavior. When hospitalized he was cyanotic, with a temperature of 104°F, pulse rate of 102, and blood pressure of 74/50. He was in severe acidosis and died five-and-one-half hours following the time of ingestion. This and other reports and unpublished case histories indicate that marked hyperpyrexia and shock are usually noted prior to death due to amphetamines. In experiments with mice given amphetamine, Hardinge and Peterson found that those whose temperatures rose above 42.4°C usually died while those whose temperatures remained below 41.7°C usually survived.[23] Zalis and associates found that dogs given amphetamines developed fever proportional to dose.[24] They found that the pulse rose at first but dropped rapidly in the terminal stage, and acidosis occurred due to the formation of lactic acid and acetic acid.

Pathologic findings in both man and animals are generally non-specific and show congestion in the lungs and other organs, including the brain. Petechial hemorrhages in various organs including the brain are frequently but not uniformly described. Only rarely is massive hemorrhage noted.[25]

Histologic examinations of the brain tissues of monkeys, dogs, and other animals have occasionally but not usually shown some cellular deterioration.[26] Zalis and collaborators report mild to moderate deterioration in the cerebral cortex of all 24 of their animals and suggest that this was due to a hypermetabolic state associated with hyperpyrexia.[24]

I have found no first-hand reports in the Western medical literature describing a histologic picture of damage to human brain cells, although Lemere[7] refers to an article by Tatetsu[27] citing histopathologic evidence (lobotomy and postmortem) of permanent organic brain damage. In two of six deaths due to amphetamine overdose reported by the San Francisco coroner, petechial hemorrhages were noted in brain tissue.

Other Effects

As in any situation in which hypodermic equipment is shared without proper sterilization, viral hepatitis is common. Evidence of hepatic damage is so common that consideration has been given to the possibility that part of the damage may be due to a direct toxic effect of amphetamines on liver.

Other forms of infection are common. Often the skin becomes shiny and delicate; small injuries may produce ulceration and healing is slow.

Jaw-grinding is often described, and, like tremors due to lesions of the basal ganglia, can be stopped briefly by conscious control. Mattson and Calverley report other dyskinesias in some individuals on therapeutic doses of d-amphetamine.[28]

RECOVERY

Although a Japanese report suggests that some high-dose amphetamine users may become chronically psychotic,[27] what has been most striking in our experience has been the slow but rather complete recovery of users who, according to their own descriptions and that of others, had become rather thoroughly disorganized and paranoid prior to their detention. Although the most florid symptoms are dissipated within a few days or weeks, some confusions, some memory loss, and some delusional ideas may remain for perhaps six to twelve months. After that time, although there may be some residual symptoms, they are slight and not disabling and are noticed primarily by the now abstinent user himself. Most commonly, exusers report slightly greater difficulty in remembering.

As a group ex-users describe being more open and talkative than they had been prior to their use of amphetamines. They like the result and declare with certainty that it is due to their experience with amphetamines.

Anyone concerned with the welfare of amphetamine users, and the users themselves, should recognize that most, if not all, can recover from even the most profound intellectual disorganization and psychosis in six months to a year of abstinence.

TREATMENT

The care of amphetamine users poses some special problems. In these are combined the problems of management of the "hooked" drug user and the paranoid personality. Although suffering with severe medical and psychiatric symptoms, amphetamine users are generally fearful of hospitalization.

Some crises may yield to phenothiazine tranquilizers or first aid, but abstinence is probably the most important therapeutic device, and that may be difficult to attain. Many users who attempt abstinence find it difficult because of the fatigue that results, extreme at first, gradually diminishing but persistent, perhaps for months.

Abstinence for many is forced by a stay in prison or commitment to a psychiatric hospital or civil addict program. No data have yet been collected to indicate the long-term value of such enforced abstinence. Certainly, many who have been incarcerated have returned to drug use upon release. Users do not readily volunteer for care, but commitment programs offer little besides enforced abstinence. Should the user be permitted to live in the limbo of his drug or forced into the limbo of an institution? Can voluntary programs be devised that are sufficiently useful and attractive that users will seek them out and persist in their program? Can commitment programs be devised that do not resemble slightly benign prisons? Or, do we just let the user seek heaven or hell on his own terms while the community offers help only on its own terms?

SUMMARY

Amphetamines taken intravenously are likely to produce dependence which has a characteristic pattern of induction into high-dose use. This results in a complex syndrome whose effects may in part be due to malnourishment, insomnia, and other drugs as well as to amphetamines. Various effects are described and discussed including the flash and euphoria, anorexia, insomnia, paranoia, violence, compulsivity, "over-amping," various other complications, and death.

Attention is drawn to the probability that virtually complete recovery is likely even for the most profoundly disorganized high-dose user, given six months to a year of abstinence. The problems of treatment and attaining abstinence are briefly considered.

ACKNOWLEDGMENT

The kind assistance of the following people is gratefully noted:

Mr. Robert Cravey, Orange County Pathologists Office; Mrs. Leah Jaffe, Smith, Kline & French Laboratories; Dr. Thomas Noguchi, Los Angeles County Coroner; Dr. David Smith, Haight-Ashbury Amphetamine Research Project; Dr. Henry Turkel, San Francisco County Coroner; Burroughs-Wellcome donated Methedrine brand of methamphetamine.

This work was supported in part by Grant D-67-10 of the Committee on Problems of Drug Dependence of the National Research Council. It was first published in the Journal of Psychedelic Drugs, volume 2, number 2, Spring, 1969.

REFERENCES

1. United Nations Narcotics Commission: Khat. *Bull Narcot 8*:6–13, 1956.
2. Connell, P. H.: The amphetamines—I. *Med World (London) 96*:18–20, 1962.
3. Kalant, O. J.: *The Amphetamines: Toxicity and Addiction.* Springfield, Ill., Charles C Thomas, 1966.
4. Council on Drugs: Abuse of the amphetamines and pharmacologically-related substances. *JAMA 183*:362–363, 1963.
5. Committee on Alcoholism and Addiction: Dependence on amphetamines and other stimulant drugs. *JAMA 197*:1023–1027, 1966.
6. Griffith, J.: A study of illicit amphetamine drug traffic in Oklahoma City. *Amer J Psychiat 123*:560–569, 1966.
7. Lemere, F.: The danger of amphetamine dependency. *Amer J Psychiat 123*:569–571, 1966.
8. Kramer, J. C., Fischman, V. S., and Littlefield, D. C.: Amphetamine abuse: Pattern and effects of high doses taken intravenously. *JAMA 201*:305–309, 1967.
9. Pickens, R., Meisch, R., and McGuire, L. E.: Methamphetamine reinforcement in rats. *Psychon Sci 8*:371–372, 1967.
10. Pickens, R., and Harris, W. C.: Self-administration of d-amphetamine by rats. *Psychopharmacologia (Berl) 12*:158–163, 1968.
11. Pasnau, R. O., et al.: The psychological effects of 205 hours of sleep deprivation. *Arch Gen Psychiat 18*:496–505, 1968.
12. Seevers, M. H., Gantz, R., and Deneau, G. A.: Amphetamine "addiction" in monkeys. Unpublished, 1958.
13. Johnson, J., and Milner, G.: Psychiatric complications of amphetamine substances. *Psychiatrica Acta Scand 42*:252–263, 1966.
14. Rockwell, D. A., and Ostwald, P. F.: Amphetamine use and abuse in psychiatric patients. *Arch Gen Psychiat 18*:612–616, 1968.
15. Leake, C. D.: *The Amphetamines—Their Actions and Uses.* Springfield, Ill., Charles C Thomas, 1958.
16. Ellinwood, E. H., Jr.: Amphetamine psychosis: I. Description of the individuals and the process. *J Nerv Ment Dis 144*:273–283, 1967.
17. Ellinwood, E. H., Jr.: Amphetamine psychosis: II. Theoretical implications. *J Neuropsychiat 4*:45–54, 1968.
18. Griffith, J. D., Cavanaugh, J., and Oates, J.: Schizophreniform psychosis induced by large-dose administration of d-amphetamine. Nashville, Vanderbilt University, no date (mimeo.).
19. Randrup, A., and Munkvad, I.: Stereotyped activities produced by amphetamine in several animal species and man. *Psychopharmacologia (Berlin) 11*:300–310, 1967.

20. Espelin, D. E., and Done, J. K.: Amphetamine poisoning: effectiveness of chlorpromazine. *New Eng J Med 278*:1361–1365, 1968.
21. Randrup, A., and Munkvad, I.: Special antagonism of amphetamine-induced abnormal behavior. *Psychopharmacologia* 7:416–422, 1965.
22. Cravey, R. H., and Baselt, R. C.: Methamphetamine poisoning. *J Forensic Sci* (in press).
23. Hardinge, M. G., and Peterson, D. I.: The effect of forced exercise on body temperature and amphetamine toxicity. *J. Pharmacol Exp Therap 145*:47–51, 1964.
24. Zalis, E. G., Lundberg, G. D., and Knutzon, R. A.: Report to Fourth International Meeting in Forensic Medicine, Copenhagen, 1966. (Reported in Ref. 22.)
25. Lloyd, J. T. A., and Walker, D. R. H.: Death after combined d-amphetamine and phenelzine. *Brit Med J 2* (5454):168–169, 1965.
26. Ehrich, W. E., Lewy, F. H., and Krumbbaar, E. B.: Experimental studies upon the toxicity of benzedrine sulphate in various animals. *Amer J Med Sci 198*:785–803, 1939.
27. Tatetsu, S.: Methamphetamine psychosis. *Folia Psychiat Neurol Jap* (Suppl.)7:377–380, 1963. (Reported in Ref. 7.)
28. Mattson, R. H. and Calverley, J. R.: Dextroamphetamine—sulfate induced dyskinesias. *JAMA 204*:400–401, 1968.

POSSIBLE DOSE-RESPONSE RELATIONSHIPS IN AMPHETAMINE PSYCHOSIS

BURTON ANGRIST, M.D., AND SAMUEL GERSHON, M.D.

Dose response relationships in amphetamine psychosis are difficult to define in such exact terms as the precise numbers of milligrams required for induction of psychotic symptoms, especially if these terms are meant to apply generally. This is because individual sensitivity to the psychotogenic effects of amphetamine appears to be quite variable.

Animal studies are of little help in this regard. Not only is it notoriously dangerous to make inferences which apply from species to species, but, more important, the psychosis *per se* can only be studied in humans.

Amphetamine has, however, been administered in psychotogenic doses under controlled experimental conditions by both Griffith's[1] and our own group;[2] these studies may be helpful in delineating general guidelines concerning the relationships between dosage and behavioral response. Although this may be true to some extent, the results as to effective psychotogenic doses were so variable that rather than giving a reliable "mean" applicable to the general population of amphetamine abusers (a criterion for inclusion as a subject in these experiments), they have, rather, served to underscore the idiosyncratic variations in individual sensitivity that emerged. Thus, in our experiments we observed subjects who consistently showed formal psychotic symptoms at cumulative doses of less than 100 mg, while others took up to 955 mg of amphetamine without showing formal signs of psychosis.[2] Whether these variations in sensitivity represent variably acquired tolerance to amphetamine's psychotogenic effects, differences in psychologic organization, or variations in innate pharmacogenetically determined vulnerability is a moot point at this time.

A second "drawback" limits the applicability of these studies to clinical situations. Administration of large doses of amphetamine is not without some danger and ethical considerations require a much more gradual "chronic" mode of administration than would occur if the users themselves controlled the drug. In these experiments, amphetamine is administered orally in doses which abusers consider quite small. Doses are given hourly at most, and their magnitude is determined by careful hourly monitoring of pulse rate, blood pressure, and temperature as well as frequent electrocardiograms. Amphetamine abusers are not so stringently concerned with the risks they incur and uniformly report more erratic usage of larger and, in some cases, frighteningly massive doses.[3,4] Moreover, the takers of the most massive doses are generally those most likely to utilize the intravenous route, the ultimate in "acute" administration.[3]

Thus, both animal studies and controlled clinical experiments are of somewhat limited value in assessing dose response relationships in terms having meaningful clinical application, leaving only the alternative of studying naturally occurring cases under "field" conditions. These conditions, however, are notoriously imprecise even in such basic matters of scientific interest as the dose actually taken. If patients have taken "diet pills" of a known type, one may, as has been done, compute individual and mean dosage required for such a crude behavioral endpoint as admission to a hospital.[5] Even in these cases, however, one is limited by the accuracy of his patient's reportage, which is, in turn, prey to such variables as faulty memory, minimization of dose in anticipation of moralistic disapproval on the part of the interviewer, or exaggeration of dose in a swashbuckling attempt to convince the examiner or, perhaps, even the patient, himself, of his ability to "handle" the drug. With black-market amphetamine powders the situation is even worse. These lend themselves to intravenous use and intravenous users tend to be even more erratic and disturbed, increasing the problems of reliability of historical data. Moreover, the users themselves generally do not know the purity and potency of the preparations they take. Roger Smith has pointed out that cutting, especially on and below the level of the "ounce dealer," can lead to enormous variations in both the adulterant and amphetamine content of a marketed "spoon" of "speed."[6]

A second serious drawback to the study of cases occurring under "field" conditions is the possibility of contamination of symptomatology through use of other drugs. Sedative drugs in particular are frequently taken to avoid the amphetamine "let down" and these may cause a "false" (if ascribed to amphetamine) clouding of consciousness.

In spite of all of these serious drawbacks, however, naturally occurring cases of amphetamine psychosis still are probably worthy of study even if the findings must be interpreted cautiously. This is particularly true of those cases in which acute administration of huge doses is reported, since these could never ethically nor responsibly be given in this way to human volunteers.

We will, then, present a series of seven such cases arranged in a continuum. The arrangement has been made to suggest, if only by implication, the qualities of the clinical states that might be seen in any one individual if route, rate of administration, and total dose were varied over a wide range. The construct being presented, then, is a hypothetical one which has all the drawbacks of being based on naturally occurring, not experimentally induced, cases of amphetamine psychosis. Moreover, it is at least in part experimentally untestable.

CASE REPORTS

Case 1

A 24-year-old male had a two-year history of heavy amphetamine use, shooting a "spoon" at a time often several times a day in the characteristic pattern of the speed "run."[3] He first came to Bellevue after injecting a spoon of street "speed." He began to feel influenced by electrical waves and forces possibly emanating from a machine operated by "people in the background" and became convinced that a police raid was imminent. To avoid prosecution, he swallowed the remaining three "spoons" in his possession and then jumped from a second-story fire escape to escape his persecutors. In the hospital, he showed auditory hallucinations which cleared after two days and ideas of influence which diminished after a week but in which he retained some belief after 15 days. Urine amphetamine output was 33.48 mg in an eight-hour specimen. Diagnostic impression after clearing was chronic paranoid schizophrenia with superimposed intoxication.

Approximately a year later this patient was caught in an actual raid in another state and swallowed what is estimated as 10 to 15 gm of amphetamine to destroy the evidence. Three hours later his status was described as follows: "They had G. in a chair up against a wall . . .

with three detectives around him. He would throw his arms up in the air and roll his head around. He was a raving maniac. I think he was so far gone he didn't realize it." He was taken to a hospital and was in shock and on a respirator seven and one-half hours after ingestion of the drug. He died 14 hours and 20 minutes post-ingestion. Quantitative assays for amphetamine were not made.

Case 2

A 27-year-old male who had used intravenous amphetamine for one and one-half years currently did so in two- to four-day "runs," injecting approximately a "spoon" per day. At his admission he had been "up" for three to four days. He was found "playing drums" on garbage cans and had attempted to tear wire grating off a fence. When approached by the police, he asked them for amphetamine. His urine (a 16-hour collection obtained approximately 24 hours after admission) contained 1.48 mg amphetamine. Diagnostic impression after clearing was chronic undifferentiated schizophrenia with superimposed intoxication.

Case 3

A 19-year-old male was found by the police, shirtless, pushing a door of an apartment demanding to be let in to see his mother. He showed complete clearing in less than a day and retrospectively stated that he had "seen" his mother enter the apartment which he took to be his sister's house in Rhode Island. Although he had used drugs of many sorts and had been in almost continuous trouble with the police since the age of 16, he had used intravenous amphetamine only a few times in the previous two to three months. He had arrived in New York City just four days prior to admission and obtained an unknown amount of amphetamine which he took in one or two injections on the day of admission. Urine amphetamine levels were not determined. Diagnostic impression was amphetamine psychosis in an antisocial personality.

Case 4

A 19-year-old male had used amphetamines orally for approximately one and one-half years (60 to 80 mg/day, every two to three days) and intravenously for three months (four to five "nickel," i.e., $5.00 bags/day).

He had not taken amphetamine for a month during which he was out of New York. On his return a week prior to admission, he began shooting larger doses (one-quarter spoonful) three to four times per day. On the day of admission, he took two such injections and had palpitations. While walking, he noted he was being followed and gave ominous referential interpretations to statements people made among themselves. He felt that other users had decided he worked for the police and that "the whole Eastside had decided to get me." He sought refuge with a girl friend in a hotel, but on looking out the window, saw hoards of people with movie cameras on the adjoining roof. He felt the hotel would shortly be "stormed," panicked, and ran about the hotel, knocking on doors at random. The police were called and he was taken to Bellevue where, on admission, he admitted to auditory as well as visual hallucinations. ("Here's 'the man,'" i.e., police, etc.) His 24-hour amphetamine excretion was 14.57 mg. Diagnostic impression was amphetamine psychosis in an antisocial personality.

Case 5

A 19-year-old single white male who had used amphetamine orally for approximately a year had taken five 20-mg Biphetamine capsules in the 12 hours prior to admission. After an argument in a Greenwich Village night club, he became apprehensive and feared that those he argued with had friends who might "get" him. Thus, any group on the street seemed ominous and suspect. He hid in a bar but asked people why they were staring at him and interpreted their expressions as ominous and inscrutable. He became more frightened and, taking refuge in a friend's hotel room, barricaded the door with a bed and got out on a third-story window ledge. He heard voices below saying, "Let's get him." He ran out of the hotel, the halls of which he recalls as a nightmare maze, and into the street with a penknife in his hand. He approached a policeman who got a taxicab to take him to a hospital. He felt pursued by a large crowd on entering the hospital. A 24-hour urine obtained approximately 18 hours after admission contained 26.76 mg of amphetamine. Diagnostic impression was amphetamine psychosis in a passive aggressive personality.

Case 6

An 18-year-old male who had used oral amphetamines for approximately three years started "snorting" "street" amphetamine four months prior to admission. In the three months prior to admission he had used the drug intravenously in doses of three to five "nickel" ($5.00) bags per day.

Over a three-day weekend he began "shooting and shooting" and took an unknown amount.

He began to feel that "everything was there for me to do something about it. It was all a big game. There were two forces, my right and left side. Right was my side because everything came more naturally to me" (the patient is right-handed). Television seemed addressed particularly to him. In a park he saw a real fire and decided it was a "signal to decide whether to take my clothes off and jump over the wall" (which would "help everybody" and "make them O.K.") or "stand by myself." In this state he asked a passing policeman to help him get to a hospital. Urinary amphetamine excretion was 2.83 mg/24 hours (collection taken 48 hours after admission). Diagnostic impression was amphetamine psychosis in an antisocial personality.

Case 7

A 22-year-old male who had used amphetamine only three times previously, "snorting" one-half of a "$5.00 bag," had, prior to this admission, "snorted" a "half spoon." His only prior drug experience was with marihuana which he had used for a year but had stopped because he found he became "too paranoid" under its influence.

For five to six hours after taking the drug, he and his girl friend (who had also taken the drug) played folk music. After six hours they noted that their voices were "shaky" and "out of tune" and that they were playing unusually softly. Both became frightened, apprehensive about their mutual relationship, and jealous. They felt "bad connotations" to everything they said to each other and attempts to analyze only made it worse. The patient then felt that he was no longer high but had become "crazy" and he and his girl friend both went to a hospital. Here he felt "totally out of reality." His thoughts took on new and profound import. For a while he found it difficult to speak. Urinary amphetamine was not determined. Diagnostic impression was anxiety reaction secondary to amphetamine.

The above cases are summarized in Table 1.

DISCUSSION

The wide range of amphetamine levels in urine is consistent with the wide variations in effective dose for psychotogenesis observed in the course of experimental studies of amphetamine psychosis.[1,2] However, this range might also have been determined by other factors such as the critical importance of the effect of urinary pH on amphetamine excretion,[7] the difficulties in obtaining reliable 24-hour urine collections from highly disturbed patients, unreliability of historical data regarding the time elapsed since the last dose or doses, and variations in time required for case-finding and starting urine collections. It seems prudent then, not to ascribe any major meaning to amphetamine levels in urine.

The cases reported have been arranged in a continuum to represent delirium and death, two primarily delirious states, paranoid hallucinatory psychoses with (a) both visual and auditory and (b) only auditory hallucinations, a paranoid psychosis without hallucinosis, and an anxiety state with diffuse symptoms. This is not to imply that these represent the only clinical presentations seen after amphetamine abuse. Both Smith and Fischer[8] and our group[5] have noted other clinical modes of presentation. Rather, this spectrum is presented to demonstrate the clinical features that might occur in any single individual under widely varying con-

Table 1 Summary of cases

Case no.	Clinical features	Dose, route, duration	Urine amphetamine	Diagnosis
1	A. Paranoid hallucinatory psychosis B. Delirium and death	1 spoon i.v., 3 spoons p.o. at once 10–15 gm at once	33.48 mg 8.0 hours Not done	Schizophrenia and amphetamine intoxication
2	Delirium	1 spoon/day for 3–4 days	1.48 mg 16.0 hours	Schizophrenia and amphetamine intoxication
3	Delirium	Dose unknown, i.v., 1 day	Not done	Amphetamine psychosis, anti-social personality
4	Paranoid hallucinatory psychosis (visual and, to a lesser extent, auditory)	¾–1 spoon/day for 1 week	14.57 mg 24.0 hours	Amphetamine psychosis, anti-social personality
5	Paranoid hallucinatory psychosis (auditory hallucinations only)	100 mg p.o. over 12 hours	26.76 mg 24.0 hours	Amphetamine psychosis, passive aggressive personality
6	Paranoid psychosis	Unknown dose i.v. over 3–4 days	2.83 mg 24.0 hours	Amphetamine psychosis, anti-social personality
7	Anxiety state	½ spoon plus 1 "$5.00 bag"	Not done	Anxiety reaction to amphetamine

ditions of dose and rate of administration. Thus, what is presented is not an exact discussion of dose-response relationships applicable on an absolute milligram for milligram scale to a given population, but rather a hypothetical spectrum which may serve as a general guide to the clinical assessment of relative level of intoxication in any given case.

This presentation has, however, raised two points worthy of more extensive discussion: (1) the occurrence of delirium after amphetamine use and (2) the possibility that variation in individual sensitivity or acquired tolerance was a factor in the clinical outcome in these cases to the same extent that they were in the formal experimental studies cited.[1,2] This latter point, in particular, is a possible criticism of the entire hypothesis presented here.

Connell, whose extensive and detailed clinical study of amphetamine psychosis makes his clinical impressions highly authoritative, noted that only three patients in his series of 42 showed delirious traits. Moreover, in two of these contamination by sedative drugs or febrile illness affected the global clinical presentation. He discussed these findings thus:

"With regard to the absence of disorientation in the present study, the writer has no doubt that amphetamine, in sufficient doses, can produce disorientation. In fact, several case histories suggested that disorientation might have been present at some time during the psychotic episode but was short-lived and no longer present when the patient eventually came under psychiatric care. The remarks made by Mayer-Gross (1951) on this point, when discussing experimental psychoses, seem apposite.

"Differences in symptoms found with various drugs have been attributed to differences in the personality of the intoxicated. This may be true for an early stage or for cases of very mild intoxication. The delirious picture, on the other hand, is probably common to all intoxicants when their effect is most severe. Between these two ends of the scale is a stage in which probably each drug shows certain special features."[9]

Our experience with regard to the occurrence of delirious states in amphetamine psychosis is similar; i.e., these seem to be rare and rather transient. We would, however, tentatively add that if delirium currently is being seen more frequently, this only underscores the fact that those investigators who report the taking of massive doses in "field" conditions are probably quite accurate.[3,4] It may well be that the style of use on the current "speed scene" is one in which even psychosis *per se* is no longer "respected" as an endpoint and that some users are now taking doses which approach the limits of organic toxicity. Kramer has described the almost playful familiarity that high-dose users show toward their psychotic symptoms.[3] If this ominous trend persists and becomes more generalized, it might lead to an increasing number of medical rather than psychiatric emergencies subsequent to use of amphetamine in these massive dosages.

Several comments can be made concerning the effects of varying sensitivity or levels of acquired tolerance on the clinical outcome. Least important, but pertinent to the specific data in this presentation, it might be noted that the mildest symptoms occurred in the naive subject, while by far the heaviest user (Case 1) died. Moreover, of the users who presented with delirium, one was a heavy habitual user while the second did not (by abusers' standards) have an extensive history of amphetamine abuse. Furthermore, of the two patients with paranoid-hallucinatory psychoses, the one with both visual and auditory hallucinations was a far heavier user than the patient with auditory hallucinations only. Thus, tolerance was certainly not the sole factor in determining clinical outcome. Nonetheless, a suggestion remains that the clinical features seen represent a response to factors of either tolerance or sensitivity or both on the one hand, and dose on the other. An expression of a ratio of the two factors may, in fact, be the case.

It is more important, however, to consider the generally held assumption that amphetamine tolerance and sensitivity, especially regarding psychosis, are interrelated. Put another way, it is often assumed that tolerance to amphetamine is an "across the board" phenomenon, and that those factors which make it possible to take massive doses without killing oneself carry over and also convey resistance to the psychosis as well. This may or may not be true.

As has been noted, in studies in which amphetamine psychosis was experimentally induced, wide differences in sensitivity were noted in both Griffith's subjects and our own. Both groups required a minimum number of drug-free days in the hospital, in part, to diminish tolerance. This took six weeks for Griffith's patients and ten days to a month for ours. Yet the same variability in sensitivity was found in both subject groups. Moreover, in our cases at least, no impressive correlation between extent of past use and sensitivity was noted. Rather, histories tended to be borne out with subjects generally becoming psychotic at approximately the dose ranges one might have predicted on the basis of what they described as their past experiences. It may be, then, that varying sensitivity is determined less by acquired tolerance than by psychologic or even innate pharmacogenetic factors.

Moreover, regardless of what has been said about the applicability of animal studies across species, the phenomenon of amphetamine tolerance has been extensively and exquisitely studied in rats by Lewander and co-workers.[10] In these studies, tolerance was induced not only by repeated administration of amphetamine itself, but also (and most interestingly from the standpoint of mechanisms of action) by pre-treatment with parahydroxy-norephedrine. Since this latter compound has been shown to be tightly bound with the storage granules of norepinephrine and to replace it at this site,[11] one might expect that pre-treatment would convey tolerance to those effects of amphetamine that are noradrenergically mediated. What was found, in fact, was that this pre-treatment could indeed induce "tolerance" to some effects of amphetamine but not to its capacity for inducing stereotyped behavior. This is consistent with a massive body of information from both pharmacologic and pharmaconeuroanatomic studies (i.e., use of drugs in addition to ablation, microinjection, and histochemical techniques) that indicate that stereotypy is largely dependent on the effects of amphetamine on dopaminergic rather than noradrenergic mechanism. This evidence has been reviewed by Randrup and Munkvad.[12]

Moreover, recent preliminary studies at our unit, utilizing differential biochemical and behavioral effects of the d and l isomers of amphetamine as delineated by Snyder et al.,[13] suggest that stereotypy is the most valid animal "model" for amphetamine psychosis and that dopaminergic mechanisms are more crucial in the pathogenesis of amphetamine psychosis than noradrenergic mechanisms.[14] This may be consistent with the possibility that tolerance to the peripheral effects of amphetamine and sensitivity to its psychotogenic effects may be separate issues mediated by noradrenergic and dopaminergic mechanisms, respectively.

REFERENCES

1. Griffith, J. J., Cavanaugh, J. H., and Oates, J. A.: Psychosis induced by the administration of d amphetamine to human volunteers. In *Psychotomimetic Drugs,* Efron, D. H. (Ed.). New York, Raven Press, 1970, p. 287.
2. Angrist, B. M., and Gershon, S.: The phenomenology of experimentally induced amphetamine psychosis. Preliminary observations. *Biol Psychiat 2*:95–107, 1970.
3. Kramer, J. C., Fischman, V. S., and Littlefield, D. C.: Amphetamine abuse pattern and effects of high doses taken intravenously. *JAMA 201*:305, 1967.
4. Smith, R. C.: The world of the Haight Ashbury speed-freak. *J Psychedelic Drugs 2*: 172, 1969.
5. Angrist, B. M., and Gershon, S.: Amphetamine abuse in New York City 1966 to 1968. *Sem Psychiat 1*:195, 1968.
6. Smith, R. C.: Traffic in speed: Illegal manufacture and distribution. *J Psychedelic Drugs 2*:30, 1969.
7. Beckett, A. H., and Rowland, M.: Urinary excretion kinetics of amphetamine in man. *J Pharm Pharmacol 17*:628, 1965.
8. Smith, D. E., and Fischer, C. M.: Acute amphetamine toxicity. *J Psychedelic Drugs 2*:49, 1969.
9. Connell, P. H.: *Amphetamine Psychosis. Maudsley Monographs, No. 5.* London, Oxford University Press, 1958, pp. 68–69.
10. Lewander, T.: Presented at Collegium Internationale Neuro-psychopharmacologicum, Prague, August, 1970 (in press).
11. Brodie, B. B., Cho, A. K., and Gessa, G. L.: Possible role of p-hydroxynorephedrine in the depletion of norepinephrine induced by d amphetamine and in tolerance to this drug. In *Amphetamine and Related Compounds,* Costa, E. and Garattini, S. (Eds.). New York, Raven Press, 1970, p. 217.
12. Randrup, A., and Munkvad, I.: Biochemical, anatomical and psychological investigations of the stereotyped behavior induced by amphetamine. In *Amphetamine and Related Compounds,* Costa, E. and Garattini, S. (Eds.). New York, Raven Press, 1970, p. 695.
13. Snyder, S. H., Taylor, K. M., Coyle, J. T., and Meyerhoff, J. L.: The role of brain dopamine in behavioral regulation and the actions of psychotropic drugs. *Amer J Psychiat 127*:199, 1970.
14. Angrist, B. M., and Gershon, S.: A pilot study of pathogenic mechanisms in amphetamine psychosis utilizing differential effects of d and l amphetamine. *Pharmakopsychiatrie Neuro-Psychopharmakologie, 4*:64–75, 1971.

A PHYSICIAN'S VIEW OF THE ADOLESCENT DRUG SCENE

DAVID E. SMITH, M.D.

Adolescent drug patterns have changed dramatically in the last ten years. Prior to the 1960's, the "drug scene" was relatively constant: "illegal" drugs were confined to certain racial, ethnic, and philosophical minorities; the white middle-class representation of the dominant culture confined itself to the traditional agents of abuse such as alcohol, cigarettes, and prescription psychoactive drugs.

In the last decade, however, there has been a dramatic change in adolescent patterns of drug use, and it is my opinion that we are still in the midst of this change. At present the only thing that we can say for sure about the adolescent drug scene is that it will change. My travels across the country have given me an opportunity to consult and talk with individuals in many different communities. It has become apparent that the drug scene is in varying phases throughout the country.

For example, as consultant to a drug center in the Hawaiian Islands, I learned that the major concern there was with solvent inhalation (e.g. paint- and glue-sniffing) particularly among the native Polynesian population.

In Los Angeles, barbiturate abuse among high school students was so widespread that a special unit was established at the county's Rancho Los Amigos Hospital, under the direction of Dr. Ruth McCormick, for the treatment of barbiturate overdose and for barbiturate-dependence withdrawal.

In certain midwestern colleges that I have visited, the drugs of choice are the psychedelic chemicals, and the culture revolving around their use is of a different nature than in other parts of the United States.

At a large urban university in Chicago, the major problem is oral amphetamine use with many of the students using amphetamine for recreation and for improved performance while studying for finals. A fairly large number were involved in low-dose maintenance patterns of amphetamine abuse with a few periodically injecting "speed." This was particularly disrupting to the university administrators because they had consistently denied the existence of a major drug problem on the campus and had made no provision in their university health services for treatment of this problem; their basic policy was expulsion for students involved with drugs. In this way, a virulent and progressively more severe drug scene remained underground on this campus.

It is difficult to make a general statement about the adolescent drug scene because of its geographic variability and constantly changing nature. When you return to your community, you must design your program to meet its particular needs. Any community-based effort, I feel, has to be balanced among *law enforcement* to control distribution and supply of drugs; *research,* even if it is nothing more than a survey to determine patterns of use in that area; *education* for prevention; and *treatment* for the young people already involved with drugs. But it is essential that these efforts be designed to meet the specific needs of the population.

Except in a few major medical centers, virtually nothing is taught about drug abuse in medical school. The average physician has little exposure to the illegal drug scene and when he is challenged with management problems in the area of drugs, he is usually poorly prepared to respond to the needs of those who turn to him for help and consultation in a crisis of this kind.

In order to understand any drug reactions, it is important to keep in mind that each drug pattern is a complex interaction of chemical, personal, and social factors. You cannot say, for example, what marihuana does in the natural drug scene without considering the personality and physical characteristics of the individual and the social environment of his drug use. This is probably one of the most important principles to carry away from my presentation. Laboratory data rarely can be translated accurately to the street drug scene because they do not include analysis of the many variables that determine individual drug reactions.

The second principle you must understand is that individuals who learn to use drugs may pass through various stages. Most adolescent drug use can be defined in three basic stages: the early experimental phase, the periodic recreational or facilitative phase, and the compulsive pattern of drug use.

An excellent review of the three basic stages is given in Frederick Meyers and associates' text, *Review of Medical Pharmacology,* and I will not elaborate on their dynamics but rather try to pinpoint the management problems associated with each phase.

EARLY EXPERIMENTAL PHASE OF DRUG ABUSE

In most areas the very young experiment with drugs for reasons that are not related to psychopathology. In the Bay area, for example, surveys done in 1970 indicated that about 50% of the high school seniors and about 12% of the eighth-graders had experimented with illegal drugs by the time of graduation from high school or grade school respectively. The vast majority had experimented with drugs out of curiosity and because of peer group pressure. They used a particular drug not because they were mentally ill or because it was forced on them by an illegal dealer, but because they had heard about drugs, they were curious, they wanted to belong to a particular group, and so on. They were using drugs primarily for social reasons; the problems that developed resulted either from something happening in their own adolescent maturation at that particular time, or from their lack of discrimination in using the drug. In an early survey done at the Haight-Ashbury Free Clinic, for example, 40% of the patient population had taken a pill at least once without knowing what the drug was. They reflected an attitude of "better living through chemistry."

I am struck by the thought that young people enter into adolescence with a general feeling that drugs are always useful, and the way to learn about a drug is to try it. When they seek help in the early stages of an acute toxic crisis, it is not because drugs are bad but because they have just gotten the "wrong" drug—some bad acid or some bad speed or some grass with something else in it—and they often tend to rationalize why they developed a particular toxic reaction.

I don't think this adolescent attitude, which reflects society's drug-oriented culture, can be emphasized enough to physicians because physicians are active contributors to this problem. Young people are bombarded with information about drugs: through television (e.g. take Compoze so you can tolerate the riot news), printed matter, and family members and others who use a variety of psychoactive drugs. Some adults feel cheated if they leave the doctor's office without a prescription and many busy doctors find it more difficult to talk a patient out of taking a drug than to give him one. I think that these and a variety of other factors contribute to a pro-chemical attitude in children entering into the period of adolescence.

I cannot overemphasize this point, for often young people in the early stages of the drug abuse problem come for help questioning not the life style of drug involvement but wondering if they just took the "wrong drug." This point can probably best be summarized by the graffiti sometimes written on our clinic wall, "Just give me Librium or give me Meth," which reflects a general attitude toward drug consumption.

In this experimental phase, young people will participate through the ingestion of a wide variety of chemicals. They may not take them more than once, and they may discard them because the effects were not pleasurable or they did not achieve the objective they sought, or they experienced some uncomfortable toxic manifestations. Young people in the early experimental stages of drug use may, for example, hear that Dr. Shiffman's asthmadore is a "good trip" or learn indirectly that there are 0.25 mg of scopolamine in Compoze and that 30 tablets yield a "good trip." They may find that the belladonna delirium is not at all pleasurable and are not anxious to repeat this particular experience. Thus, in the experimental phase, the physician will be challenged primarily by acute toxic reactions. And it is important to understand the manifestations of these acute toxic reactions. I am sure that the physician is familiar with the depressant types of reactions arising from barbiturate and heroin overdoses. He usually has less experience with hallucinogens, and often does not understand the importance of the "talk-down" (e.g. placing the individual in a quiet environment, keeping stable human points of reference from his own culture near him, and making judicious use of sedatives in the early stages). In my opinion, if the patient develops an acute toxic psychosis and is out of communication, then the administration of phenothiazines with subsequent hospitalization is necessary.

The stimulant reactions have been discussed in great detail. In terms of psychologic manifestations, acute anxiety reactions, particularly in the non-tolerant individual, may be manifest. Dr. Kramer mentioned two instances of death following amphetamine overdosage that I have seen, both in non-tolerant individuals. One patient was interesting because he had had a previous history of high-dose cyclical amphetamine abuse, had been institutionalized and released, and began to use amphetamine again. When he was in his previous high-dose speed cycle, he was regularly injecting more than 1,000 mg a day and considered this a pleasurable dosage. He experienced an orgasmic flash with each injection and had relatively few toxic side-effects until he had repeated this cycle over a long period of time, precipitating an amphetamine psychosis with the critical diagnostic signs of auditory and visual hallucinations and paranoia with ideas of reference. After this patient had been treated and released, he began using amphetamines at a lower dosage, but the first quantity he took, although only one-half of what he had taken at the peak of his speed cycle, produced a massive overdose reaction. I think that in this case the patient was lost because the physician did not realize that the thing that makes the patient "go out" is not so much amphetamine-induced convulsion but rather hyperpyrexia. This might be a critical clinical point when one is treating amphetamine overdose in the adult, or accidental poisoning of a child. The hyperpyrexia is either concomitant with or follows the convulsion and must be kept under control, because death from amphetamine poisoning usually results from hyperpyrexia and its sequelae.

Although physicians must learn how to treat acute toxic reaction to depressants, hallucinogens, and stimulants, I am sure that many are concerned with the situation in which a mother discovers that her youngster has been smoking marihuana and insists upon treatment. What does the physician treat in a case of experimental or periodic marihuana use? The young person usually has no physical problems. He may not have any psychologic problems (except for those of adolescent turmoil and family conflict), and yet the parent insists on treatment and may try to shift the burden of responsibility to the physician. It is a difficult situation, for often the person in need of treatment is the mother because of her acute anxiety. And yet the physician does not want to turn away this family, especially if their next course is to consult a law enforcement agency, because it is critical in this early stage that parent-child communication be established and maintained. Very often the physician must shift from the role of a medical doctor to that of family counselor.

One final word concerning the experimental drug-using period. An acute toxic crisis with any drug should always have at least one follow-up. For example, after a bad LSD trip there is a high incidence of flashbacks. Any drug crisis may herald the tip of a much bigger iceberg in terms of what is happening in the patient's life; even the toxic manifestations which you interpret as an experimental drug pattern may be one of compulsive use. I think that detoxification following an acute toxic episode with any drug should serve as an opportunity for some sort of aftercare, even if that aftercare consists of nothing more than a single consultation. It is improper, whether in the emergency room, in the general hospital setting, or in the physician's office, to treat a single toxic episode without providing for follow-up.

PERIODIC RECREATIONAL OR FACILITATIVE PHASE

As individuals learn to use drugs and move into the phase of periodic use for recreation, it is interesting that acute toxic reactions decrease. This is why most physicians do not see the vast majority of periodic recreational drug users. Except for an occasional overdose, they don't have problems. They have learned ways of controlling adverse manifestations and are probably part of a particular drug culture that teaches them how to avoid overdose or too deep an involvement in the drug pattern. This underscores the importance of maintaining an overall view of any drug pattern. Most of what we learn about the street drug scene in the physician's office or the clinic or hospital deals only with the acute toxic manifestation of experimental drug use or the dependence manifestations of the compulsive drug user. Thus, the broad middle category which includes most of the young people involved with drugs is often ignored.

This is a critical point because if a periodic drug user develops an acute toxic reaction, it is important that he be treated and counseled. I think that the intervention of law enforcement with its criminal arrest procedure is adverse at this point and may precipitate subsequent destructive drug use. I also find that a large number of periodic drug users mature out of this phase. At a later time they may go beyond drugs or at least reduce their drug consumption. Very often this periodic drug phase is only a phase. Too vigorous law enforcement at this point may hit the young person at a critical stage in the development of his life style and foster the more destructive aspects of drug use.

COMPULSIVE PHASE

Finally, we may see the evolution of a compulsive pattern of drug use. Whether because of psychologic or physical dependence, our major concern is that the young person may come to depend on drug use as a singular problem-solving technique. As I pointed out, the drugs of choice for compulsive use vary widely across the country. At present, in Haight-Ashbury the predominant illegal drug of choice for compulsive use is heroin. In 1969 and 1970 we have seen, for example, a dramatic increase in the use of heroin by white middle-class youth. Unfortunately, this is spreading into other parts of the country. As was mentioned earlier, drug patterns seem to be established in major drug subcultures on both coasts and then ripple out to other parts of the country at a later time. It is of great concern to me that heroin is popular today, particularly among young people who are so prone to peer group pressure.

The Haight-Ashbury heroin scene was preceded by a wave of amphetamine abuse. Not only is amphetamine abuse destructive, but its increased use often indicates the deterioration of a middle-class drug scene. The stimulants are toxic in and of themselves, and serve as the door-opener to the "downer" scene. We have unfortunately also seen a new-style junkie, a young person who participates in a broad phase of experimentation, going right to heroin without going through the speed stage.

In the major urban drug subcultures, groups other than the racial minorities appear to be in this pattern of multiple drug abuse. Thus, it is critical that each community evaluate its own drug scene and also stay abreast of what is happening in the major drug subcultures of, for example, Greenwich Village in New York and Haight-Ashbury in San Francisco. It is essential that the ripple effect—drug patterns starting in certain drug subcultures, then rippling out and involving other parts of the country at a later time—be understood. The time lag is from six to 24 months; communities not yet touched by the more severe patterns of drug abuse can predict what is headed their way and hopefully prepare a community effort to combat it. Again, any community approach to drug abuse must be balanced among law enforcement, research, education, and treatment and should attempt to predict the drug pattern that is coming their way.

THE PHYSICIAN'S ROLE IN COMMUNITY EFFORTS

It is critical that the physician know what is going on in his community and it is particularly important that he take a role in helping to resolve some of his own community's drug problems. He is already an active contributor to the drug scene, either because of misprescription or lack of understanding of what is going on. For example, it is common practice among drug users to "cash script" or talk the physician into overprescribing various substances.

I would also hope that the physician would become a positive force in the community, either by backing up treatment efforts or endorsing the total community effort. For those who want to maintain the traditional position of avoiding community social problems, I think they should understand that unless the physician does get involved, the decision may be taken out of his hands. For example, doctors have always said that they can police themselves in the use of amphetamines, but they have not. There has been massive overproduction of amphetamines, with about 50% of the excess entering the illegal drug scene. The Food and Drug Administration recently indicated that amphetamines are indicated therapeutically only for narcolepsy, hyperkinesia, and short-term diet control. That means that physicians who want to maintain their traditional practice of long-term diet control by excessive prescription of amphetamines cannot be countenanced by the Food and Drug Administration and may be subject to malpractice suits.

There are many other things that you can contribute in terms of the community drug effort, including nothing. But I want to remind you in closing that silence is a proxy vote for the status quo.

PANEL DISCUSSION—STIMULANTS

Moderator: DR. WEINER
Panel Members: DRS. ANGRIST, KRAMER, SCHWEITZER, and D. E. SMITH

Dr. Weiner: I would like to open the discussion by reinforcing Dr. Smith's comments concerning the overprescription of drugs. In our pharmacology course at Colorado, we emphasize this point most strongly. We teach our students that all drugs are potentially dangerous and one should not prescribe them without very definite indications. In the past, we have discussed this in terms of individual harm, but throughout this week I for one, and I hope that all of us, have obtained a new perspective on the general harm of over prescribing drugs; that is, there is potential harm not only to the individual but also to society, and to our youth particularly which may be even more disastrous. I am grateful to the participants for providing me with this new perspective and suggest that we start here.

Dr. John G. Cowl (Washington, D.C.): I would like to compliment the entire panel this morning, and especially Dr. Smith, for having brought drug problems back into perspective and into the area of patient-doctor relationship and patient care. Long overdue, I think it was done very well this morning. My question has to do with toxicology studies recently performed by criminal laboratories on the East Coast (I'm not quite sure of the ones on the West Coast) which showed that LSD in the street-available forms is contaminated up to 80% of the time. The frequent toxic contaminants include strychnine, and most recently animal tranquilizers in alarming concentration. In that context, Dr. Abruzzi, of Woodstock and Powder Ridge fame or notoriety, now recommends that phenothiazines *not* be used routinely to treat acute LSD "bum trips" or "bad trips," but should be used only as an absolute last resort. If they are given it should be only in the setting of an intensive care unit because of the very great possibility of orthostatic hypotension caused by phenothiazine administration. If the patient's respiratory centers have been severely depressed by the animal tranquilizers, an intramuscular or intravenous dose of phenothiazine could cause death rapidly. I was wondering what Dr. Smith, Dr. Kramer, or any member of the panel suggests concerning the management of these patients by the everyday physician.

Dr. Smith: First of all, I think it is important to understand that the quality of illegal drugs varies widely. Basically what has happened in the underground drug scene is that the sons and daughters of the middle-class have learned the same marketing techniques that are used by the pharmaceutical industry and a variety of other people. That is, they take an old drug, give it a new name, and find that people will pay for it and be very enthusiastic about using it. As a consequence, quality varies widely. For example, PCP, the veterinary anesthetic that you were talking about, appeared first in Haight-Ashbury in 1967 as the "peace pill," came out later as "hog" or THC, and continues to pop up in a variety of preparations. The quality and purity of drugs in the illegal drug scene tend to vary in proportion to demand. For example, mescaline had a "good" reputation; a lot of young people wanted to try it so the quality of mescaline for the last couple of years has been terrible. It's been, for the most part, LSD or LSD-STP or STP either pure or with amphetamine or PCP added. In our analyses recently and in some done by Dr. Joe Benforado in Madison, Wisconsin, we've seen poor quality mescaline. It is only in the last month that we've seen the first two analyses of pure mescaline. However, LSD, which is out of favor now, has the last couple of months been the highest quality seen since 1965. Again the only constant thing about the illegal marketplace is that it will change, and how it will change depends on many variables including demands.

I agree with what you say about the use of phenothiazines; we do not use them in our own center except in very rare circumstances. We generally use sedatives for the treatment of this type of drug reaction.

Dr. Angrist: To really say anything definitive about the use of phenothiazines in these situations, you need good pharmacologic studies on the interactions between phenothiazines and the agents which might have been taken. Now I don't know, perhaps some people here do, about interaction studies between phenocyclidine, which is what we are talking about in this case as the adulterant, and phenothiazines. I do know that there is a very clearly defined interaction between phenothiazines and anticholinergics, which are deliriants that have sometimes appeared on the drug scene. Gerson has established that the use of phenothiazines in anticholinergic delirium will routinely produce severe central nervous system depression. An anticholinergic delirium can readily be distinguished from a psychedelic "trip" on clinical grounds. Patients show ataxia, slurred speech, hyperreflexia, incoordination. The mental state is a delirious one with incoherence, mumbling, irrelevance, and disorientation. In short the patients appear "out of it" much like patients with delirium tremens. Anyway, this state would be an absolute contraindication to the use of phenothiazines because of the severe CNS depression that the combination induces. So I would like to underline the fact that phenothiazines must not be used in an anticholinergic delirium. I just don't know the answer about the phencyclidine.

Dr. Kramer: The LeDain Commission of Canada has recommended that laboratories be opened to drug users where they can submit samples of drugs for testing with the assurance that they will not be busted for possession of drugs. They would get a report on the components of their drugs. There might be certain advantages to this, although one of the problems is that this might imply a tacit acceptance of drug use. I think it is a subject that is open to further debate, but which deserves consideration.

Dr. Edward J. Lynn (East Lansing, Mich.): At the risk of beating a dead horse, or maybe just a stubborn one, I again would like to challenge the speakers this time, excluding Dr. Smith, with a bit of "copping out." Dr. Schweitzer suggested that one reason why amphetamine abuse will remain a problem is that it is so simple to produce that even a 9-year-old boy can prepare it in a bathtub. But we won't need bathtub stimulants as long as our pharmaceutical companies continue to produce about 8 billion amphetamine preparations every year. Now that is enough for every man, woman, and child in this country to have 40 "hits" each a year. Most users abusing amphetamines do so with ethical products, usually prescribed by their friendly family physician. Would members of the panel care to comment further in terms of steps we might take to help alleviate these situations?

Dr. Angrist: Since I now do more experimental and less clinical work, I've not kept abreast of the preparations used by patients who enter Bellevue after taking amphetamines. My experience there was from 1966 to 1969, and the scene might have changed. Also, looking at hospital admissions gives a perspective that is not really representative. I don't want to go into a long analogy, but there is a tale about blind men and an elephant in which one examined the elephant and declared that he was like a wall. Another man examined him, grabbed a tusk, and described the elephant as very much like a spear, and so on and so forth. A lot depends on the perspective from which you view the situation, and from my perspective I probably saw the craziest, the most reckless amphetamine abusers. More than half of those Bellevue admissions came in on powdered black-market preparations, and since powders tend to lend themselves so elegantly to IV use, I do think they are a critical problem. This is not to say, of course, that there is not gross over-manufacture of amphetamines, but I think the black-market preparations are potentially very dangerous. They are also very concentrated. I once experimented, not with black-market, but with laboratory amphetamine; I was in Dr. Schweitzer's lab and I said, "Let's take a small pinch as though I'm going to 'snort' it, a small, very conservative dose." Then I threw it on the scale and it rang up 30 or 40 mg. When these powders are that concentrated, it's awfully easy to get into very high-dose use.

Dr. Kramer: I think the drug companies have been very soundly beaten in the last year or two, and I think to some extent they deserve the punishment they have been getting. It has been my impression that the large drug houses have, in fact, reasonably well-controlled distribution. You very seldom find Lilly brand Seconal on the illicit market. You very seldom find Dexedrine brand amphetamine on the illicit market. What you generally find on the black market diverted from legitimate sources comes from the smaller generic drug houses. Hopefully, after some of the congressional hearings which have taken place throughout the country, some of the smaller companies may finally get the message and begin to be more careful voluntarily. I certainly think prescribing by physicians should be decreased, and I have advocated this in a paper that I have written for a book edited by Dr. David Smith. In the future it may be necessary to pass new laws if voluntary control is inadequate. The FDA has taken some action; perhaps Congress and state legislatures in a year or two will take further action if it is necessary.

The overprescribing is due in part to the fact that for years doctors thought amphetamines were perfectly safe. Pharmacists to whom I have spoken recently say that the amount of amphetamines prescribed has been reduced substantially in the last two or three years, and is down to one-half or one-third of what it was two or three years ago. This is a hopeful sign, although it certainly is not as low as it should be.

Dr. Weiner: Dr. Kramer, are you suggesting that the higher-priced trade names are less likely to get on the market? Is this a matter of economics or a matter of supply?

Dr. Kramer: The high-priced trade names are less likely to get on the market because those drug companies have been more careful about their distribution techniques.

Dr. Daniel J. Wilhelm (Port Huron, Mich.): Dr. Schweitzer noted that amphetamine secretion is enhanced by acidification of the urine. Does this principle have any clinical application in treating the acute toxic reaction?

Dr. Schweitzer: I'd rather not answer that because I am not a medical doctor. Does anyone on the panel wish to make a comment?

Dr. Angrist: Connell recommended using ammonium chloride to acidify the urine and hasten excretion, I believe; theoretically that should be useful. John Davis made an interesting point at a recent meeting, saying that in view of claimed central nervous system toxicity of amphetamine (which is really not well substantiated to date), it might well be that psychosis and organic toxicity involve different mechanisms. Theoretically then, the faster you get the amphetamine out of the central nervous system the better off you are. Now this adds some complications to the use of chlorpromazine, which Davis has shown to inhibit clearing of amphetamine from the brain, at least in animals. At any rate, Davis has also urged that ammonium chloride be used in the treatment of these cases.

Dr. Kramer: In chronic high-dose use of amphetamines, the fact that users don't eat very much would suggest to me the possibility that they may be in an acidotic state so that any attempt to change the acid-base ratio in them should be handled with great caution. You may get into trouble by acidifying such patients so much that they will turn litmus red.

Mr. Robert Winfield (Ann Arbor, Mich): My question is for Dr. Smith. In the last two days, Dr. Kaplan and Dr. Brenner underscored the anguish and alienation of the drug user and his internal discomfort. It occurred to me as they were speaking that there is a large class of drug users who are aware of this discomfort and yet continue to use drugs. Most of us in the health professions do not come into contact with these people, since they do not usually come in by their own will. On the other hand, we are in contact with the mother who brings in her son who is using marihuana, and we often find people of this class who are using much more than marihuana. I would like to ask the following: What is your projection for the future in terms of this type of drug use? And, secondly, What action would you recommend to the health professional—the physician or counselor—who comes in contact with this type of person?

Dr. Smith: Well, that's a very complex question. I think in terms of prevention, that is, primary prevention. Whatever techniques you use, whether it's drug education in the school or community-wide drug education, it really has to occur before the experimental period. Not just giving the young person information but also facilitating an attitude about the ingestion of psychoactive chemicals.

Other types of traditional educational techniques do not work once the individual has gotten into a pattern of periodic drug use. In my experience he doesn't stop unless he either finds something else that is more meaningful to him or he has a crisis. Therefore, the physician has great opportunity in the treatment of a crisis. If he establishes a meaningful relationship with his patient and does not treat him in a punitive or destructive way, and if he can establish a communicative relationship, the physician may be able to counsel him about some of the potential dangers of drug use, including arrest and the possibility of a criminal life style; this is especially true at an early stage of psychoactive periodic drug use. Beyond this I think there is little that the physician can do. The broader health team, however, can have much greater impact—particularly in the utilization of paramedical personnel, ex-addicts, and street workers—in terms of outreach. I think this middle group really responds only to their peers or individuals who are part of their own culture. Many of the community-based drug programs are gaining wider acceptance because of their utilization of outreach workers as supplements to the health team; many also use peers and assert peer group pressure not to use drugs or to at least avoid the potentially destructive patterns. I think the general philosophy of making young people part of the solution rather than part of the problem should be a dominant philosophy in this middle area. Here the physician, however, has to step down from his traditional role and acknowledge that there might be other members of the health team who can do a better job and that his responsibility in certain areas may be confined to diagnosis and coordination.

Dr. Richard G. Just (Rock Island, Ill.): I am addressing this question primarily to Dr. Kramer because he brought up the point of a possible association between a periarteritis nodosa-like syndrome and amphetamine usage. During my training in Los Angeles, many doctors and patients who had used amphetamines told me of cases of rheumatoid arthritis-like syndromes. In fact, many of them said that this syndrome was so easily reproduced that they used it in an attempt to evade induction into the armed forces. All of my questions have to do with this syndrome. First of all, are you aware that such an association in fact exists between amphetamine usage and a rheumatoid arthritis-like syndrome? If so, what is this syndrome characterized by? How common is it? Is it as predictable as I have been led to believe? Have any serologic studies been done in these cases, and if there are any studies published on these cases, can you give me the references?

Dr. Weiner: Dr. Kramer, have you an answer for that "single" question?

Dr. Kramer: No. The periarteritis nodosa-like syndrome was described at the beginning of this week, at least publicly. This was my first knowledge of it. The next morning I wrote to Dr. Citron to find out more about it, but all I know now is what I saw on TV.

Dr. Smith: I would like to make some comments on that. Just by accident I had an opportunity to talk with the investigators prior to the publication of that study, and the report suffers from the deficiencies of any street drug research. That is, the individuals involved had a history of multiple drug use, but there was a very high suspicion on the part of the clinicians that it was directly related to amphetamine. In the summer of 1967, I saw a patient, a high-dose amphetamine user, with erythema nodosum and a rheumatoid arthritis syndrome. I remember talking with Dr. Meyers about this and wondering whether there was an association. But it was absolutely impossible to prove at that time, and we didn't have the funds to do the type of work required. I really suspect that after this report there are going to be a lot of people around the country who have had clinical experience with amphetamine users who are going to say, "Hey, I remember a case like that," and you are going to see many case reports. I know that was my reaction. But proving this requires a laboratory where the variables can be controlled to insure a proper clinical study. I feel that there is an association, but I don't have the foggiest idea whether it's allergic or paradoxical or whether it's the adulterants or whatever. I do feel it is related to high-dose amphetamine use.

Dr. Angrist: I read the *Times* article that reported this, so I'm an expert, too. But what I wanted to say is that what is implied by this report is a specificity for amphetamine which is probably not true. Angiitis has been reported in persons taking drugs such as sulfonamide and penicillin. These drugs can hook up with proteins and become haptenes and lead to autoimmune reactions on that basis. Secondly, a necrotizing angiitis is well documented in association with severe hypertension. Perhaps the hypertensive effects of acute high doses of amphetamine also contribute to the condition in this way. What I mean to say is that there are well-known mechanisms that could lead to an angiitis in amphetamine abuse, but I just wonder if the phenomenon is, in fact, as specific to amphetamine as has perhaps been implied.

Dr. Eric Chivian (Washington, D.C.): I would like to address this to any member of the panel. Most of what we've heard this morning about stimulants has been concerned with the amphetamine group, although Dr. Kramer alluded to the fact that his comments could also be applied to other members of the stimulant group. I wonder if members of the panel could discuss some of the practical medical and psychologic problems that might attend other types of stimulant use, in particular cocaine, either by itself or in combination with sedative hypnotics, or opiates, because we still see a lot of cocaine use in Washington.

Dr. Kramer: Cocaine is very expensive on the illicit market so, in the past at least, it's tended to be only sporadically, rather than regularly, available. There may be some places where it's continually available; in rare instances you may find a physician or a nurse who obtains and uses it in a pattern which is similar to the amphetamine run. In those cases the intensity of the picture is much greater. I have seen only one such case and it's hard for me to draw any conclusions from that.

Dr. Smith: I'd like to supplement your observation that there does seem to be a revival in cocaine use. We've seen a lot of "speed-balling" here recently, using heroin and cocaine. It's the first time that I've had an opportunity to see a lot of cocaine use. It also seems that the same person will use amphetamine if cocaine is not around. I think that about 15% of the population we just finished studying had used amphetamine and cocaine in combination. So for some reason, cocaine is now in favor. This has been reported in San Francisco, Los Angeles, Washington, and also by some people I talked to in New York. I think the general point is, however, that cocaine toxicity is similar in pattern to amphetamine toxicity, which is the implication that I got from your presentation.

Dr. Weiner: Although it is impossible to predict this, I would guess that a mixture of cocaine and amphetamine would be particularly potent. Is there any evidence one way or the other on that?

Dr. Smith: It's additive, I would think.

Miss Jane Hassinger (Ann Arbor, Mich.): Dr. Smith, I'm very interested in your point about the necessity for follow-up, and I'm assuming that you mean psychologic follow-up as well as medical. I would like to know the design of your clinic; that is, how do you facilitate this kind of follow-up over the long-term?

Dr. Smith: Our clinic sees most of its population of 150 patients a day in crisis situations, either acute medical situations or problems of detoxification. And our after-care programs are divided essentially into two areas. First, we have a drug treatment program which focuses on the long-term management of the compulsive drug user; it utilizes not just psychologic and medical techniques, but also social techniques. John Frykman, who was the director of that program until recently, and Roger Smith, who founded the program, will be talking tomorrow, so I won't discuss that. We also have an outpatient "psych" program for the psychedelic drug user which might be applicable to many of your communities. Here we see a young person who is compulsively involved with drugs but also has a situational or personal crisis that has been aggravated by drugs. He may have psychedelic confusion following too much LSD and needs a therapist to help him sort out what is real and what is not real. I think that the techniques and approach effective with the compulsive drug user really differ from those which are useful when drugs are not the primary problem. For example, I feel that one-to-one therapy or counseling help with specific problems may be more appropriate for the adolescent crisis in which drugs are only secondary. Very often you will see an underlying thought disorder which has not been amenable to group techniques. In contrast, programs dealing with the compulsive drug user make greater utilization of group techniques for behavioral modification. They emphasize establishing relationships, and not just with the professional therapist or with peers or ex-addicts in the group. I think the move is away from the medical model toward increased emphasis on vocational rehabilitation and counseling as supplements to the psychologic processes. I would be happy to talk with you about it more, but this will be discussed in some depth tomorrow.

Dr. Robert C. Wicks (Ann Arbor, Mich.): I would like to know what happens to the compulsive drug user when he reaches the age of, say, 25 to 35; what is his life style then and what would you suggest as office treatment?

Dr. Kramer: To a large extent it depends on which group of drugs he has been using compulsively. Someone who has been involved with opiates tends to continue to be involved with them. From what I've seen and from what I've heard Dr. Smith and others say, it would appear that marihuana-psychedelic use tends to be an interlude in most people's lives which may peak around the end of high school, continue into the early college years, and fade as the person gets into his 20's. Amphetamine use also seems to have a pattern in which there is a peak and a decline. Regrettably, many of the high-dose amphetamine users we see go into some other drug use pattern. These are current observations. My own readings of historical reports suggest that drug use patterns vary in different countries, in different places, and in different times. To generalize then is difficult; all we can say is what we have seen or what we have read or heard about.

Dr. Wicks: Do users tend to be absorbed or assimilated into an establishment culture later on or do they remain a separate group of people?

Dr. Kramer: Heroin users, for example, have a distinct culture of their own, although they generally fade more easily into the landscape. They tend to keep their hair cut short and are not as obvious as some of the psychedelic drug users. Clearly, a lot of people using grass are part of the everyday scene. They may be living across the street or next door, conceivably in your own house. People who use the strong psychedelics tend to form a subculture, but there are vast differences between psychedelic subcultures and some middle-class people who occasionally drop a little acid or mescaline.

Dr. Smith: Just to supplement that, two comments. First of all, concerning the psychedelic scene I would agree with John that this appears to be a phase, but what happens after that phase can be quite varied. They may continue to use a psychedelic occasionally or not use any at all, and go on to college and become straight. They may get involved in compulsive patterns of drug use or may move beyond, like so many of the original residents of Haight-Ashbury, and get involved in an alternative life style. The communal movement, for example, represents this; a large number of the original inhabitants of the Haight are now involved in a life style beyond drugs but are certainly not back in the society they rejected. For the most part, this involvement is an alternative to drug abuse. I think the reason this is so important is that the psychedelic phase may be a crucial time, when the physician particularly and a variety of other people in the community can influence the way the young person goes. I would not rely too heavily on external symbols to define drug patterns because young people are very intelligent. One told me recently, "Well, I got into smoking a little grass, and I grew my hair longer. My mother got really upset, so I cut my hair and started dropping speed, and she thinks I'm off drugs now." If we focus on external symbols, or any superficial information, the young people will beat us at this game every time. The only way to know what's going on with a person is not through external symbols but getting to know them.

Dr. Joseph Benforado (Madison, Wis.): Dr. Kramer raised a point which is well worth discussing and perhaps the panel could direct itself to it. I think it is in the interest of society that we replace the aphorism "caveat emptor" with quality control in the illicit drug market. In Wisconsin we have been testing drug samples for ingredients such as arsenic and strychnine. However, these are not random samples. We get these drugs from physicians who provide them for us or from illicitly seized stocks. If we were to find arsenic and strychnine, we would make this knowledge available and try to insure that all these drugs were collected. On the other hand, we do run a quality control information service. I submit drugs for testing and the information comes back to me some weeks later. I use this information in retrospect to correlate the clinical symptomatology of the bad trip I have seen with what, in fact, the tablet actually contained. I'm often fooled. Kids come in on bad trips and say they took mescaline; the drug turns out, in fact, to be acid, and so on. Now I have a patient who wanted me to test a tablet of mescaline for him. The reason he wanted me to test it was that 5,000 of these tablets had just arrived in town. I said to him, "What can my testing of one tablet tell you or me about 4,000 other tablets?" Even in industry there are problems with quality control. Sound quality control is based on random sampling, and as Dr. Weiner might have read in *Science,* even reputable outfits supplying radioactive chemicals to scientists have mislabeled epinephrine and norepinephrine. And this was discovered in a scientific experiment only because the bioassays weren't turning out right. Now, the gentleman who asked me to test the tablet is a user who has solved this problem. He doesn't use a drug on himself unless he's tried it out on some other human being first.

Dr. Angrist: He's smart.

Mr. Larry Reida (Washington, D.C.): My question, directed to Dr. Kramer, relates to his comments that very few products of ethical manufacturers get into the illicit market. We've held a number of hearings across the country in the last year, and our figures indicate that 80 to 90% of all seizures of illegal amphetamines, methamphetamines, phenmetrazine, and methylphenidate come from ethical manufacturers. I also find it curious that you say there is less diversion and that doctors are prescribing less, for if you'll look at the figures manufacturers are producing more. Would you care to comment on that?

Dr. Kramer: There are a couple of questions here. I think that the rubric of "ethical manufacturers" includes the smaller generic drug houses. The differentiation I made was not between ethical and unethical, but rather between the major companies and the smaller companies or wholesalers who produce and distribute large quantities of amphetamines and barbiturates, perhaps even more than the major companies produce and distribute. My sources of information, both legal and illegal, narcotics officers as well as the consumers and dealers of these products, indicate to me that, for the most part, the "downers" and the "uppers" do not come from the major drug companies but from the smaller ones. This is the differentiation that I made. I have not previously heard your figures of total production, but I find them interesting and will accept your statement about them. I know that illicit amphetamines seem to be harder to get today in my part of the country. Barbiturates seem less readily available in Orange County than they used to be, although evidently they are as available as they were last year, if not more so, in Los Angeles County. There are variations. I hope that some of the smaller drug houses are making a greater effort to control diversion. If you have information that they are less able to control diversion now than they were one year ago, again I would accept that information. It's been my impression that they have been trying to be more careful.

I hear stories from my colleagues in the drug industry about helicopters which follow their trucks containing dangerous drugs. They have instituted a variety of other precautions but thefts take place nevertheless. I gather that special efforts are being made to check out international deliveries and would be glad to hear information to the contrary if there is such.

Dr. Smith: I'd like to supplement that by emphasizing that when we talk about diversion we must realize that diversion can take place at the manufacturer level, at the wholesaler level, at the retail pharmacist level, and at the physician level. One instance that I remember occurred in Colorado where some kids worked in the warehouse so that they could get at the drug shipments. There was a brief wave of thiopental sodium injection because they had cracked open a case and thought it was methamphetamine in vials and were using it. I think a tremendous amount of diversion occurs further on down the line and what the users happen to get hold of at the time may account for drug styles in our community.

Dr. Kramer: I would like to make one other comment. You mentioned methylphenidate. To the best of my knowledge, the only methylphenidate available is made legitimately. There has been an outbreak of methylphenidate (Ritalin) use in Seattle. Chances are there was one source of the drug responsible for its fairly wide distribution. I don't know what the drug company has done to eliminate that particular source. It is very, very difficult to eliminate sporadic diversion. In my opinion the major drug companies have generally controlled massive diversion of dangerous drugs.

Mr. Reida: Just one more question. Would you agree that it makes sense to control amphetamines, methylphenidate, and phenmetrazine in the same class, rather than take the Band-Aid approach and say just control methamphetamines? It is my feeling that if we strictly control one class, people will quickly learn that dextroamphetamine gives just as good a flash. In Congress we were bombarded with all kinds of information on how we would solve the speed problem in this country by controlling liquid injectable methamphetamine. This is what, twelve-one hundredths of 1% of what we are talking about? Would you care to comment on that?

Dr. Kramer: I agree completely that all of these drugs are in the same class and all should be equally controlled. To temper my previous remarks and to give an indication of my own view on this, I feel that if nothing else succeeds, then perhaps we ought to institute controls on the stimulants and possibly the sedatives identical to those controls now in effect for narcotics. I am perfectly willing to accept laws which will carry it that far if nothing else works. Perhaps we ought to give voluntary efforts no more than two or three years.

Dr. Alan Reed, Jr. (Milwaukee, Wis.): I have two comments and one question. The first comment relates to the type of amphetamines that are available in the illicit market. In Milwaukee I would estimate that "white crosses," which are generically-produced 10-milligram methamphetamines, comprise 80 to 90% of the illicit oral amphetamine traffic. The second comment is about cocaine. I think cocaine is one of the most rapidly rising drugs of use in the United States today, at least it certainly is in Milwaukee. I don't know, Dr. Smith, if that is your experience too? I think it would be very good if at some time during the Symposium we could put a little bit more emphasis on cocaine. Finally, there is another type of amphetamine which I have been waiting to hear about and that is methylene-dioxy-amphetamine, MDA. It may be classified as a hallucinogen and handled in the afternoon session, but I noticed that actually most of the discussion about amphetamines concerned their use in hallucinatory doses. Is MDA going to be talked about this afternoon?

Dr. Smith: Dr. Hollister, I believe, will cover it in his presentation. I think it is properly included in that category because the psychotomimetic amphetamines basically are not used in a compulsive amphetamine pattern but are used more like the psychedelics, with acute toxicity being the major problem.

Mr. Jason Janus (Chicago, Ill.): A comment about the cocaine first. I suggest that we'll get more mileage out of Freud by reviving his cocaine papers.

I wish to backtrack to the questions about treatment of amphetamine intoxication. About a year ago when I searched for something that could be used, I looked into a reference book entitled *The Hallucinogens* by Abram Hoffer and Humphrey Osmond. Abe Hoffer has worked with LSD in Saskatoon, Saskatchewan, for about 20 years, and Humphrey Osmond is at the New Jersey Neuropsychiatric Institute. I found a comment in the book concerning the use of ascorbic acid for treatment. I later asked Abe Hoffer about it and he said that he had found it in the literature. There is very little work on it, but it is believed that ascorbic acid removes methyl groups from the blood stream or from the sites of action of amphetamines. Based on this information, we started using large doses of vitamin C and found that it does alleviate the symptoms; it brings a person down from acute toxic reaction to amphetamines, in almost every case, within 25 to 40 minutes, and we find that the toxic reaction is reduced. It suggests something that might be looked into.

Dr. Angrist: I would immediately challenge methodology, placebo controls, and so on for a claim like this. The treatment of amphetamine toxicity is a difficult problem. This drug does many things, none of which, as far as I know, are affected by ascorbic acid. Its effects on central noradrenergic mechanisms might contribute to the hyperthermia seen terminally in amphetamine poisoning. In a medical crisis with hyperthermia then, one would think perhaps of a central noradrenergic blocker like chlorpromazine. There is a problem there, though, because in amphetamine poisoning hyperthermia is often part of a generalized dysautonomia in which shock occurs as well. If the patient is already in shock, then you are caught betwixt and between. I don't know about giving chlorpromazine at that point. So the medical treatment is a real problem in itself. For the psychiatric aspects, uncomplicated by medical issues, Connell has said that no treatment is indicated whatsoever, that amphetamine psychosis is a self-limited process which clears within approximately a week. Now, can this be hastened? Griffith says that chlorpromazine does produce a rapid remission of psychotic symptoms. I would think personally that haloperidol would be a logical drug to use. Anyway, I will say, while I don't know, I regard with enormous suspicion all claims to date for the psychiatric efficacy of vitamins.

Dr. Kramer: Thorazine has proven effective in the treatment of acute amphetamine overdose. I have found it to be an effective amphetamine antagonist in animals. Pre-treatment with chlorpromazine markedly raised the amphetamine and methamphetamine LD_{50}. In another experiment in which the chlorpromazine was given either 5 minutes or 25 minutes after the injection of a potentially lethal dose of methamphetamine, prompt use of chlorpromazine did appear to be effective in cutting down the mortality rate, though delayed use did not. These experiments were done with mice, and mice are not men. But it does suggest that chlorpromazine may be useful as a life-saving measure. The potential of chlorpromazine to produce or enhance shock must be considered and weighed against its value in controlling acute amphetamine toxicity.

Unidentified member of audience: In our approach to this problem, we have attempted to minimize the use of drug therapy and started out as a psychologically oriented therapeutic service. I agree, however, that chlorpromazine is a currently accepted treatment. We have learned that it is contraindicated by DOM or STP and there have been deaths in Chicago related to its use. I think Dr. Smith could probably document the report that when Thorazine was used in San Francisco to bring people down from STP in '67, five deaths occurred in a week-end of treatment.

Dr. Smith: On the follow-up of 50 hospitalized STP patients whom we saw in July 1967, we felt particularly that the high doses of phenothiazines prolonged the period of disability. We have not been able to resolve this question in a clinical situation. For example, some early cases of what we thought was STP turned out to be belladonna; the phenothiazines given for the treatment of belladonna delirium prolonged the reaction. So there was some misinformation that came out of that early period. An additive effect has certainly not been demonstrated in the laboratory, but I feel that high doses of phenothiazines do prolong the period of disability with DOM. I agree with the comments that were made over the table about demanding controls on the use of vitamins for the treatment of the acute hallucinogenic drug reactions, but I am still quite disturbed by the fact that we talk about one drug or another. I must re-emphasize my belief that the prime modalities in the treatment of the acute psychotic reaction to LSD basically involve non-drug modalities: supportive environment, the talk-down, and so on. A skilled therapist can do a much better job than this or that drug. I think that we physicians sometimes fall into a trap when we are talking about treatment and focus only on pharmacologic agents without proper understanding of the psycho-social factors involved.

Dr. Weiner: In deference to those who have been standing a rather long time at the microphones, I think we will go around just once more if each person will take only about one minute for the question and answer.

Mr. John Frykman (Carmel, Cal.): Less than a minute. I want to disagree with Dr. Kramer, too. I want to say that in my experience most of the street drugs that are sold are from the major companies. There is a certain mysticism that goes with that which makes the street dealers seek them out.

Unidentified member of audience: I have had two or three patients come in who are using around 120 milligrams of some sort of amphetamine a day. I presume that Dr. Smith has had the same kind of experience that I have where it is very difficult to get street kids into the hospital for many different kinds of reasons. I have had a difficult time coming up with a rational outpatient detoxification program. I wonder if you could tell us what you do.

Dr. Smith: I think that the challenge facing the health care delivery system relative to drug abuse is based on the fact that so many of the new generation of drug users are resistant to the medical establishment and traditional techniques of hospitalization. We have in our clinic a drug detoxification section which does a great deal of its work on an outpatient basis. This is not our after-care program, but rather outpatient detoxification, where we focus on heroin users. We have had a little experience recently with outpatient detoxification of barbiturate dependence which I am sure will throw shudders into many individuals in the audience, including Dr. Isbell whose work is the Bible in this area. But you may have a patient who is addicted to barbiturates who will not go into the hospital, or there may be no hospital beds available. In many big cities, when a drug epidemic is wide-spread, you cannot get hospital beds. You must do a lot of things on an outpatient basis, and we do a great deal of outpatient detoxification in our clinic. There is, of course, increasing utilization of the halfway or three-quarter way house. I know the drug treatment program in our clinic is soon to be involved with a three-quarter way house which will try to bridge the gap between purely outpatient care and the high cost of the often unavailable and very often unacceptable hospital bed. So, I think that there are a variety of techniques that need to be explored between pure outpatient and pure hospitalization that may be more acceptable to the patient. These houses, of course, are not hospitals and may look a lot more comfortable and familiar to them.

Mr. David Chambers (Ann Arbor, Mich.): I have a question directed to either Dr. Kramer or Dr. Smith. I was a little surprised by Dr. Kramer's statement that the whole range of amphetamines, that is, stimulants, should be rigorously controlled in terms of their distribution from the manufacturers. I was surprised and a little concerned because, if I understood earlier parts of the discussion today, it is apparently the case that the illicit manufacture of the stimulants is quite easy. Someone, I believe, referred to nine-year-olds making them in bathtubs. If that is the case, controlling even more rigorously the distribution of amphetamines manufactured by the ethical drug manufacturers, large and small, is not likely to reduce by any substantial degree the use of such drugs by young people. Even if amphetamine use is somewhat reduced, if illicit manufacture increases, do we not have greater reason to be concerned about the health implications for users of the illicitly manufactured as opposed to the licitly manufactured stimulants?

Dr. Kramer: It is quite a reasonable question, one which I ask myself. Simply because it is relatively easy to manufacture methamphetamine illicitly does not mean that we should abandon controls. Simply because it is relatively easy to get sufficient quantities of heroin into the United States to supply the demands of 100,000 to 300,000 users does not mean that we ought to decontrol or reduce our controls on opiates. I think that controlling the legal manufacture of stimulants is one part of a large procedure. Controlling certain of the precursors may also limit the production of illicit methamphetamine. I think that a combination of actions is necessary. It may be that there will be some increase in illicit production because of stricter control. The question which remains is: Will the overall problem be decreased? I believe that adequate control over legitimate production and distribution will help more than it will hurt.

Dr. Weiner: Unless most of you have taken your amphetamine this morning, I suspect you are very hungry, and I think we will close now. I want to thank both the panel and the audience for a stimulating interchange.

Afternoon Session
Wednesday, November 11, 1970

Presiding: Leo E. Hollister, M.D.

CHEMISTRY AND MEANS OF DETERMINATION OF HALLUCINOGENS AND MARIHUANA

BO HOLMSTEDT, M.D., AND ALLAN LINNARSON, D.Sc.

The hallucinogens (sometimes called phantastica or psychotomimetics) constitute a heterogeneous group of compounds of synthetic, semi-synthetic, and naturally occurring origin. It is fair to say that the majority of them are alkaloids from the plant kingdom.[1] Some are synthetic amines.[2,3] For practical reasons the hallucinogens can be placed in groups that are not chemically coherent but useful in dealing with this group of compounds. The cannabinoids are outstanding in that they are compounds without nitrogen but have a psychotomimetic action.[4,5]

The structural formulas and the pharmacology of the hallucinogens are well documented.[2,6–16]

With the rapidly changing spectrum of drugs being abused nowadays, it is sometimes necessary to identify within a short period of time the structure of completely unknown substances. The *extraction and separation procedures* for these compounds are those conventionally used for most alkaloids and are well documented in modern literature.[17–22]

Compounds can be identified by a series of methods that yield more or less positive confirmation. Reference compounds or analogues are a prerequisite for nearly all methods. The past quarter-century has seen a great change in the study of hallucinogenic drugs especially with regard to the introduction of new and more sensitive techniques, such as fluorescence in spectrophotometry. It is likely that we are now entering a period when physical-chemical techniques and the use of computers will greatly influence progress in this kind of research.[23–26]

In recent years, gas phase analysis, especially gas chromatography-mass spectrometry (GC-MS), has been successfully used for the identification of steroids, hormones, lipids, acids, and terpenes. The method has also been used for the characterization of hallucinogens and their metabolites.[24,27–30] This review will concentrate on use of gas phase analysis in the study of hallucinogens and marihuana.

DESCRIPTION OF GC-MS WITH ACCESSORIES

The technique of combining a gas chromatograph (GC) with a mass spectrometer (MS) has proved to provide a very efficient tool in the service of analytic toxicology. For a detailed description of the instrumentation and application the reader is referred to Hammar.[24]

The gas chromatograph is able to separate complex organic mixtures of low concentrations. The mass spectrometer gives structural information; it operates with sample amounts and concentrations corresponding to the gas chromatograph. The time taken to record a spectrum does not normally exceed that of the elution of a peak from the gas chromatograph.

A combination of the "high pressure" GC-system with the "low-pressure" MS-system is possible because of the various types of molecular separators.[31–33] A block diagram of the instrument used by us is shown in Figure 1.

The mass spectrometer can be regarded as a very sophisticated GC-detector and except for the scanning of spectra it can be used in various ways. The first way is to take mass spectra from the various GC-eluates in order to identify them. This may be regarded as the "normal" use of the GC-MS.

The mass spectra and fragmentation pattern form the basis for the elucidation of the structure of a drug. Comparison of the fragmentation pattern of unknown compounds with that of appropriate reference compounds provides information for determining the structure of the drug in question and its metabolites.

The mass spectrometer can also be used to monitor a certain mass fragment characteristic of an actual component or group of components eluted from the GC. In this way the mass spectrometer works as a more selective GC-detector than the conventional electron capture- and salt-detectors. The new technique of chemical ionization will probably play a role in future developments along this line.[34] With the help of a special device, three or more fragments can be monitored, which of course increases the selectivity. The slow time constant needed for the system during elution (seconds to minutes) gives a high gain and consequently a high sensitivity. The superselective detection of low concentration components in extracts from organic material makes it possible to quantitate them down to the picogram level. This technique has been called mass fragmentography.[35]

The third way in which the MS can be used is to measure very accurately the molecular or fragment masses or both, in

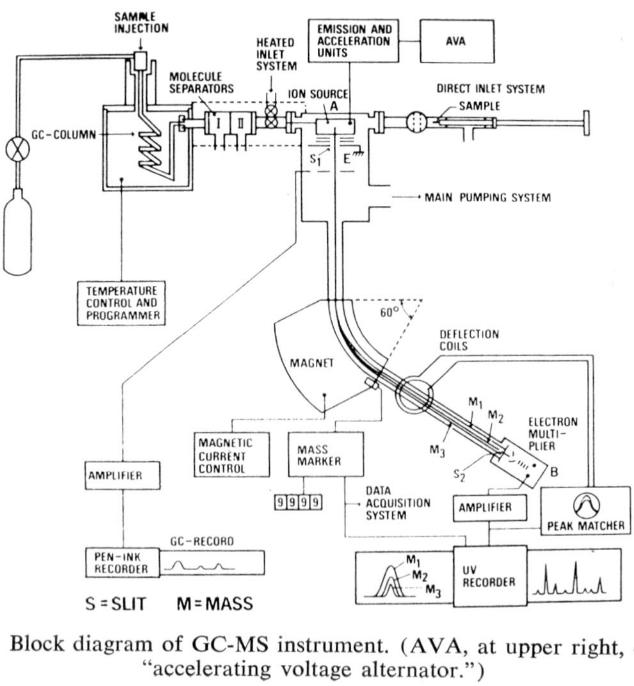

Figure 1 Block diagram of GC-MS instrument. (AVA, at upper right, designates "accelerating voltage alternator.")

order to determine their elementary composition. On a low resolution instrument this is normally done by the peak-matching technique. Quantities in the supramicrogram range are needed with this equipment when structure determination is attempted.[30] To be useful for the measurement of small GC-eluates, higher sensitivity and rapidity are required. The commercial instruments available at present are inadequate in these respects. However, a new system which fulfills the requirements of "instant peak matching" of GC eluates has recently been presented.[36]

COMPARISON OF VARIOUS ANALYTICAL TECHNIQUES

The hallucinogens are compounds which, in order to produce their action, have to be taken in quantities from grams (mescaline) to micrograms (LSD). Thus, their determination in body fluid requires a wide range of sensitivity in the methods used. One must bear in mind that the methods should be selective and applicable to the concentrations produced in the body fluids by the drug in question. In general, separation methods give little information for determining the identity of a certain component. Chromatographic methods thus are a "negative means of analysis"; i.e., certain structures can be excluded because they do not fit the measured Rf values or retention time.

Spectrometric methods give more information about structure but they are limited in their ability to separate the information obtained from several components in a mixture. Aromatic compounds, for example, in many cases absorb in the same wave-bands. This means that compounds under analysis must be as pure as possible. Several methods are compared in Table 1.

It is very difficult to find analyzing systems for biologic material without a separation method as a step in the procedure. General separation methods are gas chromatography (GC), thin layer chromatography (TLC), and liquid column chromatography. To improve both the separation and sensitivity of detection of chromatographic systems, the samples are frequently transformed into derivatives.[24(pp.55–56)] A good technique is to make a compound suitable for liquid scintillation measurements by utilizing a derivative that includes a tritium or a C^{14} atom. Trifluoroacetate derivatives of amines become sensitive to electron capture detectors and are also easier to run through the GC-column. As can be seen from Table 1, the most sensitive systems

Table 1 Approximative sensitivity in some analytical systems (uncertainty $\leqslant 10\%$)

Method	Concentration (mol/liter)	Amount (mol)
Spectrophotometry	10^{-6}	10^{-10} (0.1 ml)
Fluorometry	10^{-9}	10^{-13} (0.1 ml)
Gas chromatography		
Flame ionization detector	10^{-4}	10^{-10} (1 μl)
Phosphorus detector	10^{-7}	10^{-13} (1 μl)
Electron capture detector	10^{-7}	10^{-13} (1 μl)
Gas chromatography-mass spectrometry	10^{-5}	10^{-10} (10 μl)
Mass fragmentography	10^{-8}	10^{-14} (1 μl)
External isotopic labeling	10^{-8}	10^{-11} (2 ml)

Modified from Schill.[62]

belong to the gas chromatographic group. This is due to the highly sensitive detectors available such as the electron capture detector, the phosphorus detector (salt-detector), and the mass spectrometer. These systems also offer a high degree of selectivity.

Which system should be used for a certain analysis depends on many factors. One is the final volume of the sample. A GC requires a small volume because of the low capacity of the column. Therefore samples intended for GC-separation in most cases have to be concentrated. If evaporation of the sample to obtain a sufficient concentration for GC-analysis is impossible, a spectrometric method may be chosen. Large volumes of samples with a low concentration can, however, be separated on liquid chromatographic systems equipped with detectors of the same type used in GC.[37] The sensitivity for such systems is approximately the same as for GC.

From the discussion above, it is clear that there are tools and methods to determine very small amounts of hallucinogens in biologic material. The further development of present techniques will be in the direction of greater sensitivity, accuracy, and computer-assisted automation.

PRINCIPAL STRUCTURES OF HALLUCINOGENS

Many tabulations of hallucinogens exist. The reader is referred to Hofmann where previous reviews are fully discussed.[38] Tables 2 through 9 have been compiled in order to supply the most recent references to the analyses of the various hallucinogens and their metabolites.

The tables are intended to be used by those dealing with the analysis of hallucinogens in general and especially for those interested in determinations in body fluids.

Group 1—Compounds Related to Phenylethylamine (Table 2)

Although the members of this group are mostly primary amines, secondary amines

Table 2 Determination of phenylethylamine-related compounds (Group 1)

Substance	Method	References
Mescaline 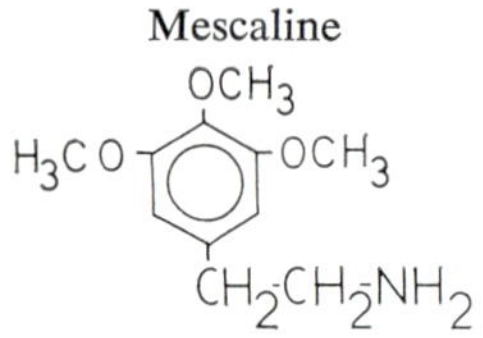	Extraction from aqueous alkaline solutions by organic solvent. Some members of the group, e.g., amphetamine, are very volatile. Steam distillation from alkaline solutions may therefore be suitable	Becket et al.: *J Pharm Pharmacol 19*:273-294, 1967[54] Bonnichsen et al.: *J Legal Med 67*: 19-26, 1970[40] Brandenberger and Hellbach: *Helv Chim Acta 50*:958-962, 1967[39] Brandenberger: Einsatz verschiedener GC-Detektoren in der toxikologischen Analyse. " 'GC-Tagung' Klinische Chemie und Pharmazeutische Industrie" Basel 3, Juli 1968[55] Campbell: *J Pharm Pharmacol 21*: 129-131, 1969[56] Gunne and Galland: *Biochem Pharmacol 16*:1374-1377, 1967[57] Holmstedt: Abuse of Central Stimulants. Symp. arr. by the Swedish Committee on International Health Relations, Stockholm Nov, 25-27, 1968[30] Schubert: *Acta Chem Scand 24*: 433-438, 1970[58] Änggård et al.: *Europ J Clin Pharmacol 3*:3, 1970[59]
Amphetamine		
Methamphetamine	TLC	
Ephedrine	UV	
Pseudonorephedrine	GC	
STP	GC-MS	
MDA		
MMDA		
Phenmetrazine (Preludin)		
Methylphenidate		

such as methamphetamine and L-ephedrine are included along with some remotely related compounds such as phenmetrazine and methylphenidate. The compounds of this group generally lend themselves to gas-chromatographic determination. It is preferable, but not necessary, to derivatize primary amines such as mescaline and amphetamine prior to gas chromatography. Amphetamine is a particularly important compound in this group and has been quantitated in body fluids by means of gas chromatography by Brandenberger[39] (as a derivative) and by Bonnichsen and associates[40] (non-derivated). For most compounds in this group mass spectra have been recorded. To our knowledge, no spectra have previously been presented for methamphet-

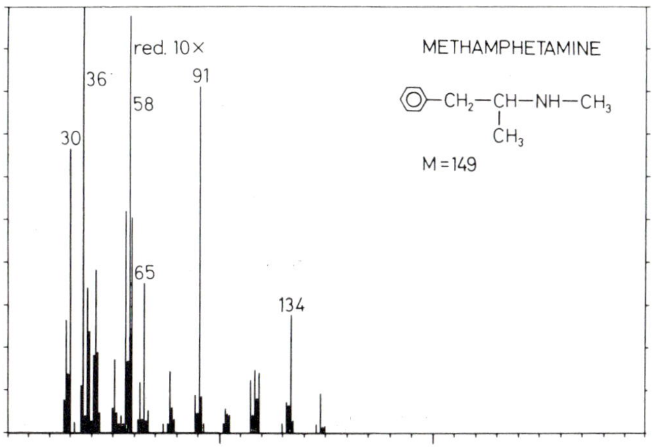

Figure 2 Methamphetamine
Instr.: LKB 9000
Inlet system: direct inlet 50-125°C
Amount: 1.5 μg
Scan: m/e 1-500/50 sec.
Ionization energy: 70 eV
Slits: source 0.05 mm
collector 0.15 m

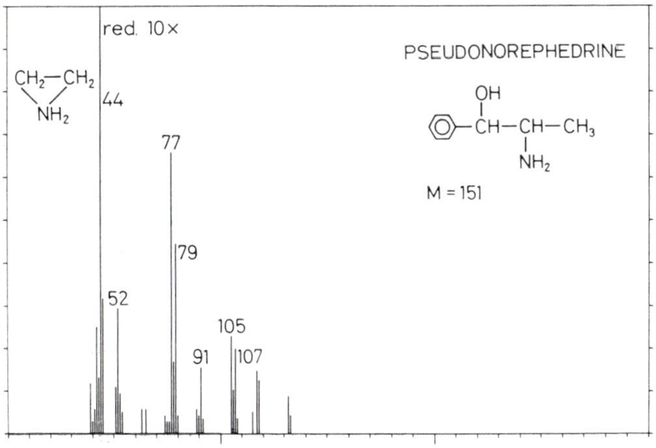

Figure 3 Pseudonorephedrine.
(Instr., inlet system, amount, scan, ionization energy, and slits—same as those of Figure 2.)

amine and pseudonorephedrine. They are therefore reproduced in Figures 2 and 3. Pseudonorephedrine is of interest because it is the psychoactive component of the drug khat (for information about this drug, see Schultes[1]). Mass spectra do exist for STP and MMDA,[30] but not for MDA which is a synthetic drug recently found to be used in some circles.[41]

Volatility and other properties make gas phase analysis an ideal means of identification and quantitation of this group of hallucinogens.

Group 2—Derivatives of Simple Tryptamines (Table 3)

Fluorometry has been the method of choice for this group of compounds since about 1950. In recent years, it has received competition from both gas chromatography and gas chromatography-mass spectrometry. Hansson has given an excellent account of all means of chemical analyses of indole alkylamines and related compounds, particularly with regard to 5-hydroxytryptamine.[42] Fluorometry is still the method of choice for quantitation of the latter but has to be carried out after column liquid chromatography. Fluorescence spectra of 5-substituted indoles when run in 3 N HCl show a shift to higher wave lengths (550 mμ). This shift is not known to occur with indoles substituted in 4, 6, or 7 position or with unsubstituted indoles.

The use of fluorometric techniques has the advantage that indoles may be detected at the 0.1 microg level in 2 ml of fluid. However, in situations in which high concentration can be obtained, gas-liquid chromatography permits not only qualitative resolution but also quantitative determination.

5-Methoxy-dimethyl tryptamine in this group is a psychotomimetic indole which to our knowledge has not been abused in modern society. It is the main component of a hallucinogenic drug used among tribes in the Amazon basin.[29]

It has been suggested that metabolism resulting in hydroxylation of the sixth position would be necessary for indoles of this type to become psychotomimetically active. This theory is no longer tenable.[12(p.286),27] Mass spectra exist for most compounds in this

Table 3 Determination of simple tryptamine derivatives (Group 2)

Substance	Method	References
Dimethyltryptamine (DMT) 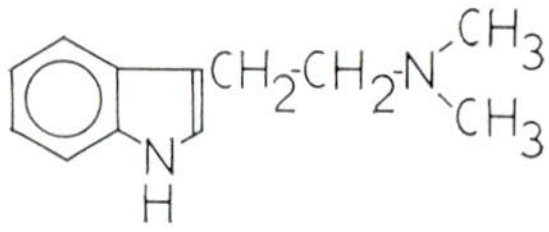 5 MeO-DMT Psilocin Psilocybin	Extraction from aqueous alkaline solutions by organic solvents TLC UV Fluorometry GC GC-MS	Ahlborg, Holmstedt, and Lindgren: *Advances Pharmacol 6 B*: 213-229, 1968[28] Agurell, Holmstedt, and Lindgren: *Biochem Pharmacol 18*: 2771-2781, 1969[46] Agurell, Holmstedt, and Lindgren: *Acta Chem Scand 23*:903-916, 1969[27] Clarke: *Isolation and Identification of Drugs,* 1969[17] Hammar, Holmstedt, Lindgren, and Tham: *Advances Pharmacol Chemother 7*:53-89, 1969[24] Hansson: *Handb. d. experim. Pharmakol. 19*:66-112, 1965[42] Holmstedt and Lindgren: *Ethnopharmacologic Search for Psychoactive Drugs,* 1967[29] Wessman et al.: *Analyt Letters 2*: 81-91, 1969[60]

group, and mass fragmentography should be an ideal method of determination. To our knowledge no quantitation of these indoles in body fluids of man has been attempted as yet with the exception of DMT and related compounds in urine.[12(p.275)] The use of these chemicals as tools in experimental psychiatry should encourage determination of the parent compounds and the metabolites in other body fluids by means of modern methods.

Group 3—Lysergic Acid Derivatives (Table 4)

In 1957 Axelrod and associates worked out a fluorometric method for the determination of LSD[43] that has since been used in studies in which the plasma levels of LSD have been correlated to behavioral tests.[44] The method is very sensitive, making it possible to record about 1 nanogram per milliliter. However, the background fluorescence presents difficulties in many cases, some of which we have had the occasion to observe in our own lab. Special precautions must be taken to avoid this.

Gas chromatographic techniques are available.[45] They are sensitive in the range of 100 nanogram, which is also in accordance with our own experience. The gas chromatographic procedure can be used in combination with mass spectrometry. A spectrum of LSD obtained in this way is presented in Figure 4. To our knowledge, no derivatization of the indole nitrogen in LSD has been attempted for analytical purpose. Heptafluorobutyryl derivatization and/or mass fragmentography should be used to increase the sensitivity and specificity of gas phase determination of LSD.

Group 4—Harmine and Derivatives (Table 5)

The compounds in this group are easily extracted and can be determined either by UV or fluorescence.[46] Many new β-carbolines have been isolated from plant material. They are also easily subjected to gas chromatography. Mass spectra have been presented in several papers. Substitution in the sixth position induces a shift in the fluorescence maximum in this group of com-

Table 4 Determination of lysergic acid derivatives (Group 3)

Substance	Method	References
LSD 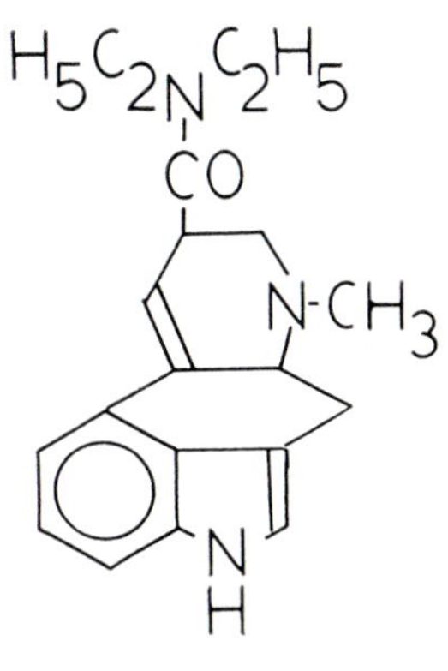	Extraction from aqueous solution by n-hydrocarbons at alkaline pH Fluorometry GC GC-MS	Aghajanian and Bing: *Clin Pharmacol 5*:611-614, 1964[44] Axelrod, Brody, Witkop, and Evarts: *Ann NY Acad Sci 66*: 435-444, 1957[43] Bellman: *J Assn Offic Anal Chem 51*:164-175, 1968[61] Katz, Tadjar, and Aufricht: *J Chromatogr 31*:545, 1967[45] Kern, Schilling, and Müller: *Gas Chromatographic Analysis of Pharmaceuticals and Drugs.* Basel, Varian Aerograph, 1967[19]

LA = d-lysergic acid amide

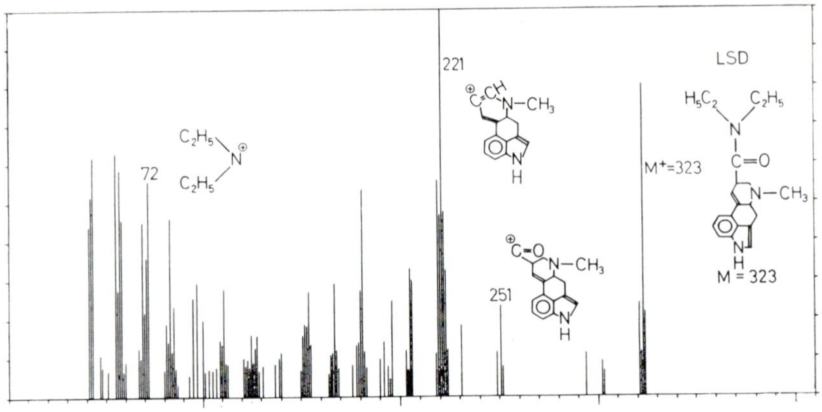

Figure 4 LSD.

Instr.: LKB 9000
Inlet system: GC
column: SE 30 5% gas chrom Q 2.7 m × 2.5 mm
temp: 260°C
Amount: 200 ng
Scan: m/e 1-500/8 sec.
Ionization energy: 70eV
Slits: source 0.01 m
collector 0.3 mm

Table 5 Determination of harmine and derivatives (Group 4)

Substance	Method	References
Harmine 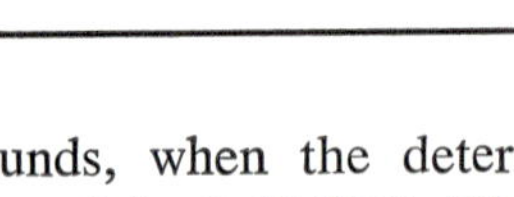 Tetrahydroharmine Harmaline	Extraction from aqueous alkaline solutions by organic solvents TLC UV Fluorescence GC GC-MS	Agurell, Holmstedt, Lindgren, and Schultes: *Acta Chem Scand 23*: 903-916, 1969[27] Clarke: *Isolation and Identification of Drugs.* London, Pharmaceutical Press, 1969[17] Hammar, Holmstedt, Lindgren, and Tham: *Advances Pharmacol Chemother* 7:53-89, 1969[24] Holmstedt and Lindgren: *Ethnopharmacologic Search for Psychoactive Drugs,* pp. 339-373,[29] 1967

pounds, when the determinations are performed in 3 N HCl. This is in accordance with what happens to simple indoles under identical conditions.

Group 5—Tropanalkaloids (Table 6)

A host of evidence of all kinds exists concerning the alkaloids in this group. For unknown reasons they are sometimes not included in tables of hallucinogenic compounds. Atropine and scopolamine are easily determined gas chromatographically. Reports of mass spectra obtained by means of direct inlet systems have been published. To our knowledge, no mass spectra obtained by means of the gas chromatograph have previously been presented. Spectra obtained in this way are shown in Figures 5 and 6. The frequent medical use of these alkaloids should prompt a quantitative determination by modern analytical methods.

Table 6 Determination of tropanalkaloids (Group 5)

Substance	Method	References
Atropine (Hyoscyamine) 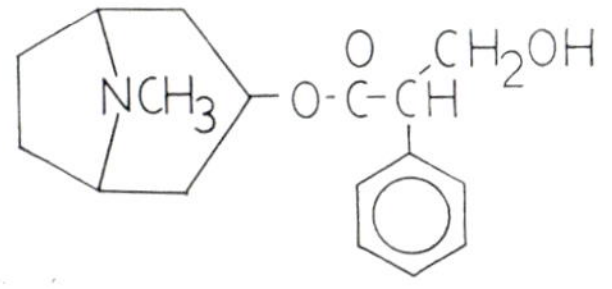 Scopolamine (Hyoscine)	Extraction from aqueous alkaline solutions by $CHCl_3$ TLC UV IR GC GC-MS	Clarke: *Isolation and Identification of Drugs.* London, Pharmaceutical Press, 1969[17] Stewart and Stolman: *Toxicology, Mechanism and Analytical Methods.* New York, Academic Press, 1960[21] Gudzinowicz and Ieller: *Drugs and Pesticides.* New York, H. Dekker, Inc. 1967[18]

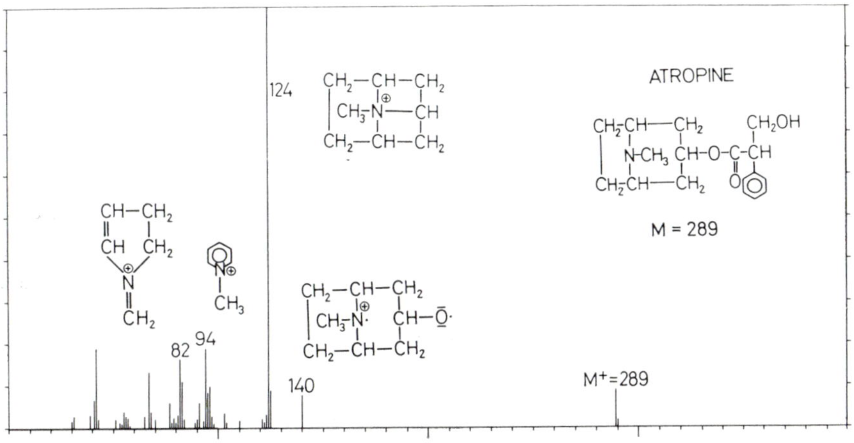

Figure 5 Atropine.

Instr.: LKB 9000
Inlet system: GC
column: SE 30 5% gas chrom Q 2.7 m × 2.5 mm
temp: 220°C
Amount: 150 ng
Scan: m/e 1-500/8 sec.
Ionization energy: 70 eV
Slits: source 0.1 mm
collector 0.3 mm

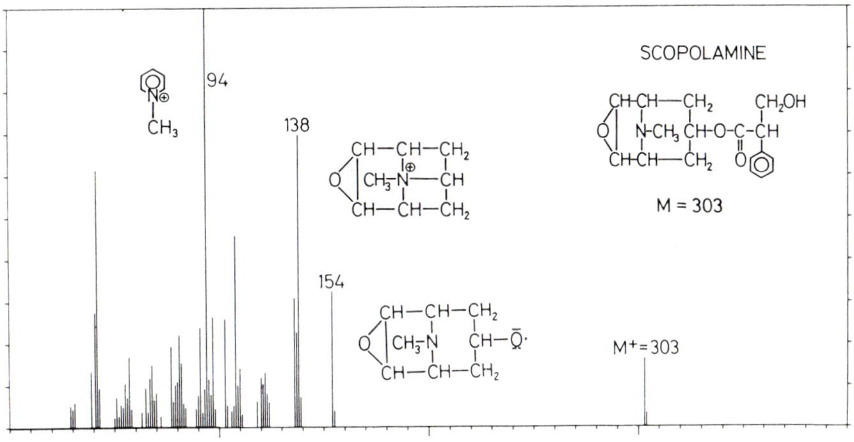

Figure 6 Scopolamine.
(Instr., inlet system, amount, scan, ionization energy, and slits—same as those of Figure 5.)

Group 6—Glycolate Esters (Table 7)

This interesting group of compounds in which are found both hallucinogens and drugs has been fully described by Abood.[3] Very little is known concerning analytical techniques that could be used with these compounds. Based upon analogy with similar compounds the technique described under "Methods" should be applicable. A mass spectrum of one of the best-known compounds of this series, ditran, is presented in Figure 7. The lack of information about tissue levels contrasts with the number of pharmacologic experiments which have been carried out with this series of substances.

Group 7—Phencyclidine Derivatives (Table 8)

Phencyclidine (Sernylan) has been known to be abused by certain groups. It is easily isolated by means of gas chromatography and its mass spectrum obtained by means of the gas chromatograph has been recorded. Peak matching was used to identify this substance.[47]

Table 7 Determination of glycolate esters (Group 6)

Substance	Method	References
Ditran 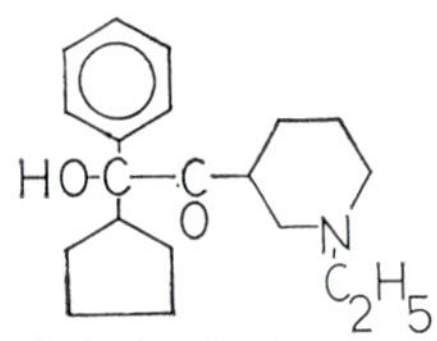 4 N-ethyl-piperidyl-benzilate	Extraction from aqueous alkaline solution by original solvent	Kuznetsov and Golikov: Synthetic atropine-like substances. U.S. Dept. of Commerce, Clearinghouse for Federal Scientific and Technical Information, Joint Publications Research Service[2]
	TLC UV GC MS	Abood: In *Drugs Affecting the Central Nervous System.* New York, Marcel Dekker, Inc., 1968, pp. 127-167[3]

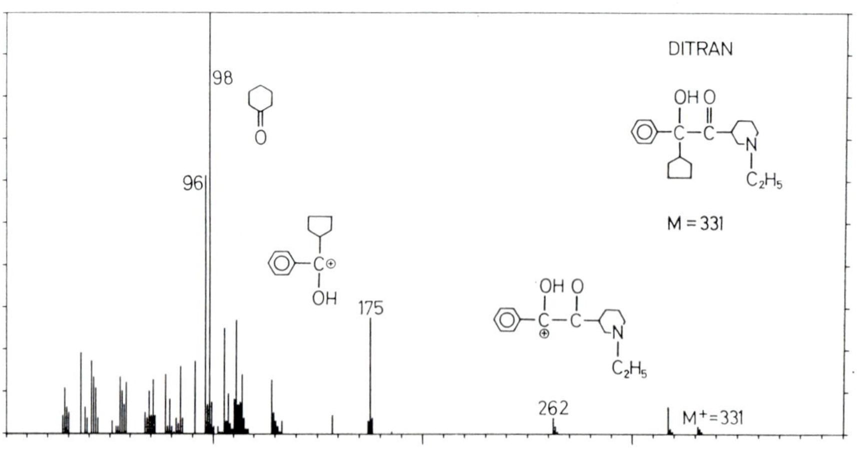

Figure 7 Ditran.
Instr.: LKB 9000
Inlet system: direct inlet 70-110°C
Amount: 2 μg
Scan: m/e 1-500/50 sec.
Ionization energy: 70eV
Slits: source 0.1 mm
collector 0.3 mm

Table 8 Determination of phencyclidine derivatives (Group 7)

Substance	Method	References
1-(1-phenylcyclohexyl)-piperidine 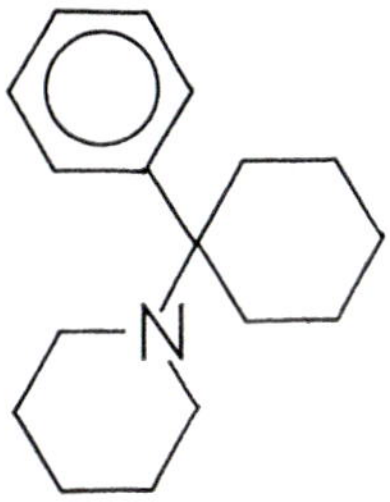	Extraction from aqueous solution at alkaline pH by organic solvents UV IR TLC GC GC-MS	Clarke: *Isolation and Identification of Drugs.* London, Pharmaceutical Press, 1969[17] Holmstedt: In *Abuse of Central Stimulants.* Stockholm, Almqvist & Wiksell, 1968[30] Lindgren, Hammar, Hessling, and Holmstedt: *Amer J Pharm 141*: 86-90, 1969[47] Usdin and Efron: *Psychotropic Drugs and Related Compounds.* Public Health Service Publication, 1967[15]

Table 9 Determination of cannabinoids (Group 8)

Substance	Method	References
Δ^1-Tetrahydrocannabinol 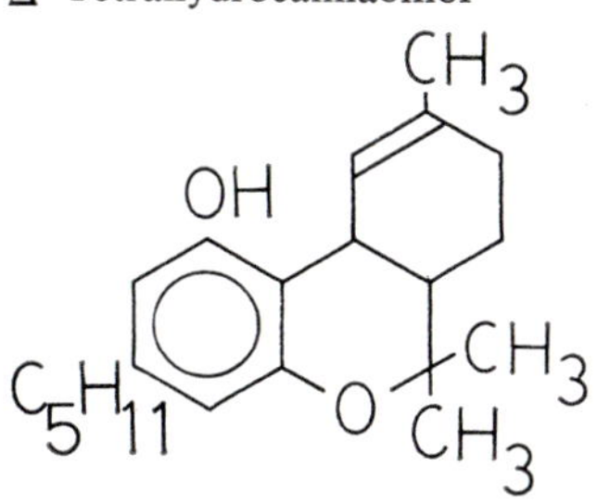 Δ1(6) Tetrahydrocannabinol 7-hydroxy-Δ^1 tetrahydrocannabinol 7-hydroxy-Δ^1 (6) tetrahydrocannabinol Cannabidiol Tetrahydrocannabidiol	Extraction from aqueous solutions at adjusted pH by n-hydrocarbons TLC GC GC-MS	Agurell: In *Botany and Chemistry of Cannabis and Its Derivatives.* London, Churchill, 1970[4] Agurell, Nilsson, Ohlsson, and Sandberg: *Biochem Pharmacol 19*:1333-1339, 1970[53] Nilsson, Agurell, Nilsson, Ohlsson, Sandberg, and Wahlqvist: *Science 168*:1228, 1970[49] Mechoulam: *Science* 168: 1159-1166, 1970[5] Foltz et al.: *Science 168*:844-845, 1970[51] Wahlqvist et al.: *Biochem Pharmacol 19*:2579- 2584, 1970[52]

Group 8—Cannabinoids (Table 9)

Knowledge of the chemistry of cannabis has advanced considerably in the past five years. The breakthrough was the elucidation of the chemical structure of the most important component, Δ^1-tetrahydrocannabinol, achieved by Gaoni, Mechoulam, and Santani in 1964.[48] Since then, numerous cannabinoids have been structurally determined.[4,5] The important cannabinoids have been synthesized, labeled radioactively, and used in metabolic experiments. Most importantly, it has been implied that hydroxylated metabolites such as 7-hydroxy-Δ^1-tetrahydrocannabinol, and 7-hydroxy-Δ^1 (6)-tetrahydrocannabinol have a psychoactive action. Gas chromatography and mass spectrometry have paved the way for the elucidation of the structure of these cannabinoids.[4,49–51] Although the cannabinoids are easily separated by gas chromatography, protein-binding makes quantitation in body fluids a problem.[52]

The mass spectra obtained by means of the gas chromatograph for Δ^1-tetrahydrocannabinol and its 7-hydroxylated metabolite are presented in Figures 8 and 9. By use of

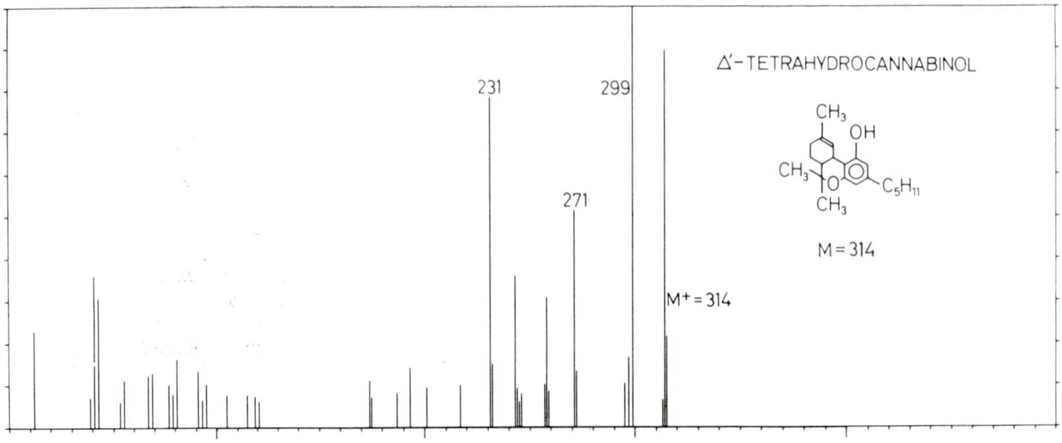

Figure 8 Δ^1-Tetrahydrocannabinol.
Instr.: LKB 9000
Inlet system: GC
column: Se 30 5% gas chrom Q
2.7 m × 2.5 mm
temp.: 200°C
Amount: 280 ng
Scan: m/e 1-500/8 sec.
Ionization energy: 70 eV
Slits: source 0.1 mm
collector 0.3 mm

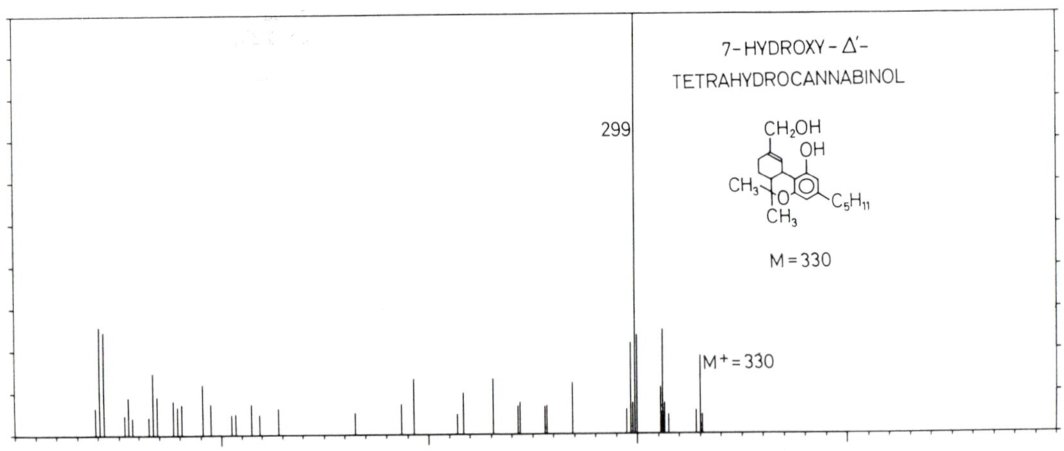

Figure 9 7-Hydroxy- Δ^1-tetrahydrocannabinol.
(Instr., inlet system, amount, scan, ionization energy, and slits—same as those for Figure 8).

mass fragmentography, Agurell and co-workers were able to quantitate as little as 8 nanograms of Δ^1-tetrahydrocannabinol.[53] The efficient use of gas phase analysis for the quantitation of the cannabinoids will have to wait until the group separation procedures in body fluids have been improved.

CONCLUSION

As in other fields of analytical chemistry, gas phase analysis has come to play an increasing role in the determination and quantification of hallucinogens. In the immediate future, methods such as gas chromatography-mass spectrometry and mass fragmentography will certainly be improved upon by various "hard- and soft-ware" developments. It is apparent that chemical ionization in mass spectrometry will play a great role in the elucidation of structures and the quantitation of hallucinogens and related compounds.

ACKNOWLEDGEMENTS

This investigation was supported by the National Institutes of Mental Health project no. MH 12007, and the National Institutes of Health, USA, project no. GM 13978; the Swedish Medical Research Council, project no. B71-40X-2629-03 and the Wallenberg Foundation. The authors want to express their thanks to E. Dufva, J.-E. Lindgren and G. Quensel for valuable help.

REFERENCES

1. Schultes, R. E.: The plant kingdom and hallucinogens. Part I, *Bull. Narcot XXI*: 3-16, 1969; Part II, *Bull Narcot XXI*: 15-27, 1969; Part III, *Bull Narcot XXII*: 25-53, 1970.
2. Kuznetsov, S. G., and Golikov, S. N.: Synthetic atropine-like substances. (USSR, *JPRS 19*: 757, 1965—OTS 63-22078.) U.S. Dept. of Commerce, Clearinghouse for Federal Scientific and Technical Information, Joint Publications Research Service.
3. Abood, L. G.: The psychotomimetic glycolate esters. In *Drugs Affecting the Central Nervous System,* Burger, A. (Ed.). New York, Marcel Dekker, Inc., 1968, pp. 127-167.
4. Agurell, S.: Chemical and pharmacological studies of cannabis. In *The Botany and Chemistry of Cannabis and Its Derivatives,* Joyce, C. R. B., and Curry, S. H. (Eds.). London, J. & A. Churchill, 1970, 217 pp.
5. Mechoulam, R.: Marihuana chemistry. *Science 168*: 1159-1166, 1970.
6. Boit, H. G.: *Ergebnisse der Alkaloid-Chemie bis 1960.* Berlin, Akademie-Verlag, 1961.
7. Bové, F. J.: The Story of Ergot. Basel, Schweiz, S. Karger, 1970.
8. Clark, W. G., and del Giudice, I.: *Principles of Psychopharmacology.* New York, Academic Press, 1970.
9. Crotchet, R., Sandison, R. A., and Walk, A.: Hallucinogenic Drugs and Their Psychotherapeutic Use. London, H. H. Lewis & Co., 1963.
10. Efron, D. H., Holmstedt, B., and Kline, N. S. (Eds.): *Ethnopharmacologic Search for Psychoactive Drugs.* Proceedings of a symposium held in San Francisco, Calif., Jan. 28-30, 1967. Public Health Service Publication No. 1645, 1967.
11. Efron, D. H. (Ed.): *Psychopharmacology. A Review of Progress, 1957-1967.* Proceedings of the Sixth Annual Meeting of the American College of Neuropsychopharmacology. San Juan, Puerto Rico, Dec. 12-13, 1967, Public Health Service Publication No. 1836, 1968.
12. Efron, D. H.: *Psychotomimetic Drugs.* New York, Raven Press, 1970.
13. Gamage, J. R., and Zerkin, E. L.: *A Comprehensive Guide to the English-Language Literature on Cannabis.* Beloit, Wisconsin, Stash Press, 1969, p. 265.
14. Hoffer, A, and Osmond, H.: *The Hallucinogens.* New York, Academic Press, 1967.
15. Usdin, E., and Efron, D. H.: *Psychotropic Drugs and Related Compounds.* U.S. Dept. of Health, Education and Welfare. Public Health Service Publication, 1967 (Suppl. 1969).
16. Usdin, E., and Efron, D. H.: *Psychotropic Drugs and Related Compounds.* U.S. Dept. of Health, Education and Welfare. Public Health Service Publication. Suppl. No. 1, Addenda and Errata, 1969, p. 148.
17. Clarke, E. G. C.: *Isolation and Identification of Drugs.* London, Pharmaceutical Press, 1969.
18. Gudzinowicz, B., and Ieller, J.: *Drugs and Pesticides.* New York, M. Dekker Inc., 1967.
19. Kern, H., Schilling, P., and Müller, S. H.: *Gas Chromatographic Analysis of Pharmaceuticals and Drugs.* Basel, Varian Aerograph, 1967.
20. Schill, G.: *Ion Pair Extraction and Its Use for the Isolation of Organic Substances.* 29th International Congress of Pharmaceutical Sciences. London, Sept. 8-12, 1969.
21. Stewart, C. P., and Stolman, A.: *Toxicology Mechanism and Analytical Methods.* New York, Academic Press, 1960.
22. Street, H. V.: Gas-liquid chromatography of submicrogram amounts of drugs. III, Analysis of alkaloids in biological media. *J Chromatogr 29*: 68-79, 1967.
23. Althaus, J. R., Biemann, K., Biller, J., Donaghue, P. F., Evans, D. A., Förster, H.-J., Hertz, H. S., Hignite, C. E., Murphy, R. C., Preti, G., and Reinhold, V.: Identification of the drug Darvon and its metabolites in the urine of a comatose patient using a gas chromatograph-mass spectrometer-computer system. *Experientia 26*: 714-717, 1970.
24. Hammar, C.-G., Holmstedt, B., Lindgren, J.-E., and Tham, R.: The combination of gas chromatography and mass spectrometry in

the identification of drugs and metabolites. *Advances Pharmacol Chemother 7*: 53-89, 1969.

25. Hites, R. A., and Biemann, K.: Computer evaluation of continuously scanned mass spectra of gas chromatographic effluents. *Anal Chem 42*: 855-860, 1970.
26. Reynolds, W. E., Bacon, V. A., Bridges, J. C., Coburn, T. C., Halpern, B., Lederberg, J., Levinthal, E. C., Steed, E., and Tucker, R. B.: A computer-operated mass spectrometer system. *Anal Chem 42*: 1122-1129, 1970.
27. Agurell, S., Holmstedt, B., Lindgren, J.-E., and Schultes, R. E.: Alkaloids in certain species of *Virola* and other South American plants of ethnopharmacologic interest. *Acta Chem Scand 23*: 903-916, 1969.
28. Ahlborg, U., Holmstedt, B., and Lindgren, J.-E.: Fate and metabolism of some hallucinogenic indolealkylamines. *Advances Pharmacol 6*: 213-229, 1969.
29. Holmstedt, B., and Lindgren, J.-E.: Chemical constituents and pharmacology of South American snuffs. *Ethnopharmacologic Search for Psychoactive Drugs,* pp. 339-373. Public Health Service Publication, No. 1645, 1967.
30. Holmstedt, B.: Gas chromatography-mass spectrometry as a tool to elucidate the structure of unknown psycho-active drugs. In *Abuse of Central Stimulants,* Sjöqvist, F., and Tottie, M. (Eds.). Stockholm, Almqvist & Wiksell, 1968, pp. 357-373.
31. Lipsky, S. R., Horvath, C. G., and McMurray, W. J.: Utilization of system employing the selective permeation of helium through a unique membrane of Teflon as an interface for gas chromatograph and mass spectrometer. *Anal Chem 38*: 1585, 1966.
32. Ryhage, R.: Use of a mass spectrometer as a detector and analyzer for effluents emerging from high temperature gas-liquid chromatography columns. *Anal Chem 36*: 759, 1964.
33. Watson, J. T., and Biemann, K.: High-resolution mass spectra of compounds emerging from a gas-chromatograph. *Anal Chem 36*: 1135, 1964.
34. Fales, H. M., Milne, G. W. A., and Veslal, M. L.: Chemical ionization mass spectrometry of complex molecules. *J Amer Chem Soc 91*: 3682, 1969.
35. Hammar, C.-G., Holmstedt, B., and Ryhage, R.: Mass fragmentography. Identification of chlorpromazine and its metabolites in human blood by a new method. *Anal Biochem 25*: 532-548, 1968.
36. Hammar, C.-G., and Hessling, R.: A new peak-matching technique: elemental analyses of compounds in submicrogram quantities without prior isolation. Short rev. at Kemistdagar i Lund, June 5-8, 1970 (in Swedish). (To be published.)
37. Haahti, E., Nikkari, T., and Kärkkäinen, J.: Design and operation of circulating chain detectors for liquid chromatography. *J Chrom Sci 4*: 12-15, 1966.
38. Hofmann, A.: In *Drugs Affecting the Central Nervous System,* Burger, A. (Ed.). New York, M. Dekker, 1968, pp. 169-235.
39. Brandenberger, H., and Hellbach, E.: Uber die gas-chromatographische Bestimmung von Amphetamin als (1-Phenyl-isopropyl)-isothiocyanat. *Helv Chim Acta 50*: 958-962, 1967.
40. Bonnichsen, R., Maehly, A. C., Mårde, Y., Ryhage, R., and Schubert, B.: Determination and identification of sympathomimetic amines in blood samples from drivers by a combination of gas chromatography and mass spectrometry. *J Legal Med* (Z Rechtsmed) *67*: 19-26, 1970.
41. Jackson, B., and Reed, A., Jr.: Another abusable amphetamine. *JAMA 211*: No. 5, s. 830/Letters to the Journal, 1970.
42. Hansson, A.: Chemical analysis of indolealkylamines and related compounds. *Handb. d. experim. Pharmakol., New Series, Vol. XIX: 5-Hydroxytryptamine and Related Indolealkylamines.* Berlin-Heidelberg-New York, Springer Verlag, 1965, pp. 6-112.
43. Axelrod, J., Brody, R. O., Witkop, B., and Evarts, E. V.: The distribution and metabolism of lysergic acid diethylamide. *Ann NY Acad Sci 66*:417-840, 1957.
44. Aghajanian, G. K., and Bing, O. H. L.: Persistence of lysergic acid diethylamide in the plasma of human subjects. *Clin Pharmacol 5*: 611-614, 1964.
45. Katz, M. A., Tadjar, G., and Aufricht, W. A.: Direct micro-analysis of LSD by gas chromatography. *J Chromatogr 31*: 545, 1967.
46. Agurell, S., Holmstedt, B. and Lindgren, J.-E.: Metabolism of 5-methoxy-N,-N dimethyltryptamine-^{14}C in the rat. *Biochem Pharmacol 18*: 2771-2781, 1969.
47. Lindgren, J.-E., Hammar, C.-G., Hessling, R., and Holmstedt, B.: The chemical identity of "Hog"—a "new" hallucinogen. *Amer J Pharm 141*: 86-90, 1969.
48. Gaoni, Y., and Mechoulam, R.: Isolation, structure and partial synthesis of an active constituent of hashish. *J Amer Chem Soc 86*: 1646, 1964.
49. Nilsson, I. M., Agurell, S., Nilsson, J. L. G., Ohlsson, A., Sandberg, F., and Wahlqvist, M.: Δ^1-Tetrahydrocannabinol: Structure of a major metabolite. *Science 168*: 1228-1229, 1970.
50. Ben-Zvi, Z., Mechoulam, S., and Burstein, J.: Identification through synthesis of an active $\Delta^{1(6)}$-tetrahydrocannabinol. *J Amer Chem Soc 92*:3468-3469, 1970.
51. Foltz, R. L., Fentiman, A. F., Jr., Leighty, E. G., Walter, J. L., Drewes, H. R., Schwartz, W. E., Page, T. F., Jr., and Truitt,

E. B., Jr.: Metabolite of (—)-trans Δ^8-tetrahydrocannabinol: identification and synthesis. *Science 168*:844-845, 1970.

52. Wahlqvist, M., Nilsson, I. M., Sandberg, F., and Agurell, S.: Binding of Δ1-tetrahydrocannabinol to human plasma proteins. *Biochem Pharmacol 19*:2579-2584, 1970.
53. Agurell, S., Nilsson, I. M., Ohlsson, A., and Sandberg, F.: On the metabolism of tritium-labelled Δ1-tetrahydrocannabinol in the rabbit. *Biochem Pharmacol 19*: 1333-1339, 1970.
54. Beckett, A. H., Tucker, G. T., and Moffat, A. C.: Routine detection and identification in urine of stimulants and other drugs some of which may be used to modify performance in sport. *J Pharm Pharmacol 19*: 273-294, 1967.
55. Brandenberger, H.: Einsatz verschiedener GC-Detektoren in der toxikologischen Analyse. " 'GC-Tagung' Klinische Chemie und Pharmazeutische Industrie" Basel, 3. Juli 1968.
56. Campbell, D. B.: A method for the measurement of therapeutic levels of (+)-amphetamine in human plasma. *J Pharm Pharmacol 21*: 129-131, 1969.
57. Gunne, L. M., and Galland, L.: Stereoselective metabolism of amphetamine. *Biochem Pharmacol 16*: 1374-1377, 1967.
58. Schubert, B.: Detection and identification of methylphenidate in human urine and blood samples. *Acta Chem Scand 24*: 433-438, 1970.
59. Änggård, E., Gunne, L.-M., Jönsson, L.-E., and Niklasson, F.: Pharmakokinetic and clinical studies on amfetamine dependent subjects. *Europ J Clin Pharmacol 3*:3, 1970.
60. Wessman, J., Moss, A. M., Horning, M. G., and Horning, E. C.: The GLC separation of indole amines and indole alcohols as heptafluorobutyryl derivatives. *Analyt Letters 2,* nr 2:81-91, 1969.
61. Bellman, S. W.: Mass spectral identification of some hallucinogenic drugs. *J Assn Offic Anal Chem 51*: 164-175, 1968.
62. Schill, G.: Kemish Läkemedelsanalys i biologiska vävnader. *Svensk Farm T 72*:975-985, 1968.

PHARMACOLOGY OF MADNESS—THE HALLUCINOGENS

EDWARD F. DOMINO, M.D.

The following quotation from de Ropp's book "Drugs and the Mind" ascribed to Théophile Gautier, a member of Le Club des Hachischins in Paris in the 1840's and colleague of Baudelaire seems appropriate:

> A certain numbness overcame me. My body seemed to dissolve and I became transparent. Within my breast I perceived the hashish I had eaten in the form of an emerald scintillating with a million points of fire. My eyelashes elongated indefinitely, unrolling themselves like threads of gold on ivory spindles which spun of their own accord with dazzling rapidity. Around me poured streams of gems of every color, in ever changing patterns like the play within a kaleidoscope. My comrades appeared to be disfigured, part men, part plants, wearing the pensive air of Ibises. So strange did they seem that I writhed with laughter in my corner and, overcome by the absurdity of the spectacle, flung my cushions in the air, making them turn and twist with the rapidity of an Indian juggler. The first attack passed and I found myself again in a normal state . . .[1]

What is happening here? What kind of pharmacology of madness is this that one can perceive such visions—turn on, then off, and remember? Positive psychologic reinforcement such as this surely will enhance drug-seeking behavior for another seemingly inexpensive yet fantastic trip! In contrast, the penalties, even if severe, seem inconsequential and bad trips apparently infrequent.

It is a curious fact that both man and animals will misuse, i.e., abuse, almost all psychoactive drugs if given the opportunity. Drugs such as the substituted phenothiazines and butyrophenones used in treating psychotic states constitute the major exceptions. Of special interest to those engaged in a biologic approach to the schizophrenias are chemicals which produce mental distortions and hallucinations. These compounds are also called psychotomimetics.

Bleuler described the primary symptoms of schizophrenia as disorders of (a) mental associations, (b) thinking and behavior, and (c) loss of affect. His system of schizophrenia has been referred to as a disturbance of the four A's: association, autism, affect, and ambivalence. The secondary symptoms are an interaction with environmental and psychogenic factors. Delusions, hallucinations, and autonomic and motor distortions result. Although other leaders of psychiatric theory have provided different classifications, the point is that most psychotomimetics really produce the secondary symptoms of schizophrenia.

It is necessary to distinguish chemicals which produce more diffuse organic brain psychopathology from those that alter thought, perception, and mood. With the possible exception of chronic amphetamine and steroid-induced psychoses, all known psychotomimetics induce toxic states that poorly mimic naturally occurring mental diseases.[2] Even so, psychotomimetics are of definite interest as possible models of the chemistry of madness. They led Gerard to speculate "that there was no crooked thought without a crooked molecule."[3] The clinical and psychiatric aspects of these compounds will be discussed in detail subsequently by Dr. Hollister.

Doctors Holmstedt and Linnarson, in the foregoing presentation, have discussed the chemistry of hallucinogens. The fact that these compounds are indolealkylamines related to serotonin (5-HT), phenethylamines related to dopamine (DA) and norepinephrine (NE), and natural or synthetic belladonna analogues related to acetylcholine (ACh) raises the question whether their pharmacology is also related to these putative neurotransmitters and/or neuromodulators. I would like to develop the well-known hypothesis that many psychotomimetics alter chemical messages in the brain by blocking them, providing false messages, reducing, and/or enhancing chemical messenger turnover.

The available literature on hallucinogens is vast. Some books on this subject include such classics as those by Moreau,[4] Lewin,[5] and Klüver.[6] More recently books by Hoffer and Osmond,[7] Hollister,[8] and the series on psychotomimetics and psychopharmacology edited by Efron and associates[9–11] have become available.

In Table 1 are listed the major hallucinogens according to chemical structure. Six important classes can be identified. One or two prototype compounds are given for each class and will be discussed separately. In addition to these, there are other compounds, such as myristicin and elemicin found in nutmeg and narcotic antagonists such as nalorphine in large dosage, which will not be discussed. In fact, the number of possible psychoactive drugs that man can discover seems limitless. The ethnopharmacology of some of these was the subject of a symposium several years ago.[9]

Table 1 Major classes of hallucinogens

Lysergic acid derivatives	*Indolealkylamines*
LSD-25	Psilocybin
	Dimethyltryptamine (DMT)
Phenethylamines	*Arylcycloalkylamines*
Mescaline	Phencyclidine
DOM (STP)	(Sernylan)
Glycolate esters	*Tetrahydrocannabinols*
Ditran	Δ^9 THC

LYSERGIC ACID AND INDOLEALKYLAMINE DERIVATIVES

Since medieval times some epidemics of St. Anthony's fire, which resulted from eating ergot-contaminated bread, were accompanied by prominent psychotic manifestations. However, it wasn't until April, 1943, that Dr. Albert Hofmann discovered the remarkable hallucinogenic properties of LSD-25, a partially synthetic amide of lysergic acid, the nucleus of many important ergot alkaloids. Very recently, Hofmann retold the fascinating story of his discovery of not only LSD-25 but also of psilocybin and psilocin (from the "sacred" Mexican mushroom, *Psilocybe mexicanna Heim*), and lysergic acid amide (from the Mexican magic drug ololiuqui).[12] Amides of lysergic acid are also found in morning glory seeds, especially *Ipomoea purpurea* and *Ipomoea violacea* (commonly called Heavenly Blue and Pearly Gates). It is of interest that one can identify an indole ring in the structure of LSD so that its derivatives, psilocybin and psilocin along with dimethyltryptamine (DMT, the business man's trip because it is short-acting), bufotenin, and harmaline, make up the major indole-containing hallucinogens related to 5-HT. Although these compounds differ considerably in the details of their pharmacology, the prototype compound LSD-25 will be discussed as the principal agent of the related lysergic acid and in-

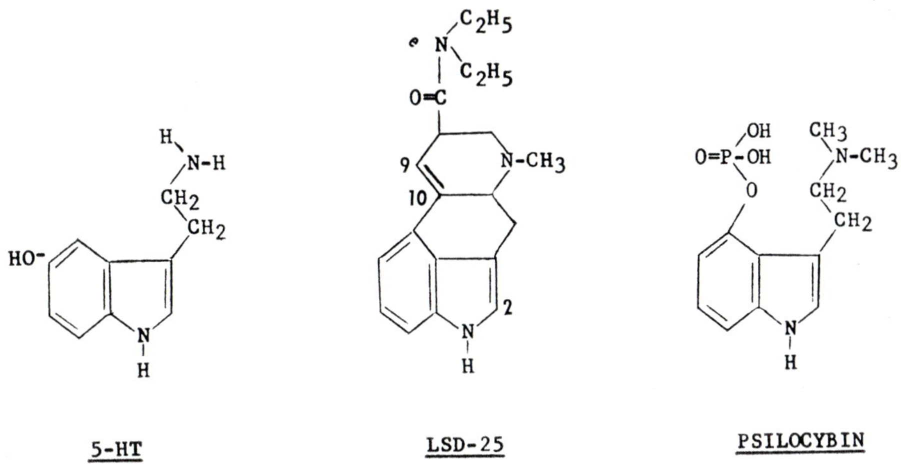

Figure 1 Structural relationship of serotonin (5-HT) to LSD-25 and Psilocybin.

dolealkylamine classes of hallucinogens. The structural relationships of these two classes to 5-HT are illustrated in Figure 1.

The 9-10 unsaturated bond in LSD-25 is important for psychic activity. Halogenation in the 2-position (2-brom LSD) markedly reduces hallucinogenic effects, but 5-HT antagonism is still present. This is an important fact which reduces the credibility of the 5-HT antagonism hypothesis of LSD action. The pharmacologic effects of LSD-25 are:

I. Central nervous system
 A. Psychic alterations including mood changes, perceptual disturbances, hallucinations, distortions of body image
 B. Sensory interactions—colors heard, sounds seen
 C. Visual after-images prolonged
 D. Initial insomnia, subsequently enhanced REM (dream) sleep greater than expected for simple deprivation
 E. Mild hyperthermia in man; marked species variation, with rabbits showing an extreme hyperthermia
 F. Increased tendon reflexes, tremor, muscle weakness
 G. Nausea
 H. Low amplitude EEG fast waves; decreased visually evoked potentials
 I. Impairment of habituation to sensory stimuli

II. Autonomic nervous system
 A. Sympathomimetic including increased blood pressure, mydriasis, tachycardia, piloerection
 B. Elicits coronary chemoreflex to decrease blood pressure
 C. Constricts smooth muscle of uterus, blood vessels, and bronchioles
 D. 5-HT agonist and antagonist

III. Sites of action
 A. Facilitates sensory input into the brain stem reticular formation
 B. Reduces transmission over some axodendritic and enhances over some axosomatic synapses
 C. Reduces frequency of discharge of 5-HT containing raphé neurons
 D. Reduces transmission at the lateral geniculate nucleus; in large doses produces functional blindness while auditory input remains intact
 E. Facilitates limbic after-discharge

IV. Mechanism of action
 A. 5-HT agonist and antagonist
 B. Increased 5-HT levels
 C. Enhanced incorporation of 5-hydroxytryptophan (5-HTP), the amino acid precursor, into 5-HT
 D. End metabolite of 5-HT, 5-HIAA is decreased, but not through MAO inhibition
 E. 5-HTP and acute reserpine increase mental effects of LSD-25. However, chronic reserpine and steroids such as progesterone decrease its effects
 F. Cholinesterase (ChE) inhibitor; relation to mental effects is unknown
 G. Tolerance and cross-tolerance to LSD-25 and hallucinogenic phenethylamines and indolealkylamines

V. Absorption and distribution
 A. Well absorbed orally
 B. Distributed throughout the body. Relatively small amounts are found in the brain, especially the hippocampus, basal ganglia, central gray, and hypothalamus

C. Plasma half-life in man about 175 minutes. Half-life in human brain unknown, but the mental effects may last 12 or more hours suggesting binding to brain substances
D. Can readily pass blood-brain, brain-blood, and blood-placental barriers

VI. Biotransformation
A. Major biotransformation occurs in the liver. The known metabolites are:
1. 2-oxy LSD
2. Hydroxylation of carboxyamide group
3. 13-hydroxy LSD
4. Glucuronides of above
5. These metabolites are not hallucinogenic

VII. Clinical uses
A. Clinical experimental use only, none widely accepted
B. Analgesic in chronic pain
C. Treatment of autistic children
D. Treatment of chronic alcoholics

VIII. Dosage
A. Orally 20 to 100 μg. Doses 10 times and larger have been taken, especially illicitly
B. Onset about two hours and duration of action about eight to twelve hours

IX. Side effects
A. Depersonalization
B. Acute panic reactions
C. Reactivation or precipitation of schizophrenia-like reactions
D. Acute paranoid reactions
E. Recurrent drug state effects in absence of drug (flashbacks). May be precipitated in former LSD users who take other psychodysleptics such as marihuana.
F. Psychotic depression
G. Suicide
H. Possible chromosomal breakage similar to stress
I. Possible birth defects if taken early in pregnancy
J. Rarely convulsions
K. Nausea and vomiting

X. Antagonism
A. Sedatives such as diazepam, etc
B. Chlorpromazine

Commentary

The how and why of LSD-induced hallucinations cannot be answered definitely because of theoretical, methodological, and manpower considerations. Investigators have provided us with a dazzling, albeit confusing, array of data. The 5-HT hypothesis for LSD-25 action was first proposed by Gaddum[13] and Woolley and Shaw.[14] Certainly 5-HT involvement in the brain is clear on the basis of biochemical[15] and electrophysiological[16] evidence. What seems certain is that LSD-25 can mimic the actions of 5-HT on some receptor sites and block its effects on others.[17] That LSD-25 acts directly on the brain is evident from iontophoretic studies of single neurons.[18] It acts qualitatively like 5-HT to produce depression when applied to lateral geniculate,[19] neocortical,[20,21] and pyriform cortex cells.[22]

Of course any chemical in large enough concentrations can affect neurons. The earlier studies of Evarts[23] and Bishop and associates[24] are therefore important in that intracarotid or intravenous LSD-25 depresses the postsynaptic "r" response in the lateral geniculate (LGN), demonstrating that LSD-25 affects the brain via a bloodborne route. While one may disagree with the precise locus of LSD-25 action at the LGN as pre- or postsynaptic, our own data with Roppolo[25] that 5-HTP, the amino acid precursor of 5-HT, has a similar action further strengthens the idea of an LSD-25 and 5-HT interaction. Evoked potential studies such as those of Marrazzi[26] on the transcallosal cortical response support the same notion. The recent work of Aghajanian and collaborators[27] that LSD-25 causes a decrease and even cessation of firing of 5-HT containing neurons in the raphé nucleus, and that 5-HTP causes a slowing but no cessation of firing of these very neurons also is in agreement. All of these findings are compatible with biochemical data that LSD-25 causes a decrease in 5-HT brain turnover.

On the other hand, 5-HTP produces EEG slow waves in the cat,[28,29] and LSD-25 produces low-voltage fast waves.[30] On this system, as well as behaviorally, these drugs have opposite effects. Bradley's data in cats that LSD-25 has alerting and excitatory effects at the level of the brain stem reticular formation are important in view of the role of the reticular formation in arousal, filtration, and integration of sensory information. Very recently, Smythies and associates[31] suggested an interaction of LSD-25 with membrane bound prostaglandin-RNA complexes causing disruption by a charge transfer process of the hydrogen bonds joining them. This would lead to membrane channel opening. 5-HT would have a similar receptor mechanism. Smythies and Antun[32] have suggested that this accounts for LSD-25 acting first as an agonist and in larger doses as an antagonist of 5-HT.

What is clear to date is that the 5-HT hypothesis of LSD action proposed 17 years ago, modified slightly, is still the best we have. However, we have a long way to go before we know why LSD-25 causes hallucinations. Furthermore, putative neurotransmitters including ACh, histamine, and probably others will have to be considered.

PHENETHYLAMINES

The eating of peyote or mescal buttons by Indians of the southwestern United States and Mexico has been the mainstay of elaborate religious ceremonies. Details of its mythology and history make fascinating reading. Klüver's book is an excellent beginning.[6] Many alkaloids are present but mescaline is said to be the active ingredient in peyote "buttons" obtained from the cactus *Lophophora williamsii*. It is a methoxylated phenethylamine derivative chemically related to amphetamine. N-Methylmescaline is also present in peyote but so little is known about its psychoactivity that it is a fruitful area for study. Mescaline and DOM are the two principal compounds whose structures are compared in Figure 2 to dopamine and amphetamine.

Many hallucinogenic phenethylamines are known. Snyder and Richelson have stressed the relationship of all of these to LSD and the other indolealkylamines.[33] The pharmacologic actions of mescaline and related compounds are very similar to LSD. Briefly, mescaline has the following actions:

I. Central nervous system
   A. Psychic alterations including increased arousal, anxiety, visual hallucinations of brightly colored geometric designs, altered color and space perception
   B. Increased tendon reflexes
II. Autonomic nervous system
   Sympathomimetic actions including mydriasis, tachycardia, etc.
III. Sites of action
   A. Stimulates brain stem activating system
   B. Additional actions not well studied
IV. Mechanism of action
   Not known. Assumed to be related to cate-

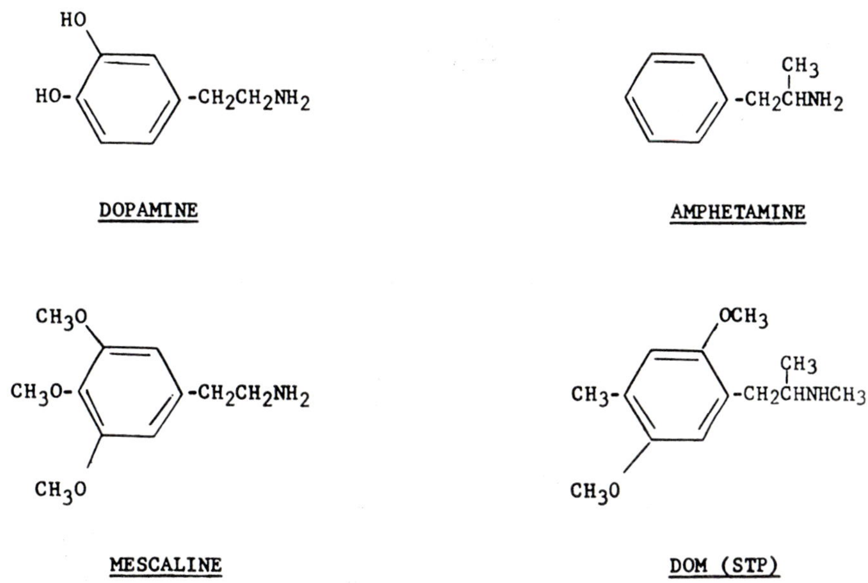

Figure 2 Structural relationship of dopamine to amphetamine, mescaline, and DOM.

cholamines such as DA or to indolealkylamines such as 5-HT. Cross-tolerance with LSD-25

V. Absorption and distribution
   A. Orally absorbed
   B. Localizes in nerve endings
   C. Duration of action—12 to 24 hours

VI. Biotransformation
   Oxidation by MAO and demethylation

VII. Clinical uses
   Experimental only

VIII. Dosage
   Orally 5 mg/kg of body weight

IX. Side effects
   A. Panic states
   B. Psychotic reactions

X. Antagonism
   A. Sedatives such as diazepam
   B. Chlorpromazine

Commentary

In spite of the fact that mescaline is one of our older hallucinogens, much less is known about its pharmacologic actions than about those of LSD-25. In fact, the evidence that mescaline is the only active hallucinogen in peyote is weak indeed. Paul has pointed out that N-methylmescaline as well as many other intriguing alkaloids are present in peyote.[34] Much more research needs to be done! The synthetic approach to the hallucinogenic phenethylamines and their relationship to other structural types has been reviewed by Shulgin,[35] Smythies and Shulgin,[36] and Snyder and Richelson.[33] A superb discussion of the quantum theory aspects of the relationship of these hallucinogens to LSD-25 can be found in the symposium chaired by Smythies.[17]

It is obvious that several synthetic phenethylamines are far more potent than mescaline. Derivatives whose action cannot be terminated by MAO are potent and long-lasting, for example DOM. At the same time, their pharmacology is altered. For example, DOM acts like amphetamine in low doses. In moderate doses it is a hallucinogen, and in large doses a convulsant. This is the reason that chlorpromazine is contraindicated as a therapeutic antagonist for DOM. The mechanism of DOM's actions is completely unknown.

Hallucinogens such as mescaline and DOM have some psychic effects in common with LSD-25, but it is also quite clear that there are considerable differences in somatic side effects, potency, and duration of action. There is considerable cross-tolerance between lysergic acid and indolealkylamine derivatives (such as LSD-25 and psilocybin) and mescaline, but not between LSD-25 and amphetamine. This would suggest that these hallucinogens are not acting like amphetamine. It is a fact that chronic amphetamine administration-induced psychosis is the best model of paranoid schizophrenia. However, current thoughts on amphetamine psychosis include the possibility of a phenolmethyltransferase forming the hallucinogenic metabolite paramethoxyamphetamine ($6 \times$ potency of mescaline), or saturating vesicular stores of catecholamines with the β-hydroxylated analogue of the normal amphetamine metabolite parahydroxyamphetamine. The latter compound has been isolated, but the former has not. Thus, in spite of much of recent information,[37] we cannot explain amphetamine psychosis, let alone that induced by mescaline or DOM.

The distribution of mescaline in the subcellular fractions of brain is of special interest. The nerve ending fraction especially takes up this drug, a finding compatible with a synaptic mechanism of action.[38]

The sad fact is that we are not sure how the phenethylamine hallucinogens really act. Our present knowledge suggests an interaction with chemical neurotransmitters, especially the biogenic amines. Mescaline acts in some ways like LSD-25 to reduce raphé neuron firing suggesting a similar mechanism of action.[27] Antagonists such as chlorpromazine further support this idea, but we have a long way to go. Some derivatives such as DOM have a different pharmacology. That these may be a "pot of gold" from a therapeutic point of view is indicated by the fact that another close analogue of DOM, DOET, has a euphorogenic effect without producing hallucinations.[39] Only research

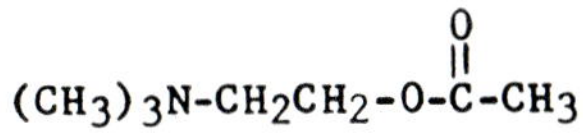

ACETYLCHOLINE

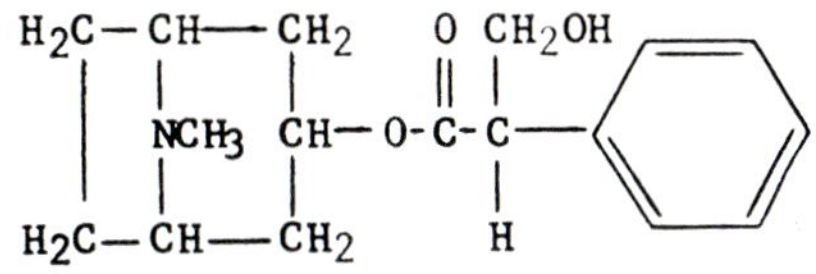

ATROPINE

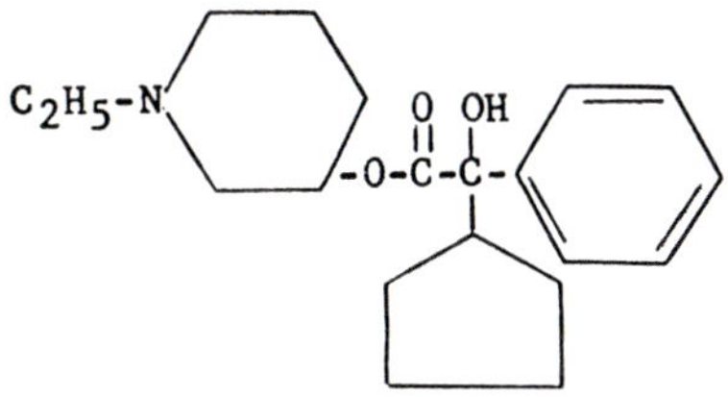

PIPERIDYL ISOMER (DITRAN)

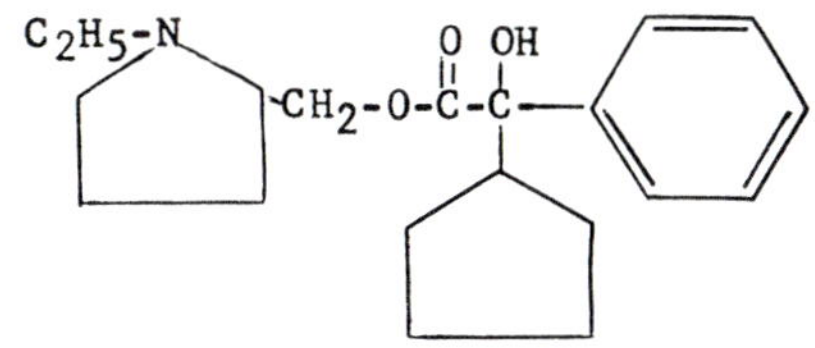

PYRROLIDYLMETHYL ISOMER (DITRAN)

Figure 3 Structural relationship of acetylcholine to atropine and Ditran.

effort and time will tell if this has therapeutic merit.

GLYCOLATE ESTERS

It is well-known since ancient times that belladonna plants produce marked mental aberrations or psychodysleptic effects, which resemble an organic brain syndrome. Drugs in this class include atropine, scopolamine, and related compounds from *Belladonna* and *Datura stramonium* (Jimson weed). About 15 years ago, several synthetic atropine derivatives known as glycolate esters were tested as possible therapeutic agents in treating patients with peptic ulcer. Surprisingly, one of these (JB-329) was shown in the early clinical trials to be a potent hallucinogen. This agent was called Ditran. It exists in two isomeric forms, about 30% the piperidyl and 70% the pyrrolidylmethyl. The structural relationship of these compounds to acetylcholine and atropine is shown in Figure 3. Abood has reviewed their chemical and pharmacologic relationships.[40,41] Their pharmacologic effects include:

I. Central nervous system
- A. Delirium similar to atropine overdosage
- B. Auditory hallucinations
- C. Marked memory loss in comparison to LSD-25
- D. Clouding of consciousness
- E. Hyperthermia
- F. Incoherence
- G. Disorientation

II. Autonomic nervous system
- A. Mydriasis and cycloplegia
- B. Tachycardia
- C. Dry mouth
- D. Antagonizes most muscarinic cholinergic actions of ACh

III. Sites of action
- A. Brain stem activating system and neocortex producing EEG slow waves
- B. Limbic system, preventing *theta* activity
- C. Peripheral muscarinic cholinergic receptor blockade

IV. Mechanism of action
- A. Central and peripheral muscarinic cholinergic blockade
- B. Membrane actions involving calcium

V. Absorption and distribution
   Well absorbed orally

VI. Biotransformation
   Unknown

VII. Clinical uses
- A. None
- B. Has been used experimentally in treatment of depression

VIII. Dosage

A. Oral 0.1 to 0.4 mg/kg of body weight
B. Duration of action related to dose, usually two to eight hours. Effects may last several days with large dosage

IX. Side effects
A. Toxic delirium
B. Hyperthermia
C. Amnesia

X. Antagonism
A compound with anticholinesterase properties, tetrahydroaminoacridine (THA) is a very effective antagonist. Surprisingly, yohimbine also antagonizes the effects of low doses of Ditran, while other cholinesterase inhibitors that penetrate well the blood-brain barrier like physostigmine and diisopropylfluorophosphate (DFP) appear to be less effective antagonists than THA. Thus the anticholinergic effects of Ditran probably are not its only mechanism of action. However, its major pharmacology is that of a centrally acting muscarinic cholinergic antagonist.

Commentary

The ditty "mad as a hen, red as a beet, dry as a bone, and hot as a cock" well-describes the "trip" with atropine overdosage or that induced by the synthetic glycolate esters. These agents tend to produce "organic brain syndrome" more so than other hallucinogens previously discussed. Auditory hallucinations are far more prominent with the glycolate esters while visual hallucinations predominate after LSD-25. Schizophrenic patients tend to have more auditory than visual hallucinations; hence, in part, the interest in glycolate esters. Because of the memory loss and delirium which these compounds induce, one can predict that they probably will never be widely abused for they seem clearly less behaviorally reinforcing. Interest in the particular "model psychosis" produced centers on the question of the site and mechanism of action of these agents. Is the action related to the cholinergic system or not? Abood stresses that these drugs bind to synaptosomes and other membranous components.[40,41] He has shown that they substitute and interfere with the regulatory role of calcium in excitable membranes. ATP interactions in membrane and calcium-phospholipid complexes with a high molecular structural protein also seem to be involved with the glycolate esters.

Yet from a behavioral and electrophysiologic point of view, these compounds can be classified as anticholinergics along with atropine and scopolamine.[42] Treatment of patients who ingest these drugs is the same as that for atropine overdosage. Tetrahydroaminoacridine seems to be a very effective antagonist. Yet why isn't physostigmine equally good? And why should yohimbine, a sympathomimetic anxiety-provoking compound, be even partially effective? It is obvious we have a lot to learn but surely one begins by examining the interaction of these drugs with central cholinergic mechanisms and ruling their use in treatment either in or out.

ARYLCYCLOALKYLAMINES

The administration of new psychoactive compounds to man, even thoroughly tested in animals, causes many surprises. Recall the discovery of Ditran. The story of phencyclidine (Sernyl or Sernylan) is similar. The drug has mixed excitant, sedative, cataleptoid-anesthetic properties in animals. It was first discovered by Chen and coworkers.[43] The class as a whole have been called cataleptoid-anesthetics by Mc Carthy who has reviewed their history and pharmacology.[44] Phencyclidine is a remarkable and entirely different kind of anesthetic when used in man than when tested in animals. In some patients an emergence delirium became very prominent and sometimes lasted for days and rarely weeks. For this reason, the drug is used as a general anesthetic only in animals. Occasionally it is taken by curiosity seekers as a psychodysleptic. It is, therefore, highly questionable if it should be classified as a hallucinogen. It is a curious fact that phencyclidine has a structure which can be related to Snyder's "A" and "C" ring analogues of LSD and amphetamine as illustrated below.[33]

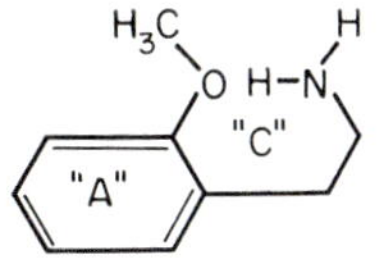

2-Methoxyamphetamine

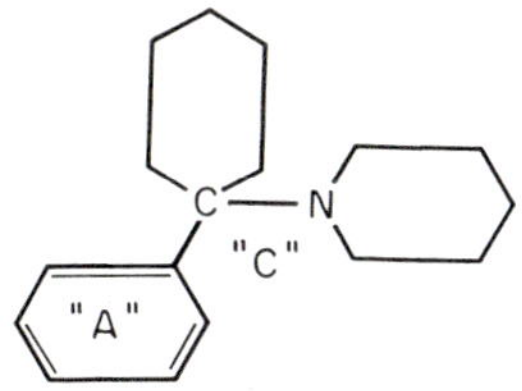

Phencyclidine

Its known pharmacology can be summarized as follows:

I. Central nervous system
   A. Small doses produce a "drunken" state with numbness of the extremities; some species are excited
   B. Moderate doses are analgesic and anesthetic
   C. Psychic state resembles crudely sensory isolation except that sensory impulses if tested electrophysiologically reach the neocortex but the neuronal signals are grossly distorted
   D. Cataleptoid motor phenomena
   E. Large doses produce convulsions
   F. Marked species differences; primates, especially man, are most susceptible

II. Autonomic nervous system
   A. Sympathomimetic
   B. Tachycardia
   C. Hypertension
   D. Potentiation of catecholamines

III. Sites of action
   A. Thalamo-neocortical system
   B. Limbic system

IV. Mechanism of action
   Unknown

V. Absorption and distribution
   All routes are effective. Rapid onset of action. Duration of action is dose-dependent; usually lasts from minutes to hours

VI. Biotransformation
   Several known metabolites, primarily hydroxylated derivatives; they are more convulsant than phencyclidine

VII. Dosage
   A. Oral—10 mg gives a slight drunken feeling
   B. I.V.—O.2 mg/kg of body weight is anesthetic

VIII. Side effects
   A. Hypertension
   B. Convulsions
   C. Emergence delirium

IX. Antagonism
   No confirmed antagonists. Succinic acid has been reported to antagonize mental effects but no antagonism demonstrable in animals.

Commentary

Perhaps more than any other class, these agents attest to the arbitrary scheme of classification of psychoactive drugs. Are these compounds, in particular phencyclidine, merely unique general anesthetics or are they also hallucinogens? Nagle has pointed out that the subjective effects of low concentrations of nitrous oxide and diethyl ether are in some ways very similar to those of hallucinogens.[45] Emergence delirium is well known and accepted as a definite anesthetic side effect. Yet when the emergence delirium persists for hours or days, we tend to classify the drug differently. No one can deny the uniqueness of the pharmacology of phencyclidine. To observe a patient under its influence is to observe a remarkable phenomenon. The group at the Lafayette Clinic, especially Luby, have contributed much to our descriptive knowledge of phencyclidine. They have pointed out that this drug in low dosage mimics the primary symptoms of schizophrenia and that schizophrenic patients are very susceptible to its effects. Other psychiatrists are less impressed with its psychotomimetic effects, but such effects cannot be denied for the drug was withdrawn from human use because of them. Phencyclidine pharmacology is just at its beginning (see review Domino[46]). We don't as yet know how to treat cases of intoxication other than through a symptomatic approach.

TETRAHYDROCANNABINOLS

There is nothing new about the psychic effects of cannabis preparations. The flowering tops of hemp have been used since 2700 B.C. In the United States cannabis is smoked usually in the form of marihuana or "grass," which should be looked upon as a dilute form of some very potent chemicals—the tetrahydrocannabinols (THC). Hashish is the resin of cannabis and contains a much higher content of THC. The resin of *Cannabis sativa* is a cesspool of organic com-

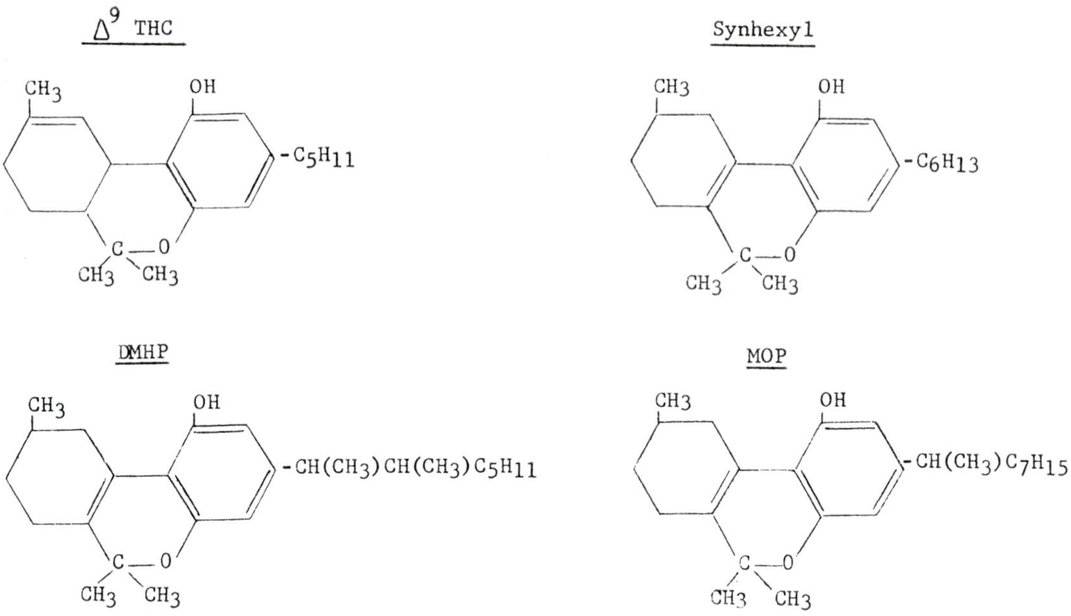

Figure 4 Structural relationships of various natural and synthetic tetrahydrocannabinols.

pounds. The most active components are the tetrahydrocannabinols. Recently, Isbell and his associates have obtained evidence that Δ9 THC is the primary active ingredient of cannabis.[47,48] There are several synthetic derivatives of which DMHP is the most potent. The chemistry is complicated by the fact that pyrolysis increases the psychic effect, indicating that many other unknown compounds may be involved in the over-all pharmacology of marihuana when it is smoked. At the present time there are several synthetic analogues of the naturally-occurring compounds as shown in Figure 4. The pharmacologic actions of these compounds are complex. They can be summarized as follows:

I. Central and peripheral nervous system
   A. Dose dependent! The psychic effects were nicely described by the French psychiatrist Moreau in 1845. This description is still valid today although often forgotten or not known
   B. General feeling of pleasure
   C. Increased excitement
   D. Distortion of time and space; time is greatly slowed
   E. Keener enjoyment of music and even noise
   F. Persistent ideas on the verge of persecution mania
   G. Altered emotion, increase of existing feelings
   H. Irresistible impulses
   I. Illusions and hallucinations with larger doses
   J. Physiologic effects include dryness of mouth, tachycardia, hypertension, slightly enhanced tendon reflexes (in contrast to LSD-25 in which tendon reflexes are greatly exaggerated), no marked pupillary changes, tremors, sedation, drowsiness, and stupor

II. Sites of action
   Mild depression of brain stem activating system causing EEG slowing and decreased arousal. Reduction of polysynaptic spinal cord reflexes

III. Mechanism of action
   Unknown. Increases slightly brain 5-HT and ACh in large soporific doses. Marked tolerance development to large doses of Δ^9 THC. Cross-tolerance to DMHP, but not LSD-25 and mescaline

IV. Absorption and distribution
   Very little known except that it is well absorbed by lungs and orally. Progress in clinical evaluation limited because of lack of suitable parenteral preparations in soluble form

V. Biotransformation
Unknown for the most part. The hydroxylated analogue of Δ^9 THC is pharmacologically active

VI. Clinical uses
None in Western medicine. Still used in India by "fakirs" and medicine men. Medical uses should be reinvestigated using Western standards

VII. Dosage
Varies widely depending on THC content. The amount of each specific THC derivative will obviously vary so that a mixture is present in the resin or "red oil"
   A. In cigarettes—50 μg/kg of Δ^9 THC = mild social "high"
   250 μg/kg of Δ^9 THC = psychotomimetic effect
   B. Orally—Δ^9 THC—5 to 30 mg

VIII. Side effects
   A. Legal penalties for non-medical use extreme
   B. Chronic use may result in an amotivational syndrome. Needs better documentation
   C. Likelihood of resins and tars causing lung disease highly probable in chronic use

IX. Marihuana: *fact vs fiction*
Any discussion of "pot," even on a purely medical and scientific basis, results in much emotion on the part of the discussant as well as the user. Note that:
   A. Pharmacologically it is not a narcotic
   B. There is usually no tendency to increase dose (except with hashish users)
   C. No physical addiction or abstinence syndrome
   D. No proven cause-effect correlation with narcotic abuse
   E. Usually, when smoked as a marihuana "joint," does not produce intense psychologic effects. Causes minimal neurologic deficit except in highest mental functions resembling a mild organic brain syndrome. Frequently the experienced user cannot tell the difference between a real "joint" and a suitable placebo of chemically extracted "grass"
   F. "Chronic dementia" with prolonged use not well substantiated. Needs much more study. However, the chronic user ("pothead") in the Middle East poses a significant social problem. These compounds would probably constitute another drug intoxication group in addition to chronic alcoholism if legalization of marihuana were to occur
   G. User more often passive than aggressive during the drug state
   H. Direct association with violent crime unsupported
   I. Contrary to widespread opinion, tolerance occurs with large amounts of active Δ^9 THC. Tolerance is not associated with physical dependence. Development of tolerance to THC derivatives is not followed by cross-tolerance to LSD-25, although an LSD "flashback" can be precipitated by marihuana smoking through an unknown psychologic mechanism

X. Antagonism
Sensory arousal and d-amphetamine may partially antagonize EEG effects, but no good antagonists are known

Commentary

The descriptive effects of cannabis have been known for centuries. Yet it is only recently that we are gaining a glimpse of its active principles. Cannabis, like any other plant, is a cesspool of chemicals, especially when smoked. Is Δ^9 THC really the primary active compound? How active are its metabolites? Does pyrolysis increase psychic activity of the pure compounds as it does the resinous extract? Does chronic use of cannabis preparations cause brain deterioration? What are other dangers? Are some of these compounds of therapeutic benefit? One could go on and on with questions! The answers lie in research which is now a fashionable area because the marihuana problem has reached epidemic proportions. One of the strange and interesting aspects of human marihuana research is that experienced marihuana users can distinguish only very poorly "active" from "inactive" marihuana cigarettes in double-blind experiments.[49] Another myth which is now being exploded is that marihuana produces no tolerance, only sensitization with repeated use. There are now many excellent studies showing profound tolerance development in animals to various THC derivatives.[50,51] The relationship of THC effects to the putative brain transmitters is still quite unclear. We don't know much about marihuana pharmacology in spite of its being around for 4,000 years. Sad isn't it, but the beginning of progress is to realize that we need it.

SUMMARY

Any review of the pharmacology of the hallucinogens is laden with difficulties. First, the available literature is vast. Secondly, it is a human failing to look at the problem on the basis of one's own very limited research and experience; and thirdly, we are just beginning to elucidate the neural and psychologic mechanisms of drug action simply because our knowledge of brain function is really quite poor. And yet there is something that must be said about the pharmacology of hallucinogens that bears on their widespread non-medical use. From both a psychiatric as well as a pharmacologic point of view, there are many chemicals that alter brain function. It is a fact that almost all drugs which affect the mind are abused. Of special interest to those engaged in a biologic approach to serious mental diseases, such as the schizophrenias, are the psychotomimetic drugs.

Although these agents mimic primarily the secondary symptoms of schizophrenia, they are still of great biologic interest. In addition, because of their abuse potential we need to know more about their sites and mechanisms of action so that we can find suitable antagonists. It seems that many of these drugs act by altering chemical messages in the brain by blocking them, providing, false messages, reducing and/or enhancing chemical messenger turnover. *From a neuropharmacologic point of view these drugs are mind distorters, not mind expanders!* We are just beginning to learn about their actions and in many cases no specific antagonists have yet been developed.

REFERENCES

1. de Ropp, R. S.: *Drugs and the Mind*. New York, St. Martin's Press, 1957, p. 310.
2. Lipton, M. A.: The relevance of chemically induced psychoses to schizophrenia. In *Psychotomimetic Drugs,* Efron, D. H. (Ed.). (Workshop Series of Pharmacology Section, NIMH No. 4.) New York, Raven Press, 1970, pp. 231–240.
3. Gerard, R. W.: Personal communication, 1970.
4. Moreau, S.: *Du Hachish et de L'Alientation Mentalle Etudes Psychologiques*. Paris, Fortin, Masson, 1845.
5. Lewin, L.: *Phantastica—Narcotic and Stimulating Drugs—Their Use and Abuse*. 1931, pp. 335. Routledge and Kegan Paul, London. Reprinted E. P. Dutton and Co., Inc., New York, 1964.
6. Klüver, H.: *Mescal and Mechanisms of Hallucinations*. Chicago, University of Chicago Press, 1966, p. 108.
7. Hoffer, A., and Osmond, H.: *The Hallucinogens*. New York, Academic Press, 1967, p. 626.
8. Hollister, L.: *Chemical Psychoses—LSD and Related Drugs*. Springfield, Ill., Charles C Thomas, 1968, p. 190.
9. Efron, D. H., Holmstedt, B., and Kline, N. S. (Eds.): *Ethnopharmacologic Search for Psychoactive Drugs*. Workshop Series of Pharmacology Section, National Institute of Mental Health. Public Health Service Publication No. 1645. Superintendent of Documents, Washington D. C., U. S. Government Printing Office, 1967, p. 468.
10. Efron, D. H., Cole, J. O., Levine, J., and Wittenborn, J. R. (Eds.): *Psychopharmacology—A Review of Progress, 1957–1967*. Public Health Service Publication No. 1836. Superintendent of Documents, Washington, D. C., U. S. Government Printing Office, 1968, p. 1342.
11. Efron, D. H. (Ed.): *Psychotomimetic Drugs*. Workshop Series of Pharmacology Section, NIMH. New York, Raven Press, 1970, p. 365.
12. Hofmann, A.: The discovery of LSD and subsequent investigations on naturally occurring hallucinogens. In *Discoveries in Biological Psychiatry,* Ayd, F. J., and Blackwell, B. (Eds.). Philadelphia, J. B. Lippincott Co., 1970, Chapter 7, pp. 91–106.
13. Gaddum, J. H.: Antagonism between lysergic acid diethylamide and 5-hydroxytryptamine. *J. Physiol 121*:15P, 1953.
14. Woolley, D. W., and Shaw, E. N.: Evidence for the participation of serotonin in mental processes. *Ann NY Acad Sci 66*:649–667, 1954.
15. Aghajanian, G. K., and Freedman, D. X.: Biochemical and morphological aspects of LSD pharmacology. In *Psychopharmacology—A Review of Progress 1957–1967,* Efron *et al.* (Eds.). Public Health Service Publication No. 1836, Washington, D.C., U. S. Government Printing Office, 1968, pp. 1185–1193.
16. Koella, W. P.: Invited discussion of Dr. J. Appel's paper: Neurophysiological effects of psychotomimetic substances (a supplemental review). In *Psychopharmacology—A Review of Progress 1957–1967,* Efron *et al.* (Eds.). Public Health Service Publication No. 1836, Washington, D.C., U. S. Government Printing Office, 1968, pp. 1223–1229.

17. Smythies, J. R. (Ed.): The mode of action of psychotomimetic drugs. *Neuroscience Res Prog Bull 8*:1–152, 1970.
18. Phillis, J. W.: *The Pharmacology of Synapses.* Int. Ser. Monographs in Pure and Applied Biology, Zoology Division. Vol 43. Oxford, Pergamon Press, 1970, p. 358.
19. Curtis, D. R., and Davis, R.: Pharmacological studies upon neurons of the lateral geniculate nucleus of the cat. *Brit J Pharmacol 18*:217–246, 1962.
20. Krnjevic, K., and Phillis, J. W.: Actions of certain amines on cerebral cortical neurons. *Brit J Pharmacol 20*:471–490, 1963a.
21. Krnjevic, K., and Phillis, J. W.: Iontophoretic studies of neurons in the mammalian cerebral cortex. *J Physiol 165*:274–304, 1963b.
22. Legge, K. F., Randic, M., and Straughan, D. W.: The pharmacology of neurons in the pyriform cortex. *Brit J Pharmacol 26*:87–107, 1966.
23. Evarts, E. V.: A review of the neurophysiological effects of lysergic acid diethylamide (LSD) and other psychotomimetic agents. *Ann NY Acad Sci 66*:479–495, 1957.
24. Bishop, P. O., Burke, W., and Hayhow, W. R.: Lysergic acid diethylamide block of lateral geniculate synapses and relief by repetitive stimulation. *Exp Neurol 1*:556–568, 1959.
25. Roppolo, J. R.: *Pharmacological Analysis of the Cat Lateral Geniculate.* PhD. Thesis, University of Michigan, 1970.
26. Marrazzi, A. S.: The effect of drugs on neurons and synapses. In *Brain Mechanisms and Drug Action,* Fields, W. S. (Ed.). Springfield, Ill., Charles C Thomas, 1957.
27. Aghajanian, G. K., Sheard, M. H., and Foote, W. E.: LSD and mescaline: Comparison of effects on single units in the midbrain raphé. In *Psychotomimetic Drugs,* Efron, D. H. (Ed.). (Workshop Series of Pharmacology Section, NIMH.) New York, Raven Press, 1970, pp. 165–176.
28. Jouvet, M.: Mechanisms of the states of sleep: a neuropharmacological approach. *Res Publ Ass Nerv Ment Dis 45*:86–126, 1967.
29. Koella W. P.: *Sleep—Its Nature and Physiological Organization.* Springfield, Ill., Charles C Thomas, 1967, p. 199.
30. Bradley, P. B., and Elkes, J.: The effects of some drugs on the electrical activity of the brain. *Brain 80*:77–117, 1957.
31. Smythies, J. R., Bennington, F., and Morin, R. D.: XVII. Specification of a possible serotonin receptor site in the brain and XVIII. A prostaglandin-RNA complex as a potential receptor site. *Neurosci Res Prog Bull 8*:117–130, 1970.
32. Smythies, J. R., and Antun, F.: The biochemistry of psychosis. *Scot Med J 15*: 34–40, 1970.
33. Snyder, S. H., and Richelson, E.: Relationship between the conformation of psychedelic drugs and their psychotropic potency. In *Psychopharmacology—A Review of Progress 1957–1967,* Efron *et al.* (Eds.). Public Health Service Publication No. 1836. Superintendent of Documents, Washington, D.C., U. S. Government Printing Office, 1968. Pp. 1199–1210.
34. Paul, A.: Personal communication, 1970.
35. Shulgin, A. T.: Chemistry and structure—activity relationships of the psychotomimetics. In *Psychotomimetic Drugs,* Efron, D. H. (Ed.). (Workshop Series of Pharmacology Section, NIMH.) New York, Raven Press, 1970, pp. 21–38.
36. Smythies, J. R., and Shulgin, A. T.: Structure-activity relationships. In *The Mode of Action of Psychotomimetic Drugs,* Smythies, J. R. (Ed.). *Neuroscience Res Prog Bull 8*:82–85, 1970.
37. Costa, E., and Garattini, S.: *International Symposium on Amphetamines and Related Compounds.* New York, Raven Press, 1970, p. 962.
38. Denber, H. C. B., and Teller, D. N.: Studies on mescaline XIX: A new theory concerning the nature of schizophrenia. *Psychosomatics 9*:145–151, 1968.
39. Snyder, S. H., Weingartner, H., and Faillace, L. A.: DOET (2,5-dimethoxy-4-ethylamphetamine) and DOM (STP) (2,5-dimethoxy-4-methylamphetamine), new psychotropic agents: Their effects in man. In *Psychotomimetic Drugs,* Efron, D. H (Ed.). (Workshop Series of Pharmacology Section, NIMH.) New York, Raven Press, 1970, p. 247–262.
40. Abood, L. G.: The psychotomimetic glycolate esters and related drugs. In *Psychopharmacology—A Review of Progress 1957–1967,* Efron *et al.* (Eds.). Public Health Service Publication No. 1836. Superintendent of Documents, Washington, D.C., U.S. Government Printing Office, 1968, pp. 683–699.
41. Abood, L. G.: Stereochemical and membrane studies with the psychotomimetic D.H. glycolate esters. In *Psychotomimetic Drugs,* Efron, D. H. (Ed.). (Workshop Series of Pharmacology Section, NIMH.) New York, Raven Press, 1970, pp. 67–74.
42. Fink, M., and Itil, T. M.: Neurophysiology of phantastica: EEG and behavioral relations in man. In *Psychopharmacology—A Review of Progress 1957–1967,* Efron *et al.* (Eds.). Public Health Service Publication No. 1836. Superintendent of Documents, Washington, D.C., U. S. Government Printing Office, 1968, pp. 1231–1239.
43. Chen, G., Ensor, C. R., Russell, D., and Bohner, B.: The pharmacology of 1-(1-

phenylecyclohexyl) piperidine HCL. *J. Pharmacol Exp Ther 127*:241–250, 1959.
44. Mc Carthy, D.: The history and development of cataleptoid-anesthetics of the phencyclidine type. (To be published.)
45. Nagle, D. R.: Anesthetic addiction and drunkenness: a contemporary and historical survey. *Int J Addict 3*:25–40, 1968.
46. Domino, E. F.: Neurobiology of phencyclidine (Sernyl), a drug with an unusual spectrum of pharmacological activity. *Int Rev Biol 6*:303–347, 1964.
47. Isbell, H., Gorodetzsky, C. W., Jasinski, D., Claussen, U., Spulak, F. V., and Korte, F.: Effects of (—)-Δ^9-trans-tetrahydrocannabinol in man. *Psychopharmacologia (Berlin) 11*: 184–188, 1967.
48. Isbell, H.: A comparison of LSD-25 with (—)-Δ^9-trans-tetrahydrocannabinol (THC) and attempted cross tolerance between LSD and THC. *Psychopharmacologia (Berlin) 14*:115–123, 1969.
49. Jones, R.: Personal communication, 1970.
50. Black, M. B., Woods, J. H., and Domino, E. F.: Some effects of (—)-Δ^9-trans-tetrahydrocannabinol and other cannabis derivatives on schedule-controlled behavior. *Pharmacologist 12*:258 (No. 2), 1970.
51. Mc Millan, D. E., Harris, L. S., Turk, R. F., and Kennedy, J. S.: Development of marked behavioral tolerance to *1* Δ^8-tetrahydrocannabinol (Δ^8-THC) in the pigeon. *Pharmacologist 12*:258 (No. 12), 1970.

CLINICAL PHARMACOLOGY OF HALLUCINOGENS AND MARIHUANA

LEO E. HOLLISTER, M.D.

A dozen years ago, hallucinogens were of interest only to psychopharmacologists, and marihuana, at least in the western world, was a material smoked by an eccentric minority. Now, "LSD," "mescaline," and "STP" are household words and "pot" or "grass" is the preferred intoxicant of millions of youth. Although these drugs were in the vanguard of the current drug movement, others, such as amphetamines, barbiturates, and heroin, have become an increasing problem. Fortunately, these latter drugs are not so widely used as those under discussion. Although the multiple aspects of these drugs cannot be covered fully within a short presentation, they have been discussed in considerable detail elsewhere.[1,2] In this presentation, some recent developments in the clinical pharmacology of hallucinogens and marihuana will be discussed.

HALLUCINOGENS

General Considerations. Drugs termed "hallucinogens" do not often produce hallucinations. Other terms have been proposed, such as "psychedelic" or "psychotomimetic," but these are equally inappropriate. Drugs of this type are taken for many reasons. According to those most widely given, the hallucinogenic drugs provide new ways of looking at the world and new insights, usually into personal problems. The former action implies varying degrees of perceptual distortion; the latter, changes in mood and increased introspection. At times, it seems as though drug users will use any drug which alters thinking or mood, including deliriants such as the synthetic central anticholinergic drug, Ditran, or organic solvents taken by inhalation. The prototypic hallucinogenic drug is lysergic acid diethylamide (LSD), because of the extent of its use, because it represents a family of drugs which are similar, and because it has been most carefully studied.

Chemical and Pharmacologic Considerations. LSD is based on the lysergic acid nucleus, within which one can discern resemblances both to the phenethylamine structure of mescaline and the indolethylamine structure of psilocybin (Figs 1 through 3). It also is readily apparent that both structures occur in brain biogenic amines, such as norepinephrine and dopamine in the case of phenethylamines, and 5-hydroxytryptamine (serotonin) in the case of the indolethylamines. Whether these chemical resemblances account for the similarity of action of the three types of drugs, or for a mechanism of action mediated by effects on brain biogenic amines is still not certain. Chemical specific-

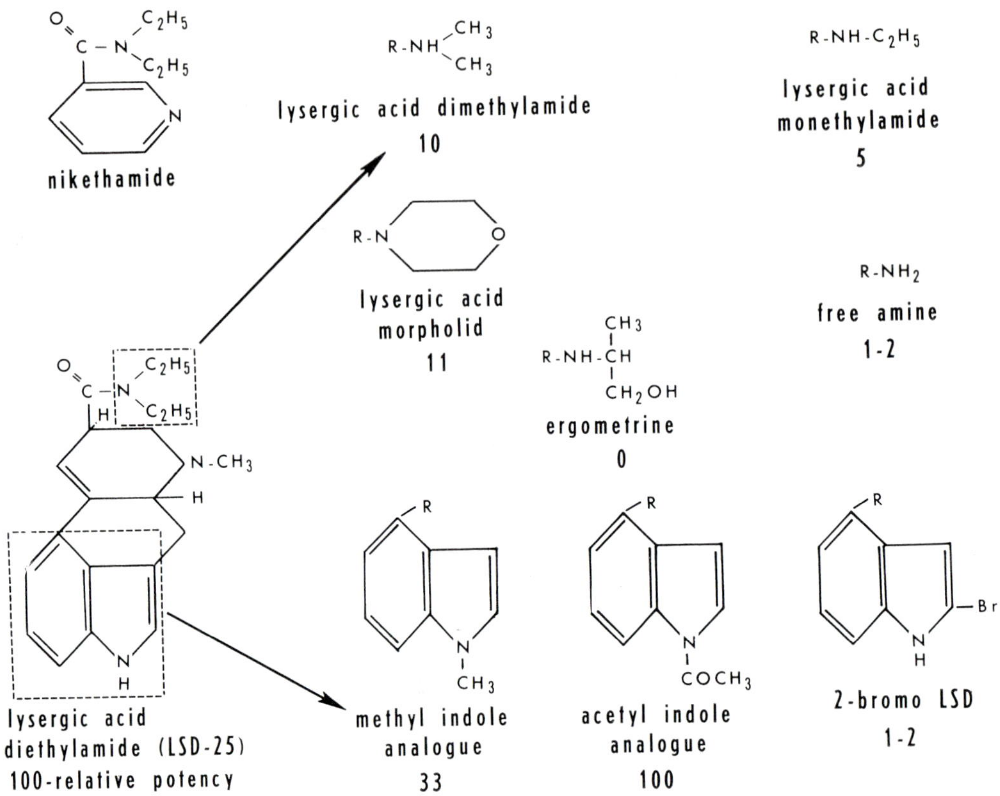

Figure 1 Structural relationships between LSD and various analogs. Numbers indicate relative psychotomimetic potency.

ity is considerable for many of these compounds, as only minor changes in the LSD structure may markedly reduce potency or activity. The same is true for ring substituents on the phenethylamines or indolealkylamines. Methoxy substituents at the 2- or 5-positions on the ring seem to be required for activity of either phenethylamines or phenylisopropylamines, the latter being represented by various hallucinogenic amphetamine analogs.[3,4] Similarly, the movement of the hydroxyl substituent from the 4-position of indolethylamine, as in psilocin, to the 5-position, as in bufotenin, markedly alters the type of activity, while movement to the 6-position inactivates the compound.

In addition to the possibility that these drugs act by virtue of their effects on biogenic amines, it has been proposed that they may act through other mechanisms.[5] The extraordinarily small quantities of LSD required to effect major mental change has suggested that the drug may interact with macromolecules at neuronal membrane surfaces, changing their levels of excitability.[6] LSD may also intercalate in the ridges of the DNA molecule.[7] Suffice it to say that no particular mechanism of action has been proposed which seems entirely satisfactory or can explain the vast differences in potency between what otherwise are essentially similar drugs. The estimated relative potency of LSD to mescaline is 4000:1, and to psilocybin approximately 250:1.

Clinical, Physiologic, and Biochemical Effects. When given in equivalent doses, LSD, mescaline, and psilocybin form a group of drugs with virtually identical clini-

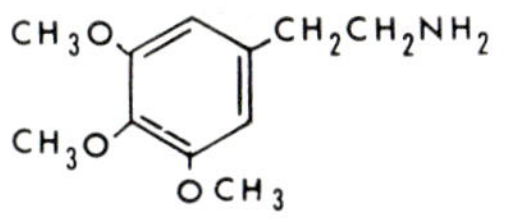

mescaline

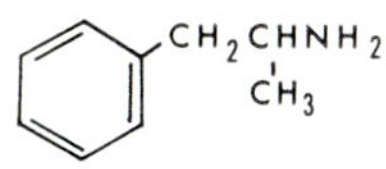

d-amphetamine

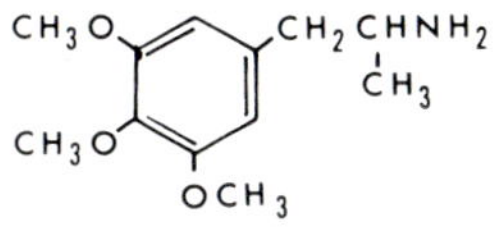

trimethoxyamphetamine

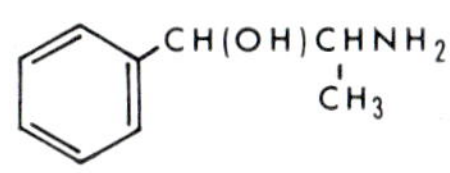

d-norpseudoephedrine

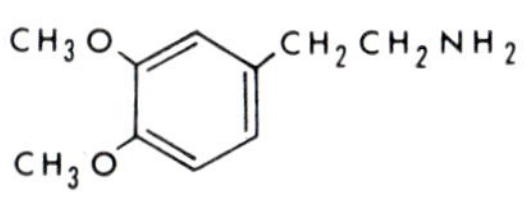

3,4-dimethoxyphenylethylamine

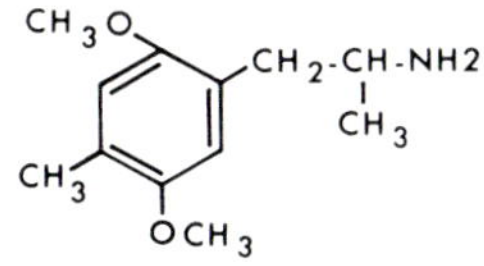

2-5 dimethoxy-4-methyl-
d-amphetamine ("STP")

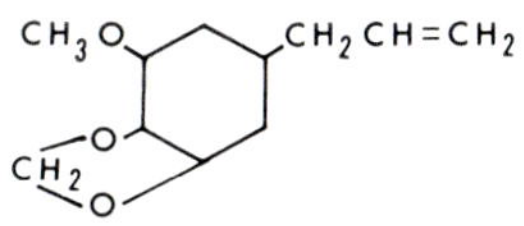

myristicin

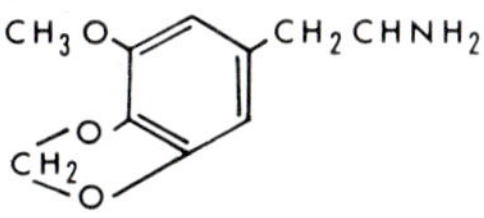

3-methoxy-
4,5-methylenedioxyamphetamine

Figure 2 Phenethylamine and phenylisopropylamine derivatives with varying degrees of psychotomimetic effect.

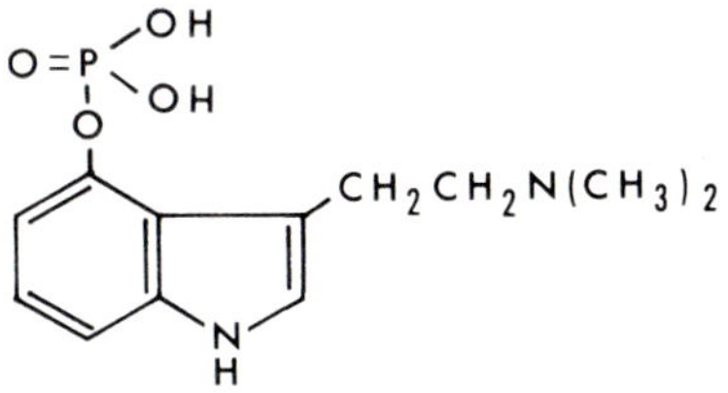

psilocybin

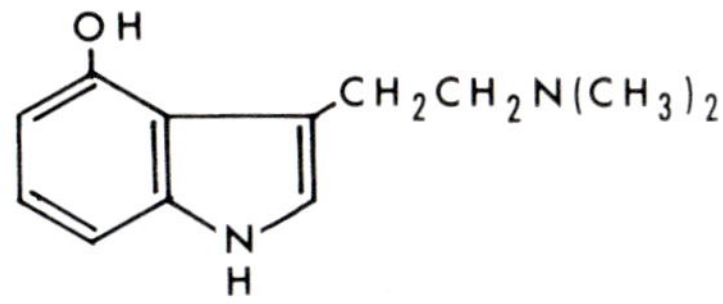

psilocin
4-hydroxydimethyltryptamine

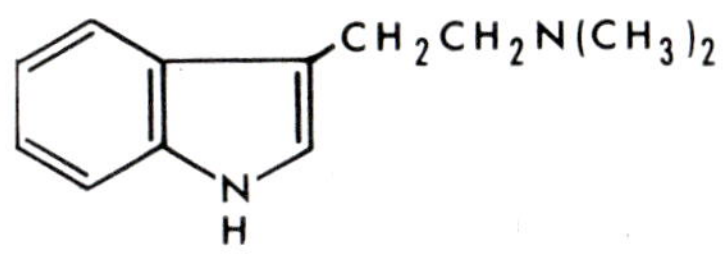

N-N dimethyltryptamine

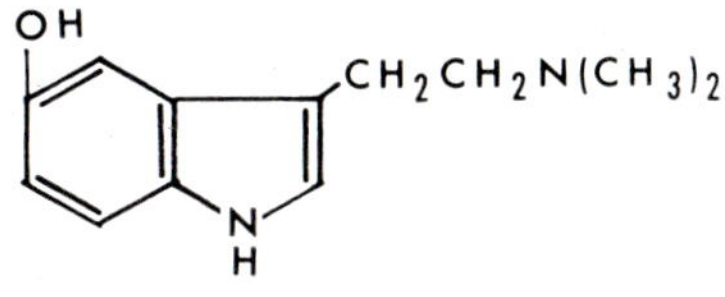

bufotenin
5-hydroxydimethyltryptamine

Figure 3 Indolethylamine derivatives. Despite N-dimethyl substitution in all, activity is changed by position of ring substituents, or their absence.

cal effects.[8] Similar effects are produced by close chemical relatives. For instance, morning glory seeds, which contain the monoethylamide of lysergic acid, or methysergide (Sansert) evoke effects similar to LSD, although both are less potent. The hallucinogenic amphetamine analogs, which are related chemically to mescaline, have very similar actions. Myristicin, the active component of nutmeg, is different enough chemically so that it is quite different clinically. The tryptamines, dimethyltryptamine and bufotenin, are related to psilocybin and to psilocin, its active metabolite, but are shorter-acting with more intense physical reactions. Accordingly, these drugs may be considered as a group consisting of members with varying degrees of similarity.

The description of the first known reaction to LSD, that unplanned experiment of Albert Hofmann, still stands as a landmark in careful, objective reporting.[9] What Hofmann described then, and what objective investigators have since repeatedly elaborated upon, have been three characteristic types of symptoms: somatic, perceptual, and psychic. In repeated laboratory experiments, our subjects have reported a basic clinical syndrome which might be described as follows:

1. Somatic symptoms—dizziness, weakness, tremors, nausea, drowsiness, paresthesias, and blurred vision.
2. Perceptual symptoms—altered shapes and colors, difficulty in focusing on objects, a sharpened sense of hearing, and rarely, synesthesias.
3. Psychic symptoms—alterations in mood (happy, sad, or irritable at varying times), tension, distorted time sense, difficulty in expressing thoughts, depersonalization, dream-like feelings, and visual hallucinations.

Physiologic effects are relatively few. Dilated pupils, hyper-reflexia, increased muscle tension, incoordination, and ataxia are common physical signs. Effects on pulse rate, respiration, and blood pressure are so variable that they probably represent varying levels of anxiety of subjects rather than true physiologic effects. Changes in appetite and salivation are inconstant, being increased in some subjects, decreased in others.

Consistent biochemical changes have been limited to elevations of free fatty acids in plasma, decreased urinary clearance of creatinine and phosphorus, and a marked decrease in free water clearance due to a strong antidiuretic effect. It should be re-emphasized that in all these respects, LSD, mescaline, and psilocybin are essentially similar in comparable doses.

The clinical syndrome tends to follow a sequential pattern with somatic symptoms presenting first, perceptual and mood changes next, and finally psychic changes, although there is considerable overlap among these phases. Between the ranges of 1 μg/kg of body weight to 16 μg/kg the severity of psychophysiologic effects of LSD in a given subject are proportional to the dose.[10] Specific types of reaction, such as paranoid ideation, are more likely a matter of personal predisposition than a function of dose.

LSD Analogs. A number of clinical comparisons have been made between LSD and closely related lysergic acid derivatives. One comparison between 2,3-dihydrolysergic acid diethylamide and LSD revealed that the former compound induced LSD-like autonomic and mental changes in man, but was less potent than LSD and its action slower in onset than that of LSD.[11] Acetylation or methylation of the LSD molecule on the pyrrole nitrogen produces two analogs, 1-acetyl lysergic acid diethylamide and 1-methyl lysergic acid diethylamide. These three compounds were tested in both patients and normal subjects.[12] About 1.5 to 2.0 times as much of the analogous drugs were required to produce essentially the same clinical reactions with the analogs as with LSD. Replacement of the diethylamine side chain of LSD with a morpholine group results in a compound which has pharmacologic and clinical effects generally similar to those of LSD; potency was probably somewhat less.[13]

Comparisons with Other Drugs. A number of systematic comparisons of LSD with other psychotomimetic drugs have revealed that the clinical syndromes from LSD closely resemble those produced by mescaline, psilocybin, and possibly some of the amphetamine analogs.[8,14–17] The ratio of dose seems to be as follows: 1.5 μg/kg of LSD is equivalent to 225 μg/kg of psilocybin, is equivalent to 5 mg/kg of mescaline. Minor differences have been described, such as a longer duration of action and more vomiting with mescaline, but in general the clinical syndromes and physiologic and biochemical effects of these drugs are quite similar.

Very early in its clinical use, tolerance to the effects of LSD was demonstrated. In chronic schizophrenics treated with the drug on a daily basis, tolerance was evident by the second day and complete on the third.[18] A period of four to six days free of LSD was necessary to reinstate the original reaction. In these patients, no cross-tolerance could be demonstrated between mescaline and LSD, although cross-tolerance was shown between LSD and lysergic acid ethylamide (LAE) and brom-lysergic acid. Cross-tolerance to other lysergic acid derivatives was confirmed by another group of investigators, although, for maximal effects, d-2-brom-lysergic acid diethylamide had to be administered in doses of 3 mg daily for five days preceding the trial with LSD.[19] This same group was able to demonstrate cross-tolerance to mescaline in patients rendered tolerant of LSD, as well as the converse situation of resistance to LSD in patients tolerant of mescaline.[20]

Metabolism of LSD in Man. Humans given doses of 2 μg/kg of body weight intravenously had blood specimens taken which were analyzed for LSD by using a spectrofluorometric technique. After equilibration had occurred in about 30 minutes, the plasma level was between 6 and 7 ng/ml. Subsequently, plasma levels gradually fell until only a small amount of LSD was still present after eight hours. The half-life of the drug in man was calculated to be 175 minutes. Performance scores on simple arithmetic tests showed considerable impairment during the period of maximal levels followed by improvement as the plasma levels of drug dropped.[21]

Therapeutic Potentials. Although hallucinogens have been proposed as a psychiatric diagnostic test, based on their alleged "mind-revealing" qualities, and although favorable results of treatment have been reported in a variety of psychiatric disorders, they have no presently established medical use. Facilitation of conventional "insight" psychotherapy with these drugs, while superficially attractive, has not stood the test of a controlled trial.[22] The intriguing claim that alcoholics can be successfully treated with minimal effort has been found wanting by controlled clinical trials.[23–25] The assertion that LSD may reduce requirements for narcotics in patients with severe pain associated with terminal malignant disease remains unproven.[26] If such treatment allowed death with more dignity than otherwise, hallucinogenic drugs would be appropriate; one would not be concerned with long-term adverse effects under such circumstances.

The initial enthusiasm for using LSD to produce a "model psychosis" was followed by disillusion and skepticism, but is currently being revived. Most clinicians would be more willing to accept the schizophreniform state following chronic abuse of amphetamines as a better model of psychosis. In any case, the LSD model psychosis has been unrewarding, either for increasing our knowledge of schizophrenia or advancing its treatment.

MARIHUANA

General Considerations. Marihuana is often classified as a weak hallucinogen, as it evokes the characteristic perceptual and psychic changes of hallucinogens only at relatively high doses. Another reason the drug is considered "weak" is that sedation limits the extent of the hallucinogenic effects of high doses. In a sense, marihuana contains its own antidote to hallucinogenic effects. Finally, marihuana preparations, being derived from a natural product, vary considerably in their potency. Most commonly available ones are weak. Perhaps the generally mild and limited intoxication that marihuana induces makes it preferred over strong intoxicants and considered innocuous by its devotees.

While marihuana has been used for several millenia and probably now by 200 million people throughout the world, its use was uncommon in western society until about a decade ago. Since then an exponential increase in its use has occurred, so that only gross approximations of its extent can be made, and these are rapidly outdated. The history of marihuana use seems to indicate that it stays in cultures once it has gained wide acceptance. Quite possibly the time is not far distant when we must decide whether or not marihuana is to be admitted to the company of social drugs.

The tall annual herb from which the drug is obtained is hemp, related to the fig-tree and the hop. Both male and female forms come to flower on separate plants which were named by Linnaeus in 1753; plants growing in different parts of the world are labeled by locality, such as *Cannabis indica* or *americana,* but botanically they are similar. The stem of the male plant is used to make the fiber known as hemp; some strains of hemp are especially good for fiber production while others are better for drug production. Despite the former notion that only the female flowers contained active material, it seems that the content of THC is about equal in plants of each sex. There are vast differences in the content of THC between the various parts of the plant; the stems contain little or none, the leaves an appreciable amount, and the flowering tops the most.

Cannabis has known many names over the years: bhang, ganja, charas, marihuana, muggles, reefer or tea. The name "marihuana" probably originated in Mexico as a combination of the proper names, "Maria" and "Juan." Hashish is an Arabic word which literally means "dried herb." Hashish is the dried resin of cannabis, which may be squeezed or extracted from the plant by a number of processes. The resin may be taken in many ways: as a candy or a sweet, an additive to a drink, or eaten or smoked directly; the leaves and flowers of the plant are dried, pulverized, and smoked, often in cigarettes, the most common mode for using marihuana in western countries.

Chemical and Pharmacologic Considerations. Of the three principal cannabinoids in marihuana, cannabidiol, tetrahydrocannabinol (THC), and cannabinol, only THC has been definitely proven to be active (Fig. 4). Two isomers of THC exist in marihuana; the Δ-1 isomer accounts for about 99% of the total THC while the Δ-6 isomer often cannot be clearly identified in chemical analyses. A number of other THC isomers and analogs have been synthesized. We have studied the effects of Δ-3-THC and its amyl side-chain analog, synhexyl (pyrahexyl). Both are active but weaker than Δ-1-THC.[27]

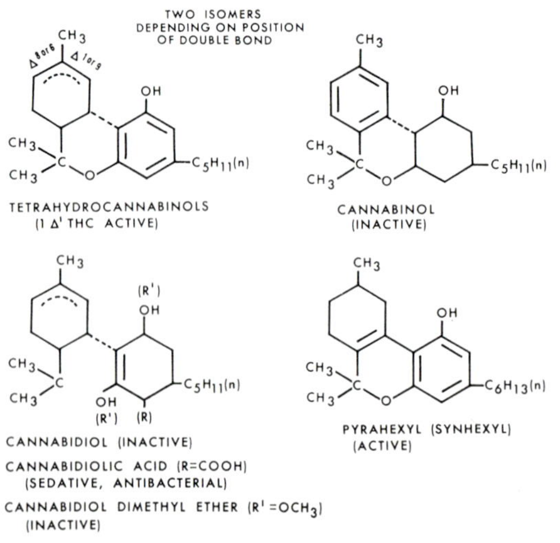

Figure 4 Three major types of cannabinoids found in marihuana as well as a synthetic Δ-3-THC analog (pyrahexyl).

Synthesis of the latter isomer became possible a few years ago; adequate amounts of the synthetic material should soon be available to satisfy research needs.

The availability of synthetic Δ-1-THC, as well as chemical techniques for quantifying its content in marihuana, has made possible for the first time pharmacologic studies which provide some but not complete precision in dose. When the material is smoked, a still uncertain and perhaps variable fraction of THC is either lost by smoke escaping into the air or exhaled incompletely absorbed from the respiratory dead space. Relatively little is lost by pyrolysis. The efficiency of delivery of a dose by smoking has been estimated from 20 to 80%, but with most experienced smokers it should approximate 50%.[28] Synthetic Δ-1-THC and marihuana extracts are also taken orally but doses equivalent in effect to those from smoking are several-fold larger.[29] It seems probable that some THC may be inactivated within the gastrointestinal tract or during its passage through the intestinal mucosa and liver. As no method for the quantitative estimation of THC levels in plasma or urine is available, the pharmacokinetics of THC is unknown.

When smoked, absorption of THC is rapid and effects appear within seconds to minutes. If marihuana is of low potency, effects may be subtle and brief.[30] Seldom do they last longer than two to three hours after a single cigarette, although users prolong effects by repeated smoking. Oral doses delay the onset of symptoms for 15 minutes to two hours. Because synthetic THC, as well as marihuana extracts, require non-polar solvents, even the administration of accurate doses for animal pharmacologic studies has been a problem, despite the greater latitude of animal experimentation.

Because of the dual route of administration, as well as the still unsettled problem of whether natural marihuana materials have activity different from that of Δ-1-THC, we shall consider each type of clinical study separately. Very likely all activity is determined by the quantity of Δ-1-THC, the only difference being in the delivery of the dose by different routes of administration.

Studies Using Synthetic Δ-1-THC Orally. Three groups have now published reports on the effects of oral doses of synthetic Δ-1-THC in man. The group at the Addiction Research Center, U.S. Public Health Service Hospital, Lexington, employed doses of 10 to 30 mg, or 120 to 480 μg/kg of body weight.[29] We used doses of 30 to 70 mg, or 341 to 946 μg/kg.[31] A group from the National Institute for Mental Health used a single dose of 20 mg, not specifying the weight of the subjects.[32] Between the three groups, however, a rather wide range of dose was explored. It should be emphasized that, even accounting for the lesser potency of the material when given orally as compared to smoking, most doses were beyond those which might be obtained from smoking an ordinary marihuana cigarette. The experimental doses might be more comparable to those obtained from smoking several cigarettes of reasonable quality, or from smoking hashish.

PHYSIOLOGIC EFFECTS. No changes in pupil size, respiratory rate, or deep tendon reflexes were observed. We found no change in oral temperature of any consequence, but the NIMH group reported a slight but consistent decline. All observers have commented on the constant increase in pulse rate, often one of the first signals of the drug effect. Blood pressure tends to fall slightly or remains unchanged; at higher doses, we observed two incidents of fainting. Conjunctival reddening is also constantly observed, and in this case indicates that this is not an artefact produced by irritation from smoke. We measured muscle strength using the finger ergograph and could demonstrate muscle weakness objectively.

PERCEPTUAL AND PSYCHIC CHANGES. Euphoria was most pronounced in our subjects, who were normal graduate students. It was less marked in the Lexington series, the members of which were imprisoned former drug addicts, or in the NIMH group of subjects who were prisoners at a correctional institution. Sleepiness was constantly observed, and often deep sleep followed the higher doses. Time sense was altered, hearing was less discriminant than normal, and vision apparently sharper than normal with many visual distortions. Depersonalization, difficulty in concentrating and thinking, and dream-like states were prominent. Many of these symptoms were similar to those produced by psychotomimetics of the LSD-mescaline-psilocybin class. On self-reporting mood scales, our subjects became more friendly initially, but less so with the passage of time; less aggressive, especially late in the course; less clear-thinking persistently; sleepy, especially after three hours; euphoric and dizzy persistently. The NIMH group found only increased sleepiness and less clear-thinking, but their comparisons were made against a placebo control rather than against baseline, and were ratings by others rather than the subjects themselves.

PSYCHOMETRIC TESTS. We used repetitive psychometric tests of arithmetic ability or drawing freehand, both of which were impaired in different ways. The arithmetic test, a familiar and simple task, showed a slowing of performance against time, with maintained accuracy. The drawing test, less familiar and more difficult, showed reduced accuracy with no slowing of performance, probably indicating some loss of finer judgment. The NIMH group found that accuracy of serial addition was impaired, but that the drug had no effect on ability to count backwards, say the alphabet, or repeat digits forwards or backwards.

BIOCHEMICAL TESTS. Our group was the only one to attempt these measures. Plasma free fatty acids remained unchanged, unlike the case with drugs of the LSD-type. Blood glucose values were also unchanged, despite the abundant literature indicating that marihuana produces hypoglycemia. The lack of change in plasma glucose values has now been verified on numerous occasions. Both creatinine and phosphorus clearance were temporarily decreased, a phenomenon which has been observed with LSD.[8]

COMPARISONS WITH OTHER MATERIALS. We compared Δ-1-THC with the synthetic Δ-3-THC homolog, synhexyl. The latter has been studied rather extensively in the past for possible clinical utility.[33] On the whole, the changes reported above were also produced by synhexyl, but the approximate ratio in potency to Δ-1-THC was 1:3, synhexyl being weaker. This drug was slower in onset of action by about one hour, but longer-lasting in equivalent doses.

The Lexington group compared the effect of taking THC orally with its effects when smoked following addition of known quantities to cigarettes. They estimate that potency is increased approximately threefold by smoking as compared with taking the same material by mouth.[29] As might be expected, effects come on much quicker, but are of briefer duration, when the material is smoked. They have compared the effects of smoked THC (75 to 225 μg/kg) with those of LSD given intramuscularly in doses of 0.5 to 1.5 μg/kg.[34] Subjective effects between the two drugs were not readily distinguished, but objective differences were marked: LSD increased body temperature, systolic and diastolic blood pressure, and deep tendon reflexes, and dilated pupils while THC had none of these effects. We made a retrospective comparison of the effects of LSD and THC taken orally and came to similar conclusions regarding the objective differences.[35] Subjectively, we thought that THC produced less total impairment with more euphoria and dream-like states than LSD at comparable doses and that, unlike the latter drug, sedation was a prominent feature with THC, most subjects falling asleep. In general, we have seen fewer psychotomimetic effects than did the Lexington group.

STUDIES USING ORAL DOSES OF MARIHUANA EXTRACTS. We have completed additional experiments using extracts of marihuana, gauging doses on the basis of THC content. We have used doses ranging from 5 to 60 mg, as well as making comparisons between these and the placebo extract. Even at the smallest dose, all subjects had appreciable clinical effects as compared with placebo; at the larger doses, psychotomimetic effects were observed. In general, one had the impression that the effects produced by these extracts were comparable to those which would have been produced by similar doses of synthetic THC.

Tests comparing the effects of 20, 40, and 60 mg THC-equivalent doses of extract with placebo indicated that while long-term memory (tested by subtraction of serial sevens) was maintained, short-term memory (tested by the span of forward and backward digits) was impaired.[36] Along with this was an impairment of a more complex task, goal-directed serial alternation, in which successive subtractions and additions had to be made to reach a specified end number. The latter task requires the retention, coordination, and serial ordering of memories relevant to a specific goal which must also be kept in mind. Disturbance of this type has been termed "temporal disintegration."[37] Both confusion and depersonalization are prominent clinical features of the intoxication from marihuana.

We also compared marihuana extracts with ethanol and dextroamphetamine in regard to their effects on mood and mental functions.[38] Marihuana and ethanol were most alike in their effects causing decreased activity and a tendency to impaired performance in certain psychometric tests as well as in the simple reaction time test. Marihuana was distinctly different from the other two drugs in regard to time estimation, resulting in longer estimates.

A group at the University of Utah has also made use of marihuana extracts for clinical studies.[39] Its members also have been impressed with the disruptive effects of marihuana on sequential thought, suggesting impairment of rapid decision-making and short-term memory. They have noted, as have others, a great variability in performance during marihuana intoxication, which may be related to the fact that subjects go "in-and-out," the effects seeming to come and go in cycles or waves.

STUDIES WITH SMOKED MARIHUANA. Because marihuana is more commonly smoked than taken orally, some investigators feel that proper studies can be done only by utilizing this particular route of administration. The major argument is that smoking native marihuana may activate other materials present in the plant which are not found in synthetic THC or extracts, or that the process of combustion may create new active materials. No proof for either assertion is at hand. The disadvantages are great: the dose, even when one knows the amount in the cigarette, is impossible to judge as variations in technique of smoking may create tremendous variations in delivery of the dose.

The first study provided marihuana in cigarettes, the putative doses being 4.5 and 18 mg, which were compared with a placebo smoke.[30] Naive smokers experienced few subjective effects, although they showed an increased heart rate and conjunctival injection. Experienced smokers of marihuana reported a typical "high," not much elaborated upon. Performance on the digit-symbol substitution test and the pursuit rotor test was unchanged. No changes in blood sugar were found. In general, the effects of drug smoked in this fashion were found to be relatively mild and innocuous, leading the investigators to take a sanguine view of the social use of the drug.

Another study tested driving skill, using a driving simulator, in subjects who were tested following smoking of two marihuana cigarettes over a 30-minute period, following consumption of walloping doses of alcohol, and following no treatment.[40] Under marihuana conditions, speedometer errors were increased, suggesting that the subjects did not monitor the speedometer as carefully as they might have normally, but driving was otherwise little impaired. As might have been expected, marked impairment was observed from the high doses of alcohol, which were intended to approximate blood levels of 100 mg per 100 ml. Such highly controversial findings have elicited criticism directed at the fact that the doses of the two drugs were disproportionate, that a dose-response curve was not obtained and that simulated driving might not be an adequate model for real life.[41] Other objections are that smoking delivers an uncertain dose of drug and that most marihuana users tend to titrate their rate of smoking to a desired clinical "high." The earlier study of smoked marihuana found that most effects dissipated by the end of one hour and that many subjects had mild effects. Although the authors were careful not to state that the marihuana did not affect one's ability to drive a car, it is unfortunate that many lawyers and courts may draw this conclusion. Sometimes it is better not to be so scientific. Since our first experiments, we have simply asked subjects, when they were "high," "Do you think you could drive a car now?" Without exception the answer has been "No!"

METABOLISM IN MAN. Very little is known at present regarding the fate of tetrahydrocannabinol in man. We have attempted, as have others with more sensitive methods (less than 600 picograms), to measure unchanged Δ-1-THC in plasma without any success. Its existence in this form must be quite brief. A relatively simple technique for measuring the excretion of metabolites of marihuana in urine used thin-layer chromatographic techniques.[42] The possible use of this test forensically is clouded by the persistence of some of the new spots following marihuana smoking or ingestion for periods far longer than the span of drug effect; we have found some present days after the last dose. Using thin-layer techniques, we have not found measurable amounts of unchanged THC in urine, even when this was the sole substance given in very large amounts. Recently a 7-hydroxy metabolite of Δ-1-THC has been described, which also has pharmacologic activity.[43] Thus, it seems possible that the drug may exert its action through an active metabolite. The prolonged latent period following ingestion of synhexyl led us to propose that it must be converted to an active metabolite. On the other hand, conversion of Δ-1-THC to any active metabolite, if this is necessary for its action, must be swift, as effects following smoking of potent preparations are noticeable within minutes.

CONCLUSIONS

The work accomplished on marihuana during the past three years has contributed little which wasn't known before. Some myths have been laid to rest; many questions are still unsettled. Among the latter, such socially important questions as whether deleterious personality changes result from chronic use of the drug cannot be studied experimentally. We shall be forced to devise scientific methods for evaluating these effects of marihuana on users who perform experiments on themselves that none of us would attempt. Whether or not the current drug revolution will wait for our answers remains to be seen.

Simultaneously with the revival of scientific interest in marihuana, research with LSD, at least in humans, has been virtually suspended. The uncertainty about the chromosomal changes which have been reported has deterred most investigators from further research until the issue becomes clear. There is some reason to believe that the same attitude prevails among the drug-taking youth and that perhaps LSD is now in a period of eclipse. Fashions in drug-taking change with remarkable speed, however, and the time may come again when LSD research will be more pressing than it is right now.

REFERENCES

1. Hollister, L. E.: *Chemical Psychoses. LSD and Related Drugs.* Springfield, Ill., Charles C Thomas, 1968, 190 pp.

2. Home Office Report by the Advisory Committee on Drug Dependence: *Cannabis.* London, Her Majesty's Stationery Office, 1968.
3. Snyder, S. H., Shulgin, A. T., Green, J. P., and Beveridge, G. L.: The mode of action of psychotomimetic drugs. Quantum theory. *Neurosciences Res Prog Bull 8* (No. 1): 90–106, 1970.
4. Hollister, L. E., Macnicol, M. F., and Gillespie, H. K.: An hallucinogenic amphetamine analog (DOM) in man. *Psychopharmacologia (Berlin) 14*:62–73, 1969.
5. Aghajanian, G. K., and Freedman, D. X.: Biochemical and morphological aspects of LSD pharmacology. In *Psychopharmacology. A Review of Progress 1957–1967,* Efron, D. H. (Ed.). Public Health Service Publication No. 1836, Washington, D.C., U. S. Govt. Printing Office, 1968, pp. 1185–1194.
6. Adey, W. R.: Neurophysiological action of LSD. In *Neurophysiological and Behavioral Aspects of Psychotropic Drugs*, Karczmar, A. G., and Koella, W. P. (Eds.). Springfield, Ill., Charles C Thomas, pp. 5–26.
7. Smythies, J. R.: The mode of action of psychotomimetic drugs. A prostaglandin-RNA complex as a potential receptor site. *Neurosciences Res Prog Bull 8* (No. 1): 123–130, 1970.
8. Hollister, L. E., and Sjoberg, B. J.: Clinical syndromes and biochemical alterations following mescaline, lysergic acid diethylamide, psilocybin and a combination of the three psychotomimetic drugs. *Compr Psychiat 5*:170–178, 1964.
9. Hofmann, A.: Chemical, pharmacological and medical aspects of psychotomimetics. *J Exp Med Sci* 5:31–51, 1961.
10. Klee, G. D., Bertino, J., Weintraub, W., and Callaway, E.: The influence of varying dosage on the effects of lysergic acid diethylamide (LSD-25) in humans. *J Nerv Ment Dis 132*:404–409, 1961.
11. Gorodetsky, C. W., and Isbell, H.: A comparison of 2,3-dihydrolysergic acid diethylamide with LSD-25. *Psychopharmacologia 6*:299–343, 1964.
12. Malitz, S., Wilkens, B., Rochrigo, W. C., and Hoch, P. H.: A clinical comparison of three related hallucinogens. *Psychiat Quart 34*:333–345, 1960.
13. Gogerty, J. H., and Dille, J. M.: Pharmacology of d-lysergic acid morpholide. *J Pharmacol Exp Ther 120*:340–348, 1957.
14. Hollister, L. E., and Hartman, A. M.: Mescaline, lysergic acid diethylamide and psilocybin: Comparison of clinical syndromes, effects on color perception and biochemical measures. *Compr Psychiat 3*:235–241, 1960.
15. Wolbach, A. B., Jr., Miner, E. J., and Isbell, H.: Comparison of psilocin with psilocybin, mescaline and LSD-25. *Psychopharmacologia 3*:219–223, 1962.
16. Hidalgo, W. T.: Psychophysiological, comparative investigation of mescaline, d-lysergic acid diethylamide and psilocybin. *Acta Med Venez 8*:56–62, 1960.
17. Snyder, S. H., Faillace, L., and Hollister, L. E.: 2,5-Dimethoxy-4-methylamphetamine (STP): A new hallucinogenic agent. *Science 158*:669–670, 1967.
18. Cholden, L. W., Kurland, A., and Savage, C.: Clinical reactions and tolerance to LSD in chronic schizophrenia. *J Nerv Ment Dis 122*:211–221, 1955.
19. Isbell, H., Miner, E. J., and Logan, C. R.: Cross-tolerance between d-2-brom-lysergic acid diethylamide (BOL-148) and the d-diethylamide of lysergic acid (LSD-25). *Psychopharmacologia 1*:109–116, 1959.
20. Wolbach, A. B., Jr., Isbell, H., and Miner, E. J.: Cross tolerance between mescaline and LSD-25 with and comparison of the mescaline and LSD reactions. *Psychopharmacologia 3*:1–14, 1962.
21. Aghajanian, G. K., and Bing, O. H. L.: Persistence of lysergic acid diethylamide in the plasma of human subjects. *Clin Pharmacol Ther 5*:611–614, 1964.
22. Hollister, L. E., Degan, R. O., and Schultz, S. D.: An experimental approach to facilitation of psychotherapy by psychotomimetic drugs. *J Ment Sci 108*:99–100, 1962.
23. Smart, R. G., Storm, T., Baker, E. F. W., and Solursh, L.: A controlled study of lysergic acid diethylamide in the treatment of alcoholism. I. The effects of drinking behavior. *Quart J Stud Alcohol 27*:469–482, 1966.
24. Hollister, L. E., Shelton, J., and Krieger, G.: A controlled comparison of lysergic acid diethylamide (LSD) and dextroamphetamine in alcoholics. *Amer J Psychiat 125*: 1352–1357, 1969.
25. Ludwig, A., Levine, J., Stark, L., and Lazar, R.: A controlled study of LSD treatment in alcoholics. *Amer J Psychiat 126*:59–69, 1969.
26. Kast, E. C., and Collins, V. J.: Study of lysergic acid diethylamide as an analgesic agent. *Anesth Analg (Cleveland) 43*:285–291, 1964.
27. Hollister, L. E.: Tetrahydrocannabinol isomers and homologues: Contrasted effects of smoking. *Nature 227*:968–969, 1970.
28. Manno, J. E., Kiplinger, G. F., Bennett, I. F., Haine, S., and Forney, R. B.: Comparative effects of smoking marihuana and placebo on human motor and mental performance. *Clin Pharmacol Ther 11*:808–815, 1970.
29. Isbell, H., Gorodetsky, G. W., Jasinski, D., Claussen, U., v. Spulak, F., and Korte, F.: Effects of (—)-Δ^9-trans-tetrahydrocanna-

binol in man. *Psychopharmacologia (Berlin) 11*:184–188, 1967.

30. Weil, A. T., Zinberg, N. E., and Nelson, J. M.: Clinical and psychological effects of marihuana in man. *Science 162*:1234–1242, 1968.
31. Hollister, L. E., Richards, R. K., and Gillespie, H. K.: Comparison of tetrahydrocannabinol and synhexyl in man. *Clin Pharmacol Ther 9*:783:791, 1968.
32. Waskow, I. E., Olsson, J. E., Salzman, C., and Katz, M. M.: Psychological effects of tetrahydrocannabinol. *Arch Gen Psychiat 22*:97–107, 1970.
33. Parker, C. S., and Wrigley, F.: Synthetic cannabis preparations in psychiatry. I. Synhexyl. *J Ment Sci 96*:276–279, 1950.
34. Isbell, H., and Jasinski, D. R.: A comparison of LSD-25 with (—)-Δ^9-transtetrahydrocannabinol (THC) and attempted cross-tolerance between LSD and THC. *Psychopharmacologia (Berlin) 14*:115–123, 1969.
35. Hollister, L. E., and Gillespie, H. K.: Similarities and differences between the effects of lysergic acid diethylamide and tetrahydrocannabinol in man. In *Drugs and Youth,* Wittenborn, J. R. *et al.* (Eds.). Springfield, Ill., Charles C Thomas, 1969, pp. 208–211.
36. Melges, F. T., Tinklenberg, J. R., Hollister, L. E., and Gillespie, H. K.: Marihuana and temporal disintegration. *Science 168*: 1118–1120, 1970.
37. Melges, F. T., Tinklenberg, J. R., Hollister, L. E., and Gillespie, H. K.: Temporal disintegration and depersonalization during marihuana intoxication. *Arch Gen Psychiat 23*:204–210, 1970.
38. Hollister, L. E., and Gillespie, H. K.: Marihuana, ethanol and dextroamphetamine. Mood and mental functions. *Arch Gen Psychiat 23*:199–203, 1970.
39. Clark, L. D., and Nakashima, E. N.: Experimental studies of marihuana. *Amer J Psychiat 125*:379–384, 1968.
40. Crancer, A., Jr., Dille, J. M., Delay, J. C., Wallace, J. E. and Haykin, M. D.: Comparison of the effects of marihuana and alcohol on simulated driving performance. *Science 164*:851, 1969.
41. Kalant, H.: Marihuana and simulated driving. *Science 166*:640, 1969.
42. Christiansen, J., and Rafaelsen, O. J.: Cannabis metabolites in urine after oral administration. *Psychopharmacologia 15*:60–63, 1969.
43. Nilsson, I. M., Agurell, S., Nilsson, J. L. G., Ohlsson, A., Sandberg, F., and Wahlquist, M.: Δ^1-Tetrahydrocannabinol: Structure of a major metabolite. *Science 168*:1228, 1970.

PATTERNS OF DRUG ABUSE—1970

SIDNEY COHEN, M.D.

If the 1960's were the decade of the rising tide of drug abuse, what can we expect of the 70's? In order to assess what is to come, we must study the trend lines of drug overuse and the shifting cultural patterns which promote our latter-day preoccupation with chemical alterations of awareness.

The drug scene of the 60's did not come from an epoch devoid of drug misuse. Every class of chemicals abused today was known and abused prior to this turbulent decade. What has changed is the slope of the curve, which now assumes a geometric configuration. What caused this acceleration in the rate of drug involvement? One of the major causes was rapid alteration in the value system of many individuals. From a lack of concern with mood-altering drugs, or even a negative attitude toward consuming illicit or illicitly obtained materials, it became not only desirable, but "in," to go on chemically propelled "trips."

THE PSYCHEDELICS

A good share of the reawakened interest in consciousness change can be traced to the popularization of the psychedelics in the early 1960's. The intrepid group at Harvard who were so overwhelmed by the enormous enlightenments of the LSD experience proceeded to proclaim that the answer to all man's hang-ups resided in an "acid" trip. So quick and magical a solution to every personal and universal ill was particularly attractive in this difficult, decadent society. The mass media, of course, spread the message far and wide.

Why the LSD state so powerfully alters sets and demolishes constancies—giving rise to abrupt shifts in attitudes, goals, and values—is not a proper part of this presentation, although it was the LSD-induced transformation of goal direction in a few people that set in motion the rising tide of mass drug abuse.

Unfortunately, what began as a loving psychedelic revolution has turned out to be considerably less. The acid-heads of yesteryear have graduated to either "speed" or "smack" or have gone over to non-chemical turns-ons, or have retreated to the less potent "pot." Their ranks have been replenished with high school and pre-high school youngsters who could not be less interested in matters mystical or transcendent.

The revolution has turned out, on further observation, to be a counter-revolution—a regression to medieval superstition, archaic myths, and ancient forms of magical thought. It is a flight from reason at a time when reason and logic have confronted us with massive and perplexing problems. The good old days and the return to nature are very attractive now. One wonders whether the good old days were that good, and whether we can return to some pristine state. Is it possible to become noble savages, or must we strive to adapt to a world in transition?

In a few short years the motives for using LSD have shifted from an all-too-sanguine desire to expand one's consciousness to an explicit expectation of pleasure now. The perennial search for a higher "high" accounts for the escalation by some "acid-heads" to "speed," "smack," and "snow."

The psychedelic prophets have muted their tune. LSD is no longer "something for everybody" nor is the worst that can happen is that "you will come back no better than you were." Now it's, "four out of five Americans shouldn't take acid." The "high" priest of drop-out has recently dropped back in, and, as is all too often the case, has swung from extreme passivity to extreme activism.

The quality control of the black market psychedelics is uniformly poor. People who prefer mescaline and Psilocybin to LSD because of their concern about their chromosomes are actually swallowing LSD. The LSD is cut and sometimes adulterated with bemusing odds and ends. The THC on the street is never THC, and even the marihuana isn't always marihuana.

Now it is pubescents who go through a romance with LSD hopefully to emerge, one day, relatively intact. The growing central nervous system is more sensitive to all psychotropic drugs and the pharmacology of juvenile drug-taking has hardly been studied. It is the developing personality which seems to need problem-solving opportunities, and a repetitive engagement with life's problems, during the process of maturation. Chemicals which avoid or blunt the encounter with problems retard personal growth.

The story with cannabis is bound to be different. The marihuana smoked in the United States is rather weak, and therefore much safer than LSD. Much of our native North American hemp is so weak that a strong imagination is needed to notice psychic effects. Even the imported material which averages 1 to 2% THC does not provide the shaking experience that LSD in ordinary amounts will induce. Unadulterated hashish is 5 to 10% THC, and it is seen in increasing quantities here.

Increasing millions of Americans have tried marihuana; a third of them continue to use it on occasion, and 5 to 10% make it a way of life. Marihuana use pervades both sexes, all classes, ages, races, and religions. It has become a symbolic gesture against the establishment. It has made the harsh laws based upon pharmacologic myths unenforceable and unenforced. Meanwhile, we witness the emergence of counterculture myths about this plant.

We have little reason to believe that the use of marihuana will phase out in the years to come. Scientific evidence concerning effects of long-term, consistent use will have to await years of further research. Even if such evidence were brought forth, we know from experience that users would not necessarily be deterred. It is naive to expect that supplies can be successfully interdicted. It is to be anticipated that large numbers of people will continue to use marihuana, some to the point where they are "stoned" during their waking hours.

Nor does it appear that this product will be legalized during the next few years. The political climate is not conducive to more than reducing the penalties for possession of small quantities to a misdemeanor. It is worth pointing out that if legalization were to occur, it would still be a misdemeanor for many of today's users: minors, those who drive while "potted," and those who are "potted" and disorderly.

THE STIMULANTS

The excessive use of amphetamines and allied substances continues to increase since the 1967 summer of love gave way to the dismal autumn of "speed" in the psychedelic subculture. There are three types of misuse, each distinguished from the others by differing personal and social implications.

First, there is the intermittent ingestion of amphetamines in average amounts to help surmount some strenuous physical or psychologic stress: an examination, a long drive without benefit of sleep, an athletic competition. Such use is generally without harm, but instances of difficulty are known even under these circumstances: impaired judgment, paranoid notions, reduced rather than enhanced mental functioning. Furthermore, athletes who have taken amphetamines in even ordinary amounts have collapsed under severe muscular activity. It appears that elevated body and ambient temperatures increase the morbidity of amphetamines.

Second, large numbers of tired, asthenic people become over-involved in the continuing use of oral amphetamines and continue to swallow them interminably. They have difficulty in "coming off" the drug and may gradually increase rather than decrease the dose. Some had been introduced to the stimulant while being treated for obesity or depression. Others have started its use through friends. One undesirable effect of continuing on amphetamines for long periods in more than the prescribed amounts is the possiblity of a thinking disorder even when daily doses of only 50 mg are consumed. This consists of ideas of reference, paranoid ideation, irritability, and impulsiveness without insight into the altered thought process. It is unknown whether the arteritis recently reported in connection with chronic high-dose amphetamine abuse is applicable to the population.

The latest (in this country) and most ominous mode of amphetamine abuse is the "speedfreak" phenomenon. Speedfreaks are people who go on binges of "snorting" or "mainlining" methamphetamine or other amphetamines to the limits of their physical endurance. The doses are unbelievable to the pharmacologist and may reach 5,000 mg a day. This will provide a series of orgiastic "flashes" which can hardly be achieved with any other drug except cocaine. The difficulties of weaning the individual off "speed" are obvious. What can be offered in the sober state to compare with the speed "rush"? In addition to the irresistible delights induced by a sudden surge of "speed" to the midbrain, the drug enhances sexual sensations and acts, as do other sympathomimetic agents, to delay orgasm or prevent it altogether. The amount of amphetamines produced in this country is staggering and requires more stringent controls. However, controlling the problem by reducing licit supplies is not so simple a maneuver. These are not difficult items to manufacture; indeed, dozens of amphetamine producers located in garages and kitchens have been discovered by enforcement officers.

There are many ways of using amphetamines—with heroin (the latter day "speedball"), with barbiturates, with LSD, and even by itself. In this period when the impact of a single drug does not seem to be a sufficient challenge, the stimulants, along with marihuana, serve as the traditional psychochemical staple.

THE SEDATIVES

Another category of drugs increasingly abused are the sleeping pills, both barbiturate and non-barbiturate sedatives. Included here should be those tranquilizers that can lead to dependence, such as meprobamate. No generation gap exists in the overuse of "downers"; young and old alike may develop a heavy habit by a gradual increase of the daily dose. These are taken, not at bedtime, but during the day to maintain an intoxicated "high."

Common as these medications are in both medical and non-medical practice, they, nevertheless, have a real potential for harm. Less than lethal amounts of sleeping capsules and alcohol can potentiate each other to produce an unexpected and unplanned demise. The barbiturates remain the favored chemical means of committing suicide. The barbiturate dependency state is also a precarious one. Orderly thinking, judgment, and performance are all impaired. Sudden withdrawal from these drugs can result in a life-endangering abstinence syndrome.

THE VOLATILE SOLVENTS

Volatile substances such as gasoline, cleaning fluid, model airplane glue, lighter fluid, fingernail polish remover, and paint thinner have been sniffed by juveniles to achieve a crude intoxication. The incidence of misuse of these substances appears to be declining slightly. Some reawakened interest in laughing gas and ether is focally noted. Modern technology has provided us with another group of abusable chemicals, the freons. These are used as carriers for insecticides, hair sprays, deodorants, and other household staples. When inhaled in sufficient quantity, they can produce a delirium by excluding the oxygen supply to the brain. They have been used by a few venturesome youths, occasionally with deadly results.

The abuse of solvents and propellants underlines the point that a few people will take anything. It is not only the emotionally disturbed who will ingest unknown materials. The intense mimicking behavior of the adolescent can result in whole neighborhoods copying some bizarre drug-taking activity of a psychopathic leader. Despite the avowed desire of the "Now" generation to do their own thing, this often consists of sedulously copying the behavior of their peers.

OPIATES

The gradual increase in the number of opiate addicts (of whom 90% use heroin as their primary drug) since World War II has changed sharply during the recent past. Not only are ghetto residents becoming more frequently involved, but for the first time increasing numbers of middle- and upper-class young people are becoming addicted to heroin. The number of opiate addicts is unknown, but 300,000 is not an excessive figure. This contrasts with the 70,000 addicts known to the authorities.

Following the opening of new consumer markets, the quality of the heroin on the street has improved. A few years ago 5% heroin was considered good heroin; now 10% heroin is not uncommonly purchased on the Eastern seaboard. This improvement in the strength of heroin accounts for a few changes in the "smack" scene. The addiction is now more difficult to lose than it was a few years ago when only the "needle" was habituating. Death from overdose can occur more frequently, especially if tolerance to heroin is lost, or if good heroin is obtained for a first fix. The withdrawal phase is becoming somewhat more intense. Until recently, withdrawal was often a negligible condition which required little or no supportive treatment. This is no longer true. The larger amount of heroin in the "bag" reflects large stockpiles that back up supplies on the black market. This can only mean that sufficient material is now in the country not only to take care of existing addicts, but to open up new markets.

The identity between pusher and user is now less distinct. Many young users are pushing to support their habit, and "turning on" their friends if business is poor. The 14-year-old junkie may have been a user for four years and a pusher for two. The lack of any moral reservations about "turning on" friends and younger brothers to the heroin way of life is notable.

The treatment of heroin users on a large scale reveals some new features. In addition to revamping the classical approach to the addict (incarceration, detoxification, discharge, relapse), and the expansion of self-help programs, additional new approaches are under trial. Methadone maintenance treatment which exchanges a controlled addiction with oral methadone for an uncontrolled addiction with injected heroin is now being tested in more than 10,000 patients. For those who cannot remain abstinent, methadone maintenance is a promising therapy. Another treatment possibility is the use of narcotic antagonists such as cyclazocine and naloxone to blockade the heroin effect, although narcotic antagonists are being tried in only a few hundred patients at this time.

SUMMARY

A number of general directions can be discerned 10 years after the current drug era started: (1) The stronger forms of a drug tend to displace weaker ones. Hashish, for example, is preferred to marihuana after a time. (2) More rapid acquisition techniques tend to be used in preference to slower forms of absorbing the drug. "Speed" may be used first in the form of tablets, later it is "snorted," and eventually it may be injected into a vein. (3) More dangerous drugs displace less dangerous ones. The escalation notion that "pot" leads to heroin is scientifically untenable. For people who remain in the drug scene over a period of time, however, it is quite likely that they will explore, and some will become locked into, more powerful drugs. The graduation from marihuana to LSD is not unusual. Neither is progression from LSD to amphetamines, nor from intravenous anything to intravenous heroin.

Might it be that the round of drug excesses is reaching a crescendo? In the last year or so a few things have happened which may indicate a "peaking out" in the not too distant future. For many drug categories the immediate rewards are great, but over time the negative aspects of a life centered on drugs are equally great. Drugs can become a drag, and if all the bridges back haven't been burned, efforts may be made to return to a sober life style. We are seeing a number of the original drug proselytizers "turning off."

The positive reinforcements to being an ex-user are more immediate. It is so "in" today that it is a career—and far better than a career of using.

Until recently, the youngster who had never used was certainly one down. Now, he too, is emerging from the depths of the one-upmanship hierarchy. He isn't quite "in," but he is beginning to be admired for being able to resist the pervasive inducements to use.

Communities are beginning to accept their responsibility for the treatment and prevention of drug abuse. No longer are they denying the existence of the problem. In many instances, local programs are innovative and are tailored to fit the real needs of the families who need help.

The wheel of change turns more rapidly these days. Given some improvement in social conditions, there is a possibility of a rollback of dangerous drug consumption. It is, perhaps, too much to expect that we will respect mind-altering chemicals as we should. But at least we can hope to condition our children to approach life without the expectation that some chemical is at hand to take care of the ordinary and extraordinary stresses of living. Then our children will have the opportunity to find values appropriate for the times, and drug-dropping will have become irrelevant.

TREATMENT OF THE COMPLICATIONS OF PSYCHEDELIC ABUSE—EDUCATION AND ALTERNATIVES

J. THOMAS UNGERLEIDER, M.D.

Another conference on drug abuse? Is this yet another look at those adverse drug reactions which are now starting to affect even the youngest members of our society? Is it but another plea to "wipe out this dreaded scourge?" More hysteria and scapegoating? Another meaningless call for further research on the dangers of drugs? There have been so many conferences on drug abuse and abusers that yet another one may seem superfluous. I hope it is not. My purpose is to suggest that we pause for a *sober* look at the drug scene today. To do that most effectively, I would like to begin by tracing the historical development of drug abuse patterns in the United States over the past few decades.

For some years prior to 1965 drug abuse patterns showed little variation. Alcohol was everybody's favorite (as it still is today); lower-socioeconomic groups were often involved with heroin; physicians and nurses had access to and often became addicted to morphine and Demerol; and marihuana use was endemic among jazz musicians.

DEVELOPMENT OF CURRENT DRUG ABUSE PATTERNS IN THE UNITED STATES

Beginning in 1965, when the mass media very seductively publicized the psychedelic drugs (particularly LSD) as happiness pills, more and more people began to experiment with these chemicals. The mass media's instant access to our youth, particularly via television, had found a ready market and, as the non-medical use of these drugs increased, adverse drug reactions soon began to be seen in various hospitals and medical facilities throughout the country. Then once again the mass media led the way, perhaps out of guilt, with an hysterical overreaction to adverse drug reactions and the occasional fatal accident that resulted from the perceptual changes caused by psychedelics, again particularly LSD. Laws were hastily passed, research decreased, crash drug abuse programs began, and fear took over. These were the days of the psychedelic proselytizers and I can recall the conversations and/or public debates (often "non-debates") with Leary, Alpert, Watts, Metzner, and the like. The psychedelic missionaries at that time were true believers who wanted everyone to share their awareness via these drugs. Everyone included their wives, children, and even family pets. They were serious and dedicated; rarely did they smoke tobacco, only marihuana, and they did not drink alcohol or ingest amphetamines or barbiturates.

Then in 1967, at about the time the flower

children were driven out of Haight-Ashbury by the police and the street gangs, we entered into the next era, the new phase of indiscriminate multiple drug abuse. The average age of the "bad trippers" dropped from 21, in our original studies, to 14 or 15. No one seemed to care what they were getting on the black market, and now drugs were used not to help change a life style but for a "high," and in order to get "stoned," which state often included delirium and occasionally loss of consciousness. "O.D." became a household word for overdose.

CURRENT STATUS AND PROBLEMS OF DRUG ABUSE

And now? Gone are the days of initial shock, and of gross denial—the "not in our school" phenomenon. The one-shot high school drug lecture as the primary educational tool for combating drug abuse in the young has faded.

And where are we today? For one thing, drug abuse now is "where the action is." For federal grants, it's where the money is; for local programs and legislative direction, it's where the interest is. Politicians admit that it's political suicide to be against drug abuse programs. Most major public officials have their advisory counsels on drugs, although their advice is often ignored. Famous entertainers who have either made movies glorifying the use of drugs or who have suffered personal drug tragedies have now become advisors to legislators, to the federal government and even to the President. Rock and roll musicians and mystics advise mayors. These advisors often command huge fees for lectures on drug abuse; they evaluate education films for national drug councils and appear in Playboy Magazine as drug panelists. In short, the public has issued a cry for instant expertise and so instant experts spring up with all sorts of unproven facts, nostrums, and panaceas mixed with some good advice.

Peer pressure to use drugs is great today; many of our young people are afraid at age 13 or 14 to admit they *haven't* taken drugs. This has become their greatest secret, much more so than guilt over sexual impulses or activity. Many young people I know who do not use drugs are repeatedly told by other youngsters that the way to become famous is to use drugs, "put them down" (i.e., discontinue usage), and then go around and give lectures on the topic as ex-addicts. This, our young people are assured, is one way that adults will listen to every word they say, send them on (traveling) trips, and render them a great deal of adulation. Several of our DARE (Drug Abuse Research and Education) teenagers who are non-users, but long-hairs, were told by some television reporters at a recent Kiwanis national convention that if they would talk about their drug abuse they would get top billing on that day's news show. When our youngsters said that they didn't use drugs, they were told that any other of their efforts, whatever they might be, were not newsworthy.

It should be obvious to us all that the effect of a drug depends on the pharmacologic action of the drug, the dose, the set (or attitude of expectation), the setting or environment in which one takes the drug, and the personality of the user. This is particularly important to keep in mind in an age when public officials and the newspapers alike proclaim that the My Lai massacre could only result from the use of marihuana, and that the Tate murders had to be due to LSD.

We keep forgetting that drug *use* and drug *abuse* are very different. Drug use, while not to be applauded, is something that is accepted in our society. It is often defined as that drug use which interferes neither with an individual's internal (intrapsychic) equilibrium nor with his external adjustment and social functioning. This includes most occasional curiosity-induced experimentation as well as recreational use even of many illicit drugs.

The problem of definition becomes difficult

when legal sanctions are imposed against the possession of a drug such as marihuana, although use and abuse can, by our definition, be differentiated. This is in contrast to a drug such as heroin, whose use and abuse are harder to differentiate.

Drug abuse treatment programs in each community proliferate rapidly; it is important, indeed vital, for each community to tackle its own drug abuse problem. But we must remember that drug abuse is a symptom—often a symptom of social pathology and unrest, but, particularly for young people, often a symptom of their own chaotic internal mental life as they struggle to develop the kinds of skills that traditionally have been equated with adult maturity. These skills include the ability to handle sexual and aggressive feelings in socially acceptable ways, to experience closeness, and to develop frustration tolerance and goal-directed behavior.

Words bandied about to describe the absence of these skills in youth include alienation, plastic, and anomie. Curiosity (engendered by adult drug abuse), rebellion, peer pressure, recreational pursuits, avoidance of internal conflicts, the boredom that results from non-challenging education and the lack of meaningful job and growth opportunities, affluence and permissiveness in upper social strata—all these are involved to some extent as the "cause" of widespread use of drugs among our youth today.

In group therapy with drug users, when the participants begin to "really" relate to each other, drugs are rarely mentioned. They no longer argue which drug should be legal and which drug shouldn't. That kind of behavior is usually employed by youngsters to test for adult reaction whether they are suitable identification figures or "hopeless moralistic squares." Thus when community programs begin to tackle the problems of drug abuse they are taking on many other social problems and should so anticipate. These other problems are often much more difficult to solve. Of course it's easier to focus on drug abuse than on poverty, or on the lack of meaningful alternatives to the use of drugs, particularly when we are anxious and frightened.

We often tend to blame others for the entire problem. How often have we heard responsibility for the drug problem placed solely at the door of rock musicians (The Beatles), liberal university professors (pot-smoking), strict laws (felony for possession), lax judges (probation for pushers), physicians (over-prescribing), pharmaceutical companies (over-manufacturing), the Communists (giant plot), the Mexican government ("on the take"), commercials ("better living through chemistry"), police (heroin pushers go free, pot-smoking students get convicted), educators (moralistic drug information), education (many drug lies), and Vietnam (of course).

DRUG ABUSE EDUCATION

Let us focus our attention on the question of *education* about drugs and drug abuse. Why educate about drugs? Where should we do it? Who shall be educated, and by whom? What shall we teach? And how?

Why education? For one thing we assume a tremendous increase in the incidence of drug abuse among young people. Of course, we really don't know how common drug abuse is because it's against the law and thus questionnaires about drug use are notably inaccurate. With peer pressure in the direction of using drugs, many young people who do not use drugs are reluctant to face ridicule by admitting it, so they say that they are users. Others who do use drugs are afraid of being placed under surveillance by police, having their school lockers searched, etc., therefore they often do not admit (on surveys) to having used drugs. All we can say is that there is a great credibility gap between the frightened school administrators (who say such things as "less than one-half of 1% of our student population has ever used any drug including marihuana") and the student body leaders, not the "dopers" (who may say "70 to 80% of us use marihuana regularly"). The truth probably lies somewhere in between.

Police arrest figures reflect only the vigor of arresting officers and their willingness to complete the paper work. (Were the 30,000 drug arrests in Los Angeles last year 1%, or 5%, or 10% of the drug-abusing population?) For example, one nightwatch of 15 police officers recently explained their working rule: if they saw less than five amphetamines or barbiturates when they stopped a car they would pretend not to see them (illegal possession of one "upper or downer" in California is a felony *arrest*). And of course, hospital and medical records reveal only that part of the drug-abusing population who have adverse reactions for which they seek help from traditional sources; many seek it from more than one source, and many from non-medical helpers such as Gurus, fellow users, etc.

The law reflects the general concern about the drug problem and many states like California require education about drug abuse. But the details of this education are not spelled out by law. The assumption is that the effects of drugs and the side-effects of their abuse should be taught in the schools. There seems little disagreement with this hypothesis, as there is with the controversy concerning sex education in the schools, for example. But it does assume that the family cannot adequately cope with the drug-taking (abusing) behavior of its youthful members. Why not? What of the parents? As parents and adults we have created one of the greatest drug-abusing societies the world has ever known: billions of amphetamines and barbiturates are manufactured each year, six to eight million people abuse alcohol, and so on. In addition, many parents themselves are frightened and anxious about their youngsters. They have raised their children in a very permissive society. They believe that to say "no" to their children means they don't love them. I see teenagers with their parents every single week, even young adults of 19, 20, and 21 who say to their parents, "I will not work, I will not go to school, I will live at home, and I demand an allowance (up to $5.00 a day) which I will use to buy drugs. If you *dare* to refuse me, I'll move out and become involved with heroin addicts and you will have failed as a parent." Blackmail? Yet 99% of the parents that I see have accepted this, and they wind up actually supplying the money which provides the drugs for their children (they often blame you, Benjamin Spock, but did you really do this to us?). So the answer to *where* we should provide education concerning drugs must include not only the classroom but the home.

Educate *whom?* Obviously our youth, in school and at home, and the parents and teachers who will do this educating. Who else needs education? Law enforcement officers, for one. Many police officers need to learn that long hair, bare feet, or a beard do not always identify drug abusers. So often "pushers" who are short-haired get away scot-free because some police are conditioned to look for "hippies." The effect on youth of this attitude of law enforcement far transcends the drug problem; when the ethos in society becomes "if you don't like a law break it" and "all the fuzz are bad," then the total system of law enforcement, of protection, and of respect for authority is in danger.

And what of the responsibilities of manufacturers and advertisers? Can these people be educated about the effects of their blandishments to "turn on"? One has only to watch the TV commercials to see that we are constantly bombarded to try such and such beer and to take "Sleep-Eze" to get to sleep at night, "Compoz" to relax through the day while at work, and "No-Doz" to stay awake. No-Doz, labeled "Exam Pill," was contained in a welcome kit, along with toiletries, that was distributed to all incoming freshmen in the fall of 1970 by a University of Michigan student bookstore. If one has a headache, it is described as "Excedrin Headache #12" by the nation's current quarterback idol as he brandishes the bottle of pills on his TV show.

Educate as to *what?* What are the *goals* of this drug abuse education? Our goal must be first to provide young people with the ability to make decisions for themselves. These decisions involve taking risks (risk-taking behavior), in this case those involved in drug use. I am convinced that young people are responsive to such risk factor information, for the most part, when it is given honestly. At least many young people now seem concerned enough about the intensity, side effects, recurrence, and unpredictability of LSD abuse to avoid that drug. And neither the physically damaging glue-sniffing nor the lethal arsenic and strychnine ingestion have ever reached major popularity despite their nice "high," which we'd call a toxic delirium. The risk factor information approach to drugs rather than a morality view should be taught at school. Morality should be taught within the home, but it is not a moral issue to say that one plant is inherently good while another plant is inherently evil. Young people consistently reject this notion. It has been a great problem when adults maintain that certain drugs such as amphetamines, barbiturates, and alcohol are "good" drugs and marihuana and other psychedelics are "bad" drugs.

Is our goal to control drug abuse or to eliminate it? Although some service organizations have set as their goal for the year the absolute elimination of drug abuse in this country, this is most unrealistic. Some people will always seek chemical relief from their troubles or for pleasure. What we can hope to do is combat some of the contributing social factors in drug abuse and thus reduce the numbers of drug abusers to about the 7% of the population—the percentage formerly estimated for our society.

But we are not just educating youth about the dangers of drugs. Many know these dangers and use drugs anyway in a form of primitive denial. Witness the recent increased use of barbiturates and heroin in affluent areas. Education must also be directed to us as adults, and must include lessons in hypocrisy ("good" vs. "bad" drugs), in respecting our young people, and in providing them *meaningful alternatives* to drugs—jobs, responsibility, and respect for themselves.

Education by whom? Drug education for youth must be by adults who are knowledgeable, who are honest, and who hold themselves up as identification models for young people. So often we adults abrogate our responsibilities and insist that only ex-users can relate to young people concerning drug abuse. Although ex-users can be most effective, many ex-users are not really ex-users at all but current users of psychedelics, who pay lip service to the negative effects but spend hours describing the joys of psychedelia. It is currently fashionable to be "against" amphetamines, barbiturates, heroin, and perhaps even LSD, but to be "for" mescaline and marihuana as the only way to endure Vietnam, pollution, and the establishment. What a tragic message for young people, at ages 13 and 14, who are struggling with their own identities, and are unable to cope. Criteria for educators might include, besides knowledge of drug abuse, the ability to be honest (to say "I don't know" when they don't know), to reveal their own attitudes, to tell why they themselves do not need a drugged way of existence, and to like young people; educators should *not* be psychedelic drug missionaries. So often, lecturing at schools, I hear that the best teachers are those who smoke marihuana and who often in very subtle and covert ways encourage young people to do likewise.

How to educate? This depends on the age of the youngster. The latency (elementary school) student, who is extremely curious and not yet rebelling, is effectively taught (not propagandized) by different techniques than is the rebellious high school student or the philosophizing college student. But the *content* of all educational programs must be scrupulously honest, including the appropriate "we don't know's."

The how of education also depends in part on the policies of each school, including recognition of such things as privileged communication and the ability to allow young people to obtain honest drug facts and not mythology from someone, perhaps an "ombudsman," who can counsel them without turning them over to the police. Since there is no teacher-student privileged communication like that between physician and patient, local law enforcement must be willing to allow such a program to take place.

TREATMENT OF DRUG ABUSERS

And what about treatment of the hallucinogenic reactions, including those due to marihuana?

The most common symptoms in patients having adverse reactions to psychedelics are auditory and visual hallucinations (including illusory phenomena); anxiety to the point of panic; depression with suicidal thoughts or attempts, or both; confusion; and paranoid thoughts. We should note that these symptoms are also frequently experienced as part of the reaction to the perceptual changes that regularly follow ingestion of LSD, and not everyone who experiences them is upset enough to seek aid. Thus, whether the acute effects from LSD are called a "freak trip," "bummer," or "freak out," as LSD users describe the adverse LSD reaction, depends not only on the perceptual alteration itself but the *user's reaction to it.*

The best way to determine who is on the drug is by questioning to secure a careful history. Prompted by a missionary zeal, the user is very prone to admit that he is "on a trip." Of course, there are some paranoid schizophrenics who complain of being poisoned with LSD by their persecutors, such as those who claimed they were being poisoned by Communists a few years ago. Upon investigation such persons are usually found to have chronic histories of mental illness; they generally know little about LSD and have not been to any of the few places where it might be placed in their drink, such as an "acid-test" party.

Treatment of the acute symptoms must be directed first toward preventing the patient from harming himself or others. We use interpersonal rather than mechanical restraint. Warm support and reassurance are usually quite effective. The patient should be encouraged to relax and not try to differentiate which perceptions are real and which are due to the drug's effect. (The guides say, "Flow with it.") Chemical agents are occasionally required as part of the treatment regimen. Chlorpromazine, Librium, and even the barbiturates have all been reported to be effective. It is also important not to intrude physically upon a calm patient. Conversely, a panicky patient cannot be allowed to remain out of control and become destructive; if talking and reassurance are not effective, physical restraint may be required eventually.

The acute side effects can be expected to wear off within a few hours or a few days. Although some patients may not recover even with treatment for a period of months, many users will return to hallucinogenic drugs despite severe reactions.

The treatment of the chronic psychologic effects of the hallucinogens, involving persistent psychoses, personality changes, and flashbacks, consists of psychotherapy with the goal of uncovering the reason why the patient has needed to maintain a drugged way of life. Conflict areas often include inability to handle sexual and aggressive feelings, dependency conflicts, and faulty self-esteem. Many persons experiencing these changes were severely disturbed before using LSD, while others appeared quite stable and normal by every indicator before the LSD experience. It is important to help the patient develop feed-back mechanisms by which *he* can measure his progress at work, in school, and even in relationships.

Because of the poor results in treating the

heavy drug user by traditional psychotherapeutic means, innovative (though often unproven) treatment regimens have proliferated. These include utilization of peer group pressure, half-way houses, live-in families, non-users as surrogate/auxiliary egos, and so forth. Some of these are covered elsewhere in this volume.

PANEL DISCUSSION—HALLUCINOGENS AND MARIHUANA

Moderator: DR. HOLLISTER

Panel Members: DRS. COHEN, DOMINO, HOLMSTEDT, and UNGERLEIDER

Dr. Hollister: To me the most interesting part of a meeting is a free exchange with the audience. I would hope that this next hour is equally interesting on both sides. To help make it so, I ask the questioners to be brief in posing their questions and the panelists to please be brief in answering.

Dr. Ben Nichols (Cleveland, O.): I address the Chairman of this symposium and the entire attendance. The consensus of the drug abuse experts today appears to be that marihuana at best is a drug that we know very little about and is at worst a drug with legal repercussions that range anywhere from seven days in jail in Nebraska to life imprisonment for dispensing to a minor in many other states, including my own. I realize that doctors don't often engage in politics or legal polemics, but, as Dr. Smith so aptly put it this morning, our silence is a vote for the status quo. This symposium, then, must not remain passive on this point. I would therefore propose that the symposium and the faculty and panel experts be polled and that a formal resolution be adopted and sent to President Nixon saying that we don't support present legal criminal actions against the possession and use of marihuana and we do support further research and education and non-punitive treatment for drug addicts.

Dr. Hollister: I'm not sure of the propriety of adoption of a formal resolution by this symposium, even if one could be obtained by a majority vote. However, I have no objection to expressing individual opinions. Your question really is: "Should pot be legal?" Dr. Domino, what about it?

Dr. Domino: What dose? I might say the same thing is true of all other drugs including LSD. What dose are you talking about? Cyanide is used as an agent to determine circulation time. A larger dose and you're in the lethal range. It depends upon the amount of THC that is contained in cannabis preparations. Although we know very little about the molecular and ultimate pharmacological mechanisms of Δ^9-THC, humanity has had 4,000 or more years with cannabis preparations. The issues are what dose, and can society control a drug by legislating dosage.

Dr. Cohen: I don't really think that the speaker has asked for legalization of marihuana. He's asked for a change in the model from a criminal one to a medical one for the possessor and user. The criminal sanctions have not worked and have led to secondary adverse effects which are very harmful for the community. I'm quite eager to see states adopt what recently has been adopted by the federal government; namely, reductions in penalties for simple possession and usage. I think we are now at a level of information about marihuana to indicate that this is certainly not a felonious item to possess for one's own use. I would hesitate to legalize until we know more about the long-range effects of consistent use. I wouldn't like to have another substance introduced into the culture through our ignorance of its adverse effects. Alcohol and tobacco fit in here, and both do really produce serious damage. Therefore, I would say, "Let's wait until the studies that are under way regarding chronic use are completed."

Dr. Hollister: Are there any additional thoughts now? I can't think of anything that has not been said. Let me remind you that the afternoon session tomorrow will consider the political and legal aspects of drug use. I'm not sure if the last questioner was aware of it, but the new law enacted within the past fortnight does include some rather interesting changes.

Dr. R. G. B. Bjornson (St. Paul, Minn.): I frankly hate to leave this question of legalization after only a rejoinder which is relative to dose. I think we might use our historical awareness of the problem with alcohol, not a perfect example but the best we have. To talk in terms of alcohol, then, it was legalized without regard to dose. The limitations that we placed on alcohol concerned such things as drunken driving, for example, when the substance was used in a dose which caused the individual to be some kind of threat to society. Now what we've done here is basically say to the individual that he may not use this in any dose, entirely on the basis of some supposed disadvantage to himself without regard to the extent to which he might under its influence threaten society. In light of this, I would like to ask what sort of educational program can we support relative to marihuana? The more we have learned about it, it seems to me, the less ground there is for supporting the law in its present form. I think this is a serious hang-up. I know it is in the educational community. We have had teachers come to us and say that the more they attend drug education programs, the less they have to tell their students because of the increasing discrepancy between whatever facts are known and the law. And incidentally, relative to facts, it seems to me that it is a little bit misleading to suggest that we don't know enough about marihuana to make some judgment about it. Now I admit we don't have detailed laboratory-type data, but we have an absolutely fantastic historical clinical trial and I don't think we should entirely throw this away.

Unidentified member of audience: Mr. Chairman, I have a question which is corollary to that. What actually do you advise for any dose? You name the dose. At what specified dose level would you support any form of legalization of marihuana today?

Dr. Hollister: Let me take a crack at this. I think the whole problem is not one of legalization, but simply making marihuana use less illegal. And there is a vast difference which people don't realize. If you make marihuana legal, then you must set up a bureaucracy to regulate the manufacture, distribution, sale, tax collection and all the rest, such as with alcohol regulation. This is not our primary goal at the moment. The primary goal at the moment is to make sure that the kids who are playing around with the stuff don't get busted and go to jail or acquire a record by being booked for a crime. These are the things that concern us. There is a very simple way to eliminate this, and that is to make it so that anyone who has marihuana, or for that matter any other drug, for personal use with no evidence of sale has committed no crime. It's that simple.

Dr. Domino: Since I raised the issue of dose, perhaps I could expand a bit further. Suppose we legalize marihuana today and say anything containing below 1% or 2% Δ^9-THC is legal. If we do that—given the American get-up-and-go system of free enterprise—we'll be able to get a "high" at a very reasonable price. Now, the question is, what's to prevent anyone from taking this material into his bathroom and adding some pure ethyl alcohol to the preparation, filtering out the weeds, and getting an extract of resinous material. If he then volatilizes off the alcohol, essentially he has hashish. If we legalize on the basis of dose then there's no question that we're going to have some entrepreneurs in our society that are going to get the stronger stuff at least through their bathroom arrangements. Society, therefore, cannot regulate these compounds on the basis of concentration. The real issue that is facing our society is "Do we really want another intoxicant?" If the answer is yes, then let's legalize marihuana. But if the answer is no, then let's not. This is the critical issue. Has alcohol done enough, or hasn't it? Do we need another intoxicant? It is in this context that we have to look, rather than at the question of dose. Consider amphetamine. Is it a hallucinogen? If the dose is large enough, you bet it is. Are we going to legalize amphetamine in low dose? Of course not.

Dr. Alan Reed, Jr. (Milwaukee, Wis.): I think the entire issue of dose is very, very faulty. The basic concept in the early days was that marihuana was one thing but hashish was something altogether different. Clinically I have not observed that. I think that if people are used to high dosage of cannabis, they can either smoke a lot of marihuana or a little bit of hashish and get to the same place. Or if they like a lower dosage, very little hashish or a little bit more marihuana. To me, it is sort of saying, "If we just make hard liquor illegal, then we won't have a drinking problem because who can get in trouble with beer?" Would there be general comment on that?

Dr. Hollister: Professor Holmstedt, haven't some of the Scandinavian countries taken a dose attitude toward alcoholic beverages?

Dr. Holmstedt: Until about 10 years ago Swedes were allowed a certain quantum of alcohol per month, but that has been abolished.

Dr. Hollister: And now we know what's happened to those Swedes! No, I was thinking particularly of Finland where they do make this distinction between what might be called fermented beverages and distilled beverages. You can get beer and wine cheaply in Finland with very little restriction. But you pay dearly for distilled spirits. This is a method by means of taxation to control the dosage form of a drug. It is still too early to see how well it works, but there are many other precedents in which either legal or tax controls have been used to regulate the dosage form of alcohol. So, it isn't entirely beside the point to talk about dose. Now I am a drinker of long-standing, and I would submit that there is a difference between beer and gin.

Dr. Reed: I have one other question for Dr. Hollister. It seems to me that you talked basically about two different groups of drugs, the one being pot and the other the LSD group. I noticed that under the LSD group you put possible implications of social problems, but under the pot group you left that category out. I wonder if that was an admission or an omission.

Dr. Hollister: No, it resulted from trying to cover a broad area in about 20 minutes. Precisely the same two slides used for LSD could be applicable to marihuana. Increasing reports of similar adverse reactions from marihuana resemble in every respect those from LSD, which are what one might expect from very high doses of marihuana.

Mr. Lawrence Sherman (Ann Arbor, Mich.): It seems to me with very few exceptions, the terms "drug use" and "drug abuse" have been used synonymously. I would like the opinion of the panel as to whether they are synonymous, or whether there are some guidelines to indicate where use becomes a medical or social problem and becomes abuse.

Dr. Hollister: Tom, I think you touched on that in your talk, didn't you?

Dr. Ungerleider: As I tried to emphasize, there is a great difference between drug use and drug abuse. Perhaps one can define the *use* of drugs as that which interferes neither with an individual's internal (or intrapsychic) equilibrium nor with one's external adjustment in behavior and social functioning. Now sometimes that's easy to measure and sometimes it's not. As a psychiatrist I often see young people who use whatever drug as a way of avoiding relating to people. They relate through the obtaining of drugs, sharing the drug experience, and enjoying the attendant secrecy and forbiddenness. It is an interaction that one does not see easily, but nonetheless it occurs. I would like to make one other comment. I think the tremendous conflict about and interest in the "legalization" of marihuana as evidenced here today by applause on both sides of the question means that we are finally aware of how we physicians have abrogated our responsibility in this area. We are outraged, and properly so, by some of the previous practices of law enforcement. I hope though that we don't throw out the baby with the bath water and let the pendulum swing too far the other way. I think it *is* meaningful to say that we don't yet have enough research to "legalize" marihuana in any form. For example, we don't have a test for marihuana in the blood, breath, or urine. We don't know what it is like to drive under the effects of marihuana (even with the Seattle study, as reported in that medical journal, *Playboy*), and this would be very helpful since kids often have accidents while under the influence. They get "stoned" driving around Disneyland before they go in, and "stop signs, parked cars, and trees" hit them. Wouldn't it be nice to obtain a bit of research on this without hysteria or panic before we give cultural sanction to this drug as we have done with alcohol? Now, this is in no way in defense of the felony for possession which has been a disaster, but it is a defense and an argument *against* making it totally legal and *for* making it a misdemeanor (which is like spitting on the sidewalk). Even the most flamboyant estimates of marihuana use (I think the most flamboyant one is that of Joel Fort, a good friend of mine, who estimates 20 million users) fall far short of the 100 million-plus users of alcohol. We just don't yet know what the liability or "risk factor" of this drug is. This is by no means my way of sitting on the fence.

Mr. Sherman: I have another question that I would like to direct to Dr. Ungerleider. One of the projects he has been involved with, which was described at the National Free Clinic Symposium at the American Psychiatric Association annual meeting and many other meetings, is called "Project Dare" in Los Angeles. As I understand it, it is an educational project which shows a beautiful interaction between physicians and young people. I think this group could benefit from a description of Project Dare.

Dr. Ungerleider: Thank you. I am planning to go into that in some detail Friday morning when we are going to have a presentation on Guidelines for the Practicing Physician. At that time I will show some slides and describe our non-using teen-age peer group. Thank you for your kind words.

Dr. Robert A. Braun (Detroit, Mich): I would like to make a brief comment regarding marihuana. I am concerned with the breakdown for the respect for laws in our society. I believe that the fact that we have more or less given up the use of the laws as they were intended regarding marihuana has a bad effect on the whole of society. I wonder whether it is preferable to have these laws and try to keep a few youngsters away from marihuana, or whether we should try to inculcate more respect by making laws which are possible to enforce. On the other hand, I agree that we don't know the long-range effects of marihuana, and I certainly would be very interested to wait until we have them. I believe from what has been said today, and from what I have read also, that the authorities agree the short-range bad effects of marihuana are negligible. I, therefore, hesitate to say that we should allow anyone and everyone to use marihuana as much as they like. But then, I always contradict myself it seems. I am told by young people that to some extent the use of marihuana is self-limiting. This has to do with the drug dose-effect relationship which Dr. Domino described. I am glad he did because in most cases, it doesn't make much sense to ignore the dosage. From what I have heard, young people smoke it slowly and the whole evening goes by. So this drug seems to be self-limiting for practical purposes. I am not talking about THC. I am talking about the practical everyday use of the drug. And finally, one more aspect on the other side. I believe that young people by using this or any other drug have a tendency to "cop-out," if I may use their term. There are many shortcomings in our society which the young people care so much about: instead of sitting in a corner and smoking marihuana I wish they would do something and join those forces who want to improve our society actively rather than withdrawing from society and smoking.

Dr. Domino: Perhaps I could begin by clarifying my own position on the issue of whether marihuana smoking is innocuous. I should preface my remarks by saying that my personal experience is limited to research studies that we conducted with co-investigators at the Lafayette Clinic in Detroit, using a rather low-grade quality of marihuana which we obtained from NIMH. Experienced marihuana smokers took it to the point of a social "high," with the agreement that they smoke until they got as high as ever in their life. We called that 10 on a scale of 0 to 10. Many of these volunteers smoked to a high of 9 or 10 using as much as eight marihuana cigarettes or about 2400 mg. I will never forget as long as I live the feelings that I had after I observed the first few subjects. I called my wife and said, "What gives with marihuana? What is the big deal? It's a big nothing." We published with Drs. Rodin and Porzak an article on this in the JAMA (*213*:300, 1970) pointing out that at the social "high" the effects of marihuana were essentially a very mild organic brain syndrome. Dr. Seevers, my chairman, said that was the worst article he ever read. I agree with him in part because the basic principle of dose-effect was not included since we were only talking about very mild stuff. At this level it causes a mild organic brain syndrome which is temporary and reversible. The second issue that Dr. Braun raised is whether marihuana smoking is self-limited. There are data on this. My own experience is that it apparently depends on the subjects you study. Most take it to the point of a social "high" and actually refuse to take more. These people are in the large majority and apparently titrate themselves by smoking to a specific end-point. A few make pigs of themselves and take marihuana until they are severely intoxicated. Thus, the idea that smoking is self-limited is not really true, at least in some hard-core users. I would object on both accounts. Marihuana is not innocuous, and it is not self-limiting in a small percentage of people.

Dr. Arnold M. Leff (Cincinnati, O.): I direct my question to Dr. Hollister. Concerning your clinical tests for marihuana on cognition, were the subjects that you used mainly people who had not tried this drug before? Also, if you have studied chronic users, do you find any differences in their reaction to the drug?

Dr. Hollister: These days, according to the ground rules, you must use non-virgins. I don't know who thought of this one. In any case, the people who volunteer for such experiments must have had some prior use. By the same token, we try to avoid the chronic drug-head and usually stipulate that they will not have had marihuana within the week preceding a drug trial, or more than several times during the preceding month. The implicit thought in your question is the one that was brought up by Weil's study suggesting a difference between naive and chronic users; I can say categorically that there is not, at least in terms of responses to adequate doses. Only when you deal with threshold doses do the elements of suggestibility, expectation, and so on enter into it, which is what he was observing. When we used Δ^1-THC in fairly substantial doses, in more than half of the subjects who were completely naive, there was never any question about an effect. It was quantitatively no different from the same doses given to experienced users.

Dr. Holmstedt: May I ask you a question? When you gave the THC, how did you give it? I never got that out of your presentation.

Dr. Hollister: You mean the synthetic Δ^1-THC? We would open the ampule on the spot, pour it in a glass, mix it with enough water to dilute the alcohol and immediately drink it. And then wash the glass pretty well because it does tend to cling to the sides.

Dr. Holmstedt: Yes, but that was an oral administration.

Dr. Hollister: We have used it in smoking, however.

Dr. Holmstedt: Comparing the inhalation to the oral. . . .?

Dr. Hollister: Well, unfortunately, we have not had a chance to use Δ^1-THC to make a direct comparison. When it is cleared again for human use, we will do this using a closed system with a micronebulizer, so you can be sure of delivering about 80% of your dose by inhalation.

Dr. Holmstedt: The reason I asked is that there is another thing besides dose and that is the means of administration. Hashish in the Orient and in India is mostly eaten, and until recently very little has been known of how much of the active substance in the marihuana cigarette is transferred into the mainstream of smoke. We now know this because the smoke can be precipitated on an electrofilter without interference to the respiration of the patient. It turns out to be approximately the same as the percentage transferred into the mainstream of smoke of alkaloids like nicotine—somewhere between 14% and 19%. What is in a marihuana cigarette comes out into the mainstream smoke. And then, as you know, we have a second step and this is the means of inhalation. By superficial inhalation from a cigarette you absorb about 5%, but if you inhale very deeply and hold your breath, about 95% is absorbed. This is basic knowledge about aerosols and respiratory physiology. The latter type of inhalation is equivalent to an intravenous injection, of course, rather than to an oral dose of a synthetic compound.

Dr. Hollister: This whole problem of administration is a sticky one because we don't know how to deliver a dose by inhalation with any exactitude. I am not sure we can do a whole lot better by oral administration, but at least whatever error is in the system should remain fairly constant.

Unidentified member of audience: My question is, what are the experts going to do about research and about reaching some solution between what I see as a gap between people who have not used the drugs, do not have the experience and are attempting to learn as much as they can in an effort to set guidelines for themselves?

I also believe that men who do have a lot of information are being moved out of their positions. Dr. Cohen is no longer with the National Institute of Mental Health, where he was doing and could do a great deal. Andy Weil who is a researcher at NIMH, or used to be, is no longer with it. Norman Zinberg does not have funds. And there are the other people who really are able to do research but are not funded. So again, my question is "What are you going to do about the situation?"

Dr. Hollister: Well, until very recently, Dr. Sidney Cohen was "Mr. Money Bags," so Sid, tell us what has been going on.

Dr. Cohen: It may well be that a place like the NIMH has not communicated well, but I am impressed with the amount of marihuana research that is going on these days. It is much greater than it was only a short time ago. At every session in which we approve supplies of drugs to investigators, increasing numbers of projects are being funded. So it is out of ignorance that you are speaking when you imply that there is little or no research going on. Much of it is in humans, not animals. A good deal is a subjective type of exploration of the boundaries of the state, and so I think that we perhaps have erred in not letting you know about the current situation with marihuana research. You have erred by not being right up on top of all the reports that are coming through now.

Dr. Hollister: Let me make a little additional comment here. Concerning some of the very important problems, ethical constraints make laboratory research impossible to tackle. The whole question of chronic brain damage with repeated doses, or the question of tolerance with repeated doses, for example, I can't tackle. I can think of many experiments that would be quite useful that I cannot do. To know what they shouldn't do is the great responsibility of people who do research in humans. So don't think we are flubbing the important questions; some of them simply can't be answered in the laboratory.

Dr. Mel. L. Barclay (Detroit, Mich.): Dr. Hollister, in your presentation you stated that there were no medical uses for cannabis. Dr. Brenner intimated, however, in his presentation that cannabis is used in other countries during labor, for instance. And Dr. Robert Walton in his book cites several anecdotal reports of cannabis use as a soporific, analgesic, and sedative. Are there any well-controlled non-anecdotal clinical studies which have been done using your purified derivative?

Dr. Hollister: No, but undoubtedly there will be a lot more exploration of possible medical uses. I alluded to a couple that we looked into, but I am sure there will be a lot more in treating depression, migraine, and other notions that have been propounded. As yet there is no established medical use. When one thinks of medical use, you also have to consider the comparison with other available drugs used for the same purposes. Any other question on the medical use?

Dr. R. G. B. Bjornson (St. Paul, Minn.): I am sincerely trying to address myself to the question of what our legal stance as a society should be. Are you, for example, as a panel willing to express your own feelings as a matter of help to those of us attending this as to whether, for example, we should remove the criminal penalties from possession and use of marihuana?

Dr. Hollister: Your question has been answered several times in the most explicit way I can imagine, but communication is a two-way street. It does seem to me that we ought to let the legal people have their day in court, and they will tomorrow.

Dr. Cohen: I think this seems to be a central issue and even though you may be right that we have dealt with it extensively, still it is a concern of many people, and maybe we ought to go further. Let me give you my personal response to your question. At present there are two studies, and others are being added which will look into the long-term effects of marihuana in countries where its use has been traditional. The hope is that we can have a look that is a little better than everything that is in the literature up to now about chronic use. Very simply, a group of users over the years is selected. A matched group of non-users is gathered from the same population. They are all examined rather thoroughly to determine whether any physical or neurological or psychiatric damage has occurred over the years to one group or the other. Now, when these studies are through, which should be within two years, and if indeed they indicate that the user group shows no substantial adverse effects over the non-user group, I would be very willing to entertain the possibility of the legalization of marihuana. I would like to wait until the results of these, to me, critical studies, are available before making that decision. That is approximately how I respond to the question of what should we do today. I would say, "Wait."

Dr. Domino: It seems to me that the present laws, even the recent version just passed by Congress, are not as good as they ought to be. I think the penalties on marihuana must be markedly reduced. But we ought to have perspective. Marihuana belongs in a class with other agents like LSD and DMT. We must consider them as a whole and not marihuana as a separate issue. There is no question that the social "high" is a minimal phenomenon. It is the big dose which causes hallucinations that I have always been concerned about. There will be escalation to big doses if we make it available for everyone. As a pharmacologist, I am unhappy with our present analgesics and convulsants, antihypertensive and pre-anesthetic medications, and I look toward possible medical uses of some of the synthetic or naturally occurring derivatives of cannabis. I know of one pharmaceutical company that is going to invest millions of dollars looking for a potential therapeutic agent among the THC derivatives. But legalization today? No, sir.

Mr. Larry Weisenthal (Ann Arbor, Mich.): As a person who has only a layman's knowledge and a layman's opinion on the scope of the problems of the social aspects of drug abuse in this country, I nevertheless have the distinct impression that the abuse of drugs such as heroin and LSD and even barbiturates and amphetamines is a much greater problem than the present abuse of cannabis-like drugs, notwithstanding the great incidence which these latter drugs are used. I would like to direct this question to Dr. Cohen. Is there any evidence that chronic use of marihuana leads to the use of more dangerous drugs, by any mechanism other than through the association of the consumer with the distributor whose merchandise often consists of very dangerous drugs in addition to marihuana? As a corollary to this, is it not conceivable that the legalization of marihuana could lead to a decrease in the overall level of drug abuse?

Dr. Cohen: I can't think that the legalization of marihuana would decrease the level of the use of marihuana or the use of other drugs of abuse. As to the escalation bit, it is certainly not a pharmacologic phenomenon. I don't know about the fact that the "pot-pusher" also has other goodies and that this is a reason for escalation. I am impressed that in surveys that have been done, there is only one kind of pot-user who seems vulnerable to escalate, and that is the "head." Here is an individual who, in almost every survey that has come forth, seems to be the person who wanders on to other types of drugs more dangerous than cannabis up to and including heroin. I might also mention that the "head" in psychological studies can be differentiated from non-users, experimenters, and social users. These latter groups cannot be differentiated from each other on psychological testing. The "head" appears to be a more disturbed individual whose needs are such that he does seek out more dangerous drugs than marihuana in his search for relief from his emotional shortcomings.

Miss Beth H. Flynn (San Luis Obispo, Cal.): I listened to a recent pediatrics tape from Audio Digest and I was led to understand from the speaker that the more an individual smoked marihuana, the less it took for him to get his "high." In talking with numerous students in our high school, I also found this to be what they felt. Is there a physiological basis for this? Also, is it true that when an individual uses marihuana several times he gets to the point where he knows what type of "high" he wants and so he is going to get it immediately?

Dr. Domino: This whole area is mixed up with a lot of folklore and nonsense. Two recent double-blind studies, one by Jones in San Francisco, California, the second by

Mano and his co-workers in Indianapolis, Indiana, have given experienced marihuana smokers THC-extracted marihuana cigarettes as opposed to active marihuana. In other words, with and without Δ^9-THC. Fact: both groups have shown that the experienced marihuana smokers do not adequately distinguish a placebo from the active marihuana cigarette. Other facts: in animals like the rat and pigeon, there have been several behavioral studies that show profound tolerance development to Δ^9-THC and cross-tolerance to synhexyl and the most potent of the Adams compounds, namely DMHP (dimethyl heptyl pyran derivative). I would like to emphasize *profound tolerance!* Dr. Meyers reported in the Journal of Psychedelic Research that chronic administration of synhexyl to man causes tolerance development. And if you read the old literature, in the 1940's Williams and his colleagues showed that when they allowed chronic marihuana smokers to smoke daily for one month there was minimal but definite evidence of tolerance development. Ergo, there appears to be little doubt that there is tolerance development to the active ingredients of cannabis if the amounts taken are high enough. We know that even with compounds such as morphine, if the dose is low enough you will not develop tolerance. It is a question of dose. So this issue of sensitization is very much involved with the question of one's personal expectations of marihuana smoking.

Dr. Cohen: Yes, expectation and conditioning are part of the so-called reverse tolerance phenomenon. There is one other factor: the user over time learns how to best inhale, a very special technique, and this will increase the effect of marihuana with less being smoked.

Dr. Rose (St. Louis, Mo.): You'll pardon me for not being as brief as I had planned, but I have been adding questions during the discussion. First of all, Dr. Domino, our department is also using that 1.5% THC and we have beaten the bushes trying to come up with people who could smoke large amounts but, unlike Dr. Meyers, we could not get anyone to smoke as much as we would have liked.

Secondly, in reference to the points that Dr. Ungerleider made earlier, I would like to endorse him wholeheartedly and expand just slightly by saying that one of the big problems of which young persons are acutely aware is the communication gap between science and themselves in the area of drugs. Often it is not deliberate lying, but rather misinformation and just as in the way that a rumor, when passed from person to person, becomes distorted so do many reputed facts about drugs. I think that we as scientists owe it to ourselves, even before we take part in something like this symposium, to be very clear about our personal prejudices being injected into what we say. Our impact as scientists rests in our credibility and that in turn has no basis other than our objectivity. I feel that some of us have been less than objective today and if the shoe fits, wear it.

In reference to the amphetamine presentation this morning, someone referred to drug deaths in Chicago. Dr. Walter Pahnke, Spring Grove State Hospital, has told me that Chicago law enforcement authorities analyzed the drug involved and it contained something like 25 μg of LSD, 20 mg of STP, 8 mg of speed and also atropine and belladonna alkaloids. Certainly chlorpromazine would not be indicated for adverse reactions to this mixture but, in fairness, broader contraindications for chlorpromazine are unwarranted. There have, of course, been pharmacologic studies on the effects of phenothiazines after pure psychoactive agents, but the point which cannot be made too often is that neither the physician nor the patient can be sure of what agent has been ingested.

It was stated that there are no medical uses for LSD. I don't want to give the impression that I necessarily advocate LSD therapy, but certainly Dr. Pahnke's experiments at Spring Grove in which he is using LSD with terminal cancer patients, alcoholics, and in a few other situations have apparently shown some positive results. I think that the possibilities for the therapeutic use of LSD are still very much in question and should not be discounted at this time.

Dr. Hollister mentioned that THC was not detectable in either urine or blood, and I would take issue with that. Dr. Agurell from the Karolinska Institut in Stockholm has reported that he has been able to find THC in blood *in vivo,* which brings me to my next statement addressed to Dr. Holmstedt. How did he do that? Perhaps through editing there is some ambiguity in the recent paper in *Biochemical Pharmacology,* and I would appreciate it if you would outline the analytical technique Dr. Agurell employed in his *in vivo* assay.

I wonder if anyone on the panel would be willing to comment on the new n-propyl THC that Dr. Gill reported recently in *Nature.* It came as a surprise to me and I would be interested to hear whatever you might have to say about it.

Also, if anyone in the audience or on the panel happens to know of any references to catnip being used as a psychotomimetic drug, I would appreciate their letting me know.

Finally, a complaint in the area of the availability of information for persons who are actually occupied in research on cannabis, as is the case with our department. We obviously are on the side of the government at least somewhat since they have given us their THC and their marihuana. But unfortunately nobody seems inclined to give us any of the Edgewood Arsenal Technical Reports on Δ^3-THC that were carried out in the late 50's and have since been declassified. That they should be declassified and released has been discussed in Senate hearings, apparently to no avail. Another unavailable document is the report on gas chromatographic studies of the cannabinoids as a function of environment and locality of parent seed being carried out by Dr. Doorenbos at the University of Mississippi. The latter apparently have been published and certainly the former have not, but we have corresponded with a number of people in an attempt to obtain these and have been unsuccessful. I think that the progress, or perhaps just the continuance, of marihuana research hinges on the free availability of information. There appears to be, I don't want to suggest it but I will, some hesitancy on the part of the government in actually making that information available. I suppose, Dr. Cohen and Dr. Hollister, since they have ins to our federal government, would best be able to comment on this point.

Dr. Hollister: If we had another hour, we would comment on all those points, but I think Ed Domino was part of the Edgewood team and he will tell you about that.

Dr. Domino: If my prejudice is showing, you bet I am prejudiced. I have five kids and I don't want any of them busted.

Dr. Rose: I didn't mean to imply that I was referring to you.

Dr. Domino: Let me go on. With regard to the last point, you are quite right. There have been a great number of basic pharmacologic and some human studies on the synthetic Δ^3-derivatives. I am very happy to report that Dr. Seevers was the project leader here at Michigan for many years on much of the animal work. Those reports have been declassified. We gave a preliminary report this August at the Pharmacology Meetings in Palo Alto, and it is available in abstract form. Dr. Hardman and I are currently putting this material into a suitable publication. The government has been very cautious and has taken a long time to release it. I think I can understand the reasons why. This is available to you as of today if you want it.

With regard to the issue of a natural endpoint for marihuana smoking, it depends upon the subjects as I said earlier. With regard to CPZ-STP interactions, it should be remembered that STP (also called DOM) represents one of the phenethylamines related to mescaline. It is an unusual compound. In low doses it is amphetamine-like and chlorpromazine as a central and peripheral adrenergic blocker is an effective antagonist.

The problem is that as the dose of STP increases, one gets other effects. Long has shown that large doses of STP cause convulsions. Chlorpromazine potentiates this aspect of STP action. So it is a double problem for the poor physician who must treat the patient without knowing either the identity or the dose ingested. Whether one should give chlorpromazine is obviously tricky. I would opt for giving a sedative like a barbiturate or a benzodiazepine.

Dr. Hollister: Can you tell us what Dr. Agurell is up to, Dr. Holmstedt?

Dr. Holmstedt: No. I do not know what he is doing, what he has said, where and when.* But as I tried to point out in my presentation, there are possibilities now for the analysis of the cannabinoids in blood, if we could get a good clean-up procedure.

Dr. Rose: How about the n-propyl derivative that Dr. Gill reports? That was my last question.

Dr. Hollister: If it just appeared in *Nature,* I doubt that anybody else but Dr. Gill knows a bloody thing about it. However, let us check with the panel. I see no one wishes to comment.

It is now 5:15. I think in all fairness to those who want to go, they may be allowed to do so. Why don't the rest of the people who have questions simply engage the panel personally rather than in the plenary session. Thank you for your kind attention.

* *Editor's Note:* Dr. Holmstedt brought this transcript to the attention of Dr. Stig Agurell in Stockholm, Sweden. Dr. Agurell made the following comments:

It appears from Dr. Rose's statement that the major obstacle is that he has not been able to identify THC in blood *in vivo.* We have made a specific statement about this[1]; comparison of our results in rabbits with the corresponding results obtained in man by Axelrod and his associates shows very similar results.[2] Axelrod also found THC in blood *in vivo.* I would assume that Dr. Rose did not use labeled THC and probably used a gas-liquid chromatographic (GLC) method for measuring levels. Not overlooking the fact that THC is very quickly distributed from blood[1] (Fig. 4),[2] it is still very difficult to use GLC to detect THC in biological extracts, for example, blood. Presumably this is due to the interreaction of THC with other co-extracted compounds. This is discussed quite thoroughly by Stone in a recent paper.[3] Also, Dr. McIsaac states: ". . . we were unable to further characterize the presence of Δ^9-THC by gas chromatography because of the interference of impurities extracted from the tissues."[4]

In summary, what has been confirmed by some groups follows:

Our results:	*Verified by:*
Presence of THC in blood *in vivo*[1]	Axelrod:[2]
Recovery of THC from plasma, light petroleum 78 ± 5%[1]	Axelrod: 95 ± 5%, heptane + 1, 5% pentanol[2]
Distribution in blood of THC, "other metabolite", more polar metabolites[1] (rabbit)	Axelrod: quite similar in man.[2]
T/2 of THC in blood–distribution phase (rabbit)[1]	Axelrod: in man about 30 min. for distribution phase, elimination phase about 56 hours.[2]
Elimination of THC-metabolites in rat[5]	Joachimoglu, Kiburis, Miras:[7]
Structure of THC metabolite[6]	Wall:[8]
THC almost completely metabolized to more polar metabolites[1]	Axelrod,[2] Hollister (see p. 326)

Thus, our major points have been confirmed. I have also recently had contact with Dr. Miras' group in Athens (Dr. Khefalas), and was informed that their material which is now in press agrees very well with our published data.

References to Editor's Note:

1. Agurell, S., Nilsson, I. M., Ohlssen, A., Sandberg, F.: On the metabolism of tritium-labelled Δ^1-tetrahydrocannabinol in the rabbit. Biochem Pharmacol 19:1333, 1970.
2. Lemberger, L., Silberstein, S. D., Axelrod, J., Kopin, I. J.: Marihuana: studies on the disposition and metabolism of Δ^9-tetrahydrocannabinol in man. Science 170:1320, 1970.
3. Stone, M. M.: An investigation into forensic-chemical problems associated with cannabis. Scientific Research on Cannabis; United Nations Document ST/SOA/SER.S/18, 12 August 1969.
4. Ho, B. T., Fritchie, G. E., Kralik, P. M., Englert, F. L., McIsaac, W. M., Heikkila, J. I.: Distribution of tritiated-1 Δ^9-tetrahydrocannabinol in rat tissues after inhalation. J Pharm Pharmacol 22:538, 1970.
5. Agurell, S., Nilsson, I. M., Ohlssen, A., Sandberg, F.: Elimination of tritium-labelled cannabinols in the rat with special reference to the development of tests for the identification of cannabis users. Biochem Pharmacol 18:1195, 1969.
6. Nilsson, I. M., Agurell, S., Nilsson, J. L. G., Ohlssen, A., Sandberg, F., Wahlquist, M.: Δ^1-tetrahydrocannabinol: structure of a major metabolite. Science 168:1228, 1970.
7. Joachimoglu, G., Kiburis, J., Miras, C.: Studies on the distribution and excretion of ^{14}C-tetrahydrocannabinol in rats. Scientific Research on Cannabis; United Nations Document ST/SOA/SER.S/15, 25 September 1967.
8. Wall, M. E., Brine, D. R., Brine, G. A.: Isolation, structure and biological activity of several metabolites of Δ^9-tetrahydrocannabinol. J Amer Chem Soc 92:3466, 1970.

Morning Session
Thursday, November 12, 1970

Presiding: John C. Pollard, M.D.

DRUG ABUSE AND DRIVING

PAUL KIELHOLZ, M.D.

Alcohol use by drivers plays an important role in a high proportion of traffic accidents in Switzerland and Sweden. A recent Swedish study indicated that 28% of the drivers involved in accidents there were under the influence of alcohol. In a Swiss study, blood samples taken from 819 automobile accident victims admitted to four hospitals revealed that 32% of these drivers were under the influence of alcohol, with 20% showing blood alcohol levels above the statutory limit of 80 mg%.[1]

Following these disclosures, we were asked by medical societies and legal authorities in both countries to investigate the effects of alcohol and drugs, alone and in combination, on driving ability. An international, interdisciplinary team was formed to consider the following questions:[2,3]

1. How is driving ability affected by therapeutic doses of the most commonly prescribed drugs?
2. How is driving ability affected by a combination of drugs and alcohol?
3. How is driving ability affected by blood alcohol levels of 80 mg% and above?

The study was carried out in a series of field tests with 320 volunteers from the Basle City Police Corps. Under double blind conditions the subjects received single doses of 10 mg and 20 mg of chlordiazepoxide (Librium), 400 and 800 mg of meprobamate (Miltown), 200 mg of phenobarbitone (Luminal), 200 mg of methyprylone (Noludar), or inert tablets. Half of each group also received enough alcohol to produce an average blood concentration of 80 mg% within one hour.

Statistical analysis of the results indicated that the two tranquilizers in therapeutic doses of 10 mg of chlordiazepoxide and 400 mg of meprobamate do not impair driving ability or potentiate the effects of alcohol. In higher doses of 20 mg of chlordiazepoxide and 800 mg of meprobamate, there was an increase in the number of driving errors which was not statistically significant. There was a significant increase in the effects of alcohol at these higher drug levels, however. The two hypnotics, in doses of 200 mg of phenobarbitone and 200 mg of methyprylone, significantly impaired driving ability and potentiated the effects of alcohol.

In test subjects receiving only alcohol and inert tablets, it was again demonstrated that a blood alcohol concentration of 80 mg% impairs driving ability to a highly significant extent.

The methods used and the most important results were presented in a short color film.

REFERENCES

1. Blood alcohol concentration in the samples was determined at the Institutes for Forensic Medicine in Basle by Professor J. Im Obersteg, M.D. and in Berne by Prof. E. Läuppi, M.D.
2. Kielholz, P., Goldberg, L., Im Obersteg, J., *et al.*: Strassenverkehr, Tranquilizer und Alkohol. *Deutsch Med Wschr 92*:1525–1531, 1967.
3. Kielholz, P., *et al.*: Fahrversuche zur Frage der Beeinträchtigung der Verkehrstüchtigkeit durch Alkohol, Tranquilizer und Hypnotika. *Deutsch Med Wschr 94*:301–306, 1969.

NEW ASPECTS OF DRUG DEPENDENCE IN WESTERN EUROPE

PAUL KIELHOLZ, M.D.

In 1968, the Council of Europe commissioned a group of specialists to investigate the extent and causes of drug abuse in Europe. The members of this study group, Dr. Kryspin-Exner, of Austria, Dr. Teigen, of Norway, and Professor Kaymakçalan, of Turkey, visited each country to gather information on the spot concerning the extent and manifestations of drug dependence. Questionnaires were sent to the national health authorities of each member state (Belgium, Cyprus, Denmark, Finland, France, Greece, Iceland, Ireland, Luxemburg, Malta, Norway, Sweden, Switzerland, Turkey, United Kingdom) of the Council[1] in order to obtain an over-all picture of the epidemiology of drug abuse.

In order to standardize the investigation, the WHO Expert Committee's definition of drug dependence (1964) was used as the basis for the study. Each country supplied information covering the six-month period from June to December of 1969. Analysis of the results showed that in the above mentioned European countries, drug abuse increased markedly since 1960. Furthermore, since 1967 the large cities and university towns of Western Europe have been inundated by a wave of hashish which spread from England by way of Denmark, Sweden, and Holland to Switzerland, thence to Germany and Austria, and finally to Italy and France.

The investigation conducted by the Council of Europe confirmed the results of several studies conducted at the national level.[7] It was found that while the preferred drug varied widely among young people, hashish was clearly in the lead in 80 to 90% in the various countries. The following principal trends were observed:

1. A growing incidence in young people.
2. A shifting to lower social levels.
3. New patterns in drug dependence (e.g., central stimulants, administered intravenously, sniffing, group phenomenon, changing of the motivation and patterns of drug dependence).
4. The rapid increase of the abuse of well-known drugs in older age groups (hypnotics, antipyretics, analgesics and central stimulants).
5. The rising frequency of multiple dependents.
6. An increasing number of women dependents.
7. A rapidly increasing problem of alcoholism.

The incidence of unrecognized drug dependence is very high and is variously estimated at between 10:1 and 50:1 in different

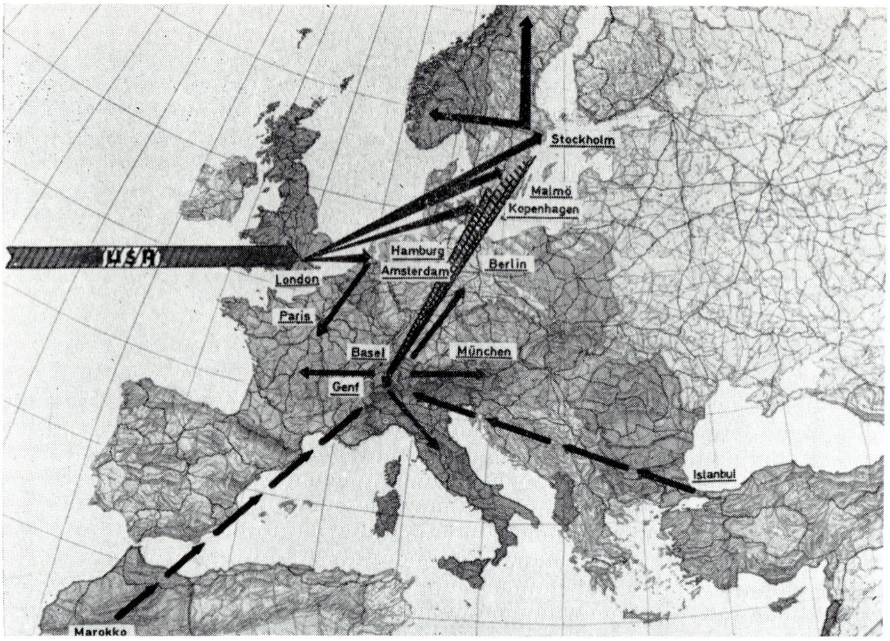

Figure 1 Map of Europe showing routes of spread of hashish use.

countries. Whereas at first cannabis smoking was observed only in the urban setting, it has recently become established in rural areas also. Investigation of groups of pupils from secondary schools and colleges of art and apprenties in Denmark, Sweden, and Switzerland has shown that 10 to 20% have experience with hashish, and that, of these, 20 to 40% gradually begin smoking it regularly.

At the same time that hashish smoking has increased among youths, the misuse of drugs, such as hypnotics, antipyretics, analgesics, and stimulants, has become more frequent among adults. As the motives for drug abuse in the young and the adult are different, two main trends of drug dependence can be distinguished among youths and young adults in the age range of 14 to 24 years, and in adults aged 25 to 45 years.

The causes of drug dependence are always multiple. In each case they are based on (1) general human disposition, (2) the actual causes, and (3) the "trigger" factors.

Drug dependence is always related to the general human desire to prevent, correct, or forget temporarily such unpleasant elements as conflicts, tension, anxiety, unhappiness, stressful situations, as well as the desire to repeat and intensify pleasures which have once been experienced.

The actual causes of dependence are to be sought in the inherited, constitutionally determined basic personality, the personal history, general sociocultural conditions, the "trigger" factors and the access to drugs. Drug dependence is founded on a triad of factors, namely, *personality, drug, and society,* and is primarily a psychologic problem.

If we study (Kielholz and Battegay[2]) first of all the heredity and previous personalities of drug-dependent individuals, we find that 60% come from families in which lability of affect, anxiety, and hypersensitivity are prominent in family history; alcoholism, abuse of drugs by the mother, suicide, and suicidal attempts are more frequent than usual. In the drug-dependent personality, isolation, hypersensitivity, and lack of self-confidence pre-

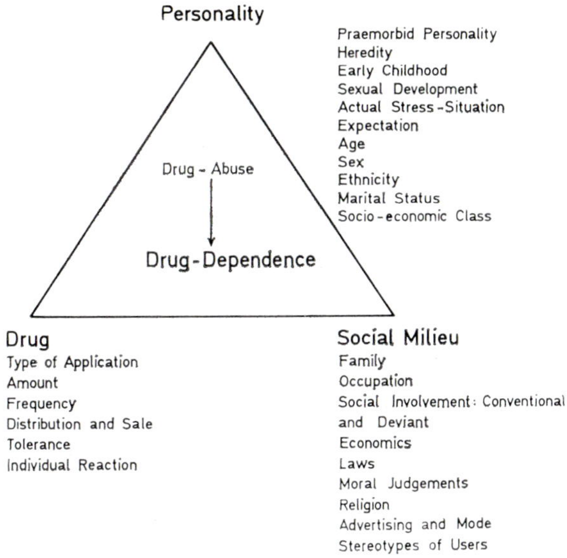

Figure 2 Triad of the development of drug dependence.

dominate. These young people are often kept under continuous emotional strain by the discrepancy between their goals and achievements, between wish and performance. These hereditary constitutional features are intensified by frustration in early childhood (Kielholz and Ladewig[3]), in particular lack of security and comfort, the inability to find roots for themselves and the absence of adequate parental example, and less frequently by sexual taboos with resulting disturbances of sexual development.

"TRIGGER" FACTORS

Drug dependence is usually based on these predisposing personality factors, intensified by neurotic patterns of development. It arises at times of increased tension as a result of conflicts and emotional stress caused by problems within the environment. Adolescents and young adults in the age range of 14 to 24 years are forced by the accelerated impulses of puberty into attitudes of violence and opposition. If they have no family ties which allow them to ventilate their emotional problems and conflicts at home, they seek others in the same situation who they feel will understand and accept them. Parents often reject these new contacts and thus reinforce the adolescent's feelings of opposition and defiance toward his home, the social establishment, the pressure for achievement and all forms of authority. These young people are nevertheless obliged to recognize that they cannot alter the structure of society fundamentally or as quickly as its use of force and so become more and more a prey to emotional tension. Amid these conflicts and tensions, they form small groups to seek drug relief for this intolerable situation.

Groups which form for hashish parties vary enormously in their motives, frequency of indulgence, and the personality structure of the individual members. From various studies it appears that 70 to 90% stop after occasional experience of hashish, while 10 to 30% continue smoking cannabis. The majority of those who give up smoking hashish after a few parties come from undisturbed families; their initial motives are mostly curiosity, peer pressures and imitation, the wish to experiment, experience, or expand their consciousness and self-awareness.

Motives of the 10 to 30% who gradually

progress to regular use of hashish are generally more profound. These young people may come from broken homes, are often undergoing a crisis in their development which is intensified by early childhood deprivation and lack of family bonds. As a result of this process, they are dominated by a sense of opposition, defiance, revolt, aggressiveness, and rejection of the established social pattern and its preoccupation with success. Often it takes 6 to 12 months for them to become dependent on the drug. At this point, they tend to withdraw from the group and indulge in their drug use regularly and alone.

So the results of any investigation will vary according to the moment at which a group has been investigated, the character of the individual structure and the milieu out of which the group has grown. Not realizing these facts has led to many misunderstandings in Europe.

Of the 10 to 30% of youngsters who gradually take up the regular use of drugs, between 30 and 90% according to various studies, transfer to other drugs, the initial drug hashish being replaced by "harder" substances. Often they become multiple abusers. This escalation or "domino-effect" has been confirmed by Buchard,[4] Goldberg,[5] Keup,[6] and Wanke.[7] The earlier the age of starting with the first drug, the more rapidly does this escalation occur. All investigations agree that drug abuse in adolescents usually has a very unfavorable prognosis, with a relapse rate of between 70% and 90%.

The causes of drug dependence suggest certain prophylactic measures and treatment:

1. Drug dependence is best prevented by an unceasing concern for a healthy harmonious family life. The best education is the good example, a figure with whom the adolescent can identify.
2. Everyone—teachers, parents, young people—should have the dangers of hashish abuse made clear to them in discussion by all available media (press, radio, television).
3. Young people themselves, especially those who have been dependent on drugs, should be involved in advising their contemporaries and in helping in education and case-finding. All efforts should be directed toward getting the drug-dependent adolescent to come voluntarily for treatment. Out-patient advisory clinics should be available for the youth who occasionally abuses hashish or other drugs; here he or his parents can obtain advice from a doctor anonymously and

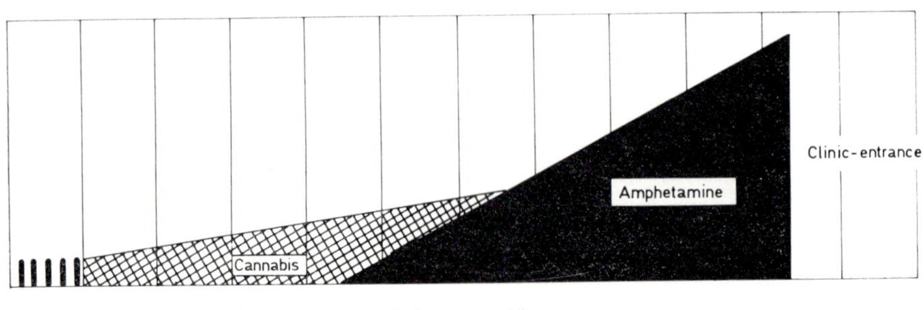

Figure 3 Development of drug dependence of the cannabis type.

without fear of criminal prosecution. Special centers for the treatment of drug dependence should be set up where withdrawal can be facilitated. After withdrawal and detoxication, the social setting, the motives for abuse and the personality defects of the abuser should be explored by appropriate diagnostic measures and followed up in individual and group psychotherapy. New human contacts and relationships are the basis for eventual gradual rehabilitation. The prognosis depends on the intensity of the social and medical follow-up, but is also determined by the environment into which the young patient is discharged.

Several aspects of drug dependence are still controversial. The World Health Organization, national health authorities and medical experts are demanding stricter control of drug production; they are proposing that substances which have a legitimate medical use but are also liable to abuse should be subject to more stringent regulations concerning their prescription, licensing of manufacture, and use. At the same time, demands are being made publicly for the free availability of drugs with no clear medical use. This situation can only be clarified by an investigation as to the degree of danger of each drug. We must determine how many of the people who take the drug regularly have become psychologically dependent on it and how many are able to stop taking it. On the basis of this dependence potential, taking into account the possible degree of medical usefulness, we shall be in a position to undertake scientifically based medical, psychosocial and medicolegal measures of prevention.

Unfortunately, the necessary medical and statistical studies in large groups of people have not yet been carried out for many substances which act on the central nervous system. Only by intensive research can the many unknown factors of drug dependence be elucidated and the free rein of fantasy, speculation and armchair theory on this important medicosocial problem be curbed.

REFERENCES

1. *Public Health Implications of Recent Developments in Drug Dependence.* Report, European Public Health Committee, Council of Europe, Strasbourg 1970.
2. Kielholz, P., and Battegay, R.: Vergleichende Untersuchungen über die Genese und den Verlauf der Drogenabhängigkeit und des Alkoholismus. *Schweiz Med Wschr 96*: 893–898, 944–949, 1967.
3. Kielholz, P., and Ladewig, D.: Ueber Drogenabhängigkeit bei Jugendlichen. *Deutsch Med Wschr 95*:101, 1970.
4. Buchard, J. M.: *Erfahrungsbericht über die Entwicklung der Rauschmittelproblematik in Hamburg seit 1868.* Kongress Deutsche Gesellschaft Psychiatrie—Nervenheilkunde, Oktober 1970, Bad Nauheim.
5. Goldberg, L.: Drug abuse in Sweden. *Bull Narcotic 20*:1, 1968, I und II/2, 1968, 9.
6. Keup, W.: *Aktuelle Probleme des Rauschmittelmissbrauchs in New York unter besonderer Berücksichtigung der Domino-Theorie.* Kongress Deutsche Gesellschaft Psychiatrie—Nervenheilkunde, Oktober 1970, Bad Nauheim.
7. Wanke, K.: *Neue Aspekte zum Suchtproblem, multifaktorielle Analysen klinischer Erfahrungen mit jungen Drogenkonsumenten.* Habilitationsschrift, Manuskript, Frankfurt, 1971.

DECREASED DRUG ABUSE WITH TRANSCENDENTAL MEDITATION—A STUDY OF 1,862 SUBJECTS*

HERBERT BENSON, M.D., AND R. KEITH WALLACE, Ph.D.
WITH THE TECHNICAL ASSISTANCE OF
ERIC C. DAHL, B.A., AND DONALD F. COOKE. B.S.

The abuse of drugs of all kinds is widespread in the United States and the extent of abuse, particularly of marihuana and hallucinogenic drugs, is growing.[1–3] It is estimated that in the United States 35 to 50% of high school and college students have tried marihuana at least once, and of these about 35% have tried marihuana more than ten times.[2] A conservative estimate of persons in the United States, both juvenile and adult, who have used marihuana is about five million and may be as high as 20 million.[1] In surveys of d-lysergic acid diethylamide (LSD) use in college populations, 5% of the students polled admitted to using LSD, with about 30% of the sample being classified as "serious" users and the remaining 70% as "experimenters."[2] Law enforcement agencies report there are approximately 65,000 active "hard" narcotic addicts in the United States. Other estimates indicate that there are 100,000 active narcotic abusers.[2] The abuse of amphetamines and barbiturates is widespread, but difficult to estimate. College surveys have indicated that more than 20% of the students have abused these drugs.[3] Stanley F. Yolles, M.D., Director of the National Institute of Mental Health summed up the situation as follows:

> The spreading of the abuse pattern into unusual and exotic drugs and the involvement of increased numbers of people have serious implications. It seems that today if a chemical can be abused, it will be. Further, it appears that stronger and more dangerous drugs tend to displace weaker drugs during this period of excessive preoccupation with mind-altering chemicals. One further identifiable ominous trend is the indulgence in drugs of abuse by younger and younger age groups.
>
> It is to be expected that the use of all sorts of drugs in the next 10 years will increase a hundred fold. It is necessary, therefore, to develop effective processes to control their abuse today.[3]

Few programs or treatments have been reported which reduce drug abuse. One apparently successful program for the rehabilita-

* Supported in part by grants from the National Heart and Lung Institute (HE 10539-05), the National Institutes of Health (SF 57-111), and from Hoffmann-LaRoche, Inc., Nutley, New Jersey, 07110.

tion of persons abusing narcotics involves the substitution of methadone.[4,5] Existing programs for the alleviation of other drug abuse usually involve education concerning the dangerous effects of drugs and sometimes provide personal counseling or psychiatric care.[6–9] The efficacy of these programs has yet to be established.

A preliminary observation suggested that the practice of Transcendental Meditation, as taught by Maharishi Mahesh Yogi, may be effective in alleviation of drug abuse.[10] The present report confirms and expands the earlier observation.

METHODS

The technique of Transcendental Meditation is reported to be an easily learned mental technique which originated in ancient Vedic tradition of India.[11,12] Practitioners are personally instructed by a teacher qualified by Maharishi Mahesh Yogi. The technique is claimed to be a spontaneous natural process, and, unlike many techniques of meditation or self-improvement, does not employ mental control, physical control, belief, suggestion, or any change in life style. It is also claimed that anyone can learn the technique in four or five instructional sessions. Practitioners are asked to abstain from drug abuse for a 15-day period prior to starting Meditation. Following instruction in Transcendental Meditation, the program involves practicing the technique twice a day for periods of 15 to 20 minutes. The program does not provide any type of personal counseling or advice about personal problems. Individuals practice the technique on their own. The only additional contact between the individual and the instructors or organization is concerned with ensuring correct practice of the technique and providing further knowledge about it.

Questionnaires were given to approximately 1,950 subjects who had been practicing Transcendental Meditation for three months or more and who were attending one of two Meditation training courses offered by the Students' International Meditation Society* in the summer of 1970. Of these, 1,862 subjects completed the questionnaire, giving information concerning age, sex, educational status, and length of time that Transcendental Meditation had been practiced. Frequency of drug use, drug-selling activity, and attitudes towards drug abuse were assessed for each of five separate time periods: (1) 6 months before starting Meditation; (2) 0 to 3 months after starting; (3) 4 to 9 months after starting; (4) 10 to 21 months after starting; and (5) 22 months or more after starting. The separate drugs and categories of drugs included in the questionnaire were (1) marihuana; (2) LSD; (3) other hallucinogens (2,5-dimethyloxy-4-methyl amphetamine [STP], N,N-dimethyltryptamine [DMT], peyote, and mescaline); (4) narcotics (heroin, opium, morphine and cocaine); (5) amphetamines; and (6) barbiturates.

Additional information was requested concerning the frequency of use of "hard liquor" and the number of packs of cigarettes smoked. Hard liquor was defined as "alcoholic beverages stronger than wine or beer."

The information on the questionnaires was analyzed on an IBM 360-65 computer. The CROSSTABS† multivariant data analysis program was utilized and all processing was done by Urban Data Processing, Inc.‡ The subjects were classified into four categories depending on the frequency of drug use: (1) non-users; (2) light users; (3) medium users; and (4) heavy users. For the subjects using marihuana, narcotics, amphetamines, barbiturates, hard liquor, and cigarettes, a "light user" indicated a frequency of three times a month or less; "medium user," once a week to six times a week; and "heavy user," once a day or more. For LSD

* National Headquarters, 1015 Gayley Avenue, Los Angeles, California 90024.

† Cambridge Computer Associates, 22 Alewife Brook Parkway, Cambridge, Massachusetts, 02138.

‡ Urban Data Processing, 552 Massachusetts Avenue, Cambridge, Massachusetts, 02139.

and other hallucinogens, "light user" indicated a frequency of less than once a month; "medium user," from one to three times a month; and "heavy user," once a week or more.

RESULTS

A total of 1,862 subjects responded to the questionnaire. There were 1,081 male subjects and 781 female subjects (Table 1). The age of the subjects ranged from 14 to 78 years and approximately half of the subjects were between the ages of 19 and 23. Most had attended college and many had college degrees (Table 2). The average length of time they had been practicing Transcendental Meditation was approximately 20 months.

Following the start of the practice of Transcendental Meditation, there was a marked decrease in the number of drug abusers for all drug categories (Tables 3 to 8). As the practice of Meditation continued,

Table 1 Age and sex of the respondents to the questionnaire

Sex		Age: 14–18 years	19–23 years	24–28 years	29–38 years	39 years and older	Totals
Male	No.	61	574	322	82	42	1081
	%	3.3	30.8	17.3	4.4	2.3	58.1
Female	No.	71	363	167	81	99	781
	%	3.8	19.5	8.9	4.4	5.3	41.9
Totals	No.	132	937	489	163	141	1862
	%	7.1	50.3	26.2	8.8	7.6	100.0

Table 2 Education of the respondents to the questionnaire

Educational experience	Less than high school	High school graduate	Some college	College graduate	Advanced college degree	Totals
Number	100	183	971	460	148	1862
Percent	5.4	9.8	52.2	24.7	7.9	100.0

Table 3 Use of marihuana and hashish before and after starting the practice of Transcendental Meditation

	Before (time in months)		After (time in months)							
	6–0		0–3		4–9		10–21		22–33	
Usage*	n	%	n	%	n	%	n	%	n	%
Heavy	417	22.4	47	2.5	39	2.1	18	1.3	1	0.1
Medium	618	33.2	190	10.2	137	7.5	65	4.6	18	2.1
Light	422	22.7	613	32.9	500	27.3	264	18.6	85	10.0
Non-user	405	21.7	1012	54.4	1154	63.1	1070	75.5	748	87.8
Totals	1862	100.0	1862	100.0	1830	100.0	1417	100.0	852	100.0

* See text for definitions.

Table 4 Use of LSD before and after starting the practice of Transcendental Meditation

	Before (time in months)		After (time in months)							
	6–0		0–3		4–9		10–21		22–33	
Usage*	n	%	n	%	n	%	n	%	n	%
Heavy	132	7.1	14	0.7	13	0.7	6	0.4	0	0.0
Medium	301	16.1	60	3.3	36	1.9	23	1.7	3	0.3
Light	467	25.1	159	8.5	151	8.3	72	5.1	23	2.7
Non-user	962	51.7	1629	87.5	1630	89.1	1316	92.8	826	97.0
Totals	1862	100.0	1862	100.0	1830	100.0	1417	100.0	852	100.0

* See text for definitions.

Table 5 Use of other hallucinogens before and after starting the practice of Transcendental Meditation

	Before (time in months)		After (time in months)							
	6–0		0–3		4–9		10–21		22–23	
Usage*	n	%	n	%	n	%	n	%	n	%
Heavy	5	0.3	4	0.2	5	0.3	3	0.2	0	0.0
Medium	56	3.0	32	1.7	30	1.7	19	1.4	0	0.0
Light	665	35.7	143	7.7	130	7.0	90	6.4	34	4.0
Non-user	1136	61.0	1683	90.4	1665	91.0	1305	92.0	818	96.0
Totals	1862	100.0	1862	100.0	1830	100.0	1417	100.0	852	100.0

* See text for definitions.

Table 6 Use of narcotics before and after starting the practice of Transcendental Meditation

	Before (time in months)		After (time in months)							
	6–0		0–3		4–9		10–21		22–33	
Usage*	n	%	n	%	n	%	n	%	n	%
Heavy	12	0.6	1	0.1	1	0.1	1	0.1	0	0.0
Medium	17	0.9	2	0.2	2	0.2	2	0.2	0	0.0
Light	286	15.4	47	2.5	39	2.1	30	2.1	10	1.2
Non-User	1547	83.1	1812	97.2	1788	97.6	1384	97.6	842	98.8
Totals	1862	100.0	1862	100.0	1830	100.0	1417	100.0	852	100.0

* See text for definitions.

Table 7 Use of amphetamines before and after starting the practice of Transcendental Meditation

	Before (time in months)		After (time in months)							
	6–0		0–3		4–9		10–21		22–33	
Usage*	n	%	n	%	n	%	n	%	n	%
Heavy	30	1.6	7	0.4	3	0.2	2	0.1	0	0.0
Medium	96	5.2	11	0.6	9	0.5	2	0.2	0	0.0
Light	470	25.2	104	5.6	79	4.3	49	3.4	10	1.2
Non-User	1266	68.0	1740	93.4	1739	95.0	1364	96.3	842	98.8
Totals	1862	100.0	1862	100.0	1830	100.0	1417	100.0	852	100.0

* See text for definitions.

Table 8 Use of barbiturates before and after starting the practice of Transcendental Meditation

	Before (time in months) 6–0		0–3		After (time in months) 4–9		10–21		22–33	
Usage*	n	%	n	%	n	%	n	%	n	%
Heavy	19	1.0	1	0.1	3	0.2	1	0.1	0	0.0
Medium	43	2.3	7	0.4	3	0.2	2	0.1	1	0.1
Light	258	13.9	56	2.9	37	2.0	27	1.9	8	1.0
Non-User	1542	82.8	1798	96.6	1787	97.6	1387	97.9	843	98.9
Totals	1862	100.0	1862	100.0	1830	100.0	1417	100.0	852	100.0

* See text for definitions.

Table 9 Use of "hard liquor" before and after starting the practice of Transcendental Meditation

	Before (time in months) 6–0		0–3		After (time in months) 4–9		10–21		22–33	
Usage*	n	%	n	%	n	%	n	%	n	%
Heavy	50	2.7	21	1.2	16	0.9	8	0.6	3	0.4
Medium	295	15.8	149	8.0	100	5.5	52	3.7	22	2.6
Light	770	41.4	646	34.7	551	30.1	365	25.8	187	21.9
Non-User	747	40.1	1046	56.1	1161	63.5	992	69.9	640	75.1
Totals	1862	100.0	1862	100.0	1830	100.0	1417	100.0	852	100.0

* See text for definitions.

the subjects progressively decreased their drug abuse until after practicing 21 months of Meditation most subjects had completely stopped abusing drugs. For example, in the six-month period before starting the practice of Meditation, about 80% of the subjects used marihuana and of those about 28% were heavy users. After practicing Transcendental Meditation six months, 37% used marihuana and of those only 6.5% were heavy users. After 21 months of the practice, only 12% continued to use marihuana and of those most were light users; only one individual was a heavy user. The decrease in abuse of LSD was even more marked. Before starting the practice of Transcendental Meditation, 48% of the subjects had used LSD, and of these subjects about 14% were heavy users. In the three months following the start of the practice of Meditation, 11% of the subjects took LSD, while after 21 months of the practice only 3% took LSD. The increase in the number of non-users after starting the practice of Meditation was similar for the other drugs: non-users of the other hallucinogens after 21 months of the practice rose from 61 to 96%; for the narcotics from 83 to 99% for the amphetamines from 70 to 99%; and for the barbiturates from 83 to 99%.

In the six-month period before starting the practice of Meditation, 60% of the subjects took hard liquor and, of these, about 4% were heavy users (Table 9). After 21 months of the practice of Meditation, approximately 25% took hard liquor and only 0.1% were heavy users. Approximately 48% smoked cigarettes before starting Meditation and 27% were heavy users (Table 10). After 21 months of practicing Meditation, 16% smoked cigarettes and only 5.8% were heavy users.

Table 10 Use of cigarettes (number of packs) before and after starting the practice of Transcendental Meditation

	Before (time in months)		After (time in months)							
	6–0		0–3		4–9		10–21		22–33	
Usage*	n	%	n	%	n	%	n	%	n	%
Heavy	503	27.0	314	16.9	222	12.2	118	8.4	49	5.7
Medium	180	9.7	165	8.9	136	7.4	86	6.0	34	4.0
Light	203	10.9	186	10.0	163	8.9	105	7.4	55	6.4
Non-User	976	52.4	1197	64.2	1309	71.5	1108	78.2	714	83.9
Totals	1862	100.0	1862	100.0	1830	100.0	1417	100.0	852	100.0

* See text for definitions.

Most subjects felt that Transcendental Meditation was instrumental in their decreasing or stopping abuse of drugs: 61.1% stated that it was extremely important; 22.8% that it was very important; 12.0% somewhat important; and 3.6% not important. Of those individuals who continued the use of drugs following starting Transcendental Meditation, 55.9% had been irregular in Meditation and 24.8% had stopped for a week or more.

Three hundred seventy-four subjects (20.1%) sold drugs before starting Meditation. Of these, 71.9% stopped and 12.5% decreased drug selling during the period 0-3 months after instruction. Among the subjects who practiced Meditation 21 months or longer and who at one time were actively involved in selling drugs, 95.9% stopped selling drugs. In addition, 997 (65.5%) had either encouraged or condoned drug abuse before starting Meditation. Over 95% of these subjects discouraged drug abuse in others after beginning the practice of Meditation.

DISCUSSION

Individuals who regularly practiced Transcendental Meditation (1) decreased or stopped abusing drugs, (2) decreased or stopped engaging in drug-selling activity, and (3) changed their attitudes in the direction of discouraging others from abusing drugs. The magnitude of these changes increased with the length of time that the individual practiced the technique. Similar decreases were noted in the use of "hard" alcoholic beverages and cigarette smoking. A high percentage of the individuals who did change their habits felt that Transcendental Meditation was very or extremely important in influencing them to change.

During Transcendental Meditation, oxygen consumption and heart rate significantly decrease, skin resistance significantly increases, and the electroencephalogram shows predominantly slow alpha wave activity with occasional theta wave activity.[13] Thus, the practice of Transcendental Meditation is physiologically distinguished from sitting quietly with eyes open or closed, from sleeping or dreaming, and from the relaxation or rest suggested by hypnosis. During Transcendental Meditation subjects report that their awareness is spontaneously drawn to "finer" or "more abstract" levels of the thinking process.

There are no simple explanations of the factors which lead to drug abuse. The types of motives which initiate and prolong drug abuse range from such things as social pressure, curiosity, desire for "kicks," rebellion against authority, escape from social and emotional problems to more philosophical motives such as self-knowledge, creativeness, spiritual enlightenment, or expansion of consciousness.[14] Student drug users are, as a group, knowledgeable about the undesirable effects of drug abuse. In general, it is not difficult for most student drug abusers to stop; the problem is to get them to want to stop. For a drug abuse program to be effective it must provide a non-chemical alternative which can fulfill at least some of the basic motivations behind student drug abuse.

Transcendental Meditation is acceptable among youthful drug abusers. It is offered as a program for personal development and is not specifically intended to be a treatment for drug abuse; the alleviation of the problems of drug abuse is merely a side effect of the practice. Thus, it may not threaten the beliefs of the committed abuser who condones the use of drugs. Since the introduction of Transcendental Meditation into the student community five years ago, more than 40,000 individuals have allegedly begun the practice.[15] Furthermore, the movement continues to grow. It is presently being presented through campus organizations at some 300 colleges and universities and is offered in the context of an accredited course at several universities.

Involvement in other kinds of self-improvement activities may also lead to decreased drug abuse. The motivation to start Meditation may have influenced the subjects to stop drug abuse. The subjects in the present study may have spontaneously stopped, continued, or increased taking drugs independently of Transcendental Meditation. However, since there are few effective programs which reduce drug abuse, Transcendental Meditation should be investigated as an alternative to drugs by a controlled, prospective study.

SUMMARY

Drug abuse is widespread and increasing in the United States, especially in student populations. However, few effective programs exist for the alleviation of drug abuse. Transcendental Meditation, a popular and easily learned mental technique which allegedly originated from the ancient Vedic tradition of India, was investigated as a possible means of decreasing drug abuse. Eighteen hundred sixty-two subjects who had practiced Transcendental Meditation at least three months formed the basis of this study. In these subjects, after starting Transcendental Meditation there was a significant decrease in the amount of drugs used or discontinuance of drug use; a decrease or cessation in engaging in drug-selling activity; and changed attitudes in the direction of discouraging others from abusing drugs. Further, the subjects decreased their consumption of "hard" alcoholic beverages and smoked fewer cigarettes. The magnitude of these changes increased with the length of time that the subject practiced Transcendental Meditation. Involvement in other types of self-improving activities may also lead to decreased drug abuse. However, since there are few effective programs which decrease drug abuse, Transcendental Meditation should be investigated as an alternative to drugs by a controlled, prospective study.

ACKNOWLEDGEMENT

We thank Miss Barbara R. Marzetta and Miss Lyne Heppner for their help in the preparation of the manuscript.

REFERENCES

1. *Resource Book for Drug Abuse Education.* National Clearinghouse for Mental Health Information, United States Department of Health Education and Welfare. Public Health Service, Health Service and Mental Health Administration, National Institute of Mental Health: 1969. Washington, D.C., Government Printing Office, 1969 (PHS Publication No. 1964), p. 25.
2. *Recent Research on Narcotics, LSD, Marijuana and Other Dangerous Drugs.* National Clearinghouse for Mental Health Information, U. S. Department of Health, Education and Welfare. Public Health Service, Health Service and Mental Health Administration, National Institute of Mental Health, 1969. Washington, D.C., Government Printing Office, 1969 (PHS Publication No. 1961), pp. 1, 2, 7, 11, 18.

3. Yolles, S. F.: Statement for Stanley F. Yolles, M.D., Director, National Institute of Mental Health, before the Subcommittee on Public Health and Welfare of the Interstate and Foreign Commerce Committee on H.R. 11701 and H.R. 13743, 1969, loose leaf, pp. 13–16.
4. Byrd, O. E.: *Medical Readings on Drug Abuse.* Reading, Mass., Addison-Wesley Publishing Co., 1970, pp. 255–257.
5. Eddy, N. B.: Methadone maintenance for the management of persons with drug dependence of the morphine type. *Drug Dependence 3*:17–26, 1970.
6. Wiesen, R. L., Wang, I. H., and Stensper, T. J.: The drug abuse program at Milwaukee County Institutions. *Wisconsin Med J 69*:141–150, 1970.
7. Murphy, B. W., Leventhal, A. M., and Balter, M. B.: Drug use on the campus: A survey of university health services and counseling centers. *J Amer Coll Health Assoc 17*:389–402, 1969.
8. Pollock, M. B.: The drug abuse problem: Some implications for health education. *J Amer Coll Health Assoc 17*:403–411, 1969.
9. Hickox, J. R.: Drug abuse education. *Texas Med 65*:31–33, 1969.
10. Benson, H.: Yoga for drug abuse. *New Eng J Med 281*:1133, 1969.
11. Maharishi Mahesh Yogi: *The Science of Being and Art of Living,* rev. ed. London, International S.R.M., 1966, pp. 50–59.
12. Maharishi Mahesh Yogi: *Maharishi Mahesh Yogi on the Bhagavad Gita: A New Translation and Commentary.* Baltimore, Penguin, 1969. (Originally published by International S.R.M., London, 1967, pp. 10–17.)
13. Wallace, R. K.: Physiological effects of transcendental meditation. *Science 167*:1751–1754, 1970.
14. Cohen, A. Y.: Inside what's happening: Sociological, psychological, and spiritual perspectives on the contemporary drug scene. *Amer J Publ Health 59*:2092–2097, 1969.
15. Jarvis, J.: Personal communication from the Students' International Meditation Society, 1015 Gayley Avenue, Los Angeles, California, 90024.

SECTION ON PSYCHIATRIC MANAGEMENT OF DRUG ABUSE

INTRODUCTORY REMARKS

JOHN C. POLLARD, M.D.

Earlier participants in this symposium have documented the fact that the use of 'hard' narcotics has been a problem in this country for several decades. At the beginning of the chapter entitled "Heroin," in their book *Drugs and Youth,* Coles, Brenner, and Meagher make the following observation:

> In the first few months of 1951, no less than 260 boys and girls, their ages ranging from 14 to 20, were admitted to Bellevue Hospital in New York City because of addiction to heroin. Other children were killed by the drug and never made it to the hospital. They were all from Harlem and they made no headlines in New York papers, let alone those in Boston or Seattle.[1]

The problem of serious drug abuse at that time tended to be confined to urban ghettos where the depressed and the oppressed used drugs mostly as anodynes. With the exception of a few physicians and some psychiatrists working in city hospitals and federal institutions, the problems of drug abuse and addiction were largely disregarded by the medical profession in general, and by psychiatry in particular. We were all aware of the chronic problem of alcohol—eight million or more alcoholics, thirty thousand road deaths directly related to alcohol use—but we considered this the uncomfortable price paid for the one socially permitted and sanctioned drug use; we also knew that an unknown number of people were dependent on amphetamine or addicted to barbiturates. In the late 50's, however, the drug abuse pattern shifted in two new and alarming directions. First of all, we began to see serious involvement among young (college age or younger), middle-class, mostly white users who were at least outwardly neither economically nor socially oppressed. Second, some but not all of the drugs were relatively new: the hallucinogens, particularly LSD, mescaline, Psilocybin, DM, DET and more recently such drugs as DOM and MDA. Marihuana, a mild hallucinogen, became increasingly popular not only among the Bohemian fringe, as many legislators would like to believe, or the morally degenerate periphery of school dropouts and reactionary activists, but, as survey after survey has indicated, by youngsters who do better than average in school and who are, if anything, more involved with vital matters. It has become a matter of great concern that these youngsters who live in constant danger of having their education and careers ruined by their use of marihuana are often indistinguishable from both parents' and society's ideal.[2]

Dr. Yolles estimates the number who have used marihuana in this country as being in the region of 10 million.[3] Dr. Louria quoted figures of up to 20 million and also estimates that 675 million marihuana cigarettes are smoked each year.[2] The Federal Bureau of Narcotics calculates that the 300 tons or so of marihuana that are intercepted by U.S. customs authorities is less than 5% of the amount that is successfully imported. One must conclude, therefore, that if millions of people consume these quantities of marihuana then either it is in fact a relatively innocuous substance or there is an epidemic of unrecognized drug toxicity in this country.

However, we know that nothing is totally innocuous and along with LSD and the more potent hallucinogens, marihuana does have its casualties. What sort of drug problems then is the psychiatrist asked to manage? First of all, the acute toxic problems that occur occasionally with all hallucinogens: panic, paranoia, prolonged effects, and, more subacute in nature, the so-called flashbacks. I have seen several cases of depression following the use of hallucinogens, a complication not usually described. The possibility of a covert or previously latent psychosis becoming manifest after the use of hallucinogens is well recognized.

Freedman[4] has suggested too many good trips as a further complication in the use of hallucinogens, sharing one of my prime personal concerns about the current drug use among younger people. Since the middle 50's, the variety and availability of drugs has increased, and it would appear that those using drugs are becoming younger and younger. The older user has some sense of identity; he is familiar with and sensitive to his aggressive and sexual drives; for him to get stoned or high or to "trip" is a totally different thing than it is for the younger adolescent. I suspect, and I emphasize that this is only a suspicion, that the use of drugs, including marihuana, beyond experimentation during the early post-pubescent years can indeed be an impediment to maturation. Marihuana is used for many reasons, but mostly because it is pleasurable; adolescence, particularly those early years, is not pleasurable. I suspect that some youngsters may use the pleasure of pot to avoid some of the pain of growing up which has to be faced, and, as Kaplan said, "(used) as a constructive instrument."[5]

Finally, I would like to add another role that I consider appropriate for the psychiatrist: namely, education, and perhaps in a way, interpretation. I do not mean by education the formal classroom instruction of school children; the latter in my opinion is far more successfully undertaken by younger educators and there is distinct tendency to a halo effect that has been described by others. I believe the psychiatrist should be involved in the education of educators, of legislators, of administrators, and of parents. Not only education about drugs, their effects and hazards, but an education concerning human needs that may possibly stem the tide of alienating forces that, in my opinion, lead to many serious casualties among young people—including the extensive crippling use of drugs. By interpretation I mean a putting in perspective, exactly what Brenner said in his paper when he remarked that we should be less urgent in seeking individual psychopathology and more helpful in putting things in terms of experience and existence. For example, if the law says that one marihuana cigarette makes the user a felon, then we can at least reassure the parents that the user is not, by the same token, necessarily mentally ill as well as criminal. It could be observed that if one has a son or daughter in a four-year college in the United States who has not used drugs, he is unusual. Should we assume, therefore, that the drug-experimenting college student is mentally ill or should we look at his use in the context of contemporary college and urban life?

Should we assume that the 16-year-old whose anxious parents bring him to the doctor because he has smoked marihuana is sick, or should we recognize that within the present societal idiom he is responding to pressures that are specific to the various resolutions he is attempting to make at his age?

There are many people, young and old, who do have serious problems with drug use and abuse over an extended period of time; what can psychiatry offer to them? Dr. Bewley said, in response to a questioner, that in his opinion psychotherapy for narcotic addicts was not successful in England. I think this experience has largely been true here, though I suspect that this may well be due to the somewhat rigid application of therapeutic models related almost exclusively to psychopathology rather than to total existence. Anyone can be taken off drugs—just see that they don't get them—but how can they be kept off? Can the psychiatrist help the drug abuser to find healthier and less destructive ways to realize . . . again as Kaplan said, "something that is in the direction of 'joy'."[5] So often those traditional therapeutic techniques have proved useless, but I believe we are at a time of great innovation in psychiatry and clinical psychology. Some of the traditional rigidity that has stultified us has been abandoned and we now have innovations both in techniques and applications: family therapy, various group therapy techniques, behavior therapy, crisis intervention, pharmacotherapy. Although we may be accused of going off like Don Quixote in all directions, this may ultimately prove more effective than perpetuating models that have proved unsuccessful in helping the many people who, indeed, need help.

REFERENCES

1. Coles, R., Brenner, J., and Meagher, D.: *Drugs and Youth*. New York, Liveright, 1970, p. 77.
2. Louria, D. B.: *The Drug Scene*. New York, McGraw Hill, Inc., 1968, p. 8.
3. Yolles, S.: In Blum, R., *et al.: Drugs II. College and High School Observations*. San Francisco, Calif., 1969, pp. 3-101.
4. Freedman, D.: "Hallucinogens." Lecture, Association for Clinical and Research Studies, Ypsilanti, September, 1970.
5. Kaplan, A.: The epistemology of drug abuse. In Zarafonetis, C. J. D.: *Proceedings of the International Conference on Drug Abuse*. Philadelphia, Lea & Febiger, 1972.

TREATMENT OF ACUTE TOXIC REACTIONS TO ABUSED DRUGS

ALFRED J. R. KOUMANS, M.D.

There seem to be two sides to the problem of treatment for acute intoxications with the drugs our youth are using. The first is clinical and I attempt to summarize known diagnostic and therapeutic information in the manual that is appended to this presentation.

The second side asks the intriguing question: How do we understand the fact that psychiatrists are being invited to inform their medical colleagues about treatment of toxic states that they are perhaps more familiar with? Clearly, considerations other than purely clinical ones are involved. Perhaps it is true to say that it would be misleading to treat the topic of acute drug toxicity as a purely clinical one, without reference to its complex phenomenology. Actually, we may see how attention to the various non-clinical issues can help formulate a reasonable approach to treatment.*

To begin with, a discussion of the treatment of acute drug toxic reactions should take two paradoxes into account:

The first paradox is contained in the following: The clinical states encountered in the intoxications with drugs that are *en vogue* with today's youth are familiar to the physician although he may not be familiar with the drugs used. Doctors who would not hesitate to treat a belladonna delirium, an alcoholic hallucinosis, a stupor induced by barbiturates, or a manic excitement often have difficulty treating not very dissimilar clinical conditions when the patient is young, long-haired, and toxic from a drug.

Behind the patient stands a culture strange and often frightening to him. *The newness in the situation is not medical, but social.* By the same token, literature concerning the subjective and objective effects of these drugs has been plentiful, but medical literature on interventions or research lags several years behind.

This immediately raises a few interesting issues. We see physicians stumped by the non-clinical elements of the situation of acute intoxication, such as demeanor and ideology of other patients. I've heard reputable doctors ask a street kid in trouble, "Why don't you take a bath and get a job instead of taking dope?" Those with or even without medical degrees, who for a variety of reasons have become familiar with what is now called the drug culture, find themselves sought after by an ever-widening circle of bewildered adults, parents, colleagues, school officials, and even the business world. These adults seem to need an introduction to this aspect of youth culture, an explanation of

* For specific recommendations, a *Manual for the Treatment of Acute Drug Reactions* is appended.

the behavior of their employees, pupils, patients, and children.

Another issue is the unspecificity of the drugs used. As a teenager I must have inhaled airplane glue while busily constructing a model of a Fokker biplane; I am sure there was nutmeg in the kitchen. In medical school we learned what we thought was all about cocaine, amphethemines, and the historical oddities like epidemics of toxic psychoses due to *Claviceps purpurea.* But not a thought ever, then, about using any of these substances for self-induced hallucinatory episodes. Today, any substance, from banana peels to explosives, that may have as a side effect disregulation of perception, will be taken, regardless of its other pharmacology, if any.

The second paradox can be illustrated with an analogy: Let us suppose that a viral epidemic had erupted five years ago that left few of our young unaffected and some with serious sequelae. In that case the medical world surely would have reacted differently and more in keeping with its humanistic and scientific traditions. Not only the medical world: Imagine the political mileage that could be gained from supporting one of our famous crash programs to study the danger of this mysterious virus that threatened the hope of our nation; it would have become a prime campaign issue. But look: Not one single candidate in the last election campaign even mentioned the urgent need for research on marihuana, the mildest of the drugs used, to outline its dangers. Today, unfortunately, this research is clearly not in the interest of politicians.

These paradoxes stimulate the psychiatrist; he has an interest in the elucidation of the seemingly irrational. He will try to come to terms with the following phenomena pertinent to the question of treatment for drug reactions.

I know of no other time in history when a distinct age group, such as adolescence or youth, has emerged with its own medical problems. The problems, I repeat, were all familiar to medicine: venereal disease, intoxications, malnutrition, psychiatric problems. New is their conjunction with an age group that, medically speaking, was the healthiest; the flower of youth between the offices of the pediatricians and internists. Adolescents until recently were allowed to do their things in schools and camps; only occasionally in a specialized psychiatric office.

The adolescent's good health shielded most practitioners from the baffling problems of adolescence; mainly, the psychologically necessary skirmishes around independence, autonomy, identity, and sexuality. A very heady mixture indeed!

Now as these difficult issues line up with social unrest, a strange yet glamorized subculture, and heavy drug use, the emotional impact on the physician or medical institution of a confrontation with sick youth is hard to face with equanimity. Nobody likes to see "wasted" that which he himself has lost—youth.

Free clinics are springing up all over the country where already overworked physicians donate their time and efforts to treat the disaffected young; at the same time a senior physician in a respectable utilization review committee questions the access to a hospital bed for a needle hepatitis victim. There is a saying in Dutch, "You'll speak the words of whoever's bread you're eating." The pharmaceutical firms lodge and dine us well. May I yet raise the question in this meeting of their responsibility, of medicine's, and of our own complicity in fostering the notion that, even for everyday frustration, anxiety, and worry, chemical remedies are appropriate? All of these inconsistencies invite serious thought along political, economic, or even philosophical lines.

Without going into these issues here, it it seems clear that treatment of the acute toxic drug reaction can be dealt with separate from this irrational scene only artificially. These issues make very clear something that has to do directly with the treatment setting.

From coast to coast, the intoxicated young have been seeking help, and sometimes finding it, in unorthodox medical settings. (The fascinating subject of how and why some physicians have responded by organizing free clinics outside the medical establishment can only be hinted at here.) More often, the young have developed helping resources within their own culture: bad-trip counsellors, whose training consisted of personal experience with some professional backing at best, or simply friends with a tablet to counteract the one they took.

The traditional medical setting of the emergency room has been neither sufficient nor effective and often counter-productive.

Experiments with free clinics, satellite doctors, mobile units, and trained or supervised peers in cities or on campuses have been more successful. These and other activities attempt to answer the question many of you have, "What can we do?" You can engage the problem this way, or stimulate similar development in your community. There still remains the question, "*How* can we do this?"

Treatment itself presupposes a physician-patient relationship. However, disturbing things have been changing in this and other relationships between representatives of the generations, whether they be in the field of education, politics, or within the confines of the family. The relationship is often a confrontation rather than an encounter.

The physician who is called to treat an acute toxic reaction may find a patient who distrusts him and who shares little of the values upon which the usual and implicit treatment contract is based. These values are more or less standard middle-class, and have something to do with a sense of pride in achieving improvement, respect for scientific authority, and, above all, a certain suspicion of the human body. That the young patient has brought the "illness" on himself in the pursuit of pleasure often repels the physician profoundly. Also, their differences can no longer be blamed on obvious factors of race or social class, as in the case of the ghetto junkie or the skid-row alcoholic. It is too close, it could be his son or daughter.

Factual ignorance and folklore about the drugs used may make some of us hesitate to apply clinical judgment, and the threat perceived in the deeply dividing differences may provoke hostility or aloofness. The adolescent has a sense of entitlement. The young drug user may confuse the issue with unrealistic expectations about the patient-doctor relationship and with his insistence on "special treatment."

What this means for the physician-patient relationship is clear: The physician is challenged to create a feeling of trust in the patient and of acceptance in himself, or else his "treatment" will be purely physical at best. He neither has to embrace his patient's values totally, nor to lecture him about his own. If he is aware of his own value-system and its relativity, he may avoid getting drawn into arguments with his patient—a futile and self-defeating exercise for both. It would help to be factual, reasonably informed as to the drug's effects, and sincere in his approach. Providing a cup of water or a gentle touch will speak louder than words, and may convey more of a feeling of genuine care and respect for the patient's integrity than a probing question or a lecture—but only if it is indeed meant and not made. The adolescent has an unfailing sense with which to detect play-acting and insincerity.

The factors that distinguish today's treatment problems of acute drug intoxications from yesterday's are in the social dimension—that includes the defiant acceptance by our young of their drugs. Where the patient-doctor confrontation does not become regressive or stereotyped, it has resulted in a reappraisal of professional as well as personal assumptions, and new alliances between the two cultures as well as a number of innovative projects and treatment methods have emerged. Perhaps this is medicine's participation in what seems to be a profound societal change.

The treatment of acute LSD intoxication may serve as an example and recapitulation of the issues outlined above as well as the consequences it has for treatment.

Psychosis following accidental poisoning with alkaloids from *Claviceps purpurea* is not new in medicine. The descriptions of subjective effects (upon which a rationale for its psychologic treatment is based) were available but not generally known. The unfortunate minority of LSD users suffering a "bad trip" finds little sympathy or credibility in society's appointed healers, the medical profession, and chooses its own: free clinics, bad-trip counsellors, and experienced friends. As a matter of fact, to be admitted to one of medicine's steel and glass temples during a derailed psychedelic experience is in itself frightening and complicates the panic and paranoia.

Finally, the physician may face the dilemmas whether to ask for parental permission to treat, and whether to report illegal drug use to the authorities; in other words, to abide by standard social and legal procedures or try to gain trust and rapport with the patient. The earlier prescriptions of medical practice now seem contraindicated.

As to specific treatment: The practice where I work varies from instant mental hospitalization (in restraints) to talking down, with, as could be expected, a variety of drugs between those extremes. (For specific recommendations I refer to the manual.)

For the treatment of the acute toxic reaction, various lay-help organizations have consolidated themselves with telephone lines and bad-trip counsellors who can be called upon to come to the scene. Usually some back-up by professional personnel as consultants or the local hospital is available. This is another instance of care provided by lay health workers as is currently developing in many other fields of medicine. Not surprisingly, the experience of these workers indicates that "talking down" the person with verbal and non-verbal means is the most elegant and least disruptive treatment. This needs time, conviction, and the willingness to engage the other on a personal level to gently guide him into a common reality. Incidentally, might this be perhaps precisely what many of our young people today are asking for?

MANUAL FOR THE TREATMENT OF ACUTE DRUG INTOXICATIONS*

ALFRED J. R. KOUMANS, M.D.

INTRODUCTION

This manual is intended for use by health professionals in the widest sense: physicians, emergency units, nurses, school infirmaries, psychologists, para-professional helpers, etc.

The rising incidence of drug† use, often of new substances and mainly among our young, has found many health professionals inadequately prepared to cope with their public and professional responsibility for treatment of drug-related conditions.

The nature of the drugs used (their mind-altering properties) as well as the unfortunate and painful rift between the "established order" and the "drug culture" makes for a complex, often conflict-ridden, social and psychological situation whenever an intoxicated drug user turns for help to an established medical facility. The acute intoxications especially present a challenge and often lead to situations where the traditional benevolent relationship between helper and patient is distorted beyond recognition by a mutual sense of bewilderment, frustration, doubts about competence, and mistrust, which may leave many patients even more alienated than they were before.

Personal convictions regarding the appearance, customs, or morals of our patients should not be allowed to jeopardize professional conduct. The best of the medical traditions—compassion, competence, and neutrality—will help us to keep our patient's well-being in our mind.

It is for these reasons that this manual puts emphasis on differential diagnosis of drug intoxications as well as on the psychological and social management of these conditions.

The manual is structured as a flow-chart of clinical decision-making with a few specific recommendations. The basic flow-chart is described in detail on pages 389 through 397.

This manual draws on the experience obtained by various people in different settings: in a free clinic for street people, a college health service, private practice, a general hospital, and was updated at the International Symposium on Drug Abuse, Ann Arbor, Nov. 1970. I am especially indebted for comments and corrections to Dr. David Lewis.

* Made available for distribution by the Massachusetts Medical Society (Nov. 1970).

† For this purpose "drugs" include the substances that influence one's level of consciousness and perception: the hallucinogens (LSD, mescaline, etc.), the amphetamines, the barbiturates, heroin derivatives, and marihuana and derivatives.

This manual is not intended to be final or complete. It will have to be continuously revised and improved. For suggestions, please contact Alfred J. R. Koumans, M.D., Rm. 11 - 203, 77 Mass. Ave., Cambridge, Mass. 02139. Tel: 864-6900 X. 2917.

GENERAL REMARKS AND RECOMMENDATIONS

1. Mixed Drugs

Many drugs available on the street have been tested and found to contain a mixture of drugs, although they are sold as a single one. The combined effects of these drugs will determine the clinical picture. Sometimes the mixture is intended to have an enhancing effect (marihuana and opium) and mask disturbing effects or mimic the real stuff (see under 2). This is a good reason for limiting the amount of drugs used in the treatment of these conditions to an absolute minimum and relying more on environmental and psychological methods of treatment that are effective and follow the dictum "Primum non nocere"—first of all do not harm (see 3).

2. Drugs Sold under a Different Name[1]

Drugs bought on the street market and tested have often been found to contain ingredients other than those that the purchaser was led to believe they contained. Most notably, "mescaline" on the recent market contained mostly no mescaline at all, but instead a mixture of amphetamines and other hallucinogens (like LSD). The illicit drug manufacturers follow the trends closely and label their product accordingly. The buyer rarely knows for sure what he buys, or the amount he gets. The treating physician should not rely too much on the information of the user but more on his clinical judgment. For those interested, the underground press (Old Mole, B.A.D., Phoenix, etc.) occasionally contains information relating to the drug market.

3. Treatment of Bad Trips

Not infrequently a psychedelic experience turns out to be a harrowing ordeal for the user. It is not known with certainty what determines this outcome although it seems that the personality, the setting, and events surrounding the use of the drug have some influence. "Bad trips" often come to medical attention. The panic of the user should be separated from the anxiety and guilt it creates in the user's environment, since one tends to reinforce the other; both will need attention.

The key is a personal approach, and understanding that the psychotic reaction is due to the patient's temporary inability to perceive reality as he does normally, to remember that his experience was caused by a drug, and to screen out the host of irrelevant sensory information to which one normally pays no attenton. The patient's state may be complicated by the fact that he is also unable to repress past experiences or memories, or to control his fantasies.

The patient is adrift; because of his extremely labile affect and suggestibility, he can easily become frightened and alarmed by what the people around him do or say, or what his physical surroundings imply to him. The same suggestibility is the physician's leverage for psychological treatment.

Those experienced in handling such cases have found the following principles most useful: (1) Establish as normal a setting as possible, keeping outside interference to a minimum. A quiet dimly lit room or an undisturbed spot on the grounds would suffice. (2) Establish *trust* with the patient. Preaching, moralizing, and punitive attitudes are not helpful, and neither are insecurity, wavering, and anxiety on the part of the helping person. (3) *Avoid discussions* or arguments about perception or ideas with the patient. *Do not probe* into his personal life or conflicts. (4) Humor works better than sarcasm or seriousness. (5) Remind the patient consistently *that his experience is the result of having taken a drug,* that he was normal before and will be normal again in a short while. (6) When the patient seems totally out of touch with reality, help him focus *attention on a real object* (a stone, a candle, a book, or your hand) and let him feel, touch, smell, and explore. This is needed to reestablish a bridge with the real world. (7) Avoid crowds, one person is enough. (8) Let the patient keep his eyes open. (9) A warm, sincere, and *appreciative attitude* on the part of the helper will be effective to talk the patient down and hold him down, even while the drug effect may persist. A bad drug experience may in this way be converted to one which has a more positive outcome.

4. Complications of Hallucinogenic Reactions (LSD, Mescaline, STP, etc.)[2]

Spontaneous Recurrence. The "flashback" may occur at any time from a few days to a year after the last hallucinogenic experience. The precipitating cause is not always clear. Frequently it is the use of a milder drug, such as marihuana. It is characterized by anxiety, perceptual distortion, and the fear of losing one's mind. The flashback seems to confirm this fear, hence the anxiety. These reactions are amenable to psychological treatment as outlined in "3. Treatment of Bad Trips." Some of these patients may require more specific psychiatric or psychological help to deal with repeated flashbacks; the terrifying idea of having indeed lost control over one's mind has driven some to suicide.

There is as yet no agreement on physical treatment, although anti-anxiety agents and anticonvulsive drugs as well as other modalities have been tried. A cautionary statement about further drug use would seem in order.

Prolonged Reaction. Any reaction lasting more than 12 hours after the ingestion of the drug and characterized by anxiety secondary to the inability to terminate the reaction needs further psychiatric treatment; it may respond to Valium.

Provocation of Psychosis. In individuals who, before taking the hallucinogen, displayed either pre-psychotic abnormalities or frank psychosis, the ingestion of the drug may precipitate a psychosis. It is unusual for a healthy normal person to remain psychotic after ingestion of a hallucinogen. The psychosis is usually of the paranoid schizophrenic or schizo-affective type. Treatment should be in an appropriate psychiatric facility.

5. Complications of Marihuana and Hashish[3,4]

After marihuana or hashish, a "flashback" from a previous drug-induced hallucinogenic experience may occasionally occur. Also, and more frightening, is the appearance at this time of a frank psychosis in persons who have shortly before used a strong hallucinogen.

It has been observed that those who react unfavorably to mild intoxication with marihuana may show personality difficulties suggestive of a pre-schizophrenic state prior to the drug use.

6. General Comments on the Differential Diagnostic Tables

The differential diagnostic tables represent highly schematic composite pictures of actual clinical conditions and are intended only as a reminder and guideline. Since the dosage of the drug and the drugs involved as well as the personality of the user will determine the clinical situation, no one actual clinical condition corresponds entirely to the picture presented in the tables' columns.

7. Disposition

Experience has shown that most patients with drug-related conditions coming to medical attention can be treated in a general hospital.

The experience of most "bad-trip counsellors" is that the "bad trip" is handled ideally in surroundings familiar to the patient (his home, that of his friends, etc.) where he is under constant care of experienced peers and if necessary, under the supervision of medical persons. If this fails, or is not available, the next resource may be an emergency room of a general hospital (which is hardly a reassuring environment for this particular type of patient, however). Yet, even severe reactions can be handled there, provided some guidelines are kept in mind (see note 3). Since "bad trips" are essentially time-limited conditions, actual admission and bedding of the patient are rarely necessary except when expedient. When admission seems indeed the last resource, a general hospital is preferable to a mental hospital. The reasons are: (1) The difference in environment; to be with "sick people" is less frightening than to be with "crazy people." (2) Social stigmatization; the patient will henceforth have to state on all employment questionnaires that he has been in a mental hospital, which can jeopardize his chances for employment, driver's license, and possible voting rights.

8. Notification[5]

There is no doubt that under present conditions, indiscriminately notifying parents and/or authorities of cases of intoxication by illegal drugs only results in further aggravation of the distrust and antagonism already present between the youthful patient, alienated or not, and his elders. Notification thus can seriously impair the doctor's accessibility to this group of patients.

Our obligation to obey the laws has to be tempered by sound medical judgment, which keeps the patient's well-being foremost.

Massachusetts law requires the following: *Acute poisoning* has to be reported by *occurrence only* (no name or other identifying data) to the Massachusetts Department of Public Health. Detailed reporting (name, address, height, weight, date of birth, color of eyes, color of hair, date treated, and name of narcotic drug) by the treating physician is required in case of *persons suffering from the chronic* use of narcotic drugs. The requirement applies only when two criteria have been met: first, the patient must be suffering from the use of narcotic drugs; and second, use of narcotic drugs must be "chronic." If the physician determines in good faith that the patient is not suffering from the use of "narcotic drugs," then the physician does not have to report under sections 210A or 212B.

The statutes themselves do not provide criteria for determining chronicity. Thus, if the physician finds that the patient is suffering from the use of narcotic drugs but the physician, acting in good faith, determines that the use of narcotic drugs is not chronic, then he is not required to report under sections 210A and 212B. Similarly, if the physician finds that the patient is suffering from the use of narcotic drugs, but the physician, acting in good faith, is unable to determine whether the use of narcotic drugs is chronic, then he is not required to report under sections 210A and 212B.

Even though a physician may not have to report under sections 210A and 212B, however, pursuant to section 187C he will still have to report "the circumstances of such poisoning" when he is "attending or treating a case of acute poisoning caused by narcotics, barbiturates or amphetamines."

The physician should realize that reporting chronic use could subject his patient to criminal prosecution.

FLOW CHART OF CLINICAL DECISION MAKING

The basic flow chart is as follows:

```
                unconscious I (p. 389)
               /
Patient is:                 uncommunicative II (p. 389)
               \           /
                conscious                  incoherent III (p. 396)
                           \              /
                            communicative
                                          \
                                           coherent IV (p. 396)
```

I. **Patient is unconscious**

1. Ask *friends* who bring the patient all relevant information (medical history? Epilepsy? Diabetes? Accidents? Drugs?, etc.).
2. *Notify* parents or authorities when necessary (see "8. Notification").
3. Obtain blood and urine *samples* for laboratory studies.
4. If history and physical examination indicate drug use, and other medical cause is absent, give appropriate antidote, or institute established routine care. (See Nos. 10 of diagnostic tables.)
5. In case of drug use causing loss of consciousness, *avoid labeling* a self-induced overdose as suicidal until you know the patient's motives. Social stigmatization by improper disposition in a mental hospital should not follow an accidental overdose. Many of these, especially with heroin, are nonintentional. The patient whose life you perhaps save will not want to see you again, and his friends will look elsewhere for help too, if you arrange for an expedient commitment to a state hospital.

II. **Patient is conscious but uncommunicative**

A. Non-verbal communication only.

1. Retain patient's *friends.*
2. Take time for patient—this will be your most effective instrument.
3. Rule out *causes other* than drugs for mute patient (speech defect, psychosis, etc. by history).
4. Try *other methods of communication*: write, whisper, and observe and respond to facial or body expressions.
5. If patient seems to be in touch with his environment: Retreat to a quiet place, preferably with his friends and gently and calmly explain his environment, its physical aspect, and its function to him. Here you assume that the patient is stricken with fear and unable to integrate his environment or your good intentions. The confusion contributes to his anxiety. ("This is a quiet room in ——— Hospital. We use it for talking with people. I am a doctor. My name is ———. You were brought here by ——— and ——— because . . . You seem to be scared—let me explain to you what we are doing and why" etc.) In other words, clear a path before you expect contact.

B. Patient seems out of touch with his environment. He takes no

Table 1 Differential diagnosis and treatment of amphetamine and cocaine toxicity, acute schizophrenia, and acute mania

	Amphetamine intoxication	Cocaine intoxication	Acute schizophrenia	Acute mania
1. History	Behavior disorders often present. Prior use of amphetamine, often in "runs" (days-weeks)→"crash"	Sniffing, injection, or swallowing of "coke" or "snow"	Often unusual, bizarre or seclusive type of people; sometimes previous psychotic episode	Variable disposition, excessive mood swings
2. Course	Depends on dose and route: oral—brief effect, no "high"; I.V.—2-4 hours "high" (1000 mg or more day). Some symptoms persist for months after abstinence	Usually shorter than with amphetamine	Gradual onset, remains longer disturbed	Gradual onset, remains longer disturbed
3. Behavior	Dose-dependent: from hyperactive to outbursts (assault, suicide). "Runs": pointless, repetitive activities	Garrulous, hyperactive	Varies (usually stable, preoccupied) can have occasional outbursts	Extremely overactive, talkative, jumpy, unable to pursue tasks
4. Symptoms & signs	Tremor, dry mouth, no appetite. Tachycardia, hypertension. Sweating, hyperreflexia, fever. "Crank-bugs," (imaginary bugs under skin), ulcers	Twitchings, no appetite, tachycardia, irregular breathing, nausea, loss of appetite. When sniffed: ulcers on nasal septum	Physical symptoms are rare, except when patient is acutely agitated. There may be dwelling on hypochondriacal complaints	Consistent with hyperactivity
5. Thought	Racing associations, in long chains, but they seem to make sense. In chronic users, becomes indistinguishable from paranoid schizophrenic thought	Same as with amphetamine. Ideas freq. of super-strength, etc. Less often paranoid	Disorder of logical thinking. Associations are loose and bizarre	Pressure of thought and speech; tangential thinking, association often in long series, comprehensible, often to sound rather than to content
6. Affect	Blunting occurs. Usually labile; anxious, tense, aggressive. Intense euphoria of "flash" lasts only minutes	Euphoric	Varies, often bland and not fitting the situation nor the items discussed—can be panic or depression	"High as a kite"

7. Delusions	Around real-life issues: e.g., violations ("police are following me"), spreading groups of foes	Ideas of grandeur	Bizarre, fantastic, religious or magical	Ideas of grandeur and omnipotence
8. Hallucinations	Can be recognised as unreal. Dreamlike, vivid, visual. When auditory: in third person ("get him")	Same as amphetamine, often tactile: "crankbugs"	Experienced as (close to) reality, mainly auditory, less often visual, olfactory, tactile	Not pronounced, in keeping with rest of syndrome
9. Diagnosis	Clinical observation. Amphetamine excretion enhanced by acid urine (No NH_4Cl if patient is already acidotic [hunger])	Clinical observation	Clinical observation	Clinical observation—patient responds better to firmness and limit-setting than do schizophrenics or patients having amphetamine toxicities
10. Therapy	Low toxicity: nothing but rest. Medium: supportive. Observe. High tox.: Admit to gen. hosp. If possible: emesis or lavage. Sedation with Thorazine *only* when amphetamine is indeed used. When in doubt, Valium or Librium. Sleep follows (often days) and should be allowed. (See NOTE in Table 2, "10. Therapy," Col. 4.) For agitation delirium: Light room, sedation	Observe and sedate if necessary	Admit to psychiatric facility. PRN phenothiazines for agitation	Admit to psychiatric facility. PRN phenothiazines for agitation
11. Complications (from amphetamines)	Hypertension (cerebral hemorrhage), sudden death, hepatitis, trauma, violence, paranoia. Malnutrition. I.V. use complications: infections, etc. High dose: convulsions, coma?, diffuse necrotizing angiitis; after run: crash with depression, suicide. "Overamping" (= overdose, exceeding tolerance): period (minutes–days) of chest pain, unconsciousness, followed by alert paralysis with racing mind. The very rare deaths directly attributed to amphetamine follow period of high fever and vascular collapse. I.V. use leads often to other I.V. drugs (heroin). Heroin often used as "downer." No evidence of brain damage after chronic use or abuse.			

Table 2 Differential diagnosis and treatment of CNS depressant intoxication and severe depression

	Heroin and other opiates	Barbiturates and other hypnotics	Depression
1. History	Previous drug use is common as is previous heroin use. Intoxication may follow (inadvertent) overdose when street sample suddenly changes in strength	Often used for suicidal motives or taken in successive doses ("losing count") as intoxication proceeds	Previous depressions are rare in young people, common in older ones (in young people, behavior disorders are common)
2. Course	Sudden onset, lasts 4 to 8 hours; methadone intoxication lasts 8 to 10 hours	Sudden onset, lasts (depending on drug used) from few hours to few days	Gradual onset, usually lasts long period
3. Behavior	From drowsiness ("nodding") to coma	From drowsiness to coma. In stimulating environment, can act as stimulant (esp. with I.V. use); violence	From general slowing of mental and motor functions to mutism and immobility. Manipulating patient (sitting him up) provokes resistance
4. Symptoms and signs	Constipation, nausea, urinary retention, constricted pupils, respiratory depression, hyperuria, needle tracks, depressed circular scars (skin popping), subcutaneous abscesses	Incoordination of movements, "drunk" (without odor of alcohol!), sub-epidermolytic bullae on extremities	Lack of appetite, insomnia: loss of weight can occur. Physical examination may reveal no abnormality
5. Thought	Preoccupation with obtaining the next dose. Thought processes obtunded, attention drifting	Respiratory depression, hyperthermia, hyporeflexia	Feelings of worthlessness, suicidal preoccupations, sin, guilt, lackluster life; thoughts may be racing or slowed down
6. Affect	The euphoria of heroin use may yield to anxiety with overdose	No typical change—yet generally dulled	Blue, depressed, etc.
7. Delusions	—	—	Of being rotten, lost, condemned, etc.

8. Hallucinations	—	—	In keeping with rest of clinical picture ("inside is rotting away")
9. Diagnosis	Opiates in urine	Serum levels of barbiturates	Clinical condition
10. Therapy	Shock: plasma expanders, I.V. pressor drugs. For pulmonary edema: O_2, diuretics. Respiratory care & opiate antagonists: levallorphan, 0.02 mg per kg body weight I.V., or nalorphine, 0.1 mg per kg body weight I.V. These drugs counteract the respiratory depression. They should not be used in excess for treatment of coma, since they are themselves respiratory depressants in high doses. If no response to this regimen occurs, mixture of drugs was probably used & diagnosis should be reviewed N.B. If no opiates in tissue, antagonists act as opiates	Respiratory care. No drugs if patient is responsive. When drug use was habitual, admit to hospital and withdraw gradually (convulsions). If patient is comatose, empty stomach of contents with endotracheal tube in place. With good renal function: forced diuresis and alkalinization to foster excretion. Dialysis in selected cases	Possible admission to psychiatric facilities **Note:** After an amphetamine high, a period of exhaustion-depression may follow ("crashing") which can be extremely uncomfortable and lasts from hours to a few days. Dangers of suicide and other drug ingestion are common in this stage and close supervision should be arranged
11. Complications	From overdose: pulmonary edema, respiratory arrest. Convulsions. From dirty needle: hepatitis, endocarditis, phlebitis, tetanus, lung abscess. Quinine-adulterants→ arrhythmias. Pulmonary edema: x-ray, fever, leukocytosis, respirations 25/min or less. Treatment: conservative, no antibiotics. If no response with 3 doses of Nalline consider brain trauma or meningitis	Convulsions (from withdrawal) From I.V. use: infections	Suicide

Table 3 Differential diagnosis and treatment of hallucinogen intoxication and hallucinosis

	Hallucinogen intoxication	Marihuana/hashish intoxication	Hallucinosis
1. History	Ingestion of LSD or hallucinogenic amines: mescaline, DOM (= STP), peyote (= Psilocybin), DMT, phenecyclidine (= Sernylan), Ditran (atropine compound) others, or mixtures	Previous drug use. Adverse reactions more common in novice or pre-psychotic individual	Preceding period of use of alcohol, followed by withdrawal. Also after use of amphetamines, cocaine, inhalation of airplane glue or solvents
2. Course	Onset in ½ hour, lasts hours, depending on drug ingested as well as on management and doses. (In general: LSD, ca 8 hrs; DMT, 1 to 2 hrs; DET, 2 to 3 hrs; STP can have prolonged effects lasting from 12 hrs to 2 days. Secondary effects may last much longer (see Item 11, below)	Onset within 20 minutes. Low dose: euphoria, sedative. High dose: hallucinogen. Duration from 2 to 5 hrs. Secondary effects: see Item 11, below	Onset gradual, usually preceded by period of apprehension. Recedes with proper care in a few hours to a week, depending on drug used—hours with glue, a week with amphetamine, days with alcohol
3. Behavior	Ranges from withdrawal to eruption. Reality is distorted, may lead to accidents (suicide, homicide)	From stimulation to sedation—hilarious, talkative, or lethargy. Speech fragmented. High dose: delirium	Unpredictable, can be violent, self-destructive (see item 8, below)
4. Symptoms and signs	Dilated pupils. Less often: tachycardia, motor restlessness, hyperreflexia, tremor, speech rambling, vertigo, headache, nausea. May be alert or dreamlike. (With peyote or organic mescaline, discomfort and GI distress.) Ditran: atropine toxicity symptoms. Sernylan: "drunk," somatic anesthesia, cataleptoid, hypertension	Dry mouth, increased appetite, conjunctival vessels dilated, bronchitis, acrid odor on breath and clothes with chronic use, tachycardia, hypotension	Confusion often absent; physical signs of withdrawal may be found (alcohol); clear sensorium is common. With cocaine: dilated pupils, hyperthermia
5. Thought	Confusion, disorientation. Distortions of perception. Loose and interrupted associations. Ineffable quality ("too many thoughts"). Ego fragmentation, hypersuggestibility, depersonalization. Unfocused attention, irrelevant stimuli given equal importance	Perceptual distortions (especially to time), depersonalization, thought sequences disrupted. May be paranoid. Increased sensory awareness and decreased inhibitions. Recent memory may be disturbed. Temporal disorganization	Sometimes disoriented. Perceptive disturbances, thoughts of persecution and danger are common

6. Affect	Panic usual, bliss rare. Can fluctuate: highly labile. Depends on environment stimuli, preexisting psychology and expectations. Highly suggestible	From euphoria to anxiety and panic	Fear, panic, violent rage. May be placid, depends on content of delusions
7. Delusions	Overlap of sensory modalities. Paranoid preoccupations. Fear of going insane, nirvana, body image distortions	Delusions rare, only with frequent or high dose. Paranoid conditions can occur	Delusions not systematized, but often consistent with real events (professional or past "criminal" record), e.g., persecutory delusions
8. Hallucinations	Frequent in all sensory modalities, usually visual, beatific, or fearful (like colorful pictures on a screen, not part of the enviroment)	Illusions (distortions or perceptions) with hashish. Hallucinations, if present, mainly visual	Most prominent feature usually vivid and auditory, at times visual and other sense modalities (with glue, visual and frightening hallucinations are common; with cocaine, tactile hallucinations may be noted)
9. Diagnosis	Clinical picture differentiate: intoxication from anticholinergics (atropine, belladonna, scopolamine): skin, mucosae dry, flushed; fixed, dilated pupils; urinary retention. Disorientation, even visual hallucinations. (These drugs may be used as adulterants.)	Clinical picture	History and clinical conditions
10. Treatment	Reassurance. If necessary, give Valium orally, 20 or 40 mg, and 10 mg prn. For Ditran: THA (tetrahydroaminacridine). For Sernylan: Do not give Thorazine. Follow-up desirable	Same as with hallucinogen intoxication. If needed, give Valium, 5-25 mg, orally	Avoid sensory and social isolation. Provide structure, consistency, company. Treat underlying organic condition, if necessary sedate by giving Thorazine orally. Rest. In extreme situations admit to psychiatric facility
11. Complications	Accidents due to poor judgment, recurrent reaction, and prolonged reactions. Triggering of psychotic reaction (see under "4: Complications of Hallucinogenic Reactions") (page 387). From Sernylan (large dose): convulsions. Fetal abnormalities: abortions (confirmed); malformations in offspring of population that uses street LSD (with pure LSD not confirmed). Chromosomal changes: conflicting studies; possibility suggested; risk-factor unknown	During or after marihuana/hashish, panic reactions and depressions may occur and persist. Reported, but extremely rare in U.S.A., is a "paranoid psychosis"—a delirious state with anxiety, confusion, hallucinations, motor restlessness. (See "5. Complications of Marihuana and Hashish" [page 387] for flashback, delayed reactions)	With alcohol, CNS and liver disease; with amphetamines, hepatitis and exhaustion syndrome; with glue, kidney and liver damage and aplastic anemia

notice of his environment, he is conscious but hardly responsive.

1. Possibilities:
   a. A psychosis (acute catatonic schizophrenia, extreme depression). See differential diagnostic Tables 1 and 2.
   b. Fugue state, trance state.
   c. Organic condition (brain tumor, pre-coma, post-ictal state).
   d. Drug reaction: hallucinogens, CNS depressant intoxication (see Tables 2 and 3)

III. **Patient is communicative but not coherent**

1. Retain his *friends* for information.
2. Try to arrange for quiet *undisturbed setting*—time to observe is paramount. Arrange for supervision of patient if you have to leave.
3. *Differentiate* basic conditions that interfere with coherent communication:

   Psychiatric conditions: schizophrenia, manic psychosis, anxiety (see Table 1).

   Neurologic conditions: CNS trauma (concussion, etc.), meningitis, post-ictal state, etc.

   Toxic conditions: diabetes, uremia, CO poisoning, alcoholic intoxication, psychedelic drugs, etc.
4. Disorders of thought (schizophrenia), mood (depression, mania), and affect (anxiety) can disrupt communication. They can, and sometimes do, coexist with drug-induced disorders. The differential diagnoses that will have to be entertained most frequently are found in Tables 1-3. They differentiate the following conditions:

   Acute schizophrenia—amphetamine intoxication—acute mania (Table 1).

   Heroin and other CNS depressant intoxication—depression (Table 2).

   Hallucinogen intoxication—marihuana intoxication—hallucinosis (Table 3).

IV. **Patient is conscious, communicative, and coherent**

1. Ask his *friends* not to leave until the patient's condition is known. You may need information only those who brought him in can give.
2. Allow friends to remain with the patient if so desired.
3. To isolate a person in a strange *environment* is the surest way to create anxiety. Anxiety is going to make your work needlessly difficult. (Remember that many hospital routines are designed for the convenience of nurses and staff rather than for the patient's comfort.)
4. Interview the patient: Who is he? Why did he come? Obtain names of parents and physicians. Reassure him that his information is confidential (only if indeed it is!) that you will not call authorities or his parents unless his medical condition requires and then only after discussing it with him. (See "8. Notification.") Without confidence, no doctor-patient relationship; without trust, no therapy.

Note: An acute drug-induced panic may be more expeditiously managed on a bench on the hospital grounds with the physician chatting with the patient and his friends, rather than in the sterile, shiny white glare of an emergency room filled with mysterious instruments, suggestive smells, and strange people.

PITFALLS AND ERRORS TO AVOID

1. Thoughtless and automatic notification of parents or authorities.
2. Isolation of patient in, for him, frightening environment (creates anxiety).
3. Committing patient to mental hospital.
4. Diagnosing overdose or accident as "suicidal."
5. Probing into conflicts; arguing; debating with intoxicated patient.
6. Substituting drugs for simple, humane approach (psychological treatment) for patient who is "tripping" on hallucinogens.

REFERENCES

1. Cheek, F. E., Newell, S., and Joffe, M.: Deceptions in the illicit drug market. *Science 167*:3922, 1970.
2. Smart, R. G., and Bateman, K.: Unfavourable reactions to LSD, a review. *Canad Med Assoc J 97*:20, 1967.
3. Weil, A. T.: Adverse reactions to marihuana. *New Eng J Med 282*:18, 1970.
4. Talbott, J. A., and Teague, J. W.: Marihuana psychosis. *JAMA 210*:299, 1969.
5. Vigoda, D. (Assistant Attorney General of Massachusetts): *Tracking the Reporting Statutes*. The Attorney Generals Newsletter on Drug Abuse: *Tracks 1*:6, 1970.
6. Abelson, P. H.: Death from heroin (editorial). *Science 168*:3937, 1970.

THE LONG-TERM MANAGEMENT OF DRUG ADDICTION WITH PARTICULAR RESPECT TO NARCOTIC ADDICTION IN YOUNG PEOPLE IN THE UNITED KINGDOM

PHILIP BOYD, M.D.

The dissimilarity in the traditional approach to the long-term treatment of narcotic addicts between the United States and the United Kingdom stems from the marked differences in the historical and epidemiologic factors of narcotic addiction in the two countries, and from the early divergence in the ethical tenets of each country's medical profession in its response to the plight and need of the addict.

The early repressive attitude which led to the troublesome years of the prohibition of alcohol in the United States also led, in 1925, to the total ban on the use of heroin, to the failure of attempts by the medical profession to develop treatment clinics, and to the view that, for all practical purposes, both the addict and the physician who attempted to treat him progressively were criminals. It was an attitude which also failed to cure the addict. During the ensuing years, apart from methadone maintenance schemes, the United States medical ethic has not substantially changed; the provision of narcotic drugs for self-administration by the addict is still regarded as unethical.

In the United Kingdom, in the earlier part of this century and in contrast with the United States, narcotic addiction was an almost negligible problem. In 1926 the British medical profession accepted the principles laid down by the Departmental Committee of the Ministry of Health in its report on Morphine and Heroin Addiction (Rolleston Report).[1] This report made it clear that heroin and other narcotics could be prescribed ethically for addicts who could not be cured after every effort had been made to effect recovery, or who could only function usefully in society when receiving a regular allowance of their addictive drug.

During the ensuing 30 years, narcotic addiction in the United Kingdom presented no special problem, and the myth of the so-called British system of treatment evolved. It was held to account for the continued absence of any serious number of narcotic addicts in the country. However, this myth was exploded in the early 1960's when it became statistically evident that a marked change had occurred, and that a new and serious problem of narcotic addiction was building up.[2] This reflected the incubation of an epidemic of general drug abuse which had been developing during the previous decade, and which had its parallel in many other western countries. The tip of this malignant iceberg manifested itself as a fresh and virulent culture of narcotic drug-taking, infecting both the United Kingdom and the United States. Its greatest impact was on the teenager and the young adult, and it changed the long-standing epidemiologic patterns of these two countries.

In the context of this new drug vogue, the freedom for British doctors to prescribe narcotics to addicts suddenly led to gross abuses. The excess of prescriptions, together with drugs easily obtained illegally from manufacturers and pharmacies, provided a black market source. In 1964 the Interdepartmental Committee on Drug Addiction was recalled to examine the new situation which had thus emerged soon after its misleading report of 1961.[3] From the new findings and recommendations in 1965,[4] the Dangerous Drugs Act of 1967 was passed to license a few special doctors only to prescribe heroin and cocaine to addicts, and to enable specific narcotic treatment centers to be established by the Department of Health. These centers were started in 1968.

The long experience of therapeutic failure associated with the ban on the prescription of narcotics to addicts in the United States, and the desire not to give rise to a vast underground system of smugglers and pushers in the United Kingdom, led to the British decision (in 1968) to continue the practice of prescribing narcotics to addicts for self-administration, with certain special reservations. The aim was first to contain and control the epidemic by a scrupulous medical supervision of the addict as well as by more strict and effective police action against those seeking purposefully to spread the "infection."

The second aim was to provide an ongoing therapeutic setting to the addict, in which he could be treated essentially as a patient, rather than as a criminal, by all the agencies concerned in his future welfare. It seemed to the medical profession in the United Kingdom that the still relatively small size of the narcotic problem there and the cohesive form of the National Health Service made this approach a potentially feasible, practical, and humane proposition. There was, however, no precedent with which to predict success for the method, and there were obvious pitfalls from the start. For example, the new law as it stood covered only the drugs heroin and cocaine; doctors could still prescribe other opiate drugs, even to addicts, if they wished to do so.

AN ADOLESCENT TREATMENT SERVICE

The details of the present British system, as it can now genuinely be called, have been described and evaluated in general terms by Dr. Bewley at the beginning of this Symposium. This presentation is concerned with the more specific problem of the teenage narcotic addict, for whom some separate provision was considered important, both in terms of the particular experience and specialized knowledge that are required in the psychiatric care of a member of this vulnerable age group, and of the need to treat addiction at its earliest stages, before it has taken on the dimension of a lifetime habit entrenched in the fully formed character of the adult.

A specially built unit was therefore opened in April 1968 in London to care for narcotic addicts being seen for the first time when they were under the age of 18 years. In that year it was estimated that 27% of all narcotic addicts were teenage and 17% were aged 18 and under.[5] The size of the clinic and other factors precluded the possibility of caring for all these youngsters, some of whom have attended adult clinics in their own areas.

While the aim of the service was to be a voluntary one, in line with the general practice of the adult clinics, the emphasis from the start has been explicitly on eventual drug withdrawal. The idea of an open-ended narcotic maintenance program for this age group has not been envisaged.

The in-patient section, for 10 boys, provides an additional facility to the out-patient section, and both are located in the one building. The out-patient clinic was designed for 50 or more youngsters of both sexes. The staff consists of two psychiatrists and a senior resident in psychiatry, a social worker, a psychologist, an occupational therapist, secretaries, and a team of nurses. The unit is located in the open grounds of a small, well-established psychiatric hospital for some 80 in-patients, which is part of and fairly close to The Middlesex Hospital, a large London University teaching hospital. The patients are referred from a variety of agencies, including general medical practitioners, the courts and probation service, other clinics, and very often by self-referral.

It should be emphasized here that the laws in England prohibit anyone from carrying on one's person either heroin or cocaine *unless* these have been personally prescribed by a specially licensed doctor. It is also illegal to carry other narcotics, cannabis, the amphetamines or like stimulant drugs, and the hallucinogens, unless these have been specifically prescribed to the person by a general doctor. However it is *not* yet illegal to carry the barbiturates and other sedatives, even without a prescription, and this fact has had unfortunate consequences. These laws are in the process of being rationalized, a program which will inevitably increase the restriction on the freedom of doctors to prescribe as they wish.

The alternative, therefore, to a youngster obtaining heroin from a licensed clinic is to seek it illegally on the black market; a dangerous and now very expensive practice. In consequence of this there is a great deal of pressure on the addict to go to a clinic, and this pressure is the pivot of the whole treatment program. Even so, there remains a small hidden section of young neophytes who avoid the clinics until they are in dire straits, for fear of discovery by their parents and other authorities.

Regular attendance at the same clinic is enforced on the one hand by a weekly or two-weekly mandatory appearance for a prescription, which provides the narcotic drug on a daily basis, and on the other by the liaison between the clinics, which prevents all but essential moves from one center to another. The whole system thus channels the addict into a regular and consistent program so long as he chooses to remain on heroin. This really also applies to the other narcotics, because there are probably no general doctors who will now risk the odium of regularly prescribing alternative narcotic drugs. (If they were to do so, the law could be suitably extended.) One loophole to all this, as stated above, is the availability on the black market of illegal supplies. So long as the clinics are available, however, and with adequate laws and police surveillance, it is likely that this supply will remain minimal.

Diagnosis and Assessment

The subject of diagnosis and assessment can only be mentioned briefly here. Before accepting a youngster as an addict a very careful evaluation is necessary. For a variety of either nefarious or psychologic reasons a false claim to addiction may frequently be made. Even injection marks and a single positive reaction to a urine test are not proof of the diagnosis. Nor are withdrawal symptoms, for two reasons; one is that these are often minimal in young people, who are usually psychologically addicted to relatively very small doses, and the other is that these somatic symptoms are quickly learned and instinctively applied when a useful gain is anticipated. Only a sophisticated experience of youngsters in this situation and a careful and extensive clinical study, together with multiple urinalyses, will give the answer. There is the danger in this respect that one clinic will be less cautious than another.

It is probable, however, that if a youngster persistently pursues the clinics to provide treatment in this context, his addiction is already psychologically predetermined and inevitable. Youngsters who continuously experiment with many drugs including narcotics may well be at such risk that, in some instances, it is advisable to prescribe a daily dose of methadone in order to bring them firmly into the discipline and stability of the unit's program, and to give them the help which they so desperately need with their over-all problems.

The assessment of the *quantity* of addictive drug required is also a matter of careful clinical judgment. Again experience alone teaches one how to quantify the youngster's need initially and this is reviewed frequently when gentle pressure is put on him to reduce his dose. Clearly there is a constant danger in this assessment either of over-prescribing and so providing an excess to be sold on the black market, or of under-prescribing and therefore of forcing the compulsive patient to seek supplementary supplies from that market, and so creating its "raison d'être." This is a tightrope which if not walked in a balanced fashion could undermine the whole system of treatment. On the whole the doctors in the clinics have learned the expertise of this balancing act remarkably well.

Treatment

Perhaps the first point that I would make about the treatment of the heroin-addicted adolescent is that, in the great majority of cases, he is basically unhappy about his predicament and he longs to be rid of it, even if at first he admits to neither of these facts. His addiction has the same quality, and usually the same meaning, as, for example, that of sexual perversion or compulsive masturbation, to which it is possibly an alternative posture. In the young, such postures are still tentative, frequently unstable, not yet deep-rooted and thus to some extent amenable to change, as maturation proceeds, either spontaneously or with appropriate psychiatric assistance. They have never been remedied by civil punishment, and are only helped by custodial care when this is used as an intermediate measure; for example, if at some stage the individual is in such personal chaos that his acts are totally out of control, and he then begins to harm and to destroy those around him, as well as himself.

The second point is that unless an addict wants the treatment offered to him and is willing to accept it, there is little prospect of his benefiting from it. To be effective the whole program, including that of drug withdrawal, must be a joint enterprise between doctor and patient. It is especially unprofitable to impose a wholly arbitrary scheme on the adolescent. Of all age groups his is the most stubborn, and that stubbornness will usually be provoked to obduracy by the insistent over-ruling of an adult, however reasoned the argument may be. It is this hardening of the heart with which one has to deal.

It is the heart therefore, not the intellect, which must be captured if one is going to be allowed by the teenager to help him, whatever may be the disturbance from which he is suffering. The first stage in the treatment of his addiction is not to remove the drug which is, in point of time, his only dependable friend, but to offer him the real understanding and acceptance which can lead to a change of motivation. Later it is possible to ask sacrifices of him, but only when he has learned to achieve some trust in the adult, and when he has reestablished a meaningful human relationship in place of the previously more dependable one with the drug.

However, more is needed than just a successful human relationship. That relationship must also be the source of greater understanding and acceptance of the self, and of new and creative ideas from within. The addict is desperately engaged in a form of self-love by means of the drug because he believes that no one else will really love him;

he enriches himself with a false and illusive treasure because he feels so poor. He is compulsively engaged in filling what he perceives to be an empty and worthless vessel, and in destroying not only the misery which this perception generates, but also those elements of inner evil which are held by him to account for his worthlessness. At the same time the social position which he adopts expresses this damaged image of himself. This is why the addict has been called depraved, and this is why he is engaged in the compulsive act of chronic suicide. Depravity and despair go hand in hand, but it is not only the former which ultimately leads to the latter. Depravity may itself be a derivative of a childhood depression in which the struggle for fortitude is abandoned when the unhappy child is thrust into the crisis of his adolescence. The basic reasons why young people use drugs in this context have been discussed in greater detail elsewhere.[6]

That famous British opium addict Thomas de Quincey, writing of his own feelings in 1822[7] said, "I hanker too much after a state of happiness both for myself and others. I cannot face misery, whether my own or not, with an eye of sufficient firmness, and I am little capable of surmounting present pain for the sake of any recessionary benefit." This was the adult statement of despair for which by then there was no remedy, but de Quincey started his addiction long before as a teenager.

With the adolescent it is *still* possible to work with the forces of maturation which, even if damaged, are still present. Provided the youngster will engage in some form of dialogue, there is reason to hope for an ultimate change of heart.

This must be a cornerstone of treatment, and we can call it psychotherapy. But it must be a therapy which is extremely painstaking, highly supportive, flexible in practice, and at times requiring active intervention as well as interpretation. The therapist must be prepared to assume a positive role and a special place in the youngster's life, and he will need a therapeutic team to work closely with him. It is clear that any program which aims to do more than contain the problem of narcotic addiction by repressive measures must be lengthy, all-embracing and dedicated. Because drug addiction is a symptom with a primary medical idiom, it is the doctor who must take over-all responsibility and lead the team, but nearly every social agency, such as the probation service and social welfare, has a vital part to play, and is incorporated into the system.

THE OUT-PATIENT PROGRAM

It is surprising how quickly the young addict adapts to his out-patient program. He is fundamentally a lonely person longing for a trustworthy friend and the clinic becomes a safe and sane port of call. It represents many things, not least the provision of structure in the place of wandering chaos.

The first lessons which he learns are those of keeping to his appointment times; of behaving with civility and in a reasonable manner; of taking care of himself and his possessions; and of keeping to his promises. In this setting he begins to learn about consistency, honesty, and loyalty in human relationships and undertakings—a lesson which he has never understood before. He also learns to accept a reliance on others without coercion or the loss of his own independence. All these are part of the attendance routine itself. In addition he learns to set aside a part of his week to think things out and to examine himself and his environment in an objective fashion. In fact he learns to be helped to help himself.

Far from being a distraction in this setting the drug itself becomes a vital therapeutic dimension along which the doctor works with his patient. It is a contract in which eventually, on the one side, the doctor offers to the patient the avenues of emancipation, and, on the other side, the youngster offers back the drug. In dynamic terms the complex meaning of the addictive behavior will sometimes become clear to the youngster, but this itself does not seem to be an essential therapeutic factor. Nevertheless it is very evident from our experience that addiction, in this age group, is far more of a psychologic event than an organic one and the ritual of the intravenous injection is at least equal in importance to the psychic experience of the drug itself.

The method of drug withdrawal employed in the case of the out-patient is usually that of gradual substitution from self-injected heroin to methadone at about equal dosage, followed by a slow reduction of the methadone to a fairly low dose level. Thus the typical patient might start on a self-administered dose of about 80 mg of heroin followed by reduction to one of about 20 or 30 mg of methadone. At that point, however, the young addict will become resistant. If the need for the intravenous ritual is very powerful it will be much harder to proceed further, and it may be necessary to remain at this stage for many months. If the need is not so powerful, it is possible to shift gradually over to the oral use of linctus methadone, perhaps giving a slightly larger dose than by injection in the initial phases. From this point, provided the youngster is making good social as well as psychologic progress, it is possible in some cases to withdraw the drug completely.

In many cases, however, it is evident that this withdrawal program is going to be long. It may require in-patient admission, and it may last well into the final stages of adolescence itself, as occurs in many other types of adolescent disorders.

Although this outline is the most typical, there are variations of it and each case is self-determined; that is to say, the addict himself is allowed a considerable choice of how and when things are done, provided it is clear that his purpose is genuine. This business of making appropriate choices is part of the whole issue of maturational learning which is fundamental to the adolescent treatment enterprise. Great patience is required to wait for the turn of the tide in this respect, but sometimes it comes with dramatic suddenness, and a patient who seemed set on a lifetime of drug taking will unexpectedly ask to have his dose cut down. Such changes require the therapist to be constantly available and alert to the mood and thoughts of his patient; if he is not, great opportunities are lost. If this choice is freely taken by the addict at a time of stability, it has the convincing sound of cure. Demands to be withdrawn in moments of crisis, or enforced withdrawal while in legal custody, are not at all as promising.

There are many problems in treating the out-patient. He continues to live in a drug-endemic environment and his family is nearly always disturbed and ambivalent about him, thus reinforcing his pathologic state. He often lives far away and resents the effort demanded of him to participate. There is little objective information about what is going on. However the youngster is usually at work under these treatment conditions and it is possible to make good use of local social agencies to support him. When he has been in trouble with the Law, the local probation officer is often a most useful ally in helping to set the social controls and disciplines which he requires and which he either does not get or will not accept from his parents.

One of the difficulties facing the clinic psychiatrist is that of liaison with the family doctor and other general practitioners. This is largely due to the fact that young people wish to keep away from the family doctor, and the adolescent addict, in particular, sees him either as an ally to authority, or as someone from whom to "con" sedative drugs. Some unsuspecting medical practitioners do in fact prescribe these drugs to teenagers in quite large quantities, seemingly unaware that they are playing directly into the hands of the potential addict. The clinical indications for giving a teenager a barbiturate or similar drug are very few.

Indeed it is the barbiturates which head

the list of narcotic alternatives, when the narcotic drugs are impossible to obtain. As the present British laws governing the abuse of these drugs are still inadequate, it is extremely easy for the youngster to buy them relatively cheaply and to crush the pill or capsule, suspend the powder in tap water, and then "mainline" it. The clinical results are very serious, and the procedure is at present undermining the national treatment program. In their search for a substitute drug, some adolescents will go even further, and all kinds of drugs have been used in this way as, for example, the antihistamines, which may be bought freely, without prescription, for combating travel sickness.

THE IN-PATIENT PROGRAM

The in-patient section of the clinic is designed to provide certain addicted adolescents with a more intense treatment situation in which all their available energy is not devoted to coping with the responsibilities of earning a living and self-dependence. It offers them a moratorium on life, a chance to go back and work through the unfinished business of childhood away from the conflictual family and social setting. It is therefore run on the lines of an adolescent therapeutic community and offers a surrogate family situation. Although it happens to be staffed with a medical team, there are no uniforms, few clinical routines, and the doors are wide open. However, there are rules of a reasonable kind that are inherent in any successful family structure, and these also include an absolute sanction against bringing drugs to the house, or against returning to it in any way intoxicated. In the belief that a youngster must learn to develop his own initiative from inner resources, however latent, there is little imposed structure, but there is a wealth of activities inside and outside to which the adolescent can attach himself and from which he can appreciate forethought and order, including his going to college or to work if these are seen to be appropriate.

It is important to provide a range of outlets whereby the youngsters are able to express their aggression, and at the same time release massive amounts of energy and feeling in constructive instead of destructive ways. It is also essential to explore and develop their talents and to arouse and widen their interests at educational, recreational, and occupational levels. Since these facts have been neglected or never developed in the past, such youngsters have almost no concept of personal contribution and creativity. Apart from exploration of talents and development of interests, the treatment includes individual and group psychotherapy; a carefully planned scheme of individual drug withdrawal, based initially on intramuscular injections of heroin and methadone given by the nurses; and the use of appropriate psychotropic medication if this is specifically indicated.

This setting is not suitable for treating the emergency medical complications of drug-taking, nor for providing shelter for the night. Nor is its immediate purpose drug withdrawal per se, although, as with the out-patient, this is implicit in his over-all program. Youngsters who are rapidly withdrawn from drugs leave the unit prematurely with a false optimism, and rapidly return to drugs. Short-term stay is usually not effective.

It has been found unprofitable in this setting to accept an adolescent for in-patient care either against his will (e.g., from the Courts) or when he wishes to please others who want this for him (e.g., his parents). It can only be effective if he wishes to come in for his own sake and if he is properly aware of the pattern and purpose of the place. It is usually wise for him to have been an out-patient for some months first, although there are occasional exceptions to this.

The aim of the unit is to offer an active group situation in which the youngster has to learn about himself by seeing his own behavior with both patients and staff reflected back to him. He has to understand that it is not enough for him to be accepted by others, and that human relationships are reciprocal.

It is only in this special environment that one can really examine the difficulties which the youngster faces in life. He himself quickly perceives the surrogate family pattern and reacts accordingly by projecting the psychopathology of his own family situation on to the staff and others. The whole interaction that ensues has to be observed in great detail if it is going to be of any therapeutic value. Both the doctors and nurses have to understand in depth what is going on. This is essential both to maintain the well-being and cohesion of the community, which the youngster is often trying to break down, as well as to demonstrate to him what he is doing to himself and to others. This is no easy task, and it requires very close liaison and communication among members of the staff. It is essential that they all know clearly what their role is meant to be so that they can apply it with mutual loyalty and consistency.

The nurse in particular finds himself at the center of many conflictual situations. Essentially his task lies in maintaining control of the group while avoiding the pitfalls of rigidity and authoritarianism on the one hand, and of unreasonable permissiveness and the possibility of being excessively manipulated on the other. He has to contend with extremes of difficult, often devious and provocative behavior at one moment, and of negativism, boredom, and lethargy at the next. He is often tempted to undermine the therapeutic purpose by being seduced into offering the false love of inappropriate concessions and favors, only to be followed later by attitudes of rejection, as a corrective measure. He is thus forced into the expected pattern so often initially set by the parents.

It has been a salutary experience for those employed in the unit to learn how easily the adult, by over-responding to his own emotional needs in this situation and by the inconsistency in his attitudes which results from this, can act adversely towards the youngster. This does not imply that the care-taking adult should not be human in his responses, but if there is going to be a corrective experience this humanity must be tempered with a self-understanding and self-discipline that others before him clearly lacked.

The nurse's task as an adult model is a vital and difficult one, and unlike that to be found elsewhere in the field of psychiatry. The traditional training of the psychiatric nurse is often inappropriate for this kind of work. The capacity to communicate with young people and the ability to offer special talents useful for the youngster to emulate in the learning program are of much greater importance than academic credentials and nursing protocol. Personal integrity, honesty, and frankness are the "sine qua non" of the worker in this field, as well as his fundamental capacity to accept the addictive behavior as a true symptom of psychopathology.

It will be obvious that not all addicted adolescents can make effective use of this form of milieu-therapy. Some do not require it and others are insufficiently stable to be able to handle it. The question has been asked many times as to what part, if any, enforced and custodial care is required in some of these cases. It is difficult to generalize on this issue. As far as the very young are concerned, there does seem to be a place for custodial care when the youngster is in such turmoil that he cannot even achieve sufficient stability under the circumstances described above. Usually these are confused youngsters in whom forces of destruction and chaos are so strong that someone else has to take over the ego-functioning role for a period of time. This form of behavior is not so unusual in the very disturbed adolescent, and it has the quality of a "pseudo-psychosis."

Some adolescents in this state may use

drugs massively and indiscriminately. In such cases there are difficult legal issues in terms of certification, and mental hospitals are seldom equipped to deal with them. Many of these adolescents eventually determine the outcome by breaking the law, and a wise judge will listen carefully to the psychiatric report if he is not to miss the cry from the youngster's heart for the right kind of *adequate* care, protection, and control. Again the difficulty here is to find the appropriate corrective setting. Punishment is not the answer to this adolescent's difficulties, and the psychiatric and social tasks as well as the time sequence involved are specialized issues which need careful consideration.

The question has also been asked as to what kind of people these narcotically addicted adolescents really are. They have all too frequently been dismissed as "psychopaths." The mixed features of their personality and mental state have been summarized elsewhere.[6] Very briefly it can be stated that at this stage these youngsters are guilt-ridden, sensitive, and extremely self-critical. They suffer from a great deal of immaturity at various levels in a largely unintegrated ego. They are profoundly affected by their emptiness and failure, and out of despair they either endorse society's contempt for them by self-denigration and self-destruction, or they display a false bravado and band themselves together as a deviant group of antipathetic social isolates who are fundamentally loved neither by themselves nor by anyone else. This is not dissimilar to the evolving pattern of adolescent sexual deviates. Their addiction may be a form of chronic suicide.

Table 1 Disposition of patients attending clinic during first year and surveyed at the end of the second year of operation

	Male	Female	Total
1. No. presenting in the first year of operation	116	14	130
2. No. not accepted due to age, wrong diagnosis, etc.	46	6	52
3. No. accepted, prescribed narcotic drugs, but no longer attending due to referral elsewhere, taken into custody, etc.	27	5	32
4. No. accepted, not prescribed drugs and no longer attending	5	3	8
5. No. accepted, not prescribed drugs, still attending	1	0	1
6. No. accepted, prescribed narcotic drugs, still attending	37	0	37

CONCLUSION

The following three tables provide a simple assessment of the patients seen during the first year of the clinic's operation.

A more detailed follow-up of all cases is being undertaken, but it is clear that a true measurement of success or failure cannot be made for several years; even then it will be difficult to judge the relevant issues because of the complex variables inherent in this kind of treatment project. Perhaps one will have to look not so much at the questionable "cure" of the individual patient as at the total and eventual impact of the national program on the broad epidemiology in the United Kingdom of this morbid psychosocial disease.

Table 2 Prescribing pattern of 37 first-year patients attending at time of survey

	Male	Female	Total	%
1. No. whose dose of narcotic drugs was slightly increased	3	0	3	=11% (items 1 and 2)
2. No. whose dose of narcotic drugs remained the same	1	0	1	
3. No. on substantially reduced dose of narcotic drugs	24	0	24	=65%
4. No. taken off all narcotic drugs	9	0	9	=24%
(Max. period drug free = 48 weeks. Min. period drug free = 3 weeks.)				

Table 3 Types of drugs prescribed to 37 first-year patients at onset of treatment and at time of survey

	Initially	At Survey
1. No. prescribed heroin alone	14	0
2. No. prescribed heroin and methadone	10	1
3. No. prescribed heroin and cocaine	2	0
4. No. prescribed methadone* alone	9	27
5. No. prescribed Ritalin alone	2	0
6. Average over-all daily dose of narcotic prescribed	83mg.	28mg.
(= 60% over-all reduction)		

* Including partial oral administration.

The Home Office statistics[8] imply that in 1969, compared with 1968, both the number and proportion of teenage narcotic addicts had slightly decreased. Those under the age of 20 years were 22% and those aged 18 or younger were 11% of the total number. This is possibly a hopeful sign, but a great deal more thinking and planning are required both in terms of the treatment and the rehabilitation of these youngsters and of all those who adventure into the dangerous game of drug experimentation.

REFERENCES

1. *Report of the Departmental Committee on Morphine and Heroin Addiction under the Ministry of Health.* London, H.M.S.O., 1926.
2. *Annual Report of H.M. Government to the United Nations,* 1964.
3. *Drug Addiction.* Report of the Interdepartmental Committee, Ministry of Health and Department of Health for Scotland. London, H.M.S.O., 1961.
4. *Drug Addiction.* Second Report of the Interdepartmental Committee, Ministry of Health and Department of Health for Scotland. London, H.M.S.O., 1965.
5. *Annual Report of H.M. Government to the United Nations,* 1968.
6. Boyd, P.: Heroin addiction in adolescents. *J Psychosomat Res 14*:295–301. Long Island City, New York, Pergamon Press, 1970.
7. De Quincey, T.: *Confessions of an English Opium Eater,* Elwin, M. (Ed.). London, Macdonald, 1956.
8. *Annual Report of H.M. Government to the United Nations,* 1969.

PANEL DISCUSSION—PSYCHIATRIC MANAGEMENT OF DRUG ABUSE

Moderator: DR. POLLARD

Panel Members: DRS. BENSON, BOYD, KOUMANS, and D. E. SMITH

Mr. Karl V. Fisher (Cincinnati, O.): My question is concerned with a minor point in the history of the development of narcotics' use in the United States. I will address it to you, Dr. Pollard, because I have heard you mention it, but it has been brought up at least two or three times by speakers here. In discussions of the sociologic development of drug use in the U.S. we are told that it began in the ghetto areas, mainly black, and only when it progressed into the white community did the recent concern result. My information is that, in 1941, when the Federal Narcotics Hospital opened in Lexington, 78% of the clients taken in were white. Figuring backward 5 to 10 years to the beginning of the problem, say back to 1931, I am wondering whether hard narcotics use really began in the ghettos since the statistics seem to suggest otherwise. What is your comment on that?

Dr. Pollard: I think maybe you are confusing treatment figures with incidence figures. I would imagine that it is still customary in this country for minority or ghetto groups in social trouble to end up in the courts; while white middle-class people in trouble go to psychiatrists. I think that is the difference, although the fact that it was predominantly a white population in Lexington in those days is surprising to me. Incidentally the average age at Lexington is 22, which is quite a change, I believe. Dr. Martin has some recent figures on the make-up of groups.

Mr. Fisher: There is the possibility of that, but when the hospital opened it was a prison in fact, so that most referrals were out of the courts.

Dr. Pollard: Yes, that's right. However, I think my impression, that narcotics use began with minority groups in the urban ghetto, is one that is pretty well established from the data we have. I see Dr. Isbell in the audience. Dr. Isbell, would you like to take this up?

Dr. Harris Isbell (Lexington, Ky.): I don't know that I have the answers but you are quite correct that during the 30's and early 40's the majority of the patients in the Lexington hospital were white. However, the majority of the patients in the Lexington hospital at that time, whether they were volunteers or prisoners, were predominantly members of what one might call minority groups. In other words, Italians were heavily represented. Jewish people were heavily represented, central Europeans were heavily represented, as were the large northern cities. Now these white addicts who predominated at the time were to an extent "ghettoized" in that they came from rather bad sections of cities like New York or Chicago, even if they were on their way up the social ladder and out to the suburbs. The other pattern, and this pattern still exists, was the southern addict who predominantly, except for the city of New Orleans, was not a resident of a big city ghetto. He was a small-town individual who used morphine rather than heroin, who obtained his drug largely by getting prescriptions from physicians or by theft and so on. Much of the population that we had in those days were individuals with long records of psychopathology. Then came the war, and remember this was at the time when opiates were the predominant drug problem. Cocaine, the sedatives, and hypnotics existed but were not yet a problem. But then came the war, and with the war a lot of things happened. In the first place the supply of narcotics was effectively cut off because shipping was controlled. The warring parties bought up all of the opium, if for no other reason than to keep each other from getting it, so the supply of narcotics in the illicit market reached a very low point. And, for a time, drug users were dependent largely on what they could get by theft. It was during this period in the war that poppy cultivation started in Mexico and opium began to come, just as heroin now comes, from Mexico to supply the market. The number of patients went down to such an extent that the hospital at Lexington was nearly closed and had to be saved by bringing in psychiatric patients from St. Elizabeth's Hospital. Then the war ended. During the war there had been, of course, a great deal of social migration. The Negro was moving from the South into the North. He moved into the places that used to be occupied by the class who were coming to Lexington, who were now on their way up; also coming in were people from the Spanish-speaking cultures of Puerto Rico and Mexico. Then very suddenly after the war, the heroin trade revived, the trade routes came back, and there was a tremendous shift in the ethnic population away from the white to Negro and Spanish-speaking persons. These were people who had migrated into a ghetto that had been vacated by whites, who we might also say had been "ghettoized," though perhaps to a lesser extent, and discriminated against. This is what seems to me to have happened. Now, in still another phase, the problem is moving out to the suburbs.

Unidentified member of audience: I have a question for Dr. Boyd. While living in Cambridge, England, I helped very briefly with a group therapy program for parents of young heroin addicts. This was run by Dr. Kenneth Meyer of Fulbourn Hospital. Group therapy for parents seemed like a very promising experimental technique and I wonder, Dr. Boyd, if you can tell me if this is widely used and if you think it is useful.

Dr. Boyd: I am not sure that I can answer that it is widely used, because I have not discussed this too much with my colleagues in the other treatment centers. We started a program of group therapy with parents and found it to be a failure. This was for two reasons. Unlike most of the treatment centers we were the only ones in the United Kingdom who treated the teen-ager specifically. We get our patients from a wide range of the country, rather than from the local area. This means that geographically it is very difficult for parents to get together. Secondly, when we did in fact send out a group psychotherapeutic team to work in one area east of London called Dagenham, where many of our addicts seemed to come from, we found other difficulties. The parents were unwilling to be insightful; unwilling to look at themselves at all. They were projecting their problems constantly onto their children and, in fact, many refused to attend after a short time. Now this may have been specific to the particular type of patients that we were getting and to the age group. Whether in a slightly older patient the parents would respond differently, I am not sure. On the whole I would say that it is very difficult to get parents of addicts to work together in this fashion.

Dr. Smith: I would like to add a comment to that. Several physicians from smaller communities have asked about handling the drug user at an earlier stage, when he is still in the recreational phase. In the area of family therapy there is one model that might be of interest to those who have asked this question. A school district in southern California developed a broad-based community effort in its approach to drug abuse; among all their educational and treatment activities, they used a special type of family group process called a child-parent contract. The school district had a tremendous drug problem, but it also had a strong interest in denying the existence of this problem for complicated reasons involving the military. Finally the drug problem became so severe that they instituted this school district program. Prior to this they had had a policy of automatic expulsion for any young person caught using a drug. However, in this particular school 60% of the children were from military families; if their children were expelled, their future promotions were in grave jeopardy. Thus, if the young person was caught in possession of the drug, the school would not expel him if the parent and child would meet twice a week with a counselor. It sounded like a good idea because, of course, the parents were panic-stricken about their careers and what was happening with their child. But it usually broke down with the contract violated not on the young person's side but on the parents' side. The parents would come to the family session and lay down what they wanted their child to be. Then the counselor would turn and the young person would lay down what he wanted his parents to be. This was where it broke down because the parents were willing to enter into counseling sessions only if the counselor would basically use their socialization techniques. But they really blew their minds if they thought that the child had an image of what he wanted the parents to be. In the vast majority of these child-parent contracts, even in the face of expulsion, the interpersonal contract was violated on the parents' side. Family therapy is possible but very difficult.

Dr. James Jackson (Muskegon, Mich.): I have a short statement. We have participated for four days in what was hailed as an international conference on drugs. I find it very disturbing that after these many days of reporting and analyzing, this conference has seemed more like a laboratory experiment rather than a serious discussion of why drug abuse has become a serious crisis and what must now be done in the light of the colossal eruption of drug use since World War II, particularly in western society and very specifically here in the United States. The very fact that this question has not been raised seriously at this conference demonstrates to me that the people here have not asked themselves the question of what role they have played in bringing this crisis about. Rather, they have accepted this phenomenon as a part of the modern way of life, just as they have accepted their role in society as a permanent part of society. For example, when the question was raised at one point whether you would prescribe for your wife or your immediate family a particular drug, the question was evaded. To me this question goes to the heart of the matter; the very ethics of the medical profession are involved here. For if a doctor will prescribe for a patient something that he or she will not prescribe for himself, his family or his close friends, then we, who are allegedly sworn to put the human being and the welfare of the patient above all other considerations, are violating our oath. What it suggests also is that we, too, are ready to subscribe to the ideology of treating the symptoms rather than the cause, which is characteristic of modern America. To what degree are we here perpetuating the myth that there is a pill for everybody to obtain eternal happiness, and that there is painlessness and freedom from suffering and responsibility for all. If we only keep our eyes glued to the TV commercials, one out of every five of which is pushing the —

Dr. Pollard: Dr. Jackson, I suspect that many of these points have been made by the presentations.

Dr. Jackson: Well, may I say just one last thing?

Dr. Pollard: Please do.

Dr. Jackson: It seems to me that the very fact that this conference is being sponsored by the drug pushers of the world, I mean the drug industry, which has made every U. S. citizen believe that the better idea in his future is a better drug, should raise some very serious questions in our minds as to whose interests this conference serves. Can we seriously condemn the pill industry for pushing this myth of instant painlessness when we ourselves have not repudiated it in our theories and in our practice? These are the questions which we must ask ourselves if we are to give some direction and some meaning to the present crisis. This duty we can no longer shirk.

Dr. Zarafonetis: I feel I must rise and correct one statement that has been made two or three times from the floor. Now, it requires money to put on a conference like this. The University does not have funds for such purposes, so we solicited many sources. Some are in industry, some are in foundations, and some are in other areas that are concerned about this major problem. These people and their organizations have contributed to make this conference possible. I wish to emphasize that considerably less than half of the total funding has come from the pharmaceutical industry. There were absolutely no strings attached by the industrial sources or anyone else. I want you all to appreciate that. No strings whatsoever. They had nothing to do with the program structure or content. They will have nothing to do with the editing or the reporting of this conference. I deny categorically the innuendo that the drug industry exerts a controlling influence on this conference.

Dr. G. William Hettler, III (Selfridge AFB, Mich.): I am in general practice in a large group, the United States Air Force. Let me say first that I took my initial training in transcendental meditation in May 1969. In view of the apparent success of transcendental meditation as a substitute for drug use or abuse as reported by Dr. Benson, I would like Dr. Smith's comments on his experience out on the West Coast with this same thing.

Dr. Smith: I think that a great deal of that type of activity occurs naturally among young people rather than within the formalized framework of any program. For example, transcendental meditation is often practiced within the framework of communes of which there are hundreds in California. On a recent trip to the Hawaiian Islands, I had the opportunity to observe commune dwellers in transcendental meditation, Yoga, and a variety of other consciousness-altering techniques. I find it difficult to conceptualize incorporation of these procedures into any inner-city program that I am familiar with that is dealing with compulsive drug users, although I know they are attempting to incorporate it into the family program at Mendocino State Hospital. I think the general philosophy of trying to find alternate ways of turning on and achieving the same objectives is the direction that we ought to go; if you can turn on to social activism, if you can turn on to working at a clinic, if you can turn on to transcendental meditation—as long as what you are turning on with is not more toxic than the drug—then it is fine with me. The general philosophy is that whatever works, if it allows the individual a better adjustment, then this is fine. I do not think it would work with the compulsive drug users that I am familiar with. I think it would work better with the psychedelic individual who has a situational crisis or some fairly long-term problems of adjustment, but who has not yet entered into a compulsive pattern of drug use. This is pretty much what is happening in the communes where kids who are alienated might have come to a Haight scene and found it far too violent and drug-involved; rather than going straight then, they move on to this alternative life style and get involved in it on their own, and they do a very good job.

Mr. Leonard Williams (Flint, Mich.): I have two comments and then a question. To the gentleman who reacted to the criticism of the drug pushers, I would be curious to know how much these same people that are sponsoring this conference have put into the rehabilitation of drug-dependent persons. My next comment has to do with the whole chauvinistic attitude of this convention. I think it is horrible that we have a situation where people have repeatedly gotten up and talked about the fact that drugs were very common in the black community for a long time, yet we have gone through damn near four days of this convention and not one black physician or one black person is represented on any of the panels. I also believe that women are very interested in their children, and yet I have not seen any female researchers take the podium. I think that this is a crime, and I am curious to know why this situation came about.

Now, to Dr. Benson: I would like to know the racial and economic make-up of the people that you talked about, and secondly, when they get out of this whole meditating bag, do they go back to drugs?

Dr. Benson: As to the latter, we don't know if the subjects go back to drugs. That is why we want to set up a prospective study. We don't know how many people started meditation and then stopped meditating, which is what you are asking. We have only looked at those that were successful.

I don't have with me the socioeconomic or racial breakdowns. There were more males than females. My impression is that there were black people involved. There were also Orientals, but I can't give you numbers now. People are entering the ghetto and attempting to offer a program of transcendental meditation for alleviation of some of the people's problems, but this is too premature to comment upon.

Dr. Boyd: Could I just take up the last questioner on the subject of race? It might console him that we seldom see black people involved in narcotics addiction in the United Kingdom. They never were, and they really still don't seem to be, which is one up to them. In fact, in all the cases that have come to my clinic, I have only had one black person. All of the rest have been white. But there are other racial origins, such as Turkish and Greek. This may refer to the specific type of drug addiction that we are treating, and in other respects in the total field, the black races may have played their part just as much as the white.

Mr. Williams: When I stated my question concerning race, I was curious to know why we here in the United States would go all the way to the United Kingdom and try to get information about how you treat people in the United Kingdom when obviously black physicians and black psychiatrists are treating people right here in our own country. And I am wondering, since last night's news reported a 5-year-old child dead from an overdose of heroin, why we don't have people who are working with the problem come to us and let us know where we can go to solve our problems in Harlem and the various ghettos, as well as the middle-class suburbs of this country. It is this country I am concerned about. In the United Kingdom I think you should take care of your own problem.

Dr. Pollard: May I attempt to respond to some of these things. It seems to me that when one has a problem, a problem that goes on, and many attempts are made at answering that problem with limited success, then you look under every possible rock, even the rocks in England. You look in every corner. You seek every possible area of expertise that you can find. It is a fact that certain clinic programs, as Dr. Bewley and Dr. Boyd described, have not been attempted in this country. I think it is very useful for us to hear about the possible pitfalls and difficulties, because this is something that we may move into even here. I feel it is very useful to learn from people in different countries what they have attempted and what successes they have had in handling their drug problems. As for the inference that this was a discriminatory group because there were no black people represented among the faculty, let me tell you this. As far as I know, the faculty was chosen in two ways. First of all, a degree of expertise; if there were black people working at the levels that most of us are working in universities and colleges, then of course they would have been invited. The other way was solicitations directly from people who felt they had something to give the Symposium. As far as I am aware, we had no solicitations from any black people to make a presentation to this Symposium, but I do feel that we can use this panel better than for a dialectic of political issues. Could we go on to the next question, please?

Unidentified member of audience: Dr. Boyd, I notice in your criteria for the treatment of adolescent addicts you said that some were rejected because they were too old. Well, how about the duration and the extent of the habit? Is that a criterion for selection? For example, I have a 17-year-old girl who has been on narcotics four years, and was on 24 bags of heroin per day when she was admitted to the hospital.

Dr. Boyd: As a matter of fact, the only criteria which we employed were that they should be under the age of 18, and that they should be bona fide addicts to narcotic drugs. It did not seem to me, and it does not seem to me after two years working in the field, that the duration of the addiction is really the paramount issue. Our experience is that within six weeks of starting on a narcotic drug the youngster can be as fully committed to his addiction as he might be after three years. What we were interested in was taking a special age group in society which we think is particularly vulnerable and perhaps particularly susceptible to treatment and to treat that age group separately.

Same questioner: The amount of drug had nothing to do with it?

Dr. Boyd: No, the amount of drug had nothing to do with it. We have seen youngsters coming to us having as much as 180 mg of heroin a day right down to 20 mg a day.

Mr. Jason Janus (Chicago, Ill.): My question is directed to Dr. Koumans. You covered treatment of toxic drug reactions, and yet I don't recall that there was any specific kind of treatment indicated for say, hallucinosis, or for acute reactions to amphetamines, barbiturates, or belladonna. I was wondering if you might discuss this specifically. As I understand it, when residents leave grad school they can pick up a cardboard-covered book which is a sort of cookbook for body mechanics. Could there possibly be a cookbook run-down of chemicals that can be used for drug reactions to bring a person down or to counter the reaction?

Dr. Koumans: I am glad that several times during the meeting this issue has been raised. We talked yesterday about advantages or disadvantages of giving, for instance, Thorazine in the treatment of either the hallucinogenic crisis or amphetamine psychosis. I did not specifically want to do that again today, since that kind of information is contained in the manual that, unfortunately, was not available here in large enough numbers; it will be available to you if you write in a few days. The few things that I picked up here that are different from the manual do not really have to disturb you because you can make those corrections yourself, after having been here this week. Rather than giving a run-down of what drugs to use for what situation, I would refer you to that list. Again I am a little uncomfortable substituting a chemical treatment for another chemical crisis, although I agree that at times you have little other choice.

Mr. Barry Adelman (Ann Arbor, Mich.): I think Dr. Pollard hit the nail on the head when he said we have to show people that there is joy in life. I think we can safely say that middle-class heroin users know the dangers of heroin. They know they can die of an overdose, and I think in a sense they are flirting with death. This may be because we live in a society which fosters death, which is death-oriented. It is hard to celebrate life when the world around you is steadily preparing for self-destruction, either through World War III or ecologic disaster. Everything seems so hopeless to them. The question then, for Dr. Pollard or anyone is, do you believe that the increased use of heroin among the white middle-class may be caused by a society which supports death, rather than celebrates life?

Dr. Pollard: I have no idea why there is an increase in the use of heroin. Sidney Cohen pointed out yesterday that it is much more readily available, but whether one is cause and one is effect I do not know. Perhaps Dr. Smith would like to respond.

Dr. Smith: Heroin is a drug of despair. In the youth culture today, there is mounting radicalism along with mounting pessimism. From this substratum of alienated youth comes violent politics in an attempt to change the system of dropping out and getting involved in more destructive patterns of drug use. I agree that alienation and the pessimism that young people express in one way or another is a major factor in the rise of middle-class heroin addiction. However, a large number of the people who are dropping out in the white middle-class heroin scene had problems that predated their heroin involvement or even their involvement with other drugs. One addict told me that the one great thing about heroin is that it is just like the "Pacific Plan," offered by a loan company in California. You take all of your little problems and put them up in one big ball and you take out one loan; and that is heroin. You shoot it up, and everything goes away. He was not talking only about problems with society, the war, and ecology; he was also talking about problems he was having in his own head. I think that you have to look at psychologic and social factors when you analyze any contemporary drug problem.

Mr. Adelman: I think the weight being placed on family background has been done at the expense of the condition the world finds itself in today.

Dr. Smith: I agree but I also think that the cultural turmoil that we are in at the present time also bears upon the family. You might say that there are societal factors that relate directly to getting involved with heroin as a drug of despair, but there are also societal factors that bear upon the breakdown of the monogamous nuclear family which has as its consequence the impairment of personality development. The vast majority of our compulsive drug users had family problems and personality problems that pre-date their involvement.

Dr. Pollard: David, you sound like a psychiatrist, and after what Dr. Kaplan said that is hardly a compliment. The trouble is that when you look into people's past histories, it is like asking if they've ever had a headache. You will always find evidence of family disruption or something that could be labeled 'family problems.'

Dr. Koumans: I want to indulge in speculation here. Since your question had to do with drug use moving into new areas, middle-class, but also middle-age, I wonder to what extent the following may contribute. You may have seen the write-up in *Life* magazine several months ago, complete with glossy photographs, of pot parties among middle-aged executives in New York City; most of the other magazines, of course, had to follow suit. I wonder to what extent the fact that this country is in a way addicted to youth—with emphasis on youth as a desirable state and trying to perpetuate it, as was also mentioned by Dr. Kaplan—to what extent does this emphasis suggest to people who are older that they try to assimilate one attribute of the youth culture, drug use, to recapture some of that lost, mystic, romantic period that they have gone through and perhaps have not been fully able to enjoy. In other words, are people approaching middle age, that very difficult time in life, somehow trying to get back to the fountain of youth? And since youth is now being equated with drug use, may this be one way they try to do it?

Dr. Fulvio Ferrari (Ann Arbor, Mich.): After three or four days, I have a growing sense of confusion from the clinical standpoint, perhaps because I am a psychiatrist. From the description of social background, statistics, symptomatology, dynamics, and so on, I see parallels between the drug user and the psychotic, even the schizophrenic with symptoms. I heard terms that I could translate into aggression, autism, and so on. Now the parallel can go even farther, historically. We used to burn schizophrenics in past centuries and now we imprison the drug users. Could this be used to help us politically not to burn our users now?

Dr. Koumans: It is dangerous to draw these kinds of analogies, I imagine, especially when you use the word schizophrenia, which is a catchall diagnosis more often than not. Personally I think that most of the noise or voices you have heard here seem to be in favor of a less legalistic or punitive approach to the problem of the drug user. I do not in any measure entertain the fantasy that there is a correlation between schizophrenia or drug use, and I certainly would have serious thoughts about using the word schizophrenia as a social denominator; such as talking about the "schizophrenic country." I think that makes no sense at all, and is a disservice to clinical medicine.

Dr. Walter D. Meester (Grand Rapids, Mich.): Dr. Boyd, one of the objectives in most of the treatment centers for heroin addicts in this country is to break the needle habit of the addict. I was a little bit surprised to hear Monday from Dr. Bewley, and today from you, that in your centers in England methadone is administered intravenously, with little being done to break the needle habit. Would you comment on this? Is it really important that we break the needle habit? Or should we perpetuate the ritual as you are doing in your clinics in England?

Dr. Boyd: Before answering your question, I would just like to comment on the previous question because I think it is relative to the same thing. When we take our patients off their narcotic drug and off their needle habit, we find two major symptom-syndrome type patterns developing. One is that of depression and the other is that of various types of neurotic, compulsive and obsessional behavior disorders. We don't see many cases of psychosis, although I did say in my talk that we do see what I call the pseudopsychosis of adolescence, the chaotic kind of state which is a multiple syndrome rather than a single entity. I would agree with Dr. Koumans here that schizophrenia as such, whatever that might be, is not really more than a small part of the whole.

Now insofar as the needle itself is concerned, I also stated that I saw this in the adolescent as comprising at least half of the addiction; i.e., the addiction to the needle is at least as great as that to the drug itself. I don't think that the drug per se really matters so very much. The narcotic addict is not different from other drug abusers because of the effect of narcotics. He is different because he has a different kind of personality problem and he chooses the narcotic simply because this is the drug which he considers most helpful in trying to resolve that problem. So essentially what we are doing is not treating the narcotic drug problem as such; we are treating the addictive personality, and in particular with the "narcotic personality" (also the "needle-personality" if you like), the needle addiction. I think this is a very important part of the whole business but you cannot remove it by coercion. It has to be something that he willfully gives up, in the same way that a sexual deviate may give up his sexual deviation. He may not be able to do entirely without it, but he may learn to control it. In the same way one can hope to achieve some success with the needle. Until one cures the needle, one will never cure narcotic addiction.

Dr. Smith: May I comment on that also? I would like to challenge that statement. I was in England this summer and had opportunity to look at some of the heroin clinics. I was very distressed that there was a growing use of intravenous methadone. I certainly think that needle freaks are one of our major concerns in the U.S., but I think we have had good luck in using oral methadone. I would like you to elaborate on the reason why, if you are going to employ methadone, it is necessary to continue to give it by injection, when we all agree that the compulsion with the needle is one of the major areas of pathology?

Dr. Boyd: Let me say first of all that although we have transferred to the use of methadone, in fact the quantity of methadone has not increased in comparison to the original figure of heroin, since our national program is to reduce the total quantity of narcotic drug used. So there has not been an increase in the use of methadone except in relation to the decrease in the use of heroin.

Why is it given by injection? Because, in fact, we are curing, or trying to cure the addiction to the needle as well as to the drug, and the same argument applies. We argue that you cannot remove the narcotic just like that and hope to cure the addict. He has to willfully give it up. Similarly, he has to willfully give up the needle. If you don't give it by needle, he will simply go to a black market, which will then thrive, and find the needle and the narcotic drug in that fashion. This is what we have seen happen when we have become tough on this problem.

Dr. Meester: Mr. Chairman, may I comment on these thoughts? During the past year, we have seen patients in our clinic who have no difficulty at all in changing from intravenous to oral medication. One thing we do find, however, is that the patient who is asked to come into the clinic every day to pick up his methadone develops a habit of coming to the clinic, and when we try to break him loose from the program he has severe difficulty in dissociating himself from the clinic.

Dr. Dennis U. Evans (Cleveland, O.): I have a practical psychiatric question which I would like to direct to the panel. I discussed it briefly last night with Dr. Koumans. At the Free Clinic in Cleveland our attempt to establish group psychotherapy for growth experiences in our outpatient population met with little success. The question is, in a Free Clinic setting, often with a transient population, can group psychotherapy have value as a growth experience, especially for the soft drug user? Could anyone comment on any successes with this population?

Dr. Koumans: I think that both David Smith and I would like to take this up. The one experience that I have is the Free Clinic in Boston, and I know enough about the Cambridge population that we can combine the two. We have tried three times so far in the 2½ years that we have existed to get a group started, on the assumption that this might be a positive growth experience. All three times we have failed, and not because there was a lack of enthusiasm on the part of the staff, psychiatrists, social workers, or patients. We feel in retrospect that the difficulty was with the approach. In the first place, the population was too transient to guarantee any long-term involvement of the same individual. Secondly, one day when I asked the people in the waiting room, "How come this group doesn't work?" they said, "Well, you know we do such a lot of rapping in our own group already, we have no need for another group."

On the other hand, if I could describe to you what the waiting room looks like, both in Joe Brenner's clinic and in ours, you would see group phenomena emerging spontaneously. Many people come in early, do not mind waiting, and after they have been served, stay on until the clinic closes. This is unusual for a doctor's waiting room. So what we have is a spontaneous, unstructured and non-ideological interaction mainly between the secretary-receptionist (who is a highly capable and extremely sensitive girl) and the social worker who is there but who is not always busy. She is usually participating in a discussion in the waiting room. The psychiatrist who is not seeing a patient at the time makes himself available and tries to get some talk or discussion started. Sometimes it falls flat. Everybody is depressed and looks into space. Sometimes it is very lively. I remember one time in particular that a discussion got started, the doctors came in, forgot treating their patients, and started talking about various issues. We were invited to join them in their commune which we did the next week-end. So there are group phenomena occurring which are not strictly classic "group therapy," but which we feel are helpful in the sense that has been pointed out several times during this Symposium; that is, there is a model of an interested mature adult that they can, and hopefully do, identify with.

Dr. Smith: I would agree with that and emphasize that we should, whenever possible, try to follow the instincts of the young people themselves. All free clinics that I have observed foster an atmosphere that facilitates these kinds of spontaneous group processes, and this is one of the major issues that was discussed in our free clinic symposium in January. If you try to institutionalize a spontaneous process, very often you end up destroying the thing that is already going on. I find that a lot of the growth activity in terms of the group process is going on already in communal settings, in free clinic waiting rooms, in much the same way that it is happening in the Boston area. If you are going to institutionalize the group process, you have to consider models of growth centers like Esalen; I think it would be very difficult to duplicate this in the inner city. For example, those programs that really focus on the use of group therapy techniques as a growth process require almost total commitment of the individual in a residential setting. Probably the best example in northern California is the Family at Mendocino State Hospital, which utilizes a large number of these techniques in encounter groups, sensitivity groups, meditation, and all the things that were discussed. But the people are there, and they are living together, and can engage in interpersonal relationships at a variety of levels. I think it is very, very difficult for a free clinic to incorporate all of these techniques when you can't possibly establish a residential or a long-term commitment on the part of the young person. If the free clinic is going to do it, then it should do it in a rural or less threatened setting. We have talked about a free clinic commune, for example, that would focus on this type of thing, working with many of the same techniques in conjunction with Mendocino.

Dr. H. DeArmond Kirkpatrick (Charlotte, N. C.): I would like to concur that the transition from the needle habit or "the needle freak" to oral methadone is made very easily, and is one step toward breaking the addict's habit of daily use of the needle.

Furthermore, Dr. Boyd, in the case of the young addict, am I to understand that you use methadone to gain initial directional control of the adolescent's already confused ego with the desired goal being complete removal of all sustaining drugs?

Dr. Boyd: Insofar as the question implies that to remove the drug compulsorily will do further harm at the initial stage to the patient, I would agree with that and say 'yes.' We leave the patient on the drug initially in order not to take away a crutch which he badly needs at a time we are initiating other forms of treatment. After that we hope that a new kind of dependency, the replacement of drug dependency by human dependency will, in fact, take over and that in good time the dependency on the therapist and on the clinic will be transferred back to society in some more mature form. That is the basis of our philosophy. As to whether one should ever use some kind of adjunctal chemotherapy, I have found that many of the adolescents I see are in a state of chaotic anxiety and that tranquilizers, the promazine or promazine-type, are effective. I try to avoid, wherever possible, giving a sedative such as a barbiturate drug for sleep. Many of these youngsters have diurnal inversion of their sleep habits and they wake at night and sleep in the day. In order to reintroduce sleeping at night, we sometimes use an antidepressant, one of the tricyclic compounds like amitriptyline, and give about 75 mg of this drug at night. It is quite a good sedative. Apart from that, I follow the philosophy that to replace drugs with drugs is on the whole unsatisfactory.

Mr. Attia Elgammal (Mt. Clemens, Mich.): Do you recommend that the drug users among students in high school and junior high school be taught separately, by a teacher who may be a stranger but who is able to deal with this kind of youngster? Do you believe that these kids should be in a separate school, that is, in a separate building? Also I have a second question for Dr. Benson. Your slide indicated that you gave the survey to roughly 1,800 subjects and your graduating class was 800. So some have moved from the area or for some other reason graduated very quickly. I am fascinated with this result, and would like to find out if we can implement this program in school physical education programs. If you have a package, I would like to know more about it.

Dr. Pollard: Let me answer the first part of your question very quickly. If we put the kids in Michigan who used drugs in separate schools, we would have them go out of the back door and back in the front door. Now, back to Dr. Benson.

Dr. Benson: The difference between the 1,862 subjects that we started with and the 852 that we ended with was simply that some subjects just did not meditate that long. The only requisition we had was that they meditate for at least three months. The difference, for example, between three months and six months was 30 people who happened to be meditating just three months, so they really were not drop-outs. It is just that they were not doing it long enough. And as to implementation of the program, perhaps we could talk after this panel about that.

Dr. Pollard: I regret that our time is limited and I must bring this discussion to a close. Thank you, Dr. Benson, Dr. Boyd, Dr. Koumans, and Dr. Smith.

Afternoon Session
Thursday, November 12, 1970

Presiding: Leo E. Hollister, M.D.

Dr. Hollister: This session will be devoted to legal matters pertaining to drug abuse. We are privileged to have a distinguished group of legal experts who embrace almost all aspects of the law: the legislative process, the judicial process, and the law enforcement process. Because we expect this to be a very provocative session, with a great many questions to be answered, we are going to ask that you write your questions on forms that will be circulated by the ushers and submitted after each speaker finishes his material. These cards should contain your question, which speaker you would like to answer it, and your name and identification that will become part of the permanent record of the discussion. I suspect that the odds of having your question answered go up in direct proportion to the legibility of your writing, so I would urge you to write your question so that it can be read.

CONGRESSIONAL REACTION TO DRUG ABUSE

JAMES H. SCHEUER, LL.B.

I am very happy to be here. I have enormously enjoyed this conference and have been enriched by it. However, I noticed on the program that I am referred to as James H. Scheuer, LL.B. Since it is a close question in my mind whether lawyers and legislators have been more productive or more harmful in the whole area of addiction, perhaps it was an act of charity on the part of those who drew up the Conference program to refer to me as both. Be that as it may, I am going to chat both as a lawyer and a legislator on the Congressional reaction to drug abuse and the role of the law in drug abuse. I shall not use my whole 35 minutes; I would probably bore you all to tears and would fall asleep myself. I am not going to give a prepared speech; I had a beautiful mimeographed speech, and if any one of you would like a copy of it, I will distribute it freely by the carload. I am just going to talk for a few minutes and then hopefully we will have a really swinging question and answer session.

The first thing that I would like to do in discussing the Congressional reaction to drug abuse is to talk about the reaction of people at large to drug abuse. Because, believe it or not, the Congress of the United States, and particularly the House of Representatives, are very representative institutions. They represent the American people with all of its warts and pimples very accurately—almost painfully accurately. Before you can really understand what the Congress is thinking and doing about drug abuse, you must understand their perception of what the public thinks about drug abuse. Now, let me give you a few facts on a survey that was published in the last month or so. The Columbia Broadcasting System conducted a survey on drug abuse. Eighty-one percent (81%) of all of the adults polled considered marihuana an addictive drug; 90% thought it dangerous to mental health; 85% thought it dangerous to physical health; 71% said the use of marihuana should be punished by incarceration, and 58% felt that the legal penalties for marihuana should be more severe than they are now. A majority of people said that using marihuana led to the use of stronger drugs—the domino theory of drug abuse. Eighty-two percent (82%) felt that the use of marihuana leads to crime and acts of violence. This is the public perception of drugs and particularly marihuana, so it is no accident that most Congressmen see things pretty much through that same prism.

Now, having said that, it is easy to understand the approach of both the Administration and the Congress to drug abuse. It has been based on the principles that drug abuse is wrong and immoral, that it leads to all kinds of terrible mental and physical effects, that it leads to crime, and above all, I suppose, that it leads to fun and pleasure. And by every criterion of our Protestant puritanical New England ethic, this is wrong and must be punished. This is the philosophy of the Congress and of the Administration, and I tell you, as a Congressman, that I consider that it is this kind of permissiveness that is ruining our country. It is these Radic-Libs who are radicalizing our criminal justice system to massage our egos and our philosophy of what is right and wrong, who are so liberal with rhetoric that in effect they have created a criminal justice system and a penal sanction that is not only ineffective, not only an exercise in frustration and futility, but is counter-productive and crimeogenic in and of itself. I suggest that I could prove to you without a shadow of a doubt that our law enforcement approach to the drug scene is of itself crimeogenic—and I consider this the extreme of permissiveness. I consider this a horrible and immoral trade-off. The majority of our population is making it an orgy of permissiveness, satisfying their own moralistic beliefs, their own religious beliefs, perhaps their own philosophy as to what is right and wrong in a whirlwind approach to the battle of narcotics, the battle over the sanity and the health and integrity of our whole young generation. This is done with a total disregard of results in a country that has always had goal-oriented strategy for everything. This disregard of results, ignoring what is good for the client population that we are concerned about in order to massage our own egos—if this isn't a type of permissiveness that should shock the country, I don't know what the word means. And we all know what the word means, don't we?

How is our law crimeogenic? Any law that does not work, in effect, reduces the respect for the law of those who know it does not work. There were over one million arrests last year for drug abuse. Any system of law that uses the arrest sanction for a kind of conduct that is broadly accepted by significant groups of the population must reduce the deterrent power of that arrest sanction. When I was a kid, we were scared sweatless of getting arrested. Kids today are not because 48% or 49% of the kids on our college campuses, according to the medical journal called *Playboy,* use marihuana frequently or occasionally. The arrest sanction cannot mean a hell of a lot to them. It degenerates the law in their eyes, it reduces the deterrent quality of the arrest sanction, and it grossly misuses police manpower.

What the people of this country really want to reduce is violent, personal crime in their neighborhood. They don't want to waste a lot of police manpower trying to regulate what we call nonvictim crime—the private, consenting relations of adults, whether it is noncrime-producing drugs, noncrime sexual relations between consenting adults, noncrime-producing loan sharking, noncrime-producing gambling. Most Americans don't care very much whether some tourist from Nebraska can find a girl in Times Square or whether a guy can place a $2 bet on the horses. Most Americans don't care if a guy gets drunk as long as he gets drunk quietly. So the spending of police manpower on nonvictim crime does not affect anyone but the person involved. This is totally crimeogenic. With limited police resources we ought to be suppressing the crime that people are really concerned about, the violent personal crime in their own neighborhoods. And there is nothing, of course, to compete with narcotics in its corrupting effects on the police establishment, except perhaps gambling which is also a nonvictim crime. If we could wipe gambling and narcotics off the law books (and I don't say that we should in the case of narcotics), we would certainly have eliminated two of the most corrupting elements of our law-enforcement profession. We have laws on our books that do not work, and we are paying a terrific trade-off for that failure.

If the harsh penalties for possession and even the harsher penalties for selling and pushing (which I favor as far as the pushers and sellers are concerned) really reduced the drug scene, I would be in favor of some of these trade-offs. I would even be in favor of the war we are waging on the Constitution. I do not like "no-knock," and I do not like preventive detention. If you told me that 150 years ago our founding fathers had no concept of the agonizing tensions and anxiety of today's urban society and that to protect people today there have to be some trade-offs between constitutional liberties and the safety of the public, maybe I would agree. I am not certain, but maybe I would. We have a law on the books that exacts tremendous trade-offs for failure, wastes and corrupts police personnel, diminishes the respect of our youth for the entire structure of our laws, diminishes the deterrent quality of the arrest sanction, and above all weakens the constitutional rights and guarantees that all Americans treasure.

So, the general attitude of the Administration and Congress, and this crosses party lines and administration lines, has been to support a hard penal sanction, without looking at the supply and demand structure that creates the problem. Indeed we have not put any significant law enforcement resources into our penal structure. We have 1300 agents in the Federal Bureau of Narcotics and Dangerous Drugs, out of four million enforcement agents in the country. If we had 10,000 to 15,000 maybe BNDD could begin to do a good job—in New York. Right now, law enforcement professionals are stopping maybe 5% to 10% of all the drugs that come into this country. They will tell you that if we had all professional enforcement officers that we ought to have, we might stop as much as 20%. So law enforcement can't be the answer. And to the extent that law enforcement is counterproductive and crimeogenic, creating more of this activity and leading to less effective enforcement of other laws, it is long since time that we have a reappraisal.

Now what is the Administration thinking about doing? What is the Congress thinking about doing, other than harsh law enforcement? I spent the month of August traveling around Europe speaking to the people at the Home Office in England about how their program is working, speaking to our Federal Bureau of Narcotics and Dangerous Drugs people in Europe, speaking in the United Nations Commission on Drug Abuse in Geneva. One million dollars is all that is being spent for education, information and protection programs for all the countries of the world. I visited the Food and Argiculture Organization in Rome, where they already have 2300 agents out in the field in the underdeveloped countries, particularly in the opium-producing countries of Turkey, Afghanistan, Burma, Laos, and South Vietnam. The Chinese professional who runs their food and plant nutrient division told me with tears in his eyes that he had no money and no authority to become active in helping opium farmers diversify into producing the melons, the winter wheat, tomatoes, and the other cash crops that Europe needs to help them get out of poppy culture. How much do you think our government is spending to control drug abuse, a problem that in New York City alone results every year in the theft of two-and-a-half *billion* dollars' worth of merchandise by narcotic addicts, and six or seven hundred *million* dollars in thefts in Washington, D.C.? How much do you think we are spending in Europe in crop diversification programs and surveillance and policing programs? In Turkey, we are spending about three million dollars, about half in policing aids and half in crop diversification. Isn't this pathetic? There has not been a reduction of one pound in the opium poppy that is coming out of Turkey. They have reduced the provinces in which they have grown it, down from 20-odd to 7. But they have increased the yield per acre so fantastically that now they are producing as much or more in seven provinces as they were producing in 25 or 30 just a couple of years ago. The Turks committed themselves to start a licensing program that the Russians and the Indians have been doing for decades. They committed themselves in the 1961 Single Convention on Narcotics to join the family of nations in strict and rigorous licensing of opium growers and they still have not done it. They committed themselves in 1967 to our ambassador to be out of poppy culture by 1970. Yet they are producing more opium poppy today than they ever did before.

As far as our efforts with the French are concerned, it pains me to report that there has not been any real impact on the refining process that converts opium to heroin in Marseilles. There has been no more than one bust of a major refining facility in France since we started working with them and that was reportedly in the last few weeks. The fact is that the poppy is grown in Turkey, comes through France, and then gets through our borders. This is quite understandable since 450,000 ships and airplanes a year arrive here from abroad. We can intercept only 10% or 20% of the drug traffic. The only result of this degree of attrition is that the organized international crime syndicates have to put in a little more at the other end of the pipeline in order to have what they need to satisfy the demand in New York. Perhaps they have to put in 10% or 20% more as a cost of doing business, knowing that it will be picked up, knowing that this is the attrition they have to expect. They are also satisfying the demand for heroin in our neighborhoods.

What kind of penal system, what kind of law enforcement system is this that imposes harsh sanctions on users without doing anything of even marginal success to stop the stuff not at our harbors or airports, but at the source in the poppy fields in Turkey? Why are we doing so incredibly little in terms of research into blocking drugs that could get the addict off this habit? We are spending about $3 million a year on blocking drugs, cyclazocine, Naloxone and the rest. Isn't this pathetic? And why is it that we cannot go with clean consciences to the nations of the world and say we must stop poppy culture now? There is no medically justifiable reason for the production of opium. Why don't we have a national program to find a nonaddictive synthetic analgesic substitute for morphine and codeine? Why aren't we in a position to say that loud and clear? There is no medically justifiable reason for growing poppy and producing opium. Isn't this a damning criticism of our national policy?

I would like to read a quotation that to my mind adequately sums up our national approach to drug abuse. It was handed to me by one of the top officials in the Home Office in England when I visited there in August. It is a quotation from John Mack, Director of the School of Social Studies, University of Glasgow, who said: "Organized crime is produced by an overworked and overridden criminal law which attempts and fails to regulate the private moral conduct of the citizen. When people are prevented by the means of the criminal law from obtaining goods and services which they do not intend to forego, the criminals will supply these goods and services under monopoly conditions at high profit to themselves. This will lead to the development of large-scale organized criminal groups which, as in the field of legitimate business, extend and diversify their operations, thus financing and promoting other criminal activities. It follows that direct action against the racketeers and mobsters by stepping up police activity, by creating special organized crime groups and similar measures, will have little or no effect on the criminal systems, and that any plan to deal with crime in America must first of all face this problem of the overreach of the criminal law, stating clearly the nature of its priorities in regard to the use of the criminal sanction and indicating what kinds of immoral and antisocial conduct should be removed from the current calendar of crime."

Now, that is as good a note as any on which to close, with but one hopeful addition. We have heard a great deal in this conference about the urgent need for characterological change, the urgent need for people who have involved themselves in the narcotic of delusion, the narcotic of self-deception, to make it in the world of reality, to equip themselves for that world by accomplishment. If they can achieve some model of success we hope that they can break away from the world of illusion and deception and failure into a world of success where they can establish their own pride and self-esteem and become effective citizens. I knew before this conference started that I was going to have the pleasure of assuring you that hope is here, hope is upon us, because when I flew out here on Monday, I read in the *New York Times* of an example that convinces me that there is hope for all of us. This is a report of the discussions of some White House spokesmen on the President's and the Vice-President's plans for the two years ahead. I quote: "These White House spokesmen pointed to the law and order issue. Whereas Mr. Nixon spent the last part of the campaign accusing the Democrats of obstructing law and order legislation and encouraging permissiveness"—now get this—"by 1972 he may be in a position to talk about his own efforts to restore domestic tranquility and stop the rise of crime." Now I suggest to you that if Mr. Nixon and Mr. Agnew can do it, if they can shake themselves from the world of rhetoric, the world of self-deception, the world of the unreal and come to realize the importance of making it by their own accomplishment, their own success. . . .

LEGAL ASPECTS OF DRUG ABUSE—A JUDICIAL VIEWPOINT

JOHN W. OLIVER, LL.B.

Six years ago I read a paper to the Section of Neurology and Psychiatry of the Medical Society in Kansas City. I gave it the title: "What Medicine Can Do for Law: Revisited Thirty-Six Years after Mr. Justice Cardozo's Address to the New York Academy of Medicine."[1] Because I was then trying to obtain the agreement of a group of physicians and other medical experts to render service for our Court, I used that part of Cardozo's 1928 address in which he said, "There are those who say that the earliest physician was the priest, just as the earliest judge was the ruler who uttered the divine command and was king and priest combined." Cardozo made the point that "our professions—yours and mine—medicine and law—have divided with the years, yet they were not very far apart at the beginning." And he added that we must become so again because he believed that "whatever enlightenment shall come (in the administration of criminal justice) will make its way, not through the unaided labors of the men of my profession, the judges and the advocates, but through the combined labors of men of many callings, and most of all your own." I am therefore grateful for the opportunity to work with you at this Symposium.

I am assigned to express "a judicial viewpoint on the legal aspects of drug abuse." The judicial view I must take is not a bright one. Indeed, the haunting words of First Corinthians XIII, 12: "For now we see through a glass, darkly. . . ." kept running through my mind as I thought about what a judge should say at this Symposium. As a point of beginning, it must be stated that judges simply do not have power to view or to deal with problems of drug abuse except as they are specifically authorized by the legislative branch of various governments. The Congress and the state legislatures have, generally speaking, proceeded in the past upon the implicit assumption that the problems of drug abuse can and will be solved, any empirical data to the contrary notwithstanding, by the passage of laws providing for heavy criminal sanctions. Authorization for, and sometimes mandatory, sentences for sales to minors of up to 40 years in prison are common; provisions for life imprisonment are not rare; and death sentences are actually authorized under particular circumstances in Georgia, Missouri, New Mexico, Oklahoma, and Texas.[2]

Generally speaking, many of the criminal laws relating to drug abuse were, and many are presently, based on the notion that the problems of drug abuse would more likely be solved if judges were deprived of their traditional power to exercise discretion on a case-by-case basis—a power accorded them as a matter of course in almost every other area of the administration of criminal justice.

The recognized movement of drug abuse from the ghettos of our inner cities to other segments of the population, including, but not limited to, students and children of affluent parents who occupy positions of importance in public and private life, has added other complicating factors to an already complicated problem. That movement, of course, has a direct relation to the sort of inquiry and research reported at conferences and symposiums such as this. But the sometimes blatant use of drugs by large numbers of alienated youth has also produced an even more avid conviction on the part of many highly vocal people who apparently are more convinced than ever before that the reason a drug abuse problem exists in the United States is either because the present laws are not harsh enough, or that the present laws may be harsh enough, but that they are not, and for no valid reason, being enforced. In spite of the fact that judges have long and regularly imposed the mandatory sentences dictated by the legislatures, judges are nevertheless accused of "coddling criminals."

I fully realize that what I have said thus far is not likely to produce much, if any, dissent at this Symposium. But that fact, if it is a fact, would not, in my judgment, improve the rather dim judicial view I am forced to take of the current drug abuse scene. And I will tell you why.

ABUSE OF ALCOHOL

The abuse of one particular drug in the United States was responsible for "two million arrests in 1965—one out of every three arrests in America" made that year.[3] The President's Commission on Law Enforcement and the Administration of Justice also stated that there were over 5,000,000 known abusers of that particular drug. And that report recognized that the present inclusion of abuse of that particular drug "within the system of criminal justice seriously burdens and distorts its operations."[3]

The President's Commission recommended that efforts to control the abuse of that drug by the attempted use of criminal sanctions should be abandoned; that communities should establish detoxification units as part of a comprehensive medical treatment program, and that appropriate recognition must be given to the fact that realistic treatment programs must include supportive residential housing facilities, such as halfway houses, community shelters, and adequate medical treatment if anyone expected any reasonable degree of success in the reintegration of former abusers back into organized society.[3]

The Commission, of course, was talking about alcohol. It fully recognized that criminal laws designed to eliminate alcoholism by prohibition of the manufacture, sale, or possession of that particular drug in any form simply have not met with substantial success and other methods must be used. That recognized failure, however, has not produced much legislation designed to attack the problem of alcoholism from any different angle which, presumptively, would be based upon assumptions quite different from the punitive criminal law approach implicit in the Volstead Act.

The October, 1970, issue of The Journal of the American Judicature Society (p. 122) carries an article by Judge Richard S. Kaplan who presides over the City Court of Gary, Indiana. Judge Kaplan describes a typical day of judicial frustration with persons charged with public intoxication. He lamented "the futility of laws which fail to recognize that chronic alcoholism is a medical and social problem and not a legal problem." He noted that in *Robinson v. California,* 370 U.S. 660 (1962), the Supreme Court of the United States had held that no state had constitutional power to pass a statute making it a crime to be addicted to use of narcotics, but that, six years later, in

Powell v. Texas, 392 U.S. 514 (1968), the Court had refused to extend the principles of *Robinson* to a chronic alcoholic.

The judge of the Gary City Court expressed his view that the Supreme Court's decision in *Powell v. Texas* "just doesn't make sense." It was his view that "In great measure, the problem [of alcoholism] can be solved [simply] by having the U.S. Supreme Court reverse the *Powell v. Texas* decision." He suggested that if the Court determined as a matter of fact that the problem of alcoholism "is not a legal problem, but is in truth a medical problem" somehow or other the Court would have a valid constitutional basis for overruling *Powell v. Texas.*

We need not reach the legal question of whether or not the latter suggestion may be sound as a matter of constitutional law. The reason for discussing the Indiana judge's unhappiness with *Powell v. Texas* and his suggestion as to how the problem of alcoholism could be solved by the Supreme Court is to focus precise attention on the fact that either a judicial or a legislative declaration that the abuse of a particular drug is a medical rather than a legal problem, does not come close to solving any of the numerous problems related to the abuse of any particular drug.

Two years before the judge of the Gary, Indiana city court protested the Supreme Court's decision in *Powell v. Texas,* the Congress had already passed the Alcoholic and Narcotic Addict Rehabilitation Amendments of 1968 to the Community Mental Health Centers Act of 1963. In what is now Section 2688E(a) of Title 42, United States Code, it was explicitly stated that "The Congress hereby finds that—(1) Alcoholism is a major health and social problem afflicting a significant proportion of the public . . . [that] (2) alcoholism treatment and control programs should whenever possible: (A) be community based [and] (B) . . . under proper medical auspices . . . [and that] (3) the handling of chronic alcoholics within the system of criminal justice perpetuates and aggravates the broad problem of alcoholism."

No one can properly complain that Congress, in contrast with the inaction of most other legislative bodies, has not legally recognized that the problem of alcoholism must be treated as a health problem. Congress has flatly stated that attempts to handle chronic alcoholics within the system of criminal justice merely "perpetuates and aggravates the broad problem of alcoholism." But, just as the passage of the Volstead Act did not solve the problem of alcoholism, the passage of the Alcoholic and Narcotic Rehabilitation Amendments of 1968, standing alone, will not solve the same problem.

The hard, cold fact of the matter is that Gary, Indiana, like most cities in the United States, is broke. State funds apparently are not available to build an adequate alcoholic detoxification facility for that city. A drive for private funds could not, as a practical matter, be sought because "unfortunately," as Judge Kaplan reported, "a drive has been going on for the past four years to raise 5½ million dollars to extend our present hospital facilities and we have (therefore) been unable to start any drive for the required treatment facility for chronic alcoholics."

The long and the short of that sad story is that the judges in the city of Gary, Indiana, and many other cities throughout the United States in similar circumstances, are going to be forced to continue to handle chronic alcoholics in exactly the way they have in the past, in spite of the fact that the Congress has fully recognized that the continuation of the existing revolving door criminal law system cannot fail to perpetuate and aggravate the broad problem of alcoholism in the United States.

It should, therefore, be clear that whether the Supreme Court overrules *Powell v. Texas* can have no real effect on the quite practical problem of how alcohol abuse is going to be treated in Gary or any place else.

ABUSE OF DRUGS

The problems concerning the abuse of alcohol are infinitely easier to solve than the problems involved in the drugs which are receiving detailed attention at this Symposium. The public has reasonably accurate information about alcohol, based primarily on its own experience. And, while there is some scientific disagreement about treatment and other facets of alcoholism, those disagreements have not produced any emotional responses in light of the experience gained by the public during and since the days of the "Noble Experiment." The reason we do not mount a concentrated attack on alcoholism is not because we do not know what to do; it is simply a matter of priorities and a willingness to pay the price.

The situation, however, in regard to the abuse of drugs other than alcohol is fundamentally different. It comes as a distinct surprise, for example, to many members of the public that two years *before* the Congress recognized that alcoholism should be treated as a medical problem, it made a similar declaration in regard to narcotic drug addiction. The Narcotic Addict Rehabilitation Act of 1966 became the law of the land four years ago. In what is now Section 3401 of Title 42, United States Code, the Congress made the following declaration of policy:

> It is the policy of the Congress that certain persons charged with or convicted of violating Federal criminal laws, who are determined to be addicted to narcotic drugs, and likely to be rehabilitated through treatment, should, in lieu of prosecution or sentencing, be civilly committed for confinement and treatment designed to effect their restoration to health, and return to society as useful members.
>
> It is the further policy of the Congress that certain persons addicted to narcotic drugs who are not charged with the commission of any offense should be afforded the opportunity, through civil commitment, for treatment, in order that they may be rehabilitated and returned to society as useful members and in order that society may be protected more effectively from crime and delinquency which result from narcotic addiction.

I think it quite unlikely that Congress would have made that declaration of policy in 1966 if it had not had the benefit of the earlier collaborative work of the American Medical Association and the American Bar Association. The Joint Committee of the AMA and the ABA, which presented an Interim Report in 1958, and a Final Report in 1959, paved the way for the White House Conference on Narcotics and Drug Abuse held in 1962. In 1963 the President's Advisory Committee on Narcotic and Drug Abuse, which grew out of the 1962 White House Conference, approved and recommended the idea of appropriate medical treatment and for the civil commitment of drug addicts. That Report concluded that the narcotic and marihuana laws passed by the Congress over the last 50 years have not solved the problem of the abuse of drugs other than alcohol.

But is there any question in anyone's mind that if our existing laws relating to drugs other than alcohol were submitted to some sort of a public referendum that the public would conclude, for example, that laws based on the rationale of the Marihuana Tax Act of 1937 should be supported more strongly than laws based on the rationale of the Narcotic Addicts Rehabilitation Act of 1965?

The reasons for the present public view are not difficult to find. The public is uninformed, if not misinformed, about the current problems of drug abuse. Nor have many members of the public had any direct experience with the drugs discussed at the Symposium. This situation is in sharp contrast to that relating to alcohol.

Few members of the public go through life without some personal experience both with alcohol and with an assorted group of alcoholics. The majority of the public also knows that they, as individuals, may drink too much on occasion but that this does not mean that they are going to end up as alcoholics.

The same situation simply does not exist

in regard to the whole new set of drugs, and some very, very old ones, such as marihuana, which, generally speaking, have been experienced only by the younger segment of our population or by persons who occupy a position in our affluent society which is, by and large, virtually unknown to those who live on the right side of the tracks.

Dr. Jonathan Cole, for example, commented in 1968 at the Rutgers Symposium on Drug Abuse that "little current information of any accuracy is available in regard to marihuana, a widely used drug."[4] Dr. Dale C. Cameron pointed out at the same meeting that the data concerning treatment for drug abuse, particularly that in regard to meaningful supervision in the community following hospitalization, are "so few that no conclusions concerning effectiveness could be drawn." Dr. Jerome Jaffe's article in the May, 1970, issue of *Psychology Today* points out that beyond almost universal agreement on the goals of treatment of compulsive narcotics users there is nothing except bitter discord.

The point is that the general public is not informed about current drug abuse by its own experience. Nor can it become informed by study of meaningful scientific data because such data, even conflicting data, are not easy to come by. But perhaps the more important point for our purposes is that those circumstances are not irreversible.

THE ROLES OF MEDICINE AND LAW IN THE CONTROL OF DRUG ABUSE

The fact that the program for this Symposium was "designed to help the practicing physician to deal more effectively with the problem of drug abuse . . . as a participant in community action programs, with school officials, the courts and representatives of social agencies" is evidence of the sort of concern which entitles one to take a brighter view of drug abuse.

The exchange and eventual publication of the data presented at this Symposium cannot help but provide a more adequate basis for public understanding.

It must not, however, be overlooked that the same data will most likely establish that there are presently but a few meaningful community action programs in actual operation in the United States. We shall also probably learn that the type of educational programs recommended at this Symposium which are in actual operation are few in number. And, perhaps most importantly, we shall be forced to face the fact that the establishment of any adequate community-based drug addict facility is subject to the same, if not greater, practical difficulties which have thus far prevented the establishment of a proper program for alcoholics in Gary, Indiana, and elsewhere.

But all that does not mean that there are no valid grounds for hope. Our professions are already very close in regard to how we, as individuals, must proceed. Mr. Justice Holmes once said that "it will be a slow business for our people to reach rational views, assuming that we are allowed to work peaceably to that end." But he never, never lost his hope that improvement could be made, provided inquiry and efforts were conducted with appropriate skepticism. Holmes used to say that one had "to get the dragon out of the cave and count his teeth" before one could take the next best step.

My judicial view and hopes for eventual improvement in the enactment and administration of laws applicable to drug abuse were enhanced by what Dr. Richard H. Blum had to say at the close of the Rutgers Drug Abuse Symposium two years ago. The thoughts he expressed illustrated how our disciplines—medicine and law—are again together as they were in the beginning. Dr. Blum expressed his faith and confidence that "the ordinary man, given simple truth about how the system of justice, treatment, or education works—both its means and its consequences—will recognize a monster when he sees one and will want to get rid of it. . . . I am confident that ordinary men will not only go along with us when we propose the abolition of monsters, but that they will demand it of us."

Dr. Blum suggested, in a manner quite similar to that of Mr. Justice Holmes, that each of us in our separate disciplines has a job "to find out where the monsters *are* and to announce their presence," and that we also have a companion job to find out where the monsters are *not*. He added that all of us must not forget that public decisions are not made solely on scientific data, and that any valid hope for real progress must rest on the concern of each interested individual "to look around us to see what is really going on and to try to make it as much better as we possibly can."

It has been my consistent experience as a judge that medicine and law can join forces in efforts to improve the administration of justice. It has been my further experience that one informed and concerned physician can do more to bring into being and to keep in operation an appropriate community action program than almost any other member of the community. I would therefore hope that each participant in this Symposium will look around his own community to see if any meaningful program is in operation. If he finds that none exists, then it is my hope that he utilize the knowledge and experience of this Symposium to become the catalytic agent for organizing and maintaining a proper program. I am certain you will be able to interest the judges in your project. I am also confident that school officials and representatives of social agencies will also lend manpower.

Where you will get the money will test your skills in grantsmanship to the breaking point. But experience again establishes that if your ideas and plans for execution are sound, and if you are concerned and tenacious enough, your project will somehow find the necessary finances.

And when that comes to pass all of us may not have to look through the glass as darkly.

I thank the University of Michigan for permitting me to join with you at this Symposium.

REFERENCES

1. Oliver, J. W.: *What Medicine Can Do for Law: Revisited Thirty-Six Years after Mr. Justice Cardozo's Address to the New York Academy of Medicine.* 37 Federal Rules Decisions 413 (1964). (Mr. Justice Cardozo's famous 1928 address "What Medicine Can Do for Law" is published in *Selected Writings of Benjamin Nathan Cardozo,* edited by Margaret E. Hall. New York, New York.
2. Eldridge, W. B.: *Narcotics and the Law,* 2nd ed. Chicago, University of Chicago Press, 1967. (Appendix B contains an excellent summary of legislation as of October, 1961.)
3. President's Commission on Law Enforcement and the Administration of Justice: *The Challenge of Crime in a Free Society.* Washington, D.C., U.S. Government Printing Office, 1967, pp. 233, 237, 238, 285.
4. Cole, J.: In *Proceedings at the Rutgers Symposium on Drug Abuse,* edited by J. R. Wittenborn. Springfield, Ill., Charles C Thomas, 1969, p. 216.

SPEED AND VIOLENCE: COMPULSIVE METHAMPHETAMINE ABUSE AND CRIMINALITY IN THE HAIGHT-ASHBURY DISTRICT

ROGER C. SMITH, D.Crim.

The Amphetamine Research Project, funded by the National Institute of Mental Health and sponsored by the Department of Pharmacology at the University of California Medical Center in San Francisco, began collecting data on patterns of high dose methamphetamine ("speed") abuse in the Haight-Ashbury district of San Francisco early in 1968. The project was located in a building which also housed the psychiatric section of the Haight-Ashbury Free Medical Clinic, a relationship which served to quickly legitimize the project's presence in the neighborhood. Soon after starting its study of questionnaires, formal taped interviews, and informal observations, the project became involved in structuring a treatment program which has since become an integral part of the Free Clinic's treatment complex. Because it combined action with research, many of the obstacles to data-collecting research in deviant subcultures were avoided. Informants agreed to participate in interviews after they had become acquainted with staff members who had provided "no strings" assistance. Because of the trusting relationships which developed, participants would often bring fellow users into the project to be interviewed, and, on several occasions, we were invited to witness the operation of illicit "speed labs" in San Francisco, a privilege granted few outsiders regardless of their intentions.

The Haight-Ashbury district was an ideal site for this research when it commenced in 1968. The pattern of drug use had shifted from a relatively benign pattern of social and ritual use of marihuana and psychedelics to a compulsive pattern involving intravenous speed. As speed became the primary drug of choice in the neighborhood, several rather dramatic changes occurred. The level of violence began to spiral as illicit speed labs moved from the area, and the patterns of distribution became increasingly rigid. During the months following the initiation of the project, specific styles and patterns of use began to appear, and alternative career routes through the speed scene became apparent.

Once the center of manufacture and distribution of speed on the West Coast, the drug scene rapidly diffused outward, setting the stage for the emergence of yet another pattern of drug use. Currently, Haight-Ashbury is experiencing a rapid rise in heroin addiction, which has now replaced speed as the drug of choice in the area.

While much is known about the effect of

amphetamines on individual behavior, little is known about the operation of the illicit marketplace to which every user must necessarily relate. Further, there have been few studies relating to the speed subculture, primarily because it is a recent phenomenon, but also because of its relative inaccessibility to most researchers.

This presentation will focus on the patterns of induction into the speed scene, the subjective experience of users in the scene, the impact of the subculture and the illicit marketplace on behavior, particularly as it relates to violence, and, finally, the divergent career routes we observed.

A great deal of statistical data were collected in the course of this 18-month project, but I will refer to the figures only generally as I attempt to paint a descriptive picture of the scene as it evolved.

CHARACTERISTICS OF THE POPULATION

Numerous articles in the popular press have described the Haight-Ashbury district as a place where thousands of young people committed to a "love ethic" congregated for a period of several years and attempted to build a counter-culture. In general, however, our observations suggest that the young people who came to the neighborhood during the 1967 "Summer of Love" and thereafter did not necessarily share values, attitudes, or expectations. Typically, the newcomer was just over twenty years old, white, had spent time in college, and usually came from a middle or upper-middle class family living in a large urban area. Researchers who studied the community agree that the only thing this heterogeneous community had in common was the use of drugs. Davis distinguished two major types in the neighborhood during the latter part of the "Summer of Love": the "head" and the "freak."[1] The "head" used drugs primarily to gain insight or new perspective into himself and his environment, while the "freak" used drugs primarily for the immediate drug effect. While the term "freak" was particularly applied to the speed user, it also applied to anyone who engaged in violent, bizarre, or destructive behavior.

Styles of drug use cannot be defined simply on the basis of the drugs used. For example, prior to the summer of 1967, the head represented the most prestigious style of drug use, mainly marihuana and the psychedelics. However, a questionnaire administered by the Haight-Ashbury Medical Clinic showed clearly that most of the heads were experimenting at the same time with a wide range of substances, including opium, oral barbiturates and amphetamines, and, to a lesser extent, heroin. Nonetheless, the primary commitment of the individual user was to the use of drugs as a means to an end; i.e., as an adjunct to a style of life which stressed exploration of self and expansion of consciousness, among other things.

Even those defined as freaks by the community, who centered their lives around the use of such drugs as speed or heroin, also used a variety of psychedelics, often in combination with other drugs. In many instances, psychedelics were ingested intravenously. Davis views the contrasting styles as ". . . the basic contrast in expressive styles extant in the American class structure; put crudely, LSD equals self-exploration, self-improvement equals middle class, while Methedrine equals body stimulation, release of aggressive impulses equals working class."[1]

While the class notion as a means of explaining differing styles of drug use may be an overstatement, nonetheless Davis clearly differentiates the two major styles observed in Haight-Ashbury and elsewhere.

Because we did not do extensive psychologic testing on the individuals who became enmeshed in the speed scene, it is difficult to determine whether those who came into Haight-Ashbury after the "Summer of Love" and became compulsively involved with speed differed significantly from those who preceded them and limited their drug use primarily to marihuana and the psychedelics. Our guess is that the social and psychologic characteristics of the young transients coming into the area did not change significantly as the patterns of drug use began to change. The individual variables which antedate involvement in the drug scene appear to be less important in determining the direction drug use will take than such immediate factors as prevailing community attitudes, peer sanctions imposed on certain kinds of social behavior, drug availability, subjective interpretations of the drug experience, the quality of social interaction in the speed scene, and, finally, the structure of the illicit marketplace to which the user must relate.

During the summer and fall of 1967, increasing numbers of young people in Haight-Ashbury began regular use of intravenous speed. Our observations during this time suggested that they were effectively isolated from other speed users except for their contacts within the marketplace. Most seemed to be fairly well integrated within the broader community. A speed culture, as such, certainly did not exist, and the individual user had few sources of information about the drug he was using, what he might expect from it, and how to deal with an adverse reaction to the drug. Consequently, many novice speed users experienced acute anxiety reactions and sought assistance from the Free Clinic. As the use of speed increased, however, the adverse reactions to the drug decreased sharply. This may have been the result of a speed culture which provided participants with guidelines for survival. Becker describes a similar development in the LSD culture:

> But individuals accumulate experience with the drug and communicate their experiences to one another. Consensus develops about the drug's subjective effects, their duration, proper dosages, predictable dangers and how they may be avoided; all these points become matters of common knowledge, validated by their acceptance in a world of users. A culture exists. When a user experiences bewildering or frightening effects, he has available to him an authoritative alternative to the lay notion that he has gone mad. Every time he uses cultural conceptions to interpret drug experiences and control his response to them, he strengthens his belief that the culture is indeed a reliable source of knowledge.[2]

Various explanations have been offered for the shift in drug-taking practices in the neighborhood. The most plausible is that the psychedelic community, which had actively proselytized their drug of choice, was overwhelmed by the sheer numbers of young people who came to the area. Without proper "instruction" in the use of the psychedelics, the experience was a frightening one for many; for others, psychedelics were just not pleasurable. Speed, on the other hand, provided immediate physical gratification, relieved depression, and required little introspection from the user. The newcomer could quickly become part of the speed scene, without having to go through the preparation and guidance which characterized early LSD use in the neighborhood.

The newcomer to an area like Haight-Ashbury in late 1967 or early 1968 had equal access to individuals in the speed or acid scenes. Many of the older "acid heads" became concerned about the motivations of the younger transients, and consequently became more selective in offering assistance, including housing. Speed users, however, seemed to care little about an individual's motivations for coming to the Haight and provided housing and speed to nearly anyone who requested them.

As the speed culture developed, "speed palaces" were formed—residences within the neighborhood where the primary activities were acquiring and shooting speed. The "speed palaces" were characterized by a high level of activity, a steady stream of visitors, some of whom might drop in for a quick visit and end up staying for months, occasional violence, and frequent arrests.

Because of the opportunity for exploitation, other groups began to attach themselves to the speed scene, including outlaw motorcycle gangs, hard-core delinquents and criminals, and young male Negroes from a nearby ghetto.

THE SPEED EXPERIENCE

Given the receptive attitudes toward drug experimentation which most young transients brought to the neighborhood with them, it was predictable that a large percentage would experiment with speed at one time or another. Whether one progressed beyond the first few experiences with the drug depended upon several factors. Because most of the residents of the Haight were middle-class whites, few had extensive experience with injectable drugs. Further, many knew the slogan "Speed Kills" and had reservations about using the drug. In a supportive environment, however, such fears are easily dissipated, and the first "fix" is often administered by a friend.

The first experience with speed is almost always described as extremely pleasurable. The initial "rush" or "flash" has been described as a full body orgasm, and some individuals claim to have reached climax shortly after injection.

For many individuals, the initial flash experience is regarded as the sought-after effect, while others prefer the long-term stimulative effects of speed. Many of the older speed users scorn those who seek only the flash, labeling them "flash freaks." For example, during the summer of 1968, the quality of street speed declined appreciably. Most users claim that the addition of ether or other adulterants to speed increases the flash, while decreasing the long-term stimulation. One "righteous speed freak" described the flash freaks as follows:

> Most of the kids that were around during the summer were flash freaks. If you gave them a crystal and ether solution that turns your stomach over, blows off your head, and five minutes later you don't even know that you even hit anything, they're perfectly happy.

In general, the attraction of speed falls into one of two categories: increased feelings of confidence and well-being, and an increased ability to perform motor tasks. Novice users in drug cultures like the Haight-Ashbury sometimes regard speed as a consciousness-expanding drug, similar to LSD. Newcomers, often reticent about making new friends, find that they are suddenly able to overcome their inhibitions and that they "can think and rap and get all of my philosophies out."

Although the heads and freaks had been fairly well integrated during the transition in drug styles, most heads viewed speed users as self-destructive and impossible to live with. Often a group of speed freaks would move in with a non-speed user, driving him from his own home within a few days. As one former speed user describes the relationship:

> You can't live with someone who's taking speed and you're not. You just can't—it's impossible, it's two different worlds. And someone who's on speed you can't take very seriously if you're not on it yourself. If you're on it, you take them desperately seriously.

Whether one continues regular speed use depends upon many factors. The excess energy, for example, may produce an acute anxiety reaction. In isolation, the novice user would probably not be able to deal with reactions like pacing and hand-wringing in a way that he could interpret as pleasurable. Within the context of a speed culture, however, several possible outlets for his energy are suggested. One novice interviewed told us that during the first two weeks of use, she learned to play the guitar, practicing fifteen to twenty hours a day, until her fingers were cut and bleeding. She became quite proficient on the guitar, a skill she retains some two years later. The user may occupy his time taking apart and reassembling complicated machinery, radios, televisions, and the like. The "knick-knack bag," a pouch which many street people use to carry drugs and cigarettes, is often filled with small objects like rocks, toys, and beads, which are taken out and fondled and polished, often for hours on end. These seemingly pointless activities take on great meaning and importance for the user; deprived of his objects, he may become anxious and upset. Several artists who had turned to speed changed their styles completely. Many began working with pen and ink, producing drawings of great detail and complexity.

The "Crash" or "Comedown"

The most uncomfortable part of the speed experience is the "crash" or "comedown." A speed run is characterized by repeated injections of increasing amounts of speed over a period of time which may last from several days to several weeks. The pleasurable phase of the speed experience commences with the flash and terminates when the stimulative effects have "peaked." When the user no longer has the feeling of "going up," he may inject more speed to avoid the unpleasantness of the comedown. While the effects of speed may last from eight to twelve hours, it is only the first few hours which are interpreted as pleasurable. During the course of an extended speed run, an individual may inject speed every three to five hours, twenty-four hours a day until the "run" terminates.

At the end of a run, an individual may crash by simply stopping the drug and waiting for the effects to wear off, or he may utilize a depressant drug, usually a barbiturate or heroin to induce sleep immediately, which effectively avoids the comedown.

The severity of the crash is related to the length of the run, the amount and quality of the drug ingested, and the physical and psychologic condition of the user. For the novice at the end of a two- or three-day run, the comedown is considered a mildly unpleasant phase, characterized by fatigue, mild depression, and, following a period of rest, by ravenous hunger. As the length of the run is extended and the physical and mental condition of the user deteriorates, the comedown is characterized by deep depression, hallucinations, and extreme fatigue. Most experienced users consider the comedown unbearable, and try to avoid it through the use of a "downer."

Compulsive Use

Although the number who progress from experimental or occasional use to compulsive use of speed is not known, our observations suggest that it is probably much higher than is the case with other substances. The reason for this may be the optimism and euphoria which characterize the speed experience. For many young people in an area like the Haight-Ashbury, speed serves to alleviate a pervading sense of depression and futility, and enables them to appear spontaneous, witty, and acceptable to others.

Other factors which seem to facilitate progression into compulsive use include: living in an area which either overtly condones drug use or applies no social or moral sanctions on users; estrangement from meaningful family or social relationships; feelings of hopelessness, despair, and depression; and an inability to relate to peers. Increasing both the frequency and amount of drug use, severing social and business relationships, and forming alliances with individuals and cliques within the speed scene may indicate the progression from occasional to compulsive use.

For most individuals, the turning point comes when they totally abandon their former life styles and assume the daily routine of the compulsive drug user: developing an identity as a speed freak; forming alliances, both social and business, with other freaks; developing a "hustle" which will support the drug habit; and adopting the rhetoric, values, and attitudes of the speed culture which allow one to justify his way of life to himself.

The most important function of the speed culture is to teach participants the art of survival. For example, the following set of instructions circulated widely in the Haight-Ashbury during the peak period of use, and was probably responsible for the survival of many vulnerable experimenters:

> 1. Use clean hypodermic paraphernalia, and do not circulate your outfit (syringe or eye-dropper and needle) among anyone. Procure it new if possible and keep it clean, if not sterile. One good system is to use permanent stainless steel needles and a glass syringe, sterilizing them in a pressure cooker after each day of use.
> 2. Take plenty of vitamins, nutritional supplements, and drink plenty of liquids. Vitamin C, ascorbic acid, is a must, as are multi-vitamins and as much good food as possible. I recommend taking desiccated liver so as to strengthen that vital organ and protect it from becoming weakened and receptive to hepatitis.
> 3. The last thing of great importance is that one should never, no matter what the reason, stay awake for longer than 72 hours, as any longer than three days awake causes irreversible brain damage. If necessary, in order to avoid this, a person is better off taking barbiturates or tranquilizers.

Within the speed scene, as with other deviant subcultures, a hierarchy of status has developed. In the heroin scene, for example, one finds a series of status positions, ranging from the "righteous dope fiend" to the "snatch-and-grab junkie" or "boot-and-shoe hype," designations based primarily on the ability of the individual to comfortably maintain his addiction through a variety of "hustles" which require varying degrees of skill, daring, or perseverance. The speed culture has begun to define, although in a less definite fashion, similar status positions based on the ability of the individual to survive.

The most general differentiation is made on the basis of the effect sought. The "flash freaks" are held in lowest esteem (flash freaks are also labeled "needle freaks" because of the pleasure which they derive from injection independent of the drug effect), while those who are committed to "good speed" and enjoy the long-term stimulative effects are simply regarded as "freaks" or "speed freaks."

Beyond these two general types, one finds a continuum, ranging from the user who buys his drug on the street whenever he can raise the necessary five or ten dollars, to the user who has a continuous supply of high-quality speed and lives rather comfortably, avoiding the hazards and dangers of street-level survival. The latter is usually one who operates in the upper strata of the speed marketplace, either as a high-level distributor or ounce dealer, who has no dealings with street freaks.

The Speed Life Style

The typical round of daily activities varies from one clique to another within the speed scene. Many individuals drift from one group to another, dropping into a speed palace and staying for varying periods of time, then moving on to another group or perhaps to another city. Since few individuals use speed in isolation, they tend to seek out others using the drug. A group of users presents a picture of frantic activity, endless talking and bickering, and occasionally, sudden outbursts of violence. Much of the talk revolves around the drug experience, tales of freaky behavior, stories about suspected "snitches" (police informants), and plans for the future. As the speed culture has developed, numerous folk heroes and villains have emerged and become an integral part of the lore of the scene.

During periods of stimulation, groups often talk for hours about their plans for the future, most of which center around drugs. For example, one user describes the kinds of fantasies which he regards as serious while stimulated, but unrealistic when "straight":

> I make lots of plans but I don't ever carry any of them out. I get into this megalomaniac

> bag about five days into it and I'll build these mountainous castles in my mind, all the far-out things I'm going to do, and all the money I'm going to make. I'll be driving a Rolls Royce and have two speed labs going at once, a heroin refining plant, my own private two-engine plane, I'll be running the Mafia, and then when I start to come down, I realize that none of this stuff is existing and that none of it is going to exist and it's like you pulled the bottom out of my brain. I feel empty and suicidal in about four to five hours.

Although most of the speed groups which form are short-lived, an occasional group comes together which survives for several months. Two such groups, calling themselves the "Methedrine Marauders" and the "Crank Commandoes," formed in the neighborhood, assumed quasi-military roles, wore "hip" uniforms, and attempted to control the market through effective organization. Both groups dissolved when a majority of the members were either arrested or killed by others in the scene.

This total identification with the speed scene has also taken on some unique dimensions. While status in most deviant cultures is related to one's abilities to survive, there are those in the speed culture who derive satisfaction from being "the biggest freak on the street"—an aspiration which is, on its face, self-destructive. One such individual, a familiar neighborhood character for several months, walked the streets daily in a long flowing robe under which he concealed several weapons, carrying a briefcase allegedly containing plastic explosives. His mission, as explained to our staff, was to seek out and destroy Lucifer, who, he explained, took many forms. (More often than not, Lucifer took the form of a uniformed policeman or a tourist who carried a camera.)

In other instances, status derived from one's ability to shoot enormous quantities of speed. One individual described, with obvious pride, having shot a full gram of speed in front of an unbelieving intern from a local hospital. Another described a public "speed shooting contest":

> This was a down to the death dope-shooting contest. One of the two of us was supposed to die when it was all over. He'd shoot a half a gram and I'd shoot a gram, and he'd shoot one and a half, and I'd shoot two grams, then he'd shoot two and a half grams and I'd have to shoot three. Nobody would back out, we'd die before we'd back out.

As mentioned previously, cliques and groups form and disintegrate rapidly in the speed scene. Often conflicts will develop over the real or imagined theft of small items. The disintegration of speed cliques appears to be related to the length of time the group has been "running" on speed, the dosage level, and the amount of directed or purposeful activity the group has to consume its energy; such as making a buy, organizing a theft, or planning to set up a speed lab. When the group has little to do in the way of purposeful activities, the potential for violence is increased greatly. Activity becomes increasingly frantic and irrational. A heavy speed user describes the torture-rape of a teen-age girl as such a situation. The girl, a heavy speed- and barbiturate-user was accused by another girl in a large group of users of stealing a barbiturate pill. The girl denied the theft, but the accusing girl began to assault her. This assault focused the attention of the entire group on the victim, and they proceeded to "train" or gang-rape her. When asked why such an event could not be stopped, our informant replied:

> But then you're on her side—out of the majority and into the minority, which is a very dangerous place. You get into something that heavy, you can't say, well, I'm going to stop playing now, I'll see you later. You can't back out, not and walk away. You have to imagine how heavy those people were into that, if you go against them you best come out shooting.

PROGRESSION WITHIN THE SPEED SCENE

During the course of our research, we followed many individuals over a period of two years, from the time they "turned on" to speed, until they either left the scene, or seemingly committed themselves to a specific career route within it.

The direction a career takes within the speed scene is related to a number of personal and situational factors, including the individual's psychologic makeup, his ability to maintain or control the extent of his drug use, his reputation with others, his ability to acquire the skills and abilities necessary for survival in this scene, and, finally, situational factors such as arrest, interest by close friends, family, or relatives, and the availability of appropriate treatment facilities.

In our evaluation of the speed culture in Haight-Ashbury and elsewhere, we have observed three distinct career patterns taken by users as they move through the speed scene: (1) continued marginal adjustment, usually terminated by arrest, hospitalization, voluntary treatment, or death; (2) elevation to the upper levels of the speed marketplace; and (3) progression to barbiturates or heroin as drugs of choice.

Marginal Existence

Most speed users are unable to rise above the level of the street scene once they have become enmeshed in it either because they lack the criminal skills necessary to succeed in the speed scene, or because they tend to lose control of the amount and duration of speed use.

The inability of most speed users to survive in the speed scene for any appreciable length of time is, in all likelihood, a reflection of their socioeconomic backgrounds; few of the individuals we interviewed had any experience in a deviant culture prior to coming to Haight-Ashbury. Most did not have the appropriate "hustling" skills, and, perhaps more importantly, did not appreciate the "rules" operating in a criminalized subculture. Most of the speed users were bizarre in appearance and behavior, and were thus regarded as a "bust" or bad risk by others involved in organized criminal activities. Few street-level speed freaks were permitted to interact with "fences," for example. A good hustler, by street definition, is someone who avoids public displays, has a variety of skills to insure a low risk and high payoff, nurtures and maintains good connections with legitimate businessmen, and can be trusted not to "snitch" or inform to the police if arrested. Ghetto hustlers learn these skills over a long period of time, but most of the middle-class youngsters in the speed scene never had to worry about street survival until their involvement in Haight-Ashbury.

The street-level speed freak exists through a series of high-risk, low-pay-off hustles, such as "working the meatrack" (male prostitution), "rip-offs" (strong-arm robbery, usually involving someone in the speed scene), "burning" (selling phony substances on the street as drugs), or dealing "nickel and dime bags" (five and ten dollar bags) of speed on the street.

The primary cause of violence on the streets is related to the practice of "burning," or selling highly adulterated drugs. Since few street users are trusted by individuals in the upper levels of the speed marketplace, they must deal in small quantities, hoping to realize enough profit to pay for the cost of their own drugs. Because of the euphoria and expansiveness which the user feels after injection, however, he may shoot all the speed which he had purchased for resale, and then find he is unable to raise enough money to purchase additional drugs. Many users will then package any substance which looks like speed and attempt to sell it on the streets. They feel that this practice is acceptable because of the transient nature of the population in Haight-Ashbury, and, in fact, some will sell only to "straights"—those who live in conventional society and only occasionally use speed. Businessmen are favorite targets for "street burns," as they rarely return to seek out the "burn artist." In the speed culture, "burning" is considered an offense which cannot go unpunished, for once one has a reputation for buying adulterated drugs without taking action, he will rarely find good speed. Thus, many of the victims of the "burn" will spend days and weeks seeking the offending dealer to either exact payment or punish him for his offense. In some cases, this involves a beating, while in others, the burn artist is either knifed or shot. In one instance, a burn artist was given the choice of being shot or taking a massive dose of LSD (ten 500 mcg capsules). He chose the latter and was discovered by the police days later, wandering the streets naked and totally incoherent.

The life of a street level speed freak is cyclic in nature. When he has speed available, he is euphoric and confident about his future. When he is unable to "score," he is depressed and often suicidal. One user, in attempting to explain why he remained in the speed scene at this level when his future appeared so dismal, stated:

> I really enjoy the whole thing—when I'm out there hustling I haven't got a thing, I'm really uptight. Part of me is uptight, the rest of me is having a good time. I'm taking all this expansive shit to everybody else. I'm building my castles about a block ahead of me and when I get there it's gone, so I build another one. But I do all right. I have my basic necessities of life and things like that. I stay a step above some of these other people who just turn into animals.

Elevation within the Marketplace

The speed marketplace, like that of marihuana and the psychedelics, is loosely organized and consists of numerous groups and individuals who are involved in the manufacture and distribution of the drug. Elevation to the upper levels of the speed marketplace is currently restricted to only a few speed users, and represents the "elite" in the vaguely defined status hierarchy of the speed culture. Prerequisite for elevation to this status is the ability to maintain one's use of drugs. While speed freaks may occasionally deal or even manufacture the drug, their tendency toward spree use is considered dangerous on this level and the freak is usually screened out. In order to be successful on this level, one must be regarded as trustworthy by all. Although it is clearly a sellers' market, one who adulterates his speed or attempts to cheat high-level dealers not only damages the reputation of the particular group, but also exposes everyone to potential violence by individuals or groups who feel they were cheated or "burned."

If one is able to maintain the level of his personal drug usage, he might participate on many different levels in the manufacture or distribution of speed. Within recent months, the individual who can purchase or steal the necessary chemical precursors of speed has become increasingly important, and he is able to command a relatively high price for his product. With few skills and minimal training, one can become an illicit chemist. For those with good reputations and far-reaching "connections," the job of distributing speed on the multiple pound level can lead to a profit of more than several thousand dollars a week.

Fortunes in the speed scene change rapidly, and even those who seem firmly established in the upper strata may become caught up in speed use, fall victim to a theft, or be arrested. Thus, the structure of the speed marketplace is in a state of constant flux with only an elite few avoiding its inherent hazards.

The speed marketplace continues to evolve and restructure itself in response to the demands of the culture and the legal restrictions with which it must contend. As it evolves and positions become more clearly defined, horizontal and vertical movement within the marketplace will become more restricted, but at this point in its development the individual speed user has numerous opportunities open to him dependent upon his abilities, reputation, and contacts within the speed world and with other groups necessary to its operation.

Progression to Other Drugs

Because it is impossible to use speed in high doses for any appreciable length of time, it is common practice to utilize a variety of "downers," such as barbiturates or heroin to counteract the stimulative effects of speed and terminate a speed "run." During the early stages of the emerging speed culture, the drug of choice for this purpose was barbiturate. For most of the middle-class participants barbiturates were familiar and did not carry the stigma of heroin. But experience with barbiturates soon leads users to experiment with heroin as an alternative, and most find this preferable. Says one heavy user:

> Barbs are what make you really mean and nasty. It's the combination of the two. Once you've got enough goofers so that you are ready to kill somebody, you have to shoot crank so that you have enough energy to do it. It's the combination of the two that really tears your head apart. But the barbs definitely overpower the speed, and the speed gets you to where you are paranoid, and when you are coming down, it is real fear, because everybody is giving up somebody else's name.

Over a period of time, the culture has come to define barbiturates as undesirable "downers" leading to irrational and often violent behavior coupled with total physical debilitation. Heroin is now regarded as a "mellow downer," one which will terminate the stimulative effects of speed, while allowing the user to function without loss of control.

While many individuals in the speed scene use heroin or barbiturates in an instrumental fashion only, drug switching has become a common practice recently. For example, it is not uncommon for one to shoot speed for several weeks, switch to intravenous barbiturates for several days, switch back to speed, then to heroin. Often these drugs are used together, the combination of heroin and amphetamine, for example, being known locally as a "San Francisco Speedball."

The shift from speed to barbiturates or heroin as primary drugs of choice usually occurs over an extended period of time, and is characterized by a marked difference in the way the individual perceives the quality of his life. In order to escape the despair and violence of the speed life, many users turn to heroin as a way to "avoid the hassles."

Although no statistics are available on the percentage of speed users who have switched to heroin as their primary drug, the Drug Treatment Program which developed from the Amphetamine Research Project reports that more than 80% of their clients are young, white heroin addicts, most of whom previously used speed compulsively.

This development in Haight-Ashbury has led the new group of addicts in two distinct directions. Many young people can justify their use of speed and rationalize their reasons for continuing in the speed culture life style, but they find it impossible to justify their entrance into a systematic criminalized scene. Heroin on the West Coast is quite expensive, and a "lightweight" habit costs at least $20 per day. Because many novices push their tolerance to heroin as rapidly as possible (a carryover from their style of speed use), one often encounters users who have gone from a small habit to one costing as much as $300 per day in a matter of six months. Our impression is that at the point where the individual realizes his situation, he will seek help if he regards the available treatment facilities as legitimate.

One problem for middle-class addicts is that they lack hustling skills. Those who attempt to survive in the heroin scene must engage in a variety of criminal activities, often without skill or the connections with legitimate businessmen without whom a good hustler could not exist. This may account for the dramatic rise in the incidence of such low-level crimes as strong-arm or armed robbery, purse-snatching, and shoplifting, all means of securing money to pay for heroin without extensive training or skills. Few "righteous dope fiends" engage in such activities because of the inherent risks and the low rewards.

It is still too early to predict with any

confidence the direction the speed scene will take in the future. One might speculate, however, that as the speed culture becomes more cohesive and increasingly criminalized, a variety of sophisticated "hustles" will develop which are suitable for this style of use and which will provide the novice heroin addict with an understanding of the techniques of successful hustling. Should this occur, the pressure to seek help now felt by many young people who have inadvertently found themselves addicted to heroin will decrease, and we may see new and distinctive routes to long-term heroin addiction.

VIOLENCE IN THE MARKETPLACE OF SPEED

The level of violence in Haight-Ashbury has been increasing since 1967 when the patterns of drug use began to change. In the first two months of 1969, for example, there were 17 murders; most of the victims were young persons identified with the neighborhood drug-using community. April saw the sensational execution of three young people known in the community as "burn artists"; they were found in a kneeling position, hands tied behind their backs, shot in the back of the head. Many other murders have occurred involving well-known drug dealers outside of the immediate neighborhood. In one bizarre case the body of a dealer was found in an adjoining county, his right arm severed from his body.

Police statistics on incidents of assault and attempted murder are poor indicators of the actual level of violence in such a neighborhood, for most of the residents have a hatred or distrust of police and will not report law violations. Most victims are unwilling to file a complaint because their own criminal activity may have precipitated the assault.

A review of some 100 acts of violence which occurred in the neighborhood over a six-month period suggests the following:

1. Acts of violence occur predominantly within a limited geographical area of high drug use.
2. Violence occurs subsequent to a business transaction centering around drugs.
3. Those involved either as assailants or victims are identified within the community as deeply committed to and involved in the use and sale of drugs.
4. The level of drug intoxication, except in clear-cut cases of mistaken identity, does not seem to be a critical variable.

Our observations suggest that most of the violence in the current speed scene results from an interaction between drug effect, lack of social controls within the subculture, and a variety of economic factors, including the way the marketplace is sustained by the individual users.

Many researchers suggest that violence in the speed scene is related primarily to the paranoid psychosis which develops after extended use of the drug.[3] The literature is replete with accounts of violent behavior by individuals using oral amphetamines in isolation, but we know little about the paranoia of individuals who use massive amounts of the drug within the context of a supportive drug culture. As mentioned previously, the speed culture helps individuals to recognize and deal with paranoia. Paranoid behavior is often described by speed users in great detail in "rap sessions," and has become an integral part of the lore of the culture. The term "creative paranoia" is used humorously by some to describe the bizarre actions they took to avert an imagined danger.

Most of the speed users interviewed carried either knives or guns. Stories circulated in the area about the availability of mortars, anti-tank guns, and similar weaponry. While this would suggest that the user arms himself in anticipation of violence, our observations suggest that arms are just part of the trappings of the speed freak. Many of the guns, for example, were without firing pins, or were for some reason inoperable. One user commented:

> The guns and knives are to bolster their ego on the paranoid trip. They want to know they have fire power if they need to use it. Most of the people that have them have ones that don't even work. They'll say, "Here, I'll sell you my .25 automatic, but it hasn't got a firing pin in it."

The usual response of the speed user when he feels paranoid is to avoid the possibility of harm to himself. One user on a lengthy run of speed secreted himself on the roof of an apartment building with an unloaded rifle for several days, watching for red Volkswagen buses which he had decided contained undercover policemen looking for him and his small quantity of speed.

In the course of our research, we interviewed many individuals who used barbiturates and amphetamines in combination. Such individuals are regarded, even by the most devoted speed user, as highly irrational, aggressive, and dangerous. One informant described a close friend who regularly used this combination:

> He's a very nice person and extremely generous; however, when he gets all jacked up and he is wired (stimulated) and doing Reds (Seconals), then he's in trouble. Because pretty quick he's got a shotgun and everybody else has got a shotgun. I've seen him out in front of 404 Oak Street in front of the freeway entrance, herding the hitch-hikers away because he's paranoid of them. At four o'clock in the afternoon, with a full-length shotgun, he's screaming 'move on, you can't stand there, move on.' That's just the way he gets.

Even the most criminalized culture requires control on the behavior of individuals. The violation of certain behavioral or conduct norms within the deviant group usually results in the imposition of sanctions by other members of the group dependent upon the degree of cohesiveness, the extent of agreement regarding appropriate behavior, and the commitment which the individual has made to the particular deviant career.

In the heroin scene, for example, patterns of dealing are relatively predictable and stable. Heroin is given in exchange for cash, the quality of the drug is reasonably assured, and there is an acknowledged interdependence between buyer and seller. In the speed marketplace, however, there are numerous sources of the drug, no regular patterns of distribution, and few standards of conduct defined for participants, who have little or no knowledge of proper conduct within a criminalized scene. Additionally, the speed scene is constantly shifting and the population turns over rapidly. Thus, the control which one sees in a stable neighborhood-based heroin scene is non-existent. Currently, there is a "game" element to the speed scene, and many of the participants view their life styles in terms of the dramatic or theatrical. There is an air of unreality about the growing brutality, and many justify the killing of fellow users in highly simplistic "cowboy ethic" ways.

An individual who compulsively abuses any drug, is committed to a deviant group, estranged and alienated from conventional society, and has few options in terms of economic survival. He must develop a method, usually illegal, for making money to continue his use of drugs, while functioning within the life style peculiar to his group. The classic example of this is the heroin addict, the majority of whom resort to a variety of money crimes in order to maintain their habits. The peculiar hustle which the addict employs depends upon the opportunities which exist within his community for illegal activities, the skill and training which the hustle requires, the commitment which is demanded, and the risks the addict is willing to assume.

A good hustler in the heroin scene usually has the ability to engage in a variety of illegal schemes where the risk is minimal and the payoff substantial. The risk of arrest increases with the frequency of the activity, as for example, shoplifting. The most apparent impediment to successful illicit activity in the speed scene is the lack of appropriate skills by the participants, most of whom are from fairly affluent, middle-class backgrounds where such skills are not learned.

Many of the hustles engaged in by heroin addicts require a certain "front," wherein one assumes a role in order to successfully execute the hustle. A good "check artist" or credit-card passer must be convincing in his appearance if he is to be successful. Few speed freaks, because of their appearance and physical demeanor, can pass checks or credit cards without arousing suspicion. Indeed, they have difficulty using legally acquired checks or credit cards because of their appearance, a situation that is often used to justify their view of the "uptight establishment."

Although burglary is the main hustle for the heroin addict, few speed users are able to burglarize because they lack skills, connections, and the knowledge of how to convert stolen goods into money. Moreover, their suspicions and paranoia make it almost impossible for them to enter a house successfully, as they often fear that someone has "tipped off" the owner who may be waiting for them.

For most young speed users, hustling involves selling small quantities of speed and other drugs on the street level—a practice which is unprofitable and dangerous, because of the practice of "ripping off" or stealing drugs from fellow users, and also because of the imminent danger of arrest.

Compulsive users of speed are the most vulnerable dealers on the scene, for they often use or give away most of their supply, and must resort to "burning" or selling adulterated drugs in order to survive. The speed culture suggests to participants that revenge is not only justified but necessary if one is "burned." Thus, many of the murders and assaults which have taken place in the speed scene involve known "burn artists," who are for the most part unsophisticated, compulsive users who have no alternatives for economic survival. One method for disposing of a burn artist who sells particularly toxic, if not lethal, substances on the street is to inject him with his own product. Many of the deaths in drug subcultures which are officially regarded as drug overdoses are, in fact, murder.

In the heroin community, the brunt of the hustling activities is borne by the non-using population. In the speed scene, most of the criminality is directed toward other members of the drug scene. The speed scene differs from other drug scenes in that the money needed to sustain the marketplace is not generated with any regularity by the individual user who is committed to the life style centering around speed use. Unable to generate money outside the community, the culture has turned on itself, creating a climate of fear, suspicion, and violence which shows little sign of abating. The only exception to the rule is the steadily growing number of previously unsophisticated young persons who become integrated into other deviant groups and learn techniques of hustling and survival from more experienced and sophisticated individuals. Should this trend continue and involve increasing numbers of criminally unsophisticated speed users, the entire complexion of the marketplace may be altered in the direction of the heroin scene; the criminalization of the speed scene would probably also lead to the development of a system where reliable and trustworthy individuals will be screened or selected out for elevation to higher status positions within the marketplace, while others will be excluded entirely. One might speculate that, with a stabilized marketplace, the level of violence will decrease significantly, while other types of secondary criminal activity will increase rapidly.

CONCLUSION

The Haight-Ashbury district since 1967 has witnessed the emergence of a unique drug scene, involving relatively unsophisticated middle-class individuals caught up in a highly criminalized situation. Lacking appropriate experiences and skills, the compulsive drug-users have turned on each other, leading to the development of a highly unstable, transient, and violent community.

As might be anticipated, society has responded to this new problem of dangerous drug abuse by increasing the legal sanctions and making possession a possible felony offense, on both the federal and state level. While the intent of this legislation is clearly to minimize the impact of the law on the possessor while attempting to control the manufacture or distribution, in practice the brunt of the law is borne by the low-level street user. Increasing the penalties has increased both profits on the upper levels of the marketplace and the violence on the street level. This would suggest that an educational and therapeutic approach may be more effective than a punitive one, both immediately and in the long run, in controlling the spread of this drug and the violence which almost certainly follows its abuse.

REFERENCES

1. Davis, Fred: Heads and freaks: patterns and meanings of drug use among hippies. J Health Soc Behav *9*:156-164, 1968.
2. Becker, Howard S.: History, culture, and subjective experience: an exploration of the social bases of drug-induced experiences. J Health Soc Behav *8*:163-176, 1967.
3. Ellinwood, Everett H.: Assault and homicide associated with amphetamine abuse. A paper presented to the American Psychiatric Association, May 1970, San Francisco, California.

THE ROLE OF INTERNATIONAL LAW IN CURBING THE ILLICIT DRUG TRAFFIC

JOHN E. INGERSOLL, A.B.

INTRODUCTION AND HISTORY

Concern about the use and abuse of narcotics and dangerous drugs has been reflected in the writings of man for thousands of years.[1] However, it has only been during the past century that meaningful efforts have been made to attempt to formulate some means of international agreement and control over the problem. As world travel facilities improved and accelerated, international communication and concern about drug abuse began to spread. The increasing magnitude of the dangerous abuse of drugs has led the world community, for the most part, to initially acknowledge the need to control these substances and seek methods to inhibit their illegal trafficking and distribution. Early bilateral agreements began to seek some international control of the trade in addicting opium.* Various governments in the world recognized and asserted that there existed a responsibility for acting in concert on an international level in order to protect individual members of each nation.[2] The United States, China, and Great Britain initiated the first multilateral convention concerned with drug abuse, called the Opium Convention, which met in Shanghai in 1909. The delegates of the thirteen nations represented had no power to draw up or sign a treaty, but met only to discuss the problem of opium use, especially in the Far East. Several resolutions were adopted, and plans were made for future conferences designed to take more definitive steps toward effective international action.[3]

The first international narcotics convention was signed at the Hague in 1912.[4] It established the basis for subsequent international agreements by enunciating a number of principles for the control of narcotic drug production and trade. It stressed that production should be limited to medical and other "legitimate" needs. Governments that became parties to the treaty were to develop permit and record-keeping requirements. Further efforts toward international control were suspended until after World War I, when the League of Nations established an Advisory Committee on Traffic in Opium and Dangerous Drugs.[5] The Committee met regularly to initiate further action on an international scale.

Subsequent international conventions expanded on the system of record-keeping for imports and exports. The Geneva Convention of 1925 resulted in the International Opium Convention of 1925,[6] which required

* For example, the Siam Opium Treaties of 1833 and 1856, the Treaty of Wang Hea of 1844, the Tientsin Treaty of 1858, and the Korean Treaty of 1882.

member nations to submit annual statistics on narcotic drug and opium production, as well as reports on the importation and exportation of these substances. The subsequent Geneva Convention of 1931 produced the International Convention for Limiting the Manufacture and Regulating the Distribution of Narcotic Drugs,[7] which created a new system of international control. Each nation was to make an annual estimate of its anticipated legitimate narcotic needs for medical and scientific purposes. An international supervisory committee would analyze the estimates and publish acceptable estimates for every nation, even those that were not party to the 1931 Convention or had not provided estimates. Imports and exports of narcotic drugs were to thereafter be limited to the set amounts, with an embargo power by the supervisory committee intended to serve as the power of enforcement, in case the limits were violated. This plan still serves as the basis of existing international control and supervision over legitimate narcotic drug distribution, and functions fairly efficiently in that capacity. However, this plan fails to even begin to affect the illicit trafficking, because it only effectively concerns itself with legitimate distribution. This weakness quickly became apparent and concerned negotiators tried to correct the problem at the 1936 conference. The resulting Convention[8] attempted to require parties to adopt specific criminal procedures and penalties for narcotic drug offenses, and it proposed a form of international jurisdiction for violations in this field. As a result, many nations retreated into the protective shell of nationalism and refused to ratify the Convention, primarily because it would seemingly create an international criminal law, and ostensibly encroach on national sovereign powers.

During these years of treaty development, the clandestine manufacture and illegal distribution of narcotic drugs spread and increased in severity. After World War II, the United Nations took over international supervisory functions and created the Commission on Narcotic Drugs.[9] The Paris Protocol of 1948[10] expanded earlier conventions so that the newly developed synthetic addiction-producing drugs could be subjected to appropriate controls. Nevertheless, the problem of illegal production continued to plague and undercut international efforts.

In 1953, another Protocol[11] was drawn up in an effort to achieve a method of limiting opium production to that amount actually needed for medical and scientific purposes. It specifically listed seven nations which would be authorized to produce opium for export. Each of these nations is supposed to enforce its own national regulatory controls so that the annual harvest will meet, but not exceed, legitimate world needs. Three of these seven nations had to accede to this Protocol before it could come into effect, and the third nation did so in 1963. However, three of the listed nations still are not party to the Protocol. Despite that fact, an attempt was made to limit world production, and the basis was built for the most recent multilateral treaty on narcotic drugs. Through the years of efforts and the different international agreements, a reasonably acceptable control of legitimate production and distribution of narcotic and addiction-producing drugs was obtained, and these have culminated in the 1961 Single Convention on Narcotic Drugs. However, problems still remain.

THE SINGLE CONVENTION OF 1961

In March of 1961, the Commission on Narcotic Drugs presented a new Convention for ratification which was intended to replace the existing treaties, expand international control somewhat, and simplify the control procedures. In December of 1964, the Single Convention on Narcotic Drugs, 1961, entered into force.[12] This Convention affirmatively outlaws the production, manufacture, and distribution of all narcotic substances for non-medical purposes (except for certain limited transitional purposes). Possession of such substances is also limited to certain persons and certain circumstances. The Single Convention divides the controlled drugs into four schedules according to specific criteria, such as the liability of the drug to abuse and its addiction-producing properties. Control of the drug depends on its classification, with varying licensing and production provisions. Import and export authorizations, labeling, and record-keeping controls have been instituted, and the estimate system for production and statistic collecting requirements have been retained. The enforcement mechanisms remain essentially the same as under treaties, relying primarily on the estimate form of production and trade control, requiring parties to take internal regulation measures and to keep detailed records. A new supervisory body was created—the International Narcotics Control Board (INCB), which supervises the administration of the treaty, administering the estimate system and the statistical returns. The Commission on Narcotic Drugs still functions, of course, to amend the schedules, research the international drug situation, and work on new treaties or protocols.

The Single Convention continues to rely on international "pressure" in order to enforce its provisions. National governments must cooperate, especially in their internal controls, in order to make the Convention effective. Economic embargo is still available as a pressure technique, but is rarely used. Despite this glaring weakness, international efforts have proved surprisingly effective at achieving control over the *legitimate* trade of narcotic drugs. International cooperation regarding import-export controls and statistical reporting has achieved sufficient regulation of the legitimate world distribution. More than 75 nations are currently parties to the Convention. The amounts of narcotic drugs legally manufactured now approximate the world's legitimate consumption.[13]

The problem which has not been solved, or even appreciably affected by international efforts, is that of the illicit production and distribution of narcotic and dangerous drugs.[14] Although the Single Convention of 1961 provides that the parties shall adopt internal measures that will ensure that certain narcotic drug offenses will be subject to criminal punishments, and provides that listed narcotic offenses will be considered extraditable crimes, these "requirements" still carry only the strength of suggestions, because they are subject to the constitutional and legal limitations of each nation's domestic law. Therefore, there is still no uniformity of criminal laws or procedure among the nations, and illicit cultivation, production, manufacture, and distribution still present the most serious problem for international narcotic and dangerous drug control. Clandestine factories and laboratories are becoming more productive as the criminal elements become better organized and more sophisticated in their international activities. The nations of the world are finding themselves faced with a growing problem of drug abuse, even in areas where such activities have never before been a problem.[14]

CONTROL OF DANGEROUS NON-NARCOTIC DRUGS

In addition to a lack of effective coordinated efforts in controlling illicit narcotic drug activities, there currently exists a vacuum internationally with respect to controls over dangerous non-narcotic drugs such as the stimulants, sedatives, tranquilizers, and hallucinogens. The misuse and abuse of these drugs, collectively known as psychotropic substances, are a growing matter of anxiety and concern for world governments and international organizations.[15] Recently, new efforts have been made to obtain international controls over these substances. In January, 1970, the United Nations Commission on Narcotic Drugs met in Geneva and worked out a Revised Draft Protocol on Psychotropic Substances.[16] This Protocol establishes five schedules of substances to be controlled, then applies progressive degrees of regulation to the manufacture, production, and international distribution of the substances. The Protocol requires parties to institute internal controls to limit the use of the substances to medical and scientific purposes, to require prescriptions for some drugs, and to tightly control production, export, and import. Parties to the Protocol must also institute strict record-keeping requirements internally as well as make required reports to the International Narcotic Control Board. Education, rehabilitation, and treatment programs are to be instituted by each nation, and all are to cooperate in combating illicit traffic. Member nations are to adopt penal measures for offenses and recognize such crimes as extraditable offenses; however, this requirement is subject to each nation's constitutional limitations. In short, the theory underlying this Protocol is quite similar to that of the Single Convention.

The representatives at the meeting of the Commission adopted the revised text of the Draft Protocol, and a plenipotentiary conference of the United Nations considered the Protocol early in 1971. The Protocol was signed by twenty nations including the United States without too much revision. Several nations have been most vigorous in their support of this new treaty because of the recent growth of abuse of psychotropic drugs within their countries.[14]

Despite the fact that this new Protocol will probably be ratified by the required number of parties and will enter into effect, and will eventually establish some international regulation over substances previously uncontrolled, the same weaknesses will be inherent in this treaty as are true for existing treaties in this field. As a general rule, international treaties only act as guidelines, because there are no effective methods of enforcement. Until the nations of the world agree to subject themselves to some real international control, or the majority of the governments begin to take effective internal measures, illicit production and trafficking of dangerous substances will continue.[17]

INTERNATIONAL POLICE COOPERATION

One important role of international law in curbing drug traffic is the cooperative efforts among the law enforcement bodies. Although there is, of course, no actual international criminal law or international police force, cooperation among the enforcement agencies of the various nations has led to some suppression of illicit trafficking of narcotic and dangerous drugs. Such cooperation is derived from shared concern and moral responsibility only, not from any binding obligations by treaty.

The International Criminal Police Organization (INTERPOL) is a prime example of police cooperation on an international level.[18] Made up of voluntary member nations that pay annual dues to support operational costs, INTERPOL functions as a central information-gathering organization and a communications facility. By maintaining information concerning known international law-violators and facilitating investigations and searches among separate nations, INTERPOL assists police efforts in tracing and capturing narcotics traffickers. Each member nation maintains a National Control Bureau (NCB), usually with radiocommunications facilities. Each NCB may deal directly with another member, or through INTERPOL headquarters in France. Cooperation is excellent, especially in tracing fugitives. Currently, there are over one hundred member nations.

INTERPOL underlines the importance of police efforts in combating international narcotics trafficking. It sends a representative to the various international meetings concerned with drug abuse and publishes position papers encouraging stronger international action in this area.[19] However, INTERPOL itself is not a law enforcement body, and has no investigative or arrest powers. Its entire function rests on the cooperation of the police organizations of each independent member. INTERPOL does not investigate possible treaty violations, nor does it directly act to enforce such treaties; however, it does facilitate investigations by its member nations of international criminal activity.

Another example of cooperation among enforcement agencies is reflected in recent informal agreements between Turkey and the United States and between France and the United States. These agreements provide for greater bilateral cooperation in law enforcement efforts. A formal written agreement between the United States and France designed to coordinate law enforcement activities was signed in Paris on February 26, 1971.

Despite this level of cooperation among nations in law enforcement activities, illicit narcotic and dangerous drug cultivation, production, and distribution continue to increase at an alarming rate. Substances grown or manufactured in one nation are smuggled to another, where they are refined in clandestine laboratories, then distributed to any number of nations for final sale to victims suffering from various stages of addiction or drug dependency. The non-medical consumption of these products continues to proliferate, and criminal activity expands proportionately on the international level. Clearly, further investigative and enforcement activities on this level are vital to combat these sophisticated criminal conspiracies. Greater cooperation among nations, more exchange of information, extensive police training and education, and speedier extradition of criminals are just a few initial improvements which must be made within the very near future.

BILATERAL TREATIES AND AGREEMENTS

Recently, a number of nations have accelerated efforts to develop stronger bilateral programs to combat illegal drug trafficking. Negotiations are resulting in new treaties, amendments of old treaties, and less formal bilateral agreements, all generally aimed at more effective cooperation in matters of criminal investigation and procedures.

Among the most important of these new treaties and amendments of existing treaties are those which modernize extradition for narcotic and psychotropic drug law offenders. Many of the existing extradition treaties do not include narcotic law offenses as extraditable crimes, much less violations of psychotropic drug laws. In view of the serious nature of these crimes and their potential for vast effect on all the societies and cultures of the world, expeditious treatment of the criminals involved is most essential. Reliable and prompt extradition of criminal fugitives coupled with just punishment will certainly act as a strong deterrent of international criminal activities.[20] When extradition of a country's own nationals is not acceptable because of constitutional or internal legal limitations, then some arrangement must be made to ensure that criminal acts will be punished. Those nations with extra-territorial jurisdiction over criminal acts committed by their own nationals must amplify their efforts to gather evidence of crimes committed abroad and bring the perpetrators to justice.

Existing extradition treaties which do not effectively apply to modern drug crimes must be updated as quickly as possible to include the sophisticated conspiracies, laboratory operations, and trafficking systems for all types of narcotics and dangerous drugs. They should be expanded to include attempted crimes as well. Loopholes in treaty arrangements too frequently permit criminal activity to continue unhampered. Returning the accused or fugitive criminal to the jurisdiction where the illegal activity took place is by far the best method of ensuring effective punishment. When extradition is absolutely impossible, some other method of law enforcement must be applied.

Treaties of mutual assistance in criminal matters are becoming more popular, especially between nations which suffer repeated criminal activity between their countries.[21] Treaties of this kind provide for reciprocal assistance between the parties in investigating and prosecuting criminal cases. Usually they facilitate the exchange of laws and information, provide access to courts and administrative bodies, and permit the taking of depositions and the use of other investigatory procedures to gather evidence. Such bilateral treaties will clearly aid in criminal investigations which involve potential defendants operating in two or more nations, especially when that activity is well organized and only conspiracy evidence is available against the higher echelon members. More bilateral treaties facilitating extraterritorial criminal investigations are desperately needed, and every concerned nation should be studying the advisability of such treaties.

In addition to treaty arrangements, many nations are making less formal bilateral agreements to deal more specifically with particular problems mutually affecting their nations. A recent example of such is the agreement between Mexico and the United States relating to the suppression of smuggling of illicit narcotics, marihuana, and other dangerous drugs over the common border.[22] The two nations agreed to work together to intensify enforcement efforts on the border. Some financial aid was granted to Mexico by the United States, and a cooperative police-training program was devised. The agreement has already resulted in increased successful enforcement activities along the border. Further high-level government consultations have been held under the continuing bilateral program,[23] and future cooperative efforts are anticipated.

Negotiations for bilateral agreements concerning narcotic and dangerous drug law enforcement are currently being held between the United States and France and between the United States and Turkey. Plans are being made to strengthen law enforcement programs through financial aid arrangements and police-training seminars.

Bilateral activities and agreements of this nature clearly will improve the cooperative efforts to combat illegal international trafficking, and therefore play a significant role in international law. However, these arrangements only involve a few countries, and the problem of illicit international drug production and trafficking is one which definitely involves the entire world community. As bilateral efforts "dry up" certain geographic areas of criminal activity, new resources will be devised and new conspiracies will develop in other nations. Until all the nations of the world decide to forthrightly attack this problem, some illicit drug trafficking will continue on an international level. Therefore, bilateral treaties and agreements represent only a limited approach to a broad problem.

UNILATERAL DEVELOPMENTS

Many nations have begun to strengthen their internal laws and enforcement systems in order to better combat the spreading drug problem. Some have even taken fairly extensive unilateral steps to stop international trafficking.

The United States has just enacted a new Federal law entitled the Comprehensive Drug Abuse Prevention and Control Act of 1970.[24] This new law emphasizes enforcement activities against the traffickers of controlled dangerous substances and tightens national regulation in the legitimate manufacturing and distribution of these substances. Additionally, the law makes provisions for expanded drug education, rehabilitation, and research activities.

The new United States law strictly regulates importation, exportation, and labeling of controlled substances, including some psychotropic drugs previously not so tightly regulated.[24(sec. 305–306, 1001–1016)] These provisions meet and exceed requirements imposed by international obligations of the United States.

The United States has taken other unilateral action toward international trafficking of dangerous drugs such as "Operation Intercept," a border search program instituted in September of 1969. This operation was designed to intensify searches of persons crossing the international borders into the United States, especially across the Mexican-United States border, in an attempt to severely cut back the illegal flow of drugs into the United States. The plan was successful in many ways, but was extremely disruptive to the normally extensive flow of persons and commerce across that border, resulting in economic disadvantages in the area, especially to Mexican border cities which are heavily dependent on the American tourist trade. Therefore, the Mexican and United States governments began a new bilateral program, designated as "Operation Cooperation." This program is intended to suppress smuggling at the border and to cooperatively investigate and eliminate organized criminal rings operating in that area.

Other nations have also adopted unilateral programs aimed at striking down international trafficking. Lebanon, for example, has been making attempts to replant all hashish (cannabis) fields with sunflowers or useful crops. India has instituted an incentive licensing program for legitimate opium growers which helps eliminate diversion of the crops into the illegal market.[14]

However, despite the laudable efforts of individual nations, the problem of illegal narcotic and dangerous drug production and distribution is an international one, and separate, divided efforts alone cannot solve it.

CONCLUSION

The spiraling increase in drug abuse is spreading throughout the world. Nations which historically have been plagued by traditional opium abuse such as by smoking preparations extracted from the opium poppy or eating the oily seeds, now are discovering a growing use of refined heroin. Other nations which have been free of any serious drug problem in the past now face a frightening misuse of stimulants and depressants by their young people.[3(p.195)] The acceptance of the new Revised Draft Protocol on Psychotropic Substances[16(pp.12–37)] by the U.N. Commission on Narcotic Drugs is a clear indication that many nations desire effective international treaties to help combat the problem.

Nevertheless, without a concerted effort to use international law as a real tool, treaties and protocols will remain ineffective and meaningless. More and better bilateral agreements, especially to improve and modernize extradition agreements, will certainly improve criminal law enforcement. Moreover, genuine application and enforcement of existing international treaties will contribute to effective control of the problem.

If the licensing, production, cultivation, and import-export control provisions of the 1961 Single Convention[12] are strongly enforced by the International Narcotics Control Board, some illicit diversion of narcotic drugs will cease. When the Board has sufficient reason to believe that any Party is violating the Convention, then it should immediately call on that Party to take remedial measures, and if such are not taken, it should seriously consider recommending an embargo against the violator. Training and improvement of member nation's law enforcement officials under U.N. supervision should be expanded. Drug abuse education and rehabilitation programs on an international level should be pursued, and statistical gathering and research efforts should be further developed.

In addition, new treaties are needed to provide a system of enforceable international controls to totally eliminate the illicit cultivation and production of narcotic and dangerous drugs. Serious consideration must be given to totally eliminating *all* opium production throughout the world, since effective synthetic substitutes can now be inexpensively produced.

The voluntary individual efforts of treaty members on a national level are no longer sufficient international requirements. More authority must be placed in an international organization to gather information, set production and distribution quotas, investigate possible violations, and take effective remedial action to control any violations. Research and education programs should be expanded and modernized, as every conceivable international tool must be applied to the problem.

Obviously, such intensified international efforts are going to require a new level of cooperation among the nations of the world. Significant funding will be necessary to institute meaningful programs. Serious enforcement efforts will have to take precedence over political differences, and genuine united participation is essential.

Significant cooperation among the international community in criminal law enforcement can reduce and perhaps eliminate illicit trafficking of dangerous substances. Improved and expedited extradition procedures, more sophisticated investigation techniques, and a united policing effort can lead to more efficient administration of justice in criminal matters.

After more than 60 years of analysis, treaty negotiations, and a great deal of primarily ineffectual rhetoric, it is definitely time to begin to make truly effective strides in combating illegal drug production and distribution. Many more delays, and the problem may conceivably begin eroding even more the national infra-structures to the point where social stability is endangered. Without question, the nations of the world must now demonstrate their sophistication and development by joining in meaningful international discipline, through law and action.

REFERENCES

1. Lewin, L.: *Phantastica, Narcotic and Stimulating Drugs, Their Use and Abuse.* Translated from the 2nd German edition by P. H. A. Wirth. New York, Dutton, 1931.
2. Resolution of the House of Commons, May 30, 1906, Parliamentary Debates, H. of C., Vol. 158, p. 494; historical materials reproduced in Lowes, P. P.: *The Genesis of International Narcotics Control.* Geneva, Droz, 1966.
3. Glatt, M. M.: The development of international control of drugs. *W. H. O. Chronicle 24* (No. 5): 189-197, 1970.
4. International Opium Convention, The Hague (Jan. 23, 1912), 38 U.S. Stat. 1912, 8 LNTS 187, 6 AJIL Supp. 177 (1912).
5. "Review of the Commission's Work during Its First Twenty Years." United Nations Economic and Social Council, Commission on Narcotic Drugs (1965), New York, United Nations Doc. E/OB/23-E/DSB/25.
6. International Opium Convention, signed at Geneva, Feb. 19, 1925, 81 LNTS 317, 23 AJIL Supp. 135 (1929).
7. International Convention for Limiting the Manufacture and Regulating the Distribution of Narcotic Drugs, signed at Geneva, July 13, 1931, 139 LNTS 301, 48 Stat. 1543, 28 AJIL Supp. 21 (1934).
8. Convention of 1936 for the Suppression of the Illicit Traffic in Dangerous Drugs, signed at Geneva, Feb. 11, 1936, 198 LNTS 229.
9. 1946 Protocol, 12 UNTS 180, TIAS 1671 and 1859.
10. Protocol signed at Paris, 1948, 44 UNTS 277, TIAS 2308.
11. Protocol for Limiting and Regulating the Cultivation of the Poppy Plant, the Production of International and Wholesale Trade in, and the Use of, Opium, signed at New York on June 23, 1953, U.N. Doc. E/N. T./8.
12. The Single Convention on Narcotic Drugs of 1961, U.N. Doc. E/Conf./31/22; TIAS 6298, 520 UNTS 204, 18 UST 1407.
13. Steinig, Leon: International Narcotics Control Board, Rapporteur, International Narcotics Control.
14. Hughes, John: *Onus for Antinarcotic Action Rests on Individual Governments.* Reprinted in remarks by the Hon. Howard W. Robison, "Curbing Narcotics Trafficking," in the Congressional Record (June 26, 1970) p. E6011.
15. "The Control of Psychotropic Substances not under International Control." United Nations, E/4606/E/CN.7/523, pp. 73-96.
16. United Nations: Commission on Narcotic Drugs, Report of the First Special Session (Jan. 12-30, 1970), Supp. No. 8, E/4785, E/CN.7/528.

17. For some United States proposals for such measures, see the Statement of John E. Ingersoll, United Stated Representative to the United Nations Commission on Narcotic Drugs, Press Release, U.S. Information Service (Sept. 28, 1970), Geneva.
18. INTERPOL: International Criminal Police Organization. *RCMP Gazette 32* (Nos. 7 & 8): 13-17, 1970.
19. The United Nations Commission on Narcotic Drugs, Int. Crim. Pol. Rev., Vol. 234 (Jan., 1970), pp. 19-21.
20. For the United States statutes which enact extradition treaties, see 18 U.S.C. 3181 et seq.
21. See generally: Whiteman, Marjorie M.: *Digest of International Law,* Vol. 6. U.S. Dept. of State Publication (Dec. 1968), pp. 183-229.
22. Agreement between the Bureau of Narcotics and Dangerous Drugs of the Department of Justice of the Government of the United States of America and the Office of the Attorney General of the Republic of the Government of the United Mexican States (Mexico City, March 5, 1970), 4 pages.
23. Report of the United States Delegation relative to the Cabinet-Level Consultations between Officials of the United States and Mexico on the Control of Narcotics, Marihuana, and Other Dangerous Drugs; Attorney General John N. Mitchell, Chairman of the Delegation; Submitted to the Secretary of State (Washington, D.C., March, 1970), 7 pages.
24. Comprehensive Drug Abuse Prevention and Control Act of 1970, Public Law 91-513, 91st Congress, H.R. 18583, 81 Stat. 1236 (Oct. 27, 1970).

DRUG ABUSE CONTROLS IN COMPARATIVE AND INTERNATIONAL LAW*

B. J. GEORGE, Jr., J.D.

Abuse of drugs, like excessive use of alcoholic beverages, has long been a major problem in the United States.[1] Indeed, the evidence is substantial that, in terms of numbers of addicts, the United States leads the world.[2] But it is also clear that every country in the world has an addiction problem large enough to cause its government concern.[3] Therefore, the process of examining the legal aspects of drug abuse can only be facilitated by looking to see how other legal systems attempt to cope with the problem of addiction and the drug traffic that feeds upon it, and how the inability to cope with international drug traffic through domestic law controls has led to increasing international efforts through treaties and cooperation in law enforcement.[4] As in any instance of comparative legal study, observation of another legal system in operation enables one to analyze his own system much more effectively.

OUTLINE OF LEGAL CONTROLS OF DRUG ABUSE IN THE UNITED STATES

There have been thus far three major periods in drug control legislation in the United States. Before the first decade of this century there were no comprehensive controls over the traffic in drugs. Aside from a few efforts at controls in the form of customs duties on importation of opium, anyone was legally free to dispense or obtain narcotics. Many persons became unwittingly addicted to patent medicines with a substantial opiate content.[5] However, the growing concern both nationally and internationally about the human cost of free use of narcotics led to major legislative regulation.

Federal Legislation

The first major federal statute was the Harrison Narcotic Law of 1914, later transferred to the Internal Revenue Code.[6] In form, control was achieved through a tax on the importation, manufacture, processing, buying, selling, dispensing, and giving away of opium, cocaine, and their various compounds, derivatives, and preparations. When the tax was paid, the container was sealed with a tax stamp. Possession of narcotics in unstamped containers would be evidence of non-payment of the tax, and therefore of a crime. All transactions in the narcotics covered by the Harrison Act were also to be based on orders written on forms supplied by the Bureau of Narcotics and Dangerous

* In this paper a "j" is employed instead of an "h" in "marijuana" to conform to the current legal and linguistic style.

Drugs. Transactions not so documented also were criminal.

Other legislation closely related to the Harrison Act included the Narcotic Drug Import and Export Act,[7] controlling the importation of crude opium and coca leaves; the Narcotics Information Act,[8] rewarding those who gave information about violations of the narcotics laws; the Narcotic Transportation Act,[9] making it criminal to use vehicles, vessels, or aircraft to transport or conceal narcotic drugs and providing for the forfeiture of means of transportation so used; and the Opium Poppy Control Act,[10] requiring a license for cultivation of the opium poppy and penalizing growth of poppy plants without a license.[11]

The Harrison Act did not cover marijuana and the so-called psychotropic drugs such as barbiturates and other central nervous system depressants, amphetamines, and LSD. Rather than amend the Harrison Act to include these drugs, Congress chose to deal with their misuse through special legislation. The oldest was the Marihuana Tax Act,[12] passed in 1937 and also found in the Internal Revenue Code because of its use of taxation as its principal control method.[13] It imposed an occupational tax on those producing, processing, or transferring any part of the cannabis plant whether growing or not, or any preparation, compound, or derivative of the plant. Transfers were taxed and were to be documented on official order forms. In form, marijuana might be lawfully prescribed by registered medical practitioners, although it has not in fact been so prescribed (aside from a few officially approved experiments to show the effects of marijuana use). Thus, with very few exceptions any transaction involving marijuana was a federal offense.

Opium, cocaine, marijuana and their derivatives have little or no modern medical use, or else are used under circumstances that permit control through taxation and registration. However, the wide array of depressant and stimulant drugs, the volume of their production, and the broad range of their legitimate use make the control devices in the narcotic and marijuana legislation inappropriate for dangerous drugs. Therefore, in 1965 Congress amended the Federal Food, Drug and Cosmetic Act to provide the necessary legal controls.[14] Enforcement of these provisions was assigned to the Department of Health, Education, and Welfare, and was achieved primarily through an administrative listing of depressant or stimulant drugs found upon investigation to be habit-forming because of their effect on the central nervous system. Because of the breadth of the legal definitions in the statute, hallucinogens, including LSD, were also brought within this statute. The chief means of control was through registration of manufacturers and dealers in the listed drugs. This system was not imposed on practicing physicians who did not in effect dispense their own drugs to patients for compensation. Simple possession by the patient was not prohibited, provided he received the drug from, or on a valid prescription provided by, a licensed practitioner.[15] Otherwise, nobody other than a registered producer, wholesale or retail dispenser, medical practitioner, researcher, government official, a supervised employee of one of these, or a common carrier or warehouseman might have any of the listed drugs for sale, delivery, or other disposal to another.[16] All violations of the act were, of course, criminal.[17]

Hospitalization of drug abusers, as contrasted with their penal incarceration, has never been a major feature of American, and particularly federal control of drug abuse. There is, however, the Narcotic Hospital Law,[18] under which federal hospitals have been established at Lexington, Kentucky, and Ft. Worth, Texas. The statute provides for voluntary commitments, but most of these apparently arise because of a pending criminal prosecution, and persons whose commitment is voluntary seem for most purposes to join the imprisoned or paroled convicts whose commitment is involuntary. Moreover, even a "voluntary" patient may be held for the maximum time the Surgeon General estimates to be necessary to effect a cure, unless he ceases to be an addict sooner[19] (hardly the epitome of voluntariness). Official United States policy has been against other than "closed" institutions for the treatment of addicts.[20]

The third era has barely begun. On October 27, 1970, the President signed into law the Comprehensive Drug Abuse Prevention and Control Act of 1970, which is a new effort to bring together all of the federal legislation controlling narcotics and dangerous drugs. Many of the provisions of the Act went into effect immediately, while for others there was a seven-month transitional period.[21] The Act is divided into two principal titles, Title II, to be known as the "Controlled Substances Act," and Title III, to be known as the "Controlled Substances Import and Export Act." Title II is in effect the regulatory statute for domestic control, while Title III governs the international traffic, to the control of which the United States is committed as a signatory to the Single Convention.[22]

The approach to control is closer to the 1965 amendments to the Federal Food, Drug and Cosmetic Act and to the Single Convention than it is to the earlier tax-controlled system. Section 101 of the Act contains the findings of Congress that although many of the drugs covered in the title "have a useful and legitimate medical purpose and are necessary to maintain the health and general welfare of the American people," the illegal flow through international and interstate commerce and the resultant illegal usage have a detrimental impact on the welfare of the people. Because of the intimate connection between interstate traffic and misuse within a state, the two aspects cannot be separated, and therefore it is necessary to control intrastate, as well as interstate, dealing in drugs. Thus, the new statute is to be the single comprehensive source of regulation.

To achieve the necessary control, the act contains schedules of controlled substances,[23] which may be added to by action of the Attorney General on the recommendation of the Secretary of Health, Education, and Welfare. Schedules I and II by and large pertain to the most dangerous substances such as heroin and opium, as well as LSD and marijuana, while Schedules III, IV and V encompass the various amphetamines, barbiturates, and other synthetic drugs in compounds and preparations. If something on these schedules contains medical ingredients not having a depressant or stimulant effect on the central nervous system and these ingredients are used in a way to vitiate the "potential for abuse" that the listed substances have, the Attorney General may by regulation except it from the schedule.[24]

All persons and firms who manufacture, distribute, or dispense a controlled substance or propose to do so must register with the Attorney General under regulations to be promulgated. The registration process is to be handled so as to keep criminal elements out and to prevent diversion of the controlled substances in question into illicit channels.[25] Various labeling and packaging requirements are also imposed.[26] Those who are registered then have imposed upon them quotas as far as Schedule I and II substances are concerned.[27] Registrants are also to maintain necessary records and make required reports,[28] and distribution must be documented on official order forms.[29] Prescriptions to the patient are also controlled, the extent of regulation depending on the schedule in which the drug in question is to be found.[30]

Substantial reliance is still placed on criminal penalties to enforce the system. The principal changes are to reduce the penalties for a first-offender possessor to the misdemeanor level and to make him eligible for probation,[31] to augment the penalties against repeated violators,[32] to define and punish heavily those who engage in a "continuing criminal enterprise,"[33] and to provide for special sentencing procedures for "dangerous special drug offenders,"[34] the latter being a special form of recidivist statute based on prior narcotics felony convictions, whether state or federal.

The basic system for controlling international traffic in Title III is similar to that in Title II, and depends on the same schedules.

The Act also contains a number of procedural revisions in matters such as search warrants,[35] administrative inspections,[36] immunity against self-incrimination,[37] and injunctive relief against violations of the statute.[38] It makes some provision for experiments in drug-abuse education[39] and special treatment regimens for addicts and drug-dependent persons,[40] authorizes the Secretary of Health, Education, and Welfare to determine appropriate methods of medical practice in the treatment of addicts and report them periodically to Congress,[41] and broadens the categories of persons eligible for treatment in Public Health Service hospitals.[42] The necessary repeals and amendments are included to make the new Act the single source of federal legal controls over the substances controlled.[43]

State Legislation

The ultimate role of state law-enforcement agencies in the control of the narcotics traffic will be determined in the future. During the years between the enactment of the Harrison Act and other federal statutes, and the new Comprehensive Drug Abuse Prevention and Control Act of 1970, much of the burden of coping with the narcotics traffic fell on the states, if for no other reason than that the federal legislation was piecemeal and not comprehensive. Most states have adopted the Uniform Narcotic Drug Act, which closely resembles the original federal legislative scheme of controlling narcotic drugs and marijuana. The principal additional areas covered are the control of premises or vehicles resorted to by addicts for the purpose of using drugs,[44] and the use of forgery, fraud, and deceit to obtain narcotics.[45] There is, however, very little uniformity when it comes to control of the so-called dangerous drugs. Although adoption of the Model State Drug Abuse Control Act has been urged at the federal level,[46] there has not yet been any great trend in that direction, so that there is great disparity in the degree to which and the methods by which dangerous drugs are controlled in the various states.[47]

Penalties in general are quite harsh, including the use of mandatory minimum terms in sale cases in some jurisdictions.[48]

LEGAL CONTROLS OF DRUG ABUSE IN OTHER NATIONS

The brief survey of American legislation offered above indicates that at least in rudimentary form there are certain basic aspects of drug abuse control, which may be summarized as criminal law penalties for unlawful traffic in drugs, hospitalization of those addicted to drugs, and administrative controls over the production and distribution of raw materials and their processed derivatives. We may also add the incidental administrative devices of addict registration and the deportation of alien addicts or drug traffickers. These aspects of control are usually much more evident in the legal systems of other advanced nations than in the United States, and the experience gained over the years in the implementation of this more comprehensive legislative coverage offers an important standard by which the efficacy of the American approach can be gauged.

Criminal Law Controls

Some sanctions against improper traffic in narcotics and dangerous drugs are found in every country's laws, even if they reach nothing more than sales outside the official control system or fraudulent prescriptions offered to circumvent it. The importance of criminal sanctions in the total system of drug control, however, turns on a number of factors, including the social or moral view of narcotics usage itself,[49] the extent to which the country in question is either a producer of the raw materials from which narcotic drugs are derived or a manufacturer of the finished drugs,[50] the extent of the local addiction problem, and the ease with which importation of raw materials or finished narcotic products into the country can be controlled. It is evident that, despite differences in the importance of these factors, in most countries, principal reliance is placed on administrative controls and on hospitalization or outpatient treatment of drug abusers, so that criminal penalties are resorted to only in the case of traffickers who fail to abide by the official regulatory scheme.

As an illustration, in Japan there are several comprehensive statutes that together regulate traffic in the important narcotics and drugs: the Narcotic Control Law,[51] the Opium Law,[52] the Cannabis Control Law,[53] and the Stimulant Drug Control Law.[54] Each is a detailed regulation of all aspects of raw material production, manufacture, and transfer of the drug in question. In contrast, however, to pre-1970 United States law, the emphasis in the Japanese legislation is on control of the offending substance through administrative measures, criminal penalty provisions being used only to aid administrative action.

Even in those countries that are not important opium- or coca bush-producing areas or do not manufacture opium derivatives such as heroin that have little or no therapeutic acceptance, the principal emphasis is on the control of addictive substances so that they are available only through legitimate channels.[55] This is true not only of countries in the civil law tradition, but of other common-law countries such as England,[56] Canada,[57] and New Zealand.[58] There, until recently at least, the emphasis has been on supervisory control of the legitimate drug trade, and penalties are used as necessary to preserve the primacy of administrative controls.

Two other features of criminal legislation in other countries are to be noted. One is that the penalties are generally much less than under American state and federal legislation. In contrast to possibilities of life imprisonment[59] and mandatory minimum terms of imprisonment[60] encountered not infrequently in the United States, the maximum terms of imprisonment are generally brief, ranging from two to ten years.[61] There are countries, such as Spain[62] and Japan,[63] in which maximum penalties approximate those in the United States, but in the case of Japan, at least, it is rare to impose sentences of more than two to three years for any crime, so that the provisions in application are not excessive.

The second feature, and one worthy of close attention, is that drug users and addicts are not themselves punished on that basis. In other words, possession and use of drugs by the ultimate addict-consumer are not brought within the coverage of the penal law. Instead, resort is had either to hospitalization or maintenance therapy. Only if the addict commits crimes such as forgery of a prescription, or some drug-motivated or -induced offense, does he come within the ambit of the criminal law. Even in these cases, he may well be dealt with on a special basis, either through application of doctrines of insanity or diminished capacity,[64] or in the disposition made of the case on sentencing, including assignment to an institution specializing in the treatment of drug abusers.

In short, criminal penalties are a universal part of all national systems of control over traffic in drugs and narcotics. However, in contrast to the United States, where in effect exclusive reliance has been placed on criminal processes to control drug abuse, most countries use criminal penalties only as a last resort when other forms of legal regulation cannot be employed effectively. In particular, they do not try to control the addict himself through the threat or imposition of harsh penalties.[65] Instead, drug abuse is dealt with administratively.

Administrative Controls over Drug Traffic

As indicated above, the effective control of drugs requires supervision of both raw materials and processed drugs from the earliest time that either are encountered in the controlling country. Some countries have long been heavy producers of raw opium, coca bushes, or cannabis. Although there may still be areas in which there is no particular concern with the use and thus the cultivation of one or more of these plants,[66] for the most part there has been an attempt to provide comprehensive control over cultivation and production. Cultivation of plants is generally dealt with either by licensing growers and limiting crops to registered tracts, or else by requiring all farmers to sell their crops to a government monopoly or to processors licensed by the government.

As will be touched on later, one of the primary problems in the international control of the drug traffic has been in confining the world rate of production to that essential to meet the legitimate medical needs of both producing and importing countries. In theory, if production is carefully controlled, there should be few if any drugs available for diversion into illicit channels. However, as long as farmers in traditional growing areas can reap substantial profits from the cultivation of the opium poppy, coca plant, or cannabis, whether disposed of through official or illicit channels, it is unlikely that they will put national and international health objectives ahead of personal gain. Therefore, in countries such as Iran, Lebanon, and Turkey, there have been official programs of subsidies to farmers to grow substitute crops that without the subsidies do not bring as high prices as the opium poppy, cannabis, or coca bush. Often the government in question will look to subsidy of its subsidy program from countries like the United States that are anxious to have the narcotics trade stopped at its source.[67]

In the non-producing countries, control begins at the stage of importation of the raw material derived from the cultivated plants, again primarily through the issuance of import permits and meticulous record-keeping of each transaction. Each stage of the manufacturing and distribution process is then subjected to like controls, either in the form of a total government monopoly on processing and manufacture, or of licensing of legitimate manufacturies.[68] The final objective is to restrict production to legitimate medical requirements and to see that the resultant drugs are administered only to bona fide users.

Ultimately, however, a system of administrative regulations of drugs is effective only if the demand by drug abusers is either eliminated through rehabilitation or met through legitimate medical channels. Therefore, most countries outside the United States have invested substantial effort in meeting drug abuse at its source, the addicted citizen.

Medical Treatment of Drug Abusers

Several variants must be taken into account in considering whether rehabilitation of drug abusers is desirable or feasible. One is the social attitude toward use of the drug in question. It may be that a strong majority views drug use as either normal or tolerable. This has been true of opium smoking in certain Asiatic and Near Eastern countries,[69] and seems to characterize the attitude of large numbers of persons throughout the world, particularly younger persons, toward use of marijuana.[70] If no medical or social problem is thought to exist, then obviously there will be little support for compulsory treatment measures.

Assuming that drug abuse is viewed as aberrant, however, two other factors are important. One is the availability of treatment facilities and the professional staffs to man them. Some countries are woefully short of medical personnel,[71] and those countries with ample resources are rarely willing to appropriate enough funds to build and staff suitable institutions. This in turn is tied to the second factor, the number of addicts identified for treatment. It is manifestly easier for Sweden to deal with an addict population of 1,000 out of a population of 8 million,[72] England[73] and Canada[74] with estimated addict populations of about 3,000, and Japan with its approximately 6,000 addicts,[75] than it is for the United States to cope with its estimated more than a quarter million drug abusers, or the state of New York to provide treatment facilities for its 40 per cent share of them.[76]

Still other matters that should be taken into account are the size of the country and its governmental structure. The Netherlands obviously has a potentially easier task of dealing with its drug abusers through medical means than might India, and a strongly controlled governmental system like that of the U.S.S.R. or the Chinese People's Republic can segregate and control addicts much more effectively than can the United States or India with a federal system in which many important control functions are decentralized and left to local initiative.

Even assuming the inevitability of variations in national law determined by the interplay of these considerations, it is nevertheless evident that, in almost every civil law country, considerable reliance is placed on civil commitment of drug abusers who are not able to function effectively in the community.[77] Sometimes the segregation of drug abusers is achieved through invocation of the ordinary civil commitment statute governing the mentally ill, while in other instances there is a special statute applicable only to drug abusers. The significant factor in either instance is that there is no criminal sanction in the background or criminal conviction that triggers the commitment process. This offers a stark contrast to civil commitment procedures in the United States, which almost always are either triggered by or closely connected with criminal prosecution of the drug abuser.[78] It seems a reasonable assumption that the civil law approach is much more conducive to successful therapy, assuming it is possible at all, than the American criminal law-oriented approach. New efforts in England and Canada fortunately seem to take the civil law approach to addict commitment procedures.[79]

Commitment of drug abusers, whether voluntary or involuntary, is an expensive approach. Moreover, to the extent that drug abuse is the product of environment, control achieved in a hospital setting may not survive a return to the home and neighborhood from which the patient came. Therefore, outpatient treatment may make better sense than hospitalization in a great many instances. Moreover, the cost to society of the illicit drug traffic is great, in that addicts commit a great many property crimes to obtain money to purchase narcotics at high prices. Providing drugs for addicts at nominal or no cost through clinics or private physicians avoids at least to a degree the indirect costs of drug abuse, although at the same time it raises at least in some countries the moral question of whether the state should underwrite drug abuse.

The Harrison Narcotic Law was interpreted by the United States Supreme Court in a way that might have left it open for doctors to prescribe maintenance doses to addicts without violating the law.[80] However, the Bureau of Narcotics reacted very strongly against the idea that maintenance doses are lawful, and the American Medical Association until recently went along with that policy rather docilely. When in 1959 a joint committee of the American Bar Association and the American Medical Association recommended that at least there be research and experimentation with maintenance doses, the result was a virulent, indeed irrational, attack by spokesmen for the Bureau of Narcotics.[81] However, overwhelmed by the sheer magnitude of its addict problems, the city of New York in recent months has been providing for methadone maintenance on a rather large scale,[82] and thus far there has been no indication of any effort on the part of the federal government to interfere on the basis of the Harrison Narcotic Act, within the provisions of which methadone clearly falls. The new federal act may also result in greater experimentation in this direction.

England is the nation with perhaps the greatest experience with controlled prescription of maintenance doses. Up to 1968, a Home Office memorandum on the duties of doctors and dentists permitted prescribing for addicts if either (a) the patient was on a withdrawal regimen preliminary to cure, (b) discontinuance of the drug would produce severe symptoms, or (c) the patient might live a "relatively normal and useful life when a certain minimum dose is regularly administered, but becomes incapable of this when the drug is discontinued." Because a few physicians (six in number) were abusing this power and supplying the illicit market in heroin and cocaine, in 1968 the Home Secretary ordered in the Dangerous Drugs (Supply to Addicts) Regulations that heroin, cocaine, and their derivatives can be supplied an addict only to relieve pain from organic disease or injury, thus precluding withdrawal or maintenance treatment. However, the Regulation does not preclude use for either purpose of morphine and appropriate medications governed by the Dangerous Drug Act.[83] Despite statements from American sources hostile to maintenance therapy suggesting that the British system is a failure and ought not be copied, the English themselves appear to be generally satisfied with the way their regulations are implemented.[84]

Meanwhile, the medical profession has reacted somewhat ambivalently to maintenance therapy. The World Health Organization Expert Committee on Drug Dependence is willing to approve methadone maintenance on an experimental basis, but believes that it must be coupled as far as possible with efforts at social rehabilitation; it has also pointed out that addicts participating in these experiments have been above average in motivation and thus are probably not typical of the addict population as a whole.[85] This will probably continue to summarize the attitude of the world medical community toward experiments in maintenance therapy, although it should be stressed again that the essentially non-punitive and regulatory approach in most legal systems other than the United States creates a relatively congenial legal atmosphere for such endeavors.

It may be that, in the years ahead, medical science will discover an opiate antagonist that can be administered orally and produce no substantial adverse side-effects. Depending on the length of time that each ingestion is effective, such a treatment method might be of great assistance in curbing heroin addiction. The principal legal problem then would be the extent to which such a medication might be administered under compulsion. However, treatments such as compulsory vaccination have been sustained by American courts,[86] so that the chances are overwhelming that legal permission will be given for such a regimen if it gains medical acceptance. Meanwhile, the medical position seems to be one of encouragement in the discovery and development of the necessary treatment.[87]

As in the instance of other public health problems, a system of in-patient or out-patient treatment requires that doctors and other health service workers report anyone who needs treatment. Such a system will not work if those reported are likely to be punished; the American emphasis on punishment of addicts to deter them has meant little or no concern with reporting requirements resting on public health considerations. In most other countries, however, both medical practitioners and governmental officials are to report addicts as a prelude to non-penal treatment.[88]

One arbitrary means of disposing of a drug abuser or addict who is an alien is to deport him as an undesirable alien, either following conviction of a narcotics offense or on the basis of his addicted condition itself. This is not an uncommon disposition of cases.[89]

In summary, the laws of all modern countries rest principally on controls over production and distribution of addictive drugs, hospitalization or other treatment of the addict, and criminal penalties to effect the primary means of control. Only in the United States have criminal penalties been relied on as the principal, indeed, almost sole means of coping with drug abuse. Whatever the country, however, it is evident that the trans-national character of the drug traffic requires something more than purely domestic law controls. It is from the awareness of this that the system of international control through treaties has come, the matter to which we now turn.

INTERNATIONAL CONTROLS OF THE DRUG TRADE

Ironically, the first treaties demanded by a western power required approval of traffic in narcotics. Despite efforts by China to control and gradually eliminate the opium trade, Britain at the conclusion of the Opium Wars in 1858 forced her, through the Treaty of Tientsin, to legalize the opium trade. In the ensuing decades, opium revenues became very important to China, Hong Kong, and India.[90] However, the same moral revulsion that led to the international abolition of slavery and traffic in women and children brought the nations of the world to the conference table to draft treaties to control the international trade in narcotics. The first international conference was convened in Shanghai in 1909, followed three years later by a second conference at The Hague which produced the International Opium Convention of 1912. This was supplemented after World War I by a second Convention of 1925, and that in turn by a Convention for Limiting the Manufacture and Regulating the Distribution of Narcotic Drugs in 1931. Supervising control bodies were created under the auspices of the League of Nations. The primary features of all these treaty systems were an assessment of the legitimate needs for the controlled narcotics, a limitation of production to the amount needed, and strict control over distribution of raw materials and manufactured drugs. Some progress was made in controlling the narcotics traffic through the invocation of these treaties by their signatories; the controls continued after World War II when the United Nations took over the functions exercised earlier by the League of Nations.[91]

The pre-World War II treaties, however, had come to show their age and their inadequacy to cover such matters as the traffic in vegetable raw material from which natural narcotic drugs are manufactured, as well as the many new synthetic narcotic drugs discovered in the interim. Therefore, in 1961 the Commission on Narcotic Drugs, which had succeeded the International Narcotics Control Board under the League, held a conference to consider the draft of a single or unified convention on narcotic drugs to replace the earlier treaties. The resulting Convention was ratified by the necessary number of nations and went into force in 1964 as the primary medium for international control of the narcotics traffic.

The basis of the control system is a classification of drugs within the coverage of the Convention into four categories listed in the Schedules annexed to it; additions are made as needed.[92] Drugs in category I are subjected to all the controls on manufacture and distribution set out in the Convention. Those in Schedule II are governed by substantially the same rules as Schedule I drugs, except for certain exemptions from record-keeping and labeling at the retail level;[93] various codeine medications are prominent in this Schedule. Schedule III includes compounds subject to even fewer restrictions, notably those governing import trade[94] and reports to the International Narcotics Control Board.[95] Most Schedule III drugs are compounds with some codeine content. The most rigidly controlled drugs are those in Schedule IV, cannabis and cannabis resin, desomorphine, heroin and ketobemidone and their salts if chemically feasible. Not only do all Schedule I restrictions apply, but signatory countries may adopt any special measures of control that they wish, including an absolute ban on production, manufacture, traffic, or use of the offending drug other than as required for medical or scientific research.

The agency responsible for administering the provisions of the Convention is the International Narcotics Control Board,[96] which is a creature of the Convention itself, and not of the United Nations. However, the U.N. Commission on Narcotic Drugs, created in 1946 under the Economic and Social Council, has certain functions to perform under the Convention,[97] the most important of which is amendment of Article 2 schedules on the recommendation of the World Health Organization or a party to the Convention.[98] Each signatory nation is to designate a national agency ("special administration") to apply the provisions of the Convention,[99] and either directly or indirectly must help to meet the operating expenses of the Board and Commission.[100] Each must also furnish annual reports on the working of the Convention in its territory, provide copies of implementing laws, respond to inquiries from the Commission about illicit drug traffic, and give the names and addresses of government authorities authorized to issue export and import documents.[101]

The initial step in the control system under the Convention is the requirement imposed on each party to file with the Board an estimate of the quantities of drugs to be consumed for medical and scientific purposes, the quantities required to manufacture other drugs and preparations whether scheduled or unscheduled, the stock of drugs to be held at the end of the report year, and the amount of drugs necessary to be added to existing stocks.[102] Each party must also furnish statistics on production or manufacture of drugs, utilization of drugs to manufacture other drugs or preparations, consumption of drugs, imports and exports of drugs and poppy straw,[103] seizure of drugs and their disposal, and stocks of drugs on hand at the end of the report year.[104] Based on these reports, no party is to manufacture or import any more drugs than will cover the quantity consumed for medical and scientific purposes, used to manufacture other drugs or preparations, exported, added to required reserves and acquired for special purposes; stocks seized in law enforcement and released for legitimate purposes must be included. Any surplus must be deducted from the following year's estimate of needs. If there is an abuse of exports, other countries may be notified of the fact, and in the event of non-cooperation an embargo is placed on exports until there is compliance.[105]

Exporting nations are to limit production of opium for international trade to legitimate world needs, and must notify the Economic and Social Council of the amounts to be produced for export if more than 5 tons, the controls exercised or proposed for production of the opium, and the countries to which exports are planned.[106] An embargo may be imposed if there are excessive exports. A producing nation may prohibit cultivation of drug-generating plants,[107] must control poppy straw to see that it is not used to produce illicit opium,[108] must similarly control coca bush cultivation[109] and cannabis growth,[110] and establish a national opium agency to supervise or monopolize opium production.[111]

To control the manufacture of drugs, each party to the Convention must either license manufacturers or do the manufacturing itself through a state monopoly.[112] Similarly, it must control both domestic[113] and international[114] trade, and must provide whatever supervision and control are required to make the system work.[115] It is not to permit possession of drugs except under legal authority.[116] Important in this regard are the requirements of prescription control, and warnings and labeling either required or recommended as part of distribution controls.[117]

The Convention does not and under traditional principles of international law cannot impose criminal responsibility directly on individuals who engage in illicit drug traffic; the only sanctions in the Convention flow against offending countries. However, each country is to enact criminal statutes enforcing the limitations imposed on the drug traffic by the Convention, punishable by imprisonment and not merely a fine.[118] Each offense committed in a different country is to be considered as a separate offense, and convictions in other countries may be taken into account in assessing penalties for recidivism. Drugs, substances, and equipment used or intended for use in the illicit drug trade are to be seized and confiscated.[119]

One of the most important tools for enforcing the criminal law against international criminal activity is extradition. However, most of the existing extradition treaties are what are called "list treaties," under which those who commit one of the crimes set out in the treaty may be extradited to the demanding country. Narcotics offenses are rarely listed in these treaties, most of which have not been revised for several decades.[120] The Convention urges signatory countries to revise their extradition treaties to include the offenses based on Article 36,[121] and, in default of that, to encourage invocation of the so-called "universality principle," under which the nation in which the offender happens to be found may prosecute him if extradition to the place where the offense actually took place is not possible and if the offender has not already been punished somewhere for the offense.[122]

In line with the prevailing doctrine outside the United States, special attention is to be given to the medical treatment, care, and rehabilitation of drug addicts.[123]

The principal deficiency of the Single Convention is that it does not apply to the so-called dangerous drugs. Because in a legal sense it came into being as a replacement and only a replacement for certain predecessor treaties that did not purport to cover barbiturates, amphetamines, and hallucinogens,[124] drugs of that character cannot be brought within its terms simply in the form of amendments to the schedules annexed to the Convention.[125] As a result, the Commission has prepared a Draft Protocol on the Control of Psychotropic Drugs Outside the Scope of the Single Convention on Narcotic Drugs of 1961, usually referred to as "The Draft Protocol on Psychotropic Substances."[126] The basic approach to control is similar to that in the Single Convention, bearing in mind the more widespread legitimate use of barbiturates and amphetamines. The WHO Expert Committee on Drug Dependence has already drafted the schedules that would be required for the implementation of the Protocol if it is adopted.[127] Adoption of the Protocol will complete the system of international control of narcotic and dangerous drugs, and will come as close to a comprehensive scheme for the control of illicit traffic in drugs as it is possible to achieve within the limits of the primacy of national sovereignty.

SUMMARY AND CONCLUSIONS

One may conclude that national legal systems for the control of drug abuse rest on three major elements: (a) control of the drug supply itself through administrative supervision, (b) rehabilitation of drug abusers or at least maintenance of their addiction other than through illicit channels, and (c) criminal penalties to enforce the basic administrative system.

A comparative study of other legal systems raises one principal question for Americans: Can any effective system for controlling drug addiction and abuse be based exclusively or even primarily on criminal statutes invoked by both regular and special law enforcement agencies, calling for generally draconic penalties against non-addict and addict traffickers alike, and embracing the addict-user in their coverage? The United States stands alone today in this approach to the control of drug abuse. Such an approach obviously rests on certain assumptions about how drug addiction develops, which may be summed up in the moral principle of free will, the exercise of which may under the deterrent theory of punishment be affected by the threat of actual infliction of severe penalties. In other countries of the world, the addict or drug abuser is viewed chiefly as a medical problem, so that criminal penalties are reserved for, or in practice applied only to, non-addicts who make possible the supply of drugs to addicts and abusers. That the free-will premise is not entirely lacking in these countries, however, is illustrated by the reluctance to approve maintenance regimens for addicts unless they are accompanied by efforts at rehabilitation.

Meanwhile, at the international level, strong support for the dichotomy between strict enforcement against non-addict traffickers and addict consumers is embodied in the Single Convention of 1961 and the activities undertaken pursuant to it by the International Narcotics Control Board and the Commission on Narcotic Drugs. There should be at least a moral obligation, if not a legal obligation, on the part of the United States as a prime mover behind and signatory of the Single Convention to bring its system into line with those of the other subscribing countries. There is some indication that this will develop, judging at least by the new federal act and New Jersey legislation[128] that substantially reduces criminal penalties for first-offender users of marijuana. It will take greater pressure from the medical profession on legislators and federal and state law enforcement agencies, however, including forthright action by the American Medical Association and other powerful medical bodies, if any breach is to be made in the half-century old emphasis on criminal deterrence as the primary if not sole means of coping with drug abuse.

However, concerned members of the medical profession still need to inquire of themselves whether they possess the ultimate answer to the drug abuse problem. The assumption common to both national legal regulation and the Single Convention which backs it up is that if the supply of narcotic and dangerous drugs is confined to licit medical and scientific channels only, and if addicts and drug abusers are treated as medical cases and not criminal defendants, the problem of drug abuse will disappear. However, there is room for legitimate doubt whether medical treatment of addiction, including psychotherapy, can in fact eliminate the causes of drug abuse or addiction. As suggested earlier, users must recognize that they are ill before treatment is possible. Abuse and addiction may be inevitable in some or many instances in every society,[129] and there will be some who will create or follow fads, during which they become "addicts."[130] Perhaps these fears are ill-founded, so that medical therapy will be the ultimate answer to addiction and abuse. But physicians, like other citizens, must be aware that if the assumptions on which a scheme of legislation is based are fallacious, that legislation is doomed to disaster.

REFERENCES

1. There is a growing recognition that addiction to alcohol and drugs presents one basic problem only, and that the etiology of drug abuse probably shares much in common with that of excessive use of alcohol. See, e.g., *WHO Expert Committee on Mental Health Fourteenth Report, Services for the Prevention and Treatment of Dependence on Alcohol and Other Drugs* (WHO Technical Report Series No. 363, 1967); Alexander: *The Criminal Responsibility of Alcoholics and Drug Addicts in Canada,* 31

Sask. B. Rev. 71 (1966). For the most part, however, traditional legal analyses treat the two problems as separate and essentially unrelated.

2. There are no reliable data available about the number of drug abusers in the United States. In part this is because criminal statistical reporting in this country is incomplete and haphazard. Also, there is no general requirement that doctors, hospitals, and clinics report cases of addiction which they uncover. Moreover, because addiction presents a spectrum and not a sharply defined category, any attempt to number addicts is probably doomed to failure. See Eldridge, W.: *Narcotics and the Law,* 2nd ed., Chicago, University of Chicago Press, 1967, pp. 66-103. However, the magnitude of the problem is indicated at least incidentally by the statistics on seizures of illicit drugs covered by the Single Convention on Narcotic Drugs of 1961. The United States seizes by far the largest amount of heroin, which of course has only illicit uses. See, e.g., *U.N. Economic and Social Council Commission on Narcotic Drugs,* Report of the Twenty-Third Session 29-40 (E/CN. 7/523/Rev. 1) (1969).
3. See, for example, the general statements on behalf of the participating nations at the commencement of the United Nations Conference for the Adoption of a Single Convention on Narcotics. I Official Records 4-16 (A/CONF.34/24, 63.XI.4) (1964).
4. Chiefly through the International Criminal Police Organization (ICPO; INTERPOL).
5. Eldridge, W.: *op. cit. supra* note 2, at 4-7.
6. 26 U.S.C. §§ 4701-4736 (1967).
7. 21 U.S.C. §§ 171-176a (1961).
8. 21 U.S.C. § 199 (1961).
9. 49 U.S.C. §§ 781-789 (1963).
10. 21 U.S.C. §§ 188-188n (1961).
11. No licenses apparently were issued under the Act, and the United States is not one of the countries formally recognized as an opium-producing nation. Current international controls are imposed through Article 24 of the Single Convention on Narcotic Drugs of 1961, discussed *infra.* See Waddel: International narcotics control, Volume number 64 *Am. J. Int'l L.* 310, 318-19 (1970).
12. Int. Rev. Code §§ 4741-4762 (1967).
13. See, in general, Rosenthal: Dangerous drug legislation in the United States: Recommendations and comments, (vol.) 45 *Texas L. Rev.* 1037, 1074-77 (1967). (See note 20.)
14. 21 U.S.C. §§ 360, 360a (Supp. 1970).
15. 21 U.S.C. § 360a(b) (Supp. 1970).
16. 21 U.S.C. § 360a(c) (Supp. 1970).
17. 21 U.S.C. § 333(a), (b) (Supp. 1970). The heaviest penalities were reserved for those 18 or over who sold or otherwise provided a depressant or stimulant drug to someone under 21.
18. 42 U.S.C. §§ 257-261a (1969). In a New York City program, see Lieberman and Brill: Rational authority and the treatment of narcotics offenders. 20 *Bull. Narcot.* No. 1, p. 33 (Jan.-March 1968).
19. 42 U.S.C. § 260(c) (1969).
20. See, e.g., the statements by the United States delegates at the United Nations Conference for the Adoption of a Single Convention of Narcotics, when the predecessor of Art. 38 of the Convention was under consideration. They strongly urged a treaty restriction to "compulsory treatment of drug addicts in closed institutions," but were overridden. I Official Records, *supra* note 3, at 103-14, 211. Cf. also Kramer: *State Versus the Addict: Uncivil Commitment,* 50 B.U.L. Rev. 1 (1970).
21. § 704
22. See text *infra* at notes 91-127.
23. § 202(c).
24. § 202(d).
25. §§ 303-304.
26. § 305.
27. § 306.
28. § 307.
29. § 308.
30. § 309.
31. § 404.
32. § 401.
33. § 408.
34. § 409.
35. § 509.
36. § 510.
37. § 514.
38. § 512.
39. § 253.
40. § 256.
41. § 4.
42. § 2(a).
43. §§ 1101, 1102.
44. Uniform Narcotic Drug Act § 13.
45. *Id.* § 17.
46. The President's Commission on Law Enforcement and Administration of Justice, The Challenge of Crime in a Free Society 220 (1967).
47. See, in general, Rosenthal: *op. cit. supra* note 13, at 1062-1074.
48. The statutes as of 1967 are summarized in W. Eldridge, *supra* note 2, at 177-231. *See also* The Challenge of Crime in a Free Society, *supra* note 46, at 222-224.
49. Note: Marihuana and the Narcotic Control Act. Vol. 3 *U. Brit. Colum. L. Rev.* 250, 252-253, 1970.
50. As an illustration of this, the points of view of Turkey and the United States are basically diametrically opposed when the question is one of control of cultivation of the opium poppy, and as opposed but the exact oppo-

site when the question is one of international control of barbiturates and amphetamines.

51. Law No. 14 of 1953, as amended.
52. Law No. 71 of 1954, as amended.
53. Law No. 124 of 1948, as amended.
54. Law No. 252 of 1951.
55. The most useful summary of statutes relating to control and treatment of drug addiction, although somewhat dated, is *Treatment of Drug Addicts: A Survey of Existing Legislation,* 13 Int'l Digest of Health Legis. 4 (1962), also available in pamphlet form from WHO.
56. A helpful summary is found in Card, *Aspects of British Drug Legislation,* 20 U. Toronto L. Rev. 88 (1970); see also Samuels: *Sentencing Drug Offenders,* [1968] Crim. L. Rev. 434.
57. See Alexander, *The Criminal Responsibility of Alcoholics and Drug Addicts in Canada,* 31 Sask. B. Rev. 71 (1966); Note, *Marihuana and the Narcotic Control Act,* 3 U. Brit. Colum. L. Rev. 250 (1970).
58. Note, *Narcotics Act 1965,* 2 N.Z. Universities L. Rev. 92 (1966).
59. E.g., Mich. Comp. L. Ann. § 335.152 (West 1967); N.Y. Pen. L. §§ 220.23, 220.44.
60. E.g., Mich Comp. L. Ann. § 335.152, 225.153 (West 1967); N.Y. Pen L. §§ 70.00(2), 220.23, 220.44.
61. For example, penalties in Austria range from one to five years, and in the case of members of criminal groups, up to ten years. In France the maximum term has been five years, although the penalty scale has been revised recently (N.Y. Times, July 2, 1970, p. 2. col. 1-2). In Italy the range is from six months to eight years, in Norway from two to four years, and in Sweden up to six years. In most instances, a fine may also be levied.
62. The most seriously punished offenses range from twelve to twenty years.
63. Violations for gain of the Narcotic Control Law are punishable by a term of from three years to life [Art. 64(2)], and of the Opium Law by any term of imprisonment of one year or more [Art. 51(2)]. Other violations of these acts, and of the Cannabis Control Act and Stimulant Drug Control Law, however, are punished generally at the level of a maximum three, seven, or ten years.
64. See Alexander, *op. cit. supra* note 57, at 82-96.
65. Cf. Packer, H.: *Limits of the Criminal Sanction.* Stanford, Calif., Stanford University Press, 1968, pp. 332-333, 337-342.
66. Cf. the problem of coca-leaf chewing, which is prevalent enough in some South American countries that it may be made the basis of a "transitional reservation" for as much as 25 years from the effective date of the Single Convention on Narcotic Drugs. Art. 49(1)(c), (2)(e). See, e.g., Commission on Narcotics Drugs Report, *supra* note 2, at 61.
67. See, for example, the praise given to the Lebanese subsidy program in the Commission on Narcotic Drugs Report, *supra* note 2, at 54-60, and 94-95, and the consternation with which the Commission reacted to indications by Iran that it was actively considering revival of licensed opium poppy production which it had suspended in 1955. *Id.* at 86-90.
68. Either alternative is acceptable under the Single Convention on Narcotic Drugs of 1961, arts. 29, 30.
69. Note, *Marihuana and the Narcotic Control Act,* 3 U. Brit. Colum L. Rev. 250, 252-253 (1970); Note, *Dangerous Drugs in Hong Kong: Recent Legislation,* 113 Sol. J. 663 (1969).
70. *Cf.* H. Packer, *op. cit. supra* note 65, at 337-341.
71. For example, in the discussions of the predecessor to Art. 38 of the Single Convention on Narcotic Drugs, the representative of Ghana pointed out that there were only 400 to 500 doctors to meet all the needs of the 7 million population of Ghana, so that compulsory treatment of drug abusers was out of the question. I Official Records, *supra* note 3, at 109.
72. *Id.* at 110.
73. Card, *op. cit. supra* note 56, at 90.
74. I Official Records, *supra* note 3, at 103-4; Alexander, *op. cit. supra* note 57, at 76.
75. Japan Ministry of Health and Welfare, A Brief Account of Narcotics Abuse and Counter-Measures in Japan 17 (1970).
76. See W. Eldridge, *supra* note 2, at 92.
77. In general, see *Treatment of Drug Addicts, supra* note 55, at 4-13; WHO Expert Committee on Mental Health Fourteenth Report, *supra* note 1, at 31-32, 33-34.
78. See, in general, W. Eldridge, *op. cit. supra* note 2, at 147-53.
79. Alexander, *op. cit. supra* note 57 at 97-99 and 100-101; Samuels, *op. cit. supra* note 56, at 444-446.
80. Linder v. United States, 268 U.S. 5 (1925); *cf.* United States v. Behrman, 258 U.S. 280 (1922), and Jin Fuey Moy v. United States, 254 U.S. 189 (1920).
81. See W. Eldridge, *op. cit. supra* note 2, at 37-41.
82. On earlier phases of the experimentation, see W. Eldridge at 157-160. See also Lieberman & Brill, *supra* note 18.
83. In general, see Card, *op. cit. supra* note 56, at 94-96.
84. *Id.* at 96.
85. WHO Expert Committee on Dependence-

Producing Drugs Fifteenth Report 9-10 (WHO Tech. Rep. Series No. 343, 1966); WHO Expert Committee on Drug Dependence Sixteenth Report 20-21 (WHO Tech. Rep. Series No. 407, 1969); WHO Expert Committee on Drug Dependence Seventeenth Report 25-26 (WHO Tech. Rep. Series No. 437, 1970).

86. E.g., Massachusetts v. Jacobsen, 197 U.S. 11 (1905); Cude v. State, 237 Ark. 927, 377 S.W.2d 816 (1964).

87. WHO Expert Committee on Dependence-Producing Drugs Fifteenth Report 9 (WHO Tech. Rep. Series No. 343, 1966); WHO Expert Committee on Drug Dependence Sixteenth Report 21 (WHO Tech. Rep. Series No. 407, 1969); WHO Expert Committee on Drug Dependence Seventeenth Report 26 (WHO Tech. Rep. Series No. 437, 1970).

88. See *Treatment of Drug Addicts, supra* note 55; A Brief Account of Narcotics Abuse and Counter-Measures in Japan 11 (1970).

89. See Samuels, *op. cit. supra* note 56, at 440; Note, *The Futile Forgiveness: Basing Deportation on an Expunged Narcotics Conviction,* 114 U. Pa. L. Rev. 372 (1970).

90. Glatt, *The Development of International Control of Drugs,* 24 WHO Chron. 189, 190 (1970).

91. In general, see Glatt, *op. cit. supra* note 90, at 190-192; Waddell, *op. cit. supra* note 11, at 312-315.

92. Arts. 2, 3.

93. Art. 30(2) and (5) do not apply to Schedule II drugs.

94. Art. 31(1)(b), (4)-(15).

95. Arts. 19, 20.

96. Arts. 5, 9-16.

97. Art. 8.

98. Procedures appear in Art. 3.

99. Art. 17.

100. Art. 6.

101. Art. 18.

102. Art. 19

103. Defined in Art. 1(r) as "all parts (except the seeds) of the opium poppy, after mowing." See, in general, *Note on Poppy Straw Submitted by the Hungarian Delegation,* in II Official Records 66 (E/CONF.34/24/Add.1; 63.XI.5).

104. Art. 20.

105. Art. 21.

106. Art. 24.

107. Art. 22.

108. Art. 25.

109. Arts. 26, 27.

110. Art. 28.

111. Art. 23.

112. Art. 29.

113. Art. 30.

114. Art. 31. Drugs in first-aid kits on international carriers are exempted from the requirements. Art. 32.

115. Art. 34.

116. Art. 33.

117. Art. 30(2)-(5).

118. Art. 36.

119. Art. 37.

120. For a list of extradition treaties now in force and their dates of execution, see Bassiouni, *International Extradition in American Practice and World Public Order,* 36 Tenn. L. Rev. 1, 3 n.5 (1968).

121. Art. 36(2)(b).

122. Art. 36(2)(a)(iv).

123. Art. 36.

124. Art. 44.

125. See Commission on Narcotic Drugs Report, *supra* note 2, at 64-71.

126. The text may be found in *id.* at 106-127, and in 8 Int'l Legal Materials 769 (1969).

127. Seventeenth Report 9-21 (WHO Tech. Rep. Series No. 437, 1970). See also Glatt, *op. cit. supra* note 90 at 194-196.

128. N.J. STATE ANN. §2A: 170-77. 9 (Supp. 1970); see also State v. Ward, 57 N.J. 75, 270 A.2d 1 (1970).

129. *Cf.* WHO Expert Committee on Dependence-Producing Drugs Fifteenth Report 13-14 (WHO Tech. Rep. Series No. 343, 1966).

130. See W. Eldridge, *op. cit. supra* note 2, at 138 n. 10, describing glue sniffing and smoking dried banana peel; some addicts reportedly have injected peanut butter or whole milk in the search for new thrills, even if potentially fatal ones.

PANEL DISCUSSION—LEGAL ASPECTS OF DRUG ABUSE

Moderator: DR. HOLLISTER

Panel Members: DR. GEORGE, MR. INGERSOLL, JUDGE OLIVER, CONGRESSMAN SCHEUER, and DR. ROGER C. SMITH

Dr. Hollister: I would like to comment on the procedure for this afternoon's session. First, the provision for submitting questions in writing rather than by verbal communication. This method is part of our general experimental and empirical approach to many problems, including drug abuse. As you can see, it is not especially innovative because these forms were printed for past Towsley conferences, and it is apparently more of a standard operating procedure than the one we have been following. In doing so, there is by no means any intention to restrict questioning; rather we thought that it would facilitate the discussion by allowing more questions to be submitted, with less dependence on the degree of aggressiveness or volubility of the questioner.

Second, concerning the exclusion of University of Michigan students from the afternoon session. I am assured by those in charge here that there has been no such exclusion, and that any registered student who wanted admission this afternoon or for any other portion of this conference was able to obtain it providing he preregistered.*

A third objection concerned the injection of politics into the discussion, but I just view this as congressional license. Now, I would like the panelists, beginning

* Editor's Note: Traditionally, students as well as faculty and other University of Michigan staff are welcome and admitted without fee to all programs in continuing medical education. This Conference was no exception. Preliminary programs were sent to deans, department chairmen, and section heads as well as to many individual faculty members. There was publicity in various local news media. The Associate Dean for Student Affairs of the Medical School noted in advance the high interest in the Conference and urged department chairmen to accommodate students and faculty members wishing to attend. The Program Planning Committee limited attendance to the seating capacity of the Rackham Auditorium, i.e., 1,138. It became evident that this number would be reached some time before the start of the Conference. Accordingly, all interested in attending were urged to register in advance and were informed that all admissions would be by identifying badge. Approximately 500 University of Michigan students registered for the Conference. Registration was closed when the number reached the capacity of the auditorium. Unfortunately, all registrants did not attend every session so that vacant seats were present during most of the presentations. To have kept every seat filled with stand-by registrants, however, was considerably beyond logistical capabilities at our disposal.

with Mr. Ingersoll, who must meet a time schedule that takes him elsewhere, to each take a single question and try as far as possible to combine it with several related ones. Mr. Ingersoll will begin with about half a dozen in a row so he can get away on time.

Mr. Ingersoll: I think you have already answered most of my questions, Mr. Chairman. They had to do with why the cards, and why no students. But I do have one or two left here. The first one that I selected is: "Why has the most recent bill sponsored by your department, and even pushed according to one of your attorneys, Mr. Sonnenreich"—I don't know what that means, we don't push much in our department—"and recently signed by President Nixon, persisted in classifying marihuana with heroin in its needs for legal restraints?" It is true that marihuana is in the same schedule as heroin for control purposes, and this is basically because it is a drug of abuse, it has a potential for abuse, and it also has no recognized medical utility in the United States. So it is classified with a group of other drugs with the same characteristics that are available only for research purposes. Now, as far as criminal sanctions are concerned, we have departed in the new bill from the old pattern. Criminal penalties imposed for trafficking or possessing marihuana are the same as those for the other non-narcotic drugs. They are significantly lower than penalties for violations involving heroin, for example. So the answer basically is that although marihuana is classified with heroin in Schedule I, that is only for regulatory purposes. The criminal sanctions are somewhat different.

Next question: "How would the effort, money, and manpower spent on the heroin problem compare to that spent on marihuana?" As far as my organization is concerned, about 13% of our operative time and investigative time is spent on marihuana cases; as I recall, 70% is spent on heroin cases and the balance on the other dangerous drugs.

"Do you believe that law enforcement should concentrate on the suppliers of abused drugs in this country rather than on harassment of the drug-dependent individual?" I do, indeed; and as far as the Federal agency is concerned, this is our practice, a practice that admittedly has been in existence only in the last two years. I might say that for the 35 or 40 years preceding that two-year period when the Federal government was involved in this, it didn't distinguish too much between trafficker and addicts or drug-dependent people. And part of the new law serves the same purpose, to distinguish between professional traffickers and those who are in possession of drugs illegally for their own use. The more severe penalties are available for traffickers, and less severe penalties for people in simple possession. In addition, on his first offense, a simple possession violator will be guilty only of a misdemeanor, and can be placed on probation. If he fulfills the condition of that probationary period successfully, his record can be expunged. He gets that opportunity once. If he comes back again, then the record can be revived for the second go-round.

An interesting question: "What is the extent of drug abuse in the Soviet Union, what are they doing to combat it, and what is their cooperative effort in the U.N., if any?" The people from the Soviet Union that I have talked to first claim that there is no significant drug abuse in the USSR, and then they immediately describe their very sophisticated up-to-date treatment programs for addicts. I might also point out that the Soviet delegates consistently take the very hardest line against marihuana, and chastise me and the United States Government for our relatively soft attitude toward marihuana. They take a very tough stance, at least before the rest of the world. We don't know too much about drug abuse in the Soviet Union beyond what they tell us. They admit to having a significant alcohol problem, and we are dimly aware of some isolated cases of drug abuse in some areas of the Soviet population.

Dr. Hollister: Congressman Scheuer has a comment.

Congressman Scheuer: Jack Ingersoll is a talented and highly professional law man. He has given us his predictably thoughtful and informed views in response to several questions from the floor. Perhaps now it might be appropriate to open up the discussion to spontaneous questions from the floor. I think it was a dubious decision on this most interesting and perhaps most controversial of all of the sessions to have closed off personal confrontation or discourse or whatever. I think both panel and audience would enjoy it much more if we reverted to the established give-and-take format. And I would only suggest in amelioration that we keep our questions short, sweet, and to the point. I am not going to answer any questions on cards. I will only answer questions from the floor.

Dr. Hollister: It seems quite obvious that there is a division of opinion here. I had in mind that if anyone really feels the need to have this kind of dialogue, then we will have it, but in fairness to the people who have taken the time to submit their questions, and who are not now standing there ready to man the mikes. I think they should also have a chance to be answered.

Congressman Scheuer: Well, they don't have a chance; I have 35 cards here.

Dr. Hollister: When we get to you, we will make the switch. Until then, are there any other questions, Mr. Ingersoll, that you feel would benefit from this kind of approach?

Mr. Ingersoll: I have to leave to catch an airplane, the only one that is available. Some of these questions are rather pedestrian, and I am willing to pass to some more interesting ones.

Dr. Hollister: Then suppose we go around the table and take the cards we think are of particular interest. After that we will open it up. Dr. George, what about you? Do you find anything interesting?

Dr. George: There are some questions relating to the Harrison Act and its significance. I think that as of May 1, 1971, the Harrison Act will be of historical interest only because the new statute goes into full effect at that time.

Another card does remind me to make a point clear. That is, the Federal statute is a matter of Federal law enforcement only. Congress makes it quite clear that it has no intention of doing what lawyers call "occupy the field"; its purpose is not to avoid state legislation. The bulk of the present state legislation was patterned on earlier federal narcotics statutes, particularly the Harrison Act and Marihuana Tax Act. The new federal legislation may conceivably bring about some reconsideration of state legislation in the years ahead. Any doctor concerned about the problem might begin to develop contacts with his local legislators.

One question inquires about using the terms "addict" and "drug abuser" interchangeably. As far as legal controls are concerned, in terms of statutory definitions, one can usually employ them interchangeably.

A question has been raised about my comment that European systems do not invoke heavy penalties against users; the questioner asks about young people being imprisoned in various countries. In these instances they have been picked up with substantial amounts of narcotics, and have been engaged in effect in international drug traffic. Of course, in every country heavy penalties are available against the professional trafficker, and I think that is frequently the reason for these prison sentences.

I have been asked to comment on the English system, legalization of marihuana and hard drugs, and any hope I have for the future. Being a pessimist, I am not sure I see any hope. My preference is for more to be done in experimenting with maintenance therapy, for no other reason than to cut down some of the cost to the community of addict-derived crimes. Congressman Scheuer has mentioned this. I think it is better to maintain a substantial percentage of addicts on free or minimal cost dosage if this reduces the collateral effects of addict crime. Then we might move toward curing addiction—if it can be cured.

Dr. Hollister: Roger, do you find any questions worth answering?

Dr. Smith: I have an important question: "Why was the format for questions from the audience changed for this afternoon?" Most people wanted to have a free and open discussion because we had the opportunity to confront a representative of the Bureau of Narcotics, Mr. Ingersoll. We have some of the best qualified people in the world in the area of drug abuse here today to describe the pharmacologic and harmful physical effects of drugs. All of us know, if we are honest with ourselves, however, that the worst thing about being a drug user in this country is the law. If we know nothing else about marihuana, we know that it hurts to go to jail. I think that we have neatly sidestepped an opportunity to question Mr. Ingersoll and I think that was really important.

Perhaps Mr. Ingersoll, or some other representative of the law-enforcement establishment, is in a better position to answer this question: "What about the drug user as a perpetrator of crime against the non-drug community?" I am wondering how we develop these criminal sub-cultures. Heroin is not that expensive; marihuana is not that expensive; none of the drugs is that expensive. Then why do we have all of this secondary criminality? Criminality is directly related to the law. Nobody takes a drug and becomes a burglar or a con-man or a whore—they do it out of necessity, once they become compulsively involved.

Some questions are related to things that I did on the streets as a researcher: "Did you know about the illegal production of speed in Berkeley and other places?" Yes, I did. I visited a number of illicit speed labs. I knew many of the chemists and top-level dealers.

"What did you do towards eliminating these individuals from our society, to prevent the further spreading of noxious drugs? In other words, what is your responsibility to report and disclose to law enforcement agencies both pushers and consumers?" When I entered the Haight-Ashbury, I was a researcher. I was not there to make moral or legal judgments. I needed access to the people who were manufacturing and dealing drugs, and I needed to come to an understanding of how those things operated. I was not there to act as a policeman. Under no circumstances would I turn in a consumer, so long as he was not hurting anyone but himself. I would make every effort, however, to put him in contact with someone who could help him.

Finally, I am going to do something that I have always wanted to hear done at a drug symposium. I have always wanted to hear a drug expert do this. The question is: "Would Dr. Philip Boyd's approach in Great Britain do away with the illicit marketplace in the United States?" My response is: "It beats the hell out of me." I wish more drug experts would say that when they don't know the answers.

Dr. Hollister: How about you, Judge Oliver?

Judge Oliver: I think the written questions have been helpful, provided the Congressman's suggestion in the aftermath is followed. I seem to detect a good many people who feel a speech coming on, and when that is the condition, the questions sometimes don't materialize. Some of the questions asked me can be answered; most of them cannot.

I would introduce the group that would not be appropriate for me to answer by two questions—one I can answer and the other I am not sure whether I can or not. Question one: "Do written questions provide free speech?" I categorically answer: "Yes." Question two: "Is the legal profession more afraid of questions than the medical profession?" I have a notion that the profession of which I am a member, which is based upon questions and the whole process of an adversary proceeding that permits not only direct examination, but cross-examination and re-direct examination, is indeed dedicated to the idea that the Socratic dialogue is a useful educational device. The legal profession believes so deeply in questions that they authorize the judiciary to put somebody in jail if he does not answer. I am not advised as to whether there is any such process among the medical profession. But on tentative judgment I would say that the legal profession is not afraid; the difficulty, it seems to me, is that without this preliminary sifting process, participants would not have asked questions, but would have been making speeches.

Another question: "As a judge and as an individual who has some knowledge about drugs, what do you personally feel should be the laws governing the sale and possession of marihuana? Please comment on the John Sinclair case." The question adds the information that "Sinclair received 10 years for possession of two marihuana cigarettes." Well, I could say that Sinclair would be 30 years better off than in my state of Missouri in Platte County last week.

I cannot answer what one might personally feel, other than saying it distresses me as a citizen and a judge not to see a greater effort on the part of individuals whose views are reflected by the question, "Can a judge reduce a sentence when a student is picked up with marihuana, even lower than the law requires?" Not unless he wants to turn in his suit as a judge; he must execute the law as written.

Still another question suggests that "the judicial branch follows the legislative mandate," and there is then an implication that there is a way to get around this, some easy solution to a quite difficult problem.

"Do you feel that youth should go to jail?" Not if this can be avoided under any and all circumstances. If the law provides, and there is no other alternative, youth is going to have to go to jail. To avoid that possibility the questioner suggests that the judge "rule against the government on constitutional issues." The constitutionality of the Harrison Narcotic Act was passed on a long time ago, and the same thing is true of every other dangerous-drug act that I know anything about. I must follow my controlling court. There are some law students here, I know, because I spoke to some. It is their job as advocates to dream up better arguments than were made in the Supreme Court of Massachusetts, which, I think, is the only jurisdiction where the question has been tested under the 14th Amendment. But the questioner goes on to suggest that if you can't beat them on constitutional issue, you "can instruct juries not to convict or give them power not to convict."

The fact of the matter is that every jury in the United States has the right not to convict and to say nothing about why it didn't convict.

Another of my questioners asks why judges do not make speeches in order to help educate the public. A judge cannot even answer a question unless he is asked to. And some questions a judge is asked are not proper for him to answer. I would like to explain that because I have questions here on the no-knock provision in the new legislation. It is not proper for me to answer that question here because it is quite likely I will have that question presented very promptly in some particular case. The answer I must give to that question now is, "I don't know." I won't know until I have the benefit of the work of counsel and afford a full hearing.

Another questioner calls for a forecast as to "whether the new legislation reducing simple possession to a misdemeanor instead of a felony will tend to reduce public fear of marihuana." I don't know. I don't know who might know. I am inclined to think that the first large scientific test which has been undertaken will provide some of the answers—if the noise and tumult of the people who believe they already know the answer do not drown out sober consideration of this question on the basis of more adequate evidence.

I would like to think it would, but I think what we really face in discussing drug abuse is one of the many numerous complicated problems with which a free society must struggle. And this process of discussion does not work unless people of intelligence rely upon reason. This is a hard thing to do when there are so many unreasonable things or at least so many things considered unreasonable by so many of us in so many different fields. I think the greatest experiment, and I hope I don't sound corny, is the one that was attempted in 1789 when we laid out the rules of the game we must follow. If we keep the faith we may struggle through; if we don't, I am not very optimistic. Reaction based upon emotion may turn our government into something that is really repressive. We must, I believe, accept the possibility that we can work out many of these problems as we obtain more information.

I think written questions have been helpful in ruling out a good many questions that need not be included in the speeches which I hope will not be made.

Dr. Hollister: Well, if the written question serves no other useful purpose than to hear Judge Oliver give that explanation of the judicial process, it was well worth it. Now, the mikes are open on the floor.

Dr. Walter D. Meester (Grand Rapids, Mich.): I had a question for Mr. Ingersoll, but he is not here now, so I will have to direct it to one of the other gentlemen, possibly to Dr. George or to Judge Oliver. In our community in the last six months, four junkies who were strung out on heroin were picked up because another junkie served as an informer; they received 20- to 21-year sentences after jury trial. Now, this leads me to two questions: First, is it common police practice throughout the United States to use the junkie informer? Is this ethical? The second question: Do these informers, the junkies that are used, generally become readdicted again in the process of informing on other junkies? And I wonder whether the Bureau of Narcotics, or anyone in the government, has set forth any guidelines concerning the use of informers; for example, should they utilize drugs such as morphine antagonists to prevent these junkies who serve as informers from becoming readdicted?

Dr. Hollister: Judge Oliver, you mentioned something about the NARA provision; would that fill the bill here?

Judge Oliver: I think NARA is hitting one particular group within the federal system. One thing that has been of great interest here has been the information I have received on the possibility of variable treatment programs, particularly after hospitalization. I have been concerned about patients committed under Title III and diverted from the criminal process; what happens to them after six months' hospitalization at Fort Worth or Lexington? It would seem clear that they should not be brought back and immediately put back on the street without any supportive half-way house or three-quarter house. The prognosis when going back immediately to the same environment is that the individual will experience relapse very rapidly. NARA recognizes the necessity of supplementation; we have several experimental projects in districts with which I am familiar which utilize community-based facilities. The utilization is dependent upon the jurisdiction of a particular state. Some states view it with favor; others do not. This depends very much on the amount of information that the particular prosecutor may have; also on the judicial attitude of the state judge.

Dr. Hollister: Roger, how about the use of police informers?

Dr. Smith: That is, of course, about the only way that police can break into the addict world. The use of undercover agents is a very risky proposition and very expensive. I think that probably quite a few of them are lost. Some find the life a little bit too tempting, others are a little bit too straight. It is not a very productive approach. The easiest approach, of course, is to pass laws like the one passed in the 1950's, I think, which provided for minimum mandatory sentences. Then you arrest some junkie on the street and charge him with a Jones-Miller violation. Instead of charging under the Harrison Act or the Marihuana Tax Act, you charge him under this act which provides for a mandatory five years and say, "If you make a deal, we will reduce the charge to a Harrison or Marihuana Tax Act." This individual is now faced with a serious dilemma. Nobody wants to go to jail for five years. On the other hand, the reality of being a junkie on the street is that if you inform to the police and other people are arrested, you put your life in very grave danger. There have been a number of killings and beatings that I am aware of in areas like the Haight-Ashbury in San Francisco and the Filmore District, an almost entirely black ghetto in San Francisco, which are related to that policy. Although it is certainly legal, the ethical question is the one that should be asked. I think it is a very destructive and highly dangerous approach to law enforcement.

Unidentified member of audience: I am a lawyer from Ann Arbor, and I also teach American history. We have been studying McCarthyism, and my students ask whether there are any similarities between today's atmosphere and that of the 50's. And I am hard put to deny that there are such similarities. I do have a question of Congressman Scheuer which deals with ethics and professionalism among the specialties. In America, I think all professionals should be made to feel very uneasy; if questions dealing with ethical topics make this audience uneasy, then I say it's all to the good. The question is: Can we really expect lawyers and doctors to endanger their status and community reputations by giving out facts or speaking out rationally in the face of ignorance? For instance, at a cocktail party, if a doctor in the community should be talking to the police chief, what is he going to do? Is he really going to endanger his reputation and status in the community? I think this is one of the basic questions. Are our professions ever going to provide leadership and rationality in America, or are we going to have another era of McCarthyism, like that of the late 40's and the 50's, with youths and blacks the scapegoats instead of communists?

Congressman Scheuer: To answer you simply, lawyers and doctors are like other people. When the chips are down, some of them will stand up and be counted and say it like they think it is, and others won't. That is true today; it has always been true. Of course, it was an eye-opener to me to come to this convention and see how the younger generation of doctors is sounding off and telling it like it is. These men don't seem to be intimidated. It is perfectly clear to me that some lawyers and some doctors are taking advantage of the public fora that are available—the press, radio, television, and community meetings—to contribute to the ferment in the discussion. Some of their colleagues are not. But it is obvious to me that those who want to participate in the evolution of our laws and of our perception of the kind of conduct that is acceptable and the way we should look upon deviant conduct and the way we ought to treat it are doing so. And this forum is a perfectly good example of it. I am proud of the way the professionals here have spoken up. I hope that it will happen more and more across the land.

Another unidentified speaker: I have a question directed to Congressman Scheuer and Judge Oliver specifically about the new law. I would like to say briefly what it is since we have been talking about this new law and nobody has really spelled it out. It contains five schedules, the so-called rationality of which I am frequently called upon as an educator to explain. I cannot do it, and I want to ask why? Why does Schedule I include LSD, heroin, marihuana, other hallucinogens? It does this on the criteria, the medical criteria, that these substances have no legitimate usage in this country at present, that they have not been demonstrated to be safe for use under medical supervision, and that they have a high abuse potential. Schedule II, which includes morphine and other drugs, has the same criteria except these drugs have been demonstrated to be safe for use under medical supervision and they do have some legitimate usage. Schedule II also includes cocaine. Schedule III includes the amphetamines and the barbiturates, and Serenyl, which, I believe, is the veterinary anesthetic we were talking about. These drugs are supposed to have less abuse potential and to be safe for use under medical supervision. While I will not argue the ethicality of using marihuana, LSD, or heroin, every pharmacology textbook I have read seems to indicate that relatively, and that is the only way you can consider this matter, they are safe for use under proper medical supervision. But we now have this written into law, not only in the federal law, but it is going to be written into all the state laws that the Justice Department can convince. Dr. Smith said the other day that if you don't get up and do something about it, they will do it for you. Well, they have already done it, as a matter of fact. The government has decreed what is safe for use under proper medical practice. I would like to know if there was any testimony on this scheduling part of the bill by doctors before it was presented. Secondly, there is a thing that I consider worse than the "no-knock" provision in the bill and that is the continual criminal conspiracy provision. This specifies that anyone convicted of a charge for which he can get a year sentence or more can, upon proper ancillary findings, be sentenced to up to 25 years in prison with no parole. All this with no jury, without being able to confront or even know who their accuser is, in a closed hearing, without the rules of evidence being followed, and on a preponderance of the evidence, not guilty beyond a reasonable doubt. This law was written as if it were an end-run around the Constitution. I would like to have the Congressman and the Judge comment upon these if they would. I think the doctors are going to have to stand up and take a stand some place.

Dr. Hollister: Let me speak first as a doctor. There were many of us who did present extensive testimony that is now part of the Congressional Record; we appeared before the congressional committees considering this bill, but I must say our educational efforts were very poor. I cannot speak for the legislative process but our efforts to try to make the law rational, at least from my point of view, were a dismal failure. But the physicians did not stand by, they fought it tooth and nail.

Congressman Scheuer: Mr. Paul Perrito, who is in the audience, is counsel to the House Crime Committee. We would like to hear from him about how Congress comes to these conclusions.

Mr. Paul Perrito: I don't think I can explain in five or ten minutes a law which covers 61 pages. Moreover, my position as counsel to the House Committee on Crime does not necessarily mean that I totally support this piece of legislation. Perhaps I can best illustrate some of the problems by telling you a story concerning the entire problem of amphetamines, methamphetamines, methylphenidate, and so on. The House Crime Committee bipartisan majority attempted to have a quota system imposed upon the production, distribution, and sale of those four substances, to be administered by the Attorney General upon the recommendation of the Secretary of Health, Education and Welfare. For months we fought; we held hearings in San Francisco; we had thousands and thousands of pages documenting promiscuous overproduction—8,750,000,000 pills, as you heard the other day, enough for 40 pills per man, woman, and child in this country. People like Jim Scheuer fought like hell for this law and yet it was cut out; and it was cut out for several reasons, many unconnected with the legislative process. For one thing, several of the major drug companies lobbied the hell out of the congressmen and put tremendous pressure on the legislative process. For another thing, the American Medical Association failed to take a definitive stand in this area. Thus, my answer is that this law, like so many other pieces of legislation, is a potpourri of the ideas and thoughts of hundreds of thousands of people, literally. The fact that a congressman voted for this does not necessarily mean that he accepts all things. If you want my personal opinion—I was a prosecutor for five years—I am displeased with this law, but it is much better than what we had. Change will come only when people in the medical and scientific professions are willing to stand up and be counted. Unfortunately their voices were not particularly powerful in Congress and the AMA failed to support critical aspects of this legislation and that hurt us grievously.

Dr. Hollister: Is it not possible, without changing the whole law, to move drugs from one schedule to another?

Mr. Perrito: Yes, there is within the law a provision for moving from one schedule to the next; it goes through the judicial process which is sanctioned by a prior administrative process. The Attorney General puts people on notice that this is going to be changed, witnesses can be heard, and, if there is an adverse ruling, one can go to the courts and the process of exhaustion of remedies is then effected. It will take a long time, but it can be done, either up or down.

Same questioner: This law, not the law itself, but the scheduling, purports to be based on medical fact and rationality. It is not, but the only people who can get up and say it is not are the doctors. I encourage you to read the law and take your stand on it because I cannot explain to my students what doctors are doing with the medical problems, or what lawyers are doing with the unconstitutional procedures.

Mr. Perrito: In response to that, the two committees that concerned themselves with the enactment of this law, namely, Interstate and Foreign Commerce and Judiciary, had thousands of pages of testimony from doctors and other people connected with the scientific community. Both houses of Congress read this and there was a substantial amount of input. Obviously, you are displeased with the scheduling. I think if Mr. Scheuer were asked directly, he would say he is displeased with it. But a congressman and a senator are faced with a terrible choice, and when I, as counsel of the Committee, was asked by the congressmen what they should do, I was faced with that same choice. It is much better than what Dr. Smith mentioned—the mandatory minimum aspect of the old Title XXI—but it is not the optimum. What has to be done is that people who are concerned must get down to Congress and put pressure on the congressmen and then they must have the guts to put pressure where it should be put, for example, on the drug companies.

Dr. Hollister: Thank you. May we have a question from the back?

Unidentified member of audience: Is it true that money allocated for the rehabilitation of drug abusers has been strictly earmarked only for heroin abuse or can you use monies allocated by the federal government to treat any drug abuser?

Mr. Perrito: That is not true. The money that has been allocated is for the broad range of drug abuse, not strictly heroin.

Dr. Hollister: Is it Act 24 money?

Congressman Scheuer: Yes, and also Poverty Program funds. There is funding, including Safe Streets Act money, for local drug abuse programs that are not specifically designated for heroin.

Unidentified member of audience: During the session of talks we heard some members of the panel mention two phrases; one was that certain drugs had no recognizable medical utility, and the other was that there was no medically justifiable reason for those drugs not being legalized. There are a number of things in this country that have no medical justification for being legal or illegal, but they happen to be legal. It seems to me that there is more than an underlying medically proven reason that marihuana or LSD is illegal in this country, but one thing that I have not heard probed is why are these drugs illegal. What are the social implications of these drugs in the minds of a gentleman like Mr. Ingersoll, who unfortunately has departed? And part of that question is: Isn't the greatest drug abuser in this country the pharmaceutical industry as a whole?

Congressman Scheuer: My bright teen-age daughters and their friends are understandably dubious about the illegality of marihuana when they see their elders drinking liquor and smoking cigarettes. Alcohol, if abused, can alter minds and create lasting physical damage. Cigarettes are proven killers. Yet alcohol and cigarettes are legal.

If we are not to create another Volstead Act era, we must step up our research on marihuana so that all reasonable men can agree on its dangers and conform our laws to the proven facts.

Dr. Robert Rose (St. Louis, Mo.): I have, I suppose, a rather philosophical question which I would like to pose for the panel, specifically the people involved with the legal aspect of marihuana. As we have all heard in some of the earlier pharmaceutical presentations, while it is not absolutely conclusive, it is fairly well accepted that THC is the active ingredient of marihuana. Yet the new law that was just signed defines possession in terms of plant material rather than something which is psychoactive. I question the rationale behind this kind of logic.

Dr. George: I would like to respond in a general way. I am troubled by this question and an earlier one because I think I hear in both of them the assumption that medical classification is firm and exclusive and that there can be no legal or legislative use of a different scheme of classification. As far as I am concerned, this is a fallacious view of what is involved in the law-making process. Without entering into the question of whether a particular drug should have been on Schedule I, II, III, IV, or V, an issue I am not equipped to resolve, remember that these schedules do not stand by themselves. Rather, they are the basis of the regulatory and licensing system that Congress has adopted in the legislation. Therefore, the principal significance of an assignment of a drug to one schedule instead of another is the degree to which it will be subjected to control. The schedule reflects a legislative judgment as to how dangerous the particular substance is in general, because the stated purpose of the statute is to control addictive substances that have no legitimate use. Therefore, if there is a feeling on the part of doctors that there are improper assignments to these schedules from a medical point of view, I simply tell you that the medical point of view was not the prevailing one. As one who has concerned himself with statute-drafting from time to time, I have to tell you that when a drafter has an objective to achieve through the legal system, he draws his own classifications with that in mind whether or not it is substantially the same as systems drawn with other (i.e., therapeutic) purposes in mind.

Questioner: I asked that question. Why keep passing the buck to the medical profession?

Dr. George: We are not passing the buck to the medical profession. For example, if the legislative concern is with the addictive tendencies of a substance, then I assume that the medical profession is a resource profession. If the legislature were drafting a building code and trying to decide what building materials fall properly in a certain schedule of permitted materials, the legislators, including the lawyer legislators, might well go to mechanical and civil engineers for suggestions about the properties of various substances.

Congressman Scheuer: In fact, historically, they have gone to the plumbing contractors and the manufacturers of the building materials to find out. They *should* go to the mechanical engineers to find out, but they don't. And that is one of the great problems with our building codes across the country.

Dr. George: What I am trying to stress is that the mere fact that a doctor thinks of a substance in a certain way does not mean that other people cannot view it in a different way. A citizen must keep this in mind as he looks at any legislation.

Another assumption I heard to a certain extent in a number of earlier questions or statements is that there is nothing worth salvaging in the existing system and that we can only wipe the slate clean, hoping then for legislation that will achieve Nirvana collectively for all of us. Again, I think that is an unrealistic position to take. As an individual I would like to see a number of things in this legislation changed. It has been said that politics is the art of the possible. Instead of condemning the whole system, I think one is much better advised to work through channels available to him. If one can bring about a little change each year, then he is ahead.

I also suggest that you watch the process of invoking "unconstitutional" as the basis for attacking policies you do not like. In fact, on the basis of actual Supreme Court decisions, most of the issues advanced by participants at this conference have been settled. For example, the use of informers has been approved; the legal issues turn on discovery of identity. As far as no-knock provisions are concerned, the Supreme Court has indicated already that it will accept a no-knock exception to the usual requirements of announcement and identification if state law permits it. Please bear in mind that in the new federal statute and others like it, prior judicial authorization is required, and that is the touchstone as far as the United States Supreme Court is concerned, as it has stated in a number of contexts. Therefore, I anticipate that the new statute will be held constitutional.

To continue, a question has been raised about the lack of jury trial on the issue of increased punishment for recidivism. This is uniformly the case under both federal and state legislation. And its constitutionality has long been sustained. Only substantive issues bearing on the ultimate question of guilt or innocence are subject to jury trial. The matter of augmented punishment on the basis of past offenses is not and never has been a jury trial matter in the history of this country.

Questioner: That is not the question. It is the continuing criminal conspiracy provision of. . .

Dr. George: Under the criminal conspiracy provision of the new law, as I read the language, all legal elements of that crime must be pleaded and proved before a jury.

Dr. Claire Ravin (Detroit, Mich.): I have been a medical examiner in Detroit for some time. I have had to testify in fatal automobile cases in which the suits were against bars where alcohol was served, but I have never been asked to testify where marihuana, hallucinogens, amphetamines, barbiturates, or narcotics were involved, or where there was a law suit against the pusher or the one who marketed the drug. Has there been such an instance?

Dr. George: On the alcoholic beverage matter, there is what is commonly called the Dram-Shop Act which creates civil liability on the part of the bar owner. I know of no corresponding legislation in the narcotics area. Nor am I aware of any effort to develop, through personal injury actions, liability based on the failure of the narcotics supplier to take appropriate care under the circumstances.

Dr. Alfred R. Heasty (Lafayette, Ind.): We have obtained some answers this week, and I think those answers suggest that we as medical men have to go home and do something about these drug problems. Now I hear a lot of argument about laws and one of the speeches this afternoon implied that we need laws that would enable us to do more. What are those laws?

Dr. Hollister: Anyone want to rise to that one?

Dr. George: To catalog changes I would like to see, the first is a restriction on the application of substantive penal statutes to an addict user. A second area is the elimination of mandatory minimum penalties and restrictions on probation eligibility. If these changes were made, then a concerned judge, even when faced with a young offender with three earlier convictions, might still conclude that probation is in order, perhaps on the condition that he submit to whatever treatment the community offers in the case. Under most of our existing legislation the judge is powerless to do this.

I think a third change is legislation that gives clear-cut approval to experimental programs directed toward controlling addiction. A very circumspect provision in Section 4 of the new Act authorizes the Secretary of Health, Education and Welfare, after consultation with the Attorney General and national organizations representative of persons with knowledge and experience with the treatment of narcotics addicts, to determine appropriate methods of professional practice in the medical treatment of various classes of narcotic addicts, and requires him to report his actions to the Congress from time to time. Of course, this offers to the medical profession an avenue it has not had before.

There are other ways we can change statutes to make it easier for doctors to do what they might wish to do as a professional matter, but until existing statutes are changed along the lines suggested, there is little room for the medical specialist to exert influence on legal process.

Dr. Jon B. Closson (Severna Park, Md.): I have a question for the Congressman. The drug industry has rightly taken a beating in this conference concerning its role in contributing to drug abuse. One of the significant underlying factors in drug abuse that has been reported so far is the advertising of drugs as the cure-all for almost any ill. Is there any thought of legislation to control this type of advertising as has been done with cigarettes?

Congressman Scheuer: I have heard of no such effort.

Dr. Hollister: Senator Moss has a bill, doesn't he?

Congressman Scheuer: I have not heard of it.

Dr. Roger Smith: There is one encouraging thing happening in medical schools. I know that it is happening in the University of California Medical Center, and I have heard that it is happening elsewhere. That is, medical students are beginning to refuse gifts which are lavished on them, saying they will make their own choice of drugs based on accurate, rational, scientifically correct studies rather than on the kind of nonsense that is put out in the advertising. I think it is a very hopeful movement which is starting in the medical schools.

Congressman Scheuer: It seems to me this is a real challenge to the organized medical profession through their teaching institutions and the AMA. I think there has to be a total revolution in the way we look at drugs and the way we prescribe drugs—and it ought best be done internally through education and the regulatory and policing capability of the medical profession. I also think that we cannot rely on the drug companies to have a different standard of ethics or propriety than any other industry. If we want to regulate or even scrutinize the way that drugs are prescribed, we should not rely on the producing industry to be the regulators or the governors. I think that we should have licensing. We should know where drugs are going. There ought to be some kind of computer bank that would record all drug prescriptions, the history of drug prescriptions by different doctors, and the history of consumption of drugs by individuals, and that computerized information should be the subject of constant scrutiny by the medical profession. The profession ought to take whatever further educational or punitive or regulatory actions that are appropriate within the profession. It seems to me that there is a gross vacuum in our knowledge of how drugs are being prescribed, how they are being consumed, and the medical results thereof. I think it is a vacuum that the medical associations with government aid, perhaps, ought to fill.

Dr. Eric Chivian (Landover, Md.): Although I very much appreciate the courage and honesty of Congressman Scheuer, I would like to take issue with one point he made in his talk. Congressman Scheuer, you pointed out the vast ignorance in this country about drug issues and quoted statistics indicating that about 80% of the country feels that marihuana is addictive, to explain why Congress passes laws that don't seem to make medical sense. I would like to take issue in that it seems to me that it is the job of congressmen not to be re-elected but to pass laws that are determined by their own conscience and by the medical and various other expert witnesses that are brought before congressional committees. It does not seem that they have done that in many cases.

Congressman Scheuer: I did not seek to justify; I only sought to explain. I said if you want to understand how and why Congress and the Administration react to the terrific emotional concern about addiction across the country, you had better first understand how the public perceives addiction. Then you can understand why Congress and the Administration react because they are perceiving it pretty much in the same way. I did not seek to say this is right; I only sought to say that that is the way it is. And to understand at least the subject which I was given to speak on—namely, Congressional reaction to drug abuse—if you want to know why Congress reacts the way it does, you had better understand why the people react the way they do.

Dr. Angrist: As I understand it, the history of the passage of this law was rather unpleasant. There was extensive medical testimony called for at first, and I would like Dr. Hollister to comment on that because he was in on the battle, I am sure. There was extensive medical opposition to the classifications as they exist but the law was then pushed through without the slightest bit of attention being paid to the medical recommendations. Then, in fact, at this point the medical community further mobilized itself and opposed it still further and again was largely unheeded. Now, what I want to say is, all right, the legislators make the laws; that is their job and they can damn well make any law they want, but how do people come off legislating the control of a highly specialized thing like drugs with gross disregard of the medical community? How do they come off doing that?

Congressman Scheuer: Can I answer that? It will be a damn sad day in this country when the Congress feels that it has an obligation to heed the advice of our nation's doctors, or our nation's lawyers, or our nation's manufacturers, or our nation's university presidents, or any other group in this country. Because not only are special groups not necessarily the best judge of the national interest, they very frequently are not the best judge of their own interests. I suggest to you that the banking industry opposed the Federal Deposit Insurance Corporation a generation ago and today they are happily advertising on the radio "all our accounts insured by FDIC." The securities industry fought to the death the requirement of full disclosure in the Securities and Exchange Act of the 30's. Yet today there is no thinking member of the securities industry who would not have a heart attack if you suggested doing away with the full disclosure requirement which is the basis of the public trust in that industry. Now doctors are human beings. In a real sense they are entrepreneurs, they are private profit makers, and they have come to the Congress with great frequency in the last generation telling us that they are against child care proposals; they are against pre-school child development programs; they are against group health insurance. There is absolutely no evidence that the medical fraternity is the best judge of what is good for the country or even what is good for the medical fraternity. On this particular case I think we were wrong not to listen to them on a very technical question. But, to quote a phrase, law-making on medical subjects is too important to be left to the doctors. . .

Unidentified member of audience: In a small group discussion a senator said that he estimated approximately 25% of his colleagues on the Hill had "turned on" at least once and that there were a couple who had used marihuana regularly. Congressman Scheuer, would you try to give your estimate?

Dr. Hollister: The question is, Congressman, "Do senators 'turn on'?"

Congressman Scheuer: I take the Fifth.

Dr. Hollister: Well, we have gone over the time in an effort to make certain that no one was cheated of free encounter. Thank you.

Evening Session
Thursday, November 12, 1970

Presiding: Chris J. D. Zarafonetis

THE INTERNATIONAL POLICE FIGHT AGAINST DRUGS

JEAN NEPOTE

In practically all countries of the world, the production, trade and possession of a certain number of substances known as narcotic drugs are regulated by law, in accordance with international conventions; the same will apply to psychotropic substances when these are brought under control by the signing of an international convention in Vienna in 1971. Any violation of these laws justifies police intervention, and the arrest and prosecution of the guilty parties.

The various penal codes define drug traffic offenses in almost identical terms. Reading out the drug legislations in force in different countries would be as boring as a bad lecture delivered by a bad lecturer. And yet, these apparently similar laws mask very different realities and I should like to dwell on this aspect of the problem first.

FORMS OF ILLICIT DRUG TRAFFIC

Illicit drug traffic takes very different forms depending on the drug involved, the part of the world where it occurs, and the stage reached in the illicit traffic process.

In Turkey, a peasant who illegally cultivates the opium poppy and sells his crop to a smuggler is breaking the law. So is a chemist in Hong Kong or Marseilles who is hired by a gang of traffickers to convert morphine base to heroin.

In the vast desert reaches of Asia, camel or mule caravans transport 500 or 600 kg of opium over hundreds of miles and are then intercepted by a police patrol. Guns are drawn, shots are fired; a few men fall and the rest escape. This too is illicit drug traffic.

To ship 100 kg of heroin concealed in the fenders of a car from Europe to America on 10 consecutive occasions is also illicit drug traffic.

The gang leader who sits in his Bronx apartment smoking cigars, organizing the traffic, and signing checks under cover of some commercial firm is a drug trafficker, as is the sordid small-time peddler who skulks along in the darkness stealthily hiding his daily supplies in doorways and alleyways. The hippie, wallowing in filth and squalor, who wanders aimlessly around the world, selling the hashish he bought in Morocco, Afghanistan, or Nepal has lately introduced a dangerous note of fantasy to the world of the more conventional trafficker.

I have listed all these examples to show the many different forms that illicit drug traffic can take and the variety of activities it covers, depending on geographic, ethnic, or even quite simply criminological factors. Obviously, the drug problem in the Chicago streets cannot be solved in the same way as the drug problem in the Asian jungle.

DIFFICULTIES IN APPREHENSION OF CRIMINAL ENGAGED IN DRUG TRAFFIC

Ordinary vs. Drug Traffic Criminal. Whatever their differences, drug offenses, when analyzed, reveal certain common factors which have an undeniable influence on police investigation techniques. The ordinary criminal, that is, one not engaged in drug traffic, seeks to harm his victim, by physical attack, by stealing his possessions, by damaging his reputation or by "getting at him" in his private life. The victim can report the crime to the police, describe what he saw, and supply details which will help the police in their investigations. It may even be that the victim knows the identity of the person who committed the offense. In drug cases, whether in Tokyo, London, or New York, there is no victim in the legal sense of the word, only in the moral or medical sense. In other words, there is no complainant; nobody ever complains to the police that they have bought a few grams of opium or smoked a few joints. Contrary to other forms of crime, all those involved in illicit drug traffic are accomplices and on the same side. They are all against the police.

Again, compared with ordinary crime, international drug trafficking offenses leave few clues to help the police in their investigations. In a burglary case, there may be fingerprints, toolmarks, or objects or clothing left by the criminal at the crime scene. There is very often material evidence, a lead to work on. In illicit drug traffic, there is obviously no material evidence that the police technicians can analyze to help the investigating officers. It is, however, true that drugs can be analyzed. Analysis can confirm whether or not a suspect substance is a drug, and if so, which drug, but I do not think it can do much more. It may be able to tell the police that the opium seized in a particular case came from Burma or India, but what contribution will such a vague piece of information really make to investigations?

Traffickers have so little fear of the information the police can obtain from drugs analysis that they do not hesitate to put veritable "trademarks" on the wrappers. The morphine slabs bearing the trademark "999" are made in the Yunnan jungle. The small bags of hashish marked with a cedar are from Lebanon. Traffickers even go so far as to guarantee the customers satisfaction and warn them against imitations! We do no better for Camembert or champagne!

No victims. No clues.

Difficulties Due to International Character of Drug Traffic. To present an idea of the difficulties due to the international character of drug traffic, I should like to tell the story behind the dose of adulterated heroin which some addict is at this very moment administering to himself somewhere in the United States. A year ago perhaps, a peasant in Anatolia, Turkey, planted opium poppy seeds; he is authorized to grow the opium poppy, but controls are lax and inadequate. He declares to the authorities that he harvested 3 kg of opium an acre. In fact, he harvested 7 kg; he also has some clandestine plantations and some friends who are smugglers. One night, a "friend" comes to his house to collect 50 kg of illicit opium. This drug then travels 50 or 100 miles before arriving in the suburbs of Ismir or Istanbul. It is taken to a cellar boasting a few pots and pans. This is what is grandly known as a "clandestine laboratory." It is here that the opium is converted into 5 or 6 kg of medium-quality morphine. The operation is carried out rather like French cooking; all you need is a good recipe and the knack. Two or three months go by and one day this morphine is packed into a carefully prepared hiding-place in the petrol tank of a car, together with morphine from other sources. The trip across Europe begins—from Istanbul, the drug crosses Greece or Bulgaria, Yugoslavia, Austria, Switzerland,

and, finally, after three or four days, reaches Marseilles on the French Mediterranean coast. What about border controls? Millions of people cross frontiers every day and the customs and immigration authorities have only two or three seconds to check a car and its passengers. In other words, the risks are minimal! Then, taking a thousand precautions, the traffickers deliver the drug to a small house in the country. One day, a car draws up to the house and a man steps out; all he knows is that his skill as a chemist is needed. He has been provided with the ingredients and the premises; 100 kg of morphine base have been collected. His job is to convert them into heroin, a process which takes a couple of weeks.

During this time, the group of traffickers who brought the morphine from Turkey and arranged for it to be converted to heroin have recruited chance acquaintances to transport the drug from Europe to America for $1,000 a kilogram. The mission of the chance acquaintances is very simple. They are to ask no questions. They are given a plane or boat ticket and a false-bottomed suitcase and then—trusting to providence—they are off, bound for the New World. If they are arrested, which is unlikely, they will say that someone called "Duval" asked them to deliver the cases to a Mr. Smith 4,000 miles away; apart from that they have seen or heard nothing. As for the heroin, they had no idea what was in the suitcase. In New York, if the couriers are not arrested, the heroin is taken over by another team that divides it into small quantities of 500 or 1,000 grams. At a certain stage along the line, lactose is added, partly to dilute it, but mainly to increase its volume and consequently push up the profits. The drug is passed on from hand to hand until it finally ends up in those small twists of paper which will satisfy the cravings of our unfortunate addict.

The international nature of illicit drug traffic could be illustrated by many other examples—morphine slabs made in the forests of the Yunnan are shipped to Hong Kong via Bangkok and Singapore; hashish grown in Pakistan, Nepal, or Nigeria finds its way to England; amphetamines manufactured in Italy or Germany travel up to Sweden and then down to West African ports and from there to Central Africa; the cocaine in Detroit comes from Bolivia.

The imagination of traffickers knows no bounds. To transport drugs, toys, African drums, goat skins, car bodies, refrigerators, tennis rackets, ski poles, canned food, and, in fact, almost anything can be used.

Where does the all-powerful drug king so dear to the press come in all this? I should like here and now to say quite definitely that there is no such person. We are not up against a single king, but a whole series of little kings; each has his own sidekicks, his own hirelings, and his own customers. It is these little kings of the underworld who transport drugs from one end of the world to the other. In the United States, they have tried to stake out the boundaries of their kingdoms, to divide up the territory between them. The sinister unwritten law of organized crime has enabled them to run the black market and to defend their interests, with machine guns if necessary.

On the other side of the world, in the large cities in Asia, crime is organized through the Triad secret societies. Elsewhere, the structure of the criminal underworld is less clearly defined.

INTERNATIONAL COOPERATION IN THE CONTROL OF DRUG TRAFFIC

No victims, no clues, no respect for boundaries—these then are the characteristics of illicit drug traffic. The fact that there are no complainants, no victims, and no clues means that in all cases the police have always to act on their own initiative and provide the necessary impetus themselves; they cannot wait upon events but must anticipate them. Consequently, all law enforcement agents involved in combating drug traffic must be highly dedicated and determined men.

The conspiracy of silence which investigating officers come up against obliges the police to use very special techniques. The use of undercover agents, first introduced in the United States, has been one of the more successful. The law enforcement agencies must be able to infiltrate secret drug circles, obtain reliable information, and be prepared to wait, sometimes months, before closing in.

The international nature of drug offenses has made international cooperation essential and I am proud to say that the police are real pioneers in this field, for the foundations of such cooperation were laid almost 50 years ago, when the International Criminal Police Organization, known as Interpol, was set up.

As early as 1926, three years after its founding, our organization was recommending a series of measures to combat illicit drug traffic through cooperation.

It is commonly supposed that Interpol is a James Bond-type organization, made up of glamorous ace detectives who spend their time roaming the world, tracking down criminals. This is a childish fantasy, something dreamed up by thriller writers. Police cooperation can only exist on the basis that governs all relations between countries, that is, the principle of national sovereignty. This means that each country uses its *own* police force, and its *own* body of laws. How can there be international detectives when there is no international criminal law and no international criminal court? An international court responsible for trying certain crimes clearly beyond the competence of a single nation could possibly serve a useful purpose. But the time is not yet ripe and furthermore this is not our present concern.

One of Interpol's main merits is to have found the common denominator on which police cooperation could be based: the desire to combat ordinary law crime. It has found wide support because no present-day civilization tolerates murder, rape, or theft. Narcotic drugs offenses are obviously ordinary law crimes and therefore come within the scope of our organization's activities.

Interpol's second merit is to have provided permanent and official machinery for this international police cooperation. How? Firstly, by asking each country to appoint a law enforcement agency to function at national level as permanent Interpol correspondent; secondly, by setting up, at international level, headquarters which coordinate the activities of affiliated countries and centralize information that may be drawn on by all; thirdly, by instituting rapid and reliable communications systems between these units.

This is the network which Interpol has patiently built up over the last 50 years to combat ordinary law crime. Today it covers 107 countries throughout the world and includes, of course, the United States. I will not present the full list, but perhaps it will be of interest that our newest member is Burundi.

This permanent machinery for cooperation offers vast possibilities: exchange of criminal records; checking of suspects' whereabouts, movements, and activities; interrogations; and arrests for subsequent extradition. The list is endless.

At this point, I will quote a few figures to give an idea of the results obtained and the extent of this cooperation. Last year, our radio network broadcast 140,000 messages. Some 650,000 criminals are listed in our General Secretariat records; they have all been engaged in international criminal activities. Statistics covering only 17 countries show that 980 criminals were arrested for subsequent extradition. These same countries exchanged over 200,000 items of information.

The fight against illicit drug traffic obviously occupies a large place in this international police cooperation. In fact, there is no fear of unemployment in this field!

No day goes by without the Interpol General Secretariat receiving reports of international drug cases; last year we recorded over 3,000 such cases, all of them with international ramifications.

A good example of this cooperation is the recent arrests in Switzerland and France of three important American traffickers wanted by the United States' authorities. There are a thousand such examples.

But it is obvious that the efficiency of international law enforcement depends on the efficiency of the individual law enforcement agencies in each country. Here, we have to admit that things are not always as we should like them to be. Not all the ships flying the Interpol flag sail at the same speed. Differences in the levels of development, equipment, and efficiency of various countries are found in all sectors of human activity. Why should the police be any different?

In one country there is a shortage of manpower, in another of equipment. In one, the police officer does not consider drug trafficking reprehensible, but rather a commercial activity with an essential part to play in the society in which he lives. In another country, the police are preoccupied with other problems or else are reluctant to launch campaigns to eliminate drug traffic which could lead to mass rebellion from whole sectors of the community.

PROBLEMS IN CONTROL OF INTERNATIONAL DRUG TRAFFIC

As you can see, the police problem, at international level, is infinitely complex. For example, it is fair to say that if there were no opium in Turkey, there would be no heroin in New York. But it is also fair to say that if there were no addicts in New York, there would be no smugglers in Istanbul.

Let us suppose that tomorrow we could stamp out illicit opium crops in Turkey; this would be a great step forward, but we must realize that the traffickers would simply turn to Afghanistan or the Yunnan for their supplies. To illustrate the point, I can do no better than describe how the situation has developed in Iran.

In 1950, the cultivation of the opium poppy was legal in this country and there were two to three million opium addicts. Opium addiction was part of the Iranian way of life. But a courageous and radical decision was taken and opium production and abuse were declared illegal by decree.

Twenty years later, Iran was forced to admit that opium was entering the country by the ton from its neighbors; the number of opium addicts had dropped considerably but heroin addiction had risen instead. Discouraged, the Iranian Government reauthorized the cultivation of the opium poppy on state farms.

In the fight against illicit drug traffic, any positive step or omission on the part of one government, or police force, has repercussions on the situation in other countries. Interdependency is such that, to be effective, action must be multilateral. There is not and there cannot be any isolated action. This is the very philosophy behind our organization.

But the problems arising from the international nature of drug traffic cannot explain all the difficulties facing the police, and, in the field of investigation techniques, there is another question which has to be asked: As the law stands today, do the police have sufficient freedom for maneuver to deal with these more sophisticated and vile forms of organized crime, or are they hemmed in by too literal interpretations of the law which deprive them of adequate means to combat the drug problem?

How, for instance, is it possible to put an end to illicit traffic if a gang leader is arrested and released on bail after 20 minutes' interrogation and if he can continue his activities until he comes up for trial a year later? Am I right in believing that heroin found on a suspect cannot be admitted in evidence against him unless certain procedural niceties were respected when he was searched?

In the United States, professional drug traffickers tend to be given very heavy sentences. This is all very well, but what is this policy worth if the more dangerous, clever and skillful criminals can outdo the police and defy justice by taking shelter behind procedural niceties or articles of the law designed to protect law-abiding citizens?

You would hardly go elephant-hunting armed only with an airgun; if you did, you wouldn't expect to come back alive. Perhaps this hunting metaphor will provide food for thought.

I realize that I am raising the fundamental and agonizing question of the balance between individual rights and the protection of society. People always tend to fear the powers of the police and are afraid they will abuse them. But surely the dictatorship exercised in a certain section of society by drug traffickers is worse and more to be feared than the action—however vigorous—of police and judicial authorities who are answerable to the public. If you have any doubts, just think of the victims who die of drug overdoses every day.

I personally am convinced that a happy medium can be found between the respect for individual freedoms and vigorous police action against professional criminals. I refuse to accept that any new powers granted the police will result automatically in a restriction of the rights of law-abiding citizens who have nothing to fear from the police.

Having thus stressed some of the difficulties encountered by the American police in the execution of their duty, I am all the more pleased to be able to pay tribute publicly to their work, at local, state, and federal level. In particular, I should like to mention the enterprising and dedicated men at the Bureau of Narcotics and Dangerous Drugs, who are a credit to the United States administration. It is a pleasure to be able to say so here tonight, in the presence of their chief and my friend, Mr. Ingersoll. I am sure he will be happy to associate himself with me to pay tribute to the Bureau of Customs which is working shoulder to shoulder with the BNDD for the suppression of drug smuggling.

CONCLUSION

In spite of our collective efforts, the general situation is becoming worse every year. But we are only too aware of this unprecedented crime wave for me to dwell on it at this time. Nine hundred tons of opium are harvested legally every year and are sufficient to meet industrial and medical requirements. One thousand two hundred tons find their way onto the illicit market, according to the recent estimates of the United Nations Commission on Narcotic Drugs. According to our own calculations, the police manage to seize approximately 10% of the drugs channeled into illicit traffic.

These statistics demonstrate much better than any long speech could that the police cannot hope to eliminate drug abuse and solve the drug problem unaided. They can contain it, and make drug trafficking a risky business, but they cannot possibly hope to deal with it effectively in the face of the proportions it has taken.

I have said that it is a multinational problem; it is also a multiprofessional problem. The United Nations are fully aware of this and, at the prompting of the American government, have launched a coordinated campaign in which the police, of course, have their part to play. Police action is obviously necessary; it must have a place in any program to combat drugs. But law enforcement agencies cannot be expected to bear the sole burden.

I am firmly convinced that the best—the only—solution to the drug problem calls for the combined efforts of all doctors, educators, social workers, economists, parents, and citizens. In certain underdeveloped areas of the world, centuries-old customs have to be transformed, agriculture modernized, and economic development stepped up. Elsewhere, the concept of freedom of expression may have to be modified for it is certainly illogical that, in a country where drugs are illegal, books extolling the so-called virtues of drugs can be circulated in the thousands.

On the judicial level, measures other than imprisonment will no doubt also have to be used against traffickers: withdrawal of passports, seaman's discharge books, and pilot's licenses; a ban on practicing certain professions or occupations, etc. A whole arsenal of measures which would paralyze traffickers' activities should be available to the courts. No doubt, the indiscriminate marketing of too many pharmaceutical products will have to be brought under control also.

In many countries, certain young people on the fringe of society will have to be reintegrated and society itself will have to set a more worthy example, in particular, through the behavior of its leaders. It is an enormous task which must be tackled with enthusiasm. Young people, who sometimes complain that they have no ideal to work for, have one here ready made for them.

Technological progress has increased man's capacity, whether he be a law-abiding citizen or a criminal, ten- or a hundred-fold. Even so, it is wrong and dangerous to think that in modern society everybody is entitled to make unrestricted use of technological facilities for any purpose whatsoever. In other words, it is wrong to believe that in a technological society everything can be tolerated and allowed.

I refuse to believe that drug abuse is an inescapable form of pollution in highly industrialized societies. I hope and believe that society and family structures are still sufficiently strong to resist the corrosive effects of drug abuse. I hope and believe that the moral values of society as a whole will strengthen any weakening in individual moral standards and counterbalance the fallacious teachings of a few irresponsible philosophers.

Morning Session
Friday, November 13, 1970

Presiding: Neal A. Vanselow, M.D.

GUIDELINES FOR THE PRACTICING PHYSICIAN

J. THOMAS UNGERLEIDER, M.D.

Although the problem of drug abuse has become a grave national concern, many physicians still approach the matter reluctantly—if at all. For many doctors drug abuse lies in largely uncharted areas, where traditional methods and training are often inadequate. This presentation will offer some guidelines for the physician in treating drug abuse, and will attempt to help him make the decision to become "involved" in the first place.

As physicians, unfortunately, our role in the history of drug abuse has been inglorious to date; we have been too much involved in the creation of the problem. Young people from families who traditionally have called the physician for more than their health problems too often grow up to become drug abusers. Magical belief in the prescription as a panacea can have tragic ramifications, and there is reason to believe that we have oversold our product.

I witnessed the results of this oversell, in purest form, at a large army basic training center in northern California. There, physicians who worked through the night in the base emergency room would often be reported by the patients (servicemen and dependents) to the Inspector General for refusing to prescribe tetracycline or some other antibiotic for a simple cold. Such was the belief in the magic of the prescription.

Even more frequently, however, the medical profession has been accused of over-prescribing drugs such as the amphetamines, barbiturates, and tranquilizers. The pharmaceutical companies have been accused of both over-production and poor control of distribution. Men in white coats constantly inform us, via television commercials, that anyone can go through life without being angry, depressed, or frustrated if he uses Compoz, No-Doz, Sleep-Eze, and Nytol. Even our (Excedrin) headaches are now numbered. No wonder our young people have followed suit and adopted the motto, "Better living through chemistry."

THE PHYSICIAN'S ROLE IN TREATMENT OF DRUG ABUSE

The physician can play a crucial role, not only in helping to treat the drug abuser or in seeing that an appropriate referral is made, but also in the *prevention* of drug abuse. For treatment is all too unreliable; effective programs and techniques are sadly lacking and results of rehabilitation are poor at best. The best way to combat drug abuse is to remove the consumer demand. Only when, and if, our young people no longer desire or need a chemical way of life will the drug problem recede.

The physician is most effective in prevention through education of his patients. The goal is to provide information to young people concerning the risks of various kinds of drug use. There has been so much hysteria that young people usually distrust adults and rely only on their peers for drug information. Much of this information is mixed with mythology and the dangerously false "facts" which abound on the street.

Prevention also derives from physicians who hold themselves up as figures with whom young people can identify. The physician is still one of the most respected of adults to young people who desperately want to identify with adults who are neither "square" nor drug abusers. Much of their testing and probing of the doctor is designed to serve this end. When healthy identifications take place the teenager will emulate and "improve upon" the life style and goals of the identification figures. He will find his own meaningful alternatives to drugs; in the language of youth, he will "find his bag" or "do his thing."

But when the physician is reluctant to become involved, then the so-called "hippie therapists" take over. This is a tragedy because so often these therapists are not only drug users but psychedelic missionaries. It is fashionable now to be against "speed" and the other amphetamines, the barbiturates or "downers," heroin, and even LSD because it's unpredictable. But so often 13- and 14-year-olds (who are desperately struggling with their identity problems as they try to form healthy kinds of relationships and develop effective maturational patterns) are told by "hippie" therapists that they must take mescaline, or at least marihuana and hashish in order to be able to endure smog, the establishment, and Vietnam. This "prescription" does incalculable harm to these young people.

Special Problems in the Treatment of Drug Abuse

There are some particularly difficult problems in treating the youthful drug abuser. One has to do with privileged communication, particularly for pediatricians and practitioners of adolescent medicine. What does one do when an adolescent patient asks a question about drugs? Does the doctor give the information without determining whether the youngster is using drugs? In other words does he *not* take a careful history? And there is often an intense conflict between privileged communication and the matter of revealing information about the youngster's drug use to the parents. Doctors are often reluctant to even mention the drug issue to teenagers, thus investing the whole idea of drugs and drug abuse with an aura of forbidden secrecy. When even discussion is taboo, then it must be very exciting and pleasurable indeed.

Further difficulty lies in the fact that physicians, like most other adults, are frightened and anxious about the drug problem and often burdened by mythology and prejudice. Our DARE (Drug Abuse Research and Education) teenagers put on an award-winning scientific exhibit called the "Happening" at the 1970 A.M.A. annual meeting in Chicago. Some of the physicians in attendance refused to let their children listen to the rock music which was being played on tape at the exhibit as an educational tool to indicate the connection between rock music and the drug scene. I would suggest that it is only by understanding all facets of the drug problem—including the language of drugs—that any adult can effectively handle the drug issue with a teenager.

Then, too, as doctors we usually anticipate respect and perhaps even reverence from our grateful patients. This is not true in the treatment of drug abuse. Young people have read more on this subject than many doctors, although much of what they read is mythology. They never say, "O wise physician, tell me about the harm of marihuana," but rather, "Alcohol is worse than pot." Then they stick out their chins and wait provocatively for a lecture on morality, which they invariably "tune out." It is obvious to today's teenager that one drug is not inherently *good* (from a moral standpoint) while another drug is *bad*. Chemicals have effects and side effects. That is what can be taught—but honestly. To impart the morality of drug taking, as well as the morality of all behavior, is a parental function.

In addition, many physicians themselves were brought up in the era of "the big lie" promulgated years ago by federal narcotics authorities, who decided apparently that the end (removal of drug abuse) justified the means (tell people anything at all about the dangers of drugs). This attitude has come back to haunt us, so scrupulous honesty is vital. Young people respect a straightforward "we just don't know." The decision concerning any kind of risk-taking behavior, here the risk involved in using drugs, is a very personal one which must be made by the youngster himself. Hopefully the physician will be able to provide the risk-factor information (when it is known) as the factual basis for these decisions.

Discussion of peer group pressures with the youngster is vital too, because in order to gain acceptance many youngsters will adopt ways of behavior, including drug-taking, which are really alien to them.

What is required then in order to be effective with young people? Obviously the physician must like youngsters and, although not necessarily approving their long hair or "psychedelic" clothing, he must be able to distinguish these from drug abuse, a feat often difficult for law enforcement officers. We have already noted that the physician must emphasize honesty and be neither a drug missionary nor a drug moralist. Tragically I have heard physicians defend alcohol, smog, cigarette-smoking, and even war as all being "good for you" in comparison to the ultimate evil in the world today, the smoking of marihuana cigarettes. And this is no subtle plea, on my part, for the promulgation of marihuana use or legalization, for we see too many young people with side effects from this drug as more potent forms of "grass" appear and use grows among younger people.

Most importantly, the physician must not be so committed to the concepts either of personal freedom or of obedience to authority that he misses clues as to what the teenager is seeking. The need may be for more parental interest and expanded responsibility and privilege, or it may be for more parental control and less permissiveness.

DIAGNOSING DRUG ABUSE

For those of us willing to become involved and who are not too frightened to take a good history, some clues to diagnosis may be helpful. Drug abuse is a symptom, a piece of behavior, that must be scrutinized carefully. A vital differentiation must be made between *drug use* and *drug abuse*. Drug use is usually defined as that form of drug-taking which neither interferes with a patient's intrapsychic or internal equilibrium nor his social effectiveness in his daily life, in his job, and in relationships with other people.

First, the physician must know what drugs are being used. Much that is sold "on the street" is spurious and does not contain what is claimed; nonetheless one can usually obtain helpful information by a careful history of what the patient thinks he is using. The route of administration is important; parenteral use of any medication is more serious prognostically than oral ingestion. It is vital to ascertain the pattern of drug use. Does the patient use drugs in the morning before school or all day long, or only in the evening or on weekends? Is the use continuous or in sprees? Is it to alleviate depression, avoid responsibilities, alert significant adults to the problem, or is it in lieu of relationships? Does the youngster seek a delirious reaction (as with barbiturates) or an increase in sensory bombardment (as with psychedelics); is his drug use meant to find happiness and pleasure or a new religious way of life? Is his life style altered to attain the goal?

The physician must also obtain a detailed

history of concomitant legal problems, including recent arrests and/or probation. It is important to determine exactly who in the community knows what about the drug behavior. In other words, what specifically is known to parents, school, and peers? These details, when pieced together, will usually indicate whether drug use or abuse is present.

A further reason for a careful history is that many young people are seeking desperately to have someone put controls on them. Their parents are often too frightened and anxious, even when the first marihuana is found in the home, to do anything about it. This is why so many young people *manage to get arrested,* especially if the schools also ignore the problem. As a general rule, those who use drugs for recreation or relaxation rather than abuse rarely get arrested, unless they are picked up by an undercover agent who occasionally infiltrates a college campus.

A detailed history of impairment of function—at work, at school, while driving—is necessary for a definite diagnosis of drug abuse. Some young people have accidents if they drive while "stoned" and many others drop out of school or are unable to work. They will usually cover this loss of function with complex rationalizations about why school is not meaningful or how the job was just "not my thing." One cannot assume, however, that concomitant behaviors are necessarily causally related, and that drug use produced the school drop-out. A complete history will be most helpful, if the patient trusts the physician enough to provide it, in this determination. With some youngsters it is possible to correlate several cycles of occupational, social and educational endeavor with abstinence from drugs and loss of these interests with return to drug use.

We may conclude then that most young people use drugs for some combination of reasons perhaps including curiosity, peer group pressure, the search for pleasure, rebellion against an affluent and often permissive society, the lack of meaningful alternatives in terms of jobs and education, and in order to avoid anxiety, frustration, anger and their own confused sexual feelings.

GUIDELINES FOR TREATMENT

After diagnosing either drug use or drug abuse the doctor must decide whether to tell the parents, an act to which the young person usually acquiesces. If the patient is using drugs rather than abusing them, which most often means the recreational or occasional use of marihuana with no progression to hashish, mescaline or other drugs, then there is little that the doctor must do. He should make the patient aware of the possible legal consequences of his behavior, and suggest some feedback mechanisms (observing ego or personality on the part of the patient so that if the patient later becomes functionally impaired he himself will consider the possible role of drugs). The likelihood of "successful drug use" (which is nothing to be applauded either) increases with the age of the patient but depends, of course, on which drugs are being used and for what reasons. Young people usually become involved with more than one drug. If drug use is seen as a cry for help, then the doctor must intervene with the parents. Of course it is important that the physician not tacitly condone even successful drug use, remembering that alcohol too is not only illegal for, but often harmful to, the teenager.

One of the purposes of bringing the parents in to discuss these problems is to alleviate their anxiety and help them to recognize the problem in order to intervene effectively. Very frequently the parental setting of limits and expression of interest will be all the youngster needs to stop his drug experimentation. But if all intervention ultimately fails, the parent also has the right to refuse to tolerate drug use while the youngster lives at home. Leaving home may involve another facility if the teenager is young, but if he is over 18 it may only be a matter of insisting that he move out to make it on his own.

Parents can also help provide meaningful

alternatives for their children by helping them work, by giving them responsibility, and by neither loading them with material possessions nor replacing their interest with a permissiveness that is all too often combined with emotional neglect. The young person rebelling through drugs is not a very effective rebel; perhaps the physician can help channel this energy into more effective expression. The teenager must know, without rationalizing, whether he is able to set a goal and stick to it. The goal, of course, must be his and not his parents'. When he has difficulty in being creative or independent or sustaining relationships, then he should be able to realize that something is wrong. If he is "into drugs," it is hoped that he will realize the connection.

It is sometimes difficult to decide when to refer the patient for psychiatric care or other specialized help. Certainly it is more difficult for any physician to ascertain the *etiology* of the drug use than to decide whether it has progressed into drug abuse. When abuse is apparent and simple confrontation and involvement of parents do not markedly alter the situation, then referral is indicated.

When performance at school, behind the wheel, or on the job is impaired, parents must be urged to step in and place firm controls on the youngster. Often the physician will be consulted only after the patient has been arrested. Then the question of probation with psychiatric care vs. forestry or youth camp becomes a matter for his judgment. It should be remembered that some young people do need an environment where they can live drug-free for a time. Some of the younger teenagers who are heavily into drugs do very well at the forestry camps maintained by many states and counties. Other voluntary live-in treatment facilities like the Manhattan Project (by the Salvation Army in Los Angeles) and Synanon are excellent. Unfortunately many young people refuse to enter and many parents are reluctant that they go to these facilities.

The physician must help dispel the idea prevalent today that if you say no to your child you do not love him. When a 19-, 20-, or 21-year-old repeatedly says he will not work or go to school and demands to live at home and receive $5 a day as an allowance *for drugs* from his parents (and his parents acquiesce, as 99% of those I see have done), then we must step in and show how destructive permissiveness can be, and how love and lack of limit-setting are not the same thing.

If a drug-induced psychosis or a schizophrenic reaction concomitant with the use of drugs is detected, then hospitalization is indicated. Hospitalization is also indicated for withdrawal of a physiologic addiction to barbiturates, particularly from the secobarbital or "reds" so popular today. The withdrawal cannot be achieved by a "crash pad," cold-turkey type of approach, as can be done with heroin, but must be gradual and on an in-patient basis to avoid seizures.

If the patient is on heroin the physician should know of the methadone maintenance programs which are recent additions to our therapeutic armamentarium in the State of California and throughout the United States. The requirements for acceptance into this type of treatment program vary, but usually require that the patient be an older, confirmed heroin user who has been addicted for some time and has failed in other treatment modalities.

There are other drug-related areas where we physicians have been inactive, with often tragic results for our young people. One concerns our stand on drug abuse legislation directly relating to patient care. For example, when we wish to use methadone in a hospital like UCLA to withdraw someone from heroin over a several-day period (not to be confused with the daily continuous use of methadone in the maintenance programs), we must report the patient to a *police facility*. This requirement prevents many heroin addicts from seeking treatment, especially if they have not had prior convictions. I receive numerous calls from heroin addicts who must subsequently travel to another state for heroin withdrawal with methadone. California is, to my knowledge, the only state in the country which has such a reporting requirement.

Of course what happens is that many hospitals and physicians ignore the law and do not report such patients. The situation has degenerated to the point where many of the staff working in these hospitals ridicule those who attempt to abide by these laws. Whenever I mention this law and the necessity for changing it in drug abuse lectures throughout the state, professionals respond with comments such as, "Who ever heard of following the law? Break it, we all do."

The tragedy here is that young people quickly receive the message from us as adults in general and physicians in particular that if we don't like any law, all we have to do is break it. How can we then wonder why our young people grow up with disrespect for law and order? The paradox is that law-enforcement officials, with whom I have also worked closely, are so swamped with the problem of drug abuse (60,000 arrests in Los Angeles alone last year, reflecting probably no more than 5% or 10% of the total drug abusers in the area) that they do not even follow up on names of these drug users who are turned in by the hospitals. There is a similar law also usually ignored for hospitals to report needle marks on emergency room patients to the police.

We as physicians must help to make *effective* drug laws, ones that are meaningful rather than permissive, enforceable, and which do not discourage young people from coming to reputable medical facilities. It is no wonder that our young people flee to volunteer medical facilities that have been springing up throughout the state. There are too many hospital emergency rooms where people with long hair and different dress are treated with contempt and hostility, another area where we abrogate our responsibility.

Physicians could also help by supporting the establishment of laboratory facilities to analyze street drugs. It would be extremely helpful in terms of prompt and effective treatment to identify what's being sold on the black market. It might also deter many young people if they know that they are not receiving the drug they think they are getting. But this is now done only by the police laboratories—which will not analyze unless they have a name for an arrest.

SUMMARY

This communication has been a plea—to us as responsible adults and as physicians—to assume our responsibilities in the area of drug abuse. We are in a unique position to make a vital contribution toward solving this serious health problem if we will but become involved.

Note: The Special Drug Issue (April 20, 1970) of *Medical Economics,* entitled "The Pursuit of High," should be read by all to whom this article is pertinent.

DISCUSSION AND QUESTIONS—GUIDELINES FOR THE PRACTICING PHYSICIAN

Dr. Ungerleider: I would like an open discussion on how the practicing physician might use the guidelines developed by those who are involved in this area. Dr. Pollard has agreed to join me and to respond to your comments or questions, and I hope that others in the audience will feel free to help us as resource people.

Mr. Robert Winfield (Ann Arbor, Mich.): I started the medical student drug education program here. When I go out and give honest factual drug education without a moralistic point of view, I do not know if I am promoting drug use or not promoting drug use. Drawing on your work in this area, what is your opinion of this? Do you really think drug education has a role in putting another perspective into drug use?

Dr. Ungerleider: I feel very strongly that drug education is probably the only way to significantly reduce drug abuse. If the course provides education in more than the pharmacologic effects of drugs, it becomes education into a way of life, a life style. The effective treatment programs usually depend almost completely upon the charisma of the person running it. So you represent an adult who has found a way, whose hair may not be as short as some others, but who has found his way. I think if you listen with a third ear, as we call it, you will find the kids asking things about you and your life, and in this way I think you can make tremendous headway. I don't think the kids are going to say, "Thank you. I've decided not to abuse drugs." Or, "Thank you. I've decided not to use them." But I think one can follow and note all kinds of results if one listens carefully.

Dr. Pollard: I think this question is often brought up in relation to our small organization here, Drug Help, Inc., which is educational as well as helpful. Some people argue that when these kids come in high, stoned, freaked, on a bummer, you should give them a lecture on the evils of drugs and the bad things they are doing to themselves. My reply is it would be analogous to delivering a lecture on morality to a man you are treating for gonorrhea. At that time our goal is to help the kid. As we discussed previously, follow-up is not always very successful. And a follow-up is of value only as information. It is a great mistake to treat our young people as though they are stupid and talk down to them. If you can put the facts before them and, as we noted in previous sections and as Tom has said, they find an alternative for turn-ons, then we are going in the right direction. But the moralistic attitudinal things are a complete turn-off, and if anything, I think, would send them right back to the drug again.

Dr. Mohammed Shaalan (Ann Arbor, Mich.): Dr. Ungerleider's question about reporting the heroin user to the police brings to mind the issue of allegiance to God vs. Caesar: the police ask you to report heroin and your own code as a physician says that you should keep this as privileged communication. So I wonder about the message that we are giving here to the kids. Is it that my allegiance is first to the police and second to my profession, or vice versa? I think the key to understanding much of this rebellion lies in the fact that these young people feel there is a certain point beyond which they cannot change the system and are faced with this immovable rock, represented by the 80% of Congress who believe unfounded things about marihuana. The only alternative they have is not whether they should obey the law or not, but how not to get caught, how to avoid the law, and how to have their own code of conscience. What I am asking is, should physicians give a picture of allegiance to their own morals or mode of conscience when it is opposed to the police, even if it means breaking the law from a strictly legal point of view?

Dr. Ungerleider: The concern I have about the message of allegiance here is that if it is O.K. to break the law, then we must connect this with a group that grows up and has no respect for any laws. I think it is very directly connected. And in addition, we are seeing a group that won't avail themselves of traditional medical facilities. They are fleeing traditional medical facilities like never before in our history. What applies somewhat to each physician is even more vital when you consider a hospital. Hospital policy should be: "People in trouble, come here. We'll help you." Now, you say, "Why not? What is going to happen?" I will give you an example.

We have something called methadone maintenance programs in California as everywhere else. John Kramer may have talked about it, because it happened with him. It happened a couple of times in California. There was a fellow who was on probation from one of the jails, one of the correctional institutes. He knew he couldn't stay off heroin so he went into the methadone maintenance program. When this was found out, he was locked up again. So, trying to sneak in and doing something without changing the law had a rebound effect, a negative effect.

Dr. Pollard: I would like to make quick comment on that. First of all, I think the law may be somewhat different in California. As far as I know, in Michigan you do not have to report heroin use. The only thing you have to report is mayhem, a gunshot wound, stabbing and the like. You cannot use methadone as a maintenance. You can use it as for withdrawal but not as maintenance medication.

One thing I want to disagree with, Tom, and this comes under the idea of treating kids as if they are stupid. I think it is fine to say I am a law-and-order man; the law should be obeyed. But this patently is not true. It is not an insignificant fact that doctors' tax records are audited more than any other professional groups? They know perfectly well that we play games with the law. I was rather late for the dinner last night, so I went 45 in a 35-mile-an-hour zone. I think we have to recognize that there are in fact good and bad laws. Look at the epilepsy laws. There are still many states in the United States that prohibit people who have epilepsy from marrying, and if they do marry, their children are illegitimate and there is no estate. Did you know about that? What about the New England "blue" laws? What about consenting sexual behavior between people? The legislation on how we go to bed with our wives, mistresses, and what have you. There are lots of lousy laws. So let's be honest with kids about that.

Dr. Daniel Wilhelm (Port Huron, Mich.): I think this area of privileged communication is something that a lot of physicians have hang-ups on and worry a lot about. A law has passed the Michigan State House allowing privileged communication and treatment of minors for drug abuse on their consent alone. This law is in the State Senate and I don't know what action is going to be taken on it. In our community we are very concerned about this. A lot of physicians would not undertake the treatment of minors for this problem. I wrote my own insurance carrier on this matter and got what I felt was a very enlightened response. They stated that they felt this was such a social problem that some "trail-blazing" needed to be done and they would stand behind me if I was willing to do so. I think this is revealing to a lot of people.

Dr. Ungerleider: At the free clinics in California, where you can treat an "emancipated" minor, they will say to the young people who come in and really don't want their parents to know, "Well, we can only treat you at 15 or over if you are an emancipated minor. We, of course, have no way of checking, but if you sign this form and say you are emancipated, we will treat you." And that is the way they get around it there in their treatment.

Dr. Edgar J. Geist, Jr. (Rochester, Mich.): John, I would like to question you again about the validity of chemical analysis to determine whether a kid is taking drugs or not. I know about acid and pot, but I am concerned about making, if you want to call it, the diagnosis or differentiation of drug-induced abnormal behavior. When we get them into the emergency room, we do bloods and urines on them and we come out with a lot of positive tests. I am anxious about the validity of these tests aside from acid and pot. I would like to pin you down, John, as to the validity of the other drugs that the kids are taking; that is, the evaluation of the chemical analysis of them.

Dr. Pollard: I like to be pinned down, but I am afraid the toxicology that is available in this neighborhood is such that I do not even have to ask myself that question. We don't have full-range toxicological analysis. Maybe, Tom, you have. I know at Bellevue they have and if the pharmacologists in the audience would like to respond to Dr. Geist as to the validity of the toxicological analysis, please do.

Dr. Ungerleider: I have nothing to say on that. Dr. Meyers, would you want to add something?

Dr. Meyers: What does it matter? By the time you get the results, you have to have already acted.

Dr. Geist: I am talking about after-care. That is the important thing.

Dr. Meyers: I don't know of any persistent effect that I would be concerned about other than effect on life style. I think organic toxicity can probably be ignored here.

Dr. Geist: Thank you.

Dr. Albert I. Aronoff (Cincinnati, O.): I will soon become a senior citizen but am still young at heart and working with the Cincinnati Free Clinic. What impressed me most through the 40-some presentations and eight or nine panel discussions that we have had so far is the obvious hostility and undercurrent of hostility between some of the straight physicians and the new style physicians. I cannot see how the former can go back and treat the new style patients, the so-called drug abusers, unless they do something about their own emotional feelings. A vote was asked for from one group that the panel be polled as to how they felt about decriminalization of marihuana. Now this is a big hangup of people in the drug scene. We are always taught through medical school to take a history. And the most important two items that come up in every history of a drug abuser are being "busted" and Vietnam. I understand that the panel cannot take a poll of themselves, many of them have vested interests, and I am not condemning them for not doing it, but I know when I return to Cincinnati, I will be asked how other physicians feel about this. And I would appreciate. . . .

Dr. Vanselow: Doctor, excuse me. Just before the intermission, we are going to ask the audience to indicate their feeling about the legalization of marihuana. This is not going to be a resolution or a proclamation. It is just going to be a poll. We are going to do that later this morning.

Dr. Aronoff: This would answer my question. Thank you.

Dr. Fulfio Ferrari (Ann Arbor, Mich.): In response to Dr. Ungerleider's invitation to contribute, I would like to relate an episode which may be of practical use. All week I have been haunted by a memory. About four or five months ago, I was asleep and the telephone rang at my bedside. When I answered, a very young voice asked me, "Are you a doctor?" "Yes. Dr. Ferrari speaking." "Are you a psychiatrist?" "Yes." And the discussion went on from then in a way that I interpreted with my 4 A.M. mentality as a crank call. I am wondering how many of you have heard at 4 o'clock in the morning a call from a young voice that sounds like a crank call, but maybe it isn't. Maybe we had better listen to the next call. That is all.

Dr. Vanselow: Does someone want to comment on this? Next question.

Dr. Thomas Pollard (Danville, Ill.): First of all, Dr. Ungerleider, what is the mailing address of Project Dare? I think some people might want to communicate with them. I have two other questions. You mentioned Project Dare having a house. Would you talk a little about what goes on there? And, the other question: What effect, if any, does LSD have on the reproductive system and on a pregnancy?

Dr. Ungerleider: The address of Project Dare is c/o UCLA, Neuropsychiatric Institute in Los Angeles.

Nothing goes on at the house right now since they were evicted because they played such loud music when they practiced.

I'll defer the question on pregnancy to someone who feels he can really answer it.

Dr. Joseph Benforado (Madison, Wis.): I am asked this question every day by coeds on the University of Wisconsin campus. And I must say, if you look at all of the experimental work in terms of its implications to human beings, the one thing that you come away with is the following: in other systems, namely lower animals, you can experimentally, under good controlled conditions, produce fetal deformities. Now, the action is very hot here and some laboratories do not reproduce the results. But if you look at any competent investigator who has published, he can show you that under his conditions you can reproduce this in lower forms. This has no direct relation to the human being at present. It only says that in one system it is possible. And from what we know about drugs and what we don't know about drugs in human beings, I think this evidence alone suggests that we are going to have to wait a little while until humanity does its experiment and then we will find whether in some percentage of human beings this same thing obtains. So I tell the girls that if I were in the age group where I could become pregnant, if I were a female, I would not expose myself to LSD for the following reason: By the time you know you are pregnant, if you have exposed yourself to LSD the animal experiments suggest that you may already be in trouble. So you have to wait, and the girl says, "Well, Doctor, how long will it take to know?" "Fifteen, 20, 25 years." "God damn it, Doctor, I can't wait that long." And the only answer there is, "Well, that is your problem, not mine." And I think that is the only tenable thing in the area. We just cannot say, and I think girls ought to wait.

Dr. Ungerleider: I think there is one thing we can say with certainty. If you are a pregnant rat, you certainly should not let anybody inject LSD.

Dr. Benforado: No question about it.

Dr. Pollard: Wouldn't it be wise to add also that these fetal abnormalities are probably related to a direct effect on the uterine muscle and the vasculature rather than to a genetic effect?

Dr. Benforado: How do you know that, sir?

Dr. Pollard: For the simple reason that in studies where LSD has been given later, after the various developmental processes have taken place, it has also resulted in abortion.

Dr. Benforado: I think the answer there again is that we really don't know. And recently a letter in *Science* from a competent investigator at the University of Wisconsin chided one of his colleagues who said, "I could not reproduce your experiments." And he said "Of course, you are not using the same rats that I am using." And I think that is the answer. There is so much variability here that regardless of what the theory is, you have to stay with the facts and the only facts you know are that in the lower forms you can do it. As to whether it happens in the human, we really don't know.

Dr. Vanselow: Thank you. I think we will have to stop the question period at this point. Actually for two reasons: First, we are running out of time and secondly, we did want to take the poll that I mentioned. Dr. Ungerleider has some movies which he will be showing during the break. Do you want to make a comment or two about the movies, Dr. Ungerleider?

Dr. Ungerleider: Yes, I am just going to show a six-minute excerpt of the new four-in-one series called "The Psychiatrists" which is currently being screened. This particular episode concerns a methadone maintenance program that is sabotaged by a California senator. Any relation to people living or dead is coincidental. It does not have a happy ending, and is nothing like the Beverly Hillbillies. You may rather go out and have coffee, but I am going to show it in a couple of minutes.

Dr. Vanselow: Dr. David Lewis, who is here from the Harvard Medical School, has requested that we take a poll regarding the legalization of marihuana. I want to emphasize that this is not a proclamation. We are not going to send this to anybody. We are just going to see how people vote. So I will call on Dr. Lewis to come up and briefly explain this to you.

Dr. Lewis: Thank you. This poll is simply an outgrowth, as was said before, of the rather heated response between the panel and one of the questioners about opinions on marihuana. It is not meant to sway the group in any one fashion or another, but just to determine the range of opinion and put it up as a mirror so that we can see what we believe collectively. As has already been stated, it is not a resolution and is not for the press. It is of interest only to see how we as a group collectively and as a community that has spent a week here feel about the issue. Moreover, I would not even use the word "legalization" of marihuana. Let's just respond to our feelings and our opinion about the new Federal law; let's make it specific regarding the possession penalties of marihuana which, I understand, go up to a year's imprisonment for first offense. It is a misdemeanor, and probation and parole are possible. And for a second offense the penalty may be up to two years. Both misdemeanor and felony penalties are classified in the criminal statutes. The question is, "What is your opinion of the possession penalty in this new Federal law that is about to take effect?" We have four choices: A. Not severe enough. B. Just right. C. Too severe, and D. Should be abolished. "Should be abolished" means that we believe that no criminal sanctions should be used for controlling the possession of marihuana.

We have some cards at the back of the hall. As you leave for intermission, you may pick up one of these cards and express your opinion on this issue. The poll will be tabulated during lunch period and given to Dr. Zarafonetis. He can announce the results this afternoon. [Editor's note: The results of the poll are given in my concluding comments, p. 604.]

AN INNOVATIVE APPROACH TO SO-CALLED DRUG EDUCATION: "TRIGGER FILMS FOR HEALTH" DEVELOPED AT THE UNIVERSITY OF MICHIGAN

STANLEY H. SCHUMAN, M.D., Dr. P.H.

Previous speakers in this symposium have indicated the crucial role of drug education in reaching the growing numbers of youngsters in the population who are at risk of experimenting with a wide variety of drugs. Important as are the treatment and rehabilitation aspects of drug abuse, the preventive value of reaching the beginner cannot be underestimated.

Unfortunately, many of our current efforts at drug education miss their mark of helping a youngster cope intelligently with an ambivalent society—one that uses, abuses, seeks, glamorizes and abhors drugs all at the same time. By emphasizing the so-called facts of drug use and abuse, the educators may run aground on certain areas of incomplete or controversial scientific data (i.e., LSD and birth defects). By emphasizing the dire legal, medical and moral consequences of drug use, educators risk diversion of their message into debates on the legal issues of marihuana and the inconsistent behavior of adults who abuse alcohol, tobacco, and tranquilizers. By emphasizing pharmacology, the educator runs the risk of running out of time and information to cover the numerous classes of drugs and substances, and the numerous combinations thereof, which may interest the experimentally inclined youngster. Finally by emphasizing fear, disability, and death, the educator may be relatively ineffectual because of the "risk-discounting" and "temporary social system" defenses described by Seymour Fiddle in his work at Exodus House among heroin addicts from Harlem.[1]

Fortified by several years of research with young drivers, we reasoned that risky behavior behind the wheel might be similar to drug-seeking behavior, and we wondered whether drug education might be supplemented by a new and different approach.[2] Why not stimulate youngsters to examine the root causes of drug behavior and examine their own feelings and impulses in small group discussions with a skilled discussion leader? Sharing of feelings, facts, and value systems by the open-ended discussion technique takes a little longer, but educators know that the gains of self-discovery and the peer-sharing process are rarely equaled by the didactic approach. During the stages of early and mid-adolescence, the average youngster is resentful of adult-imposed standards, rules, and advice; often, however, he will exceed adult ethics if given the freedom to arrive at his own conclusions.

The "trigger film" concept was originally

conceived at the University of Michigan Television Center in 1967 when the Highway Safety Research Institute requested films as part of a research program to change driving behavior among youthful drivers. "Trigger films" were the answer. They are brief, evocative, attractive films which establish an emotional situation involving a young man and his car, and then stop abruptly, triggering response and discussion among the target audience. "Trigger Films for Young Drivers" have been widely accepted by their young audiences and studies of the films' effectiveness are being conducted. They have been honored in international film competitions, having won major awards in the CINE, U.S. Industrial, and Venice Film Festivals.[3]

Designing drug-oriented "trigger films" for teenagers presented challenges of another dimension. Could a theme as subtle, diverse, and emotional as drug use be translated effectively into visual terms? How could the drug scene be suggested without resorting to stereotyped images such as needles, pills, joints, and bad trips? Themes for the films were chosen from the clinical and educational literature quite empirically: (1) rebellion against authority, especially at home, (2) peer pressures to gain status or conformity, and (3) feelings of loneliness, despair, or boredom.

Three films on these themes have been produced. Ordinarily the films are not shown to youngsters in sequence. Rather, each film is shown once or two or more times as desired by the group, and the remainder of the hour is devoted to searching questions, argument, and discussion under the guidance of a skilled discussion leader. The first of the films, "Linda," presents a mother and daughter in conflict over the radio playing too loudly and suggests drug abuse through Linda's detachment, her flushed face, and continuing music after the radio has been removed. Incidentally, field tests of this particular film during the summer of 1970 in nine schools in southeast Michigan revealed a student-teacher perception gap (judged by the written mention of drugs by viewers of the ambiguous film) of considerable dimensions. Under standardized conditions of viewing and written response, one out of three (32.9%) of the 428 students suspected that Linda was under the influence of drugs, compared to only one out of eight (12.5%) of the 72 teachers and counselors in the same schools.[4] The second film, "Door," suggests peer pressures at a tempting teenage party-within-a-party. What is going on behind the door? Who is invited and why? The third film, "Window," suggests the loneliness, boredom, and depression which might lead to sensation-seeking. A boy with a guitar is in his room. How does he feel? Who are the people he sees out of his window? Why does he suddenly grab some money and leave his room? What happens next?

The "Trigger Films for Health" were written and produced by Alfred Slote and photographed by William LaCrosse, both of the Television Center. They were developed for the University of Michigan School of Public Health with the support of a Public Health Service Grant and are being field-tested in elementary and secondary schools by the author.

REFERENCES

1. Fiddle, S.: Some speculations on risk discounting among young ghetto heroin users. In *Social Health Papers No. 5: An Examination of the Risk-Taking Behavior of Youth,* National Leadership Conference of the American Social Health Association, November, 1969, New York City, pp. 16-27.
2. Schuman, S. H., and Pelz, D. C.: A field trial of young drivers: an epidemiological approach in Michigan. *Arch Environ Health 21*:462-467, 1970.
3. Miller, E. J.: The trigger film triggers response. *Audiovisual Instruction 13*:876-878, 1968.
4. Schuman, S. H.: Drug perception and the student-teacher gap: Reactions of 428 students and 72 teachers to an experimental film on drugs. *JAMA 216*:659–666, 1971.

THE PHYSICIAN AND COMMUNITY ACTION

FREDERICK H. MEYERS, M.D.

Four years ago a prescient member of the hip community accurately predicted the events of the summer of 1967 in San Francisco and persuaded some physicians and others to prepare to work outside of the conventional structure to meet the general medical needs of a youthful, transient population. The activities have continued and broadened, but no one, in this or any other city, can claim to have slowed the evolution of more and more destructive patterns of drug use or stayed the development of ghettos for youthful drug users; nor has anyone developed treatment programs for drug users that are both effective and large enough to have impact. Indeed, if you attended the meeting of free clinics last night, you probably concluded that there is limited interest in supporting people in their present life style with general assistance.

My introduction must, therefore, be cautionary and very complex.

This panel will not consider general medical care, which is provided and accepted in this group much as it is in any other poor population. Treatment here means assistance offered to people who wish to change; to change a habit or to change or develop a life style. There are a variety of problems relating to drugs and to maturation, but, at this time, your facility will be swamped with compulsive heroin users.

When the misnamed "free" (instead of "freeman's") clinics began, the populations of the high-risk neighborhoods were young, mobile and unfamiliar with the facilities of their city of brief residence, a situation that changed rapidly after Labor Day 1967. There is a myth that the people presently caught up in drug use won't accept care from established agencies. No one who has seen how the street people use the welfare system will continue to accept that myth.

I would certainly like to see them get first-class help at a well-financed agency, and I have already said that they will accept such help. What then is the problem? Establishment agencies are not unacceptable; they are simply inadequate in size and ineffective. I will first give you two reasons why this is so and then provide a few guides derived from our admittedly not completely successful efforts to provide counsel for troubled young people, drug users and other deviant sub-cultures.

The first reason why some new type of agency must be developed is that conventional psychiatry and psychiatrists are inadequate in this area. I do not suggest that all psychiatrists fit a stereotype and that my conclusions apply to all. In fact, in my opinion, the most useful remarks in this meeting to date were made by Dr. Pollard.

However, like the rest of us, psychiatrists do react as members of a group. Most are trained to be passive, non-involved and non-directive, characteristics which repel most of the younger patients. Psychiatrists tend to exert a low level of therapeutic pressure, in contrast to the amount of daily treatment time as described by the representative of a therapeutic community who will speak in a minute. Psychiatrists tend, until they have had special experience that overrides their training, to classify drug dependence with other "character disorders" and assume that their therapeutic results will be poor. Actually, the medical model for drug abuse which assumes a psychiatrically "diseased" individual in whom drug use is symptomatic is very doubtful. The most widespread problems today are better viewed in sociologic or anthropologic context. And the most destructive problem (compulsive heroin use) stigmatizes the individual's personality no more (and no less) than cigarette smoking, adolescent masturbation, gluttony, or other compulsion.

Whatever your theoretical position, in assisting the people we are talking about you will find that earthy efforts at environmental manipulation require more time than most of us feel we can expend.

The above argument is important as it influences the design of a treatment effort, but whether or not "shrink" is a term of disapprobation is in practice an unimportant question because of the inadequacy of the total resources. Not only established psychiatric treatment resources, but the total general medical effort is quantitatively inadequate to treat the current problems. Unfortunately, the total educational and social effort appears to be qualitatively inadequate to prevent the problems.

With virtually the entire emerging generation alienated and with potentially or overtly destructive habits of drug use found even below the junior high school level, the number of young people soliciting help in their efforts to alter or establish life styles is so large that a huge number of therapists must be recruited. The professions are not now meeting the traditional demands upon them; how can they expect to add this function to their obligation? But can non-professionals take up the slack? They certainly have not thus far, but we have no choice other than to keep trying.

In judging the programs to be presented, I will be influenced by their adherence to the following precepts:

1. *Gain Leverage, Don't Give up Responsibility*

The goal of the professional therapist should be to gain leverage, to amplify his competence, not to give up responsibility. I do not equate "M.D." and "professional," but some trained and experienced therapist must remain in control.

The non-professional therapists working in the program must be trained. They must learn a few elementary concepts (transference turns out to be of critical importance) and appreciate the influence of their own personality and motives.

Volunteers (whether professionals or untrained) must be assumed to be present for the wrong reasons. They can be useful only if they accept training and refrain from "politicalizing" your program.

Ex-addicts (not to be confused with maintained addicts) enjoy an undeserved special status today. They require the same training as any other prospective worker. Most new programs keep their first success around; indeed, he is usually a success because the program is his new passion. He has value as a display for new patients, but he is at risk because you are keeping him in contact with his drug.

2. *Gain Community Support*

When we talk about the "physician and community action" we are referring to the straight community that must someday provide adequate financial support and personal participation.

It is a mistake to say that the size and nature of the problem force us to work through "community-based" programs. The size of the problem requires support from the general community for its management.

If, as we have insisted above, treatment means working with people who have solicited our help, the patient by asking for help is denying a sense of community with others with the same problem.

"Community-based" may have meaning if one wishes to begin work in the black community (although I have no shortage of black patients in my lab on the 12th floor of an imposing complex of buildings). Otherwise, it means control and destruction of a program by people with problems close to those under treatment, or it is a euphemism for improvised, non-professional care, that is, band-aid medicine in play-store clinics.

3. *No Zealotry*

The people on this panel will present experience rather than models. None of the programs are polished and their practices will be modified by data from their own and other programs. It follows, then, that we want various and different programs to continue operating. It is also obvious that we should be gratified by the same tolerance from other programs. Such tolerance is not always forthcoming, especially when competition for funds is a factor. I see no results that can justify the religiosity projected by some programs.

4. *Audit Function*

To keep zealotry within limits, to protect the program and its clients from the senseless repetition of an ineffective treatment regimen, and to convince and assist other programs, treatment results should be evaluated as in other areas of medicine. No standardized criteria of success or failure are imposed, but there is some obligation to honestly collect and present information. There are many ways in which a program can harm clients and adverse results should be included.

If this is done and if the program has an organization that provides control, continuity and evaluation beyond that provided by the possibly inconstant interest of one or two individuals, the clients and the supporting community will be protected against two major risks: continued provision of ineffective treatment because it gratifies the therapist, and continued provision of ineffective treatment because the vested interest of the workers in the program precludes their honestly evaluating it. This vested interest need not be financial although for the less well trained person it may be. For the professional, it may not be financial but reflects the efforts of a "careerist," as Friedman names him, to convert a problem into a career that will rescue him from obscurity.

COMMUNITY APPROACHES TO DRUG TREATMENT

ROGER C. SMITH, D.Crim.

In communities throughout the country, groups are being formed to cope with the rising incidence of drug abuse, particularly among the young. While many groups have fallen back on a more traditional medical-psychiatric model, others have attempted to establish programs which speak directly to the express needs of the drug-abusing population. While the strategies of the new programs may vary considerably from one community to another, most share a common philosophical understanding of the role which drugs play in all of our lives, and are concerned with how we, as a community, can most effectively, rationally, and humanely deal with compulsive or destructive patterns of drug abuse. Further, most have rejected the arbitrary legal definition of drugs as either "good" or "bad," focusing instead on individual motivations for use, style in which drugs are taken, and the context of use.

As long as we were able to see drug abuse in "we-they" terms, we could cling to one of two views of the abuser: either as a criminal who had to be isolated for long periods of time for his protection as well as our own; or as a patient deserving medical and psychiatric attention, but within the confines of a locked and supervised ward. Efforts to rehabilitate abusers in this context, with the aim of returning them to "conventional" society, have produced minimal results and have led to the widely held belief that drug abuse is virtually untreatable. Failure within this system is considered patient failure, with little thought given to the possibility that it may, in fact, represent treatment failure.

The changing attitudes of many professionals and thoughtful laymen toward drug use and abuse may be traced to the wholesale involvement of middle-class young people in the "hip" drug scenes—from Haight-Ashbury to the East Village of New York—during the early and middle 1960's. The public imagery that has surrounded illicit drug users since the late 1920's has been questioned and we have come to recognize that all sanctioned drugs are not safe for all people, and some illicit drugs are not dangerous for all people.

At the same time, many people have begun to understand more fully the individual and social consequences of attempting to control individual behavior solely through the imposition of harsh criminal penalties. As increasing numbers of children from substantial homes are caught up in the criminal justice system, the problem of criminalization has become more relevant and immediate. Pressures to change our legal stance toward narcotics and dangerous drugs began as the problem spread from the highly publicized centers of youthful drug use to the previously unaffected neighborhoods and suburbs, and eventually into smaller and more isolated communities.

In this presentation, I present one approach with which a community might face its drug problems. The approach recognizes the varieties of styles in drug-taking, and is not focused exclusively on one particular facet of the drug-abuse problem. One of its basic aims is to make treatment accessible and relevant to those in the early stages of drug abuse and to forestall their involvement in a more destructive and criminalized drug scene. Underlying this approach is the notion that every community has the capacity to mobilize its resources in such a way as to be able to deal effectively, rationally, and humanely with the needs of the citizenry.

In describing the principles and techniques underlying community-based programs, I will refer to three programs which I have helped to establish in three separate communities. The first was a specialized drug program, executed in the Haight-Ashbury district of San Francisco, which grew out of the expressed needs of the thousands of transient young people who came to the San Francisco area during the late 1960's. The second, based in Pittsburg, California, a highly industrialized town of approximately 25,000 with a sizable percentage of minority persons, is directed to a varied population of users, ranging from a considerable number of heroin and barbiturate addicts to youthful experimenters with myriad legal and illegal substances. Finally, I will make reference to Marin Open House, a county-wide drug education, prevention, and treatment program, based in an affluent, predominantly white area immediately north of San Francisco. These programs are by no means unique, but they incorporate many of the attitudes, procedures, and components found in community-based programs developing across the country.

ORGANIZATION

Many communities, confronted with a confusing and a frightening rise in drug use, will turn to "drug experts," both professional and non-professional, who are able to sell a particular approach to drug treatment in which the community is only the passive recipient. Hundreds of hastily organized "single-solution" programs have developed in communities within recent years, most of them short-lived and ineffective.

The director of a community-based program is not necessarily a "drug expert." His primary function is to bring together a variety of divergent elements within the community whose support is essential if such a program is to function effectively. The initial organizational task involves legitimizing the program to both the community at large and the drug-using community. If such a program is to operate honestly, one must deal directly with the hysteria and fear which often cloud the issue of drug use. In seeking broad community support, one can speak directly to the self-interests of a majority of business, professional, and lay groups within the community. For example, in most communities where compulsive drug abuse has increased arithmetically, the home burglary rate usually increases geometrically, at great cost not only to the home-owner, but also to the tax-payer who must ultimately pay for additional police services, courts, probation, incarceration, and parole.

Not only must the program be legitimate in the eyes of the broader community, but it must also be regarded as a trustworthy program by those it intends to serve. For that reason, it is often better to establish a private non-profit organization rather than affiliate with an established agency which may present a negative image to the drug-using population. Trusting and open relationships with the drug-using community are essential in the organizational stages if one is to anticipate program priorities. In many instances, there is a great difference between the perceived needs of a community and the real needs. For example, the primary motivation for establishing a program in Pittsburg was related to the relatively benign use of marihuana by middle-class youngsters, many of whom began to challenge their parents' style of life. Few people were initially concerned about the more serious population of heroin addicts and spree users of barbiturates.

In the early organizational stages, a variety of talent within the community can be utilized to obtain necessary physical facilities, establish a sound financial base, make necessary contacts with medical and social agencies, and handle the legal problems of incorporation or contractual agreements with public and private agencies.

PROGRAM COMPONENTS

Prior to the actual establishment of both the Pittsburg Drug Therapy Center and the Marin Open House, the specific goals and objectives of the program were clearly outlined to the community. Some of the basic goals presented to the community in seeking their assistance and support were: (1) to provide students and other young people with rational, scientifically accurate and believable information on the whole range of drugs, both legal and illegal, which are currently being used in the community. The aim is to provide experimenters with facts upon which to base their decision to use or not use drugs, rather than relying on peer pressure and street myths about drugs alone. (2) To inform parents and others about the drugs which they use as opposed to the drugs their children use, and to assist them in dealing with drug-related problems within their own families. (3) To provide a resource for school officials, law enforcement officers, probation and parole agents, and various other persons within the helping professions, both in terms of education and as a referral source for clients. (4) To avert the possible tragic consequences of drug experimentation or overdoses by making available on a 24-hour basis a crisis intervention center, which may be staffed by professional and lay volunteers who are trusted by drug users, have available to them accurate information, and are aware of medical and psychiatric facilities in the community. (5) To offer viable, constructive, and realistic alternatives to individuals who are either destructively or compulsively using drugs, through the provision of a wide range of services that would include assistance during withdrawal, individual and group counseling, referral, and possibly residential care. (6) To provide long-term residential care, when indicated, for individuals who are unable to withdraw from drugs in their present environment. (7) To provide out-patient detoxification, counseling, and crisis intervention to the heroin addict community. (8) To provide a variety of vocational services, including job counseling, liaison with local industry and business, vocational rehabilitation, educational testing, and contacts with educational or vocational schools. (9) Finally, to work closely with the "significant others," to help individuals who have been compulsively involved in drugs and therefore alienated from the broad community.

Depending on the specific styles of use within the community, the program components discussed below should be developed.

Crisis Intervention Center. A crisis intervention program is designed to assist those individuals who are experiencing an acute drug reaction. For the most part, such individuals will be relatively naive and have little experience with the drug that they are using. Occasionally, however, a more experienced drug user will ingest a substance of higher quality than he is accustomed to and find himself in need of immediate assistance. Most calls coming into crisis centers involve acute anxiety reactions related to the use of psychedelic drugs, and problems of heroin and barbiturate overdose. To operate a crisis center effectively, it must be available to callers on a 24-hour-per-day basis. Many programs will have telephones manned by staff and volunteers around the clock, while in other programs where emergency calls are only sporadic, an answering service or phone diverter may be used. Additionally, many centers have developed "calm rooms" where an individual who is acutely anxious may be isolated and given supportive help. This approach is most effective with the psychedelic "freak-out." It is also important that physicians be on call to render emergency assistance when needed, or to make referrals to appropriate medical or psychiatric facilities. It is important that when referrals are made to hospitals or clinics that they are aware of the program's existence and philosophy, and have the appropriate skills and attitudes.

Detoxification. Within recent years, we have challenged the notion that withdrawal from a narcotic can be accomplished only in a closed hospital ward. Many of the heroin addicts currently being seen in community clinics have a relatively short history of addiction; with proper assistance and support they can be withdrawn over a period of approximately a week on an out-patient basis. Since few community programs are licensed to use narcotics in the process of a withdrawal, many have had to resort to the use of non-triplicate, non-narcotic drugs for symptomatic treatment. Few community programs are equipped to handle the rather difficult task of barbiturate withdrawal, and in most instances the addicted individuals are referred to a medical facility for in-patient care.

When an individual is heavily involved in a drug scene, and perhaps living in a home where others are actively using the drug, it becomes necessary to remove him temporarily to accomplish withdrawal. Again, this may be accomplished within a hospital, but it can be done very effectively in a non-medical setting, with 24-hour supervision, and medication administered on a "house-call" basis by a physician.

To avoid the "addiction-withdrawal-readdiction" syndrome, it is necessary to provide a variety of services during the period of withdrawal, including supportive counseling, family therapy, and vocational and legal services.

While many individuals are able to alter their destructive patterns of drug abuse on an out-patient basis, others require a more intensive residential program. The type of residential care varies greatly. For those who have a fairly extensive history of involvement in a criminalized drug scene, where their lives revolve around the use of and obtaining their drug of choice, it is often appropriate to make a referral to an agency which can provide long-term live-in care. The Pittsburg and Marin programs frequently refer individuals to a therapeutic community within a state mental hospital, or to such self-help organizations as Synanon.

In other instances, more specialized residential facilities may be indicated. In the Pittsburg program, for example, a 51-room hotel, entirely refurnished by individuals and groups within the community, now serves as a residential center, providing care for up to six months. At Marin Open House, a rural commune built and run by youthful drug users will be operating a farming program to supply organically grown foods to the program. In the Haight-Ashbury, a series of small urban communes have been developed which in many instances have become autonomous.

Development of Viable Alternatives to Drug Use. Traditionally, the goal of most drug programs has been to prepare the individual for reentry to conventional society. This, however, is a goal which needs to be closely examined. In many instances, drug abuse represents an attempt by the individual to retreat from conventional society and to develop an entirely new life-style. This is particularly true when working with the more affluent middle-class young people. While it is important that a comprehensive program include such services as vocational training and placement, it is also important that the program staff understand that the individual has the right to select his own style of living. It is not a program's prerogative, for example, to suggest to a transient drug user that transiency is an unacceptable form of life. Programs which deal with young people involved in the "hip community" should focus instead on helping the individual to involve himself in a creative and non-destructive life-style.

Many programs are also incorporating the techniques and approaches currently being utilized by the human potential movement, including such activities as meditation, yoga, sensory awareness, massage, Gestalt therapy, and encounter groups. At Marin Open House, for example, such techniques are utilized as an alternative way to get "high."

Ancillary Services. Depending upon the nature of the drug population served, it may be necessary to provide a range of emergency services, including food, housing, money, transportation, legal services, temporary employment, or simply connections to helping people within the community. Drug programs can also be of great assistance to the user by intervening at critical points with courts, probation and parole officers, schools, businesses, social agencies, and parents.

While such services may be carried on by the program, it is more likely that there are agencies which can do a more adequate job if they are sympathetic to the goals of the program. For that reason, it is important that community programs establish a link to legal services, welfare, emergency housing and food programs, and with the various agencies within the criminal justice system.

Information Center. A community program can be a valuable resource for many professional individuals and agencies within the community, in terms of providing up-to-date, reliable, and scientifically accurate information on drugs. Further, it can serve as a resource for adults who may or may not have children involved in the drug scene, in helping them understand how and why people use drugs.

In areas such as the Haight-Ashbury district, new and often unidentified drugs appear on the street regularly. It is important that any drug treatment facility have access to laboratory facilities where compounds may be analyzed and the information rapidly made available to the drug-using population. Such a service will often prevent a flurry of acute drug crises if information on potentially harmful substances can be rapidly disseminated throughout the drug culture.

Drug Education in the Schools. The most productive approach to drug education in the schools is a matter that has been sharply debated in recent years. Traditionally, we have resorted to "scare techniques," relying on the threat of arrest and long jail sentences or adverse reactions to drugs as a method of deterring students from experimentation. However, in recent years, most young people have been able to obtain scientifically reliable information on their own, and, as a consequence, much of the information which is presented by "experts" is regarded as fraudulent. Likewise, the drug films which have been extensively employed in school programs are for the most part unrealistic, and often regarded as ludicrous by students. Many such films are outdated, fail to address the concerns of the student, and are often law-enforcement- or punishment-oriented.

As interest in the drug scene has increased nationally, many large industries and businesses have begun to produce "drug packages," pamphlets, and a variety of educational gimmicks, most of which are expensive and of questionable value.

Likewise, the drug "spectacular" which has become the standard approach to drug education in the schools must also be called to question. Normally, drug "experts" are assembled to speak to large numbers of students and present divergent views on the pharmacology, psychology, and legal aspects of drug abuse. For the most part, such presentations tend to be confusing to students, and often end up in debate between the participants. The members of the audience are then left to pick and choose a point of view with which they happen to agree.

A community-based drug facility has a responsibility to provide the schools with a resource which they can count on in developing an appropriate drug curriculum. Two approaches to the problem of drug education appear to be worth further exploration. First, many schools are developing curriculums centering around drug abuse which are an integral part of the school curriculum, grades kindergarten through twelve. Normally, they are a part of an on-going health or family living series, or are incorporated into one of the basic sciences. Drug information is presented in much the same way that any information relating to physical health is presented.

The second approach, which appears to have great promise, is a much more individualized approach, utilizing the skills of teachers who are trusted and respected by students. Such teachers can discuss in small groups the various questions which students have regarding drugs, present factual information regarding drugs, and allow for a free, open, and honest exchange of ideas. Additionally, such teachers and/or counselors should be readily available to individual students who wish to discuss specific questions in depth, or wish to share drug-related problems with a concerned and sympathetic adult. This approach is hindered in some states by laws which require teachers to report any student involvement in drugs to the administration, which is then required to relay this information to the police. In California, for example, students are well aware of the fact that they share their drug experiences with teachers or counselors at their risk, and, conversely, teachers are aware that their positions could be jeopardized should they fail to relay this information to administrators.

Case Finding. One of the major strengths of the community-based approach, derived from its relationship to the drug-using community, is its ability to prevent the young drug user from progressing into a more destructive or criminalized drug scene by intervening during the early stages of drug involvement. This is often possible because of information given to staff members by other clients and because the facility is regarded as trustworthy and reliable by the drug-using community. In some instances, however, individuals in need of assistance for drug-related problems are either hesitant or unable to seek assistance. In such cases, it is possible for agency staff to seek out those persons and attempt to involve them in a therapeutic program. It is particularly important that the community program have a good working relationship with the police department, the juvenile court, juvenile and adult probation and parole agencies. In Pittsburg, for example, many young persons who might otherwise be arrested and processed through the criminal justice system are instead referred by police to the drug therapy center for assistance. Likewise, probation and parole agencies prefer to keep young people in the community rather than attempt to find appropriate treatment within the confines of a juvenile institution. Many juvenile court judges have come to recognize within recent years that institutionalization is, for many young people, a sufficiently negative experience to assure continued and progressively more destructive involvement in a drug scene. Many judges are sympathetic to the notion that such problems can be dealt with most effectively on the community level and are willing to use such a program as an alternative to incarceration.

Incidentally, such intervention on behalf of drug-users caught up in the criminal justice system is tremendously important in terms of legitimizing the program's function in the eyes of the drug-using community.

SUMMARY

Many have questioned the necessity to establish another specialized helping agency in communities where a variety of social, medical, psychiatric, and vocational agencies are already established and operating. Perhaps the major reason for such agencies' existence can be found in the attitudes of the community toward the drug abuser. Despite the changing view of the nature of drug abuse, many established agencies continue to be judgmental and moralistic in their attitudes; because drug abuse is a patient-initiated and patient-sustained problem, they do not regard it as a legitimate problem which they are obliged to treat.

As middle-class drug use increased in the early and middle 1960's, many individuals sought assistance from more traditional health and social agencies in the community, and found them totally unresponsive to the problems of drug abuse. Discussions with many such agencies have suggested that for the most part they have become rigid, bureaucratized, self-serving, and unable to respond to the changing human needs within the community.

Finally, many well-meaning and sympathetic agencies continue to regard drug use and abuse as a special form of pathology for which they are totally unprepared. While the medical problems associated with drug abuse are, in fact, a specialty within medicine, the problems of alienation, compulsive behavior, and loneliness are not. Were such agency personnel to attempt to deal with drug users in the same fashion that they deal with other individuals in need of assistance, they would quickly realize that drug abuse is not particularly a unique human phenomenon.

A well-organized and administered community-based drug education, prevention, and treatment program can provide valuable services to a community in many areas. To the drug-using community, it provides timely and appropriate service, it is accessible and non-threatening and will assist many individuals through a period of their lives when the choices are either personal growth and fulfillment or a disastrous involvement in a criminalized drug subculture.

Such a program may also be regarded as an asset to a community on a strictly monetary basis. With each addict taken from the streets, the cost to citizens in terms of criminal activity decreases, as do the costs of the entire criminal justice system.

Whether such agencies continue to function in the specialized role of drug treatment or not, it is probable that their impact will be considerable. In many communities where such programs have operated successfully, they have been a catalyst for bringing together divergent individuals and groups to work on common social problems. The approach has also called to question the whole legal approach to controlling individual behavior. Finally, and perhaps most important, this approach may have great impact on existing and established social service agencies in the community. Many agencies have already begun to adopt some of the therapeutic approaches which have been utilized in creative drug programs within recent years, and many are beginning to realize that they can no longer hide behind the slick professionalism which has been their protection for too many years.

THE OPERATION OF A COMMUNITY-BASED TREATMENT PROGRAM—USES OF PROFESSIONAL AND NON-PROFESSIONAL THERAPISTS

JOHN H. FRYKMAN, B.D.

I should like to share with you some of the things we've learned from our operation of a community-based program for the treatment of drug abuse.

Let me begin by stating my feeling that any drug treatment program should operate on an ecologic basis rather than an interdisciplinary one. By this I mean that there should be a team of people working together to help people solve their drug abuse problems, a team which provides a life support system which allows the drug user to begin to take responsibility for himself. In this system there is little or no need for titles; each worker contributes in his way to the life and operation of the treatment facility, making use not only of his own particular skills as a physician, counselor, or therapist, but also sharing in the responsibilities of the others. This is quite different from the interdisciplinary approach, where each discipline does its own task, and little else: the doctor gives pills; the counselor counsels, the nurse takes temperatures.

The outpatient treatment program should operate as a free-flow, off-the-street, therapeutic community in action. As a person steps through the door, he should feel that he is beginning to participate in a community which is going to make it possible for him to alter his life style. In such a system, each person extends the ability and the affect of the other persons in the team. The ex-addict helps the doctor to work more easily as a physician; the doctor helps the counselor understand more adequately the physical problems which are involved. No person is confined to one particular precise function. It seems to me that as the use of "physicians' assistants" increases in our country, they will play an increasingly important role in drug treatment programs.

In terms of the physical arrangement of the treatment facility, this means that there are no desks and offices as in a typical setting, but small groupings of comfortable chairs. There are no small cubby-holes where people go off unseen behind plastic and glass to counsel in some mystical way. Everything is out front. In fact, in most counseling situations more than one person becomes involved. If a patient-therapist is talking to someone and needs the advice of a physician, he is free to call across the room, "John, will you come and help with this problem." In this way all concerned receive continuous and constructive feed-back on what is taking place.

This kind of community-based treatment program is, in fact, organized around the needs of the people who come for help, not the needs of the people who work there. The latter is too often the case in so many establishment as well as non-establishment programs; endless, fruitless hours are spent in arguing who is in charge of what, who gets which office, who is to represent the program to the public, and so forth, *ad infinitum.*

To show how important these principles are consider the following examples:

1. In Alviso, California, a health clinic was set up for Mexican-American farm workers. The most significant barriers which they had to hurdle before moving on to effective work were: (a) The physicians working there wanted all patients to come into the clinic to receive care; other workers knew that a large number of the population were unwilling to do this. Solution: a cadre of medical aides was trained to go out into the community. (b) In order to be totally accepted in that particular community, everyone who worked for the clinic would have to be able to speak Spanish; there was great weeping and wailing and gnashing of teeth before all agreed that this had to be done.

2. On a lighter note, to show how the physical plant of the clinic can sometimes get in the way, a beautiful new OEO Health Center was erected in East Palo Alto to meet the needs of the predominantly black population. The clinic had a full staff, also predominantly black, and yet no one would come and use the facility. There was no trust in the community that the people would receive better care there than they had received from the established institutions of the past. A friend of mine was called in as a consultant to suggest ways of encouraging people to come and use the facilities. As he walked in the front door, the first thing the consultant noticed was a large copy machine which was working away, "clunk, clunk, spiiit, clunk." It was his guess that if he could change the image of the machine, people would begin to come in. He hung a simple sign on the front of the machine, "Out of Order," changed the game, humanized the facility by so doing, and people began to come in. Enough about facilities.

In terms of supervision and coordination, the first principle of a community-based program is that almost never should a physician be the head of the team. Doctors are viewed with great suspicion by the drug-using community: they have been turned in by doctors; doctors have told them that they can do things for them which they can't; they have been put in the "one-down" position by physicians too many times; and, probably most common of all, preached to about how dangerous drug use is. Believe me, drug users know well the dangers of a compulsive style of drug use.

The second principle of a community-based program states that there should be shared responsibility, with sole authority never resting with one man. Authoritarian models are only latter-day versions of the "man," already too well known to the drug user. The staff of the program should be involved in the decision-making process with respect to policy and changes in direction with specified individuals assuming responsibility for certain sections of the program. The focus should be primarily on the treatment of people, not on research, statistics gathering, or good feelings among staff.

The third principle is that there ought to be a stated policy about the way that money is to be used in the program—the money already in hand as well as money that might come in the future from a new grant, or government contract. Staff members can too easily fall to haggling about where the funds are to be used if these priorities have not been worked out in advance.

According to the fourth principle, the staff must determine whether they will recruit people who have specific qualifications for a certain job or whether the program is going to employ whoever is available, training them to do the work. Many programs have gotten into deep trouble because they have taken on staff people who were available and who would work for little because of either lack of policy or funds. Sometimes, of course, available people have worked out splendidly.

The fifth principle is that each facility should develop a therapeutic design which it intends to follow consistently. Many programs have been wrecked by letting everyone "do his own thing" in the context of what is supposed to be treatment. A multi-modality approach is probably the best approach for a *community,* but there should never be a multi-modality approach within a particular *agency;* this is tantamount to institutional schizophrenia, and such agencies need to be treated as what they are, crazy. I'm not suggesting any particular modality at this point, only that each plan be thought through. Please remember that if the form of treatment is not acceptable to the person coming for help, no matter how well you have thought it through or how successfully you think it will work, you won't have anyone to treat. Just as an aside, please don't use Rorschach tests, MMPI's, or CPI's to evaluate the people coming for help; from my experience, they are not very helpful and in many instances get in the way of a therapeutic process.

From time to time invite a capable, nonbiased person to come and spend a few days observing your operation. Let him evaluate what you are doing. It is easy to lose sight of problems within your own organization, especially when you are close to them and have a deep personal commitment to the place where you spend a major portion of your time.

One final word with respect to organization, coordination, and supervision: it is relatively easy for a program to become too professional and to be taken over by the professionals. Many of you will remember what happened when alcoholism was finally recognized as a disease: when the professionals came in and took over, the alcoholics received less and less care. Professional energy went into statistics-keeping, research, hospitals and other centers, with less and less actual care being given directly to alcoholics. This has been well documented by Dr. Sanford J. Feinglass, formerly with the National Institute of Mental Health and now Co-Director of the National Drug Abuse Training Center in Hayward, California. I fear the same outcome as the danger of drugs becomes a more popular issue.

In terms of training, it's my opinion that all levels of staff should be included in both professional and sub-professional training programs. Expose the staff to a variety of approaches: transactional analysis, operant conditioning, Gestalt, primal, intervention therapy, brief therapy, chemotherapy, and so on. Whatever the facility's therapeutic design, the staff should be continuously exposed to new input. They are in it together, and should learn to draw from a variety of sources to make their own approach much more effective. This procedure also exposes the persons whom you bring in to do the training to your program, and valuable feedback can often result. In terms of the way people relate to those who come in for help, a team approach seems best to me. Always counsel, rap, discuss, intervene in tandem, with one person assuming the major responsibility for what is happening and the others giving him valuable feedback and criticism after the session is over. This has proven to be an effective training technique in the situations of which I have been a part.

Another technique, in terms of staff involvement, is to have the member list all of the problems which they see in the particular treatment program. Then begin systematically to find solutions to the problems as part of a group staff process.

Other group experiences have proved valuable to me as training tools as well: reversing roles of specific staff members and letting them act them out; making use of psychodrama; having one member work with a patient while the rest of the staff observes, discussing the case after the person has gone; and making constructive use of the wide variety of other group methods which are available such as encounter, attack, Synanon Game, sensory awareness, dance groups, music groups, and so on.

The "Plunge Experience" has proven a helpful training technique for many who are new to the drug scene and to this particular subculture. It consists simply of putting a person out on the street for a 12- to 24-hour period, with little money and no identification; he is asked to survive (within a certain circumscribed geographical area) for the period of the "plunge." He looks for work, tries to find food and a place to stay. It's amazing how quickly certain prejudices break down after such an encounter.

When using volunteers, it is a good idea to establish a regular system for training and be ready to drop any volunteers who are not willing to conform to the system. The program should require a certain number of hours of training before a volunteer begins to work and a certain number of hours of continuing training each month as they serve.

The volunteers should understand the therapeutic design of the treatment program and be willing to support it, whether it agrees with their own views or not. It should be made very clear to them that they are there to serve the needs of the people who come to be helped, not their own. They are part of a healing team and are respected as such.

It is usually a good idea not to have volunteers answering the telephone. They are not always able to handle the calls, and must spend their time tracking down one of the staff who can, a demoralizing waste of time for everyone.

Finally, it is my opinion that volunteers should not share in policy-making along with the rest of the staff since their own personal needs are met in many other ways through their participation in the program. Many programs have been in deep trouble with their community and the people they serve over this very issue.

In conclusion, we ought to remember that everything that we do setting up community-based drug treatment programs is experimental. It is easy for us to take ourselves too seriously; we need to do a good job, a responsible job, but we ought to realize that most of the people who come to us for help are going to get better or worse no matter what we do. Because our part in it is so infinitesimal, we ought to strive for that unique spark that can provide the catalyst for change in the people who come to us. This cannot be accomplished by playing it safe. Crazy problems cannot be solved by rational solutions, sometimes they require crazy solutions. Let us be free enough to dare to be wrong as we strive to do what is right for those who need our concern and help. We are treating people, not diseases; not dope fiends, not someone whose chemistry is altered, but people who have the same basic needs, feelings, thoughts, desires, and aspirations as the rest of us. If we focus on the people rather than the disease, we'll go a long way toward contributing something to the solution of the problem of drug abuse.

The following guidelines, adapted from my book, *A New Connection,* published as a cooperative effort by the Center for Drug Information, Research and Education, Hayward, California, and Scrimshaw Press, 149-9th Street, San Francisco, California, are included in the hope that they will provide practical, pragmatic, and useful directions for both professional and nonprofessional workers in the field of drug abuse.

IN GENERAL . . .

There is no *one* right way to conceptualize a drug treatment program. Each community must evaluate its own local situation: the staff that is available, their particular skills, the facilities and, most importantly, the needs of the community.

It is important, however, that each drug treatment program strive to provide a model that differs from those which have for so long dominated the health care scene. The interdisciplinary approach only encourages doctors, social workers, group therapists, employment counselors, and others to apply their particular Band-Aids to particular sore spots. I prefer the alternative ecological conceptualization—where everyone is part of a system of care which supports and stimulates life, where titles have little meaning and where professionals, paraprofessionals, and lay people assist each other as well as those who come for help.

A DRUG TREATMENT CENTER SHOULD BE . . .

— a friendly, warm, human, relaxed place, a sanctuary from the streets;
— a place where people receive help, whether they are able to pay for it or not;
— willing to follow through with the person on all his needs, not only stopping the use of drugs;
— guided by the needs of those who come for help, not by the needs of those who dispense help.

YOUR TREATMENT CENTER MIGHT INCLUDE

— around-the-clock emergency drug information and referral service (especially important in communities with many young and naive drug users; anxious parents will also make good use of such a service);
— psychiatric and medical referral service;
— detoxification facilities, residential and out-patient;
— group living houses, therapeutic communities to provide longer-term care;
— individual counseling;
— group therapy;
— drug information and education service (may be provided through lectures, forums, and publications);
— training institutes helpful to other agencies (social workers, probation and parole officers, teachers, etc.);
— employment, training, educational, and vocational counseling for persons in the program;
— street workers, who can circulate throughout the community noticing problems and helping people;
— outposts in removed, idyllic settings for marathons, training sessions, creative activities, special programs, staff planning and evaluation;
— the "night rap" (see page 538);
— developing "home" industries (arts and crafts, restaurants, shops) as a "first step" towards employment opportunities;
— cooperation with other programs;
— contracts with government agencies (such as NARA) and state.

These are but some of the possibilities. The direction and scope of any drug treatment program will ultimately depend on the needs it serves and the imagination of those who run it, and, regrettably, on the limitations imposed by the available resources.

WHAT TO DO WHEN THE PATIENT COMES . . .

Giving Information:

— Whenever giving information over the telephone about drug-related emergencies, be sure to stress the need to get the person to a hospital or physician if there are serious problems. If the caller is fearful that he might be turned in to the police or abused by the hospital staff, offer to send someone to accompany him and help make the bridge to the institution. If he still refuses, give whatever information you can to help him cope with the situation. If you do not feel that you can help, try to keep him on the phone while you seek advice from someone who can.

— When a person comes in for help, treat him with respect and avoid anything that might make him defensive. Many people will check a number of times to see what kind of "vibes" you give out before asking for help with a problem. When someone comes into the center, don't taunt or patronize ("I'm Mr. Jones, the intake worker; what do you want?"); instead greet people casually and offer something of yourself ("Hi, how are you? . . . I'm John . . . Would you like a cup of coffee?").

— People often get "up-tight" if they must wait too long to see someone. Casual rapping not only eases waiting, but loosens people up so that they can "get into it" more quickly when they see a counselor.

Securing Information:

— In getting information—for treatment forms, records, whatever—the most important thing is to approach the individual as a person. Your patients are not freaks, misfits, weak-spined, or anything else. They are individuals with the same kinds of feelings as anyone else, especially the need for approval and acceptance.

— In the first meeting, don't start filling out a form. Talk to the person; make brief notes if you like; draw him out. You can transfer your notes and impressions to the form after the person has left. Place the form in a folder which you can use to collect further information in the future. At a later date the individual might be encouraged to write some things himself—his feelings about the way he has been treated, criticisms about you or the program, things which he'd like you to know and which you haven't been able to draw out.

— Never give the person a form to fill out by himself; this only puts him into an institutional box with which he is probably very familiar, and about which he is probably very defensive.

— You can secure most of the factual information you need without being an interrogator, by talking with the patient about himself and his life ("Have you been living around here long?"). People usually like to talk about themselves, even when their lives have been troubled. Don't ask "When is your birthday?" but rather, "When were you born?" This allows him to relate an experience rather than give you a statistic. Instead of "What is your address?" try, "Where are you living now, Peter?" By asking questions in a way which communicates your interest in him, you will facilitate the establishment of a relationship with him. If he resists, or you feel that he doesn't want to give the information, forget it; if getting the information interferes with your potential to help him, you don't need it.

— The main information you should try to get in the first interview is:

1. The person's name, where he lives, who else is in his family, how old they are (should take only one or two minutes).

2. What's the problem? If you talked to him on the phone before he came in and he told you the problem, pretend that it didn't happen, get him to say it over again. Try to clarify what the problem is as he sees it, not as you see it; this can give you a clue to where intervention might be made.

3. Why now? The problem has been going on for some time; why has he decided that this is the time to seek help? This may often point to the direction needed for a particular individual.

4. How has he been trying to deal with the problem? This may suggest the best place to begin; often the difficulty lies in the method with which the user tries to solve his problem (i.e., shooting dope), rather than the problem itself.

5. Set an initial goal, either in your own mind or stated directly to the person, in order to evaluate what's happening. If the drug user doesn't achieve the goal, it's not always "his fault"; often the goals are inappropriate or unrealistic.

6. In my opinion, the initial interview should be kept as short as possible, usually not more than an hour and sometimes as short as thirty minutes. There is real danger, in a long initial interview, that the user may begin to identify his dependency with the treatment program, rather than take responsibility for himself.

— Again, the needs of the person coming for help will determine what happens, and we need to be continuously aware of that fact. If the system of treatment is unacceptable to the person coming for help, no matter how well conceived or how well trained the staff is, it is to no avail.

— Some programs make it a point to gather certain kinds of information for research purposes, or for use in justifying the program when making application for funds. When you inquire about the chronology of drug use and ask what drug he started with, the patient is very likely to answer "Marihuana." This is in many instances an unreliable answer, and warrants further clarifying questions: "Do you mean that you never tried alcohol, cigarettes, or any kind of pills before you tried grass?"

— Getting this kind of information may or not be helpful to the person coming for help. If obtaining this information gets in the way of establishing or continuing a relationship with him, give up on it quickly.

VOLUNTEERS AND STAFF

Duties:

— The staff member who first identifies with a person who comes in for help should be the bridge to all the services this person needs—medical, social, housing, job, school, whatever. Generally he should not procure these services for the person, but help him to obtain them himself.

Education and Feedback:

— Staff and volunteers should take the initiative in educating themselves about drug use and abuse. We keep a resource notebook in our center where interesting articles can be shared; everyone is encouraged to bring things to be included in it. Up-to-date lists of good reference books can also be kept in the notebook.

— We also keep a log book with two basic sections: a log of all incoming and outgoing telephone calls complete with appropriate messages and actions taken, and a reference section with referral information on other agencies and programs (employment, medical, training, housing, legal aid) locally available to persons who come for assistance.

Shared Responsibilities:

We try to develop a feeling among staff and volunteers that they share responsibility for what happens in the center. This would include:

— keeping the place neat and clean;

— discouraging people from using the place as a social center, to avoid scaring off someone who comes for help by what seems like an "in" crowd laughing and having fun together;

— contributing to the resource notebook by writing down questions, answers, descriptions, raps, theorizations, poems, songs, suggestions, anything that seems appropriate;

— following through with the people who come in, whether they are drug people, anxious parents or spouses, or non-drug people seeking help;

— making the proper referral whenever possible, preferably to a particular person in an agency rather than to the agency in general, even setting up appointments for people and, if possible accompanying them;

— answering the telephone with care; because many important initial contacts and crisis calls come by way of the telephone, we try to have staff rather than volunteers on the phone whenever possible—otherwise the volunteer spends his time running down someone who can handle the problem;

— keeping tabs on what goes on, noticing the interplay among people so that unfortunate things can be averted (like rip-offs) and opportunities can be seized (as when a person first indicates that he is ready to seek help);

— counseling in teams, with one person taking the major responsibility and the other giving feedback later; this provides excellent on-the-job training for the counselors and eliminates the phony one-to-one mystique for all concerned.

RESIDENTIAL FACILITIES

Short-Term Detoxification Houses

Almost all detoxification facilities currently in operation are either too expensive or unsuccessful, or both. Most are conducted in a hospital setting with high-priced and largely unnecessary professional personnel. This approach tends to reinforce a person's feeling of hopelessness (someone is taking charge of him while he is getting "clean"), and he returns to shooting dope soon after leaving the detoxification facility.

A residential detoxification house should exist to turn people on to an opening experience. Much of what has happened to them in the past has consisted of "take your medicine, lie around," an unnecessary and counterproductive approach. Instead of making detoxification the symbolic killing of the old "dope fiend" person, we see it as the symbolic re-birth of a new person, one who is capable of a totally new life style following the detoxification experience.

Rather than the feeling of a hospital, we would opt for the ambience of a country club. Excellent medical care is provided, but the rest should be a "turn on." Things that might be available include a juke-box, pool table, pin-ball machine, color TV, quiet room, steam bath, sauna bath, masseur, swimming pool, solarium garden, coffee shop, private rooms, books, games, magazines, sun roof, Polaroid camera, tape recorder, light show machine—whatever else can be obtained to provide opportunities to thrust people into a new awareness of themselves and the world. Local facilities will in most instances be able to provide only a few such things; but no matter how meager the resources, it can be a turning-on experience, one which communicates support—a commitment to the person that says to him, "You are of value, even if you're unable to grasp that for yourself." When setting up a residential detoxification facility, remember that . . .

1. Detoxification is a limited experience, not long-term treatment. The maximum stay should be two weeks: one week to overcome the effects of withdrawal, another week to make some plans and get them moving.
2. People brought into the house should be allowed to come and go as they please. This is not a police operation; these people are making their own choice to do something about their lives. If someone uses drugs and returns to the house, or uses drugs in the house, the staff people working with him should decide whether or not he is allowed to remain. Make every attempt to anticipate this kind of problem; don't let the person take you by surprise, but be ready for him with a thought-through response. ("When you went home tonight, we were almost positive that you were going to shoot up. We thought you might consider staying in the house for a couple of days, just so you can get hold of yourself.") Obviously there will be problems which you cannot anticipate, but keep in mind that hard-and-fast legalistic rules are soon regretted.

3. Medication used in withdrawal should be handled by a physician who practices "out of his bag," on a regular house call, once a day reviewing drug-related and other medical problems which might present themselves.
4. Persons in withdrawal should participate at every level in the life of the house—helping to prepare meals, doing housework, painting, redecorating, setting up special activities (using musical skills, for instance). We have already mentioned the importance of "doing," of action, for those giving up a drug habit.
5. While a person is in residence, he should be continually exposed to the possibilities open to him in the rest of the program such as group work and private counseling.
6. There should be an "Isolation Room," where a person learns to live with his feelings of loneliness. Let him begin by spending two or three minutes a day in isolation, working up to a couple of hours if possible. This is the best way for someone to learn that it is all right to be alone, that loneliness is only a state of mind, not a sickness.
7. Try to use as many different types of people as possible on the staff: square, straight, hip, professional, para-professional, non-professional. It is surprising to the junkie to find this many people interested in him, it gives him a chance to respond and to feel that he is able to relate to people who are not junkies.
8. Underestimate what will be possible; don't pretend you can do more than you can. Say to the person, "I'm not really too experienced; I don't know whether I'll be able to help. But if you're willing, let's give it a try."
9. Tell the person that you'll take the full responsibility, but that you'll need his cooperation. ("You look like you're in bad shape, but if you're truly making a choice, we'll do whatever we can to help.") In this way you give him a chance to fight for himself.
10. Let him know that you are trying to force him into doing it himself. ("You know, we've instructed the staff to leave you alone, to not have much to do with you. Sometimes they are going to ignore you, just so you can see if you can handle rejection.") This helps him to experience rejection not as a defeat, but as something that happens normally in the course of living.
11. To the person who is continually trying to blame a bad "head" for his problem ("I just need to get my head straight"), don't deny his feeling, but simply point out the hopelessness of it. ("O.K., but you have to realize that it will take several years of intensive therapy to straighten it out, at about $25 per hour. We don't do that kind of thing in our program, but we would be happy to give you a referral.") This helps him to realize that he'd better "get with it," that he can't face up to such a plan financially or otherwise.
12. Help a person with his re-entry into his old neighborhood; send him back to it even before the detoxification is complete. Before he goes, however,

draw out from him all the possible details of who he might see, what they might say or do, what possibly could happen, and how he would have formerly responded. Ask him to try to get one of his old acquaintances to turn him on for free; they probably won't, and even if they offer, his way of approaching it will have been changed because of your participation in the planning of the excursion, and he'll refuse himself. He needs to learn very early in the detoxification process that he can control himself, regardless of the neighbohood in which he lives.

13. Individuals coming to the house may own their records (sell them to them for a penny, if necessary), and can take them when they leave if they so desire. In this way, there cannot be any question that the records might be subpoenaed by a court (self-incrimination laws).

14. A follow-up system is extremely important. The staff must determine what has happened to a person one month after he has left the Detox House. No follow-up should be contemplated for those who leave before the detoxification procedure is completed.

15. Outpatient detoxification might well be operated out of a detoxification residence. Those who live at home could participate in other activities of the inpatient facility, at least on a daytime basis. The "Night Rap" could also be conducted at the residential facility.

In short, the detoxification experience needs to turn a person on to himself, to his own ability to make decisions about himself, to his own choice about what he is and has the possibility to be if he takes action on behalf of himself.

Most important, get everyone tuned in to the major goal in detoxification: getting results and changing behavior. Don't get side-tracked into trying to understand junkies better or yourselves and your mistakes better. Organize what you do around getting results, without tying your own personality to a "success" motive. Results mean that people are changing, not necessarily staying off drugs.

Long-Term Therapeutic Communities

Group-living houses which are part of a drug treatment program should not be referred to as "families," for this implies a dependent relationship which we strive to avoid; nor should they be called "communes," for this denotes a financial independence which is not possible; nor are they "cooperatives," for this denies that they are primarily places where therapy takes place. We therefore refer to our group living houses as "therapeutic communities."

In our program to date we have had experience sponsoring three separate therapeutic communities, from which we have learned a great deal.

People don't generally have difficulties with themselves or with things; they experience difficulties in their relationships with others. In working with the complexities of a residence where ten to twenty people have shared a drug problem and probably not much else, don't try to change the people; change instead the system in which they are living, the structure in which they relate.

Give the person who comes into the community (usually after a period of detoxification) a clear understanding of what is expected of him; it will add a secure dimension to his beginning experience in the "house." The "Residence Agreement" is useful in this regard.

There should be a staff coordinator who follows through with the people in the residence, who acts as their primary personal counselor, mediator and group facilitator, and who provides the link to the rest of the program.

The group living together should be in charge of what happens in the house: making the day-to-day rules, sharing responsibility for the smooth operation of the community, dealing with problems, and facilitating secure feelings among themselves so that each of them can leave as soon as he is able to take charge of himself.

Make clear to a person how long he can expect to stay in the community. From our experience, six months should be the maximum. This can be extended if it seems appropriate, but extensions should be a group decision between the coordinator and other residents.

If the community structure is tight, the residents remain under the same type of discipline as when they were using drugs, in jail, or in an institutionalized hospital setting; they are neither responsible nor in charge. The looser the structure a community can keep, the more likely it is to foster progress in solving problems and in developing individual problem-solving patterns among the residents. Again, there should be a genuine sense of the community determining together the structure within which they will live. It is a good idea to encourage as many opportunities for group experiences as possible—going to ball games, lectures, anything that seems appropriate.

When serious problems develop—such as a person using drugs, getting ill, stealing, being arrested—the entire community should convene to handle the situation, rather than having a rigid "answer" imposed from above.

When a person is working, he can contribute part of his income toward the expenses of running the house. In our experience, people usually feel that one-third of their income represents a fair amount to contribute.

The wider the diversity of persons living in the residence, the better the chance for long-term results. A homogeneous group tends to reinforce old models of relating to and dealing with problems. A broad spectrum of people provides a better opportunity to break through to new understandings of self and of living within the context of a community where all the members share some responsibility for each other.

Night Rap

One of the most difficult times for the addict in withdrawal is the late evening hours. He is alone, unable to sleep, anxious, likely to walk the streets, and tempted to "fix"; this is when he most needs a supportive environment. For this purpose we have put into practice (at the suggestion of one of our staff, who is himself a former heroin addict, Art Harper) an "All-Night Rap" program. Staff people are on duty at our center every night from ten o'clock until six in the morning. To prevent it from becoming a night-time drop-in center (which is another type of program), only those going through withdrawal—identified by special cards for the purpose—are allowed to enter. They are engaged in conversation by the staff, discuss their problems, allowed to sleep, and are given an alternative to going back to the use of their drug.

PROGRAM OPERATIONS GUIDELINES AND EXAMPLES OF FORMS

The guidelines and forms on the following pages will further clarify the organization and operation of a drug treatment program.

Drug Treatment Program
144 Main Street
Headtown, U.S.A.

GUIDELINES FOR CENTER OPERATIONS

1. A sign shall be posted in the Drug Treatment Center, stating: NO HOLDING, USING, OR DEALING. ANY OF THESE COULD CLOSE OUR PROGRAM.

2. A biographical sketch shall be kept in the files on all employees and staff people, including letters of recommendation, if appropriate. A duplicate copy is sent to the sponsoring corporation.

3. If and when the facilities of the program are over-taxed, reasonable limits should be set concerning the number of people receiving treatment. The number of admissions should be determined prior to the establishment of residential programs.

4. An advisory committee will be established to evaluate and make suggestions concerning the operation of the program. The committee should include professionals, former drug users, volunteers, staff, and members of the community. There shall be at least one liaison person representing the program to its sponsoring corporation.

5. A position paper outlining the program's philosophy and plan of treatment shall be kept on file in each center. This would also include the standard forms used by the program for medical histories, referrals, volunteer applications, etc.

6. The staff will communicate through its relationship with those who come into the office that the center is not a place for lounging, loitering, or rapping. Outside of the framework of actually working with people who have come for help, this kind of activity is disruptive to the program.

7. Volunteers should be screened according to their ability to work within the program. Before taking on responsibilities, they are required to participate in our groups and attend at least one training session. Volunteers need to be stable, mature people, able to respond to any crises that may occur.

8. When the office is opened, a staff member should be present to assume responsibility for what takes place. Except under unusual circumstances, the center should not be left in the care of a volunteer.

9. Emergency procedures, which are clearly defined for volunteers, will be kept up-to-date.

10. Regular training sessions will be set up for volunteers: to train them initially, to keep them up-to-date, and to give volunteers a chance to share their experiences with the staff.

TREATMENT RECORD
DRUG TREATMENT PROGRAM

144 Main Street
Headtown, U.S.A.

Name ______________________________ Date __________
Last First Middle

Address ______________________________ Phone __________

Birthdate ____/____/____ Birthplace ______________________

Education Level ______________ Race ________ Occupation __________

How Long in Area? __________ How Long at Present Address? __________

Referred by __________ Address ______________ Phone __________

Welfare? ________ Worker ________ Welfare or Insurance Identification Number __________

Medical History (Hepatitis, Heart, Lung, Epilepsy, etc.): ______________________

____

____

General Physical Appearance: ______________________________

____

Psychiatric History: ______________________________

____

Previous & Current Medical Treatment (What, Where, When): ______________

____

____

Chronological History of Drug Use (What Drugs, When, Frequency, & Amounts)

1st Drug	2nd Drug	3rd Drug	4th Drug	5th Drug	6th Drug

Present Drug Problem: Type __________ Frequency ________ Amount ________

Main Complaints: ______________________________

____

Action Taken: ______________________________ Interviewer __________

DRUG TREATMENT PROGRAM

144 Main Street
Headtown, U.S.A.

REFERRAL FOR SERVICE

Referred to (Agency): ______________________________

Address: ______________________________

Re: ______________________________
Last First Middle

Address: ______________________________

Reason for Referral: ______________________________

____

____

____

____

Referred by: ______________________________ Phone: ____________

I give my consent for release of information on myself to the Drug Treatment Program

____
Patient

Please send us information concerning your treatment of the above patient. We appreciate your help with this case.

John Q. Anyone
Director

DRUG TREATMENT PROGRAM

144 Main Street
Headtown, U.S.A.

APPLICATION TO VOLUNTEER

This information is for our confidential files only.

Name ____
Last First Middle

Address ______________________________ Phone ____________

Age __________ Education (circle one): High School 1 2 3 4

College 1 2 3 4

Graduate School ____________________

Major ____________________

Do you type? __________ WPM __________

Would you have time for treatment-oriented activities & outings? __________

Would you have an available car? __________

What kinds of activities are you most interested in? ____________________

____

How would you describe your own drug experience, if any.

Describe any experience you have had with nursing, social work, drug-abuse treatment, alcohol-abuse treatment, social problems, counseling, or encounter groups.

What are your feelings on drug abuse; how would you define drug abuse; and how does that relate to your volunteering your services to our program?

RESIDENCE AGREEMENT
Group Living Houses

DRUG TREATMENT PROGRAM
Headtown, U.S.A.

Name ______________________ Date __________ Rent __________

Room Assignment __________ Occupation ______________________

Friend or Relative (for reference) ______________________
Name

____
Relationship Address Phone

____

This is a treatment program facility. Because of this and the fact that the house could be closed for improper operation, we need to have you make a few simple agreements with us.

1. There will be no use, holding, or trafficking of dangerous drugs, narcotics, paraphernalia, or anything else you might suspect to be illegal.
2. Weapons will not be permitted. This is a house of peace and healing, a sanctuary, not a place of destruction and violence.
3. The privacy of the house will be upheld; no one will be allowed to visit or intrude without the permission of the community, the coordinator, and the Director of the Drug Treatment Program.
4. Crashers and friends are not to be allowed to stay overnight in the house, and are to visit only with the permission of the Drug Treatment Program coordinator.
5. No one is to sit around the house all day. You will be required to work, to be in a school or training program, or to be looking for work or training.
6. Living expenses such as rent, food, cleaning supplies, household items, etc., will be shared by the entire community.
7. Responsibilities for maintaining the house (cooking, cleaning, housework, etc.) will be shared.
8. Group therapy sessions will take place once weekly. Community meetings will also take place at least once weekly (to assign responsibilities, take care of housekeeping matters, etc.).
9. The respect of personal property will be upheld at all times.
10. Both couples and single individuals will be allowed to live in the house. When possible, a couple will be allowed to share the same room.
11. Any changes in this agreement will be worked out by the community, the coordinator, and the program director.

The Group House is here to help you change your style of life, to give you a chance to turn on to happiness and security without the use of drugs. We are only helpers; the main job is up to you. We ask your full cooperation. Anyone not living up to this agreement makes the choice himself to leave the Group Living House immediately. Any instances of breaking the agreement will be reviewed at the community meetings. Ultimate responsibility will rest with the coordinator and the Director of the Drug Treatment Program.

Signed ______________________

Date ______________________

APPROACHES TO DRUG PROBLEMS WITHIN THE SCHOOL SETTING

ELGIE L. BELLIZIO, A.B.

Today's drug scene presents a challenge to education which is crucial and unparalleled in its history. It is different in so many ways that schools generally have been ineffective in dealing with it realistically. Most schools and society in general are willing to deal with drug abuse only as an isolated problem; a problem, in their view, that can best be solved by traditional structured educational approach.

In the past few years, tremendous effort has been exerted in the preventive approach to drug abuse. It has become fashionable at all levels of education, government, and law enforcement to advocate education of the young as the real answer. The questions are: How do you do it? How do you get bright, young, and often very bitter students to consider the drug scene? How do you engage young people in a dialogue that might help them develop acceptable alternatives to drug use?

The first step for a school district to take in dealing with the problem is for its leadership to recognize that there is in fact a problem. Approximately three years ago it became apparent that the normal devices used to handle student behavior problems simply were not effective in dealing with the student drug scene in the Salinas Union High School District. School suspensions, expulsions, devices of punishment and rejection, were neither appropriate nor effective in dealing with the drug user or potential drug user.

With the recognition of the need for special study and work in drug abuse prevention, the next step was to communicate the urgency of this need to the professional staff, the students, and the community at large. The formidable consequences of drug abuse and of our failure to prevent it were underscored by its increased incidence in our school population.

It was fairly easy to convince people of the inability of students to learn when they are destroying themselves with drugs. In the normal budget processes of our district, the number-one priority in the expenditure of funds is for the protection of health and safety. In spite of agonizing salary deliberations which threatened to deprive us of all funds, we were able to budget enough money to get a program off the ground. The commitment of the Board of Trustees was and is total; the commitment of the staff is sometimes less than total and frequently requires new shots of zeal. The question from the staff, "What does this have to do with teaching?" must be answered frequently and with conviction.

THE SALINAS AREA

The Salinas Union High School District, located in the heart of Monterey County, California, comprises six schools with an enrollment of more than 8400 students in grades seven through twelve. The school population is 69% white, 24% Mexican-American, 2.5% Oriental, 1.5% Negro, and 3% non-white and American Indian.

The district provides a comprehensive educational program designed to meet the needs of all students. Individual schools, teachers, and students are encouraged to experiment with new approaches, methods, and materials. The district provides a full range of special services and classes including guidance counseling, testing, EMR classes, educationally handicapped classes, speech therapy, psychologic services, adaptive physical education, home teaching, English as a second language, and remedial reading.

The use of drugs, narcotics, and other harmful substances has increased in recent years especially among teenagers and young adults to the point where the Monterey County Public Health Department now considers it to be the most significant health problem in the county. It is virtually impossible to estimate the number of persons in the Salinas area who are using drugs in dangerous ways. Statistics tend to be unreliable and at best only reflect a part of the whole. However, from data gathered in a county survey, from research reported in the professional literature, from surveys done in individual schools, and empirical observation and expert opinion of persons closely allied with youth and the drug scene, we have concluded the following: (1) Many young people use potentially harmful substances for experimental, social, and recreational purposes, sometimes without apparent harm and sometimes with serious results. (2) Drug misuse occurs in all neighborhoods and among all subcultures in our area. (3) Drug use in the Salinas area is probably influenced by the drug subcultures in San Francisco, Santa Cruz, and on the Monterey Peninsula, including the Big Sur area.

THE SALINAS PROGRAM

The Board of Trustees of the Salinas Union High School District, and its administrative staff had the foresight and courage to admit that there were students in the schools who had used or were still using drugs. In order to do something about the situation, educational administrators, counselors, teachers, students, parents, and the community worked together to develop a program as responsive as possible to the student's needs. The need of the student is foremost and board policy and administrative regulations were developed to meet these needs (see Appendices 1 and 2). We were faced with few alternatives except punitive ones, which we recognize as necessary in some circumstances but which may serve as the basis for further alienating, disorienting and embittering experiences. We are dealing with young people who are feeling lonely or isolated, worthless or bored and frustrated. Drugs are a symptom as well as an "effective" way of coping with life. It is only one of many solutions that may be selected to improve the quality of life.

In 1968, the district started the drug control program for its students. After the budget was arranged, I was selected to direct the program (see Appendix 3). The administration felt that I had the respect of the parent community, the professional staff and the young people with whom I had contact. I was released from my teaching assignment to devote full time to studying the drug scene. I was given time and money to go anywhere to gather information on the vastly complex social problem of drug use. The training lasted several months before I began intensive discussions of these problems in the small group that I found to be most effective in both communicating and gaining a point of view for the staff.

It is extremely important that any program focus on the art of the possible. In developing the Salinas program, the staff approached the problem as an outgrowth of our society—our educational, political and economic system as well as our method of raising children. We see drug use as a symptom of the alienation of youth and we direct our efforts toward the modification or elimination of those aspects of community life which contribute most to this alienation. The district's program is also designed to deal with the immediate problems of students involved in drug abuse, including crisis situations involving students and their relationship with their parents (see Appendices 4 and 5). Although information concerning the pharmacology of drugs is supplied to the students, the district holds that the problem of drug abuse will not be appreciably affected by any factual program forced on all students.

The over-all objective of drug education programs is the reduction of drug abuse among all people. We feel that this objective will not be attained merely by instituting instruction about drugs in our schools. Its attainment demands a commitment on the part of community, parents, school personnel, and responsible students to attack underlying causes of drug abuse as these are identified in a given situation. Thus, the Salinas program attempts to focus public knowledge on the underlying aspects of the drug problem, and to assist our students in developing the pride and self-concept which are inconsistent with drug abuse. We attempt to act constructively when real issues arise in students' lives, but we also try to provide students with the leadership they need to deal with their own problems. Our program also works to improve the capacity of the schools to teach problem-solving and the constructive selection of life styles. The public must be made aware of the underlying causes of the drug problem so that it can react realistically and focus its resources effectively.

To meet the basic objectives of the program, we emphasize education and counseling. If the goal of society's effort in this area is control and prevention rather than retribution, education and counseling offer a real hope. Some areas contributing to the alienation of youth which have been revealed by our students are: school failure, not only of the student to existing curricula but of the curricula to meet the needs of our youth, the aspirations of youth in conflict with those of their parents and other adults, and finally the unwillingness of youth to conform to the behavioral expectations of the community.

We are doing many things differently in our schools in regard to drugs. Drug abuse education must be comprehensive, relevant, and diffuse. In response to the different situations we've encountered, we have had to change our policies and views toward students who become involved with drugs in and out of school. The traditional punishment model is not the answer. There are no drug problems—there are only people problems and we must be willing to deal with them on this level. The needs of the student are the most important needs to be cared for. The problem for most students cannot be solved by removing them from school and the community. Nor can such problems be dealt with by looking at the failures of the past. We meet a student at the point of immediate need and try to help him go forward from there. An immediate, non-judgmental, non-punitive approach is used. We are not permissive about drug use, and at the same time we realize that it is not a criminal problem, but a medical-sociologic-psychologic problem. Teachers and counselors working with our young people are trying to create a climate in which authentic relationships can develop if we come together as people, rather than maintain our distance as adults, teachers, or whatever.

Teachers and counselors participate with the student as he tries to work his way through his problems. The student must be shown positive support—and positive support is not necessarily agreement. It is important that the teacher be aware of the fact that each student is unique; in recognizing this, he must be careful not to try to solve the problem, but to help the student come to solve his problem himself. Counseling students with any type of problem is extremely time-consuming and relationships cannot be forced. The teacher must wait patiently until the student-user feels that he wants to make the contact. When contact is made it is emphasized that all confidences are a sacred trust. If students are afraid that they will be turned over to authorities, the gap widens and communication becomes impossible. Once a student feels comfortable and non-threatened he will usually be receptive to education and counseling. In order to assure this, we obtained a ruling from the County Counsel concerning the privileged nature of our communications with our students (see Appendix 6).

SERVICES RENDERED BY DISTRICT PERSONNEL

1. Provide an educational program for students at the seventh-, ninth-, tenth-, and twelfth-grade levels.
2. Provide 24-hour emergency service seven days a week to students and parents.
3. Provide counseling to students and their families.
4. Conduct classroom discussion on request of teachers or students.
5. Conduct parent education classes.
6. Assist religious and community agencies with in-service training programs and materials.
7. Maintain a speaker's bureau and research library open to the public.
8. Act as a community resource for information relating to use and misuse of drugs and narcotics, treatment of drug users, and rehabilitation.

We meet our students and their parents anywhere they are most comfortable. The school is not the sole source, but rather a part of the life-support system in meeting the needs of its students. Teachers and counselors go into the community, the home, juvenile hall, court, or wherever the student or parent requests.

In the development of our approach we consulted many parts of the community as to their feelings concerning drug abuse and the school's role in handling students who become involved. One of the most important aspects of our investigation was our involvement with our students. Students in all schools were given a part in creating their own drug education program by telling us how they felt the program should be approached. They were very honest, open, and eager to describe to us their conception of a helpful program. This is their scene; to work with it, one must work with them (see Appendix 7).

A lot of time was taken to learn the students' opinion of drugs. And, most of all, we attempted to communicate on their level, for students have their own knowledge and opinion of drugs and will resist the moralizations of elders. Students were given time and the opportunity to honestly evaluate the program as it was being developed. Their comments and feelings have given us faith in the direction of our approach to this vastly complex social problem.

THE TEACHER'S ROLE

The primary task of the classroom teacher in our program is to study young people and the development of the human personality rather than to study drugs. The teacher has to know something about drugs, but if drug use is symptomatic it's more important to focus on the reasons for use than on the drug itself. Thus, we emphasize patterns of use and factors other than the properties of the drug to explain the varying vulnerability of individuals or members of different groups. We discuss individual factors, or, if you prefer, psychiatric factors, by pointing out what it is in their personality or personality of other people in their group that makes this or that drug appealing to them. We feel that drugs are less important than what it is about that individual that compels him to use them. We point out that drugs may be meeting a need of that person that cannot be met in any other way that is apparent to him. We attempt to make our students aware that they are going to have to live with their anxieties the rest of their lives; we try to help them handle these anxieties. We define anxiety and discuss its role in the development of the human personality: in our discussions we learn what is meant by neurotic anxiety and inappropriate anxiety. This allows us to talk about compulsion as a mechanism for handling anxiety and for suppressing and relieving anxiety. Unless this material is included in our discussion sessions we may run the risk of inadvertently becoming proselytizers for drugs.

We believe in an educational program and feel that it should be directed at the student when he is forming his attitudes. We do not believe that a prepared program or a formal program is very useful. Our students have told us that there is no movie worth showing, no program worth buying from outside agencies.

Our method of sitting down with our students in relatively small groups to exchange information has proven to be a very effective method, according to the evaluation of the students. Young people today are not incapable of making moral or ethical judgments, but they will not accept the teacher, doctor, lawyer, or law enforcement officer who provides the information on a platter or as a canned lecture without feeling and real concern for people.

An experimental program is being tried in the junior high schools at the seventh-grade level. Each class is organized into small groups (10 to 15 students) for approximately a week. The program has a number of objectives: to discuss problems students encounter in their daily lives and methods for dealing with these problems, to foster the feeling of dignity within the student, to develop pride in the student's self-image, and to encourage respect for a healthy mind and body. Concerted effort is made to answer questions and listen to the student's views rather than to lecture or belabor adult opinions. Instruction focuses more on what is happening each day in adolescents' and children's lives than on the pharmacologic properties of a variety of drugs. Most importantly, we try to help our students to understand human motivation and relationships, individual and social responsibilities, and to develop decision-making skills which will enable them to resist the temptation to experiment with drugs or to use them for recreational purposes.

The program at the senior high school level is conducted by a specially trained teacher at the ninth- and twelfth-grade levels. In addition, all tenth-grade students enrolled in driver education receive an instructional unit on the relationship of drugs and narcotics to driving. The driver-education teachers have had several hours in-service training on the problems of drug abuse. The teacher specialist is also available to teachers and students on a request basis. The number of referrals being received by the teacher-specialists and counselors indicates that we are relating to our students and are able to provide a service for them. The presentations in the classroom are accepted by our students. They want good information about chemicals and will listen and discuss them with the teacher who is interested in them. The teacher must be honest, open, and willing to be challenged.

CONCLUSIONS

As seen from this brief description, the Salinas program is one of both general and specific education. The public must be made aware of the underlying causes of drug abuse so that it can react realistically and focus its resources effectively. The individual must receive specific assistance which can lead him to the personal attitudes inconsistent with drug use; if needed, he must be helped through the physical and mental results of drug misuse.

The organized effort to reach these objectives belongs in the school system, supplemented by a non-school local agency. The school system provides an existing communications network to and from young people; the non-school agency can reach beyond youth and can stir more public involvement than the school system alone.

The Salinas Union High School District has taken the lead in cooperation with the Salinas Police Department, Monterey County Probation Department, Juvenile Hall, Health Department, and other public agencies in developing an educational approach to drug use and abuse. Through the efforts of the teacher-specialist, the district has carried on an extensive program, including in-service training for all staff members, counselors, and administrators; training workshops for juvenile judges, probation officers, juvenile hall supervisors, and public health personnel; counseling with individual students and parents; classroom presentations; and parent-education classes. Hundreds of speeches have been given to the PTA, service, church, youth, and parent groups, and many other organizations.

In addition, the district has also taken the leadership in the development of the Salinas Area Youth Drug Information/Crisis Center (see Appendix 8). Through a Joint Powers Agreement, a number of agencies in the Salinas area have joined together to establish the Center. It is the core of a community-wide effort to deal realistically with the underlying causes of drug misuse by all people and to develop new lines of communication among students, parents, schools, and the community. Costs of operating the Center on a 24-hour a day basis are borne by the participating agencies. The function of the Center is to provide help for people of all ages in situations resulting from the misuse of drugs or other personal problems and to offer counseling and referral services with regard to medical, psychologic, legal, and employment problems. It is our hope that the Center will give the individual a feeling of taking charge of himself—that he really controls the way his life functions in the midst of his environment, no matter how "fouled up" his environment is. We hope also that it will be a place where a person comes and feels himself part of a community where people become involved together in a life support system for those who need help. The Center is a challenging undertaking for the city of Salinas, its residents, its schools, and other service agencies. We must stimulate changes in attitude and behavior as a prelude to the reduction of drug abuse among youth. If our community, our schools, our classrooms, and our homes are places where the emotional, physical, intellectual, and spiritual aspects of our children are encouraged to flourish through human involvement—then we can accomplish our goal, but we must DARE TO CARE!

From my counseling, and the impressions of teachers in our junior high schools, it has become evident that in the area of drug-abuse education we really do not need to deal extensively with the pharmacology of drugs. The solution of drug abuse seems to be in the area of developing solutions to "people" problems. The home, the community, and the schools contribute to these problems for youth and thus to youths' alienation. Our concern now is to examine these social institutions in an attempt to expose the aspects which are alienating to youth and which lead students into the use of drugs or some other form of behavior which is life-threatening.

I would like to point out the contribution of the schools to the alienation of their students. The first thing we are faced with, in the public secondary schools at least, is the grading system. Dr. Glasser, in his book,

"Schools without Failure," probably best describes for us the effect of a grading system which requires that there be losers. The use by teachers of a grading system to enforce behavioral conformity instead of as an indicator of proficiency in a subject has been a serious problem in secondary schools and a major source of student frustration. Irrelevant rules are another major source of concern. The automatic penalty for rule infractions and rigid programs for regimented physical education are other problem sources. The arbitrary suspension of students without due process, which might very well result in the student dropping out or quitting school is another. There is glaring need for major curriculum reform. Because of the impetus given by this approach to the drug problem in our district, we have embarked on a strategy of change in the educational program in order to become more responsive to the individual needs of students. We are trying to get people to care about people and to give a damn about kids; we are prepared to eliminate a few institutional sacred cows to accomplish this.

APPENDICES

1. Proposed Board Policy on Drugs
2. Proposed Administrative Regulations on Drugs
3. Teacher Specialist on Drug Abuse—Job Description
4. Procedures Followed in Handling Students Who Are Observed as Being under the Influence of Drugs While at School
5. Questions: Role of the School in Handling Students Involved in Drugs (used as a basis of discussion as an aid in developing board policy)
6. County Counsel Opinion—Privileged Nature of Communications
7. Teaching Aids Developed by Students and Others in the Community for Use in the Classroom and with Parent Groups
8. Joint Powers Agreement—Development of Crisis Center

Appendix 1—Proposed Board Policy on Drugs

The Salinas Union High School District will:

1. Provide for its students a forceful and well-organized instructional program in regard to the effects of drug use with the intention of preventing students from becoming involved with drugs.
2. Cooperate with community agencies to help students who have become involved with the use of drugs.
3. Cooperate in the enforcement of all laws regarding drug use and abuse.
4. Focus the resources of the school upon the solution of students' problems which have been brought on by drug use.

Appendix 2—Proposed Administrative Regulations on Drugs

An administrative case study will be made on each student charged with the use of drugs, utilizing medical help, the counseling of the teacher specialist on juvenile drug abuse, and conferences with parents and other relevant persons or agencies, to make a determination as to the appropriateness of the student's continuance in school or of other action to be taken.

Alternative lines of action in handling cases might be (these are not to be considered as all-inclusive):

1. Continuance of student in school if the Principal is satisfied that the student's presence is not inimicable to the welfare of other students and is also in his best interests.
2. Continuance of student in school dependent upon his following recommended medical or psychiatric therapy.
3. Placement of student in home-study program.
4. Major adjustments of student's school program to involve work experience, adult school, or other.
5. Suspension, expulsion, or exclusion of student from school.

Appendix 3—Teacher Specialist on Juvenile Drug Abuse

Appointment, Term, and Remuneration. The Teacher Specialist on Juvenile Drug Abuse shall be appointed for a term of one year upon the recommendation of the Assistant Superintendent—Instruction and Personnel, concurrence of the Superintendent, and by action of the Board of Trustees. Remuneration shall be by placement on the teacher salary schedule, with compensation beyond the regular teachers' work year on a pro rata share of base teacher's salary.

Responsibility. Under the direction of the Assistant Superintendent—Instruction and Personnel the Teacher Specialist on Juvenile Drug Abuse will develop and implement a comprehensive program of information designed to combat the juvenile drug problem in this area. He will be responsible for developing and implementing this program consistent with the laws of California and the policies of the Salinas Union High School District.

Duties:

1. To become involved in activities which will provide training appropriate to prepare him to become the district specialist in matters pertaining to drug abuse.
2. To work cooperatively with selected teachers in the development of instructional units pertaining to drug abuse.
3. To provide in-service training to teachers who will be involved in teaching the units developed.
4. To cooperate with all agencies involved in matters pertaining to drug abuse and to provide liaison between the district and these agencies.
5. To serve as consultant to all members of the staff in matters pertaining to juvenile drug abuse.
6. To counsel students who have questions relative to drugs and narcotics.
7. To develop programs of parent and adult education in the area of juvenile drug abuse.
8. To work in the development of community support for the District's efforts in juvenile drug-abuse education by making presentations before organizations and public bodies.
9. To teach classes within his areas of competence as assigned by the administration.
10. Other duties as assigned.

Appendix 4—Procedures Followed in Handling Students Who Are Observed as Being under the Influence of Drugs While at School

The highly individualized nature of this problem requires the exercise of discretion.

Lines of action to take when students are seen by the administration and thought to be under the influence of drugs are as follows:

1. The student is placed in a non-threatening setting in as non-traumatic a manner as possible. All school nurses and counselors are trained to handle this type of situation. Students specially trained by district personnel and with an understanding of the problem are invaluable and are utilized in many situations under the supervision of the nurse or counselor.
2. Someone is assigned to be with the student to talk him down, if appropriate, or for his own safety. In most cases very little more than acceptance and reassurance are necessary.
3. When it is reasonably obvious that the parent might be notified of the student's problem without compounding the trauma, the parent might then be requested to pick up his or her child. If it is necessary to take the student to the hospital the parent must be notified. We want the student back in school as soon as possible (as soon as he desires to return) so that follow-up counseling may begin.

The resolution of the student's involvement in drugs usually will require the participation and cooperation of the parents, and it is our goal to secure this cooperation at the earliest feasible time consistent with the best interest of the student. However, it is necessary to recognize that there might be times when parental involvement may not be in the best interest of the student. Counseling at this point by understanding counselors is a must. The student must remain in the system if any kind of follow-up is to have any effective meaning to the student.

Repeated episodes of drug abuse on the part of individuals are dealt with by a case-study process involving a broad spectrum of professional services, both school related and non-school related. Recommendations to the Board of Trustees for exclusion of students from school are preceded by recommendations from the case-study panel and in-put from a due process administrative hearing.

Appendix 5—Role of the School in Handling Students Involved in Drugs

1. a. Must drug or narcotic offenses be committed while the student is under the supervision of the school?
   b. What is the school's position when offenses are committed off campus and not under school supervision?
2. How should a student be handled at school who is referred to the office by school personnel for being under the influence or in possession of a drug or narcotic?
   a. Should the parents be notified by the school?
   b. Are there situations in which the parents may not be notified?
   c. Should juvenile authorities be notified?
   d. Should the school treat the student?
3. How should a student who voluntarily reports to the office and admits use of a drug or narcotic and requests assistance with his problem be handled?
   a. Should the parents be contacted if the student objects?
   b. Should juvenile authorities be notified?
4. What standards of evidence should the school use in determining whether a student has wrongfully used, sold, or been in possession of prescribed drugs or narcotics?
   a. Police record
   b. Official charge
   c. Should the school take action without due process?
      —Punitive suspension
      —Protective suspension
5. Should the school recommend professional counseling for students identified as using drugs?
   a. With parent's consent
   b. Without parent's consent
   c. What agencies?
      —Law enforcement
      —Probation
      —County medical department
      —Specialist employed by the school district
6. Should such counseling become a condition for remaining in school?
7. Should school personnel contribute to the rehabilitation of students who have a problem with drug use?
8. Should the school provide group counseling sessions for students who are known to be using drugs?
9. Should counselors and other certificated personnel report information received in confidence from students to:
   a. Parents
   b. Police
   c. Other authorities
10. Source of the problem most frequently is in the home. What should be the role of the school in attacking this problem?
11. Should an instructional unit on drugs be offered?
    a. What should be its content?
    b. Where should it be taught?
    c. Who should teach it?

Appendix 6

Mr. Robert P. Binns
Superintendent
Salinas Union High School District
431 West Alisal Street
Salinas, California 93901

Re: Schools—Counselors—Privileged Nature of Communications

Dear Mr. Binns:

You have pointed out in your letter of August 26, 1968, that high school counselors and other certificated personnel frequently receive in confidence information from students about law violations and other indiscretions. You ask whether there is any legal reason why counselors and other certificated personnel need report such information to parents, police, or other authorities.

There appears to be no *legal* obligation which requires that parents, police or other authorities be informed of the commission of criminal offenses or "indiscretions" by students who have related such acts in confidence to their counselor or teacher.

Section 130001 of the Education Code provides that any principal, teacher, employee, or school officer of any elementary or secondary school who refuses or wilfully neglects to "make such reports as are required by law" is guilty of a misdemeanor. But the reports contemplated in this section are those relating to school functions and duties, e.g., annual reports (section 13562 et seq.); section 6710 (principal's report on students coming within the 24-hour elementary schools provisions); section 11906 (report to parent by supervisor of health of visual defect of pupil).

Although communications between students and counselors are not privileged in the legal sense—thus, a counselor could be compelled in a judicial proceeding to disclose the information related to him by the student—there is no legal requirement which obliges a counselor or other certificated personnel to disclose to parents or other authorities information given in confidence by students concerning criminal offenses or indiscretions.

California has no "misprision of felony" statute which requires any person having knowledge of the commission of a felony to reveal that fact to law enforcement authorities under pain of criminal prosecution. California has a misprision of treason offense (section 38, Penal Code). Obviously, such law has little application here.

A distinction should be made between one's having knowledge of the commission of a crime and one's qualifying as a principal to a crime. The former does not constitute an offense; the latter does when advice and encouragement have been given in the commission of an offense, when the person committing the crime has been aided or abetted, when a person counsels, advises, or encourages children under the age of 14 to commit any crime, and when certain other unusual situations arise (cf. section 31, Penal Code). Every person who, after a felony has been committed, harbors, conceals or aids the person committing the crime, *with the intent that said person may avoid or escape from arrest, trial, conviction, or punishment,* having knowledge that such person has committed a felony or has been charged with a felony or convicted thereof is an accessory to the crime and is subject to criminal prosecution (section 32, Penal Code). Thus, mere knowledge of the commission of a crime does not constitute an offense when not reported to the authorities unless the information is withheld with the intent that the student avoid or escape from arrest, trial, etc.

Certainly, the student should be encouraged to tell his parents or authorities in the case of serious crimes, e.g., drug use or addiction, and a situation could easily present itself involving the best interests of the student where the counselor would inform the parents or law enforcement officials despite the confidential relationship between him and the student. As to a legal requirement that he do so, however, none appears to exist, and failure to report such information cannot be prosecuted as a crime or as a breach of some legal duty. The welfare of the child is, of course, paramount and as long as the counselor acts in good faith and exercises reasonable care in serving that interest, he should suffer no liability for protecting the confidential relationship between him and the student.

Very truly yours,

WILLIAM H. STOFFERS

Paul R. DeLay, Deputy

County Counsel of Monterey County
Salinas, California

PRD:jg

Appendix 7A—Drug Reaction Cycle—Marijuana

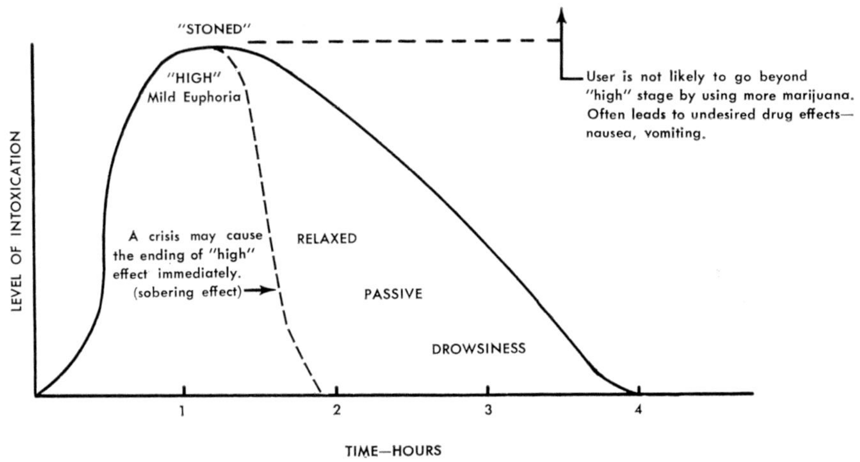

REACTION OF DRUG

Due to
1. Chemical effect
2. Subjective effects
   A. Setting
   B. Set
   C. Expectations
   D. Experience
   E. Potency

→

POSSIBLE PHYSIOLOGICAL EFFECTS

Time-space distortion—spaced
Tunnel vision—increased concentration
  decreased peripheral vision
Heightened sense perception—sound, taste, tactile
Inability to coordinate movement
Hallucinations
Increased appetite
Reddening of eyes—inflammation of mucous membrane
Harsh cough—burning sensation
Dryness of mouth and eyelids

→

POSSIBLE NEGATIVE EFFECTS

Confusion
Anxiety
Panic   Toxic dose
Paranoia
Impairment of judgment or memory
Lessens inhibitions
Creates false reality based on user's wants,
  motivations or set
Distorting agent on future psychological
  development of adolescent user
Predisposed people may have temporary
  psychoses
Exposure to drug culture
Diversion of energy and money
Conflict with the law

DEPENDENCE

No physical dependence
Tolerance—variable
Yes—psychological
No abstinence syndrome
No positive evidence of
  organic damage

CLAIMED PLEASURABLE EFFECTS

Freedom from pressures—relaxed
Increased sensual feelings
  pleasurable feelings towards sex
  and people (socialize)

NO NO's

Driving   Study
Converse intelligently
Fine motor control
Critical decisions

CAN DO'S

Music   Eating   Sex
Group ritual—sociability

Appendix 7B—Drug Reaction Cycle—LSD

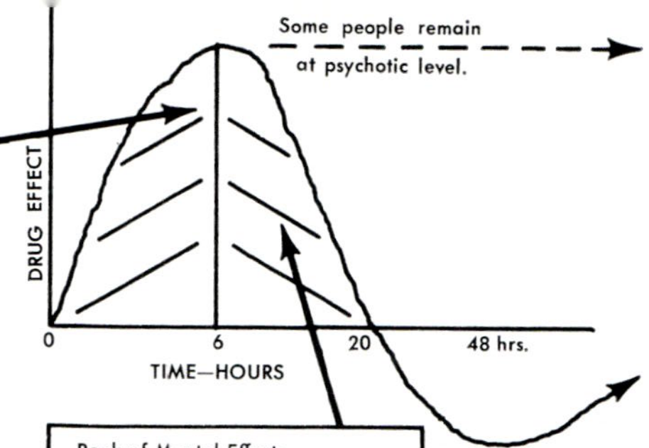

USUAL DOSAGE: 100-250 Mcg.
Colorless, Odorless, Tasteless
No test yet developed to detect inside the body.
Readily absorbed from the intestinal tract.
Doses as little as 25 mcg. (.000025 gram) are capable of producing psychological effects in susceptible individuals.

Start of mental effects.
Dangerous period—unknown effect.
Mild euphoria—may change to sadness.
Internal TV show.
Gales of laughter or tears.
Belligerency may occur.
Communications may be difficult.
May assume unusual postures.
  Crouching on chair
  Lying on the floor

Peak of Mental Effects
  Accidental deaths
  Depersonalization
  God-like ways
  Heightened self-centeredness
  Sense of being detached

MENTAL HANGOVER
Depression
Mild fatigue
Some degree of insomnia
May feel bad with the world

CLAIMED PLEASURABLE EFFECTS

Self meaning and mind expanding.
Makes one real.
Increases sensory awareness.
Kicks.

POSSIBLE PHYSIOLOGICAL EFFECTS

Average latent period of 35-40 minutes follows ingestion of an average dose. Onset may be noted within 15 minutes in some—others delayed an hour.

ACTS AS A SYMPATHOMIMETIC AGENT:
Mimics the effects that are brought about when the sympathetic nervous system is stimulated.
- —Pupils dilate
- —Blood pressure rises slightly
- —Pulse may quicken

Ingestion may be followed by unpleasant symptoms:
- —Nausea, vomiting
- —Sweating
- —Chills, flushes
- —Tingling
- —Tremors
- —Dry mouth
- —Irregular breathing
- —Visual distortion
- —Increase in body temperature

TOLERANCE

High potential for tolerance.
Tolerance develops rapidly—user rarely uses LSD more than twice weekly.
No physical dependency.
Yes—psychological dependence.
No lethal dosage known in humans.

POSSIBLE PSYCHOLOGICAL OR BEHAVIORIAL EFFECTS

After the sympathetic response, the first thing the individual usually notices are illusions (objects in his sensory environment changing shape and color).
Feeling of depersonalization and loss of body image.
Great anxiety.
Change in perception—increased sensory awareness, colors brighter, etc.
Hallucinations rare—usually pseudo-hallucinations.
Synesthesia.
Time sense frequently affected—past, present and future may merge.
Altered speech patterns.
Panic reaction to bad trip.
Belief in magic.
Impaired intellectual functions—cognitive functioning or value judgment formation resulting in confusion, inappropriate association of thoughts, and stickiness in thinking.
No loss of consciousness—usually remembers most of what happens.
Suspiciousness or hostility may develop.
Sense of intimate communication with God.

POSSIBLE ADVERSE EFFECTS—PSYCHOTIC IN NATURE

ACUTE IMMEDIATE EFFECTS
  Acute panic state—individual either tries to flee the situation or actually lashes out. Feels drug get away and can't control psychological effects.

  Can show very poor judgment—can fly feeling, etc. Feeling of invincibility. "It doesn't matter if my body dies, my spirit will live." Inadvertent suicide.

CHRONIC PSYCHOLOGICAL AFTER EFFECTS
  Long term schizophrenic reaction: this usually occurs in people who were disturbed prior to use of LSD.

  Long term depression: Recurrent panic reactions; usually occur in an acute panic reaction.

  Long term perceptual and recurrent hallucinosis or "flash back."

POSSIBLE NEGATIVE EFFECTS

No documented evidence of brain damage. Some persistent alterations in E.E.G. patterns in chronic users that might suggest chronic brain damage.

No documented proof of human birth defects. Some reported chromosomal alterations in peripheral white blood cells.

Serious long term disruption in personality.
Flash-back—recurrence of acute reaction.
  Long term depression.
  Long term perceptual alterations.
Acute panic reaction.
Conflict with the law.
Diversion of time and money.
May precipitate or intensify an already existing psychosis.

If a person is given LSD under properly supervised circumstances and is screened medically, the drug may be relatively safe; but where a person who has not been screened medically and psychiatrically takes LSD of unknown purity, in unknown quantity, in unsupervised circumstances, IT CAN BE AN EXTREMELY DANGEROUS DRUG.

Appendix 7C—Compulsive Use Cycle—"Speed" (Methamphetamine) (Central Nervous System Stimulant)

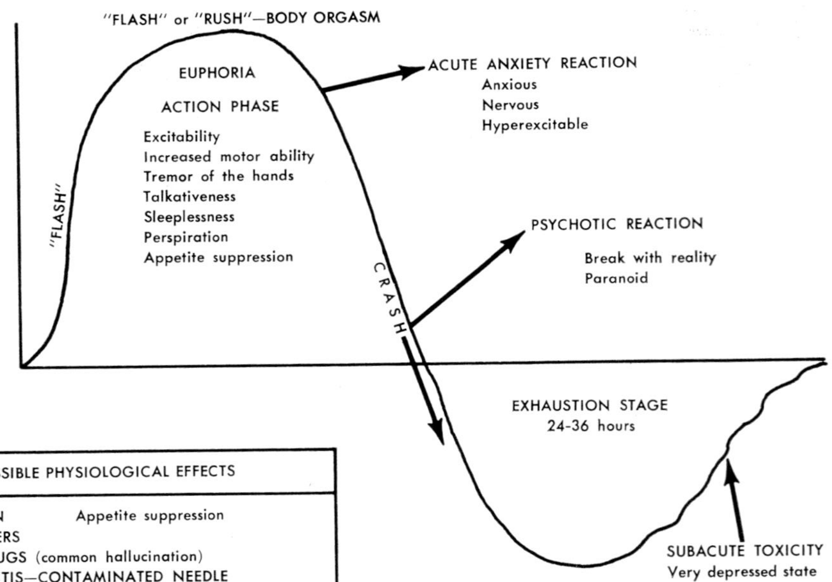

HIGH DOSE USE: "SPEED BINGE"—1 to 15 grams a day. (AVERAGE MEDICAL DOSE IS 5 to 15 Mg.) User repeats injection 2 to 10 times a day. The cycle is repeated again and again. With each "HIT" a "FLASH" occurs (body orgasm). User may stay high for several days. Between "HITS" user is EUPHORIC, HYPERACTIVE, HYPEREXCITABLE. May last 4 to 5 days. User does not sleep and rarely eats. This action phase will eventually end. May stop voluntarily because of FATIGUE, CONFUSION, PARANOIA, PANIC, OR QUANTITY OF DRUG MAY NOT PRODUCE DESIRED EFFECT.

CLAIMED PLEASURABLE EFFECTS

High effect or "flash"—body orgasm
Increased alertness and deeds.
Increased ability to communicate.
Creativity.
Kicks.

DRUG REACTION DUE TO:

1. Chemical effect
2. Subjective effect
   - A. Set
   - B. Setting
   - C. Expectations
   - D. Experience
   - E. Potency—purity

ACTION-REACTION CYCLE (IV)

ACTION STAGE—hyperexcitability is followed by REACTION STAGE of extreme exhaustion:
May sleep 24-48 hours
Awakes—May be psychologically depressed.
—may be quite hungry.
—may enter into a prolonged subacute phase of extreme psychological depression.
Depression may be so severe user may start another "BINGE."

POSSIBLE PHYSIOLOGICAL EFFECTS

MALNUTRITION — Appetite suppression
SKIN DISORDERS
CRANK BUGS (common hallucination)
SERUM HEPATITIS—CONTAMINATED NEEDLE
TACHYCARDIA—INTERPRETED AS HEART ATTACK
ACUTE RESPIRATORY DISTRESS—LIFE THREATENING CHOKING.
ACUTE ABDOMINAL PAINS
LOSS OF BODY RESISTANCE TO DISEASE—BODY MASKS NORMAL SIGNS OF ILLNESS.
FATIGUE
GASTROINTESTINAL SYMPTOMS—NAUSEA, VOMITING, CRAMPS.
AUDITORY AND VISUAL HALLUCINATIONS.
PROLONGED DEPRESSIVE AND HALLUCINATORY REACTION MONTHS AFTER USE.
REDUCED ABILITY TO PERFORM SEXUALLY AFTER EXTENDED USE.
THERE MAY BE SOME INDICATION THAT IT CAN PRODUCE PERMANENT, IRREVERSIBLE ORGANIC BRAIN DAMAGE.

DEPENDENCE

PHYSICAL * POSSIBLE
TOLERANCE POTENTIAL HIGH (IV RAPIDLY)
PSYCHOLOGICAL * HIGH POTENTIAL

POSSIBLE NEGATIVE EFFECTS

MAJOR BREAK WITH REALITY—TOXIC PSYCHOSIS—MAY NEED HOSPITALIZATION.
PARANOID OR SCHIZOPHRENIC REACTION.
SUICIDE POTENTIAL HIGH.
TENDENCY TO DRIVE BODY BEYOND MENTAL AND PHYSICAL CAPABILITIES.
LIKELY TO OVERMEDICATE.
MEANINGFUL SOCIAL LIFE DIFFICULT.
FEAR AND DISTRUST OF OTHERS MAY ALIENATE FRIENDS, EMPLOYERS, TEACHERS.
INCREASED ACTIVITY WITHOUT JUDGMENT AND CONSIDERATION OF WHAT GOES INTO THIS ACTION.
CONFUSION
IRRITABILITY
HOSTILITY
CONFLICT WITH THE LAW
EXPOSURE TO DRUG CULTURE
DIVERSION OF ENERGY AND MONEY

Appendix 8—Joint Powers Agreement (Establishing the Salinas Area Youth Drug Information/Crisis Center)

THIS AGREEMENT is entered into pursuant to statutory provisions of the State of California, namely, Title 1, Division 7, Chapter 5, Article 1, Government Code and provides for the joint exercise of powers among the following parties:

The County of Monterey
The City of Salinas, California
The Salinas Union High School District
The Hartnell Joint Junior College District
The Salinas City School District
Alisal Union Elementary School District
The Monterey County Board of Education
and such other parties as may be eligible and who apply for participation in this Agreement as is provided hereinafter.

WITNESSETH:

WHEREAS, it is to the mutual benefit of the parties hereto and in the best public interest that said parties join together to establish an agency to accomplish the purposes hereinafter set forth; and

WHEREAS, the parties hereto find and determine that there is a serious growing problem concerning the use among the youth of the community of narcotics, marijuana, and dangerous drugs (hereinafter called "drugs"); and

WHEREAS, the use of such drugs poses one of the most serious social and moral problems confronting society; and

WHEREAS, the use of such drugs often results in anti-social behavior reflected in criminal activities, alienation of families, and degradation of the person; and

WHEREAS, the alarming increase in the use of drugs has centered particularly among the young, for whom the parties to this Agreement recognize a responsibility and entertain a strong desire to lead and educate as useful citizens; and

WHEREAS, the professional workers on probation, police department, county health department, school districts, and in the field of medicine attest to the immediacy of the need for a concerted community effort to identify and alleviate some of the causes inherent in drug use; and

WHEREAS, the 1969 Grand Jury of Monterey County found that there is a need to establish in Monterey County a coordinated, supervised, directed, and County-supported program of activities designed to ameliorate the drug abuse problem; to provide assistance to those who are addicted to or dependent upon drugs; to provide a source of information and support for parents, friends, and relatives of persons with drug abuse problems; to assist teachers, medical personnel, and others who have responsibilities for dealing with the drug abuse problem; to evaluate the effectiveness of the various measures employed in the County program; and

WHEREAS, the use of narcotics, drugs, and other harmful substances has increased in the course of recent years, particularly on the part of teen-agers and young adults, to the point that the Monterey County Public Health Department now considers it to be the most significant health problem in the County;

Under a directive of the Board of Supervisors, the Assistant Director of Public Health conducted a survey of drug usage among the County's students in grades 7 through 12. Based on in-depth interviews with a representative 5% sample of the 22,000 students in these grades conducted during the 1968-69 school year, the following findings resulted:

—76% of Monterey County junior and senior high school students have never used drugs;

—24% of County junior and senior high school students have used drugs of some type at least once;

—7% of County junior and senior high school students use drugs of some type with some regularity;

—The percent of County junior and senior high school students that have never used drugs varies inversely

—with grade from 92% in the 7th grade to 59% in the 12th grade.

—There is some variation among schools regarding the incidence of drug usage. In general, the urban area schools tend to have more drug usage among students than do the schools in the rural areas.

and,

WHEREAS, the parties find and determine that the development of a forceful and hopefully effective recreational and educational program is necessary to counter the growing pattern of drug use; and

WHEREAS, the State Legislature, under Part 3, Division 12, Chapter 6, Education Code, has authorized public authorities including counties, cities, and school districts to organize, promote and conduct programs and to establish centers contributing to the physical, mental or moral development of the community.

NOW THEREFORE, the parties mutually agree:

1. TERM OF AGREEMENT. The Agreement shall be effective, and the Center shall be established, upon the execution hereof by the parties as attested by the signatures and dates of execution hereof, and shall continue in effect until terminated as provided herein. Additional eligible entities may, from time to time, become party to this Agreement upon the unanimous consent of the Board of Directors.

The Salinas area shall be defined as that geographical area encompassed by *that portion* of the Hartnell Joint Junior College District *within Monterey County*.

2. NAME OF AGENCY. The parties hereby agree that an Agency, wholly separate and apart from the entities of the parties, be and is hereby created under the aforesaid provisions of law, and shall hereafter be designated as the Salinas Area Youth Drug Information/Crisis Center (herein referred to as "Center").

3. CONTROL OF AGENCY. The Center shall be under the control of the Salinas Area Youth Drug Information/Crisis Center Board of Directors, hereinafter referred to as "The Board."

3.1 The Board of Directors shall consist of one representative and one alternate appointed by the governing body of each participating public agency.

3.1.1. Alternate will act on behalf of the agency he represents only in the absence of the appointed representative.

4. NOTICES. The Board, by resolution, shall designate a specific location at which it will receive notices, correspondence, and other communications and shall designate one of its members or its secretary as an officer for the purpose of receiving service on behalf of the Center.

5. MEETINGS. The Board may hold special meetings as it may determine and shall hold regular meetings at least once every two months. The date, hour and place for each such regular meeting shall be fixed annually by resolution of said Board, which resolution shall be publicly posted for two weeks on the bulletin board regularly used for official notices by the Center.

5.1 An Annual Meeting of the Board shall be held in __________ for the purpose of selection of officers and the conduct of whatever other business of an organizational nature is necessary, in addition to consideration of the regularly scheduled items.

5.2 The Secretary of the Board shall cause to be kept minutes of its meetings, both regular and special, and shall promptly transmit to each of the parties hereto, and to such other persons as the Board may determine, true and correct copies of the minutes of such meetings, which shall be primarily, a reflection of the actions taken by the Board.

6. VOTING. The presence of a majority of the membership shall be required in order to constitute a quorum necessary for the transaction of the business of the Board. No action of the Board shall be valid unless a majority of the membership concur therein by their votes.

7. OFFICERS AND EMPLOYEES. The Board shall elect a chairman and a vice-chairman from its members upon the occasion of its Annual Meeting.

7.1 The Board shall appoint the Executive Director of the Center, who shall act as executive and administrative officer of the Center, shall serve as secretary of the Board and who shall perform such other and further duties as may be determined by the Board. The Executive Director shall be an ex officio member of the Board, but shall have no voting power on the Board.

7.2 The Board, for the achievement of the purposes herein, shall have the power to appoint any other officer, or officers, as it may deem necessary and appropriate, to fix their duties, and to fix and pay their compensation.

7.3 The Board shall have the power to fix the duties, to fix and pay the compensation, and to regulate the reimbursement of expenses of all employees of the Center.

7.4 Nothing shall prohibit the Board, acting through its Executive Director, with approval of the Board, and within the constraints of the regularly adopted budget, per Section 11 hereof, from contracting for services of employees of parties hereto, or of other institutions and organizations, and reimbursing them for the costs, including salaries and fringe benefits, involved in providing such services.

7.5 The members of the Board shall serve without compensation.

8. SCOPE OF POWER.

8.1 The Center shall be a community youth agency which shall serve as an organization and place where young persons or members of their family of the Salinas area as defined herein and such other persons as may be authorized by the Board may seek comprehensive services regarding drug related problems. The Center shall develop and implement a vigorous total community awareness program to provide accurate information on drug usage to adults and youths. The Center will assist teachers, administrators, religious and community agencies with inservice training programs and materials. The Board shall have the power and authority to exercise any power common to the parties hereto and to exercise any power set forth in California Government Code Section 6508, provided that the same are for the furtherance of the objectives of this Agreement as contained herein and in the recitals set forth above, and may to the extent permissible thereunder enter into contracts in its own name with persons and with public or private agencies, boards, and other entities all subject to the terms and conditions of this Agreement. In the absence of its consent, no party to this Agreement, appointing power or members of the Board shall be bound to provide any sum of money, property, or service of any kind to the Center. The Board shall have no power or authority to bind any of the parties to this Agreement to any debt, liability or obligation in the absence of express written authorization from the party to be bound.

8.2 The Board shall adopt appropriate rules not inconsistent herewith for the orderly transaction of the business of the Center.

8.3 The Board shall have no power or authority to incur any obligation in excess of the amount appropriated to its use by a funding source.

9. ADVISORY COMMITTEES. The Board may from time to time establish or abolish one or more advisory committees, including youth advisory committees, to perform such functions as the Board may determine.

10. INSURANCE. The Center shall insure itself, the Board and the signatory parties, in a manner which the Board may determine, from loss, liability and claims arising out of or in any way connected with the performance of this Agreement.

11. FUNDS AND EXPENDITURES.

11.1 The City of Salinas is designated as the Fiscal Agent for the purpose of receiving and disbursing funds for the operation of the Center and its program pursuant to budget approved by the Board and shall at all times maintain a complete and accurate system of accounting for such funds consistent with established accounting procedures for the City of Salinas. The City's responsibility in this regard is limited to jurisdiction of funds of the Center provided by participating agencies in addition to grants of Federal, State and County governments. The City of Salinas in no way assumes financial obligation for the operation of the Center other than that agreed to by the City as a participating agency.

11.2 Each and every expenditure of Center funds shall be authorized by the Board through adoption of a budget. The budget shall not be adopted over the objection of the legislative body of any party hereto who provides financial support to the agencv. The Board will file annually with participating agencies a complete statement of estimated annual income and expenditures made during the prior fiscal year, in addition to a copy of projected budget for the forthcoming fiscal year (July-June). Said statement and projected budget shall be filed with the parties hereto at least 45 days prior to date said budget is to be considered for adoption by the Board.

11.3 Expenses for operation of the Center and its program not borne by Federal, State or County financial assistance to the program shall be borne by participating agencies on the basis of a formula approved by the legislative body of such participating agencies. Notice shall be sent by April 1st of each year to participating agencies advising what financial assistance, if any, will be needed from the agencies to carry on the program of the Center.

11.4 The Board shall have no power or authority to assess the parties to this Agreement for contribution of any kind whatsoever unless specifically approved by the legislative body of the participating agencies.

12. WITHDRAWAL OF PARTIES. The powers and authority of the Board shall continue until termination of this Agreement. Upon sixty (60) days written notice, any party may withdraw from its status as a party to this Agreement, provided that at such time said party has either discharged, or has arranged to the satisfaction of the remaining parties for the discharge of any pending obligations it may have assumed hereunder, and further provided that written notice of intention to so withdraw has been served. A party to this Agreement will incur no liability for any obligation incurred after the date of submission of the notice of withdrawal provided for herein, provided withdrawal is accomplished at the end of the sixty (60) day period.

13. DISPOSITION OF PROPERTY, EQUIPMENT AND/OR FUNDS. In the event of the complete rescission, or other final termination of this Agreement by the agencies then a party thereto, property, equipment and/or funds, following discharge of all obligations due by the Center, shall be disposed of in the following manner:

13.1 Property shall be disposed of by sale or other disposition according to law, and the proceeds, in cash or by their fair market value in kind at the time of rescission or other final termination as aforesaid, distributed to the signatories in proportion to the contribution made by them.

13.2 Equipment shall be disposed of per 13.1 above, although specific items, if required by a particular donor, shall be returned to said donor.

13.3 In accordance with Government Code 6512, surplus money, other than that obtained per 13.1 and 13.2 above, shall be returned in proportion to the contributions made by any and all donors or grantees whose funds might be within the budget of the Center at the time of the rescission.

14. AGREEMENTS. This Agreement may be amended at any time by the unanimous vote of all parties hereto.

15. SEVERABILITY. Should any portion, term, condition, or provision of this Agreement be decided by a court of competent jurisdiction to be illegal or in conflict with any law of the State of California, or to be otherwise rendered unenforceable or ineffectual, the validity of the remaining portions, terms, conditions and provisions shall not be affected thereby.

IN WITNESS WHEREOF, the parties hereto have caused this Agreement to be duly executed as set forth hereinbelow:

On the __________ day of _______________ 19__ by the COUNTY OF MONTEREY

On the __________ day of _______________ 19__ by the CITY OF SALINAS

On the __________ day of _______________ 19__ by the SALINAS UNION HIGH SCHOOL DISTRICT

On the __________ day of _______________ 19__ by the HARTNELL JOINT JUNIOR COLLEGE DISTRICT

On the __________ day of _______________ 19__ by the SALINAS CITY SCHOOL DISTRICT

On the __________ day of _______________ 19__ by the MONTEREY COUNTY BOARD OF EDUCATION

On the __________ day of _______________ 19__ by the ALISAL UNION ELEMENTARY SCHOOL DISTRICT

Appendix 8 (Continued)—Salinas Area Youth Drug Information/Crisis Center—Purpose

I. The Salinas Area Youth Drug Information/Crisis Center is a community youth agency which will serve as an organization and a place where young persons, or members of their family, may seek comprehensive services regarding drug related problems. It exists to give immediate help either directly through its trained personnel or by referral to another appropriate agency or professional person.

II. The Center serves to INFORM and COUNSEL young people and adults concerning drugs and related problems rather than MORALIZE OR CRITICIZE.

III. All contacts of any nature with the Center are confidential without fear of police or other reprisal.

IV. The Center is a community project, a public service to the citizens of the Salinas Area. There are NO charges for services rendered.

V. Progress in our community concerning social problems, such as drug abuse, will be achieved only when our citizens, both young and old, can evaluate these problems with understanding and knowledge, rather than with misinformation, speculation, fear and superstition.

**Appendix 8 (Continued)—Salinas Area Youth Drug Information/Crisis Center—
Organizational Chart**

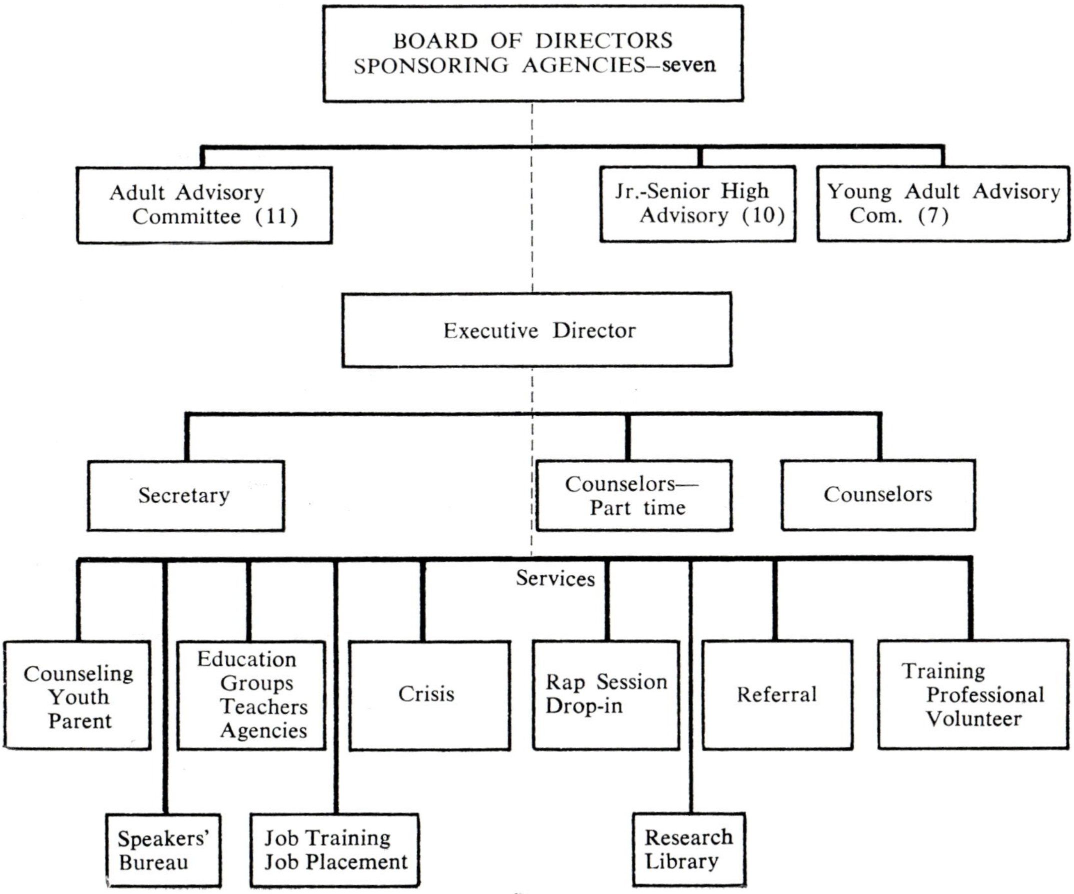

Appendix 8 (Continued)—Salinas Area Youth Drug Information/Crisis Center—Fiscal Channelization

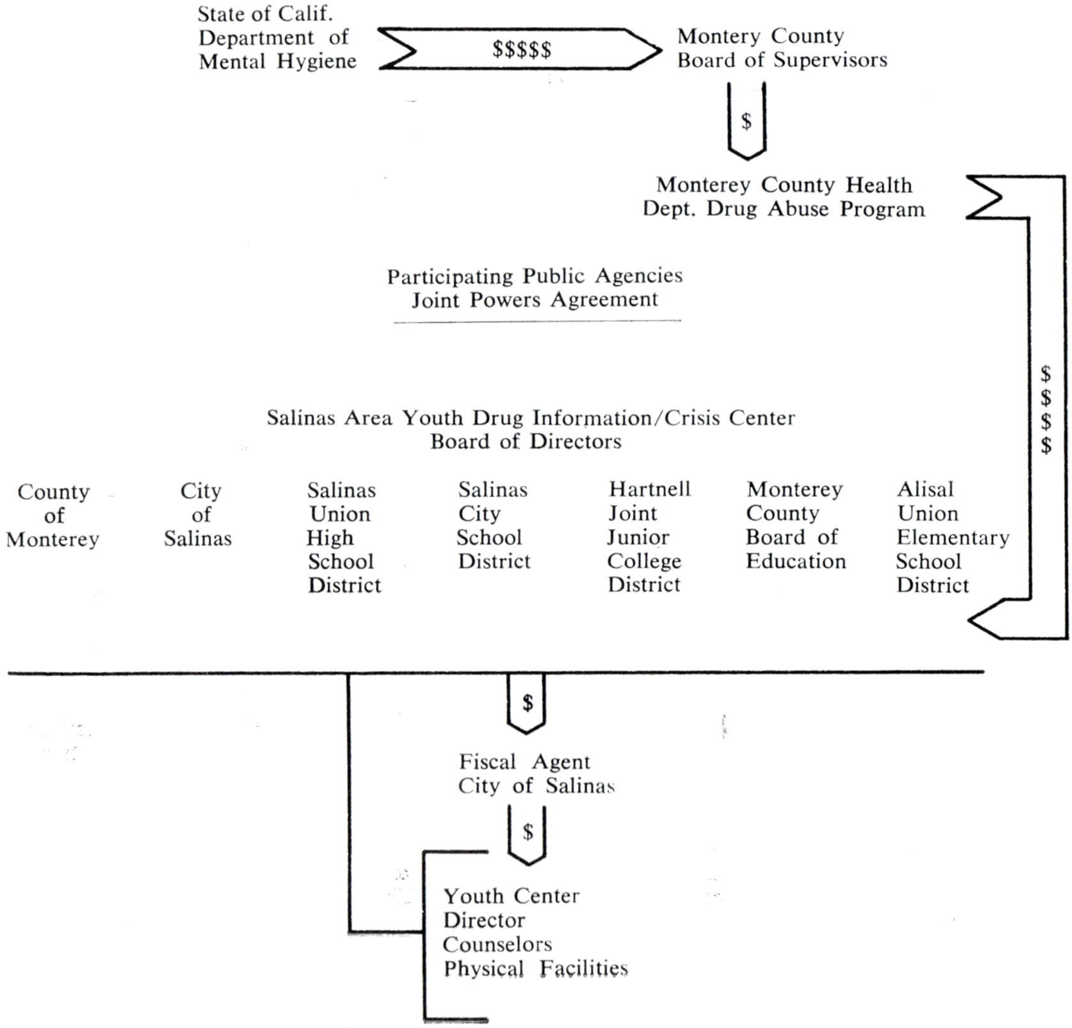

THE THERAPEUTIC COMMUNITY: THE ODYSSEY HOUSE CONCEPT

CHARLES C. ROHRS, M.D., JAMES P. MURPHY, AND JUDIANNE DENSEN-GERBER, M.D.

The way one chooses to attack a problem is patterned greatly by the concept one has of that problem; this concept is a product of both experience in that particular area and the knowledge about it that has been made available. Implicit in this definition is the idea that the concept is continuously subjected to change as more data are made available.

Such was the evolution of the physician's approach to the treatment of fever. When fever was thought of as a disease, treatment was aimed at reducing the temperature. As we became more sophisticated and knowledgeable in clinical science it was apparent that there were many causes of fever and an understanding of these various causes allowed for a more effective treatment method. Thus, fever was recognized as the symptom rather than the disease and the way was made clear to work with more fundamental problems masquerading under the same symptom.

A similar evolution has occurred in the area of dependency diseases, especially those manifesting as drug abuse and drug addiction. Presently there are three major modalities in the treatment of drug users, in addition to the traditional custodial and punitive methods of correctional institutions. These are best categorized as social, medical, and psychiatric. The social or lay agencies are one of two types. First are those in which the emphasis is on crisis intervention, environmental change, and vocational rehabilitation. It is hoped here that massive supportive aids will enable patients to live without drugs.

The second type is represented by the Synanon, Daytop, Phoenix House model where rigid behavioral controls in structured settings are imposed. Here the ex-addict, and the ex-addict alone, is believed capable of understanding the problems that an addict faces. It is believed that attitudes will change through forced behavioral conditioning in this type of therapeutic community setting. Professionals, and indeed most "straights," are considered valueless to the rehabilitation process. Educational and intellectual expertise are considered of limited value and psychodynamics are unimportant. A basic commitment to function is emphasized but little encouragement is given to return to the corrupt society at large. However, this new environment, in spite of its limitations, is drug-free.

The medical approach is basically a

chemotherapeutic method where greatest attention is paid to the symptom of this disease, the drug itself. Although this best represents the medical profession's traditional method of problem-solving—a drug remedy for every malady—in reality it only applies to the chronic long-term opiate addict. As the drug abuse problem evolves into the present-day pattern of multiple-drug usage among individual addicts, ever younger drug abusers (10% of our adolescent heroin abusers are under age 14), and antisocial behavior patterns in "functioning" individuals that antedate drug abuse, this method appears to be less and less of a panacea.

The psychiatric method attempts to deal with the underlying personality. Here the basic attempt is to effect attitudinal change and/or personality reconstruction in an environment where behavior can be regulated. Traditional Freudian theory and practice have been most effective in dealing with the psychoneurotic patient. Psychotics and persons with character disorders remained, for the large part, unaffected. Chemotherapy with the major tranquilizers has, in recent years, allowed many psychotics to return to and function in society. The character disorders, alone, have remained unamenable to known treatment. And the great majority of drug-dependent individuals are in this category of sociopathic personalities. To these individuals, to whom conning, gaming, exploitation, and manipulation are a way of life, one-to-one therapy between patient and therapist is a lost cause, with most patients seeming to gain pathologic satisfaction from thwarting their therapist. They choose to self-medicate themselves rather than cope with their anxieties and frustrations. In most sociopathic persons, productive functioning is absent; when they are capable of functioning, most exhibit antisocial, especially criminal, behavior. Only about 20 years ago did treatment utilizing massive supportive peer group therapy and/or living in a sheltered environment—such as A.A. or Synanon—begin.

Four years ago Odyssey House attempted to amalgamate the structured setting of the therapeutic community typified by Synanon and its marked effect on behavior, with the more traditional psychiatric thinking and its emphasis on attitudinal changes. This all occurs in a drug-free environment.

The proven effectiveness of the therapeutic community is based on the concept that the drug-dependent person is not a chronically sick individual in need of a crutch, but, rather a person whose way of life or orientation toward functioning is sick or distorted. The patient must be shown that his behavior is a result of free choice. No matter how deviant his past behavior has been, he has the innate capacity to function positively when properly challenged.

Acceptance by their peers constitutes the prime force available to the professional and ex-addict attempting to treat a group of sociopaths. Once deviant behavior is no longer rewarded by his group the addict has two choices: he may persist, which is grounds for exclusion by the group; or he may orient himself towards positive, affirmative attitudes about treatment and accept the judgment of his peers as to his therapist's qualifications. Although the professional is a participant-guide in therapy settings, it is the patient-participants who are primarily responsible for the success or failure of these endeavors. The same is true for the living-in arrangements. The peer group relations permit the residents to relate to each other and the staff without fear of reprisal. They can reward the motivated patient and punish the inevitable manipulation and testing. They come quickly to understand that it is only they who stand to benefit from treatment or to lose if they game.

Residents have assumed sole responsibility for enforcing the rules that they have instituted. This permits the professional staff to devote full time to treatment rather than enforcing police or security measures. It also helps to avoid much of the frustration and depression often seen in professionals who treat addicts, since they are now free to utilize their training to the maximum.

The program is divided into three basic phases: motivation; treatment; re-entry. In the motivation period a patient must demonstrate that he has some understanding of the concept of a therapeutic community and a commitment to abide by it while in residential therapy. This means that he accepts the doctor-patient relationship, he is willing to look into himself for the answers to his problems, he will accept responsibility for himself and his actions, and he will submit to the authority and discipline of the community.

The emphasis in therapy is on present behavior, interpersonal relationships, and attitudes. The use of the past to excuse current behavior or to avoid present responsibilities is frowned upon. The individual is forced to accept the unchangeable reality of the past and to understand that nothing he may do can ever change that. He is forced to think in terms of "what can be done now" and to examine the resources present in him which he can use to cope effectively from this time on. In other words, he is challenged to accept the past and begin anew today.

It is inevitable that previous behavior patterns will interfere with an individual's functioning in the therapeutic community. This is combated by permitting group activity only, in both therapy and work. This insistence on group situations exclusively subjects each resident to the constant scrutiny of, and open confrontation by, his peers. No single resident can play one staff member against another or against the community. This forces openness in therapy, honesty in interpersonal relationships, and correctness of behavior. The only demands met by the staff are those of guidance and therapy. The professional is no longer a symbol of authority, police, or pill-pusher. There is a constant refusal "to do for"—only to help them explore themselves and a group identity. It cannot be too greatly emphasized how little help the ex-addicts require. They develop pride and "grow behind" doing for themselves. Whatever is accomplished has much more meaning and value under these circumstances, and affords the ex-addict the stimulus he needs for required maturational growth. Thus, many of the basic pathologic features in the addict—his loneliness, isolation, dependency, and low self-esteem—are overcome.

Another basic facet of drug-using pathology is dealt with effectively in this setting. Most patients have a great contempt for authority and have submitted to it only when they have seen personal value for themselves. Within the structure of the therapeutic community they must accept willingly the authority of others, especially ex-addicts. With time and improvement under treatment they gradually reach the point where they must exercise more and more authority through the responsible positions they assume. To utilize the basic tool they once held in such disdain in itself has great therapeutic significance. Of course, any wielding of authority must be carried out fairly. With its abuse the patient must stand up to the confrontation of the community.

Because of the severity of the psychopathology of drug addiction and its attendant social disruption in the lives of the addicts, it is felt that an in-residence setting is a prerequisite for successful treatment and that psychiatric intervention be intensive. All activities within a facility are considered to be part of the therapy which is directed to returning the ex-addict to normal living patterns.

The transition from the protection of the residential unit with its structured environment to functioning in the community at large is accomplished in gradual stages. The ex-addict must spend increasingly more time outside of the facility. He is expected to live out and to assume authority not only over others within the therapeutic community, but eventually, and with more difficulty, over himself in the world at large. He acts upon and effects his own reintegration plan. He may start school or formal vocational pursuits. He must obtain the equivalency of a high school diploma. He must hold employment outside the field of drug addiction. He must think out his relationships with his family.

When sociopathic behavior patterns are lifted, problems of a neurotic nature are unmasked in about 30% of the patients. Individual therapy is permissible and desirable at this time. The patients now have the anxiety required to respond to treatment, they can form the necessary transference relationships, and they can control their behavior sufficiently to keep appointments.

If, after handling this difficult period successfully and demonstrating an ability to function maturely and independently, the patient should choose to work in the therapeutic community, his appointment as an ex-addict staff member may be considered.

The major point of departure of Odyssey House from the other forms of the therapeutic community is the integration of the professional and the ex-addict. This is vitally important in two respects: the development of attitudinal change and the success of re-entry. The ex-addict who has successfully undergone the process of behavioral and attitudinal change is preeminently qualified to motivate addicts to seek to help themselves and to supervise and effect the process of behavioral change. The professional is usually incapable of relating to the raw addict in a way that provides a sufficiently credible role-model to induce sustained motivation. However, the professional's expertise allows him to concentrate on the attitudinal and emotional problems which the ex-addict is still working out for himself.

This illustrates the concept of paramedical bridges at work. The ex-addict induction worker is a role-model to the using-addict seeking admission. During the treatment phase the ex-addict acts as a bridge between the patient and the professional therapist in the group. In the re-entry phase the professional serves as a bridge back to normal society by demonstrating meaningful constructive methods of problem-solving. In the beginning of the therapeutic process the patient listens to and relates almost exclusively to the paramedical worker. As the patient proceeds in rehabilitation he is better able to relate to others and, in the final stages, he is no longer bound to his addict identity but is able to relate as an equal to non-addicted persons.

In addition to his value as a therapist and a professional role-model, the physician has a responsibility to identify each individual's psychopathology. Irreparable damage occurs in therapeutic communities when the fundamental distinction between the schizophrenic and the sociopath is not made. Stressful anxiety and healthy guilt feelings are the tools by which a sociopath improves. These techniques must be altered and made more flexible to help the individual with true psychosis.

The Odyssey House techniques, although effective, have been difficult for many professionals to accept. They require that the senior patient population be accorded authority and responsibility as well as respect. It is even more difficult for some to accept the ex-addict therapist or administrator as a peer. The ex-addict must develop respect for himself and for others; this can be accomplished only through a supervised structure of which he is an integral part. The goal is positive, fully functioning, independent people. It is difficult for many professionals to survive in a therapeutic community environment because acceptance and status are accorded by the patients to a professional on the basis of his functioning as a real, warm, and capable person—and not on title per se. Neither patient nor therapist can demand respect; they can only earn it.

The concept of earning position, respect, and emotional growth through positive social interaction is deeply ingrained in all the work. Residents are continually evaluated and re-evaluated by each other and by the staff. It is partially on the basis of these evaluations that a resident will move from one level of the program to another. This creates a hierarchy which provides additional incentive for growth and is in accordance with socially acceptable norms. The desire of an individual resident to win the approval of his peers and favorable recognition from an authority figure has become clearly demonstrable.

During the past four years, Odyssey House has admitted 2500 patients, of whom 50% were referred through the courts. One thousand of these patients were adolescents between 9 and 18 years of age. Ninety per cent have used heroin. There are no restrictions on admission and no waiting list. Since the program is voluntary and each individual must demonstrate adequate motivation in order to remain in the community, approximately 45% leave or are expelled within the first week. Of the remainder, 72% complete the program which ranges from 12 to 18 months. Of the initial 41 graduates during the first three years, 96% remain drug-free on the basis of urine testing and present level of function. Of the nearly 30% who drop out during the course of treatment, about 50% are re-addicted, 20% remain drug-free, and 30% are lost to follow up and are presumably back on drugs.

PANEL DISCUSSION—THE PHYSICIAN AND COMMUNITY ACTION

Moderator: DR. MEYERS
Panel Members: MR. BELLIZIO, MR. FRYKMAN, and DRS. ROHRS and ROGER C. SMITH

Dr. John G. Cowl (Towson, Md.): I would like to direct my comments to Dr. Rohrs, the only speaker in the field of therapeutic communities, because I feel that if I were a general practitioner in Kansas, for instance, I would come away from his talk with little knowledge and even less enthusiasm for them. Dr. Rohrs made some specific references which were, in fact, untrue and I think correction is needed.

In Part II of his grouping of three different modalities of therapeutic communities, Dr. Rohrs listed Synanon as the forerunner, along with Daytop and Phoenix House. He suggested that only Group 1 had vocational facilities which stressed that aspect of rehabilitation. For Dr. Rohrs information, Synanon Industries, a multi-million dollar corporation, offers all aspects of vocational rehabilitation. Furthermore, members of Synanon are required to participate; they have no choice and must start within seven days of entry into the Synanon Foundation.

Secondly, Dr. Rohrs said that Synanon and other places were very nonprofessional. I am a psychiatric resident and formerly provided free medical services at Synanon. I had the opportunity of remaining in Synanon as a psychiatric resident, possibly under the direction of Dr. Wallerstein, Chief of Psychiatry at Mt. Zion in San Francisco. Further, there are many physicians in San Francisco who work with groups like Synanon. Dr. Rohrs said Synanon was antiprofessional and relied exclusively upon ex-drug users. Well, I am not an ex-drug user, nor is their Chief of Public Relations, Mike Kaiser, nor is their Chief of Dietary Control, Dan Silverman, nor is the President's secretary, who happens to be his wife, Terry Hurst; nor are countless other people such as the Medical Director, Carl Deisler, who lives in Synanon, and a number of other physicians who have given up their private practices and entered Synanon.

Dr. Rohrs also mentioned that the typical therapeutic community did not push people back into the community and implied that this was one of the virtues of Odyssey House. Specifically, Daytop has a very comprehensive program for putting people back into the community. Of the thousands who have come through Synanon, many have returned to the community. Furthermore, in Oakland, where Synanon has a very large facility, they have gone into the community providing excess food to the Black Panther Party and to the community, and have set up a low-priced warehouse supermarket for the ghetto dwellers. They have opened their doors and medical clinics to all members of the community. They also have given clothing to those same ghetto dwellers. I cannot imagine a program sending more back into the community than that.

Dr. Meyers: To save time, can you perhaps limit yourself to the question whether Synanon strives for re-entry or not? We are more or less familiar with the program, but time is a factor.

Dr. Cowl: Those are major points—raw facts which disagree with Dr. Rohrs'. I wonder why he chose to represent them in that fashion.

Dr. Rohrs: I cannot disagree with a single thing you have said because it was absolutely true. What I tried to say at the beginning was that therapeutic communities evolve and change as do all other things. They are continually amalgamating several different types of modalities. In general, I was speaking about the prototypical therapeutic community of 1958 and 1960, just as I was when I spoke of a social agency used to provide only supportive services. If I left you with another impression, I apologize. I did not mean to suggest that one excludes the other by any means. If I left that impression of Synanon or Daytop or any other, I certainly apologize. What is true is that these particular communities have developed treatment methods that were not part of their fundamental treatment in the early stages; this also occurred in the evolution of psychiatric hospitals.

Dr. Cowl: The Symposium is not concerned with aged aphorisms. The question is, "What is going on today?" I wish you as well as the other panel members would address yourselves to what is really going on today in the drug scene.

Dr. Meyers: I think it is a matter for each individual here to decide whether giving up a practice and entering a sheltered environment is optimal. I live in San Francisco, not in Kansas and I am familiar with Synanon Industries. Most people who participate still live at Synanon, so keep this as an open question in your mind. As I said at the very beginning, there is a hell of a lot of zealotry in this area and I certainly am not free of that taint. I also said at the very beginning that if anyone can accept Synanon, then for him it is the best program, but none of these programs need be mutually exclusive. There is no one optimal program. The Catholic Church is very good for people who can accept it. But there are a lot of people who cannot, and the same thing is true for many other treatment modalities.

Unidentified member of audience: I have talked to people like Sam England, Efraim Ramirez in New York, and others who have started some of these programs. There seems to be some antagonism, but it is based upon the therapeutic result. They all agree that they are working toward the end result of a positively functioning individual. I do see, however, that there is a difference between the therapeutic community that Charles Rohrs speaks of and the one that John Frykman presented. If I could give some information that might clarify these differences, I believe they can be reconciled. First, as I understand them, the psychodynamics that they are based on have to do with what Erich Fromm and others have seen as the basic experience of the person, whether he was emotionally accepted or rejected by the mother. In our programs, for example, those from Jewish families who were positively supported children and permissively raised seem to fit into one kind of program. Another group of drug abusers grew up with negative sanctions in an authoritarian atmosphere; they fit into another program. We found that a program with rigid structure is best suited for people who were permissively raised and have a positive image of themselves but can't appreciate the consequences of their actions. A Synanon-type program works well for them.

The other type of program is an open community in which the people learn by achievement to accept themselves. They arrive at optimal functioning for themselves when they learn through experiences of group encounters and achievement that they can stand on their own. Then they can accept responsibility for themselves and function optimally in the community in many different ways. Some people coming out of our program have gone into the media, for example, and a number of our people have become executive leader types in their respective fields.

Dr. Meyers: We agree that junkies can make it, I don't think it is necessary to belabor that.

Same speaker: The point I am making is that the same agency or community group can work both programs. A child coming in can be entered into the open one; if he presents a problem and it doesn't work for him, he can switch over to a more rigid or sanctioning type program. Indeed, the same agency can do this. That is the point I am making.

Dr. Meyers: Fine. That is certainly a valid idea.

Dr. Ben Nichols (Cleveland, O.): I have questions for John and Roger. Since June of this year, we have been operating in an outpatient facility that has a well-oiled medical, dental, psychiatric, and referral machinery that eats up a lot of time, space, money, and what-have-you. We have about 250 professional volunteers and we have seen about 4,000 patients. My duties and responsibilities at the clinic include recruiting and orienting nonprofessional volunteers. I am particularly concerned with the community excitement. My questions to John are: What can I expect from nonprofessional volunteers? How much training can they be subjected to and how much quality can you expect from them? Can they be efficiently and adequately trained to do paramedical jobs of counseling?

My other question is, Can the medical, dental, psychiatric and what-have-you machinery operate in the same building? We have a three-story house and a basement. Along with some other programs that I would like to start, we have a workshop and a kitchen. I would like to get a newspaper going, perhaps a day-care center, and so on. Can we effectively run these two seemingly different kinds of programs?

And finally, a question for both John and Roger. Roger, I believe, made the statement that he did not feel these kinds of programs needed charismatic people any more. I thought you might discuss that between the two of you.

Mr. Frykman: Well, I might say that Roger is very charismatic—to start off with—and I think that contributes to what he does. But I don't think it is necessary either.

As for what you can expect from nonprofessional people, as much as they are comfortable doing when they are given adequate training. For instance, right now, I have four high school students who are sitting in on all the counseling I do. I don't do any counseling by myself unless it happens to be at a family's home. And the students participate. They are not just there as oddballs on the side of the road. We have a continuous process of training with them and they are a part of interventions. I am into a thing called brief therapy, too, which we do not have time to talk about. When I see a person with a problem, I do not meet with him more than ten times. That is known from the beginning. Together we are going to see progress and change in that period, or we won't see it at all. These four high school students participate in that kind of aggressive intervention-type therapy. In terms of operating things in the same building, I think you can do that as long as all the things going on keep to business. If the building becomes a social center, you cannot do it.

Dr. Roger Smith: I would like to describe briefly how I use volunteers in Marin County. Marin County is a very affluent and predominantly white area that in the last few years has seen a radical increase in the use of drugs. We are beginning to see some people who formerly lived in the Haight-Ashbury but now live in communes in this area. When we started the program, we had all kinds of people who wanted to volunteer for us. We did not know what to do with them, so we decided that all the people who came to the program would have to involve themselves in groups if they wanted to work with people. In the groups, initially, we explained the nature of the program and what they could expect and then went into the classical encounter group type of thing, asking them to really risk something of themselves, to share with us who they were. We did the same with them. The emphasis in Open Houses was on getting people "high." Because of the nature of the population we deal with, our emphasis is on developing alternatives to the chemical high. I think it is time to discuss that. I think it is great when people are high. We should find some good ways to get high, teach our volunteers to get high. Once people have been able to risk something of themselves in the groups, we utilize a team approach with a staff person and a volunteer. We work jointly with the patient. We now have someone in from the community other than from the drug program. Since we serve the entire county, we have a lot of space to cover. If someone comes in, say from Novato, we will send him to a volunteer who is from Novato. The relationship is an open one; the relationship of equals in one sense. It is a kind of big brother approach. I am there; you can count on me. If you feel a "yen" at 10 o'clock at night or 3 o'clock in the morning, call me. If the volunteer gets depressed and needs some help, he can call on his friend. They have that kind of relationship. We also feel that if a volunteer is fairly together personally, he also has some community resources that perhaps the drug abuser or young person does not have. He has access to schools, to businesses, to industry, to training programs, to people and to situations. That essentially is how we feel volunteers can be most helpful. I think it takes a real commitment on the part of the volunteers. It is not something you do so that you can talk about it at cocktail parties, which happens with many volunteer "types" who plague drug programs.

Dr. Meyers: This may not be handled intuitively. There are data on this. It goes back to volunteers as subjects in experiments and I think you have to anticipate that most of them will be unsatisfactory.

Dr. Nichols: Let me add one distressing detail. A lot of the street people, the people who are into grass and acid come to me and say, "I cannot get involved in the clinic because there are too many junkies around and too much bad pot." It is that kind of thing that concerns me.

Mr. Frykman: Then I would say don't encourage them. You do not want people who are trying to get at it themselves or who want to solve their own deficiencies through your program.

Rev. Larry Kettlehake (Westlake, O.): I am a Lutheran minister still in the church. I felt paranoiac until I noted John Frykman was here. There are a few observations I would like to make. First, I hope the scientists here continue to be existentialists because I think it is very important; secondly, I hope especially the psychiatrists will become more eclectic in their approach than they have been; and thirdly I would like to see doctors, patients, and ministers, in their minds at least, not think of the people before them as patients but as people. These are the kinds of things that bother me as an outsider in a sense. I throw them out for what they are worth. Thanks.

Mr. Bixley (Bowling Green, O.): I am in health and education and have a question for Mr. Bellizio regarding the way he handles questions from students about marihuana—acceptability, unacceptability, and so forth.

Mr. Bellizio: One of the amazing things about our kids is that very rarely does marihuana come up, and I do not really know why. I guess maybe it is because we have gone into the thing together. In all honesty, I have never been asked to compare marihuana to alcohol. We have never had that hang-up with the thousands of kids we have had the chance to sit down with; we just do not get into that sort of thing for some reason. If a question does come up, we will give the best information we have. Although I anticipated it, it just has not become an issue at all.

Dr. Roger Smith: If I may add to that, I think that is one of the first things young people will do in an attempt to find out whether you are for real. And if you do a toe dance around the issue of marihuana, you are discredited. That's all, and generally that is why it is the first thing that comes up.

Mr. Bellizio: I let them know we have the information. I have tapes I let them listen to, as a Dr. Jones has been doing recently in the Atlantic Reporter. I let them know exactly what we have. We share with them everything that we have in terms of their making a decision about what they want to do with the drug.

Congr. James Scheuer: Mr. Bellizio, I very much enjoyed your remarks, but I was a little concerned that you said you would not accept any federal dollars. It seems to me that with the scope and dimensions of the problem of addiction, the federal government ought to be helping local communities. If there are any hang-ups that you have about accepting federal dollars, it seems to me that there ought to be some discourse. Maybe somebody is doing something wrong. Maybe Washington is doing something wrong. But the resources are there and it seems a shame for a model program, such as yours seems to be, to hold itself out self-righteously and not accept these federal dollars. The federal government ought to be helping and if we are not helping right, we ought to change the way that we perceive ourselves as being of assistance. I would like to hear why you do not want federal dollars and what you think the powers-that-be in Washington ought to do differently in order to make those federal dollars more attractive to you.

Mr. Bellizio: I think federal dollars should be used in more important areas, like improving the environment. To me, making sense with the kids in our community requires a commitment on the part of the community. As soon as we go looking for an easy solution which is wrapped in federal dollars, it gives somebody a chance to go out the back door, and I don't want that. I want the people to become shareholders. As soon as it becomes a federal dollar, it just seems to detract from what we are trying to do and then we start gearing everything to the federal dollar. The next year we don't know whether it is going to come or not. I want the community to make a long-term commitment to people-to-people problems. It does not make sense to go after the federal dollar. Our superintendent is interested in it and so is our city manager. I am not. They are going to have to do some pretty strong talking to convince me that we ought to. Maybe I would be interested if we could provide better services which we could not do locally. But I think the community has to pick up the costs. That is where it has to go and it is the only way it makes any sense to me. That, I guess, is my hang-up.

Dr. Roger Smith: I want the Congressman to know that I am not proud and would be happy to accept any federal money he wants to send my way.

Mr. Frykman: I would just like to say one thing. The drug treatment program where I worked was loosely associated with the Haight-Ashbury Medical Clinic. It was almost destroyed by the manipulations and the kind of things that we had to get into just trying to get some portion of the federal money. It was a very demoralizing thing for our staff and a difficult situation, trying to make a program creditable in order to get federal dollars.

Dr. Meyers: On the other hand, John, your salary for the first three months came out of a federal research grant from which I stole the dollars. I don't really think that federal funds bring trouble with them if your application is honest. Of course, you may not get your grant renewed. I do not now have a grant, but I have not had to pay back the money.

I think that as a temporary employee of The University of Michigan I am going to have to stop this discussion. I see the man coming with the hook—

Dr. Zarafonetis: I am sorry to interrupt but our time is up. I want to thank Dr. Meyers and the panelists for a most interesting session.

Afternoon Session
Friday, November 13, 1970

Presiding: Chris J. D. Zarafonetis, M.D.

MEDICAL COMPLICATIONS OF ILLICIT DRUG USE

DONALD B. LOURIA, M.D.

The striking increase in prevalence of drug abuse in the United States during the last five years has resulted in a marked rise in the number of cases in which serious medical complications have followed the use of drugs for hedonistic purposes. In this review, I shall discuss some of the more interesting complications reported.

HEROIN

The major medical complications of heroin abuse are listed in Table 1.

Overdose. Administration of an amount of heroin in excess of that to which the body is acclimated results in overdose (OD). There appear to be at least three different overdose syndromes. First, profound depression of respiration may occur, resulting in a breathing rate of less than 10 per minute. In some cases, apnea may result. Second, cardiac arrest or conduction aberrations may occur, probably not due to the heroin, but rather to the quinine frequently used as an adulterant for the heroin.[1] Third, lung congestion may supervene rapidly after heroin administration.[2] This is not a new observation nor is it confined to heroin. Pulmonary edema secondary to morphine was reported in 1880 by Osler,[3] several years prior to the introduction of diacetyl morphine (heroin). Although in Osler's patient the edema appeared to be confined to one side, ordinarily there are diffuse rales and clear evidence on chest x-ray of bilateral involvement (Fig. 1). Elevation of body temperature and leukocytosis are frequently found; on occasion the former may reach 104 to 105° F and the latter 20,000 to 30,000/mm^3. Neither necessarily connotes superinfection. Indeed, ordinarily overdose should not be treated with antibiotics, and if infection is thought to accompany overdose, a single antimicrobial should be given in order to avoid the substantial risk of superinfection incurred by the administration of two or more antibiotics simultaneously.[4]

Table 1 The major medical complications of heroin abuse

Overdose
Endocarditis
Tetanus
Hepatitis
Cellulitis and/or thrombophlebitis
Lung abscess

Treatment consists of supportive therapy, oxygen administration, and, if respirations are markedly slowed, administration of nalorphine (3 to 10 mg) or levallorphan (1 mg). Nalorphine and levallorphan are narcotic antagonists which will usually produce prompt increase in the breathing rate. If the breathing rate does not increase sub-

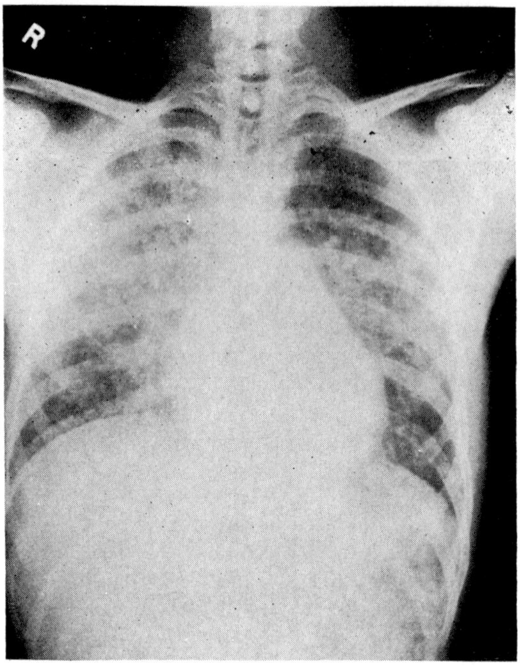

Figure 1 Heroin-induced pulmonary edema in a 20-year-old man.

stantially after the first dose, the antagonist can be given two more times at 15- to 30-minute intervals. If no respiratory response is observed after three doses, the diagnosis must be considered in doubt, and an assiduous search made for other causes of respiratory depression, such as brain trauma.

Usually the diagnosis is easy to make. The patient generally shows evidence of needle punctures over peripheral veins, is young and apparently fundamentally healthy, and has markedly constricted pupils. If pulmonary edema is present, the breathing rate is unpredictable; often it is normal or slowed even in the presence of a diffuse pulmonary infiltrate that should activate Herring-Breuer receptors and induce tachypnea.[2,5] The disparity between roentgenologic findings and anticipated breathing rate is often virtually diagnostic of opiate overdose.

Although the lungs are markedly congested at autopsy in those patients found dead on the street or in those few who fail to respond to therapy, there is no evidence that the edema is due to cardiac failure. Steinberg and Karliner found the central venous pressure elevated in two of seven patients,[5] suggesting that the edema could be at least part cardiogenic in origin in some subjects, but in the overwhelming majority there is no reason to treat for congestive heart failure. Lung function has not been adequately studied during the acute phases of overdose, but Karliner, Steinberg, and Williams[6] have evaluated it in five patients two to twelve days after the acute episode of overdose. They found a decreased vital capacity and total lung capacity as well as diminished diffusion capacity, and these abnormalities appeared to persist long after clinical recovery.

Overdose exacts an appalling toll in lives among the addict population. In New York City, most of the heroin-related deaths are due to overdose. Between 1950 and 1965, according to data from the Office of the Chief Medical Examiner, the number of deaths due to narcotism ranged between 250 and 380. Thereafter, the number has increased progressively so that, in 1969, 900 persons died of narcotic-related causes; most of the deaths were due to overdose. Of the 900, 224 were teenagers, and 55 were 16 years of age or younger.

Endocarditis. The following three case histories, of patients treated recently at the New Jersey College of Medicine and Dentistry, illustrate the endocardial problems arising as a consequence of opiate injection.

CASE 1. A 28-year-old man was hospitalized because of a seven-day history of right-sided pleuritic chest pain, accompanied by shaking chills and fever. He had been addicted to drugs for 14 years, and, at the time of admission, was taking an enormous amount of heroin and cocaine intravenously. Physical examination showed rales over the right lower lobe; sputum examination revealed gram-positive cocci in clumps; and chest x-ray showed a right lower lobe infiltrate. A diagnosis of right-sided staphylococcal pneumonia was made and anti-staphylococcal antibiotic therapy instituted, but over the ensuing week bilateral pleuritic pain appeared, hemoptysis became a prominent manifestation, electrocardiograms suggested acute

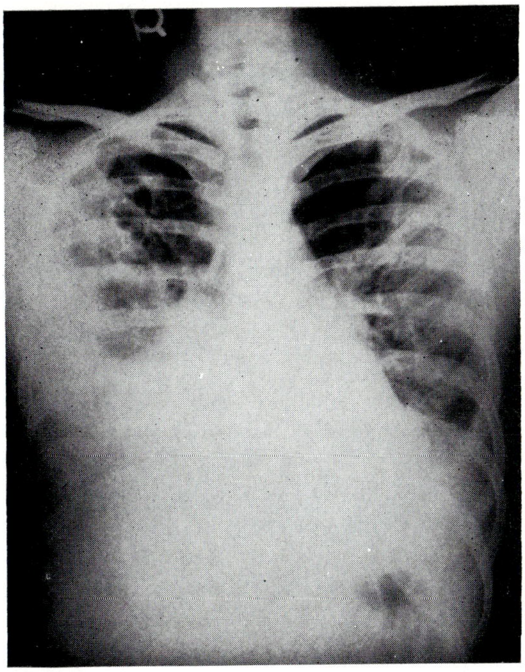

Figure 2 Bilateral embolic infiltrates in addict with tricuspid staphylococcal endocarditis.

cor pulmonale, and x-rays showed a bilateral infiltrate (Fig. 2). Even though no murmur suggestive of tricuspid endocarditis was heard, the sequence of events clearly indicated that this was present, an impression strengthened by positive sputum and blood cultures showing *Staphylococcus aureus*. Despite vigorous therapy directed against the staphylococcus, intermittent hemoptysis, pleuritic chest pain, splenic pain, and hematuria persisted for approximately 6 weeks. In the absence of positive blood cultures in either standard or hypertonic media, it was felt that the symptoms and signs represented recurrent non-septic emboli, and heparin was administered, following which the evidence of embolization gradually abated.

CASE 2. A 36-year-old heroin addict of 20 years' duration was hospitalized because of sweats, fever, shortness of breath, and swelling of his knees and ankles. Physical examination showed fundal hemorrhages and exudates as well as a prominent murmur of aortic insufficiency. Previous hospital examination had shown no murmurs and there was no history of underlying heart disease. Because the patient was an addict, a diagnosis of acute staphylococcal endocarditis was made and appropriate anti-staphylococcal therapy initiated. Multiple blood cultures showed a conventional alpha streptococcus sensitive to penicillin.

CASE 3. A 50-year-old man with a 32-year history of heroin abuse suddenly developed pain involving his left posterior thigh, accompanied by fever. Physical examination showed petechiae and two murmurs—a loud pansystolic apical murmur accompanied by a thrill and a decrescendo diastolic murmur heard best at the base and left sternal border. Blood cultures all showed *Candida guilliermondii*. The patient improved transiently during treatment with 5-fluorocytosine and blood cultures became sterile. Subsequently, blood cultures again showed *C. guilliermondii* and treatment with amphotericin-B was started. Despite this, relentless congestive heart failure ensued, followed by evidence of an acute myocardial infarction, presumably embolic in origin.

At autopsy only the mitral valve was involved and there was no evidence of underlying valvular disease.

COMMENT. The first patient had staphylococcal endocarditis involving the tricuspid valve. In a recent review, Ramsey, Gunnar, and Tobin noted that 35 of 93 cases of endocarditis in opiate users were due to staphylococci, and the tricuspid valve was the site of the disease in 27 of 57 heroin users.[7] If the tricuspid valve is involved, the offending organism is almost always a staphylococcus.[2,7] In these cases the valvular involvement may be manifest first, followed by pulmonary emboli, or sequential pulmonary embolization may be the first evidence of tricuspid valve mischief.[2,8] Often there is no murmur or the murmur appears to be clinically insignificant. It is well to note that even if the disease is limited to the tricuspid valve, there may be emboli to organs such as the kidneys, spleen, and brain. In any patient suffering discrete lesions in disparate areas of the lung, the possibility of tricuspid endocarditis should be considered.

The second patient had a *Streptococcus viridans* infection on an apparently normal aortic valve. This is unusual in two respects. In the first place, the alpha streptococcus does not ordinarily attack normal valves. Second, the addict acquires staphylococcal, gram-negative and fungal endocardial infections, but alpha streptococcal endocarditis is infrequent in this setting. This contrasts with subacute endocarditis in the non-addict, in whom alpha streptococci account for at least 60% of the infections. In one series of 48 collected cases of endocarditis in addicts, only one was attributed to this organism.[2] In a more recent series, nine of 93 were due to *Streptococcus viridans*[7]; in three of these cases, there was no evidence of underlying valvular heart disease.

The third patient suffered from infection with an ordinarily non-pathogenic species of Candida. In 10 of the 93 cases reported by Ramsey, Gunnar, and Tobin, a Candida species was the offending organism[7]; in five of the patients, *C. parapsilosis* was isolated, although this, like *C. guilliermondii,* is ordinarily a non-pathogen. Although underlying valvular disease was present in most of the patients, this is not always so; in three of the 10 patients there was no underlying abnormality, and in two of these infection was due to a supposedly non-pathogenic species of Candida. In case 3 described above there was no demonstrable underlying defect, and this was also true in another patient I saw in whom *C. parapsilosis* affected a previously normal aortic valve. Thus, in at least four cases, alleged non-pathogenic species *C. guilliermondii* and *C. parapsilosis* not only caused the endocarditis, but also attacked previously normal heart valves.

The mortality rate from endocarditis arising in addicts is very high; among those infected with Candida species the disease is almost always fatal. There is very little reason to hope that 5-fluorocytosine will be efficacious; indeed it is not as effective as amphotericin-B in these patients, and even amphotericin appears to exert only a transitory beneficial effect in most instances.

Hepatitis. In more than half the hepatitis patients seen at Bellevue Hospital in New York City the infection was attributed to use of heroin.[2] Several studies show that even if there is no clinical evidence of hepatic derangement, up to 80% of habitual users have hepatitis, evidenced by abnormalities detected on liver biopsy and routine liver chemistry profiles.[9,10] Many heroin users who acquire overt hepatitis are SH antigen-positive but surprisingly only about 2 to 3% of addicts who do not have clinical evidence of hepatitis were found to be SH antigen-positive.[11] Clearly, a great deal more data are required on the status of the liver in anicteric individuals who take heroin regularly by vein.

Miscellaneous. Tetanus is a frequent complication of addiction in users who are no longer able to inject the drug by vein and have resorted to "skin popping"—the subcutaneous injection of heroin, usually into the superficial areas of the thighs or arms. Scarring, cellulitis, and abscesses frequently ensue, providing an ideal milieu for *C. tetanus.* Unlike the other medical complications of heroin use, tetanus is found more often in women than in men even though women constitute less than 25% of the addict population. This is usually ascribed to the fact that the men are far more likely to have served in the Armed Forces and undergone compulsory immunization. The mortality from tetanus is high in the addict population, approximately 70% dying from the disease or from the complications of medical therapy.

Richter and Rosenberg[12] reported thoracic cord destruction leading to profound, sometimes fatal, paralysis in five patients; cord involvement occurred shortly after injection of heroin, but the mechanisms underlying the defect remain obscure.

There is some as yet unsatisfactory evidence that heroin abusers may be more susceptible to upper respiratory tract-acquired bacterial pneumonia, but these data are difficult to interpret in view of the lack of adequate controls who are matched for age and similarity of life style. Clearly, the addict is more prone than is the normal person to develop embolic pneumonia either from the tricuspid valve or from peripheral abscesses and thrombophlebitis; if the infection comes from peripheral sources, the organisms implicated most frequently are *Streptococcus pyogenes* and *Staphylococcus aureus*.[13]

Pulmonary granulomatosis has been described in a small number of addicts in whom the opiate was admixed with a second substance containing talc.[14] In these the talc induced diffuse granulomas leading to pulmonary hypertension. Cotton fiber emboli, due to drawing the heroin through a cotton plug, have been considered a cause of pulmonary infiltrates in addicts, but this entity has not been documented conclusively.

STIMULANTS

The main medical complications of stimulant abuse are listed in Table 2.

Sudden death was described by Kalant in six of 54 amphetamine abusers.[15] This clearly occurs in a small number of cases. Amphetamines in large dosage may cause serious arrhythmias, intracerebral hemorrhage, severe hypertension, convulsions, striking pyrexia, and circulatory collapse. In children, these adverse effects usually arise from ingestion of a substantial amount of central stimulants, whereas in adults these severe toxic effects ordinarily follow the intravenous administration. But this is not always so. For example, Smith[16] reported a patient, a 19-year-old man, who, during police pursuit, swallowed the drugs he had in his possession and then experienced convulsions, fever of 108°F, and cardiac arrhythmias. At autopsy, blood and tissues showed only methamphetamine hydrochloride. There was no evidence of recent or past injection of drugs by vein.

Goodman and Becker[17] reported intracranial hemorrhage in an 18-year-old following ingestion of 50 mg of amphetamine.

Recently, Citron and associates[18] described necrotizing angiitis arising in 14 young drug users who generally used multiple drugs, especially intravenous methamphetamine. This is a fascinating finding that has given rise to two postulates as to mechanisms. First, the repeated injection of foreign protein might induce a host antibody response which in some unexplained fashion is directed against blood vessels. Second, the angiitis might be a response to microbial antigens. Heroin addicts show serologic reactivity to a variety of antigens[19] and have substantial titers of circulating opsonins against both gram-positive and gram-negative microorganisms.[20] Conceivably, the angiitis could also represent a reaction involving an antibody response to SH antigen.

Table 2 The major complications of intravenous stimulant abuse

Hypertension
Cerebral hemorrhage
Sudden death
Hepatitis
Trauma
Violence
Paranoia
Convulsions
Coma

Other complications of abuse of central stimulants (including methamphetamine, methylphenidate, and phenmetrazine) taken intravenously include hepatitis and malnutrition. The former is probably a consequence of unhygienic techniques used during injection, but a direct toxic effect of methamphetamine has been postulated, although not adequately documented. The life style of the binge user of intravenous methamphetamine and the drug itself are together responsible for the malnutrition and debility, which in turn may predispose the habitué to the frequently observed subcutaneous abscesses. The heavy intravenous user takes the drug every four to six hours for days or weeks at a time, eating little and sleeping for only very brief periods. Debilitation resembling that found in World War II concentration camps can occur with astounding rapidity.

Trauma and violence also characterize the "speed" scene. The former often occurs because of the frenzied and uninhibited physical activities induced by the drug in the individual taking large amounts. The violence is usually a consequence of the paranoia that characterizes central stimulant abuse. In non-violent societies, such as that in Sweden, central stimulant-induced paranoia most often produces ludicrous but non-violent behavior; but in the United States high-dose methamphetamine abuse has been associated with rape, mayhem, and murder.

An additional problem results from the severe depression often accompanying the withdrawal from the drug. During this period, suicide may be attempted. Consequently, close observation is mandatory if this complication is to be obviated.

Ginzberg, Hertzman, and Schmidt-Nowara[21] recently reported an interesting case in a 21-year-old man who ingested 2.2 gm of amphetamine and subsequently developed increased fibrinolytic activity, marked hyperthermia, and reversible renal failure. Although the renal failure might have been due to the amphetamine-induced hyperpyrexia, it could also conceivably have occurred as a consequence of a transient period of post-hospitalization hypotension, resulting in renal ischemia. If renal failure does occur following amphetamine intoxication, dialysis may be beneficial.

The treatment of the patient with acute medical complications (hyperpyrexia, hypertension, agitation, arrhythmias, and coma) consists of the usual supportive measures plus administration of chlorpromazine. Espelin and Done[22] recommend an initial dose in children of 1 mg/kg of body weight given intramuscularly. In adults up to 100 mg can be given initially parenterally, followed by smaller amounts administered every three to six hours as necessary. Short-acting barbiturates are alternative therapy, but may accentuate the post-stimulatory depression, and are currently not considered the agents of choice. It is important to realize that if there is a possibility that the patient ingested or injected an atropine-like agent, use of the phenothiazines is contraindicated; in such cases, diazepam appears to be effective, and diazepoxide a reasonable alternative agent.[23–25]

Informal data from Japan, Sweden, and the United States suggest that an organic mental syndrome can result from the heavy use of central stimulants. How often this occurs, and how frequently it remits upon discontinuing use of the drug remain to be determined.

HALLUCINOGENS

Lysergic Acid Diethylamide (LSD)

Although the major adverse effects of LSD relate to psychiatric abnormalities and recurrences of bad experiences, the most acrimonious debate surrounding the drug concerns its alleged ability to produce chromosomal damage and fetal malformation. Fetal malformation clearly does occur in experimental animals, but the data in man remain conflicting. In a few cases, malformed infants have been born of mothers who used LSD early in and/or prior to pregnancy. Six cases are listed in Table 3. The number of cases is small, the malformations could have been coincidental rather

Table 3 Reports of infant abnormality following maternal LSD use

Author	Defect	LSD use by parents
Charakushansky et al.[26]	Limb	No details
Hecht et al.[27]	Limb	Both parents—mother during pregnancy
Zellweger et al.[28]	Limb—right leg	Both parents—mother during pregnancy
Eller and Morton[29]	Bony thorax—spine	Mother at conception
Hsu et al.[30]	D trisomy	Mother, father pre-conception
Assemany et al.[31]	Limb	Mother—unknown amounts

Table 4 LSD and chromosomal damage—positive studies

Author	Patient population	Type LSD	Dosage	Defects	Comment
Cohen et al.[84]	Illicit users =18	Street	300-600 μg	Breaks, dicentrics, rearrangements, exchanges—15/18	Most used other drugs; also 2/4 exposed in utero had increased breaks
Egozcue et al.[35]	Illicit users = 46	Street	110-1000 μg	Breaks, dicentrics, rearrangements, exchanges—26/50 completely outside normal range	Not dosage-related; also, 3/4 exposed in utero had increased breaks
Nielsen et al.[36]	Psychiatric patients = 5	Pure	- - - -	Gaps, breaks, hyperdiploid	
Jacobsen[37]	Pregnant women	Street	- - - -	Details not available	
Zellweger et al.[28]	Medically treated = 10 Street users = 2	Pure Street	- - - -	Breaks, exchanges, tetraploids	
Cohen et al.[38]	Children of users = 13 (Exposed in utero = 9 Exposed prior to conception = 4)	Street	- - - -	Increased breakage, dicentrics, quadriradials	
Hungerford et al.[39]	Psychiatric patients = 4 Prospective = 3	Pure	200 μg	Transient breaks and dicentrics	Breaks disappeared rapidly

- - - - = not stated

Table 5 LSD and chromosomal damage—negative studies

Author	Patient population	Type LSD	Dosage/ ingestion	Comments
Corey et al.[40]	21: 10 = prospective 11 = retrospective	Pure laboratory	200-600 μg (prospective)	
Dorrance et al.[41]	Street users = 14	Street	- - - -	
Bender and Siva Sanker[42]	Schizophrenic children = 7	Pure	100-150 μg	Could have repaired
Sparkes et al.[43]	8: 4 = medically treated 4 = street users	Pure Street	250-500 μg 250-1000 μg	Could have repaired Took other drugs but used LSD mostly
Hultén et al.[44]	Illicit users = 2 Illicit users = 1 with testicular biopsy	Street Street	100-800 μg 800-1000 μg	
Lucas and Lehrnbecher[45]	20—no details			
Aase et al.[46]	10 children—mothers took LSD during pregnancy	Street	Dose not stated	
Judd et al.[47]	9 continuing users 8 prior users 8 controls	Street Street	275-625 μg "normal" dose 150-300 μg "normal" dose	Heavy users Not using for 15-47 months

Table 6 LSD and chromosomal damage—negative studies that are statistically disputed

Author	Rebuttal authors	Patient population	Type LSD	Dosage/ingestion
Loughman et al.[48]	Slatis[49]	8 users	Street	300-4000 μg
Tijo et al.[50]	Markowitz and Klotz[51]	32 given pure LSD	Laboratory	50-450 μg
		5 illicit users given pure LSD	Laboratory and street	100-150 μg
		8 studied retrospectively	Laboratory	

than caused by LSD, and the purity of the LSD was not documented. Additionally, some of the mothers were multiple drug users.

A recent study[32] covering 127 infants is more impressive, showing a substantial increase in birth defects, but another study of 121 infants showed no such increase.[33] Interestingly, in both reports there appeared to be an increased risk of spontaneous abortion among LSD-using women.

The data on chromosome damage are equally conflicting. Some of the more important studies are summarized in Tables 4, 5, and 6. The majority of studies now show no effects or only transient chromosomal breaks, but some well-conducted and adequately controlled investigations show a striking increase in both chromosomal breaks and more serious abnormalities including dicentrics, tetraploids, and exchange figures. Resolution of the discrepancies is not easy, but a growing consensus is that the role of multi-drug use and adulterants admixed with the LSD have not yet been adequately explored. The experimental data *in vitro* and *in vivo* indicate that LSD can cause serious chromosomal aberrations, but it may well be that in man at the dosage and frequency used, LSD rarely produces these abnormalities unless its effects on chromosomes are facilitated by other agents taken concomitantly. There are also two case reports of leukemia arising in LSD users[52,53] and perhaps another dozen similar cases of leukemia or lymphoma as yet unreported.

It seems clear that many persons using LSD escape genetic harm, and it is equally clear that in view of our incomplete knowledge of LSD's mutagenic and teratogenic effects, any young person experimenting with the drug has to accept the potential genetic risks. It will probably take two generations to fully sort them out; in the interim it is folly to disregard the possibility of this complication. It seems to me it is the obligation of the scientific community to make it absolutely clear to the public that there is a big difference between "inadequately documented" (or "not proved") and "clearly disproved."

The other major medical complications of LSD use relate to trauma secondary to suicidal attempts or drug-induced misinterpretation of the environment; thus, for example, those who blithely walk out of windows in order to immerse themselves in the ineffable beauty outside created by LSD illusions are clearly misinterpreting the realities of their ambient environment and may suffer serious trauma depending on the height of the window. Convulsions have been reported to occur following LSD use,[54] but these are a rare occurrence.

Other Hallucinogens

There is not much known about the medical complications of use of mescaline, dimethyltryptamine, or Psilocybin.

The proprietary medication, Asthmador, when ingested in teaspoon or tablespoon amounts, can induce profound mentation change associated on occasion with impressive hyperpyrexia and even convulsions.[55] The toxicity of 3-methoxy-4, 5-methylene-dioxy-amphetamine (DOM, STP) remains unclear. Laboratory studies do not show atropine-like activities,[56] but clinical experience has, at least according to some observers; dry mouth, as well as difficulty with vision, swallowing, and breathing have all been observed.[57] Reportedly, lethal respiratory paralysis has occurred, and phenothiazines are thought by some to accentuate the adverse mental effects and to even induce lethal cardiovascular collapse.[57] The discrepancy between these clinical observations and carefully controlled laboratory studies has not been explained. It is possible that some of the apparent accentuation of the mental aberrations by phenothiazines may represent direct toxic effects of large amounts of the phenothiazines themselves.

Even though the data are divergent, phenothiazines should be avoided if STP ingestion is suspected, especially since diazepam appears to be a perfectly effective alternative.

MISCELLANEOUS

Various Substances Taken by Inhalation

Inhalation of glue, gasoline, hair shampoos, deodorants, foot medicaments, Freon, and decongestants has produced serious medical complications in a small number of cases. If the substance is inhaled in a plastic container pulled over the head, suffocation has followed in some persons who lost consciousness.[58] Similarly, spraying Freon into the mouth apparently "froze" the upper respiratory tract in some instances. Glue-sniffing has been associated with lethal, acute cerebral edema and with profound bone marrow depression attributed to the benzene base in the glues.[59,60] Recently, Bass summarized 110 cases of sudden deaths due to glue-sniffing and adduced considerable evidence to suggest that cardiac arrhythmias played a prominent role in the deaths; in many cases, the individual would sniff the substance, become agitated, run a modest distance, then collapse and die.[61]

Combinations and Mixtures

As the drug scene becomes increasingly more complex, multi-drug use is becoming the norm, and often on chemical analysis the drug is found to be composed of either an entirely different substance from that alleged or a mixture of pharmacologic agents.[62,63] LSD may be admixed with mescaline, STP, phencyclidine, methamphetamine, or strychnine. Several cases of strychnine poisoning have occurred. The mescaline being sold today is very often not mescaline at all but LSD, STP, phencyclidine, or inactive materials. Even the marihuana may be admixed with dimethyltryptamine or an atropine-like agent. And at times "marihuana" contains no tetrahydrocannabinol at all; instead what is sold as marihuana is in actuality one of a variety of other substances, some of which are potentially poisonous.

The deceptions in the illicit market are so numerous and unpredictable that the risk to the user is markedly increased, and the physician must often treat the patient suboptimally because he does not know the drug utilized. For this reason, it would be well to have available to every community a regional diagnostic laboratory the major function of which is to advise both potential users and physicians of the true nature of the drugs being purveyed in that community. Such information, obtained in a fashion guaranteed to protect the provider of the sample from surveillance, arrest, or harassment, should not only be publicized in the local press, but should be immediately available for use on a local hot line that is, at least to some extent, supervised by those knowledgeable in the areas of medicine and pharmacology.

REFERENCES

1. Lupovich, P., Pilewski, R., Sapira, J. D., and Juselius, R.: Cardiotoxicity of quinine as adulterant in drugs. *JAMA 212*:1216, 1970.
2. Louria, D. B., Hensle, T., and Rose, J.: The major medical complications of heroin addiction. *Ann Intern Med 67*:1, 1967.
3. Osler, W.: Oedema of the left lung in morphia poisoning. *Montreal Gen Hosp Rep 1*:291, 1880.
4. Louria, D. B., and Kaminski, T.: The effects of four antimicrobial regimens on sputum superinfection in hospitalized patients. *Amer Rev Resp Dis 85*:649, 1962.
5. Steinberg, A. D., and Karliner, J. S.: The clinical spectrum of heroin pulmonary edema. *Arch Intern Med 122*:122, 1968.
6. Karliner, J. S., Steinberg, A. D., and Williams, M. H.: Lung function after pulmonary edema associated with heroin overdose. *Arch Intern Med 124*:350, 1969.
7. Ramsey, R. G., Gunnar, R. M., and Tobin, J. R., Jr.: Endocarditis in the drug addict. *Amer J Cardiol 25*:608, 1970.
8. Collins, W., Romansky, M., and Vlalukin, L.: Clinical pattern of endocarditis in heroin addicts. Abstracts of papers read before the Seventh Interscience Conference on Antimicrobial Agents and Chemotherapy, Chicago, Oct. 25-27, 1967.
9. Norris, R. F., and Potter, H. P., Jr.: Hepatic

inflammation in narcotic addicts: Viral hepatitis a possible cause. *Arch Environ Health 2*:662, 1965.
10. Chapple, P. A. L.: Cannabis: A toxic and dangerous substance: A study of 80 takers. *Brit J Addict 61*:269, 1966.
11. Cherubin, C. E., Hargrove, R. L., and Prince, A. M.: The serum hepatitis related antigen (SH) in illicit drug users. *Amer J Epidem 91*:510, 1970.
12. Richter, R. W., and Rosenberg, R. N.: Transverse myelitis associated with heroin addiction. *JAMA 206*:1255, 1968.
13. Briggs, J. H., McKerron, C. C., Souhami, R. L., Taylor, D. J. E., and Andrews, H.: Severe systemic infections complicating mainline heroin addiction. *Lancet 2*:1227, 1967.
14. Wendt, V. E., Puro, H. E., Shapiro, J., Mathews, W., and Wolf, P. L.: Angiothrombotic pulmonary hypertension in addicts. "Blue velvet" addiction. *JAMA 188*: 755, 1964.
15. Kalant, O. J.: *The Amphetamines.* Toronto, University of Toronto Press, 1966.
16. Smith, D. E.: The characteristics of dependence in high dose methamphetamine abuse. *Int J Addictions 4*:453, 1969.
17. Goodman, S. J., and Becker, D. P.: Intracranial hemorrhage associated with amphetamine abuse. *JAMA 212*:480, 1970.
18. Citron, B. P., Halpern, M., McCarron, M., Lundberg, G. D., McCormick, R., Pincus, I. J., Tatter, D., and Haverback, B. J.: Necrotizing angiitis in drug addicts. *New Eng J Med 283*:1003, 1970.
19. Cherubin, C. E., Hargrove, R. L., and Prince, A. M.: The serum hepatitis related antigen (SH) in illicit drug users. *Amer J Epidem 91*:510, 1970.
20. Nickerson, D. S., Williams, R. C., Jr., Boxmeyer, M., and Quie, P. G.: Increased opsonic capacity of serum in chronic heroin addiction. *Ann Intern Med 72*:671, 1970.
21. Ginzberg, M. D., Hertzman, M., and Schmidt-Nowara, W. W.: Amphetamine intoxication with coagulopathy, hyperthermia and reversible renal failure. *Ann Intern Med 73*:81, 1970.
22. Espelin, D. E., and Done, A. K.: Amphetamine poisoning: Effectiveness of chlorpromazine. *New Eng J Med 278*:1361, 1968.
23. Solursh, L. P., and Clement, W. R.: Use of diazepam in hallucinogenic drug crises. *JAMA 205*:644, 1968.
24. Dimijian, G. G., and Radelat, F. A.: Evaluation and treatment of the suspected drug user in the emergency room. *Arch Intern Med 125*:162, 1970.
25. Taylor, R. L., Maurer, J. I., and Tinklenberg, J. R.: Managing bad trips on the drug scene. *JAMA 213*:422, 1970.
26. Charakushansky, G., Neu, R. L., and Gardner, L. I.: Lysergide and cannabis as possible teratogens in man. *Lancet 1*:150, 1969.
27. Hecht, F., Beals, R. K., Lees, M. H., Jolly, H., and Roberts, P.: Lysergic acid diethylamide and cannabis as possible teratogens in man. *Lancet 2*:1087, 1968.
28. Zellweger, H., McDonald, J. S., and Abbo, G.: Is lysergic acid diethylamide a teratogen? *Lancet 2*:1066, 1967.
29. Eller, J. L., and Morton, J. M.: Bizarre deformities in offspring of user of lysergic acid diethylamide. *New Eng J Med 283*: 395, 1970.
30. Hsu, L. Y., Strauss, L., and Hirschhorn, K.: Chromosome abnormality in offspring of LSD user. D trisomy with D-D translocation. *JAMA 211*:987, 1970.
31. Assemany, S. R., Neu, R. L., and Gardner, L. I.: Deformity in a child whose mother took LSD. *Lancet 1*:1290, 1970.
32. Berlin, C. M., and Jacobsen, C. B.: Personal communication.
33. McGlothlin, W. H., Sparkes, R. S., and Arnold, D. O.: Effect of LSD in human pregnancy. *JAMA 212*:1483, 1970.
34. Cohen, M. M., Hirschhorn, K., and Frosch, W. A.: *In vivo* and *in vitro* chromosomal damage induced by LSD-25. *New Eng J Med 277*:1043, 1967.
35. Egozcue, J., Irwin, S., and Maruffo, C. A.: Chromosomal damage in LSD users. *JAMA 204*:214, 1968.
36. Nielsen, J., Friedrich, U., and Tsuboi, T.: Chromosome abnormalities and psychotropic drugs. *Nature 218*:488, 1968.
37. Jacobsen, C. B.: Personal communication.
38. Cohen, M. M., Hirschhorn, K., Verbo, S., Frosch, W. A., and Groeschel, M.: The effect of LSD-25 on the chromosomes of children exposed in utero. *Ped Res 2*:486, 1968.
39. Hungerford, D. A., Taylor, K. M., Shagass, C., La Badie, G. U., Balaban, G. B., and Paton, G. R.: Cytogenic effects of LSD-25 therapy in man. *JAMA 206*:2287, 1968.
40. Corey, M. J., Andrews, J. C., McLeod, M. J., MacLean, J. R., and Wilby, W. E.: Chromosome studies on patients and cells treated with LSD. *New Eng J Med 282*: 939, 1970.
41. Dorrance, D., Janiger, O., and Teplitz, R. L.: *In vivo* effects of illicit hallucinogens in human lymphocyte chromosomes. *JAMA 212*:1488, 1970.
42. Bender, L., and Siva Sanker, D. V.: Chromosome damage not found in leukocytes of children treated with LSD-25. *Science 159*: 749, 1968.
43. Sparkes, R. S., Melnyk, J., and Bozzetti, L. P.: Chromosomal effect *in vivo* of exposure to lysergic acid diethylamide. *Science 160*:1343, 1968.

44. Hultén, M., Lindsten, J., Lidberg, L., and Ekelund, H.: Studies on mitotic and meiotic chromosomes in subjects exposed to LSD. *Ann Génét 11*:201, 1968.
45. Lucas, G. J., and Lehrnbecher, W.: Chromosomal effects of LSD-25, a controversy? *J Arkansas Med Soc 66*:348, 1970.
46. Aase, J. M., Laestadius, N., and Smith, D. W.: Children of mothers who took LSD in pregnancy. *Lancet 2*:100, 1970.
47. Judd, L. L., Brandcamp, W. W., and McGlothlin, W. H.: Comparison of the chromosomal patterns obtained from groups of continued users, former users and non-users of LSD-25. *Amer J Psychiat 126*:626, 1969-70.
48. Loughman, W. D., Sargent, T. W., and Israelstam, D. M.: Leukocytes of humans exposed to lysergic acid diethylamide: Lack of chromosomal damage. *Science 158*:508, 1967.
49. Slatis, H. M.: Chromosome damage by LSD. *Science 159*:1492, 1968.
50. Tijo, J., Pahnke, M. N., and Kurland, A. A.: LSD and chromosomes. A controlled experiment. *JAMA 21*:849, 1969.
51. Markowitz, E., and Klotz, J. H:. LSD and chromosomes. *JAMA 211*:1699, 1970.
52. Grossbard, L., Rosen, D., McGilvray, E., de Capoa, A., Miller, O., and Bank, A.: Acute leukemia with ph[1]-like chromosome in an LSD user. *JAMA 205*:791, 1968.
53. Garson, R. M., and Robson, M. K.: Studies in a patient with acute leukemia after lysergide treatment. *Brit Med J 2*:800, 1969.
54. Fisher, D. D., and Ungerleider, J. T.: Grand mal seizures following ingestion of LSD. *Calif Med 106*:210, 1967.
55. Di Giacomo, J. U.: Toxic effect of stramonium simulating LSD trip. *JAMA 204*: 265, 1968.
56. Hollister, L. E., MacNicol, M. F., and Gillespie, H. K.: An hallucinogenic amphetamine analog (DOM) in man. *Psychopharmacologia 14*:62, 1969.
57. Solursh, L. P., and Clement, W. R.: Hallucinogenic drug abuse. *Canad Med Assoc J 98*:407, 1968.
58. Press, H. E., and Done, A. K.: Solvent sniffing. Physiologic effects and community control measures for intoxication from intentional inhalation of organic solvents. *Pediatrics 39*:451, 611, 1967.
59. Winek, C. L., Collom, W. D., and Wecht, C. H.: Fatal benzene exposure by glue sniffing. *Lancet 1*:683, 1967.
60. Powars, D.: Aplastic anemia secondary to glue sniffing. *New Eng J Med 273*:700, 1965.
61. Bass, M.: Sudden sniffing death. *JAMA 212*: 2075, 1970.
62. Marshman, J. A., and Gibbons, R. J.: The credibility gap in the illicit drug market. *Addictions 6*:22, 1969.
63. Cheek, F. E., Newell, S. and Joffe, M.: Deceptions in the illicit drug market. *Science 167*:1276, 1970.

SYMPOSIUM SUMMARY

HENRY BRILL, M.D.

In the summary which follows I shall neither attempt to compress the proceedings into abbreviated form nor to reorganize the material into a synthesis of all that was said, although the latter might be considered an ideal form of review. Instead I shall summarize my own reactions, based on a re-reading of the text of the papers and of the panel presentation, to both the formal and informal aspects of the Symposium.

My first impression, and one which deepened as the meetings went on, was that the structure, organization and mood of this Symposium were characterized by great openness and by a willingness to subject every point of view to critical re-examination. Nothing was considered beyond or above question; considerable soul-searching, self-criticism, and mutual criticism took place and was especially apparent in the active participation of the audience.

Emotions ran high on many questions, reflecting the way in which the use of drugs has become entangled in this country with the related social and political problems of poverty, the war in Vietnam and racism. All of these issues were raised at various times, creating strong reactions which can be recognized in the written record only in muted form because the non-verbal elements of the interchanges do not appear on the printed page. Nevertheless, the debate remained at all times within the acceptable bounds of academic discussion; this was a credit to the understanding and restraint of the audience on the one hand, and to the considerable parliamentary skill of the panel chairmen on the other.

The elements of controversy also helped to broaden the focus of the discussion to include an unusually wide range of topics which ran from Kaplan's epistemology at one extreme to Holmstedt's analytical chemistry at the other. Brenner spoke of special areas of sociology and George of comparative law; Nichols covered the genetic implications of drug abuse and Fraser its long medical history. Important new data were presented by Martin, while Maynert, Elliott and Hollister outlined the complex pharmacology of narcotics and sedative-hypnotic drugs.

It is doubtful whether any one participant was in a position to comprehend fully all of the disciplines which were invoked; I for one was not. Yet each paper was given close attention, with every avenue of discussion considered relevant if it offered a new way of getting at the problems. This same broad approach could also be seen as a sort of flanking maneuver in logic, one which led to a great deal of constructive digression. The Symposium was also marked by a considerable amount of consensus with respect to

N.B. Since all references are to papers in this Symposium the usual bibliographical references are omitted the vast body of information needed for medical practice in this field, an attitude which was consistent with its title and announced purpose.

Let us now re-examine this purpose as stated on the inside cover of the printed program: "This symposium is designed to help the practicing physician deal more effectively with the problem of drug abuse, not merely in the context of the individual physician-patient relationship but also as a participant in community action programs with school officials, the courts, and representatives of social agencies. . . ."

In this statement one can identify two trends of major importance to medical practice. First, the role of the physician in the field of drug dependence is no longer limited to situations where he takes primary responsibility, that is, situations governed by the physician-patient relationship. Now his role has been expanded and requires that he also participate in action programs where the lead may be taken by the courts, the schools, or social agencies. The second trend is implicit in the reference to community action programs; drug dependence today is no longer the exclusive responsibility of distant governmental agencies. It now requires local initiative by community groups. This is a radical change from the situation of only a few years ago when two federal hospitals and a few state facilities constituted the treatment resources for the entire country. In part this may be an expression of the general trend toward locally based organizations such as regional medical centers and community mental health facilities for the delivery of all health care.

In the area of community control of drug abuse D. Smith and S. Cohen described the rapid escalation in the number of cases and the diffusion of drug dependence problems to communities throughout the country. Bellizio pointed out that by its nature, prevention must be a local matter as is the treatment of drug-related emergencies as described by Angrist and Ungerleider. Fink and Jaffe outlined the treatment of non-emergency drug dependence problems with pharmacologically-based techniques, suggesting that the changes in the delivery of health care services result from the nature of the drug problem itself. In its current form this was seen as significantly new and different even though Fraser was able to trace the problem back historically for many hundreds of years and Isbell described significant drug addiction in the U.S. in the 1800's.

In spite of its long history there was, as has already been stated, general agreement that the current drug situation is new in many important respects. Some of the dependence-producing substances are of recent development (the amphetamines for example, as outlined by Schweitzer). Cameron noted the relatively recent shift from ingestion and smoking of drugs to injection; perhaps the most striking new aspect of drug dependence is its growing internationalization, a phenomenon described by Cameron, Nepote, Kielholz, and others.

These papers make it clear that whatever the nature of the new problem it is worldwide and not limited or indigenous to the United States. Thus, we can learn much from other countries whose experience is often strikingly similar to our own and often closely related in time. Cameron outlined the international dimensions and Nepote described some of the realities of international control, in the process deftly divesting the topic of some of the unreal aspects created for the world of fiction. His account was rounded out by Ingersoll who spoke from an American viewpoint and by George who dealt with comparative law. The spectacular story of Japan's drug experience was described by Kato and that of Sweden by Goldberg, both countries having had major problems with stimulant abuse by young persons. Ideström reported a study from Sweden and another from Norway indicating extensive abuse of sedative-hypnotic drugs and benzodiazepines by adults, especially alcoholics. Together with the European account by Kielholz, these confirm the impression that the similarity of experience of various countries is more important than the difference. They also indicate that there has been a significant surge of drug abuse among young people throughout the world since World

War II. The British situation, which had had extensive American attention for some 15 years or more, was described by Bewley and by Boyd who felt that the recently opened heroin maintenance clinics were effective to the extent that the number of new cases was now sharply diminished. Although the British policy of containment showed signs of success, it is too early to be sure and both speakers cautioned against uncritical projection of the British experience onto the American scene.

In the course of these presentations it was natural to think in terms of some sort of epidemiologic model to describe the post-World War II spread of drug dependence. Although the terms "epidemic" and "pandemic" were repeatedly used during the Symposium, no one described an organized epidemiology similar to those developed for contagious disease. Demographic and other descriptive epidemiological elements were described in Cohen's paper and the dynamics of spread were suggested by D. Smith, who spoke of a "ripple effect," the spreading out of a drug-taking pattern from a geographical center. No more formal account was offered as to the mechanism of this spread. Since it is presumably due to the transmission of an idea rather than to a "contagium vivum" the term epidemic is used in a different sense than that of bacterial contagion. This aspect is unexplored as yet, as is the relation of these so-called drug epidemics to each other. Some relationship seemed to be assumed by all speakers, but its nature was not elucidated.

The development of drug dependence has been studied throughout the medical and social sciences—in psychopharmacology, clinical psychiatry, and clinical medicine as well as in the fields of genetics, chemistry, criminology, sociology, and law. It is significant that for the most part the Symposium papers were confined far less than usual to any one of these disciplines and were characteristically multidisciplinary in nature. Thus, Fraser included a considerable amount of clinical and pharmacological material in his paper on history while Schweitzer included historical data in his paper on the chemistry of amphetamines. Brenner dealt with questions of law and ethics in his paper on sociology and in general each of the participants seemed interested in information from any source provided only that it related to the problems of drug dependence, drug-dependent persons, and associated social problems.

If a multidisciplinary viewpoint is characteristic of the papers in this field so also is a tendency to differences of opinion which reflect varying basic judgments and philosophies of life as much as interpretations of fact. It was in relation to the law and law enforcement that some of the sharpest controversy developed, and the issue raised most vehemently was whether marihuana should be released from legal controls since everyone seemed to agree that penalties should be mitigated markedly. Brenner attacked the marihuana laws vigorously, as did Scheuer, while Oliver took an intermediate position and included alcohol with other dependence-producing substances in his paper. In discussing the use of medical data for legislative purposes he quoted a statement to the effect that public decisions are not made solely on scientific data, and it is significant that this principle was fully supported by other members of the legal profession when it was invoked by another speaker to help explain why the laws did not accept medical classification of drugs but made certain important modifications. Needless to say this provoked lively and even heated reactions from physicians in the audience, as did a number of other issues involving medico-legal matters and public policy. This was one of the sessions when the skills of the chairman were well tested.

In contrast to this, consensus if not unanimity appeared to prevail on a principle which has been widely enunciated, namely, that when drug dependence or abuse does create a problem it is a medical problem, and the individual should be treated accordingly. On closer examination, however, this statement proved to require considerable interpretation if it was not to be misunderstood. The term "medical problem" to many of the Symposium participants meant "health problem" (as contrasted to a legal, law enforcement, or correctional problem) rather than a problem primarily for doctors. This point was made repeatedly with respect to treatment as well as to preventive education and community organization.

Frykman described a rather complete service including a "crash pad," or emergency unit which used medical services only as a back-up and R. Smith described a different type which was also not organized along traditional professional lines. The use of transcendental meditation as outlined by Benson was entirely nonmedical and was thought to offer a significant alternative to drug use, while Rohrs gave an account of the Odyssey House group in New York which operates essentially as a self-help unit under medical supervision.

In general the tendency was to reduce authoritarian and establishment-type elements to a minimum, and to create an egalitarian atmosphere emphasizing self-help and mutual trust. Pollard described the psychiatric management of drug abuse while Schuman introduced a new technique which studied small groups and their reactions to films. Self-help was repeatedly stressed in descriptions of "crash pad" or emergency treatment units. At other points questions were raised directly or by implication about the role of the physician in counseling the drug experimenter and in managing the long-term rehabilitative efforts of the addict. Questions were raised also about the physician's place in the organization and administration of community efforts, a complex matter well detailed by Meyers.

Yet other papers seemed to point in a somewhat different direction. Ungerleider spoke of potential dangers in the hippie therapist who may also be a psychedelic missionary while Oliver indicated that medical leadership was expected. At many points in the Symposium the medical profession was criticized for not becoming more actively involved. Society requires the participation of the medical profession and of the other health professions even though at this time the exact role or roles which they may play still remains to be defined. Perhaps the simplest formula to describe the present attitude as it appeared in the Symposium was, "not by doctors alone, but not without them either." In the meantime one had the impression that the roles of various workers in the field were being determined to a considerable degree by personal interests and personality rather than traditional professional qualifications.

While all this points to a large area of overlap in the activities of medical and other groups, the Symposium also identified certain areas of primary and basic medical responsibility especially with respect to the treatment of serious physical and mental complications; this responsibility will inevitably bring an increasing number of physicians into contact with drug cases. Moreover, one cannot expect that such cases will be pre-selected or pre-sorted before they appear in the consulting room, the emergency service, or the hospital ward. The considerable number of physicians who can expect to come into contact with drug dependence problems must have as broad a knowledge as possible of the entire range of their manifestations, from the most benign and innocuous to the most severe, in order to diagnose and evaluate each case for treatment or suitable referral. This is a complex matter and requires an understanding not only of the action of individual drugs but also of their likely combinations, and their interactions with the personality of the drug taker and his environment. Several papers pointed out that although the crash pad can do a creditable job of caring for the average bad trip, these facilities have the option of referring to hospitals those patients that they cannot handle and medical groups must be prepared to accept them. Medical facilities must also accept patients who come directly and these often require rapid decision and immediate medical action. For the physician who faces such problems the highly practical paper by Koumans on medical emergencies and their differential diagnosis is of great value, as is the paper on barbiturates by Essig, Kramer's discussion of stimulants, and Angrist's description of stimulant-induced hyperpyrexia.

Seevers' paper on classification of drugs and definition of terms offers a clear conceptualization of the problem while the evaluation of cases is simplified by D. Smith's formulation of three phases or stages of drug dependence, namely experimentation, regular use, and compulsive use, each with its own management and treatment.

Laboratory testing of urine has become a valuable resource in the last few years, as Parker described in a very practical paper. The non-psychiatric complications of the drug life are essentially medical problems and many of these were described by Louria, Kramer and others. Pulmonary and hepatic infections led the list, which also included over-dose, tetanus, and endocarditis. Finally, the pharmacologically based treatments using methadone and narcotic antagonists were described by Jaffe and by Fink.

In summary, all of the commonly used methods for prevention of drug dependence, for its evaluation and treatment were described in the Symposium papers and discussed in the round-table sessions. With respect to virtually all of these topics there were differences of opinion and as one would expect these differences tended to receive the most attention. Nevertheless, there were important areas of consensus which should not be lost to sight. Let us now examine some of them.

Judging from what was said and not challenged, it appears that on an operational level many communities in the United States are developing a new pattern of response to local drug dependence problems. Incidentally, this pattern is somewhat different from that reported by other countries, even though their experience appears to resemble the American one quite closely. The emerging pattern of social response in the U.S. places increasing emphasis on prevention through education even though there is controversy and doubt about what should be taught, to whom and by whom. Treatment is considered a community-based activity, with a network or pattern of services which may include "a hot line" or continuously available telephone emergency contact to provide immediate help for many patients and to secure necessary assistance for others; a center for emergency care under non-professional or professional auspices; full hospital back-up or support service with laboratory and emergency care; counselling in the non-medical problems of young people; methadone maintenance; and residential facilities for long term rehabilitation. All such programs in a given community tended to function within the framework of at least an informal cooperative understanding, and it was not unusual for several of them to be under the aegis of a single agency. Leadership and initiative might be medical or non-medical, depending on local circumstances, but such services usually had some constructive medical liaison, and it was assumed that physicians were to become involved when called upon. Many of the authors also indicated that useful liaison with the courts and with law enforcement had been achieved, and it would seem that this is another element of the system of community-based drug services.

Another point of general consensus was that the social and recreational use of drugs does exact a price among users and much of the Symposium effort was devoted to descriptions of ways and means of reducing that price to a minimum. It seems clearly a medical responsibility to participate in these efforts for preventive education, diagnosis and evaluation, or treatment, and this participation requires a continual updating of technical information because of its rapidly changing nature.

Finally, I would like to draw attention to a topic that had relatively little attention at this Symposium or any other that I have attended, namely what is to be done about the large numbers of intractable and fully addicted users who do not respond to any of the treatments now available. For them the chief hope is that they may be helped to survive until they in fact become reachable by treatment. This represents an unsolved problem, as is the fact that even for the patients now entering treatment the available methods are time-consuming, exacting, uncertain, and far from standardized. Finally, no one was able to describe a method or approach which had been successful in reducing the overall burden of drug dependence in the United States, although the situation was reported as relatively stationary in Britain and in Japan where at least two epidemics had been suppressed. The limited effectiveness of current methods of prevention and treatment conveys what may be a major message of the Symposium, namely, that much of the real hope for the future must remain in research, which appears to be active as never before in the drug field. Virtually every paper dealt with some aspect of advancing knowledge and even in the controversy and debate one could see the formulation of testable hypotheses and stimulus for further investigations. As an additional benefit it is reasonable to expect that advances in the drug dependence field will have wider relevance and cast important new light on other areas of psychic function and malfunction. This was clearly implied by Angrist and Weiner and openly stated in Domino's paper which bore the intriguing title of "The Pharmacology of Madness."

In closing I wish to say that it was particularly appropriate that the International Symposium on Drug Abuse was held at the University of Michigan in Ann Arbor where Dr. Maurice Seevers has for more than 25 years directed a pharmacology research laboratory of outstanding international reputation. Major contributions to the scientific understanding of drug dependence and abuse were made here under his guidance.

Finally, I know I speak for all the participants when I say that special recognition goes to Dr. Chris Zarafonetis for the planning and organization of this Symposium, which Dean Gronvall told us had its inception in conversations held with Dr. Zarafonetis some 2½ years ago. All of us owe him a debt of gratitude for the thought and effort which he put into this important undertaking.

CLOSING REMARKS

CHRIS J. D. ZARAFONETIS, M.D.

First, I wish to thank Dr. Brill for his splendid summary of this conference. It was the most demanding task of the entire program, and we are all grateful to Dr. Brill for his perception and skill in highlighting the most important elements of the meeting and in indicating the major implications of the respective presentations.

In my opening remarks, I stated my hope that our attitudes were not fixed in advance of the discussions and that an open and receptive posture would be retained throughout the sessions. I think this was achieved quite well in view of the large and diverse audience which has been in attendance this week. Dr. Brill suggested that as a group we were probably not far from a consensus, at least with respect to certain of the issues that were discussed. In this connection, I will shortly give the results of the informal poll taken this morning regarding the penalties for possession of marihuana in the new Federal law. Before doing so, however, I wish to cite a reference which I believe has great relevance to some of the problems that surfaced, especially during that portion of the program having to do with the legislative, judicial, legal, and law-enforcement aspects of drug abuse. The reference to which I refer deals with the question: "How do scientists reach a decision?" The question was asked by Mr. Emilio Q Daddario at the time he was Chairman of the House Science and Astronautics Sub-committee on Science, Research and Development (1963-1970). Mr. Daddario apparently hoped that the answer to his question would provide a rational basis for the evaluation of legislation dealing with scientific problems. Accordingly a special study was commissioned which was performed by the Science Policy Research Division, Legislative Reference Service of the Library of Congress. The study consisted of an analysis of some 14 science-related legislative actions which had been taken by Congress in recent years.[1] Briefly, some of the findings in the repcrt may be noted. It became quite apparent that few scientists are politicians and few politicians are scientists. Moreover, it was observed that whenever there was a political issue associated with the scientific problem, the political aspects always overrode the scientific. The Library of Congress document warned against the tendency of lawmakers to total up the number of scientists on either side of an issue. Scientists do not decide a scientific question by voting on it; they decide by reaching a consensus. This takes longer than taking a simple head count. Finally, it was stressed that urgency associated with the timing of the political decision process may require decisions without first having achieved scientific consensus. In other words, if scientists haven't reached agreement because of their political hang-ups or a genuine lack of data, the politicians may just have to decide without them. I believe that there is great relevance of these and other observations made in the document to the legislative issues such as have arisen in the area of drug abuse.

Now, as to results of the poll which Dr. Lewis requested of us this morning. First, it should be understood that while there were more than 1100 registrations for this conference, only a portion have attended the entire program. Moreover, a formal poll was not considered to be an appropriate function for this meeting for reasons which should be self-evident. With these points in mind, then, it will be recalled that Dr. Lewis asked our group opinion regarding the Federal penalties in the new law with regard to possession of marihuana. The categories for your opinions were: A, not severe enough; B, just right; C, too severe; and D, this element of the law should be abolished. There were 358 who chose to turn in a "ballot," and there was one who voted "not severe enough"; 39 that it was "just right"; 159 that it was "too severe"; and 159 that "it should be abolished." Thus, should we manipulate the data, 318 believe it is too severe or should be abolished. That would perhaps be a consensus—of those voting—but not necessarily the position of all the others who attended some portion of the conference.

I will now declare the formal aspects of this conference adjourned *sine die*.

REFERENCE

1. Technical Information for Congress. Report to the Subcommittee on Science, Research, and Development of the Committee on Science and Astronautics, U.S. House of Representatives, Ninety-first Congress, First Session. Prepared by The Science Policy Research Division, Legislative Reference Service, Library of Congress. Serial A. April 25, 1969. U.S. Government Printing Office, Washington, D.C. 1969.

INDEX

Page numbers followed by f refer to figures, those by t to tables.